Electrical, Optical, and Magnetic Properties of Organic Solid-State Materials IV

MATERIALS RESEARCH SOCIETY
SYMPOSIUM PROCEEDINGS VOLUME 488

Electrical, Optical, and Magnetic Properties of Organic Solid-State Materials IV

Symposium held December 1–5, 1997, Boston, Massachusetts, U.S.A.

EDITORS:

John R. Reynolds
University of Florida
Gainesville, Florida, U.S.A.

Alex K-Y. Jen
Northeastern University
Boston, Massachusetts, U.S.A.

Michael F. Rubner
Massachusetts Institute of Technology
Cambridge, Massachusetts, U.S.A.

Long Y. Chiang
National Taiwan University
Taipei, Taiwan.

Larry R. Dalton
University of Southern California
Los Angeles, California, U.S.A.

Materials Research Society
Warrendale, Pennsylvania

CAMBRIDGE UNIVERSITY PRESS
Cambridge, New York, Melbourne, Madrid, Cape Town,
Singapore, São Paulo, Delhi, Mexico City

Cambridge University Press
32 Avenue of the Americas, New York NY 10013-2473, USA

Published in the United States of America by Cambridge University Press, New York

www.cambridge.org
Information on this title: www.cambridge.org/9781107413450

Materials Research Society
506 Keystone Drive, Warrendale, PA 15086
http://www.mrs.org

First published 1998
First paperback edition 2013

Single article reprints from this publication are available through
University Microfilms Inc., 300 North Zeeb Road, Ann Arbor, MI 48106

CODEN: MRSPDH

ISBN 978-1-107-41345-0 Paperback

CONTENTS

*Invited Paper

*Invited Paper

PART II: PHOTONIC MATERIALS AND DEVICES

*Invited Paper

*Invited Paper

*Invited Paper

*Invited Paper

PART V: ORGANIC METALS AND MAGNETIC MATERIALS

PART VI: POSTER PRESENTATIONS

*Invited Paper

PREFACE

The presentations of this symposium again attest to the fact that research involving electrical, optical, and magnetic properties of organic solid-state materials continues to grow both in scope and technological importance. Early studies of charge transport in conducting polymers have evolved from the elucidation of fundamental structure/function relationships to applications as batteries, simple electrical devices such as diodes, chemical sensors, antistatic coatings, microwave and millimeter wave-absorbing materials, and photochromic devices. A particularly exciting evolution has been the discovery and development of organic light-emitting diodes (OLEDs) which appear to be nearing commercialization in an amazingly short period of time. This application is of particular interest because both electrical and optical properties must be considered, and these have been important parallel themes of the conference. Moreover, nanostructure control is important for OLEDs, and nanoscale architectural engineering has been an increasingly important theme of the conference. Indeed, not only has the study of conjugated (quasidelocalized) electrons in organic solid-state materials resulted in interesting physical properties and device applications, but the desire to exploit these properties has promoted the development of new synthesis and processing methodologies to achieve special nanoscale and microscale structures.

John R. Reynolds
Alex K-Y. Jen
Michael F. Rubner
Long Y. Chiang
Larry R. Dalton

January 1998

ACKNOWLEDGMENTS

The symposium chairs would like to acknowledge the contributing and invited authors for the outstanding quality of their presentations and proceedings manuscripts. The invited speakers included:

Jean-Luc Bredas	Alan MacDiarmid
Michael Canva	Seth Marder
Arthur Epstein	Tobin Marks
Susan Ermer	Fotios Papadimitrakopoulos
Steven Forrest	Joseph Perry
Richard Friend	Paras Prasad
W. Haase	Hiroyuki Sasabe
Aaron Harper	Samuel Stupp
Alan Heeger	Timothy Swager
Chain-Shu Hsu	Daniel Talham
Samson Jenekhe	Michael Therien
Patrick Kinlen	Mark Thompson
Bernard Kippelen	Sukant Tripathy
Hilary Lackritz	

In addition, the symposium organizers are indebted to the session chairs for their efforts in overseeing the sessions, guiding subsequent discussions, and reviewing the manuscripts.

The symposium chairs wish to express their appreciation to the following organizations which provided financial support and enabled us to present the "Electrical, Optical, and Magnetic Properties of Organic Solid State Materials" symposium.

International Specialty Products
National Taiwan University
Taiwan Four Pillars
Taiwan President Enterprises
United Microelectronics Corporation

A special thanks is reserved for the Materials Research Society and staff, as well as the 1997 MRS Fall Meeting chairs, for their development of an outstanding conference.

MATERIALS RESEARCH SOCIETY SYMPOSIUM PROCEEDINGS

MATERIALS RESEARCH SOCIETY SYMPOSIUM PROCEEDINGS

Prior Materials Research Society Symposium Proceedings available by contacting Materials Research Society

Part I

Organic Light-Emitting Materials and Devices

OPTICALLY AND ELECTRICALLY EXCITED SEMICONDUCTING POLYMER STRUCTURES

H. BECKER*, S. E. BURNS,* G. J. DENTON*, N. T. HARRISON, N. TESSLER*, T. D. WILKINSON**, R. H. FRIEND*
*Cavendish Laboratory, Madingley Road, Cambridge CB3 0HE, UK, rhf10@cam.ac.uk
**Department of Engineering, University of Cambridge, Trumpington Street, Cambridge CB2 1PZ, UK

ABSTRACT

We report recent advances in the development of conjugated polymer device structures. We have fabricated vertical microcavity structures in which the optical path distance is tunable through control of the orientation of a liquid crystal layer, and demonstrate cavity tuning over 59 nm. We show that this structure allows convenient matching of cavity modes to the free-space emission spectrum from poly(; henylene vinylene), PPV. We have investigated the operation of electrically-excited polymer LEDs with short pulse operation, and have achieved current densities above 10^3 A/cm^2 and peak brightnesses of up to 5 x 10^6 cd/m^2. We have measured the optical absorption due to the injected carriers under these drive conditions, and find an absorption band near 1.6 eV which we attribute to the presence of polarons. This allows measurement of the injected charge density, from which we deduce that carrier mobilities are strongly enhanced at high fields, reaching values of 4 x 10^{-2} cm^2/Vsec for PPV diodes.

INTRODUCTION

Since the first report of electroluminescence in poly(p-phenylenevinylene) (PPV) [1] there has been increasing interest in conjugated polymer light-emitting diodes (LED's) both from basic physics and applications perspectives.[2] Recent reports of lasing [3] and line narrowing [4] from optically-pumped solid-state conjugated polymer structures have raised the possibility of the construction of an electrically-pumped polymer laser diode. However, there remain formidable technical challenges to be overcome before this can be realised, and in several areas, it is first necessary to carry out measurements of fundamental physical properties of the materials used and of the characteristics of the optical structures used.

We note that although in-plane guided-wave modes provide a convenient structure for achieving relatively low gain or lasing thresholds [4, 5], such structures require substantial modification to enable electrical excitation. The vertical microcavity structure formed by placing thin layers of semiconductor between mirrors [3] is however easily modified to allow electrical excitation, and we consider that it is this structure which provides the best information for the study of conditions necessary for optical gain. In our earlier work [3], we used a semitransparent silver mirror on top of the polymer layer, and found that thresholds for lasing were at very high excitation densities, substantially above the threshold for significant exciton-exciton collisions and quenching, which we find to be at an excitation density of 10^{18} cm^{-3} [6]. The presence of the silver mirror is the cause of significant losses in the cavity, due to the non-ideal juxtaposition of the node in t!.e EM field close to the mirror [7] and the excited semiconductor; this suppresses both spontaneous and stimulated emission within 50 nm of the

3

mirror, and also quenches excitation by energy transfer to surface plasmon modes in the silver layer [8]. We have recently developed microcavities which use only dielectric stack mirrors, using evaporated layers of ZnS and MgF_2 deposited on top of the PPV layer [9], and we have reported that these structures show much lower lasing thresholds.

We report here two recent studies we have conducted which advance our understanding of the issues that determine conditions for lasing. In the first, we summarise results obtained on microcavities which are arranged to be wavelength-tunable, by introduction of a liquid crystal layer within the cavity [10]. In the second, we report the response of polymer LEDs driven to high current densities [11].

RESULTS

Tunable Microcavity Structure

The planar microcavity is a Fabry-Perot resonator with a mirror separation of the order of the optical wavelength which contains a photon emitting medium. The value of this structure is that the spontaneous emission rate and the emission spectrum of an optical emitter can be modified, since the cavity enhances the rate of emission at the resonance wavelengths of the cavity and suppresses the rate of emission at other wavelengths [12-14]. Structures such as the one investigated here affect the total emission rate only marginally. However, the spectral and spatial dependence of the radiative rate is still strongly modified by the microcavity. At the resonance wavelengths of the cavity the radiative rate for emission in the forward direction can easily be enhanced by more than two orders in magnitude, and large enhancements are found experimentally [15]. In the same fashion it will be suppressed at other wavelengths. This leads to a spatial and spectral redistribution of emission as a result of channeling of radiation power into the modes for which the radiative rate is enhanced. For an efficient broad band emitter inside a tunable microcavity this effect can be used to control the spectral and spatial distribution of emission without the losses of radiative power associated with filters.

Three factors determine the effective length, L_{eff}, and thus the resonance wavelengths of a microcavity, the separation of the mirrors ($L_{mirror\ separation}$), the phase change on reflection of light from the mirrors ($L_{phase\ change}$) and the refractive indices (n) of the materials inside the cavity:

$$L_{eff} = n \cdot L_{mirror\ separation} + L_{phase\ change} \qquad (1)$$

and

$$\lambda_{Res} = \frac{2L_{eff}}{q} \qquad (2)$$

where λ_{Res} are the resonance wavelengths and q is an integer. In our devices, changes in the effective refractive index of the materials inside the cavity are used to control the effective length of the microcavity. Although many materials display refractive index changes in response to applied electric fields, the magnitude of these changes is too small to shift significantly the resonance wavelengths of a microcavity. We have therefore used highly birefringent liquid crystalline materials. Liquid crystalline materials have previously been used to tune passive Fabry-Perot structures, performing as wavelength tuning filters [16]. Liquid crystals commonly exhibit large optical anisotropies. The orientation of the molecules and thus the effective refractive index inside the cavity can be controlled by application of an AC electric field. We used here a positive uniaxial nematic liquid crystal, Merck Ltd Licrilite® BL048,

4

which has a low refractive index n_o ($n_0=1.5277$) in the two dimensions perpendicular to its optic axis and a high refractive index n_e ($n_e=1.7904$) along the optic axis. Consequently the two normal modes of polarization in the liquid crystal will sample the same refractive index n_0 if the light propagates along the optic axis but two different refractive indices n_0 and $n(\theta)$ if the light propagates in any other direction, with $n(\theta)$ depending on the angle between the optic axis and the direction of light propagation and $n_0 < n(\theta) < n_e$. Because we are not changing the refractive index of a medium, but rather the orientation of molecules, only one polarization samples a refractive index change.

The microcavity cells used in our experiment are shown in figure 1. Transparent indium tin oxide (ITO) electrodes with sheet resistance 30 ohms/square were used for application of an ac electric field. Each electrode was coated with a 12 layer dielectric stack mirror (alternating layers of MgF_2 and ZnS) with a reflection band (higher than 97% reflectivity) from 490-650nm and a peak reflectivity of about 99%. Poly(p-phenylenevinylene) (PPV) precursor was spun onto one mirror at 2000 rpm for 60 seconds and then baked at 200°C for 12hrs under vacuum to form an 80 nm PPV layer. An alignment layer of SN7212 polyimide was spun onto the other mirror at 3000 rpm for 40 seconds and then baked at 220°C for 3hrs. The alignment layer was then buffed at 500 rpm to give the rubbed alignment for the liquid crystal. The cells were assembled with UV glue. A spacing of around 2 µm was obtained by using 2 µm epostar spacers or due to the mirror roughness and residual dust. The cells were then vacuum-filled with BL048 nematic liquid crystal before sealing.

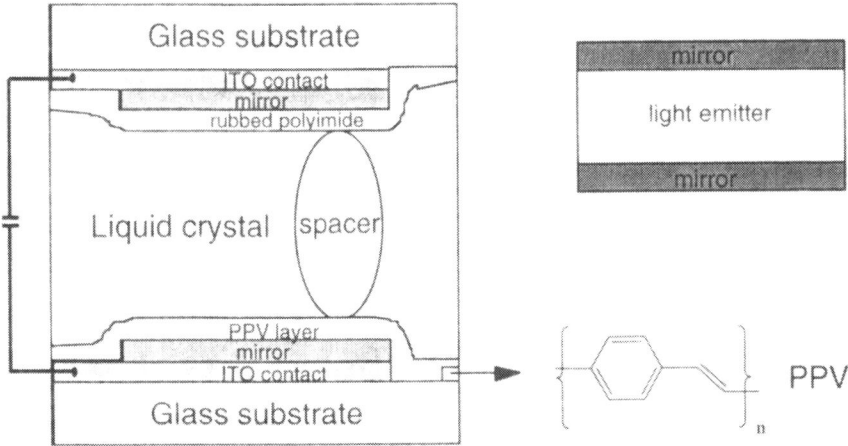

Figure 1 Schematic diagram of a microcavity (right) and of the device structure used in our experiments. Both glass substrates were coated with ITO layers and dielectric stack mirrors (alternating layers of MgF and ZnS). Before assembling the cell one substrate was coated with an 80 nm thick poly(p-phenylenevinylene) (PPV) film and the other with a polyimide film that was then rubbed. The chemical structure of PPV is shown in the bottom right.

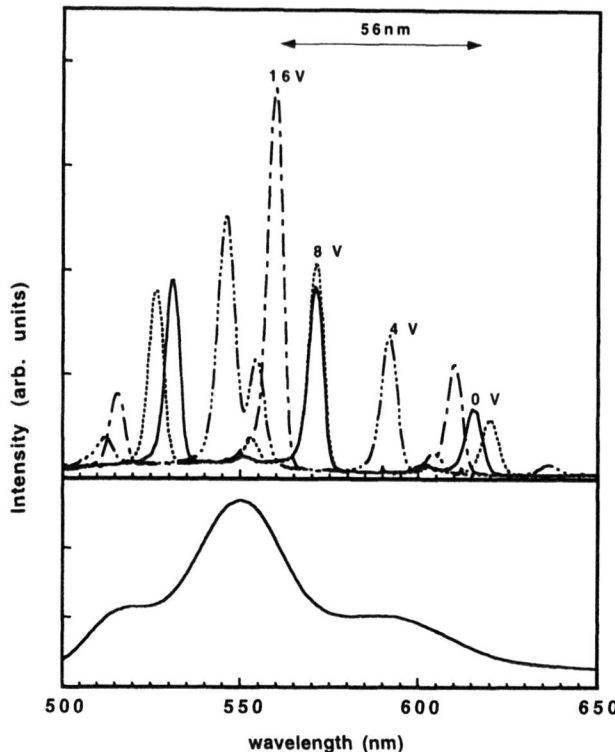

Figure 2 Upper panel: PL emission from the microcavity device with dielectric stack mirrors and no spacers as a function of applied ac bias. A polarizer transmitting light linearly polarized along the rubbing direction of the polyimide was placed in front of the detector. Lower panel: Free space PL emission spectrum of poly(p-phenylenevinylene) (PPV).

A 458 nm laser served as the photoluminescence (PL) excitation source and the photoluminescence was measured using a CCD array spectrometer (Oriel Instaspec IV). For some experiments a polarizer was placed in front of the device, so as to pass light linearly polarized along the rubbing direction of the polyimide. An ac 1kHz square wave electric field was applied across the cavity to change the orientation of the liquid crystal molecules and thus the cavity resonances. PPV and related copolymers can show efficient luminescence, with PL efficiencies up to 80% and is widely used as the emissive layer in polymer light-emitting diodes [2] and in optically pumped lasers [3]. It has a broad emission spectrum which conveniently spans 3 cavity modes for the devices investigated here. A broad emission spectrum is also a precondition for the tunability of the cavity as the emission wavelengths of the cavity can only be selected from within the natural emission spectrum of the emitter. The free-space emission spectrum of PPV is shown in the lower panel in figure 2; its structure is due to vibronic coupling to the excitonic emission process.

The polarized PL in the forward direction from a microcavity device is shown in the upper panel in figure 2 for applied electric fields between 0V and 16V in amplitude. Note that the cavity modes are narrow (6 nm FWHM) and can be tuned in wavelength. The 12th mode moved from 616 nm at 0V to 560 nm at 16V. The effective refractive index n(θ) of the liquid crystal layer decreases as a function of bias thus shifting the resonance wavelengths. The shift of 56 nm in resonance wavelength corresponds to a change in refractive index from 1.73 to 1.55 (including the refractive indices of all layers in the cavity) or a change in the optical thickness of the cavity from 3.3 μm to 3.7 μm. At 4V a relatively high fraction of the modes corresponding to the ordinary refractive index, n_0, appear in the spectrum. This is caused by the fact that at low fields and at intermediate angles the forces on the liquid crystal molecules due to the alignment layer and the electric field are not high enough to keep them oriented in the plane perpendicular to the mirrors and along the rubbing direction. The light transmitted by the polarizer has therefore sampled both refractive indices, n_0 and n(θ).

We have demonstrated a shift in resonance wavelengths of up to 56 nm in our microcavity devices. This change in wavelength spans the range from green to blue or red to green. We consider that with greater birefringence and improved alignment of the liquid crystal molecules shifts across the visible spectrum are accessible. There are, however, engineering limits to the wavelength changes obtainable from microcavity devices. Equation 3 describes the shift in resonance wavelength as a function of the optical anisotropy Δn assuming that the thickness of the emissive layer is small compared to the liquid crystalline layer:

$$\Delta\lambda = \lambda_0 \cdot \frac{\Delta n}{n} \qquad (3)$$

where n is the refractive index for which the resonance wavelength of the cavity is λ_0. Using BL048 a better alignment of the liquid crystal molecules could yield changes in the refractive index close to the birefringence of the molecules allowing shifts around 90 nm in the visible. However, for thicker microcavities the mode spacing decreases. Increasing Δn optimises Δλ only to the point where the mode spacing becomes equal to Δλ. That is the case for:

$$d_{microcavity} = \frac{\lambda_0 \left(\dfrac{n}{\Delta n} + 1 \right)}{2n} \qquad (4)$$

For BL048 and a wavelength shift of 90 nm in the visible this means that in order to achieve a mode spacing of 90 nm the mirror separation of the cavity has to be 1.23 μm.

In summary, To conclude, we have built a microcavity structure whose resonance wavelengths can be tuned by application of an external control voltage. We have achieved shifts of 56 nm in wavelength. In contrast to filters, microcavities achieve tuning of the wavelength of emission by channelling of radiation power into the resonant modes. We have presented a device that absorbs light at short wavelengths (blue, UV light) and emits light in part of a spectrum where there is control of the emission wavelength. The potential applications for such a structure are numerous. Mirror design, as here, can allow efficient absorption from a shorter wavelength. This allows the structure to be used as an efficient and tunable wavelength converter. Appropriate excitation might be provided by blue nitride LED's or UV fluorescent lamps. In order to built microcavities that will be tunable over the whole visible spectrum the challenge will be to find materials with a large $\Delta n/n_0$ and to fabricate thinner cells. The losses associated with the polarizer could be avoided by orientating the dipole emitters in a way that they emit polarized light or incorporating the polarizer into the cavity so that only polarized emission is supported by the cavity. Our experiments have also shed light on the nature of the energy transfer into the resonant modes in microcavities. We expect this structure to be a useful tool in future microcavity experiments.

High Current Densities in Pulsed-Drive Polymer LEDs

We have optimised the structure of the standard PPV LEDs to allow high current operation using short drive pulses, by working to reduce the series resistance associated with the indium-tin oxide electrode. LED's were fabricated on indium-tin oxide (ITO) coated glass substrates, ITO serving as the hole injecting electrode. Insulating strips (typically 100 nm SiO_2) were deposited onto the ITO in areas where contact was to be made to the cathode. Two LED structures were examined: (A) single layer PPV ; and (B) double-layer devices as described in ref. 2 with a conducting polymer between ITO and PPV, and a PPV copolymer as the emissive layer. The same PPV and PPV copolymer has recently been used to demonstrate optically-pumped lasing[3]. Both PPV and the PPV-copolymer are spin-coated as precursor polymers and thermally converted to the final form.[1, 3] The emissive polymer layers were typically 100-120 nm thick after conversion, and both device structures had calcium cathodes which were thermally evaporated under vacuum of around 6×10^{-6} mbar. In order to be able to measure intrinsic device characteristics and avoid problems associated with series resistance due to the ITO a 100nm Al layer was deposited underneath the insulating strips and the distance between the Al and the active pixel was kept to less than 0.5mm (typically 0.2mm). To avoid series resistance within the pixel, its dimensions were kept below $1 mm^2$. Device storage, and, where possible, device processing was performed in a dry nitrogen filled glove-box. We note that these diodes show exceptionally low turn-on voltages, and show high current densities at relatively low forward voltages [2].

Electrical excitation of the devices was provided by a home-made voltage pulse generator with a rise time of ~10 ns operating at a repetition rate of typically 45 Hz, with pulse width chosen to avoid heating problems, typically 100-200 ns. The voltage drop across the LED was measured with a voltage probe, and the current flow with an inductively-coupled current probe. Average light output was measured using a calibrated Si photodiode, and the temporal evolution of the luminance with an avalanche photodiode (APD). The current probe and APD have rise times of ~10ns. Voltage, current and light pulses were recorded using a 400 MHz bandwidth digitising oscilloscope. All measurements were performed in vacuum.

Figure 3 shows the J-V characteristics of a ~140nm thick device of structure A. We note that the device can support current densities up to $1 kAcm^{-2}$. The log-log scale extends over 5 decades and shows that at low current densities the current rises rapidly, approximately as V^n where n = 9. However, it levels off, towards a V^2 dependence at high current densities, as would be expected for a space-charge limited current with a constant carrier mobility (e.g. under conditions where one charge carrier predominates). At these high current densities the luminous efficiency remains high. At 600 A/cm^2 the peak brightness reaches 10^6 cd/m^2 for device A and 5×10^6 cd/m^2 for device B. Device B shows a maximum efficiency of 2.1 cd/A at 4 A/cm2, comparable to literature values for this device structure [2] and this value falls to 0.8 cd/A at 600 A/cm^2.

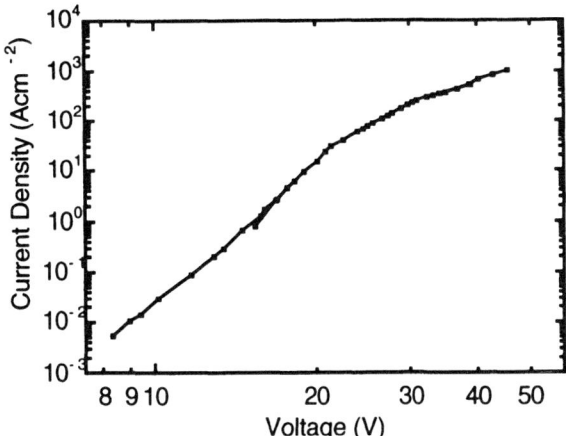

Figure 3 Current density as a function of voltage for a ~140nm thick single-layer PPV LED. The voltage is corrected to account for the residual series contact resistance

These results are unexpected because charge carriers are known to have very low mobilities in polymer LEDs of this type, less than $10^{-6}cm^2/Vsec$, so that such current densities should not be achievable. Low carrier mobility imposes high density of charge carriers to obtain high current densities. The electric field due to this space charge density should then limit the achievable current density (though high tip current densities are reported in STM experiments [17]). The high current densities measured here point towards a strongly increased mobility for at least one charge carrier (presumed here to be the holes in view of the higher hole mobility in PPV at low fields [18]). This is consistent with low-field measurements reported by Blom et. al.[19]. The maintenance of a high device efficiency to such high peak brightness is a significant result because there are many factors which might have been expected to have reduced the efficiency very strongly. Among these we list (i) the effects of exciton quenching by electronic charges, seen for photoinduced charge [20] and field-effect charge [21, 22], which is significant when the space charge density is such that diffusing excitons are likely to make collisions with such charges (above about 10^{18} cm^{-3} for PPV [6, 23]); (ii) electron-hole recombination takes place in the bulk for this type of diode, and is sensitive to charge carrier densities and mobilities [24].

It is difficult to obtain carrier mobilities from current-voltage characteristics. We note in particular that the effects of charge compensation for bipolar currents are hard to find evidence for, and indeed hard to model [25]. We note also that where the mobility varies with field, the modelling of current-voltage characteristics is complex even for unipolar currents [26]. We have therefore developed further an optical technique which allows direct measurement of the areal concentration of injected charge, which we had previously used for field-effect diodes [21, 27, 28], and which we and others have extended to LEDs [29, 30]. Electronic excitations in conjugated polymers are considered to be relatively localised (to individual polymer chains) and they therefore couple to the local chain geometry; this is established for both charged and neutral excitations through evidence for vibronic contributions to electronic transitions. This causes rearrangement of oscillator strength, to generate sub-gap absorption, and is generally termed polaron formation. For the case of PPV, absorption due to charged excitations is found in a band centred near 1.6 eV, as is seen in measurements of transient optically-pumped absorption [6, 31].

Techniques developed earlier [29] have been extended to be able to sample absorption changes with 10 ns resolution. The PPV LED was driven as described above. The current-induced optical absorption was measured as the proportional change in reflection of a probe optical beam which was directed through the ITO and PPV layers and reflected of the Ca/Al electrode. The reflected light was measured using a fast photodiode with better than 10ns response time, which was passed through a 300 MHz bandwidth amplifier and recorded with a 400 MHz bandwidth digitising oscilloscope. It was necessary to use a range of IR diode lasers to provide probe beams with sufficient intensity, but measurements were still restricted to the high excitation regime (above 10 A/cm^2).

Figure 4 shows the temporal response of the diode current, light emission and change in reflection (absorption) at 780nm (~1.6 eV). The diode was driven at 35 V with the voltage waveform shown in the figure. Note that the current and light output follow one another, except that there is a negative current transient as the voltage drive is taken to zero. We consider that this is due to the sweep-out of charge present in the polymer during the 'driven' state. This transient decays over about 150ns; from a crude analysis of the integrated charge that flows during this transient, we consider that the transit time for carriers in the 'driven' phase is no more than 10ns, consistent with the rapid rise in both current and voltage as the drive voltage is raised from zero at the start of the pulse. The absorption, shown as fractional change in reflection, tracks the rise in current, but has a longer decay. This decay is on many time-scales; the initial fall in the absorption has a typical time constant of ~150ns and is similar to the negative current measured in the external circuit. This is consistent with the physical picture of charge carriers being swept from the device in the reverse direction, by the internal field once the applied voltage is turned off. There is also a much longer-lived component which is seen as the baseline offset after the drive pulse (but which is not present on the leading edge, which is 22 ms after the previous pulse).

The detailed identification of the nature of the charged excited states responsible for this transient absorption is not central to the present discussion. Its spectral response [11] is similar to that of the short-time photo-induced absorption for PPV [6, 32]. Knowledge of the optical cross-section per injected charge allows conversion of the reflection change into a direct measure of the areal density of injected charge. This cross-section can be calculated quantitatively from measurements on metal-insulator-semiconductor field-effect diodes, and for PPV we have used the value measured on a PPV derivative, poly(2,5-dimethoxyphenylenevinylene) by Harrison et al.[28]. Figure 5 shows the current dependence of the proportional change in reflection and the evaluated charge density. At low currents the carrier density is rising steeply but, it saturates towards a value of 2 x 10^{17}cm^{-3}. The reduced slope of the carrier density with current density is a direct measure of the enhanced mobility at high fields. If we assume, for simplicity, that the mobility is constant across the device we can derive two limiting expressions for the effective mobility, μ, as μ_{ohm}=J/qEN for ohmic conduction and μ_{SCL}=2eJ/(N^{2}q^{2}L) for space charge limited current, where J is current density, q the electronic charge, E the applied field, N the carrier concentration, L the device thickness, and e the dielectric constant of the polymer layer. The 'ohmic' limit covers the situation where space-charge effects are unimportant (as would be the case for balanced electron and hole

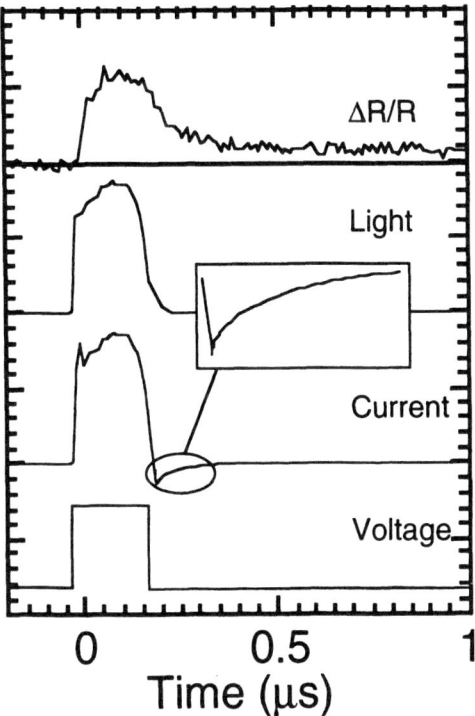

Figure 4   Temporal response of diode current, light emission and change in reflection (absorption) at 780nm (~1.6eV). The diode was driven at 35 V with the voltage waveform shown. Shown in the inset is an enlargement of the current transient seen as the drive voltage is switched to zero. Measurements are shown here for a pulse width of 200ns; this was found to be sufficiently long to allow "steady state" operation of the LED, as is evident in the figure.

currents with little recombination), and the space-charge limit is appropriate when there is a unipolar current (no screening by one charge of the other). In view of other aspects of device operation, particularly the constancy of efficiency at high drive currents, we consider that the space-charge limit is appropriate here [24] but note that the 'ohmic' limit represents a lower-limit on carrier mobility. Both mobility values are extremely high, in the 10^{-2}cm^2/Vs region; for the SCLC limit, this rises as high as 4×10^{-2} cm^2/Vsec.

The analysis we have performed to extract the carrier mobilities from the optical and electrical characteristics of the diodes takes as input parameters the average injected charge density, N, and the current density. It is therefore much less model-dependent than analyses based on current-voltage measurements. We note however that our analysis does not include the field-dependence of the mobility through the device, as discussed for example by Young [26]. We can compare the results we have determined from our method against the simple application of Child's law for unipolar current flow, and find reasonable agreement. For example, at a current density of 600 A/cm^2 Child's law gives a lower mobility (3×10^{-3} cm^2/Vsec) and higher hole concentration (6×10^{17} cm^{-3}). We note that it will be necessary to include the effects of bipolar currents (the significant electroluminescence is evidence for this) before such variations can be modelled quantitatively.

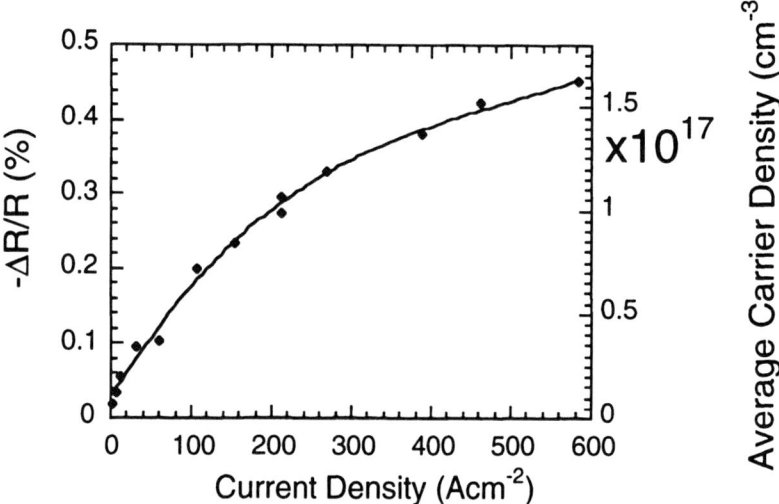

Figure 5   The relative change in reflection (left ordinate) and the deduced carrier density (right ordinate) as a function of injected current density. The right ordinate was calculated assuming an absorption cross section of 10^{-15}cm^2.

Although it is conventional to associate changes in mobility with the applied electric field, we note that there is a strong dependence of mobility on the concentration of carriers, as is established from measurements of the field-effect mobility on doped semiconducting polymers [33]. This behaviour is understood in terms of the filling of valence band states so as to raise the density of states at the Fermi energy. In the context of studies on polymer EL diodes, the high current densities achieved in the present experiments produce relatively high injection densities of carriers, which we measure here to be as large as 10^{17} cm^{-3}. However, this density is still low in the context of these studies of field-effect mobilities; at 10^{17} cm^{-3} the expected mobility is no higher than 10^{-6} cm^2/Vsec [33].

These results have implications for the possible development of organic semiconductor injection lasers. Although the high peak brightness and high peak current densities that we have reported here are encouraging, we note that there are considerable barriers still to overcome. Based on the measured light output from the diodes and assuming a radiative lifetime for the excitons of 1ns it is possible to estimate the exciton density at high currents (10^{14}cm^{-3}). We note that this value is several orders of magnitude lower than is currently required to achieve lasing in microcavity structures which have metallic mirrors [3] (and which might therefore function as injection electrodes in electrically-excited structures). We also note that at this level of exciton density, the density of injected charges is at least a factor of 1000 times larger (see figure 5). It will therefore be necessary to work very carefully to arrange that the charge-induced absorption which we have measured here does not overlap with the luminescence (gain) region and suppress the gain produced by the relatively small number of excitons. Designs that transfer energy to lower-energy emitters, as recently reported by Berrgren et al.[34] will need to be selected very carefully to avoid such problems. Based on the above we conclude that, at present, it is device efficiency rather than peak current density which is limiting laser action.

We mention finally that our finding of enhanced carrier mobility and rapid LED switching, both on and off, is very encouraging in the context of the development of passive-matrix polymer LED displays, for which LED switching speed determines how many rows can be addressed.

ACKNOWLEDGEMENTS

We acknowledge the help of A.R. Beadle and C.J. Moss in building vital parts of the experimental apparatus. We thank K. Pichler, W. R. Salaneck and W. A. Crossland for their support and advice for this work. We thank Cambridge Display Technology (CDT) for the provision of the PPV precursor polymer, and the Engineering and Physical Sciences Research Council for support.

REFERENCES

1. J. H. Burroughes, D. D. C. Bradley, A. R. Brown, R. N. Marks, K. Mackay, R. H. Friend, P. L. Burns and A. B. Holmes, Nature **347**, 539 (1990) .
2. J. C. Carter, I. Grizzi, S. K. Heeks, D. J. Lacey, S. G. Latham, P. G. May, O. R. delospanos, K. Pichler, C. R. Towns and H. F. Wittmann, Appl. Phys. Lett. **71**, 34 (1997) .
3. N. Tessler, G. J. Denton and R. H. Friend, Nature **382**, 695 (1996) .
4. F. Hide, M. A. Diazgarcia, B. J. Schwartz, M. R. Andersson, Q. B. Pei and A. J. Heeger, Science **273**, 1833 (1996) .
5. G. J. Denton, N. Tessler, M. A. Stevens and R. H. Friend., Adv. Mater. **9**, 547 (1997).
6. G. J. Denton, N. Tessler, N. T. Harrison and R. H. Friend, Phys. Rev. Lett. **78**, 733 (1997) .
7. H. Becker, S. E. Burns, N. Tessler and R. H. Friend, J. Appl. Phys. **81**, 2825 (1997) .
8. H. Becker, S. E. Burns and R. H. Friend, Phys. Rev. B **56**, 1893 (1997) .
9. G. J. Denton, N. Tessler, M. A. Stevens, S. E. Burns, N. T. Harrison and R. H. Friend, Proceedings of SPIE. July/August San Diego, 1997), .
10. H. Becker, T. D. Wilkinson and R. H. Friend, Appl. Phys. Lett., (in press) .
11. N. Tessler, N. T. Harrison and R. H. Friend, Adv. Mater., (in press) .
12. K. H. Drexhage, in *Prog. Opt.* E. Wolf, Eds. (North-Holland, 1974), vol. XII, pp. 163.
13. T. Tsutsui, C. Adachi, S. Saito, M. Watanabe and M. Koishi, Chem. Phys. Lett. **182**, 143 (1991) .
14. A. Dodabalapur, L. J. Rothberg, T. M. Miller and E. W. Kwock, Appl. Phys. Lett. **64**, 2486 (1994) .
15. J. Grüner, F. Cacialli and R. H. Friend, J. Appl. Phys. **80**, 207 (1996) .

16. G. D. Sharp, K. M. Johnson and D. Doroski, Optics Lett. **15**, 523 (1990) .

17. L. J. Rothberg, M. Yan, F. Papadimitrakopoulos, M. E. Galvin, E. W. Kwock and T. M. Miller, Synth. Met. **80**, 41 (1996) .

18. D. D. C. Bradley and R. H. Friend, J. Phys.:Condens Matter **1**, 3671 (1989) .

19. M. G. Harrison, K. E. Ziemelis, R. H. Friend, P. L. Burn and A. B. Holmes, Synth. Met. **55**, 218 (1993) .

20. P. W. M. Blom, M. J. M. Dejong and J. J. M. Vleggaar, Appl. Phys. Lett. **68**, 3308 (1996) .

21. D. G. Lidzey, D. D. C. Bradley, S. F. Alvarado and P. F. Seidler, Nature **386**, 135 (1997) .

22. P. W. M. Blom, M. J. de Long and M. G. van Munster, Phys. Rev. B **55**, 1 (1997) .

23. K. E. Ziemelis, A. T. Hussain, D. D. C. Bradley, R. H. Friend, J. Rühe and G. Wegner, Phys. Rev. Lett. **66**, 2231 (1991) .

24. P. Dyreklev, O. Inganäs, J. Paloheimo and H. Stubb, J. Appl. Phys. **71**, 2816 (1992) .

25. J. J. M. Halls, K. Pichler, R. H. Friend, S. C. Moratti and A. B. Holmes, Appl. Phys. Lett. **68**, 3120 (1996) .

26. H. Vestweber, H. Bässler, J. Grüner and R. H. Friend, Chem. Phys. Lett. **256**, 37 (1996) .

27. N. C. Greenham and R. H. Friend, in *Solid State Physics* H. Ehrenreich, F. A. Spaepen, Eds. (Academic Press, 1995), vol. 49, pp. 2.

28. R. A. Young, Phil. Mag. Lett. **70**, 331 (1994) .

29. J. H. Burroughes, C. A. Jones and R. H. Friend, Nature **335**, 137 (1988) .

30. A. R. Brown, K. Pichler, N. C. Greenham, D. D. C. Bradley, R. H. Friend and A. B. Holmes, Chem. Phys. Lett. **210**, 61 (1993) .

31. M. Redecker and H. Bassler, Appl. Phys. Lett. **69**, 70 (1996) .

32. M. Yan, L. J. Rothberg, F. Papadimitrakopoulos, M. E. Galvin and T. M. Miller, Phys. Rev. Lett. **72**, 1104 (1994) .

33. J. W. P. Hsu, M. Yan, T. M. Jedju, L. J. Rothberg and B. R. Hsieh, Phys. Rev. B - Cond. Matt. **49**, 712 (1994) .

34. H. Bässler, G. Schonherr, M. Abkowitz and D. M. Pai, Phys. Rev. B - Cond. Matt. **26**, 3105 (1982) .

35. C. P. Jarrett, A. R. Brown, R. H. Friend and D. M. de Leeuw, J. Appl. Phys. **77**, 6289 (1995) .

36. M. Berggren, A. Dodabalapur, R. E. Slusher and Z. Bao, Nature **389**, 466 (1997) .

TRANSIENT ELECTROLUMINESCENCE FROM PPV UNDER STRONG VOLTAGE PULSES

V. SAVVATE'EV*, R. POGREB*#, D. DAVIDOV*, H. CHAYET*,
R. NEUMANN& and Y. AVNY@

*Racah Institute of Physics, The Hebrew University of Jerusalem, ISRAEL 91904
#Research Institute, College of Judea and Samaria, Ariel, ISRAEL 44837
&Casali Institute of Applied Chemistry, The Hebrew University of Jerusalem, ISRAEL 91904
@Department of Organic Chemistry, The Hebrew University of Jerusalem, ISRAEL 91904

ABSTRACT

We have studied the transient electroluminescence (EL) from poly(phenylenevinylene), PPV, as a function of electric field under strong electric pulses up to fields of $E \approx 10^9$ V/m with emphasis on (a) the time delay, τ, between the electric pulse and the onset of the EL emission pulse and (b) the EL intensity as a function of the field. A monotonic decrease of τ with increasing E is explained by an increase of the carriers mobility according to Frenkel-Poole model. The EL intensity at high fields is proportional to E^3 suggesting that the contacts at the polymer-metal electrode interfaces are practically ohmic. We demonstrate significantly improved brightness, peak power and lifetimes for polymer-based light emitting diodes working under such a pulsed mode.

INTRODUCTION

The recent observations of optically pumped lasing from PPV polymer and the prospects for electrically pumped lasing makes transient EL measurements under strong electric fields both timely and interesting area of research [1-3]. Recently, we have reported [4,5] on pulsed electroluminescence (EL) measurements in light emitting diodes (LED) based on thin films (~500Å thickness) of PPV sandwiched between Al and ITO electrodes under short (250 nsec) and strong voltage pulses (up to 500V or fields of 10^{10} volts/meter). We demonstrated the existence of a delay time, τ, between the voltage pulses and the onset of the EL pulse which depended on the strength of the electric pulses. We observed also two regimes in the operation of the LED depending on the pulsed current density. In the first regime, up to a current density of 60 A/cm^2, the EL pulses are strongly enhanced by a factor of at least 10^4 above that achieved in stably operating DC devices, and follow the DC EL characteristics of PPV with green emission centered at 530 nm. Above this current density, in the second regime, we have observed an additional uv-violet emission (centered at 390 nm, ~3.17 eV) which increased exponentially with the height of the voltage pulses. This exponential increase was taken as evidence that hot electrons were responsible for the uv-violet emission. We proposed a possible explanation for the appearance of the uv-violet emission based on "hot" carriers in strong electric fields which partially suppress the formation of singlet exitons and enhance the probability for direct LUMO-HOMO inter-band radiative transition.

The goal of the present work is to gain a better understanding of the enhanced transient green EL emission in the first regime and especially the transition to the regime of highly enhanced electroluminescence. Therefore, we have measured the EL intensity, the current density, and the delay time as a function of the electric field pulse height. We have used relatively thick devices in order to reduce the device capacitance and the effect of the Al/PPV [3] interface on the transport properties of the devices.

EXPERIMENTAL AND RESULTS

We have studied LED devices of ~2000Å thick PPV films sandwiched between ITO coated glass and Al electrodes; the device active area is 20 mm².

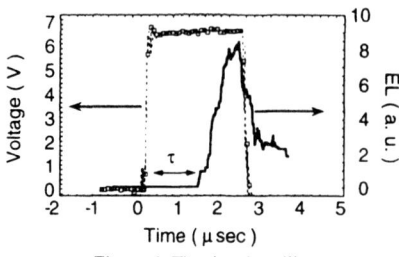

Figure 1. The signal oscillogram.

The pulser output load, connected in parallel with the device, was 10-100 Ω; device capacitance 1 nF, so, the RC time constant was less than 100 nsec. We have used pulses of 1-10 μsec duration in a single pulse mode or in the repetitive mode. The experimental setup is described elsewhere [4,5]. During measurements the devices were placed in a vacuum cell at room temperature. Figure 1 exhibits an oscillogram of a voltage pulse and the EL signal. A single voltage pulse is followed with a pulse of EL. Note the time delay τ of the EL after the onset of a voltage pulse. This time delay is field dependent and decreases with the increase of the applied electric field. Figure 2 demonstrates the field dependence of τ up to fields of 10^8 V/m. Above such fields τ is of the order or smaller than the RC time constant of the circuit and can not be measured using the existing setup.

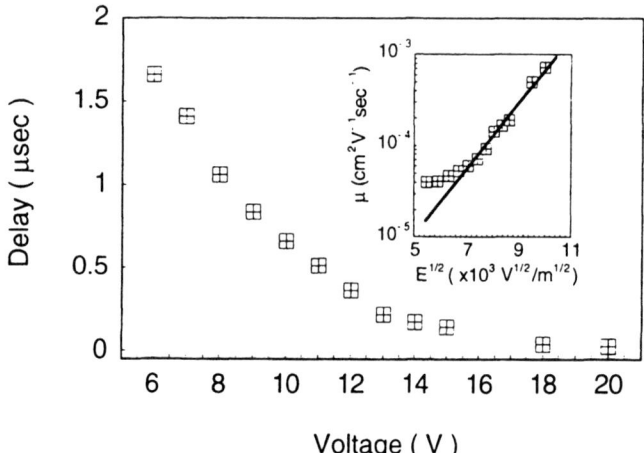

Figure 2. The time delay, τ, as a function of the voltage pulse height. Insert: a fit to the Frenkel-Poole model using the parameters: $\mu(0)=8.2 \times 10^{-7}$ cm²/V sec, $\gamma=6.4 \times 10^{-4}$ V$^{-0.5}$cm$^{0.5}$. Above 20V τ is limited by the RC time constant of the device.

The delay time, τ, is related to the field dependent mobility [5], μ(E), according to the relation

$$\tau = \frac{D}{\mu(E)E} \tag{1}$$

where D is the film thickness. According to the Frenkel-Poole model :

$$\mu = \mu(0) \times \exp\left(\gamma \sqrt{E}\right) \tag{2}$$

where $\mu(0)$ is a low-field mobility and the coefficient γ is given as :

$$\gamma = \frac{\left(e^3 / \pi\varepsilon_o\varepsilon\right)^{1/2}}{kT} \tag{3}$$

where k is Boltzmann constant, T is the absolute temperature, e is the electron charge, ε_o is permeability of vacuum and ε is the dielectric constant of a polymer. The Frenkel-Poole mechanism has been observed previously in organic semiconductors under low electric field [7-14]. The solid line in the insert of Figure 2 indicates a good fit to the Frenkel--Poole model. The fit yields the effective dielectric constant ε =10±2. As a rule, the effective dielectric constant determined from the Frenkel-Poole fit differs from that obtained from an independent impedance measurements. For example Karg et al. [9] obtained ε =22 for PPV from the field dependence of mobility and ε =4 from impedance data.

Figure 3 shows the peak EL intensity (per pulse) and the brightness, measured with 1:1000 duty cycle (averaging of 1000 EL pulses of 1 μsec duration per second). The overall increase in EL intensity is approximately 4 orders of magnitude with the fourfold increase in electric field.

There is a clear crossover from the exponential dependence (or at least high power law $EL \propto E^n$, $n \geq 8$) at low fields to a low-power law, $EL \propto E^n$, $n \approx 3$, at high fields. Also the functional dependence of the EL on the current density, J, shows a crossover behaviour (see insert to Figure 3). Note that J is smaller by almost an order of magnitude compared to that in our previous paper [4] due to the higher film thickness. The emission spectra are almost identical to those obtained in PPV under low DC currents. Hence, the transition from the exponential to power-law behavior in Figure 3 is related only to the change in the electron injection mechanism.

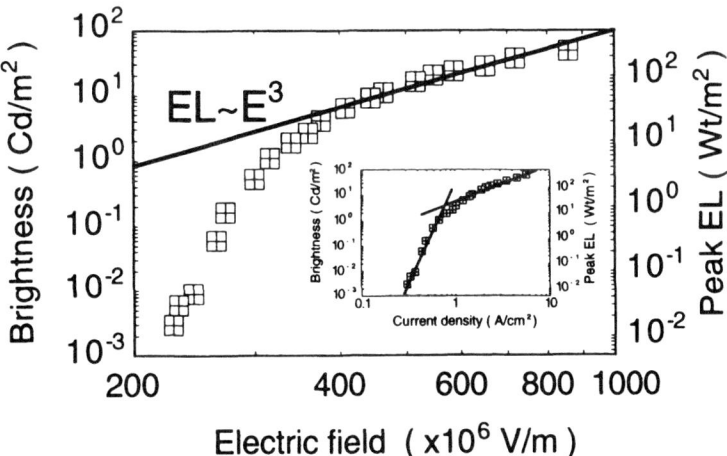

Figure 3. The EL peak power (per pulse) and brightness versus electric field demonstrating a crossover. Insert: EL brightness and peak power versus the current density through the device. The solid lines are for eye guidance to emphasize the crossover.

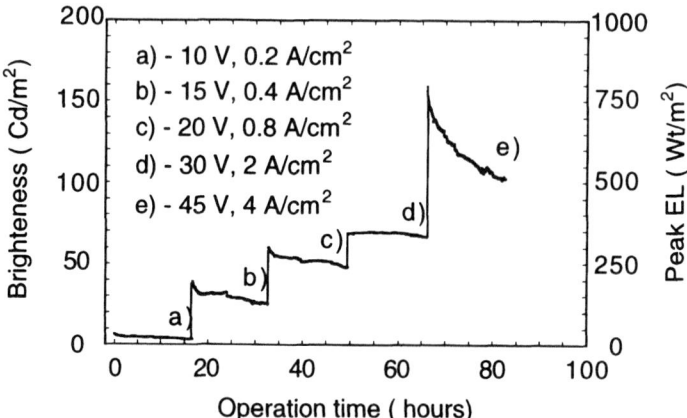

Figure 4. Brightness and peak power as a function of time. Note the high brightness achieved under low (1:1000) duty cycle. The sample was stressed with 1000 pulses of 1μsec duration for more than 80 hours - total of 300,000,000 pulses.

Figure 4 exhibits the brightness and the peak EL as a function of operation time under different pulse voltage heights and respective transient currents. The device could be operated for relatively long time without degradation under 1:1000 duty cycle with a 30V pulse height and 2 A/cm^2 transient current density to give an average brightness of ~70 Cd/m^2. We emphasize that under DC voltage and such brightness the life-time of such a device is only few minutes.

DISCUSSION AND CONCLUSIONS

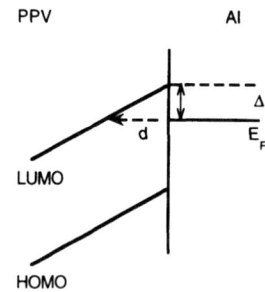

Figure 5. Band diagram. The band offset Δ and the applied field E determine the barrier width d~Δ/E.

Probably our most exciting observation is the dramatic increase of the EL intensity and the crossover behaviour in Figure 3. We suggest a very simple model based on band structure concepts to explain these features. The hole-injecting contact at the ITO/PPV interface behaves like an ohmic contact at all electric fields. The hole current density, j_p is space-charge limited in the filled-traps regime ($j_p \sim E^2$) [15-18]. In contrast, the electron injection at the Al/PPV interface strongly depends on the energy barrier at this interface as shown in Figure 5. At low fields the electron injection current, j_n, is due to field-assisted tunneling across the barrier and depends exponentially on the external electric field and the thickness of the barrier. However, at high electric fields, above 4x10^8 V/m, the interface energy barrier narrows down to a nanometer, which is of the order of intermolecular spacing (upper limit estimate for the barrier width, d, can be obtained from the ratio Δ/E where Δ is the energy offset at the interface, see Figure 5). For such narrow energy barrier there is a significant overlap of the electronic wave-functions and the electrons injection is no longer governed by tunneling. It is virtually ohmic, namely j_n is proportional to E since the electrons injected into PPV are effectively screened out with a more mobile and abundant holes. The EL intensity is due to exciton recombination and is proportional to the product:

$$EL \propto j_p \times j_n. \tag{4}$$

Consequently, the EL intensity versus E at low fields is expected to be a product of an exponential function of E and a power law function of E. At high fields, when both contacts are ohmic, the EL intensity should be roughly proportional to E^3. These simple considerations are remarkably consistent with our experimental observations, namely, at high electric fields the EL intensity versus field obeys a power law $EL \propto E^3$ (see Figure 3). Similar considerations indicate that the EL intensity versus the current density at high fields exhibits a low power law. Indeed, this is observed experimentally (insert to Figure 3).

The demonstration of ohmic contact at the metal/PPV interface at high fields is a most interesting feature. The ohmic contact makes the use of highly reactive metallic electrodes for matching the LUMO band of organic semiconductor unnecessary. Indeed, we found that replacing the Al electrode by a MgAg electrode did not lead to any significant increase of the transient EL intensity under strong voltage pulses, although under DC conditions at low fields the use of a MgAg electrode significantly improved the brightness.

In conclusion

(a) We have demonstrated that the mobility follows the predictions of Frenkel-Poole model. At fields of the order of $E \approx 10^8$ V/m the mobility reaches values of 10^{-3} cm^2/Vsec. Extrapolation of the Frenkel-Poole fit (see inset to Figure 2) to a higher field of $E \approx 10^9$ V/m gives mobility of the order of 10 -100 cm^2/Vsec. This, however, should be considered with much caution.

(b) We have demonstrated a transition in the injection mechanism at electric fields above 4×10^8 v/m. This is attributed to narrowing of the width of the energy barrier at the Al/PPV interface leading to ohmic contact. In our opinion this is an important finding which alleviates the seemingly unsolvable contradictory requirement of strong injection and chemical stability at polymer-metal interface. The analysis of the group in Tucson [19], under similar situation, also predicts a transition in the EL versus field at above 10^8 V/m.

(c) The high peak radiation power created by strong electric pulses allows to achieve significant EL brightness with duty cycles as low as 1:1000. We attribute the high brightness to higher mobility and to the appearance of the ohmic contacts at high fields. Low heating of the device under these conditions allows operation for at least several days at brightnesses at which their DC counterparts decay in minutes. The damage mechanism is also different: the device lifetime is no longer limited by bubble formation [20] but rather by damage to the ITO.

ACKNOWLEDGEMENTS

The authors are grateful to Boris Laikhtman and Aharon Yakimov for insightful discussions. We thank Galina Perepelitsa and Iris Benjamin for substance preparation. This work was partially supported by VW Foundation and the Israeli Ministry of Science and Arts. R. Pogreb acknowledges the support from the Israeli Ministry of Absorption.

REFERENCES

1. N. Tessler, G. J. Denton and R. Friend, Nature **382**, 695 (1996).
2. M. A. Diaz-Garsia, F. Hide, B. J. Schwartz, M. R. Andersson, Q. Pei and A. J. Heeger, Proc. ICSM-96, Salt-Lake City, Utah, Synth. Met. **84**, 455 (1997).
3. S. V. Frolov, W. Gellerman, Z. V. Vardeny, M. Ozaki and K. Yoshino, Proc. ICSM-96, Salt-Lake City, Utah, Synth. Met. **84**, 471 (1997) .
4. H. Chayet, R. Pogreb, and D. Davidov, Phys. Rev. B **56**, R12702 (1997) .
5. H. Chayet, R.M. Pogreb, D.Davidov, Proceedings of SPIE **3148** (1997).
6. J. L. Bredas, Adv. Mater. **7**, 263 (1995) .
7. D. Pai, J. Chem. Phys. **52**, 2285 (1970).
8. W. D. Gill, J. Appl. Phys. **43**, 5033 (1972).
9. S. Karg, V. Dyakonov, M. Meier, W. Reiß and G. Paasch, Synth. Met. **67** (1994) 165.
10. H. Meyer, D. Haarer, H. Naarmann and H. H. Hörhold, Phys. Rev. B **52**, 2587 (1995).
11. P. W. Blom, M. J. M. de Jong and M. G. van Munster, Phys. Rev. B **55**, R656 (1997).
12. L. B. Lin, S. A. Jenekhe, R. H. Young and P. M. Borsenberger, Appl. Phys. Lett. **70**, 2052(1997).
13. A. J. Pal, R. Osterbacka, K. M. Kallman and H. Stubb, Appl. Phys. Lett. **71**, 228 (1997).
14. E. Lebedev, Th. Dittrich, V. Petrova-Koch, S. Karg and W. Brutting, Appl. Phys. Lett. **71**, 2686 (1997).
15. V. N. Savvate'ev, M. Tarabia, H. Chayet et al., Proc. ICSM-96, Salt-Lake City, Utah, Synth. Met. **85**, 1269 (1997).
16. P. W. Blom, M. J. de Jong and J. J. M. Vleggaar, Appl. Phys. Lett. **68**, 3308 (1996).
17. S. Karg, M. Meier and W. Reiß, J. Appl. Phys. **82**, 1951 (1997).
18. I. Musa, S. J. Higgins and W. Ecclestone, J. Appl. Phys. **81**, 2288 (1997).
19. Y. Kawabe, G. E. Jabbour, S. E. Shaheen, B. Kippelen and N. Peghambarian, Appl. Phys. Lett. **71**, 1290 (1997).
20. V. N. Savvate'ev, A. V. Yakimov, R.M. Pogreb, D. Davidov, R. Neumann and Y. Avny, Appl. Phys. Lett. **71**, 3344 (1997)

Light-Emitting and Hole-Transporting Polymers for LEDs

Toshihiro Ohnishi, Shuji Doi, Masato Ueda, Fumi Yamaguchi, Takanobu Noguchi

Tsukuba Research Laboratory, Sumitomo Chemical Co. Ltd.,
6 Kitahara, Tsukuba, Ibaraki, 300-32 Japan, ohnishi@tuc.sumitomo-chem.co.jp

ABSTRACT

Various copolymers of arylene vinylenes, having strong fluorescence, showed predominantly the emission in the multi-layer device using an electron-transporting material(ETM) such as tris(8-quinolinolato)aluminum(Alq$_3$). The emission from Alq$_3$ was suppressed due to the high hole-injection barrier from the copolymers to ETM in spite of low or no barriers of electron injection from ETM to the copolymers.

We have successfully prepared highly hole-transporting polysilane having a triphenylamine group as a side chain(TPA-PS). The hole mobility as high as 10^{-3}cm^2/Vs is attributable to the intermolecular hopping process facilitated by the interaction between the polysilane backbone and the triphenylamine group. The polysilane is effectively used as a hole transporting material. The bilayer LED device consisting of TPA-PS and Alq$_3$ showed high luminance (2000cd/m^2) and high efficiency (4cd/A).

INTRODUCTION

Organic and polymer light-emitting diodes, O- and P-LEDs, have attracted much attention as an accessible flat panel display[1,2]. Many P-LEDs have been fabricated by using poly(phenylene vinylene),PPV, and its derivatives because of a wide structural variety[3]. The device performance of the P-LED was, however, insufficient for industrial requirements, compared with that of O-LED. Improving the performance, especially the efficiency, we have prepared highly luminous copolymers of various arylene vinylenes and a bilayer device, using the copolymers and tris(8-quinolinolate) aluminum, Alq$_3$, as an electron-transporting material,ETM, [4]. From the studies on the relation between the emitting properties of the polymers and the devices, the device efficiency depends not only on the fluorescence intensity of the polymers but also on the charge-injecting efficiency[5].

The hole- and electron- injections should be taken into consideration to elucidate the injection process at the polymer-ETM interface, because both carriers recombine to generate an exciton. It is, therefore, important to estimate barrier heights of hole- and electron- injections at the interface. The poly(arylene vinylene) derivatives can be used for this purpose, because their different energy levels related to the composition. Although it is a good ETM, Alq$_3$ cannot be used for this purpose, because it shows an emission in the same wavelength region as the copolymers. Bis(2-hydroxyphenylbenzoxazolate) zinc(ZnBox)[6] , having blue emission and electron transporting properties, is useful to estimate the barrier heights. When ZnBox is used in a bilayer device with the copolymers, it is easy to obtain the precise emission spectra from copolymers seperately from that of ZnBox.

Polysilane, such as poly(methylphenylsilane), is known as a good hole-transporting material, showing a hole mobility as high as 10^{-4}cm^2/Vs, and has been used in LEDs as a hole-transporting layer[7]. Some authors have successfully prepared a new polysilane, showing a hole mobility higher than 10^{-3}cm^2/Vs[8]. The polysilane consists of a silicone main chain and a triphenyl amine side group.

In this paper we will discuss the charge injection to poly(arylene vinylenes) and the hole transporting of polysilane in LEDs.

EXPERIMENT

Random copolymers having arylene vinylene units A and B, or A and C, were prepared by the Wittig reaction, as reported previously[5]. The number-average molecular weights, measured by GPC using polystyrene as a standard, were 5×10^3 - 1×10^4 for all copolymers.

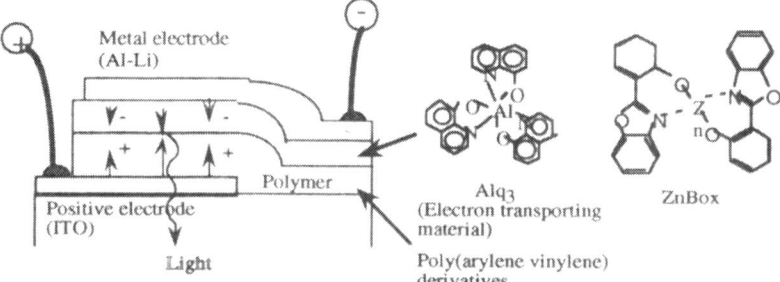

R:C_8H_{17}

Fig.1 The chemical structure of arylene vinylene units.

Alq$_3$ was commercially obtained and purified by vacuum sublimation. A bis(2-hydroxyphenyl benzoxazolate) zinc(ZnBox) was synthesized using zinc acetate and 2-(2-hydroxyphenyl) benzoxazol in the same manner reported[6] and purified by sublimation.

A polysilane having triphenylamine group(TPA-PS) as a side chain was obtained by the Kipping reaction of methyl(4-(N,N-diphenyl)aminophenyl)dichlorosilane and Na[8]. The hole mobility of the polysilane was obtained with a conventional time-of-flight method using Se as a carrier generating material.

Fig.2 Kipping reaction of methyl(4-(N,N-diphenyl)aminophenyl)dichlorosilane

The P-LED device, consisting of a transparent electrode, a light-emitting polymer film, an electron-transporting or hole-blocking layer, and a negative electrode, as shown in Figure 2, was fabricated as follows: a polymer film, 50 to 100 nm in thickness, was spin-cast from toluene solution onto an indium tin oxide, ITO, transparent electrode. A film of Alq$_3$,or ZnBox, used as an electron-transporting layer, was then deposited onto the polymer film, followed by

Metal electrode (Al-Li)

Polymer

Positive electrode (ITO)

Light

Alq3
(Electron transporting material)

ZnBox

Poly(arylene vinylene) derivatives

Fig.3 Schematic device structure of polymer LEDs and electron-transporting materials.

co-evaporating Li-Al alloy as a negative electrode at 10^{-6} Torr. A buffer or additional layer[9] was used between the ITO electrode and the light-emitting polymer film in the case of a three-layer device. The LEDs, using TPA-PS as a hole transporting material and Alq_3 as a light-emitter, were fabricated in a manner similar to that described above.

Luminescence spectra of devices using copolymers were measured with a Hitachi 850 fluorescence spectrophotometer. Luminance and current-voltage curves were obtained with a TOPCON BM-8 luminance meter, a Keithley 199 digital multimeter and a Takasago GP050-2 DC voltage source, which were controlled with a personal computer.

RESULTS AND DISCUSSION

<u>Electroluminescence of P-LED fabricated using copolymers.</u>

Figure 4 shows the luminance-voltage curve of the multi-layer device using the A/C(1/1) copolymer, using ZnBox, together with that of Alq_3. A green emission was observed from both devices. The luminance of the device increased with the voltage, and reached 16,000 cd/m^2 of the maximum luminance before the break-down of the device, while that of Alq_3 reached $25,000cd/m^2$. The turn-on voltage of the device using ZnBox was higher than that of Alq_3. These results suggest that ZnBox needs a higher voltage to operate the device than Alq_3, although it works well as a electron-transporting material.

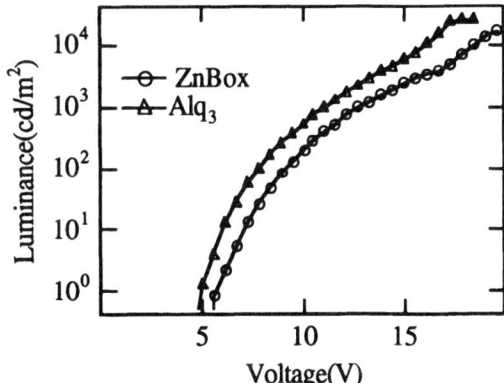

Fig.4 Luminance-current density curve for P-LED using the A/C copolymer(A:C=1:1) and ZnBox together with that of Alq_3.

Figure 5 shows the electroluminescence, EL, spectra of the devices using A/C copolymers together with those of Alq_3. The emission from the polymers was seen predominantly for the random copolymers, while the spectrum of homopolymer of unit C is observed in the same region as that of Alq_3. Consequently it is not clear that emission only from the homopolymer of C is observed.

To separate the emission of the polymer from Alq_3, the P-LEDs were fabricated using ZnBox. The devices showed basically the same spectra, as those using the polymers and Alq_3, as shown in Fig.6. In the case of the device using homopolymer of the unit C, the emission from ZnBox was observed as a shoulder in the EL spectrum. Although the spectrum of the device is in fairly good agreement with that of Alq_3, besides the shoulder arising from ZnBox, the EL intensity in the 500-550nm is a little smaller than that of the device using the polymer and Alq$_3$. This result means that the emission from homopolymer C is predominant in the both devices, and a small emission from ZnBox and Alq$_3$ is observed. Similar results were obtained for the copolymer of A and B units.

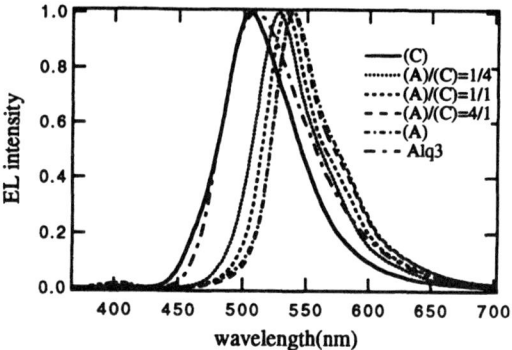

Fig5. EL spectra of P-LED using Alq$_3$ and A/C copolymers having various monomer ratios

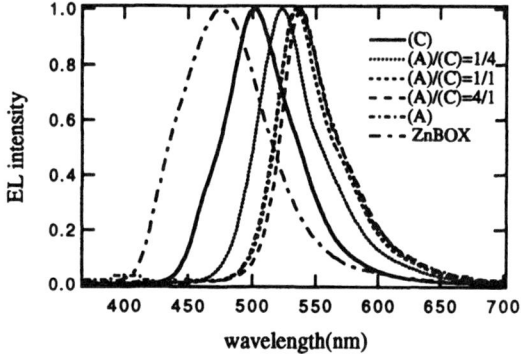

Fig6.EL spectra of P-LED using ZnBox and A/C copolymers having various monomer ratios.

The emission mechanism can be illustrated by using the energy diagram of the copolymers and ETMs, as shown in Fig.7, where the ionization potentials and electron affinities were obtained from redox potential and absorption edge wavelength of the material. The barrier height of the hole injection from the polymer to ETM is about 0.2-0.3eV for all polymers. The hole-injection is an uphill process in the device. The electron injection from ETM to the polymer is downhill, except for homopolymers of B and C. Thus, the recombination of holes and electrons injected from the electrodes takes place mainly inside the polymers, because of the energy barrier for electron injection at the polymer-Alq$_3$ interface. In the case of homopolymer B and C, the electron injection from ETM to the polymer becomes difficult, due to the high barrier and consequently the device efficiency decreased[5]. The high barrier of electron injection results in electron accumulation inside of the ETM layer, in recombination of holes and electrons and in emission from ETM.

The high performance of the device mentioned above resulted from strong fluorescence of the polymers and hole accumulation near the interface between the polymers and Alq$_3$.

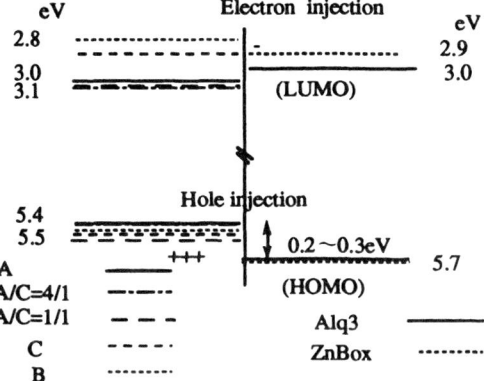

Fig.7.Schematic energy diagram of the copolymers , Alq$_3$, and ZnBox

Hole-transporting polymer

The polysilane having a triphenylamine group showed a high hole mobility, which depended on the molecular weight of the polysilane, as shown in Fig.8. The mobility reached over $10^{-3} cm^2/Vs$, which is the highest known so far among the polysilanes. Conventional poly(methylphenylsilane) showed $10^{-4} cm^2/Vs$ even in the high molecular weight region. This result implies that the interaction between polysilane main chain and triphenylamine effectively facilitates the intermolecular hopping of the holes.

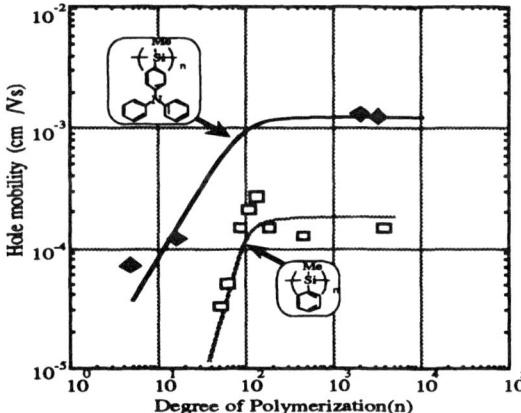

Fig.8 molecular weight dependence of hole mobility of polysilane

Figure 9 shows current-voltage and luminance-voltage curves of the LED using the polysilane and Alq$_3$, together with those of polyvinylcarbazole,PVCz. The LED using polysilane showed a lower turn-on voltage(4V) and a higher efficiency (4cd/A) than that using PVCz. This results are attributable to the high hole mobility of the polysilane. The current density and resulting luminance were, however, suppressed in the high current density region. The reason of this result is not clear at present.

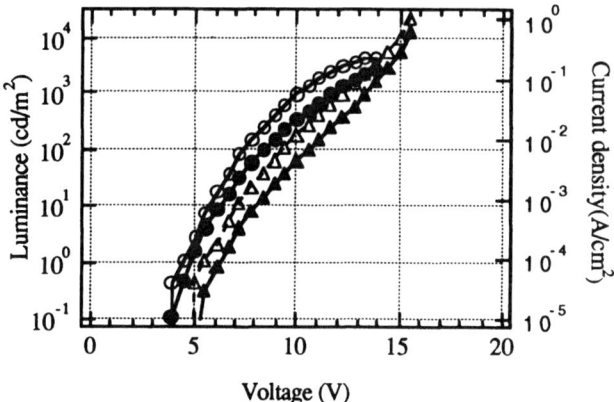

Voltage (V)

Fig.9. Luminance-current density-voltage curves for the LED using TPA-PS($\bigcirc$,$\bullet$) and PVCz $(\triangle,\blacktriangle)$.$\bigcirc$,$\triangle$:Luminance-voltage, and $\bullet$,$\blacktriangle$: current density-voltage curves.

CONCLUSIONS

The poly(arylene vinylene) derivatives are effectively used as a light-emitting material to give highly luminous P-LED devices. When Alq_3 is used as an electron-transporting material in the multi-layer device, the emission from the polymers is predominant, due to the energy barrier for hole-injection, and the device efficiency depends on the barrier height of electron injection from Alq_3 to the polymers.

The polysilane having triphenylamine works as a hole-transporting layer in the LED and gives a high efficiency in low current density region.

REFERENCES

1. C.W.Tang and S.A.VanSlyke, Appl. Phys. Lett., **51**, 915, 1987.
2. J.H.Burroughes, D.D.C.Bradley, A.R.Brown, R.N.Marks, K.Mackay, R.H.Friend, P.L.Burns, and A.B.Holmes, Nature, **347**, 539, 1990.
3. P.L.Burn, A.B.Holmes, A. Kraft, A.R.Brown, D.D.C.Bradley, and R.H.Friend, MRS Symposium Proceedings, **247**, 647, 1992.
4. T. Ohnishi, T. Noguchi, and S.Doi, Japanese Patent 5-247460,1993.
5. T. Ohnishi, S. Doi, Y. Tsuchida, and T. Noguchi, Materials and Measurements in Molecular Electronics, Springer Proceedings in Physics, **81**, 245, 1996.
6. N.Nakamura, S. Wakabayashi, K. Miyairi, and T. Fujii, Chem.Lett., 1994,1741.
7. Kido, K. Nagai, Y. Okamoto, and T. Skotheim, Chem.Lett., 1991,1267.
8. M.Ueda, F.Yamaguchi, Y. Fujii, I.Yahagi, M. Sasaki, T. Osada, M. Kitano, and Y.Abe, Europian patent 0654495A1
9. S. Doi, M. Kumwabara, T.Noguchi, and T. Ohnishi, Synth. Met., **57**, 4174,1993.

WHITE LIGHT EMITTING ELECTROLUMINESCENCE DEVICES WITH ORGANIC OLIGOMERS

F.MEGHDADI, S.TASCH, W.GRAUPNER, S.E.DÖTTINGER*, M.HANACK*
M.GRAUPNER**, A.HERMETTER** and G.LEISING
Institut für Festkörperphysik, Technische Universität Graz, Petersgasse 16,
A-8010 Graz, Austria
*Institut für Organische Chemie, Eberhard Karls Universität Tübingen, Auf der Morgenstelle
18, Germany
** Institut für Biochemie und Lebensmittelchemie, Technische Universität Graz, Graz, Austria

ABSTRACT

We report on the properties of white light emitting electroluminescence (EL) devices fabricated by combining blue and orange light emitting organic materials, parahexaphenyl (PHP) and a new isothianaphthen-S-oxide (ITSO) oligomer. The EL spectrum of this oligomer blend covers the range from 400 to 650 nm. When the concentration ratios of these materials are suitably adjusted a pure white light emission (CIE coordinates of x=0.333 and y= 0.338) and a high external EL quantum efficiency of 1.2% photons/electron can be obtained. The mechanism of the white light generation is ascribed to an internal self-absorption of part of the blue PHP light emission by ITSO and a subsequent photoluminescence emission process of ITSO. The results of photoinduced absorption, which is sensitive to charged species and triplet states of these guest-host systems, are discussed.

INTRODUCTION

Organic light emitting diodes based on the PHP oligomer have great potential for application in several optoelectronic devices and for organic light-emitting diodes (OLEDs) to be used in large area displays in particular [1-3]. The intensive, deep blue light emission of PHP can be converted into other visible light due to the fact that its emission spectrum is located on the high energy side of the visible spectrum [4,5]. In a bilayer configuration using PHP and the organic-metal complex Gaq3 as the light-emitting layers, it possible to switch color emission from green to blue depending on the polarity of the applied bias voltage [6]. Recently we have been successful in the realization of multicolor LEDs red, green, blue (RGB) by a new external color conversion technique based on PHP OLEDs [4].

One special goal is to realize efficient white light emission based on OLEDs, which is necessary for many applications such as backlighting of liquid crystal displays and producing full color displays using suitable color filters. There have been several concepts to achieve white light emitting devices based on thermally vacuum deposited organic materials [8-15]; i) using multilayer LEDs consisting of two or more active layer, ii) by spectral filtering of the emission light of a suitable emitter, iii) by superposition of red green and blue LEDs. Recently, we showed how to improve the performance OLEDs by an internal conversion technique using an active layer based on a blend of two conjugated polymers [7]. The white emission in this polymers blend is generated by an excitation energy transfer from the host polymer into lower lying emitting states of the guest polymer [7].

In this article we present the fabrication of efficient white EL devices using a blend of a PHP and an ITSO oligomer as the active layer, which is produced by co-evaporation of the

27

oligomers. A pure white EL light emission is obtained by introducing only 4 % ITSO into the PHP oligomer. The combination of two emission spectra, blue light from PHP and orange light from ITSO results in efficient white OLEDs. The EL efficiency of this OLED is 1.25%. The generation of white light in mixed single layer EL is due to an internal color conversion method. The ITSO oligomer, which strongly absorbs in the blue spectral range, is excited when illuminated with the PHP EL emission and subsequently emits PL light at longer wavelengths. A part of the blue PHP emission therefore is absorbed and transformed in the orange spectral range, so that a pure white light emission can be realized, if the concentration ratio of these two oligomers is suitably adjusted.

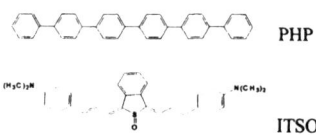

Fig. 1 : Chemical structure of PHP and ITSO oligomers.

EXPERIMENT

Fig.1. shows the chemical structure of PHP and ITSO oligomers used in this study. The synthesis of pure PHP powder and ITSO (1,3-Bis-[1-(dimethylamino) phenylenevinylene-methyliden]-1,3-dihydro-isothianaphthence-2-oxide are presented in Refs. [16-18]. The homogeneous thin films of PHP and ITSO on the Indium-tin oxide (ITO) surface were produced by a co-deposition technique using two individual Knudsen cells under high vacuum conditions ($<10^{-6}$ mbar). High purity aluminum as a stable metal was used as the top electrode. The film thickness, which is controlled by quartz microbalance for all samples, was 2000Å. The concentration of ITSO in PHP, determined by IR-transmission using an co-deposited films of both oligomers on Si substrate. The concentration of ITSO relative to that of PHP was varied in a range from 4% to 40 %. The EL spectra were recorded with CCD spectrometer 77400 L.O.T.-ORIEL. Electrical characteristics and external EL quantum efficiency (emitted photons per injected electrons) were measured simultanously with a Keithley 236 and 180. The photoluminescence quantum yield was measured in an integrating sphere. The excitation from a 1000 W Xe lamp, dispersed by an excitation monochromator, is guided into the sphere via a glass fiber. A second fiber is used to direct the light from the sphere into the emission monochromator. The photoinduced absorption (PIA) experiments were performed with mounted samples in an optically accessible liquid nitrogen cryostat within a conventional transmission set up [19].

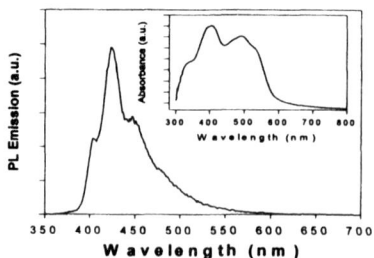

Fig. 2: PHP PL emission and (inset) ITSO absorption spectrum.

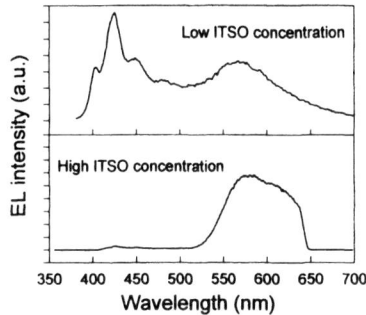

Fig. 3: The orange and white EL emission spectra for co- deposited PHP and ITSO EL devices with high (40%) and low (4%) ITSO concentration .

RESULTS

In order to illustrate the efficient energy transfer expected from PHP to ITSO we show the PHP emission and ITSO absorbance in Fig.2. The resulting EL spectra of the OLEDs consisting of blend with 4% and 40% ITSO in PHP, are shown in Fig. 3. For the blend with 4% ITSO, the EL emission consists of the blue light from PHP and orange light from ITSO. The combination of the two emission spectra produces efficient white emission with a relatively high external EL quantum efficiency around 1.2%. The CIE coordinate of the emitted light is calculated to be x=0.332 and y=0.338. The generation of white light is based on a self-absorption and PL emission process of ITSO, as presented above. It is very unpropable that the ITSO contributes directly to the EL emission, since in single layer devices of ITSO no light emission was observed. In addition ITSO is subject to strong concentration quenching as demonstrated in Figure 4. In Figure 4 we show the PL decay behavior of the spectrally integrated emission of PHP/ITSO layers with increasing ITSO concentration. It is clearly seen that the pure PHP and the low concentration ITSO/PHP layer show a decay time of about 0.9 ns while for higher ITSO concentrations we observe a faster decay, i.e. lower quantum yield.

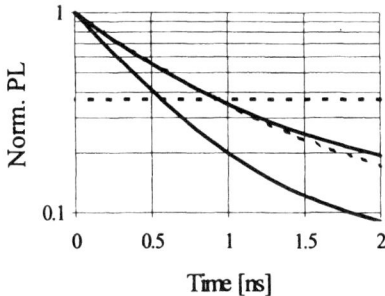

Fig. 4: Decay of the integrated PL emission of PHP/ITSO layers. The PL becomes increasingy faster as the ITSO concentration increases. The curves correspond to pure PHP, white PL emission of a PHP/ITSO layer and 40 % ITSO; excitation was at 350 nm; the horizontal dotted line indicates 1/e, i.e. the lifetime for monoexponential decay.

For concentration of ITSO above 40%, the blue emission is completely absorbed by ITSO and therefore completely internally converted into orange light. A similar color conversion is also observed in the PL of the PHP/ITSO blend. The PL external quantum efficiencies for 4% and 40% ITSO blends are found to be 27% and 4%, respectively. The pure PHP film has high PL efficiency of 30% [20], whereas pure ITSO has a PL efficiency of about 7%. In solid solution this value slightly increases. The reduced efficiency for higher ITSO concentrations is therefore due to concentration quenching of the ITSO emission within the coevaporated layer. We want to emphasize that the mechanism of the white light generation is not due to internal excitation (and charge transfer), which was observed in other white OLEDs (compare [7]).

Fig . 5. shows the typical I-V-electric field characteristics of the white OLEDs with 4% and the orange OLED with 40% ITSO in the blend. The onset of the light emission for this white OLED occurs for electric fields in the range of 1 MV/cm. This onset value and also the external EL quantum efficiency are only slightly lower than that observed for pure PHP OLEDs. In contrast the EL efficiency and stability of the OLEDs strongly decreases for increasing ITSO concentration in the blend, whereas the onset electric field increases. Also the magnitude of the current passing through the device at a certain voltage (i.e. field) decreases with increasing ITSO concentration in the blend. This shows, that the charge transport properties of ITSO are inferior compared to PHP. Therefore it is not surprising that for pure ITSO OLEDs no light emission was observed.

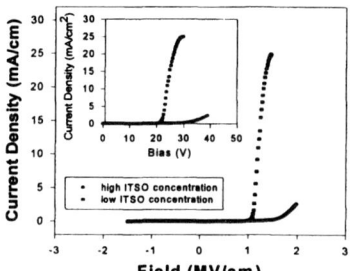

Fig 5: Comparison of the I-V-electric fieldcharacteristics of single PHP device with low and high ITSO concentration EL devices.

In order to investigate the photoexcitations present in ITSO we also have performed photoinduced absorption experiments which probe the absorption from excited states in the oligomer layer - for experimental details see ref. [19]. The results are shown in Figure 6 together with the absorption of the ITSO. Judging from the spectral signature of a narrow (≈ 0.1 eV) peak centered at 1.33 eV and a broader feature located at about 1.6 eV we can conclude that either (1) a triplet at 1.33 eV with vibronic progressions at higher energies is observed or (2) both triplets at 1.33 eV and polarons at 1.6 eV are detected. Chemical doping experiments will clarify this question.

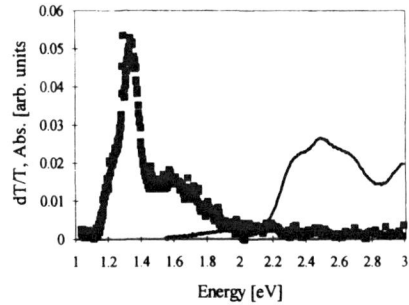

Fig 6: Photoinduced absorption (squares) and absorption (solid line) of an ITSO layer .

CONCLUSION

Bright, efficient white light emitting devices have been successfully fabricated from blend single layer of PHP and ITSO oligomers by thermal co-evaporation. Efficient white emission light is generated for low ITSO concentration due to the combination of blue and orange emitted light, and covered the wide range of visible spectrum with high external EL quantum efficiency around 1.25%.

ACKNOWLEDGEMENT

This work was supported by the Österreichische Nationalbank (Project Nr. 6433) and the SFB Elektroaktive Stoffe.

REFERENCES

1. W.Graupner, G.Grem, F.Meghdadi, C.Paar, G.Leising, Mol. Cryst. Liq. Cyst., 265, 54 (1994).
2. M. Era, T.Tsutsui, and S.Saito, Appl. Phys. Lett., 67, 2436 (1995).
3. F. Meghdadi, G.Leising, W.Fischer, F.Stelzer, Synth. Metals, 76, 113 (1996).
4. S.Tasch, C.Brandstätter, F.Meghdadi, G.Leising, L.Athouel, G.Froyer, Adv.Mat., 9, 33 (1997).
5. G.Leising, S.Tasch, F.Meghdadi, L.Athouel, G.Froyer, U.Scherf, Synth. Met.,81 185 (1996).
6. F.Meghdadi, E.Eder, F.Stelzer, G.Leising, Conference Proceeding ANTEC 97, 2, 1291 (1997).
7. S.Tasch, J.W.E.List, O.Ekström, W.Graupner, P.Schlichting , U.Rohr, Y.Geerts, U.Scherf, K.Müllen, G.Leising, Appl. Phys. Lett., 71, 2883 (1997)
8. G.E.Jonson, K.M.McGrane, SPIE, San Jose, California, (1993).
9. J.Kido, K.Hongawa, K.Okuyama, K.Nagai, Appl. Phys. Lett., 81, 564 (1994).
10. M.Berggren, G.Gustafsson, O.Inganäs, M.R.Anderson, T.Hjertberg, O.Wennerström, J. Appl. Phys., 76, 7530 (1994).
11. J.Kido, M.Kaimura, K.Nagai, Science, 267, 1332 (1995).

12. J.Kido, W.Ikeda, M.Kaimura, K.Nagai, Jpn. J. Appl. Phys., **35**, L394 (1996).
13. M.Matsura, H.Tokailin, M.Eida, C.Hosokawa, Y.Hyronaka, K.Kosumoto, Proc. Asia Display, **95,** 269 (1995).
14. C.C.Wu, J.C.Strum, R.A.Register, Appl.Phys.Letter, **69**, 3117(1996).
15. R.H.Jordan, A.Dodabalapur, M.Strukelj, T.M.Miller, Appl. Phys. Lett., **68**, 1192.(1996).
16. P. Kovacic and R.M. Lange, J.Org.Chem., **29**, 2416 (1964).
17. G.Froyer, Y.Plous, E.Dall´arche, C.Chevrot, A.Sivoe, Fr.Patent No. 9008091 (1990).
18. M.P.Cava, N.Pollack, O.A.Mamer, J.J.Mitchell, J. Org. Chem., **36**, 3932 (1971).
19. W.Graupner, S.Eder, K.Petritsch, G.Leising, Synth. Met., **84**, 507 (1997).
20. J.Stampfl, S.Tasch, G.Leizing, Synth. Met., **71**, 2125 (1995).

PLASTIC MULTILAYERED MOLECULAR ORGANIC LIGHT EMITTING DIODES

George M. Daly,[†] Hideyuki Murata,[‡] Charles D. Merritt and Zakya H. Kafafi[*]
United States Naval Research Laboratory
Washington, DC 20375

Hiroshi Inada and Yasuhiko Shirota
Department of Applied Chemistry, Faculty of Engineering, Osaka University,
Yamadaoka, Suita, Osaka 565, Japan

ABSTRACT

Enhanced performance has been observed for plastic molecular organic light emitting diodes (MOLEDs) consisting of two to four organic layers sequentially vacuum vapor deposited onto patterned indium-tin oxide (ITO) on polyester films. For all device structures studied, the performance of plastic diodes is comparable to or better than their analogs on glass substrates. At 100 A/m², a luminous power efficiency of 4.4 lm/W and external quantum yield of 2.7% are measured for a device structure consisting of two hole transport layers, a doped emitting layer and an electron transport layer on a polyester substrate. The same device made on a silica substrate has a luminous power efficiency of 3.5 lm/W and external quantum yield of 2.3%. Electrical and optical performance for comparable device structures has been characterized by current-voltage-luminance measurements and electroluminescence spectra collected normal to the emitting surface. In addition, an integrating sphere was used to collect the total light emitted and to determine the optical output coupling on glass versus plastic substrates.

INTRODUCTION

Emissive flat panel displays (FPDs) fabricated on a plastic rather than a glass back-plane represent the ultimate achievement in thin, lightweight devices with the added advantage that they are flexible and less costly to manufacture than their glass counterparts. Plastic FPDs are particularly suited to markets where flexibility and/or safety issues are important such as military and commercial head mounted and hand-held gear, and automobile and entertainment displays. Recently, it has been demonstrated that light emitting diodes (LEDs) made from polymer[1,2] (PLEDs) and molecular organic[3-5] (MOLEDs) materials can be fabricated on plastic substrates. The realization of flexible displays is possible through low temperature deposition processes for transparent indium-tin-oxide (ITO) electrodes[6] and the use of 'soft' organic materials as active device components. Amorphous organic films can be deposited onto any substrate, with no requirement for lattice matching or substrate orientation. Light emitting diodes made from molecular organic thin films have shown higher efficiencies compared to PLEDs. This is due largely to more efficient luminescent molecules and in some cases better carrier transport

---

[†] American Society for Engineering Education Postdoctoral Fellow
[‡] Sachs Freeman Associates, Inc., Largo MD 20774
[*] Author to whom correspondence should be addressed.

properties. In the present study, we evaluate the performance of MOLEDs fabricated on ITO/polyester (PET) and compare them to their analogs on ITO/silica (glass) substrates.

EXPERIMENTAL SECTION

Four MOLED devices (2 mm x 2 mm) per substrate are made in a high vacuum system (base pressure $\sim 5 \times 10^{-8}$ Torr) equipped with thermal evaporation sources. Up to 56 devices can be fabricated in a single deposition sequence. Deposition rates of the organic layers, monitored by quartz crystal microbalances, are typically in the range of 2-4 nm/min. The cathode materials, Mg:Ag (12-14:1 by weight), are codeposited through a shadow mask at a rate of 4 nm/min. Voltage-current-luminance measurements are made in a dry nitrogen atmosphere inside a glove box using a Keithley Model 236 Source Measure Unit and a Minolta LS-110 Luminance Meter. The MOLED light output, collected by a fiber optic coupled to a monochromator/CCD setup, is used to measure the electroluminescence spectrum. The same collection fiber can also be connected to an integrating sphere for measurements of the total light emission of the MOLED.

Four MOLED device configurations were made for this study. A typical control device consists of a 50 nm TPD (4,4'-bis(3-methylphenylphenylamino)biphenyl) hole transport layer (HTL) and a 50 nm Alq$_3$ (tris(8-quinolinolato)aluminum) electron transport (ETL)/emitting (EML) layer heterojunction sandwiched between an ITO anode and a magnesium-silver (MgAg) alloy cathode (Figure 1, device A). Device B incorporates a double HTL consisting of 40 nm of the starburst compound m-MTDATA (4, 4', 4''-tris(3-methylphenylphenylamino)-triphenylamine)[7] and 10 nm of TPD with the Alq$_3$ layer (Figure 1, device B). Devices A' and B' are the same as devices A and B, respectively, with the exception that the 50 nm Alq$_3$ film is replaced with a 25 nm 1% molar diethylquinacridone (DEQ) in Alq$_3$: 25 nm Alq$_3$ bilayer (Figure 1, devices A' and B').

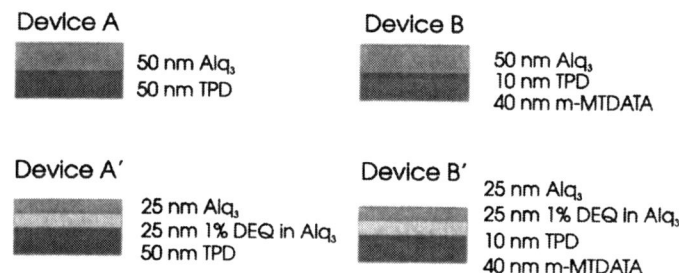

Figure 1 - Device configurations used in this study. A $\equiv$ 50 nm TPD : 50 nm Alq$_3$, B $\equiv$ 40 nm m-MTDATA : 10 nm TPD : 50 nm Alq$_3$, A' $\equiv$ 50 nm TPD : 25 nm 1% molar DEQ in Alq$_3$: 25 nm Alq$_3$, B' $\equiv$ 40 nm m-MTDATA: 10 nm TPD : 25 nm 1% molar DEQ in Alq$_3$: 25 nm Alq$_3$

RESULTS AND DISCUSSION

ITO and Substrate Characterization

Table I shows a summary of selected properties for the ITO/PET and ITO/glass substrates. Using Atomic Force Microscopy (AFM), we have shown that the ITO on PET surface is considerably rougher than the ITO on glass.[5] The ITO/glass substrates have a grain diameter of 50-75 nm compared to a grain diameter of 150-250 nm for the ITO/PET substrates. In the

ITO/PET substrates typical peak-to-valley heights are on the order of 20 nm. Figure 2 shows the x-ray diffraction (XRD) spectra measured for the ITO substrates. The ITO on glass is polycrystalline with the predominant reflection along the (400) plane. No characteristic peaks for ITO are observed in the XRD spectrum for the PET substrates indicating that the ITO is amorphous. The intense peak observed (Figure 2b) at $2\theta \approx 26°$ is assigned to PET.

Table I - ITO and Substrate Properties

	ITO/Glass	**ITO/PET**	**Method**
ITO Sheet Resistance	< 10 Ω /sq.	60 Ω /sq.	Manufacturer
ITO Thickness	160 nm	180 nm	Profilometry
ITO Grain Size	50-75 nm	150-250 nm	AFM
ITO Roughness (RMS)	3.7 nm	5.7 nm	AFM
ITO Refractive Index	2.03	1.97	Ellipsometry
Substrate Refractive Index	1.46	1.65	Near IR
Substrate Thickness	1000 μm	175 μm	Calipers
ITO/Substrate Transmittance (@ 550 nm)	~90%	~70%	UV-VIS

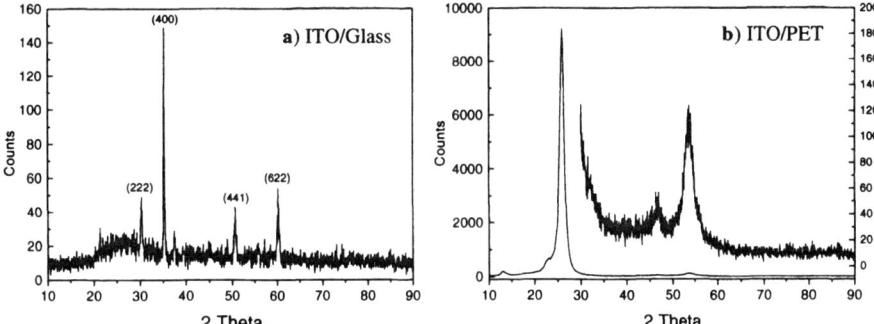

Figure 2 - XRD Spectra of a) ITO/Glass and b) ITO/PET substrates.

Refractive indices were determined using cylindrical ellipsometry (Gartner Inc.) and interference fringe spacings observed in the near infrared. In determining the index of refraction of ITO via ellipsometry, the circularly polarized incident beam was aligned along the edge of the substrate to remove the contribution from the back reflection of the glass substrate. In the measurement, both the thickness and refractive index are determined from a two parameter fit of the degree of rotation of the polarization and the phase difference of the reflected beam. Predicted thicknesses were in agreement with thicknesses measured with a scanning profilometer. The ITO on glass showed a slightly higher index of refraction compared to that of ITO on PET. Overall, the properties of ITO on glass are superior to those of ITO on PET with a lower sheet resistance and greater optical transparency in the visible region.

Device Optimization

Figure 3a shows the current density-luminance-voltage (JLV) responses for devices incorporating a *m*-MTDATA/TPD double HTL (device B). For devices (A) with a single TPD

HTL a current density of 100 A/m^2 is produced at ~8.8 V while in devices incorporating a double HTL (B), the same current density is achieved at a voltage of ~7.5 V. The reduction in driving voltage is attributed to the increased efficiency of hole injection from ITO into the m-MTDATA layer due to the smaller energy barrier between the work function of ITO and the ionization potential of m-MTDATA.[8]

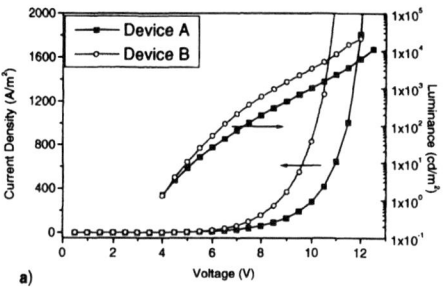 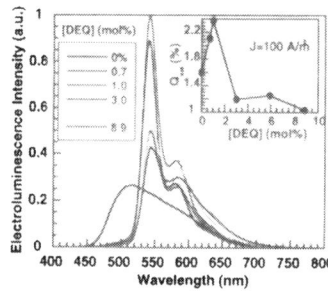

a)   b)

Figure 3 - a) Current density versus voltage for device A and B on ITO/glass. b) Electroluminescence spectra for devices A (0% DEQ) and A' (0.7-8.9% DEQ). Inset shows the dependence of external quantum efficiency on DEQ concentration (mol %).

Figure 3b shows the electroluminescence (EL) spectra for devices with DEQ doped into the Alq$_3$ emitter layer (device A'). In the absence of DEQ doping the EL spectrum is similar to the observed photoluminescence (PL) spectrum of Alq$_3$. With DEQ doping, the EL spectrum is characteristic of the PL spectrum for liquid solutions of DEQ indicating complete energy transfer from Alq$_3$ to DEQ. The inset to Figure 3b shows that the device external quantum efficiency is maximized at near 1% DEQ doping. Quenching of electroluminescence occurs at higher DEQ doping due to molecular aggregates formed by plane to plane stretching.[9]

ITO/Glass versus ITO/PET Device Performance

Figure 4a shows a plot of current density (J) versus voltage (V) for device A on both substrate types. The similar JV behavior for device A on glass and plastic suggests that device electrical performance is not affected by the differences in the ITO sheet resistance or surface morphology. Device performance, in regards to ITO, depends upon the ease of hole injection into the TPD HTL and should be more sensitive to surface states at the ITO/TPD interface. Both ITO substrate surfaces are oxygen plasma ashed prior to deposition, which eliminates residual surface impurities and increases the ITO work function, thus lowering the energy barrier between ITO and TPD. Figure 4a shows a plot of luminance (L) versus current density (J) for device A. This plot shows that the ratio of the luminance to the input current density is greater for device A fabricated on an ITO/PET substrate compared to its analog on glass. This result is surprising given that the light is attenuated more by the ITO/PET substrates (Table I). Similar results have been obtained for all device configurations (Figure 1) in this study. For the optimized device structure (B') on ITO/PET, we report an average luminous power efficiency of 4.4 lm/W and an external quantum efficiency of 2.7% (at 100 A/m^2) compared to 3.5 lm/W and 2.3% for device B' on ITO/glass.

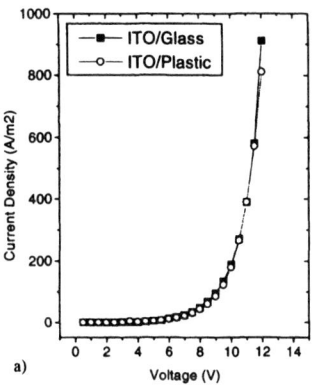 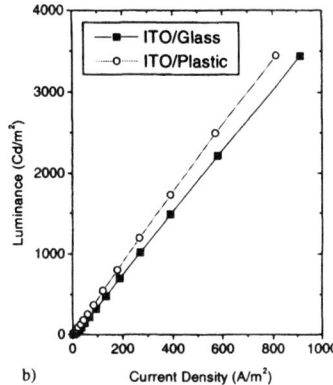

Figure 4 - a) Current density (A/m²) versus voltage for device A. b) Luminance (cd/m²) versus current density for device A.

Evidence for Optical Interference and Waveguiding

Figure 5a compares the photopic eye response curve and the measured electroluminescence (EL) spectra for device structure A on glass and plastic substrates collected normal to the MOLED surface. The Alq_3 EL spectrum for devices on ITO/PET exhibits a red shift (~ 20 nm) compared to that on ITO/glass, counter to that expected based on the transmittance curves[5] for the respective substrates. The graphically determined (Figure 5) degree of overlap between the Alq_3 EL spectra and the photopic eye response curve predicts a 14% enhancement in the measured luminance for Alq_3 based MOLEDs made on ITO/PET. This enhancement is significant but not enough to account for the overall increase in the observed luminous efficiency. This result suggests that optical interference effects, such as waveguiding, may play an important role in the device luminous power efficiency.

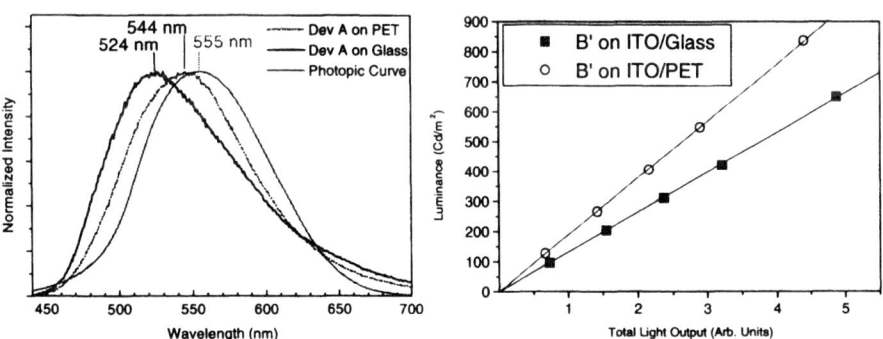

Figure 5 - a) Electroluminescence (EL) spectra of Alq_3 for device A on ITO/glass (524 nm peak max) and ITO/PET (545 nm peak max) compared to the photopic eye response curve (555 nm peak max). b) Plot of luminance collected normal to the device surface versus the total light output measured from the device in an integrating sphere.

To investigate waveguiding phenomena, radiant power measurements have been carried out using an integrating sphere coupled to a photodiode and compared to the luminance measured normal to the MOLED surface. The measured current from the photodiode coupled to the integrating sphere is proportional the total light emitted from the device. Using the same device at a fixed current density, the ratio of the luminance to the photodiode current provides a measure of the amount of light that is forward scattered to the total amount of light emitted from the device. Figure 5b shows a plot of the ratio of the forward scattered light to the total light emitted from device B' operated with increased current density for both substrate types. The slope (light output coupling ratio) of each respective line in Figure 5b gives the relative ratio of light forward scattered to the total light emitted from all sides of the diode. The greater optical output coupling ratio for the ITO/plastic devices indicates that more light (~ 40%) is forward scattered than in the ITO/glass devices. Since the calculated emission patterns for MOLEDs shows a maximum emission intensity at a viewing angle of 0° (normal to the MOLED surface) and then decreases as the viewing angle is increased,[10] the total light output of a MOLED can be approximated by the light emitted normal to the MOLED surface and the light that is waveguided. The ratio of the slopes indicates that more light is forward scattered in the PET devices than their glass counterparts. The enhanced performance in ITO/PET MOLEDs can be attributed to decreased waveguiding in the PET devices, which is not surprising given the morphology of the polyester surfaces. The rougher ITO and PET (compared to glass) surfaces result in increased light scattering and reduced waveguiding.

CONCLUSIONS

In conclusion, we have fabricated optimized molecular organic light emitting diodes on ITO/PET substrates that demonstrate enhanced performance when compared to their ITO/glass analogs. The enhanced luminous power efficiency of the devices fabricated on ITO/PET substrates has been attributed to optical interference and waveguiding effects.

ACKNOWLEDGMENTS

We thank Dr. J. M. Pond for use of patterning facilities, Dr. D. B. Chrisey for use of the x-ray diffractometer and Dr. C. Dulcey for use of the ellipsometer, all at NRL. We also thank the Office of Naval Research (ONR) and the Defense Advanced Research Projects Agency (DARPA) for partial funding for this work.

REFERENCES

[1] G. Gustaffson, G. M. Treacy, Y. Cao, F. Klavertter, N. Colaneri and A. J. Heeger, Synth. Metals 57, 4123 (1993).
[2] P. Yam, Sci. Am. 273, 83 (1995).
[3] G. Gu, P. E. Burrows, S. Venkatesh and S. R. Forrest, Optics Lett. 22, 172 (1997).
[4] G. Gu, Z. Shen, P. E. Burrows, and S. R. Forrest, Adv. Mater. 9, 725 (1997).
[5] G. M. Daly, H. Murata, C. D. Merrit, Z. H. Kafafi, H. Inada and Y. Shirota, SPIE Proc. 3148, 393 (1997).
[6] G. L. Harding and B. Window, Solar Energy Mater. 20, 367 (1990).
[7] Y. Shirota, Y. Kuwabara, H. Inada, T. Wakimoto, H. Nakada, Y. Yonemoto, S. Kawami and K. Imal, Appl. Phys. Lett. 65, 807 (1994).
[8] Y. Hamada, T. Sano, K. Shibata and K. Kuroki, Jpn. J. Appl. Phys. 34, L824 (1995).
[9] H. Murata, C. D. Merritt and Z. H. Kafafi in Science and Technology of Polymers and Advanced Materials, (Plenum Press, New York, 1998), p. xx.
[10] S. Saito, T. Tsutsui, M. Era, N. Takada, C. Adachi, Y. Hamada and T. Wakeo, SPIE Proc. 1910, 212 (1993).

ADVANCED PHOTOREFRACTIVE AND LIGHT-EMITTING ORGANIC MATERIALS

N. PEYGHAMBARIAN, B. KIPPELEN
Optical Sciences Center, The University of Arizona, Tucson, AZ 85721

ABSTRACT

The tailoring of organic molecules and polymers has enabled the recent development of multifunctional materials such as photorefractive polymers and organic electroluminescent materials. This paper presents recent advances in both areas.

INTRODUCTION

Photonic technologies are at the base of current high bandwidth telecommunication systems and are being widely used for storage applications. The development of high-performance optical materials plays a major role in enabling future optical technologies. Organic materials such as polymers combine ease of processing, low cost, light weight, and high performance. They look, therefore, particularly promising for optoelectronic applications. Due to the structural flexibility of organics, multifunctional materials can be designed and synthesized. Among these multifunctional materials, photorefractive [1] and light-emitting polymers [2-3] have emerged recently.

Organic photorefractive materials combine photosensitivity, transport, and electro-optic properties [4][5]. They are used to store and retrieve optical information in real-time. The encoding is achieved through reversible refractive index changes, leading to an efficient optical reconstruction of the stored information [6]. These encoding properties make them suitable for many optical applications including holographic storage [7], optical correlation [8], time-average interferometry [9], phase conjugation [10], time-gating holography [11], and image processing.

Electroluminescent polymers combine transport and light-emitting properties. In contrast to organic molecules dissolved in a solvent (like in dye lasers) or in an inert polymer binder or sol-gel (like in solid state dye lasers), conjugated polymers do not exhibit any strong quenching of their light-emitting properties at high concentration or even in neat films. Due to their semiconducting properties, they have transport properties that make them suitable for electroluminescent devices. In contrast to inorganic semiconductors, the emission properties of organics can be tuned through the entire visible part of the spectrum. These materials are therefore suitable for full-color displays. With the recent evidence of stimulated emission [12-16] and optical gain in conjugated polymers and organic molecules, electrically injected organic lasers are under investigation.

This paper is divided into two sections. In the first part, we will report some recent advances in two different classes of organic photorefractive materials: polymers and polymer dispersed liquid crystals [17]. In polymers, the dynamic range could be improved by a factor of four compared with previous materials by using chromophores that were designed according to a new design rationale [18-20]. New polymers with high dynamic range in the near IR are discussed. We also report on new guest/host polymer composites that are thermally stable and where the dopant chromophore has a triple functionality. Then we describe a new class of organic photorefractive materials: photorefractive polymer dispersed liquid crystals (PDLC) [17].

In the second part of this paper, we report on several optically pumped laser structures using a conjugated polymer in devices where feedback is provided in different ways.

Mat. Res. Soc. Symp. Proc. Vol. 488 © 1998 Materials Research Society

PHOTOREFRACTIVE POLYMERS

New polymers with improved dynamic range and near IR spectral sensitivity

In low glass transition temperature (T_g) photorefractive polymer composites, both electro-optic and orientational birefringence contribute to the total refractive index modulation [18,6]. For such materials, a figure of merit *FOM* for the design of chromophores for photorefractive applications is [18-20]:

$$FOM = A(T)\,\Delta\alpha\,\mu^2 + \beta\mu \qquad (1)$$

where $A(T) = 2/9kT$ is a numerical scaling factor. Recently, we used the BLA (Bond Length Alternation) model to define new design guidelines of chromophores for photorefractive applications [20,21]. Within that model, molecular quantities such as the dipole moment, the polarizability, as well as the hyperpolarizability can be correlated with the degree of ground-state polarization [22,23]. Donor-acceptor substituted molecules with a π-electron conjugation path have a ground-state structure that can be viewed as a linear combination of two limiting resonance forms: a neutral form (Fig. 1a) and a charge-separated form (Fig. 1b). The relative

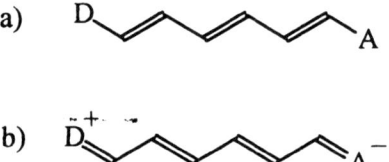

Fig. 1: Two limiting charge transfer resonance forms of a donor-acceptor polyene molecule: the neutral form a) and the charge separated form b).

contribution of these two forms in the ground state can be correlated to the values of BLA or BOA (Bond Order Alternation), where BOA is the difference in the π-bond order between adjacent carbon-carbon bonds. BOA and BLA is usually varied by changing the strength of the donor and acceptor substituents, or by changing the properties of the surrounding medium such as its polarity. In model calculations, an internal field can be used to vary BLA or BOA. The calculated values of the dipole moment μ, the linear polarizability α_{zz} along the axis of the molecule, and the first hyperpolarizability β as a function of applied field are plotted in Fig. 2 (a)-(c) for the molecule $(CH_3)_2N-(CH=CH)_4CHO$. The calculations indicate that that $\alpha_{zz} \gg \alpha_{xx}, \alpha_{yy}$, thus $\Delta\alpha \approx \alpha_{zz}$. From these results, the figure of merit *FOM* can be calculated as a function of BOA. Previous studies [19] suggested that an optimal *FOM* was obtained at the point where β vanishes and where the polarizability anisotropy is maximized. In contrast, our study [20] that takes into account the linear increase of μ as a function of BOA over the region considered, indicates that *FOM* is optimized for BOA where β is roughly maximized in amplitude with a negative sign as clearly shown in Fig. 2d.

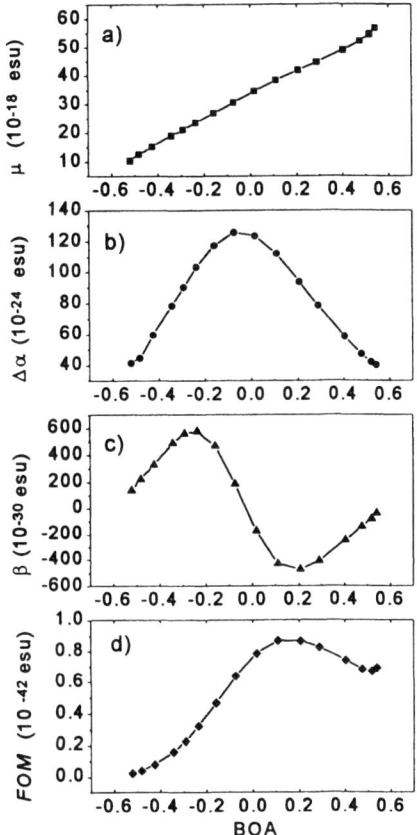

Fig. 2: The dipole moment (a), polarizability anisotropy Δα (b), first hyperpolarizability β (c), and figure of merit FOM (d), as a function of bond order alternation for the molecule (CH₃)₂ N(CH=CH)₄ CHO.

To explore this chromophore design rationale, we have focused our studies on linear molecules such as polyenes [24], rather than on chromophores that contain benzene rings such as DMNPAA for instance (DMNPAA was used as the dopant chromophore in the first photorefractive polymer to show maximum internal diffraction efficiency). This is because polyenes exhibit a considerable charge transfer that is confined along the quasi one-dimensional π-conjugated bridge, providing a large $\Delta\alpha$. In addition, polyenes can have an important charge separation in the ground state that provides large dipole moment μ. To comply with the practical requirements of a well-performing PR polymer, we synthesized the polyene molecule DHADC-MPN (2-N,N-dihexylamino-7-dicyanomethylidenyl-3,4,5,6,10-pentahydronaphthalene). The hexyl groups help impart solubility to this highly dipolar molecule. The incorporation of the polyene into the fused ring systems enhances thermal and photochemical stability. The molecule was used as a dopant molecule in mixtures of PVK and ECZ. Sensitivity in the visible (633 nm) was provided by (TNF). By using the sensitizer (2,4,7-trinitro-9-fluorenylidene)malonitrile (TNFDM), the spectral response of the photosensitivity could be extended to the near infrared (λ = 830 nm). The spectral response of the new sensitizer in PVK/ECZ is shown in Fig. 3. The photorefractive properties, in particular the dynamic range, or Δn, were tested by four-wave mixing experiments in the tilted geometry described in [6]. The

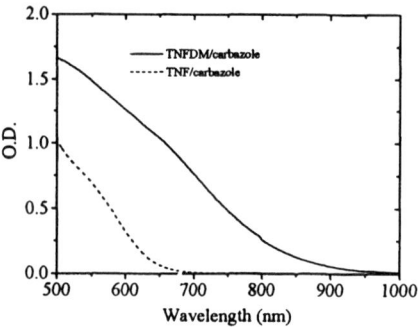

Fig. 3: Linear optical absorption spectrum of the charge transfer complexes of TNF/carbazole in PMMA doped with ECZ, and TNFDM/carbazole in PVK.

thickness of all the samples was 105 μm and the polymer was sandwiched between two transparent indium-tin oxide (ITO) electrodes. Two-beam coupling, i.e., energy exchange between the two interfering laser beams, was observed in all the photorefractive composites presented here and confirmed the photorefractive nature of the optical encoding. The normalized diffraction efficiency (corrected for absorption and reflection losses and for small electro-absorption effects) of a composite DHADC-MPN:PVK:ECZ:TNF (40:39:19:2 wt. %) as a function of applied field is shown in Fig. 4a. For comparison, we also plotted the normalized diffraction efficiency of a composite DMNPAA:PVK:ECZ:TNF (40:39:19:2 wt. %). In both samples the oscillatory behavior of the diffraction efficiency η is in agreement with Kogelnik's coupled-wave theory [25]:

$$\eta \propto \sin^2 \left[\frac{G \pi \Delta nd}{\lambda} \right] \tag{2}$$

where G is a geometrical factor that depends on the polarization of the beams and the experimental geometry [18,5], Δn is the refractive index modulation amplitude, d the thickness of the grating, and λ the wavelength of the light. The composite based on DHADC-MPN exhibits a first maximum at an applied field $E_{\pi/2} = 30$ V/μm, to be compared with $E_{\pi/2} = 65$ V/μm for the DMNPAA-based composite (Fig. 4a). At this first maximum of the diffraction efficiency, the value of the argument of the *sin* function in Kogelnik's expression for the diffraction efficiency (Eq. 2) is $\pi/2$. This reduction in $E_{\pi/2}$ by a factor larger than 2 is indicative of the large efficiency of DHADC-MPN based polymers. Calculations of Δn using Kogelnik's expression (Eq. 2) show that Δn in the DHADC-MPN composite is over four times higher than in a DMNPAA composite with the same doping level (wt. %). When using TNFDM as a sensitizer, the spectral sensitivity is extended to the near infrared. In the normalized diffraction efficiency of a composite of DHADC-MPN:PVK:ECZ:TNFDM (25:49:24:2 wt. %) (Fig. 4b) maximum diffraction is observed at $E_{\pi/2} = 59$ V/μm at 830 nm. The real diffraction efficiency η_{max} at $E_{\pi/2}$ is 30% for that composite. This value can be further optimized by reducing the sensitizer concentration, that is reducing the absorption of the sample at 830 nm, and can reach $\eta_{max} = 74\%$ in a composite of DHADC-MPN:PVK:ECZ:TNFDM (25:49:25:1 wt. %).

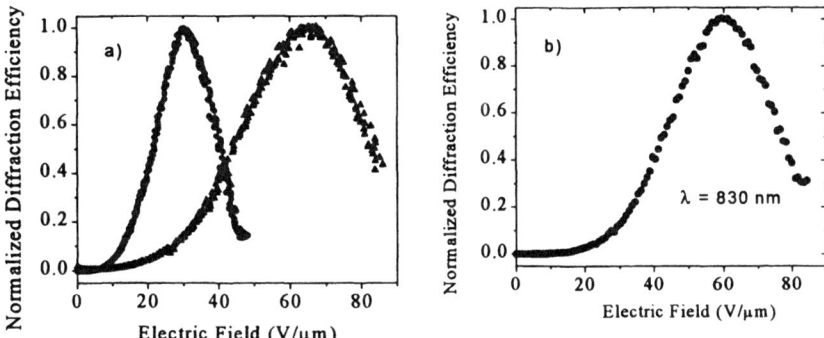

Fig. 4: (a) Normalized diffraction efficiency versus field measured at 633 nm in a composite DMNPAA:PVK:ECZ:TNF (40:39:19:2 wt. %) (circles) and in a composite DHADC-MPN:PVK:ECZ:TNF (40:39:19:2 wt. %) (squares). (b) Normalized diffraction efficiency versus field measured at 830 nm in a composite DHADC-MPN:PVK:ECZ:TNFDM (25:49:24:2 wt. %).

These materials are, to the best of our knowledge, the first efficient photorefractive polymers in the near infrared. These results suggest that our proposed design rationale [20,21], that is to synthesize molecules that combine high μ and high $\Delta\alpha$, provides an efficient route to optimize the performance of low glass transition temperature photorefractive polymers. The photorefractive polymer composites with high Δn and spectral sensitivity in the near infrared developed here offer new opportunities for numerous photonic applications. As a sensitive holographic recording medium with high dynamic range, we used them for instance, for imaging through scattering media [11,24]. Of particular importance is their compatibility with the emission of high quality GaAs semiconductor laser diodes and commercial solid-state femtosecond lasers, such as Ti:Sapphire lasers. More importantly, their near infrared spectral response is compatible with the transparency of biological tissues (700-900 nm) and therefore this imaging technique can be extended to medical imaging.

Thermally stable guest/host photorefractive polymers

Early highly efficient guest/host photorefractive polymer composites had limited phase stability and consequently shelf lifetimes of a few months at room temperature [6]. By tailoring the structure of the new chromophore DHADC-MPN, we could improve the shelf lifetime by two orders of magnitude compared with DMNPAA-based materials [24]. This stability was studied by performing accelerated aging tests at high temperature using the optical scattering set-up described in [26]. At 85° C, samples of DHADC-MPN:PVK:ECZ with composition 39:41:20 wt. % have a lifetime of 220 min before phase separation gradually starts to degrade the optical quality of the samples. From these results, a shelf lifetime of several years can be deduced at room temperature.

The finite phase stability of polymer composites can be related to the large difference in polarity between the dopant molecule (the guest) and the matrix (the host). This is particularly the case for PVK-based materials where the relatively apolar PVK matrix is doped with chromophores that have a high dipole moment. To improve the compatibility between guest and host, we doped the chromophore DHADC-MPN is the more polar matrix of poly(methyl methacrylate-co-tricyclodecyl methacrylate-co-N-cyclohexyl maleimide-co-benzyl methacrylate) (PTCB). PTCB is a birefringent-free acrylic resin from Hitachi Chemical co. Diphenylisophthalate (DIP) was used as a plasticizer to ensure free orientational mobility of the chromophore at room temperature. Composites of DHADC-MPN:PTCB:DIP did not exhibit any signs of phase separation even after 6 days at 70 °C and can be considered as thermally stable. After solvent evaporation, PVK-based composites are opaque and must be heated above the melting point of the dopant chromophore and rapidly cooled down to room temperature to insure optical clarity. In contrast, PTCB-based samples are optically transparent after the solvent used to combine the different compounds is evaporated.

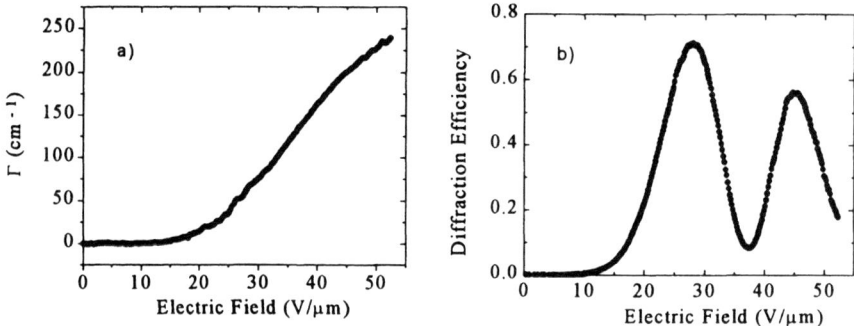

Fig. 5: a) Gain coefficient and b) diffraction efficiency versus field measured in DHADC-MPN:PTCB:DIP:TNFDM with composition 37.6:49.7:12.5:0.18 wt. %

We performed four-wave mixing and two-beam coupling experiments in thermally stable composites of DHADC-MPN:PTCB:DIP in the tilted configuration described previously [6] ($2\theta = 20°$ and $\psi = 60°$). PTCB was purified three times by precipitation in ethanol from a chloroform solution. The compounds were dissolved in chloroform and filtered. The solvent was then evaporated on a rotary evaporator and the mixture was dried for 2 hours at 75 °C and 1000 Pa. The composite was then homogenized by heating it to 150 °C for 10-15 min. Finally the composite was sandwiched between two ITO-coated glass slides and calibrated glass beads of 105 μm diameter were used to control the thickness. The laser source was a HeNe Laser (633 nm). With s-polarized writing beams and a p-polarized reading beam, samples of DHADC-MPN:PTCB:DIP with composition 37.4:50.3:12.3 wt. % showed a maximum diffraction efficiency of $\eta = 78\%$ at an applied field of $E = 36$ V/μm. A net gain coefficient of $\Gamma = 196$ cm^{-1} was observed at $E = 50$ V/μm with p-polarized beams. In these samples, the chromophore DHADC-MPN provides three different functionalities: photosensitivity, transport, and refractive index changes. Adding small amounts of TNFDM to these composites results in a broad absorption band from 600 to 1100 nm that can be attributed to the formation of a charge-transfer complex between TNFDM and DHADC-MPN. Samples of DHADC-MPN:PTCB:DIP:TNFDM with composition 37.6:49.7:12.5:0.18 wt. % exhibit a diffraction efficiency of η = 71% at an applied field of E = 28 V/μm and a net gain coefficient of Γ = 202 cm^{-1} at E = 50 V/μm (Fig. 5). To the best of our knowledge, these samples are the most efficient photorefractive polymers to date, and in addition they are thermally stable.

<u>Photorefractive polymer dispersed liquid crystals</u>

As shown in previous sections, photorefractive polymers with a low T_g, exhibit strong electric-field induced orientational effects that are responsible for their high efficiency. However, a drawback of these materials is the high electric field (30-100 V/μm) that needs to be applied in order to orient these molecules efficiently. Another obvious way to induce a molecular reorientation with an electric field involves the use of materials with a dielectric anisotropy, such as liquid crystals. In contrast to dipoles dispersed in amorphous polymers, liquid crystal molecules in mesophases can be reoriented with much lower electric fields (few V/cm). However, due to the coherence length in bulk nematic liquid crystals, the resolution of these materials is limited and their photorefractive performance is low for small values (< 5 μm) of the grating spacing [27,28].

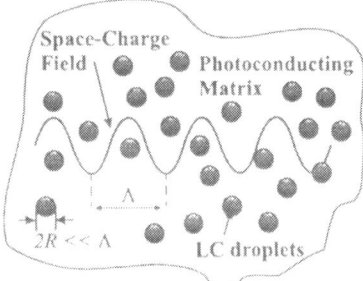

Fig. 6: Schematics and design principle of photorefractive polymer dispersed liquid crystals

To combine the high resolution of photorefractive polymers, and the high refractive index changes associated with field-induced reorientation of nematic liquid crystals, we developed photorefractive polymer dispersed liquid crystals [17]. Such materials are prepared by dispersing liquid crystal domains of almost spherical shape in a photoconducting solid organic polymer matrix. The sensitizing, photoconducting, and trapping properties necessary for the build-up of a space-charge field are provided by the polymer matrix and the refractive index changes are due to the liquid crystal droplets as illustrated in Fig. 6. In addition to the photorefractive properties, these new materials present also the characteristic field-dependent transmission properties of PDLCs [29]. For our experiments we used poly-methylmethacrylate (PMMA) for the polymer matrix, doped with ECZ for hole transport, TNF for charge generation, and the eutectic nematic mixture E49 (purchased from Merck) for the orientational properties. For this study, we prepared samples with the following composition: PMMA:E49:ECZ:TNF (45:33:21:1 wt. %). For an applied field of 22 V/μm, an external diffraction efficiency of 8% was observed in 53 μm-thick samples for a grating spacing of 4.5 μm. The corresponding internal diffraction efficiency defined as the ratio between the diffracted and transmitted beam intensities was 40%. This corresponds to a refractive index modulation amplitude of $\Delta n = 2 \times 10^{-3}$. That value is five times higher than the value of $\Delta n = 3.8 \times 10^{-4}$ measured with the same applied field in the highly efficient photorefractive polymers DMNPAA:PVK:ECZ:TNF [6] and illustrates the high efficiency of these new materials. The speed of these materials is rather slow at this stage (min.) due to the non optimized transport properties of the PMMA matrix doped with ECZ. However, due to the structural flexibility of organic photoconductors and liquid crystals, materials with further optimized properties can be expected. This new class of materials combines photorefractive properties with the field-dependent transmission properties of PDLCs, which makes them intriguing materials for new optical applications.

INTEGRATED LASER STRUCTURES BASED ON CONJUGATED POLYMERS

Conjugated polymers and organic molecules have gained increasing attention in recent years due to their promise in the fabrication of full-color emitting displays [30]. Electroluminescent devices based on thin films of evaporated organic molecules [31], conjugated polymers [2,3], oligomers [32], or molecularly doped polymers [33], have been extensively investigated because of their low operating voltage, red-green-blue output-color and tunability, light weight, high brightness [34], mechanical flexibility, and ease of fabrication. Recently, photoluminescence line narrowing and stimulated emission have been reported in a variety of conjugated polymers [12-16, 35]. Recently, we have reported photoluminescence line narrowing, stimulated emission, and gain coefficients of up to 10^4 cm^{-1} in thin solid-state films of the polymer poly(2,5-bis(2'-ethyl-hexyloxy)-1,4-phenylenevinylene) (BEH:PPV) [16]. This large light amplification per unit length allows laser action to occur in devices where feedback is provided in different ways. In this paper we investigate laser emission in three different structures: a vertical cavity surface emitting laser, and two different microring resonators.

Experiment

Films of 100 nm to 500 nm thickness of the semiconducting conjugated polymer BEH:PPV were spin-coated from xylene-solution onto different substrates. To realize a high quality vertical surface emitting resonator, dielectric mirrors were produced by electron beam deposition that contained a stack of 21 alternating quarter-wave layers of low (SiO_2) and high (TiO_2) refractive index materials. These mirrors were characterized by low losses and extremely high reflectivity (>99 %) in the wavelength region of optical gain in the polymer. By spin-coating the polymer onto one of the dielectric mirrors and properly adjusting the second mirror we produced a low loss, high Q planar microcavity of about 9 μm optical thickness (inset of Fig. 7a). The two microring resonators were obtained as follows: the first microring was formed by coating an optical fiber of 56 μm diameter with the polymer; the second ring resonator was formed by coating a Si_3N_4 waveguide ring structure with a ring diameter of 35 μm on SiO_2 with a 200 nm thick polymer film. In this devices, the light output was observed from a channel waveguide which was designed to have a small coupling (< 5 %) to the ring resonator.

To excite the polymer films in the vertical cavity laser we use an amplified CPM laser system providing intense and tunable pulses of 100 fs duration. The central wavelength of the light pulses of 555 nm was exciting BEH:PPV in its π-π* absorption band. For the waveguide ring resonator, laser emission was obtained under quasi-stationary excitation conditions using 7 ns pulses from a frequency doubled Nd:YLF laser at 524 nm. The microring fiber device was excited by the second harmonic of a Q-switched Nd:YAG laser focused by a cylindrical lens from the side of the fiber. The repetition rate was 2 ~ 10 Hz, the pulse width was ~ 10 ns, and pulse energies up to 10 μJ were used. The optical signals were detected by an optical multichannel analyzer with a spectral resolution < 0.3 nm. All experiments were carried out in vacuum and at room temperature.

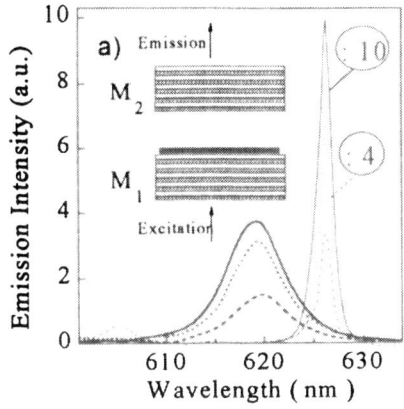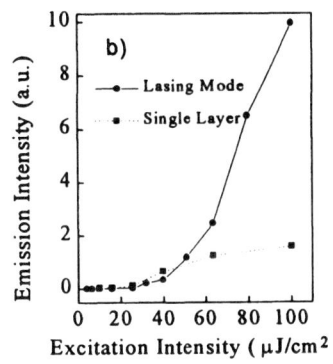

Fig. 7: a) (a) Emission spectra of a single BEH:PPV layer (thick lines) and a planar polymer cavity of 9 μm optical thickness (thin lines). The excitation intensities are 40 μJ/cm² (dashed lines), 63 μJ/cm² (dotted lines), and 100 μJ/cm² (solid lines), respectively. In the inset the structure is shown schematically. (b) Spectrally integrated single layer emission intensity and lasing cavity mode intensity as a function of pump laser fluence.

Vertical cavity surface emitting polymer laser

Fig. 7a shows the emission spectrum of a polymer in the vertical cavity and the emission of a single BEH:PPV layer without a cavity for three pump intensities. For all these intensities, the excitation is high enough to obtain optical gain and stimulated emission in the polymer layer. The emission spectrum of the polymer layer without cavity shows a narrow emission band that is typical for excitation above the threshold for stimulated emission. In the resonator configuration the emission characteristics change dramatically. Two narrow longitudinal cavity modes at 605 nm and 626 nm overlap with the polymer emission. For excitation below 40 μJ/cm² the cavity emission is much weaker than the single layer emission because even small material losses do not allow the emitted photons to leave the high finesse cavity. However, at about 50 μJ/cm² excitation intensity a clear laser threshold is reached and the gain per cavity roundtrip is larger than the losses. Above this threshold the cavity emission is much stronger than the single layer emission. Fig. 7b shows that the cavity output above the threshold grows linearly with the excitation, a characteristic behavior for laser emission. A striking evidence for laser action is the directionality of the laser emission: the laser emission is concentrated in a cone of less than 3° half-apex angle. Here, we collected the emission through various apertures and showed that the emission cone was almost diffraction-limited. A high degree of polarization than 50:1 and parallel to the exciting laser was measured [36].

Whispering gallery mode polymer devices

Microcavities are easy to integrate and have the advantage that high Q values can be obtained even in very small and controllable mode volumes in which the current densities necessary for electrical pumping can in principle be achieved. Here, we demonstrate whispering gallery modes in two different microring configurations under optical excitation: a coated fiber ring resonator [37] and a Si_3N_4 waveguide ring resonator.

The structure of the fiber microring resonator is shown in the inset of Fig. 8. This kind of structure forms a high quality cavity for whispering gallery modes [38-41]. Light output from the fiber was detected as light scattered from the side. Fig. 8 shows the spectrum of the emission for three excitation intensities. Whispering gallery modes clearly dominate the spectra above 100 kW/cm^2. The mode separation of 1.5 nm agrees well with the separation of the eigenvalues of this ring laser of 56 μm diameter. The inset of Fig. 8 shows the threshold behavior of the emission intensity as a function of excitation. For the highest intensities gain saturation sets in.

Fig. 9 shows the emission spectrum of the Si_3N_4 waveguide ring structure. The spectra show the ring cavity modes. The inset of Fig. 9 shows the spectrally integrated output as a function of excitation intensity. Under 2.2 MW/cm^2 no emission could be detected. Above the intensity threshold of 2.2 MW/cm^2 the normalized integrated emission intensity increases linearly with the excitation until saturation is observed above 4 MW/cm^2. Despite its slightly higher threshold intensity, this structure is of particular interest since it can be easily integrated and the emitted light can be coupled from the waveguide into an optical fiber.

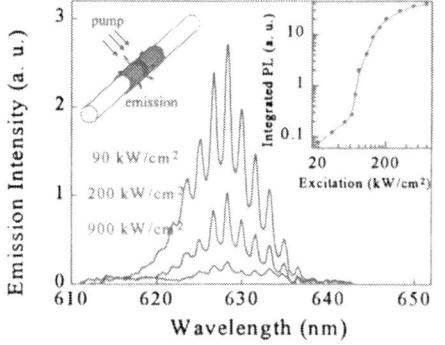

Fig. 8: Emission spectra from a polymer microring fiber for three different excitation intensities. Inset: Integrated emission intensity as a function of excitation.

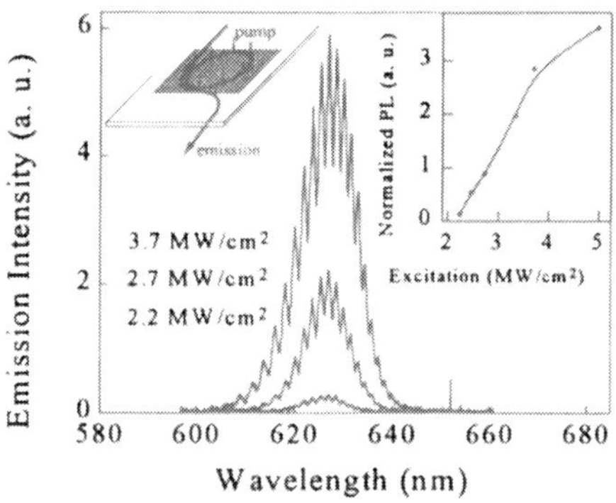

Fig. 9: Emission spectra of a waveguide ring resonator for different excitation levels. Inset: Normalized integrated output of the channel waveguide as a function of optical excitation of the polymer film.

SUMMARY

In summary, we have significantly improved the performance of photorefractive polymers in the visible and have extended their spectral sensitivity to the near infrared. In addition to high performance, PVK-based polymers doped with DHADC-MPN show shelf lifetimes of several years at room temperature. When PTCB is used as a matrix, we have shown that thermally stable polymers with net gain coefficients in excess of $\Gamma = 200$ cm^{-1} at an applied field of 50 V/μm can be obtained. We have developed a new class of organic photorefractive materials: photorefractive polymer dispersed liquid crystals. They show high refractive index changes at lower fields compared with polymers and are a promising new class of materials. In the area of light emitting organic materials, we have demonstrated laser action from a high finesse vertical cavity surface emitting structure containing a conjugated polymer. Finally, we have observed whispering gallery modes from microring devices.

ACKNOWLEDGMENTS

This work was supported by the US Office of Naval Research (ONR) through the MURI Center CAMP, by the US National Science Foundation (NSF), by the USAF Office of Scientific Research (AFOSR), by an international CNRS/NSF travel grant, and by a NATO travel grant. The authors would like to acknowledge collaboration with Drs. S. R. Marder and H. Röckel from the California Institute of Technology, Dr. A. Golemme from Università Della Calabria, Rende, Italy, their colleagues at The University of Arizona, Drs. E. Hendricks, Ch. Spiegelberg, A. Schülzgen, S. Honkanen, S. B. Mendes, Y. Kawabe, M. -F. Nabor, E. A. Mash, P. M. Allemand, Mr. J. L. Maldonado, J. Herlocker, M. M. Morrell, Prof. M. Kuwata-Gonokami from the University of Tokyo, and Prof. M. Leppihalme from VTT Electronics, Finland.

REFERENCES

1. S. Ducharme, J. C. Scott, R. J. Twieg, and W. E. Moerner, Phys. Rev. Lett. **66**, 1846 (1991).
2. J. H. Burroughes, D. D. C. Bradley, A. R. Brown, R. N. Marks, K. Mackay, R. H. Friend, P. L. Burns, and A. B. Holmes, Nature **347**, 539-541 (1990).
3. D. Braun, A. J. Heeger, Appl. Phys. Lett. **58**, 1982 (1991).
4. W. E. Moerner and S. M. Silence, Chem. Rev. **94**, 127-155 (1994).
5. B. Kippelen, K. Meerholz, and N. Peyghambarian, in Nonlinear optics of organic molecules and polymers, edited by H. S. Nalwa and S. Miyata (CRS, Boca Raton, 1997), p. 465-513.
6. K. Meerholz, B. L. Volodin, Sandalphon, B. Kippelen, and N. Peyghambarian, Nature **371**, 497 (1994).
7. P. M. Lundquist, C. Poga, R. G. DeVoe, Y. Jia, W. E. Moerner, M.-P. Bernal, H. Coufal, R. K. Grygier, J. A. Hoffnagle, C. M. Jefferson, R. M. Macfarlane, R. M. Shelby, and G. T. Sincerbox, Opt. Lett. **21**, 890-892 (1996).
8. B. L. Volodin, B. Kippelen, K. Meerholz, B. Javidi, and N. Peyghambarian, Nature **383**, 58-60 (1996).
9. B. L. Volodin, Sandalphon, K. Meerholz, B. Kippelen, N. Kukhtarev, and N. Peyghambarian, Opt. Eng. **34**, 2213-2223 (1995).
10. A. Grunnet-Jepsen, C. L. Thompson, and W. E. Moerner, Science **277**, 549-552 (1997).
11. D. D. Steele, B. L. Volodin, O. Savina, B. Kippelen, N. Peyghambarian, H. Röckel, S. R. Marder, to appear in *Opt. Lett.* (1997)
12. N. Tessler, G. J. Denton, and R. H. Friend, Nature **382**, 695-697 (1996).
13. F. Hide, B. J. Schwartz, M. A. Diaz-Garcia, and A. J. Heeger, Chem. Phys. Lett. **256**, 424-430 (1996).
14. G. J. Denton, N. Tessler, M. A. Stevens, and R. H. Friend, Adv. Mater. **9**, 547-551 (1997).
15. S. V. Frolov, W. Gellermann, M. Ozaki, K. Yoshino, and Z. V. Vardeny, Phys. Rev. Lett. **78**, 729-732 (1997).
16. Ch. Spiegelberg, A. Schülzgen, P. M. Allemand, B. Kippelen, N. Peyghambarian, to appear in Phys. Stat. Sol. (b).
17. A. Golemme, B. L. Volodin, B. Kippelen, and N. Peyghambarian, Opt. Lett. **22**, 1226-1228 (1997).
18. W. E. Moerner, S. M. Silence, F. Hache, and G. C. Bjorklund, J. Opt. Soc. Am. B **11**, 320 (1994).
19. R. Wortmann, C. Poga, R. J. Twieg, C. Geletneky, C. R. Moylan, P. M. Lundquist, R. G. DeVoe, P. M. Cotts, H. Horn, J. E. Rice, and D. M. Burland, **105**, 10637-10647 (1996).

20. B. Kippelen, F. Meyers, N. Peyghambarian, and S. R. Marder, J. Am. Chem. Soc. **119**, 4559-4560 (1997).
21. S. R. Marder, B. Kippelen, A. K.-Y. Jen, and N. Peyghambarian, Nature **388**, 845-851 (1997).
22. S. R. Marder, D. N. Beratan, and L.-T. Cheng, Science **252**, 103-106 (1991).
23. F. Meyers, S. R. Marder, B. M. Pierce, and J. L. Brédas, J. Am. Chem. Soc. **116**, 10703-10714 (1994).
24. B. Kippelen, S. R. Marder, E. Hendrickx, J. L. Maldonado, G. Guillemet, B. L. Volodin, D. D. Steele, Y. Enami, Sandalphon, Y. J. Yao, J. F. Wang, H. Röckel, L. Erskine, and N. Peyghambarian, to appear in *Science* (1998)
25. H. Kogelnik, Bell. Syst. Tech. J. **48**, 2909-2947 (1969).
26. E. Hendrickx, B. L. Volodin, D. D. Steele, J. L. Maldonado, J. F. Wang, B. Kippelen, and N. Peyghambarian, Appl. Phys. Lett. **71**, 1159-1161 (1997).
27. I. C. Khoo, H. Li, and Y. Liang, Opt. Lett. **19**, 1723-1725 (1994).
28. G. P. Wiederrecht, B. A. Yoon, and M. R. Wasielewski, Science **270**, 1794-1797 (1995).
29. P. S. Drzaic, Liquid crystal dispersions (World Scientific, Singapore, 1995).
30. S. A. Jenekhe, K. J. Wynne, Eds. Photonic and optoelectronic polymers, ACS Symposium Series 672 (ACS, Washington, DC, 1997).
31. C. W. Tang and S. A. VanSlyke, Appl. Phys. Lett. **51**, 913-915 (1987).
32. W. Tachelet, S. Jacobs, H. Ndayikengurukiye, H. J. Geise, and J. Gruner, Appl. Phys. Lett. **64**, 2364-2366 (1994).
33. J. Kido, M. Kohda, K. Okuyama, and K. Nagai, Appl. Phys. Lett. **61**, 761-763 (1992).
34. G. E. Jabbour, Y. Kawabe, S. E. Shaheen, J.-F. Wang, M. M. Morrell, B. Kippelen, and N. Peyghambarian, Appl. Phys. Lett. **71**, 1762-1764 (1997).
35. H. J. Brouwer, V. V. Krasnikov, A. Hilberer, and G. Hadziioannou, Adv. Mater. **8**, 935-937 (1996).
36. A. Schülzgen, Ch. Spiegelberg, , M. M. Morrell, S. B. Mendes, M. F. Nabor, E. A. Mash, P. M. Allemand, B. Kippelen, N. Peyghambarian, to appear in *Appl. Phys. Lett.*
37. Y. Kawabe, Ch. Spiegelberg, A. Schülzgen, M. F. Nabor, E. A. Mash, P. M. Allemand, M. Kuwata-Gonokami, K. Takeda, B. Kippelen, and Peyghambarian, to appear in *Appl. Phys. Lett.*
38. S. L. McCall, A. F. J. Levi, R. E. Slusher, S. J. Peartson, and R. A. Logan, Appl. Phys. Lett. **60**, 289 (1992).
39. J. C. Knight, H. S. T. Driver, R. J. Hutcheon, and G. N. Robertson, Opt. Lett. **17**, 1280 (1992).
40. M. Kuwata-Gonokami, R. H. Jordan, A. Dodabalapur, H. K. Katz, M. L. Schilling, R. E. Slusher, and S. Ozawa, Opt. Lett. **20**, 2093 (1995).
41. S. Frolov, M. Shkunov, Z. V. Vardeny, K. Yoshino, Phys. Rev. B, **56**, R4364-R4366 (1997).

SYNTHESIS AND ELECTROLUMINESCENCE STUDY OF A NOVEL COPOLYMER : POLY(PHENYLENE VINYLENE-CO-QUINOLINE VINYLENE)

Ramesh K. Kasim, Bela Derecskei[†], Martin Pomerantz[†] and Ronald L. Elsenbaumer
Materials Science & Engineering Program
[†]Department of Chemistry & Biochemistry
University of Texas at Arlington, Arlington, TX 76019

Abstract

High-efficiency polymer light-emitting diodes (LEDs) are often fabricated using multi-layered structures with separate carrier transport and light emission layers. Recently, we reported on the synthesis and electroluminescence (EL) characteristics of poly(2,6-quinoline vinylene) (PQV) and its potential for use as an electron transport layer in poly(phenylene vinylene) (PPV) LEDs. To take advantage of the high emission efficiency of PPV and electron accepting ability of PQV, a copolymer of PPV and PQV, poly(phenylene vinylene-co-quinoline vinylene) (PPVQV) was synthesized via the precursor polymer route and converted to the conjugated form by thermal elimination. When used as the emissive layer with Indium-Tin Oxide (ITO) and aluminum as positive and negative electrodes respectively, PPVQV emitted blue light at an onset electric field of 1.05×10^6 V/cm and emission efficiency of 0.08%. Improved efficiencies of the order of 0.15% were obtained when blends of copolymer with PPV were used in conjunction with PPV in a multi-layered structure. Along with copolymer chemical characterization data, results from EL studies on single and multi-layered devices are discussed. We also report on a simple and cost-effective chemical deposition of silver for the negative electrode in polymer LEDs.

1. INTRODUCTION

After the observation of EL in PPV, a number of conjugated homopolymers, copolymers and polymer blends have been investigated to obtain light emission over the entire visible range [1-3]. Imbalances in injection and transport properties of charge carriers inside the active polymer are considered to be partly responsible for low efficiencies in polymer LEDs. Strategies to overcome these drawbacks have included the use of multi-layer structures with separate transport layers such as PBD dispersed in PMMA, polyquinoxaline, etc. [4,5]. Other methods involve incorporation of electron withdrawing groups such as cyano- or electron deficient nitrogen atoms in the repeat unit of the emissive polymers to improve the electron accepting property [6,7]. Polymers based on the pyridine, quinoline or quinoxaline ring systems fall into this class. The n-type electrical conductivity of this class of polymers is said to originate from the π-deficient nature of the ring system [8].

EL of quinoline based polymers was first investigated by Parker et al [9]. Blue emission was reported from devices in which the active layer consisted of rigid phenylquinoline units spaced by a flexible non-conjugated ether segment containing the hexafluoroisopropylidene group. Internal quantum efficiencies were reported to be 4% indicating the great potential for use of quinoline based polymers in LEDs. Later, electron injection and transport properties of poly(phenylquinoline) (PPQ) were put to use by blending acid protonated PPQ with the emissive polymer in MEH-PPV LEDs [10]. The improvement in emission efficiency was attributed to enhanced electron injection, facilitated by higher electron affinity of PPQ. The complexation-mediated solubility of PPQ with protonic acid was reported to yield films that did not show EL when used as active layer in LEDs. Thus the protonated PPQ could only serve the role of electron transporting medium.

Recently, we reported the synthesis of PQV by direct polymerization of bis(chloromethyl) quinoline and via the xanthate precursor route. This precursor polymer was converted into the conjugated form by thermal elimination under vacuum [11]. Upon doping with sodium naphthalide, PQV turned from dark red to blue showing the potential for n-dopability. Polymer LEDs fabricated using precursor PQV showed broad emission with the peak wavelength in the yellow region.

Literature shows that PPV and its derivatives have yielded the best EL efficiencies so far, despite the associated problems. To utilize the electron transport and EL properties of PQV in conjunction with the hole transport and light emission characteristics of PPV, it was proposed to prepare a copolymer of PPV and PQV *via* the precursor route. The nonbonding electrons on nitrogen act as shallow traps and prevent positive charge carriers from migrating to the cathode without radiative recombination [12]. Such a copolymer, PPVQV, will have great potential for applications in polymer LEDs.

2. EXPERIMENTAL

Scheme 1

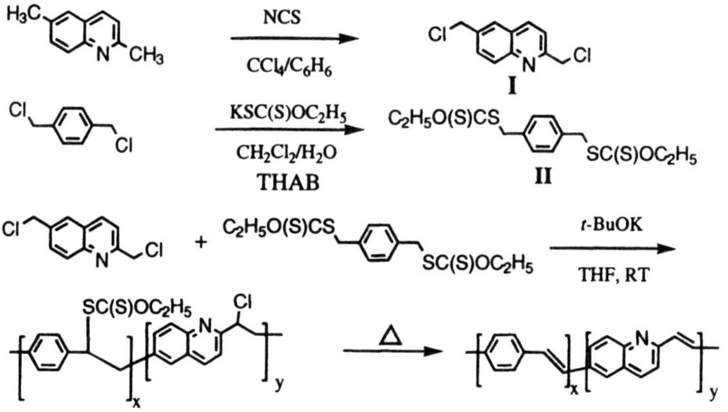

2,6-Bis(chloromethyl) quinoline (I): Monomer I, 2,6-bis(chloromethyl)quinoline (BCMQ) was synthesized from DMQ by adapting a procedure published by Lenz et al [13]. 4 g of DMQ was dissolved in 150 mL of a CCl_4 and C_6H_6 mixture (2:1). To this, was added 10 g of NCS and 200 mg of dibenzoyl peroxide. The mixture was re-fluxed for 18 hours under nitrogen and it changed in color to yellow. It was cooled to room temperature and the succinimide precipitate was filtered and washed with CCl_4 to re-dissolve BCMQ that had also precipitated. The combined filtrates were evaporated to obtain a yellow solid wetted with a red oil. The mixture was washed with a minimum amount of cold methanol to remove the red oil by-product and this produced a light yellow solid. The solid was found to contain some succinimide which was removed by stirring with water to obtain pale yellow solid. The solid was re-crystallized from ethanol to obtain pure BCMQ with a yield of about 22%. ^1H NMR ($CDCl_3$) δ: 4.77 (s, CH_2), 4.84 (s, CH_2), 7.60-8.20

(m, aryl-H). ^{13}C NMR (CDCl$_3$) δ: 45.84, 47.2 (CH$_2$), 121.02, 127.08, 127.17, 129.98, 130.37, 136.13, 137.33, 147.16, 157.36 (aryl). Anal. Calcd for C$_{11}$H$_9$Cl$_2$N: C, 58.43; H, 4.01; N, 6.19; Cl, 31.36. Found: C, 58.57; H, 3.86; N, 6.25; Cl, 31.07. m.p. = 120 °C.

α,α'-Xylenediyl bis(ethoxycarbodithioate) (II): The xanthate monomer for PPV was prepared according to a published procedure with slight modification [14]. 10 g of α,α-dichloro-p-xylene was dissolved in 200 mL of methylene chloride. 22.8 g of O-ethylxanthic acid, potassium salt and 1 g of a phase transfer catalyst i.e. tetrahexyl ammonium bromide were dissolved in 100 mL of deionized water. The mixture was stirred vigorously to ensure thorough mixing of aqueous and organic phases. After 12 hours, the organic phase was separated and the solvent was evaporated to obtain a white crystalline solid. The yield was about 80%.

Copolymerization of I and II: 400 mg of bis(chloromethyl) quinoline and 612 mg of xanthate monomer of PPV were dissolved in 25 mL of dry THF. 500 mg of potassium-t-butoxide was dissolved in 50 mL of dry THF and both solutions were purged with nitrogen for half an hour prior to mixing. The base was slowly added to the monomer solution at room temperature under nitrogen. The reaction mixture initially turned light yellow to green and finally red with some precipitate formed. The mixture was stirred for 12 hours, solvent was removed on a rotary evaporator and the precipitated polymer was washed with methanol, water, ethanol and acetone. The PPVQV precursor thus prepared was found to be soluble in common organic solvents such as THF, chloroform etc. GPC (Polystyrene standards) analysis gave a number average molecular weight (M$_n$) of about 4100 with a polydispersity (M$_w$/M$_n$) of 4.3. UV-vis$_{(HCOOH)}$: λ$_{max}$= 294, 336 and 392 nm with onset band edge at 478 nm corresponding to a band-gap of about 2.59 eV.

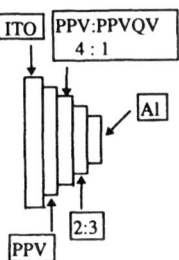

Fig. 1 Schematic diagram of graded structure.

To fabricate LEDs, ITO coated glass plates were cleaned in an ultrasonic bath using organic solvents and de-ionized water. Simple photolithography and etching were performed to pattern substrates with a 4 mm wide ITO strip. PPV, PPVQV precursors and PPV/PPVQV blend solutions were prepared by dissolving in THF at concentrations of 5 mg/mL. The blends were prepared with a PPV/PPVQV ratio of 4/1 and 2/3 to vary the concentration of copolymer from ITO to Al as shown in Fig. 1. Polymer solutions were filtered prior to casting in an Ar-filled glove box. Thermal elimination of the films was carried out by heating at 170 °C for 2 h under high vacuum. It was followed by deposition of Al or Ag for the negative electrode either by conventional evaporation under vacuum or by chemical deposition of Ag on a lab bench without any special precautions.

Chemical deposition of Silver: Chemical deposition of silver was developed on the lines of Brashear's process used for silvering mirrors [15]. It involves chemical deposition of a silver mirror by using a two-part solution. In this process, it was observed that it is necessary to age the reducing solution at least for a week prior to its use. The silver nitrate solution is generally prepared when used.

Reducing Solution (Aged). An 80 g portion of pure sugar was dissolved in 700 mL of distilled water. To this solution were added 175 mL of ethanol and 3 mL of concentrated nitric acid. Water was added to make up a solution of 1 L. This solution was then stored for at least one week at room temperature.

Silver Nitrate Solution (Fresh). A 2 g sample of $AgNO_3$ was dissolved in 5 mL of water. A potassium hydroxide solution was prepared separately by dissolving 1 g of KOH in 5 mL of water. Dilute aqueous ammonia was added drop-wise to the $AgNO_3$ solution and the solution turned dark brown. Ammonia was continually added slowly until the precipitate was nearly but not quite all redissolved. At this stage, the KOH solution was poured into the $AgNO_3$ solution. Again ammonia was added little by little to redissolve the precipitate that formed upon adding KOH. It is necessary to make sure that excess ammonia is not added when redissolving the precipitate, as it influences the quality of silver films deposited. The solution was filtered to remove particles of undissolved precipitate and the filtered solution exhibited a brown tint.

Deposition Method. A quantity of reducing solution equal to about one-fourth of the fresh silver nitrate solution prepared was measured out. The polymer sample was masked with a film of Teflon to allow silver deposition only in a specified region. Samples were then laid out "face up" at the bottom of a glass beaker and a quantity of the fresh silver nitrate solution was poured into the beaker to cover the polymer sample. The measured amount of reducing solution was added to it and the mixture was stirred with a glass rod taking care not to scratch the sample surface. The solution turned black to brown and finally a light gray, indicating completion of the deposition. The sample was taken out, washed with water followed by rinse in 2-propanol, and dried by blowing with compressed nitrogen. The silver metal film surface was gently wiped with a tuft of cotton to remove sediment, and the rinse and drying steps were repeated to obtain a shiny silver surface. Samples were then dried under high vacuum for at least 8 h to remove all moisture.

3.   Results and Discussion

The PPVQV used in this study was prepared by polymerizing bis(chloromethyl) quinoline and α,α'-Xylenediyl bis(ethoxycarbodithioate) with a little more than 1 eq. base as shown in Scheme 1. Synthesis of PQV via the xanthate precursor showed that it is difficult to completely eliminate the xanthate group attached to the quinoline unit even after heating it under vacuum for 12 h at 300 °C. To avoid such difficulties, the chloro monomer was chosen for PQV and the xanthate monomer for PPV to retain the advantages associated with the precursor processing.

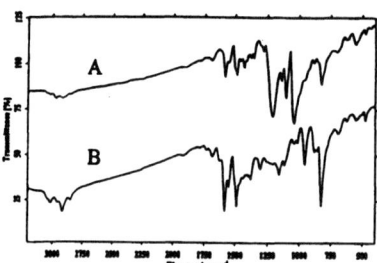

Fig. 2 IR spectra of PPVQV copolymer (A) before, (B) after elimination

Fig. 2 shows IR spectra of the copolymer before and after thermal elimination. The strong peak at 1044 cm^{-1} in the precursor spectrum corresponds to the xanthate stretch and disappears in post-elimination spectra showing that conversion proceeds via the elimination of the xanthic acid. The 732 cm^{-1} peak, which corresponds to the C-Cl stretch was observed to be weak in precursor as it is a better leaving group. A strong, new peak at 960 cm^{-1} in the post-elimination spectra corresponds to trans-vinylene C-H out of plane bending. Formic acid solution of the precursor copolymer showed

λ_{max} at 294, 336 and 392 nm, while PQV showed a single peak at 450 nm. PPV was found to be insoluble in formic acid.

Fig. 3a illustrates the EL spectrum of PPVQV with emission peaks at 425 and 445 nm. The emission spectra was blue shifted with respect to both PQV and PPV indicating a shortened conjugation length in the copolymer. Fig. 3b and 3c compare the I-V curves of ITO/PPVQV/Al and ITO/PQV/Al devices. It is clear that the copolymer yields a better diode characteristic with smaller leakage currents at lower voltages before the light emission occurs. This can be attributed to balanced charge injection into the device due to the p-type and n-type nature of the phenylene and quinoline moieties present in the copolymer. To further improve efficiency, devices were constructed with homopolymer/blend configuration as shown in Fig. 1. The devices showed dominant emission peak corresponding to PPV at 530 nm and higher efficiencies approximately by an order of magnitude.

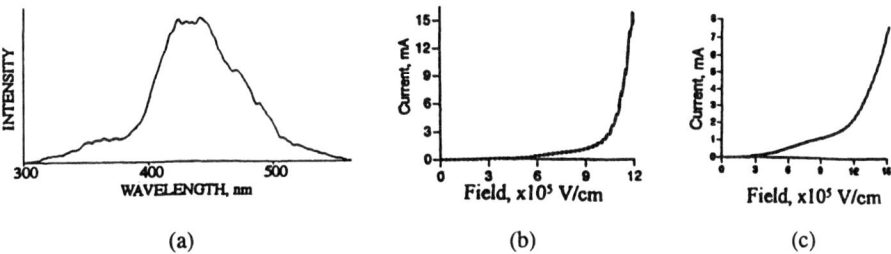

(a)	(b)	(c)

Fig. 3(a) EL spectrum and (b) I-V curve of ITO/PPVQV/Al, (c) I-V curve of ITO/PQV/Al

Chemical Deposition. Physical metal deposition methods are cost intensive processes requiring high-cost capital equipment and have limits on shape/size of the substrates that can be used for deposition. Wet metal deposition processes have an advantage in this regard and deposition can be on any size/shape of substrate without limitation. The chemistry involved in the deposition of silver is based upon Tollen's test used for identifying aldehydes. The role of the nitric acid added to the sugar solution is to oxidize the terminal CH_2OH groups in the sugar to form aldehydes. Upon addition of the ammonical silver nitrate solution, the aldehyde groups are oxidized to carboxylic acids while the Ag^+ is reduced to give a deposition of silver metal. Reaction continues until all the Ag^+ is consumed and colloidal silver is deposited on all surfaces that were in contact with the reaction mixture. Scheme 2 shows the pertinent reactions.

Scheme 2

$$AgNO_3 + 3NH_4OH \longrightarrow [Ag(NH_3)_2]OH + NH_4NO_3 + 2H_2O$$

$$2[Ag(NH_3)_2]OH + R\text{-}CHO \longrightarrow 2Ag^0 + RCOONH_4 + H_2O + 3NH_3$$

Reaction conditions were observed to exert a great influence on the quality of films deposited. Strong reduction conditions which gave fast reaction rates resulted in gray or brown spongy films of poor quality, while slower reaction rates under mild reduction conditions yielded consistent films of good quality. Under the optical microscope, the polymer side of the silver film appeared very smooth and was as shiny and reflecting as vacuum deposited silver films. The films were observed to be continuous and contain agglomerations of silver grains, which is a characteristic of chemically deposited films from colloidal solutions.

The shape of I-V curves recorded for these devices was similar to those reported for LEDs with vacuum deposited electrodes except for difference in onset voltage. The light onset, for PPV LEDs with chemically deposited electrodes, was observed at applied fields of 1.4×10^6 V/cm, which is slightly higher than that reported for LEDs with vacuum evaporated Ag electrodes. At 10 V applied voltage, the light emitted by the device could easily be seen in the dark and devices emitted for approximately 2 h in air. Although the light emission was uniform over the entire device area, LEDs with chemically deposited electrodes showed an order of magnitude lower efficiency compared to LEDs fabricated in our laboratory with vacuum deposited silver. Lower efficiencies for these devices were not surprising since the deposition was carried out on a lab bench without any special precautions. It is known that the crystallinity of metallic films can be improved by annealing after the films are deposited. Ample room exists for further improvement in efficiency by subjecting the devices to annealing under vacuum before operation.

We like to acknowledge the financial support from Texas Advanced Technology Program under Grant No. 003656-009.

REFERENCES

1. J.H. Burroughes, D.D.C. Bradley, A.R. Brown, R.N. Marks, K. Mackay, R.H. Friend, P.L. Burn and A.B. Holmes, Nature 347, 539 (1990).
2. I. Sokolik, Z. Yang, F.E. Karasz and D.C. Morton, J. Appl. Phys. 74, 3584 (1993).
3. C. Zhang, H. von Seggern, K. Pakbaz, B. Kraabel, H.W. Schmidt and A.J. Heeger, Synth. Met. 62, 35 (1994).
4. A.R. Brown, J.H. Burroughes, N.C. Greenham, R.H. Friend, D.D.C. Bradley, P.L. Burn, A. Kraft and A.B. Holmes, Appl. Phys. Lett. 61, 2793 (1992).
5. T. Fukuda, T. Kanbara, T. Yamamoto, K. Ishikawa, H. Takeze and A. Fukuda, Synth. Met. 85, 1195 (1997).
6. N.C. Greenham, S.C. Moratti, D.D.C. Bradley, R.H. Friend and A.B. Holmes, Nature 365, 628 (1993).
7. J. Tian, C-C. Wu, M.E. Thompson, J.C. Sturm, R.A. Register, M.J. Marsella and T.M. Swager, Polym. Prepr. (Am. Chem. Soc., Div. Polym. Chem.) 35 (1), 275 (1994).
8. T. Kanbara, N. Saito, T. Yamamoto and K. Kubota, Macromolecules 24, 5883 (1991).
9. I.D. Parker, Q.Pei and M. Marocco, Appl. Phys. Lett. 65, 1272 (1994).
10. F. Hide, C.Y. Yang and A.J. Heeger, Synth. Met. 85, 1355 (1997).
11. R.K. Kasim, S. Satyanarayana and R.L. Elsenbaumer, MRS Abstracts, Spring '97 Meeting, San Francisco, CA.
12. H. Chayet, E.Z. Faraggi, H. Hong, V.N. Savvate'ev, R. Neumann, Y. Avny and D. Davidov, Synth. Met. 84, 621 (1997).
13. S. Antoun, D.R. Gagnon, F.E. Karasz and R.W. Lenz, J. Polym. Sci., Polym. Lett. 24, 503 (1986).
14. S. Son, A. Dodabalapur, A.J. Lovinger and M.E. Galvin, Science 269, 376 (1995).
15. Laboratory Arts and Recipes. Handbook of Chemistry and Physics, Ed. by C.D. Hodgman, R.C. Weast and S.M. Selby, 38th ed., Chemical Rubber Publishing Co., Cleveland, 1956, p 3047.

LECs MADE OF m-LPPP

F. P. WENZL [1], S. TASCH [1], J. GAO [2], L. HOLZER[1], U. SCHERF [3], A.J. HEEGER [2] and G. LEISING [1]

[1] Institut für Festkörperphysik, Technische Universität Graz, Petersgasse 16, A-8010 Graz, Austria
[2] Institute for Polymers and Organic Solids, University of California, Santa Barbara, Santa Barbara,California 93106-5090, USA
[3] Max-Planck-Institut für Polymerforschung, Ackermannweg 10, D-55128, Mainz, Germany

Abstract

We report the application of the blue light emitting conjugated polymer m-LPPP (methyl substituted laddertype poly(paraphenylene)) in light emitting electrochemical cells. The active layer of the LEC consists of a blend of m-LPPP with the ionically conductive polymer PEO and LiCF₃SO₃ as ionic salt. Investigations of different concentrations of PEO and salt showed that the best LECs made of m-LPPP up to now where realised with an active layer consisting of a blend of m-LPPP:PEO:salt in the range of 20:10:3. In this case we are able to realise LECs with response times in the range of 30 µs. The I/U characteristics show low turn on voltages both for current and electroluminescence, but only in the case of ITO biased as a cathode. The initial electroluminescence spectra are quite the same as those for LEDs made of m-LPPP but turn into green after some time of operation.

Introduction

Nowadays highly efficient light-emitting diodes based on conjugated polymers (PLEDs) can be realised over the whole visible spectral range.[1-5]There are a lot of possible applications for PLEDs, maybe the most attractive is their use in self-emissive flat panel displays. While low band band gap polymers show low turn on voltages, this is not the case for wide band gap ones, due to the fact, that for these polymers the charge injection is more difficult because of the higher energetic barriers at the interface polymer/electrode.

However, in order to realise multicolor flat panel displays the fabrication of blue PLEDs of sufficient lifetimes is absolutely necessary. Moreover, one of the technologically most promising ways to produce multicolor displays is based on blue PLEDs, where the red and green light emission are realised by an external color conversion technique [6-8].

Recently, a new type of polymer light-emitting device, the polymer light-emitting electrochemical cell (LEC) has been presented[9-10], which has the advantage, that the EL onset occurs for *all* emission colors at a bias voltage close to E_g/e [11] where E_g means the single particle energy gap and e is the elementary charge. The different performance of LECs compared to PLEDs is due to the fact that the charge carrier injection in these devices relies on different physical mechanisms. In contrast to PLEDs, in LECs a *pin*-junction is formed during operation[10].The formation of the *pin*-junction is not an instantaneous process but depends on the mobility of the ions in the blend of the ionically conductive polymer, the light-emitting polymer and the ionic salt. Longer response times are therefore necessary for LECs compared to PLEDs[12]. The response time strongly depends on the morphology of the polymer blend network of the LEC. For that reason the pulse response times for LECs based on MEH-PPV can be improved from seconds to ms by improving the morphology of the blend[12].

We demonstrate the realisation of efficient blue and blue-green LECs based on a blend of the conjugated polymer m-LPPP (methyl substituted laddertype poly(paraphenylene) and the ionic conductor PEO poly(ethyleneoxide) complexed with a Li salt, in this case LiCF₃SO₃ (lithium trifluoromethanesulfonate, Li triflate for short), a well known solid state electrolyte. By optimising the ratio of these components a phase morphology in the blend is obtained, that allows to realise response times below 30µs. This fast response makes LECs based on m-LPPP also attractive for applications in self-emissive displays.

Experimental

The LEC devices were fabricated on indium-tin oxide (ITO) coated glass substrates onto which the m-LPPP: PEO (molecular weight 5000000 or 100000, purchased   from Aldrich) : LiCF$_3$SO$_3$ (purchased from Aldrich) blend was spincoated from a cyclohexanone solution (see Ref.12) which results in an active layer of a thickness of about 200 nm. The film was annealed afterwards at a temperature of 60° C under argon atmosphere for half a day before Al or Au as top electrodes were evaporated in high vacuum to realise a thickness of about 150 nm. All measurements were done in argon atmosphere. The EL spectra were recorded with a CCD-spectrometer (Oriel 77400 and Princeton Instruments CCD-576). For the pulsed measurements the excitation of the LECs was performed with a 1kHz (duty cycle: 1ms on / 1ms off) square wave of a HP-pulse generator and the detection was provided by a photo-diode in combination with an oscilloscope. The absorption of the LEC during operation was determined in a Perkin Elmer λ9-spectrometer.

Results and Discussion

The application of m-LPPP in LECs yields bright and stable devices. The brightest and most efficient LECs up to now, were obtained by using a blend of m-LPPP, PEO, molecularly weight of 5000000 and LiCF$_3$SO$_3$ in a weight ratio of 20:10:3. These LECs yield luminance values of around 250cd/m^2 (at operating voltages of -10V (means ITO as cathode, see below)) and external EL quantum efficiencies of around 0.3 %. This value of the EL efficiency of the LEC should be compared to that of an ITO/m-LPPP/Al - LED, which is usually between 0.1 and 1%.[13]

The main advantage of the LECs based on m-LPPP compared to PLEDs based on m-LPPP is the much lower threshold voltage, which is necessary for the onset of the electroluminescence. The onset of the current and the electroluminescence for the LECs based on m-LPPP can be detected at a voltage of about -2.7 V (s. Fig.1), which matches well with the value of the onset of the absorption spectrum (which is around 0.5 eV lower than the band gap[14] divided by e . This should be compared to the onset of a PLED of pure m-LPPP in a single layer configuration of about 12 V for an active layer of about 200 nm (which is the thickness as the active layer of the LEC) and also the incline of the current for driving voltages above the onset is less steep compared to that of the LEC. [13]

The doping of the active layer in LECs based on m-LPPP during operation can be observed by a change of the absorption properties of the active layer, which is visible by a change of the appearance color of the layer from yellow to grey. In the absorption spectrum (s. Fig.2) a sub bandgap absorption with peaks at 1.9 and 2.1 eV evolves under operation (see Ref.10). This absorption is due to an electronic transition between polaron levels, which are generated by doping of m-LPPP. The same polaronic peaks are therefore also visible in the absorption spectrum of m-LPPP films, when they are chemically doped with conventional dopands such as FeCl$_3$ (s. Fig.2) in a nitromethane solution followed by carefully washing of these films in pure nitromethane. After the bias of LEC is set to 0 V again, the polaronic absorption peaks in the spectrum vanish and the layer returns to its original yellow color within seconds. Simultaneously, a slowly decreasing ionic discharging short circuit current in the opposite direction of the initial current flow can be observed.

The film quality and the phase morphology of the films cast from the conjugated polymer: PEO: LiCF$_3$SO$_3$ solution is of great importance for the performance of the LECs.

In order to obtain LECs with low response times, the phase morphology of the active layer in the LEC and the ionic salt concentration have to be optimised. As it has been demonstrated recently[12] the lowest response times can be expected for a phase morphology of the blend, where the conjugated and the ion-conductive polymer form a fine, interpenetrating network.

As mentioned above, we obtained the most efficient LECs with lowest response times for a ratio of m-LPPP:PEO: LiCF$_3$SO$_3$ of 20:10:3. These blends form a very nice and fine interpenetrating network, even without using any kind of additive, as it was confirmed by AFM studies. The exact dependence of the response times and the efficiencies of the LECs made of m-LPPP on the dimensions of the phase morphology is currently analysed. These response times for the optimised LEC based on m-LPPP are significantly faster than it has been reported for other LECs at room temperature up to now [12] (e.g.

in LECs based on poly(fluorenes) type materials the maximum brightness is achieved after more than 100 min[15]).

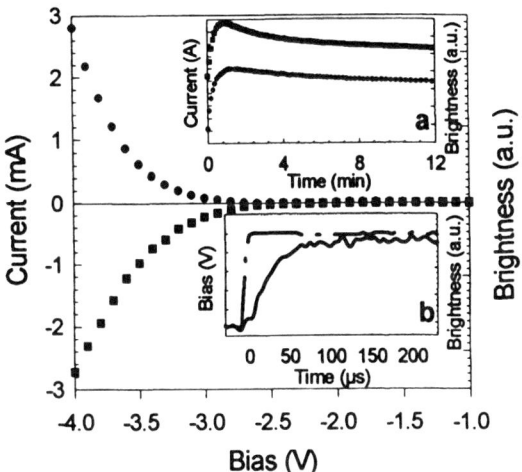

Figure 1: Current/bias and brightness/bias characteristic of an ITO/m-LPPP+PEO+Li triflate/Al LEC (circles... brightness, squares... current) for a weight ratio of 20: 10: 3. (inset: time behaviour of the current and the EL of uncharged LEC **a**) at a stress voltage of –6V **b**) under pulsed excitation)

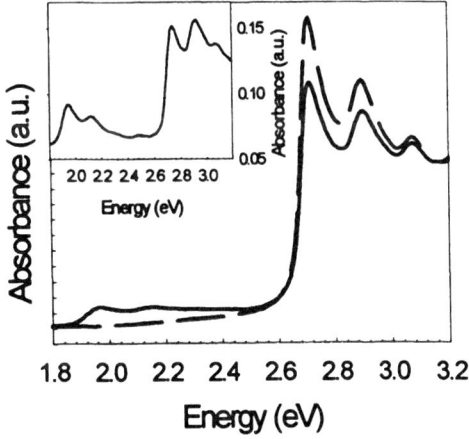

Figure 2: Absorbance spectrum of a LEC based on m-LPPP at 0V (dashed) and under operation (inset: absorbance spectrum of a m-LPPP film on glass, which is doped with FeCl₃)

Once the *pin*-junction is formed in m-LPPP based LECs, the response time is several orders of magnitude lower. In the optimised LECs based on m-LPPP, the turn-on response time (defined as the time until 90% of the maximum EL brightness value is reached) under pulse operation is only between 30-40 μs (which is already close to the time resolution of the used electronic equipment). This value is significantly below the reported value for the turn on response (≈30ms) of LECs at room temperature. Similar response times have only been achieved in 'frozen junction' LECs based on MEH-PPV.

For this optimised ratio ot the components, we were also able to detect light emission from LECs at turn on voltages *lower* than $E_{g,s}/e$, especially in the case of gold as counterelectrode, a behaviour which also was reported for the case of a fine interpenetrating network. In figure 3

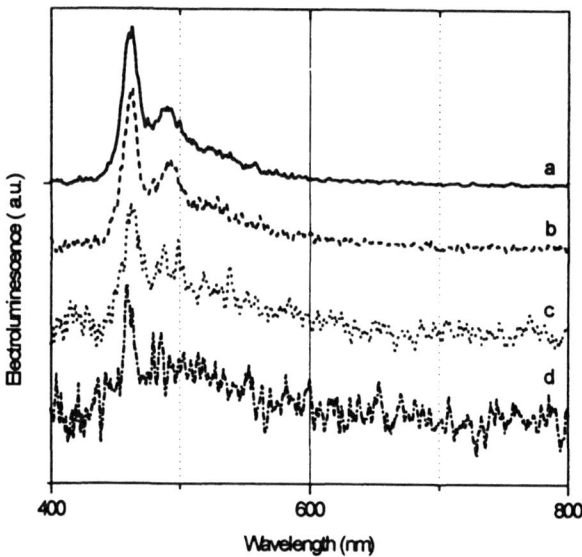

Figure 3: Electroluminescence-Spectra for a ITO/m-LPPP+PEO+li triflate/Au device
for the following voltages: from top to bottom: **a**)-3V, **b**)-2.6V,**c**)- 2.3V,**d**)- 2.2V

the electroluminescence spectra for turn on voltages lower than $E_{g,s}/e$ are also presented.

The EL emission color of a fresh LEC based on the optimised blend changes from blue to green within the first minutes of operation. In the EL spectrum this color change is visible in a decrease of the two dominant peaks at 460 and 490 with respect to peaks at 530 and 560nm (s. Fig. 4).

This change of the emission color is mainly due to interaction of m-LPPP with the polar environment, as it has also been observed for m-LPPP in other polar matrices. Similar effects have also been observed in LECs based on other conjugated polymers.[15]

Due to the creation of ohmic contacts on both electrodes by electrochemical doping, the I/U characteristics of LECs should show a symmetric behaviour. Blends of m- LPPP, PEO and $LiCF_3SO_3$ don´t show such a symmetric behaviour. They show turn on voltages close to the value of Eg,s /e and even lower always when the bottom electrode (usually ITO) is contacted as a cathode. Note, that for this poling direction (which is called the reverse direction in PLEDs) the charge carrier injection is supposed to be more difficult since the nominal potential barriers at the interface polymer/electrode are increased compared to the forward poling direction [16].

This asymmetric behaviour is intensivly investigated at the moment.

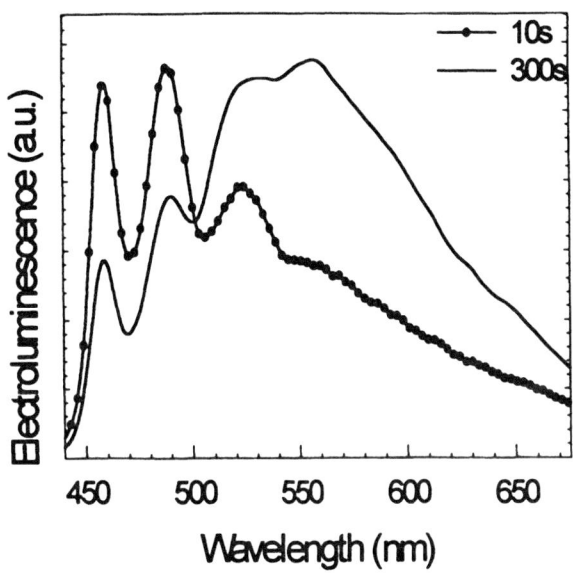

Figure 4: Change of Electroluminescence- spectra for a ITO/m-LPPP+PEO+Li triflate/Al
LEC operated at - 4V after 10s and after 300s

Conclusion

We have demonstrated, that blue and green light-emitting LECs can be fabricated with m-LPPP as
the light-emitting polymer. These LECs based on m-LPPP are of high interest due the high brightness
and the high EL-efficiency, which can be achieved, and especially since response times in the
microsecond regime were realised by optimising the phase morphology of the active blend layer.

ACKNOWLEDGEMENTS

The authors would like to thank Dr. Gang Yu for helpful and exciting discussions and Dipl.-Ing.
Norbert Koch for performing atomic force microscopy on our samples. The financial support of this
work the *SFB Elektroaktive Stoffe* is gratefully acknowledged.

REFERENCES

1. J. H. Burroughes, D. D. C. Bradley, A. R. Brown, R. N. Marks, K. Mackay, R. H. Friend, P. L.
Burn, A. Kraft, A. B. Holmes, Nature **347**, 539 (1990).
2. D. Braun, A. J. Heeger, Appl. Phys. Lett. **68**, 1982 (1991).
3. G. Grem, G. Leditzky, B. Ullrich, G. Leising, Adv. Mater. **4**, 36 (1992).
4. N. C. Greenham, S. C. Moratti, D. D. C. Bradley, R. H. Friend, A. B. Holmes, Nature **365**, 628
(1993).
5. Y. Ohmori, M. Uchido, K. Muro, and K. Yoshino, Jpn. J. Appl. Phys. **30** , L1938 (1991).
6. C. W. Tang, D. J. Williams, J. C. Chang, US-Patent 5,294,870 (1994).

7. M. Matsura, H. Tokailin, M. Eida, C. Hosokawa, Y. Hironaka and T. Kusumoto, Proc. Asia Diplay '95, 269 (1995).
8. S. Tasch, C. Brandstätter, F. Meghdadi, G. Leising, G. Froyer, L. Athouel, Adv. Mater. 9,33(1997)
9. Q. Pei, G. Yu, C. Zang, Y. Yang, A. J. Heeger, Science 269, 1086 (1995).
10. Q. Pei, Y. Yang, G. Yu, C. Zhang, and A.J. Heeger, J. Am. Chem. Soc. 118, 3922 (1996).
11. Q. Pei, Y. Yang, Synth. Met. 80, 131 (1996).
12. Y. Cao, G. Yu, A.J. Heeger, C.Y. Yang, Appl. Phys. Lett. 68, 3218 (1996).
13. S. Tasch, A. Niko, G. Leising, U. Scherf, Appl. Phys. Lett. 68, 1090 (1996)
14. G. Meinhard, A. Horvath, G. Weiser, G. Leising, Syth. Met. 84, 669 (1997)
15. Y. Yang, Q. Pei,J. Appl. Phys. 81,3294 (1997)
16. S. Tasch, J. Gao, F.P. Wenzl, U. Scherf, K. Müllen, G. Leising, A.J. Heeger (unpublished)

ELECTROCHEMICALLY BASED LIGHT EMITTING DEVICES FROM SEQUENTIALLY ADSORBED MULTILAYERS OF A POLYMERIC RUTHENIUM (II) COMPLEX AND POLY(ACRYLIC ACID)

Aiping Wu, Jin-kyu Lee*, Michael F. Rubner
Department of Materials Science and Engineering, Massachusetts Institute of Technology, Cambridge, MA 02139, waiping@mit.edu
*Department of Chemistry, Massachusetts Institute of Technology, Cambridge, MA 02139.

ABSTRACT

We have investigated a new light emitting material, $Ru(bpy)_3^{2+}$ polyester for fabricating electrochemically based solid state light emitting devices using the layer-by-layer sequential adsorption processing technique. By controlling the deposition conditions such as the pH of the $Ru(bpy)_3^{2+}$ polyester and poly(acrylic acid) (PAA) solutions, we systematically altered the layer thickness and bilayer composition to obtain multilayers that contain different amounts of $Ru(bpy)_3^{2+}$ polyester (from 46% to 70%). Differences in the $Ru(bpy)_3^{2+}$ polyester composition, in turn, influence the device performance dramatically.

INTRODUCTION

Over the past few years, we have demonstrated that the layer-by-layer sequential adsorption processing scheme offers exciting new possibilities in the fabrication of various thin film opto-electronic devices. Examples of devices that we have fabricated include light emitting diodes based on conjugated polymers such as poly(p-phenylene vinylene) PPV [1, 2], photovoltaic devices based on conjugated polymer/C60 hetero-junctions and electrochemical light emitting devices based on active ruthenium (II) complexes [3, 5].

The layer-by-layer sequential adsorption process involves the alternate spontaneous adsorption of oppositely charged polymers from their dilute aqueous solutions. The driving force for such multilayer assembly is the electrostatic interactions developed between the oppositely charged moieties of the polymer chains. Consequently, the charge density of the adsorbing polymer and the previously adsorbed polymer are crucial in the multilayer adsorption process. Fundamentally, molecular parameters such as polymer pKa, the ionizable functional group type and the polymer molecular weight as well as polyelectrolyte solution parameters such as solution concentration, pH and ionic strength play an important role in determining the structural features of the resultant multilayers. Structural features such as the bilayer thickness and the composition of the polycation/polyanion building block and the interpenetration between the polycation/polyanion layers are key factors in determining the performance of devices. From this point of view, the layer-by-layer sequential adsorption processing technique provides a possibility to optimize device characteristics by simply controlling the thin film processing parameters as mentioned above.

In this paper, we focus on light emitting devices made from sequentially adsorbed multilayers of a $Ru(bpy)_3^{2+}$ polyester and poly(acrylic acid). As will be presented, by controlling the thin film processing conditions such as solution pH, we can systematically alter the bilayer thickness and the composition of the $Ru(bpy)_3^{2+}$ polyester/PAA building block. This in turn has made it possible to dramatically alter and optimize the device performance.

EXPERIMENT

The chemical structure of the polymeric ruthenium (II) complex, $Ru(bpy)_3^{2+}$ polyester 1, is shown in Scheme 1. A detailed description of the synthesis and properties of this material are described in another paper [4]. The structures of the polyelectrolytes, poly(allylamine hydrochloride) (PAH) 2 and poly(acrylic acid) (PAA) 3, used for sequential adsorption processing are also shown in Scheme 1. Both the PAH (MW=50 - 65K) and PAA (MW=50K, 25% aqueous solution) are commercially available.

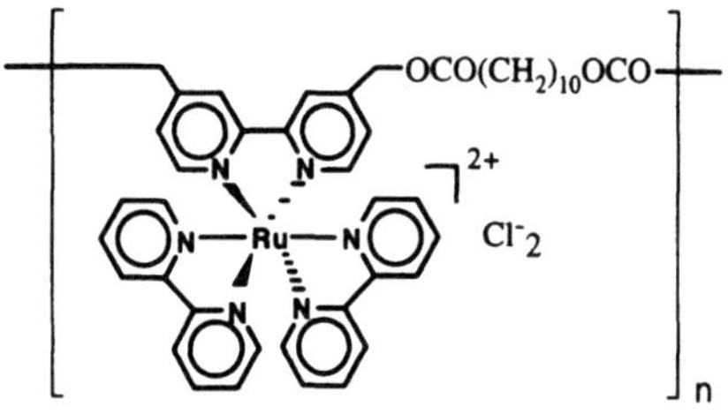

1 Ru(bpy)$_3^{2+}$ polyester

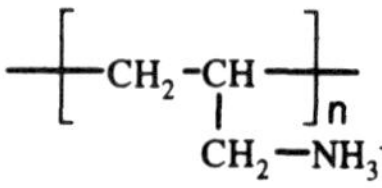

2 poly(allylamine hydrochloride) PAH

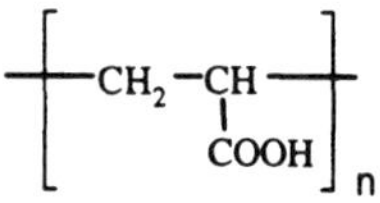

3 poly(acrylic acid) PAA

Scheme 1 Chemical structures of the polyelectrolytes used for sequential adsorption processing.

Solutions of the polycations and the polyanion were prepared by dissolving polymers in Millipore water to a concentration of 1×10^{-2} M. Solution pH was adjusted by adding hydrochloric acid HCl or sodium hydroxide NaOH. The rinsing water was prepared by adjusting the Millipore water to the same pH as that of the polyelectrolyte solutions. Prior to multilayer adsorption, the polyelectrolyte solutions were filtered through a 0.45 μm filter.

Sequentially absorbed multi-bilayers of Ru(bpy)$_3^{+2}$ polyester and polyanion PAA were made by the automatic alternate dipping of the substrates in solutions. Each step for adsorption of the polycation or the polyanion in their solutions takes 10 minutes. Following the adsorption process, the substrate was rinsed by pH adjusted Millipore water three times. A single bilayer of PAH/PAA (pH=3.5 for both) was hand dipped on the substrate prior to multilayer deposition. This single bilayer worked as a preparatory layer for subsequent multilayer growth.

Light emitting devices were fabricated in a sandwich architecture. The transparent ITO electrode was patterned on a 1 x 2 inch glass substrate with four 2 mm wide lines. After sequential adsorption of a PAH/PAA single layer and the Ru(bpy)$_3^{2+}$ polyester/PAA multilayers on the ITO substrate, the multilayers were thermally treated at 110 °C under vacuum for 12 hours. Immediately after annealing, aluminum electrodes were thermally evaporated onto the multilayers through a mask to give 2 mm wide lines perpendicular to the ITO lines. The measurement of device characteristics was conducted inside a nitrogen glove box. Devices were tested automatically using a programmed current-voltage and light-voltage setup, which includes a Keithley programmable voltage source, a Hewlett-Packard multimeter and a calibrated silicon photodiode. A step pulsing method was used to quick test devices, wherein voltage was applied from zero to the desired voltage and held for 5-10 seconds before being turned off.

RESULTS

In order to determine the influence of thin film processing conditions such as the solution pH on the device characteristics of the sequentially adsorbed Ru(bpy)$_3^{2+}$ polyester/PAA devices, we conducted multilayer assembly using dipping solutions with different pH levels. For this purpose, the concentration of all solutions was kept at 0.01 M and no salt was added to any solution. The pH of the rinsing water was adjusted to the same pH as that of the previous dipping solution. The pH of the Ru(bpy)$_3^{2+}$ polyester or PAA dipping solution was systematically varied from 2.5 to 4.5, while the pH of the oppositely charged dipping solution was kept at 3.5.

The linear plots in Figures 1a and 1b, for the case that the pH of the $Ru(bpy)_3^{2+}$ polyester varied, show that the layer-by-layer sequential adsorption process proceeds in a linear and reproducible manner. Similar linear plots (data not shown) were also obtained in the case that the pH of the PAA varied. These figures also show that both the visible absorbance per bilayer and the thickness per bilayer of the $Ru(bpy)_3^{2+}$ polyester/PAA multilayers are strongly influenced by the pH of the dipping solutions. As shown in Figure 1, both the absorption and the total thickness tend to increase with increasing the pH of the $Ru(bpy)_3^{+2}$ polyester solution. On the other hand, both the absorption and the total thickness tend to increase with decreasing the pH of the PAA solution (data not shown). It should be mentioned that the absorption and thickness data shown in this work were obtained from one substrate. The deposition was interrupted by drying with compressed air for corresponding measurements. For comparison, we also assembled multilayers at the same dipping condition but without the interruption. Absorption data indicate no significant difference between these two methods.

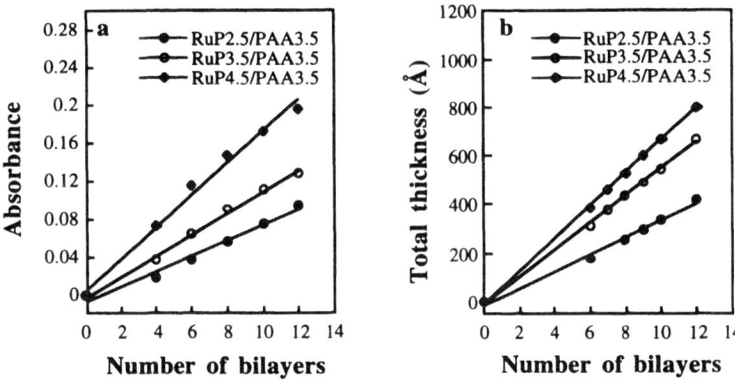

Figure 1   pH dependent sequential adsorption of the $Ru(bpy)_3^{2+}$ polyester and PAA multilayer system. a) visible absorbance of the $Ru(bpy)_3^{2+}$ polyester measured at 460 nm (maximum) versus number of bilayers, b) thickness of the $Ru(bpy)_3^{2+}$ polyester/PAA multilayers versus number of bilayers.

In order to further analyze the multilayer structure, we examined the incremental thickness contribution of the $Ru(bpy)_3^{2+}$ polyester and PAA in each bilayer building block. Figures 2a and 2b show that the thickness contribution of both the $Ru(bpy)_3^{2+}$ polyester and PAA layer systematically changes depending on the pH of the dipping solutions. For instance, the incremental thickness of PAA decreases from 54 Å to 7Å as the dipping solution pH increases from 2.5 to 4.5 (Figure 2a), while the incremental thickness of the $Ru(bpy)_3^{2+}$ polyester increases from 18 Å to 40 Å as the dipping solution pH increases from 2.5 to 4.5 (Figure 2b). Also note that the thickness of the adsorbed PAA layer does not depend on the thickness of the previously adsorbed $Ru(bpy)_3^{2+}$ polyester layer, whereas the thickness of the adsorbed $Ru(bpy)_3^{2+}$ polyester layer does depend on the thickness of the previously adsorbed PAA layer. For instance, at a constant pH of 3.5, the incremental thickness of the $Ru(bpy)_3^{2+}$ polyester decreases from 63 Å to 18 Å (Figure 2a), but the incremental thickness of PAA remains nearly constant (Figure 2b). It should be mentioned that we have strong evidence that the oppositely charged polymers layers (the Ru(bpy)$_3^{2+}$ polyester and PAA layers) are interpenetrated. Thus, the blocks shown in Figure 2 simply represent the incremental thickness increase that is observed when a layer of the Ru(bpy)$_3^{2+}$ polyester or PAA is added to the film.

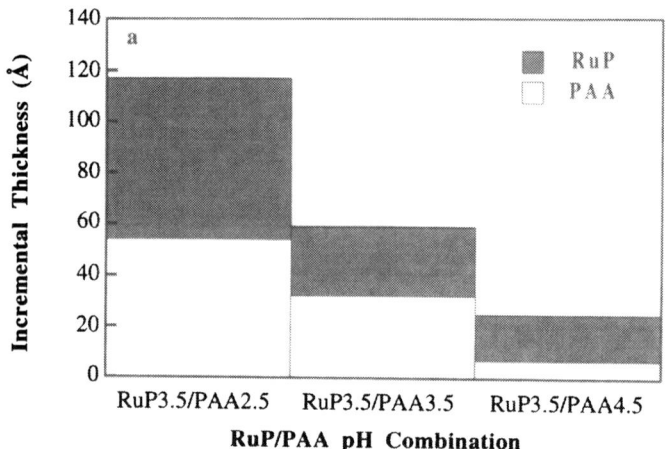

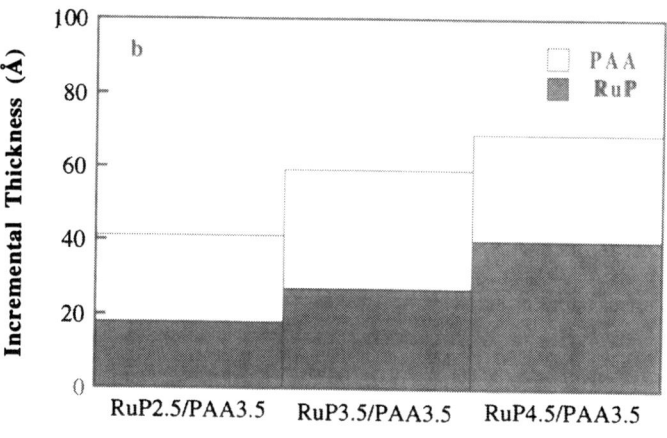

Figure 2 Incremental thicknesses of the Ru(bpy)$_3^{2+}$ polyester and PAA in the bilayer building block. a). the pH of Ru(bpy)$_3^{2+}$ polyester is kept at 3.5, the pH of PAA varies from 2.5 to 3.5 and 4.5, b). the pH of PAA is kept at 3.5, the pH of Ru(bpy)$_3^{2+}$ polyester varies from 2.5 to 3.5 and 4.5.

To understand the effect that pH has on layer organization, we need to consider the key factors in the deposition process. In general, the thickness of an adsorbed polymer in a multilayer deposition process depends on a number of different factors including the linear charge density of the adsorbing polymer chain, the ionic strength of the dipping solution and the surface charge density of the previously adsorbed polymer. In our case, since there is no salt added to the dipping solutions, the charge density of the adsorbing polymer and the surface charge density of the previously adsorbed polymer determine layer thickness. The variation of charge density (degree of ionization of carboxylic acid groups for PAA) in this case is achieved by the pH adjustment. In the pH range used (2.5 to 4.5), the charge density the $Ru(bpy)_3^{2+}$ polyester remains constant. The weak polyacid PAA (pKa is around 4), on the other hand, has a charge density that is very sensitive to pH with lower pH values creating a less charged polymer. It is generally understood that the polymer chains of weak polyacids like PAA tend to adsorb as thin layers with flat conformations when they are highly charged (high pH) and as thick layers with loopy conformations when they are less charged (low pH). In addition, it is also known that the amount of polyelectrolyte adsorbed to a surface increases with increasing surface charge density since more material is needed to balance and overcompensate the higher charge density of the oppositely charged surface. The decrease in the incremental thickness of PAA (Figure 2a) indicates that the thickness contribution of PAA is controlled by the pH of the PAA solution. Whereas, the increase in the incremental thickness of the $Ru(bpy)_3^{2+}$ polyester (Figure 2b) indicates that the thickness contribution of the $Ru(bpy)_3^{2+}$ polyester is controlled by the surface charge density of the previous adsorbed PAA layer, which, in turn, is determined by the pH of the $Ru(bpy)_3^{2+}$ polyester solution. The nearly constant thickness of PAA at the same pH of PAA solution (Figure 2b) suggests that the thickness contribution of the PAA depends on the pH of the PAA solution and does not depend on the thickness of the previously adsorbed $Ru(bpy)_3^{2+}$ polyester layer. In contrast, the thickness variation of the $Ru(bpy)_3^{2+}$ polyester at constant pH of the $Ru(bpy)_3^{2+}$ polyester solution (Figure 2a) suggests that the thickness contribution of the $Ru(bpy)_3^{2+}$ polyester depends on the thickness of the previously adsorbed PAA layer. Thus, a higher adsorbed amount of the $Ru(bpy)_3^{2+}$ polyester occurs with a thicker previously adsorbed PAA layer.

Having understood what it takes to vary the composition of the $Ru(bpy)_3^{+2}$ polyester/PAA bilayer, we chose several different conditions for making light emitting devices. The main motive of this study is to improve light output by changing the composition of $Ru(bpy)_3^{+2}$ polyester in the bilayer building block. Three multilayer systems which contain 46%, 58% and 70% $Ru(bpy)_3^{+2}$ polyester in the bilayer combination have been studied.

Table 1 Device performance measured by the step pulsing method.

Devices	*Composition of RuP	Forward bias		Reverse bias	
		**Light	**Efficiency	**Light	**Efficiency
RuP3.5/PAA3.5	46 %	1337	3.2%	721	3.2%
RuP4.5/PAA3.5	58 %	898	2.1%	193	1.2%
RuP3.5/PAA4.5	70 %	184	0.01%	23	0.006%

* Based on thickness contribution of the $Ru(bpy)_3^{2+}$ polyester in the bilayer building block.
** Maximum values. Light: nW, Efficiency: external quantum efficiency (photons/electron)

Table 1 summarizes the device performance of these three devices, which have a total thickness of the $Ru(bpy)_3^{2+}$ polyester/PAA multilayers of around 1100 Å. Remarkably, the device performance changes dramatically with small changes in the composition of the bilayer. In the device that has 46 % of $Ru(bpy)_3^{2+}$ polyester, the highest device efficiency of 3.2 % is obtained and the device behaves almost symmetrically in the forward and reverse biases. The efficiency we achieved is comparable to some of the best devices based on conjugated polymers [6, 7]. Notably, increases in the $Ru(bpy)_3^{2+}$ polyester composition tend to decrease the device efficiency and result in asymmetric device behavior in the two biases.

It should be mentioned that Murray's group has extensively studied the dynamics of Ru(II) type redox materials. Recently, they were successful at operating a solid-state ECL device based on a Ru(II) complex by establishing serial, frozen concentration gradients of the Ru^{3+}/Ru^{2+} and Ru^{1+}/Ru^{2+} species [8]. Based on their work and ours, it is possible to identify some of the key elements that control device performance in electrochemical systems of this type. The three key steps are: (1) the redistribution of counter-ions to establish a concentration gradient of mixed valent states of ruthenium redox species, Ru^{3+} and Ru^{1+}; (2) transition to an electron hopping transport dominated mode; (3) the bimolecular reaction of neighboring Ru^{3+} and Ru^{1+} states to generate excited Ru^{2+*}. Detailed studies are now underway to determine how these steps are influenced by the change of the $Ru(bpy)_3^{2+}$ polyester composition. Currently, we believe that changing the composition of the $Ru(bpy)_3^{2+}$ polyester influences the relative rates of electron transport by the Ru^{3+}/Ru^{2+} and Ru^{2+}/Ru^{1+} species thereby controlling the proximity of the recombination zone to the electrodes.

CONCLUSIONS

In conclusion, we have demonstrated that by controlling thin film processing conditions such as solution pH, we can systematically alter the bilayer thickness and the composition of the sequentially adsorbed polycation/polyanion building block. This in turn has made it possible to dramatically alter the device performance. This approach is not only effective in the $Ru(bpy)_3^{2+}$ polyester/PAA system as shown here, but also applicable to other systems such as the PPV/PAA system.

ACKNOWLEDGMENTS

This work was supported in part by the Office of Naval Research, the National Science Foundation and the MIT MRSEC program of the National Science Foundation (award number DMR-9400334).

REFERENCES

1. A. C. Fou, O. Onitsuka, M. Ferrari, M. F. Rubber and B. Heist, J. Apple. Phys. **79**, p. 7501 (1996).
2. O. Onitsuka, A. C. Fou, M. Ferreira, B. Hsieh and M. F. Rubner, J. Appl. Phys. **80**, p. 4067 (1996).
3. J-K. Lee, D. S. Yoo, E. S. Handy and M. F. Rubner, Appl. Phys. Lett. **69**, p. 1686 (1996).
4. J-K. Lee, D. Yoo and M. F. Rubner, Chem. Mater. **9(8)**, p. 1710 (1997).
5. A. Wu, J-K. Lee and M. F. Rubner, Thin Solid Films, accepted.
6. J. H. Burroughes, D. D. C. Bradley, A. R. Brown, R. N. Marks, K. Mackay, R. H. Friend, P. L. Burns and A. B. Holmes, Nature, **347**, p. 539 (1990); N. C. Greenham, S. C. Moratti, D. D. C. Bradley, R. H. Friend and A. B. Holmes, Nature, **365**, p. 628 (1993);
7. Q. Pei, Y. Yang, G. Yu, C. Zhang and A. J. Heeger, J. Am. Chem. Soc. **118**, p. 3922 (1996
8. K. M. Maness, R. H. Terrill, T. J. Meyer, R. W. Murray and R. M. Wightman, J. Am. Chem. Soc. **118**, p. 10609 (1996). K. M. Maness, H. Masui, R. M. Wightman and R. W. Murray, J. Am. Chem. Soc. **119**, p. 3987 (1997).

IMPROVED ELECTRON INJECTION IN ORGANIC LIGHT EMITTING DEVICES BY APPLYING THIN INSULATING LAYERS

R. PAIRLEITNER*, S. TASCH*, G. LEISING*, U. SCHERF**
*Institut für Festkörperphysik, Technische Universität Graz, Petersgasse 16, A-8010 Graz, Austria
**Max-Planck-Institut für Polymerforschung, Ackermannweg 10, D-55128, Germany

ABSTRACT

We present the fabrication and characterization of organic light emitting devices (OLEDs) using thin insulating layers for improved electron injection. The OLEDs are constructed with an ITO anode and an aluminum cathode. For the active layer we use either ladder-type Polyparaphenylene (m-LPPP) or Parahexaphenyl (PHP). A thin film of an insulating material is applied between the active layer and the cathode, in order to achieve a better tunnel injection due to a higher electric field at the interface. We compared different insulating materials with various thickness. The best results are obtained by using a LiF-layer with a thickness between 10 Å and 15 Å. Thereby the onset voltage decreases and the current density in the device increases significantly.

INTRODUCTION

After the application of conjugated polymers as organic semiconductors in organic light-emitting devices (OLEDs) had been initially demonstrated [1,2], poly(paraphenylene) (PPP), was the first simple conjugated polymer to produce visible blue light, when used as the active layer [3]. The realization of blue OLEDs is a necessary step for the possible application of the polymers in multicolor screens.

To realize OLEDs with operation lifetimes sufficient for industrial application, is much easier with small band gap conjugated polymers, because for these polymers the mismatch between the energy bands of the polymers relative to the electrodes are much smaller than for wide band gap polymer. The charge injection in OLEDs depends exponentially on the energetic barrier height at the interface between the polymer and the electrode. Thus, lower operation voltages are necessary for OLEDs with a small band gap than for high band gap polymers, so that it is easier to realize red-orange and also green OLEDs with long operation lifetimes than it is the case for blue ones.

One possibility to decrease the onset voltages (i.e. electric fields) of OLEDs is to insert a thin insulating layer between the active layer and the cathode, as it has been shown recently [4-9]. Suitable insulating layers for this purpose can be made of organic (e.g. Poly(methyl-methacrylate) [4] and some inorganic materials, as SiO_2 and LiF [5-9]. Although this effect has been observed by several groups using various active layers, the fundamental mechanism and the influence of the specific material parameters of the insulating layers onto the device performance are still not clarified [8].

In this article we investigate the influence of various thin insulating layers on the device performance of blue OLEDs based on Parahexaphenyl (PHP), which are produced in a one step UHV-vacuum deposition process, and of OLEDs based on a derivative of ladder-type Polyparaphenylene (m-LPPP) as shown in Figure 1. The device perfomance of the OLEDs is presented and the influence on the elctroluminescence (EL) efficiency is discussed.

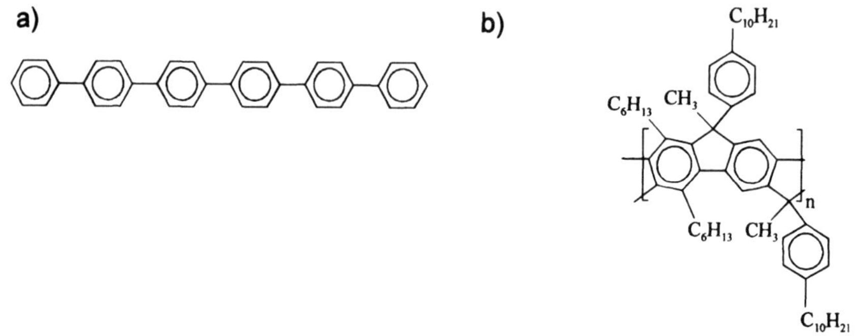

a)   b)

Figure 1: Chemical Structures of the used organic materials: a) para-hexaphenyl (PHP) and b) methylated ladder-type poly(para-phenylen) (m-LPPP)

EXPERIMENT

The EL devices were produced with thoroughly cleaned Indium Tin Oxide (ITO) (work function $\Phi \approx 5.1$ eV) coated glass substrates, serving as the hole injection electrode, onto which the active organic layer was applied. In case of m-LPPP the polymer film was spin cast from oxygen-free m-LPPP solutions in an Argon Box. The m-LPPP was dissolved in dry toluene by stirring at 60°C for several hours. Finally, the cast film has been annealed in the Argon Box at a temperature at about 60°C for at least a half day. The thickness of the m-LPPP layer was determined via absorption measurement, with a Perkin Elmer λ9 spectrophotometer. In case of PHP, the film was thermally evaporated at a pressure below $1 \cdot 10^{-7}$ mbar, with a thickness of 150 nm. The thickness was measured with a quartz crystal thickness monitor.

We made two different series of PHP devices. One with a low deposition rate of approximately 0.2 Å/s and the other with a high deposition rate of 1.6 Å/s. For a low deposition rate, it has been shown, that the PHP molecules are arranged nearly perpendicular to the substrate surface. The angle between the substrate normal and the axis of the PHP is around 20° [10]. Whereas for high deposition rates the PHP molecules are lying parallel to the substrate. The orientation of the PHP molecules effects the device characteristic due to the high anisotropy of the electronically and optical properties [11-13].

The insulating layers were evaporated by an electron beam evaporation source. After that the Al electrode ($\Phi \approx 3.9$ eV) was thermally evaporated in the same vacuum chamber. The I/V-characteristics were measured with a KEITHLEY source - measure unit. The external EL quantum efficiency was measured in an integrating sphere with a calibrated photo-diode. The devices were operated with a series resistance of 995 Ω. All measurements were performed in air at room temperature.

RESULTS

When LiF is arranged between the active polymer layer and the cathode a significant decrease of the onset electric field for the current is obtained (we define the current onset as the point, where the current exceeds a value of 0.1 mA). The thickness of the LiF-layer varies from 6 Å to 20 Å. For LiF-thickness below 10 Å no improvement of the device performance was observed. We speculate this is due to the thin LiF layers growing as islands and inhomogeneous LiF-layers.

70

The m-LPPP OLEDs with a LiF thickness of 10 Å and 14 Å, respectively are characterized by a relatively low current onset at about 0.9 MV/cm and 0.8 MV/cm. Whereas, the onset of the current for the device with a 20 Å thick LiF layer is 1.4 MV/cm. In comparison, the m-LPPP single layer device has a current onset at about 1 MV/cm [14].

The m-LPPP OLEDs based on a 10 Å LiF layer shows a very high current density, indicating a very efficient electron injection. However, the m-LPPP devices using LiF as electron injection layer are characterized by a poor EL quantum efficiency. The EL efficiency diminishes with increasing thickness of the LiF layer up to 14 Å and then increases again for higher LiF-thickness. The turn-on field for the electroluminescence increases with increasing LiF thickness, as it is shown in Figure 2. Whereas, the onset for the m-LPPP device with 10 Å thick LiF-layer is only slightly higher than for the single layer device (1 MV/cm) it raises up to a value of 1.5 MV/cm for a 20 Å thick LiF-layer.

When we use other insulating layers as MgO or SiO_2, different results are observed as compared to the LiF results. The I-V and EL-voltage characteristics of these devices are shown in Figure 3. The MgO device (thickness $d_{MgO} = 11$ Å) shows a similar good electron injection into the active layer (compare current onset and current density) as the device with LiF, but a EL quantum efficiency more than 10 times higher than the LiF device. However, this quantum efficiency is a factor 2 lower than the efficiency of the single layer device. In contrast to the MgO device and the LiF device respectively, the device with SiO_2 ($d_{SiO2} = 15$ Å) shows a significantly higher onset field (1.3 MV/cm), and an EL efficiency that is a factor of 4 lower than the single layer device. Although the onset is higher than for the single layer device the slope of the SiO_2 device is clearly steeper. These results clearly give proof that the effect of decreasing the onset electric field strongly depends on the material parameters and the ionic character of the binding and dielectric constant of the thin film.

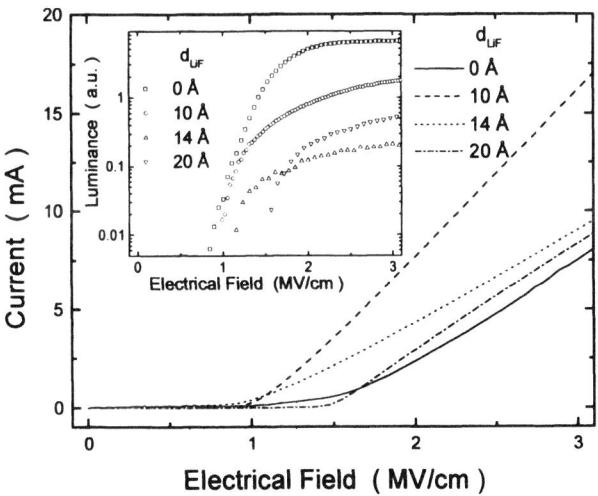

Figure 2: Current versus Electrical Field characteristics and EL characteristics (inset) of ITO/m-LPPP/LiF/Al devices (Thickness of the m-LPPP layers: 82 nm for single layer device; 94 nm for 10 Å LiF; 65 nm for 14 Å LiF; 64 nm for 20 Å LiF)

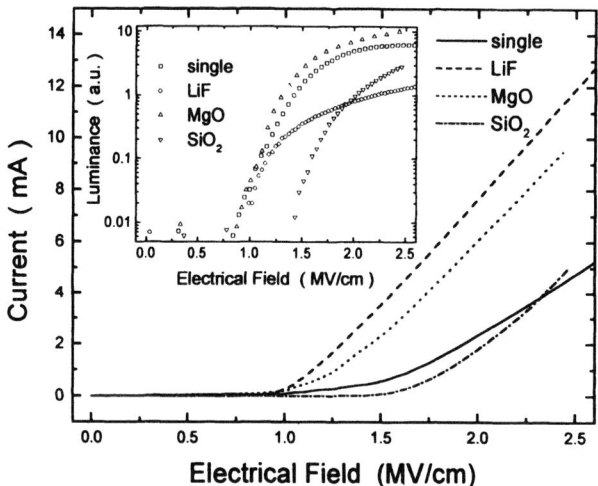

Figure 3: Current versus Electrical Field characteristics and EL characteristics (inset) of ITO/m-LPPP/Al devices with different insulating layers between m-LPPP and Al (Thickness of the layers: 95 nm for m-LPPP single layer device; 94 nm m-LPPP / 10 Å LiF; 121 nm m-LPPP / 11 Å MgO; 122 nm m-LPPP / 15 Å SiO₂)

The influence of these insulating layers on PHP OLEDs is significantly different compared to m-LPPP OLEDs and shows a strong dependence on the orientation of the PHP layer, which can be controlled by the deposition parameters (as the deposition rate [15]).

For OLEDs where PHP is produced with a high deposition rate, the onset of the OLEDs cannot be significantly improved with LiF layers of any certain thickness. The PHP single layer device show an extremely high EL quantum efficiency of around 2% photons/electron. The single layer device and the device with a 11 Å thick LiF layer exhibit a similar characteristic with a very low onset voltage (7 V) and a moderate current density. But the EL efficiency for the 11 Å LiF device is 3 times lower than that of the single layer device. The devices with thicker LiF layers show a significantly higher onset and only a slightly higher EL efficiency than the PHP device containing the 11 Å thick LiF layer.

For PHP devices with a low deposition rate, a completely different behavior is observed. The onset voltage for the OLEDs can be decreased compared to single layer devices, if a LiF layer of suitable thickness is applied between the active layer and the cathode. For a 12 Å thick LiF layer the onset voltage can be decreased to 6 V (d_{PHP} = 150 nm) compared to an onset voltage of 9 V for the PHP single layer. The onset voltage for pure PHP OLEDs, where the chains are lying perpendicular to the plane of the substrate, occurs at higher voltages, since the charge transport in this case is much worse then for PHP molecules which are lying in the substrate plane. In contrast to the m-LPPP and to PHP devices, from a fast deposition rate, the slowly deposited PHP devices shows a significant increase in the EL quantum efficiency. The quantum efficiency for a 12 Å thick LiF layer is a factor of 2 higher compared to the identically prepared single layer PHP device.

However, the stability of these devices, where PHP has been produced by a low deposition rate, is lower than the PHP fabricated with a fast deposition rate.

The performance of the OLEDs with other insulating materials as MgO or SiO$_2$ based on PHP produced with a low deposition rate can *not* be significantly improved:

(i)   the current onset of the PHP devices containing a 10 Å thick MgO layer occurs at 9 V and is the same as for identically prepared single layer devices, but the incline of the current above the onset voltage is much steeper than for the single layer OLEDs. The EL efficiency is by a factor of 2 lower than the EL efficiency of the single layer device.

(ii)   For the SiO$_2$ device (d_{SiO2} = 10 Å) the onset occurs at higher voltages of around 13 V and the EL efficiency is 4 times smaller than that of the single layer device.

DISCUSSION

Thin insulating layers applied between the active PHP and m-LPPP layer and the Al-cathode can be used to decrease the onset electric field of these OLEDs, as it has been observed and discussed for OLEDs based on other active materials [5-9]. The effect of the insulating layers on the EL quantum efficiency strongly depends on the active layer, especially on ordering phenomenon in the active layer, onto which they are applied. An increase of the EL quantum efficiency was only achieved, if the polymer or oligomer chains are oriented perpendicular to the plane of the substrate as it is realized for PHP layers deposited with a low deposition rate. This fact can be attributed to following scenario:

Holes injected into the active layer accumulate at the high potential barrier at the interface to the insulating layer. The emission process therefore also occurs adjacent to this interface within the active layer. The distance between the emissive zone and the cathode is only slightly higher than the thickness of the insulating layer, which in order to achieve a reduction of the onset electric field should be in the range of 10 Å. Within such small distances from the metal surface, the emitting properties of the active layer is strongly influenced due to electric dipole interaction processes [16] resulting in a strong decrease of the radiative emission. Such quenching effects near metal electrodes have been observed in the photoluminescence and electroluminescence processes of other organic materials [17,18]. One has to keep in mind that the maximum quenching process occurs, if the emitting material is oriented *parallel* to the metal layer and is lower for any other orientation. In case of the PHP produced with a high deposition rate and also for m-LPPP, the oligomer or polymer chains are oriented preferentially in the plane of the substrate [14,15], so that the quenching near the metal electrode will be strong. In the case of PHP fabricated with a low deposition rate, the oligomer chains are lying *perpendicular* to the plane of the substrate, so that quenching effects occurring near the interface to the metal electrode in the OLEDs are much weaker compared to the case where the chains are oriented *parallel* to the plane of the substrate.

The reason for the difference in the behavior of the various insulating layers on the device performance is currently under investigation.

CONCLUSIONS

We have shown that with the implementation of thin insulating layers between the emitting layer and the Al cathode the electron injection process in organic light emitting devices can be strongly influenced. By the application of a LiF layer with a thickness between 10 Å and 15 Å the electron injection in m-LPPP - and PHP - based OLEDs can be significantly increased. However, depending on the orientation of the organic molecules quenching processes which are occurring at the interface are drastically decreasing the EL quantum efficiency. The best improvement, referring to electron injection *and* EL quantum efficiency, can be obtained for the application of LiF on PHP which is arranged perpendicular to the substrate plane.

ACKNOWLEDGMENTS

The financial support of this research work by the Spezialforschungsbereich 'Elektroaktive Stoffe' is gratefully acknowledged.

REFERENCES

1. J. H. Burroughes, D. D. C. Bradley, A. R. Brown, R. N. Marks, K. Mackay, R. H. Friend, P. L. Burn, A. Kraft, A. B. Holmes, Nature **347**, 539 (1990).

2. Y. Ohmori, M. Uchida, K. Muro, K. Yoshino, Jpn. Appl. Phys. **30**, L1941 (1991)

3. G. Grem, G. Leditzky, B. Ullrich, G. Leising, Adv. Mater. **4**, 36 (1992)

4. H. H. Kim, T. M. Miller, E. H. Westerwick, Y. O. Kim, E. W. Kwock, M. D. Morris and M. Cerullo, J. of Lightwave Techn. **12**, 2107 (1994).

5. L. S. Hung, C. W. Tang and G. M. Mason, Appl. Phys. Lett. **70**, 152 (1997).

6. Y.E. Kim, H. Park and J.-J. Kim, Appl. Phys. Lett. **69**, 680 (1996).

7. R. N Marks, F. Biscarini, T. Virgili, M. Muccini, R. Zamboni and C. Taliani, Phil. Trans. R. Soc. Lond. A 355, 763 (1997).

8. H. Tang, F. Li and J. Shinar, Appl. Phys. Lett. **71**, 2560 (1997).

9. G. E. Jabbour, Y. Kawabe, S. E. Shaheen, J. F. Wang, M. M. Morrell, B. Kippelen and N. Peyghambarian, Appl. Phys. Lett.**71**, 1762 (1997).

10. L. Athouel, G. Froyer, M. T. Riou, Synth. Met. **57**, 4734 (1993).

11. C. Ambrosch-Draxl, J. A. Majewski, P. Vogl, G. Leising, Phys. Rev. B **51** , 9668 (1995).

12. A. Niko, F. Meghdadi, C. Ambrosch-Draxl, P. Vogl, G. Leising, Synth. Met. **76**, 177 (1996).

13. F. Meghdadi, G. Leising, W. Fischer, F. Stelzer, Synth. Met. **76**, 113 (1996).

14. S. Tasch, A. Niko, G. Leising, U. Scherf, Appl. Phys. Lett. **68**, 1090 (1996).

15. R. Resel, N. Koch, F. Meghdadi, G. Leising, W. Unzog, K. Reichmann, Thin Solid Films **305**, 232 (1997).

16. K. H. Drexhage, J. of Luminescence 1,2, 693 (1970).

17. V.-E. Choong, Y. Park and Y. Gao, T. Wehrmeister and K. Müllen, B. R. Hsieh, C. W. Tang, J. Vac. Sci. Technol. A 15(3), 1745 (1997)

18. H. Becker, S. E. Burns and R.H. Friend, Phys. Rev. B **56**, 1893 (1997).

CONTROL OF LIGHT-EMITTING POLYMER DEVICES USING POLYMER/POLYMER INTERFACES

A. J. Epstein [a,b], Y. Z. Wang [a], D. D. Gebler [a], D. K. Fu [c], T. M. Swager [c]
[a] Department of Physics, The Ohio State University, Columbus, OH 43210-1106
[b] Department of Chemistry, The Ohio State University, Columbus, OH 43210-1185
[c] Department of Chemistry, Massachusetts Institute of Technology, Cambridge, MA 02139-4307

ABSTRACT

We present the use of polymer/polymer interfaces to control light-emitting polymer devices. Bilayer devices utilizing poly(9-vinyl carbazole) (PVK) as a hole transporting/electron blocking polymer together with a pyridine containing electron transporting layer show dramatically improved efficiency and brightness as compared to single layer devices. This is attributed to charge confinement and exciplex emission at the PVK/emitting polymer interface. The introduction of emeraldine base (EB) form of polyaniline (PAN) on both side of the emitting layer enables the device to work under both forward and reverse bias, as well as in AC modes. Interfaces play an important role in the operation of these devices. Furthermore, when the EB is replaced by sulfonated polyaniline (SPAN) on the cathode side and the emitting layer is properly modified to balance electron and hole transport, the device generates different colors of light, red under forward bias and green under reverse bias.

INTRODUCTION

Electroluminescence (EL) combined with other unique properties of polymers, such as solution processibility, band gap tunability and mechanical flexibility, make conjugated polymers excellent candidates for low cost large area display applications [1-4]. Most "conventional" polymer light-emitting devices have been shown to be tunneling diodes and can be operated only under forward DC bias [5]. Among the most important limitations associated with many of the "conventional" polymer light-emitting diodes (LEDs) are poor stability and shelf lifetime. The double charge injection mechanism of the "conventional" polymer LEDs requires the matching of the cathode (anode) work function to the corresponding LUMO (HOMO) level of the polymer in order to achieve efficient charge injection. The relatively low electron affinity of most conjugated polymers requires metals with very low work functions to achieve efficient electron injection.

Pyridine-containing conjugated polymers are promising candidates for light-emitting devices [6]. The pyridine containing polymers are highly luminescent, especially the copolymers. The high electron affinity of pyridine based polymers enables the use of relatively stable metals such as Al, Au, or ITO and doped polyaniline as electrodes. Taking advantages of the better electron transport properties of the pyridine-containing polymers, we fabricated bilayer devices utilizing poly(9-vinyl carbazole) (PVK) as hole transporting/electron blocking polymer, which improves the device efficiency and brightness significantly due to the charge confinement and exciplex emission at the PVK/emitting polymer interface [7]. By inserting a layer of emeraldine base (EB) form of polyaniline on both side of the emitting polymer, we fabricated symmetrically configured AC light-emitting (SCALE) devices [8] which work under both forward and reverse DC bias as well as in AC modes. When we replace the EB layer on the cathode side of the SCALE device with sulfonated polyaniline (SPAN) and modify the emitting layer to appropriately balance

Mat. Res. Soc. Symp. Proc. Vol. 488 © 1998 Materials Research Society

electron and hole transport, color variable bipolar/AC light-emitting devices are fabricated that can generate different colors of light depending on the polarity of the driving voltage, red under forward bias and green under reverse bias [9]. Figure 1 shows the repeat units of the pyridine-containing polymers and other polymers that were used in these devices.

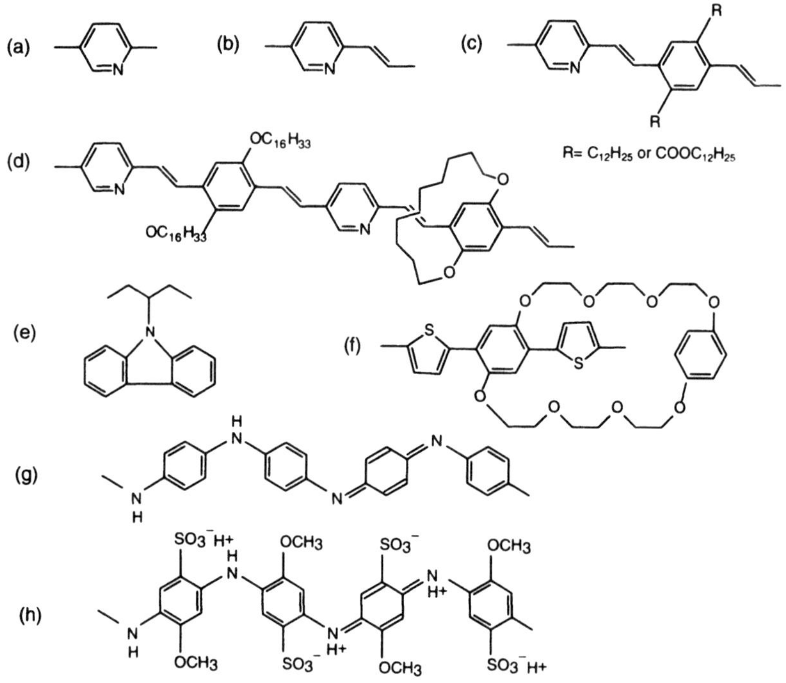

Fig. 1 Repeat units of the pyridine-containing polymers and other polymers used in the study. (a) poly(p-pyridine) (PPy); (b) poly(p-pyridyl vinylene) (PPPyV); (c) copolymer of PPV and PPyV (PPyVP(R)₂V); (d) wrapped copolymer of pyridyl vinylene and phenylene vinylene (@PPyVPV); (e) poly(9-vinyl carbazole) (PVK); (f) wrapped copolymer of dithienylene and phenylene (@PTP); (g) emeraldine base (EB) form of polyaniline; (h) sulfonated polyaniline (SPAN).

EXPERIMENT

For bilayer devices, the PVK layer was spin coated onto the ITO substrate from solution in tetrahydrofuran (THF) (~10 mg/ml) at ~3000 rpm. The emitting layer was then spin coated on top of the PVK layer from appropriate solutions (typically xylenes). As the solvent for the emitting layer does not dissolve PVK, no significant intermixing of the two polymers is expected. The SCALE devices were fabricated by spin-casting a solution of EB in N-methyl pyrrolidinone (NMP) (~5 mg/ml) onto an ITO substrate. After drying in dynamic vacuum, a layer of the emitting polymer was spin-coated on the EB surface. Another EB layer was similarly coated on top of the emitting layer. The spinning speed for all layers is ~ 2000 rpm. The color variable/AC light-emitting devices were similarly fabricated as the SCALE devices except that the top EB layer was replaced by SPAN cast from aqueous solution and the emitting layer was comprised of a blend of polymers. All the spin coating procedures were carried out inside a class 100 cleanroom in air. The top metal electrodes were deposited by vacuum evaporation at a pressure below 10^{-6} torr. To prevent damage to the polymers, the substrates were mounted on a cold-water cooled surface during the evaporation.

Absorption spectra were measured on spin-cast films using a Perkin-Elmer Lambda 19 UV/VIS/NIR spectrometer. Photoluminescence (PL) and EL were measured using a PTI fluorometer (model QM-1). The current-voltage characteristics were measured simultaneously with EL using two Keithley model 195A multimeters while dc voltage was applied by a HP model 6218A DC power supply. We note that all the device testing procedures were performed in air on as-made devices without any encapsulation.

RESULTS AND DISCUSSION

The performance of a bilayer devices using PVK as the hole transporting/electron blocking layer improves dramatically as compared to single layer devices. The internal quantum efficiency and brightness of the bilayer devices increase two to three orders of magnitude, reaching ~0.5% and over 300 cd/m^2, respectively. Figure 2 compares the electroluminescence-voltage and electroluminescence-current characteristics for a single layer device and a bilayer device using the wrapped copolymer as the emitting layer. PVK enhances the transport of holes injected from the anode and blocks the transport of electrons injected from the cathode such that the electrons accumulate at the PVK/copolymer interface. In addition, the PVK layer removes the recombination zone from the vicinity of the electrode so that the radiative recombination is protected against the non-radiative quenching at the electrode/polymer interfaces.

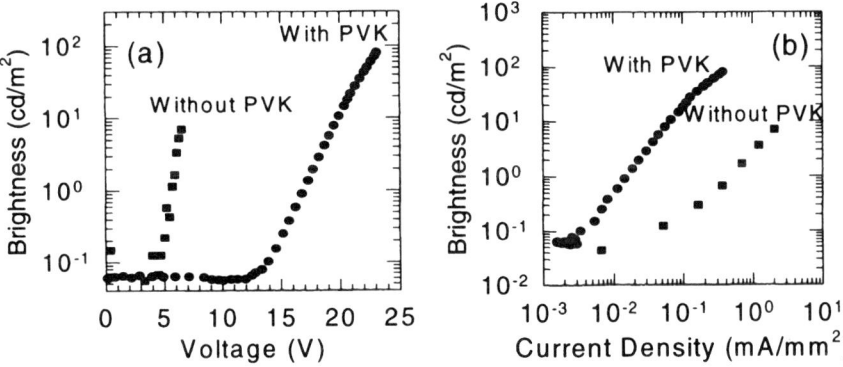

Fig. 2 Comparison of (a) brightness-voltage and (b) brightness-current characteristics for a single layer device (square) and a bilayer device (circle).

Quenching at the electrodes has been reported for phenylene vinylene oligomers by Park *et al.*[10] and in MEH-CN-PPV by Becker *et al.* [11]. Becker *et al.* also note that the removal of the recombination zone away from the metal electrode to a polymer/polymer interface is crucial to increasing the EL efficiency in their PPV/MEH-CN-PPV bilayer devices [11] In order to probe the effect of the electrodes on the luminescence properties of our single and bilayer films a number of samples were fabricated on ITO. Electrodes were evaporated on single layers of PVK and wrapped copolymer @PPyVPV. On each film the PL was measured in the vicinity of and far from the electrode with the exciting light entering the sample from the ITO side and recording the EL from the ITO side. The insets of Fig. 3 show the device configurations. The relative PL intensities of PVK at these two locations for 330 nm excitation wavelength (Fig. 3a) shows that the PL increases modestly when the excitation light is incident on the electrode. In the absence of quenching an increase in PL intensity is expected for two reasons: reflection of the excitation light by the Al and reflection of the photoluminescence by the Al. The small increase for PVK indicates significant quenching near the Al electrode. Figure 3b shows the same experiment for a single layer of wrapped copolymer @PPyVPV using a 450 nm wavelength excitation. The nearly two-fold decrease in PL intensity indicates extensive quenching by the Al electrode for this polymer.

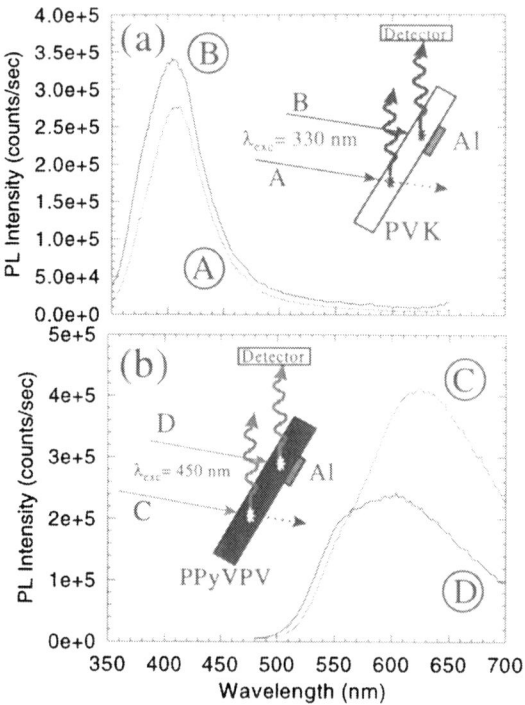

Fig. 3 PL of films of PVK (A,B) and PPyVPV (C,D) for the configurations shown schematically in the insets. (a) 330 nm excitation and (b) 450 nm excitation.

For bilayers in which the luminescence occurs primarily at the interface the PL and EL intensity should be unaffected by the presence of the electrode, i.e., no quenching should occur. Figure 4b shows the PL intensity for a PVK/@PPyVPV bilayer with a 330 nm excitation (experimental configuration shown in Fig. 4a). When the excitation wavelength is incident on the evaporated Al electrode the PL intensity increases dramatically. The increase is attributed to the reflection of the luminescence and the excitation light implying that the emission is primarily occurring at the polymer/polymer interface. Figure 4c shows the PL intensity of the bilayer when the lower energy excitation is used (450 nm). In this case the PL intensity is also increased, but the emission is shifted away from the single layer emission to the exciplex (interface) emission, when the exciting light is incident on the reflecting electrode. Although the low energy excitation is used, the emission at the interface is observed because the wrapped copolymer @PPyVPV PL is quenched at the electrode and the interface emission is enhanced. Energy is transferred from excitons in the wrapped copolymer @PPyVPV to exciplexes at the interface. The diffusion coefficient for excitons in the @PPyVPV copolymer can be estimated from the diffusion distance ~10^{-6} cm) and the lifetime of the exciton in the copolymer (~1 ns) [12], giving a diffusion coefficient of ~ 10^{-3} cm^2/s. Thus, exciplex emission in the bilayer devices substantially reduces the effects of quenching by electrodes in single layer devices.

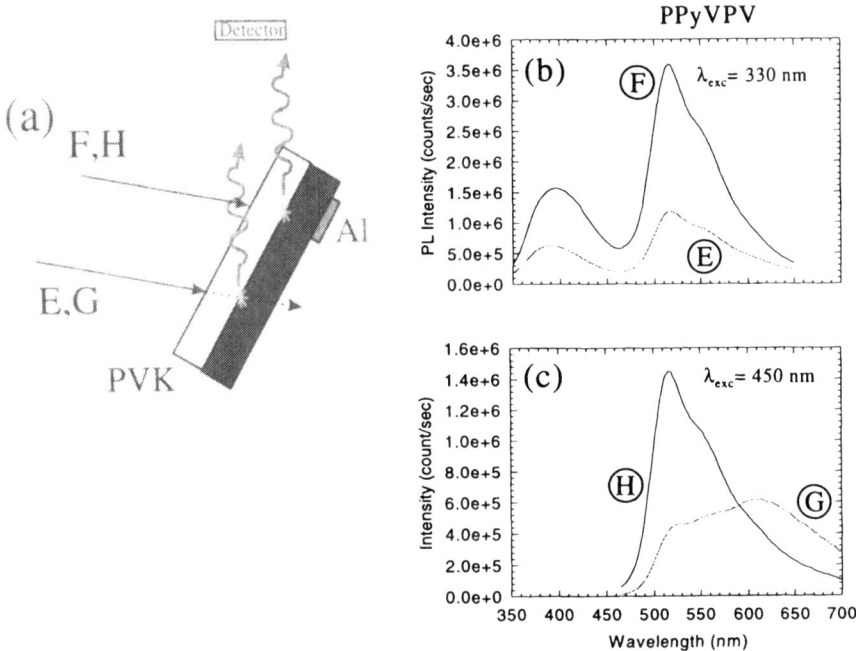

Fig. 4 PL of bilayer films of PVK/PPyVPV with the configuration shown schematically in (a). The excitation energies were (b) 330 nm (E,F) and (c) 450 nm (G,H).

An unwanted side effect of using the PVK layer is that it increases the device operating voltage substantially. One effective way to reduce the device turn on voltage is to use a high surface area network electrode [13]. The rough electrode will create a non-uniform high electric field that enhances the charge injection. This technique has been successfully applied to PPV based single layer devices [14].

The light generation in the bilayer device is attributed to decay of exciplexes formed at the PVK/copolymer interface. Figure 5 compares the PL of pure wrapped copolymer, pure PVK, and bilayer of PVK/copolymer, as well as the EL spectra of the bilayer device. The PL of PVK film excited at 3.6 eV has an emission peak at 3.06 eV. The PL of pure wrapped copolymer film excited at 2.65 eV peaks at 2.03 eV with a shoulder at 2.25 eV. When the PVK/copolymer bilayer film is excited at 3.6 eV, a new peak appears at 2.38 eV with another peak at 2.23 eV. The new peak is assigned to exciplex emission at the PVK/copolymer interface, supported by the optical absorption and photoluminescence excitation (PLE) measurements. The EL of the bilayer device follows closely with the PL of the bilayer film in accord with the EL emission originating primarily from the exciplex formed at the PVK/copolymer interface.

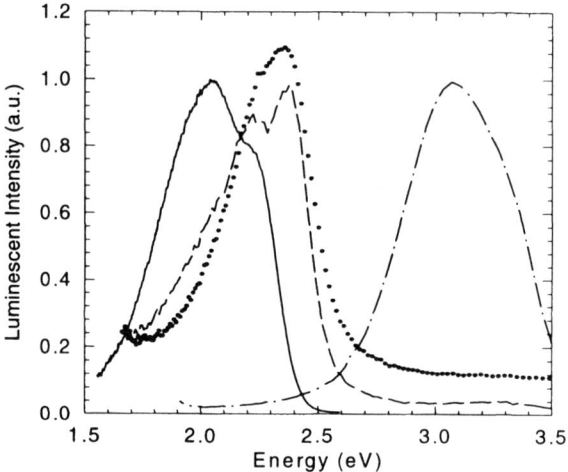

Fig. 5 Normalized PL spectra of pure wrapped copolymer film (solid line) excited at 2.65 eV, pure PVK film (dash-dotted line) excited at 3.6 eV, bilayer of PVK/copolymer (dashed line) excited at 3.6 eV, and the EL of the bilayer device (filled circle). The EL spectra are offset for clarity.

Two novel device configurations that can be operated in both forward and reverse DC bias as well as in AC modes are the SCALE devices and color variable bipolar/AC light-emitting devices. The SCALE devices consist of a light-emitting layer sandwiched between two "insulating" redox polymer layers, then placed between two electrodes. The SCALE device configuration is quite general, it can be applied to a variety of emitting and redox polymers, as well as electrodes. For the SCALE device presented here, PPy was used as the emissive layer; the emeraldine base (EB) form of polyaniline was utilized as the redox material; ITO and Al were used as electrodes. Figure 6 inset shows schematically the structure of such SCALE devices. They emit light under both forward and reverse DC bias as well as under AC driving voltage. Under low frequency AC (sinusoidal) driving voltage, nearly uniform light pulses with twice the driving frequency were observed, Figure 6. This unusual behavior is attributed to the effects of charge accumulation at the polymer/polymer interfaces [8]. We propose the following mechanism for the SCALE device operation. Under low bias voltages, electrons and holes can be injected from the electrodes into the quinoid and benzenoid levels of EB and form negative and positive polarons, respectively. These polarons transport to the EB/PPy interfaces via a hopping mechanism and populate the EB/PPy interfaces at the polaron levels. Because the polarons levels are within the π-π^* band gap of EB, and are also likely within the band gap of PPy, the barriers for charge injection from electrodes to polaron levels of EB are significantly reduced as compared to injection directly to conduction and valence bands of PPy. Within this model, the limiting barriers for charge injection are changed from the electrode/polymer contacts as proposed for conventional polymer LEDs to the polymer/polymer (EB/PPy) interfaces of the SCALE devices. When the applied electric field is high enough, the stored charges begin to tunnel into the conduction and valence bands of PPy. When they meet, the injected charges may form intrachain excitons and decay radiatively to emit photons or follow other nonradiative decay paths. If the charge injection is not balanced, as is the case for most polymer LEDs, the excess charge carriers may migrate through the PPy layer without decaying. Most of these charges will be trapped in the opposite PPy/EB interface. When the bias voltage is reversed, the shallow trapped charges will be released from the interfaces and contribute to the recombination current. The deep trapped charges which act as quenchers or injection limiters in DC devices will be neutralized. We point out that the use of stable high work function metals, such as Au, as electrodes to inject both electrons and holes for the SCALE devices may reduce the problems of aging of contacts of polymer light-emitting devices. Also, continuous reversal of the sign of the driving voltage under AC operation may reduce degradation.

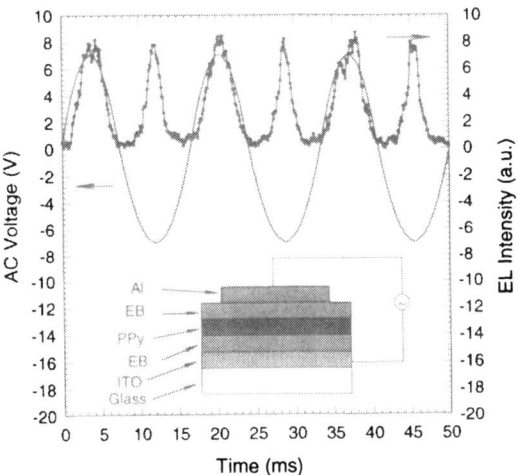

Fig. 6 EL intensity as a function of time for a SCALE device driven by a 60 Hz sinusoidal voltage. Inset: schematic device structure of the SCALE device.

We have reported a new approach to the color variable polymer light-emitting devices based on SCALE device structure: color variable bipolar/AC light-emitting devices [9]. In this approach we replace the EB layer on the cathode side of the SCALE device structure with sulfonated polyaniline (SPAN). Because the SPAN is capable of modifying the emission properties of certain pyridine containing polymers at the interface such that the interface emits different colors of light than the bulk does, the device is able to generate a different color of light under forward and reverse bias. In this approach, the colors of light are controlled by selecting the desired emission locations which in turn are controlled by the polarity of driving voltage and the charge injection and transport properties of the polymer layers. Since motion of ionic species is not required for device operation a relatively fast time response is achieved, allowing the colors to be switched rapidly.

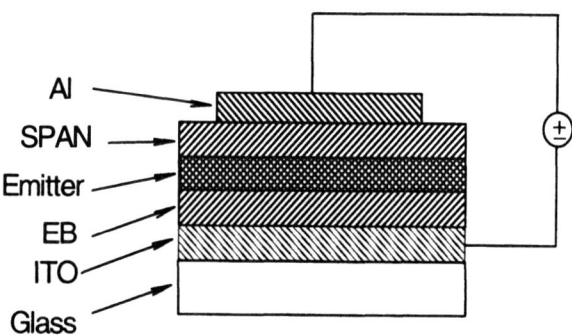

Fig. 7 Schematic structure of the color variable light-emitting devices

For the devices presented here, the wrapped copolymer @PPyVPV, and a wrapped copolymer of polythiophene and polyphenylene derivative, @PTP, were used as the emitting materials; SPAN and EB were used as the redox materials; ITO and Al were used as electrodes. Figure 7 shows the schematic diagram of the device structure of the color variable bipolar/AC light-emitting devices. Figure 8 shows the typical current-voltage and brightness-voltage characteristics of the color variable devices. The devices have typical turn on voltages of ~4-8 V depending upon film thickness and work equally well under both polarities of driving voltage with different colors of light being emitted. The light appeared red and green to the eye under forward and reverse bias, respectively, and was clearly visible under normal indoor lighting. Internal device efficiencies of 0.1% photons/electron has been achieved for the initial devices. Figure 9 shows the EL spectra of the color variable device under forward and reverse bias. For comparison, the EL spectra for single layer @PPyVPV and @PTP are also shown. The CIE chromaticity x,y coordinates of the two spectra are calculated to be (0.654,0.345) and (0.471,0.519), respectively, showing both colors to be relatively pure (see Fig. 8 inset). The colors of the devices can be switched rapidly, up to ~20 kHz, depending upon device impedance and geometry.

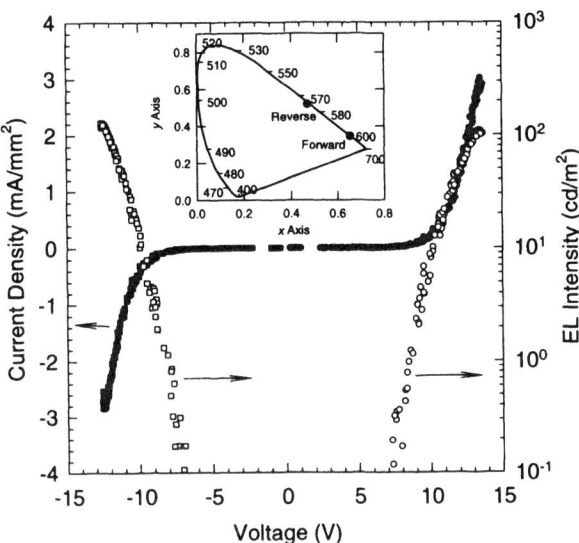

Fig. 8 Current-voltage and brightness-voltage characteristics of a color variable light-emitting device under forward and reverse bias conditions. Inset shows the colors of the device under forward and reverse bias in the CIE chromaticity diagram.

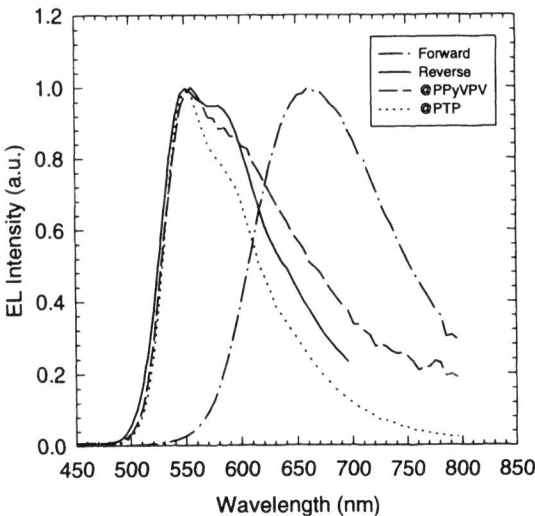

Fig. 9 EL spectra of the color variable device under forward and reverse bias conditions. For comparison, the EL spectra for single layer PPyVPV and PTP devices also are shown.

The EL spectrum under forward bias is substantially different from that of the single layer devices of either @PPyVPV or @PTP, suggesting that the light is generated from the interface between the emitter blend and either EB or SPAN under forward bias. To further clarify this, we fabricated the following devices: ITO/@PPyVPV/Al; ITO/SPAN/@PPyVPV/Al; ITO/@PPyVPV/SPAN/Al; ITO/SPAN/@PPyVPV/SPAN/Al and similar devices replacing SPAN with EB. Only ITO/@PPyVPV/SPAN/Al and ITO/SPAN/@PPyVPV/SPAN/Al show dramatically redshifted EL. Similar studies using @PTP as the emitting layer show that the emission properties of @PTP are not affected significantly by the presence of the SPAN layer.

The EL spectrum of the color variable device under reverse bias are similar to those of the single layer @PPyVPV and @PTP devices implying that the light is generated either in the bulk of the emitting polymer or at the EB interface. The EB layer functions as a charge injection enhancement layer, playing a similar role as it does in SCALE devices reported earlier [8]. Under reverse bias, the SPAN layer on the cathode side plays a similar role.

It is noted that when the blend in the color variable devices is replaced by pure @PPyVPV polymers, the devices emit red light in forward bias and red-orange light in reverse bias. This indicates that under reverse bias the light is still generated near @PPyVPV/SPAN interfaces. This can be understood in terms of different electron and hole transport properties of the @PPyVPV and @PTP polymers. It is known that most conjugated polymers, including poly(phenylene vinylene) (PPV), poly(p-phenylene) (PPP) and polythiophene, have better hole than electron transport properties [3]. The addition of a high electronegativity unit, pyridine, to the backbone is expected to improve the electron transport properties. Therefore, under reverse bias the light is still generated near the SPAN interface for the ITO/EB/@PPyVPV/SPAN/Al device. By adding @PTP, which is expected to have better hole transport properties, to @PPyVPV, the overall hole transport properties of the blend is improved, and hence the light is generated away from the SPAN interface for the ITO/EB/Blend/SPAN/Al device under reverse bias, emitting green light. The mechanism for the SPAN layer changing the emission properties of the @PPyVPV polymer is attributed to the formation of new emissive species due to protonation of the pyridyl units by SPAN. The new species are emissive and emit redshifted light as compared to the non-protonated polymer.

SUMMARY AND CONCLUSION

In summary, we have presented a number of DC and AC light-emitting devices based on the role of polymer/polymer interfaces. The high electron affinity of pyridine based polymers enables the use of relatively stable metals such as Al as efficient electron injecting contacts. Taking advantages of the better electron transport properties of the pyridine-containing polymers, we have fabricated bilayer devices utilizing PVK as hole transporting/electron blocking polymer. The bilayer device structure improves the device quantum efficiency and brightness significantly due to the charge confinement and exciplex emission at the PVK/emitting polymer interface. The fabrication of novel devices such as SCALE devices and color variable bipolar/AC light emitting devices opens new possibilities for device design and operation.

ACKNOWLEDGMENT

This work was supported in part by Office of Naval Research.

REFERENCES

[1]   C. W. Tang and S.A. VanSlyke, Appl. Phys. Lett. **51**, 913 (1987).
[2]   J. H. Burroughes, D. D. C. Bradley, A. R. Brown, R. N. Marks, K. Mackay, R. H. Friend, P. L. Burns, and A. B. Holmes, Nature **347**, 539 (1990).
[3]   D. D. C. Bradley, Synth. Met. **54**, 401 (1993); J. Kido, Trends in Polymer Science **2**, 350 (1994).
[4]   See, for example, Proc. Int. Confs. Sci. Tech. Synth. Met., Seoul, Korea, July 21-29, 1994 (Synth. Met. **69-71**,1995); Snowbird, Utah, July 28 - Aug. 2, 1996 (Synth. Met. **84-86**, 1997); J. W. Blatchford and A. J. Epstein, Am. J. Phys. **64**, 120 (1996), and references therein.
[5]   I. D. Parker, J. Appl. Phys. **75**, 1656 (1994).
[6]   A. J. Epstein, J. W. Blatchford, Y. Z. Wang, S. W. Jessen, D. D. Gebler, L. B. Lin, T. L. Gustafson, H. L. Wang, Y. W. Park, T. M. Swager, and A. G. MacDiarmid, Synth. Met. **78**, 253 (1996).
[7]   D. D. Gebler, Y. Z. Wang, J. W. Blatchford, S. W. Jessen, D. K. Fu, T. M. Swager, A. G. MacDiarmid, and A. J. Epstein, Appl. Phys. Lett. **70**, 1644 (1997).
[8]   Y.Z. Wang, D.D. Gebler, L.B. Lin, J.W. Blatchford, S.W. Jessen, H.L. Wang, and A.J. Epstein, Appl. Phys. Lett. **68**, 894 (1996).
[9]   Y. Z. Wang, D. D. Gebler, D. K. Fu, T. M. Swager, and A. J. Epstein, Appl. Phys. Lett. **70**, 3215 (1997).
[10] Y. Park, V. -E. Choong, B. R. Hsieh, C. W. Tang, and Y. Gao, Phys. Rev. Lett. **78**, 3955 (1997)
[11] H. Becker, S. E. Burns, and R. H. Friend, Phys. Rev. B **56**, 1893 (1997).
[12] J. W. Blatchford, S. W. Jessen, L. B. Lin, T. L. Gustafson, H. L. Wang, T. M. Swager, A. G. MacDiarmid, and A. J. Epstein, Phys. Rev. B **54**, 9180 (1996).
[13] Y. Z. Wang, D. D. Gebler, D. K. Fu, T. M. Swager, and A. J. Epstein, Synth. Met. **85**, 1179 (1997).
[14] Y. Yang, E. Westerweele, C. Zhang, P. Smith, and A. J. Heeger, J. Appl. Phys. **77**, 694 (1995).

NEW LUMINESCENT PPV DERIVATIVES FOR LED APPLICATIONS

B.S. Chuah[*], F. Cacialli[**], J. E. Davies[***], N. Feeder[***], R. H. Friend[**], A. B. Holmes[*], E. A. Marseglia[**], S. C. Moratti[*], J.-L. Brédas[****], D.A. dos Santos[****]

[*]Melville Laboratory for Polymer Synthesis, Department of Chemistry, University of Cambridge, Pembroke Street, Cambridge, CB2 3RA U.K.
[**]Cavendish Laboratory, Department of Physics, University of Cambridge, Madingley Road, CB3 0HE U.K.
[***]University Chemical Laboratory, Department of Chemistry, University of Cambridge, Lensfield Road, Cambridge, CB2 1EW, U.K.
[****]Centre de Recherche en Electronique et Photonique Moléculaires, Université de Mons-Hainaut, Place du Parc, 20, B-7000 Mons, Belgium

ABSTRACT

A new poly(phenylenevinylene) [PPV] derivative **8** with dialkoxy substituents at the 2,3-positions of the phenylene ring of the polymer backbone has been prepared. This exhibited significantly different properties compared with the typical 2,5-dialkoxy substituted PPV derivatives. The polymer is not only significantly blue-shifted in its optical and luminescence properties but also has high photoluminescence (PL) and electroluminescence (EL) efficiencies. Investigation of the model oligomer **11** offers insight into this interesting behavior.

INTRODUCTION

The 2,5-dialkoxy substituted PPVs have been extensively explored. The orange-red luminescent poly(2-methoxy-5-ethylhexyloxy)-1,4-phenylenevinylene (MEH-PPV) has dominated research, owing to its reliable PL (15-20%) and EL device efficiency [1, 2]. Other PPV analogs with different substituents such as long chain alkoxy-[3], silyl [4], alkylthio- [5] and alkyl have since been investigated. An area less often studied is the effect of substitution patterns on the phenylene ring. We were interested in the effect of 2,3-dialkoxy substitution on the properties of PPV derivatives [6].

We present in this paper the synthesis of the poly(2,3-dibutoxy-1,4-phenylenevinylene) [PDB-PPV] **8** having the unconventionally positioned dialkoxy substituents on the phenylene ring of the PPV backbone, using the Gilch dehydrohalogenation [7] route. The characterization and observation of significantly blue-shifted emission properties in both PL and EL devices are reported, together with supporting information on the crystal structure of the model oligomer **11**.

RESULTS

Synthesis of the Polymer PDB-PPV 5

The monomer was synthesized using the method outlined in Scheme 1. Alkylation of the bis phenol **3** using the Mannich reagent derived from **1** and **2** afforded the morpholino derivative **4** [8]. This underwent transfer hydrogenolysis [9] using Pearlman's catalyst to yield the dimethyl catechol **5**. The catechol derivative was O-alkylated to produce the dibutoxy-xylene **6** which underwent side chain bromination [10] using N-bromosuccinimide in tetrachloromethane to give the bis(bromomethyl) monomer **7**.

The polymerization of the bis(bromomethyl) benzene **7** was carried out in tetrahydrofuran (THF) at room temperature using excess potassium t-butoxide to ensure a high degree of dehydrohalogenation. It was found (elemental analysis) that six equivalents of potassium tert-butoxide yielded <0.1% bromine in the resulting polymer.

87

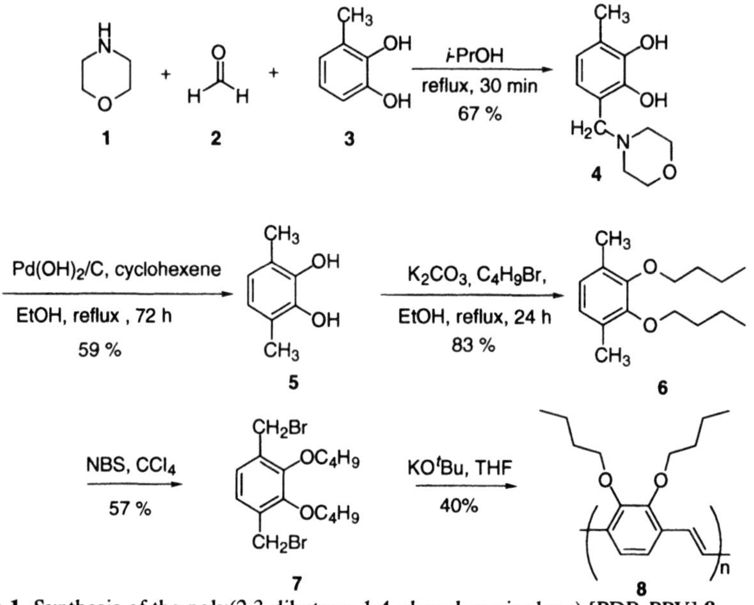

Scheme 1 Synthesis of the poly(2,3-dibutoxy-1,4-phenylenevinylene) [PDB-PPV] **8**

Characterization and Optoelectronic Properties of PDB-PPV **8**

The polymer **8** is bright yellow and is highly luminescent. The polymerization route afforded a polymer with high molecular weight and a polydispersity typical of a dehydrohalogenation route. Despite the high molecular weight the alkyl chains of PDB-PPV enable solubility in organic solvents such as chloroform and tetrachloroethane, thus rendering **8** solution-processible and easy to handle in device fabrication. The polymer **8** also possesses good film-forming properties.

The solid film of PDB-PPV **5** absorbs in the UV-VIS at a maximum of 454 nm, exhibiting similar optical properties to its parent polymer, poly(phenylenevinylene) [PPV] [11] (Fig 1). This is in contrast to a typical 2,5-dialkoxy PPV, e.g. MEH-PPV, which absorbs in the red region (λ_{max} ca. 500 nm). The unexpected behavior is also reflected in the photoluminescence properties, where the emission falls in the yellow-green region (Fig 2) with a PL emission maximum ca. 519 nm (2.25 eV). The emission is significantly blue-shifted compared with that of MEH-PPV [PL maximum at ca. 589 nm (2.0 eV)] and is almost identical to the emission obtained from PPV. The absence of any vibrational features can be attributed to the disorder effects resulting from the introduction of two alkoxy groups.

The EL emission of polymer **8** in a single layer EL device (ITO/**8**/Al or Ca) is observed at voltages above 7 V with the expected spectral distribution similar to the PL emission. The typical turn-on voltages in these devices range from 10-15 V. The single-layer devices operate at internal quantum efficiencies of 0.15% for Ca electrodes while Al cathodes gave typical efficiencies of ca 0.06%.

Device performance improved significantly with an additional layer of hole-transporting material sandwiched between the polymer **8** and the ITO electrode. Using PPV as the second layer, the bilayer device internal efficiency increased to 1.5% (Ca) with a brightness in excess of 4,500 cd/m^2 at an operating electric field of ca. 1 MV/cm. The turn-on voltage was reduced to

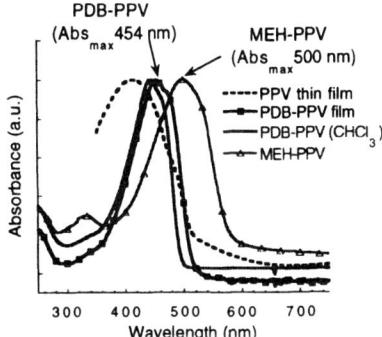

Fig 1 Absorption spectrum of PDB-PPV compared with films of PPV and MEH-PPV

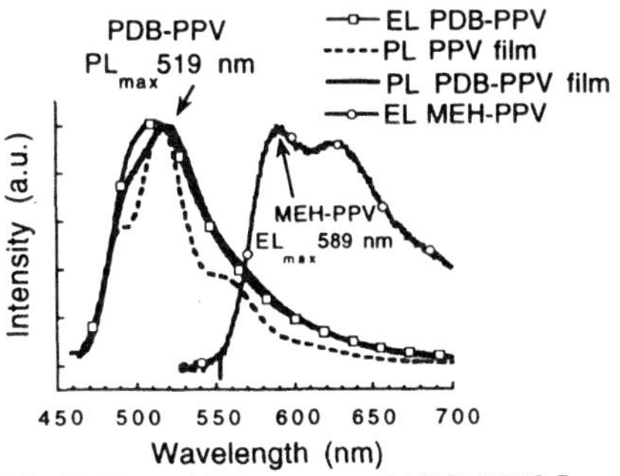

Fig 2 PL and EL spectra of PDB-PPV **5** compared with PPV and MEH-PPV

9 V. A second bilayer device using poly(vinylcarbazole) as the hole-transporting layer produced maximum efficiencies of ~0.6% with Al as cathode.

Fig 3 Plot of current density *vs.* applied voltage for a bilayer EL device (ITO/PPV/PDB-PPV/Ca)

<u>Synthesis and Characterization of the Model Oligomer **11**</u>

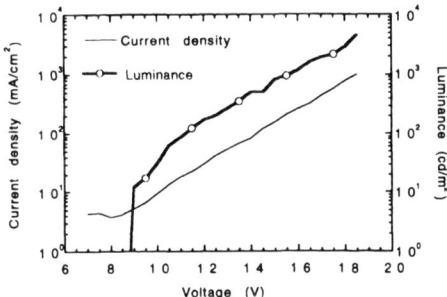

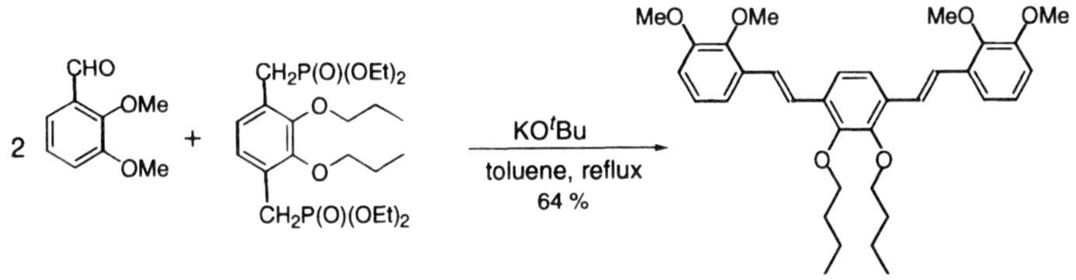

9   **10**   **11**
Scheme 2 Synthesis of the model oligomer **11**

The model distyrylbenzene **11**, a PPV trimer with a similar dialkoxy substitution pattern on the phenylene rings to that of the polymer PDB-PPV **8**, was selected to allow further insight into the origin of the blue-shifted luminescence. The highly fluorescent oligomer was synthesized using the Horner-Wittig-Emmons reaction [12] of the dimethoxybenzaldehyde **9** and the dibutoxybenzyl bisphosphonate **10** (Scheme 2).

Evidence for the origin of the blue-shifted emission and possible high fluorescence efficiency comes from the single crystal X-ray structure of **11** which shows twisting of the aromatic rings to a significant extent (Fig 4). This twisting results in the disruption of the effective conjugation length of the polymer backbone. The localization of the conjugation could also be due to the cisoid-like disposition of the styryl substituents in **11**. The benzene rings are twisted at a 30° angle from one another in a regular pattern along the oligomer backbone which is believed to lead to an interruption in the polymer backbone conjugation after every 4 benzene rings. The steric bulk of the butoxy chains could also cause significant disruption of the effective orbital overlap of the lone electron pairs of the oxygen with the aromatic rings. Results of Austin-Model 1 (AM1) geometry optimization and valence effective Hamiltonian (VEH) band-structure calculations are consistent with this hypothesis.

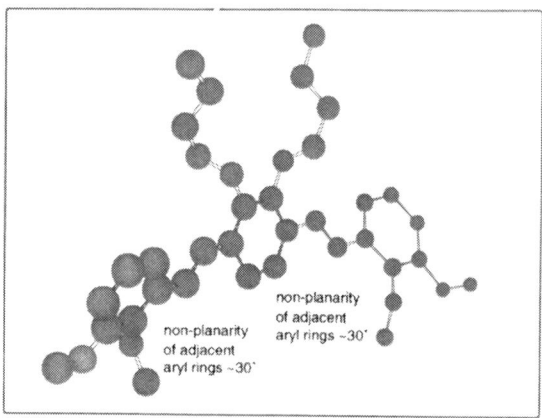

Fig 4 X-ray crystal structure of the model oligomer **11** showing the twist of the backbone
A film of the oligomer **8** also showed surprisingly efficient PL emission (*ca.* 80%) in the blue-green region (Fig 5).

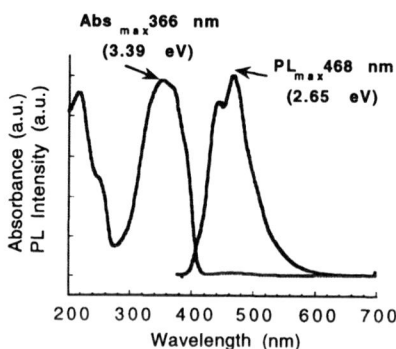

Fig 5 UV-VIS absorption and PL emission spectra of a thin film of the oligomer **11**

CONCLUSION

The synthesis of a new PPV derivative **8** with a 2,3-dialkoxy-substitution pattern has afforded a solution-processible polymer with highly efficient luminescent properties. The polymer exhibits an unexpected blue-shifted optical and emission behavior, showing that the 2,3-disubstitution pattern on the aryl ring can be utilized as an effective tool for tuning the color of emission. This surprising behavior has been attributed to the helical-like twist in the polymer backbone as well as to the steric bulk of the dialkoxy substituents. Both these factors disrupt the conjugation of the PPV backbone, thus causing the blue-shift.

EXPERIMENTAL

Preparation of polymer 8

A degassed solution of potassium tert-butoxide (0.41 g, 3.6 mmol) in dry THF (16 cm^3) was slowly added to a degassed solution of the monomer **7** (0.24 g, 0.6 mmol) in dry THF (16 cm^3) over 20 min at room temperature. The reaction mixture was stirred under nitrogen for 18 h, after which it was poured into methanol. The resultant yellow precipitate was collected by filtration and reprecipitated into methanol. The polymer **8** was collected as a bright yellow solid (60 mg, 40 %). Found: C, 76.6; H, 9.1; Br, 0 ; [C$_{16}$H$_{22}$O$_2$]$_n$ requires C, 78.0; H, 9.0; Br, 0 %). GPC assay in CHCl$_3$ vs polystyrene standards revealed M_n *ca.* 10^5; λ_{max}(CHCl$_3$)/nm 444 (2.29 eV); λ_{max}(film)/nm 454 (2.73 eV); PL emission (max)/nm 519 (2.25 eV); PL efficiency 40 %; EL emission (max)/eV 2.28.

Fabrication of a bilayer EL device with polymer 8

A film of PPV (*ca.* 100 nm in thickness) was prepared by spin-coating a solution of a sulfonium precursor polymer [13] onto a glass plate coated with indium tin oxide (ITO) and thermally converting the film at 250 °C under vacuum for 10h. A 1% (w/v) solution of the polymer **8** in chloroform was spin-coated (1600 rpm) on top of the converted PPV film to give a uniform film about 90 nm thick. The deposition of aluminum or calcium on the coated film surface was carried out under vacuum inside a metal evaporator to produce metal layers typically 500-1000 Å in thickness. The active device area was typically 2 mm^2.

ACKNOWLEDGMENTS

We thank the EPSRC (U.K.), the Cambridge Commonwealth Trust and the Committee of Vice Chancellors and Principals (ORS studentship to BSC), the Royal Society (University Research Fellowship to FC), the Commission of the European Community (support under the TMR network 'SELOA') and Cambridge Display Technology for financial support.

REFERENCES

1.	D. Braun and A.J. Heeger, Appl. Phys. Lett. **58**, 1982 (1991).
2.	N.C. Greenham, I.D.W. Samuel, G.R. Hayes, R.T. Philips, Y.A.R.R. Kessener, S.C. Moratti, A.B. Holmes and R.H. Friend, Chem. Phys. Lett. **241**, 89 (1995).
3.	J. Salbeck, Ber. Bunsenges. Phys. Chem. **100**, 1666 (1996).
4.	D.-H. Hwang, S.T. Kim, H.K. Shim, A.B. Holmes, S.C. Moratti and R.H. Friend, J. Chem. Soc., Chem. Commun. 2241 (1996).
5.	C.-B. Yoon, I.-N. Kang and H.-K. Shim, J. Polym. Sci. Pt. A 2253 (1997).
6.	B.R. Hsieh, H. Antoniadis, D.C. Bland and W.A. Feld, Adv. Mater. **7**, 36 (1995).
7.	H.G. Gilch and W.L. Wheelwright, J. Polym. Sci. A-1 **4**, 1337 (1966).

8.	R.C. Helgeson, T.L. Tarnowski, J.M. Timko and D.J. Cram, J. Am. Chem. Soc. **99**, 6411 (1977).
9.	S. Hanessian, T.J. Liak and B. Vanasse, Synthesis **118**, 396 (1981).
10.	G.-J.M. Gruter, O.S. Akkerman and F. Bickelhaupt, J. Org. Chem. **59**, 4473 (1994).
11.	J.H. Burroughes, D.C.C. Bradley, A.R. Brown, R.N. Marks, K. Mackay, R.H. Friend, P.L. Burn and A.B. Holmes, Nature **347**, 539 (1990).
12.	A. Lux, A.B. Holmes, R. Cervini, J.E. Davies, S.C. Moratti, J. Grüner, F. Cacialli and R.H. Friend, Synth. Met. **84**, 293 (1997).
13.	P.L. Burn, D.D.C. Bradley, R.H. Friend, D.A. Halliday, A.B. Holmes, R.W. Jackson and A. Kraft, J. Chem. Soc., Perkin Trans 1 3225 (1992).

ALTERNATING-CURRENT LIGHT-EMITTING DIODES AND THEIR TRANSIENT CHARACTERISTICS: RESPONSE TIME AND CARRIER TRANSPORT

A. J. PAL [*,**], T. ÖSTERGÅRD [*], R. ÖSTERBACKA [*], K-M. KÄLLMAN, H. STUBB [*]
[*] Åbo Akademi University, Department of Physics, Porthansgatan 3, FIN-20500 ÅBO (Finland), henrik.stubb@abo.fi
[**] Indian Association for the Cultivation of Science, Department of Solid State Physics, Jadavpur, Calcutta 700 032 (India)

ABSTRACT

We report the fabrication and characterization of alternating current light-emitting diodes (LEDs) with quinquethiophene as the emitting material. We have obtained equal electroluminescence intensity in both bias sections. From the frequency response of the LEDs, we have estimated the device response times and compared them with the response times obtained from the transient response of dc LEDs. Langmuir-Blodgett film deposition technique has been employed to control the thickness of the emitting layer on the molecular scale. We have shown that the response times originate from the accumulation rather than the transit of charge carriers. We have compared the photo- and electroluminescence spectra of QT LEDs.

INTRODUCTION

Most of the organic light-emitting diodes (LEDs) can only operate under forward dc bias, because of the tunneling mechanism involved in these devices [1]. To study the response characteristics of LEDs basically two methods have been used. The transient electroluminescence (EL) characteristics of simple LED structures showed a time lag between the sharp rise of a square-wave voltage pulse and the first appearance of the EL signal [2-5]. This time lag can be considered as the response time of LEDs.

Response times can also be estimated from alternating current (ac) LEDs. These have recently attracted much attention because of their unique mode of operation [6-8]. Symmetrically configured ac LEDs consist of an active electroluminescent layer sandwiched between insulating layers. In earlier reports [9,10], we have described the use of Langmuir-Blodgett (LB) films of the emitting polymer and the insulating materials to fabricate ac LEDs. From the frequency dependence of ac LEDs, we have estimated response times.

The LB film deposition technique has the unique advantage of controlling the film thickness on the molecular scale. With this advantage, we have fabricated dc and ac LEDs with a precise thickness of the emitting material. This has enabled us to compare the response times obtained from both measurements. The role of interfaces has been emphasized and the operation mechanism of ac LEDs has been discussed.

EXPERIMENT

Symmetrically configured ac LEDs consist of an active EL layer sandwiched between two insulating layers. The active layer in the present work was quinquethiophene (QT), an oligomer of polythiophene. Polymethyl methacrylate (PMMA) was used for the insulators. QT was synthesized according to methods described elsewhere [11, 12]. The purity was checked by measuring the melting point (254 °C) and the IR, UV-Visible absorption, photoluminescence and excitation spectra [12]. We have deposited LB films of QT mixed with arachidic acid (molar ratio 60%/40%) at a surface pressure of 30 mN/m. We have used a value of 2.5 nm as the average thickness of a monolayer [11]. Isotactic PMMA (molecular weight = 100,000) was purchased from Polyscience Inc. and used as received. PMMA LB films were deposited at a surface pressure of 15 mN/m and the thickness of each monolayer was reported to be 1 nm [13]. Each LED typically had an area of 10 mm^2 which was defined by a mechanical mask during the aluminum (Al) evaporation. Indium tin oxide (ITO) coated glass was used as the other electrode. All measurements were carried out in nitrogen atmosphere at room temperature. For the measurements of the ITO/PMMA/QT/PMMA/Al LED-structures, ac sinusoidal voltages were applied from an EICO 377 audio generator. The electroluminescence (EL) intensity was measured using a RCA 6342A photomultiplier and a Tektronix 2230 storage oscilloscope. For transient EL measurements, single square-wave voltage pulses were applied from a Hewlett-Packard HP 214B pulse generator. Photoluminescence (PL) and EL spectra were recorded by a Perkin-Elmer LS-50 Luminescence Spectrometer.

RESULTS

We have studied the EL intensity of the ac LEDs operated under sinusoidal drive voltages of different frequencies and amplitudes. Figures 1(a) and 1(b) show typical EL and applied bias signals in 12 layer and 26 layer devices, respectively. The EL signal has almost equal intensity in forward (ITO positive) and reverse bias. In a polymer ac LED using emeraldine base polyaniline as the insulator, we have obtained [9] stronger EL in the forward bias compared to that in the reverse bias. The difference in relative intensities can reflect the degree of charge accumulation at the interfaces between emitting and insulating layers [14].

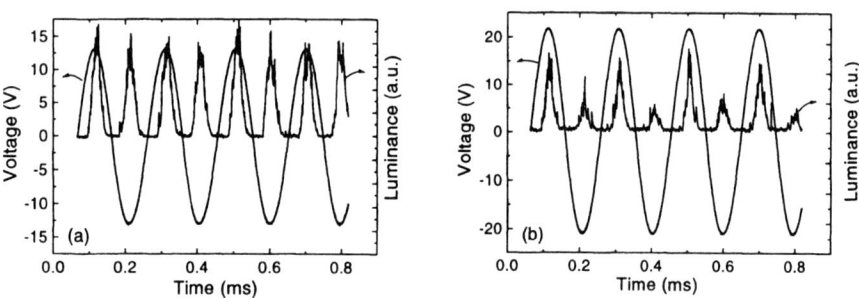

FIG. 1. EL and sinusoidal ac voltage as a function of time of (a) 12 layer QT device driven by 5120 Hz and 9.9 V rms and (b) 26 layer QT device driven by 5120 Hz and 15.6 V rms.

The frequency dependence of the light output from the QT ac LEDs followed our previous results [9]. The EL emission and applied bias voltage are in perfect phase at low frequencies but at higher frequencies, EL starts to lag the applied voltage. Figures 2(a) and 2(b) show the frequency response of the EL intensity during forward and reverse bias from 12 layer and 26 layer devices, respectively, driven by different voltages. At low frequencies the EL intensities for both bias directions remain constant. As the frequency increases beyond a certain value, the EL intensity starts to decrease. If the applied voltage across the structure is increased, the devices remain active to higher frequencies. The -3 dB frequency (where the EL intensity has decreased to -3 dB) also increases with the applied voltage. This trequency can be taken as the limiting frequency of operation of the ac LEDs. We can thus also estimate the response times of the ac LEDs as the duration of voltage half-cycle at the frequency f_0.

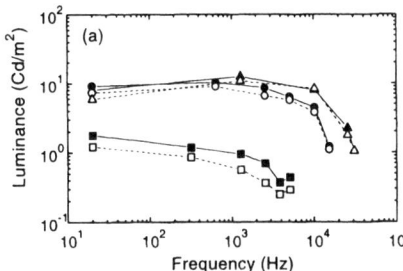 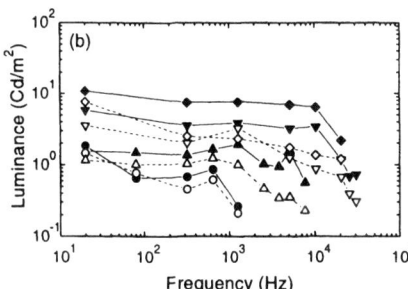

FIG. 2. Frequency dependence of EL intensity during forward bias (filled symbols) and reverse bias (open symbols) for a device with (a) 12 layers and (b) 26 layers of QT driven by different voltages. The open-circuit rms voltages are 7.9 V (square), 9.9 V (circle), 12.7 V (up-triangle), 15.5 V (down-triangle), and 18.4 V (diamond). The lines are only to guide the eye.

The response times can also be estimated from the transient EL measurements of the dc structures (ITO/QT/Al). We have studied the transient response of EL from 13 and 27 layer QT LEDs operated under single square-wave voltage pulses of different amplitudes. Figures 3(a) and 3(b) show the typical light output from a 13 layer QT device with voltage pulses of 4.0 V and 8.0 V, respectively. It is clear from the figures that the EL emission lags the sharp rise of the voltage

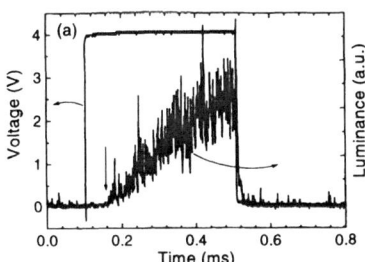 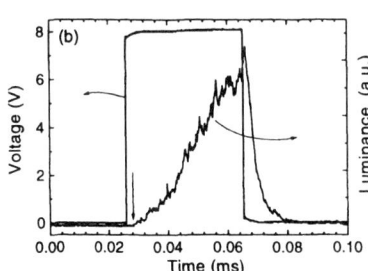

FIG. 3 EL response of ITO/QT/Al LEDs from 13 layers of QT under a (a) 4.0 V and (b) 8.0 V rectangular voltage pulse.

pulse and the delay time is shorter when the applied bias is higher. A similar dependence has also been observed in structures with 27 layers of QT.

We have compared the response time obtained from both measurements. The dc transient measurements should yield a shorter response time compared to ac LEDs, since in the former case the time required for the first appearance of EL is considered. In ac LED measurements on the other hand, the time taken for EL to reach 3 dB has been considered. The response times of dc and ac LEDs can be compared, primarily because the emitting layer thickness in both devices has been controlled on the molecular scale by the LB deposition technique. Figure 4 shows the response times versus applied field across the device for a variety of structures. The plots for each type of measurements show a single behavior for different thicknesses of LEDs. This shows that the response time is not directly related to the transit time of the charge carriers through the emitting layer. Moreover, considering the hole mobility values in QT obtained from field-effect measurements (1×10^{-5} cm^2/Vs [15]), the transit time for the holes in the 67 nm (= 27 layers) emitting layer will be less than a microsecond. On the other hand, we have obtained response times which are orders of magnitude higher than the transit time. Such unusually long response times have recently been explained to be due to accumulation of charges to build up an internal field [16]. In LEDs, holes are injected and move through the emitting layer with a high mobility. They build up a space charge near the Al electrode and modify the barrier for the electrons. At high fields, electrons are injected and light emission occurs. The time delay is thus due to accumulation of charge carriers rather than due to their transit times through the emitting layer.

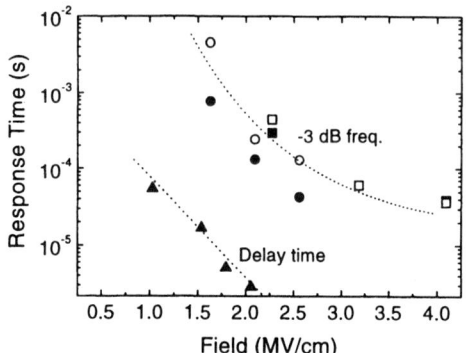

FIG. 4. Field dependence of response time from ac LED measurements (circles and squares) and transient EL measurements (triangles). Circles and squares represent response times from 26 and 12 layer ac LED measurements, respectively. Forward and reverse bias results are shown by filled and open symbols, respectively. For the dc transient case, up triangles represent response times for 13 a layer structure.

We have recorded EL spectra of PMMA/QT/PMMA ac devices and compared them with the photoluminescence (PL) spectra of QT LB films. Figure 5 shows the EL spectrum of a typical QT ac device along with the PL spectrum. EL is broad compared to the PL spectrum especially towards the low energy region. EL also lacks the structures of PL.

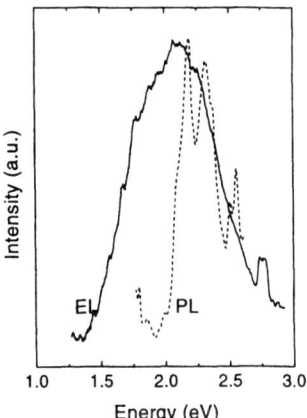

FIG. 5. EL and PL (dashed line) spectra of QT LB films. The former was measured using a 800 Hz sinusoidal voltage (14.2 V rms) across a 26 layer QT LED structure.

CONCLUSIONS

We have fabricated ac LEDs with QT as emitting layer and PMMA as insulating material. We have obtained equal EL emission in both bias sections. From the frequency response of the LEDs, we have estimated response times of the devices. We have also studied the transient response of dc LEDs. LB deposition technique has been employed to control the thickness of the emitting layer on the molecular scale. This has enabled us to compare the response times obtained from both types of measurements. We have shown that the response times originate from the accumulation rather than transit of charge carriers. We have compared the PL and EL spectra of QT LEDs.

ACKNOWLEDGMENTS

Quinquethiophene was synthesized by E. Vuorimaa and ITO substrates were provided by Planar International Ltd. TÖ and RÖ participate in the Graduate School of Materials Research, Turku, Finland. This work was financed by Technology Development Centre (TEKES), Finland, Academy of Finland, and Swedish Academy of Engineering Sciences in Finland. AJP was on leave from Indian Association for the Cultivation of Science, and his visit was arranged with Academy of Finland.

REFERENCES

1. D. R. Baigent, N. C. Greenham, J. Grüner, R.N. Marks, R. H. Friend, S.C. Moratti, and A. B. Holmes, Synth. Met. **67**, 3 (1994), and references therein.

2. C. Hosogawa, T. Tokailin, H. Higashi, and T. Kusumoto, Appl. Phys. Lett. **60**, 1220 (1992).

3. S. Karg, V. Dyakonov, M. Meier, W. Reiß, and G. Paasch, Synth. Met. **67**, 165 (1994).

4. P. Delannoy, G. Horowitz, H. Bouchriha, F. Deloffre, J.-L. Fave, F. Garnier, R. Hajlaoui, M. Heyman, F. Kouki, J.-L. Monge, P. Valat, V. Wintgens, and A. Yassar, Synth. Met. **67**, 197 (1994).

5. Y.H. Tak, J. Pommerehne, H. Vestweber, R. Sander, H. Bässler, and H. Hörhold, Appl. Phys. Lett. **69**, 1291 (1996).

6. Z. Yang, B. Hu, and F.E. Karasz, Macromolecules **28**, 6151 (1995).

7. Y.Z. Wang, D.D. Gebler, L.B. Lin, J.W. Blatchford, S.W. Jessen, H.L. Wang, and A.J. Epstein, Appl. Phys. Lett. **68**, 894 (1996).

8. H.L. Wang, A.G. MacDiarmid, Y.Z. Wang, D.D. Gebler, A.J. Epstein, Synth. Met. **78**, 33 (1996).

9. A.J. Pal, R. Österbacka, K.-M. Källman, and H. Stubb, Appl. Phys. Lett. **70**, 2022 (1997).

10. R. Österbacka, A.J. Pal, K.-M. Källman, and H. Stubb, J. Appl. Phys. in press (1998).

11. S. Tasaka, H. E. Katz, R.S. Hutton, J. Orenstein, G. H. Fredrickson, and T.T. Wang, Synth. Met. **16**, 17 (1986).

12. E. Vuorimaa, P. Yli-Lahti, M. Ikonen, and H. Lemmetyinen, Thin Solid Films **190**, 175 (1990).

13. R.H.G. Brinkhuis and A.J. Schouten, Macromolecules **25**, 2717 (1992).

14. T. Östergård, A. J. Pal, and H. Stubb, J. Appl. Phys. in press (1998).

15. J. Paloheimo, P. Kuivalainen, H. Stubb, E. Vuorimaa, and P. Yli-Lahti, Appl. Phys. Lett. **56**, 1157 (1990).

16. J. Pommerehne, H. Vestweber, Y.H. Tak, and H. Bässler, Synth. Met. **76**, 67 (1996).

DYNAMICS OF POLARONS IN GUEST-HOST-SYSTEM POLYMER LIGHT EMITTING DEVICES

E.J.W. LIST[a], J. PARTEE[b], W. GRAUPNER[a], J. SHINAR[b] and G. LEISING[a]

a Institut für Festkörperphysik, Technische Universität Graz, A-8010 Graz, Austria

b Ames Laboratory-U.S. DOE, Iowa State University, Ames, Iowa 50011

ABSTRACT

The conjugated ladder-type poly(paraphenylene) is an attractive material for blue polymer light emitting devices (PLED). Blending the active layer with small amounts of a red emitting guest polymer, the emission shifts from blue to red with increasing guest concentration due to efficient excitation energy transfer. The results of photoluminescence detected magnetic resonance, electroluminescence detected magnetic resonance measurements and current detected magnetic resonance measurements on PLEDs based on 0.05w% - 2w% red emitting poly(perylene-co-diethynylbenzene) (PPDB) in the active layer of the PLED are presented and discussed.

INTRODUCTION

One of the advantages of using conjugated polymers as electroluminescent emitters is the possibility to vary the degree of conjugation by *chemical tailoring* [1,2] which results in a change of the emission color[3,4]. Blending a certain amount of the organic red dye PPDB into films of the blue emitting methyl-substituted ladder-type poly(*para*-phenylene) (m-LPPP) results in white light emitting devices [5,6]. In order to improve the performance of these PLEDs there is considerable interest in the role of the non-emissive states like polarons in the emission processes. To probe these states we use electroluminescence (ELDMR) and photoluminescence detected magnetic resonance (PLDMR) to compare the dynamics of the involved polarons in both emission creating processes. Both methods are sensitive to states with lifetimes in the range of ~10 microseconds to 30 milliseconds, while the decay time of electroluminescence (EL) and photoluminescence (PL) is in the nanosecond range [7].

EXPERIMENT

The synthesis of the utilized polymers m-LPPP and PPDB is described in [8] and [9], respectively (see Figure 1 (a), (b)).

The PLEDs were fabricated in a single layer configuration, containing both the host and the guest polymer, using Indium-Tin-Oxide (ITO) on glass as the hole injection electrode. The polymer layer ($\approx$100nm) was spin-coated from a cosolution of m-LPPP/PPDB onto the ITO coated glass substrate of dimensions 1 mm x 4.5 mm x 12 mm. As the electron injecting electrode Al was evaporated onto the polymer film so that an active device area of ~6 mm^2 was obtained.

For preparing the films for PLDMR measurements, the polymer powders were dissolved in toluene and filled in electron spin resonance quartz-tubes. The toluene was evaporated leaving thin polymer films on the inner tube walls, which were sealed under vacuum. The sealed sample tubes were placed in the quartz dewar of an Oxford Instruments He gas-flow cryostat, enabling temperature control from 4-300 K, inside an optically accessible X-Band microwave cavity.

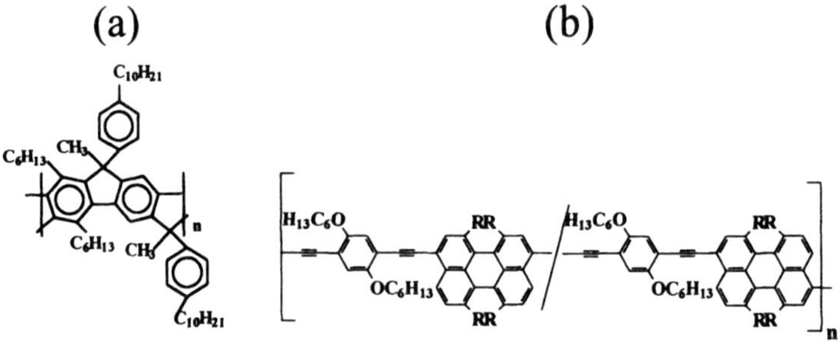

Figure 1: Chemical structure of m-LPPP (n≈25) (a) and PPDB (b) (R=phenoxy-t-butyl) The „/" in figure (b) denotes the random nature of the location of the single bond connecting the perylene to the ethynylene unit

The PL was excited by 457.9 nm and 515 nm photons from a Pockels-cell stabilized Ar^+ laser (30 mW). The changes in the EL and PL intensity, induced by the X-band microwaves at the applied DC magnetic field were detected by connecting the photodiode output to a lock-in amplifier referenced to the microwave chopping frequency v_C (400Hz $\leq v_C \leq$ 900Hz). In the case of PLDMR the laser line was blocked by an appropriate cut off filter. For CDMR measurements the PLEDs were driven by a constant bias voltage and the device current as well as the induced change was measured.

RESULTS AND ANALYSIS

The Guest Host System

The electric and optical properties of the guest-host system containing the conjugated polymers m-LPPP and PPDB have been thoroughly studied [4,5,6,7]. The absorption and PL spectra of m-LPPP and PPDB at room temperature are shown in Figure 2. The EL emission spectrum of m-LPPP (Figure 2 A (b)), the polymer which accomplishes the charge transport and the primary EL generation process in the PLEDs, [10] is dominated by a maximum at 461nm, a vibronic replica at 491nm and a broad shoulder at 530nm [4]. It is homologous to the absorption spectrum (Figure 2 A (a)) and the dominant PL maximum is only very slightly Stokes shifted (≈ 0.04eV). The PL spectrum of the polymer PPDB in the solid state (Figure 2 A (d)) is dominated by a broad maximum at 595nm with a vibronic shoulder at 650nm where the absorption spectrum (Figure 2 A (c)) is homologous to the PL emission spectrum. The PL spectrum of PPDB films is shifted towards lower energies compared with the PL spectrum of PPDB in a liquid toluene (Figure 2 A (e)) or solid PMMA solution.

The PL and the EL emission spectra obtained from the polymer-blends (see Figure 2 B) are composed of the superposition of the m-LPPP emission spectrum and the emission spectrum of PPDB in solution, where the contribution of each polymer to the spectrum is determined by the concentration of the blend. The photophysics of the of guest-host system after the primary excitation is governed by a very efficient excitation energy transfer (EET), involving the diffusion of an electron hole pair from the higher energy gap m-LPPP to the lower-gap PPDB [4]. As clearly observed in Figure 2 B the contribution of the PPDB emission to the obtained spectrum of the blend is dependent on the bias voltage. We found that the integral PPDB emission increases linearly with the bias voltage independent of the concentration of PPDB in the blend.

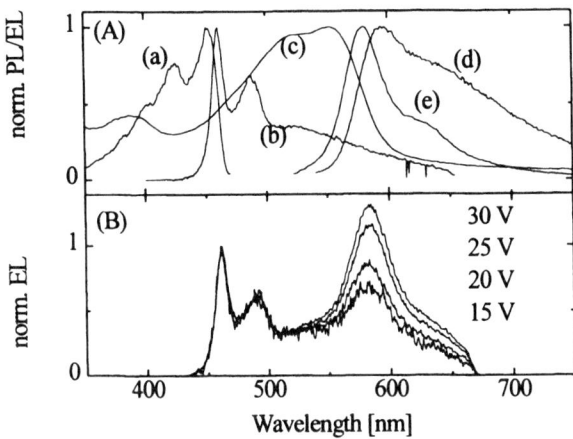

Figure 2: (A) absorption of m-LPPP (a), PL of m-LPPP (b), absorption of PPDB (c), PL of PPDB in solid state (d), PL of PPDB in solution (e), (B) (normalized) EL emission spectrum of a PLED containing 0.2w% PPDB recorded for different bias voltages of 15V-30V.

The PLDMR Resonance

The full-field PLDMR pattern of films of m-LPPP containing different concentrations of PPDB at ~20 K are shown in Figure 3 (B), the corresponding half-field pattern are shown in Figure 3 (A). The features H_{1-6} in the full-field triplet pattern, which is distributed between H_1 and H_2 [11], permit the determination of the zero field splitting (ZFS) parameters D and E using:

$$H_{1,2} \cong \frac{1}{g\mu_\beta}(h\nu \pm D) \tag{1}$$

$$H_{3,4} \cong \frac{1}{g\mu_\beta}\left(h\nu \pm \left(\frac{D+3E}{2}\right)\right) \tag{2}$$

$$H_{5,6} \cong \frac{1}{g\mu_\beta}\left(h\nu \pm \left(\frac{D-3E}{2}\right)\right) \tag{3}$$

where h is Planck's constant, g the g-factor of the free electron and μ_β the Bohr-magneton. From D and E one can calculate the spectral features H_{min}, H_{Sh}, H_{Si}, H_{max} observed in the half-field powder pattern using:

$$H_{min} \cong \frac{1}{2g\mu_\beta}\sqrt{(h\nu)^2 - \frac{4}{3}(D^2 + 3E^2)} \tag{4}$$

$$H_{Si} \cong \frac{1}{2g\mu_\beta}\sqrt{(h\nu)^2 - (D+E)^2} \tag{5}$$

$$H_{Sh} \cong \frac{1}{2g\mu_\beta}\sqrt{(h\nu)^2 - (D-E)^2} \tag{6}$$

$$H_{max} \cong \frac{1}{2g\mu_\beta}\sqrt{(h\nu)^2 - 4E^2} \tag{7}$$

with the half field resonance distributed between H_{min} and H_{max}.

Furthermore from D and E one can determine the deviation from spherical symmetry of the wave function and r_{ub} in Å, the upper limit of its extent is approximately given by: $r_{ub} \approx 30(D/2g\mu_\beta)^{-1/3}$ with $D/g\beta$ is in Gauss [12].

The forgoing analysis of the full-field patterns shown in Figure 3 (B) yields a $D/g\mu_\beta \sim 685$ G and $E/g\mu_\beta \sim 16$ G from the resonance spectrum of m-LPPP [13] and $D/g\mu_\beta \sim 430$ G and $E/g\mu_\beta \sim 67$ G from the samples containing pure PPDB. In the spectra of the blends containing m-LPPP and different amounts of PPDB one finds that the spectral features of PPDB dominate already in the sample with 0.2 w% over the m-LPPP features and are about at the same position as observed for pure PPDB. The upper limit of the triplet wavefunction of PPDB was determined to be $r_{ub}=$ 4.0 Å where m-LPPP yielded $r_{ub}= 3.4$ Å [13].

Since the analysis of the full-field pattern of pure PPDB Figure 3 (a) yielded $D/g\mu_\beta= 430$ G and $E/g\mu_\beta=67$ G the foregoing features of the half field resonance should be at $H_{min} = 1648$ G, H_{si} $= 1658$ G, $H_{sh} =1650$ G, $H_{max} =1666$ G. As clearly seen from Figure 3 (A) (a)-(d), the measured pattern is in excellent good agreement with these values. With $D/g\mu_\beta=685$ G and $E/g\mu_\beta=16$ G for pure m-LPPP (Figure 3 (B) (e)) one obtains the half field resonance at $H_{min} = 1620$ G, $H_{sh} = 1631$ G, $H_{si} =1634$ G, $H_{max} =1668$ G which is in agreement with the spectra in Figure 3 (A) (b)-(e).

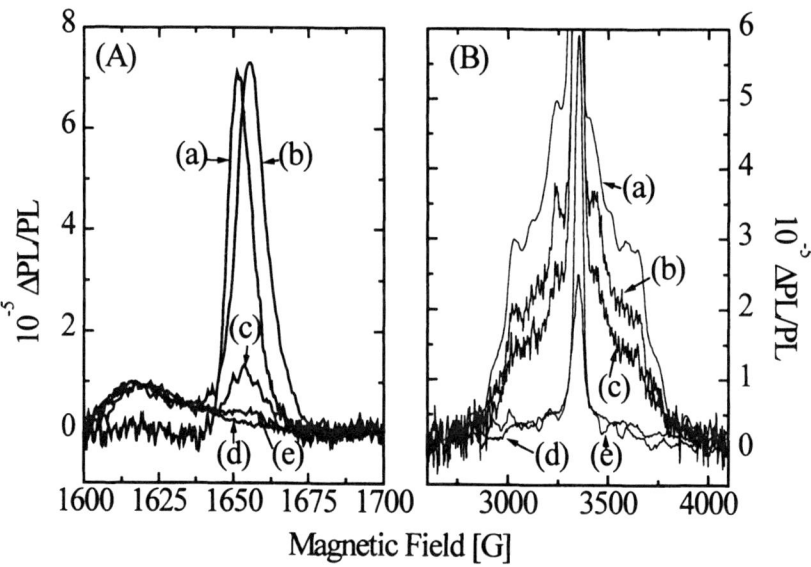

Figure 3: (A) Half field triplet powder pattern of a PPDB film (a), of a m-LPPP film with 2.0w% PPDB (b), of a m-LPPP film with 0.7w% PPDB (c), of a m-LPPP film with 0.05w% PPDB (d) and of a m-LPPP film (e), $\lambda_{EXC}= 458$ nm for (b,c,d,e) and $\lambda_{EXC}= 514$ nm for (a), laser power 30 mW, microwave power 810 mW. (Plots b-e) are normalized on m-LPPP maximum at approximately 1614 G. Temperature 20K.

(B) Full field triplet powder pattern (a-e) as in (A), $\lambda_{EXC}= 458$ nm for (b,c,d,e) and $\lambda_{EXC}= 514$ nm for (a), laser power 30 mW, microwave power 810 mW for all (a) is magnified by 100% for better visibility.

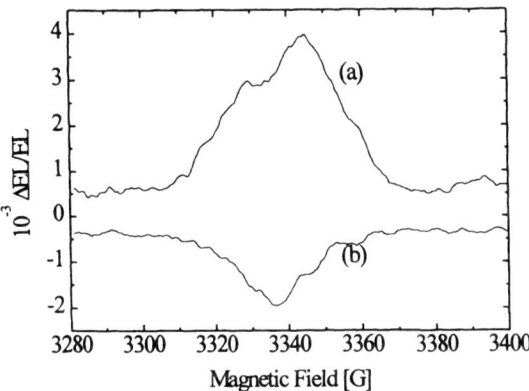

Magnetic Field [G]

Figure 4: ELDMR spectrum of a m-LPPP PLED at T = 15 K (a) and a PLED containing 0.7w% PPDB in the m-LPPP layer at T = 20 K (b), microwave power 810 mW.

The ELDMR and CDMR Polaron Resonance

The full field polaron ELDMR spectra at ~20 K of the single layer devices with pure m-LPPP and one containing 0.7 w% PPDB in the active m-LPPP layer are shown in Figure 4. The device fabricated of pure m-LPPP shows a broad enhancing ELDMR signal (Figure 4 (a)) with a full width at half maximum ($\Delta H_{1/2}$.) of ~ 40 G where the PLED containing PPDB shows a quenching resonance. This change in the sign of the ELDMR signal is not observed in the CDMR spectra. Both PLEDs, the one containing pure m-LPPP and the one consisting of a blend of both polymers show a positive CDMR signal [14].

CONCLUSION

We observed PL and EL-enhancing magnetic resonance for m-LPPP as previously described [13,15] as well as in PPDB films and m-LPPP/PPDB blends of different concentrations and photoexcited at different wavelengths [14]. This positive resonance can be assigned to magnetic resonance enhancement of the recombination of long lived interchain or intrachain interconjugation-segment polarons pairs [13,16,17,18]. The reduction of their steady state population enhances the PL and EL as there is strong evidence that these polaron pairs are quenching sites for singlet excitons.

From the lineshape analysis of the polaron PLDMR signal [14] we found that the PLDMR signal is mainly due to the recombination of one kind of polaron pair a rather close intrachain polaron pair. The analysis of the triplet full and half field powder pattern of different concentrations of the blend shows that the spectral features H_{1-6} and H_{min}, H_{Sh}, H_{Si}, H_{max} stemming from PPDB govern the PLDMR spectra. Furthermore one finds the triplet wavefunction of PPDB to be more extended compared with the one of m-LPPP and different in shape, where we suggest that it is disklike and located on the perylene unit, with an extent of r_{ub}= 4.0 Å.

Studying the voltage dependence of blend PLEDs one finds a linearly increasing contribution of the PPDB emission to the total blend spectrum, which can be assigned to changes in the charge transport process, or a field dependency of the EET mechanism.

The ELDMR signal of a PLEDs containing only m-LPPP in the active layer shows an enhancing resonance where the one containing PPDB (0.7%) shows a quenching resonance.

These observed ELDMR resonances are more difficult to analyze due to fact that the emissive species in the PL are created via photogeneration, whereas in EL they are created via charge carrier injection and recombination. In the case of the blend PLEDs we have also to consider the presence of the EET process and compare the spectra with CDMR spectra of pristine m-LPPP and blend PLEDs, where we observe an enhancing resonance for both types. Various processes may be responsible for these features, which are currently under investigation. However, the different behavior of the ELDMR spectra compared with the PLDMR and CDMR spectra as well as the bias voltage dependency of the blend PLEDs underline the different dynamics of excited species in the PL and in the EL of the PPDB/m-LPPP blends involving an EET processes.

ACKNOWLEDGEMENTS

The authors wish to thank P. Schlichting, U. Rohr, Y. Geerts, U. Scherf and K. Müllen for providing the polymer powders. E.J.W.L. gratefully acknowledges the partial financial support of the Bundesministerium für Wissenschaft und Verkehr for the work at the Ames Lab. Ames Lab is operated by the Iowa State University for the U.S. Department of Energy under Contract No. W-7405-Eng-82. The work in Graz was partially supported by the SFB *Elektroaktive Stoffe*.

REFERENCES

[1] O. Lhost , J.L. Bredas, S.C. Graham, R.H. Friend, D.D.C. Bradley, *Synth. Met.* **55-57**, 4290 (1993).

[2] F. Meyers, A. J. Heeger, J.L. Bredas, *Synth. Met.* **55-57**, 4308 (1993).

[3] G. Grem, G. Leditzky, B. Ullrich, G. Leising, *Adv. Mater.* **4**, 36 (1992).

[4] S. Tasch , E.J.W. List, C. Hochfilzer, G. Leising, P. Schlichting, U. Rohr ,Y. Geerts, U. Scherf, K. Müllen, *Phys. Rev. B* **56**, 4479 (1997).

[5] S. Tasch, E.J.W. List, O. Ekström, W. Graupner, G. Leising, P. Schlichting, U. Rohr, Y. Geerts, U. Scherf, K. Müllen, *Appl. Phys. Lett.* **71**, 2883 (1997).

[6] E.J.W. List, S. Tasch, C. Hochfilzer, G. Leising, P. Schlichting, U. Rohr Y. Geerts, U. Scherf, K. Müllen, *Optical Materials* (in print).

[7] M. Wohlgenannt, W. Graupner, S. Tasch, E.J.W. List, G. Leising, M. Graupner, A. Hermetter, P. Schlichting, Y. Geerts, U. Scherf, K. Müllen, *Chemical Physics*(in print).

[8] U. Scherf, K. Müllen, *Makromol. Chem., Rap. Commun.* **12**, 489 (1991).

[9] H. Quante, P. Schlichting, U. Rohr, Y. Geerts, K. Müllen, *Macromol. Chem.* **197**, 4029, (1996).

[10] S. Tasch, G. Kranzelbinder, G. Leising, U. Scherf, *Phys. Rev. B* **55**, 1 (1997).

[11] Z.V. Vardeny, X. Wei, *Handbook of Conducting Polymers* (in print).

[12] M. Bennati, A. Grupp, M. Mehring, P.Bäuerle, *J. Chem. Phys.* **100**, (1996).

[13] E.J.W. List, J. Partee, W. Graupner, J. Shinar, G. Leising, submitted to *Phys. Rev. B*

[14] E.J.W. List, J. Partee, W. Graupner, J. Shinar, G. Leising, unpublished results

[15] W. Graupner, J. Partee, J. Shinar, G. Leising, U. Scherf, *Phys. Rev. Lett.* **77**, 2033 (1996).

[16] J. Shinar, *Synth. Met.* **78**, 277 (1996).

[17] E.L. Frankevich, A.A. Lymarev, I.Sokolik, F.E. Karasz, S. Blumstengel, R.H. Baughman, and H.H. Hörhhold, *Phys.Rev.B* **46** ,9320 (1992).

[18] M.Yan, L.J. Rothberg, F. Papadimitrakopoulos, M.E. Galvin, and T.M. .Miller, *Phys. Rev. Lett.* **72**, 1104 (1994).

POLYMERIC ZINC-BISQUINOLINE BASED
SELF-ASSEMBLED LIGHT EMITTING DIODES

F. PAPADIMITRAKOPOULOS*, D. L. THOMSEN, III, K. A. HIGGINSON
Department of Chemistry, Polymer Science Program, Nanomaterials Optoelectronics Laboratory,
Institute of Materials Science, University of Connecticut, Storrs, CT 06269-3136

ABSTRACT

Bifunctional 8,8'-dihydroxy-5,5'-biquinoline (bisquinoline) is reactively self-assembled with diethyl zinc to form a linear coordination polymer. The potential of this method to produce insoluble and intractable structures of controllable supramolecular architecture suitable for semiconducting applications has stimulated an in-depth investigation of the growth mechanism of these polymeric chelates. These films were characterized by FTIR, UV/VIS and photoluminescence spectroscopy. The film growth on glass or indium-tin oxide (ITO) coated substrates was monitored by UV/VIS spectroscopy and ellipsometry. FTIR spectroscopy indicates that the self-assembled films are polymeric in nature. Single layer light emitting diodes exhibited an orange electroluminescence, consistent with the corresponding photoluminescence spectrum.

INTRODUCTION

Metal chelates of 8-hydroxyquinoline (8-Hq), and in particular aluminum tris(8-Hq) (Alq3), have received considerable attention toward the development of organic light emitting diodes (OLEDs).[1,2] In contrast to traditional inorganic light-emitting devices, OLEDs based on sublimed organic glasses or semicrystalline films are naturally limited by morphological changes at elevated temperatures and high current densities.[3-5] In order to suppress crystallization and densification in OLEDs, materials have been engineered with increased molecular asymmetry, hyperbranching, ring-puckering and entropic disorder via molecular doping.[6,7] While there is considerable difficulty in achieving stable disordered systems with these molecular compounds, polymeric materials offer a more traditional route to inhibit large morphological reorganizations.

The incorporation of polymers in light emitting devices (LEDs) is currently challenged by purity and processability issues. The ability to purify macromolecules devoid of defects requires tight control of side reactions.[8] Furthermore, when traditional thin-film spin or dip casting techniques are utilized to fabricate multilayer LEDs, interlayer solubility, thickness variations, and pinhole defects may commonly occur. Self-assembly of thin films has introduced ways to improve film uniformity and layer architecture at the molecular level.[9] Rubner et al.[10] have constructed pinhole-free poly(p-phenylenevinylene)-based LEDs as thin as 300 Å. Although the versatility of a polycation/polyanion layering approach shows increasing promise as an alternative fabrication method, it is inapplicable for molecular-based devices which lack ionic functionalities, such as Alq3.

Self assemblies of small molecules on glass and metal surfaces have been demonstrated in numerous motifs with alkanol and alkanoic acids,[11] alkylsilanes,[12] disulfides and thiols,[13] and isonitriles.[14] Zirconium phosphonates represent an interesting class of self assemblies. The aqueous or highly polar (e.g., dimethyl sulfoxide (DMSO)) solvents necessary for solubilizing the $ZrOCl_2$ and the bifunctional phosphonic acid terminated organics result in heavily hydrated assemblies which develop pores upon drying, further compromising film quality.[15,16] Transition metal coordination with bifunctional isonitriles in polar solvents has also been used to fabricate multilayer self-assemblies.[14] Currently, only a limited number of coordinating transition metals have been suitable for multilayer fabrication, and the substantial film shrinkage upon drying (ca. 35%) are limiting factors in device fabrication.

---

* To whom correspondance should be addressed.

The fabrication of polymer analogs of 8-hydroxyquinoline-based metal chelates (such as Alq$_3$, etc.) for electroluminescence applications has been a challenging task. These metal chelate polymers are nontraditional polymers and are usually associated with considerable handling difficulties. Their major intricacy arises from complexation-decomplexation dynamics, which are very sensitive to the pH and ionic strength of the solute.[17] For linear metal chelate polymers, solubilization typically occurs only in polar aprotic solvents,[18] which are difficult to remove from spun films. The insoluble and intractable nature of these polymers makes them amenable to the self-assembly growth which is the subject of this paper. Single layer light emitting diodes of these polymers were constructed via alternative dipping of hydroxy-functionalized ITO substrates in solutions of bisquinoline and ZnEt$_2$ separated each time by a washing step to remove excess reactants.

EXPERIMENTAL

Monomer Synthesis: The synthesis of bisquinoline (**I**) was done following the method of Archer *et* al.[19] Prior to its use, (**I**) was purified by sublimation to yield a white powder with a m.p. 280-300 °C (dec.). The room temperature ^{1}H-NMR spectrum of (**I**) in d^6-DMSO showed peaks (ppm) at 9.98(br s, 2H, OH), 8.87(d, 2H, J = 2.77 Hz), 7.64(d, 2H, J = 8.24 Hz), 7.42(m, 4H), and 7.21(d, 2H, J = 7.75 Hz). The FTIR of (**I**) is shown if Figure 3(a).

Zinc-Bisquinoline Powder Preparation: Zinc-bisquinoline powder (**II**) was prepared by adding 1:1 equiv. of bisquinoline (0.347 mmole) and ZnCl$_2$ in a 10 ml glacial acetic acid, 10 ml concentrated HCl, and 10 ml distilled water and then adding 4 M ammonia dropwise until precipitation occurred at pH 3.5. The powder was collected via gravity filtration, washed with distilled water, and then vacuum dried overnight at 56 °C to produce a 100% yield. It was then soxhlet extracted overnight in THF and vacuum dried at 56 °C for three hours, with a final yield of 75% yield. FTIR is shown in Figure 3(b).

Polymer Synthesis by Self Assembly: Zn(Et)$_2$ was purchased from Aldrich. THF was purchased from J. T. Baker and refluxed over Na with benzophenone under nitrogen and distilled. Glass and quartz substrates were hydroxy-functionalized according to the method described elsewhere.[20] Silicon substrates were hydroxy-functionalized with concentrated NaOH solution, washed with distilled water, and air dried. Solutions of bisquinoline and Zn(Et)$_2$ were made from THF at 10^{-5} to 10^{-3} molar concentration . Hydroxy-functionalized surfaces were first dipped in the Zn(Et)$_2$ solution for two minutes, followed by a rinse in the THF bath for an equal amount of time. The substrate was then dipped into the bisquinoline solution for two minutes followed by another THF rinse to form a polymer repeat unit of zinc-bisquinoline. In this fashion, the film thickness increases by cycling through the dipping sequence.

Instrumentation: Room temperature ^{1}H NMR was performed on a Bruker 500 MHz spectrometer.
    FTIR was performed on a Nicolet 60sx instrument using a DTGS detector. 32 scans were accumulated at 4 cm^{-1} resolution. 1% KBr powder samples of bisquinoline (**I**) and zinc-bisquinoline (**II**) pellets were used. Double polished silicon (*100*) wafer was utilized to grow self-assembled zinc-bisquinoline (**III**). These substrates were transparent above 1200 cm^{-1}, and their spectra were taken by orienting them at 45° angle with respect to the incident beam to minimize thickness induced interference fringes (see Figure 3).
    UV/VIS Absorption Spectroscopy of (**III**) on glass was performed on a Perkin-Elmer UV/VIS/NIR Spectrophotometer Lambda 900. Spectra were collected from 750-250 nm with an interval of -0.50 nm at speed of 250 nm/min.
    The photoluminescence (PL) emission spectra were done using a Perkin Elmer LS50B. A zinc-bisquinoline molecular assembly on a silicon (*100*) wafer was done with an excitation wavelength of 400nm, 100nm/min. scan rate, 10nm excitation slit width, and 20nm emission slit

width. The PL of powder zinc bis (8-hydroxyquinoline) (Znq$_2$) and powder zinc-bisquinoline in a KBr pellet were recorded with a 400nm excitation wavelength.

Film thicknesses were determined using a Tencor Alpha Step 200 after scratching the slides with a fresh razor. Measurements for (III) on silicon and glass were done with a 10 mg weighted stylus for 400 μm sweep at 8 sec scan.

Ellipsometric measurements were performed on a WVASE 32 M-44 spectroscopic ellipsometer manufactured by J. A. Woollman Co., scanned between 600 - 1000 nm at incident angles of 70, 72, 74, 76 and 78°. Scanned intervals of 10 nm at 20 revolutions per measurement were collected. The thicknesses were determined by fitting a CAUCHY model using the optical constants of A = 1.72, B = 0.044 and C = -0.0015.

X-ray diffraction (XRD) was performed on a Norelco/Phillips diffractometer using Cu K_α radiation (λ=1.5418 Å). Zinc-bisquinoline powder (II) was packed on a glass slide for measurement of its XRD spectrum, followed by subtraction of the substrate background.

RESULTS AND DISCUSSION

Quinoline chelate based coordination polymers were studied extensively in the early 1960s by Philips et al.[17] and by Berg et al.[21] The chelation of zinc and other divalent cations with bisquinoline (I) was shown to be an 1:1 metal/ligand ratio (see Scheme 1). The good thermal stability of such compounds stimulated additional investigation by Archer et al.[19] On the other hand, processing difficulties were demonstrated with reduced solubility and purification of bisquinoline derivatives and related metal chelates. The introduction of high-vacuum sublimed bisquinoline (I) to dilute solutions of Zn(Et)$_2$ provided insoluble oligomers of poly(Zn-bisquinoline) (II) of high purity (see Scheme 1).

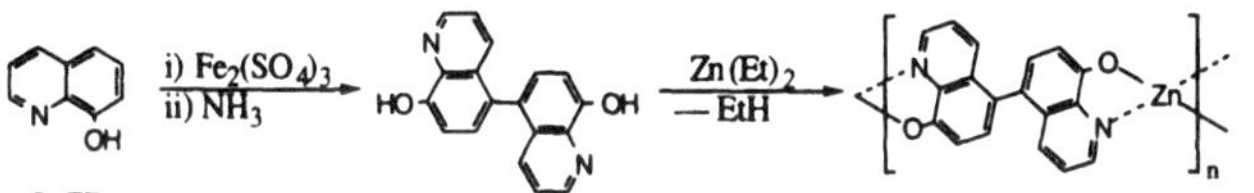

Scheme 1. Synthesis of bisquinoline (I) and its zinc-chelate polymer (II)

The natural limitations of solution polymerization to achieve high molecular weight were circumvented by sequential layering of the polymer repeat units based on a reactive self-assembled motif (see Scheme 2). Self-assembled polymeric Zn-bisquinoline films were grown on hydroxy-functionalized glass and silicon substrates. Highly pure Zn(Et)$_2$ and bisquinoline were diluted to 10^{-3} and 10^{-5} M in dry THF. Coating was initiated by dipping first into a THF solution of Zn(Et)$_2$ followed by a rinse in the THF bath (which is constantly refluxed and circulated over Na), and then dipping in a THF solution of bisquinoline. Both dips were limited to equal times of one to two minutes. The layer thickness progressively increased by repeated cycling through the dipping sequence (see Figure 1).

A crucial point in the fabrication of this polymer by self-assembly is the insolubility of (II).[21] The layered film growth by repeated cycling through the dipping sequence was confirmed by the steady increase in absorbance due to the formation of zinc-bisquinoline (UV maximum at 396.6 nm, see Figure 1). The film thickness also increased as measured by profilometry. Superb film uniformity was readily achieved over the entire dipped substrate and no major scattering from large film inhomogeneities was observed. The UV/VIS and film thickness measurements showed that the layering efficiency increased after four dipping cycles. This could be attributed to initial anchoring difficulties on the substrate surface.

A more accurate determination of film thickness was obtained by spectroscopic ellipsometry. Figure 2 illustrates the ellipsometrically determined film thicknesses as a function of dip cycle. The two markedly different regimes with growth rates of 8.3 and 26.5 Å/dip cycle (d.c.) respectively, indicate a complex film growth. Films thicker than 100 Å maintain the 26.5 Å/d.c. growth rate as far as we have measured (c.a. 1,250 Å). The expected growth for the molecular repeat of Zn-bisquinoline is 12.3 Å per dip cycle. Preliminary results from quartz microbalance measurements suggest that this complex growth pattern is related to film densification. However, considerable work is still needed to understand this behavior. Above the seventh dip-cycle the films attain a very high film uniformity, as indicated by atomic force microscopy, with mean square surface roughness below 10 Å.[22] Such film uniformity demonstrates the suitability of this thin-film growth for semiconductor applications. Further investigation into molecular anisotropy could shed additional light to the molecular mechanisms that cause such complex growth.

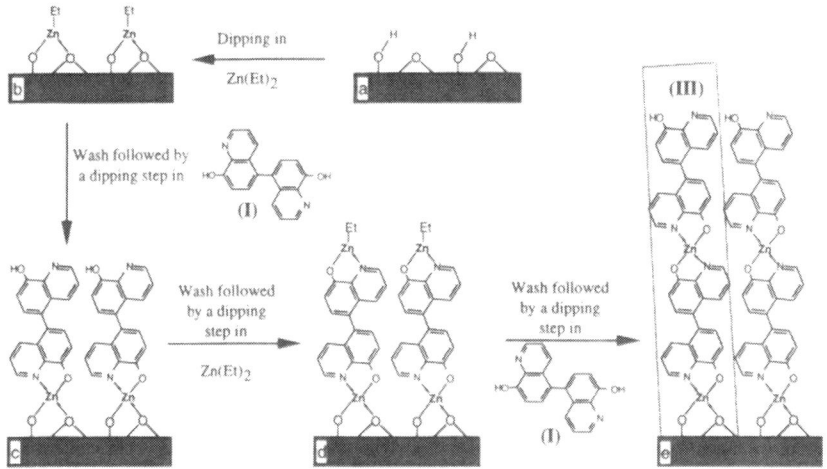

Scheme 2. Methodology for growing poly(Zn-bisquinoline) **(III)** via reactive self-assembly.

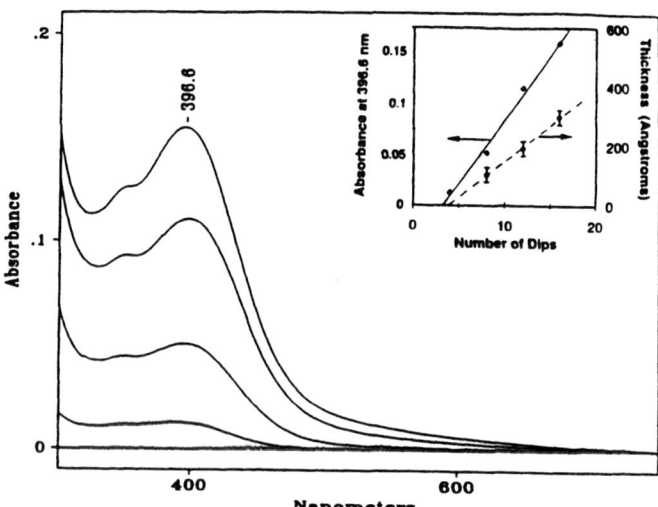

Figure 1. UV/VIS absorption vs. number of complete dip cycles. Inset depicts the film thickness and absorbance at 396.6 nm.

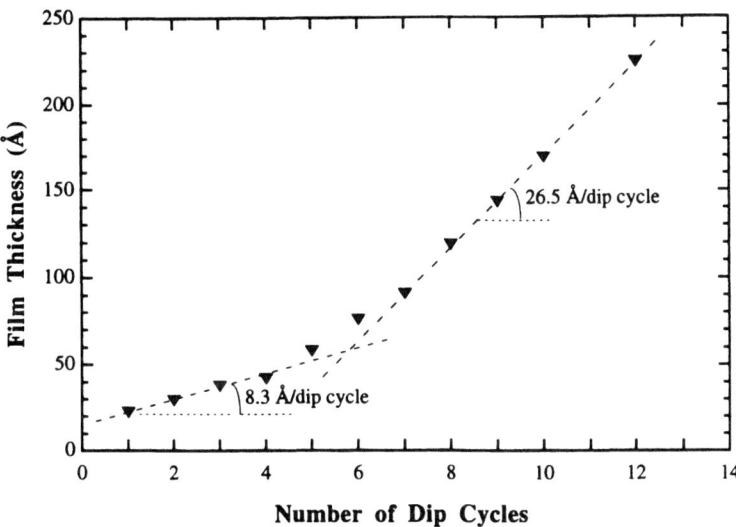

Figure 2. Ellipsometrically determined film thickness growth of (**II**) as a function of successive dip cycles.

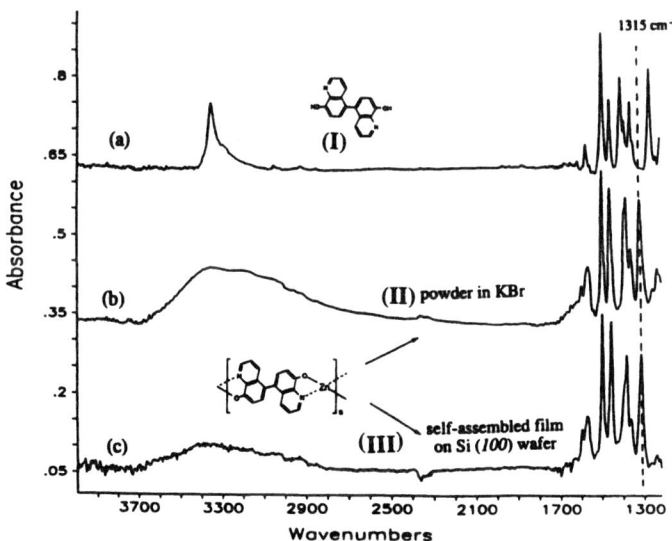

Figure 3. FTIR spectra of (a) bisquinoline (**I**) in KBr, (b) powder form of poly(Zn-bisquinoline) (**II**) in KBr and (c) self-assembled film of (**III**) on a (*100*) silicon wafer.

The FTIR spectra of bisquinoline (**I**) and zinc bisquinoline powder (**II**), is illustrated in Figure 3(a) and 3(b) respectively. Evidence of chelation is shown by comparing the changes in vibrational frequencies of (**I**) with the powder (**II**) and the self-assembly (**III**). The FTIR of zinc-bisquinoline (**III**) layered on a silicon wafer was compared to powdered samples of bisquinoline (**I**) and zinc-bisquinoline (**II**) and showed a noticeable decrease in the broad hydroxyl stretch at 3400 cm^{-1} (see Figure 3). This decrease was attributed to disappearance of terminal hydroxyl groups that strongly absorb at 3355 cm^{-1} for the unchelated bisquinoline (free OH stretch) versus the broad OH band (hydrogen-bonded OH stretch) based on terminal end-groups. The peak at 1315 cm^{-1} which is indicative of chelation in the layer assembly (**III**) is present in the powder sample (**II**) as well, but is not present in unchelated bisquinoline (**I**). Bisquinoline (**I**) has a peak at 1285 cm^{-1} for a phenolic C-O stretch, not found in the chelated samples. The FTIR spectra provide sufficient evidence that chelation is the primary mechanism contributing to self-assembly.

The photoluminescence (PL) of the zinc-bisquinoline molecular assembly on a silicon wafer shows a broad emission centered at 590 nm (see Figure 4). This spectrum is significantly

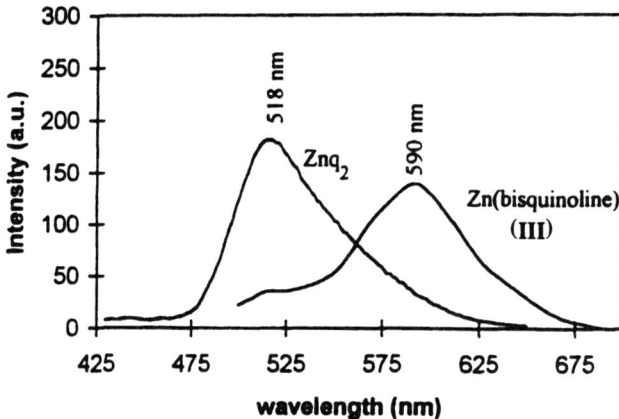

Figure 4. Photoluminescence spectra of molecular zinc-bis-(8-hydroxyquinoline) powder and zinc-bisquinoline (**III**) self-assembled on a silicon wafer.

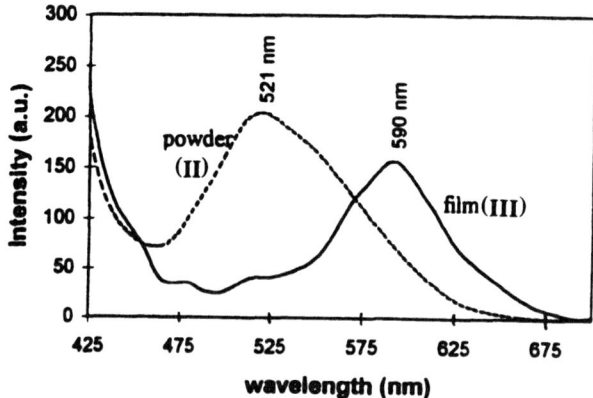

Figure 5. Photoluminescence spectra of zinc-bisquinoline powder (**II**) and self-assembled film (**III**) on a silicon wafer.

red-shifted with respect to the zinc bis-(8-hydroxyquinoline) (Znq_2) photoluminescence emission spectrum which has a sharper emission maximum at 518nm. This is attributed to the increased conjugation for the coordination polymer versus its low molecular weight analog of Znq_2. The difference in conjugation length between the precipitated zinc-bisquinoline powder (II) and self-assembled films of (III) is further demonstrated in Figure 5. The red-shifted shoulder in the PL emission of (II) is indicative of the significantly lower conjugation of the self-assembled films of (III).

The crystalline packing arrangement of poly(zinc-bisquinoline) has been a topic of considerable interest. Up to the present, evidence of crystalline order has been witnessed only from the solution-synthesized oligomers of poly(zinc-bisquinoline) (II) (see Figure 6). This is in good agreement with the behavior of its low molecular analog (zinc-bis-(8-hydroxyquinoline)) where its dihydrate forms highly ordered single crystals.[21] Self-assembled poly(zinc-bisquinoline) (III) on the other hand, has shown a lack of crystalline order. The incommensurate substrates, chain misalignment and end-group impurities could be a few of the reasons why clear evidence of crystalline order has not been witnessed in (III). However, the potential of achieving highly ordered self-assembled films is significant, considering the dense packing and high refractive index (n = 1.68 at 633 nm) of (III).[22]

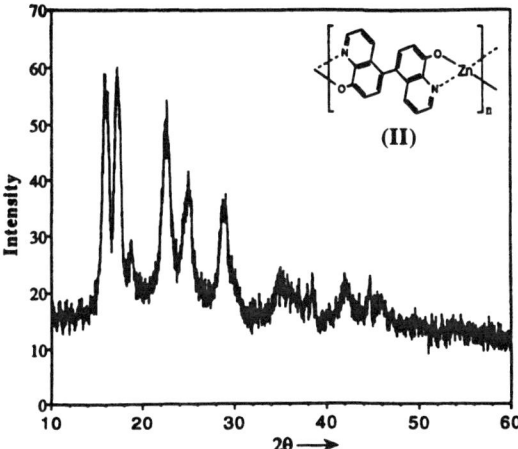

Figure 6. Wide-angle X-ray diffraction of the solution synthesized oligomers of poly(zinc-bisquinoline) (II).

The insoluble and intractable nature of these chelates, along with their improved processability, film forming quality and increased purity enables their usage in a variety of semiconducting applications.[22,23] Single layer LEDs were fabricated by evaporating Mg contacts on self-assembled films of poly(Zn-bisquinoline) (III) grown on ITO substrates. Figure 7 illustrates the electrical characteristics of two such devices with thicknesses of 400 Å (24 dips) and ca. 1200 Å (40 dips), along with their visual responses (the 40 layer device became visible at approximately 20 V). The electroluminescence response of these devices produced an orange emission shown in Figure 8. The broad orange emission of these devices, peaking at 588 nm is slightly red shifted from the photoluminescence (PL) spectrum of self-assembled zinc-bisquinoline (III) which peaks at 582 nm when excited by the 345 nm UV line of He-Cd laser. The nature of the smaller peak at 425 nm at the PL spectrum is presently unknown. At low voltage the electrical characteristics of both 400 Å and 1200 Å devices where shown single carrier transport followed by minority carrier injection and light emission at 6 and 10.6 V respectively. Ongoing investigation of the transport mechanism of these devices point towards that of their low molecular analogs of Alq_3 and

Znq$_2$.[1,24] The unbalanced carrier injection device presently limits the device efficiency and lifetime which have yet to be optimized.

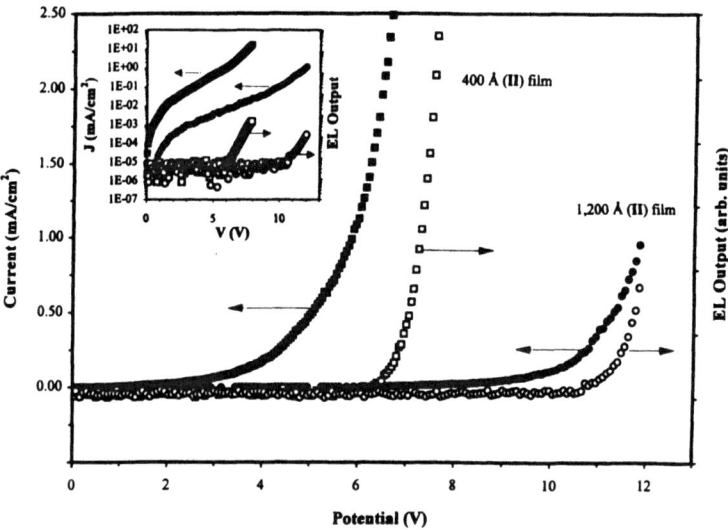

Figure 7. Typical current-voltage and luminance response of a 400 Å and a 1200 Å single layer light emitting diode (+)/ITO/(III)/Mg/(-) of (III).

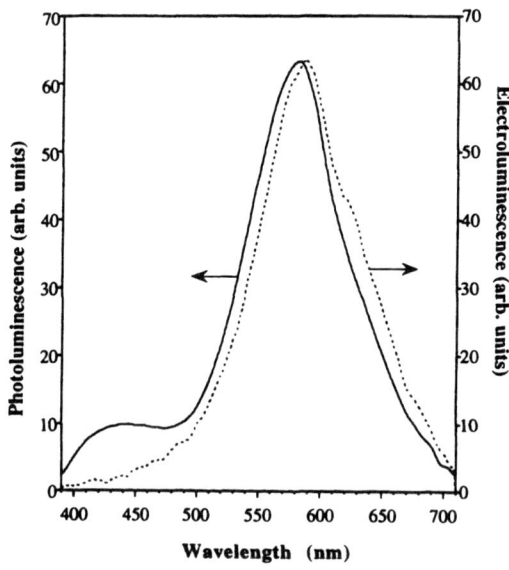

Figure 8. Photoluminescence and electroluminescence response of self-assembled poly(Zn-bisquinoline) (III).

CONCLUSIONS

Light emitting diodes (LED) were constructed from a novel reactive self-assembly technique developed for the synthesis of polymeric metal chelate derivatives of 8-hydroxyquinoline. Reactive self-assembly of diethyl zinc and bisquinoline onto hydroxy-functionalized ITO coated substrates resulted in thin, homogeneous films of coordination polymers in a timely fashion. These insoluble and intractable chelates are suitable for semiconducting applications, where controlled multilayer design and film forming uniformity are essential requirements. This approach exhibits a strong potential for the fabrication of electroluminescent (EL) devices as compared to traditional vacuum deposition and spin coating techniques.

ACKNOWLEDGMENTS

The authors would like to thank Dr. Y. Lvov for helping with the QCM measurements. Financial support from NSF CAREER Grant DMR-9702220 and the Critical Technologies Program through the Institute of Materials Science, University of Connecticut are greatly appreciated.

REFERENCES

1. Tang, C. W.; VanSlyke, S. A.; Chen, C. H. *J. Appl. Phys.* **1989**, *65*, 3610.
2. Sheats, J. R.; Antoniadis, H.; Hueschen, M.; leonard, W.; Miller, J.; Moon, R.; Roitman, D.; Stocking, A. *Science* **1996**, *273*, 884.
3. Do, L.-M.; Han, E. M.; Niidome, Y.; Fujihira, M.; Kanno, T.; Yoshida, S.; Maeda, A.; Ikushima, A. J. *J. Appl. Phys.* **1994**, *76*, 5118.
4. Han, E.-M.; Do, L.-M.; Yamamoto, N.; Fujihira, M. *Thin Solid Films* **1996**, *273*, 202.
5. Higginson, K. A.; Zhang, X.-M.; Papadimitrakopoulos, F. *Chem. Mater.* **1997**, Accepted for Publication.
6. Shirota, Y.; Kuwabara, Y.; Inada, H.; Wakimoto, T.; Nakada, H.; Yonemoto, Y.; Kawami, S.; Imai, K. *Appl. Phys. Lett.* **1994**, *65*, 807.
7. Chen, C. H.; Shi, J.; Tang, C. W. *A.C.S. Polymer Preprints* **1997**, *38*, 317.
8. Papadimitrakopoulos, F.; Yang, M.; Rothberg, L. J.; Katz, H. E.; Chandross, E. A.; Galvin, M. E. *Mol. Cryst. Liq. Cryst.* **1994**, *256*, 663.
9. Decher, G.; Hong, J. D.; Schmitt, J. *Thin Solid Films* **1992**, *210/211*, 831.
10. Fou, A. C.; Onitsuka, O.; Ferreira, M.; Rubner, M. F.; Hsiesh, B. R. *J. Appl. Phys.* **1996**, *79*, 7501.
11. Allara, D. L.; Nuzzo, R. G. *Langmuir* **1985**, *1*, 52.
12. Sagiv, J. *J. Am. Chem. Soc.* **1980**, *102*, 92.
13. Nuzzo, R. G.; Allara, D. L. *J. Am. Chem. Soc.* **1983**, *105*, 4481.
14. Ansell, M. A.; Zeppenfeld, A. C.; Yoshimoto, K.; Cogan, E. B.; Page, C. J. *Chem. Mater.* **1996**, *8*, 591.
15. Katz, H. E. *Chem. Mater.* **1994**, *6*, 2227.
16. Feng, S.; Bein, T. *Nature* **1994**, *368*, 834.
17. Phillips, J. P.; Deye, J. F.; Leach, T. *Analytica Chimica Acta* **1960**, *23*, 131.
18. Shi, S. Q.; Tempe, F. S. *US Patent* **1996**, 5.
19. Archer, R. D.; Hardiman, C. J.; Kim, K. S.; Grandbois, E. R.; Goldstein, M. In *Metal Containing Polymer Systems*; J. E. Sheats, C. E. J. Carraher and C. U. J. Pittman, Ed.; Plenum Press: NY, 1985; pp 355.
20. Ferreira, M.; Rubner, M. F. *Macromolecules* **1995**, *28*, 7107.
21. Berg, E. W.; Alam, A. *Anal. Chim. Acta* **1962**, *27*, 454.
22. Thomsen, D. L.; Phely-Bobin, T.; Papadimitrakopoulos, F. *Mat. Res. Soc. Symp. Ser.* **1997**, *within this volume*,
23. Thomsen, D. L.; Higginson, K. A.; Papadimitrakopoulos, F. *ACS Polymer Preprints* **1997**, *32*, 353.
24. Hamada, Y.; Sano, T.; Fujita, M.; Fujii, T.; Nishio, Y.; Shibata, K. *Jpn. J. Appl. Phys.* **1993**, *32*, L514.

POLYMER LAYER ORDERING OF POLYANILINE DERIVATIVES IN PLED DEVICES: SURFACE ADSORPTION AND CHARACTERIZATION

R. C. ADVINCULA*[1], W. KNOLL**, C. W. FRANK*
D. ROITMAN***, R. MOON***, J. SHEATS***
*Department of Chemical Engineering and Center for Polymer Interfaces and Macromoleculer Assemblies (CPIMA), Stanford University, Stanford, CA 94305
*[1]currently with the Department of Chemistry, University of Alabama at Birmingham, Birmingham, AL 35294
**Max Planck Institute for Polymer Research, Mainz, Germany D-55021
***Hewlett-Packard, Solid State Laboratories, Palo Alto, CA 94304

ABSTRACT

The fabrication and characterization of polyaniline (PANI) derivatives deposited on ITO coated glass is investigated as possible hole injection layers for MEH-PPV based polymer light emitting diode (PLED) devices. This involved multilayer ordering by the alternate polyelectrolyte adsorption of polyaniline and sulfonated poyaniline with an oppositely charged polyelectrolyte from solution. A combination of spectroscopic and microscopic techniques was utilized to determine the layer ordering, film structure, morphology, and homogeneity. The deposition process generally showed a linear behavior for all pairs as shown by ellipsometry and UV-vis spectroscopy. However, surface plasmon spectroscopy (SPS) and AFM revealed that thicker films are accompanied by increased surface roughness regardless of concentration. Comparison in performance was made between bare ITO and PANI or SPANI coated devices. Initial investigations of PLED performance showed significant improvements in lifetime and efficiency compared to bare ITO.

INTRODUCTION

The optimization of polymer light emitting diode (PLED) device performance requires the proper combination of materials and processing protocols. The goal being to increase the quantum efficiency, lifetime, lower the operating voltage, and increase the luminance levels. A common processing protocol involving the ITO/MEH-PPV/Ca system [1], has been widely investigated by the UNIAX, IBM, and HP groups. These generally showed poor behavior due to oxidation or catastrophic failures. In contrast, a device formed by the incorporation of a polyaniline complex (PANI-CSA or Panaqua™) by spin coating, prior to MEH-PPV has been shown to have dramatic improvements in both efficiency and lifetime. [2] These layers are typically several hundred up to a thousand Å's thick. However, at very thin films of less than two hundred Å, the formation of pinhole defects is evident. Investigations at these thicknesses are believed to be important in determining the mechanisms of device performance and failure. [3] A question arises as to the critical thickness by which these layers play an important role in the hole injection process. Thinner layers are also believed to result in lower turn-on voltages.[3] Rubner et. al. has initially reported the formation of heterostuctures of polyaniline and sulfonated polyaniline with a number of other conducting polymers using alternate polyelectrolyte adsorption.[4]

In this work, we initially report our investigations on the layer ordering, characteristics, and film properties of polyelectrolyte adsorbed polyaniline (PANI) and sulfonated polyaniline (SPANI) layers on the ITO/ MEH-PPV/Ca device configuration. A number of surface sensitive

115

spectroscopic and microscopic techniques such as surface plasmon spectroscopy, ellipsometry, UV-vis spectroscopy, atomic force microscopy (AFM) and x-ray reflectivity were applied. The absence of pinhole defects is critical in determining the performance of these devices. Initial lifetime and efficiency measurements for fabricated devices showed dramatic improvements for these devices compared to bare ITO. A more detailed investigation will be described in a future publication, to include the critical parameters of the device performance and its relation to layer ordering, e.g. thickness, polymer pairs, annealing, etc. [5]

EXPERIMENT

Polyaniline (PANI) and sulfonated polyaniline (SPANI) were synthesized according to the procedure of McDiarmid et. al.[6] and Epstein et. al.[7] respectively. Dilute aqueous solutions (20 mg/ml) of the PANI and SPANI derivatives were prepared according to the procedure reported by Rubner:[8] It is important that the right solvent and pH be utilized as solubility limits the MW fraction that dissolves. MW for both PANI and SPANI was estimated between 10,000-50,000 based from the literature procedure. FT-IR measurements indicate between 40 to 50 % doping for the PANI and SPANI polymers in the bulk.

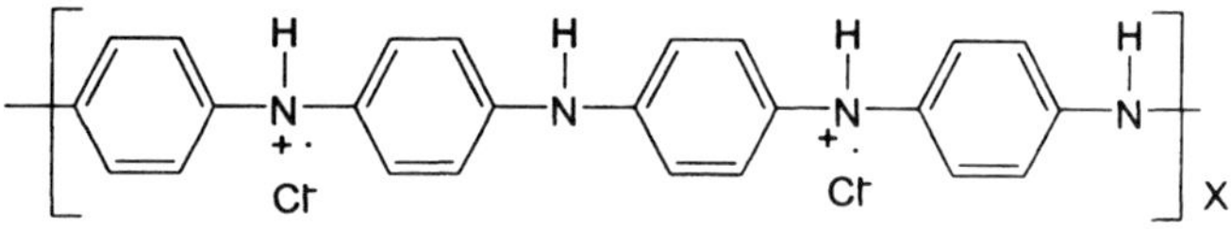

emeraldine, HCl (42%-doped)

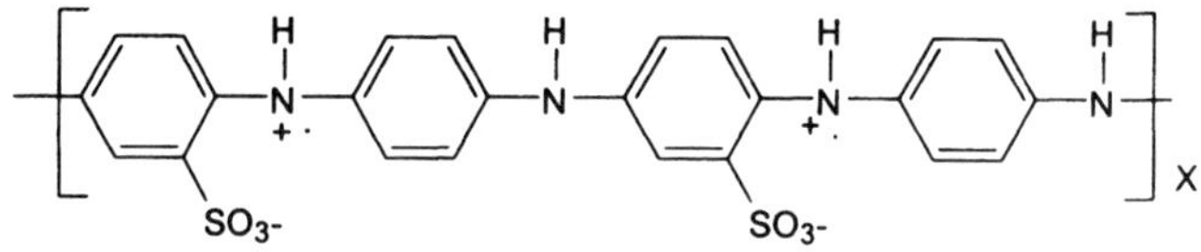

SPANI - (1 structure)

Figure 1. Representative chemical structures of SPANI (-) and PANI(+) polyelectrolytes
(depending on doping, protonation, and sulfonations) used for the alternate deposition process.

Dilute aqueous solutions of other commercially available (Aldrich) polyelectrolytes were prepared for poly(allyl amine)HCl (PAH)(+) MW= 28,000, poly(sodium-4-styrene sulfonate) (PSS)(-), MW=70,000, poly(diallyldimethyl ammonium chloride) PDADMAC (+), low MW (unspecified at 20% solution). Poly(2-methoxy,5-(2'-ethylhexyloxy)-1,4–phenylene-vinylene) (MEH-PPV) in xylene solution was obtained from UNIAX corporation.

The alternate polyelectrolyte layer deposition approach was used to fabricate the films. This involved the initial preparation of a uniformly charged surface on the ITO surface using (N-2-aminoethyl-3-aminopropyl) triethoxy silane (APS) and acidified with dilute acid (0.01 M HCl) prior to use. The deposition of the polyelectrolytes were done by dipping the substrates alternately between the oppositely charged solutions (e.g. SPANI and PDADMAC) using concentrations of 0.02 to 0.0002 monoMolar (mM). Rinsing was utilized in-between each adsorption. This procedure was iterated up to the desired number of layers. Both manual and automated deposition (Carl Zeiss Programmable Stainer) methods were used.

Solution and film substrate Uv-vis spectroscopy was done using a HP8452A UV-Vis spectrophotometer. FT-IR measurements (ATR and transmission) were done on a BIORAD Digilab FTS-60A. The *in situ* adsorption process was monitored by surface plasmon spectroscopy (SPS). This set-up based on the Kretschmann configuration has been previously described elsewhere.[9] Ellipsometry was done using a Geartner L116C ellipsometer. Atomic force microscopy (AFM) with both tapping and contact mode (Digital Instruments, Nanoscope III) was used to investigate the morphological and film structure at the immediate surface. Preliminary X-ray reflectivity measurements were done with the Rigaku θ–θ diffractometer with an 18kW rotating anode X-ray source. Device fabrication and testing methods have been reported elsewhere.[10]

RESULTS AND DISCUSSION

The polymer properties of the synthesized PANI and SPANI were initially characterized by UV and FT-IR measurements. Doping, protonation, and structural properties were consistent with previous investigations.[6,7] It was only possible to form dilute aqueous solutions as precipitation occurred with higher concentrations. By using the right pH (pH=3-4), the solution stability of an amphoteric polymer like SPANI can be controlled. Optically visible uniform films were formed both by manual and automated deposition up to 30 pair-layers. As shown in Figure 2, the linearity of the absorbance rise was best after 5-8 double layers and more.

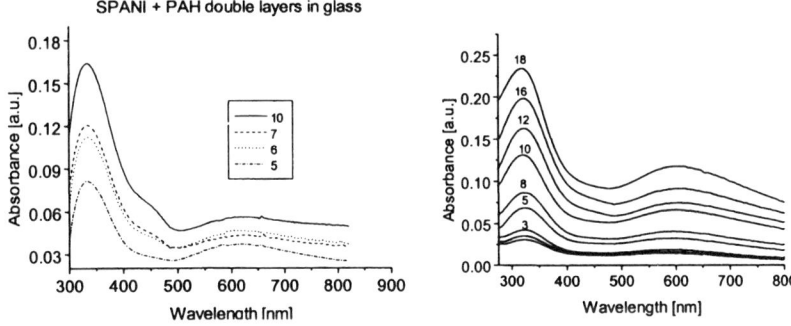

Figure 2. UV-Vis spectra of SPANI/PAH on glass substrate as a function of increasing layers.

In most instances, this is due to any non-uniform surface coverage on the initially charged surface (APS) which, tends to "smooth" out with the addition of subsequent layers. A good estimation of the amount of polymer deposited per increase in absorbance can be estimated using Beer's law (0.004/20 Angstrom at 630 nm). This was useful as a rough guide for comparison with subsequent depositions. The values are summarized in Table 1. The layer thicknesses for each double layer were verified by ellipsometry for the dry film, likewise giving an average value for each pair. The thickness also increased linearly with increasing layer number for all pairs. Previous attempts by Rubner et. al. to deposit the polymers at different pH conditions showed that lower pH's tended to give thicker layers.[11] This suggested that the amphoteric nature of the polymers could probably play a larger role in the level of segmental electrostatic repulsion which favors greater polymer adsorption. This has been found to be the case especially with SPANI layers. The SPANI polyelectrolyte had a complex solution behavior and therefore the deposition to much thicker layers is not possible without using fresh solution.

Table I. Comparison of thickness and absorbance properties for each polymer pair as investigated by the three techniques. Linear behavior was found in all cases.

Polymer Pair 0.001 monoMolar	Ellipsometric Thickness (Ave. per 2 layers) in Angstroms	SPS (aqueous) *in situ* Thickness (Ave. per 2 layers) in Angstroms	UV-vis Absorbance at 630 nm. (Ave. per 2 layers)
SPANI/PAH	18	26	0.004
SPANI/PANI	28	42	0.008
SPANI/PDADMAC	21	38	0.004
PANI/PSS	19	22	0.003

SPS could also be utilized to determine the thickness values for the dried films. In this case, it was used to measure *in-situ* adsorption properties of the polyelectrolytes on the substrate surface. As shown in the SPS curves (Figure 3), the adsorption properties followed a progressive shift of the reflectivity minima to higher angles indicating a linear increase in thickness. The average pair-layer thickness values obtained were larger than that of a dried film as measured by ellipsometry. Broadening of the plasmon curve at much thicker layers, as shown for the SPANI/PAH system, indicate increasing overall inhomogeneity for the films. This means that the adsorption process involves increments on the measured thickness and refractive index, which becomes increasingly out of resonance. This has generally been attributed to the formation of a "rougher" layer surface.[12] It is not clear though how the local concentration of unadsorbed polyelectrolytes can affect the actual film refractive index. Further measurements are in progress to verify these results and to include determining the kinetics and equilibrium of adsorption.

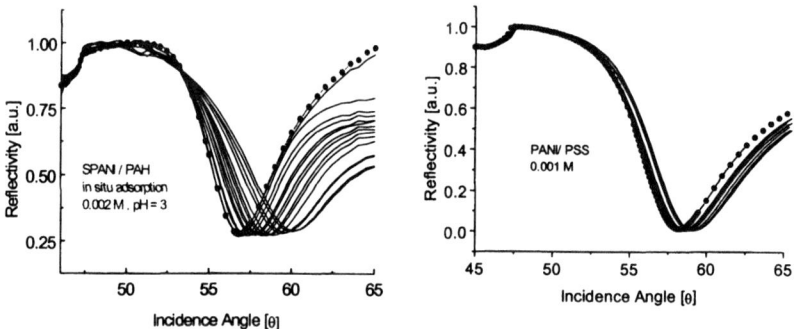

Figure 3. SPS curves of the pairs SPANI/PAH and PANI/PSS with the same solution conditions

In-situ deposition measurements with a quartz crystal microbalance apparatus has shown a linear increase in the mass deposited per layer consistent with the measured thickness per each pair.[13] Careful measurements of the resistance change per pair are currently in progress to determine possible correlation with the viscoelastic and mechanical properties of the polyelectrolyte films.

X-ray reflectometry on films deposited to the gold-coated SPS substrates showed that roughness indeed increased with thickness as observed from the plasmon curve broadening. Based on a 3-layer model fit,[14] this roughness is of the order of 1.8 nm or one pair-layer polyelectrolyte thickness. This was observed for a 20.7 nm (10 double layer) thick film with an average pair-layer thickness of 2.07 nm. These values are again consistent with that observed from SPS and ellipsometry. However, it is expected that thicker layers would result in greater surface roughness.

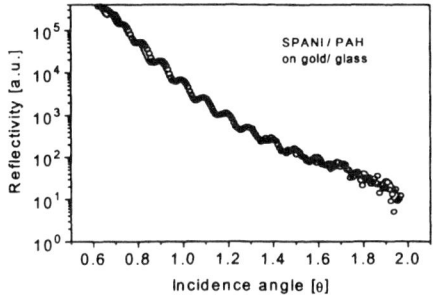

Figure 4. X-ray reflectometry for the SPS-gold coated substrates with SPANI/PAH layers.

AFM and optical microscopy showed the absence of pinhole defects on the SPANI/PAH covered ITO substrate surface. Large area scans showed effectively complete coverage of the rougher ITO Surface (2.1 nm rms). However the mean roughness for the film changes with the polymer pairs and solution conditions used to adsorb the film (1.5- 2.4 nm rms). In general, this is of the order of a pair-layer thickness, consistent with X-ray reflectometry. Again thicker layers are expected to give greater surface roughness. No preferential anisotropy is observed.

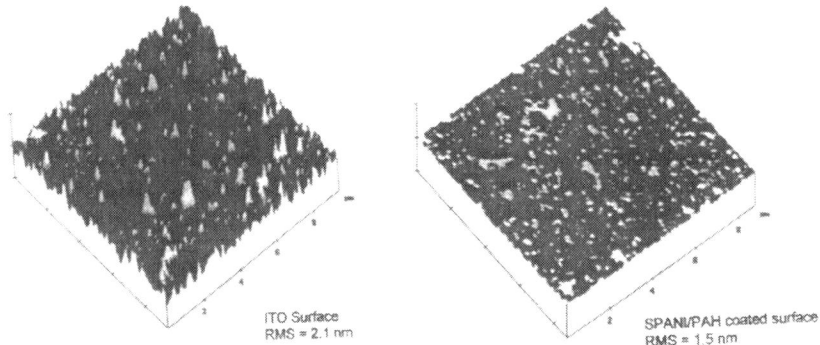

Figure 4. AFM (tapping) mode micrographs comparing the bare ITO and the SPANI/PAH coated substrates (10 x 10 μm). Roughness varies with polymer pairs and solution conditions.

Device Performance was investigated by constant current measurements (J-V) and (I-V) at defined stress periods. Shown in Table II are the results of bias voltages and efficiencies for MEH-PPV based devices made from ITO coated with different polymer pairs . The improvement in efficiencies over bare ITO is evident. The J-V and I-V plots showed smoother curves with lower operating voltages. Lifetimes showed a typical increase up to several hundred hours and showed better average stability than devices prepared using bare ITO at constant current stress measurements.

Table II. Summary of initial device performance measurements for 10 pair-layer coated devices.

Polyelectrolyte Film	Bias V	Luminance /Area	Luminance/ current
PDADAMAC: SPANI	5.9 V	281 Cd/ m^2	1.4 Cd/ A
PAH: SPANI	4.5 V	193 Cd/ m^2	1 Cd/ A
PSS: PANI	3.4 V	130 Cd/ m^2	0.65 Cd/ A
ITO	6.8 V	77 Cd/ m^2	0.4 Cd/ A

Recent studies by the HP group indicate that surface passivation is important in avoiding early catasthropic failure in these devices.[15] In this case, the SPANI and PANI layers not only act as efficient hole injecting transport layers but as a type of "limiting resistor" which tends to smooth out the imperfections on the electrode contacts and emission pathways.

CONCLUSION:

The alternate polyelectrolyte deposition technique provides a uniformly deposited film, which is characterized by the absence of pinhole defects. The layer ordering and roughness are dependent on the solution properties and characteristic polymer pair. PLED devices made from modified ITO substrates showed better performance characteristics over bare ITO. The versatility of the technique and the mechanisms of device improvement will be the focus of future research.

REFERENCES:

[1] Y.Yang and A.J. Heeger J. Appl. Phys. **64**, 1245 (1994).
[2] J.C. Scott, S. Carter, S. Karg, and M. Angelopoulos Synth. Metals. **85**, 1197 (1997).
[3] I.D. Parker J. Appl. Phys. **75**, 1656 (1994).
[4] M. Ferreira, J. Cheung, M. Rubner Thin Solid Films **244**, 772 (1994).
[5] R. Advincula, W. Knoll, C. Frank, D. Roitman, J. Sheats to be submitted to Chem. Materials
[6] A. MacDiarmid, J. Chiang, A. Richter, N. Somasiri in Conducting Polymers ed. L. Alcacer (D. Reidel Publishing, New York, 1987) p. 105-120.
[7] A.J. Epstein and J. Yue, J. Am. Chem. Soc. **112**, 2800 (1990).
[8] J. H. Cheung, W. Stockton, M. F. Rubner, Macromolecules 30, 2712 (1997).
[9] B. Rothenhausler, C. Duschl, and W. Knoll, Thin Solid Films **159**, 323 (1988).
[10] D. Roitman, J. Sheats, H. Antoniadis, M. Hueschen, Proc. Int. SAMPE Technical Conference **27**, 681 (1995).
[11] M. Ferreira and M.F. Rubner Macromolecules **28**, 7107 (1995).
[12] R. Advincula, E. Aust, W. Meyer, W. Knoll, Langmuir, **12**, 3536 (1996).
[13] C. E. Reed, K.K. Kanazawa, and J. H. Kaufman, J. Appl. Phys. **68**, 1993 (1990).
[14] S. K. Sinha, Physica B, **173**, 25 (1991).
[15] H. Antoniadis, M.R. Hueschen, J. McElvain, J.N. Miller, R. Moon, D. Roitman, and J. Sheats, Polymer Preprints **38**, 382 (1997).

INTERNAL FIELD DISTRIBUTION IN ORGANIC LIGHT EMITTING DIODES WITH DOUBLE LAYER STRUCTURE

C. HOCHFILZER*, T. JOST*, A. NIKO*, W. GRAUPNER*, G.LEISING*, C. W. TANG**, E. FORSYTHE***, Y. GAO***
*Institute for Solid State Physics, University of Technology Graz, 8010 Graz, Austria
**Imaging Research and Advanced Development, Eastman Kodak Co., Rochester, NY 14650
***Department of Physics and Astronomy, University of Rochester, NY14627

ABSTRACT

Double layer organic light emitting devices (OLED) are constructed by evaporating tris(8-hydroxy) quinoline aluminum (Alq$_3$) on a spin cast thin film of a methyl substituted ladder type poly-para-phenylene (m-LPPP). A thick layer of Mg:Ag is used as the cathode material. These organic materials are very suitable for application in OLEDs both, as transporting materials as well as active layers. Alq$_3$ predominantly transports electrons while m-LPPP is a conjugated polymer having higher hole mobilities. Due to these transport properties the formation and radiative recombination of the excitons in ITO/m-LPPP/Alq$_3$/Mg:Ag devices occur close to the m-LPPP/Alq$_3$ interface. We compare the device performance of OLEDs with varying Alq$_3$ layer thickness (0, 50, 150, 300, 500Å) and constant m-LPPP layer thickness (900Å). A difference in the device parameters and performance as a function of the Alq$_3$ layer thickness is observed. We analyze these results with respect to the internal electric field distribution of the double layer devices derived from electroabsorption measurements.

INTRODUCTION

The first reports on efficient electroluminescence emission (EL) from an organic multilayer device using tris(8-hydroxy) quinoline aluminum (Alq$_3$) as the emissive material [1][2] have stimulated intensive research in the field of fluorescent molecules with the focus on future applications in display technology [3][4]. Also, the successful use of fluorescent polymers has enlarged the selection of materials which can be used in organic light emitting devices (OLEDs) [5]. Among the conjugated polymers, poly (para-phenylene) (PPP) [6] and its derivatives are very suitable materials for EL applications. One of these derivatives is methyl substituted ladder-type poly (para-phenylene) (m-LPPP), which shows bright EL emission in the blue-green region of the visible spectrum [7].

Neither Alq$_3$ nor m-LPPP forms an efficient bipolar charge transport layer. Alq$_3$ is known as a better electron transport material [1] whereas m-LPPP is a predominantly hole transporting material [7]. Highly efficient OLEDs with these materials typically incorporate separate transport and emission layers. We present Alq$_3$ layer thickness dependent EL and electroabsorption (EA) measurements on these bi-layer OLEDs. This gives us insight into the field distribution in each layer.

EXPERIMENT

Methylated poly-para-phenylene (m-LPPP) and a metal chelate complex, tris(8-hydroxy)quinoline aluminum (Alq$_3$) were used as the organic compounds for these double layer devices. The chemical structures of these materials are shown in Figure 1. The synthesis routes for m-LPPP and Alq₃ can be found in Ref. [8] and Ref. [9], respectively. Figure 1 shows the structure of the OLEDs investigated in this work. The substrate is an indium tin oxide (ITO) coated glass plate where ITO is the transparent anode material with a sheet resistance of about 300Ohm/sq. Prior to the organic deposition, the substrates were thoroughly cleaned by scrubbing, sonication, vapor degreasing and irradiation in a UV-ozone chamber. Following this step a 900Å thick homogenous m-LPPP film was spin-cast on top of the substrate using spectroscopic clean toluene as the solvent. Both deposition and the following thermal annealing process at 60°C were performed under argon atmosphere in a dust free glove box. The film thickness was measured with an Alpha-step profilometer and UV/vis transmission spectroscopy using Lambert's law.

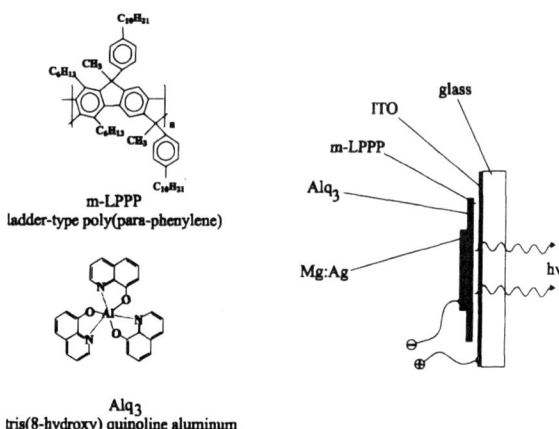

Figure 1: Chemical structure of the organic materials and schematic diagram of the device structure

Thin layers of Alq₃ (50, 150, 300, 500Å) were deposited on top of the m-LPPP layer in a vacuum chamber (about 10^{-5} Torr) by vapor deposition using resistively heated tantalum boats. The deposition rate was 4Å/s at a distance between substrate and evaporation source of about 30cm. The cathode material, a magnesium-silver compound, was evaporated on top of the organic layer without breaking the vacuum. The Mg:Ag (10:1) electrode was deposited using separately controlled sources. The deposition rate for Mg and Ag was typically 10 and 1Å/s, respectively. The active area of the devices, defined by the overlap of the ITO and the Mg:Ag electrodes was $0.1cm^2$. The OLEDs were completed with encapsulation in a dry argon glove box. All OLEDs investigated in this work were produced in one pump down.

The EL spectra were measured using a Photo Research PR650 spectrophotometer. Electroabsorption (EA) measurements were performed using the device shown in Figure 1 in a double pass transmission geometry: the collimated beam from an incandescent light source was passed through glass, ITO, mLPPP and Alq₃, reflected by the elctrode and guided through a grating monochromator to be detected by a Si-photodiode equipped with a built in preamplifier. The electric field in the organic layer was generated using the output of a function generator applied to the ITO and Mg:Ag electrodes. The driving voltage amplitude was 15 V and we typically used 1 kHz. The synchronization output of the function generator was used as a reference for a lock-in amplifier which picked up the synchronous signal from the Si-photodiode. All experiments were performed in air and at room temperature.

Figure 2 shows the EL spectra of bi-layer OLEDs with equal m-LPPP layer thickness (900Å) and different Alq_3 layer thickness d_{Alq3} (150Å and 500Å). The two peaks at 2.7 and 2.5eV arise from the m-LPPP emission whereas the broader feature at 2.3eV comes from the Alq_3 layer. The photoluminescence (PL) spectrum of a pure Alq_3 film is shown for comparison. For an Alq_3 layer thickness of 150Å the EL spectrum is dominated by the m-LPPP emission. As the Alq_3 thickness increases, the Alq_3 emission becomes more and more prominent relative to the m-LPPP emission [10]. The region where the EL occurs is called the emissive zone and is governed by the location where the positive and negative charge carriers recombine. This in turn depends on the charge and electric field distribution within the device which was investigated by electroabsorption.

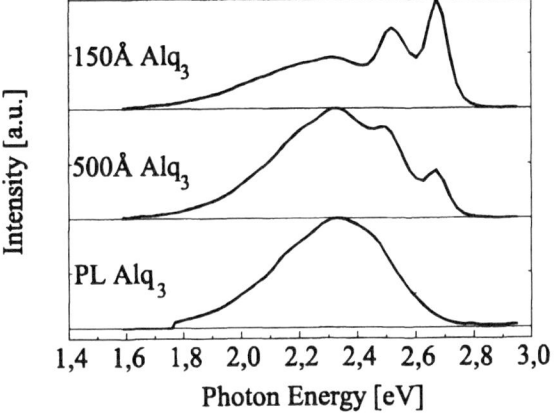

Figure 2: Electroluminescence spectra of ITO/m-LPPP/Alq_3/Mg:Ag OLEDs with constant m-LPPP (900Å) and different Alq_3 (150, 500Å) layer thickness (top, middle) at 100mA/cm² driving current density and the photoluminescence spectrum of pure Alq_3 (bottom)

We have applied EA technique to these devices to gain more information about the internal electric field distribution under applied forward and reverse bias. The EA response is proportional to the square of the applied electric field. In the presence of a dc electric field superposed by ac bias an electroabsorption signal is identified spectroscopically both at the fundamental and the second harmonic frequency of the applied ac bias whereas if there is no static electric field the response is detected only at the second harmonic frequency of the applied ac bias. Figure 3 shows a typical EA spectrum for a bi-layer OLED. This response is coming from m-LPPP which has a prominent feature at 2.7eV [11]. The EA response coming from Alq_3 is negligible because the absorption coefficient of Alq_3 is considerably smaller than in m-LPPP in this energy region - even at 3 eV the values are 20 000 cm⁻¹ vs. 80 000 cm⁻¹ [12][7]. Additionally, the spectral position of EA response coming from Alq_3 is expected higher in energy since the absorption onset is at around 2.8eV and therefore well above that from m-LPPP. Hence the changes in the EA response are proportional to the dc field in the m-LPPP layer. Another indication for the signal being due to m-LPPP exclusively is the fact that the EA spectra are identical for m-LPPP single-layer devices and for the Alq_3/m-LPPP bilayer devices. See references [13][14] for further details on this technique.

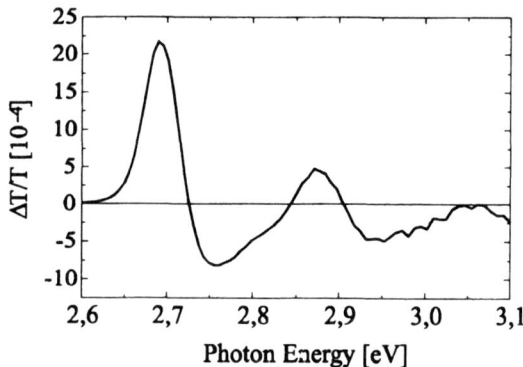

Figure 3 Typical electroabsorption spectrum of a bi-layer ITO/m-LPPP/Alq$_3$/Mg:Ag OLED under dc bias at the fundamental frequency of the applied ac bias

RESULTS

The magnitude of the ac and dc electric field at the m-LPPP layer of the ITO/m-LPPP/Alq$_3$/Mg:Ag bi-layer OLEDs is determined by measuring the EA response at both the fundamental and the second harmonic frequency of the ac bias and comparing the signal amplitudes to reference spectra acquired with the ITO/m-LPPP/Mg:Ag single layer OLED. Figure 4 shows the dc electric field distribution for bi-layer OLEDs with varying Alq$_3$ layer thickness under different bias. A linear dependence of the electric field on the applied bias voltage in forward as well as in reverse direction was observed.

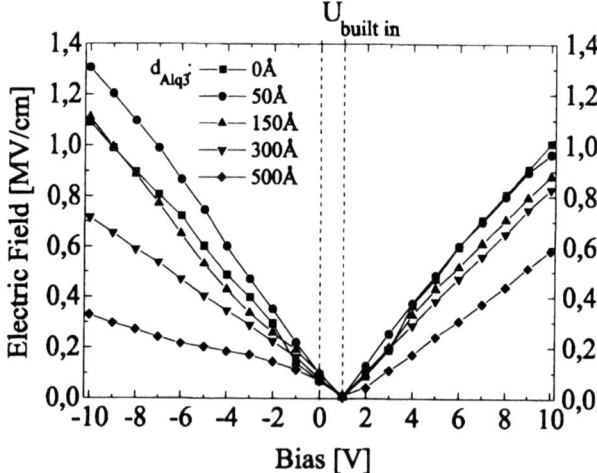

Figure 4: Electric field at the m-LPPP layer as a function of dc bias voltage for ITO/m-LPPP/Alq$_3$/Mg:Ag OLEDs with different Alq$_3$ layer thickness d$_{Alq3}$; the built in potential U$_{built\,in}$ is also shown

The Fermi energies of the anode and cathode metal are 4.7eV (for ITO) and 3.7eV (for Mg:Ag), respectively. Therefore we expect the electric field in the m-LPPP layer to vanish when we apply a forward bias of 1V. At this potential the built in field of the OLED arrangement is compensated. The experimental results (see Figure 4) show a built in potential of 1V which is in good agreement with our predicted value. The Fermi levels of the metals are aligned without applied bias.

The absolute magnitude of the electric field in m-LPPP under forward bias decreases with growing Alq_3 layer thickness under constant voltage. Considering a model of two series capacitors to describe the bi-layer OLED structure, we have analyzed the electric field in m-LPPP under forward bias in more detail [15]. Injected holes traverse the m-LPPP layer - a predominantly hole transporting conjugated polymer - and surmount the relatively small barrier between m-LPPP and Alq_3 (0.2eV). Alq_3 is a predominantly electron transporting material [1]. Therefore, the holes will accumulate within the Alq_3 layer close to the m-LPPP/Alq_3 interface. The electrons injected into Alq_3 are more efficiently blocked at the m-LPPP/Alq_3 barrier with a height of 0.6eV. Hence, the recombination zone is in the Alq_3 layer close to the m-LPPP/Alq_3 interface. A portion of the holes may drift to the Alq_3/Mg:Ag interface under the applied bias field and leave the OLED when the Alq_3 layer thickness is very low. For increasing Alq_3 layer thickness the hole distribution is shifted away from the cathode and hence the internal field in the m-LPPP layer will be reduced.

When centered at the determined built in field as described above, a symmetric behavior is observed for the electric field at the m-LPPP single layer OLED in forward and reverse bias (filled squares in Figure 4).

The magnitude of the electric field for varying Alq_3 layer thickness under reverse bias shows a different behavior. For thin Alq_3 layers of 50Å thickness the electric field in m-LPPP increases compared to the single layer OLED, whereas the electric field in m-LPPP decreases as the Alq_3 layer gets thicker (150, 300, 500 Å). We explain this with a different effect. No EL emission can be detected under reverse bias indicating that the opposite charge carriers do not meet each other. The electrons are injected from ITO into the m-LPPP and will be trapped close to the ITO/m-LPPP interface. The holes are injected form Mg:Ag into Alq_3 and will be trapped in the Alq_3 layer close to the Alq_3/Mg:Ag interface.

Consider thicker Alq_3 films, we can visualize this system as a charge distribution over greater total thickness of the multilayer. Hence, the internal field would be lower compared to a single layer OLED. We neglect charge accumulation at the m-LPPP/Alq_3 interface because in this case the energy "ramps" at the interface are no longer barriers. The field increase for low Alq_3 layer thickness may result from the hole accumulation in the Alq_3 layer (in this case closer to the m-LPPP/Alq_3 interface).

We do not observe symmetry in the electric field for forward and reverse bias of a bi-layer OLED for one particular Alq_3 thickness as we have two unrelated processes. However for a single layer device this symmetry is observed.

CONCLUSIONS

We have shown a correlation between the Alq_3 layer thickness and the device EL spectrum for bi-layer OLEDs which results in an increasing Alq_3 emission relative to the m-LPPP emission. Information was gained how the internal field is built up for different Alq_3 thickness' in forward as well as in reverse bias. In forward bias we attribute the response to the charge accumulation at the emissive zone close to the m-LPPP/Alq_3 interface. In reverse bias the field distribution is governed by the transport properties of the organic materials.

ACKNOWLEDGMENT

The authors thank L. Nugyen and John Burtis from Eastman Kodak for device preparation and device characterization. We gratefully acknowledge the supply of polymer powder by U. Scherf from the MPI Mainz. This work was partially sponsored by the *Bundesministerium für Wissenschaft und Verkehr 'Kurzfristige wissenschaftliche Auslandsaufenthalte'*, the *SFB Elektroaktive Stoffe* and the NSF DMR-9612370. The support form Nokia Telecommunications Austria GmbH is also acknowledged.

REFERENCES

1. C. W. Tang, S. A. VanSlyke, Appl. Phys. Lett. 51, 913 (1987)
2. C. W. Tang, S. A. VanSlyke, C. H. Chen, J. Appl. Phys. 65, 3610 (1989)
3. S. Tasch, C. Brandstätter, F. Meghdadi, G. Leising, G. Froyer, L. Athouel, Adv. Mater. 9, 33 (1997)
4. C.W. Tang, D.J. Williams, J. C. Chang, US-Patent 5,294,870, (1994)
5. J. H. Burroughes, D. D. C. Bradley, A. R. Brown, R. N. Marks, K. Mackay, R. H. Friend, P. L. Burn A. Kraft, A. B. Holmes, Nature 347, 530 (1990)
6. G. Grem, G. Leditzky, B. Ullrich, G. Leising, Adv. Mater. 4, 36 (1992)
7. S. Tasch, A. Niko, G. Leising, U. Scherf, Appl. Phys. Lett. 68, 1090 (1996).
8. U. Scherf, A. Bohnen, K. Müllen, Makromol. Chem. **193**, p. 1127 (1992).
9. D. C. Freeman, C. E. White, J. A. Chem. Soc. **78**, p. 2,68 (1956)
10. C. Hochfilzer, G. Leising, E. Forstythe,Y. Gao,C. W. Tang, submitted to Appl. Phys. Lett.
11. G. Meinhardt, A. Horvath, G. Weiser, G. Leising, Synth. Met. **84**, p. 669 (1997).
12. D. Z. Garbuzov, V. Bulovic, P. E. Burrrows, S. R. Forrest, Chem. Phys. Lett. 249, 433 (1996).
13. D. E. Aspnes, J. E. Rowe, Phys. Rev. B **5**, p. 4022 (1972)
14. I. H. Campbell, M. D. Joswick, I. D. Parker, Appl. Phys. Lett. **67**, p. 3171 (1995)
15. A. Niko, C. Hochfilzer, T. Jost, F. Meghdadi, W. Graupner, G. Leising, this proceedings

PHOTOEXCITATIONS OF OLIGOTHIOPHENES WITH REDUCED CONFORMATIONAL MOBILITY

L. ROSSI [a,b], G. BONGIOVANNI [a,c], C. BOTTA [d], G. CERULLO [e], G. LANZANI [a,f], A. MURA [a,c], F. SANNICOLO' [g], T. BENINCORI [g], R. TUBINO [a,h]
(a) Istituto Nazionale per la Fisica della Materia (Italy).
(b) Dipartimento di Fisica "A. Volta," Università di Pavia, Pavia (Italy).
(c) Dipartimento di Scienze Fisiche, Università di Cagliari, Cagliari (Italy).
(d) Istituto di Chimica delle Macromolecole, CNR, Milano (Italy).
(e) Istituto di Matematica e Fisica, Università di Sassari, Sassari (Italy).
(f) CEQSE-CNR Dipartimento di Fisica, Politecnico di Milano, Milano (Italy).
(g) Dipartimento di Chimica Organica ed Industriale, Università di Milano, Milano (Italy).
(h) Dipartimento di Scienza dei Materiali, Università di Milano, Milano (Italy).

ABSTRACT

We present a comprehensive study of the optical and electronic properties of a series of oligothiophenes in which one or two inter-ring torsional angles have been blocked by chemical bridging. These give us the possibility to investigate the role of the conformational mobility and of the coupling with the inter-ring torsion on the deactivation process of the singlet excited states. We find that both the radiative and the non-radiative deactivation channels are affected by the inter-ring bridging.

INTRODUCTION

The class of thiophene oligomers (nT) have attracted considerable attention as model compounds for a better understanding of the electronic and optical properties of the parent polymer polythiophene [1] and because thin films of nT have found applications as active layers in field effect transistors [2-4] and light emitting diodes [5]. The optical properties of nT can be easily tuned by changing the conjugation length of the chain or via the chemical substitution of hydrogen atoms by many different groups.

With the intent of controlling the molecular conformation and enhance the solubility, we have synthesized and studied a series of quaterthiophenes in which one or two inter-ring torsional angles have been blocked by chemical bridging. Two types of adjacent ring bridgings have been realized one rigid, which employs a CH_2 group, corresponding to a planar inter ring configuration and the other flexible, with a CH_2OCH_2 group, in which a torsional angle up to $20°$ is expected.

Both cw and time resolved spectroscopical techniques have been applied on bridged nT in solutions. The comparison with the parent unsubstituted oligomer T4 provides information on the role of the reduced conformational mobility.

EXPERIMENT

The chemical structures of the four studied quaterthiophenes are shown in Fig. 1. The chemical bridging have been realized via a CH_2 unit in T4B and via one or two CH_2OCH_2

groups in T4BO and T4BBO, respectively. All these compounds have been synthesized via the chemical route reported in Ref. 6.

The absorption spectra of oligomeric solutions were recorded with a Cary 2400 spectrophotometer at room temperature. Photoluminescence (PL) spectra were obtained using a double monochromator equipped with a photomultiplier and exciting with the 350-363 nm lines of an Ar^+ laser. PL quantum yields (QY) of oligomers solutions in tetrahydrofuran (THF) were determined at concentration of about 9.1 x 10^{-6} mol/L (T4), 10^{-5} mol/L (T4B), 4.3 x 10^{-5} mol/L (T4BO) and 4.6 x 10^{-5} mol/L (T4BBO) by using quinine sulfate solutions as a reference.

T4

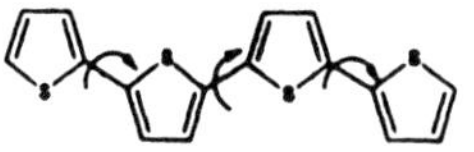

T4B

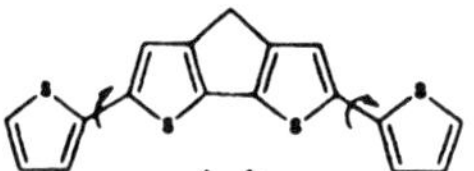

T4BO

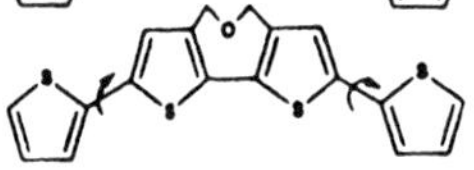

T4BBO

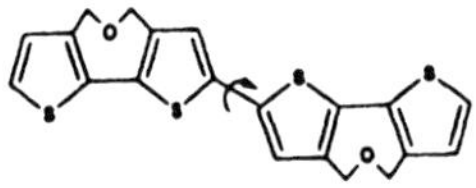

Fig. 1 Chemical structures of quaterthiophenes.

PL time resolved measurements on solutions were performed by exciting with the second harmonic of a pulse compressed Nd:YAG laser. Temporal dispersion of the PL signal was achieved through a Hamamatsu optical sampling oscilloscope, which combines a high time resolution (≈ 20 ps) with a high dynamical range ($\approx 10^5$). All the oligomers were dissolved in THF.

Transient transmission difference spectra ($\Delta T/T$) have been measured in the range from 400 to 770 nm using a Ti:Sapphire laser system. The excitation pulses at 390 nm were generated by second harmonic of the fundamental beam and had a duration of 180 fs. To probe the photoinduced transmission changes we used a supercontinuum pulse generated white light. Transmission difference spectra were obtained by subtracting pump-on and pump-off data collected by the same silicon diode array. The pump beam was focused to a spot size of about 100 µm and linearly polarized at the magic angle. All the oligomers were dissolved in THF.

RESULTS

In Fig. 2 the absorption and PL spectra of the oligomers in diluted THF solutions are reported. The broad and structureless absorption of T4 is indicative of the multiplicity of conformations assumed in the electronic ground state by the unsubstituted thiophene oligomers in good solvent solutions. Upon single and double bridging of the thiophene rings the absorption spectra slightly narrow and some structure appears, because of the partial reduction of the degree of freedom for inter-ring torsions. A significant narrowing could be obtained only in a completely rigid structure, where a localization of the singlet states can take place.

A red shift of both the absorption (λ_{Abs}) and emission (λ_{PL}) are observed. From T4 to T4BBO these shifts can be evaluated from the values reported in Table 1 and are of about 180 meV in the case of the absorption and of 100 meV for PL. In a simple one dimensional free electron model the main parameter is the effective conjugation length (cl_{eff}), which describes the π-electron delocalization along the molecular backbone. Single and double bridgings seem to slightly increase the cl_{eff} values thus inducing a red shift of the structures. Besides, PL spectra show well resolved emission peaks whose energy separation ($\approx$150 meV) roughly corresponds to the vibrational ring mode more coupled with the electronic π–π* transition, while absorption spectra are structureless.

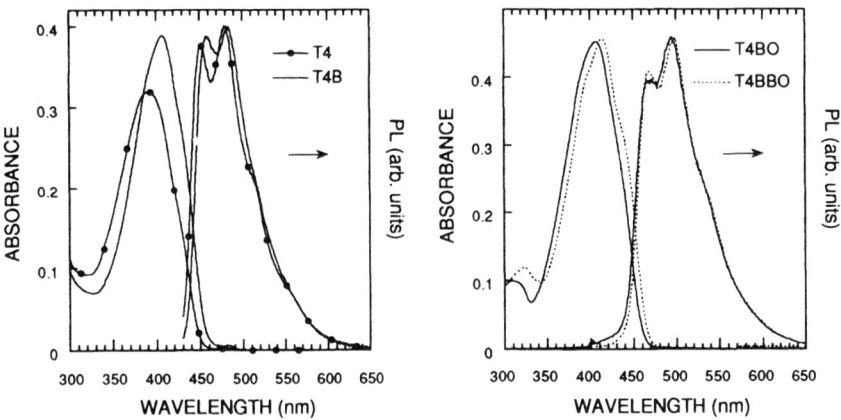

Fig. 2 Absorbance and photoluminescence spectra of quaterthiophenes in solution at room temperature.

	λ_{Abs} (nm)	$\lambda_{PL}{}^a$ (nm)	$\Phi_F{}^b$	τ_F (ns)	K_R (ns^{-1})	K_{NR} (ns^{-1})
T4	390	450	0.17	0.410	0.415	2.02
T4B	407	459	0.24	0.630	0.380	1.21
T4BO	406	467	0.11	0.280	0.39	3.18
T4BBO	413	468	0.10	0.380	0.26	2.37

[a] Short wavelength maximum

Table 1 Photophysical data of the oligomers in THF solution, at room temperature

The reason for these differences between absorption and PL is that the structure of the ground state is benzenoid so we expect that at room temperature several non planar conformers are present: the inter ring rotational vibration give rise to loss of π–π* conjugation decreasing the effective conjugation length. In contrast, the structure of the first excited state is more quinoid, so with double bond character of the ring connecting bonds which leads to a more planar configuration. The double bond linking the thiophene rings in the quinoid structure partially inhibits free inter-ring rotations narrowing the distributions of rotational conformers [7].

The temporal dynamics of PL up to 2 ns for all the oligomers show that the formation times of the emitting states are in the picosecond or subpicosecond range. The kinetics are exponential with time constant τ_F reported in Table 1. Radiative (K_R) and non radiative rates (K_{NR}) are also shown. K_R have been calculated using the following expression $K_R = \Phi_F/\tau_F$ where Φ_F is the QY of the PL. The literature data on T4 report Φ_F values ranging from 0.11 to 0.2 [8-10]. The higher value of the Φ_F in T4B could be due to two different concomitants effects: an increase in the efficiency of the radiative decay channels or a decrease in non radiative channels: the former effect is related to a static property, like a change in the oscillator strength of the electronic transition involved, while the latter is affected by dynamical processes such inter system crossing (ISC) or internal conversion (IC). If we look at the non radiative decay rates the effect of the reduced conformational mobility is detected by the decrease of K_{NR} by going from T4 to T4B and from T4BO to T4BBO. By looking at Table 1 we can therefore conclude that non radiative decay is responsible for the observed changes in the optical response. In general, the non-radiative decay channels consist in the IC in which the excitation energy is dissipated through the coupling with internal vibration and the ISC process which populates the non radiative triplet state. In the oligothiophene series the heavy atom effect due to the presence of the sulphur heteroatom is responsible for the efficient ISC. The high torsional mobility peculiar of the unsubstituted system can enhance this effect. Consistently with the latter point, it has been recently reported that triplet yields in thiophene oligomer can be reduced in polymer matrix, where the inter-ring torsional motion is suppressed, this can hold even for T4B where this motion is partially contrasted. It is also commonly assumed that K_{IC} is a relative small fraction of K_{ISC} and is constant over the oligothiophenes family.

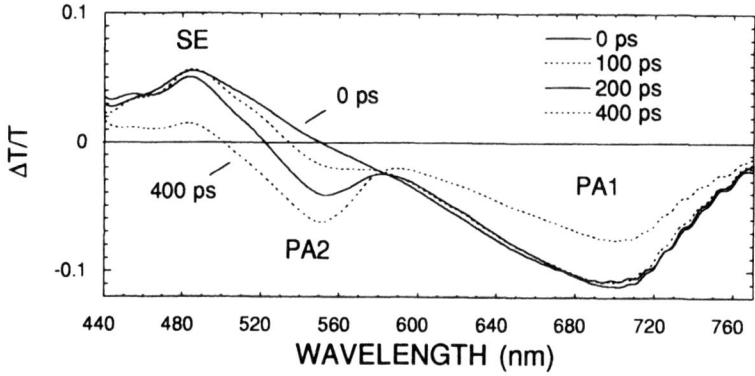

Fig. 3 Transient transmission difference ($\Delta T/T$) spectra of T4B at different delays.

An excellent tool to study both radiative and non radiative processes is time resolved pump and probe experiments. Transient transmission difference spectra ($\Delta T/T$) of T4B in solution are shown in Fig. 3. At 0 ps pump-probe delay we see a positive ΔT between 440 nm and 550 nm, which is to be assigned to stimulated emission (SE), and a negative ΔT, i.e. photoinduced absorption (PA1), at longer wavelengths. Upon increasing the probe delay with respect to excitation, a new negative band appears at 550 nm (PA2), while the initial SE and PA1 decay.

At 10 ps all the spectra show very similar features corresponding to SE and PA1. To show the evolution of the PA spectra for the four compounds we report spectra at 400 ps delay times in Fig. 4. SE and PA1 are reduced for all the compounds, while PA2 shows a peculiar behaviour: it presents a doublet for T4 and T4BBO and a single band for T4B and T4BO: the low energy (LE) peak disappears when the central inter-ring rotation is inhibited by the presence of a chemical bridging. Because PA1 and fluorescence display the same dynamics, they can be assigned to transitions from a common state, the excited singlet state. The long living PA2 band can be reasonably identified with optical transition within the triplet manifold based on comparison with previously reported data on shorter oligomers [11].

In general, the high energy (HE) peak could be assigned to a vibrational replica of the LE peak, to the coexistence of different conformers whose geometrical configuration minimize the total energy of the chain or to transitions to different excited triplet states (T_n). The possibility to selectively detect the HE peak in T4B and T4BO suggests to disregard the former hypothesis, while the latter can be excluded on the basis of theoretical calculations on different nT which prove that only one transition $T_1 \rightarrow T_n$ should be expected [12]. A comparison between T4B and T4BO indicates that there is an effect of the oxygen which is only a conformational effect.

The strong influence of the central inter ring rotation on the properties of these compounds is consistent with the fact that the triplet wavefunction is strongly confined in the central part of the oligothiophene and its extension does not exceed two rings. We can therefore suppose the formation of different conformers: 1) two quasi planar conformers in which the sulphur atoms of the central inter-ring bond are arranged in a anti (A) or syn (S) configurations, 2) a highly distorted conformer in which the central inter-ring angle has a very high value (D). Since, the formation of the most probable conformer, namely the A [13], is not allowed both in T4B and in T4BO, where only the HE peak have been detected, we can conclude that this peak should be related to the formation of S or D conformers.

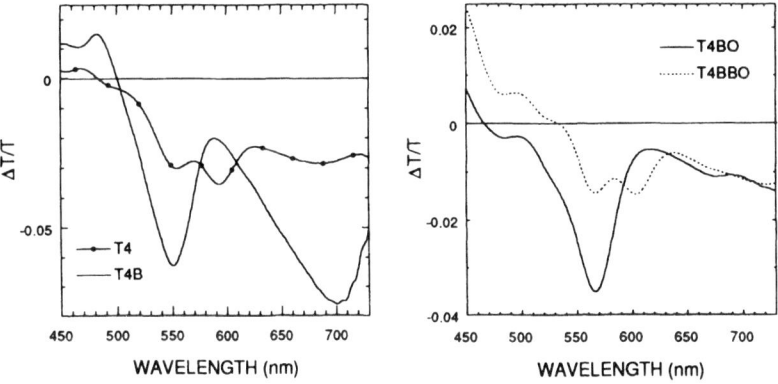

Fig. 4 Transient transmission difference of quaterthiophenes in solution measured at 400 ps delay time.

CONCLUSIONS

We have studied the photophysics of different bridged and non bridged quaterthiophenes in solution. In particular, we have estimated the effect of the multiple bridging on the peak energy of absorption and PL which is on the order of 100 meV. The reduction of the rotational degree of freedom affects the dynamics of deactivations of the excited states by reducing the cross section to the triplet manifold and therefore increasing the PL quantum yield.

REFERENCES

1. D. Fichou, G. Horovitz, B. Xu, and F. Garnier, Synth. Met **39**, p. 243 (1990).

2. H. Koezura, A. Tsumura, H. Fuchigami and K. Kuramoto, *Appl. Phys. Lett.*, **62** p1794 (1992).

3. F. Garnier, R. Hajlaoui, A. Yassar and P. Srivastava, *Science*, **265** p1684 (1994).

4. A. Dodabalapur, L. Torsi and H.E. Katz, *Science*, **268** p270 (1995).

5. D.D.C. Bradley, *Synth. Met.*, **54** p401 (1993).

6. M.E. Brenna, Ph. D. Thesis, University of Milan, Italy (1993).

7. H. Akimichi, K. Waragai, S. Hotta, H. Kano, and H. Sakaki, Appl. Phys. Lett. **58**, p. 1500 (1991).

8. H. Chosrovian, S. Rentsch, D. Grebner, D.U. Dahm, and E. Birckner, Synth.Met **60**, p. 23 (1993).

9. J.P. Reyftmann, J. Kagan, R. Santus, and P. Moliere, Photochem. Photobiol. **41**, p. 1 (1985).

10. R. S. Becker, J. Seixas de Melo, A.L. Maçanita, and F. Elisei, Pure & Appl. Chem. **67**, p. 9 (1995).

11. D.V. Lap, D. Grebner, S. Rentsch, and H. Naarman, Chem. Phys. Lett. **211**, p. 135 (1993).

12. D. Beljonne, J. Cornil, J.L. Brédas, and R.H. Friend, Synth. Met. **76**, p. 61 (1996).

13. G. Barbarella, M. Zambianchi, A. Bongini, and L. Antolini, Adv. Mater. **5**, p. 834 (1993).

SELF-ASSEMBLED MULTILAYERS AND PHOTOLUMINESCENCE PROPERTIES OF A NEW WATER-SOLUBLE POLY(PARA-PHENYLENE)

Xiaobo Shi, DeQuan Li, M. Lütt, M. R. Fitzsimmons, and G. P. Van Patten

Chemical Science and Technology Division(CST-4) and Manuel Lujan Jr. Neutron Scattering Center, Los Alamos National Laboratory, Los Alamos, NM 87545

ABSTRACT

This paper reports the synthesis and characterizations of a new water-soluble poly(para-phenylene) (PPP) and its applications in preparing self-assembled multi-layer films. This new water-soluble conducting polymer was prepared through the sulfonation reaction of poly(p-quarterphenylene-2,2'-dicarboxylic acid). The incorporation of sulfonate groups has dramatically improved PPP's solubility in water at a wide pH range, whereas previous PPP is only slightly soluble in basic solutions. Dilute aqueous solutions of this polymer with acidic, neutral or basic pH emit brilliant blue light while irradiated with UV light. The sulfonated PPP emits from 350 nm to 455 nm with a maximum intensity at 380 nm. Self-assembled multilayers of this sulfonated PPP were constructed with a positively charged polymer poly(diallyl dimethyl ammonium chloride) and characterized with various surface analyses. Conductive (RuO_2 and ITO), semiconductive (Si wafer), and non-conductive (SiO_2) substrates were used in the preparation of self-assembled multilayers. Electrical, optical and structural properties of these novel self-assembled thin films will be discussed.

INTRODUCTION

Conducting polymers are a novel class of conjugated materials which combine the electronic and optical properties of semiconductors and the processabilty of conventional polymers[1]. Conducting polymers, such as PPP or (poly(para-phenylene)), PPV or (poly(phenylene vinylene), and polythiophene, typically possess the delocalized π electrons in their polymeric backbones. These aromatic, rigid-rod polymers play an important role in some important technologies including smart and high performance molecular engineered materials[2] and polymer optoelectronic devices[3]. The organic electroluminescence device (LED) based on these conjugated conducting polymers have been studied extensively in the last decade. The blue-emission LEDs based on PPP was reported by Leising et al. in 1992[4]. Later on, Fou *et al.* reported the incorporation of self-assembled multilayer technology into the fabrication of LED's[5]. In general, the aforementioned unmodified PPP and cured PPV polymers are not soluble in water. To avoid using toxic and corrosive solvents, Novak *et al.* reported the first example of water soluble poly(para-phenylene) derivative[6] in 1991 which provides a new route to synthesize and process rigid-rod polymers in water. The molecular structure of this water-soluble poly(p-quaterphenylene-2,2'-dicarboxylic acid is illustrated in Figure 1(a). This polymer, however, is only slightly soluble in water in its salt form.

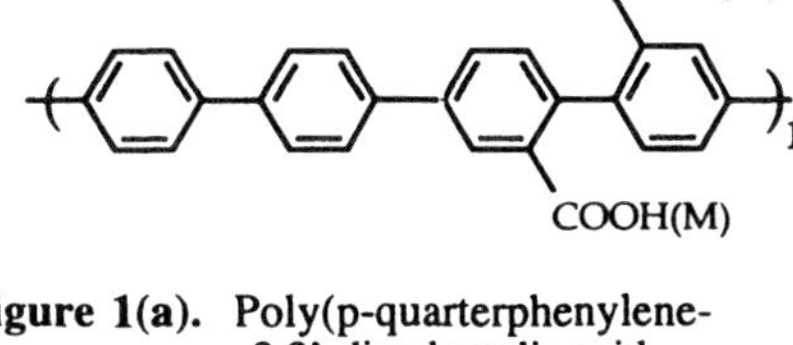

Figure 1(a). Poly(p-quarterphenylene-2,2'-dicarboxylic acid

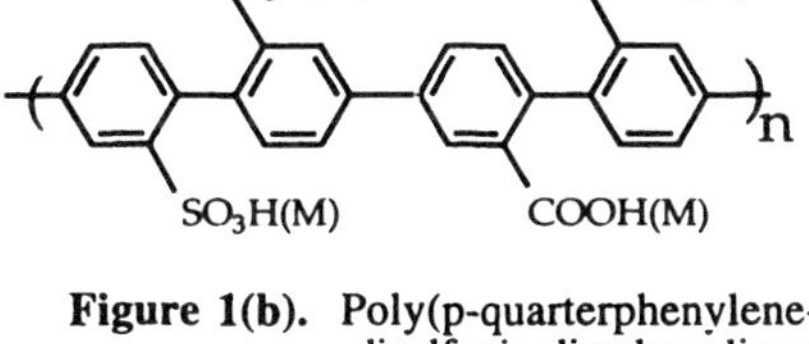

Figure 1(b). Poly(p-quarterphenylene-disulfonic-dicarboxylic acid

The mixture of dimethylformamide/aqueous solvent(25/75) was needed in order to prepare concentrated polymer solution for processing (>5% w/v). Thus, there is a need to prepare highly water-soluble, rigid-rod conducting polymers which can be processed entirely under environmentally benign conditions. Another important consideration is that the charged groups on the repeating unit of the polymer backbone can serve as anchor points in self-assembled thin film growth or LED's fabrications. It is anticipated that the increase in the number of charges on each monomeric unit of the polymer will provide stronger cation-anion attractions in the self-assembly process. In this paper, we report a new PPP polymer—poly(p-quaterphenylene-disulfonic-dicarboxylic acid (Figure 1(b)) through sulfonation of poly(p-quaterphenylene-2,2'-dicarboxylic acid). The sulfonated PPPso$_3^-$ has enhanced solubility in aqueous solution and photoluminescent properties in both solution and solid state films. Self-assembled multilayer films of this new PPP are prepared and the structure of the films is characterized with X-ray reflectivity measurements.

EXPERIMENT

The modified PPP bearing both sulfonate and carboxylate functional groups was prepared by the following route. In a typical experiment, the golden color powder of poly(p-quaterphenylene-2,2'-dicarboxylic acid) was slowly added into an aqueous solution containing 23% SO$_3$/H$_2$SO$_4$ (the ratio of polymer/SO$_3$ is 1:2.3) at 0°C degree under vigorous stirring. Upon the completion of the addition, the reaction solution was stirred at 0°C degree for an hour, and then was heated at 70°C for 7 hours. The product mixture was cooled down to room temperature and was neutralized by adding dilute KOH solution. The solvent in the mixture was removed under vacuum condition. The dried crude product was purified on a column packed with Florisil. Methanol was selected as eluent to collect the product. Final product was collected as yellow color powder.

Monosulfonated triphenylphosphine and its palladium catalyst, which are needed in the cross coupling polymerization, were synthesized according to the literature[7-8], while the monomers of 4,4'-dibromicdiphenic acid and bis(boronic acid) diphenyl were also synthesized and characterized by the reported methods[9-10]. The precursor polymer, poly(p-quaterphenylene-2,2'-dicarboxylic acid) used for sulfonation reaction, was obtained using the procedures reported by Novak et al.[6]. The purified product was characterized by 200 MHz ^{1}H NMR(D$_2$O) (Gemini NMR spectrometer) and IR (KBr pellet) (Bio-Rad-40). Absorption spectra were recorded using a Perkin-Elmer Lambda-19 UV/VIS/NIR spectrometer. Photoluminescence(PL) measurements on (un)sulfonated PPP solutions and self-assembled films were obtained using a SPEX Fluorolog Fluorometer.

The negatively charged PPPso$_3^-$ polymer was assembled with a positive charged polymer poly(diallyl dimethyl ammonium chloride) or PDDA to form multilayers. The self-assembled multilayer of PDDA/PPPso$_3^-$ films and spin casted films were formed on various substrates; these include conductive RuO$_2$ and ITO, semiconducting Si wafer, and nonconductive quartz or glass substrates. The conductivity measurements were performed on a 100 PDDA/PPP bilayer film self-assembled on RuO$_2$ substrate. Self-assembled multilayer PDDA/PPP films on quartz and glass substrates were also characterized by UV-Visible absorption and photoluminescence measurements. Self-assembled PDDA/PPP multilayer films on Si(100) substrates were used for the X-ray reflectivity study which was carried out with an 18-kW rotating anode X-ray generator with a copper target and a four-circle diffractometer.

RESULTS AND DISCUSSION

The new sulfonated poly(para-phenylene) PPPso$_3^-$ exhibits a high solubility in water at a wide range of pH including acidic, neutral, or basic conditions. It can be easily dissolved in water at pH values ranged from 1 to 14. This high solubility can be attributed to the increased numbers of negative charges on the repeating unit (possibly four negative charges on each repeating unit) of the sulfonated and carboxylated poly(para-phenylene) polymer. This highly water-soluble polymer was characterized by a number of techniques to confirm its composition and rod-like structure. ^{1}H NMR spectrum(D$_2$O) showed broad resonances ranged from 6.6 to 8.2 PPM, which can be attributed to the polymeric effect. The resolved vibronic features in the IR spectrum (KBr pellet) displays expected bands at 1608, 1440, and 3448 cm⁻¹ for -COOH vibrations, bands at 1440 and 1196 cm⁻¹ for -SO₃H vibrations, bands at 3060, 1130 and 1046 cm⁻¹ for Ar-H vibrations, and bands at 809 and 690 cm⁻¹ for phenyl rings in PPP backbone.

The aqueous solution of sulfonated poly(para-phenylene) emits very brilliant blue light when irradiated by UV light. The emission spectrum of sulfonated PPP aqueous solution is compared with the emission spectrum for PPPco₂⁻ solution in Figure 2 with the same concentration (10⁻³M). Figure 3 displayed the photoluminescent spectra of a 100-bilayer-film of self-assembled PPP and PDDA. It is noticed that the introduction of -SO₃H(M) functional groups onto the polymer backbone caused the blue-shift for either absorption or photoluminescence spectrum.

This phenomenon can be explained with either electron withdrawing effect or interruption of π-conjugation. It is more likely that the band gaps between conducting band and valence band is increased by the introduction of negatively charged electron withdrawing group such as -SO₃H(M). Comparing the blue shift effect for aqueous polymer solution, the effect is relatively small to the modified PPP/PDDA bilayer film. The explanation might be weak spatial interactions among the PPP layers through the PDDA linker. As shown in Figure 2, the caboxylaed PPP emits blue light with a maximum intensity at about 430 nm, but the modified PPPso₃⁻ with both sulfonate and carboxylate groups emits very brilliant blue light with maximum intensity at 380 nm and covered a narrow spectrum range from 355 to 450 nm. The emission spectrum of the 100-bilayer PDDA/PPP films has the maximum intensity at 411 nm (Figure 3) with a shoulder at about 430 nm.

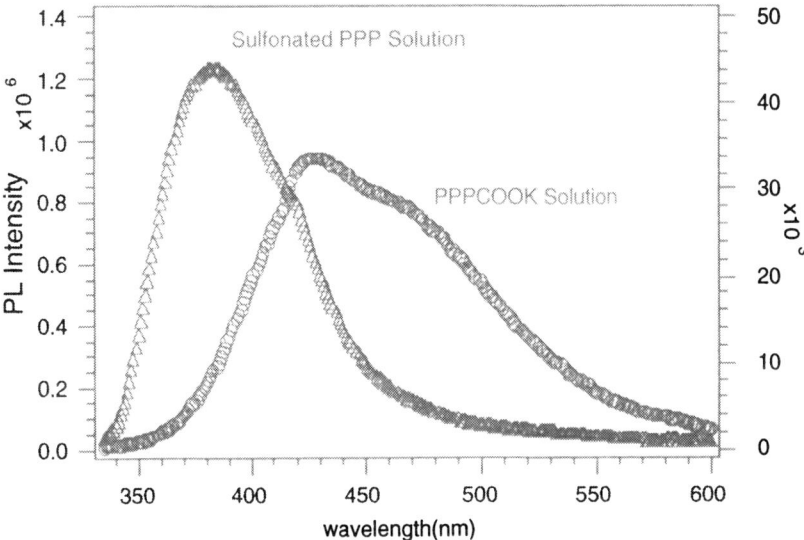

Figure 2. The solution emission spectra of poly(p-quarterphenylene- 2,2'-dicarboxylic acid (referred as PPPco₂⁻) and Poly(p-quarterphenylene- disulfonic-dicarboxylic acid(referred as PPPso₃⁻).

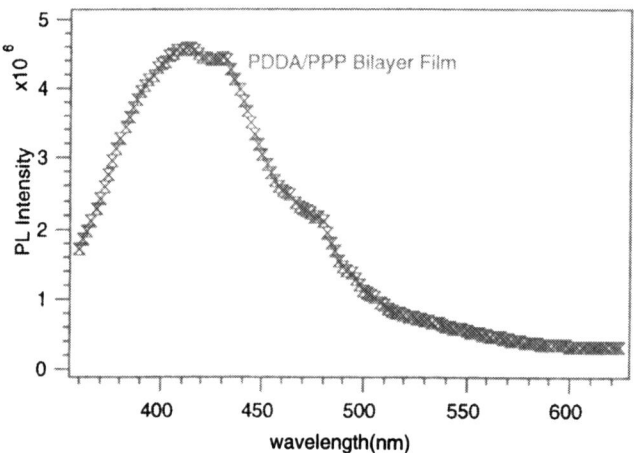

Figure 3. The emission spectrum of the 100-bilayer PDDA/PPP film

In an attempt to build high quality multilayer films, the modified PPP bearing carboxylate groups was used first. However, it failed to yield good quality films due to the use of mixture aqueous and organic solvent. The possible reasons for this failure might be that the organic solvent such as DMF may cause desorption of the deposited bilayers. The sulfonated PPP, however, is very soluble in water and can be successfully employed to build up multilayer films. Several factors were considered to play important roles in such kind of self-assembly process. First, the number of net negative charges on each repeating structural unit in the polymer was doubled, the increased negative charges provided stronger chemical forces for this conducting polymer to bind with positively charged polymer linkers. Second, the incorporation of more negatively charged functional group dramatically increased the solubility of sulfonated PPP in water at ambient pHs so that the use of polar organic solvent in self-assembly was eliminated avoiding desorption of the films from the substrate surface.

Figure 4 shows the electron-depth-density profile of a bilayer film of PDDA/PPP which was derived from the analysis of the X-ray reflectivity measurement shown in the figure inset. The reflectivity is the ratio of the intensities of the X-rays reflected from the sample to that which illuminate the sample. Such measurements can be fitted to calculations of reflectivity profiles from model systems (such as that shown in figure 4) using the Parratt-formalism [11]. The best fitting model is one of a substrate terminated with a native oxide of a thickness of 7.1 Å and a roughness of 2.4 Å (rms). The modeled PDDA layer was 17.4 Å thick and had a surface roughness of 1.2 Å. These values are similar to those of a pervious study of PDDA and NiPc [12]. The sulfonated PPP layer deposited on the PDDA layer was 12.5 Å thick and has a slightly larger roughness (2.3 Å) than the PDDA-layer.

The X-ray study was sensitive to a scattering contrast between the PDDA and sulfonated PPP layers, whereas previously no scattering contrast was determined between PDDA and NiPc [12]. The difference in the scattering contribution in the present system suggests that the single layers which comprise the bilayer of PDDA and PPP are distinct (not intermixed).

PDDA/PPP

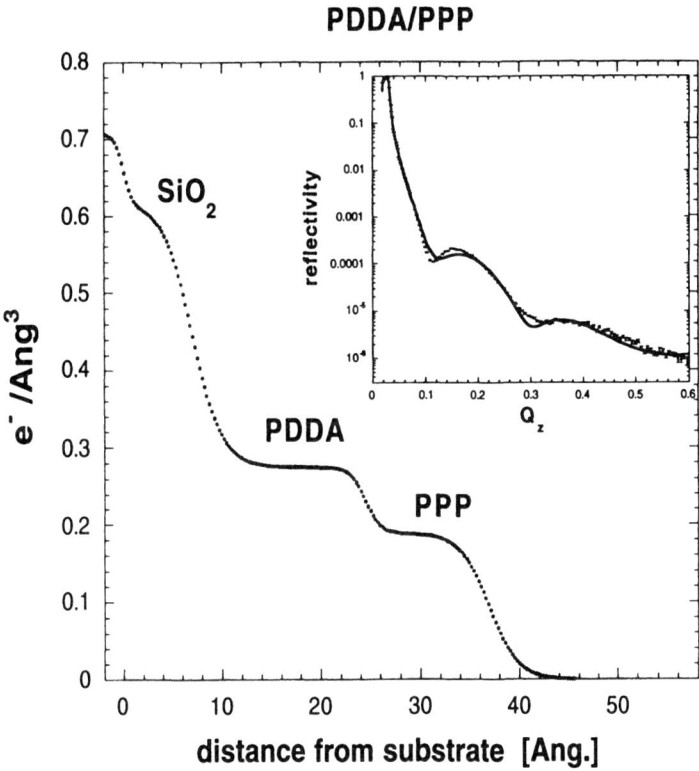

Figure 4 The electron density profile of a PDDA/PPP bilayer film on Si(100) derived from an X-ray reflectivity measurement (shown in the inset)

CONCLUSION

Through the chemical modification of poly(p-quarterphenylene-2,2'-dicarboxylic acid), a newly prepared Poly(p-quarterphenylene-disulfonic-dicarboxylic acid) shows a promising solubility in water at a wide range pH values (ranged from 1 to 14), which can be self-assembled at an environmentally benign condition. The brilliant blue luminescence was observed when the aqueous Poly(p-quarterphenylene-disulfonic-dicarboxylic acid) was irradiated using UV-light. The incorporation of electron withdrawing group -SO$_3$H(M) into the rigid polymer backbone caused the blue shift in both absorption and emission spectrum while compared to poly(p-quarterphenylene-2,2'-dicarboxylic acid). Photoluminescence was also observed when a self-assembled multilayer PDDA/PPP film or a spin coated film was exposed to the UV-light. Finally, the thin-film structure of a PDDA/PPP bilayer was determined by X-ray reflectivity.

ACKNOWLEDGMENTS

We gratefully acknowledge for the financial support of this work provided by Los Alamos National Laboratory Directed Research and Development. Part of this study was supported by the U.S. Department of Energy, BES-DMS, under Contract No. W-7405-Eng-36.

REFERENCES

1. J. W. Blatchford and A. J. Epstein, Am. J. Phys. 64, p. 120(1996).

2. Q. Pei, G. Yu, C. Zhang, Y. Yang and A. J. Heeger, Science 269, p. 1,086(1995).

3. J.J.M. Halls, C.A. Walsh, N.C. Greenham, E.A. Marseglia, R.H. Friend, S.C. Moratti and A.B. Holmes, Nature 376, p. 498(1995).

4. G. Leising, S. Tasch, F. Meghdadi, L. Athouel, G. Froyer and U. Sherf, Synthetic Metal 81, p. 185(1996).

5. A. C. Fou, O. Onitsuka, M. Ferreira, M.F. Rubner and B.R. Hsieh, J. Appl. Phys. 79, p. 1,316(1995).

6. Thomas I. Wallow and Bruce M. Novak, J. Am. Che. Soc. 113, p. 7411(1991).

7. S. Ahrland, J. Chatt, N.R. Davies and A.A. Williams, J. Chem. Soc. p. 276(1958).

8. A.L. Casalnuovo and J.C. Calabrese, J. Am. Chem. Soc. 112, p. 4,324(1990).

9. E. Balogh-Hergovich, G. Speier and Z. Tyeklar, Synthesis p. 731(1982).

10. G.C. Coutts, H.R. Goldschmid and O.C. Musgrave, J. Am. Soc. C. p. 488(1970).

11. L.G. Parratt, Phys. Rev. 95, p. 359(1954).

12. M. Lütt, M.R. Fitzsimmons, D.Q. Li, accepted for publication in J. Chem. Phys. (1997)

Part II

Photonic Materials and Devices

PHOTOFABRICATION OF SURFACE RELIEF GRATINGS USING POST FUNCTIONALIZED AZO POLYMERS

SUKANT K. TRIPATHY*, JAYANT KUMAR[#], DONG-YU KIM*, XINLI JIANG[#], XIAOGONG WANG*, LIAN LI**, MONGKOL SUKWATTANASINITT*, DANIEL J. SANDMAN*
Center for Advanced Materials, Departments of Chemistry* and Physics[#], University of Massachusetts Lowell, Lowell, MA 01854
Molecular Technologies Inc.**, Westford, MA 01886

ABSTRACT

A series of azobenzene functionalized polymers has been synthesized by post polymerization azo coupling reaction. Photo-fabrication of surface relief gratings were studied on the polymer films. Epoxy based azo polymers were prepared by post azo coupling reaction to form polymers containing donor-acceptor type azo chromophores. The azo chromophores were designed to contain ionizable groups to impart self-assembling and photoprocessing capabilities to the polymers. The polymers containing 4-(4-(carboxylic acid)phenylazo)aniline chromophores can be directly photofabricated to form surface relief gratings with large surface modulations. Charge interactions had a strong influence on the details of the writing process. A new soluble polydiacetylene, post-functionalized with azobenzene groups was also prepared. Large amplitude surface gratings could be fabricated on this polydiacetylene film as well.

INTRODUCTION

Polymers containing aromatic azo units have attracted considerable interest in various research fields.[1-8] The mesogenic azobenzene groups have widely been used in the design of liquid crystal polymer (LCP) studies.[1] The highly conjugated azobenzene groups can act as efficient nonlinear optical (NLO) chromophores. A variety of NLO azo polymers have been developed in the last decade.[2-5] Optically induced trans-cis isomerization of aromatic azo groups is being investigated for diverse potential applications.[6-8] Upon exposure to light, azobenzene groups undergo reversible photoisomerization between the generally more stable trans form and the less stable cis form.[9-10] The rate of isomerization of the azo groups depend on the structure of azo compounds and polymer matrices incorporating the aromatic azo chromophores.[11,12] This photo-induced geometrical or conformational change can significantly influence the bulk and surface properties of the polymer.[6-8] Therefore, this photoisomerization of the azobenzenes in polymer matrices and its related property changes are being actively explored for potential applications, such as in optical switching and optical information storage.[13-16]

Upon exposure to a polarized light reorientation of the azobenzene groups in polymer matrices can be induced through repeated trans-cis photoisomerization and subsequent cis-trans relaxation of the azobenzene groups.[17-20] Reversible holographic phase gratings have been demonstrated using this optically induced orientation or birefringence.[19,20] Recently, our group[21-26] and several other groups[27-29] have shown that large surface modulation can be photo-induced on azobenzene polymer films. The surface relief gratings were formed upon exposure to an interference pattern of Ar+ laser beam at modest intensities without any subsequent processing steps. We have found that this process is due to the large scale macromolecular translational motion, which may be attributed to the photoinduced trans-cis-trans isomerization cycles of azobenzene chromophores and the forces experienced by the chromophore dipoles in the electric field gradient of the superimposed standing wave pattern. At the present stage, the photoprocessibility of different types of polymer matrices containing azobenzene moieties and the photo-response behavior of different systems are being pursued to fully understand the structure-property correlation and to develop versatile new materials.

Recently we have synthesized a series of polymers functionalized with aromatic azo groups by post polymerization azo coupling reaction for the study of their NLO properties and surface relief grating formation.[30-32] Post polymerization azo coupling reaction can provide a convenient way to synthesize various azo functionalized polymers. It is often the simplest path to prepare azo functionalized polymers. The extent of the functionalization also could be easily controlled using post azo-coupling reactions. In addition, versatile functional groups can be incorporated on the azobenzene groups. Polymers containing azo chromophores with ionizable groups could also be synthesized. Under appropriate pH condition the polymers can be ionized into azo polyions in water. By a layer-by-layer dipping process, along with an oppositely charged polyelectrolyte, the azobenzene polyelectrolyte can be assembled into multilayer films. A number of important potential applications are expected in these self assembled multilayer films.

In this paper, we report the photo-fabrication of surface gratings on various azo polymers prepared by post polymerization azo-coupling reactions. A series of epoxy based azopolymers with various substituents were prepared and the structure-property relationship in photo-fabrication is discussed. We have also prepared a novel soluble polydiacetylene (PDA) functionalized with azobenzene groups in the side chain. Surface grating formation in this conjugated polymer was also demonstrated.

EXPERIMENTAL SECTION

Materials

The chemical structures of the epoxy based polymers post-functionalized with various azobenzene groups are shown in Fig. 1. The synthesis and structure characterization of the epoxy based polymers were reported elsewhere.[33]

Azobenzene functionalized PDA was prepared by post polymerization azo coupling reaction on the precursor PDA. The precursor PDA was reacted with the diazonium salt of p-aminobenzoic acid in N,N'-dimethylformamide (DMF) to give a quantitative yield (Fig. 2). The detailed synthesis of the precursor and functionalized PDAs are reported elsewhere.[34]

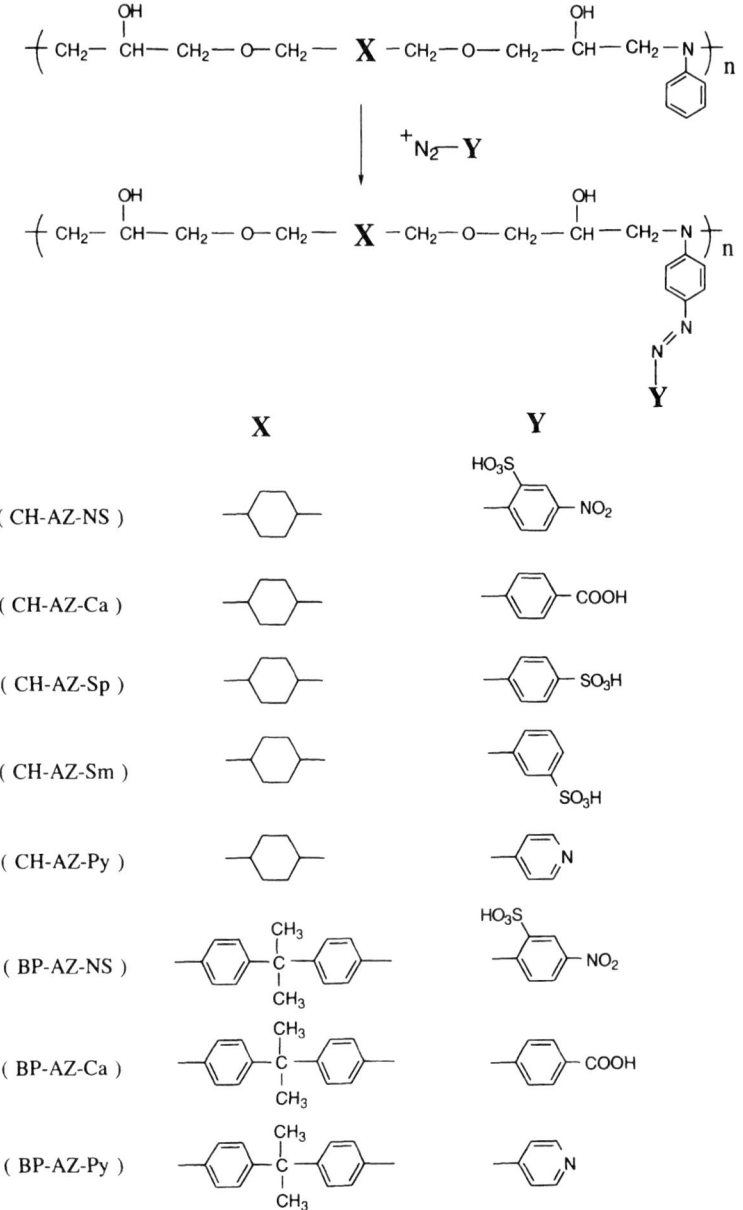

Figure 1. Synthetic scheme and chemical structures of epoxy-based polymers post-functionalized with various azobenzenes.

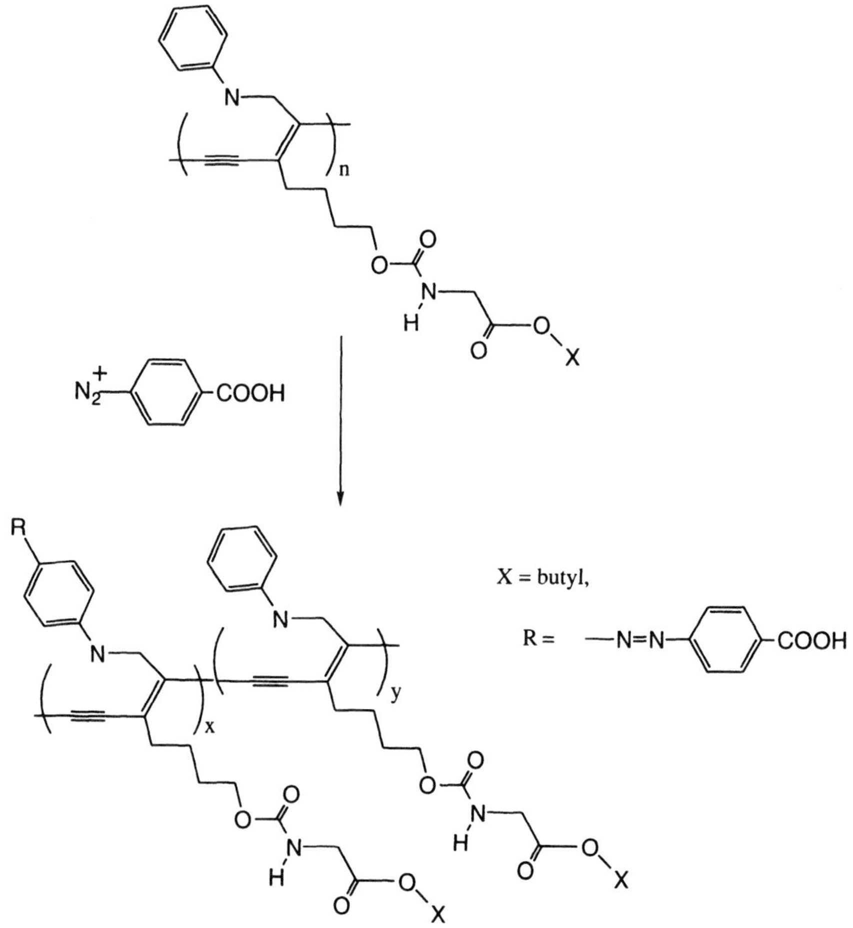

Figure 2. Synthetic scheme of the polydiacetylene post-functionalized with azobenzene groups.

Polymer Film Preparation

The homogeneous solutions of polymers in spectroscopic grade DMF were filtered through 0.2 μm membranes. The solutions were spin-coated onto glass slides. The film thickness was controlled to be 0.4-1 μm by adjusting the solution concentration and the spin speed. The spin-coated films were dried under vacuum for 48 h at 40-50 °C and were stored in a desiccator for further measurements.

The experimental setup for the grating formation has been reported elsewhere[22,23]. A linearly polarized laser beam at 488 nm from an Ar^+ laser was used as the recording source. The polarized laser beam passes through a halfwave plate, and is then expanded and collimated. Half of the collimated beam is incident on the film directly. The other half of the beam is reflected onto the film from a mirror. The polarization condition of the recording beam can be adjusted by rotating the halfwave plate. The angle between the polarization with respect to s-polarization, was selected to be 45° for optimal recording[23]. The intensity of the recording beams ranged from 50 to 100 mW/cm^2. The incident angle θ of the recording beam was adjusted to control grating spacing. The diffraction efficiency of the first order diffracted beam from the gratings in transmission mode was probed with an unpolarized low power He-Ne laser beam at 633 nm.

RESULTS AND DISCUSSION

The precursor polymers were functionalized to incorporate different azobenzene chromophores by post azo coupling reaction (Fig. 1). Optical and thermal properties of these azobenzene polymers are listed in Table 1. The first part of the polymer nomenclature is an abbreviation to distinguish between precursor polymers from 1,4-cyclohexanedimethanol diglycidyl ether (CH) and diglycidyl ether of bisphenol-A (BP). The subsequent parts refer to different conjugation bridges, electron acceptors and ionizable groups of the chromophores.

Table 1. Optical and thermal properties of azobenzene post functionalized polymers.

Polymer	D.F. [a] (%±3%)	λ_{max}[b] (sol.)	λ_{max} (film)	T_g (°C)
CH-AZ-NS	100	482	508	144
CH-AZ-Ca	100	437	428	90
CH-AZ-Sp	69	429	514	98
CH-AZ-Sm	54	424	464	143
CH-AZ-Py	100	440	437	88
BP-AZ-NS	100	483		105
BP-AZ-Ca	100	438		105

a. Degree of functionalization measured by ^{1}H NMR. b. In DMF solution.

The solubility characteristics of the synthesized polymers were greatly influenced by the chromophore and backbone structures. All of the polymers discussed above were able to form homogeneous solution in polar organic solvents such as DMF. CH-AN based azo polymers containing acid groups such as CH-AZ-NS and CH-AZ-Ca are highly soluble in aqueous alkaline medium. CH-AZ-PY is soluble in aqueous acid solution. BP-AN based azo polymers containing acid groups such as BP-AZ-NS are highly soluble in polar organic solvent, but can hardly be dissolved even in strong alkaline aqueous solution. BP-AZ-PY containing 4-(4-pyridylazo)aniline chromophores is not soluble in any solvent tested in our laboratory. The insolubility of BP-AZ-PY is presumed due to the interaction between the pyridine nitrogen and the hydroxyl groups on the main chain.

Thermal behavior of the precursor and azo polyelectrolytes were studied by using differential scanning calorimetry (DSC). All the polymers exhibit thermal behavior typical of amorphous polymers. The T_gs for various polymers determined by DSC are listed in Table 1. The T_g of the precursor polymer CH-AN is relatively low (41 °C) due to the flexible backbone. The T_gs of the functionalized polymers are much higher than the precursor polymer and are highly dependent on the structure of the azo chromophores. CH-AZ-CA and CH-AZ-PY show T_gs of about 90 °C and the increase in T_g is attributed to significant increase both in the size and the dipole moment of the side groups. CH-AZ-NS and CH-AZ-Sm show much higher T_gs (144°C, 143°C). As shown by the following UV-Vis spectroscopic studies, the polyelectrolytes containing sulfonic groups exhibit strong inter- and intramolecular interactions in the solid state.

Absorption maxima (λ_{max}) of the polymers from UV-Vis spectra are listed in Table 1. The λ_{max} values of CH-AZ-NS and CH-AZ-Sp shift from 482 and 441 nm to 507 and 514 nm respectively in case of spin coated films. CH-AZ-Sm shows a more complex spectrum. λ_{max} of 464 nm and a second absorption band can be seen at a longer wavelength. It is believed that the reason for the color change is similar to halochromism effect of aminoazobenzene caused by protonation of the ß-Nitrogen atom of the azo group. Although λ_{max} of CH-AZ-PY remain almost unchanged, a broad shoulder can be seen in the range from 480 to 540 nm. One possible reason for the spectral change is caused by the pyridine ring and hydroxyl group on the main chain. In the system mentioned above, the proton donating and accepting processes accompanied by local charge separation result in strong inter or intramolecular interaction.

Polymer Photoprocessing

Spin coated thin films of newly synthesized azo polymers were used to carry out photoprocessing experiments. Surface relief gratings with designed dimension and patterns can be formed on the film surfaces. The grating formation efficiency depends on the type of polymer and also on the recording conditions. The experimental setup and recording conditions have been previously reported and also described briefly in the experimental part of this paper.

The polymer films were exposed to an interference pattern formed by two polarized Ar+ laser beams at a wavelength of 488 nm. Polymers undergo photodriven transport and a surface relief pattern is formed. The surface modulation profile is dose dependent and the periodicity is determined by the writing angle and wavelength. It was found that these azo polymers show significantly different photoprocessibility depending upon chromophore structures. The polymers containing 4-(4-(carboxylic acid) phenylazo)aniline chromophores such as CH-AZ-CA and BP-AZ-CA form surface relief gratings with large surface modulations and high diffraction efficiencies. As CH-AZ-CA and BP-AZ-CA possess polymer backbones with quite different rigidities, the similar recording efficiency for both polymers imply that the azo chromophores play the defining role in the grating formation process. Chromophore densities being similar the two polymers are subjected to similar forces and are equivalently plasticized in the writing process. The polymer containing 4-(2-(sulfonic acid)-4-nitrophenylazo) aniline and 4-(pyridineazo)aniline chromophores such as CH-AZ-NS and CH-AZ-PY can form surface relief grating with low efficiency. In case of CH-AZ-Sp and CH-AZ-Sm, surface grating was hardly observed under the same recording condition. It is believed that the driving force, resulting from localized variations of magnitude and polarization of the resultant electric field in the film, are counterbalanced by the strong intermolecular interaction mentioned above. Electric field induced poling experiment of these NLO polymers also proved this strong intermolecular interaction present in sulfonic acid functionalized azo polymer systems. While the side chain azo polymers CH-AZ-CA, BP-AZ-CA and CH-AZ-PY could be effectively poled, poling was inefficient for sulfonic acid functionalized azo polymer systems. We also investigated the surface grating formation on the guest-host system. A PMMA film doped with an azo dye (Disperse Orange 3) with 20 wt% dye content was prepared and exposed to an identical exposure condition. Amplitude of surface modulation was much smaller (<100 Å) than those of functionalized polymer films.

The atomic force microscopy (AFM) images of the surface relief gratings formed on CH-AZ-CA and BP-AZ-CA films are shown in Figure 3, and Figure 4, respectively. Regularly spaced sinusoidal surface relief structures can be seen in both cases. The spacings depend on the period of interference pattern, which can be adjusted in range between 0.3 and 5 μm by changing the angle (2θ) between the two writing beams. The depth of the gratings formed in typical cases were in the range of 1000 to 3000 Å. Large surface modulation (>6000 Å) and high diffraction efficiency (>40%) were obtained under optimized condition. The gratings were stable when stored below the polymer T_g and could be erased by heating the polymer above T_g. One of the interesting properties of the azo polyelectrolytes such as CH-AZ-CA is the unique ability to form both surface relief grating upon exposure to an interference pattern of laser beam and formation of multilayer structures by a layer-by-layer deposition process.[32]

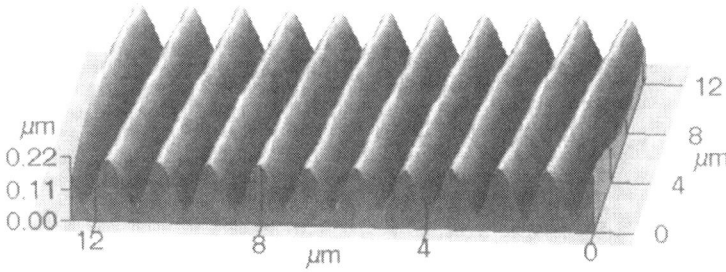

Figure 3. Atomic force microscopy image of surface relief gratings on the CH-AZ-CA film.

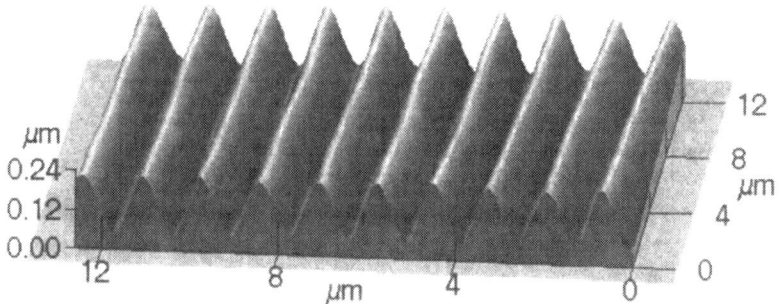

Figure 4. Atomic force microscopy image of surface relief gratings on the BP-AZ-CA.

We also investigated surface grating formation in a PDA system which was post functionalized with azobenezene groups by a post azo coupling reaction. The precursor PDA was reacted with the diazonium salt of p-aminobenzoic acid in DMF to give a quantitative yield of the azobenzene functionalized PDA (Fig. 2). The polymer was found by ^{1}H NMR to have 60% of the aromatic rings functionalized. No T_g was observed from DSC studies of this polymer. Exposure of a spin-coated film to an interference pattern from two polarized argon-ion laser beams at 488 nm leads to formation of a surface grating. As shown in Figure 5, the surface relief grating shows very regularly spaced surface structure similar to other epoxy based systems with depth modulation of over 800 Å. The grating spacing can be adjusted by changing the angle between the two writing beams.

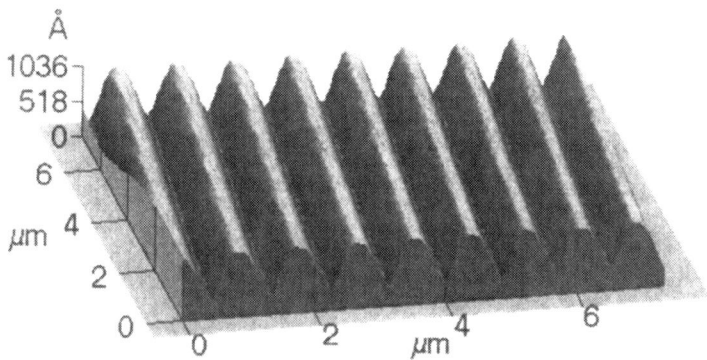

Figure 5. AFM 3-D image of the surface grating fabricated on an azobenzene functionalized
PDA film.

Even though the surface grating formation was less efficient than other epoxy based systems such as CH-AZ-CA and BP-AZ-CA, it is very interesting to observe such a large and smooth surface modulation in a conjugated polymer system which is generally believed to possess a very rigid polymer backbone. This provides the first example of such laser-induced grating formation in a conjugated polymer lacking a glass transition.

CONCLUSIONS

In summary, two epoxy based precursor polymers have been functionalized to form a series of side-chain azo polymers which contain ionizable groups on the chromophores. The CH-AN based azo polymers show typical behaviors of a polyelectrolyte. Surface relief gratings with large surface modulation were obtained on the polymer film containing 4-(4-(carboxylic acid)phenylazo)aniline chromophores. In case of the polymers containing azobenzenes with sulfonyl groups surface modulation could not be observed due to the strong intermolecular interactions. Surprisingly, large photo induced surface modulation was also observed on a conjugated polydiacetylene polymer functionalized with azobenzene side groups. We have demonstrated post azo coupling reaction is a convenient and versatile synthetic procedure for the preparation of various kinds of azo polymers. Due to the presence of ionic groups in our azo polymers self assembled polymer multilayers can be readily fabricated and characterization of optical, electronic and other interesting properties are being explored in our group.

ACKNOWLEDGMENTS

Financial support from ONR and ONR MURI program and discussions with the MURI group are gratefully acknowledged.

REFERENCES

1. McArdle, C. B. *Side Chain Liquid Crystal Polymers*, Plenum and Hall, Glasgow, 1989.
2. Prasad, P.; Williams, D. *Introduction to Nonlinear Optical Effects in Molecules and Polymers*; John Wiley and Sons: New York, 1991.
3. Burland, D. M.; Miller, R. D.; Walsh, C. A. *Chemical Reviews* **1994**, *94*, 31.
4. Dalton, L. R.; Harper, A. W.; Wu, B.; Ghosn, R.; Laquindanum, J; Liang, Z.; Hubbel, A.; Xu, C. *Adv. Mater.* **1995**, *7*, 519.
5. Marks, T. J. and Ratner, M. A. *Angew. Chem. Int. Ed. Engl.* **1995**, *34*, 155.
6. Kumar, G. S.; Neckers, D. C. *Chem. Rev.* **1989**, *89*, 1915.
7. Xie, S.; Natansohn, A.; Rochon, P. *Chem. Mater.* **1993**, *5*, 403.
8. Kumar, G. S. *Azo Functional Polymers: Functional Group Approach in Macromolecular Design* , Technomic Publishing Company Inc.: Lancaster Basel, 1993.
9. Neckers, D. C. *Machanistic Organic Photochemistry*, Reinhold: New York, 1967.

10. Zollinger, H. *Azo and Diazochemistry*, Interscience: New York, 1961.

11. Rau, H. *Photochemistry and Photophysics*, Rabek, J. F., Ed; Vol. II, CRC Press: Boca Raton FL, 1990.

12. *Mechanisms of Photophysical Processes and Photochemical Reactions in Polymers*, Rabek, J. F., Ed.; John Wiley&Sons: Chichester, 1987.

13. Ikeda, T.; Tsutsumi, O. Science **1995**, *268*, 1873.

14. Seki, T.; Sakuragi, M.; Kawanishi, Y.; Suzuki, Y.; Tamaki, T.; Fukuda, R.-I.; Ichimura, K. *Langmuir* **1993**, *9*, 211.

15. Tomita, H.; Kudo, K.; Ichimura, K. Liquid Crystals **1996**, *20*, 171.

16. Chen A. G.-S.; Brady, D. J. *Optics Letters* **1992**, *17*, 1231.

17. Todorov, T.; Nikolova, L.; Tomova, N. *Appl. Opt.* **1984**, *23*, 4588.

18. Ebralidze, T.; Mumladze, A. *App. Opt.* **1990**, *29*, 446.

19. Eich, M.; Wendorff, J. H.; Reck, B. Ringsdorf, H. *Makromol. Chem. Rapid Commun.*, **1987**, *8*, 59.

20. Eich, M.; Wendorff, J. H. *J. Opt. Soc. Am., B: Opt. Phys.*, **1990**, *7*, 1428.

21. Kim, D. Y.; Li, L.; Kumar, J.; Tripathy, S. K. *Appl. Phys. Lett.* **1995**, *66*, 1166.

22. Kim, D. Y.; Li, L.; Jiang, X. L.; Shivshankar, V.; Kumar, J.; Tripathy, S. K. *Macromolecules*, **1995**, *28*, 8835.

23. Jiang, X. L.; Li, L.; Kim, D. Y.; Shivshankar, V.; Kumar, J.; Tripathy, S. K. *Appl. Phys. Lett.* **1996**, *68*, 2618.

24. Tripathy, S. K.; Kim, D. Y.; Jiang, X. L.; Li, L.; Lee, T. S.; Wang, X.; Kumar, J. *SPIE Proceedings* for 1st International Conference on Interactive Paper, Vol. 3227, 176, 1997.

25. Kim, D. Y.; Jiang, X. L.; Lee, T. S.; Li, L.; Kumar, J.; Tripathy, S. K. *Macromol. Symp.*, **1997**, *116*, 127.

26. Lee, T. S.; Kim, D. Y.; Jiang, X. L.; Li, L.; Kumar, J.; Tripathy, S. K. *Macromol. Chem. Phys.* **1997**, *198*, 2279, .

27. Rochon, P.; J. Batalla, E.; Natansohn, A. *Appl. Phys. Lett.* **1995**, *66*, 136.

28. Barrett, C.; Natansohn, A.; Rochon, P. *J. Phys. Chem.* **1996**, *100*, 8836.

29. Ramanujam, P. S.; Holme, N. C. R.; Hvilsted, S. *Appl. Phys. Lett.* **1996**, *68*, 1329.

30. Wang, X. G.; Chen, J.; Marturunkakul, S.; Li, L.; Kumar, J.; Tripathy, S. K. *Chem. Mater.* **1997**, *9* , 45.

31. Wang, X. G.; Li, L.; Chen, J.; Marturunkakul, S.; Kumar, J.; Tripathy, S. K. *Macromolecules.* **1997**, *30* , 219.

32. Wang, X. G.; Balasubramanian, S.; Li, L.; Jiang, X. L.; Sandman, D. J.; Rubner, M. F.; Kumar, J.; Tripathy, S. K. *Macromol. Rapid Commun.* **1997**, *18*, 451.

33. Wang, X. G.; Balasubramanian, S.; Li, L.; Kumar, J.; Tripathy, S. K. *Chem. Mater.* **1997**, submitted.

34. Sukwattanasinitt, M.; Wang, X. G.; Li, L.; Rubner, M. F.; Kumar, J.; Tripathy, S. K.; Sandman, D. J. *Chem. Mater.* **1997**, submitted.

HIGH E-O COEFFICIENT POLYMERS BASED ON A CHROMOPHORE CONTAINING ISOPHORONE MOIETY FOR SECOND-ORDER NONLINEAR OPTICS

Jinghong Chen, Jingsong Zhu, Galina Todorova, Mingqian He, Larry R. Dalton

Loker Hydrocarbon Research Institute, Department of Chemistry,
University of Southern California, Los Angeles, CA 90089-1062

Sean M. Garner, Antao Chen, Sang-Shin Lee,
Vadim Chuyanov, William H. Steier

Center for Photonic Technology, Department of Electrical Engineering,
University of Southern California, Los Angeles, CA 90089-0483

ABSTRACT

A high-$\mu\beta$ chromophore APII utilizing isophorone as π-conjugation bridge was processed into both PMMA guest host and crosslinked (XL) polyurethane network with various loading densities. High electro-optic coefficients, $r_{33} = 30$ pm/V (in PMMA) and $r_{33} = 32$ pm/V (in polyurethane) at 1.06 μm were obtained. Alignment temporal stability ranged from 90 to 120°C for XL polymer network. There is virtually no intrinsic absorption loss at 1.3 μm (solution measurement). Also noteworthy is that high optimum loading densities of this chromophore are attainable in both cases without detectable chromophore aggregation due to intermolecular electrostatic interactions.

INTRODUCTION

Organic second-order nonlinear optical (NLO) materials have been intensely pursued for the past two decades. The superior nonlinearity, along with the inherent ultrafast response and large laser damage threshold, suggests that organic NLO materials are ideal for electro-optic and telecommunication applications.[1] Recently, remarkable progress has been made in the design and synthesis of chromophores with exceptionally high-$\mu\beta$ values.[2] Nevertheless, translation of the microscopic nonlinearity to large macroscopic nonlinearity (r_{33}) is not an easy task.[3] In addition, for a material to be practically used for electro-optic device applications, several other criteria must be fulfilled: 1) good photo-, electro-, and chemical stabilities of chromophores, 2) good processibility, 3) high temporal stability of chromophore alignment, 4) low optical loss at the operating wavelength. Unfortunately, a material that meets all of the addressed device-quality criteria has not yet been be realized.

Recently, it has been shown that very large molecular nonlinearities can be achieved by using extended polyene bridge systems.[4] In particular, the chromophore E,E,E-1-(4-dimethylaminophenyl)-1,3-pentadien-5-ylidene 3-phenyl-5-isoxazolone has a high $\mu\beta$ value (3156×10^{-48} esu at 1.907 μm) and low absorption maximum (562 nm in chloroform). Comparing with this chromophore, most of the large bulk nonlinearities materials possess long wavelength absorption (λ_{max}=650-850 nm) which in many cases extended into the near infrared and contributed to unacceptably large intrinsic optical losses at operating wave length.[5] These favorable features make this isoxazolone chromophore worthy of further exploration. However, the unsubstituted polyene segment of such chromophores render them thermally, photochemically, and chemically unstable.[6] In this paper, we demonstrate the improvement of the above mentioned literature chromophore by introducing a six-membered aliphatic ring (isophorone moiety) into the polyene bridge unit. Both unfunctionalized and hydroxyl-functionalized chromophores have been synthesized and incorporated into PMMA composite and rigid three dimensional polyurethane thermosetting network respectively.

EXPERIMENT

Synthesis of Chromophores

The synthesis and the chemical structures of the unfunctionalized chromophore E,E,E - 1-(4-diethylaminophenyl)-1,3-(2,2-dimethylpropane-3,5-diyl)pentadien-5-ylidene 3-phenyl-5-isoxazolone (denoted APII-1) and dihydroxyl-functionalized chromophore (denoted APII-2) are shown in scheme I.[7] Pure *trans* isomer of the key intermediate **7** was used in the last step of APII-2 synthesis, and different amount of base was used to manipulate the ratio of trans/cis isomers which has significant impact on the bulk nonlinearity. ^1H and ^{13}C spectra were taken using a Bruker-250 FT-NMR spectrometer operating at 250 MHz. Elemental analysis were performed by Atlantic Microlab Inc. FT-IR spectra were obtained using a Perkin-Elmer 1760 FTIR spectrophotometer. Absorption maxima were determined by a Perkin-Elmer Lambda-4C UV/Vis spectrophotometer. The chromophore thermostability was measured by Perkin-Elmer DSC7 at a scan rate of 5 °C/min, under nitrogen atmosphere. HPLC has been used to determine the *trans/cis* isomer ratio.

Scheme I. Synthetic Scheme of APII

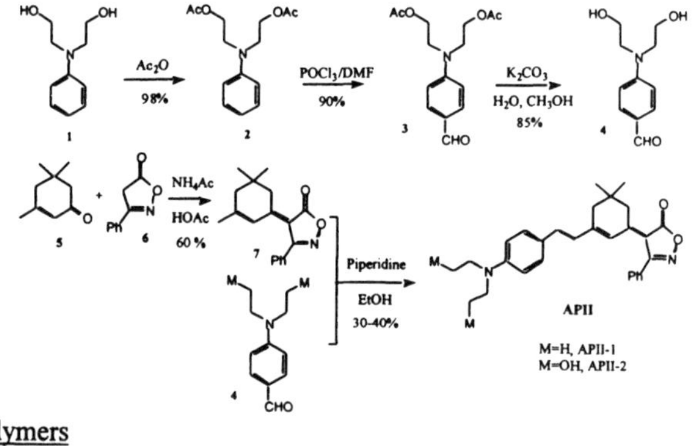

Synthesis of Polymers

Polymer composites of this chromophore were prepared by co-dissolving with PMMA in dioxane. Loading densities were 10, 20, 25, 30, 45, 55, and 75 percent by weight. After filtered through a 0.2 μm microfilter, excellent-quality thin films were spin cast onto ITO-coated glass substrates. In all cases, there was no observable phase separation, as determined by microscopy and DSC. After drying of the films *in vacuo* for two days, noncentrosymmetry was induced in the films by corona poling at the temperature which corresponded to the largest thermally-dependent second-harmonic signal. Electro-optic coefficients, refractive indices, and film thicknesses were determined by the attenuated total reflectance (ATR) technique at

1.064 μm.[8] Film thicknesses (*ca.* 3 mm) were confirmed by a Sloan Dektak IIA surface profilometer. Optical loss data were acquired by a dipping technique.[9]

There are two steps in the synthesis of thermosetting polyurethane (Scheme II). Anhydrous dioxane, tolylenediisocyanate (TDI), triethanolamine (TEA) were purchased from Aldrich. In order to add the correct amount of the TDI at appropriate time, FTIR was used to determine the unreacted amount of isocyanate groups (absorption peak at ca. 2200 cm^{-1}) in the prepolymer. DSC was utilized to figure out the suitable polymerization temperature. A mixture of the APII-2 chromophore and TDI in dioxane was heated at 80°C for 20-40 min., then a dioxane solution of TEA was added and cured for another 10-20 min. The film processing and r_{33} measurement methods are the same as above.

Scheme II. Synthesis of APII-2 Polyurethane Thermoset

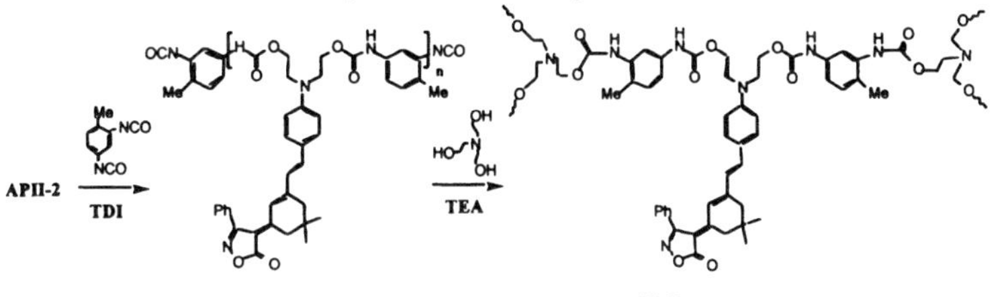

RESULTS AND DISCUSSION

In this paper, we demonstrate the improvement of the above mentioned literature chromophore by introducing a six-membered aliphatic ring (isophorone moiety) into the polyene bridge unit. The results are several-fold. First is the provision of a configuration-locked geometry of part of the polyene segment to elevate the thermal and chemical stability of the polyene segment. Second is that the planarizing effect (restriction of rotation around certain single bonds) of the isophorone moiety serves to maximize electron correlation (orbital overlap) of relevant polyene π-orbitals, and should enhance the molecular nonlinearity. Third is that derivatization of chromophore with inert steric substituents will effectively inhibit close approach along chromophore minor axes, thus minimizing chromophore aggregation. As results, both large electro-optic coefficients and low optical loss were achieved.

Guest-Host System

APII-1 exhibits high thermal stability, the decomposition temperature of this chromophore is 235 °C. APII-1 has excellent solubility in common organic solvents. The absorption maximum is 568 nm in chloroform, which is slightly higher than the corresponding unsubstituted polyene chromophore. High-resolution absorption measurements of a 10 (wt/vol)% solution of APII-1 in chloroform shows that the chromophore has no intrinsic optical (absorptive) loss at 1.3 μm (Figure I). The intrinsic nonlinearity of this chromophore is expected to be greater than that of the corresponding unsubstituted polyene chromophore. However, due to the existence of isomerization at the bridge-acceptor double bond, the definite μβ value is unpredictable. From the HPLC data, the *trans/cis* isomer ratio is around 2:1. EFISH and HRS characterization of APII-1 are underway.

Figure I. UV-Vis-NIR of 10 *wt./v.*% APII-1 in chloroform

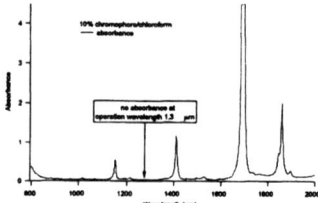

Figure II indicated the change of r_{33} with respect to different chromophore weight loading density in PMMA. The optimum loading density (45 *wt.*%) is higher comparing with conventional high $\mu\beta$-chromophores. This observation indicates that the deleterious intermolecular electrostatic interactions are less pronounced in this system. Consequently, light scattering optical loss is also greatly reduced. Optical loss of a slab waveguide at 1.3 μm was measured to be 0.96 dB/cm by the liquid outcoupling method (Figure III), which could be further decreased by more careful processing of the material. At the optimum poling temperature of 38 °C, the 45 *wt.*% composite film of APII-1 in PMMA gave an electro-optic coefficient $r_{33} = 30$ pm/V at 1.06 μm. At such a low poling temperature, significant relaxation of dipolar alignment is expected and the achievable electro-optic coefficient should be higher. This material has been successfully applied in a novel device: an electrooptic modulator with a constantly applied bias electric field. At 1.31μm, the lowest Vπ of 1.57 V was achieved from EO polymer and the corresponding r_{33} is 70 pm/v.[12]

Figure II. Electro-optic coefficient *vs. wt.*% loading densities (1eq. of base, 30min. of reaction)

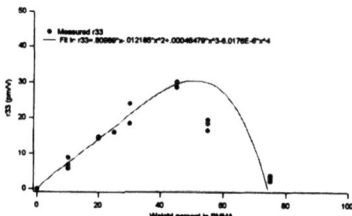

Figure III. The optical loss of APII-1 on the thin film (in PMMA host)

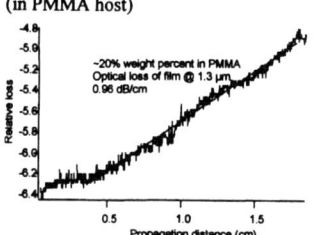

Polyurethane Thermosetting System

For practical applications, it is desired that NLO polymers exhibit long term alignment stability (around 100 °C) at operating temperatures. Due to the intrinsic limitations of guest host polymer system, the induced acentric ordering of chromophore dipole in the matrix rapidly relaxes back to the thermodynamically more stabilized random state, thus ends up with diminished bulk nonlinearity. This poses the most critical challenge of utilizing guest host system for practical device application. To preserve the induced acentric dipole alignment, it is desirable to covalently attach chromophores with polymers.

In this paper, we report the synthesis and incorporation of dihydroxyl-functionalized APII-1 (denoted APII-2) into rigid three dimensional polyurethane thermosetting network. This polymer system has been demonstrated to have the advantages of mild reaction condition, good processability, and relatively high long term temporal stability.[10] The electro-optic coefficient (r_{33}) was largely dependent on chromophore loading density as indicated in Table I.

The highest r_{33} (27 pm/v @ 1.06 μm) corresponded to the optimum loading density 40 wt.% (sample from 1 equiv. of base and 40 min. of reaction). Electro-optic coefficients vs. different weight loading densities, as well as their respective thermal stability and optical loss, are also listed in Table I. Comparing with other XL polymer systems, the optimum loading density of APII-2 has also been dramatically elevated. Such a characteristic is mainly attributed to the introduction of isophorone moiety to the chromophore architecture which increases the average molecular separation distance thus leading to lessened chromophore aggregation.[11]

Table I. Electro-optic coefficients (r_{33}) of APII-2 in polyurethane thermoset (1eq. of base, 40 min.of reaction)

wt% loading	r_{33} (pm/v)	thermal stability (°C)	optical loss (dB/cm)
20	11.2	90	2.3
30	21.0	80	—
40	27.0	95	1.5
50	23.2	120	5.0

Table II. Electro-optic coefficient vs. the amount of the catalyst and the reaction time.

base amount	reaction time	r_{33} (pm/v)
1 eq.	40 min	27
0.25 eq.	40 min	32
1 eq.	60min	15

In the current study, the further increased bulk nonlinearity r_{33} from polyurethane network is contributed to the increased trans/cis isomer ratio. According to our NMR and HPLC analysis, the final APII-2 product contains mixture of trans and cis isomers. Because of their different π-conjugation configurations, electron transfer efficiencies of these two isomers also differ from each other. In general, the trans isomer has much larger efficiency than that of cis isomer.[6b] In the synthesis of APII-2, although pure trans isomer 5 was used in the next Knoevenagel condensation, the keto/enol equilibrium still competes with condensation reaction under basic condition. Therefore the final product APII-2 was expected to be a mixture. However, it is possible to optimize the ratio of trans/cis forms by carefully controlled catalyst amount and reaction time. According to our experimental data, the lowered trans/cis isomer ratio correlated to larger amount of catalyst and longer reaction time (Table II).

Thermal stability of both chromophore and its noncentrosymmetric alignment in bulk materials were studied. The chromophore exhibits high thermal stability, the decomposition temperature is 235 °C. The temporal stability of acentric alignment is listed in Table I. At optimum loading density, thermal stability is as high as 95 °C.

Besides the large nonlinearity, optical loss is another critical concern regarding to practical device application. The film absorption of 30 wt.% thermoset APII-2 is 560nm. Its $λ_{max}$ lies in the deep blue region. In contrast to other high μβ chromophores, this chromophore will less likely be resonantly enhanced, hence the r_{33} values at operating wave length (1.3 μm) are expected to be very close to the obtained data at 1.06μm. Compared with the optical loss in the guest-host system (0.96dB/cm@1.3μm), the loss in the thermoset (40 wt.% @1.3μm) is slightly higher. Noteworthy, the sample at lower loading density (20 wt.%) exhibited higher optical loss (2.3 dB/cm) than that of higher loading density (40 wt.%). In order to determine the cause of optical loss, we undertaken IR absorption measurements of the polymer film, and the film virtually exhibited no absorption at 1.3 μm. The possibility of intrinsic optical loss can be eliminated. Therefore, the optical loss is attributed to the inhomogeneity of the thin film. Optimization of polymer processing (the polymer synthesis, the spinning and the corona poling) is required to minimized the optical loss. The large nonlinearity along with other properties have well qualified this material for device application such as Mach-Zehnder modulator.

CONCLUSIONS

The APII chromophores possess the major requirements for electro-optic applications, as mentioned in the introduction. High optimum loading densities of both composite and covalently attached thermoset materials indicate that isophorone is an efficient moiety in preventing chromophore aggregation. By systematic optimization of key structural elements of the chromophore, and by careful film processing, large electro-optic coefficients and low optical loss were obtained. Auxiliary chromophore characteristics such as solubility, thermal, chemical, and photochemical stabilities are all sufficient for materials processing, device fabrication, and performance tolerances. APII-1 and APII-2 have been successfully used in a new electrooptic modulator (with a constantly applied bias electric field) and the conventional Mach-Zehnder modulator respectively. Low Vπ values were obtained. Current work involves trifunctionalizing the chromophore on both donor and acceptor ends for the purpose of further increasing the alignment temporal stability.

ACKNOWLEDGMENTS

This project is supported by the Air Force Office of Scientific Research (contracts F-49620-94-1-0323, F-49620-94-1-0201).

REFERENCES

1. Prasad, P. N.; Williams, D. J. Introduction to Nonlinear Optical Effects in Molecules and Polymers; John Wiley & Sons: New York, 1991.
2. Ahlheim, M.; Barzoukas, M.; Bedworth, P.V.; Blanchard-Desce, M.; Fort, A.; Hu, Z.-Y.; Marder, S.R.; Perry, J. W.; Runser, C.; Staehelin, M.; Zysset, B. *Science* **271**, 335(1996).
3. Zhu, J.; He, M.; Harper, A. W.; Sun, S. S.; Dalton, L. R.; Garner, S. M.; Steier, W. H.; *Polymer Preprints* **38**(1), 973 (1997).
4. Marder, S.R.; Cheng, L.-T.; Tiemann, B.G.; Friedli, A.C.; Blanchard-Desce, M.; Perry, J.W.; Skinhoj, J. *Science* **263**, 511(1994).
5. a) Boldt, P.; Bourhill, G.; Brauchle, C.; Jim, Y.; Kammler, R.; Muller, C.; Rase, J.; Wichern, J. *Chem. Commun.* 793(1996). b) Sun, S.-S.; Harper, A. W.; Dalton, L. R.; Garner, S. M.; Chen, A.; Yacoubian, A.; Steier, W. H. *Mat. Res. Soc. Symp. Proc.* **413**, 263(1996).
6. a) Shu, C.-F.; Tsai, W.J.; Chen, J.-Y.; Jen, A.K.-Y.; Chen, T.-A. *Chem. Commun.*, 2279(1996). b) Cabrera, I.; Althoff, O.; Man, H.-T.; Yoon, H.N. *Adv. Mater.* **6**, 43(1996).
7. Brooker, L.G.S.; Craig, A.C.; Heseltine, D.W.; Jenkins, P.W.; Lincoln, L.L. *J. Am. Chem. Soc.* **87**, 2443(1965).
8. Dentan, V.; Lévy, Y.; Dumont, M.; Robin, P.; *Optics Communications* **69**, 379, (1989).
9. Teng, C.; *Applied Optics*, **32**, 1051, (1993).
10. Mao, S. S. H.; Ra Y. S.; He M.; Zhu, J.; Zhang, C.; Harper, A.; Dalton, L. R.; Garner, S.; Steier, W. H.; *PMSE*, **77**(2), 564, (1997).
11. Dalton, L. R.; Harper, A. W.; Chen, J.; Sun, S.; Mao, S.S.H.; Garner, S.; Chen, A.; Steier, W. H.; *SPIE*, **3007**, 313, (1997).
12. Chen, A.; Chuyanov, V.; Zhang, H.; Garner, S. M.; Steier, W. H.; Chen, J.; Zhu, J.; He, M.; Mao, S. S. H.; Dalton, L. R. *Optics Letters*, submitted, 1997.

PASSIVE POLARIZATION ROTATORS FOR NONLINEAR OPTICAL DEVICES

T.C. KOWALCZYK, L-Y LIU, H.S. LACKRITZ
Gemfire Corporation, Palo Alto, CA 94303

ABSTRACT

Initial experiments of a passive polymeric TE-to-TM polarization converter that utilizes laterally offset upper and lower poling electrodes to induce a 45° optical axis in the core of a nonlinear optical polymer waveguide are reported. By optimizing the device length, we have demonstrated polarization conversion efficiencies greater than 95% in devices as short as 240 um with negligible insertion loss.

INTRODUCTION

Polymers are seeing increasing   use in passive and active waveguiding applications for integrated optics technology.   Polymeric devices have advantages because of their ease of manufacture, low cost, and compatibility with other processing needs and components. Nonlinear optical polymers have already been demonstrated as efficient modulators, couplers, and frequency converters [1,2]. More recently, polymeric polarization converters and rotators have been proposed and demonstrated [3,4]. Unlike their inorganic counterpart, polymeric polarization converters can operate in both active and passive modes. In particular, active polarization converters are interesting for optical communications as large band-width modulators and passive polarization converters for simplifying packaging requirements. In the later case, an integrated polarization converter simplifies attachment of a laser diode (primarily TE polarized output) to an electro-optic waveguide device that requires TM polarization. The devices are passive in the sense that applied voltage is unnecessary for device operation after poling.

The polarization converter described in this study is shown in figure 1. The essential elements of the polymer polarization rotator are its: birefringence, optic axis orientation, and length. In polymer polarization converters, the orientation of the optical axis in the waveguide is determined by the poling electrode geometry (lateral offset). In the interesting case of a polymer half wave-plate, the optical axis is oriented at a 45° angle with respect to the input polarization state. The resulting output polarization state is then rotated by 90°, producing an orthogonal polarization state compared to that of the input light. In the special case of a half-wave plate, power oscillates between the TE and TM modes as described by the coupled mode equations:

$$I_{TE}(L) = I_o \cos^2(\Delta\phi L/2)$$

$$I_{TM}(L) = I_o \sin^2(\Delta\phi L/2)$$

$$(1)$$

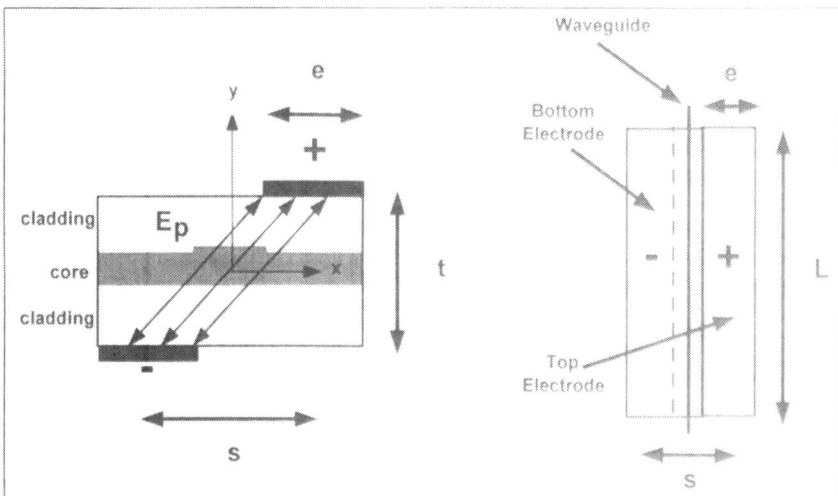

Figure 1. Cross-sectional (left) and top view (right) of the polymer half-wave plateTE-TM polarization converter showing device fabrication parameters e, t, s, L, and E_p. (electrode width, polymer stack thickness, lateral offset, length, and poling field, respectively)

where the phase difference is defined as $\Delta\phi = 2\pi/\lambda \ \Delta N$, and ΔN is the waveguide mode birefringence. Unlike active polarization conversion methods, the polymer half wave-plate requires a phase-mismatch between the propagating modes. The amount of phase mis-match determines the correct device length for polarization conversion. For a half-wave plate the phase between the fast and slow axis must be a multiple of π. The minimum propagation length for polarization conversion is then defined as the polarization beat length

$$L_B = \lambda / 2 \ \Delta N \qquad (2)$$

In typical poled polymers the induced birefringence can be as high as ($\Delta n = 0.02$) so that the beat length and corresponding electrode length must be accurately controlled. However this also provides a means for changing or adjusting the beat length during processing, so that the index difference is small ($\Delta n \sim 0.005$). In this case the beat length is longer and the conversion efficiency does not critically depend on electrode fabrication tolerances. We used equations 1 and 2 to determine the device's sensitivity to fabrication parameters.

As mentioned above, the refractive index and birefringence of a NLO polymer can be tuned by appropriately adjusting the poling field strength during processing. We calculated the refractive index assuming the chromophores do not interact with each other or the surrounding polymer environment. Under these assumptions an oriented gas model accurately describes the induced orientational order in a polymer matrix. Within these approximations, we calculated the refractive index of a nonlinear optical polymer as a function of poling field by taking the thermodyanmic average of the linear polarizability [5,6]. The resulting refractive indices are:

$$n_{TM} = n_o + 2/3 \, \delta \, E_p^{\,2}$$

$$n_{TE} = n_o - 1/3 \, \delta \, E_p^{\,2}$$

(3)

where n_o is the unpoled refractive index, δ is a constant containing orientational averages and material parameters, and E_p is the poling field. In the following sections we will discuss how these fabrication parameters affect device performance.

EXPERIMENT

Because it is necessary for a device to be minimally affected by fabrication uncertainties, we first determined how fabrication tolerances affect device performance. To investigate how the parameter space defined by our fabrication parameters affected the poling field direction or optical axis in the polymer stack we used a finite difference algorithm to numerically solve Poisson's equation in two dimensions. Equipotential surfaces and electric field lines were extracted from the simulations and average electric field angles were tabulated. Figure 2 illustrates one such electric field contour plot. In this figure the black bars represent the laterally offset poling electrodes, arrows indicate the calculated direction of the electric field at a specific location, and the box in the center outlines the waveguide region.

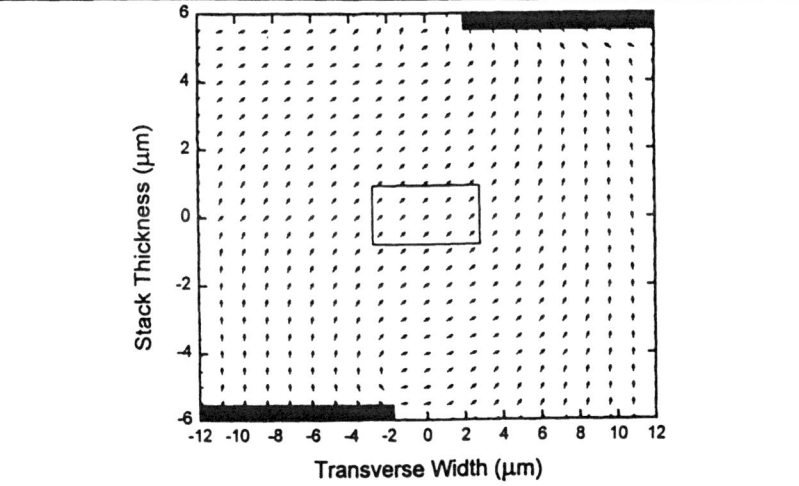

Figure 2. Electric field contour plot in a polymer stack with laterally offset poling electrodes above and below the polymer stack. The center box defines the waveguide region, and shows that poling field in the core is at a 45° with respect to the x- and y-axis.

A conducting boundary condition was used to limit the space to a manageable number of grid points so that reasonable convergence times were obtained. The mesh size for the calculation grid was 0.25 μm by 0.25 μm. In these calculations the polymer stack is introduced as a dielectric with a dielectric constant ε = 2.2 embedded in a semi-infinite medium of air ε =1. Indeed, figure 2 shows that 45° oriented optic axis is obtainable with the proper lateral poling electrode offset.

We modeled the electrode width and the electrode center-to-center separation (using a constant stack height of 12 μm) to examine how these parameters change the poling field direction in the core. Figure 3 shows the simulation results as data points that represent the electric field angle as determined by averaging over the 2 μm x 6 μm waveguide area. Our modeling showed that many electrode width (8.0-14.0 μm) and separation pairs produced a 45° tilt in the optical axis in the waveguide core. It is also apparent from figure 3 that the slopes of the lines of electric field direction plotted against electrode separation are nearly constant as are the standard deviations of field directions for data points contained on the same line. As expected electric field angles became increasingly horizontal with increasing electrode separation. Standard deviations in the calculated electric field orientations are listed on the graph (in brackets next to each line) and vary between 4.0° and 5.4°. More importantly, the data in figure 3 indicates that processing deviations resulting in 1.0 μm of electrode center to center separation misalignment result in 2.5 degrees of change in the average poling field angle.

It was necessary to understand how stack thickness variations affect device performance too. We modeled how stack thickness variations affect the average electric field angle in the core of the waveguide as a function electrode center-to-center separation (assuming an electrode width of 10.0 μm). Two trends were apparent from the modeling: First, the slope of each line increased as the total stack height was reduced and Second, the standard deviation in the average electric field direction in the core increased as the stack height was reduced. Using a stack height of 12.0 and assuming an electrode separation of 17 μm we predict a 2.4 degree shift in average electric field angle per micron of stack thickness change. Because stack height reproducibility can be controlled to better than +/-1%, this variable is not critical for device efficiency.

The electrode length must be precisely controlled to have efficient TE-TM conversion, particularly for highly birefringent materials. To minimize the device's critical dependence on length we calculated the conversion efficiency as a function device length using birefringence as an adjustable parameter. A birefringence of Δn = 0.005 resulted in a device design that could vary by +/- 22 μm in length and still produce a 90% conversion efficiency, while Δn = 0.01 requires a device length control to +/- 12 μm. Clearly, small Δn values ease fabrication tolerances and minimize any effective electrode length contributions. For this reason small poling fields or equivalently small birefringence values are optimum for TE-TM conversion in this geometry.

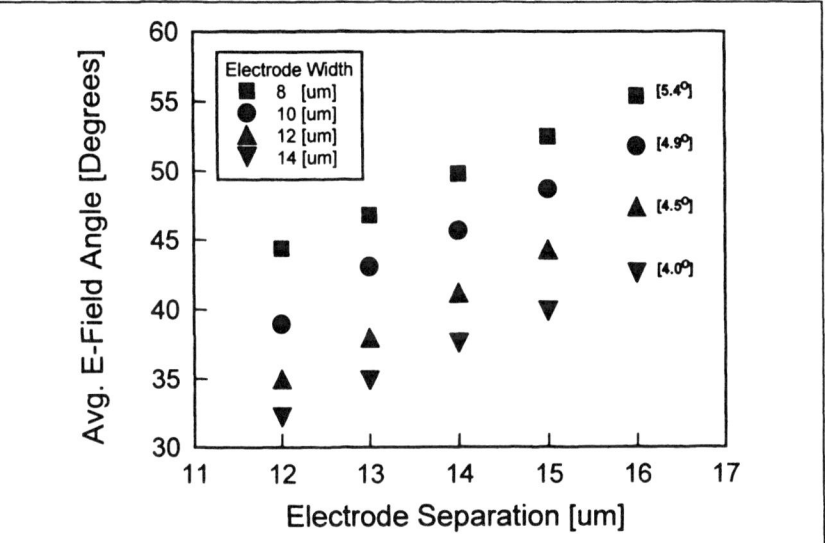

Figure 3. Average electric field angle in the waveguide core calculated as a function of electrode width and separation. Bracketed numbers on the graph represent the standard deviation in the average electric field angle. (See figure 1 for a description of these parameters.)

Another requirement for efficient TE-TM conversion is spatial mode overlap of the TE and TM modes. In a typical EO devices, large poling fields are desirable to produce large EO coefficients. Large poling fields also produce larger birefringences which in turn changes the modal profile distribution and limits the power that can be exchanged during the polarization conversion. There are two effects that contribute to insertion loss. First, scattering loss from the index and spatial mode discontinuities as light enters the device and second is spatial overlap of the TE-TM mode profiles in the device during conversion. Calculations of the mode profile before and in the device yield overlaps greater than 99% and indicate that these losses will be negligible. For the second contribution, again the overlap integrals are greater than 98% indicating that we expect only minor losses in the device. The large index difference between the core and cladding (n_{core}-$n_{cladding}$ ~ 0.05 for our waveguides) produces strongly confined modes with spatial mode profiles that remain relatively unchanged after poling. Figure 4 shows the guided mode profile for an unpoled waveguide and a waveguide poled at 200 V/µm (TE-TM converter). The overlap integral calculated from these fields was 99%. From these calculations we expect the total insertion losses to be near 6%. Overall these insertion losses are quite low and therefore negligible compared to loss mechanisms such as material absorption and coupling (fiber-to-device) losses.

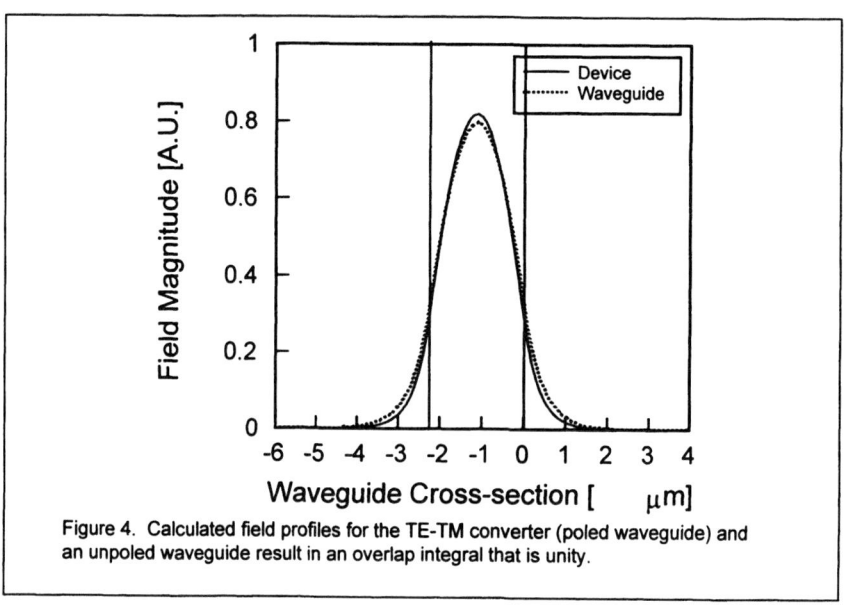

Figure 4. Calculated field profiles for the TE-TM converter (poled waveguide) and an unpoled waveguide result in an overlap integral that is unity.

We fabricated a series of devices that allowed us to probe the dependence of TE-TM conversion efficiency on fabrication parameters. Specifically, each sample wafer contained TE-TM converters with varying length (75 μm to 440 μm) and multiple lateral electrode offsets (15, 17, 19, 25 μm). Samples were fabricated using the following procedure. Glass substrates were pre-cleaned to eliminate surface contaminants. A gold lower electrode was then evaporated onto the glass substrate and coated with photo-resist. After photo-lithography and development the excess metal was removed with a gold etchant. The lower cladding layer (uv-curing acrylate) was spun to a thickness of 5.0 μm and cured. The core layer (DR1-MMA commercially available from IBM) was spun to a thickness of 2.0 μm and then reactive ion etched to produce a 0.3-0.4 μm raised rib. The core layer was chosen because of its appreciable nonlinearity and adequate optical transparency in the infrared. A top cladding layer was spun to match the thickness of the lower cladding layer and uv-cured. Finally, the upper electrodes were deposited and the device poled. Typical poling fields varied between 50 and 250 V/μm at poling temperatures of 130° C. After poling, the sample end-faces were cut and polished to allow fiber butt-coupling. Samples were placed on a 3-axis high precision translation stage while a 6.0 μm core pig-tailed IR laser was butt-coupled to the waveguide. The polarization extinction ratio of the laser was greater than 100:1 with typical throughput powers of 1.0 mW. A microscope objective collected light output from the waveguide and passed it through a linear polarizer. The power was measured for each polarization state and the ratio of TM component divided by the sum of TE and TM components was recorded as the conversion efficiency.

RESULTS

Figure 5. shows data collected from a sample poled with a field of 100 V/μm using a standard time and temperature ramp poling profile (poling temperature 130°C) and plotted as a function of propagation length and fit to a theoretical curve containing adjustable parameters for the birefringence and poling field angle is shown in figure 5. A reasonably qualitative agreement between theory and experiment was obtained using Δn = .0075 and an electric field poling direction of 45°. These parameters are in agreement with electric field angles predicted from static field calculations in equation 3.

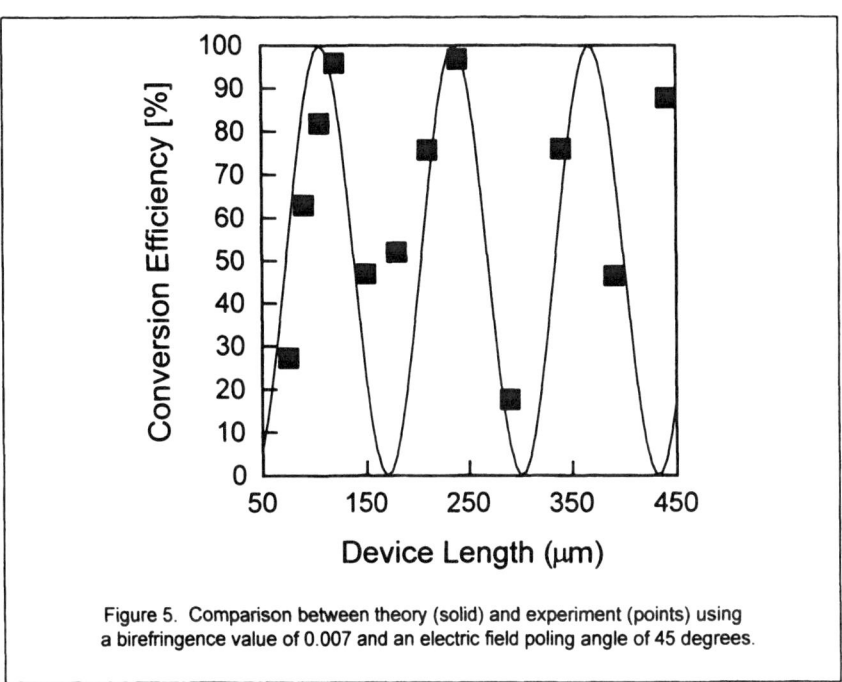

Figure 5. Comparison between theory (solid) and experiment (points) using a birefringence value of 0.007 and an electric field poling angle of 45 degrees.

The slight discrepancy may be attributed to differences between the static field modeling and the actual dynamic system one encounters during electric field poling or deviations in the electrode center-to-center separation. As figure 5 shows, there are conversion efficiencies (data points) that do not lie on the theoretical curve. These differences may be attributed to local poling field variations. It is also interesting to note that the birefringence obtained from fitting was in agreement with samples poled with the same electric field magnitude (0.007) as determined from equations 3 and verified experimentally. Overall the largest conversion efficiency we measured was 96% and 97% with device lengths of 120 μm and 240 μm, respectively.

CONCLUSIONS

We have demonstrated a passive polymeric polarization conversion efficiency of 97% with a 240 um device length while maintaining low insertion losses. Our experimental results were consistent with theoretical predictions allowing us to correlate processing variables (poling field and device length) with device performance.

ACKNOWLEDGEMENTS

This work was funded by DARPA under contract DAAH01-96-C-R218.

REFERENCES

1. D. Chen, H.R. Fetterman, A. Chen, W.H. Steier, and L. R. Dalton, Appl. Phys. Lett. **70**, (1997).

2. T.A. Tumolillo and P.R. Ashley, Appl. Phys. Lett. **62**, pp.3068-3070 (1993).

3. W-Y Hwang, J-J Kim, T. Zyung, M-C Oh, S-Y Shin, IEEE J. Quantum Electron. **QE-32**, pp.1054-1062 (1996).

4. M-C Oh, S-Y Shin, W-Y Hwang, J-J Kim, Appl. Phys. Lett. **67**, pp. 1821-1823 (1995).

5. K.D. Singer, M.G. Kuzyk, and J.E. Sohn, J. Opt. Soc. Am. B **4**, 968 (1987).

6. C.C. Teng, M.A. Mortazavi, and G.K. Boudoughian, Appl. Phys. Lett. **66**, 667 (1995).

MOLECULAR SECOND-ORDER OPTICAL NONLINEARITY OF PUSH-PULL BISDITHIOLENE NICKEL COMPLEXES

Chin-Ti Chen*, Sish-Yuan Liao, Kuan-Jiuh Lin, and Long-Li Lai
Institute of Chemistry, Academia Sinica, Taipei, Taiwan 11529

ABSTRACT

The synthesis and characterization of donor-acceptor substituted unsymmetrical bisdithiolene nickel complexes are described for the first time. X-ray single crystal data indicate that the complexes exist with two types of bonding structures, namely, the π-localized and π-delocalized structures. The relation between bonding structures and the molecular second-order nonlinear optical properties, i.e., solvatochromism, dipole moment, and molecular first hyperpolarizability is discussed.

INTRODUCTION

Bisdithiolene d^8 metal complexes have been used as Q-switch dyes for near-IR lasers[1] as well as molecular magnets, conductors, or superconductors[2,3]. Such complexes can also show mesogenic behaviors and are potential for dichroic dyes in liquid crystal displays[4] Recently, a series of symmetrical complexes have been reported for third-order optical nonlinearities and considered to be promising candidates for all-optical signal-processing devices[5]. Despite diversified applications, bisdithiolene d^8 metal complexes have not been examined for their second-order nonlinear optical (NLO) properties. Very recently, negative molecular first hyperpolarizability (β) were observed for a series of mix-liganded d^8 metal complex, M(diimine)(dithiolate)[6]. Solvatochromism and intramolecular charge-transfer (ICT) wise, anionic unsymmetrical bisdithiolene d^8 metal complex has been shown with similar behavior to M(diimine)(dithiolate)[7,8]. Accordingly, here we report the preliminary results from our synthesis and characterization donor-acceptor unsymmetrically substituted bisdithiolene nickel (II) complexes. The characterization of the complexes will be focused on two types of bonding structures accompanying with X-ray single crystal structures, solvatochromic behavior, dipole moments of ground-state as well as excited-state. At the end, we will examine β values of the complexes within the context of two-state model assumption.

EXPERIMENT

Compounds Ni(tfd)$_2$, bis(3,3',4,4',5,5'-hexamethoxydithiobenzil)nickel(II), and 1,4-dialkylpiperazine-2,3-dione were prepared according to literature methods[9-11]. Reagent (diamine)(maleonitrildithiolato)Ni(II) was synthesized by mixing disodium maleonitrildithiolate[9] and nickel(II) chloride in concentrated ammonia solution. The synthetic schemes for **1-7** are shown in Scheme 1.

Scheme 1   a) ex. P_4S_{10}, CH_2Cl_2, reflux 4h, RT, 12 h, R = 2-ethylhexyl 69%, R = cyclododecyl, 72%; b) CH_2Cl_2/CH_3CN, reflux 10 min, RT 24 h, R = 2-ethylhexyl, 59%, R = cyclododecyl, 60%; c) CH_2Cl_2, RT 3 h, 62% d) CH_2Cl_2, RT 3 h, 89%. e) CH_2Cl_2, RT 3 h, 56%.

Compounds **1-7** gave ^{1}H NMR, FTIR, UV-Vis, and FABMS fully consistent with the proposed structure and satisfactory elemental analyses were obtained for complexes **5-9**. Transition dipole moments (μ_{eg}) were calculated from the area of the absorption band, taking twice the numerical integral of the low-energy half of the absorption band. Ground-state dipole moment were determined from an analysis of the solution dielectric constant and refractive index versus concentration (10^{-3}-10^{-5} M) of the solute, using chloroform as the solvent except dichloromethane was used for **9**. All solvents used in μ_g measurement were dried over 4-Å molecular sieves. The equation adopted for calculating μ_g was based on that of Osipov-Onsager for the solute dissolved in polar solvents[12,13]. Excited-state dipole moments (μ_{es}) of **3** and **5-9** were determined using solvatochromic method as previously described[14]. The dipole moment change, $\mu_e - \mu_g$, was calculated from the coefficient, obtained from a least-squares fitting of McRae expression to the λ_{max} of ICT band in 10-20 different solvents. The most uncertain quantity of solvatochromic method is the poorly defined Onsager radius a. We took the sum of spherical radius of solute and solvent for a [15]. The spherical radii of solutes were calculated from measured density values of compounds 3 and 5-9.

RESULTS AND DISCUSSION

X-ray single crystal structures [16] of 1, 4, and 6 were obtained in order to elucidate differences in ground-state bonding structures, i.e., 3-5 of one type and 6-7 of the other. Selected bond distances are listed in Table 1.

Table 1   Selected bond distances (Å) of 1, 4, and 6 from X-ray single crystal structures

donor—C_1—S_1 S_3—C_3—acceptor
C_2—S_2 Ni S_4—C_4
donor—   acceptor

	C_1-C_2	C_1-S_1 C_2-S_2	Ni-S_1 Ni-S_2	Ni-S_3 Ni-S_4	C_3-S_3 C_4-S_4	C_3-C_4
1	1.522(9)	1.654(4)	—	—	—	—
4	1.34(2)	1.661(10) 1.688(11)	2.178(3) 2.151(3)	2.145(3) 2.145(3)	1.745(12) 1.718(11)	1.50(2)
6	1.410(5)	1.713(4) 1.696(4)	2.124(1) 2.113(1)	2.122(1) 2.118(1)	1.717(4) 1.706(4)	1.366(6)

For the central Ni(S$_2$C$_2$)$_2$ moiety of 4, all bond distances of 4 are consistent with the fact that nickel is coordinated unsymmetrically by one dithiodiketone and one dithiolate ligands. In contrast, differences of bond lengths between donor and acceptor sides of 6 are less substantial. Complex 6 has a long C—C bond (1.41 Å) on the donor side which is not as long as that of 4 ; the short C—C bond (1.37 Å) on the acceptor side which is not as short as that of 4. The comparison of C—S bonds leads to the same conclusion. In fact, the average C—C and C—S bond length of 6 are very close to 1.38 and 1.71 Å of Ni(tfd)$_2$, respectively, a dithiolene complex with π-delocalized central Ni(S$_2$C$_2$)$_2$ moiety [17,18]. In addition, compound 6 has a shorter average Ni—S distance than that of 4. This is the evidence further support that 6 has a more π-delocalized central Ni(S$_2$C$_2$)$_2$ moiety than that of 4. The structure data suggest that ground-state bonding structures of 6 and 4 are different and best described as those of 6B and 5A in Scheme 2, respectively.

For complexes 3, 5, and 6, measurements of the ground-state and excited-state dipole moments provide further insight into the electronic structure of these molecules. The measured ground-state dipole moment (μ_g) of 6 is ~ 6 debye and is considerably smaller than those of 3 and 5 (Table 2) Both theoretical and experimental evidence have demonstrated that the Ni(S$_2$C$_2$)$_2$ central unit are highly delocalized (B in Scheme 2) for symmetrical complexes [1]. However, one of resonance forms, **A** or **C** depicted in Scheme 2, may prevail in the ground-state structure if the complex becomes unsymmetrical with different substituents.

Scheme 2   Major canonical resonance structures (**A, B, C**) of **5** and **6** with the indication of major components of dipoles. The square frames include the most important resonance forms. For **6B**, due to the nearly symmetrical π-delocalized nature of $Ni(S_2C_2)_2$ moiety, major component of μ_g is from methoxy donors to trifluoromethyl acceptors. The μ_g of **5A** is mainly composed of two dipoles with the same direction. One dipole, presumable the weak one, exists between the donor and acceptor. The other dipole, presumable the strong one, originate from the charge separated metal cation and the dithiolate ligand.

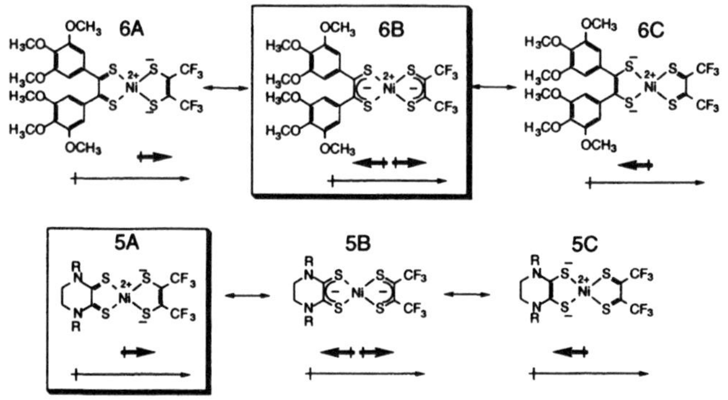

Table 2. Optical data of complexes **3** and **5-9**.

Compound	λ_{max} [nm]	$\varepsilon \times 10^{-3}$ [$M^{-1}cm^{-1}$]	μ_{eg}[a]	d [g/mL]	a [Å]	μ_g[a]	μ_e-μ_g[b]	μ_e[c]	$\beta_0 \times 10^{30}$[d] [esu]
3	829	9.8	3.74	1.23	7.26	14	-12	2	-43
5	840	8.7	3.59	1.34	7.37	8	-12	-4	-48
6	842	19.6	5.91	1.65	7.13	5	1~9	6~14	5~81
7	1099	21.3	5.85	1.53	7.00	13	4	17	70
8	562	6.8	3.27	2.12	6.19	9	-9	0	-11
	563[e]	7.0[e]	—	—	—	9[e]	—	—	-15[e]
9	582	7.2	3.20	2.08	6.13	9	-12	-3	-15
	583[e]	7.0[e]	—	—	—	9[e]	—	—	-16[e]

[a] debye, $\pm \approx$ 10%. [b,c] debye, $\pm \approx$ 20-30%. [d] zero-frequency term of β, $\pm \approx$ 20-30%. [e] reference 6.

With ground-state bonding structure (based on X-ray crystal structures), the constitution of μ_g of **5** (as well as **3** and **4**) and **6** (similarly **7**) can then be analyzed (see caption of Scheme 2). The dipole analysis of **5** and **6** are consistent with observed solvatochromic behavior: strong negative solvatochromism was observed for **3** and **5** but moderate solvatochromic shift with positive in sign for **6** and **7** [19]. For **5** (so is **3** and **4**), with the smaller dipole component remains mostly unchanged, the larger dipole component is greatly reduced upon excitation. The excitation here has been referred to be mainly the strong ligand (dithiolate) to ligand (dithiodiketone) charge transfer (LLCT)[7]. Since the direction of strong LLCT is critically opposite to that of the ground-state dipole, **5** has a strong negative solvatochromic behavior. However, with nearly symmetrical Ni(S_2C_2)$_2$ bonding structure, it is the weak ICT, from the donor through delocalized central Ni(S_2C_2)$_2$ moiety to the acceptor parallel to the dipole direction, which gives **6** a small and positive solvatochromism. With stronger dialkylamino donor, **7** shows considerable solvatochromic shifts but still positive in sign. Based upon these data we hypothesize that the strong near-IR absorption ($\varepsilon \sim 2 \times 10^4$) of **6** and **7** is mainly π—π^* original with limited ICT character like those of symmetrical complexes. From X-ray crystal structure, the weak ICT of **6** can be attributed to the twisting angle between the donor substituted phenyl rings and the central Ni(S_2C_2) plane. Having a stronger dialkylamino donor, **7** has a μ_g and β considerably larger than that of **6**. From our measurement and calculation, μ_e of **3** and **5** is ca. 2 and -6 debye, respectively (Table 2). The μ_e of **3** and **5** indicates that excited-state dipole is either considerably smaller or with opposite direction to the ground-state dipole, which is accordant with the observed strong negative solvatochromism. On the other hand, μ_e of and **7** is ca. 4 debye larger than μ_g, and is consistent with the observed moderate but positive solvatochromic shifts of **7**. Regarding **6**, its weak solvatochromic behavior decrease the accuracy of determined μ_e values, although the sign of μ_e-μ_g, like that of **7**, is always positive.

Finally, with λ_{max}, μ_{eg}, μ_e - μ_g, μ_e available, we performed a rough calculation of β for **3**, **5**, **6**, and **7** based on the two-state model for β of NLO chromophores with low-lying strong ICT transition[20-22]. As shown in Table 1, negative β values were obtained for both **3** and **5** but positive β values were found for **6** and **7**. Altogether, our calculated β values are in general consistent with solvatochromic shifts, dipole moment changes, and differences in proposed bonding structures of **3**, **5**, **6**, and **7**. For reference, Table 1 also includes M(diimine)(dithiolate), **8** and **9** [6], with known β values from EFISH measurement [6] as well as the calculated values from this work. Undoubtedly, solvatochromic measurement provides decided sign but not the magnitude of β values. Nevertheless, for **8** and **9**, it is interesting to note that β values from EFISH and solvatochromic method are comparable. More accurate β values of **3**, **5-7** will be determined by EFISH and that will be reported in due course.

ACKNOWLEDGMENTS

This research was supported by the National Science Council of Taiwan (Grant No. NSC 86-2113-M-001-016). We thank Professor Ulrich Mueller-Westerhoff for helpful discussions about the synthesis. Valuable opinions from Dr. Fabienne Meyers are also acknowledged.

REFERENCES

1. U. T. Mueller-Westerhoff, B. Vance, and D. I. Yoon, Tetrahedron **47**, 909 (1991).

2. P.Cassoux, L. Valade, H. Kobayashi, A. Kobayashi, R. A. Clark, and A. E. Underhill, Coord. Chem. Rev. **110**, 115 (1991).

3. J. S. Miller, and A. J. Epstein, Angew. Chem. Int. Ed. Engl. **33**, 385 (1994).

4. A. -M. Giroud-Godquin, and P. M. Maitlis, Angew. Chem. Int. Ed. Engl. **30**, 375 (1994).

5. C. A. S. Hill, A. Charlton, A. E. Underhill, S. N. Oliver, S. Kershaw, R. J. Manning, and B. J. Ainslie, J. Mater. Chem. **4**, 1233 (1994).

6. S. D. Cummings, L. -T. Cheng, and R. Eisenberg, Chem. Mater. **9**, 195 (1997).

7. A. Vogler and H. Kunkely, Angew. Chem. Int. Ed. Engl. **21**, 77 (1982).

8. T. R. Miller and I. G. Dance, J. Am. Chem. Soc. **95**, 6970 (1973).

9. A. Davison and R. H. Holm, Inorg. Synth. **10**, 8 (1967).

10. G. N. Schrauzer, V. P. Mayweg, J. Am. Chem. Soc. **87**, 1483 (1965).

11. U. T. Mueller-Westerhoff, M. Zhou, J. Org. Chem. **59**, 4988 (1994).

12. T. Thami, P. Bassoul, M. A. Petit, J. Simon, A. Fort, M. Barzoukas, A. Villaeys, J. Am. Chem. Soc. **114**, 915 (1992).

13. K. L. Kott, C. M. Whitaker, R. J. McMahon, Chem. Mater. **7**, 426 (1995).

14. M. S. Paley, J. M. Harris, H. Looser, J. C. Baumert, G. C Bjorklund, D. Jundt, R. J. Tweig, J. Org. Chem. **54**, 3774 (1988).

15. Ch. Bosshard, G. Knopfle, P. Pretre, P. Günter, J. Appl. Phys. **71**, 1594 (1992).

16. X-ray single crystal data of **1**: Monoclinic, space group $C\ 2/c$, $a = 25.124(4)$, $b = 6.925(2)$, $c = 16.424(4)$ Å, $\beta = 92.30(2)°$, $V = 2855(1)$ Å^3 $\rho = 1.109$ gcm^{-3}, $Z = 8$, $R = 0.087$, $wR(F^2) = 0.238$ (146 variables); **4**: Monoclinic, space group $P\ 2_1/n$, $a = 7.048(2)$, $b = 27.927(10)$, $c = 17.880(5)$ Å, $\beta = 93.09(3)°$, $V = 3514(2)$ Å^3 $\rho = 1.278$ gcm^{-3}, $Z = 2$, $R = 0.069$, $wR(F^2) = 0.191$ (370 variables); **6**: Triclinic, space group $P\ 1$, $a = 8.555(6)$, $b = 11.834(5)$, $c = 14.783(3)$ Å, $\alpha = 99.48(2)°$, $\beta = 94.38(4)°$, $\gamma = 107.05(4)°$, $V = 1390(1)$ Å^3 $\rho = 1.679$ gcm^{-3}, $Z = 2$, $R = 0.041$, $wR(F^2) = 0.108$ (370 variables).

17 R. D. Schmitt, R. M. Wing, A. H. Maki, J. Am. Chem. Soc.**91**, 4393 (1969).

18. Z. S. Herman, R. F. Kirchner, G. H. Leow, U. T. Mueller-Westerhoff, A. Nazzal, M. C. Zerner, Inorg. Chem. **21**, 46 (1982).

19. In dichloromethane, acetone, and dimethylsulfoxide, the absorption λ_{max} of **3** is 812, 763, and 751 nm, respectively; λ_{max} of **5** is 830, 798, and 779 nm, respectively; λ_{max} of **6** is 845, 848, and 917 nm, respectively; λ_{max} of **7** is 1161, 1168, and 1225 nm, respectively.

20. J. L. Oudar, D. S. Chemla, J. Chem. Phys. **66**, 2664 (1977).

21 J. L. Oudar, J. Chem. Phys. **67**, 446 (1977).

22. D. R. Kanis, M. A. Ratner, T. J. Marks, Chem. Rev. **94**, 195 (1994).

EXCITON BANDWIDTH AND COUPLING TO INTRAMOLECULAR PHONONS IN PTCDA

M.H. HENNESSY*, Z.G. SOOS*, V. BULOVIC**, and S.R. FORREST**
* Department of Chemistry, Princeton University, Princeton, NJ 08544, mhh@princeton.edu
**Department of Electrical Engineering, Princeton University, Princeton, NJ 08544

ABSTRACT

A Holstein model is introduced for PTCDA stacks forming a one-dimensional exciton system and analyzed using Merrifield's variational method. The bandwidth and hopping integral V = 0.15 eV are obtained by comparing the absorption, emission, and fluorescence excitation of PTCDA films to solution spectra.

INTRODUCTION

The striking structural and electronic properties of perylenetetracarboxylic dianhydride (PTCDA, Fig 1) have recently been reviewed by Forrest [1]. PTCDAs crystallize face-to-face with strong π-overlap indicated by the spacing c = 3.38 Å along the stack, while interstack contacts are van der Waals or longer. The spacing c is uniform within each stack, the z axis. Two PTCDA per unit cell in crystals or ordered films correspond to parallel stacks with long axes in Fig. 1 nearly perpendicular to each other and to the stack. The intense π-π^* transition is long-axis polarized. To first approximation, crystals contain uniform PTCDA stacks that are textbook realizations of one-dimensional excitons with dispersion relation

$$E(k) = E_M + I + 2V \cos k. \tag{1}$$

E_M is the gas-phase excitation energy, I is the gas-to-solid shift of a localized excitation, k is the wavevector in the first Brillouin zone, $-\pi < kc \le \pi$, and $V = \langle M^*M|H|MM^* \rangle$ is the transfer integral for motion along the chain. Since c is less than the 3.5 Å separation of π-clouds in van der Waals contact, 4V is expected to be comparable to the bandwidths 4t ~ 0.5-1.0 eV of ion-radicals that also form face-to-face stacks [2]. Although PTCDA has an exceptionally close π-contacts for closed-shell molecules, its spectra have been analyzed [1] without invoking finite V. The purpose of this paper is to examine bandwidth contributions to solid-state spectra. Since molecular transition dipoles are normal to the stack, we have V > 0 and k = 0 at the top of the band.

PTCDA is an ideal system for other reasons than its uniform isolated stacks. E_M is strongly coupled to single vibration and is sufficiently below other excited states for a vibronic progression. The simulated absorption and fluorescence in Fig. 1 closely follow PTCDA spectra in methylene chloride solution [3]. The spacing $\hbar\omega = 0.18$ eV is typical of polyenes and polyacetylene and represents out-of-phase C-C and C=C stretches [4]. The same a_g mode is the effective conjugation coordinate discussed by Zerbi and coworkers in conjugated polymers [5]. The relative intensities of the 0-n vibronics go as $g^{2n}/n!$, where g = 0.82 is the dimensionless electron-phonon (e-ph) coupling constant used in Fig 1. The emission has characteristically smaller quanta ~0.16 eV and a Stokes shift of ~0.07 eV. Perylene has almost unit quantum yield for fluorescence and PTCDA's yield is over 50 %. At this resolution PTCDA mimics a diatomic and has negligible e-ph coupling to any other vibration.

The Holstein model [6] for polarons is a chain with linear e-ph coupling to a molecular harmonic oscillator. We take $E_M + I$ as the reference energy and model each PTCDA stack as

171

$$H = \sum_{p=1}^{N} V(a_{p+1}^{+}a_{p} + a_{p}^{+}a_{p+1}) + g\hbar\omega(b_{p}^{+} + b_{p})a_{p}^{+}a_{p} + \sum_{n=1}^{N} \hbar\omega(b_{n}^{+}b_{n} + 1/2) \qquad (2)$$

where the bosons b_n^+ create a vibrational quantum in the nth molecule and the fermions a_n^+ create a molecular excitation. As in the polaron problem, we seek solutions of (2) with a single electronic excitation. In contrast to previous applications, the parameters $\hbar\omega = 0.18$ eV and $g = 0.82$ are accurately known from solution data and only the hopping V is adjustable in crystals. The structural, electronic, and vibrational simplicity of PTCDA make it arguably the best realization of Holstein's one-dimension chain of H_2 molecules.

In this paper, we analyze PTCDA spectra in terms of (2) to obtain the exciton bandwidth and to identify phonon-assisted processes. An exciton band and V > 0 rationalize a red-shifted emission and low (~1%) quantum yield in films, since the lowest state at $k = \pi$ has vanishing transition dipole with the ground state. Crystal selection rules are regained in phonon-assisted processed. Finite V accounts naturally for features that were previously treated phenomenologically and relates PTCDA to polaron theory. The idealized model (2) omits important PTCDA features, notably charge transfer (CT) states, and complete solution for arbitrary V, g, and ω is still pending [7]. Since polaron discussions have focused on transport [6-8], PTCDA spectra are interesting new applications of a familiar solid-state model.

EXCITON-PHONON COUPLING IN PTCDA STACKS

The time scales of nuclear and electronic motions must be considered for transport and are comparable when $V \sim \hbar\omega$. In the Condon approximation, absorption or emission is vertical and the nuclei are fixed. Excitations are localized for V = 0, with excited-state potential $\hbar\omega(\lambda - g)^2$ where λ is the dimensionless displacement along the coupled a_g mode. The displacements λ_j extend to nearby molecules for finite V and are the variational parameters proposed by Merrifield [9] to solve (2) with cyclic boundary conditions and any V, g, and ω. The trial function for k is

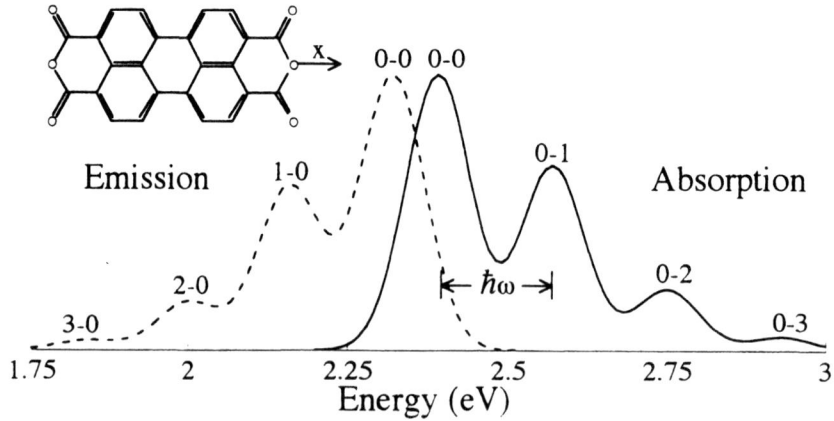

Fig. 1. Franck-Condon progression of PTCDA in methylene chloride, for e-ph coupling constant g = 0.82 and Gaussians with FWHM of 0.05 eV.

$$|k, \lambda(k)\rangle = N^{-1/2} \sum_p e^{ikp} |p, \lambda(k)\rangle$$

$$|p, \lambda(k)\rangle = a_p^+ \exp[-\sum_j (\tfrac{1}{2}\lambda_{j-p}^2 + \lambda_{j-p} b_p^+)]|0,0\rangle \quad . \tag{3}$$

Excitation at site p induces displacements λ_{j-p} in nearby sites; $|0,0\rangle$ is the electronic and vibrational vacuum; the virtual phonons in the normalized state $|p,\lambda(k)\rangle$ depend on k.

We present a real-space version of the Merrifield solution that gives the same result at the band edges. The expectation value of (2) with respect to $|k,\lambda(k)\rangle$ is

$$E_k(V, g) = 2VS_k \cos k - g^2 + (\lambda_0 - g)^2 + \sum_{p \neq 0} \lambda_p^2 \equiv A_k + \Lambda_k^2 - 2g\lambda_0 \tag{4}$$

where λ_0 is the displacement of the excited molecule and the overlap S_k is

$$S_k = \exp\{-\sum_{j=-N/2}^{N/2-1} [\lambda_{j+1}(k) - \lambda_j(k)]^2 / 2\} \quad . \tag{5}$$

To find $\lambda_j = \lambda_{-j}$, we minimize E_k and obtain linear equations for even N

$$(\lambda_0 - g) = (\lambda_0 - \lambda_1)A_k$$

$$2\lambda_j = (2\lambda_j - \lambda_{j+1} - \lambda_{j-1})A_k \quad , \quad j = 1, 2, \dots, \tfrac{1}{2}N - 1 \tag{6}$$

$$\lambda_{N/2} = (\lambda_{N/2} - \lambda_{N/2-1})A_k \quad .$$

The quadratic dependence of S_k on displacement differences ensures that each partial derivative occurs twice, with opposite signs. The A_k thus cancel on summing over (6) and give the sum rule,

$$g = \sum_j \lambda_j(k) = \lambda_0 + 2\sum_{j=1} \lambda_j \tag{7}$$

for any k. The net displacement is conserved under linear e-ph coupling to degenerate modes. Similar results hold for odd N.

To solve (6), we eliminate $\lambda_{N/2}$ in the last equation, $\lambda_{N/2-1}$ in the j = N/2-1 equation, and so on, to generate a continued fraction. The same result follows by assuming λ_j to decrease as $\lambda_0 x^j$. We obtain

$$\lambda_0 = g(1 - 2A_k)^{-1/2}$$

$$\lambda_{\pm j} = \lambda_0 x_k^j, \quad x_k \equiv -2A_k / (1 + \sqrt{1 - 2A_k})^2 \tag{8}$$

when N is large enough to have undistorted sites far from the excited molecule. The displacements are in-phase for $A_k = 2VS_k \cos k < 0$, out-of-phase for $A_k > 0$, and reduce to the molecular value at $A_{\pi/2} = 0$. We sum the geometric series for λ_j in S_k, Λ_k and E_k to find

$$S_k = \exp(-g^2 \Delta_k), \quad \Delta_k \equiv (1 - 2A_k)^{-3/2}$$

$$\Lambda_k^2 = g^2(1 - A_k)\Delta_k \tag{9}$$

$$E_k = 2VS_k(1 - g^2 \Delta_k)\cos k + g^2(\Delta_k - 2\Delta_k^{1/3})\hbar\omega \quad .$$

These equations are identical to Merrifield's and give the same results at $k = 0$ and π. For general k, Merrifield [9] considered virtual phonons with finite momentum, with phase $\Phi \neq 0$ that changes Δ_k in (9). Since $A_k = 2VS_k\cos k$ depends on S_k, the first equation is a self-consistency condition.

The exciton bandwidth, $E_0 - E_\pi$, is shown in Fig. 2 in units of $\hbar\omega$ for $g = 0.82$. Coupling to a_g modes reduces the width from the band value of $4V$ in (1). The sum over molecular displacements, Λ_k in (4), is larger (smaller) than g at the top (bottom) of the band; the $k = 0$ and π values of $(\Lambda_k/g)^2$ in Fig. 2 follow from (9) and change the Franck-Condon factors in films. Increased $k = 0$ displacements generates a broader vibronic progression in absorption, while reduced $k = \pi$ displacements indicate narrower emission. The choice of V in (2) simultaneously fixes the bandwidth and the vibronic progression in absorption and emission.

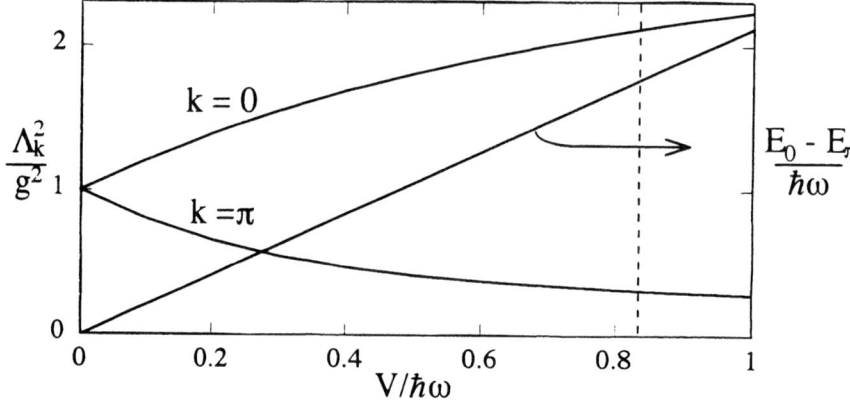

Fig. 2. Relative displacements Λ_k^2/g^2 and bandwidth $E_0 - E_\pi$ as a function of exciton hopping V, from eq. (9); the dashed line at $V = 0.15$ eV is discussed in the text.

PTCDA SPECTRA AND PHONON-ASSISTED PROCESSES

Figure 3 summarizes the absorption and emission of PTCDA films and the quantum yield for fluorescence as a function of excitation energy [10]. In contrast to solution, the absorption and emission have different vibronic progressions and an apparent Stokes shift of ~0.5 eV. To assign the 0-0 absorption line, we need the shift I in (1) of a localized excitation from solution to film. The previous choice [1] of $I = V = 0$ was used to identify the CT state at 2.23 eV and excitons starting at 2.39 eV. Alternatively, finite I and V can be chosen to place 0-0 at 2.23 eV and admixtures of CT states are then needed for electroabsorption data. The 2.23 eV polarization is along the stack for a CT state, normal to it for an exciton. Polarization data will guide the proper choice. In either case, the relative 0-n intensities in films indicate displacement $\Lambda_0 > g$.

The quantum yield is reduced by more than an order of magnitude between solution and films [10]. Moreover, in contrast to the intense 0-0 fluorescence in solution, the 0-0 of films at 1.85 eV is weak and 1-0 emission at 1.70 eV dominates. Reduced e-ph coupling in emission supports $\Lambda_\pi < g$ in films. The enhanced quantum yield for fluorescence in Fig. 3 for excitation at 2.11 eV suggests a 0-1 line at $E_\pi + \hbar\omega$ that is fully consistent with $E_\pi = 1.91$ eV within the 0.07 Stokes shift in solution and ± 0.01 eV experimental resolution.

Having identified the bottom of the band, we find an exciton bandwidth of 0.42 eV if the 2.23 eV line is taken as 0-0 and obtain V = 0.15 eV, Λ_0 = 1.19, and Λ_π = 0.46 from (9) for the solution parameters g = 0.82 and $\hbar\omega$ = 0.18 eV. The alternative 0-0 assignment yields larger V and Λ_0, but smaller Λ_π. In either case, the variational result (9) indicates in detail how solution spectra evolve with V in a PTCDA stack. The origin of these changes has not been considered previously. The calculated Λ_0 = 1.19 accounts for more intense 0-1 than 0-0. The full vibronic progression in Fig. 3 requires larger Λ_0 ~1.3 for 0-0 at 2.23 eV, or the present Λ_0 for 0-0 at 2.39 eV and larger V that increases Λ_0. Neither choice is quantitative and CT contributions cannot be neglected.

The transition dipole for π-π^* excitation of each PTCDA in the stack is

$$\langle e|\mu(\lambda)|g\rangle \;=\; \mu_e + \left(\frac{\partial\mu}{\partial\lambda}\right)_e \lambda \;=\; \mu_e + \frac{\mu_e'\hbar}{\sqrt{2m\omega}}(b^+ + b) \tag{10}$$

where μ_e is along the long axis, the dipole derivative μ_e' represents variations along the a_g mode, and the reduced mass m for CC stretches is 6 amu. Pariser-Parr-Pople calculations [11] for the related π-system of octatetraene indicate that μ_e' is a 5-10% correction, an unusually large value that reflects strong e-ph coupling to the extended conjugation coordinate. The molecular correction is nevertheless < 1% for dipole-allowed processes. The transition moment of a PTCDA stack is

$$\mu \;=\; \sum_n \langle e|\mu(\lambda)|g\rangle(a_n^+ + a_n) \;=\; \mu_e\sqrt{N}\delta_{k0} + \frac{\mu_e'\hbar}{\sqrt{2m\omega}}\sum_q (a_{-q}^+ + a_q)(b_q^+ + b_{-q}) \tag{11}$$

where we have introduced Einstein phonons b_q^+. Absorption requires k = 0 and gives the standard 0-0 oscillator strength, $N\mu_e^2\exp(-\Lambda_0^2)$. The second term also conserves the crystal momentum.

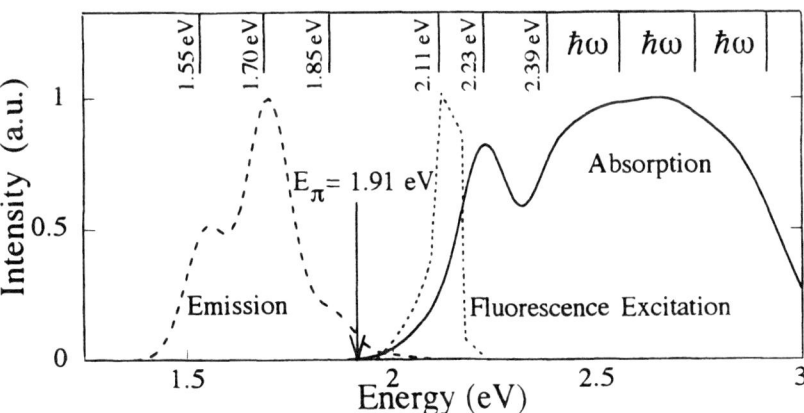

Fig. 3. Experimental [10] absorption, emission, and fluorescence excitation spectra of PTCDA films, with measured lines indicated at the top; the bottom of the exciton band is E_π.

The normalized excited states of a stack are the $|k,\lambda(k)\rangle$ functions defined in (3), with allowed transitions to $k = 0$. To obtain μ_e contributions to 0-1, we expand the exponential operator in (3) and find, after some straightforward algebra, the one-phonon part

$$|0,\lambda(0),1\rangle = \frac{\exp(-\Lambda_0^2)}{\sqrt{N}}\sum_q \lambda_q(0)b_q^+ a_{-q}^+|0,0\rangle \quad , \tag{12}$$

where we have introduce q-space displacements

$$\lambda_q(k) \equiv \sum_n e^{iqn}\lambda_n(k) = \frac{g(1-x_k^2)}{\sqrt{1-2A_k(1+x_k^2-2x_k\cos q)}}$$
$$\Lambda_k^2 = N^{-1}\sum_q \lambda_q(k)^2 \tag{13}$$

using A_k and x_k from (4) and (8), respectively. As seen from (12), the total momentum vanishes, but the exciton can have any q.

The total intensity on one-phonon states follows from the second sum in (13) and is Λ_0^2 times the 0-0 intensity. The normal Franck-Condon 0-1 is distributed over the exciton band, starting at $E_\pi + \hbar\omega$. The exciton density of states, $(dE/dk)^{-1}$, follows from (9); some 25% of the states are within 0.03 eV of E_π, where $\lambda_\pi(0)$ is the largest. The assignment of the 2.11 eV line in Fig. 3 as phonon-assisted absorption from $k = \pi$ is supported by its estimated intensity of ~30% compared to 0-0. About half of the 0-1 intensity is at $E_0 + \hbar\omega$ because the band is nearly flat at k = 0. The remaining 20% gives a background contribution. Similar treatment of 0-n transitions yields multiphonon processes whose total q is opposite to the exciton's wavevector. The lower resolution of absorption in films follows immediately, although as already noted the overall width of the vibronic progression requires larger $\Lambda_0 \sim 1.3$.

Emission originates from the vibrational ground state of $k = \pi$ excitons and 0-0 is forbidden by crystal selection rules. The small intensity in Fig. 3 at 1.85 eV for 500 Å thick films may be due to the surface, an interpretation supported by increased intensity in multiple quantum wells with PTCDA stacks as short as three in 10 Å layers. The $|\pi,\lambda(\pi)\rangle$ state in (3) has displaced oscillators relative to the $|0,0\rangle$ ground state. The expansion of the excited-state vibration in the ground state basis corresponds to $-\lambda_j$ for all j and is the simplest approximation in which frequency changes [6] due to V are neglected. The 1-0 emission parallels the development above

$$|\pi,\lambda(\pi),1\rangle = -\frac{\exp(-\Lambda_\pi^2)}{\sqrt{N}}\sum_q \lambda_q(\pi)b_q^+ a_{-q+\pi}^+|0,0\rangle \tag{14}$$

for total wavevector π. A k = 0 transition becomes allowed at $q = \pi$ and generates a phonon in the ground-state potential. The 1-0 emission at $E_\pi - \hbar\omega$ has relative intensity $[\exp(-\Lambda_\pi^2)]\Lambda_\pi^2/10 \sim 0.02$ if a tenth of the states are within kT of $k = \pi$; this is the strongest emission in Fig. 3. The high quantum efficiency for fluorescence at 2.11 eV excitation fits in well with a 0-1 transition in the k = π well.

The dipole derivative also leads to ~1% emission for 1-0. Using the second term of (11), we annihilate a $k = \pi$ and conserve momentum by creating a ground-state vibration. The Franck-Condon factor is $\exp(-\Lambda_\pi^2) \sim 1$ and the $(\mu_e')^2$ factor is now small. The relative contributions of

176

dipoles and dipole derivatives require further study. We are also seeking more accurate excited states than $|k,\lambda(k)\rangle$ and corrections to phonon frequencies due to V. The Holstein model (2) accounts qualitatively for both energy and intensity changes due to PTCDA stacking.

DISCUSSION

The identification of CT states in organic molecular solids or conjugated polymers has motivated much research. Such states are crucial for charge separation and thus for current efforts towards electroluminescent devices or electrically pumped lasers [1]. CT states in polyacene crystals [12] are typically masked by strong molecular absorption, as can be expected for small intermolecular overlap. Strong overlap along PTCDA stacks is favorable for CT and unequivocal evidence for their participation has been accumulated [1], notably from polarized electroabsorption. We omitted CT states in order to focus on excitons and electron-phonon coupling in the familiar Holstein model (2). Even then, the regime when V and $\hbar\omega$ are comparable requires further analysis and many spectroscopic consequences remain to be considered.

Some qualitative implications of CT participation are readily found. We start with the ground-state geometry (all $\lambda_n = 0$) and consider $k = 0$ states with adjacent PTCDA ions

$$|CT \pm\rangle = (2N)^{-1/2} \sum_n \left(|M_n^+ M_{n+1}^-\rangle \pm |M_n^- M_{n+1}^+\rangle \right) \tag{15}$$

and ground-state PTCDA at the remaining sites. Mulliken complexes mix $|CT+\rangle$ and $|0,0\rangle$, thereby generating transitions polarized along the stack. We discount this mechanism in PTCDA stacks because the HOMO and LUMO hardly overlap in the actual structure. The high hole mobility of PTCDA indicates significant LUMO-LUMO overlap, as we indeed find in AM1 calculations. We take the transfer integral t_h for hole motion to be proportion to this overlap and electron transfer t_e to the HOMO-HOMO overlap. This specifies the excited-state mixing

$$\langle CT + |H| k = 0, \lambda(0) = 0 \rangle = \sqrt{2}(t_h + t_e) \tag{16}$$

between $|CT+\rangle$ and the $k = 0$ exciton (4) with no displacement. There is no mixing with $|CT-\rangle$. The calculated overlaps add at $k = 0$. The corresponding $k = \pi$ states lead to small $(t_h - t_e)$ with $|CT-\rangle$. We consequently expect much larger CT contributions in absorption than emission because the $k = 0$ excited states are closer in energy and mix more strongly.

The mixing (16) strongly enhances the polarizability along the stack, where the molecular polarizability is small, and as proposed by Petelenz [13] can be comparable to the in-plane polarizability of conjugated systems. The long-axis polarizability of PTCDA is 340 Å^3 according to STO-3G calculations [14], close to the 300 Å^3 measured in electroabsorption [15] from the shift of the 2.23 eV line in Fig. 3. Reasonable transfer integrals and a CT state within ~0.2 eV of the exciton rationalize the data, although in a clearly phenomenological manner. Since the mixing (16) borrows intensity from the $k = 0$ exciton, it conserves oscillator strength and redistributes it to higher energy. The same result can be achieved by increasing Λ_0, which also shifts oscillator strength to higher energy. The PTCDA absorption in Fig. 3 requires $\Lambda_0 \sim 1.3$ rather than the 1.19 implied by V = 0.15 eV.

In summary, we have located the bottom $(k = \pi)$ of the exciton band to be around 1.91 eV by using a Holstein model with hopping V to analyze changes in absorption and emission between

PTCDA in solution and films. The exciton bandwidth is at least 0.32 eV, with V at least 0.15 eV. CT states must be included to deal with the wealth of PTCDA spectra [1]. Their strong mixing with $k = 0$ excitons is under investigation through the polarization of the 2.33 and 2.39 eV lines in Fig. 3. The surprising applicability of a diatomic Holstein chain to PTCDA stacks is due to strong coupling to a single molecular vibration, the out-of-phase C-C and C=C stretch encountered in polyenes and polyacetylene, and a single low-lying electronic state.

ACKNOWLEDGMENTS

We gratefully acknowledge support of this research by the National Science Foundation through DMR-9530116 and the MRSEC program under DMR-9400362.

REFERENCES

1. S.R. Forrest, Chem. Rev. **97**, 1793-1896 (1997).

2. Z.G. Soos and D.J. Klein, in <u>Molecular Association</u>, Vol. 1, edited by R. Foster (Academic, New York, 1975) p. 1-109.

3. U. Gomez, M. Leonhardt, H. Port and H.C. Wolf, Chem. Phys. Lett. **268**, 1 (1997).

4. Z.G. Soos, M.H. Hennessy and G. Wen, Chem. Phys. Lett. **274**, 189 (1997).

5. M. Gussoni, C. Castiglioni and G. Zerbi, in <u>Spectroscopy of Advanced Materials</u>, Vol. 19, edited by R.J.H. Clark and R.E. Hester (Wiley, New York, 1991) p. 251-353.

6. T. Holstein, Ann. Physics **8**, 325, 343 (1959); D. Emin, Adv. Phys. **22**, 57 (1973).

7. Y. Zhao, D.W. Brown and K. Lindenberg, J. Chem. Phys. **106**, 5622 (1997); D. Feinberg, S. Ciuchi, and F. de Pasquale, Int. J. Mod. Phys. B **4**, 1317 (1990); G. Wellein, H, Röder and H. Fehske, Phys. Rev. B **53**, 9666 (1996).

8. R. Silbey and R.W. Munn, J. Chem. Phys. **72**, 2763 (1980); V.M. Kenkre, J.D. Anderson, D.H. Dunlap and C.B. Duke, Phys. Rev. Lett. **62**, 1165 (1989).

9. R.E. Merrifield, J. Chem. Phys. **40**, 4450 (1964).

10. V. Bulovic, P.E. Burrows, S.R. Forrest, J.A. Cronin and M.E. Thompson, Chem. Phys. **210**, 1 (1996).

11. M.H. Hennessy and Z.G. Soos, unpublished.

12. P.J. Bounds, W. Siebrand, I. Eisenstein, R.W. Munn and P. Petelenz, Chem. Phys. **95**, 197 (1985).

13. P. Petelenz, Chem. Phys. Lett. **215**, 607 (1993).

14. Z.G. Soos, M.H. Hennessy and G. Wen, Chem. Phys. (in press).

15. E.I. Haskal, Z. Shen, P.E. Burrows and S.R. Forrest, Phys. Rev. B **51**, 4449 (1995).

ADVANTAGES OF MODAL DISPERSION PHASE-MATCHING AND MATERIALS REQUIREMENTS FOR POLYMERIC DEVICES USING EFFICIENT SECOND HARMONIC GENERATION AT TELECOMMUNICATION WAVELENGTH

Matthias L. JÄGER, Vincent RICCI, Wook-Rae CHO,
Michael T. G. CANVA and George I. STEGEMAN
Center for Research and Education in Optics and Lasers – CREOL,
University of Central Florida - UCF,
P.O. Box 162700 - 4000 Central Florida Boulevard
Orlando FL 32816-2700 - USA

ABSTRACT

Modal Dispersion Phase Matching appears to be currently much better adapted to parametric mixing in polymeric material waveguides than Quasi Phase Matching. For second harmonic generation at telecommunication wavelengths, using organic materials should allow better performance than with ferroelectric crystals. Promising results are expected in view of theoretical expectations and continuously improving experimental past and current results.

INTRODUCTION

Polymers have recently penetrated photonics in a number of areas. Notable examples are electro-optic modulators operating around 100 GHz, photorefractive media with record writing efficiencies, and now electroluminescent and even lasing devices. Parametric mixing, most specifically second harmonic generation was initially tried as early as 1990 with polymers for doubling diode lasers into the blue. However, for this application, the transparency-nonlinearity tradeoffs proved problematic and, although there were sporadic reports of second harmonic generation, this type of parametric mixing in poled polymers was effectively abandoned.

For incident wavelengths in the communication bands, however, and most specifically at 1550 nm, the transparency-nonlinearity tradeoffs are very favorable for poled polymers. In this wavelength range there are many applications ranging from frequency shifting in Wavelength Division Multiplexing (WDM), to cascading for all-optical signal processing and to tunable frequency generation via optical parametric generation and amplification. The primary prerequisite for these applications is to obtain very efficient SHG. Quasi-phase-matched $LiNbO_3$ channel waveguides are currently the material of choice for these applications but this kind of device is near its theoretical limits, which could be by far surpassed by doped polymer based components. This is the main motivation for our investigation in this field.

In this paper, we shall first briefly recall the basic equations governing second harmonic generation (SHG) and the associated figures of merit (FOMs). Two phase-matching techniques were considered as potentially useful, namely Quasi Phase Matching (QPM) and Modal Dispersion Phase Matching (MDPM). Then, we shall discuss how poled polymer devices were realized and characterized. These devices were based on the well-known stilbene DANS, (4-dimethylamino-4'-nitrostilbene) and an azobenzene DR (disperse red) chromophores. Finally, we will speculate about the ultimate potential of poled polymers for parametric optical processes at telecommunications wavelengths, and the materials challenges that must be met for these applications to become a reality.

179

THEORETICAL EXPECTATIONS

Second-Harmonic Generation

The SHG process is usually studied first in the limit of low conversion for the fundamental beam. In a channel waveguide capable of two-dimensional confinement of the optical beams, the wavevector detuning is defined as $\Delta\beta = \beta(2\omega) - 2\beta(\omega) = 4\pi[N(2\omega)-N(\omega)]/\lambda$, where N is the waveguide effective index for a given propagation mode and associated wavevector β. The following general expression gives the second harmonic power $P_2(L)$ as a function of device length L and fundamental input power $P_1(0)$:

$$P_2(L) = \frac{\varepsilon_0^2 \omega^2 P_1^2(0) d_0^2 L^2}{4} \left| \frac{1}{L} \int_0^L dz\, d(z) e^{i\Delta\beta z} \right|^2 \left| \iint dx dy\, d_{ijk}(x,y) e_{\omega j} e_{\omega k} e_{2\omega i}^* \right|^2 \qquad \text{Eq. 1}$$

It can be seen that $P_2(L)$ is proportional to the product of two integrals: the longitudinal phase-matching integral and the transverse spatial overlap integral, both squared.

The first integral is easily integrated into a sinc^2 function of the phase-mismatch $\Delta\beta L/2$:

$$\left| \frac{1}{L} \int_0^L dz\, d(z) e^{i\Delta\beta z} \right|^2 = \left(\frac{\sin(\Delta\beta L/2)}{\Delta\beta L/2} \right)^2 = \mathrm{sinc}^2(\Delta\beta L/2) \qquad \text{Eq. 2}$$

The value of the second integral depends on the overlap in the nonlinear active region of the waveguide $(d_{ijk}(x,y) \neq 0)$ of the transverse electric field profiles, e_ω, for the fundamental mode and, $e_{2\omega}$, for the second harmonic mode. The repercussions of this "overlap integral" will be discussed in more detail later.

The SHG performance of a device is usually characterized by the following three figures of merit, (FOM):

(1) $\eta = P_2/(P_1 L)^2$ in % / W / cm^2, normalizes the performance of a device to its length L;

(2) $\eta_0 = P_2/(P_1 L)^2$, is the extrapolated version of η for L $\rightarrow$ 0;

(3) $\eta' = P_2/P_1^2$, in % / W, is the net conversion efficiency.

The first two FOMs are equal in the absence of losses, and both increase as the square of the nonlinear coefficient. The third FOM is the real measurement of the device efficiency. It increases as the square of the ratio of the nonlinearity to the losses and optimizing η' is the ultimate goal for a device. We would like to introduce a fourth figure of merit, η_0', as a rough, but yet useful (over)estimate of the largest value achievable by η':

(4) $\eta'_0 = \eta_0 \times L^2_{eff}$, where L_{eff} is the effective optimum length of the device, mainly limited by propagation losses and phase-matching limitation, as defined further.

Also, in order to judge the potential for SHG of a material with nonlinearity $d_0 = d^{(2)}$ and a specific phase-matching geometry, one often uses an idealized (and unachievable) figure of merit, η_{max} which assumes perfect phase-matching, no loss and a constant field distribution over the effective area A of the waveguide (It allows a quick estimate without having to calculate the exact field distributions).

$$\eta_{max} = \frac{2\omega^2 d_0^2}{\varepsilon_0 c^3 A N^3} = \frac{8\pi^2}{\varepsilon_0 c} \frac{d_0^2}{\lambda^2 A N^3} . \qquad \text{Eq. 3}$$

Depending on the phase-matching technique employed, in practice, one or both integral terms will not be optimized, resulting in reduced efficiencies. One should note that this figure of merit shows a $1/\lambda^4$ dependence since the optical beam in a waveguide is confined to a distance of the order of the wavelength ($A \approx \lambda^2$). Considering that the spectral dispersion of the $d^{(2)}$ coefficient would lead to a even sharper decrease of the FOM with increasing wavelength, this strong wavelength dependence has to be kept in mind when comparing the FOMs achieved at different wavelengths, especially since most of the early work on SHG was targeted for different applications at shorter wavelengths.

Phase-Matching Techniques

To ensure optimum transfer of energy from the fundamental to the second harmonic, the phase of the radiating dipoles must be matched to the propagating SH field. One needs to either match the effective speeds of the wavefronts, or, at least, to adjust the sign of the nonlinearity in order to eliminate destructive interference in the phase-matching integral. Different phase-matching techniques can be used to achieve these goals : we have found that modal dispersion phase-matching (MDPM) and quasi-phase-matching (QPM) are the most suitable for our purpose, i.e. $\chi^{(2)}$ SHG with ultimate application to $\chi^{(2)}:\chi^{(2)}$ cascading behavior.

Quasi Phase Matching - QPM

QPM is certainly the most effective phase matching technique used to date to produce efficient SHG in ferroelectric materials such as $LiNbO_3$ and $LiTaO_3$ [1, 2, 3]. Destructive interference between the propagating second harmonic beam and the locally generated field is eliminated by adapting the nonlinearity to their dephasing $\Delta\psi$, i.e. the sign of the nonlinearity is reversed (+/-)each time $\Delta\psi$ changes by an additional π. The coherence lengths L_c are typically of the order of a few microns to a few tens of micron, and the realization of QPM devices is technically feasible using lithographic techniques. However, any modulation of the refractive index associated with the nonlinearity modulation should be avoided since an index periodicity is appropriate for out-coupling the guided waves into radiation fields, thus increasing the guided wave losses. As pointed out previously, this technological challenge for the process has been met quite successfully in the case of ferroelectric crystals, resulting in propagation losses well below 1 dB/cm. Similar research and development has being carried on with doped polymers [4, 5, 6, 7, 8]. Although promising losses, of about 5 dB/cm for the fundamental wave, have been reported, the FOM's are not yet competitive [9].

Modal Dispersion Phase Matching - MDPM

MDPM is particularly well suited for SHG in poled polymer waveguides because it utilizes some unique properties of polymer processing [10, 11].

In a guided wave geometry, light of a fixed frequency may propagate in different modes, each characterized by a different propagation wavevector β and/or state of polarization (and hence by a different velocity and a different spatial field distribution). Each mode is thus associated with a different effective index (higher order modes have lower effective indices). This is shown in Figure 1, where an example, for both ω and 2ω, of the effective indices for the first few lowest order transverse magnetic TM modes is represented as a function of the core waveguide thickness. TM modes are used here because their polarization direction is appropriate for parallel plate poling of the chromophores that are oriented normal to the surfaces of the device. On this graph it can be seen that building a structure of about 1.0 μm (or 2.25 μm) core thickness will phase-match the TM_{00} mode at ω with the TM_{01} (or TM_{02}) mode at 2ω. Precision realization of waveguide dimensions is required. Given a fixed nonlinearity, the change in sign of the second harmonic field across the transverse dimension may result in canceling contributions in the overlap integral. This can be compensated by designing a core consisting of several layers with the proper alternating nonlinearities corresponding to the sign reversal of the high order mode in the transverse directions (+/0 or +/-), as done in QPM for the longitudinal dimension.

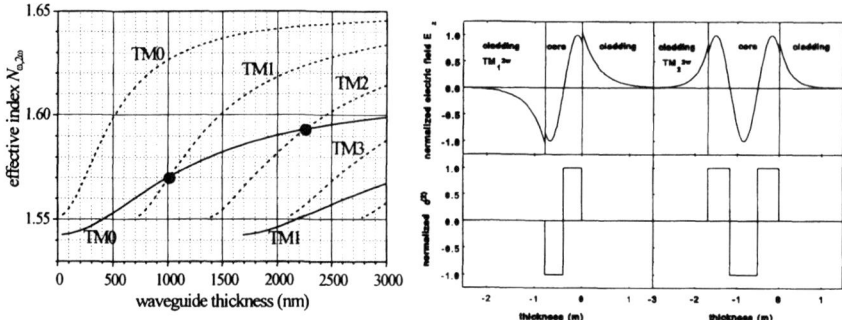

Figure 1 (Left): Typical effective index variations as a function of core thickness for the first few TM order modes at the fundamental and second harmonic frequencies. The data used for this graph correspond to a PC/DANS/PC structure used at the fundamental wavelength λ = 1550 nm.

Figure 2 (Right): Electric field distribution of the fundamental and second harmonic of the $TM_{00} \rightarrow TM_{01}$ and TM_{02} phase-matching conditions shown in Figure 1. The π phase change in adjacent lobes of high order modes can be compensated for by using successive layers of different nonlinearity, as explained in the text.

As we shall see, the apparent drawbacks to this MDPM technique are not really important for our applications. First, the mode profiles. A second harmonic output profile consisting of several lobes is not well suited for "source" applications that usually require single-lobed beams. However, for both of our projected applications, it is only the structure of the fundamental (cascading) or difference frequency (WDM frequency shifting) output beam that is important, and these can be arranged to be in the lowest order mode. Second, one will note that the spatial overlap between the modes is not optimal due to the presence of zero(s) in the second harmonic field in region(s) where the corresponding fundamental field is strong. In fact, this loss in efficiency is of the order of, or less than the reduction in effective nonlinearity obtained with using only the first Fourier component of the nonlinear grating in QPM.

Actually, overall, MDPM can be more efficient than QPM. This is illustrated in Figure 3 where the FOMs η_0, in % / W cm^2 are presented for different structures. First, the phase-matching of the lowest order modes $TM_{00}(\omega)$ and $TM_{00}(2\omega)$ are plotted as a function of core thickness, (a) in the perfect phase matched case and (b) in the QPM case, smaller by a factor $4/\pi^2$ taking into account the impact of the grating structure along the propagation axis; $\blacktriangledown$ and $\blacktriangle$ correspond to the highest efficiency and to the most stable (minimum) grating period respectively. Second, the efficiency of the two MDPM structures matching $TM_{00}(\omega)$ to $TM_{01}(2\omega)$, and $TM_{02}(2\omega)$ respectively, are indicated by two points ($\blacksquare$, $\bullet$). Also represented is the structure with a core made of alternately nonlinear and linear material ("+/0" instead of "+/-"). In Figure 4, is plotted the corresponding ultimate FOM η'_0, in % / W, which basically affects the previous shape by normalizing η_0 by h^2, h being the thickness of the nonlinear layer. Clearly, MDPM is as efficient as QPM, and furthermore the highest efficiency is calculated for the "MDPM $TM_{00} \rightarrow TM_{01}$" structure.

What is also illustrated is the fact that as long as the absorption of the nonlinear chromophore is the dominant mechanism for the beam propagation losses, it is not worthwhile making complex, oppositely poled structures. This will only increases η, not the true efficiency η'_0 ! Structures much easier to fabricate made of a waveguiding core with alternated nonlinear absorbing and transparent linear layers (+/0) will be as efficient, in % /W.

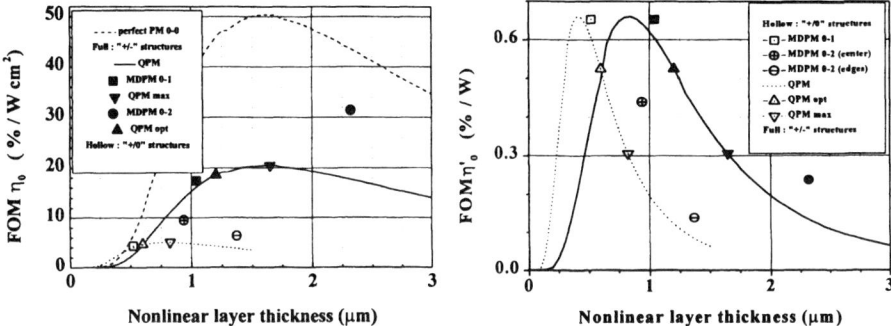

Figure 3 (Left) : FOM η_0 , in % / W cm^2, for (a) perfect phase-matched structures, (b) QPM structures and (c) MDPM structures as a function of core thickness. As explained in the text, the corresponding device parameters are assumed for DANS (nonlinearity of 6 pm/V) and PC polymers. The structure "+/0" (hollow and dot) is calculated assuming a high index polymer perfectly matched to the DANS.

Figure 4 (Right) : FOM η'_0 , in % / W, as a function of nonlinear material thickness, for the same cases as in Figure 3. The normalization assumed a propagation loss due to absorption of 40 dB/cm at the harmonic and 5 dB/cm at the fundamental for a h=2 μm thick layer of DANS.

The FOMs plotted in the above graphs are calculated for a channel waveguide geometry with a weak lateral confinement in the horizontal direction (the index difference is only about 0.03-0.05 on the lateral direction) and a strong confinement in the vertical direction. As a result, we used the effective index method which allows a separation of the two transverse dimensions in the calculation of the field profiles involved in the overlap integral. First, we calculated the effective indices of the modes corresponding to vertical confinement together with their vertical field profiles. An integration of these profiles along the vertical dimension can be used to achieve a partial normalization. However, in order to be complete, the normalization must be done also along the other channel dimension. Instead of directly calculating the impact of the weak lateral confinement on the field profile distribution, we calculated the effective widths of the channel modes and used these effective widths to complete the normalization of the fields.

Effect of losses

Losses are the key problems leading to a decrease in the overall SHG efficiency. The losses may be due to absorption, scattering, disordering, impurities, clustering etc. Loss has been very successfully addressed in the case of passive devices made from polymers with electronic transitions in the UV leading to propagation losses well below one dB/cm in the near infrared. However, it is still a problem that is far from solved in the case of nonlinear polymer systems where the electronic transitions occur in the visible. Research is still very active in this area and any progress is very beneficial to the nonlinear optics community. In such systems losses due to absorption are always much higher at the lowest wavelength, which is nearest to the main absorption band. Scattering losses also decrease with increasing wavelength. The material loss coefficients are quoted in units of cm^{-1}, or in more device-oriented papers in dB/cm (1 cm^{-1} $\approx$ 4.34 dB/cm), with $\alpha(\omega) < \alpha(2\omega)$. Typically $\alpha(\omega)$ is a few cm^{-1} and $\alpha(2\omega)$ can be as high as a few tens of cm^{-1}, which is just too large. Qualitatively, the impact of losses is to reduce the available pump power at the fundamental frequency, and to attenuate the second harmonic that is generated too far away from the output face of the device. Quantitatively, this can be taken into account in some FOMs in the low conversion limit. Writing $\Delta\alpha=\alpha(2\omega)/2 - \alpha(\omega)$, leads to the expression:

$$P_2 = 2\eta_0 P_1^2 e^{-(\alpha(\omega)+\alpha(2\omega)/2)L} \frac{\cosh(\Delta\alpha L) - \cos(\Delta\beta L)}{(\Delta\alpha)^2 + (\Delta\beta)^2}$$

Eq. 4

From this equation it can be shown that there is an optimal length, L_{opt}, at which the net conversion efficiency P_2/P_1^2 is maximized:

$$L_{opt} = \frac{\ln(\alpha(2\omega) / 2\alpha(\omega))}{\alpha(2\omega)/2 - \alpha(\omega)} \quad .$$

Eq. 5

In the event that these propagation losses are small, L_{eff} is limited by L_{pm}, the distance over which the phase-matching can be technologically maintained:

$$\frac{1}{L_{eff}} \approx \frac{1}{L_{pm}} + \alpha(\omega) + \frac{\alpha(2\omega)}{2} \quad .$$

Eq. 6

We shall see in the experimental section how important those parameter are.

EXPERIMENTS

In this part we will summarize our latest efforts to enhance the efficiency of SHG in poled polymer waveguides. It has already been increased by several orders of magnitude over the past years [12, 13, 14, 15, 16, 17], bringing it, at least in terms of some figure of merit to a level comparable to the best ferroelectric based devices. The nonlinear polymers we used in this study were based on the well-known stilbene DANS, (4-dimethylamino-4'-nitrostilbene) and an azobenzene DR (disperse red) chromophores [Figure 5].

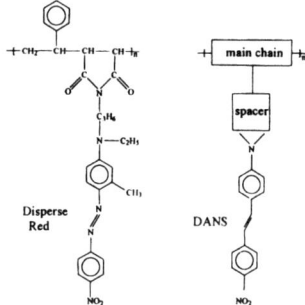

Figure 5 : The nonlinear polymers used in this study were based on (right) the well-known stilbene DANS, (4-dimethylamino-4'-nitrostilbene) and a (left) azobenzene DR (disperse red) chromophores. These raw materials were supplied respectively by Sandoz-OptoElectronics and Akzo-Nobel.

Device Realizations

The realization of our SHG devices typically needed 4 steps after material characterization and structure design: spin coating of the different layers, poling of the chromophores, channel waveguide formation, and dicing of the endfaces.

The different layers were deposited by spincoating. The substrate was placed on a rotation stage that could be spun at high speeds (typically between 2,000 and 5,000 rpm). The polymeric solution was dropped onto the substrate that was held in place by a low vacuum chuck. The spinning process spread the polymer solution evenly across the substrate surface. The thickness was controlled by rotation speed and viscosity (i.e. dilution). Finally, the sample was cured to remove the residual solvent in an oven, typically just below the glass transition temperature of the polymer film. Cleanliness of all the components is critical to obtaining high quality samples. This is especially true for poled polymer devices because the poling process does not tolerate impurities in the polymer layers. They can lead to dielectric breakdown at high poling fields. Therefore, all fabrication was performed in a cleanroom environment, class 10,000 or better.

To orient the chromophores, we have mainly used parallel plate poling which makes it possible to fabricate and test hundreds of high quality NLO waveguides on a single substrate. For the top electrode, a thin metal film (50 nm aluminum or gold) was deposited on the polymer stack, and was removed by chemical wet etching after the poling. The details of the bottom electrode were determined by the device requirements. For specially shaped electrodes, a thin aluminum layer was deposited first on fused silica substrates and then photolithographically patterned. For devices requiring transparent electrodes, indium tin oxide (ITO) coated glass was employed. All other samples were prepared on silicon wafers that have a sufficiently high conductivity for the poling process. With well-characterized materials, the in-plane poling technique was also used because it allows poling with a much higher local electric field, producing a larger degree orientation and thus a larger optical nonlinearity. However it yields only a few samples to study per substrate. In the case of the DANS based side chain polymer, the glass transition temperature T_g is 142 $^{\circ}$C and the poling was performed at 137 $^{\circ}$C. In the case of the DR based material, two different chromophore loadings of the polymer were available with different T_gs, 164 $^{\circ}$C and 137 $^{\circ}$C. In a structure with alternating layers of high and low T_g,, the stack is poled first around the high T_g and then reverse poled around the lower T_g in which only the lower T_g layers are reoriented [15, 18]. Using this technique one can build structures with core active layers of opposite nonlinearity, optimizing the spatial transverse overlap in the MDPM case.

To transform our planar waveguiding structure into channel waveguides, we exposed them to UV-blue irradiation through a mask which was opaque in the channel regions and transparent elsewhere. The UV-blue radiation photochemically bleaches the organic dye doped material, thus reducing its refractive index in the near IR band.

Endfire coupling was chosen to ensure that the coupling of the fundamental field into and out of the waveguides is compatible with commercial fiber coupled devices. It was implemented by dicing the samples with a diamond-grinding wheel, a technique commonly used for semiconductor processing. This technique was optimized for application to polymer samples in collaboration with Thermocarbon, Inc. [19], making it possible to produce good and reproducible endfaces and coupling efficiencies for a variety of substrates [17]. To fully characterize the devices, samples of different length, typically between 1 and 10 mm, were cut in this way.

SHG results

QPM devices
Some effort was first invested into a number of variations of the classical QPM structures, trying to realize alternating longitudinal +/0 structures by bleaching the dye in the active area, or by modulating the nonlinearity using digitized electrode patterns [8]. In the case of the bleached grating structures (+/0) which typically also exhibit strong index modulation, the losses at the fundamental were of order 25 dB/cm due mainly to the grating out-coupling of the fundamental guided field into radiation fields. The periodic electric field poling (combination of digital and planar electrodes) also led to poor performance based on net SHG efficiency, $\eta' < 0.035$ % / W. The losses were still important, at the harmonic wavelength ($\alpha(2\omega) > 30$ dB/cm) and to a lesser degree at the fundamental one, ($\alpha(\omega) > 7$ dB/cm). These samples exhibited a poor figure of merit, $\eta < 0.05$ % / W cm^2, best case, primarily because the modulation in the nonlinearity achieved via quasi phase matching was at best $d^{(2)} \approx 0.7$ pm/V. Furthermore, when using the patterned electrodes on the upper surface, electrostrictive forces caused sinking of the periodic electrodes into the soft polymer stack during the poling process which led to a thickness

modulation of the films. These effects could be minimized but not totally avoided and were responsible for the extra propagation losses due to coupling to radiation fields. Overall, efficient QPM structures were found to be technically very challenging to realize with poled polymer materials. This caused us to consider the MDPM structures whose realization does not require poled, small structure patterning, but "only" the spincoating of layers with precisely controlled thickness.

MDPM devices

In the case of modal dispersion phase-matched devices, the experimental results have been very encouraging. The first samples were prepared with only one core layer of DANS to phase-match the fundamental TM_{00} with the $TM0_2$ mode of the harmonic. Although this modal combination did not produce an efficient transverse field overlap, these devices already exhibited more than an order of magnitude higher SHG efficiency and figure of merit than any of the previous QPM samples, i.e. 1 % / W / cm^2 [13].

To improve the spatial transverse overlap integral, we then used the high index, transparent polyetherimide polymer together with the DANS material to fabricate a two-layer core waveguide, phase-matching the TM_{00} mode at ω with the TM_{01} mode at 2ω. Again the figure of merit η jumped by an order of magnitude to reach 14 % / W cm^2 [14]. This maximum figure of merit was obtained with a short sample, 1 mm long, but it was demonstrated for longer samples that phase-matching could be maintained over 7 mm. Nevertheless, we estimate that the DANS nonlinearity in these structure was only 6 pm/V. We attribute this to the electrical conductivity mismatches between the layers.

The next step was to optimize the spatial overlap integral by using only active materials in the core of the waveguide, with successive, oppositely poled layers. This was done using DR based polymers with two different loadings of the chromophore and thus two different T_gs. The experimental SHG tuning data for samples 1 and 10 mm long are given in Figure 6 together with the theoretical curves. Note the good agreement between theory (which includes measured losses) and experiment in terms of the spectral dependence of the efficiency and the bandwidth narrowing, even for such long devices. This suggests that the main limitation is waveguide losses rather than waveguide inhomogeneity. Unfortunately, because of the severe absorptive losses at the second harmonic for these DR based devices relative to the previously discussed DANS ones, the figure of merit η obtained for a 1 mm long sample was still only 14 % / W cm^2 [15]. However, taking into account the high losses, a value η_0 of about 44 % / W cm^2 can be extrapolated, as shown in Figure 7. By comparison, the extrapolated value in the case of the previous DANS sample was only $\eta_0 = 18.6$ % / W cm^2. Altogether, those former samples, with an optimum length of 2 to 3 mm, were more efficient than the latter ones. This illustrates the critical importance of losses: they limit the useful device length and decrease the efficiency of the conversion process

To fully understand the SHG data, the experimental loss in the DANS polymer is given in Figure 8. (The absorption losses of the DR based samples were even higher around 800 nm.) This absorption was measured as film transmission losses near the absorption band and as waveguided propagation losses in the so-called "transparent" near-IR band and communications wavelengths. In this case the absorption cannot be fitted to a simple two level model which leads to a gross overestimate of the absorption coefficient in the tail of the absorption line. It is more accurately modeled by a Voigt profile taking into account the vibronic contributions, fitting the data with a dispersion model based on inhomogeneous broadening [12]. The peaks observed in the IR are essentially due to C-H vibrations band overtones.

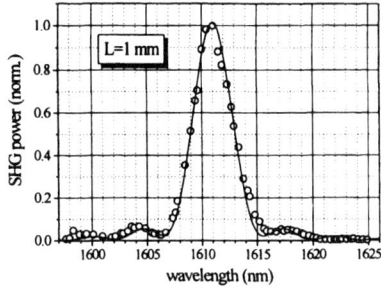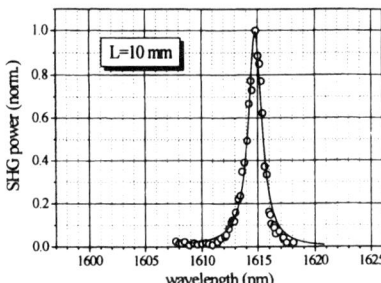

Figure 6 : SHG tuning curves, experimentally generated from a DR based device with three oppositely poled "-/+/-" layers, for (left) 1 mm and (right) 10 mm long samples.

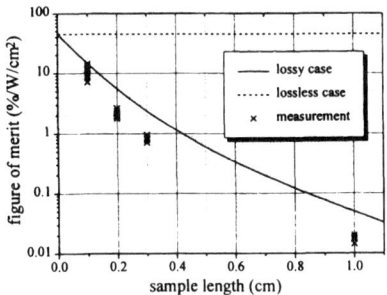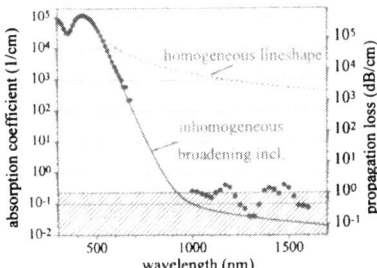

Figure 7 (Left) : FOM η as a function of length for the DR based devices. The strong decrease observed is due to the high chromophore absorption at the second harmonic.

Figure 8 (Right) : Absorption losses of DANS-SCP which were measured as film transmission in the visible and waveguide propagation losses in the IR. The were fitted using a Voigt profile. Note that although the absorption peaks around 430 nm, it decreases sharply, on the log scale, until 1 μm, and is still noticeably appreciable around 775 nm.

DISCUSSION

The progress that has been made in the field of "second harmonic generation at telecommunications wavelengths with poled polymer waveguides" is dramatic. This "young" field has gained several orders of magnitude in performance during the past years, especially the MDPM devices when compared to the QPM ones [17]. This is fairly natural if one considers the respective challenges in realizing multi-layer systems versus creating structures with strong nonlinear modulation while still avoiding linear losses.

However, this polymeric material family is still far from being used at its maximum potential and a lot can still be gained in SHG efficiency by developing and using chromophore doped polymers optimized for this specific IR application. As we shall see now, polymer devices, using their very favorable trade-off of optical nonlinearity versus absorption losses in the near infrared, have the potential of achieving far better performance.

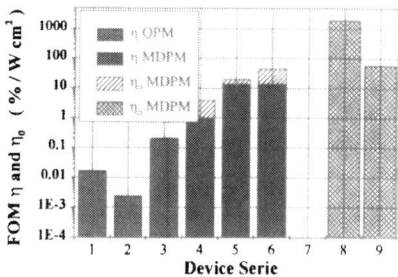 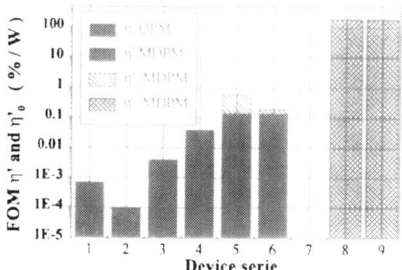

Figure 9 : Summary of the evolution of the FOMs for our series of samples; (1-3) are QPM structured devices using (1) periodic bleaching, (2) periodic poling with digitized electrodes on the top, (3) periodic poling with digitized electrodes on the bottom, (4-6) MDPM structured devices matching the $TM_{00}(\omega)$ to, (4) $TM_{02}(2\omega)$ using one core layer, (5) $TM_{01}(2\omega)$ using two core layer – nonlinear and linear "+/0", (6) $TM_{02}(2\omega)$ using three nonlinear core layers "-/+/-". Theoretical projections for (8) DANS polymer poled to 60 pm/V, and (9) new chromophore characterized by smaller non linearity, 10 pm/V, and lower losses, 2 dB/cm.

Material Requirements

The most important material requirement is to obtain low propagation losses and high nonlinearity simultaneously. The emphasis is therefore on low absorption in the red and near infrared regions of the visible spectrum (currently around 750 nm for operation at 1.55 µm and later at 650 nm for operation at 1.3 µm). Also needed is a high value for the product $\mu\beta$ of the molecular dipole moment μ (facilitating a high degree of molecular alignment by poling) and the hyperpolarisability β (needed for an efficient nonlinearity). What can be expected in terms of an increase in the net SHG efficiency (% / W) by improvements in these polymer characteristics is illustrated in Figure 10. The actual best experimental performance to date is given by curve (a) with parameters corresponding to our DANS device. Other curves (b, c, d) are given for a similar device with different material characteristics : (b) corresponds to lowering the losses; (c) to increasing the nonlinearity, and (d) to achieving both optimizations simultaneously. Curves (e) and (f) correspond to existing and theoretically optimized $LiNbO_3$ devices. Figure 11 shows that such high nonlinearity, as used in preceding calculations, have already been achieved with DANS polymer, using in-plane poling and DC electric field of 300 V/µm [12].

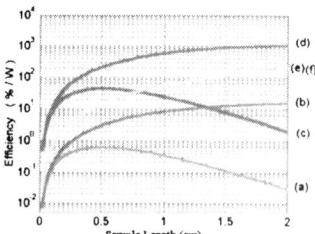 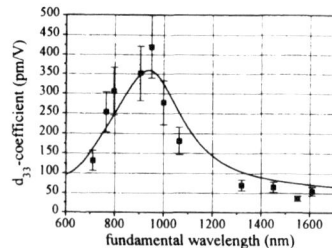

Figure 10 (left) : FOM η', as a function of length, for different sample characteristics based on one of our DANS device – MDPM $TM_{00} \rightarrow TM_{01}$; (a) existing device with $\alpha(\omega)$=5 dB/cm, $\alpha(2\omega)$=20 dB/cm, $d^{(2)}_{eff}$=6 pm/V; (b) hypothetical device optimized for low loss with $\alpha(\omega)$=1 dB/cm, $\alpha(2\omega)$=5 dB/cm, $d^{(2)}_{eff}$=6 pm/V; (c) hypothetical device optimized for nonlinearity with $\alpha(\omega)$=5 dB/cm, $\alpha(2\omega)$=20 dB/cm, $d^{(2)}_{eff}$=50 pm/V; (d) proposed device optimized for loss and nonlinearity $\alpha(\omega)$=1 dB/cm,

$\alpha(2\omega)$=5 dB/cm, $d^{(2)}_{eff}$=50 pm/V. For LiNbO$_3$: (e) existing device with $\alpha(\omega)$=0.4 dB/cm, $\alpha(2\omega)$=0.6 dB/cm, $d^{(2)}_{eff}$=30 pm/V at theoretical maximum; (f) hypothetical device optimized for loss with $\alpha(\omega)$=0.1 dB/cm, $\alpha(2\omega)$=0.2 dB/cm and $d^{(2)}_{eff}$=30 pm/V, indistinguishable from (e).

Figure 11 (right) : Nonlinear d_{33} coefficient of a DANS sample, in-plane poled using a field of 300 V/μm [12], showing that 50 pm/V is a reasonable value in the precedent figure.

A useful summary of the potential of polymeric doubling devices is given in Figure 12. In this "SHG losses - $d^{(2)}_{eff}$" plot, the lines of equal SHG efficiency, ranging from 0.1 to 1,000 % / W have been graphed using the structure corresponding to the DANS device which gave our best net efficiency. It is indicated by " O " on the graph. It is clear that gaining one order of magnitude in transparency and nonlinearity, which is achievable with polymeric materials, will allow us to gain about four orders of magnitude in net SHG efficiency, from about 1 to 10,000 % / W ! Such efficiencies would be several orders of magnitude greater than any other device based on inorganics. A second example based on the new material 4 [N-(2-hydroxyethyl)-N-(methyl)aminophenyl]-4'-(6-hydroxyhexyl sulfonyl) stilbene, named APSS (■ on the graph) [20] which exhibits a nonlinearity similar to DANS but in which losses are far less, should already give us an enhancement of nearly two orders of magnitude in net SHG efficiency. Harmonic loss is a critical issue!

Another area in which material development is important is to improve the availability of a wide range of inactive transparent polymers with different T_g. This will allow better compatibility of the conductivities for the active and passive polymers, and hence lead to better poling and nonlinearities. The optical dispersion and electric conductivity of these materials also needs to be well documented.

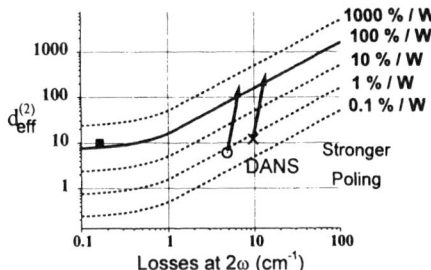

Figure 12 : "Iso-optimum efficiency" lines as a function of second harmonic propagation losses and effective nonlinearity. The corresponding device parameters are those of the MDPM TM$_{00}$→TM$_{01}$ using DANS, polyetherimide and PC polymers (see main text). Interesting points are highlighted : (O)   actual DANS based devices – projection using higher poling of DANS and (■) new "transparent" nlo chromophore.

SUMMARY AND CONCLUSIONS

Modal Dispersion Phase Matching (MDPM) is the technique of choice for waveguided SHG geometries. It takes advantage of the possibility of choosing waveguide structure parameters so that the fundamental spatial mode of a given frequency has the same effective optical index as a higher spatial mode of its second harmonic. In the case of polymeric waveguides, the devices are made of several layers deposited on a substrate. Technically, the phase matching is realized by precise control of the core layer(s) thickness and cm long phase-matched lengths have been experimentally demonstrated. To optimize the spatial mode overlap efficiencies, and to take into account that the field of adjacent lobes of high order spatial modes are π out of phase, the guiding part can be made of several nonlinear layers of opposite sign – or more simply of nonlinear and linear layers. It is worth pointing out that such structure can be as, or more efficient for a given material than those based on the classical Quasi Phase Matching (QPM) technique.

For polymeric materials, it is much easier to realize MDPM structures than the QPM ones. The former structures have proven experimentally to be far more efficient than the latter ones, with figures of merits now of a few 10s % / W cm^2, comparable to those obtained with the best ferroelectric based devices. However, the strong absorption tail of the dye we used, stilbene DANS or azo dye DR, induced large propagation losses at the second harmonic. Hence the maximum net efficiencies are obtained for samples a few mm long and are only of the order of 0.1 to 1 % / W . These optimum efficiencies scale as the square of the ratio of the effective nonlinearity to the propagation loss.

Optimization of the transparency of the material by a more suitable choice of the chromophores is currently being pursued. Furthermore, previous results were obtained with nonlinear coefficients of about 5 pm/V induced by 100 V/μm DC electric fields in a parallel plate poling geometry. Using the same material and in plane poling of a conductivity matched structure with a 300 V/μm DC electric field should allow us to induced nonlinearities of more than 50 pm/V. Preliminary results using such efficiently poled modal dispersion phase matched structures exhibited side band resonance as efficient as the previous main band resonance, indicating a very high net efficiency, at least one order of magnitude greater. Unfortunately due to a higher film refractive index than anticipated, the phase-matched wavelength was outside the tuning band of our color center laser. New phase matched structures are currently being processed.

Further progress in this field requires the synthesis and characterization of a large number of optically linear and nonlinear polymers. Their global potential for SHG, and simultaneously for other active operations such as electro-optics, etc, is being assessed. Polymer chemists should be able to make great contributions by modifying polymer structures to optimize their characteristics. Taking advantage of the best materials available and improving their operating characteristics (higher macroscopic nonlinearity, lower propagation losses at the second harmonic and fundamental wavelengths, optimized phase-matched guiding structures...) will allow the field to move on to potential applications. The ultimate pay-off should be the demonstration of the most efficient waveguided SHG devices to date at telecommunication wavelengths. New applications such as cascading of $\chi^{(2)}$ nonlinearities for optical processing and wavelength shifting for wavelength multiplexing communications will then be feasible. However, for this to happen, strong collaborative efforts between optical physicists and organic chemists are needed in order to develop new, targeted polymeric materials.

ACKNOWLEDGMENTS

The authors wish to thank the Heinrich Hertz Institute for preparation and characterization of part of the devices (DR based), Akzo, SANDOZ and MOEC for providing the raw materials, and also gratefully acknowledge US Air Force for financial support.

M. Canva is a scientific visitor from the French CNRS and gratefully acknowledges the French DGA for support through contract n° ERE 96-1101.

REFERENCES

1 S. Somekh and A. Yariv, *"Phase-matchable nonlinear optical interactions in periodic thin films."*, Appl. Phys. Lett., **21**, 140-141. (1972).

2 T. Suhara, T. Morimoto and H. Nishihara, *"Optical second-harmonic generation by quasi-phase-matching in channel waveguide structure using organic molecular crystal."*, IEEE Photon. Technol. Lett., **5**, 934-937, (1993).

3 M. Bortz, M. Fujimura and M. Fejer, *"Increased acceptance bandwidth for quasi-phase-matched second-harmonic generation in LiNbO₃ waveguides"*, Electron. Lett., **30**, 34-35, (1994).

4 Y. Azumai, M. Kishimoto, I. Seo and H. Sato, *"Enhanced SHG power using periodic poling of vinylidene cyanide / vinyl acetate copolymer."*, IEEE J. Quantum Electron., **30**, 1924-1933, (1994).

5 G. Khanarian, R. Norwood, D. Haas, B. Feuer and D. Karim, *"Phase-matched second-harmonic generation in a polymer waveguide."*, Appl. Phys. Lett., **57**, 977-979, (1990).

6 R. Norwood and G. Khanarian, *"Quasi-phase-matched frequency doubling over 5 mm in a periodically poled polymer waveguide"*, Electron. Lett., **26**, 2105-2106, (1990).

7 G. Rikken, C. Seppen, S. Nijhuis and E. Meijer, *"Poled Polymers for frequency doubling of diode lasers."*, Appl. Phys. Lett., **58**, 435-437, (1990).

8 M. Jäger, G. Stegeman, W. Brinker, S. Yilmaz, S. Bauer, W. Horsthuis and G. Möhlmann, *"Comparison of quasi-phase-matching geometries for second harmonic generation in poled polymer channel waveguides at 1.5 μm"*, Appl. Phys. Lett., **68**, 1183-1185 (1996).

9 Y. Shuto, T. Watanabe, S. Tomaru, I. Yokohama, M. Hikita and M. Amano, *"Quasi-phase-matched second-harmonic generation in diazo-dye-substituted polymer channel waveguides."*, IEEE J. of Quantum Electron., **33**, 349-357, (1997).

10 M. Flörsheimer, M. Küpfer, Ch. Bosshard, H. Looser and P. Günter, *"Phase-matched optical second-harmonic generation in Langmuir-Blodgett film waveguides by mode conversion"*, Adv. Mater. Commun., **4**, 795-798, (1992).

11 H. Ito and H. Inaba, *"Efficient phase-matched second-harmonic generation method in four-layered optical-waveguide structure"*, Opt. Lett. **2**, 139-141, (1978).

12 A. Otomo, M. Jäger, G. Stegeman, M. Flipse and M. Diemeer, *"Key trade-offs for second-harmonic generation in poled polymers"*, Appl. Phys. Lett., **69**, 1991-1993 (1996).

13 M. Jäger, G. Stegeman, G. Möhlmann, M. Flipse and M. Diemeer, *"Second harmonic generation in polymeric channel waveguides using modal dispersion"*, Electron. Lett., **32**, 2009-2010 (1996).

14 M. Jäger, G. Stegeman, M. Flipse, M. Diemeer and G. Möhlmann, *"Modal dispersion phase matching over 7 mm length in overdamped polymeric channel waveguides"*, Appl. Phys. Lett., **69**, 4139 (1996).

15 W. Wirges, S. Yilmaz, W. Brinker, S. Bauer-Gogonea, S. Bauer, M. Jäger, G. Stegeman, M. Ahlheim, M. Stähelin, B. Zysset, F. Lehr, M. Diemeer and M. Flipse, *"Polymer waveguides with inverted nonlinear layers for optimizing the overlap integral for modal dispersion phase-matched second-harmonic generation"*, Appl. Phys. Lett., **70**, 3347-3349 (1996).

16 M. Jäger, G. Stegeman, S. Yilmaz, W. Wirges, W. Brinker, S. Bauer-Gogonea, S. Bauer, M. Ahlheim, M. Stähelin, B. Zysset, F. Lehr, M. Diemeer and M. Flipse, *"Poling and characterization of polymer waveguides for modal dispersion phase-matched second harmonic generation"*, JOSA B, special feature on NLO materials, accepted for publication (1997).

17 M. Jäger, *"Efficient second harmonic generation in poled polymer waveguides using copropagating geometries"*, Dissertation, defended July 1997, CREOL – Univ. of Central Florida (1997).

18 These DR1 doped polymer based samples were prepared at the Heinrich Hertz Institute fur Nachrichtentechnik in Berlin.

19 Thermocarbon, Inc., specialized in diamond grinding technology for efficient dicing is located in Central Florida.

20 C. K. Park, J. Zieba, C. F. Zhao, B. Swedek, W. M. K. P. Wijekoon and P. N. Prasad, *"Highly cross-linked polyurethane with enhanced stability of second-order nonlinear properties"*, Macromolecules, **28**, 3713, (1995).

Recent Progress of Electro-optic Polymers for Device Applications

Alex K-Y. Jen[1], Qing Yang[1], Seth R. Marder[2], Larry R. Dalton[3] and Ching-Fong Shu[4]
1.Department of Chemistry, Northeastern University, Boston, MA 02115, ajen@lynx.neu.edu
2.Molecular Materials Resource Center, The Beckman Institute,California Institute of Technology, Mail Stop 139-74, Pasadena, CA 91125
3.Department of Chemistry, Loker Hydrocarbon Research Institute, University of Southern California, Los Angeles, CA 90089-1661
4.Department of Applied Chemistry, National Chiao-Tung University, 1001 Ta Hsueh Road, Hsin-Chu, Taiwan 30035

Introduction

Electro-optic (E-O) polymers have drawn great interest in recent years because of their potential applications in photonics devices such as high speed modulators and switches, optical data storage and information processing[1-2]. In order to have suitable materials for device fabrication, it is essential to design and develop polymeric material systems (active and passive polymers) with matched refractive indices, large E-O coefficients, good temporal and photochemical stability.[3-8] The E-O response of an active polymer commonly arises from the electric field induced alignment of its second-order nonlinear optical (NLO) chromophore, either doped as a guest/host system or covalently bonded as a side-chain. Because of the strong interaction among the electric dipoles, the poled structure is in a meta-stable state; the poled NLO chromophores which possess large dipole moment will tend to relax back to the randomly oriented state. As a result, the stability of the poled structure strongly depends on the rigidity of the overall material system. As it might be expected, the continuous increases of the rigidity and T_g of poled polymers imposes constraints on the selection of suitable chromophores that can survive the high-temperature poling and processing conditions. To circumvent this problem, we have developed a series of chromophores that possess conformation-locked geometry and perfluoro-dicyanovinyl-substituted electron-accepting group which demonstrate both good thermal stabilty and nonlinearity. This paper provides a brief review of these highly efficient and thermally stable chromophores and polymers for device applications.

Synthesis of NLO Chromophores with Enhanced Thermal Stability and Nonlinearity by Using a Conformation-locked *trans*-Polyene Approach

Recently, it has been shown that very large nonlinearities can be achieved by employing heteroaromatic rings which have lower aromatic stabilization energy upon charge separation,[9-10] or by using extended polyene π-bridge systems that have strong electron acceptors.[11] Incorporation of these chromophores into polymer matrices as guest/host and their subsequent alignment by an electric field have demonstrated very large electro-optic coefficients (r_{33}=45-55 pmV^{-1}) at a wavelength of 1.3 μm[12-13]. These values are significantly larger than the value of 31 pm/V for commercially available LiNbO₃. However, the long-term alignment stability of these E-O polymer systems was limited due to the low inherent thermal stability (< 175 °C) of these chromophores that prevents them to be processed and poled at higher temperatures. We have recently developed a facile approach to synthesize NLO chromophores with both enhanced nonlinearity and thermal stability. This synthetic method combines the advantages of using a thiophene ring and a triene as efficient conjugating moities for easier charge separation, and a 2,2'-dimethyl-propyl group connected 6-membered ring system that provides a conformation-locked geometry of *trans*- triene to prevent the thermally induced *cis-trans* isomerization. This allows us to synthesize a series of NLO chromophores with broad variations of electron acceptors to fine tune linear absorption property (λ_{max}) and βμ.

Comound $\underline{3}$ with a conformation-locked *trans*-triene bridge possesses very good thermal

Mat. Res. Soc. Symp. Proc. Vol. 488 © 1998 Materials Research Society

stbility at 225 °C. Furthermore, when **3** was doped (15 wt %) in a polyquinoline PQ-100 host and poled at 210 °C with a poling field of 1.4 MVcm^{-1}, the resulting polymer had an electro-optic coefficient, r_{33}, of 27 pm/V measured at 1.3 μm[14] consistent with the large molecular nonlinearity. The thermal stability of the poled polymer was demonstrated by heating the poled sample in a oven at 100 °C for over 1000 h. The E-O activity of the sample shows an initial drop to 75 % of its original value within 80 h then remains unchanged during this period of time.

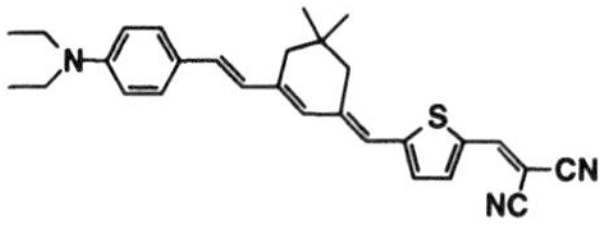

1

λ_{max}= 559 nm

βμ= 3450 x10^{-48} esu

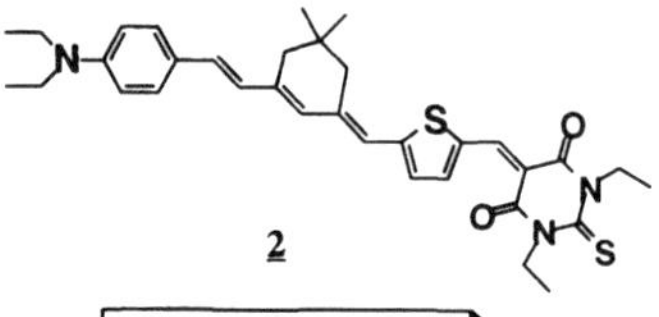

2

λ_{max}= 611 nm

βμ= 5380 x10^{-48} esu

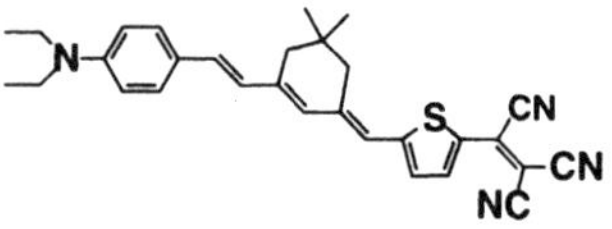

3

λ_{max}= 684 nm

βμ=13000 x10^{-48} esu

Using the conformation-locked bridge approach described above, we have developed a series of donor-acceptor substituted chromophores **4-6** with completely ring-locked polyene bridges. By quenching the lithiated product of 4-bromo-N,N-diethylamino aniline with 3-ethoxy-2-cyclohexenone and 7-methoxy-3,4,4a,5,6-pentahydronaphthalen-2-one, then condensation with malononitrile and 3-phenyl-5-isoxazolone gave compounds **4-6**, respectively. These chromophores possess high nonlinearity, good thermal, chemical and photochemical stability. These chromophores will be appropriately functionalized for attachment onto high temperature polymers for the evaluation of their E-O properties.

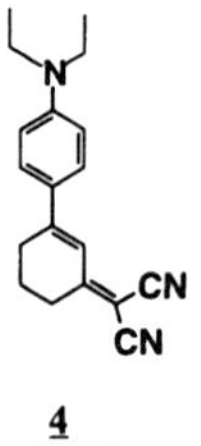

4

λ_{max}=461 nm

βμ=875 x10^{-48} esu

T_d= 240°C

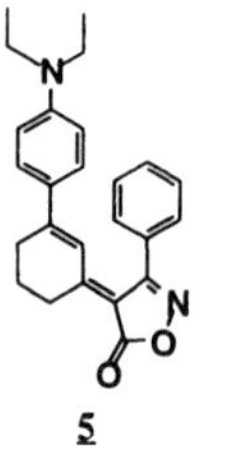

5

λ_{max}=502 nm

βμ=1050 x10^{-48} esu

T_d= 250°C

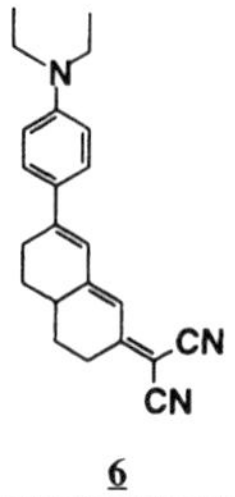

6

λ_{max}= 484 nm

βμ=1500 x10^{-48} esu

T_d= 270°C

Synthesis of Highly Nonlinear, Thermally Stable Perfluoropropyldicyanovinyl-substituted NLO Chromophores with Blue-shifted Absorptions

Earlier research has clearly shown that large molecular nonlinearities ($\beta\mu$) can be achieved by combining easily polarizable thiophene moiety and tricyanovinyl electron-acceptor in donor-acceptor substituted stilbenes. Although these chromophores exhibit impressive $\beta\mu$ values, they possess long-wavelength absorption (λ_{max}=600-750 nm) which contribute significantly to unacceptably high intrinsic optical losses for E-O devices operated at telecommunication wavelengths (1300 and 1550 nm). In addition, the α-CN in the tricyanovinyl group is very sensitive to nucleophiles at elevated temperatures which prevents these NLO chromophores to be used in high temperature, more reactive polymer matrices. Recently, Dalton's group reported the result from the study of a NLO chromophore that dispersed in poly methylmethacrylate (PMMA) which possesses pentadecafluoroheptyl-dicyanovinyl group as an electron-acceptor.[15] The preliminary results indicate that the chromophore has excellent solubility, high thermal stability, and a reasonable E-O coefficient from the high electric field poling study. These attributes, in conjunction with a blue-shifted absorption peak (ca. 50 nm lower than the corresponding tricyanovinyl substituted chromophore) make materials based on these perfluoroalkyl-dicyanovinyl acceptor very attractive for practical applications.

In order to develop NLO chromophores with optimized nonlinear, thermal, and linear absorption properties, we have incorporated a similar perfluoroheptyl-dicyanovinyl group into a series of dialkylamino- and diarylamino-substituted NLO dyes (Scheme 1). The results from Table I demonstrate the blue-shifted absorption characteristics compared to their tricyanovinyl-substituted counterparts. These chromophores also possess very good thermal stability (>250 °C) and solubility.

Table I
Comparison of λ_{max} for Heptafluoropropyl-dicyanovinyl- and Tricyanovinyl-substituted NLO Chromophores

	λ_{max} (nm) in dioxane		λ_{max} (nm) in dioxane
Et$_2$N—C$_6$H$_4$—C(CF$_2$CF$_2$CF$_3$)=C(CN)(NC)	489	Et$_2$N—C$_6$H$_4$—C(CN)=C(CN)(NC)	506
Ph$_2$N—C$_6$H$_4$—C(CF$_2$CF$_2$CF$_3$)=C(CN)(NC)	479	Ph$_2$N—C$_6$H$_4$—C(CN)=C(CN)(NC)	510
Et$_2$N—C$_6$H$_4$—CH=CH—(thiophene)—C(CF$_2$CF$_2$CF$_3$)=C(CN)(NC)	604	Et$_2$N—C$_6$H$_4$—CH=CH—(thiophene)—C(CN)=C(CN)(NC)	640
Ph$_2$N—C$_6$H$_4$—(thiophene)—C(CF$_2$CF$_2$CF$_3$)=C(CN)(NC)	541	Ph$_2$N—C$_6$H$_4$—(thiophene)—C(CN)=C(CN)(NC)	578
Ph$_2$N—C$_6$H$_4$—CH=CH—(thiophene)—C(CF$_2$CF$_2$CF$_3$)=C(CN)(NC)	561	Ph$_2$N—C$_6$H$_4$—CH=CH—(thiophene)—C(CN)=C(CN)(NC)	601

Synthesis and Characterization of High Temperature E-O Polyquinolines

Recently, we have explored a new optical polymer system, polyquinolines, for second-order NLO applications. The studies of their electro-optic properties have shown very promising results for both guest/host[12] and side-chain polyquinoline systems.[15] Our previous approach to NLO side-chain polyquinolines includes the synthesis for polyquinoline precursor polymers, and a post-tricyanovinylation of these polyquinolines to activate the side-chain NLO chromophores.[15] However, this approach is limited to the synthesis of side-chain polyquinolines with only tricyanovinyl containing chromophores. Here, we describe a generally applicable synthetic approach to side-chain NLO polyquinolines with a broad variety of high temperature NLO chromophores covalently attached onto the polymer backbones. The side-chain polyquinolines were synthesized by the direct polymerization of chromophore-containing bis(ketomethylene) monomers and bis(o-amino ketone) monomers. Their physical properties including electro-optic coefficients, thermal stability, temporal stability of the poling-induced polar order, etc. are also reported.

Thermal and chemical stability of the chromophores are critical for the development of high temperature side-chain NLO polymers. Recently, a new class of high temperature second-order nonlinear optical chromophores were developed through the replacement of the most reactive cyano-group in the tricyanovinyl-substituted derivatives with donor-substituted aryl units.[16] Excellent tradeoffs among optical, chemical, and thermal properties have been achieved in this class of nonlinear optical chromophores. The NLO side-chain polyquinolines were synthesized via polymerizations of monomers containing these chromophores with 4,4'-diamino-3,3'-dibenzoyldiphenyl ether. A high quality fluorinated polyquinoline moiety was introduced into the backbone of the side-chain co-polyquinolines in order to achieve the desired physical properties and the flexibility of adjusting the chromophore contents (**Scheme 1**). The stuctures and the purities of the side-chain polyquinolines were verified by ^{1}H NMR spectroscopy and elemental analysis. This simple synthesis is applicable not only to NLO polyquinolines with the side-chain chromophores described above, but also to the side-chain polyquinolines with many different chromophores, such as disperse red and DCM type of chromophores. Moreover, this methodology allows us to synthesize NLO side-chain polyquinolines with broad variations of polymer backbone to fine-tune the physical properties of the polymeric materials for device applications.

These side-chain polyquinolines also possess high glass transition temperature (T_g) and excellent thermal stability. Preliminary results through high electric field poling showed that these NLO side-chain polyquinolines exhibited large electro-optic coefficient (r_{33}) values (15-20 pmV^{-1} at 0.83 μm) and good temporal stability of dipole alignment at 100 °C.

In summary, we have developed efficient approaches to synthesize both highly efficient NLO chromophores and side-chain E-O polyquinolines. These methodologies offer great flexibility in choosing chromophores and the backbone structures of polyquinolines to fine-tune the physical properties of the polymeric materials for device applications. We are continuing in examining several aspects which may further improve the properties of the side-chain NLO polyquinolines.

Acknowledgments

Financial support from the AFOSR (F49620-97-1-0124) and the Center for Advanced Multifunctional Polymers and Assemblies, ONR (N00014-95-1-1319) are gratefully acknowledged.

Scheme 1

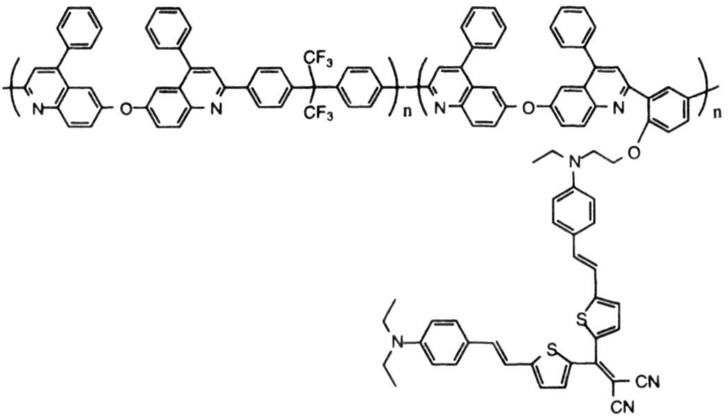

References

1. a) Marder, S. R., Kippelen, B., Jen, A. K-Y. Jen, and Pheyghambarian, N., *Nature*, **1997**, *388*, 845.
    b)Lindsay, G. A., Singer, K. D. Eds. *Polymers for Second-order Nonlinear Optics*; ACS Symposium Series No. 601; American Chemical Society: Washington, DC, 1995.
2. Dalton, L. R., Harper, A. W., Ghosn, R., Steier, H. W., Ziari, M., Fetterman, H., Shi, Y., Mustacich, R. V., Jen, A. K-Y., Shea, K. J. *Chem. Mater.* **1995**, *7*, 1060.
3. Wu, J. W., Valley, J. F., Ermer, S., Binkley, E. S., Kenney, J. T., Lipscomb, G. F., Lytel, R. *Appl. Phys. Lett.* **1991**, *58*, 225.
4. Twieg, R. J., Lee, V. Y., Miller, R. D., Moylan, C. R., Volksen, W., Knoesen, A., Hill, and R. H., Yankelevich, *Technical Digest*, **1995**, *21*, 279.
5. Verbiest, T.; Burland, D. M.; Jurich, M. C.; Lee, V. Y.; Miller, R. D.; Volksen, W. *Science* **1995**, *268*, 1604.
6. Yu, D.; Gharavi, A.; Yu, L.; *Macromolecules* **1995**, *28*, 784.
7. Chen, T.-A.; Jen, A. K-Y.; Cai, Y. M. *J. Am. Chem. Soc.* **1995**, *117*, 7295.
8. Wong, K. Y., Jen, A. K-Y., *J. Apply. Phys.* **1994**, *75*, 3308.
9. Dirk, C. W., Katz, H. E., Schilling, M. L., King, L. A.,*Chem. Mater.* **1990**, *2*, 700.
10. (a) Jen, A. K-Y., Rao, V. P., Wong, K. Y., Drost, K. J. *J. Chem. Soc. Chem. Commun.* **1993**, 90. (b) Jen, A. K-Y., Wong, K. Y., Rao, V. P., Drost, K. J. Cai, Y., *J. Electronic Mater.* **1993**, 1118.
11. Marder, S. R., Cheng, L-T. Tiemann, B. G., Friedli, A. C. Blanchard-Desce, M., Perry, J. W., Skindhoj, J. *Science* **1994**, *263*, 511.
12. Cai, Y. M.; Jen, A. K-Y. *Appl. Phys. Lett.* **1995**, *67*, 299.
13. Ahlheim, M., Barzoukas, M., Bedworth, P. V. Hu, J. Y., Marder, S. R. Perry, J.

W. Fort, Runser, A., Stahelin, C. M. Zysset, B.*Science*, **1996**, *271*, 335.

14.   a) Hsu, C. F., Tsai, W. J., Chen, J. Y., Jen, A. K-Y., Zhang, Y., Chen, T. A., *Chem. Commun.*, **1996**, 2279.

       b) Hsu, C. F., Tsai, W. J. and Jen, A. K-Y., *Tetrahedron Lett.*, **1996**, *37(39)*, 7055.

15   Wang F., Harper, A. W., He, M., Dalton, L. R., Garner, S. M., Yacoubian, A., and Steier, W.H., Polymer Preprints, **1997**, 38(1), 971.

16.   Chen, T.-A.; Jen, A. K-Y.; Cai, Y. M. *Chem. Mater.* **1996**, *8*, 607.

17.   Rao, V. P.; Jen, A. K-Y.; Cai, Y. M. *J. Chem. Soc., Chem. Commun.* **1996**, 1237.

CHARACTERIZING AND CIRCUMVENTING INTERMOLECULAR ELECTROSTATIC INTERACTIONS IN HIGHLY ELECTRO-OPTIC POLYMERS

AARON W. HARPER,[1] JINGSONG ZHU,[1] MINGQIAN HE,[1] LARRY R. DALTON,[1]* SEAN M. GARNER,[2] WILLIAM H. STEIER[2]

[1]Departments of Chemistry and of Materials Science and Engineering, University of Southern California, Los Angeles, CA 90089-1661
[2]Department of Electrical Engineering-Electrophysics, University of Southern California, Los Angeles, CA 90089-0483

Abstract

In general, polymers possessing non-resonant electro-optic activities exceeding 20 pm/V require chromophores with strong electron withdrawing groups (cyanovinyls, carbon acid moieties, etc.) as well as highly polarizable bridges. Although much progress has been made on designing and preparing materials with molecular "electro-optic" activities, their incorporation into polymers to show comparably large bulk electro-optic activities has been met with little success. We report here the mature of the difficulty of the translation of microscopic to macroscopic electro-optic activity. The optimization of molecular activity increases intermolecular electrostatic interactions between chromophores, and these interactions impede induction of polar acentric order in the polymers. Theoretical analysis of the problem is presented, as well as one example of a material that is designed to circumvent these interactions. The resulting material possesses electro-optic coefficients as high as 29 pm/V and optical losses as low as 1.5 dB/cm.

Introduction

For many electro-optic devices to reach commercial viability, their drive voltages (V_π) must be in the digital range. This in turn necessitates electro-optic coefficients on the order of 30 pm/V for organic polymeric materials. It is generally regarded that conventional electro-optic chromophores such as stilbene and azobenzene type chromophores possess insufficiently large molecular first hyperpolarizabilities to achieve this goal. In recent years tremendous progress has been made in optimizing the nonlinear optical response of organic molecules for electro-optic application, such that when incorporated into polymers in identical manner as with conventional materials electro-optic coefficients far exceeding 30 pm/V may be expected.*(1)* Unfortunately, incorporation of these chromophores into polymers to show the expected large electro-optic responses have been generally met with disappointingly poor results. We have recently shown that this is primarily due to intermolecular electrostatic interactions between chromophores.*(2)* In this communication we report our recent progress in the theoretical understanding of this phenomenon, its effect on the poling behavior of chromophores, and present results on a representative material that has been designed to circumvent these interactions.

Experimental

General. The polymer DH-TCV chromophore was prepared as shown in figure 1. The thermosetting polyurethanes were prepared as shown in figure 2. Full details of the synthesis of the chromophore and polymer will be published elsewhere. Optical spectra were obtained from a Perkin-Elmer Lambda-4C UV-Vis spectrophotometer or a Hitachi U-2000 spectrophotometer. Dipole moments were calculated from the concentration dependence of dielectric constant of solutions containing the chromophores, where 1,4-dioxane was used as the solvent and dielectric constants were measured at 100 kHz. Ionization potentials were estimated from the ionization potential-optical bandgap correlation of Tang *(3) et al.*, using absorption maxima in methanol. First hyperpolarizabilities were measured by the electric field-induced second-harmonic generation method,*(4)* in dioxane, and at a fundamental wavelength of 1907 nm. Static hyperpolarizabilities were calculated from the frequency-dependent values using the standard dispersion-correction equation.*(5)* Linear polarizabilities were determined from the concentration-dependence of the refractive index of the guest-host polymer films, as *per* the method of Cheng,*(6) et al.*

Preparation and processing of polymer films. The appropriate amounts of prepolymer and crosslinker were dissolved in 1,4-dioxane at a concentration of 7-10% (wt/vol), depending upon the viscosity of the resulting solutions. All solutions were tested for conductivity to ensure that they were free of ionic contamination. Solutions were then filtered through a 0.2 μm syringe filter. The solutions were heated at 80°C until spin cast films were of sufficient quality. The solutions were spin coated onto precleaned (water, methanol, acetone) indium-tin-oxide coated fused silica substrates. The films were then dried *in vacuo* for several hours to overnight. In all cases, there were no observable phase separation, as determined by optical microscopy and differential scanning calorimetry. Film thicknesses were measured by attenuated total reflectance (ATR) spectroscopy as well as by surface profilometry using a Sloan Dektak II surface profilometer. The films were corona-poled (6-8 kV potential applied to the corona needle at a tip-to-plane distance of 2 cm) at the temperature which corresponded to the largest thermally-dependent second-harmonic signal. After poling, the film were rinsed with deionized water to remove residual surface charges, and ATR measurements were performed within fifteen minutes. Electro-optic coefficients were measured at 1.06 μm by the ATR method modified for the determination of electro-optic activity.*(7)* Optical losses were measured at 1.3 μm by the waveguide immersion technique.*(8)*

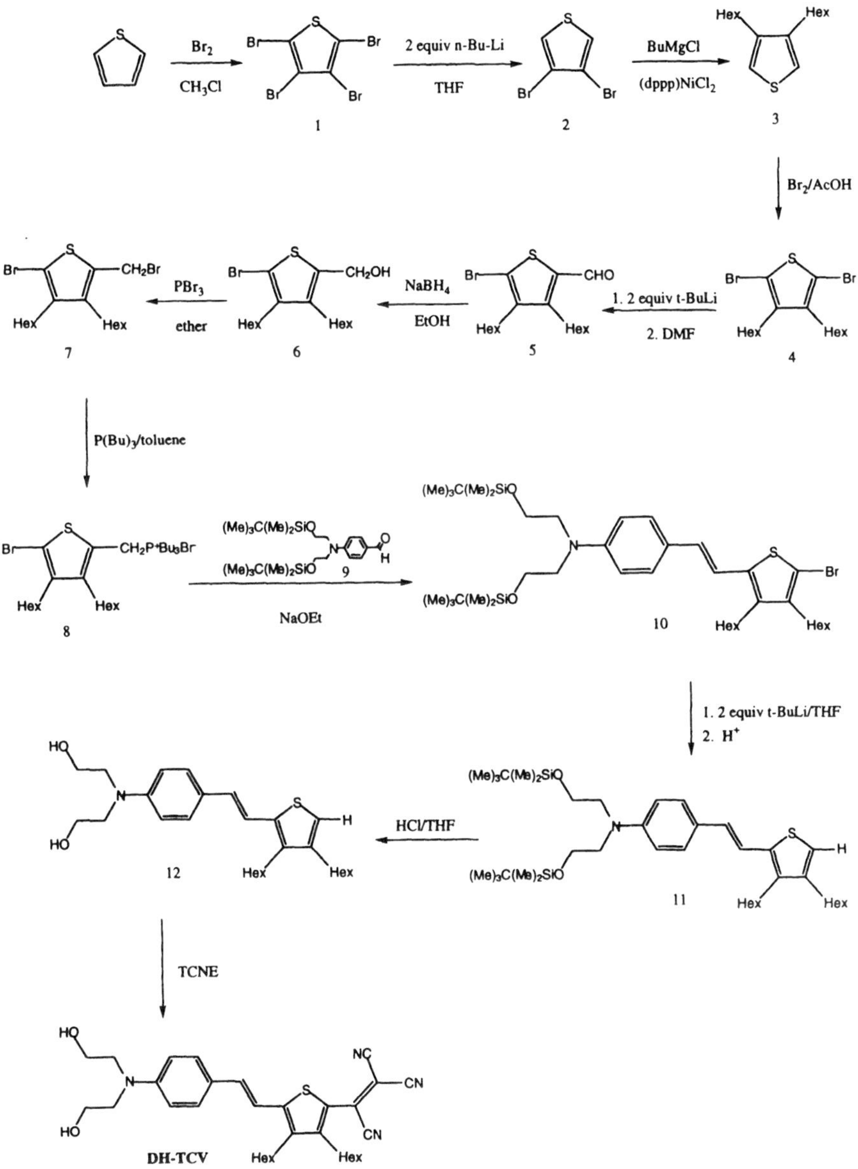

Figure 1. Synthetic scheme of chromophore DH-TCV.

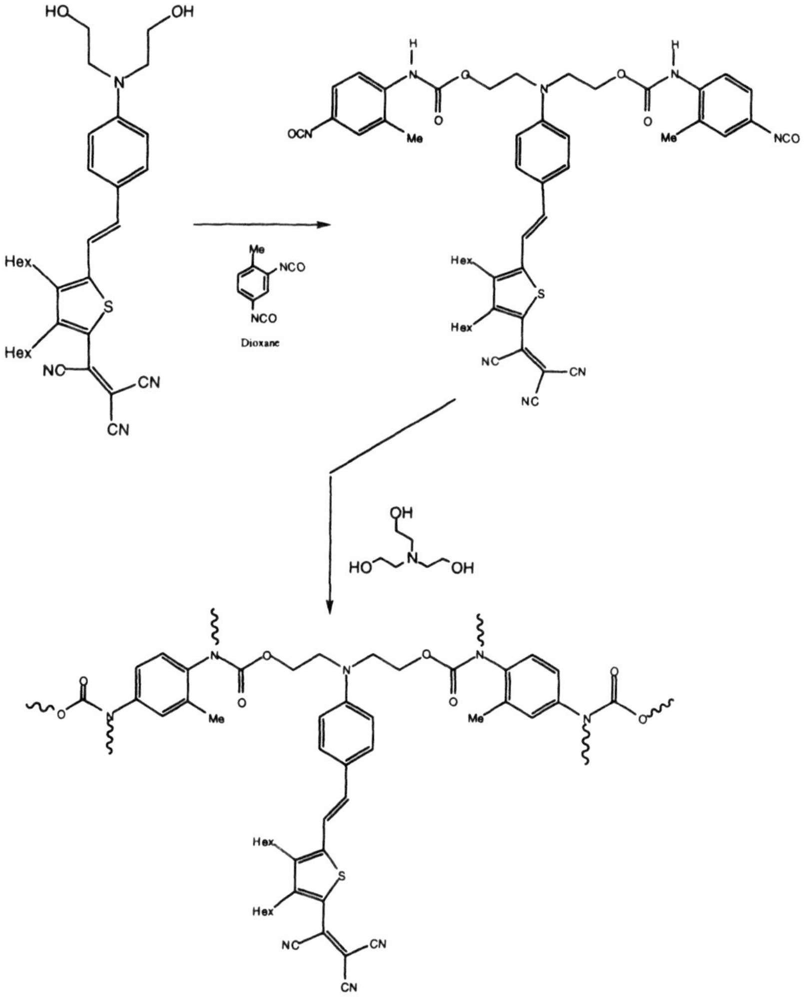

Figure 2. Preparation of thermosetting polyurethanes containing DH-TCV.

Results and Discussion

Role of intermolecular electrostatic interactions. To determine the impact on electrostatic interactions between chromophores, we have focused on the series of chromophores shown in figure 3. As can be seen, optimization of molecular optical nonlinearity results in the concurrent increase in the ground-state dipole moment and linear polarizability. To first order, the strengths of the interactions may be taken as the crystallization potential energy

$$W = -\frac{1}{R^6}\left[\frac{2\mu^4}{3kT(4\pi\varepsilon_0)^2} + \frac{2\mu^2\alpha}{(4\pi\varepsilon_0)} + \frac{3}{4}\alpha^2 I\right] \tag{1}$$

where R is the average interchromophore separation distance, and the other variables have their usual meanings. The terms in brackets denote (from left to right) the orientation, induction, and dispersion interactions, respectively. Using the values given in figure 3, equation 1 shows that the interaction energies between pairs of traditional chromophores (i.e., DMNA, DANS, and DR) are below dipole alignment energies and Boltzmann thermal

Compound	Structure	μ (Debye)	α (10^{-23} cm^3)	I (10^{-19} J)	β_0 (10^{-30} esu)
DMNA	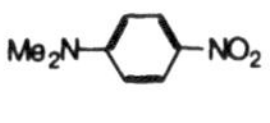 Me$_2$N—⟨ ⟩—NO$_2$	6.4	2.2	9.04	9.7
DANS	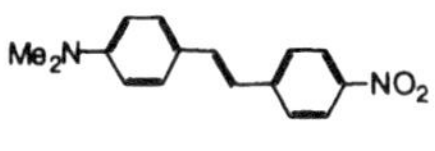 Me$_2$N—⟨ ⟩—⟨ ⟩—NO$_2$	6.6	3.4	8.70	55
DR	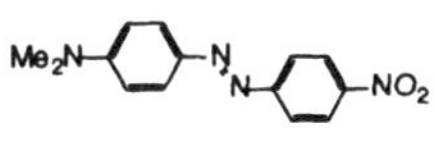 Me$_2$N—⟨ ⟩—N=N—⟨ ⟩—NO$_2$	7.0	3.8	8.26	55
ISX	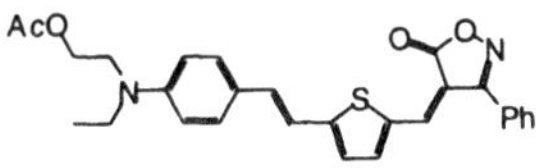 AcO... Ph	8.0	7.6	7.90	153
FDCV	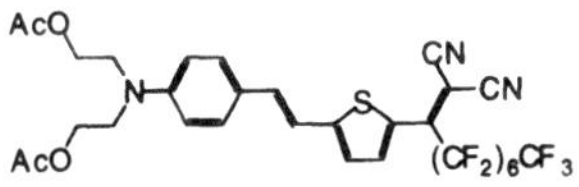 AcO... CN, CN, (CF$_2$)$_6$CF$_3$	9.0	10.3	7.77	350

Figure 3. Chromophores used in the theoretical study.

energies at interchromophore separation distances encountered in typical polymeric electro-optic materials containing these chromophores. On the other hand, intermolecular electrostatic interaction energies for the high-β chromophores (i.e., ISX and FDCV) are significantly greater than dipole alignment and thermal energies at the same separation distances. Modification of the theory (isotropic gas model) to include electrostatic intermolecular interactions leads to the following equation describing the polar order parameter $<\cos^3 \theta>$

$$\left\langle \cos^3 \theta \right\rangle = \frac{\mu f E}{5kT}\left[1 - L^2\left(\frac{W}{kT}\right)\right] \tag{2}$$

where L is the Langevin function, $L(x) = \coth(x) - 1/x$, the argument being the reduced crystallization potential energy. This modified expression for the polar order parameter may be substituted into equation 3, which will allow for the calculation of effective electro-optic coefficients as a function of chromophore concentration,

$$r_{eff} = \frac{2NF\beta}{n^4}\left\langle \cos^3 \theta \right\rangle \tag{3}$$

Plots of order parameter versus separation distance versus interchromophore separation distance, and the more familiar effective electro-optic coefficient versus chromophore concentration are shown in figure 4. From figure 4(a), it can be seen that as the chromphore's ground state dipole moment increases, greater potential polar order may be achieved, as expected. What is remarkable, however, is that attenuation of polar order occur as the interchromophore separation distance reaches a critical value, and that attenuation is rapid over a rather narrow range of separation distances. While the onset of attenuation of polar order does not occur for traditional chromophores until rather small values of R, attenuation occurs at relatively large separation distances (corresponding to rather low chromophore concentrations) for the high-β chromophores. The impact of this trend may be readily seen in figure 4(b), in that instead of the monotonic increase in electro-optic coefficient as chromophore concentration increases, there is a critical concentration (depending upon the chromophore) at which further increase in concentration results in a *decrease* in electro-optic coefficient. The inflections in the curves of figure 4(b) correspond to the chromophore concentration at which the relative strengths of the dipole alignment interaction, Boltzmann thermal randomization interaction, and collective intermolecular electrostatic interactions are in balance according to equation 2. At this point, increases in chromophore concentration result in the dominance of the electrostatic interactions over the other two interactions, and hence tend to drive the system toward centrosymmetric aggregation (either or both transient or nontransient) of chromophores. In this case, birefringence may still be observed, which denotes *axial* ordering of chromphores, although little or no *polar* order may be present.

In the simplest case, the apparent course of action would be to increase the average separation distance between chromphores. Keeping in mind that as two chromophores approach each other, intermolecular electrostatic interactions result in mutual attraction of the chromphores, and so that interacting chromophores may in many cases be better described by a separation distance corresponding to the relative strengths of electrostatic attraction and electrostatic repulsion (often derived from the hard-sphere approximation). Values of R in

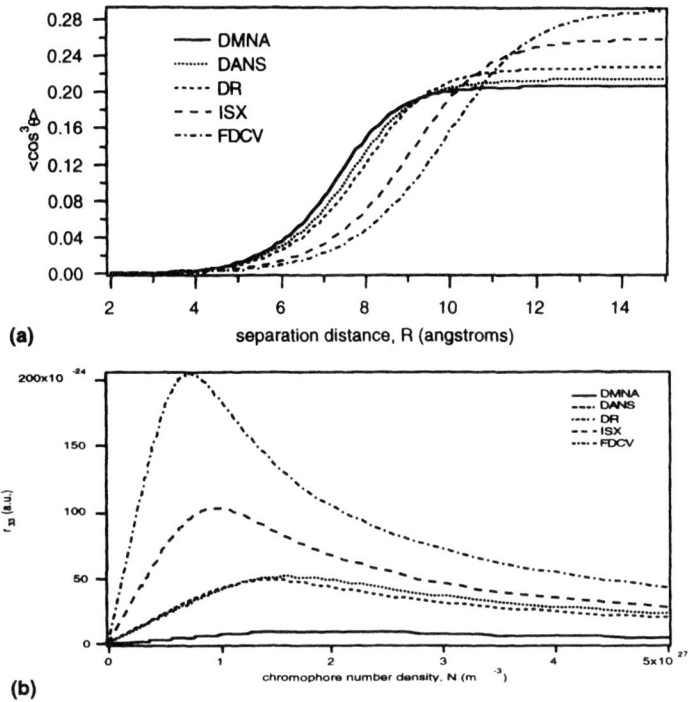

Figure 4. (a) Calculated polar order parameters versus intermolecular separation distance, and (b) calculated electro-optic coefficient versus chromophore concentration, for various electro-optic chromophores of figure 3. Calculations were performed for $T=100°C$ and $fE=2.5 \times 10^8$ V/m.

these cases are on the order of 3~5Å, despite average values of R dictated by chromophore concentration. At such close interchromophore separation distances, the impact of electrostatic interactions are severe. The simplest prescription, then, would be to increase the minimum chromophore separation distance in these "aggregates." Realizing that the approach of chromphores occurs normal to the molecular plane, derivatizing chromphores by placing bulky groups normal to the chromophore faces will increase the minimum values of R for the dimer ensembles, and result in greater poling efficiencies of materials containing the derivatized chromophores over the underivatized analogs. Provided that enough bulk is placed between two chromophores, even modest increases in chromophore separation may result in significant increases in effective electro-optic coefficients for the derivatized materials. Several examples of this prediction has been confirmed in the recent literature.*(9,10)*

Derivatization of a high-β chromophore to realize high electro-optic activities. The parent chromophore (underivatized version of DH-TCV) has a reported *(11)* molecular nonlinearity of $\mu\beta_{1907nm}$=6200x10^{-48} esu. Despite the high molecular "electro-optic" activity, comparably large bulk electro-optic activity has not been realized. As this chromophore is well known, we chose this parent chromophore are the subject of our focus. The synthesis of the derivatized chromophore, denoted DH-TCV, is shown in figure 1. AM1-calculated models of this chromophore indicate that the hexyl groups occupy space above and below the plane of the molecule, without distorting the π-framework of the molecule. Incorporation of this chromophore into the well-known polyurethane thermosetting system *(12)* (figure 2) resulted in materials that are viable candidates for device application. Optical spectra of several films of these materials are shown in figure 5. Although the absorption maxima of

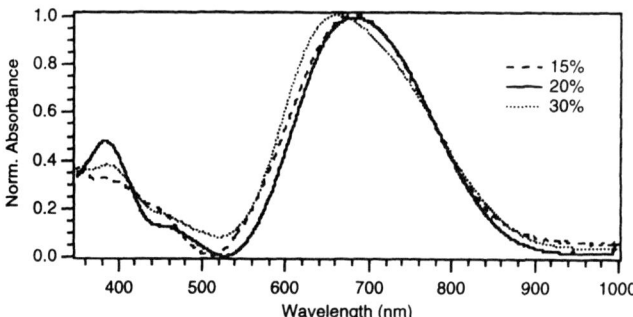

Figure 5. Optical spectra of PU-DH-TCV films.

these materials is at circa 685 nm, the absorption falls to zero before the measurement wavelength of 1064 nm. While this does not exclude all possibility of resonance enhancement of electro-optic data, such enhancement will not contribute greatly to the activity observed. Figure 6 shows the chromophore concentration-dependent electro-optic coefficients obtained for these materials. Interesting to note is that such loading densities of materials containing the underivatized chromophore resulted in intractable prepolymers.   In accord with the expectations given earlier, the plot of electro-optic coefficient versus loading density goes through a maximum, which we contend reflects the onset of the dominance of intermolecular electrostatic interactions above the corresponding loading density.

Dynamic thermal stability of nonlinear optical activity for one of these materials is shown in figure 7. As can be seen, the stability of polar alignment depends on the thermal treatment of the material during its curing stage. The optimized protocol resulted in a material with a dynamic stability of about 100°C, which is slightly greater than the stabilities obtained from the well-known PU-DR19 thermoset polyurethane.*(12)*

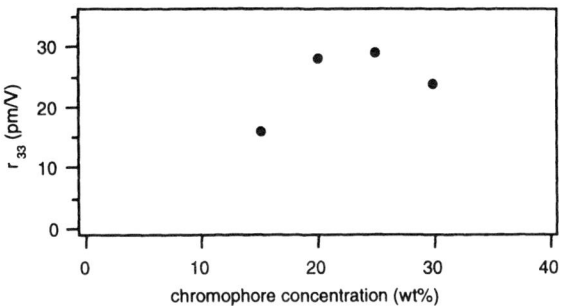

Figure 6. Electro-optic coefficient versus chromophore loading density for PU-DH-TCV materials.

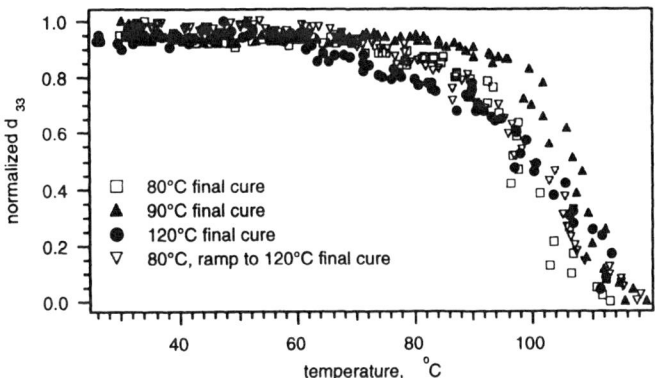

Figure 7. Dynamic thermal stability of nonlinear optical activity for PU-DH-TCV (30 wt%).

Conclusion

We have characterized various intermolecular electrostatic interactions for several representative electro-optic chromophores, using traditional London theory. Results showed that while conventional chromophores such as DANS and Disperse Red type chromophores possess interaction strengths below poling field strengths and thermal energies encountered during typical poling protocols, so-called high-β chromophores possess interaction energies significantly exceeding the dipole alignment and thermal energies. The impact of these interactions on the induction of polar order has been shown to possess strong dependence on the average separation distance between chromophores. By extension, the expected electro-optic coefficients as a function of chromophore number density was shown to go through a maximum. The inflection of such curves corresponds to the case where electrostatic intermolecular interactions exceed reduced dipole alignment energies. Furthermore, the role of molecular shape on the induction of electro-optic activity has been shown, in that prolate ellipsoidal shaped chromphores are expected to be more problematic than spherical shaped chromophores, relative to induction of polar order and electro-optic activity in materials containing these chromophores. Finally, measures to circumvent or reduce the level of electrostatic interactions between chromophores by chemical derivatization have led to materials possessing electro-optic activities on the order of 30 pm/V or greater (29 pm/V in the example shown here), and optical losses typically below 2 dB/cm (and as low as 0.75 dB/cm *(10)*), which are significant improvements over materials containing underivatized chromphores.

Acknowledgment

We would like to thank Professor Bruce H. Robinson (University of Washington, Seattle) for assisting in calculations involving chromophore shape dependencies on electro-optic activities. Financial support was provided by the Air Force Office of Scientific Research (F49620-94-1-0201, 94-1-0312, 95-1-0450, 96-1-0035, 97-1-0307, and 97-1-0491) and by the National Science Foundation (DMR-9528021).

References

1. Dalton, L.R.; Harper, A.W.; Ghosn, R.; Steier, W.H.; Ziari, M.; Fetterman, H.R.; Shi, Y.; Mustacich, R.V. *Chem. Mater.* **7**, 1060(1995).
2. Dalton, L.R.; Harper, A.W.; Robinson, B.H. *Proc. Natl. Acad. Sci. USA* **94**, 4842(1997).
3. Tang, T.B.; Seki, K.; Inokuchi, H. *J. Appl. Phys.* **59**, 5(1986).
4. Oudar, J.L.; Chemla, D.S. *J. Chem. Phys.* **66**, 2664(1977).
5. Lequan, M.; Lequan, R.M.; Ching, K.C.; Barzoukas, M.; Fort, A.; Lahoucine, H.; Bravic, G.; Chasseau, D.; Gaultier, J. *J. Mater. Chem.* **2**, 719(1992).
6. Cheng, L.T.; Tam, W.; Stevenson, G.R.; Meredith, G.; Rikken, G.; Spangler, C.W.; Marder, S.R. *J. Phys. Chem.* **95**, 10631(1991).
7. Chen, A.; Chuyanov, V.; Garner, S.; Steier, W.H.; Dalton, L.R., in *Organic Thin Films for Photonic Applications,* Vol. **14**, OSA Technical Digest Series (Optical Society of America, Washington DC, 1997), 158.
8. Teng, C. *Applied Optics* **32**, 1051(1993).

9. Harper, A.W.; He, M.; Wang, F.; Chen, J.; Zhu, J.; Sun, S.S.; Dalton, L.R.; Chen, A.; Garner, S.M.; Yacoubian, A.; Steier, W.H.; Chen, D.; Fetterman, H.R., in *Organic Thin Films for Photonic Applications,* Vol. **14**, OSA Technical Digest Series (Optical Society of America, Washington DC, 1997), 232.

10. Wang, F.; Ren, A.S.; He, M.; Harper, A.W.; Dalton, L.R.; Garner, S.M.; Zhang, H.; Chen, A.; Steier, W.H. *Proc. Polym. Mater. Sci. Eng.* **78**, *in press.*

11. Jen, A.K.-Y.; Cai, Y.; Bedworth, P.V.; Marder, S.R. *Adv. Mater.* **9**, 132(1997).

12. Chen, M.; Dalton, L.R.; Yu, L.P.; Shi, Y.Q.; Steier, W.H. *Macromolecules* **25**, 4032(1992).

ELECTRONIC PROPERTIES OF A NOVEL CLASS OF CONJUGATED SYSTEMS : TRANSITION METAL SUBSTITUTED OLIGOTHIOPHENES

L. LANCELLOTTI*, R. TUBINO*, S. LUZZATI**, E. LICANDRO***, S. MAIORANA***, A. PAPAGNI***
*Dipartimento di Scienza dei Materiali e INFM, Universita' di Milano, Milano, Italy
**Istituto di Chimica delle Macromolecole e INIMITER, CNR, Milano
***Dipartimento di Chimica Organica e Industriale, Universita' di Milano, Milano

ABSTRACT

We report on a spectroscopical study of a novel class of push-pull molecules containing a Chromium or Tungsten atom connected to an oligothiophene through a carbenic bond.
The electronic coupling between the π electrons of the conjugated system and the d electrons of the transition metal has been monitored through absorption and Raman spectra.
This interaction between d and π electrons leads to a red shift of the thienylene π-π* absorption band and to the appearance of a new strong metal to ligand charge transfer band at lower energies. In this latter transition the electrical dipole reverses its direction upon photoexcitation from the ground to the first excited state, thus accounting for the enhanced non linear optical response of these molecules. Solvatochromic effect has been used to estimate the second order molecular hyperpolarizability.

INTRODUCTION

The incorporation of a transition metal in a conjugated structure is particularly interesting because enhanced nonlinear activity is expected due to the the induced asymmetry in the charge distribution [1].
Two key ingredients are necessary to achieve large second order molecular hyperpolarizabilities, namely i) low-lying electronic transitions possessing large oscillator strengths (μ_{eg}^2) with ii) associated large changes of the dipole moment between the ground and the excited state (μ_e-μ_g) [2]:

$$\beta \propto \mu_{eg} (\mu_e\text{-}\mu_g) \tag{1}$$

In consideration of Eq. 1 chemists have syntetized a variety of strongly absorbing conjugated structures with attached electron-donors and acceptors in order to produce asymmetric charge distributions and this approach has been successful in engineering molecules with enhanced second order nonlinearities.

In this context investigation on organometallic systems have been greatly intensified [3]. Incorporation of metal in a conjugated structure introduces new flexibility in the molecular design because transition metals can exhibit different oxidation states and a variety of ligand environments. Moreover the high polarizability of the d electrons, ensures nonlinear activity for conjugated molecules to which transition metal compounds are attached.
In this paper we report on the spectroscopical manifestation arising from the interaction between the electrons of a conjugated system and the d electrons of a transition metal, as well as an estimate of the second order molecular hyperpolarizability using the solvatochromic effect. In particular, among the stable and isolable metal complexes, our attention has been focussed on the Fischer-type carbene complexes [4], which are organometallic compounds containing a VI group transition metal (Cr, Mo, W) of general structure depicted in Scheme 1.
Because of the six dative bonds the metal atom in the ground state is negatively charged so there is a large dipole moment from the transition metal to the carbenic bond. This situation is expected to be reversed in the first excited state which usually corresponds to the transition from an occupied d orbital of the metal to a p_z empty carbene orbital followed by the charge delocalization throughout the oligothiophenic system.

The nature of this low-lying Charge Transfer (CT) transition and the modification of the π electronic structure in the oligothiophene system induced by the metal substitution have been studied by using optical and Raman spectroscopy.

Scheme 1

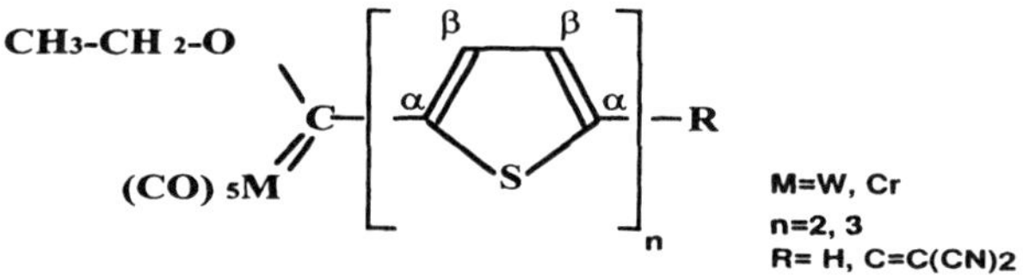

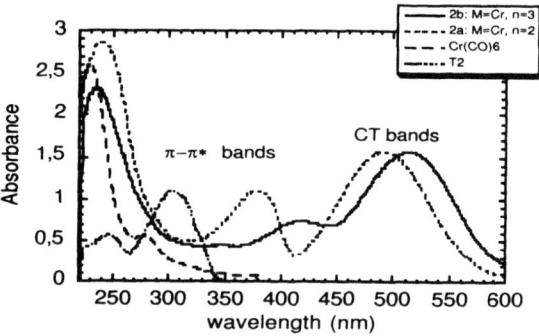

EXPERIMENTAL

The synthesis of the metal substitued oligothiophenes studied in the present work (see Scheme 1) has been described in detail in previous publications [5].
The compounds have been dispersed in Kbr pellets for the IR absorption measurements while UV-Vis absorption , Raman measurements have been carried out in solution. Dichloromethane has been used as a solvent for the Uv-Vis absorption spectra and carbondisulfide for Raman spectra.
The UV-Vis absorption spectra have been measured with a Cary 2300 spectrophotometer. A Bruker FTIR IFS66 has been used for IR absorption and for the Raman spectra obtained with a 1064 nm Nd-Yag laser excitation.The Raman spectra with excitation in the visible region have been obtained with the laser lines of an Argon-ion laser, using a flat field triple monochromator with an intensified diode array detector (Jasco TRS300 monochromator and 1420 EG&G detector).

RESULTS

Electronic absorption spectra

Figure 1 shows the electronic absorption spectra of the organometallic complex with M=Cr and n=2,3 together with those of the Cr(CO)$_6$ and of the bithiophene (T$_2$).

Fig. 1 Absorption spectra of : M= Cr, n= 2, 3 complexes; T$_2$.; Cr(CO)$_6$.

The comparison between these spectra shows that the formation of the organometallic molecule is monitored by the appearance of a new absorption band at 482 nm. This band, which is assigned to the metal to ligand charge transfer transition (MLCT) from the Chromium to the conjugated system, is the origin of the intense colour of the complex ($Cr(CO)_6$ and T2 are colourless). The electronic spectrum shows also an increase of the conjugation length, with respect to the oligothiophene, detected from the red shift of the π-π^* transition from 310 nm to 385nm in the case of the bithiophene complex.

The high-energy part of the spectra, which is not particularly interesting in this context , results from the overlap of many ligand to ligand transitions.

Fig. 1 also shows the effect of the increasing the number of thiophene rings on the electronic absorption spectrum. Both the π-π^* and the Charge Transfer transitions undergo a batochromic shift from the bithiophene to the terthiophene complex. This is undubtedly due to the conjugation, which increases the stability (thus lowering the energy) of the π^* state.

Finally the substitution of the chromium with the tungsten atom has little effect on the electronic absorption spectrum of the organometallic complex. A slight red shift of the electronic transitions occur in the cyano-substituted compound.

Raman spectra

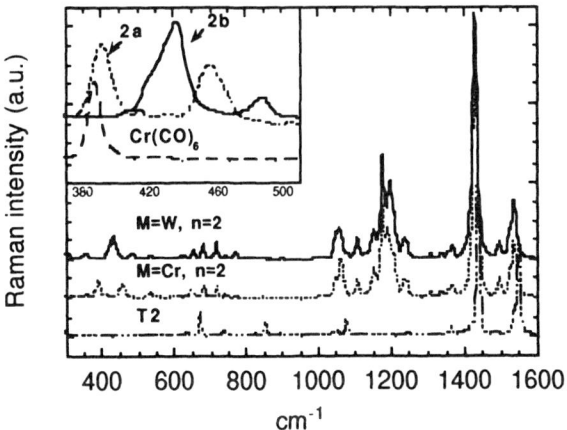

Fig. 2 Raman spectra of 2a : M= Cr, n=2; 2c : M=W, n=2; T_2. Inset : low frequency spectra of $Cr(CO)_6$ and the metal complexes 2a, 2c. Excitation wavelength : 1064 nm.

Fig. 2 shows the off-resonance Raman spectra of T2 and of the organometallic compounds with n=2 and Me=Cr,W. The spectra are normalized to the intensity of the dominating C=C stretching mode of the thienylene fragment (at 1440-1430 cm^{-1}). There are a number of new bands in the organometallic compounds which are related to the metal-carbonyl fragment as well as to the OEt group. The spectral features in the 350-550 cm^{-1} region are mostly due to metal-carbonyl modes while the C-C stretching modes of the OEt group appear in the 1200 cm^{-1} region.

The spectra exhibit several bands which can be ascribed to the break of symmetry of the metal carbonyl fragment and of the thienylene molecule due to the formation of the organometallic compound. Such interpretation is supported by the presence of many modes which are both Raman and IR active. The reduced symmetry of the thienylenic fragment is indicated in the Raman spectra by the presence of a band at 1539 cm^{-1}, corresponding to a mode which is only IR active in unsubstituted T2 (the B_u antisymmetric C=C stretching T2 mode [6]

The inset of Fig. 2 shows the metal-carbonyl stretching region of the n=2 organometallic compounds. For Cr, two main bands are observed at 393 cm^{-1} and at 465 cm^{-1} which are assigned respectively to Cr-CO stretching and to Cr-carbene stretching modes [7]. The blue shift of these bands in the W complex indicates higher force constants and thus a higher strength of the W-C bond with respect to the Cr-C bond. This accounts for the higher stability of W complex.

The red shift of the electronic absorption bands upon complex formation is responsible for the preresonance enhancement, with the 1064 nm excitation, of the C=C stretching mode of the thienylenic fragment which has been inferred by normalizing the Raman bands to the intensities of the C-H stretching modes, which are not coupled to the electronic transitions.

Figure 3 compares the Raman spectra obtained at 1064 nm and 514 nm respectively; normalized to the thienylene symmetric C=C stretching band intensity.

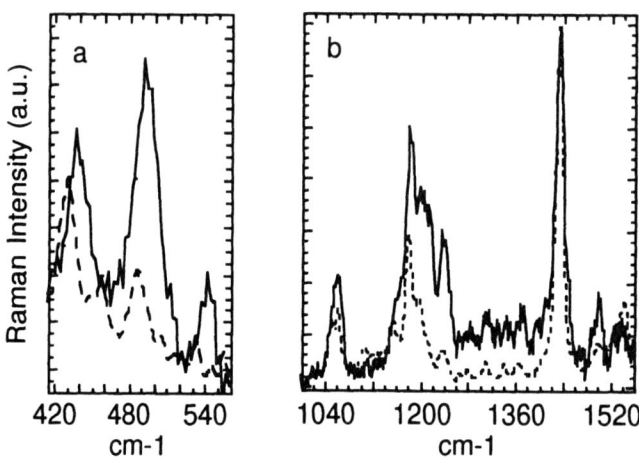

Fig. 3 Resonance (solid line) and off-Resonance (dotted line) Raman spectrum of M=Cr; n=3.

Fig. 3/a shows the region of the modes involving the metal atom. It can be seen that the mode which is mostly coupled to the CT transition is the W-carbene stretching at 493 cm^{-1}. This is consistent to the fact that the atoms mostly involved in this vibration are respectively the donor and acceptor of the charge transfer process. Fig. 3/b shows the 1150-1550 cm^{-1} region. It can be seen that, upon excitation under the CT transition, some thienylene bands and notably the C-H bending at 1240 cm^{-1} and the 1499 cm^{-1} antisymmetric C=C stretching, are enhanced. The broad band around 1200 cm^{-1} is also enhanced probably because of a contribution of the thienylene C-C interring stretching mode [8]. The coupling of these modes to the CT transition is probably due to the π delocalization of the charge along the conjugated system.

Second order molecular polarizability

On the basis of the previously discussed results which have shown the reversal of the dipole moment occur upon excitation in the CT band, Eq. 1 predicts that metal substituted oligothiophenes should exhibit a good second order non linear response.

Assuming a simple two state model the hyperpolarizability component along the molecular axis x is given by :

$$\beta xxx\,(2\omega) = \frac{6\pi^2 \mu_{eg}^2 \left(\mu_e - \mu_g\right)\omega_{eg}^2}{h^2\left(\omega_{eg}^2 - \omega^2\right)\left(\omega_{eg}^2 - 4\omega^2\right)} \qquad (2)$$

where ω_{eg} is the frequency of the charge transfer transition. The various term appearing in Eq. 2 have been evaluated as follows: ω_{eg} and μ_{eg} are measured from the electronic absorption spectrum of a solution of known concentration; μ_g has been measured using an impedance bridge connected with a cylindrical capacitor in which the space between the electrodes (kept at a distance of one mm   apart with teflon spacers) is filled with the solution. From   the Guggenheim- Debye equation [9], a value μ_g= 16.51 x 10^{-30} C · m has been obtained.

One of the experimental techniques to evaluate μ_e is to detect the changes of the transition energy upon changing the solvent polarity (solvatochromism) [10,11]. These changes, which are not related to conformational effect, arise from the interaction between the electrostatic reaction field (**F**) created by the polar molecules of the solvent and the molecular dipole in the excited state (μ_e) The physical origin of this effect is related to the fact that the solvent molecules cannot rearrange, in 10^{-15} s, in response to the change of the molecular dipole produced by the transition and a shift in absorption energy occur ($\Delta E= -(\mu_e - \mu_g) \bullet F$).

Assuming that the dipole moments of the ground and excited states are parallel and neglecting the dipole induced from the solvent molecules the solvatochromic shift can be written as (McRae equation [12]):

$$\omega_{eg}^s - \omega_{eg}^0 = A\left(\frac{n^2-1}{2n^2+1}\right) + B\left(\frac{\varepsilon-1}{\varepsilon+2} - \frac{n^2-1}{n^2+2}\right) \qquad (3)$$

$$B = \frac{\mu_g\left(\mu_g - \mu_e\right)}{e_0 h a^3} \qquad (4)$$

where ω_{eg}^s is the transition frequency in a given solvent and ω_{eg}^0 is the transition frequency in vacuum; ω_{eg}^0, A and B can be determined by measuring ω_{eg}^s in various solvents of known dielectric constant and refractive index.

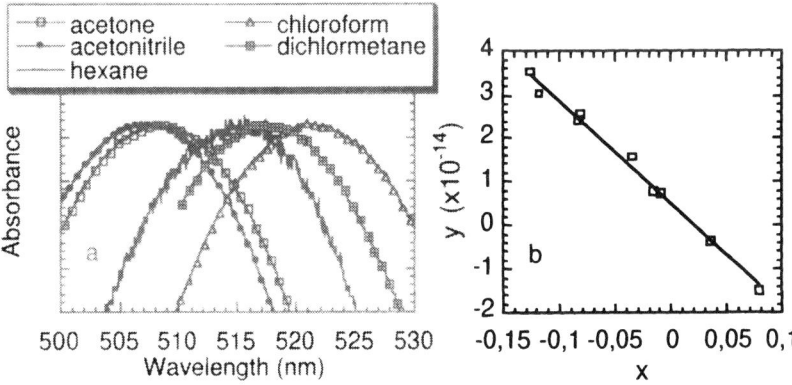

Fig. 4 Solvatochromic effect in metal-substitued oligothiophenes. a: CT absorption maxima in solvent of various polarity; b: fit to the linearized McRae equation.

The results are shown in Fig. 4a which depicts the solvatochromic shift in solvents of various polarity. Fig. 4b shows the excellent fit of the experimental data to Eq. 3, where the variables y and x are properly chosen in order to obtain a linear relation. In order to determine μ_e from the fitting of the MacRae equation, the radius a of the spherical cavity occupied by the molecule in the dielectric has to be known. A common used assumption is $a = 0.7 L$, where L is the total length of the molecule. This procedure yields $\mu_g - \mu_e = 39.74 \times 10^{-30}$ C·m. The data previously obtained inserted in Eq. 2 yield a $\beta_{xxx} = 7 \times 10^{-28}$ esu at 1064 nm and 1×10^{-28} esu at 1300 nm. This indicates that these molecules show a significant non linear second order response with a strong enhancement at 1064 nm due to preresonant conditions ($\omega \approx \omega_{eg}/2$).

CONCLUSIONS

The possibility of linking transition metal atoms with a conjugated oligothiophene system offers new possibilities of molecular design and applications in the crowded research area of the thiophene-based materials for non linear optics. In fact, by changing the transition metal, its oxidation state, the thiophene substituents and the coordination pattern, the second order non linear optical response can be tuned.
A key role in determining the optical (both linear and non linear) properties of this novel and promising class of organometallic compounds is played by the interaction between the π electrons of the conjugated system and the d electrons of the transition metal atom. This interaction, resulting from the large overlap between the two electronic systems through the carbenic link, has two main effects namely the appearance of a strong low-lying charge transfer transition and the disruption of the symmetry of the electronic charge distribution in the oligothiophenic system. Both these effects, which have been investigated by using various spectroscopical techniques, appear to be essential in enhancing the molecular hyperpolarizability which has been measured using the solvatochromic effect.

REFERENCES

1. Ch. Bosshard, K. Sutter, Ph. Pretre, J.Hulliger, M. Florshemeier, P.Kaatz e P. Gunter in Advances in nonlinear optics-Vol.1. Organic nonlinear optical materials Gordon and Breach Publishers, 1995.
2. T. Yoshimura, App. Phys. Lett. 55, 534 (1989).
3. N. Long, Angew. Chem. Int. Ed. Engl. 34, 21 (1995).
4. E.O. Fischer et al., Transition Metal Carbene Complexes, P. Seyferth Ed.,1983.
5. S. Maiorana, A. Papagni, E. Licandro, A. Parsoons, K.Clay, S. Houbrechts and W. Porzio Gazz. Chim. It. 125, 377 (1995).
6. G. Louarn, J.P. Buisson, S.Lefrant e D. Fichou, J. Phys. Chem.99, 11399 (1995).
7. D. M. Adams, Metal-Ligand and related vibrations, Edward Arnold Ltd., London, 1967.
8. Y. Furukawa, M. Akimoto and I. Harada, Synth. Met. 18, 151 (1987).
9. E.A. Guggenheim , Trans.Faraday Soc. 45, 714 (1949).
10. L. Cheng, W. Tam, S. H. Stevenson, G. R. Meredith, G. Rikken and S.R. Marder, J. Phys. Chem. 95, 10631 (1991).
11. M.S.Paley, J.M. Harris, H. Looser,J.C. Baumert, G.C.Bjorklund, D. Jundt and R.J. Twieg, J.Org. Chem. 54, 3374 (1989).
12. E.G. McRae, J. Phys. Chem. 61,562 (1957).

NEW PHOTOPOLYMERS BASED ON TWO-PHOTON ABSORBING CHROMOPHORES AND APPLICATION TO THREE-DIMENSIONAL MICROFABRICATION AND OPTICAL STORAGE

B.H. CUMPSTON[†,§], J.E. EHRLICH[†], L.L. ERSKINE[§], A.A. HEIKAL[†,§], Z.-Y. HU[§], I.-Y.S. LEE[†,§], M.D. LEVIN[§], S.R. MARDER[†,§], D.J. McCORD[§], J.W. PERRY[†,§], H. RÖCKEL[§], M. RUMI[†,§], AND X.-L. WU[†,§]

† Jet Propulsion Laboratory, California Institute of Technology, Pasadena, CA 91109
§ Beckman Institute, California Institute of Technology, Pasadena, CA 91125

ABSTRACT

Molecules exhibiting strong two-photon absorption hold great potential for a wide range of applications including two-photon fluorescence imaging, three-dimensional (3D) optical data storage, and 3D microfabrication. We have observed two-photon absorptivities as high as 1500×10^{-50} cm^4 s/photon in bis-donor diphenylpolyene derivatives that are correlated to simultaneous charge transfer from the end groups to the polyene bridge in the molecule. Many of these molecules are also excellent photoexcitable electron donors that can initiate charge-transfer reactions with acrylate monomers. Marcus theory is used to describe the efficiency of these charge-transfer reactions. Polymerization rates have also been measured and we show that these two-photon chromophores display increased sensitivity and recording speed over conventional UV photo-initiators. The fabrication of complex, three-dimensional structures by two-photon polymerization is demonstrated and discussed in the context of advanced photonic applications.

INTRODUCTION

Many applications in imaging, data storage, and lithography can benefit from the non-linear process of two-photon absorption. The advantages lie chiefly with the small excitation volumes that can be produced and the ability to excite in the near-infrared where many organic compounds have no absorption. Low absorption minimizes unwanted photochemical processes and optical fatigue upon optical readout. For instance, two-photon excitation allows one to polymerize very small volumes near the focus of a laser beam creating a change in refractive index. This index change allows for data storage densities of 10^{12} bits/cm^3 [1]. The greatest advantage in the use of two-photon excitation comes from an extremely small excitation depth in the longitudinal direction. Two-photon absorption falls off as z^4 in this direction since the process depends on the square of the excitation intensity. Confinement is smaller than that achieved by single-photon excitation in the lateral dimension as well, but only by a factor of $1/\sqrt{2}$.

The exceptional volumetric confinement afforded by two-photon excitation also presents unique opportunities in three-dimensional microlithography or microfabrication. This technique has been previously presented by Webb and coworkers [2, 3]. The Webb group has demonstrated 0.5 μm trenches with an aspect ratio of 8 in polyimide films as well as trenches with an undercut profile. They have also used two-photon polymerization of acrylate systems to write letters with linewidths on the order of 1 μm. Maruo *et al.* have fabricated a helical structure with a diameter of 6 μm and a pitch of 10 μm [4].

Unfortunately, little work has been done to synthesize, by design, molecules with large two-photon absorptivities. Photopolymerization by the two-photon process, as discussed above, has generally used commercially available one-photon photopolymer systems, with no effort to optimize the materials system. Recognizing that the class of bis-donor-substituted molecules with extended conjugation should have strong two-photon absorption, our group has synthesized and characterized many compounds of this type. Structure/property studies have allowed for a greater understanding of the chemical properties leading to strong two-photon absorption. Also, control of the chemical structure of these compounds affords wavelength tunability of the two-photon absorption maximum. Compounds with absorption maxima ranging from 530 nm to 980 nm with two-photon absorptivities, δ, of up to 1500×10^{-50} cm^4 s/photon have been characterized. This value of δ is unprecedentedly high. Many of these chromophores are also good electron donors and can initiate radical polymerization of acrylic monomers by photo-induced electron transfer. Excited-state charge transfer rates have been determined for a series of bis-donor diphenyl polyenes and triarylamines. In order to explore the efficiency of polymerization initiated by these molecules, rates of polymerization have also been measured. Photopolymer systems incorporating these chromophores have been developed and used to demonstrate 3D microfabrication and optical data storage potential.

EXPERIMENTAL

Two photon absorptivities, δ, were measured by ns and/or ps two-photon fluorescence. Measurements were made in toluene at a dye concentration of 10^{-4} M. Excitation intensity was maintained at a level where the fluorescence signal was proportion to I^2, where I is the incident intensity, ensuring that no higher order processes were taking place. The total fluorescence signal was collected by a PMT which was filtered to eliminate scattered light from the excitation laser.

Photo-induced electron transfer rates were determined for several two-photon chromophores in SR9008, a triacrylate monomer supplied by Sartomer Corp. Rates were measured by both fluorescence intensity quenching and fluorescence lifetime quenching methods. Steady state intensity quenching measurements were made using a Spex Fluorolog fluorimeter. Fluorescence lifetimes were measured using ps excitation from a Nd:YAG-pumped dye laser and a time-correlated single photon counting detection system. Stern-Volmer analysis of these data was used to extract the electron transfer rate constants.

For polymerization rate measurements, solutions of chromophores were prepared in neat SR9008 triacrylate monomer. The dye solution was placed between two glass windows separated by a Viton O-ring and held in place by two binder clips. For two-photon excitation, the solution was irradiated using a 5 ns pulsed dye laser operating at 20 Hz or a 5 ns Nd:YAG-pumped optical parametric oscillator (OPO) with a repetition rate of 10 Hz. Single photon (UV) photo-initiation was performed using an 8 ns, 10 Hz, frequency tripled Nd:YAG laser at 355 nm. Exposures were performed in a dose array pattern with time increasing along one axis of the array and power along the other. After exposure, the unpolymerized solution was washed away using tetrahydrofuran. The polymerized columns were coated with a thin layer of Au/Pd alloy and the height and diameter of the columns were measured using scanning electron microscopy (SEM). The rate of polymerization was then estimated as the slope of a plot of volume of polymer versus exposure time.

Microfabrication was performed in solid films consisting of 30% by weight polymer binder (75% polystyrene:25% polyacrylonitrile copolymer) (PSAN), 69.9% by weight reactive monomer, and 0.1% by weight two-photon dye. Solutions of this composition were prepared in dioxane such that the PSAN concentration was 150 mg/ml to obtain the proper viscosity to cast films having a thickness of about 120 μm. Exposure was performed using a two-photon microscope incorporating a Ti:Sapphire laser operating at 75 MHz with a pulsewidth of about 150 fs. The excitation light was focused through a high-power oil-immersion objective with a numerical aperture of 1.4. X-Y-Z control of the sample was achieved through use of a computer controlled stage. After exposure, the unpolymerized film was washed away with dimethylformamide and the features were characterized using SEM.

RESULTS AND DISCUSSION

Two Photon Absorptivity

To assess the strength of the two-photon absorption, the two-photon absorptivity, δ, must be experimentally determined. This is typically done by non-linear transmission (NLT) or two-photon fluorescence (TPF) techniques. Our group has determined that the value of δ measured by NLT depends on the pulsewidth of the laser used in the experiment. Nanosecond pulses lead to excited-state absorption resulting in an effective absorptivity, δ_{eff}, which is a combination of two-photon and excited-state absorption. TPF, however, given an accurately measured reference compound such as rhodamineB or fluoroscein [5], results in a measure of δ that is independent of the excitation pulsewidth. All values of δ reported in this paper were measured by TPF, taking care to ensure that the fluorescence signal was proportional to I^2. As shown in Table 1, compound I, a bis-donor-substituted stilbene, has a two-photon absorptivity, δ, of 240×10^{-50} cm^4 s/photon at 605 nm, its two-photon absorption maximum, $\lambda^{(2)}_{max}$. In comparison, unsubstituted stilbene has a reported two-photon absorptivity of only 12×10^{-50} cm^4 s/photon [6]. We reasoned that the increased strength of the two-photon absorption must be associated with the redistribution of charge from the ends of the molecule to the π-conjugated bridge upon excitation. Quantum chemical calculations confirm this notion and show that this charge redistribution is correlated to an increase in the transition dipole moments controlling the strength of the two-photon absorption [7]. The two-photon absorption properties of these molecules can be tuned by changing the nature of the donor groups and altering the conjugation length. For instance, increasing the number of polyene units (I through V) shifts the two-photon absorption maximum further to the red. Also, changing donors to diphenylamino groups (VI) red-shifts the two-photon absorption maximum by 85 nm relative to dialkylaminostilbene (I). These variations do not result in much change of the absorptivity, δ.

Phenylene vinylene derivatives (VII through X) allow for longer wavelength absorption maxima and increased absorption strength. A dramatic increase in the value of δ is obtained by substituting the central phenyl ring with electron-withdrawing cyano groups (IX), resulting in a donor-acceptor-donor chromophore. Acceptor-donor-acceptor molecules, such as X, with very strong thiobarbituate acceptors, also have large two-photon absorptivities at long wavelengths.

Electron Transfer from Two-Photon Chromophores

In addition to having very strong two-photon absorptivities, many of the compounds in Table 1 also have relatively low oxidation potentials. Two-photon excitation of many of these dyes can result in photo-induced electron transfer to relatively weak electron acceptors such as acrylates. This is especially true for compounds I - IV, where the fluorescence intensity and lifetimes are significantly quenched in the presence of alkoxylated triacrylate monomers. The electron transfer that takes place in the chromophore/triacrylate system was observed to result in radical polymerization of the

Compound/ID Number	$\lambda^{(2)}_{max}$ (nm)	$\delta * 10^{50}$ (cm^{4}s/photon)
n=1 **I**	605	240
n=2 **II**	645	240
n=3 **III**	695	270
n=4 **IV**	695	220
n=5 **V**	730	170
n=1 **VI**	690	190
VII	730	740
VIII	770	800
IX	835	1560
R=C$_{12}$H$_{25}$ **X**	980	870

Table 1. Two-photon chromophores with two-photon absorption maxima, $\lambda^{(2)}_{max}$, and two-photon absorptivities, δ

monomer. Due to the multiple reactive functionality of the triacrylate, the polymer that results is reasonably crosslinked.

For compounds **V** and **VII** - **X**, the electronic excitation energy of the chromophore has decreased to the point that it is energetically unfavorable for electron transfer to the triacrylate monomer to compete with fluorescence. Compound **VI** has a sufficient electronic excitation energy (comparable to **III**), but has a much higher oxidation potential due to the replacement of alkyl groups on the amine with phenyl groups so that it too is not highly quenched by SR9008.

Excited-state electron transfer data for these two-photon chromophores in the SR9008 triacrylate monomer are summarized in Figure 1. This Marcus plot shows the relationship between the electron transfer rate constant, k_q, and the thermodynamic driving force, ΔG, for charge transfer to take place. In Figure 1, values of k_q obtained by both intensity and lifetime quenching methods are shown. According to Marcus theory [8], data presented in this form should fit a parabola, and the maximum of the parabola represents the condition where there is no energy barrier to electron transfer. In Figure 1, the parabolic fit is made to the lifetime quenching data. The fluorescence intensity quenching measurements may also include contributions from static quenching processes. It is obvious that compounds **I** and **II** are nearest the top of the parabola and should, of the compounds studied, be the most efficient initiators of polymerization in SR9008.

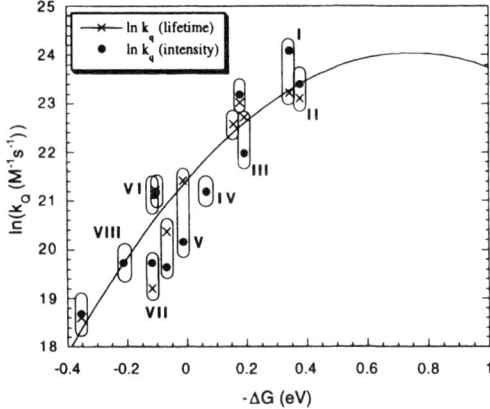

Figure 1. Marcus plot for excited-state electron transfer between dyes **I** - **VIII** and SR9008. Quenching rate constants determined by both intensity (solid circles) and lifetime (crosses) quenching measurements are given. Unlabeled points are two-photon dyes not discussed in this paper. The parabolic fit is made to the lifetime quenching data (crosses).

Photopolymerization Kinetics

Because many of the two-photon chromophores discussed above can efficiently transfer electrons to acrylates, they are able to initiate polymerization of acrylic monomers. In order to assess the relative efficiencies of initiation, it is necessary to determine the rate of acrylate polymerization. Radical polymerization kinetics show that the rate of polymerization, R_p, is given as [9]:

$$R_p = k_p[M]\left(\frac{R_i}{2k_t}\right)^{1/2}$$ (Equation 1)

In Equation 1, k_p and k_t are the polymerization and termination rate constants which are determined by the monomer system, $[M]$ is the monomer concentration, and R_i is the rate of initiation. The rate of initiation can be derived for a two-photon process and is given by:

$$R_i = \Phi_i(hv)^{-2}\delta[C]I_0^2$$ (Equation 2)

In Equation 2, Φ_i is the quantum efficiency of initiation, hv is the photon energy, δ is the two-photon absorptivity, $[C]$ is the concentration of the two-photon chromophore, and I_0 is the incident intensity. The electron transfer rate constant k_q is included in Φ_i. From Equations 1 and 2 it is clear that the rate of polymerization depends linearly on the incident intensity for two-photon initiation. For the case of one-photon initiation, the polymerization rate depends on the square root of the intensity.

The rate of polymerization was experimentally measured for several two-photon chromophores by the method discussed in the experimental section. Figure 2 shows the rate for SR9008, initiated by compound **I**, versus intensity for two-photon (600 nm) excitation (left panel), and versus the square root of intensity for one-photon (355 nm) excitation (right panel).

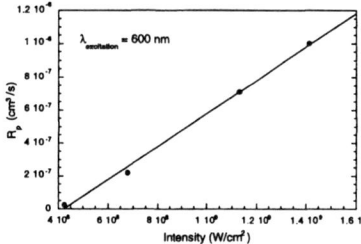 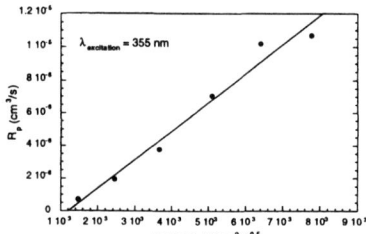

Figure 2. (Left) Rate of polymerization, R_p, as a function of I for SR9008 initiated by **I** at 600 nm. (Right) Rate of polymerization, R_p, as a function of $I^{0.5}$ for SR9008 initiated by **I** at 355 nm.

Figure 2 shows good agreement with the kinetics outlined in Equations 1 and 2. Linear behavior of R_p versus intensity for two-photon initiation was also observed for compounds **I, II** and two bis-diphenylaminobiphenyl-based chromophores. The rate of polymerization initiated by two-photon excitation has not been measured for the other compounds in Table 1.

To gauge the efficiency of SR9008 polymerization by compounds **I** and **II**, their rates of polymerization and threshold energies for polymerization were compared to values for several conventional UV photo-initiators. Benzil, benzophenone, dl-camphorquinone, 2-methyl-4'(methylthio)-2-morpholino propiophenone, and 2-isopropyl-9H-thioxanthen-9-one were among the UV photo-initiators studied. All measurements were performed at 600 nm, the $\lambda^{(2)}_{max}$ of **I**. The morpholine initiator performed best under these conditions, displaying a polymerization rate of 3×10^{-7} cm^3/s and a threshold energy of 1 mJ. Compound **II** had a polymerization rate of 7×10^{-7} cm^3/s and a threshold of 0.2 mJ, and compound **I** had a rate of 30×10^{-7} cm^3/s and a threshold of 0.3 mJ. The polymerization rate of **I** was three times higher and the threshold energy three times lower than that of the best UV photo-initiator studied. This increased sensitivity results in faster recording times for data storage.

Interestingly, compounds **V** - **VIII** initiate polymerization in SR9008 despite the fact that their electron transfer rates to the monomer are not very high. It is believed that higher order processes may be involved in these cases, particularly for compounds **V** and **VII** - **X** which have two-photon absorption maxima at wavelengths greater than 700 nm. Two photon excitation followed quickly by the absorption of one or two more photons, at sufficiently high intensities, may lead to photo-ionization of the chromophore, leading to polymerization that is initiated by free electrons or by the radical cation produced by photo-ionization. Higher order processes were evident in the one-photon initiated polymerization of SR9008 by compound **VII**. Figure 3 shows the rate of polymerization versus intensity for the SR9008/**VII** system. At low incident intensities, the rate increases as the square root of the intensity as it should for one-photon initiation. But at higher incident intensities, the order of the rate dependence on intensity increases, consistent with an absorption of at least two photons.

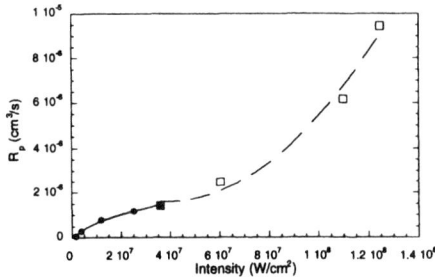

Figure 3. Single-photon rate of polymerization of SR9008 initiated by **VII** at low intensity (solid circles) and high intensity (open boxes). The order of the relationship changes at higher incident intensity. The solid line is a fit to $I^{0.5}$ and the dashed line is present to guide the eye.

3D Microfabrication and Data Storage

A photopolymer system was developed that consists of a polymer binder, multifunctional acrylate monomers, and a two-photon chromophore. The composition is detailed in the experimental section. The monomeric portion is made up of 50% SR9008 and 50% SR368, a second commercially available triacrylate monomer. SR9008 is chosen because it easily reduced by the two-photon chromophores discussed in this paper and has good adhesion to glass substrates. SR368 is chosen because of its good mechanical properties. Compound **VIII** was used in these photopolymer films and was excited at 800 nm, near $\lambda^{(2)}_{max}$ (see Table 1).

By controlling the position of the sample under the microscope objective, it is possible to define 2- and 3-dimensional objects. As an example, photonic bandgap (PBG) structures have many applications in photonics, including low-loss mirrors and waveguides. However PBGs rely on a periodic, high dielectric contrast to be effective

[10]. Although acrylate polymers do not provide the requisite contrast, it is conceivable to produce a mold by the two-photon polymerization technique which can then be used to form a high dielectric refractory part, for instance. Figure 4 demonstrates that it is possible to produce a well-defined, periodic structure using the two-photon polymerization process. In this case, the PBG form has 3 μm line widths and a periodicity of 10 μm. The true advantage that the two-photon process affords here is excellent control of polymerization normal to the substrate.

Of course there are applications in which the polymer itself has the required properties and no additional fabrication steps are necessary. Optical data storage is an application that takes full advantage of two-photon polymerization. Because of the strong 3D confinement of the two-photon excitation, we have shown that it is possible to write bits of information into the photopolymer film that are approximately 1 μm in diameter. Furthermore, individual planes of these bits are resolvable by two-photon microscopy when the planes are spaced less than 10 μm apart. The ability to write information in three dimensions will lead to increased data capacity in compact disc-like media.

The features that we have photopolymerized by two-photon absorption are comparable in size to those reported earlier [1-4], although the fabricated 3D structures are greater in complexity. However, because of the improved sensitivity of the two-photon initiators used here over conventional materials, the speed at which these features can be produced is enhanced. This is particularly important in optical data storage where large numbers of bits need to be recorded quickly.

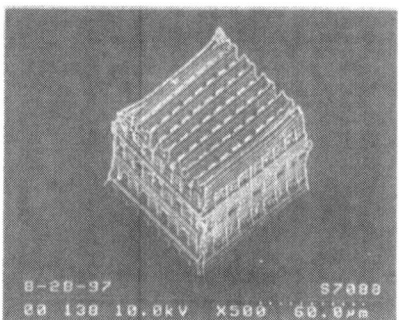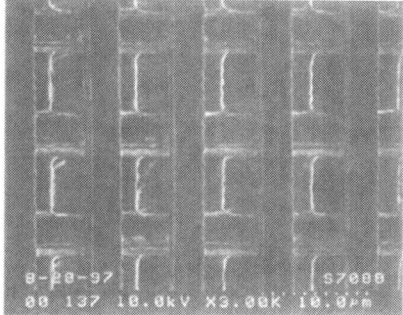

Figure 4. SEM micrographs of PBG form. (Left) Complete form as seen from above. (Right) Close-up looking down on top of the structure.

CONCLUSIONS

We have reported a class of organic molecules which have unprecedented two-photon absorption strengths and have demonstrated a broad range of operating wavelengths within this class. In addition to their non-linear properties, many of these compounds also have low oxidation potentials and can undergo excited-state charge transfer reactions with acrylates. Electron transfer rate constants were experimentally determined for many two-photon chromophores and show excellent agreement with Marcus theory.

Photo-induced electron transfer in the dye/acrylate system leads to polymerization of the acrylate. The use of multifunctional acrylate monomers results in the formation of a cross-linked, mechanically stable polymer. Although certain two-photon chromophores do not undergo electron transfer with acrylates, they can, at sufficiently high intensities, still initiate polymerization by higher order non-linear processes.

Rates of polymerization have been measured and it was shown that the rate allows one to discern the order of the initiation, i.e. one-, two-, or multi-photon initiation. The dependence of the rate was clearly demonstrated for compound **I**. The rates of polymerization obtained using bis-donor-substituted molecules with extended conjugation were as much as 10 times higher than the conventional single-photon initiators studied and also had lower threshold intensities for polymerization. This increased sensitivity leads to faster optical data recording and 3D microfabrication. It was also shown, for compound **VII**, that multi-photon processes become more dominant at higher incident intensities, as expected.

By designing a photopolymer system that incorporates one of the two-photon dyes, we have demonstrated the fabrication of complex three-dimensional structures produced by two-photon initiated photopolymerization. Clearly, there is a nearly limitless range of opportunities for this method of polymerization.

ACKNOWLEDGMENTS

This work was performed in part at the Jet Propulsion Laboratory (JPL), California Institute of Technology as part of its Center for Space Microelectronics Technology and was supported in part by the JPL Director's Research and Development Fund and the Ballistic Missile Defense Organization, Innovative Science and Technology Office, through an agreement with the National Aeronautics and Space Administration. Support at Caltech from the Office of Naval Research through the Center for Advanced Multifunctional Nonlinear Optical Polymers and Molecular Assemblies, the National Science Foundation, the Office of Naval Research, and the Air Force Office of Scientific Research is gratefully acknowledged. We thank Sartomer, Inc. for providing us with various acrylate monomer materials used in this study.

REFERENCES

1. J.H. Strickler and W.W. Webb, *Optics Letters* , **16**, 1780 (1991).
2. J.H. Strickler and W.W. Webb, *SPIE Proceedings*, **1398**, 107 (1990).
3. E.S. Wu, J.H. Strickler, W.R. Harrell, and W.W. Webb, *SPIE Proceedings*, **1674**, 776 (1992).
4. S. Maruo, O. Nakamura, and S. Katawa, *Optics Letters*, **22**, 132 (1997).
5. C. Xu and W.W. Webb, *J. Opt. Soc. Am. B*, **13**, 481 (1996).
6. R.J.M. Anderson, G.R. Holton, and W.M. McClain, *J. Chem. Phys.*, **70**, 4310 (1979).
7. Manuscript in preparation.
8. J.R. Bolton and M.D. Archer, "Basic Electron-Transfer Theory", in *Electron Transfer in Inorganic, Organic, and Biological Systems*, J.R. Bolton, N. Mataga, and G. McLendon, eds., Advances in Chemistry Series, **228**, American Chemical Society, Washington, D.C. (1991).
9. G. Odian, Principles of Polymerization, John Wiley & Sons, New York (1981).
10. J.D. Joannopoulos, P.R. Villeneuve, and S. Fan, *Nature*, **386**, 143 (1997).

CONJUGATED POLYMER BASED NANOCOMPOSITES FOR PHOTONICS APPLICATIONS

P. N. Prasad, N. Deepak Kumar, Manjari Lal, Mukesh P. Joshi

Photonics Research Laboratory, Department of Chemistry
State University of New York At Buffalo
Buffalo, NY 14260-3000

Abstract

Nanoscale synthesis and processing provides a novel approach for making a new generation of nanocomposite materials with exceptional optical and electrical properties that are needed for the development of new technologies. This presentation will focus on the preparation of nanocomposites made of Poly (para-phenylene vinylene) (PPV) with other polymers, inorganic glasses and semiconductors. We will present a new approach of nanoscale polymerization for making processable monodispersed oligomeric species of PPV which uses the base catalyzed polymerization of PPV monomer within the cavity of a reverse micelle nanoreactor. Application of this approach of fabricating novel materials for a variety of applications in photonics will also be discussed. In addition, we will discuss fabrication of bulk nanocomposites of PPV and silica by in-situ polymerization of monomer within a porous glass and their lasing properties.

INTRODUCTION

Multifunctional polymers that simultaneously exhibit more than one property are a new generation of materials which hold considerable promise for numerous applications in the field of electronics and photonics. Polymer light emitting diodes, polymer lasers and photorefractive polymers are some significant examples [1-3]. Nanosize manipulation of molecular architecture and their morphology provides a powerful approach to control electronic and optical properties of a material as well as its processability. This is why the design and processing of nanostructured materials has emerged as a frontier area of materials research. Poly(p-phenylene vinylene) is a promising π-conjugated polymer for applications in photonics due to its large non-resonant optical nonlinearity. The large photoluminescence quantum efficiency of PPV and the discovery of electroluminescence in PPV and its derivatives generated a great deal of interest in this class of materials. Recently, researchers have demonstrated optical amplification in PPV and its analogs in solution and in pure thin film when excited with an appropriate light source.

In this paper we present novel approach for synthesis and processing of nanosized oligomeric species of PPV through reverse micelle technique as well as the fabrication of nanocomposites of PPV with other polymeric systems, inorganic glasses and semiconductors. Applications of nanocomposites in polymer lasing and elecroluminescence are also discussed.

NANOSCALE OLIGOMERIC PPV VIA REVERSE MICELLE TECHNIQUE
1. Synthesis

Reverse micellar solutions were prepared by dissolving appropriate amounts of sodium bis(2-ethylhexyl)sulfosuccinate(AOT, a surfactant) in isooctane (2,2,4 trimethylpentane) and adding specific volumes of aqueous solution of the base (NaOH) to achieve the desired W_0 value. The W_0 value, defined as polar fluid-to-surfactant molar ratio, can be varied to obtain the maximum size of reverse micelle cavity. The synthesis of the PPV polymer involved a base catalyzed polymerization of the monomer, xylene bis(tetrahydrothiophenium chloride), which in the form of an ethanolic solution was mixed with the continuous phase i.e. isooctane. As a result of the interdroplet interaction arising due to the Brownian motion of the reverse micellar droplets, the monomer slowly diffuses inside the base catalyzed core and polymerization takes place inside the cavity. The polymerization reaction scheme is shown below.

The size of the reverse micellar cavity is governed by the amount of the aqueous solution of the base. The degree of unsaturation (chain length) can be controlled by varying the droplet size W_0 [4]. The PPV oligomer was extracted from the reverse micellar system by the addition of water and ethanol in 1:1 volume ratio.

2.Characterization and Discussion

Figure 1 shows the uv-vis spectra obtained from the in-situ synthesis of the PPV oligomer formed in reverse micelles having various droplet (W_0) sizes. As observed at low droplet size, the uv-vis spectra show a structured absorption band around 380 nm with the absorption edge at 440 nm indicating that at low droplet size the polymer has very few repeating units. With increase in the droplet size, the intensity of these bands increases and broadens and at $W_0=20$ the absorption edge is located at 500 nm; this is indicative of a more conjugated system. In all the cases, the retention of structure in the absorption spectra is due to the conjugated sequences resulting from the chains which are essentially oligomeric[5]. Also, with the increasing cavity size (W_0), a shift in the absorption edge towards longer wavelengths was observed. This is in exact correlation with the data obtained by laser light scattering (LLS) measurements of different droplet sizes as shown in Figure 2. Since in reverse micelles the size of the aqueous droplet is directly related to the amount of electrolyte dissolved in the polar core[6], the LLS data reveals that the particle size increases from 9nm to 74nm as W_0 is increased from 5 to 20.

Figure 1 also shows the uv-vis spectra of the PPV film made by extracting the polymer from the reverse micellar system. The spectra of the extracted polymer ($W_0=20$) show a bathochromic (red) shift suggesting that the polymeric chains once removed from the reverse micelle cavity are free to link. This is perhaps, due to the presence of the reactive end groups inside the micelles which form large aggregates by interacting with other polymer chains. This process is prevented due to the encapsulation within the surfactant head groups. This feature can also be observed from the LLS data which show that the particle (polymer) size increases from 74nm to 135nm when the polymer extracted from the reverse micellar system. Also, the loss

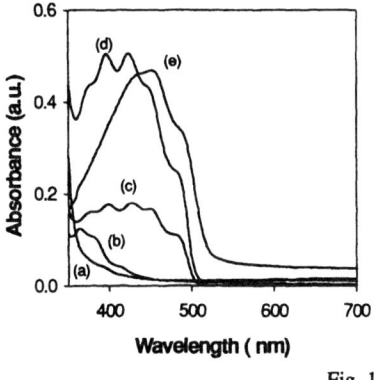

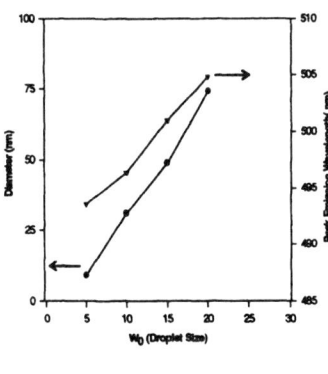

Fig. 1   Fig. 2

Figure 1: UV-vis spectra of poly (p-phenylene vinylene) synthesized within the reverse micellar cavities of varying droplet size, (a) $W_0 = 5$, (b) $W_0 = 10$, $W_0 = 15$, (d) $W_0 = 20$; (e) PPV extracted from the micellar system ($W_0 = 20$).

Figure 2 : Size of the polymeric nanoparticles within reverse miceller system having various droplet size: $W_0 = 5,10,15,20$ as measured by laser light scattering method and their corresponding peak emission wavelengths.

of structured sequence in the absorption spectra of the extracted polymer indicates that the polymeric chains are no longer oligomeric, rather they have grown or become interconnected to form long chain polymers. The extracted PPV can be dispersed in any solvent (polar or nonpolar). Thus we have produced a form of PPV that can be used for making composites with other polymeric systems like PMMA and PVK.

The degree of different conjugation of PPV polymer is clearly observed in their fluorescence spectrum (fig. 2). We observe that as the W_0 increases from 5 to 20, the fluorescence emission peak also increases from 493 nm to 505 nm, respectively.

COMPOSITE OF PPV WITH SEMICONDUCTOR AND SILICA GLASS
1. PPV/CdS heterojunction LEDs

High fluorescence efficiency and intrinsic hole conductivity of PPV and its derivatives are frequently utilized in polymer LEDs [7,8]. Efficiencies of single layer (ITO/PPV/Al or Ca) PPV based polymer LEDs are typically low (< 1%)[9]. This is mainly due to poor electron transport properties of PPV and inefficient charge injection across PPV/Al interface. To improve upon this we introduced a thin film of inorganic semiconductor, CdS, between PPV and metal layer. The double layer heterojunction LED was fabricated as follows: a water soluble precursor of PPV was spin coated onto a ITO coated glass plate. The film was heated to 250 ^{0}C under nitrogen for full conversion to PPV polymer. A mixture containing equimolar amounts of cadmium acetate and thiourea in methoxyethanol was then spin coated onto PPV layer. The resultant composite film was heated in air to 150 ^{0}C to form CdS thin film. Finally, Al electrode was vacuum deposited onto CdS layer. The schematic geometry of ITO/PPV/CdS/Al multilayer heterostructure is shown in Figure 3. For such a double layer heterojunction LED we achieved an overall quantum efficiency of about 1% and 1200 hrs of continuous operation. The luminescence intensity was 150 cd/m²[10]. The I-V and EL- V characteristics of ITO/PPV/CdS/Al system

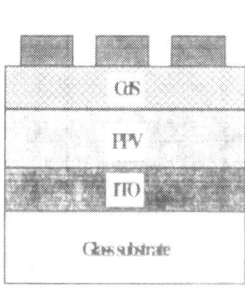

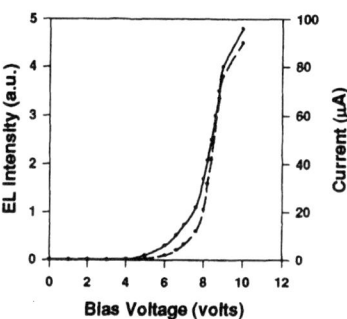

Fig.3   Fig. 4

under forward bias are shown in Figure 4. The enhancement in electroluminescence efficiency in such organic-inorganic heterojunction LED can be explained using the energy level scheme, shown in Figure 5, for the PPV polymer LED with and without CdS layer[10]. For the pure PPV layer sandwiched between ITO and Al electrodes, the energy barrier for electron injection (ΔE) is about 1.6 eV, while the energy barrier for hole injection (ΔE^+) is only 0.3 eV. The strong hole

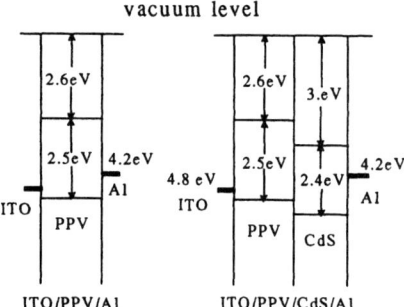

Fig.5

injection coupled with the poor mobility of electrons in the polymer makes the electron-hole recombination possible only at the PPV/metal interface which can degrade the junction. The presence of CdS layer improves the electron injection into the device due to the higher electron affinity of cadmium sulphide (3 eV) compared to pure PPV (2.6eV). With aluminum as the negative electrode having a work function of 4.2 eV the barrier for electron injection is reduced to

1.2 eV and, hence, more electrons are injected into the device than in the case of a single layer system. The CdS layer also helps in blocking the flow of the injected holes towards the negative electrode and the electrons are blocked by the PPV layer from flowing to the positive electrode. The accumulation of positive charges at the PPV/CdS interface increases the field across the CdS layer and, consequently, further improves the injection of electrons. The mismatch of energy levels at the PPV/CdS interface also helps in localizing the electrons and holes near the junction[11], thereby increasing the probability of charge carrier recombination near the PPV/CdS interface. This enhances the efficiency and lifetime of the device.

2. Fabrication and lasing in PPV:Silica nanocomposite

The large photoluminescence quantum efficiency of PPV and the discovery of electroluminescence in PPV and its derivatives generated a great deal of interest in this class of materials. Recently, researchers have demonstrated optical amplification in PPV and its analogs in solution[12] and in pure thin film[13] when excited with an appropriate light source.

The optical losses in pure PPV are very high due to its large absorption and scattering. This problem is very severe for applications where thicker samples are required. On the other hand, a nanocomposite of PPV with an inorganic material (glasses) have superior optical quality and, therefore, could be useful for a number of applications in photonics. The first report on such a composite of PPV/silica using sol-gel process was published by Wung et al[14], who reported the preparation of films of thickness in the micron range. Since then there have also been other reports on nonlinear optical properties and fluorescence studies from these composites.

We have fabricated PPV/SiO$_2$ bulk conjugated polymer-glass composite of a PPV monomer within the pores of a Vycor® glass (30% pore volume) by the in situ base-catalyzed polymerization reaction and studied their lasing properties. Base-catalyzed polymerization reaction eliminates the need for extended heat treatment under vacuum for full conversion of precursor to PPV and allows to eliminate up to 99% of the polymerization byproduct (tetrahydrothiophene). Utilizing this approach, we have been able to prepare very high quality PPV/silica composite films of thickness up to a few millimeters. The extremely small pore size (30 Å) of the Vycor® glass enables nanometer size distribution of the PPV network within the glass. The nanocomposite prepared by polymerization of PPV on the surface of the pores also results in the polymer topology that has reduced interchain coupling. This provides us with an opportunity to examine the role of such interactions that affect the optical gain and, thus, lasing. We have achieved solid state cavity lasing from a PPV/SiO$_2$ nanocomposite bulk sample fabricated using this approach.

Clean, porous Vycor® glass samples of thickness 1mm and surface area of 1cm^2 were dipped in a 25% solution of tetrabutylammonium hydroxide for 4 hours to make the pores within the glass strongly basic (high pH); the pores were later dried under vacuum. This was followed by the impregnation of the glass with a 25% ethanol solution of the PPV monomer salt. The polymerization reaction was then carried out by keeping the monomer solution containing the glasses at 60^{0}C in an oven for 2 days. The formation of PPV within the glass pores was evident from the greenish-yellow coloration of the glass samples. The glass samples were then removed from the solution, thoroughly washed and dried. The amount of PPV in the glass samples was about 10% by weight. The dry samples were then heated under flowing nitrogen at different temperatures to study the effect of heat treatment. The color of the samples varied from bright greenish yellow to deep yellow with the increase in processing temperature. After the heat treatment, the remaining pores of the samples were filled with methyl methacrylate (MMA) monomer and polymerized in-situ by using benzoyl peroxide as the initiator. Finally, the samples were polished to achieve high optical quality, characterized by means of UV-visible and photoluminescence measurements and used for lasing studies. Figure 6 shows the absorption spectra of a 100 μm thick PPV-glass composite processed at 150°C. The spectra showed an absorption maximum around 375 nm as opposed to 410 nm obtained for a pure PPV thin film. This blue shift may be due to a partial conversion and hence a reduction in the effective conjugation length of the PPV formed within the pores. The nature of the absorption spectra is very similar to that reported for PPV/SiO₂ composites made by the sol-gel process[15]. We observe a red shift in the absorption edge of the spectra when samples are heated beyond 200°C prior to the impregnation of PMMA. Figure 7 shows the photoluminescence (PL) spectra of PPV/SiO₂ samples

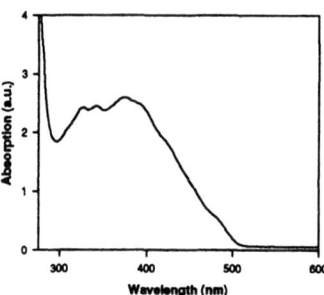

Figure 6.   Absorption spectrum of a 100 μm thick PPV/SiO₂ composite processed at 150°C.

processed at 100, 150, 200 and 250°C. The PL spectra of all samples showed well resolved vibronic structure. We observed a reduction in the intensity of PL emission from samples processed at higher temperatures. This is consistent with the report by Herold et al[16]. The reduction in the PL intensity may be due to the partial oxidation of samples when processed at higher temperatures.

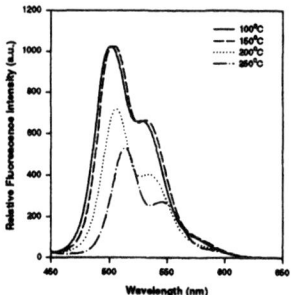

Figure 7.   Photoluminescence (PL) spectra of PPV/SiO₂ samples processed at 100, 150, 200 and 250°C.

There is also a red shift in the emission peaks for samples processed at higher temperatures due to the increase in conjugation length of the PPV network. It was found that only the samples processed up to 150°C, which do not possess extensive conjugation, exhibited a net gain and, hence, lasing. A resonant cavity was formed using a 100 % reflecting concave mirror as an end mirror, and a 50% reflecting dichroic plane mirror as an output coupler. The sample was transversely pumped at 415 nm using the frequency doubled output from an optical parametric oscillator (OPOTEK, Inc.) focused on the sample by a cylindrical lens. The optical parametric oscillator (OPO) was a broadband pump source with a linewidth of about 4 nm and a repetition rate of 10 Hz. When the cavity was properly aligned, intense green lasing was observed. The beam was collimated and observed on a screen in the near-field and the far-field regimes to determine beam divergence. The beam divergence was determined to be 3.85 mrad while the diffraction limited divergence was calculated as 0.32 mrad.

Figure 8 shows the fluorescence and lasing spectra of the PPV/SiO_2 sample. It can be seen from the spectra that the lasing occurred at 529.4 nm, on the lower vibronic energy peak in the

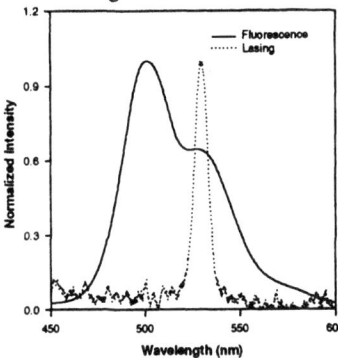

Figure 8. Fluorescence and lasing spectra obtained from the PPV/SiO_2 composite.

fluorescence spectrum. The sharp decrease in FWHM of the spectrum from fluorescence (49nm) to lasing (8nm) is a clear evidence of gain narrowing. The emission was due to cavity lasing as opposed to amplified spontaneous emission (ASE) as evidenced by the fact that the intensity of this highly directional emission dropped to zero when the end mirror was blocked or removed[17].

A characteristic reduction of the laser pulse duration as compared to the pump pulse is seen in figure 9. The FWHM of the temporal profile of the pump pulse was 2.2 ns while for the lasing output pulse, it was 1.55 ns. This pulse narrowing is due to the existence of a pump threshold intensity above which there is a net gain in the system.

By varying the pump intensity from zero to 22μJ and measuring the corresponding lasing output intensity, it was possible to determine the pump threshold and the optical efficiency of the system. Figure 10 shows the output laser pulse energy as a function of the pump pulse energy. From the figure, the pump threshold intensity is seen to be about 4 μJ. The optical efficiency of lasing at 22 μJ pump energy, was about 11.4%.

Studies were conducted to determine the operating lifetime of the laser medium. The medium was excited by 20 μJ pulse at a repetition rate of 3 Hz, and the output power was monitored continuously. The output power dropped to 50% of its initial value after approximately 8000 shots. We presume that the lifetime was limited by photooxidation of the PPV. The thermal conductivity of SiO_2 is typically over three orders of magnitude larger than most polymers. Therefore, it is expected that the SiO_2/PPV composite will have a much higher thermal conductivity and hence superior heat dissipation capabilities. This fact combined with the high thermal stability of PPV reduces the risk of thermal degradation of the lasing medium.

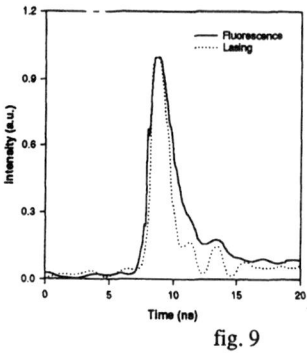
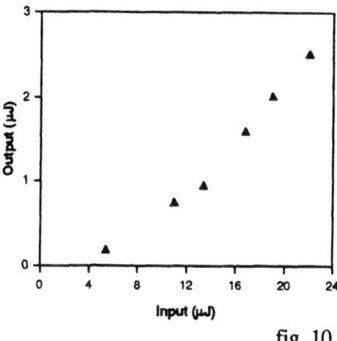

fig. 9   fig. 10

Figure 9.   Temporal profile of the pump pulse and the laser output pulse from the PPV/SiO$_2$ composite

Figure 10   Laser output energy per pulse plotted as a function of the pump energy for the PPV/SiO$_2$ composite.

In conclusion, we have fabricated high optical quality PPV/SiO$_2$ composite monoliths using a base-catalyzed reaction of the PPV salt monomer within the pores of a Vycor® glass and demonstrated highly efficient lasing from these composites. The described method opens up a new avenue to fabricate optically active materials.

Acknowledgment

We thank Dr. J. Swiatkiewicz and Dr. G. S. He for valuable discussions. This research was supported by the Directorate of Chemistry and Life Sciences of the Air Force Office of Scientific Research through MURI grant #150-7252A-000507 and F49620-96-1-0035 administered by the University of Southern California.

References

1. D. Braun, A. Heeger, *Appl. Phys. Lett.* **58**, 1982(1991).
2. Y. Ohmori, M. Uchida, K. Muro, K. Yoshino, *Solid State Commun.* **80**, 605(1991).
3. Y. Ohmori, M. Uchida, K. Yoshino, *Jpn. J . Appl. Phys.* **30**, L1941(1991).
4. Manjari Lal, N. Deepak Kumar, Mukesh P. Joshi and P. N. Prasad, *Chem Mater.*, accepted for publication.
5. I. Murase, T. Ohnishi, T. Noguchi, M. Hirooka, *Poly. Commun.* **25**, 327(1985); F. E. Karasz, J. D. Capistran, D. R. Gagnon, R. W. Lenz, *Mol. Cryst. Liq. Cryst.* **118**, 327(1985); D. D. C. Bradley, *J.Phys.* **D. 20**, 1387(1987).
6. M. P. Pileni, *J. Phys. Chem.* **97**, 6961(1993).
7. D. Braun, G. Gustafsson, D McBranch and A. J. Heeger, *J. Appl. Phys* **72**, 564(1992).
8. J. Kido, M. Kohda, K. Okuyama and K. Nagai, *Appl. Phys. Lett.* **61**, 761(1992).
9. V. L. Colvin, M. C. Schlamp, and A. P. Alivisatos, *Nature* (London) **370**, 354(1994).
10. N. Deepak Kumar, Mukesh P. Joshi, Christopher S. Friend, P. N. Prasad and Ryszard Burzynski, *Appl. Phys. Lett.* **71**,1388 (1997).

11. C. W. Tang, Information Display **12**, 16 (1996).

12. D. Moses, *Appl. Phys. Lett.* **60**, 3215 (1992).

13. F. Hide, B. J. Schwartz, M. A. Díaz-García, and A. J. Heeger, *Chem. Phys. Lett.* **256**, 424 (1996).

14. C. J. Wung, Y. Pang, P. N. Prasad, and F. E. Karasz, *Polymer* **32**, 605 (1991).

15. E. Z. Faraggi, Y. Sorek, O. Levi, Y. Avny, D. Davidov, R. Neumann, and R.Reisfeld, *Advanced Materials* **8**, 833 (1996).

16. M. Herold, J. Gmeiner, W. Reib, and M. Schwoerer, *Synthetic metals* **76**, 109 (1996).

17. N. Deepak Kumar, J. D. Bhawalkar, and P. N. Prasad, *Appl. Opt.*, accepted for publication.

OPTICAL LIMITING PROCESSES IN DERIVATIZED FULLERENES AND PORPHYRINS/PHTHALOCYANINES

R. KOHLMAN*, V. KLIMOV*, X. SHI*, M. GRIGOROVA*, B. R. MATTES*, D. MCBRANCH*, H. WANG**, F. WUDL**,J.-L. NOGUÉS*** and W. MORESHEAD***
*Chemical Science and Technology Division, Los Alamos National Laboratory, Los Alamos, NM 87545
**University of California, Los Angeles, CA 90095
***GELTECH, Inc., Orlando, FL 32826

ABSTRACT

We review our results from spectral studies of the ultrafast excited-state absorption in fullerenes and derivatized fullerenes. These results allow determination of both the spectral response of reverse saturable absorption (RSA) nonlinearities such as optical limiting (OL) in fullerenes, and the dynamical response for different morphologies. We have investigated the effects of thin film and various sol-gel glass environments on the nanosecond OL and femtosecond dynamics of derivatized fullerenes. These data provide evidence of decay pathways which compete with the intersystem crossing to a triplet from the initial singlet states. With appropriate processing, however, the OL response of derivatized-fullerene sol-gel glasses can be enhanced to approach that of the same molecule in solution, while significantly enhancing the optical damage threshold. The optical limiting of these derivatized fullerenes is compared with that of various porphyrin and phthalocyanine molecules.

INTRODUCTION

Optical power limiting (OL) occurs when the optical transmission of a material decreases with increasing laser fluence [1]. This property is desirable for providing protection to optical sensors (such as human eyes) from the high fluence output of modern pulsed lasers. One mechanism for obtaining OL is reverse saturable absorption (RSA), which occurs when excited states formed through optical pumping of the ground state have a higher absorption cross-section than the ground state [1, 2]. The fullerene C_{60} has demonstrated broadband optical limiting [2, 3, 4] due to broad singlet and triplet excited-state absorptions that extend throughout the visible and near infrared, in regions where the ground-state absorption cross-section (σ_0) is small.[3, 5, 6, 7] Limiting for nanosecond and longer pulses has been attributed to rapid (650 ps)[6, 8] intersystem crossing to a triplet excited-state which has an absorption cross-section, $\sigma_T > \sigma_0$ over a broad wavelength range and a lifetime of microseconds[7] to milliseconds[6] depending on the exposure to quenchers such as oxygen.[7] Selective derivatization across a bond joining two hexagons in the C_{60} molecule (to form a 6,6 fullerene adduct) allows these materials to be optimized for solubility and consequent ease of processing [9]. In fact, sol-gel glasses with a wide range of linear transmissions [10, 11] can be obtained due to enhanced solubility. This 6,6 derivatization also extends the OL further into the infrared as a consequence of additional ground-state absorption at long wavelengths.[3, 12]

To extend the maximum laser pulse energy that the optical limiter can sustain before catastrophic optical damage as well as to improve the ease of incorporating the limiter into an optical system, many researchers have sought to disperse limiting materials into a solid-state matrix. [10, 11, 13] However, early studies [10, 12] showed that the induced absorptions have much faster decay dynamics in the solid state. Consequently, the optical limiting response of fullerenes incorporated into sol-gel glasses was weaker than for the same molecules in solution.[10, 11, 12] It is important to determine whether this reduction in OL response in the solid state can be overcome with materials processing. Optical limiting due to RSA has also been reported in solutions of heavy-metal substituted phthalocyanines and porphyrins, with significantly higher ratios of the excited-state to ground-state absorption cross sections ($\sigma^*/\sigma_0 \sim 10$-30) near 532 nm.[14, 15] It is useful to compare the behavior of solid-state dispersions of derivatized fullerenes with heavy-metal substituted porphyrins and phthalocyanines to determine the strengths and weaknesses of both classes of materials for solid-state limiters.

In this paper, we present transient absorption (TA) and optical limiting spectra for representative 6,6 functionalized fullerenes in solution and sol-gel glass matrices. We compare these results with the optical limiting for heavy-metal substituted porphyrins and phthalocyanines. We demonstrate that the essential dynamics of 6,6 substituted fullerenes are similar to those of C_{60}, but with a broadening of the singlet and triplet excited-state transitions. OL spectra measured at multiple wavelengths for these derivatized fullerenes confirm the broadband nature of the limiting. σ^*/σ_0 increases at long wavelengths. Though the solid-state dispersions of fullerenes show faster relaxation dynamics and weaker OL, we demonstrate, by careful processing of sol-gel glasses, that the OL response of 6,6 mono-adducts in a sol-gel matrix can approach that of the fullerene derivatives in solution, while maintaining a high optical damage threshold. The reduction in OL performance for fullerene sol-gel glasses is attributed to rapid relaxation to a trap level which does not contribute to OL. The heavy-metal substituted porphyrin and phthalocyanine sol-gel glasses undergo irreversible damage likely due to demetallation at relatively low pulse energies (substantially below the damage threshold of the sol-gel glass), restricting their use in applications to low input-pulse energies.

EXPERIMENTAL

C_{60}-pyrrolidine(3)-5-benzyloxy-1H-pyrrolo[2,3-C]-pyridine (CPBPP) [12] and the 6,6 derivative 1-(3-methoxycarbonyl)propyl)-1-phenyl-[6,6]-C_{61} (PCBM) [9] were synthesized as published previously. Silica gels containing PCBM were prepared by mixing into a precursor sol, using either o-dichlorobenzene (DCB), tetrahydrofuran (THF), or toluene as co-solvents, as previously described.[10] These gels are termed "pre-doped" gels. "Post-doped" sol-gel samples were prepared using porous glass (GELTECH) with 75 Å nominal pore size produced by their proprietary process. Treated porous glass samples were placed in a solution containing the fullerene derivatives in various concentrations, and the solvent was dried to trap the fullerenes within the porous matrix. Zinc *meso*-tetra(p-methoxyphenyl)tetrabenzporphyrin (Zn-TMPTBP) was prepared by M. Nakashima at the U.S. Army Natick Research Laboratory, and purified before use using gel chromatography. Metal substituted phthalocyanine (Pc) compounds were obtained commercially (Aldrich).

Time-resolved excited-state absorption spectra were measured using a femtosecond (fs) pump-probe technique.[16] Broadband optical limiting was measured using the output of an optical parametric oscillator (CASIX OPO BBO-3B with 6 ns pulses from 400-2200 nm) pumped by the frequency-tripled output (355 nm) of a Nd:YAG laser (Quanta Ray GCR-3).[12] Each point represents the average of 20 - 100 laser shots. For TA and OL measurements of 6,6 mono-adducts and neat C_{60} in solution, 2 mm path length cuvettes were used. For TA measurements, the optical densities were adjusted to be near unity at 400 nm.

RESULTS AND DISCUSSION

Previously, we reported fs transient absorption and nanosecond optical limiting spectra for C_{60} in solution.[3, 8, 10, 12] We recently have studied 6,6 derivatization of C_{60} for improved solubility, as well as the resulting effects on the optical properties. The TA measurements for C_{60} showed excited-state absorption peaks attributed to the lowest excited singlet state at $\sim$975 nm and $\sim$500 nm as well as absorption attributed to the lowest excited triplet state at $\sim$750 nm and $\sim$530 nm. The singlet decay and triplet growth show complimentary dynamics with an intersystem crossing time ranging in the literature from 650 ps[5] - 1.2 ns

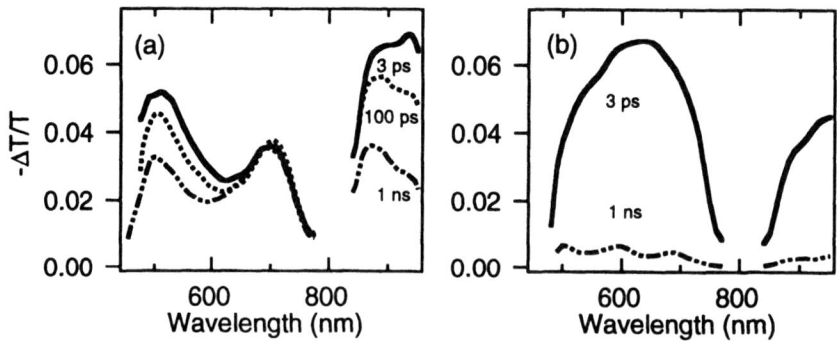

Figure 1. -ΔT/T spectra (a) at delay times of 3 ps (solid line), 100 ps (dotted line), and 1 ns (dot-dashed line) for CPBPP in THF. (b) for PCBM in a pre-doped sol-gel glass.

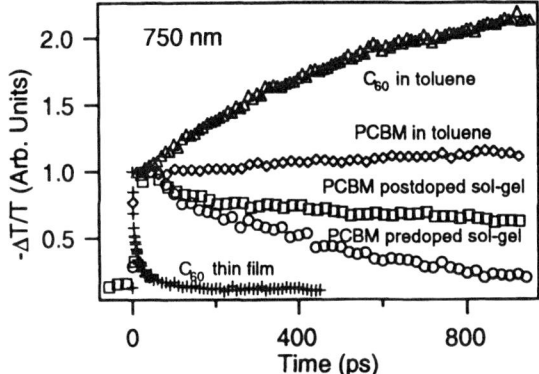

Figure 2. Comparison of normalized dynamics at 750 nm (triplet state) for C_{60} in toluene (triangles) and thin film (crosses) with PCBM in toluene (diamonds), predoped sol-gel glass (circles), and postdoped sol-gel glass (squares).

[6]. Our multispectral analysis of the dynamics yielded an intersystem crossing time of ~650 ps.[8] The peaks in the transient absorption spectrum have been compared with the general energy level diagram obtained from linear absorption.[12]

Figure 1 shows the TA spectra for CPBPP in toluene at delay times of 3 ps, 100 ps, and 1 ns compared with those for PCBM in a sol-gel glass matrix measured using chirp free TA spectroscopy. The spectra for CPBPP show the same qualitative features reported earlier for C_{60}[3, 5, 6, 7, 8], including singlet absorptions at ~500 nm and ~900 nm at early times. However, the triplet transition near ~750 nm has been substantially broadened so that it does not show clear growth as in C_{60}. These results are similar to the spectral dynamics observed in PCBM and other 6,6 derivatives.[3, 17] For PCBM in the predoped sol-gel glass, the spectrum is broadened further, likely from interactions in the solid state. Also, the decay dynamics are clearly much faster than for CPBPP in toluene.

Figure 2 shows the normalized dynamics at 750 nm for C_{60} and PCBM in both solution and solid-state matrices. The clear growth of the triplet is observed for C_{60} in toluene. In contrast, the initially-excited singlet states are rapidly quenched for C_{60} thin films so that the triplet state is not formed.[10, 12] Some authors have suggested the importance of exciton-exciton annihilation in C_{60} thin films [18]. This provides one explanation for the striong and rapid deactivation pathway for the singlet which occurs when there is aggregation of the C_{60}. Upon derivatization, there is less prominent growth at 750 nm for PCBM in toluene, consistent with the TA

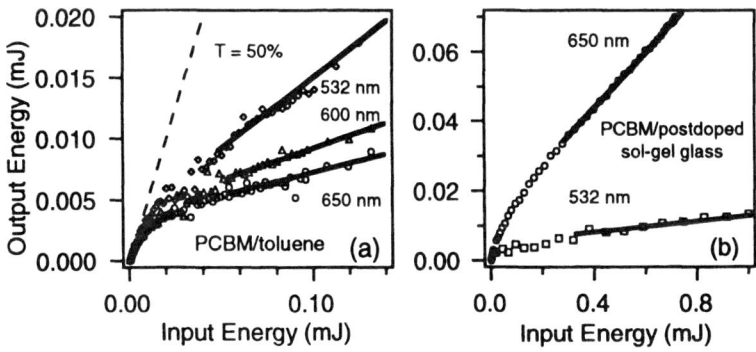

Figure 3. OL versus wavelength for PCBM in (a) toluene and (b) predoped sol-gel glass.

Table I: Estimate of σ^*/σ_0 vs wavelength for PCBM in toluene, pre- and postdoped sol-gel glasses. The damage threshold is also shown for comparison.

Matrix	Wavelength (nm)	T_0	T_{sat}	σ^*/σ_0	E_{dam} (mJ)
toluene	532	0.52	0.11	3.3	0.36
	600	0.51	0.054	4.4	
	650	0.51	0.036	5.0	0.34
postdoped	532	0.13	0.009	2.4	2.25
sol-gel	650	0.52	0.081	3.9	1.4
predoped					
sol-gel	650	0.53	0.34	1.7	0.8

spectral data. Compared with thin films of C_{60}, in which there is rapid quenching of the singlet (on ps or sub-ps time scales), PCBM sol-gels have much slower dynamics, indicating molecular dispersions of the PCBM.[10] However, the dynamics of the PCBM sol-gels indicate some quenching or competing pathways to the triplet, as there is no clear growth of the triplet state as seen for both PCBM and C_{60} in solution [8, 10]. These quenching pathways lead to a substantially reduced quantum yield for intersystem crossing. The quenching of the intersystem crossing for PCBM in postdoped sol-gel glasses is not as strong as in predoped sol-gel glasses.

For 6,6 fullerene derivatives such as PCBM and CPBPP, strong excited-state absorption and weak ground-state absorption extend throughout the visible and near-infrared, indicating that both materials should be broadband optical limiters. Figure 3 shows the wavelength-dependent optical limiting for PCBM in toluene and in a postdoped sol-gel glass. For PCBM in toluene, whose initial transmission was adjusted to 50 % at each wavelength, the output energy clamps at lower values with increasing wavelength between 532 nm and 650 nm. This behavior was reported previously for neat C_{60} and other 6,6 mono-adducts.[3, 12] This demonstrates the similar broadband OL response of many 6,6 mono-adducts. Since the initial transmission in the sol-gel glasses cannot be adjusted at various wavelengths, we gauge whether this same broadband OL response occurs in sol-gel glasses by estimating the ratio of the excited-state to ground-state absorption cross-section ratio (σ^*/σ_0), a figure of merit for the OL response. To estimate σ^*/σ_0, we use the expression [1]

$$\sigma^*/\sigma_0 \sim lnT_{sat}/lnT_0, \qquad (1)$$

where T_{sat} is the high fluence transmittance (obtained from optical limiting measurement) and T_0 the linear transmittance. The estimates for σ^*/σ_0 for PCBM both in solution and postdoped sol-gel glass are in Table I. $\sigma^*/\sigma_0 \sim 3$ is obtained at 532 nm for PCBM in toluene, in agreement with past estimates for C_{60}.[14] For PCBM in toluene, σ^*/σ_0 grows at long wavelengths, consistent with the improved OL response at longer wavelength, Fig. 3. For the postdoped sol-gel glass, σ^*/σ_0 also increases at longer wavelengths. This indicates that the broadband OL response is preserved in solid-state dispersions.

In past reports, PCBM pre-doped sol-gels have demonstrated much less effective limiting than solutions[10]. In Figure 4 (a), we show the optical limiting at 650 nm of PCBM in toluene (2 mm path length), in a pre-doped sol-gel (1.4 mm thick), and in a post-doped sol-gel (2 mm thick) all prepared to have a linear absorption ~50%. Comparison of σ^*/σ_0 at 650 nm for PCBM in toluene and predoped sol-gel glasses indicates a factor of nearly 3 reduction in σ^*/σ_0 for the predoped sol-gel compared with solution (see Table I). For the post-doped PCBM sol-gel in Figure 4(a), $\sigma^*/\sigma_0 \sim 3.9$, indicating that the OL response in postdoped sol-gel glasses can approach that in solution. A definite advantage of incorporating PCBM into the sol-gel matrix involves the increase of the energy at which the optical limiter suffers irreversible damage (the damage threshold). The damage threshold for PCBM in the sol-gel matrices is substantially higher than for PCBM in toluene (see Table I).

It is interesting to compare the respective behaviors for the dynamics (Fig. 2) and the optical limiting (Fig. 4(a)) for PCBM in toluene, pre- and postdoped sol-gel glasses. For PCBM in toluene, the dynamics show a weak growth due to intersystem crossing to the triplet; this same material has the highest σ^*/σ_0 for PCBM. For the predoped sol-gel glass, the dynamics show the fastest decay (strongest quenching of the triplet) as well as the lowest σ^*/σ_0 reported here for PCBM. This indicates that the triplet state is quenched by more rapid population of a state which does not contribute to the optical limiting. We have been able to explain the reduction in OL in sol-gel glasses with preliminary modeling of the optical limiting

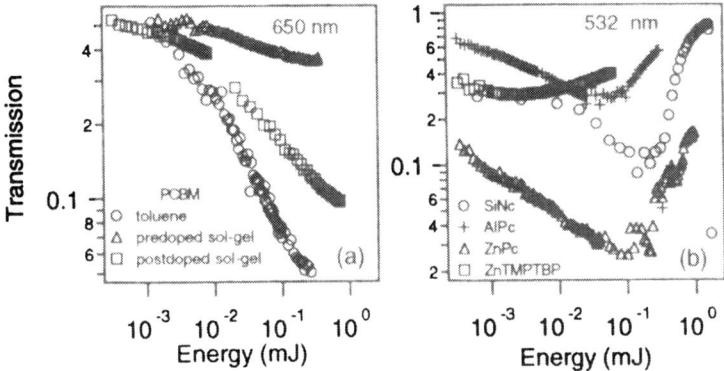

Figure 4. (a) OL at 650 nm for PCBM in toluene, pre- and post-doped sol-gel glasses. (b) OL at 532 nm for sol-gel glasses with ZnTMPTBP (squares), AlPc (crosses), SiNc (circles), and ZnPc (triangles)

of fullerenes in sol-gel glasses using a five level RSA model[1] with an additional state to which the singlet decays. We assumed that this state does not contribute to the RSA. The fast dynamics measured at 750 nm for sol-gel glasses is attributed in this model to quenching of the intersystem crossing due to rapid population of the trap state.

Fig. 4(b) shows the optical limiting response of sol-gel glasses doped with members of the porphyrin and phthalocyanine families, both of which have demonstrated superior OL response compared with fullerenes at 532 nm in solution.[14, 15] We specifically show results for Zn-TMPTBP, Aluminum and Zinc substituted phthalocyanine (Pc), and Silicon Naphthalocyanine (SiNc). Studies of these materials in solution indicate $\sigma^*/\sigma_0 \sim$ 10-30 at 532 nm.[14, 15] In sol-gel matrices, these materials have problems with photostability. The optical limiting response for each of these materials, Fig. 4(b), has an irreversible photobleaching which occurs for pulse energies in the range of 10^{-6}-10^{-4} Joules/pulse depending on the central metal. This photobleaching is likely due to photon induced demetallation.[13] This photobleaching occurs well below the damage threshold of the sol-gel glass, and so does not allow a significant improvement of either the OL response or optical damage threshold over the same materials in solution.

CONCLUSIONS

Transient absorption measurements indicate that the essential singlet/triplet dynamics of C_{60} are preserved in 6,6 derivatives such as CPBPP and PCBM, though the spectral features are broadened. Upon incorporation of PCBM into a sol-gel glass, the singlet/triplet features are further broadened and the relaxation dynamics at 750 nm are more rapid than for solution samples. The broadband nature of the optical limiting of PCBM, including growth of σ^*/σ_0 at long wavelengths, is preserved in the sol-gel matrix, though the magnitude of the OL response is reduced in comparison with PCBM in solution. The reduction of the OL along with the more rapid dynamics at 750 nm in the sol-gel matrix compared with solutions suggests the presence of states which quench the intersystem crossing and do not contribute to optical limiting. The OL response of PCBM sol-gel glasses can be made to approach that of PCBM in toluene by appropriate processing. Sol-gel glasses which incorporate heavy-metal-substituted porphyrins and phthalocyanines suffer from irreversible bleaching at low input-pulse energies, likely due to demetallation.

ACKNOWLEDGEMENTS

This work was supported in part by LANL Directed Research and Development funds and Collaborative University of California/ Los Alamos Research funds, under the auspices of the U.S. DOE. Additional funding was provided by the U.S. Army Small Business Technology Transfer program, under contract DAAH04-96-C-0077.

REFERENCES

1. L. Tutt and T. Boggess, "A review of optical limiting mechanisms and devices using organics, fullerenes, semiconductors, and other materials," *Prog. Quant. Elect.*, vol. 17, pp. 299–338, 1993.

2. L. Tutt and A. Kost, "Optical limiting performance of c_{60} and c_{70} solutions," *Nature*, vol. 356, pp. 225–226, 1992.

3. L. Smilowitz, D. McBranch, V. Klimov, J. Robinson, A. Koskelo, M. Grigorova, B. Mattes, H. Wang, and F. Wudl, "Enhanced optical limiting in derivatized fullerenes," *Opt. Lett.*, vol. 21, p. 922, 1996.

4. D. McLean, R. Sutherland, M. Brant, D. Brandelik, P. Fleitz, and T. Pottenger, "Nonlinear absorption study of a c_{60}-toluene solution," *Opt. Lett.*, vol. 18, pp. 858–860, 1993.

5. R. J. Sension, C. M. Phillips, A. Z. Szarka, W. J. Romanov, A. R. McGhie, J. P. M. Jr., A. B. S. III, and R. M. Hochstrasser, "Transient absorption studies of c_{60} in solution," *J. Phys. Chem.*, vol. 95, p. 6075, 1991.

6. T. W. Ebbesen, K. Tanigaki, and S. Kuroshima, "Excited-state properties of c_{60}," *Chem. Phys. Lett.*, vol. 181, p. 501, 1991.

7. J. Arbogast, A. Darmanyan, C. Foote, Y. Rubin, F. Diederich, M. Alvarez, S. Anz, and R. Whetten, "Photophysical properties of c_{60}," *J. Phys. Chem.*, vol. 95, pp. 11–12, 1991.

8. V. Klimov, D. McBranch, L. Smilowitz, J. Robinson, B. R. Mattes, A. Koskelo, H. Wang, and F. Wudl, "Femtosecond to nanosecond dynamics of c_{60}: implications for excited-state nonlinearities," *Res. Chem. Intermed., special issue*, vol. 23, pp. 587–600, 1997.

9. J. C. Hummelen, B. W. Knight, F. Lepec, F. Wudl, J. Yao, and C. L. Wilkins, "Preparation and characterization of fulleroid and methanofullerene derivatives," *J. Org. Chem.*, vol. 60, p. 532, 1995.

10. D. McBranch, V. Klimov, L. Smilowitz, M. Grigorova, B. R. Mattes, J. Robinson, A. Koskelo, H. Wang, and F. Wudl, "Femtosecond excited-state absorption dynamics and optical limiting in fullerene solutions, sol-gel glasses, and thin films," *SPIE Proceedings, Fullerenes and Photonics III*, vol. 2854, p. 140, 1996.

11. M. Maggini, G. Scorrano, M. Prato, G. Brusatin, P. Innocenzi, M. Guglielmi, A. Renier, R. Signorini, M. Meneghetti, and R. Bozio, "C_{60} derivatives in sol-gel glasses," *R. Adv. Mater.*, vol. 7, pp. 404–406, 1995.

12. R. Kohlman, V. Klimov, M. Grigorova, X. Shi, B. R. Mattes, D. McBranch, H. Wang, F. Wudl, J.-P. Nogués, and W. Moreshead, "Ultrafast and nonlinear optical characterization of optical limiting processes in fullerenes," *SPIE Proceedings, Fullerenes and Photonics IV*, p. in press, 1997.

13. M. Brunel, F. Chaput, S. Vinogradov, C. B, M. Canva, J. P. Boilot, and A. Brun, "Reverse saturable absorption in palladium and zinc tetraphenyltetrabenzoporphyrin doped xerogels," *Chem. Phys.*, vol. 218, pp. 301–307, 1997.

14. J. Perry, K. Mansour, I.-Y. Lee, X.-L. Wu, P. Bedworth, C.-T. Chen, D. Ng, S. Marder, P. Miles, T. Wada, M. Tian, and H. Sasabe, "Organic optical limiter with a strong nonlinear absorptive response," *Science*, vol. 273, pp. 1533–1536, 1996.

15. T. Xia, D. J. Hagan, A. Dogariu, A. A. Said, and E. W. V. Stryland, "Optimization of optical limiting devices based on excited-state absorption," *Appl. Opt.*, vol. 36, no. 18, pp. 4110–4122, 1997.

16. V. Klimov and D. McBranch, "Femtosecond high-sensitivity, chirp-free transient absorption spectroscopy," *Opt. Lett., submitted.*

17. D. Guldi and K.-D. Asmus, "Photophysical properties of mono- and multiply-functionalized fullerene derivatives," *J. Phys. Chem.*, vol. 101, pp. 1472–1481, 1997.

18. S. L. Dexheimer, W. A. Varecka, C. V. Shank, D. Mittelman, and A. Zettl, "Nonexponential relaxation in solid C_{60} via time-dependent singlet exciton annihilation," *Chem. Phys. Lett.*, vol. 235, p. 552, 1995.

SYNTHESIS AND NONLINEARITY OF TRIENE CHROMOPHORES CONTAINING THE CYCLOHEXENE RING STRUCTURE

SUSAN ERMER,* STEVEN M. LOVEJOY,* DORIS S. LEUNG,* HOPE WARREN,* CHRISTOPHER R. MOYLAN**† AND ROBERT J. TWIEG**‡

*Lockheed Martin Advanced Technology Center, O/H1-32, B/204, 3251 Hanover Street, Palo Alto, CA 94304
**IBM Almaden Research Center, 650 Harry Road, San Jose, CA 95120-6099

ABSTRACT

A series of conjugated donor-acceptor trienes in which the central double bond is incorporated into an unsaturated isophorone, verbenone or chromone ring has been synthesized. In each case, the donor group consists of an amine and an aromatic or heterocyclic ring system, and the acceptor is the dicyanomethylidene group. The nonlinear optical properties of each of the compounds has been measured and correlated with its structure. The dipole moments and molecular hyperpolarizabilities of these compounds, like those of other conjugated polyenes, are large enough to be used as the active components of electro-optic polymers. Unlike other donor-acceptor polyenes, however, these compounds exhibit the thermal stability required for such applications.

INTRODUCTION

A major challenge in the design of nonlinear optical chromophores for electro-optic (EO) devices is simultaneously achieving acceptable thermal stability, transparency, nonlinearity, and processability in one compound. Thermal stability is typically enhanced by incorporating aromatic rings into the delocalized bridge between the electron donating and accepting groups, but it has been shown that such substitution produces excited states that feature quinoidal bonding patterns in those rings and concomitant loss of aromaticity.[1] The first excited state in such compounds is thus energetically disfavored, decreasing the hyperpolarizability β. The nonlinearity can be enhanced to some extent by incorporating quinoidal and aromatic rings into the bridge in pairs, so that the net aromatic stabilization energy upon excitation is close to zero, but the thermal stabilities of such compounds will be compromised. This nonlinearity-stability tradeoff can be evaded to some extent by using aromatic substitution upstream of the electron donor group only.[2] This substitution pattern is especially effective when combined with cyanomethylene acceptor groups.[3] Nevertheless, the largest measured molecular hyperpolarizabilities have been found in long conjugated polyene systems with few or no aromatic rings between the donor and the acceptor.[4]

---

† Current address: Lightwave Microsystems Corporation, 2950 Scott Boulevard, Santa Clara, CA 95054

‡ Current address: Department of Chemistry, Kent State University, Kent, OH 44242

Unfortunately, such compounds decompose far below the 300 °C required for incorporation into real devices[5] and are sufficiently red-shifted as to fail the transparency requirements. One class of compounds has been found to evade the nonlinearity-transparency tradeoff somewhat by featuring two low-lying excited states that contribute to β in an additive manner,[6] but other systems with that desirable quality remain to be found. Accordingly, it remains of great interest to determine whether the thermal stabilities of unsaturated delocalized systems are irretrievably low, or whether structural modifications can increase decomposition temperatures without decreasing hyperpolarizabilities proportionally.

EXPERIMENTAL RESULTS

Synthesis

The work described in this paper was initially motivated by success using the laser dye 4-(dicyanomethylene)-2-methyl-6-(p-dimethylaminostyryl)-4H-pyran (DCM) in the development of an all-polyimide integrated process for electro-optic waveguides in 1991.[7] In an effort to extend this work, we embarked upon the design, synthesis, and characterization of a series of chromophores with related molecular structures and improved characteristics. Like DCM, all of these chromophores incorporate the dicyanomethylidene [>C=C(CN)$_2$] acceptor group.

The synthesis of donor-acceptor analogues of DCM is complicated by the symmetrical nature of the dicyanomethylidenepyran reactant which becomes the acceptor part of the molecule.[8] Reaction of the donor-containing aldehyde with this reactant in a one-to-one stoichiometric ratio tends to produce mixtures (see Figure 1) in which a considerable portion of the product is the doubly condensed product, requiring difficult chromatographic separation.

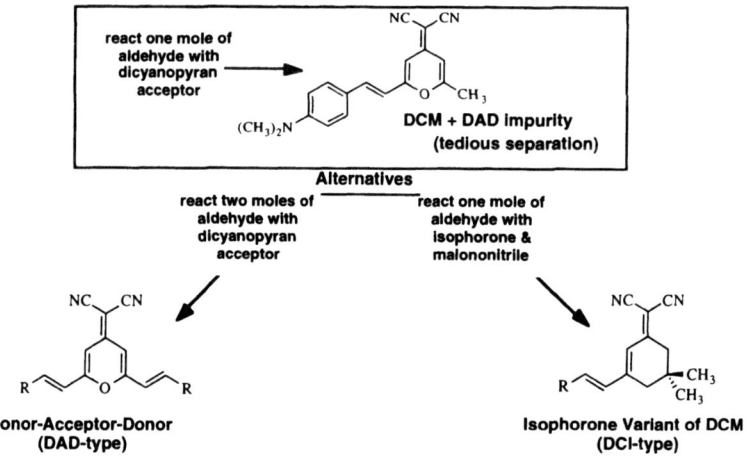

Figure 1. Two categories of DCM-related compounds: Donor-Acceptor-Donor (DAD)series (left) and Donor-Acceptor Isophorone variants of DCM (DCI) (right)

We identified two methods to avoid the tedious separation of reaction mixtures related to DCM. The first strategy, as depicted in Figure 2, is the use of an excess of donor aldehyde to push the reaction entirely to the doubly-condensed product. This produces the donor-acceptor-donor (DAD) series of compounds described previously.[9,10] The DAD compounds are of interest as EO chromophores for both their thermal stability at temperatures over 300 °C and an advantageous nonlinearity-transparency tradeoff. The usual combination of high thermal stability and photobleachability of one of the DAD compounds (DADC) has allowed the use of high processing temperatures in the fabrication of a demonstration Mach-Zehnder modulator.[11]

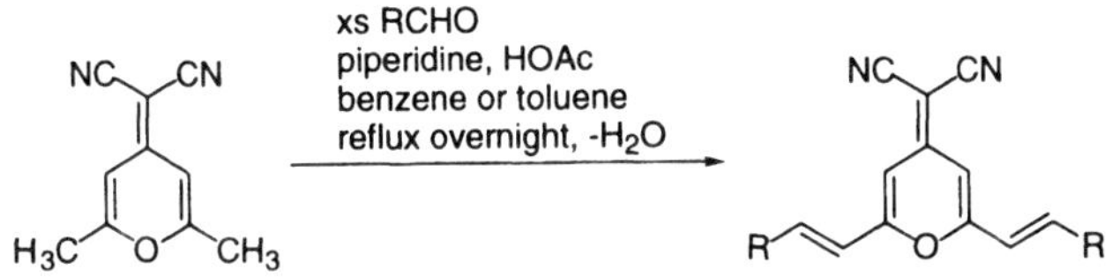

Figure 2. Preparative scheme for Donor-Acceptor-Donor (DAD) compounds

To further our studies we wished to also prepare donor-acceptor analogues of DCM. In view of the synthetic difficulties discussed above in obtaining pure mono-condensed products, the second strategy we employed was to seek compounds chemically similar to 2,6-dimethyl-4H-pyranone as starting materials which could not undergo multiple condensations. Isophorone, verbenone, and chromone are such materials. As shown in Figure 3, isophorone can be condensed with malononitrile to give an intermediate dicyanomethylidene isophorone derivative, and this can be further reacted with a donor-containing aldehyde to produce isophorone-based analogues of the type 1. Alternatively, isophorone, malononitrile, and the donor-containing aldehyde can be condensed in a "one-pot" manner as described by Lemke[12,13,14] to give the desired NLO chromophore.

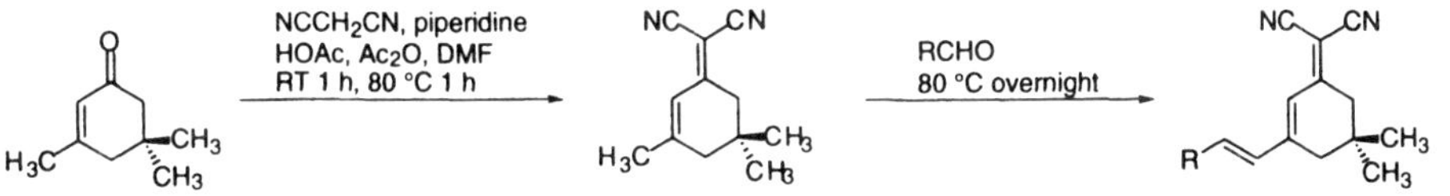

Figure 3. Preparation of Donor-Acceptor compound through stepwise synthetic method

Structures and Physical Properties

A series of compounds of type 1 was prepared by employing different donor-containing aldehydes in the synthesis. Table I gives their structures. The nonlinear optical properties of 1a in solution[15] and in poled polymers[16,17] have been investigated by others. Substitution of (1S)-(-)-verbenone for isophorone in this synthesis gives chiral analogues of the type 2. Table II shows the structures of the two compounds in this series. For a given donor the physical properties of the type 2 compounds are quite similar to those of type 1, but their solubilities in polar solvents commonly used to cast polyimide films is markedly improved. A third class of compounds can be made by substituting 2-methylchrom-4-one for isophorone in the synthesis yielding compounds of the type 3 (see Table III). The right-hand columns of the tables give melting point (mp), absorption maximum (λ_{max}), and thermal weight loss temperature onset (T_o) as determined by thermogravimetric analysis (TGA) in nitrogen for each of the analogues synthesized. The compounds exhibit intense absorption maxima in the wavelength range of 470 to 545 nm when measured in N-methylpyrrolidinone (NMP).

Table I. Compounds prepared with isophorone

Compound	R =	melting point	λ_{max}	T_o
1a	—⟨⟩—N(CH3)2	226-228	519	286
1b	—⟨⟩—N(C4H9)2	119-120	539	309
1c	(carbazole, CH2CH3)	195-196	474	343
1d	(indole, CH3)	234.5-235.5	498	326
1e	—⟨⟩—N(CH3)2 (naphthalene)	209-211	479	334
1f	—⟨⟩—N(CH2CH3)(CH2CH2OH)	187.5-188.5	534	317
1g	(thiophene-pyrrolidine)	>200[b]	618	N/A
1h	(thiophene-morpholine)	>235[b]	556	328
1i	—⟨⟩—N(Ph)2	188-189	496	346
1j	—⟨⟩—N(p-CH3Ph)2	216-217	510	344

246

Table II. Compounds prepared with verbenone

Compound	R=	melting point	λ_{max}	T_o
2a	—⟨benzene⟩—N(CH$_3$)$_2$	193-194	525	289
2b	(carbazole) N—CH$_2$CH$_3$	197.2-198	477	389

Table III. Compounds prepared with chromone

Compound	R=	melting point	λ_{max}	T_o
3a	—⟨benzene⟩—N(CH$_3$)$_2$	267-269	530	361
3b	—⟨benzene⟩—N⟨CH$_2$CH$_3$ / CH$_2$CH$_2$OH⟩	210-211.5	544	370

[a]Melting points (mp) and onset temperatures of weight loss (T_o) in N_2 are given in °C, wavelengths are given in nm, and N-methylpyrrolidinone (NMP) was used as the solvent for the λ_{max} measurements.
[b]Sublimes.

Attachable Chromophores

The list of compounds in the previous tables includes two (**1f** and **3b**) which have been specifically designed for covalent attachment to polymers via a *hydroxyl*-group appended to the donor amine. Many researchers have pursued covalent attachment of standard donor-acceptor chromophores, such as azo dyes, to polymers such as polyimides in order to improve EO polymer stability [18]. The attachment approach has the dual advantages of circumventing out-diffusion and improving chromophore loading levels. These improved loading levels will yield higher activity electro-optic polymers, improving device performance. Guest-host systems have provided a way for us to rapidly screen a large variety of chromophores with a modest expenditure of effort. Once a chromophore with optimized properties for a given device application is developed, covalent attachment is a logical next step for enhancement of loading level and long-term stability of the poled state.

Thermal Stability

Relative thermal stability as reported in the tables above was determined using thermogravimetric analysis (TGA) under nitrogen atmosphere at a heating rate of 20 °C/minute. The results obtained for a selected few of the analogues are shown in Figure 4. Plot of thermogravimetric analysis results for selected compounds in a N2 atmosphere with a heating rate of 20 °C/minute. Weight loss in this experiment may be due to either sublimation or decomposition of the chromophore. As can be seen, **1a** and **2a** lose weight at slightly lower temperatures than does DCM. Higher weight loss temperatures are seen in the compounds in which either the nitrogen donor group (**1c** and **1d**) or the dicyanomethylidene acceptor group (**3a**) is part of a fused ring system. The compound which survives to highest temperature is **1c**, in which the donor amino group is part of a carbazole ring system. This is consistent with our results for the DAD series, in which the carbazole-containing compound DADC loses less than 2% of its weight when heated to 400 °C at a rate of 20 °C/minute.

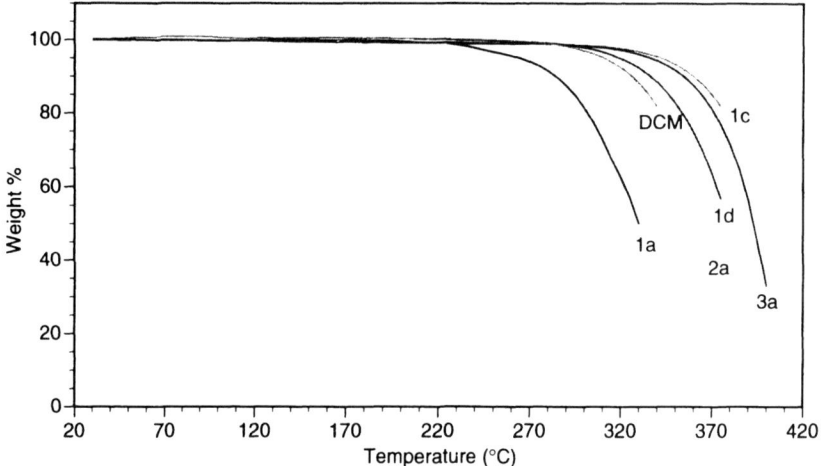

Figure 4.

EFISH Results

Electric field-induced second harmonic (EFISH) generation experiments were performed on the isophorone-series as previously described[19] in order to determine the quantity $\mu\beta$ at the experimental wavelength of 1907 nm (the Stokes Raman line obtained from the YAG fundamental of 1064 nm in a hydrogen cell). The second harmonic signals from chromophore solutions in chloroform were referenced to a quartz standard[20] ($d_{11} = 6.7 \times 10^{-10}$ esu), and the B convention of Willetts et al.[21] was used. The dielectric constants of the solutions were measured and used to determine dipole moments independently as previously described.[19] The resulting value of $\beta(1907$ nm) was extrapolated to zero frequency (β_0) using the two-level model of Oudar and Chemla,[22] and the absorption maximum was determined by UV/visible spectroscopy in chloroform solution. The EO figure of merit[20] for applications at 1300 nm (referred to here as the nonlinearity) was determined by extrapolating β_0 to 1300 nm using the dispersion relationship for the Pockels effect under the two-level model, multiplying by the dipole moment, and dividing by the molecular weight (MW).

Table IV lists the spectral properties (λ_{max} and oscillator strength, f), dipole moments, nonlinear optical properties (β and β_0), EO figures of merit, and thermal decomposition temperatures (T_d) for each of the isophorone derivatives. Calculations of β_0 were made using the two-level model. This set of decomposition temperatures were measured by differential scanning calorimetry (DSC) in air, scanning at 20 °C/minute. Extrapolation of the exotherm onset back to the baseline yielded T_d. Not surprisingly, the decomposition temperature values obtained by DSC in air differ somewhat from the values reported in the previous tables, which were obtained from the TGA onset as measured in a nitrogen atmosphere.

Table IV. EFISH experimental data for isophorone-based compounds

Compound	λ_{max}	f	μ	β	β_0	$\mu\beta_0$/MW	$\mu\beta_{1300}$/MW	T_d
1a	508	0.83	8.7	246	164	4.50	5.94	307
1b	530	0.84	9.1	277	177	4.02	5.45	268
1c	464	0.62	8.3	146	105	2.22	2.79	322
1d	482	0.74	9.0	114	79	2.17	2.78	317
1e	478	0.68	6.8	146	103	1.91	2.43	315
1f	506	0.78	8.4	219	146	3.39	4.68	306
1h	530	0.80	7.8	213	136	2.91	3.95	290
1i	502	0.63	7.1	242	163	2.61	3.42	313
1j	520	0.74	7.3	352	229	3.56	4.75	367

[a]Wavelengths are given in nm, dipole moments in debye, hyperpolarizabilities in units of 10^{-30} esu, and temperatures in °C. All spectral and EFISH data were taken in chloroform solution.

Nonlinearity-Transparency Tradeoff

The degree to which a series of similar compounds suffers from the tradeoff between nonlinearity and transparency is typically quantified by determining the slope of a log-log plot of β vs. λ_{max}. We have created such a plot, fit a straight line to the plot, and determined the slope of the best fit line. This was done so we could compare the nonlinearity/transparency tradeoff with that observed for other classes of chromophores which have been similarly analyzed.[23] Most classes of compounds examined to date have exhibited slopes of approximately 6 (6.5 ± 1.4 for triphenylimidazoles[19], 6.0 ± 0.7 and 7.1 ± 1.1 for tolanes[24,25], 6.5 ± 1.0 for pyrazoles[26], 6.2 ± 0.7 for thiazoles[27], and 6.1 ± 0.7 for coumarins[28]). Such a plot for the isophorone series is shown in Figure 5. The slope of this plot is 5.2 ± 1.7, equal to the typical value within experimental error.

Therefore, the improved thermal stability in these compounds has not led to any decrease in transparency (at least as measured by λ_{max}) for a given nonlinearity.

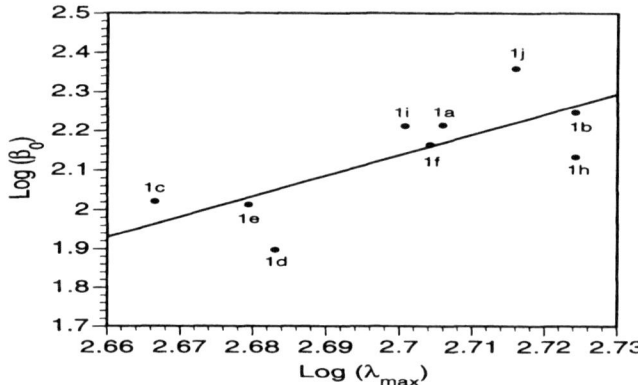

Figure 5. Log(λ_{max}) versus log(β_0) plot for isophorone-type compounds

Discussion

It has been noted previously[20] that chromophores with nonlinearities ($\mu\beta_{1300}$/MW) above 3.0 and thermal decomposition temperatures above 300 °C are quite rare. The most striking aspect of the data in Table IV is that six of the nine compounds meet the nonlinearity criterion, seven meet the thermal stability criterion, and four meet both. Partial protection of the triene bridge with the isophorone ring has afforded a surprising amount of thermal stability, while allowing the β values to remain substantial. The oscillator strengths also remain high (0.74 on average) as do the dipole moments (8 D).

The improvement in thermal stability over simple extended single-double bond alternant systems is consistent with expectations. Protection of polyene chains in chromophores by means of aliphatic rings was previously proposed by Cabrera and co-workers.[29] The compounds described in that work exhibited appreciable values of β, as anticipated, but no thermal stability information was provided. Shu, Jen, and co-workers[30,31] have also reported the incorporation of isophorone rings into longer configuration-locked triene nonlinear optical chromophores and showed this to give an enhancement of their thermal stabilities over the corresponding simple triene compounds. We are currently preparing attachable versions of some of these longer chromophores for use in both guest-host and covalently attached polymer systems for electro-optic devices.

SUMMARY

A series of nonlinear optical chromophores with dicyanomethylidene acceptor groups has been synthesized and characterized. Nine of these compounds include an isophorone cyclohexene ring as part of the delocalized bridge segment. Not only are the dipole moments and hyperpolarizabilities encouragingly large, but their thermal decomposition temperatures are substantially higher than one would expect from unsubstituted trienes. The synthesis of these isophorone derivatives is facile compared to that of the previously studied dicyanomethylidenepyran series, because the asymmetric structure of the parent isophorone allows only one condensation reaction rather than two, thus obviating the need for a difficult chromatographic separation. The increase in hyperpolarizability with absorption maximum is the same as for most other classes of chromophores. It has recently been shown[32] that compounds of this type form glasses which can be used for holographic optical storage, suggesting that materials of very high chromophore concentration and good optical quality can be prepared. The successful inclusion of hydroxyl attachment groups in the amino moiety of the chromophore enables the chromophore to be covalently incorporated into a wide variety of polymers.

EXPERIMENTAL SECTION

This section includes the synthetic methodology used for at least one compound from each of the classes listed in Tables I, II and III. We provide details for the syntheses of the remaining compounds together with complete characterizations in a recent paper.[33]

All organic chemicals were purchased from Aldrich Chemical Company except as noted and were used as received. Compound **1a (DCI)** and 3-dicyanomethylidene-1,5,5-trimethylcyclohex-1-ene were synthesized following the procedures of Lemke.[13] The latter compound was also prepared by the condensation of isophorone with malononitrile using ammonium acetate and the azeotropic removal of water.[34] Compounds **1b (DCIB)** and **2a (DCV)** were also reported earlier.[35] Melting points (mp) were determined in capillary tubes on a Mel-Temp II capillary melting point apparatus utilizing a digital thermometer and are uncorrected. Proton nuclear magnetic resonance (^{1}H NMR) spectra were obtained at room temperature (RT) on a Varian XL-300 NMR spectrometer using deuteriochloroform ($CDCl_3$) as solvent and tetramethylsilane (TMS) as an internal reference unless otherwise indicated. Chemical shifts are reported in parts per million (ppm). Fourier transform infrared (FT-IR) spectra were measured on a KVB/Analect RFX300 FT-IR spectrometer. Ultraviolet-visible (UV-vis) spectra were obtained on a Varian Cary 5E spectrometer. Elemental microanalyses were performed by Desert Analytics (Tucson, AZ).

Compound 1c (DCIC). A solution of isophorone (8.25 mL, 55 mmol), malononitrile (3.15 mL, 50 mmol), piperidine (0.90 mL, 9.1 mmol), glacial acetic acid (0.20 mL, 3.5 mmol), and acetic anhydride (0.10 g, 1.0 mmol) in N,N-dimethylformamide (25 mL) was stirred at RT for 1 h, then 80 °C for 1 h. After adding 9-ethyl-3-carbazolecarboxaldehyde (5.60 g, 25 mmol), the reaction solution was stirred an additional 16 h at 80 °C. Cooling to RT followed by roto-evaporation of the solvent gave a red oil which was redissolved in ethyl acetate. This solution was washed with water, dried with anhydrous sodium sulfate, filtered, and roto-evaporated. Flash chromatography (85/15 hexane/ethyl acetate) and recrystallization from hexane/ethyl acetate gave the product as fine red needles, 1.08 g (11%): mp 195-196 °C; ^{1}H NMR ($CDCl_3$) δ 1.09 (s, 6H), 1.46 (t, 3H, $J = 7$ Hz), 2.50 and 2.56 (two s, 2H each), 4.38 (q, 2H, $J= 7$ Hz), 6.82 (s, 1H), 7.04 (d, 1H, $J = 15$ Hz), 7.25-8.23 (m, 8H); FT-IR (KBr) 2210 cm^{-1} (CN stretch); UV-vis λ_{max} (NMP) 474 nm ($\varepsilon = 2.91$ X 10^4); Anal. Calcd for $C_{27}H_{25}N_3$: C, 82.83; H, 6.43; N, 10.73. Found: C, 82.50; H, 6.42; N, 10.51.

Compound 2b (DCVC). A solution of (1S)-(-)-verbenone (8.26 g, 55.0 mmol), malononitrile (3.15 mL, 50 mmol), glacial acetic acid (0.2 mL), acetic anhydride (0.5 mL), and piperidine (0.9 mL) in N,N-dimethylformamide (25 mL) was stirred at RT for 1 h, then 80 °C for 1 h. After adding 9-ethyl-3-carbazolecarboxaldehyde (5.60 g, 25 mmol) and stirring 3.5 h at 80 °C, the reaction solution was cooled to RT, and the solvent was roto-evaporated. The residue was flash chromatographed (4/1 hexane/ethyl acetate) and recrystallized twice from toluene to give the product as a dark orange solid, 1.95 g (19%): mp 197.2-198 °C; [^1]H NMR (CDCl$_3$) δ 0.94 (s, 3H), 1.45 (t, 3H, J = 7.3 Hz), 1.64 (s, 3H), 1.93 (d, 1H, J = 9.2 Hz), 2.91 (ddt, 1H, J = 9.2, 5.5, and 5.5 Hz), 3.23 (dd, 1H, J = 5.5 and 5.5 Hz), 3.36 (ddd, 1H, J= 5.5, 5.5, and 1.6 Hz), 4.37 (q, 2H, J = 7.3 Hz), 6.66 (s, 1H), 7.03 (d, 1H, J = 15.8 Hz), 7.22-8.26 (m, 8H); FT-IR (KBr) 2217 cm^{-1} (CN stretch); UV-vis (NMP) λ$_{max}$ 477 nm (ε = 3.09 X 10^4); Anal. Calcd for C$_{28}$H$_{25}$N$_3$: C, 83.34; H, 6.24; N, 10.41. Found: C, 83.64; H, 6.49; N, 10.24.

4-Dicyanomethylidene-2-methylchromene. Malononitrile (0.79 g, 12 mmol) and 2-methylchrom-4-one[36] (1.73 g, 10.8 mmol) were refluxed together overnight in acetic anhydride (10 mL). The reaction mixture was cooled to RT and filtered. The crude solid was washed with acetic anhydride, then treated with boiling water (20 mL) and refiltered. Flash chromatography (65/35 dichloromethane/hexane) and recrystallization from ethyl acetate/hexane gave the pure product as pale yellow crystals, 0.80 g (36%): mp 185-186 °C (lit.[37,38] 332-334 °C, 194-196 °C); [^1]H NMR (CDCl$_3$) δ 2.44 (s, 3H), 6.72 (s, 1H), 7.42-7.49 (m, 2H), 7.68-7.75 (m, 1H), 8.92 (dd, 1H, J = 1.4 and 8.6 Hz); FT-IR (KBr) 2212 and 2200 cm^{-1} (CN stretch); Anal. Calcd for C$_{13}$H$_8$N$_2$O: C, 74.99; H, 3.87; N, 13.45. Found: C, 75.03; H, 3.75; N, 13.34.

Compound 3b (BDCMOH). A solution of 4-(ethyl(2-hydroxyethyl)amino)benzaldehyde (0.77 g, 3.98 mmol), 4-dicyanomethylidene-2-methylchromene (0.80 g, 3.84 mmol), piperidine (1 mL), and glacial acetic acid (0.5 mL) in toluene (125 mL) was refluxed overnight under argon using a Dean-Stark trap. The reaction mixture was cooled to RT and filtered. The crude solid was washed with toluene and hexane, then flash chromatographed using ethyl acetate/dichloromethane 1/9 as eluent. The purified material was recrystallized from acetonitrile to give the product as dark blue needles, 1.18 g (80%): mp 210-211.5 °C; [^1]H NMR (CDCl$_3$) δ 1.24 (t, 3H, J = 6.9 Hz), 3.50-3.60 (m, 4H), 3.87 (br s, 2H), 6.56 (d, 1H, J = 15.5 Hz), 6.75-6.90 (m, 3H), 7.38-7.59 (m, 5H), 7.71 (t, 1H, J = 7.71 Hz), 8.89 (dd, 1H, J = 1.2 and 8.5 Hz); UV-vis λ$_{max}$ (NMP) 544 nm (ε = 5.17 X 10^4); FT-IR (KBr) 2189 and 2204 cm^{-1} (CN stretch); Anal. Calcd for C$_{24}$H$_{21}$N$_3$O$_2$: C, 75.18; H, 5.52; N, 10.96. Found: C, 75.13; H, 5.57; N, 10.86.

ACKNOWLEDGEMENTS

We appreciate the contributions of our colleagues at the Lockheed Martin Missiles and Space Advanced Technology Center involved in the development of electro-optic materials and devices: W.W. Anderson, D.G. Girton, J. Marley, M.M. Steiner, R. Taylor, and T.E. Van Eck. We thank S. Marder, S. Gilmour, and P. Bedworth[†] of CIT-JPL for helpful chemical discussions. This work was partially supported by Lockheed Martin Independent Research and Development Funds. We also appreciate the support of J. Zetts and K. Hopkins of the Wright Laboratory

---

[†] Current address: Lockheed Martin Advanced Technology Center, Palo Alto, CA 94304

Materials (WL/MLPO) Directorate for development of improved materials which will enhance performance and utilization of active polymer EO devices. Work at IBM was performed under partial support from the National Institutes of Standards and Technology.

REFERENCES

[1]Marder, S.R.; Beratan, D.N.; Cheng, L.-T. *Science* **1991**, *252*, 103.

[2]Moylan, C.R.; Twieg, R.J.; Lee, V.Y; Swanson, S.A.; Betterton, K.M.; Miller, R.D. *J. Am. Chem. Soc.* **1993**, *115*, 12599.

[3]Moylan, C.R.; Miller, R.D.; Twieg, R.J.; Lee, V.Y.; McComb, I.-H.; Ermer, S.; Lovejoy, S.M.; Leung, D.S. *Proc. SPIE* **1995**, *2527*, 150.

[4]Marder, S.R.; Cheng, L.-T.; Tiemann, B.G.; Friedli, A.C.; Blanchard-Desce, M.; Perry, J.W.; Skindhøj, J. *Science* **1994**, *263*, 511.

[5]Lytel, R.; Lipscomb, G.F.; Kenney, J.T.; Binkley, E.S in *Polymers for Lightwave and Integrated Optics: Technology and Applications*; Hornak, L.A., Ed.; Marcel Dekker: New York, 1992, pp 433-472.

[6]Moylan, C.R.; Ermer, S.; Lovejoy, S.M.; McComb, I.-H.; Leung, D.S.; Wortmann, R.; Krdmer, P.; Twieg, R.J., *J. Am. Chem. Soc.* **1996**, 118, 12950-12955.

[7]Ermer, S.; Valley, J.F.; Lytel, R.; Lipscomb, G.F.; Van Eck, T.E.; Girton, D.G. *Appl. Phys. Lett.* **1992**, *61*, 2272.

[8]Ermer, S.; Valley, J.F.; Lytel, R.; Lipscomb, G.F.; Van Eck, T.E.; Girton, D.G.; Leung, D.S.; Lovejoy, S.M. *Proc. SPIE* **1993**, *1853*, 183.

[9]Ermer, S.; Leung, D.S.; Lovejoy, S.M.; Valley, J.F.; Stiller, M. *Organic Thin Films for Photonics Applications Technical Digest* **1993**, *17*, 50.

[10]Ermer, S; Lovejoy, S; Leung, D.S.; in *Polymers for Second-Order Nonlinear Optics,* ACS Symposium Series 601, (American Chemical Society, Washington, D.C., **1995**)

[11]Girton, D.G.; Ermer, S.; Valley, J.F.; Van Eck, T.E. *Organic Thin Films for Photonics Applications Technical Digest* **1993**, *17*, 70.

[12]Lemke, R. *Chem. Ber.* **1970**, *103*, 1894.

[13]Lemke, R. *Synthesis* **1974**, 359.

[14]Lemke, R. Ger. Offen. 2 345 189, 1975; *Chem. Abstr.* **1975**, *83*, 29885s.

[15]Mignani, G.; Soula, G.; Meyrueix, R. Fr. Demande FR 2 636 441, 1990; *Chem. Abstr.* **1990**, *113*, 181123n.

[16]Gadret, G.; Kajzar, F.; Raimond, P. *Proc. SPIE* **1991**, *1560*, 226.

[17]Man, H.T.; Shu, C.F.; Althoff, O.; McCulloch, I.A.; Polis, D.; Yoon, H.N. *J. Appl. Poly. Sci.* **1994**, *53*, 641.

[18] Some examples of work in this area include: a) Lin, J.T.; Hubbard, M.A.; Marks, T.J.; Lin, W.; Wong, G.K. *Chem. Mater.* **1992**,*4*, 1148-1150; b) Zysset, B.; Ahlheim,. M.; Stähelin, M.; Lehr, F.; Prêtre, P.; Kaatz, P.; Günter, P., in *Nonlinear Optical Properties of Organic Materials VI*, Möhlmann, G.R., Ed.; Proc. SPIE 2025; Society of Photo-Optical Instrumentation Engineers: Bellingham, WA, 1993; pp 70-77; c). Meyrueix, R.; Tapolsky, G.; Dickens, M.; Lecomte, J.-P. In *Nonlinear Optical Properties of Organic Materials VI*, Möhlmann, G.R., Ed.; Proc. SPIE 2025; Society of Photo-Optical Instrumentation Engineers: Bellingham, WA, 1993; pp 117-128; d). Becker, M.W.; Sapochak, L.S.; Ghosen, R.; Xu, C.; Dalton, L.R.; Shi, Y.; Steier, W.H.; Jen, A.K.-Y. *Chem. Mater.* **1994**, *6*, 104-106; e) Sotoyama, W.; Tatsuura, S.; Yoshimura, T. *Appl. Phys. Lett.* **1994**, *64*, 2197-2199; f). Jen, A.K.-Y.; Drost, K.J.; Rao, V.P.; Cai, Y.; Liu, Y.-J.; Mininni, R.M.; Kenney, J.T.; Binkley, E.S.; Marder, S.R.; Dalton, L.R.; Xu, C. *ACS Polymer Preprints* **1994**, *35(2)*, 130-131; g). Yu, D.; Yu, L.*ACS Polymer Preprints* **1994**, *35(2)*, 132-133.

[19]Moylan, C.R.; Miller, R.D.; Twieg, R.J.; Betterton, K.M.; Lee, V.Y.; Matray, T.J.; Nguyen, C. *Chem. Mater.* **1993**, *5*, 1499.

[20]Moylan, C.R.; Miller, R.D.; Twieg, R.J.; Lee, V.Y. in *Polymers for Second-Order Nonlinear Optics*; Lindsay, G.A.; Singer, K.D., Eds.; ACS Symposium Series 601; American Chemical Society: Washington, DC, 1995, pp 66-81.

[21]Willetts, A.; Rice, J.E.; Burland, D.M.; Shelton, D.P. *J. Chem. Phys.* **1992**, *97*, 7590.

[22]Oudar, J.L; Chemla, D.S. *J. Chem. Phys.* **1977**, *66*, 2664.

[23]Cheng, L.-T.; Tam, W.; Feiring, A.; Rikken, G.L.J.A. *Proc. SPIE* **1990**, *1337*, 203.

[24]Cheng, L.-T.; Tam, W.; Marder, S.R.; Stiegman, A.E.; Rikken, G.; Spangler, C.W. *J. Phys. Chem.* **1991**, *95*, 10643.

[25]Moylan, C.R.; Walsh, C.A. *Nonlin. Opt.* **1993**, *6*, 113.

[26]Miller, R.D.; Moylan, C.R.; Reiser, O.; Walsh, C.A. *Chem. Mater.* **1993**, *5*, 625.

[27]Miller, R.D.; Lee, V.Y.; Moylan, C.R. *Chem. Mater.* **1994**, *6*, 1023.

[28]Moylan, C.R. *J. Phys. Chem.* **1994**, *98*, 13513.

[29]Cabrera, I.; Althoff, O.; Man, H.-T.; Yoon, H.N. *Adv. Mater.* **1994**, *6*, 43.

[30]Shu, C.-F.; Tsai, W.-J.; Jen, A.K.-Y. *Tetrahedron Lett.* **1996**, *37*, 7055.

[31]Shu, C.-F.; Tsai, W.J.; Chen, J.-Y.; Jen, A.K.-Y.; Zhang, Y.; Chen, T.-A. *Chem. Commun.* **1996**, 2279.

[32]Wortmann, R.; Lundquist, P.M.; Twieg, R.J.; Geletneky, C.; Moylan, C.R.; Jia, Y.; DeVoe, R.G.; Burland, D.M.; Bernal, M.-P.; Coufal, H.; Grygier, R.K.; Hoffnagle, J.A.; Jefferson, C.M.; Macfarlane, R.M.; Shelby, R.M.; Sincerbox, G.T. *Appl. Phys. Lett.* **1996**, *69*, 1657.

[33] Ermer, S.; Lovejoy, S.M.; Leung, D.S.; Warren, H.; Moylan, C.R.;R.D.; Twieg, R.J.; *Chem. Mater.*, **1997**, 9, 1437

[34]Bardakos, V.; Sandris, C. *Org. Magn. Res.* **1981**, *15*, 339.

[35]Ermer, S.; Lovejoy, S.M.; Leung, D.S. *Mat. Res. Soc. Symp. Proc.* **1995**, *392*, 3.

[36]Badcock, G.G.; Dean, F.M.; Robertson, A.; Whalley, W.B. *J. Chem. Soc.* **1950**, 903.

[37]Zeid, I.; El-Bary, H.A.; Yassin, S.; Zahran, M. *Liebigs Ann. Chem.* **1984**, 186.

[38]Eiden, F.; Kochs, I. *Planta Med.* **1967**, *15*, 81.

MULTIFUNCTIONAL HYPER-STRUCTURED MOLECULES

T. WADA*,†, Y. ZHANG*, T. AOYAMA†, Y. KUBO‡ and H. SASABE*,†
*Core Research for Evolutional Science and Technology (CREST), †Frontier Research
Program, The Institute of Physical and Chemical Research (RIKEN), JST, Wako, Saitama
351-0198, Japan, tatsuow@postman.riken.go.jp
‡Dept. of Applied Chemistry, Saitama University, Urawa, Saitama 338, Japan

ABSTRACT

To fill the gap between molecular design and the architecture of three-dimensional
functional structures, we propose novel hyper-structured molecules (HSMs) based on well-
defined and topologically controlled molecular systems. To this end we have developed
carbazole dendrimers, trimers, cyclic oligomers and chromogenic calix[4]arenes as HSMs.
Photorefractivity was selected as the primary target function of these HSMs. Oligomers
developed in our laboratory exhibit intrinsic photocarrier generation, transport, electro-optic,
film-forming and poling properties. These multifunctional properties allow us to demonstrate
optical image processing using optical phase conjugation. The topological shapes of
indoaniline-derived calix[4]arenes were studied by hyper-Rayleigh scattering. The two
indoaniline moieties in calix[4]arene derivatives were pre-aligned so as to enhance the net
molecular hyperpolarizability. Besides dendric oligomers, cyclic oligomers can be used as a
molecular platform which allows molecular level tuning of shape, size and topology for superior
opto-electronic functions.

HYPER-STRUCTURED MOLECULES

In biological molecular systems higher order structures of protein provide the special
coordination of active sites. Consequently efficient energy and/or electron transfer processes are
achieved. These biological functional structures have been built based on the specific interaction
of functional groups in amino acid residues in proteins. There have been many efforts to
synthesize molecular systems which mimic the biological systems. Even though we can control
the primary structures and tacticities of polypeptides by synthetic chemistry, there is a large gap
between molecular design and three-dimensional (3-D) functional structures, as illustrated in
Figure 1.

Organic molecules can order themselves and condense to form higher structures in a
specific manner. While the opto-electronic properties of the condensed state can be derived from
those of the isolated molecules, the organization of molecules also effects the properties of the
condensed state. Therefore we have to incorporate not only the organizing parameters but also
opto-electronic functions into the molecular building blocks in order to construct a functional 3-
D structure. Taking these ideas into account, we have utilized not a point-like molecule but
oligomers which offer more flexibility to build the 3-D functional structure.

To fill this gap between molecular design and the architecture of 3-D functional
structures, we propose hyper-structured molecules (HSMs) in which molecules themselves have
well-defined and specific structures [1,2]. In other words, the molecular design of HSMs is
directly related to the realization of the specific 3-D structure without molecular architecture.
Furthermore, functions in the molecular level can be expected in HSMs based on their specific
3-D structure as molecules with "topologically well-defined structure". HSMs have at least
more than two molecular units which must be coordinated or organized for particular purposes to
create a specific structure using various bond arrangements including covalent bonds.

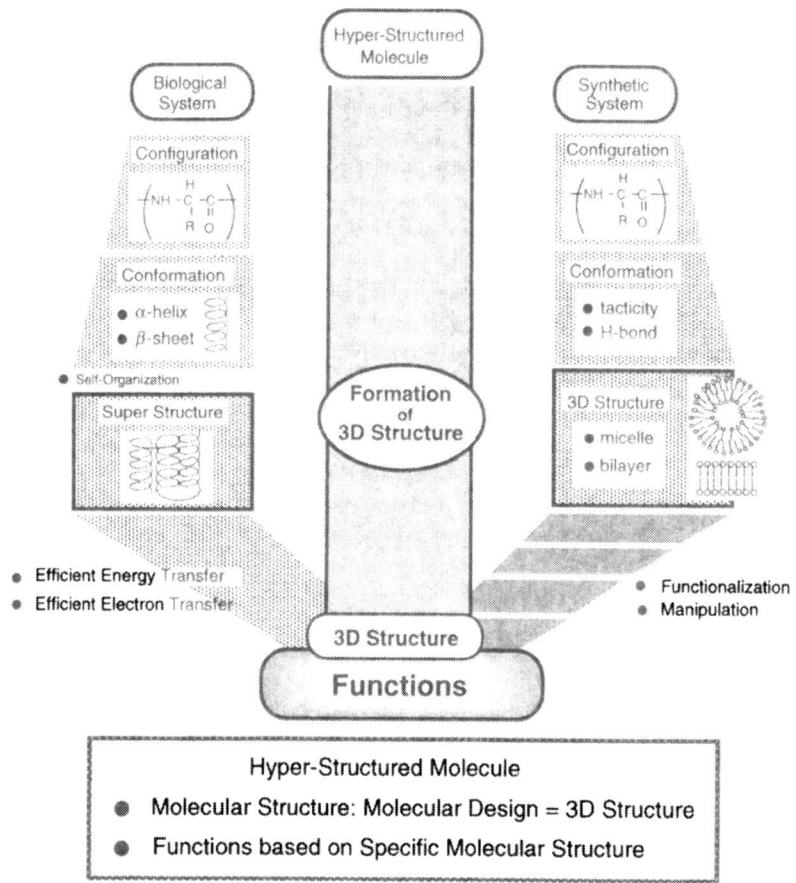

Figure 1. Approach to hyper-structured molecules.

One of the most interesting and important features in the dendrimer structure is based on their divergent structure, that is, the number of nodes increases depending on the generation as shown in Figure 2. In contrast to linear chain oligomers, the large number of terminal positions can be obtained. By introducing electroactive groups into the terminal positions of dendrimers, we can modify the total features of these dendrimers chemically and physically. While there is a structural freedom in linear chain oligomers, molecular shape of dendrimers can be controlled by the structure of the spacer units and the generation. Using rigid spacer groups, we can obtain stretched linear oligomers, however, these rod-like molecules show poor processability. On the other hand, it was reported that higher generation dendrimers are quite soluble in common organic solvents. In the case of cyclic oligomers, opto-electronic functional groups can be introduced as a member of a ring. Interaction through space and through bonds among these functional groups depends on the ring size and the sequence of ring members. In addition to ring size, the topological shape of a ring can be controlled using specific cyclic compounds such as chromogenic calixarenes. Besides the feasibility of chemical modification, interesting physical properties can be expected in these oligomeric systems.

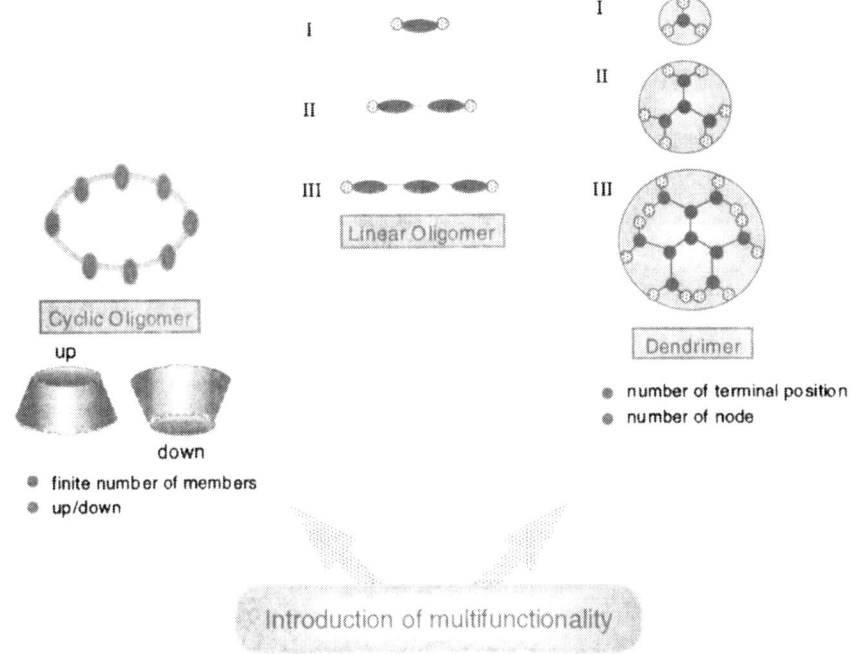

Figure 2. Generation of dendrimers, linear and cyclic oligomers.

MOLECULAR BUILDING BLOCKS

As the basic building block for these HSMs we selected an acceptor-substituted carbazole. Carbazole derivatives are well-studied hole-transporting molecules and their photocarrier generation efficiency can be sensitized by the formation of a charge-transfer complex. The carbazole molecule has an isoelectronic structure of diphenylamine. Therefore, the introduction of acceptor groups in the 3 and/or 6-position induces intramolecular charge transfer. The results of photoconductive and electro-optic experiments confirm the dual-functionality of acceptor-substituted carbazoles [3]. Using acceptor-substituted carbazole as a building block, we have developed dendrimers (Cz Dendrimer (A)), conjugated trimer (Cz Trimer (A)), and cyclic oligomers as shown in Figure 3. Based on the opto-electronic properties of these building blocks, we can expect dual-function of these oligomers.

The primary target of these HSMs is the synthesis of materials which display strong photorefractive properties. Although diffraction of light by index gratings is a bulk response, photorefractive effects require multifunctional responses in the materials [4]. This is one of the interesting and challenging targets for the demonstration of multifunctional performance of HSMs. The requirements for photorefractivity in organic materials are usually achieved by using multicomponents, each of which fulfill a single requirement. Recently remarkable progress has been made to improve the photorefractive performance in multicomponent photorefractive polymers by addition of a plasticizer which reduces the glass transition temperature (T_g) [5]. Although large photoinduced refractive index changes have been obtained in low-T_g multicomponent polymers, there is a limitation of maximum concentration of chromophores because of phase separation and crystallization. In order to overcome this problem, we have developed multifunctional chromophores which enable us to obtain "monolithic" photorefractive materials [6]. A single component of a multifunctional chromophore fulfills all the requirements for photorefractivity. It should be noted that an alternative to our approach is through fully functionalized polymers in which multicomponent moieties were incorporated into polymer side chains or main chains [7]. These fully functionalized polymers, including monolithic polymers have been developed as a macroscopic material, successfully suppress phase separation and increase the density of functional groups. However, they have distributions of molecular weight, number and size of free volume, and molecular structures. On the other hand, the oligomeric molecules have a perfectly defined structure.

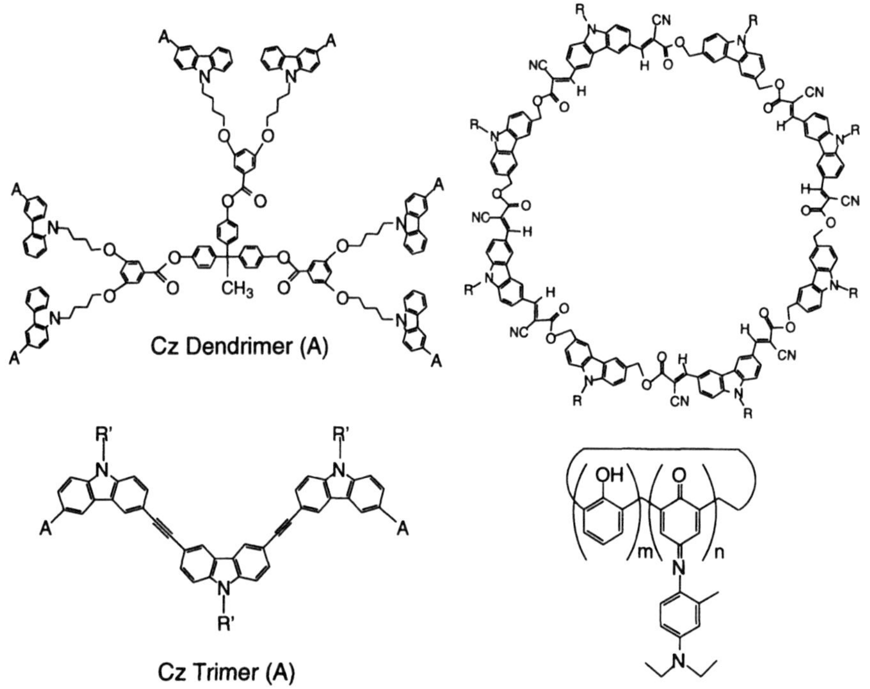

Figure 3. Hyper-structured molecules.

CARBAZOLE DENDRIMER AND TRIMER

Since carbazole dendrimers and conjugated trimers have a highly branched structure with long spacer groups, they have film-forming properties and show a glass-rubber transition. The T_g changes depending on the length of spacer groups and chromophore structures. Typical differential scanning calorimetry (DSC) traces of carbazole dendrimers with different acceptor groups (A) are shown in Figure 4. Amorphous solid films can be formed by spin coating from a solution, and poled at an elevated temperature above T_g. The electric field-induced alignment of the chromophores was confirmed by a second-harmonic generation (SHG) measurement.

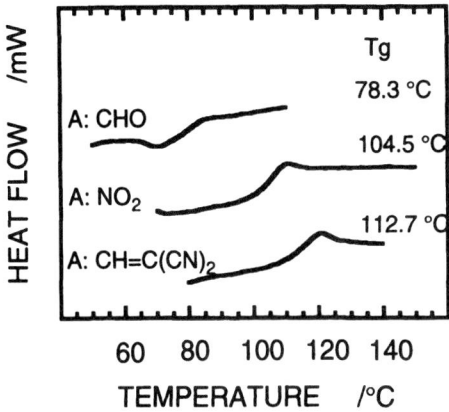

Figure 4. DSC traces for Cz Dendrimer (A).

Due to the attachment of a long aliphatic chain (R'=$C_{14}H_{29}$ in Figure 3), the carbazole trimer is amorphous and has a T_g of 20 °C. The second order nonlinear coefficient (d_{33}) at a fundamental wavelength of 1064 nm monotonically increased with an applied electric field.

Typical values of the electro-optic coefficient (r_{33}) and photoconductive sensitivity (σ/I) for carbazole trimers are also summarized in Table 1. In this trimer, carbazole rings are linked by an acetylene bond, and peripheral carbazoles are substituted with a nitro group. Acceptor-substituted carbazole moieties provide photocarrier generation and an electro-optic response. The central carbazole moiety has hole-transporting properties, and forms charge-transfer complexes with acceptor molecules such as 2,4,7-trinitro-9-fluorenone (TNF) [8]. A conjugated structure prefers to exhibit a relatively high photocarrier mobility [9]. In order to obtain a conjugated carbazole trimer with low T_g with film-forming properties a long aliphatic group was introduced at the 9-position of each carbazole ring.

Table I   Electro-optic coefficient (r_{33}) and photoconductive sensitivity (σ/I) for Cz Trimer (A).

A	r_{33} (pm/V)	σ/I (Scm^{-1}/Wcm^2)
$CH=C(CN)_2$	1.12* (30V/μm)	1.22 x 10^{-10} * (39V/μm) @633nm
NO_2	—	1.2 x 10^{-11} (39V/μm) @532nm

* doped with 0.06 wt% of 2,4,7-trinitro-9-fluorenone (TNF)

The photorefractive properties were characterized by the techniques of two-beam coupling and four-wave mixing. In the two-beam coupling measurement an asymmetric energy exchange between the two beams was observed when the electric field was applied. At an applied electric field of 33 V/μm, a photorefractive gain of 35.0 cm^{-1} was obtained for Cz Trimer (NO_2). Four-wave mixing was used to obtain phase conjugation. A strong electric field dependence of the diffraction efficiency was observed. At an applied electric field of 33 V/μm, a diffraction efficiency of 13 % was obtained and a photoinduced refractive index change of 3.6 x 10 $^{-4}$ was calculated. These efficient photorefractive effects allow us to demonstrate optical image processing. Figure 5 shows the phase conjugation via four-wave mixing in Cz Trimer (NO$_2$). The original image was distorted by the insertion of a phase aberrator. The wave front of this distorted image (Figure 5 (a)) was reconstructed via optical phase conjugation as shown in Figure 5 (b) [10].   In summary, we have developed "monolithic photorefractive materials" using multifunctional chromophores and demonstrated their capability to perform image reconstruction using low power laser sources.

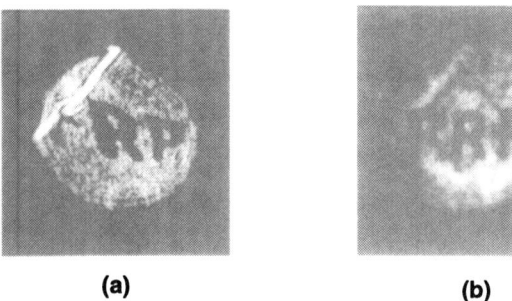

(a)   (b)

Figure 5.   Optical phase conjugation using four-wave mixing: (a) distorted image; (b) conjugated image.

CYCLIC OLIGOMERS

As a cyclic oligomer, two types of molecular systems were studied: carbazole cyclic oligomer and indoaniline-derived calix[4]arene. Recently we have developed the efficient synthesis of a novel carbazole cyclic oligomer and a main-chain polymer via the Knoevenagel condensation as shown in Scheme 1 [11]. The carbazole cyclic tetramer could be obtained in a high yield by a one-stage Knoevenagel condensation of 3,6-diformyl-9-heptylcarbazole and 3,6-bis(cyanoacetoxymethyl)-9-heptyl-carbazole in tetrahydrofuran (THF) without the use of the high dilution principle. In the polycondensation of diformyl and cyanoacetate, there are several interesting issues: the main products could be obtained in a polymer or a cyclic oligomer by controlling the reaction conditions. The cyclic oligomer was always obtained in THF reaction solution as a major product in a high yield. Size-exclusive chromatography shows a sharp peak which means that the cyclic oligomer has monodispersion of molecular weight, and only one kind of macrocycle was obtained from this condensation reaction. The carbazole main-chain polymer with very large molecular weight also could be obtained in a high yield by a two-stage Knoevenagel polycondensation. The polycondensation in THF followed by solid state polycondensation after removal of THF yielded the polymer as a main product.

We also found Knoevenagel condensation of 3,6-diformylcarbazole and bis(cyanoacetate)s can yield cyclic oligomers with a ring size which can be determined from the bis(cyanoacetate) structures. In the case of condensation of 3,6-diformylcarbazole and 1,4-bis(hydroxymethyl)benzene in THF solution, a yellow cyclic dimer was obtained as a major product in 78 % yield (as shown in Scheme 1). It was also found that the carbazole main-chain polymer could be obtained in 94 % by a two-stage Knoevenagel polycondensation [12]. The monomers and linear oligomers yielded by the polycondensation are soluble in THF. This condensation is a reversible reaction in the presence of a base and is the reason why we can not obtain carbazole main-chain polymer from the solution condensation. On the other hand, the cyclic oligomer is insoluble in THF. Once the cyclic oligomer was produced, the cyclic oligomer precipitated from the reaction solution. Therefore, the chemical equilibrium must be favorable for the formation of the cyclic oligomer in THF solution. This is the reason why the cyclic oligomer can be obtained in a high yield. In general, the production of high molecular weight polymers is often accompanied by the formation of cyclic oligomers with various ring sizes in polycondensation reaction [13]. However, in our case, only one size of the cyclic oligomer was obtained.

The cyclic oligomer has alternating units of acceptor-substituted and electron donative carbazole moieties connected through 3,6-linkages. Here we can expect efficient energy transfer in a cyclic oligomer. Figure 6 shows the excitation spectra at ~450 nm for carbazole tetramer and monomer solutions. Emission from the carbazole moieties in the cyclic oligomer excited at 300-350 nm was quenched and acceptor-substituted carbazole moieties played as an energy acceptor. Therefore relative excitation yields increased in cyclic oligomer at 300-350 in comparison to those of a carbazole monomer as shown in Figure 6.

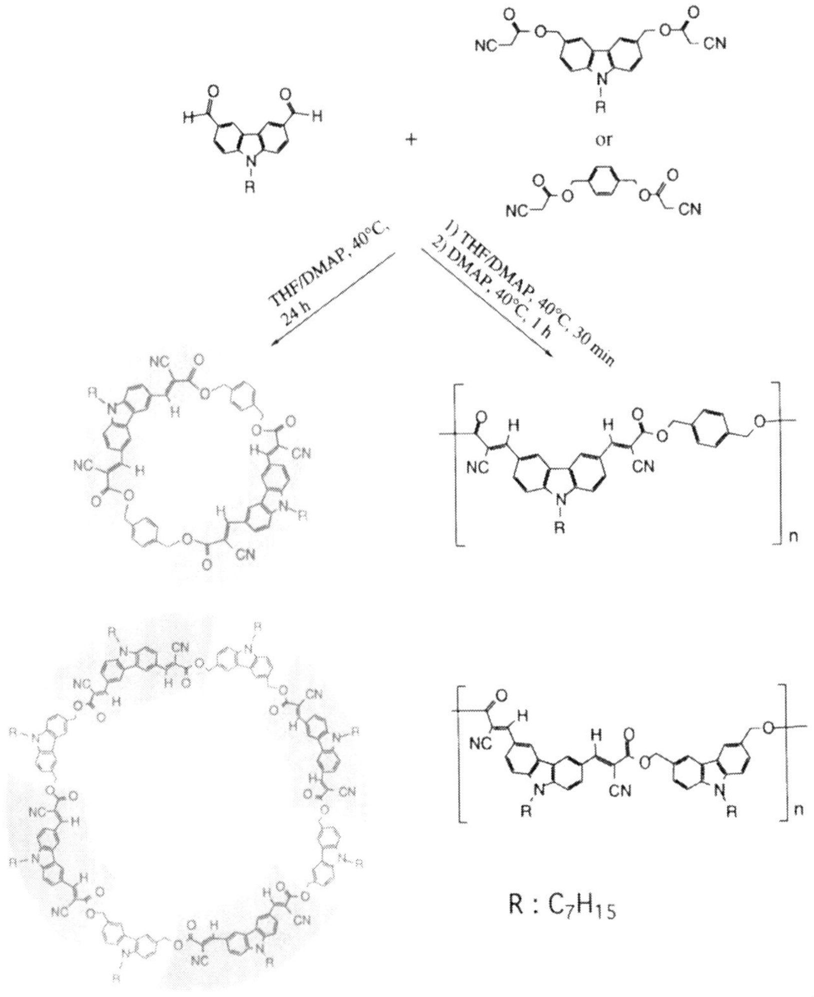

Scheme 1. Synthesis of carbazole cyclic oligomer and main-chain polymer.

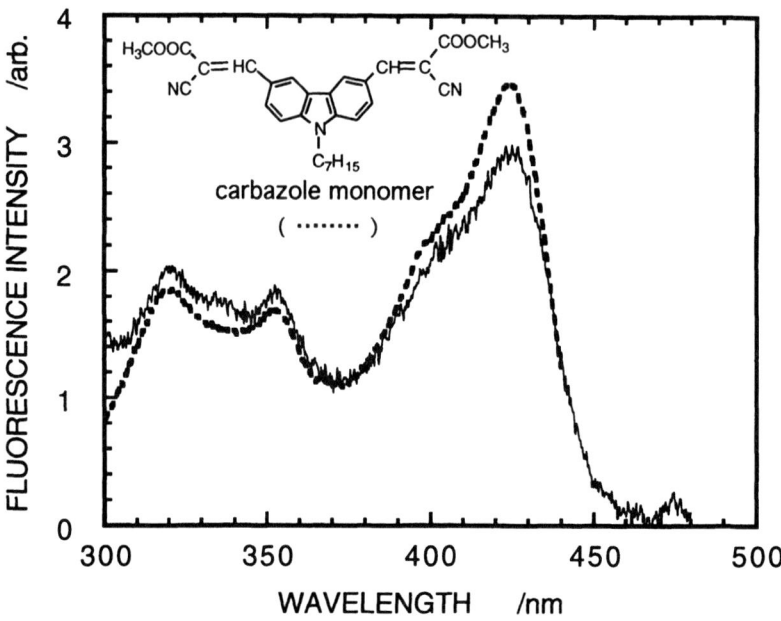

Figure 6. Excitation spectra for carbazole cyclic oligomer (————) and monomer
(-------) solutions.

INDOANILINE-DERIVED CALIX[4]ARENES

As a chromogenic calix[4]arene, we describe the indoaniline-derived calix[4]arene with different numbers of indoaniline moiety. Originally this chromogenic calix[4]arene was developed as an artificial receptor: host-guest interaction induced the change of absorption for sensing by a chromophore group due to a perturbation of the electronic state [14]. Synthesis is quite straightforward except for separation of number of isomers. Calix[4]arenes with two indoaniline groups (Figure 7) have four different conformers: cone, partial cone and their alternates. In order to estimate the conformation of calixarenes, we used the hyper-Rayleigh scattering (HRS) technique. Hyper-polarizability (β) is related to the difference between the ground state and excited state dipole moment, and oscillator strength. To obtain a compound with a large β value, we must optimize each parameters for β. Indoaniline compound has a larger β value than that of donor-acceptor stilbene due to the contribution of mixed benzene and quinoid structure in the ground state [15]. Since HRS signal at a fundamental wavelength of 1064 nm located in the absorption band of indoaniline derivatives, one may argue the multi-photon emission besides HRS. A time-resolved subpicosecond HRS study however showed that we could distinguish HRS signals from the multi-photon fluorescence. The β value of 870 x 10^{-30} esu for indoaniline was obtained. On the other hand, that of 1,3-bis(indoaniline)-derived calix[4]arene was 1300 x 10^{-30} esu. Enhancement factor was less than 2, however, two indoaniline sites were pre-aligned in the solution. Using calix[4]arene as a molecular platform, we can also control the topological shape of cyclic oligomers.

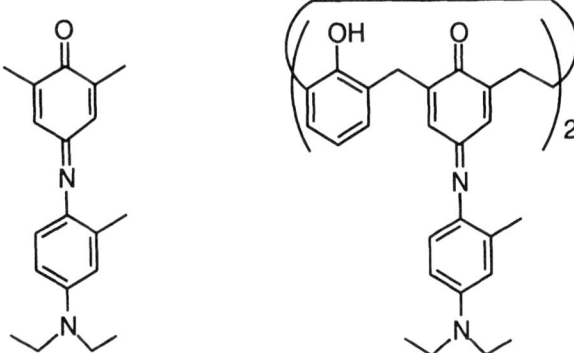

Figure 7. Chemical structures of indoaniline and 1,3-bis(indoaniline)-derived calix[4]arene.

CONCLUSION

We have proposed and demonstrated the novel idea of hyper-structured molecules in which the molecular design is directly related to the realization of the specific 3-D structure. As HSMs, we have developed carbazole dendrimers and trimers which exhibit multifunctional properties. These systems allow us to obtain monolithic photorefractive materials. We believe the approach using HSMs opens new possibilities for photonic applications.

REFERENCES

1. H. Sasabe, Hyper Structred Molecules I-Chemistry, Physics and Applications, H. Sasabe (Ed), Gordon and Breach Sci. Pub., in press

2. T. Wada, Y. Zhang and H. Sasabe, Hyper Structred Molecules II-Chemistry, Physics and Applications, H. Sasabe (Ed), Gordon and Breach Sci. Pub., in press

3. T. Aoyama, T. Wada, Y.-D. Zhang, N. Saito, H. Sasabe and K. Sasaki, Mol. Cryst. Liq. Cryst., **295**, 63 (1997).

4. W. E. Moerner and S. M. Silence, Chem. Rev., **94**, 127 (1994).

5. K. Meerholz, B. L. Volodin, Sandalphon, B. Kippelen and N. Peyghambarian, Nature, **371**, 497 (1994).

6. T. Wada, L. Wang, Y. Zhang, M. Tian and H. Sasabe, Nonlinear Optics, **15**, 103 (1996).

7. L. Yu, W. Chan, Z. Bao and X. F. Cao, Macromolecules, **26**, 2216 (1992).

8. L. Wang, Y. Zhang, T. Wada and H. Sasabe, Appl. Phys. Lett., **69**, 728 (1996).

9. C. Beginn, J. V. Grazulevicius and P. Strohriegel, Macromol. Chem. Phys., **195**, 2353 (1994).

10. T. Wada, Y. Zhang, T. Aoyama and H. Sasabe, Proc. Japan Acad., **73**, 165 (1997).

11. Y. Zhang, T. Wada and H. Sasabe, Tetrahedron Lett., **37**, 5909 (1996).

12. Y. Zhang, L. Wang, T. Wada and H. Sasabe, Macromolecules, **29**, 1569 (1996).

13. G. Montaudo, E. J. Scamporrrino, C. Puglisi, D. Vitalini, Polym. Sci., Part A: Polym. Chem., **25**, 1653 (1987).

14. Y. Kubo, S. Tokita, Y. Kojima, Y. Osano and T. Matsuzaki, J. Org. Chem., **61**, 3758 (1996).

15. S. R. Marder, D. N. Beratan, L.-T. Cheng, Science, **252**, 103 (1991).

SYNTHESIS OF A NEW CLASS OF POLYPHENYLATED PHTHALOCYANINES AND METALLOPHTHALOCYANINES

CHRISTOPHER J. WALSH AND BRAJA K. MANDAL*

Department of Biological, Chemical, and Physical Sciences, Illinois Institute of Technology, Chicago, IL 60616

ABSTRACT

We wish to report the synthesis of a new class of peripherally modified, polyphenylated phthalocyanines (Pcs) and metallophthalocyanines (MPcs). These compounds were synthesized by cyclic tetramerization of functionalized pentaphenylbenzophthalonitriles. The phthalonitriles were made in several steps via Diels-Alder cycloaddition between phenylethynylphthalonitrile and substituted tetracyclones. The polyphenylated design leads to enhanced solubility. Thin films can be cast from solutions of the butoxy and dodecaoxy substituted polyphenylated Pcs and MPcs. Cyclodehydrogenation of the Pcs was studied in an attempt to produce planar systems. Preliminary results indicate the pentaphenyl substituent is too bulky to planarize under the conditions employed.

INTRODUCTION

The phthalocyanines and metallophthalocyanines continue to play an important role in the development of new materials for electrical, optical and magnetic applications.[1] In the past decades, a number of synthetic strategies have been developed to obtain novel Pc an MPc derivatives. New materials based on these derivatives can be classified into three categories: (i) axially modified MPc derivatives, some of which are polymerized through either covalent or coordinative bonds[2], (ii) symmetrical and peripherally, disubstituted isomers for uncrosslinked ladder-type polymers[2a], (iii) peripherally modified derivatives for which many different applications, including the development of liquid crystalline mesophases[3] and non-linear optics and photonics[4], have been reported. In the past several years, this laboratory has investigated novel axially and peripherally modified Pcs and MPcs mainly for their use in optical applications such as optical switching via non-resonant third-order nonlinear absorption and optical limiting.[5]

A recent development in peripheral modification chemistry has been the use of the Pc macrocycle as a core subunit around which a dendritic macromolecule was built.[6] Preliminary investigations have shown that it is possible to alter the aggregating properties of the core macrocycle. Recently, this laboratory has taken initiatives to develop novel polyaromatic, star-branched, peripherally modified Pcs and MPcs for NLO applications. This paper will report on the progress of this work, namely on the synthesis of polyphenylated Pcs and MPcs.

Although this is a general approach, polyphenylated MPcs should exhibit properties suitable for optical limiting applications. The use of MPcs as optical limiters has been developed[7] and the important issues have been outlined as follows: fast reverse saturable absorption through fast intersystem crossing rates (from singlet to triplet excited states), high thermal and laser damage thresholds and high transmittance in the working wavelength region at low light fluence. The polyphenylated design would improve the processability (i.e. solubility), while increasing the molecular weight and maintaining its high thermal stability. It is well known that polyphenylenes have high performance characteristics. The ground state absorption of the MPc would increase much in the working range (532nm), and the intersystem crossing rates can be controlled by the appropriate choice of metal.

Mat. Res. Soc. Symp. Proc. Vol. 488 © 1998 Materials Research Society

EXPERIMENTAL

Typical Diels -Alder Cycloaddition Reaction to form **3c**

Into a 50 mL drying ampul (ChemGlass) was added 2.00 g (8.81 mmoles) of **1** and 4.65 g (8.81 mmoles) of **2c** along with 3 drops of cyclohexylbenzene. The ampul was sealed under vacuum after flushing with dry nitrogen gas. The tube was placed into a 215°C furnace for 6 hours. Upon cooling, the tube was broken and the mixture was deposited onto silica gel using dichloromethane. The solid was loaded onto a silica gel column and purified using flash chromatography and toluene/hexanes (1:1) as the eluent. The fractions containing purified **3c**, detected and analyzed by TLC, were collected and the solvent evaporated *in vacuo*. Recrystallization from methylcyclohexane gave the title compound as a white solid, melting point 205°C, yield 65% (4.17 g).

Typical Cyclization Reaction of **3c** to form **4b**

Into a 5 mL drying ampul (ChemGlass) was added 53.1 mg (0.073 mmoles) of **3c** along with 4.9 mg of hydroquinone. The ampul was flushed with nitrogen and sealed under vacuum. The ampul was placed in a 225°C furnace for 48 hours. Upon cooling, the ampul was broken and the contents were deposited onto silica gel using dichloromethane. The solid was loaded onto a silica gel column and purified using flash chromatography with dichloromethane as eluent. The green fractions were collected and combined and the solvent removed *in vacuo*. This produced a waxy green solid, 11mg (0.004 mmoles), 20%.

RESULTS AND DISCUSSION

The syntheses of polyaromatic systems has been well investigated, and the use of the Diels-Alder cycloaddition reaction between alkyne and cyclopentadieneone derivatives has been shown to be a powerful tool for building high molecular weight, fully aromatic compounds and polymers.[8] The current approach, outlined in Scheme I, consists of a Diels-Alder reaction between 4-phenylethynylphthalonitrile, **1**, and 2,3,4,5-tetraphenylcyclopentadiene-1-one (tetracyclone), **2a**, to produce 4-(2,3,4,5,6-pentaphenyl)-phthalonitrile, **3a**. This compound could then be forced to cyclictetramerize to produce the Pc of interest, **4a**. This has proven to be possible and the methodology was extended to include alkoxy substituted derivatives such as **4b-f**.

Scheme I

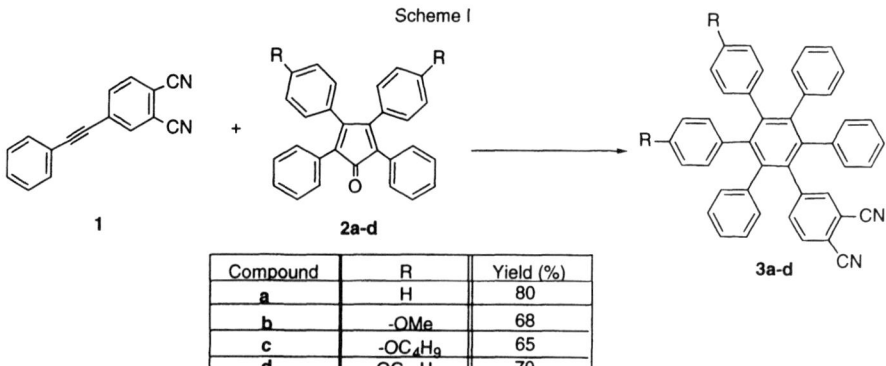

Compound	R	Yield (%)
a	H	80
b	-OMe	68
c	$-OC_4H_9$	65
d	$-OC_{12}H_{25}$	70

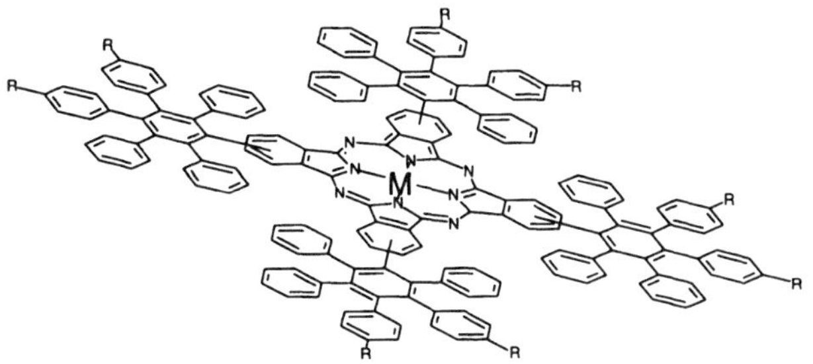

4a-f

Compound	R	M	Yield (%)
4a	H	2H	10
b	-OC$_4$H$_9$	2H	20
c	-OC$_4$H$_9$	Ni	37
d	-OC$_4$H$_9$	Pb	41
e	-OC$_{12}$H$_{25}$	2H	30
f	-OC$_{12}$H$_{25}$	Pb	45

The synthesis of **1** begins with 4-nitrophthalonitrile as shown in Scheme II. Reduction[9] followed by diazotization and iodide displacement gave 4-iodophthalonitrile, **5**, which was catalytically coupled with phenylacetylene by Pd and Cu co-catalysts to prepare the Diels-Alder adduct **1**.[10] The cycloaddition reaction between **1** and **2** was selective in that the alkyne acted as the dienophile and not the nitrile. This is significant in that both possibilities have been reported at the reaction temperature of 215°C. The cycloaddition was performed neat, plasticized by cyclohexylbenzene, in a sealed tube, and gave the adduct **3a**. In order to synthesize alkoxy substituted pentaphenylbenzophthalonitriles, such as **3b-d**, the substituted tetracyclones **2b-d** were used. These compounds were synthesized according to Scheme III, by cyclocondensation of the 1,2-diketones **6a-c** and diphenyl acetone, **7**.

Scheme II

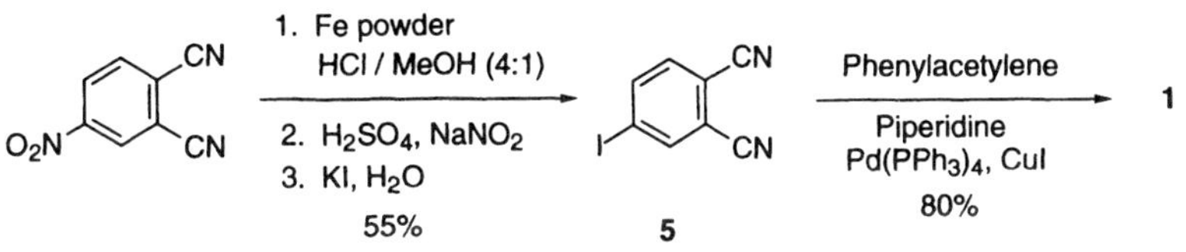

Scheme III

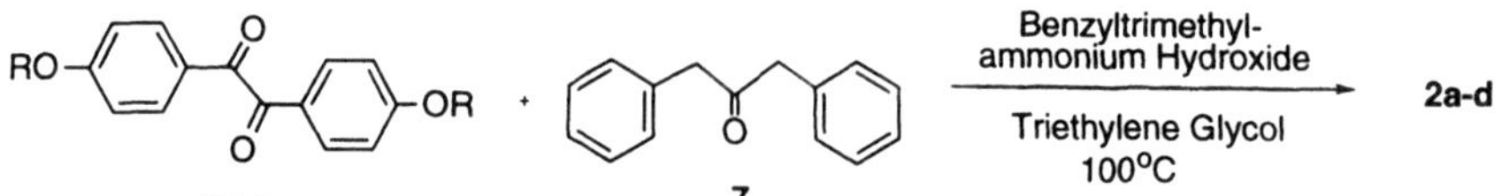

RO—[benzil structure]—OR   +   [dibenzyl ketone structure 7]

Benzyltrimethyl-
ammonium Hydroxide
———————————→   **2a-d**
Triethylene Glycol
100°C

6a-c   **7**

Compound	R
6a	-CH$_3$
6b	-C$_4$H$_9$
6c	-C$_{12}$H$_{25}$

Tetracyclones **2b-c** were both readily prepared via the method of Fieser.[11] Basic catalyzed condensation in triethylene glycol (TEG), at 100°C gave **2b-c** in 85% and 65% yields respectively in the presence of a basic phase-transfer catalyst, benzyltrimethylammonium hydroxide. However, compatibility problems between the diketone **6c** and TEG due to poor solubility forced a change in solvent. After extensive research compound **2d** was obtained in the highest yield of 58% using cyclohexanol as the solvent and a full equivalent of base.

The diketones **6b-c** were available via two different routes as shown in Scheme IV. In the first method, starting from 4,4'-dimethoxybenzil, deprotection using pyridinium hydrochloride and etherification gave alkoxy derivatives **6b-c** in excellent yields (75%-85%). In the second method, starting from 4-iodophenol, etherification followed by lithiation, via metal-halogen exchange, and reaction with N,N'-dimethylpiperazine-2,3-dione (DMPD)[12] gave **6b-c** in 70% and 30% respectively. The low yield for **6c** reflects the low reactivity of the lithiated species with DMPD, due to insolubility of the alkoxylated carbanion in THF at -78°C. The cycloaddition reaction between **1** and **2b-d** yielded the dinitriles **3b-d** in 68%, 65% and 70% yields, respectively.

Scheme IV

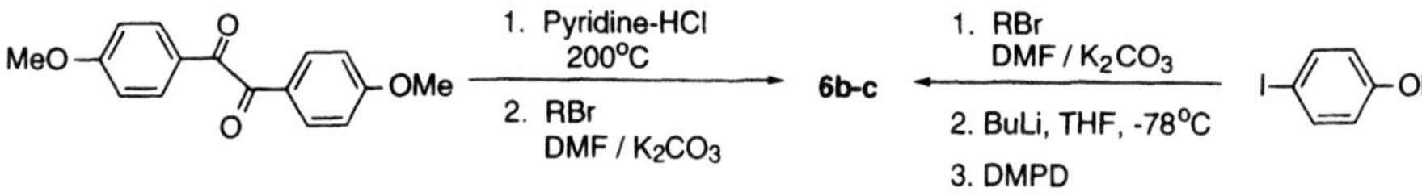

MeO—[benzil structure]—OMe

1. Pyridine-HCl
200°C
———————————→   **6b-c**

1. RBr
DMF / K$_2$CO$_3$
←———————————

2. RBr
DMF / K$_2$CO$_3$

2. BuLi, THF, -78°C

3. DMPD

I—[phenol structure]—OH

Cyclictetramerizations of the dinitriles had to be performed as a melt and proved to be difficult probably due to the steric crowding of the penta-aryl substituent. Reaction of **3a-d** in the presence of hydroquinone[13] produced the Pcs in 10%, 10%, 20% and 30% yields respectively. In the first series, **a** and **b**, the reaction temperature had to be maintained near 300°C due to the high melting points of the dinitriles (305°C and 265°C). When the melting points were lowered significantly as in **3c-d** (205°C and 180°C), less decomposition and more Pc formation was observed when the reaction was run at 225°C. Attempts to cyclize dinitriles and the corresponding isoindolines (even in the presence of metal salts) in solution was fruitless. Reasons for this are unclear, but solubility was a factor for the isoindoline of **3a** which was neither soluble in n-octanol or dimethylamino ethanol. The metallated derivatives were also produced neat, in the presence of metal salts, and in most cases gave better yields (30%-40%).

The polyphenylated Pcs and MPcs have different physical properties, but all maintain similar electronic properties. All the materials synthesized are soluble in common organic solvents such as aromatic hydrocarbons like toluene and xylene, chlorinated hydrocarbons like chloroform and dichloromethane, and cyclic ethers such as THF and 1,4-dioxane. The unsubstituted polyphenylated Pc is crystalline and doesn't exhibit a melting point below 400°C. However, the butoxy substituted macromolecule is much more amorphous and forms films easily from solution.

Recently, there have been reports on the synthesis and planarization of dendritic polyphenylenes.[14] It was of interest to us whether or not the polyphenylated Pcs and MPcs could be planarized. This would produce large, conjugated molecules which may show interesting properties such as near-IR electronic absorption and discotic liquid crystallinity. Attempts were made to planarize the pentaphenylbenzo substituents on the metal-free butoxy substituted derivative, 4d. However, initial attempts have not been successful. The planarization reactions were run photochemically with and without iodine catalysts and were monitored by a UV-Vis spectrophotometer. Only slow decomposition of the Pc was observed. Attempts to planarize the dinitrile precursor, before cyclictetramerization to the macrocycle are in progress.

The electronic absorptions of 4b and d are shown in Figure 1. The metal free derivative shows a Q-band λ_{max} at 712 nm. This absorption is red-shifted about 40 nm compared to the unsubstituted Pc. Alkoxy substitution of the pentaphenylbenzo substituent has no additional effect on the λ_{max} of the Q-band as compared to the unsubstituted derivative, 4a. Insertion of a metal changes the position of both the Q-band and B-band. The Pb derivative, 4d, has the farthest red-shifted Q-band, placed at 724 nm, and B-band, placed at 356 nm. However, as is typical with PbPcs, the compound 4d decomposes rapidly over time (24 hours) in chlorinated and nucleophilic solvents (CHCl$_3$ and 1,4-dioxane) when stored in the light. It is interesting to note that decomposition occurred much more slowly in toluene (several days). All the Pcs and MPcs have the typical window in which there is little or no ground state absorption between 500-540 nm. The studies on the NLO properties are in progress and will be reported elsewhere.

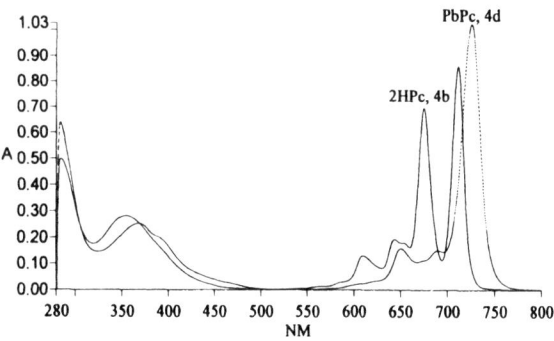

Figure 1. Ground state electronic spectra of 4b and d.

ACKNOWLEDGMENTS

This work is supported in part by the U.S. Army Research Office under the grant #DAAH04-96-I-0130 and by the IIT-ERIF grant.

REFERENCES

(1) For general references see: (a) Phthalocyanines Properties and Applications edited by C. C. Leznoff and A. B. P. Lever (VCH, New York, 1989). (b) Phthalocyanine Research and Applications edited by A. L. Thomas (CRC Press, New York, 1990).

(2) (a) M. Hanack, M. Lang, Adv. Materials 6(11), 819 (1994). (b) M. Hanack, S. Deger, A. Lange, Coord. Chem. Rev. 83, 115 (1988). (c) B. K. Mandal, A. K. Sinha, M. Kamath, Polymer Journal 27, 529 (1995).

(3) (a) J. F. Pol, E. Neelemen, J. C. Miltenburg, J. W. Zwikker, R. J. M. Nolte, W. Drenth, Macromolecules 23, 155 (1990). (b) M. J. Cook, D. A. Mayes, R. H. Poynter, J. Mater. Chem. 5, 2233 (1995).

(4) (a) Nonlinear Optics of Organic Molecules and Polymers edited by H. S. Nalwa, S. Miyata (CRC Press, Boca Raton, FL, 1997). (b) J. W. Perry, K. Mansour, I. Y. S. Lee, X. L. Wu, P. V. Bedworth, C. T. Chen, D. Ng, S. R. Marder, P. Miles, T. Wada, M. Tian, H. Sasabe, Science 273, 1533 (1996). (c) P. V. Bedworth, J. W. Perry, S. R. Marder, Chem. Comm. 1353, (1997). (d) J. S. Shirk, J. R. Lindle, F. J. Bartoli, M. E. J. Boyle, Phys. Chem. 96, 5847, (1992). (e) R. D. George, A. W. Snow, J. S. Shirk, S. R. Flom, R. G. S. Pong, Polym. Prep. 35(2), 236 (1994).

(5) B. K. Mandal, B. Bihari, A. K. Sinha, M. Kamath, Appl. Phys. Lett. 66(8), 932 (1995).

(6) (a) M. Kimura, K. Nakada, Y. Yamaguchi, K. Hanabusa, H. Shirai, N. Kobayashi, Chem. Comm. 1215 (1997). (b) B. K. Mandal, A. K. Sinha, X. Li, C. J. Walsh, D. Lamb, O. J. Smith, in Electrical, Optical and Magnetic Properties of Organic Solid State Materials III edited by A. K-Y. Jen, C. Y-C. Lee, L. R. Dalton, M. F. Rubner, G. E. Wnek, L. Y. Chiang (Mater. Res. Soc. 413, Pittsburgh, PA, 1996) pp. 431.

(7) see J. W. Perry, ref. 4a, pp. 813-839.

(8) (a) U. Kumar, T. X. Neenan, Macromolecules 28, 124 (1995). (b) Y. Sakaguchi, F. W. Harris, Polymer Journal 24(10), 1147 (1992). (d) for a general review see M. A. Ogliaruso, M. G. Romanelli, E. I. Becker, Chem. Rev. 65, 261 (1965).

(9) J. Griffiths, B. Roozpeikar, J. Chem. Soc. Perkin Trans. 1, 42 (1976).

(10) M. Alami, F. Ferri, G. Linsrumelle, Tetrahedron Lett. 34(40), 6403 (1993).

(11) Organic Experiments edited by L. F. Fieser (D. C. Heath and Co., Boston, 1964) pp. 303.

(12) U. T. Mueller-Westerhoff, M. Zhou, J. Org. Chem. 59, 4988 (1994).

(13) A. N. Snow, N. L. Jarvis, J. Amer. Chem. Soc. 106, 4706 (1984).

(14) F. Morgenroth, E. Reuther, K. Muellen, Angew. Chem. Int. Ed. Engl. 36, 632 (1997).

STUDY ON THE RELATIONS OF STRUCTURE AND THE ULTRAFAST OPTICAL KERR EFFECT OF POLYNITRILES

Q. WANG[1], Y. Y. LI[1], W. W. PEI[1], Y. K. HE[1], C. Q. LUO[1], H. Y. CHEN[1*], L. LIN[2], C. F. WANG[2], Y. H. ZOU[2]
[1]Peking Univ., College of Chemistry and Molecular Engineering, Beijing 100871, P. R. China
[2]Peking Univ., Dept. of Physics, Mesoscopic Physics Lab, Beijing 100871, P. R. China

ABSTRACT

A series of polybenzonitrile derivatives, prepared by plasma glow discharge processes or chemical polymerization, were characterized by FT-IR, UV-Vis, GPC and VPO etc. The second-order nonlinear optical hyperpolarizability γ of various benzonitrile monomers and their polymers were measured by the femtosecond time-resolved optical Kerr effect technique, and the structure-property relationships were discussed. The power law dependence of γ on the average-polymerization-degree of several polybenzonitrile derivatives were also reported.

INTRODUCTION

Organic conjugated polymers have attracted much attention because of their optical nonlinearity, fast response, relatively low cost, ease of fabrication and integration into devices[1-2]. Many contributions concerned on the third-order polarizability and structure correlation study of some typical conjugated polymers, such as polyacetylene, polythiophene, polyaniline etc.[3-6], manifested that the molecular geometry, electronic state, the donor or acceptor substitutents and conjugated chain length affected the third-order nonlinear susceptibility of the materials. These polymers are all of carbon-carbon or heteroaromatic conjugation system. We have reported a novel conjugated system, polynitriles with unique third order nonlinear optical properties in the previous papers [7, 8]. In this paper we report the further study on the synthesis of various polybenzonitrile derivatives, and the relationship between molecular structure and optical nonlinearity of polybenzonitriles.

EXPERIMENTS

Benzonitrile(BN, >99%) and p-tolunitrile(TN, >98%) were purified by vacuum distillation just before use. The p-aminobenzonitrile(ABN, >98%) and p-nitrobenzonitrile(NBN, >99%) monomers were recrystallized from ethanol, and p-hydroxybenzonitrile(HBN, >97%) recrystallized from hot water. 3,4-Dimethoxybenzonitrle (DMBN)was synthesized according to literature[9].

Plasma polymerization of Benzonitriles (BN, ABN, NBN, NBN and TN) was carried out by radio-frequency plasma glow discharge. The process was described in literature[8]. PTN was dissolved in chloroform and reprecipitated by anhydrous ether. PHBN was dissolved in tetrahydrofuran and precipitated by anhydrous ether. Both PTN and PHBN were extracted with anhydrous ether to remove unreacted monomers. Poly(3,4-dimethoxybenzonitrile) (PDMBN) were prepared by chemical polymerization. The monomer and boron-trifluoride-anisole complex catalyst were sealed in a glass tube under vacuum, polymerized at 100°C for 24 hours. The polymers prepared by plasma method had a large molecular weight distribution from 2 to 5.1, so the polymers were fractionated by Gel Permeation Chromatography(WATERS 208 GPC analyzer) with dimethylformamide or tetrahydrofuran. The average number molecular weight of selected polymer fractions were measured by KNAUER-1000 Vapor Pressure Osmometer at 90°C. The polymers were characterized with Nicolet Magna FT-IR 750 infrared analyzer and Shimadu-250 UV-Vis spectrometer. The optical Kerr effect measurements were performed by using ultrashort laser pulses generated from an Satori Model 774 ultrafast dye laser. The detailed procedures have been described in Ref[10].

RESULTS

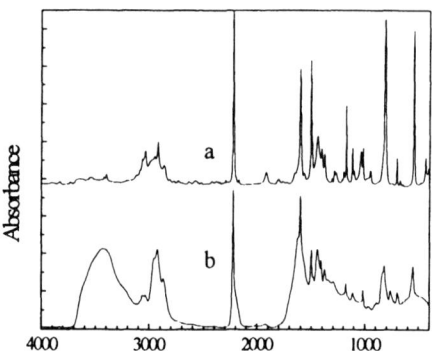

Figure 1. IR spectra of a) tolunitrile monomer, and b) polytolunitrile.

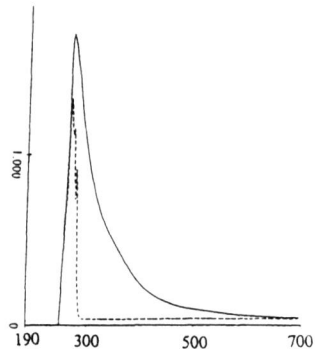

Figure 2. UV-Vis spectra of tolunitrile monomer (dash line), and polytolunitrile(solid line).

PTN, PBN, PABN, PHBN, and PNBN were prepared by plasma polymerization at various discharge power from 30 to 80W. Figure 1 shows the infrared spectra of p-tolunitrile monomer and its polymer. There is a strong broad band at $1640cm^{-1}$ in polytolunitrile attributing to the vibration of C=N double bond. From the UV-Vis spectra shown in Figure 2, a broad absorption band that extends to the visible light region in polytolunitrile can be observed, again revealing the formation of conjugated $-(C=N)_m-$ in PTN. For other polymers mentioned above, including PDMBN synthesized by chemical polymerization, similar results were obtained. Furthermore, by plasma polymerization, relatively high molecular weight up to 10^4 for polybenzonitriles could be obtained, and polymer thin films with good optical properties were prepared by deposition on glass or CaF_2 substrates.

The dipole moment, Π_n^m, the lowest intense absorption wavelength λ_{max} of monomers, the third-order nonlinear susceptibility $\chi^{(3)}$ and the second-order hyperpolarizability γ data of various benzonitrile monomers and their polymers are listed in Table 1, where various benzonitrile monomers are arranged in the order with decreasing π conjugation extent. The order of γ values for monomers follows as: $HBN(\Pi_9^{10})>ABN(\Pi_9^{10})>TN(\Pi_8^8) >NBN (\Pi_{11}^{12})>BN(\Pi_8^8)$. The para-donor substituted benzonitrile derivatives , such as HBN, ABN and TN with larger dipole moment than NBN and BN, possess γ values of one order higher than NBN and BN. But NBN monomer with dipole moment much lower than BN, has larger γ value than BN, implying that the state of conjugation influence is more important. Further, this is also in the case of HBN, TN and ABN, where γ values did not increase with increasing dipole moment. Similar influence of charge transfer and conjugation extension on the third-order nonlinear susceptibility had also been reported on the donor-acceptor capped cyanine[11] as well as the para-substituted benzenes and stilbezene[12].

Table 1. Structure-property data of various benzonitrile monomers and polymers[a].

Structure of monomer (abbreviation)	dipole moment of monomer[b]	Π_n^m of monomer[b]	λ_{max}[b] (nm)	$\chi^{(3)}$ $(10^{-15}$esu) of monomer	γ $(10^{-33}$esu) of monomer	$\chi^{(3)}$ $(10^{-13}$esu) polybenzonitrile(m=6)	γ $(10^{-31}$esu) polybenzonitrile(m=6)
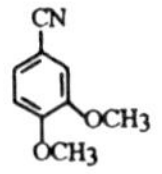 DMBN	3.419	Π_{10}^{12}	291	/	/	164[c]	216[c]
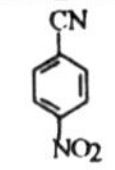 NBN	1.208	Π_{11}^{12}	258	4.4	5.6	1.9	2.4
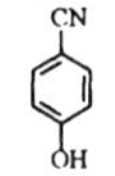 HBN	4.354	Π_9^{10}	283	23	38	4.2	5.3
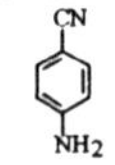 ABN	4.610	Π_9^{10}	268	11	14	1.6	2.1
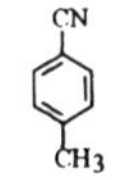 TN	4.190	Π_8^8	277	32	31	0.82[c]	1.0[c]
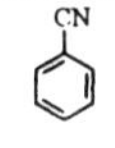 BN	4.091	Π_8^8	275	3.0	3.8	1.2	1.6

a. The samples for optical Kerr effect measurement were kept at 4.0×10^{-4}M in DMF.
b. The dipole moment of monomers were calculated with the PCM software; Π_n^m stands for n atoms with m π electrons involves in conjugation; λ_{max} is the lowest intense absorption wavelength of different monomers detected by UV-Vis analysis.
c. The average polymerization degree m for the PDMBN sample was 9.1, for PTN sample was 5.3.

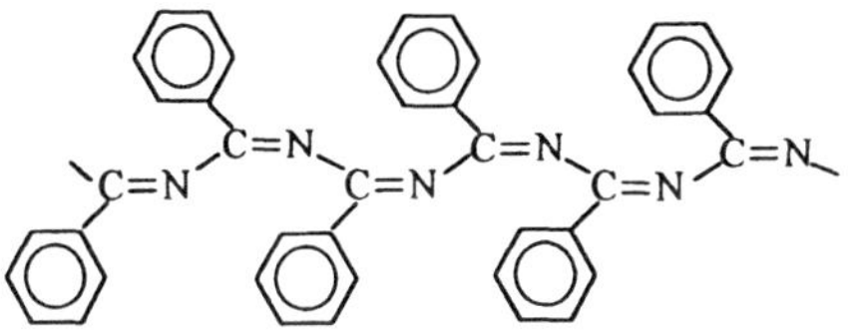

Fig. 3. Illustration of the optimal conformation of PBN.

As described above, para-donor or acceptor substituents generally increase the third-order nonlinear response of benzonitrile monomers. In case of their corresponding polymers, great increase in the $\chi^{(3)}$ and γ are also observed, perhaps due to conjugated -(C=N)$_m$- formation in polybenzonitriles. The conformational optimization on polybenzonitrile done with cerius 2 package gives the preferential lowest energy conformation with the benzene ring distributed at the opposite sides of the cis-(C=N)$_n$- chain alternatively as shown in Fig. 3, thus the spatial resistance in the para-substituted polybenzonitrile derivatives became lowest and even be neglected. The aromatic π electrons will conjugated with main chain π electrons, thus influence the charge density on the C and N atoms along the conjugated chains, therefore lead to the decrease in energy band gap between ground state and excited state, and enhance the nonlinear response of polymers. From the comparison of various polybenzonitrile derivatives with the same average-polymerization-degree(m=6), a stronger dependence of the γ values on the side group's conjugation extent for polybenzonitriles was found. The γ value of various polybenzonitrile derivatives decrease in the following order: PDMBN>PHBN>PNBN>PABN>PBN>PTN. PDMPN polymer has absorption value at 647nm (the wavelength of the pump laser) at least eight times as high as those of other polymers detected by UV-Vis analysis, so that the resonant response may occur in PDMBN during the optical Kerr effect examination, and thus results in very large γ value.

Table 2. Comparison of the power law dependence of γ value
on the chain length of various polybenzonitriles.

Polymer[a]	PBN	PTN	PHBN	PABN	PNBN
μ[b]	2.4	0.75	1.4	1.45	2.2
SPG[c]	<12	>14	>16	>16	<14

a. The structures of different polymers are shown in Table 1.
b. The power law dependence of γ on the average-polymerization-degree m is described as $\gamma = am^\mu$, where a is a constant, μ is the linear fit of the log-log plot of γ vs m value.
c. SPG(Saturation Polymerization Degree) means that when the repeated -(RC=N)- units amount to m, the increase of γ slows down and approaches to saturation value.

Further study of the conjugated chain length influence on the second-order hyperpolarizability of some polybenzonitrile derivatives, including PBN, PABN, PNBN, PTN and PHBN, indicates that the power value μ is greatly influenced by the conjugation state as shown in Table 2. The energy band gap between the HOMO and LUMO of the polymer decreases evidently as new - (RC=N)- unit with larger conjugation degree adding to the main chain. The saturation γ value for various polymers, except PTN, PABN and PHBN, appeared at about 14 average-polymerization-degree. The γ value of HBN monomer was much higher than that of NBN monomer, the PHBN polymer also have large γ value than PNBN polymer, but the increasing rate in γ value of PHBN polymer is lower than that of PNBN polymer. The methyl group on PTN interacts with benzene ring by superconjugation, and this may cause the low γ and μ value of PTN polymer.

CONCLUSIONS

High molecular weight polybenzonitrile derivatives have been prepared by plasma polymerization technique. Infrared spectra and UV-Vis analysis confirm the formation of large π

conjugated -$(C=N)_m$- main chain. The OKE results indicated that 1) the para-donor or acceptor substitution on benzonitrile monomer improves its γ value, 2) the large conjugated side groups along the polynitrile chain enhance the γ value of polybenzonitrile derivatives, and 3) the power value of various polybenzonitrile derivatives depend upon the state of conjugation system, and the γ value reach to saturation after about 14 -$(C=N)$- units.

ACKNOWLEDGMENT

The authors are grateful to the financial support from the National Science Foundation of China.

REFERENCES

1. D. S. Chemla and I. Zyss, Nonlinear Optical Properties of Organic Molecules and Crystals, Academic, New York, Vol. 1 and 2, 1988.

2. P. N. Prasad and D. R. Ulrich, Nonlinear Optical Effects in Organic Polymers, Plenum, New York, 1987.

3. H. Nakanishi, Macromolecules, **21,** p. 1238 (1988).

4. A. Drury and A. P. Duvey, in Nonlinear Optics, edited by T. Kobayashi, Gordon and Breach Publishers, France, **10,** p. 139 (1995).

5. M. Ando and H. Matsuda, Polym. J., **25(4),** p. 417 (1993).

6. J. L. Bredas, C. Adant, P. Tackx and A. Persoons, Chem. Rev., **94,** p. 243 (1994).

7. X. Y. Zhao, Q. Xu, Y. K. He, H. Y. Chen, D. Qiang, C. F. Wang, X. C. Ai, Z. J. Xia and Y. H. Zou, Chinese Chem. Lett., **7,** p. 71 (1996).

8. H. Y. Chen, X. Y. Zhao, Y. K. He, Y. H. Zou and M. Chen, in ISPC-13 13th International Symposium on Plasma Chemistry, Peking University Press, Beijing, Vol. III, p. 1342, 1997.

9. J. I. Wang and Z. T. Huang, Acta Polymerica Sinica, **2,** p. 108 (1981).

10. C. F. Wang, X. Y. Zhao, H. Y. Chen, X. C. Ai, Z. J. Xia and Y. H. Zou, Appl. Phys. B, **64,** p. 45 (1997).

11. S. R. Marder and J. W. Perry, J. Am. Chem. Soc., **115,** p. 2524 (1993).

12. L. -T. Cheng, W. Tam, G. R. Meredith, G. L. J. A. Rikken and E. W. Meijier, in Nonlinear Optical Properties of Organic Materials II, (Proc. SPIE -Int. Soc. Opt. Eng., 1989, Vol 1147), p. 61-72.

RADIATIVE DECAY OF EXCITONS IN MODEL AGGREGATES OF π-CONJUGATED OLIGOMERS

ERIC S. MANAS AND FRANK C. SPANO
Department of Chemistry, Temple University, Philadelphia, PA 19122
spano@fspa.chem.temple.edu

ABSTRACT

Spontaneous emission from exciton states in an aggregate of π-conjugated oligomers is studied theoretically. Each oligomer is taken as a ring of N carbon atoms and is treated using a PPP Hamiltonian. Coulombic interactions between rings are treated to first order. The radiative decay rate γ from an exciton state in an aggregate of M aligned oligomers is superradiant, being M times faster than the decay rate of an isolated oligomer exciton. Inter-oligomer interactions have little effect on the exciton size and energy when the oligomer size N is large compared to the inter-oligomer spacing. However, when N is small, both the exciton size and energy are strongly affected by these interactions, leading to a markedly different N dependence for γ.

INTRODUCTION

Over the past decade or so, a great deal of effort has been directed towards a better understanding of the emissive properties of conjugated polymers in films and aggregates. Detailed analysis of the time and frequency resolved fluorescence decay has revealed a rapid (less than several picoseconds) energy transfer from short (high energy) segments to long (low energy) segments in amorphous polyphenylvinylene (PPV) films [1,2]. Very recently, several groups have reported cooperative radiative emission - superradiance or superfluorescence - from polymer films [3,4]. Pumping thin films of PPV derivatives above a certain threshold leads to narrowing of the emission spectrum as well as a shortening of the fluorescence lifetime. There is some speculation that such cooperativity may arise from aggregated domains within the film, where the short-range order enhances inter-polymer cooperativity.

Thus far, there have been relatively few detailed theoretical investigations of the effects of intermolecular interactions on polymer emission [5-7]. In this communication we present a theory of exciton spontaneous emission in a small aggregate of π-conjugated oligomers, treating each oligomer with the Pariser-Parr-Pople (PPP) Hamiltonian and inter-oligomer interactions to first order. This work is an extension of a previous paper [7], where we studied radiative decay in aggregates of oligomers modeled using a Su-Schrieffer-Heeger Hamiltonian.

AGGREGATE HAMILTONIAN

The model aggregate consists of M π-conjugated rings each containing $N = 4s + 2$ ($s = 1, 2 \ldots$) carbon atoms with average C-C bond-length a. The ring planes are all normal to the aggregate (z) axis, with distance d between adjacent rings. We further assume the more thermodynamically stable in-phase bond alignment, and periodic boundary conditions along the aggregate (z) axis. Finally, the aggregate dimensions are taken to be smaller than an optical wavelength, $Md, Na/\pi << \lambda$. The aggregate Hamiltonian can be written as:

$$H = \sum_n (H_0^{(n)} + H_{e-e}^{(n)}) + \frac{1}{2} \sum_m \sum_{n \neq m} H_{e-e}^{(m,n)}, \tag{1}$$

where $H_0^{(n)}$ is the one electron Hamiltonian for the nth ring, $H_{e-e}^{(n)}$ represents electron-electron interactions within each ring, and $H_{e-e}^{(m,n)}$ represents the interaction between rings m and n. The first sum in Eq. (1) yields a Pariser-Parr-Pople (PPP) Hamiltonian, with:

$$H_0^{(n)} = -\sum_\sigma \sum_p t[1 + (-1)^p \delta](c_{np,\sigma}^\dagger c_{np+1,\sigma} + c_{np+1,\sigma}^\dagger c_{np,\sigma}) \tag{2a}$$

$$H_{e-e}^{(n)} = \sum_p U\rho_{np}(\rho_{np}-1)/2 + \sum_p \sum_{p'\neq p} V(R_{|p-p'|}^0)(1-\rho_{np})(1-\rho_{np'})/2, \tag{2b}$$

while $H_{e-e}^{(m,n)}$ is given by:

$$H_{e-e}^{(m,n)} = \sum_p \sum_{p'} V(R_{|p-p'|}^{|m-n|})(1-\rho_{mp})(1-\rho_{np'}), \tag{2c}$$

where $c_{np,\sigma}^\dagger(c_{np,\sigma})$ creates (destroys) a spin σ electron in the p_z orbital of the pth carbon atom residing in the nth ring, and $\rho_{np} \equiv \sum_\sigma c_{np,\sigma}^\dagger c_{np,\sigma}$ is the electron density operator. The nearest neighbor resonance integral is $-t(1\pm\delta)$ for a double (+) or single (-) bond. In the calculations which follow we take $t = 2.4$ eV which best describes trans-polyacetylene (PA). For the long range interaction potential, we assume the simple form:

$$V(R_{|p-p'|}^{|m-n|}) = V/R_{|p-p'|}^{|m-n|} \qquad (p \neq p'), \tag{3}$$

with $V = e^2/4\pi\varepsilon a$, where ε is the permittivity of the material and $-e$ is the electron charge. $R_{|p-p'|}^{|m-n|}$ is the distance (in units of a) between carbon atom p of ring n, and carbon atom p' of ring m. The on site (Hubbard) Coulombic interaction is given by $U \equiv V(0)$. In all calculations that follow, we assume $U = 2V = 4t$ and $\delta = 0.1$. These values, which are close to those commonly used with the Ohno potential, yield a single polymer band gap of approximately 2 eV within our model. In addition, we assume an inter-ring spacing $d = 3a$, the approximate inter-chain contact for PA.

The above Hamiltonian is diagonalized as follows. First we obtain the eigenstates of the one electron Hamiltonian $H_0^{(n)}$, which includes a ground state represented by a filled valence band, and the set of single electron-hole (e-h) pair excited states $|k_1, k_2 >_n$, obtained by annihilating an electron with wavenumber k_2 in the valence band, and creating and electron with wavenumber k_1 in the conduction band. We then define an effective Hamiltonian, such that the subspace containing the ground state and all single e-h pair eigenstates of $H_0^{(n)}$ is disconnected (to first order) from the space containing higher energy multiple e-h pair states. Within this approximation, Coulombic interactions result in the first order mixing of the single e-h pair states. The cylindrical symmetry of the effective Hamiltonian dictates that the one e-h pair states will mix only if $k_3 - k_4 = k_1 - k_2$. This condition allows the effective Hamiltonian to be factored into blocks, with each block defined by the value of $k_1 - k_2$. In addition, taking advantage of the translational symmetry along the aggregate (z) axis, we write H_{eff} in a Frenkel exciton basis defined by $|k, k+\alpha l; K> = M^{-1/2}\sum_n e^{iKn}|k, k+\alpha l>_n$, where $\alpha = \pm1, \pm2...$ and $l \equiv 2\pi/N$. This further block diagonalizes H_{eff} into (K, α) subblocks with dimension $N/2 \times N/2$, each containing Frenkel exciton states with common wave number K along the aggregate axis. The eigenstates of H_{eff} are found by diagonalizing the (K, α) subblocks, which for the general case is accomplished numerically. The matrix elements of H_{eff} in the (K, α) subblock are:

$$< k, k+\alpha l; K|H_{eff}|k', k'+\alpha l; K > = e(k, k+\alpha l)\delta_{k,k'} + {}_n< k, k+\alpha l \| H_{e-e}^{(n)}|k', k'+\alpha l >_n$$

$$+2\sum_{j=1}^{(M-1)/2}\cos[Kj] \ {}_m< k, k+\alpha l \| H_{e-e}^{(m,m+j)}|k', k'+\alpha l >_{m+j} \tag{4}$$

where $e(k,k') = \varepsilon(k,\delta) + \varepsilon(k',\delta)$, with $\varepsilon(k,\delta) = 2t(\cos^2 k + \delta^2 \sin^2 k)^{1/2}$. The optically allowed subblock is $(K = 0, \alpha \pm 1)$, so we only need diagonalize this portion of the Hamiltonian to evaluate the radiative decay rates. In Eq.(4), the first term represents the non-interacting electron energy, while the second term contains the effects of intra-oligomer Coulombic interactions, which mix the states $|k, k+l\rangle$ on the same oligomer, leading to the formation of Wannier excitons. We find that for $U = 2V = 4t$ and $\delta = 0.1$ for instance, there are three intra-oligomer Wannier exciton states significantly bound relative to a continuum of free e-h pair states in the large N limit [8,9]. We label these states by their symmetry with respect to reflection through a bond center (A or B), and with respect to charge conjugation (+/-). The two optically allowed excitons are the antisymmetric states $1B^-$ and $2B^-$, while the symmetric exciton state $2A^+$ is optically dark. The ground state is $1A^+$. The third term in Eq.(4) contains the effects of inter-oligomer Coulombic interactions, which depend on V, as well as N and d. These interactions will delocalize the Wannier exciton states over the oligomer aggregate, creating Wannier-Frenkel excitons with unique radiative properties.

RADIATIVE EMISSION

If the dimensions of a molecule are much smaller than an optical wavelength, the radiative or spontaneous emission rate of an excited state is given by the Einstein A coefficient, which scales as the square of the transition dipole moment and the cube of the excited state energy. For an excited state $|i\rangle$ with energy $\hbar\omega_i$ and transition dipole moment $\bar{\mu}_i$, the spontaneous radiative decay rate is (assuming any states degenerate with $|i\rangle$ have transition dipole moments orthogonal to $\bar{\mu}_i$):

$$\gamma_i = \frac{|< G|\bar{\mu}|i >|^2 \, \omega_i^3}{3\pi\varepsilon_0 \hbar \, c^3} \tag{5}$$

We can thus use Eq. (5) to evaluate the radiative rates of the $(K = 0, \alpha \pm 1)$ subblocks, as long as we restrict the aggregate dimensions to be much smaller than an optical wavelength, $(Md, Na/\pi \ll \lambda)$. In what follows, we will consider the radiative rates of the $1B^-$ and $2B^-$ exciton states, since these carry most of the oscillator strength.

The effects of aggregation on the radiation rates are twofold. First, the K=0 exciton transition dipole moments scale as $\sqrt{M}$ times the single oligomer dipole moments, so it is readily seen from Eq. (5) that both γ_{1B^-} and γ_{2B^-} will be enhanced by a factor of M, and thus the radiative decay from these states will be superradiant. Secondly, inter-oligomer Coulombic interactions will mix the B^- symmetry states on different oligomers, leading to shifts in the exciton energies and oscillator strengths, which in turn leads to a different N dependence for γ_{1B^-} and γ_{2B^-} in an aggregate relative to a single oligomer. It is to this effect that we now turn our attention.

First we will consider the effects of aggregation on the exciton energies. In Figs. 1a and b, the energies of the $1B^-$ and $2B^-$ excitons, as well as the $2A^+$ exciton, are shown as a function of N, for the case of interacting and non-interacting oligomers, respectively. Interactions cause the $1B^-$ and $2B^-$ excitons to blueshift, as can be more clearly seen in Fig 1c, which shows the blueshift of these states, Δ_{1B^-} and Δ_{2B^-}, as a function of N. Note that these blueshifts become negligible in the large N limit, a result shown in a previous study for aggregates of bond alternating Hückel oligomers [7,10]. The blueshift is greatest for the $1B^-$ exciton, which carries the greatest oscillator strength. In fact, for $14 < N < 58$, the $1B^-$ state is blueshifted *above* the $2A^+$ state. Although these blueshifts tend to increase the radiative rates γ_{1B^-} and γ_{2B^-}, it is important to note that the binding energy of the exciton is simultaneously decreased, and hence the exciton oscillator strengths go down as a result of enhanced mixing with higher energy e-h pair states.

It is instructive to consider how the exciton size r, rather than the actual oscillator strength,

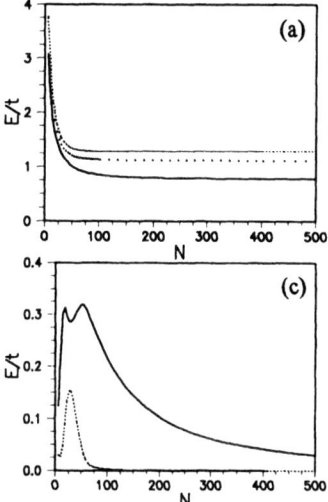

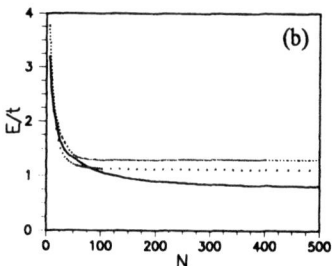

Figure 1 The energies of the 1B⁻ (solid), 2B⁻(dashed), and 2A⁺ (dotted) excitons as a function of N, for the case of interacting (a) and non-interacting (b) oligomers. (c) shows the blueshift of the 1B⁻ (solid) and 2B⁻ (dashed) excitons caused by inter-oligomer interactions. In all aggregate calculations in this and subsequent Figures, $U=2V=4t$, $\delta=0.1$, $d=3a$, and each molecule interacts with 20 nearest neighbors.

depends on the oligomer size N. This provides a natural length scale, a property of the exciton, with which to compare N. Wannier excitons can be expanded in terms of valence and conduction band Wannier functions centered at each of the $N/2$ double bonds of a given oligomer [9], which defines the oligomer unit cell. We can then define r as the root mean square separation (in unit cells) between the electron and hole. When r is small, there will be a greater probability of finding the electron and hole at the same unit cell, and to a good approximation, the exciton oscillator strength will be proportional to this probability, due to greater likelihood of e-h recombination. This approximation neglects two-center-corrections in which, for example, the transition dipole moment operator connects a valence Wannier function centered at one unit cell to a conduction Wannier function centered at another unit cell [11]. Once the exciton size has converged with N, the transition dipole moment to the Wannier exciton states scale as $\sqrt{N}$ due to delocalization of the e-h center of mass. Thus, the oscillator strength will increase linearly with N in the large N limit.

Figures 2a and b show the sizes r_{1B^-} and r_{2B^-} of the 1B⁻ and 2B⁻ excitons, respectively, for the case of interacting and non-interacting oligomers. The most pronounced effect of aggregation is on the radius of the 1B⁻ exciton. For the range $14 < N < 58$, when the blueshift of this state is the largest, mixing with higher energy e-h pairs occurs, resulting in a large increase in the exciton size and subsequent decrease in oscillator strength. The size of the 2B⁻ exciton is

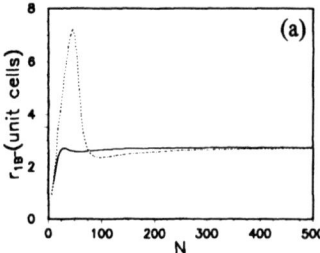

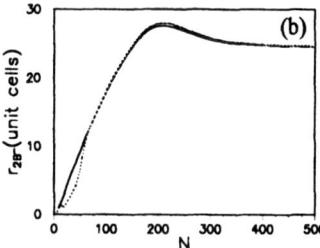

Figure 2 The sizes of the 1B⁻ (a) and 2B⁻ (b) excitons as a function of N, for the case of interacting (dashed) and non-interacting (solid) oligomers.

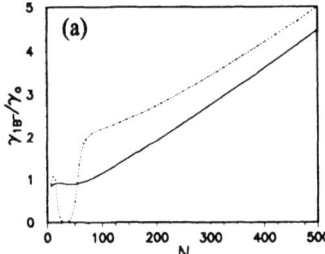 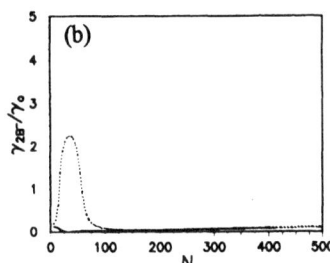

Figure 3 The radiative rates of the 1B⁻ (a) and 2B⁻ (b) excitons as a function of N, for the case of interacting (dashed) and non-interacting (solid) oligomers. The rates are measured relative to γ_0, which is the radiative rate of a single ethylene unit. Although not shown, the interacting value is also enhanced by a factor of M.

simultaneously decreased, due to its mixing with the 1B⁻ exciton. In the large N limit, inter-oligomer Coulombic interactions essentially have no effect on the exciton sizes, so their values in both the interacting and non-interacting cases converge to the same value.

Figures 3a and b show the radiative rates, γ_{1B^-} and γ_{2B^-}, of the 1B⁻ and 2B⁻ excitons, respectively, for the case of interacting and non-interacting oligomers. The dependences of the radiative rates on N are determined by a balance between the exciton oscillator strengths and energies. It is helpful to consider this dependence in two regions: (1) when both the exciton energy and size has converged, and (2) when the exciton size and energy both depend significantly on N. In region 1, it is clear from Figures 1-3 that, once the exciton energies and sizes have converged, the radiative rates scale linearly with N, due to the $\sqrt{N}$ dependence of the dipole matrix element. In addition, as the blueshifts Δ_{1B^-} and Δ_{2B^-} become negligible, the interacting and non-interacting values of γ_{1B^-} and γ_{2B^-} approach each other. This is most clearly seen in Figure 3b. The slight difference between the interacting and non-interacting values of γ_{1B^-} is due to the blueshift Δ_{1B^-}, which has not yet converged with N. In region 2, for the case of non-interacting oligomers, r_{1B^-} and r_{2B^-} are approximately independent of N until N becomes small, at which point they sharply decrease. This is accompanied by a rise in the non-interacting values of the exciton energies ε_{1B^-} and ε_{2B^-}. However, γ_{1B^-} and γ_{2B^-} undergo only a small increase with decreasing N, due to confinement of the e-h center of mass. For the case of interacting oligomers, the most prominent effect occurs when Δ_{1B^-} and Δ_{2B^-} become significant. When this occurs the mixing of the 1B⁻ and 2B⁻ excitons with each other and with higher energy states increases, resulting in a sharp increase in r_{1B^-} accompanied by a significant decrease in r_{2B^-}. This effect on the exciton sizes dominates, and the radiative rate γ_{1B^-} dips sharply, while γ_{2B^-} is strongly enhanced. However, as N decreases further, r_{1B^-} decreases and r_{2B^-} increases, so the radiative rates γ_{1B^-} and γ_{2B^-} increase and decrease, respectively.

CONCLUSIONS

Since the transition dipole moments of the K=0 Frenkel exciton states scale as $\sqrt{M}$ times the single oligomer dipole moments, the radiative decay rate γ from these states in an aggregate of M aligned oligomers will clearly be superradiant, being M times faster than the decay rate of an isolated oligomer exciton. This is true even when the inter-oligomer Coulombic interactions are negligible, as when $N a >> d$, since the eigenstates will remain delocalized excitons. Of course any amount of disorder will limit the degree of delocalization. In the case of site disorder for instance, a spread σ of site energies will destroy superradiance whenever $\sigma \gg \Delta$, the exciton blue shift.

Once the energies and sizes of the intra-oligomer Wannier excitons have converged with oligomer size N, the radiative rates will also scale linearly with N, due to greater delocalization of the e-h center of mass. Inter-oligomer interactions have little effect on the exciton radius and energy when the oligomer size N is large. However, when N is small, both the exciton size and energy are strongly affected by inter-oligomer interactions. Thus there is a significant departure from the linear dependence of γ on N in an aggregate relative to a single oligomer.

Enhanced radiative rates will always compete with non-radiative decay routes to determine the fluorescence of a system. This is the principle behind Kasha's rule, which states that fluorescence will be observed from the lowest excited singlet state of a molecular aggregate. In our model system, although aggregation will enhance the radiative decay of the K=0 Frenkel exciton states by a factor of M, non-radiative decay to the lowest energy K=π exciton, which is forbidden from the ground state, will inhibit strong fluorescence. There are several ways to enhance the fluorescence quantum yield. First, as N becomes large, the K=0 and K=π excitons become nearly degenerate, making it more likely to observe superradiant decay from the lowest K=0 exciton. Disruption of symmetry by including disorder, site or otherwise, is likely to destroy the degree of coherence, as mentioned above. However, disorder introduced by varying the oligomer orientations can also distribute oscillator strength to lower energy states, which may enhance fluorescence efficiency [5]. There also exists the possibility of having an out of phase double bond alignment, such that the double bonds of one oligomer overlap the single bonds of its neighbors, which expands the aggregate unit cell to include two oligomers. In this case, the lowest excited state may be strongly allowed. Finally, we point out that polymers in an oriented film may have a lesser degree of conformational disorder than in unoriented films or solutions. Thus, the center of mass of the e-h pair will be delocalized over a much greater range along the individual chains in the oriented film, leading to enhanced radiative rates for these excitons.

ACKNOWLEDGEMENTS

We would like to acknowledge Temple University for support of this research.

REFERENCES

[1]   R. Kersting, B. Mollay, M. Rusch, J. Wenisch, G. Leising and H.F. Kauffmann, J. Chem. Phys. **106**, 2850 (1997).

[2]   R. Kersting, U. Lemmer, R. F. Mahrt, K. Leo, H. Kurz, H. Bässler and E. O. Göbel, Phys. Rev. Lett. **70**, 3820 (1993).

[3]   S. V. Frolov, W. Gellermann, M. Ozaki, K. Yoshino and Z. V. Vardeny, Phys. Rev. Lett. **78**, 729 (1997).

[4]   H. J. Brouwer, V. V. Krasnikov, A. Hilberer and G. Hadziioannou, Adv. Mat. **8**, 935 (1996).

[5]   J. Cornil, A.J. Heeger and J. L. Bredas, Chem. Phys. Lett. **272**, 463 (1997).

[6]   M. H. Hennessy and Z. G. Soos, Synth. Met. **85**, 1051 (1997)

[7]   M. J. McIntire, E. S. Manas, and F. C. Spano, J. Chem. Phys., **107**, 8152 (1997).

[8]   S. Abe, J. Yu, and W. P. Su, Phys. Rev. B **45**, 8264 (1992).

[9]   D. Yaron and R. Silbey, Phys. Rev. B **45**, 11655 (1992).

[10]   Z.G. Soos, G. W. Hayden, P. C. M. McWilliams and S. Etemad, J. Chem. Phys. **93**, 7439 (1990).

[11]   R. S. Knox, Theory of Excitons, Academic Press, New York, 1963, pp. 112-136.

SYNTHESIS OF NEW MULTIFUNCTIONAL POLYMERS INCORPORATING BIS-(DIPHENYLAMINO)DIPHENYLPOLYENE MOIETIES

C.W. SPANGLER, T. FAIRCLOTH, E.H. ELANDALOUSSI, B. REEVES
Department of Chemistry and Biochemistry, Montana State University, Bozeman, MT 59717,
uchcs@earth.oscs.montana.edu

ABSTRACT

A series of bis-(diphenylamino)diphenylpolyenes containing up to eight double bonds has been synthesized, both as parent model compounds, and functionalized for attachment as pendant chromophores. Oxidative doping of these new materials yields exceptionally stable bipolaron-like dications, even at the triene level. These new materials show promise as bimechanistic optical limiters by reverse saturable absorption and by 2-photon absorption.

INTRODUCTION

Over the past eight years our research group has studied the formation of polaron-like radical cations and bipolaron-like dications in diphenyl and dithienyl polyenes under oxidative doping conditions[1-5]. In π-conjugated electroactive polymers with nondegenerate ground states, such as polythiophene, poly(p-phenylene vinylene) (PPV) and poly[2,5-thienylene vinylene] (PTV), bipolarons are generally recognized as the dominant charge carriers[6]. Upon doping, both the fully conjugated polymers and the model oligomers exhibit dramatic changes in the UV-VIS spectra, with large red shifts in oscillator strength. Most recently we have been able to correlate these shifts with bipolaronic enhancement of the third-order optical nonlinearity[7,8], as predicted by theory[9,10]. The bipolaron-like dications obtained from dithienylpolyenes are among the most nonlinear small molecules yet observed, with γ, the third order hyperpolarizability approaching 10^{-30} esu[11]. These materials also possess outstanding processibility, and can be designed as guest-host system in a variety of different polymer hosts, incorporated as pendant chromophores in poly(methyl methacrylate) (PMMA), or as formal polyester, polyamide or polycarbonate repeat units[12,13]. Whether incorporated into composites, as pendants, or into the polymer main chain, we have shown that each and every polyene repeat unit can be oxidized to the dication form[14]. In this study, we present preliminary results on a new polyene system which may prove to be even more interesting and versatile than the dithienylpolyenes.

SYNTHESIS AND OXIDATIVE DOPING OF BIS-(DIPHENYLAMINO)-DIPHENYLPOLYENES

A series of bis-(diphenylamino)diphenylpolyenes were synthesized by Wittig and Horner-Emmons-Wadsworth methodology developed in our lab over the past ten years[5,15]. Condensation of diphenylaminopolyenes with bis-Wittig reagents allowed the synthesis of polyenes with n=3-8. Bis-(diphenylamino)stilbene was prepared directly from diphenylaminobenzaldehyde by McMurry Coupling. The diene was prepared by standard Wittig condensation. These procedures are outlined in Schemes 1 and 2. All polyenes up to n=8 could be purified by column chromatography, or by recrystallization from a variety of solvents.

OXIDATIVE DOPING OF BIS-(DIPHENYLAMINO)DIPHENYLPOLYENES

Oxidative doping to bipolaron-like dications was accomplished by addition of $SbCl_5$ to solutions of the polyenes in CH_2Cl_2. Exceptionally stable bipolarons were formed for n=3-8, and the n=3 sample is the first stable trienic bipolar ever observed. For comparison, bis-(N,N-dimethylamino)diphenylhexatriene does not form a stable charge state when similarly doped, but rather decomposes rapidly. Similar observations have been made in our lab for other substituted diphenylhexatrienes and dithienylhexatrienes. We have summarized these results in Table 1,

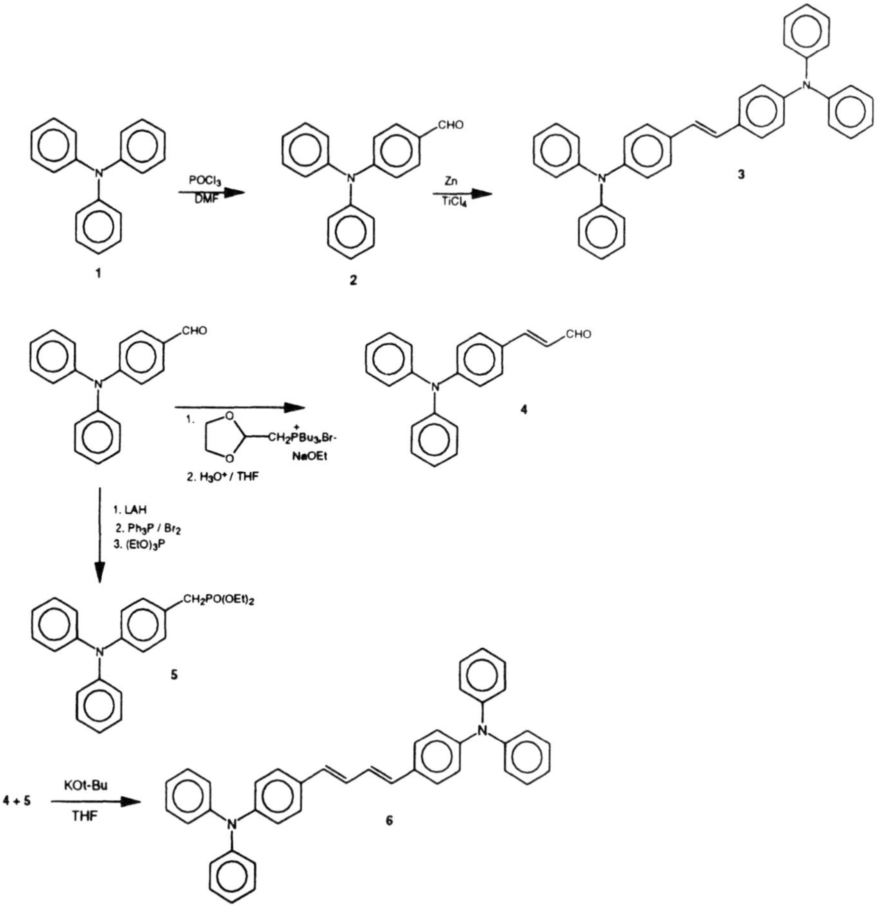

Scheme 1. Preparation of bis-(diphenylamino)stilbene and bis-(diphenylamino)-diphenylbutadiene

comparing these results to those previously obtained for the bis-(N,N-dimethylamino)diphenyl polyene series[1,15].

Table 1. Oxidative doping of 4', 4''-disubstituted diphenylpolyenes

n	x	λ_{max} (π-π^*) neutral (nm)[a]	λ_{max} BP (nm)[a]
3	Ph$_2$N	424	730
4	Ph$_2$N	438	770
5	Ph$_2$N	449	805
6	Ph$_2$N	462	848
7	Ph$_2$N	464	892
8	Ph$_2$N	485	922
3	Me$_2$N	410	b
4	Me$_2$N	425	b
5	Me$_2$N	445	714
6	Me$_2$N	458	752

[a] 10^{-5}M solutions in CH$_2$Cl$_2$; [b] not stable

INCORPORATION OF BIS-(DIPHENYLAMINO)DIPHENYLPOLYENE PENDANT CHROMOPHORES ON PMMA

Bis-(diphenylamino)diphenylpolyenes functionalized for attachment to PMMA can be readily synthesized, as outlined in Scheme 3, followed by attachment to poly(methacroyl chloride). Quenching of the residual COCl groups with excess MeOH yields the substituted PMMA. This approach eliminates variations in polymer properties due to differing chain lengths attached to monomer (methyl methacrylate) and then polymerized. We also believe that this approach gives a more uniform, though random, distribution of the pendant groups on the chain. This procedure is outlined in Scheme 4.

POTENTIAL APPLICATIONS

We have recently shown, in collaboration with coworkers at Laser Photonics Technology, that polyene type molecules may function as optical power limiters via reverse saturable absorption based on photogeneration of highly stable bipolarons[16]. It has also recently been shown by Perry and coworkers that bis-(diphenylamino)stilbene is an extremely good 2-photon absorber[17]. In addition, we are currently studying 2-photon absorption in dithienylpolyenes in collaboration with the Laser Hardening Group at Wright Labs (MLPJ), and have recently shown the 1,8-dithienyl-1,3,5,7-octatetraene has a very large 2-photon cross-section[18]. The bis-diphenylamino)diphenylpolyenes reported in this study thus represent a potentially new class of optical power limiting materials, whether in solution, formulated as guest-host composites or as functionalized polymers (e.g. PMMA). These materials have the unique capability of providing optical limiting and eye protection via two distinctly different mechanisms: (1) at the higher end of the visible energy range by photogeneration of highly absorbing bipolarons (RSA), and (2) via two-photon absorption (TPA) at the lower end of the visible energy range. We believe that this is the first such broad-band bimechanistic optical limiting that has been proposed, and the materials

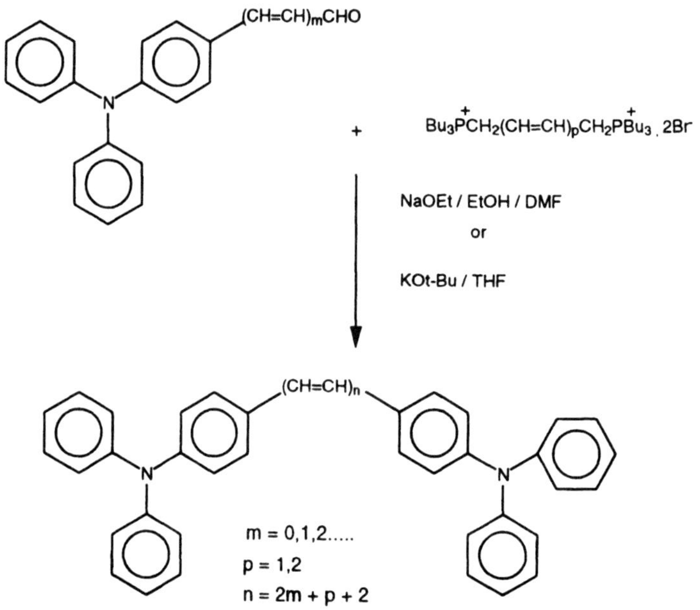

Scheme 2. Preparation of bis-(diphenylamino)diphenylpolyenes

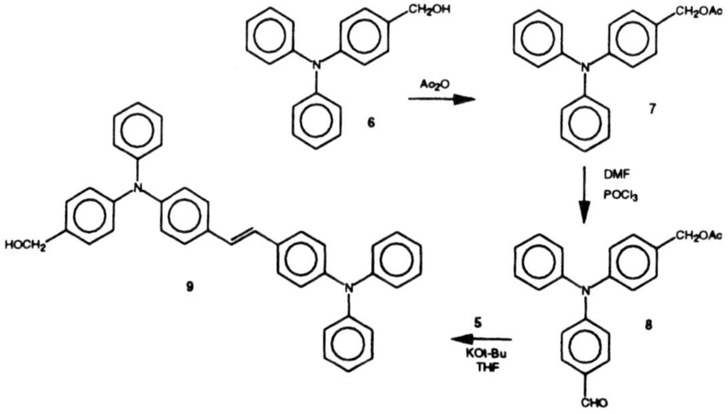

Scheme 3. Preparation of a functionalized bis-(diphenylamino)stilbene

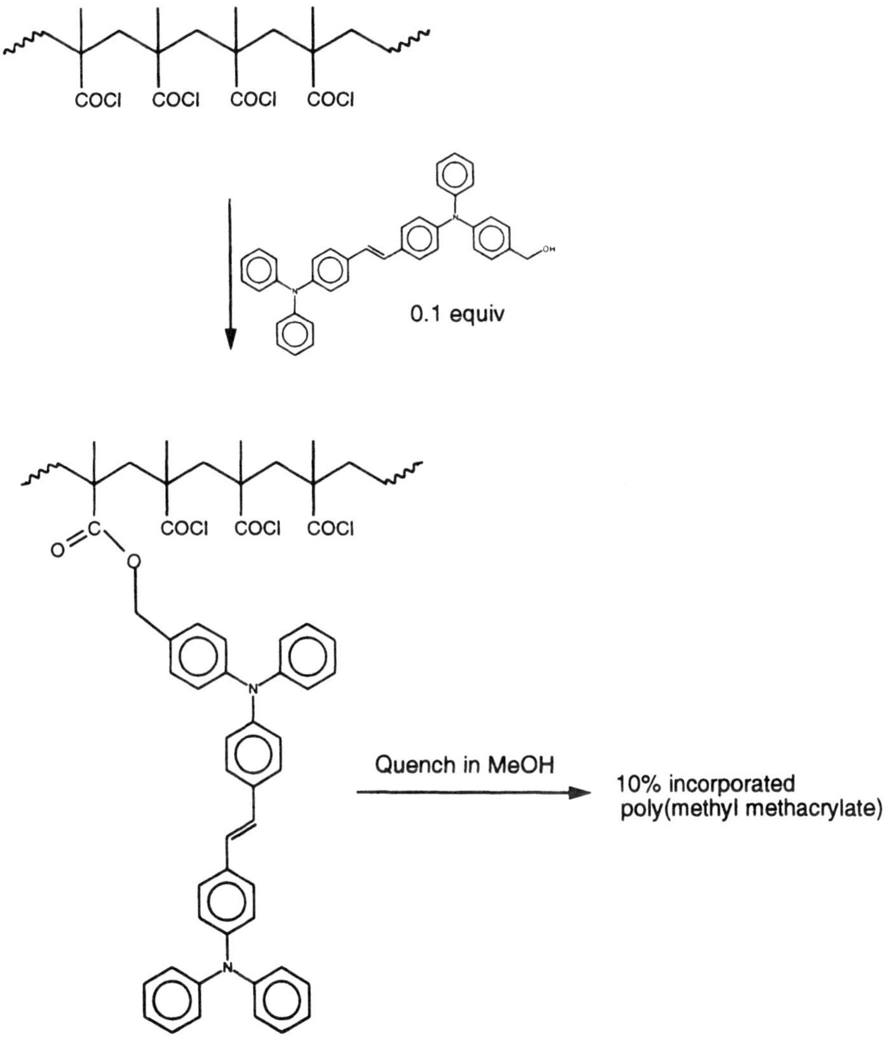

Scheme 4. Attachment of a bis-(diphenylamino)diphenylpolyene moiety to PMMA

described in this paper are currently being studied in collaboration with both Wright Labs and Laser Photonics Technology to confirm the efficacy of this approach to the design of practical limiters for eye protection in the visible region, and sensor protection in the NIR.

CONCLUSIONS

A synthetic approach to a new class of polyenes that can be readily functionalized for polymer attachment has been described. These new materials form exceptionally stable bipolarons when oxidatively doped, whose absorption characteristics may be used to model the RSA of the corresponding photogenerated species. These polyenes also possess the capability of having large 2-photon cross-sections, and as such may provide bimechanistic optical power limiting.

ACKNOWLEDGMENTS

The authors gratefully acknowledge financial support for this research by The Air Force Office of Scientific Research under Grants # F49620-96-1-0440 and F49620-97-1-0413 and also acknowledge helpful advice and discussion with the following: Martin Casstevens, Ryszard Burzynski (Laser Photonics Technology), Paras Prasad (Photonics Research Center (SUNY-Buffalo), and L. Natarajan and Tom Cooper (Wright Labs-MLPJ).

REFERENCES

1. C.W. Spangler, L.S. Sapochak and B.D. Gates in Organic Materials for Non-Linear Optics, edited by R. Hahn and D. Bloor (Roy. Soc. Chem. Spec. Pub. # 69, London, 1989) pp. 57-62.

2. C.W. Spangler and R. Rathunde, J. Chem. Soc. Chem. Commun. 1969, 26.

3. C.W. Spangler and T.J. Hall, Synthetic Met. 44, 85 (1991).

4. C.W. Spangler and P.-K. Liu, J. Chem. Soc. Perk. Trans. 1 1992, 1959.

5. C.W. Spangler and M.Q. He, J. Chem. Soc. Perk. Trans. 2 1995, 715.

6. J.E. Fromer and R.R. Chance in Electrical and Electronic Properties of Polymers, A State of the Art Compendium, edited by J.I. Kroschwitz (John Wiley Publishers, New York, 1998) pp. 56-101.

7. C.W. Spangler, M.Q. He, J. Laquindanum, L. Dalton, J.P. Partanen and R. Hellwarth in Electrical, Optical, and Magnetic Properties of Organic Solid State Materials, edited by A. Garito, A.K.-Y. Jen, C.Y.-C. Lee and L.R. Dalton (Mater. Res. Soc. Proc. 328, Pittsburgh, PA, 1994) pp. 655-660.

8. E.G. Nickel, C.W. Spangler, N. Tang, R. Hellwarth and L. Dalton, Nonlinear Optics, 6, 135 (1993).

9. C.P. deMelo and R. Silbey, Chem. Phys. Lett. 140, 537 (1987).

10. C.P. deMelo and R. Silbey, Chem. Phys. Lett. 88, 2567 (1988).

11. N.Tang, J. Partanen, R. Hellwarth, J. Laquindanum, L. Dalton, C.W. Spangler and M.Q. He in Nonlinear Properties of Organic Materials VII, edited by G.R. Möhlmann (Proc. SPIE, Bellingham, WA, 1994) pp. 186-192.

12. C.W. Spangler, K.O. Havelka, P.-K. Liu, E.G. Nickel, D. Polis, L.S. Sapochak and L.R. Dalton in Frontiers of Polymer Research, edited by P.N. Prasad and J.K. Nigam (Plenum Press, New York, 1991) pp. 149-156.

13. C.W. Spangler, P.-K. Liu, T.J. Hall, D. Polis, L. Sapochak and L.R. Dalton, Polymer, 33, 3937 (1992).

14. C.W. Spangler and M.Q. He in Electrical, Optical, and Magnetic Properties of Organic Solid State Materials III, edited by A.K.-Y. Jen, C.Y.-C. Lee, L.R. Dalton, M.F. Rubner, G.E. Wnek and L.Y. Chiang (Mater. Res. Soc. Proc. 413, Pittsburgh, PA, 1996). Pp. 221-229.

15. C.W. Spangler, R.K. McCoy, A.A. Dembek, L.S. Sapochak and B.D. Gates, J. Chem. Soc. Perk. Trans. 1 1989, 151.

16. M.K. Casstevens, R. Burzynski, J.F. Weibel, C. Spangler, G. He and P.N. Prasad in Nonlinear Optical Liquids and Power Limiters, (Proc. SPIE, Bellingham, WA, 1997) pp. 152-159.

17. D. Beljone, J.-L. Brédas, B.H. Cumpston, J. Erlich, L. Erskine, A.A. Heikal, Z.-Y. Hu, T. Kogej, I.-Y.S. Lee, S.R. Marder, J.W. Perry, H. Röckel, S. Thayumanavan and X.-L. Wu, Polym. Preprints, 38(2), 497 (1997).

18. L. Natarajan, A. Sowards, N. Tang, P. Fleitz, R. Sutherland, T. Cooper and C. Spangler in Materials for Optical Limiting II (Mater. Res. Soc. Proc. 479, Pittsburgh, PA, 1997) (in press).

Part III

Conducting and Electroactive Polymers and Materials

ELECTRONIC STRUCTURE AND OPTICAL RESPONSE OF ELECTROLUMINESCENT CONJUGATED POLYMERS

J.L. BREDAS[1], J. CORNIL[1,2], D. BELJONNE[1], D.A. DOS SANTOS[1], Z. SHUAI[1], R. SILBEY[2]
[1]Service de Chimie des Matériaux Nouveaux, Centre de Recherche en Electronique et Photonique Moléculaires, Université de Mons-Hainaut, Place du Parc 20, B-7000 Mons, Belgium; [2]Department of Chemistry, Massachusetts Institute of Technology, Cambridge MA 02139, USA.

ABSTRACT

In this contribution, we investigate by means of correlated quantum-chemical calculations the influence of intermolecular interactions on the absorption and emission properties of conjugated chains. Various strategies are suggested to avoid a substantial decrease in fluorescence quantum yield in condensed media. Finally, the reliability of our theoretical approach is validated by showing the remarkable agreement obtained between the experimental data and the calculated optical properties of clusters formed by sexithienyl molecules.

INTRODUCTION

Since the first report of electroluminescence from polyparaphenylene vinylene, PPV [1], interest in conjugated polymers has been renewed and has opened the way to the fabrication of electro-optic devices exploiting their semiconducting properties and high luminescence yield. Efficient organic light-emitting diodes (LED's) [2] are now available in the whole UV-visible range [3-5] and the commercialization stage is approaching fast. Conjugated polymers are also used as active element in light-electrochemical cells [6], photodiodes [7,8], and solid-state lasers [9-11]. In spite of outstanding achievements in the field of device fabrication, a full understanding of the intrinsic electronic and optical properties of the conjugated chains is still far from being reached [12]. This is partly due to the fact that the experimental results strongly depend on the purity and/or morphology of the samples. In this context, the theoretical techniques of quantum-chemistry can prove to be useful in order to better apprehend the electronic and optical properties of the chains, for instance to characterize the nature of the excited states playing an essential role in the electro-optic devices.

According to a large number of experimental studies, the most stable photogenerated species in the lowest excited states of conjugated chains are electron-hole pairs bound by Coulomb attraction and associated to a local deformation of the backbone, i.e., polaron-excitons [13]. A good insight into the properties of these species can be provided by quantum-chemical calculations performed on isolated chains, as illustrated for instance by some of our recent studies focusing on the lowest excited state in PPV oligomers (i.e., the state acting as source of light emission in LED's and lasers) [14-16] The choice of finite-size systems to modelize the properties of actual polymer samples can be justified by the often reduced degree of delocalization along the chains, resulting from the presence of physical or chemical impurities [17].

Many recent experimental studies have, however, clearly demonstrated that interchain effects can play an important role [18-23]. Among the various manifestations of interchain couplings in the solid state, we observe: (i) a significant increase in radiative lifetime without any substantial loss in quantum efficiency [20]; (ii) a lack of correlation between the dynamics of the luminescence and that of the fast photoinduced peaks, which is not observed for the isolated chains [18,23]; and (iii) a significant decrease in fluorescence quantum yield [19]. These studies indicate that interchain species are present in condensed media even though their nature and origin is still unclear. It is thus of interest to try and uncover by means of quantum-chemical calculations the nature of photogenerated interchain species and to rationalize the somewhat contradictory experimental observations. Deeper understanding of the interchain effects in the field of organic LED's and solid-state lasers should also provide new guidelines towards the design of devices with enhanced efficiencies.

In this contribution, we describe the influence of intermolecular interactions on the absorption and emission characteristics of conjugated chains, on the basis of correlated theoretical calculations performed on complexes formed by PPV oligomers; their number, size, and relative orientations are varied in a systematic way. We also validate the reliability of our theoretical approach by showing the remarkablement agreement obtained between the calculated optical properties of clusters of sexithienyl molecules and the corresponding experimental measurements obtained for the sexithienyl crystal [24].

THEORETICAL METHODOLOGY

The ground-state geometry of isolated and interacting PPV oligomers assumed to be planar are optimized by means of the semiempirical Hartree-Fock Austin Model 1 (AM1) technique [25]. We determine the equilibrium geometry in the lowest excited state of isolated and interacting PPV chains by coupling the AM1 Hamiltonian to a limited Configuration Interaction (AM1/CI) scheme. In the case of sexithienyl molecules, the geometry used is that provided by X-ray data [26].

On the basis of the optimized ground-state geometries, we simulate the absorption spectra by combining the semiempirical Hartree-Fock Intermediate Neglect of Differential Overlap (INDO) Hamiltonian to a Single Configuration Interaction scheme, as previously described [15]. In order to avoid any size inconsistency, we systematically involve all the occupied and unoccupied π-levels when generating the singly excited configurations. The emission spectra are simulated on the basis of a very similar approach, except that the starting geometry is the one optimized for the lowest excited state in its relaxed conformation [15].

INTERCHAIN INTERACTIONS

Absorption properties of highly symmetric complexes

We first focus on cofacial dimers formed by stilbene molecules; in such conformations, the amplitude of interchain interactions is expected to be maximized [27]. For large interchain separations (8 Å $\leq$ R $\leq$ 30 Å), the LCAO coefficients of a given molecular orbital are localized on a single chain, as intuitively expected. The lowest excited state of these dimers results from a destructive interaction of the two intrachain transition dipole moments, whereas a constructive interaction prevails for the second excited state; this result is fully consistent with the molecular exciton model developed by Kasha [28]. In this weak coupling regime, the calculated optical properties can thus be rationalized by a simple dipole-dipole interaction model. The energy splitting between the lowest two excited states calculated increases when the chain separation is lowered. In this range of interchain distances, the CI description of the lowest excited state indicates that the probability to find the exciton on one chain is the same for the two stilbene units. However, due to low energy transfer rates between adjacent molecules, the exciton is localized at any given time on a single chain; as a result, the luminescence signal is characteristic of an isolated molecule and is not quenched by the interchain effects.

In a regime of strong interaction between the chains, there also occurs no optical coupling between the ground state and the lowest excited state. The absence of coupling, however, has a different origin. Indeed, below 7 Å, the LCAO coefficients start to delocalize over the two chains; the wavefunctions become entirely symmetric below 5 Å due to an efficient exchange of electrons between the chains; this delocalization of the wavefunction is not taken into account in the molecular exciton model, which therefore becomes unreliable at short chain separations. Analysis of the one-electron structure of the complexes indicates that the HOMO (Highest Occupied Molecular Orbital) and LUMO (Lowest Unoccupied Molecular Orbital) have the same parity, as is also the case for the HOMO-1 and LUMO+1 levels, see Figure 1. The lowest optical transition is described by a mixing of the H→L and H-1→L+1 excitations, and is forbidden according to the selection rules; the whole intensity becomes concentrated in the second excited state originating from the interaction between the H→L+1 and H-1→L transitions. The energy splitting calculated between the two excited states continuously increases when going from 30 Å down to 3.5 Å, as illustrated in Figure 2. We stress that, at short distances, the splitting is not symmetric with respect to the optical transition calculated for the isolated stilbene molecule. This is rationalized in perturbation theory by the fact that a symmetric splitting is provided by the first-order correction whereas an identical stabilization of the two states is induced by the second-order correction [29].

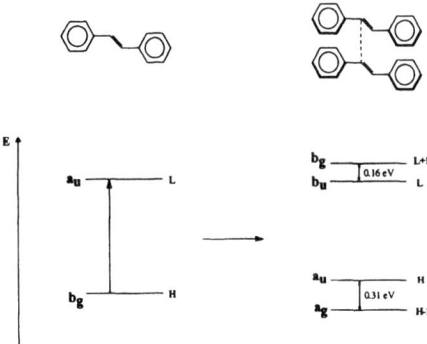

Figure 1: Schematic representation of the one-electron structure of a single stilbene molecule and that of a cofacial dimer formed by two chains separated by 4 Å. The INDO-calculated energy split of the HOMO and LUMO levels when going from the isolated molecule to the dimer are also given.

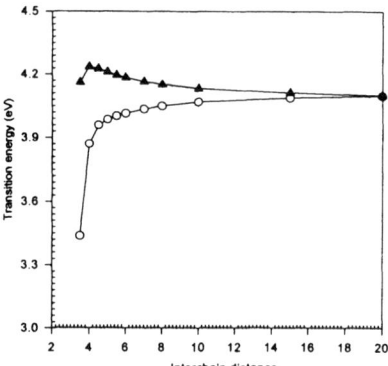

Figure 2: INDO/SCI-calculated transition energies of the lowest two optical transitions of a cofacial dimer formed by two stilbene molecules as a function of interchain distance. The horizontal dashed line refers to the transition energy of the isolated molecule. Note that the upper value reported at 3.5 Å corresponds to the transition to the fifth excited state, which provides the lowest intense absorption feature.

The one-electron structure of the cofacial dimer also reveals that the splitting of the HOMO and LUMO levels are not equal (0.31 eV and 0.16 eV at 4 Å, respectively). The lack of electron-hole symmetry reflects changes in the strength of the overlap between the levels; the smaller LUMO splitting can be rationalized by the fact that there are more nodal surfaces in the LUMO wavefunction than in the HOMO. Indeed, the overlap of the molecular orbitals is made up of a sum of many atomic overlaps with a product of LCAO coefficients; the larger the number of nodal surfaces, the larger the possibility of cancellation of terms, due to the signs of the products of coefficients. By extrapolation, in a one-dimensional stack of interacting chains, the valence band can be expected to be wider than the conduction band. The present results have strong relevances with regard to charge transport properties in the solid state, since they are consistent with the typical observation that in conjugated polymers holes are more mobile than electrons.

We now address the way the properties calculated for highly symmetric conformations are affected by the size of the interacting unit. To do so, we have built complexes formed by two PPV oligomers containing up to 7 phenylene rings and separated by 4 and 6 Å, respectively. In general, the trends prevailing for interacting stilbenes hold true for the longer chains. However, a striking difference is that the splitting between the two lowest excited states calculated at a given separation distance, tends to decrease nonlinearly with chain length. It is worth stressing that this evolution contrasts with the molecular exciton model which predicts the splitting to be proportional to the oscillator strength (hence, to increase with chain length). Though reduced, the amplitude of the splitting is expected to be significant for chains having a size corresponding to the typical effective conjugation lengths encountered in polymer samples. For polymers that can be considered as having very long conjugation lengths, such as MEH-PPV in highly drawn and oriented polyethylene [30], the splitting could become vanishingly small. In this case, the interaction would lead to a red-shift of the lowest optical transition, as suggested by recent highly-correlated calculations performed on interacting polyenes [29]; accordingly, the luminescence would not be quenched by the mechanism operating in finite-size systems.

At this stage, we can draw several important conclusions regarding the absorption and emission properties of cofacial dimers formed by two identical PPV chains:
(i) The lowest symmetry-allowed optical transition involves the second excited state (or a higher-lying state at short interchain distances) and is blue-shifted with respect to the isolated molecule. Such an evolution has been observed experimentally for stilbene molecules [31] and for other conjugated systems [32,33]. We also stress that the blue-shift expected from our calculations could be masked by additional effects such as conformational changes upon interaction.
(ii) Due to the absence of optical coupling between the ground and the lowest excited states, intermolecular interactions are expected to strongly affect the luminescence efficiency, however more so in molecular materials than in polymers since the amplitude of the interchain effects decreases with increasing chain length.

Photoluminescence properties of highly symmetric complexes

In the previous discussion, we have not considered the possible influence of the large electron-phonon coupling that is a trademark of conjugated systems; for instance, the equilibrium geometry in the lowest excited state can differ significantly from that in the ground state [15,34]. It is therefore of interest to investigate the role played by relaxation effects in the lowest excited state, assuming there is enough time for them to occur.

As expected, the AM1/CI-optimized geometry of a cofacial dimer formed by two largely separated stilbene molecules indicates that one chain possesses the typical ground-state geometry of an isolated unit while the other strand presents an equilibrium geometry characteristic of a single stilbene molecule in its lowest excited state. When the interchain distance is lowered, the lattice deformations remain highly asymmetric. Even though the chain initially in the ground-state geometry becomes increasingly affected by the intermolecular interactions when the chain separation decreases, the major relaxation phenomena always take place over a single chain. The mostly intrachain character of the exciton can be understood as driven by the Coulomb attraction between the electron and the hole, which is maximized when the two charge carriers are on the same chain. Since there is an equal probability to find the relaxed exciton on each of the two chains, it is worth stressing that the most general solution should be described beyond the Born-Oppenheimer approximation as a linear combination of the asymmetric states ($\psi_a = \phi_{1e}(R_e)\phi_{2g}(R_g)$ and $\psi_b = \phi_{1g}(R_g)\phi_{2e}(R_e)$, where we have explicitly noted that the equilibrium geometry of the two molecules is different); however, when the dimer is put in a condensed phase, there can occur different solvent shifts of the two states and the correct description is expected to be ψ_a and ψ_b or not [$\psi_a \pm \psi_b / \sqrt{2}$] [27].

The self-localization of the electron-hole pair is also supported by a detailed analysis of the wavefunction of the lowest excited state of a cofacial dimer formed by two five-ring PPV oligomers. In analogy to what we have described elsewhere [35], we have computed the probability amplitude $|\psi(x_e, x_h = 16, \text{chain I})|$ to find the electron on a given site x_e when fixing the hole on site 16 in the middle of chain I of the dimer (see Figure 3). The corresponding wavefunction presents a Gaussian lineshape centered on the position of the hole. The probability of finding the electron on chain II, *i.e.*, of creating an interchain exciton, is only 6.7% when the two oligomers are separated by 4 Å; this probability evolves exponentially with the intermolecular distance R going from 28.1% at 3.5 Å down to 0.6% at 4.5 Å (inset of Figure 3). We emphasize that charge-transfer excited states are observed at higher energies, the lowest of those being 0.63 eV above the lowest excited state for the two chains separated by 4 Å; in these states, there is a large probability to find the electron on chain II when the hole is fixed on chain I.

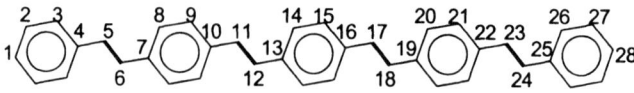

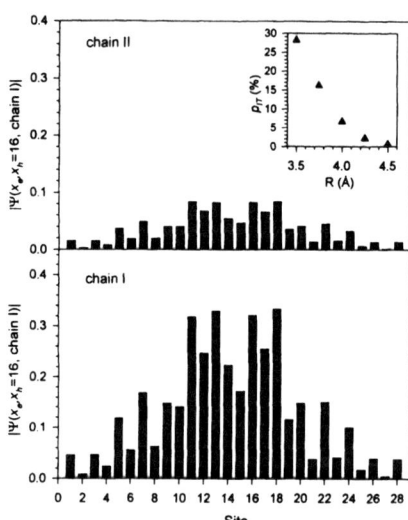

Figure 3: INDO/SCI simulation of the first singlet excited state wavefunction in a cofacial dimer formed by two five-ring PPV oligomers, $|\psi(x_e, x_h=16, \text{chain I})|$, assuming an intermolecular distance of 4 Å. $|\psi(x_e, x_h=16, \text{chain I})|$ represents the probability amplitude to find an electron on a given site x_e assuming the hole is centered on site 16 of chain I of the dimer (the other chain being referred to as chain II). The site labelling is represented in the top panel. The inset shows the evolution of the probability to create an interchain exciton, p_{IT}, as a function of the intermolecular distance R. The R values are given in Å.

The above results indicate that the selection rules are relaxed when the geometry modifications taking place upon photoexcitation are considered. Although the transition dipole moment between the ground state and the lowest excited state remains small, the luminescence is no longer entirely quenched by the interchain interactions and is red-shifted with respect to the emission of the single molecule; this is illustrated in Figure 4 where we have reported the INDO/SCI-simulated absorption and photoluminescence spectra of a cofacial dimer formed by two stilbene molecules separated by a huge distance and by 4 Å. In the regime of strong interactions, the results also show that the shorter the interchain distance, the larger the red-shift of the emission peak; the optical coupling between the ground state and the lowest excited state is simultaneously reduced due to the fact that the wavefunctions become increasingly symmetric for low chain separations. The red-shift and the attenuation of the emission signal predicted by our calculations appear to be fully consistent with recent experimental measurements probing the luminescence properties of PPV chains as a function of an external applied pressure, *i.e.*, more or less as a function of interchain distance [36]. It is worth stressing that the same trends prevail when increasing the number of units in interaction [27].

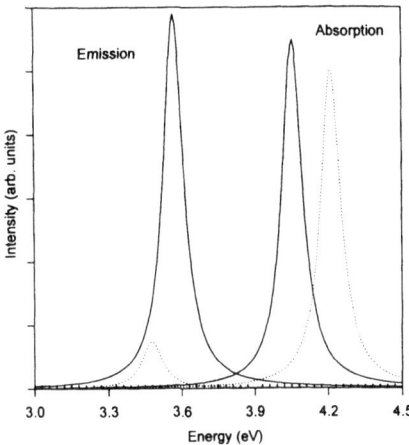

Figure 4: INDO/SCI-simulated absorption and emission spectra of two stilbene molecules with a huge interchain distance (solid lines) and those of a cofacial dimer formed by two stilbene chains separated by 4 Å (dotted lines).

The main conclusion is that interchain couplings have not necessarily to be associated to a substantial decrease of the luminescence quantum yield. As a matter of fact, a decrease in the radiative decay rate increases the relative importance of multiple nonradiative decay routes; oppositely, the effectiveness of these channels can be strongly limited by the confinement of the excitons due for instance to the formation of excimers or to the migration of the photogenerated species towards regions where the chains are strongly interacting (*i.e.*, the low-energy domain of the samples), as suggested by recent experimental studies [21,22]. The actual luminescence thus results from the balance of these contributions. In this context, the significant increase in the radiative lifetime of the luminescence observed in recent studies [20] can be attributed to the joint influence of a confinement process and of the weak optical coupling between the ground state and the lowest excited state.

Influence of the relative orientations of the interacting molecules

In order to address the possible influence of positional disorder, we have chosen to analyze the way basic operations such as translations and rotations affect the properties calculated for highly symmetric configurations. This approach could provide guidelines to prevent the loss of significant optical coupling between the ground state and the lowest excited state, and hence the quenching of luminescence in the solid state.

We start from a cluster formed by two or three interacting stilbene units separated by 4 Å. The translation of any unit along the chain axis or in a direction tranverse to the chain axis is a very inefficient process to recover a significant optical coupling. Referring to the dipole-dipole approximation, this is explained by the fact that the relative orientations of the transition dipole moments (which are polarized along the chain axis) are not changed. In all cases, the lowest optical transition is weakly coupled to the ground state and the energy splitting between the lowest two excited states is directly related to the overlap between the wavefunctions of the two units.

The second set of operations consists in rotating the molecules; here, starting from a cofacial dimer constituted by two stilbene units separated by 4 Å, we have rotated one molecule: (i) in the stacking plane around an axis perpendicular to the molecular planes and passing through the center of the vinylene linkage; and (ii) around its long axis, thus breaking the parallelism between the molecular planes. In the latter case, the calculated properties are very similar to those obtained for the cofacial dimer because the relative orientations of the transition dipole moments are still not modified. In the first case, we observe that increasing the angle between the chain axes leads to a reduction of the splitting between the lowest excited states, due to a decrease in overlap efficiency, as well as a gradual transfer, from the second excited state to the lowest one, of the electronic coupling with the ground state. In such conformations, the interaction between two chains thus leads not only to an intense absorption band blue-shifted with respect to the single molecule but also to the appearance of a weak red-shifted peak (*i.e.*, a red tail in highly disordered samples). The calculations also teach us that a supramolecular architecture characterized by a perpendicular orientation of the long chain axes turns out to be an efficient configuration to reduce luminescence quenching in the solid state; this is due to the fact that the lowest two excited states are then degenerate and that the whole oscillator strength involved in absorption is conserved in emission. Such perpendicular orientations are encountered for instance in spiro-conjugated molecules [37].

In summary, over a large range of interchain distances, the most stable species photogenerated in the lowest excited state of clusters formed by identical molecules are excitons with a mostly intrachain character. The calculations also show that polaron-pairs, also referred to as interchain excitons, (corresponding to a positive polaron on one chain in Coulomb attraction with a negative polaron on an adjacent chain) can be generated in charge-transfer excited states; the latter are found at energies higher than that of the lowest intrachain excited state. Since the average separation between the electron and the hole in a polaron-pair is expected not to be significantly different from that in an intrachain exciton, the polarization effects should not strongly stabilize the energy of the charge-transfer excited states with respect to that of the lowest excited state.

Several strategies that could be followed to prevent a substantial decrease in luminescence efficiency in condensed media emerge from the above results. Approaches that can be suggested are: (i) to minimize the splitting between the optically-forbidden and optically-allowed excited states; this can be done either by using long chains with highly delocalized excitations or by separating the chains with the help of bulky substituents; (ii) to reduce the diffusivity of the excitons to luminescence-quenching traps, by forming aggregation zones acting as low-energy centers from which emission is allowed.

SEXITHIENYL CLUSTERS

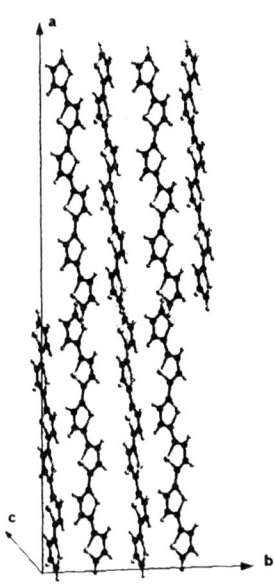

We now turn to an investigation of the optical properties of clusters of α-sexithienyl molecules adopting the herringbone structure, as determined from X-ray diffraction (see Figure 5) [26]. The theoretical results are confronted to relevant experimental data, which allows us to assess the reliability of our theoretical approach. Various clusters containing from 2 to 6 units are considered. We have first estimated the amplitude of the Davydov splitting as a function of configuration (this splitting is defined as the difference between the energy of the lowest excited state and that of the higher-lying state strongly coupled to the ground state [38]).

Figure 5: Sketch of the relative orientations of the sexithienyl molecules in the crystal and representation of the crystal axes a, b, and c discussed in the text. Note that the layers are stacked along the a axis.

When considering sexithienyl molecules located in the same layer (*i.e.*, no interaction between molecules aligned along the chain axis), the results indicate that the Davydov splitting is growing with the size of the cluster but tends to saturate around 0.5 eV in the largest systems we have considered. For a cofacial dimer, the energy difference is very sensitive to the interchain separation; it evolves from 0.2 to 1.0 eV when going from 4 to 3.3 Å. In all cases, the lowest excited state is only weakly coupled to the ground state due to the lack of significant angles between the transition dipole moments of the units in the herringbone structure. Note that the amplitude of the Davydov splitting is largely dominated by intralayer interactions (where the typical interchain distances are around 3.5-4 Å) and is little affected by the interactions between adjacent layers, which are separated by ~20 Å. We also observe that the lowest charge-transfer excited state is located above the high-lying Davydov state, as suggested by the previous calculations and confirmed by electroabsorption spectra recorded on a sexithienyl crystal [39].

The dominant features of the experimental optical spectrum of the sexithienyl crystal is an absorption onset at 2.28 eV, polarized along the b axis, and an intense vibronic progression at higher energy with the 0-0 peak at 2.60 eV [24], polarized along the c axis. We report in Figure 7 the theoretical absorption spectrum of a complex containing six units on top of each other within the same layer. The main goal of this simulation is to reproduce the overall shape and the main absorption characteristics of the sexithienyl crystal; consequently, we have not taken into account the Herzberg-Teller coupling (*i.e.*, vibronic borrowing) providing intensity to the low-lying excited states. Our approach is based on the displaced harmonic oscillator model where the Franck-Condon factors for a given mode are applied only to the states significantly coupled to the ground state and expressed for intrachain [14] and interchain excitons by [40]:

$$F_{0-v} = \frac{e^{-\frac{b_0^2}{4}} b_0^v}{\sqrt{2^v v!}} \qquad\qquad F_{0-v} = \frac{1}{\sqrt{v+1}} \sum_{i=0}^{v} \frac{e^{-\frac{b_+^2}{4}} b_+^i}{\sqrt{2^i \, i \, !}} \frac{e^{-\frac{b_-^2}{4}} b_-^{v-i}}{\sqrt{2^{v-i}(v-i)!}}$$

In these expressions, υ is the vibrational quantum number in the excited state, b_0 represents the displacement parameter associated with the mode for a neutral excitation ($b_0^2/2=S$ is the corresponding Huang-Rhys factor); b_+ and b_- are the parameters for the positive and negative polarons forming the charge transfer excitons. The model is based on the use of the single dominant mode at 0.18 eV observed in Raman spectroscopy [41]; the displacement parameters are taken as adjustable parameters chosen to match the experimental vibronic progressions (b_0=1.18, b_+=b_-=0.55 in our case); and the line broadening is fitted to the experimental resolution.

The INDO/SCI-simulated absorption spectrum reported in Figure 6 is characterized by a weak absorption at 2.24 eV polarized along the b axis and by a strong vibronic progression with the 0-0 feature at 2.58 eV, polarized along the c axis. These results are in remarkable agreement with the corresponding experimental data; they also provide a Davydov splitting of 0.35 eV between the onset of the first band and the 0-0 peak of the intense vibronic progression, while a value of 0.31 eV is obtained experimentally [24]. Our results cast doubt on other recent experimental estimates of the Davydov splitting, > 1 eV [42].

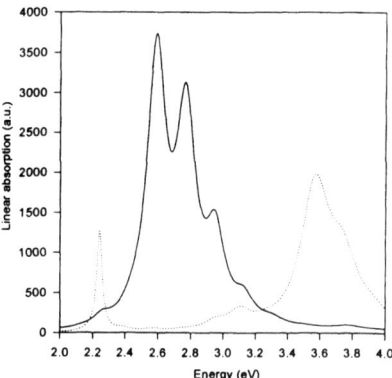

Figure 6: INDO/SCI-simulated absorption spectra of a cluster containing six molecules on top of each other within the same layer, for polarizations along the b (dotted line) and c (solid line) axes.

CONCLUSIONS

In this contribution, we have shown that the changes occurring in the absorption and luminescence properties when going from an isolated molecule to a cluster containing several units in interaction can be unraveled with the help of quantum-chemical calculations. The reliability of our theoretical approach was further established by the remarkable agreement observed between the calculated properties of clusters formed by sexithienyl molecules and the corresponding experimental data; this brings much confidence on the use of an INDO/SCI approach to shed light into the absorption properties of interacting systems, provided that the relevant excited states are suitably described by single excitations. The interest of such calculations is reinforced by the fast saturation of the calculated properties when increasing the size of the clusters and their applicability to a large class of compounds.

ACKNOWLEDGEMENTS

This work has been partly supported by the Belgian Prime Minister Office of Science Policy (SSTC) "Pôle d'Attraction Interuniversitaire en Chimie Supramoléculaire et Catalyse Supramoléculaire (PAI 4/11)", the Government of the Region of Wallonia and FEDER (Objectif 1-Hainaut), the Belgian National Fund for Scientific Research FNRS/FRFC, and an IBM Academic Joint Study. JC is Chargé de Recherches and DB is Chercheur Qualifié of the FNRS. We acknowledge stimulating discussions and collaborations with R.H Friend, A.J. Heeger, M. Muccini, and C. Taliani.

REFERENCES

[1]   J.H. Burroughes, D.D.C. Bradley, A.R. Brown, R.N. Marks, K. Mackay, R.H. Friend, P.L. Burn, and A.B. Holmes, Nature 357, 477 (1990).
[2]   R.W. Gymer, Endeavour 20, 115 (1996).
[3]   M. Berggren, O. Inganäs, G. Gustafsson, J. Rasmusson, M.R. Andersson, T. Hjertberg, O. Wennerström, Nature 372, 444 (1994).
[4]   R.E. Gill, G.G. Malliaras, J. Wildeman, and G. Hadziioannou, Adv. Mater. 4, 36 (1992).
[5]   S. Tasch, C. Brandstätter, F. Meghdadi, G. Leising, G. Froyer, and L. Athouel, Adv.Mater. 9, 33 (1997).
[6]   Q. Pei, C. Yu, Y. Zhang, and A.J. Heeger, Science 269, 1086 (1995).
[7]   N.S. Sariciftci, L. Smilowitz, A.J. Heeger, and F. Wudl, Science 258, 1474 (1992).
[8]   J.J.M. Halls, C.A. Walsh, N.C. Greenham, E.A. Marseglia, R.H. Friend, S.C. Moratti, and A.B. Holmes, Nature 376, 498 (1995).
[9]   F. Hide, M.A. Diaz-Garcia, B. Schwartz, M.R. Andersson, Q. Pei, and A.J. Heeger, Science 1833, 273 (1996).
[10]   N. Tessler, G.J. Denton, and R.H. Friend, Nature 382, 695 (1996).
[11]   S.V. Frolov, W. Gellermann, M. Ozaki, K. Yoshino, Z.V. Vardeny, Phys. Rev. Lett. 78, 729 (1997).
[12]   Primary Photoexcitations in Conjugated Polymers: Molecular Exciton versus Semiconductor Band Model, edited by N.S. Sariciftci (World Scientific, Singapore), in press.
[13]   R.H. Friend, G.J. Denton, J.J.M. Halls, N.T. Harrison, A.B. Holmes, A. Köhler, A. Lux, S.C. Moratti, K. Pichler, N. Tessler, K. Towns, and H.F. Wittmann, Solid State Commun. 102, 249 (1997), and references therein.
[14]   J. Cornil, D. Beljonne, Z. Shuai, T.W. Hagler, I.H. Campbell, K. Müllen, C.W. Spangler,

D.D.C. Bradley, and J.L. Brédas, Chem. Phys. Lett. **247**, 425 (1995).

[15] J. Cornil, D. Beljonne, C.M. Heller, I.H. Campbell, B.K. Laurich, D.L. Smith, D.D.C. Bradley, K. Müllen, and J.L. Brédas, Chem. Phys. Lett. **278**, 139 (1997).

[16] D. Beljonne, Z. Shuai, R.H. Friend, and J.L. Brédas, J. Chem. Phys. **102**, 2042 (1995).

[17] Electronic Materials - The Oligomer Approach, Edited by G. Wegner and K. Müllen (VCH, Weinheim), in press.

[18] B. Schwartz, F. Hide, M.R. Andersson, and A.J. Heeger, Chem. Phys. Lett. **265**, 327 (1997).

[19] L.J. Rothberg, M. Yan, F. Papadimitrakopoulos, M.E. Galvin, E.K. Kwock, and T.M. Miller, Synth. Met. **80**, 41 (1996).

[20] I.D.W. Samuel, G. Rumbles, and C.J. Collison, Phys. Rev. B **52**, 11573 (1995).

[21] J.W. Blatchford, S.W Jessen, L.B. Lin, J.J. Lih, T.L. Gustafsson, A.J. Epstein, D.K. Fu, M.J. Marsella, T.M. Swager, A.G. MacDiarmid, and H. Hamaguchi, Phys. Rev. Lett. **76**, 1513 (1996).

[22] U. Lemmer, S. Heun, R.F. Mahrt, U. Scherf, M. Hopmeier, U. Siegner, E.O. Göbel, K. Müllen, and H. Bässler, Chem. Phys. Lett. **240**, 373 (1995).

[23] S.V. Frolov, W. Gellermann, Z.V. Vardeny, M. Ozaki, and K. Yoshino, Synth. Met. **84**, 493 (1997).

[24] M. Muccini, E. Lunedei, and C. Taliani, private communication.

[25] M.J.S. Dewar, E.G. Zoebisch, E.F. Healy, J.J.P. Stewart, J. Am. Chem. Soc. **107**, 3902 (1985).

[26] G. Horowitz, B. Bachet, A. Yassar, P. Lang, F. Demanze, J.L. Fave, and F. Garnier, Chem. Mater. **7**, 1337 (1995).

[27] J. Cornil, D.A. dos Santos, X. Crispin, R. Silbey, and J.L. Brédas, submitted for publication.

[28] M. Kasha, Radiation Research **20**, 55 (1963).

[29] Z.G. Soos, G.W. Hayden, P.C.M. McWilliams, and S.J. Etemad, J. Chem. Phys. **93**, 7349 (1990).

[30] T.W. Hagler, K.J. Pakbaz, K.F. Vos, and A.J. Heeger, Phys. Rev. B **44**, 945 (1991).

[31] F.D. Lewis, T. Wu, E.L. Burch, D.M. Bassani, J.S. Yang, S. Schneider, W. Jäger, and R.L. Letsinger, J. Am. Chem. Soc. **117**, 8785 (1995).

[32] K. Liang, M.S. Farahat, J. Perlstein, K.Y. Law, and D.G. Whitten, J. Am. Chem. Soc. **119**, 830 (1997).

[33] J. Gierschner, H.J. Egelhaaf, and D. Oelrug, Synth. Met. **84**, 529 (1997).

[34] J.L. Brédas and G.B. Street, Acc. Chem. Res. **18**, 309 (1985).

[35] A. Köhler, D.A. dos Santos, D. Beljonne, Z. Shuai, J.L. Brédas, R.H. Friend, S.C. Moratti, and A.B. Holmes, Synth. Met. **84**, 675 (1997).

[36] S. Webster and D.N. Batchelder, Polymer **54**, 13713 (1996).

[37] N. Johansson, D.A. dos Santos, S. Guo, J. Cornil, M. Fahlman, J. Salbeck, H. Schenk, H. Arwin, J.L. Brédas, and W.R. Salaneck, J. Chem. Phys. **107**, 2542 (1997).

[38] D. Beljonne et al., to be published.

[39] L.M. Blinov, S.P. Palto, G. Ruani, C. Taliani, A.A. Tevosov, S.G. Yudin, and R. Zamboni, Chem. Phys. Lett. **232**, 401 (1995).

[40] P. Petelenz and V.H. Smith Jr., Chem. Phys. **131**, 409 (1989).

[41] G. Bongiovanni, C. Botta, J. Cornil, J.L. Brédas, D.R. Ferro, A. Mura, A. Piaggi, and R. Tubino, Chem. Phys. Lett. **278**, 146 (1997).

[42] F. Deloffre, F. Garnier, P. Srivastava, A. Yassar, and J.L. Fave, Synth. Met. **67**, 223 (1994).

AN EMULSION POLYMERIZATION PROCESS FOR SOLUBLE AND ELECTRICALLY CONDUCTIVE POLYANILINE

P. J. KINLEN*, Y.DING, C.R. GRAHAM, J. LIU AND E.E. REMSEN
Monsanto Company, St. Louis, Missouri 63167

ABSTRACT

A new emulsion process has been developed for the direct synthesis of the emeraldine salt of polyaniline (PANI) that is soluble in organic solvents. The process entails forming an emulsion composed of water, a water soluble organic solvent (e.g., 2-butoxyethanol), a water insoluble organic acid (e.g., dinonylnaphthalene sulfonic acid) and aniline. Aniline is protonated by the organic acid to form a salt which partitions into the organic phase. As oxidant (ammonium peroxydisulfate) is added, PANI salt forms in the organic phase and remains soluble. As the reaction proceeds, the reaction mixture changes from an emulsion to a two phase system, the soluble PANI remaining in the organic phase. With dinonylnaphthalene sulfonic acid (DNNSA) as the organic acid, the resulting product is truly soluble in organic solvents such as xylene and toluene(not a dispersion), of high molecular weight (M_w >22,000), film forming and miscible with many polymers such as polyurethanes, epoxies and phenoxy resins. As cast, the polyaniline film is only moderately conductive, (10^{-5} S/cm), however treatment of the film with surfactants such as benzyltriethylammonium chloride (BTEAC) or low molecular weight alcohols and ketones such as methanol and acetone increases the conductivity 2-3 orders of magnitude.

INTRODUCTION

Polyaniline (PANI), an inherently conducting polymer (ICP), has attracted considerable attention because of its ease of preparation and environmental stability. PANI is unique among ICP's in that its conductivity can be reversibly controlled either electrochemically (by oxidation reduction) or chemically (by protonation/deprotonation).

Synthesis of polyaniline is commonly performed by chemical oxidative polymerization in an aqueous solution (see e.g., Cao, et al.[1]). The method involves combining water, aniline, a protonic acid and an oxidizing agent and allowing the mixture to react while maintaining the reaction mixture at a constant low temperature (normally about 5 °C). After a period of several hours, the conducting emeraldine salt of polyaniline precipitates. Material synthesized by this approach is predominately amorphous, intractable and insoluble in most organic solvents[2].

Emulsion polymerization processes for preparing polyaniline salts have been reported[5,3,4,5,6]. In these disclosures, aniline, a protonic acid and an oxidant are combined with a mixture of water, a non-polar or weakly polar liquid, e.g., xylene, chloroform or toluene, all of which are sparingly soluble or insoluble in water. In order to form an emulsion in the above systems, protonic acids were employed that have substantial emulsifying properties in weakly polar liquids, e.g. dodecylbenzene sulfonic acid. In addition, the products of the above reactions were not isolatable directly, since the polyaniline salt remained entrained in the emulsion at completion of the reaction along with by-products of the reaction. In most cases, the product was isolated by breaking the emulsion (adding acetone, e.g.) and collecting the precipitated polyaniline salt, filtering and washing.

This paper describes a new emulsion polymerization process for polyaniline which yields a

305

soluble, conducting emeraldine salt directly without the need for a post-doping process step. The reaction is unique since the emulsion flocculates during the course of reaction to form a two phase system, the polyaniline remaining as a soluble component in the organic phase. Ammonium sulfate, a by-product of the reaction, remains in the aqueous phase which is readily removed from the organic layer.

EXPERIMENTAL SECTION

Preparation of Polyaniline Dinonylnaphthalene Sulfonic Acid Salt

0.1 mole DNNSA (50% w/w solution in 2-butoxyethanol from King Industries) was mixed with 0.06 moles aniline (Aldrich) and 200 ml water to form a milky white emulsion. The emulsion was chilled to 5°C, mechanically stirred and blanketed with nitrogen. Ammonium peroxydisulfate (0.074 moles in 40 ml water) was added dropwise to the mixture over a period of approximately one hour. During this period the emulsion changed color from white to amber. The reaction was allowed to proceed for 17 hours, during which time the emulsion separated into a green 2-butoxyethanol phase and a colorless aqueous phase. The progress of the synthesis was monitored by pH, OCP (open circuit potential, mV) and temperature.

The organic phase was washed three times with 100 ml portions of water, leaving a dark green, highly concentrated polyaniline phase in 2-butoxyethanol. This concentrate was readily soluble in xylene from which thin films were cast. Addition of acetone to a portion of the above concentrate resulted in the precipitation of the polyaniline salt as a green powder. After thorough washing of the powder with acetone and drying, elemental analysis indicated the expected stochiometric ratio of sulfonic acid to aniline of 1:2.

Using the same process polyaniline DNNSA salts were prepared using acid/aniline molar ratios of 1:1, 1:2, and 1:5 while the peroxydisulfate/aniline mole ratio was kept constant at 1.23. Attempts at synthesizing the polyaniline DNNSA salt employing either kerosene or xylene as the organic solvent yielded no product. UV/VIS/NIR measurements were made on a Cary 5E spectrophotometer.

Conductivity Measurements

A 6 mil wet coating of PANI-DNNSA concentrate (ca. 50% solids in xylene) was applied across a pattern of four gold contacts using a 0.006" draw down blade, 3.8 mm wide. The gold stripes sputtered onto a PET film were 6 mm wide by 50 mm long by 0.5 microns thick and spaced 6mm apart. After drying overnight at 80°C/27" Hg vacuum, the sample was stored in a desicator over P_2O_5. Resistance measurements between two adjacent gold contacts was taken using a Keithly Model 2001 Multimeter in the two point resistant mode. The contact resistance was judged to be minor, since the resistance through the coating between adjacent gold bars was additive.

Molecular Weight Determinations

Molecular weight distribution averages determined by size-exclusion chromatography (SEC). Chromatograms were obtained with two SEC systems: a model 150-CV SEC/viscometry (SEC/VISC) system (Waters Chromatography Inc.) and a multi-component SEC system (Waters Chromatography Inc.) assembled from a model 590 pump, a model 712 autoinjector, a model 410 differential refractive index detector, and a model TCH column heater. Both SEC systems were operated at 45°C and employed a bank of two styragel SEC columns (Waters Chromatography Inc.) with mean permeabilities 10^5 and 10^3 Å was employed. UV-grade N-methylpyrrolidinone (Burdick & Jackson Co.) modified with 0.02M NH_4HCO_2 (Fluka Chemical Co.) was used as the mobile phase and polymer solvent. A flow rate setting of 0.5 mL/min was employed. The actual flow rate was determined gravimetrically before the start of the analysis.

The preparation of polyaniline salt solutions in NMP/0.02M NH_4HCO_2 deprotonated the polymer and produced the emeraldine base. The presence of the dinonylnaphthalene sulfonic acid dopant in the solutions necessitated a determination of emeralidine base concentration in order to obtain accurate SEC/VISC-based polymer molecular weights. This was accomplished by the measurement of the emeralidine base's visible absorbance at 636 nm and the use of a literature value[7] of 7.7 x 10^6 M^{-1} cm^{-1} for the polymer's molar absorptivity.

Calibration of the SEC was performed with a set of 12 nearly monodisperse polystyrene standards (Toya Soda Inc.) ranging in molecular weight from 1.1 x 10^6 to 2,698. Intrinsic viscosities of the polystyrene calibrants were measured using the SEC's viscometeric detector. These values provided the Mark-Houwink expression for polystyrene in NMP/0.02M NH_4HCO_2 at 45°C needed to calibrate the SEC according to universal calibration:

$$[\eta](dL/g) = 1.947 \times 10^{-4} \, M^{0.66}$$

A linear least squares fitting was used to generate a universal calibration curve or a polystyrene-based molecular weight calibration curve.

Calculations of SEC/VISC-based molecular weight distribution employed previously described methodology[8]. Mark-Houwink constants for polyaniline were determined from the set molecular weight distribution averages and intrinsic viscosities calculated for individuals data points of SEC/VISC chromatograms.

Solubility Measurements

The polyaniline salt was prepared as described above using a 1:2 mole ratio of aniline to DNNSA. A 10 to 100 mg sample of the PANI-DNNSA salt (60% w/w in 2-butoxyethanol) was mixed with a test solvent (0.5 to 1.5 grams). Solubility was determined immediately after preparation.

RESULTS AND DISCUSSION

Emulsion Process

The emulsion polymerization process, illustrated in Figure 1, utilizes a water soluble organic solvent (2-butoxyethanol) that facilitates the polymerization of polyaniline salts of hydrophobic organic acids such as DNNSA.

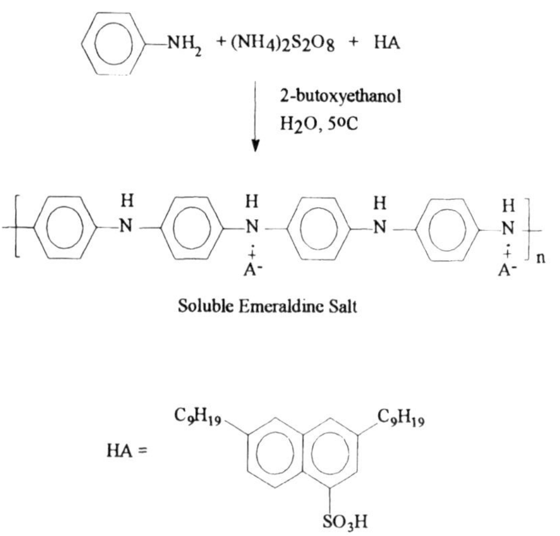

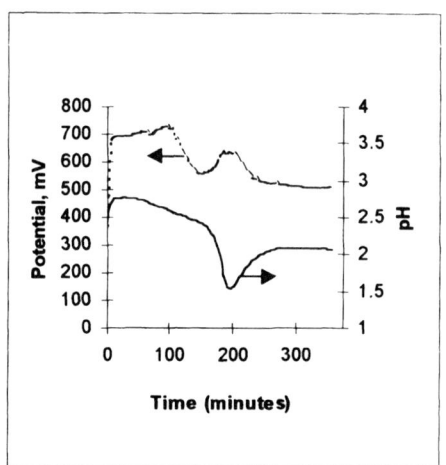

(top reaction scheme area — within the figure)

Figure 1.Emulsion polymerization scheme

The synthesis was monitored monitored by pH, OCP[9,10,11], and temperature, as well as color changes (see Figures 2 and 3).

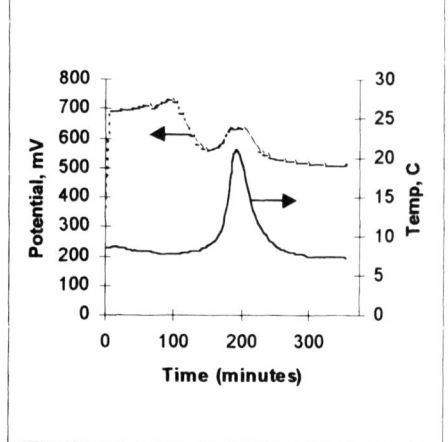

Figure 2. pH and OCP vs. time profile

Figure 3. Temperature vs. time profile

Addition of ammonium persulfate oxidant to the aniline/DNNSA/2-butoxyethanol/water mixture causes the OCP to increase to about 700 mV (this potential is approximately that of the ammonium peroxydisulfate solution). Prior to addition of the oxidant, dynamic light scattering analysis of emulsion size indicated a mean hydrodynamic diameter of 210 nm for emulsion particles. As the reaction proceeds, the emulsion flocculates forming a two phase system. After approximately 100 minutes into the reaction, the reaction the mixture begins to turn green, and the OCP drops to about 550mV, suggesting that the ammonium persulfate has been consumed and a new oxidizing species is formed. As discussed above, pernigraniline is believed to be an intermediate present at this point, however, sampling of the reaction mixture at this time showed very little polymer; the small amount of polymer found had a similar molecular weight distribution to that of the final product (coatings of this material exhibited very low conductivity). Sampling and warming to room temperature at this time resulted in precipitation of black solids from the colorless aqueous phase, indicating a decomposition of some intermediate species of limited stability (quinone imines, e.g.). As seen from Figure 2, the pH shows a steady decline during the first phase of the reaction due to sulfuric acid generation.

Our observations of the PANI-DNNSA synthesis are generally consistent with the mechanisms described in the literature. For example, a similar OCP vs. time curve is observed, an induction period exists and the polymerization becomes autocatalytic. However, there are important differences because of the presence of butylcellosolve (which is miscible in both aqueous and organic phases); and the dopant DNNSA which is hydrophobic and insoluble in water.

A pictorial representation of the reaction sequence is presented in Figure 4.

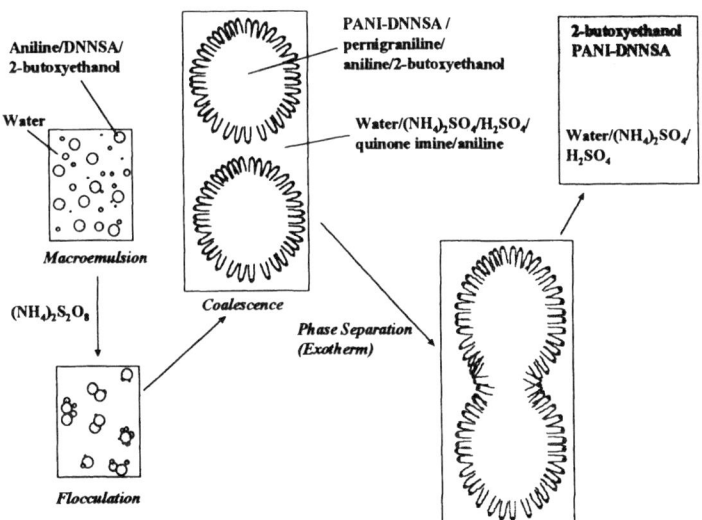

Figure 4. Pictorial representation of PANI-DNNSA emulsion reaction sequence

At the onset of the reaction essentially all of the aniline is in the butylcellosolve phase as the DNNSA salt. At 1/22,800 dilution of the standard emulsion laser light scattering shows that the primary emulsion particle has a mean hydrodynamic diameter of 150 nm and ranges in diameter from about 115 nm to 210 nm. As ammonium persulfate is added to the emulsion it flocculates and coalesces as the ion content of the aqueous phase increases. We conjecture that at this point pernigraniline begins to form in the 2-butoxyethanol phase as indicated by the appearance of quinone imine in the aqueous phase.

The observation of a long induction period (hours compared to minutes when using water soluble dopants such as chloride, sulfate or p-toluene sulfonic acid) may be explained by the slow diffusion of peroxydisulfate into the organic phase.

Molecular Weight

The SEC/VISC chromatograms for a deprotonated polyaniline salts were typically unimodal and nearly baseline resolution of the PANI and its sulfonic acid component was found as shown in the representative data in Figure 5.

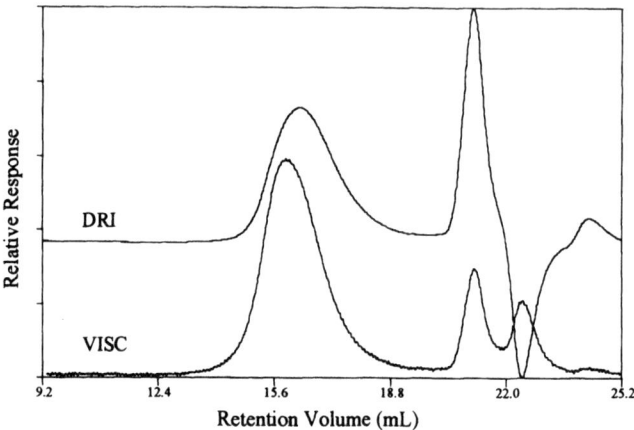

Figure 5. Size exclusion chromatogram of PANI with viscometric (VISC) and differential refractive index (DRI) detection

Table 1 lists the SEC/VISC molecular weights of several of the polyaniline salts synthesized as well as Versicon. The sulfonic acid components separated from the polyaniline peaks and were not included in the molecular weight calculations. In general the polyaniline salts tested produced broad size exclusion chromatograms, with M_w/M_n (polydispersity) >1.5. The Mark-Houwink (M-H) plot for PANI-DNNSA (1:2) is linear with alpha = 0.671 and logK = -3.146.

Table 1. Molecular Weights of Polyaniline Salts

Polymer	Mole Ratio of DNNSA to Aniline	M_w(SEC/Viscosity)	M_w (Polystyrene-equivalent)
PANI-DNNSA	1.7	31,250	70,400
PANI-DNNSA	0.5	25,300	63,300
PANI-DNNSA	0.2	5,685	16,750
Versicon™	----	19,600	55,800

Table 2 compares the M-H behavior of the PANI-DNNSA polymers synthesized as well as Versicon PANI. These values are in the range expected for typical linear random coil polymers[12].

Table 2. Comparison of M-H Behavior for Polyanilines

Polymer	Mole Ratio of DNNSA to Aniline	Alpha	log K
PANI-DNNSA	1.7	0.671	-3.146
PANI-DNNSA	0.5	0.667	-3.05
Versicon™	----	0.667	-2.997
Average		0.668	-3.059

The accuracy of polystyrene-equivalent molecular weights for polyaniline was evaluated by comparing these values against those determined by SEC/viscometry. The latter approach has the important advantage of using universal calibration and measured intrinsic viscosities to correct the SEC calibration for the difference polystyrene and polyaniline hydrodynamic volume.

The comparison between polystyrene-equivalent and SEC/viscosity weight-averages molecular weight, summarized in Table 1, points out the overestimation of molecular weight resulting from polystyrene-equivalent calibration. These results are in agreement with previous conclusions based on the comparison of light scattering and SEC molecular weights[16] which indicated that the polystyrene-equivalent SEC overestimates the true molecular weight by approximately a factor of 2.

Table 3 lists the solubility the PANI-DNNSA salt (2:1 molar ratio of sulfonic acid to aniline) in a variety of solvents.

Table 3. Solubility of PANI-DNNSA (1.7:1 DNNSA: Aniline ratio)

Solvent	Initial Solubility	Dielectric Constant	Solubility Parameter
Hexane	S	1.9	14.9, p
Cyclohexane	S	2.0	16.8, p
1,4 dioxane	S	2.2	20.5, m
Xylene	S	2.4	18.0, p
Trichloroethylene	S	3.4	18.8, p
Chloroform	S	4.8	19.0, p
Butylacetate	S	5.1	17.4, m
n-decylalcohol	S[a]	7.9	---
2-butoxyethanol	S[a]	9.4	---
1-octanol	S[a]	10.3	21.1, s
m-cresol	S	12.4	20.9, s
4-methyl-2-pentanone	S	13.1	17.2, m
1-pentanol	S	15.1	---
2-methoxyethanol	SS	17.2	---
1-butanol	S[a]	17.8	23.3, s
Methylethylketone	S	18.6	19.0, m
Isopropanol	SS	20.1	23.5, s
Acetone	I	21.0	20.3, m
N-methyl-2-pyrrolidinone	S	32.2	23.1, m
Acetonitrile[b]	SS	36.6	24.3, p
N,N-dimethylformamide	S	38.2	24.8, m
Dimethylsulfoxide	SS	47.2	24.6, m
Water	I	80.1	47.9, s
Propylene carbonate	I	---	27.2, m
Diethylene glycol	I	---	18.0, m
Diethyleneglycol monoethylether	SS	---	20.9, m
Diethylene glycol monobutylether	S	---	19.4, m

[a] Polymer phase settled out within 24 hours of preparation; [b] Solution turned brown

The rating of solubility was designated as follows: S, solubility equal to or greater than 1% by weight; SS, solubility less than 1% by weight and greater than about 0%; and I, insoluble (approximately 0% soluble). Also listed in Table 3 are solvent dielectric constants and Hildebrand solubility parameters[13].

As seen in Table 3, good solvents for PANI-DNNSA include: hexane, cyclohexane, xylene, dioxane, trichloroethylene, chloroform, butylacetate, m-cresol, 4-methyl-2-pentanone, methylethylketone, N-methyl-2-pyrrolidinone and N,N-dimethylformamide. It should be noted, however that solubilities of the acetone precipitated DNNSA salts were found to be very low in solvents that are excellent for the unprecipitated polymer (e.g., xylene and toluene). It thus appears that once the polymer chains entangle and/or crystallize that it is very difficult to resolubilize them.

Electrical Conductivity/Spectroscopy of PANI-DNNSA

Films of PANI-DNNSA ca. 0.15 mm thick cast from xylene solutions and dried at 70°C at 10-20 mm Hg for 7 hours yielded conductivities ranging from 6.5×10^{-6} to 63×10^{-6} S/cm. The UV/Vis/Near IR spectra of a dried PANI-DNNSA film is shown in Figure 6, curve A.

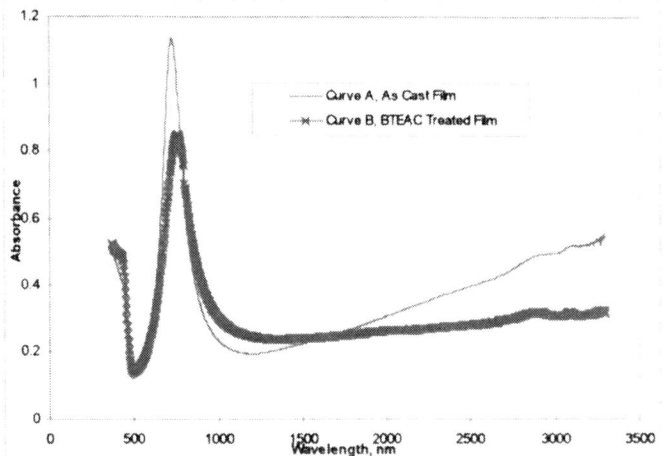

Figure 6. UV/Visible/NIR spectra of PANI-DNNSA films (curve A "as cast" film, curve B treated with 0.5 M BTEAC)

A strong polaron absorption at 720 nm is observed with a "free carrier tail" commencing at ca. 1000 nm. Similar spectra have been observed for PANI-camphorsulfonic acid exposed to m-cresol[14]. Curiously, in the case of PANI-DNNSA, the free carrier tail is present without secondary doping. After treatment of the PANI-DNNSA films in 0.05 or 0.5 M BTEAC solutions at 22 or 58°C, conductivities were found to increase by 3 to 4 orders of magnitude as shown in Table 4. Figure 6, curve B shows a decrease in the polaron absorption band at 720 nm with a concomitant decrease or "flattening" of the free carrier tail absorption.

A pronounced humidity dependence of the BTEAC treated film was noted. Thorough drying of the BTEAC treated films caused the conductivity to decrease significantly (although final values were still 10x greater than that of the original untreated samples). In contrast to the methanol treated films discussed below, the BTEAC treated films remained flexible and soluble in xylene.

No extraction of DNNSA into the BTEAC solutions was detected. We believe the BTEAC treatment results in morphological change from discontinuous (PANI-DNNSA) conducting domains to multiple connected pathways of PANI-DNNSA in an amorphous dopant(DNNSA/BTEAC) matrix[15].

It was also discovered that treatment of PANI-DNNSA films with methanol or acetone caused a large increase in conductivity, up to 5 orders of magnitude. The effect was also observed with MEK, 2-propanol and ethanol. No effects were noted with water or heptane treatment. The rate at which resistance dropped was very rapid, and substantial increases in conductivity were observed with as little as 5 seconds contact with the solvent. During the

Table 4. Effect of Surfactant Teatment on PANI-DNNSA Conductivity

	0.05 M BTEAC 22°C (30 s)	0.05 M BTEAC 58°C (10 min)	0.5 M BTEAC 22°C (30 s)	0.5 M BTEAC 58°C (10 min)
Conductivity Before Treatment (S/cm)x10^{-6}	6.5	12.0	8.1	63.0
Conductivity After Treatment and Drying at 22 °C (S/cm)x10^{-6}	18,000	130,000	40,000	610,000
Conductivity After Treatment and Drying at 70°C Under 10-50 mmHg Vacuum (S/cm)x10^{-6}	320	160	1400	2,500
Fold Increase in Conductivity	49	13	170	40

treatment process, the coating underwent shrinkage and became insoluble in solvents like dichloromethane and xylene (these solvents readily dissolve dried PANI-DNNSA coatings as well as the BTEAC treated coatings discussed above). The coatings were soluble in NMP, and no loss in molecular weight was noted. Quantitative extraction experiments in conjunction with HPLC showed that the treatment procedure extracts excess DNNSA from the films, leaving the stoichiometric equivalent of DNNSA behind (i.e., the mole ratio of DNNSA to aniline decreases from 1.4 to 0.5). As in the case of the BTEAC treated film, the UV/Vis/Near IR spectra exhibits both a decrease in the polaron absorption and free carrier absorption (Figure 7). Preliminary XRD analysis shows an increase in crystallinity for both the methanol and BTEAC treated films[16].

SUMMARY

A new emulsion process has been discovered for the direct synthesis of the emeraldine salt of polyaniline (PANI) that is soluble in organic solvents. As the reaction proceeds, the reaction mixture changes from an emulsion to a two phase system, the soluble PANI remaining in the organic phase. Thus, organically soluble polyaniline is formed directly. SEC/viscometry shows that the polymer behaves as a non-branched "stiff" random coil. Thin films of the are readily cast from xylene and have relatively low conductivities due to aggregation of conducting domains.

Treatment of thin films of PANI-DNNSA with quaternary ammonium salt solutions changes the morphology to one of multiply connected conducting pathways and results in a 3-4 order of magnitude increase in conductivity. Treatment of the films with acetone or methanol also results in an increase in conductivity, however the mechanism is through densification and crystallization of the film by removal of dopant.

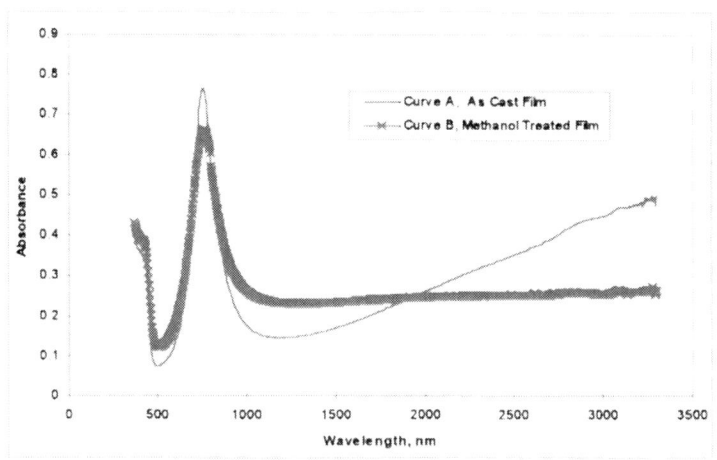

Figure 7. UV/Visible/NIR spectra of PANI-DNNSA films (curve A "as cast" film, curve B treated with methanol)

REFERENCES

1.   Y.Cao, A. Andreatta, A.J. Heeger and P. Smith, *Polymer, 30*, 2305(1989).

2.   B.K. Annis, A.H. Narten, A.G. MacDiarmid, and A.F. Richter, *Synthetic Metals, 22*, 191 (1986).

3.   Y. Cao and J.-E. Osterholm, WO94/03528, 1994.

4.   Y. Cao and J.-E. Osterholm, *U.S. Patent No.* 5,324,453, 1994.

5.   J.-E. Osterholm, Y. Cao, F. Klavetter and P. Smith, *Synth. Met., 55*, 1034 (1993).

6.   J.-E. Osterholm, et al., *Polymer, 35*, 2902 (1994).

7.   Y. Wei, K.F. Hsueh, and G-W. Jang, *Macromolecules 27*, 518 (1994).

8.   C-Y. Kuo, T. Provder, and M. E. Koehler, *J. Liq. Chromatogr., 13*, 3177 (1990).

9.   S.K. Manohar, A.G. MacDiarmid, and A.J. Epstein, *Synth. Met., 41*, 711 (1991).

10.   Y. Wei, K.F. Hsueh and G. Jang, *Polymer, 35,* 3572 (1994).

11.   Y. Wei, K.F. Hsueh and G. Jang, *Polymer Preprints, 35,* 267 (1994).

12.   F.W. Billmeyer, *Textbook of Polymer Science, Third Ed.,* John Wiley and Sons, New York, 1984, p. 211.

13.   E.A. Grulke, *Polymer Handbook,* 3d Ed., Brandrup and Immergut, Eds., Wiley & Sons, New York, pp. 519-517, 1989.

14.   Y. Min, Y Xia, A. G. MacDiarmid and A.J. Epstein, *Synth. Met.,* 69, 159 (1995).

15.   J. Liu and P.J. Kinlen, *Proceedings 1997 MRS Fall Meeting.*

16.   B.G. Frushour and P.J. Kinlen, to be published.

MODULATION OF THE ELECTRONIC PROPERTIES IN POLYDITHIENOTHIOPHENE MATERIALS

C. ARBIZZANI *, M. CATELLANI **, S. LUZZATI **, M. MASTRAGOSTINO ***
* Università di Bologna, Dip. di Chimica 'G. Ciamician', via Selmi 2, 40126 Bologna, Italy
** Ist. di Chimica delle Macromolecole, INMITER - CNR, via Bassini 15, 20133 Milano, Italy
*** Università di Palermo, Dip. di Chimica Fisica, via Archirafi 26, 90123 Palermo, Italy

ABSTRACT

Conjugated polymers with narrow band gap are promising candidates for symmetric electrochemical devices, such as electrochromics and redox supercapacitors, which involve the p- and n-doped state of the polymer. Polydithienothiophene materials which are prepared from isomer monomers show this characteristic. The copolymerisation of two dithienothiophene isomers leads to materials whose electronic properties depend on the monomer ratio. Optical and electrochemical characterizations of different copolymers are reported and discussed.

INTRODUCTION

Conjugated polymeric materials with narrow band gap have been receiving considerable attention due to their capacity to be both n- and p- doped, along with a suitable transparency in the doped state [1]. This opens the possibility of application in symmetric electrochemical devices such as electrochromics and redox supercapacitors. We have recently described a series of narrow band gap isomeric polydithienotiophenes, which can undergo p- and n-doping processes [2-4].

A powerful strategy for chemical modification of the conjugated materials and for electronic property control is the copolymerisation of different monomers. This strategy produced thiophene-based materials with tunable optical properties [5], narrow band gap [6], and which are stable in both the neutral and doped states [7].

The copolymerisation of two dithienothiophene isomers in well defined ratios leads to materials with properties depending on the chemical composition of the chain. Synthesis of the copolymers and their preliminary optical and electrochemical characterization are presented.

EXPERIMENT

Dithienothiophene polymers and copolymers were electrochemically grown on transparent ITO or TO glass and on Pt electrodes in galvanostatic conditions at 1mA cm^{-2}. The electrosyntheses were performed at room temperature in acetonitrile - 0.1 M tetrabutylammonium tetrafluoroborate (Et$_4$NBF$_4$) - 0.015 M monomer (or monomers).

The electrochemical measurements were performed with EG&G PAR M270A potentiostat/galvanostat and with a Solartron 1255 frequency response analyzer in propylene carbonate (PC) - 1 M Et$_4$NBF$_4$ in a MBraun Labmaster 130 dry-box (O$_2$ and H$_2$O contents < 1ppm). All chemicals were reagent-grade products, dried and purified before use. The capacitance values were evaluated by impedance spectroscopy after charging the electrode for 150 s at the peak potential of the voltammetric waves in the p- and n-doping potential domains, and by maintaining this potential during the impedance measurement. The reference electrode was an Ag wire (+100 mV vs saturated calomel electrode, SCE).

The UV-Vis spectra were recorded using a Varian Cary 2400 spectrophotometer.

RESULTS

Dithienothiophene (DTT) molecules (Scheme 1) consist of three heteroaromatic fused rings and have a high π-delocalisation. The four DTT isomers have different condensation types: i) in dithieno[3,2-b:2',3'-d]thiophene (DTT0) the lateral thiophene moieties are fused in αβ positions to the central ring; ii) in dithieno[3,4-b:3',4'-d]thiophene (DTT1) the condensations are in β positions; iii) in dithieno[3,4-b:3',2'-d] thiophene (DTT2) and dithieno[3,4-b:2',3'-d]thiophene (DTT3) both αβ and β condensations are present.

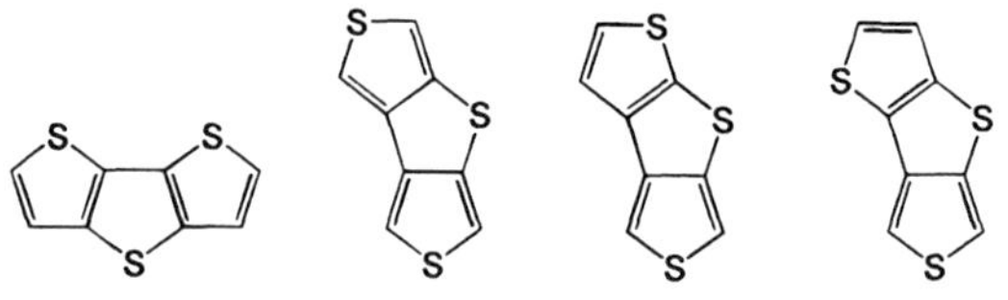

Scheme 1   DTT0   DTT1   DTT2   DTT3

The different chemical geometry of the four DTT monomers suggests different types of polymeric enchainments. DTT0 is expected to polymerize on the lateral thiophene rings, so that all the three fused rings take part in the conjugated chain (5,5' enchainment). Conformational calculations [8] suggested that DTT1 polymerization affects only one lateral ring, hence only one thiophene should lie in the conjugated backbone (2,5 enchainment). In DTT2 and DTT3 polymerization, too, only the ring with β condensation is probably involved (2,5 enchainment).

The electropolymerization of DTT1, DTT2 and DTT3 gave polymers with extended conjugation and narrow gap between HOMO and LUMO bands. In contrast, DTT0 yielded a polymer with energy gap not significantly lower than that of conventional polythiophene. A comparison of the electronic properties of the four PDTTs is in Table I. In addition, the electrochemical characterization of the polymers demonstrated that, unlike PDTT0, PDTT1, PDTT2 and PDTT3 are promising candidates for symmetric electrochemical devices [2-4].

Table I. Electronic absorption maximum and energy gap (onset of the absorption) of PDTTs grown on transparent electrodes.

Polymer	Absorption max (eV)	Energy gap (eV)
PDTT0	2.53	1.68
PDTT1	2.06	1.15
PDTT2	1.91	1.12
PDTT3	1.70	1.05

With the aim of tuning both the optical and the electrochemical properties of PDTTs we designed a family of dithienothiophene-based copolymers. The DTT copolymers are easy to prepare as the oxidation potentials of the four molecules are not significantly different as shown in Figure 1. We performed the electrosynthesis of these copolymers in galvanostatic conditions using different ratios of isomeric DTTs (co-monomer concentration of 0.015 M). The copolymer chemical composition probably matches the monomer ratio at the polymerization start.

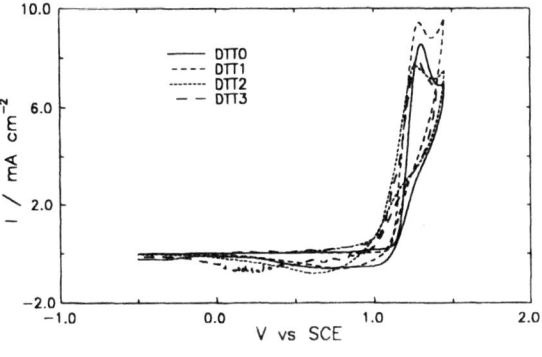

Figure 1. First cyclic voltammetry (CV) at 50 mV s^{-1} on Pt electrodes in the electrosynthesis solutions of DTT0, DTT1, DTT2, DTT3.

The copolymerisation of DTT0 with one of the other isomers gave three copolymer series, P(DTT0-DTT1), P(DTT0-DTT2) and P(DTT0-DTT3), whose optical and redox properties are intermediate between those of PDTT0 and those of the low energy gap PDTT used in each series.

The electronic structure of the copolymers is determined by two components: the contribution of the high energy gap PDTT0 and that of the low energy gap PDTT. Figure 2 shows the evolution of the absorption maximum and of the energy gap with the DTT1, DTT2 or DTT3 percentage in the starting co-monomer mixture for the three copolymer series. The increase of DTT1, DTT2 or DTT3 produces a shift of both $\pi-\pi^*$ absorption maximum and band gap to lower energies, but these variations are not a linear function of the co-monomer ratio.

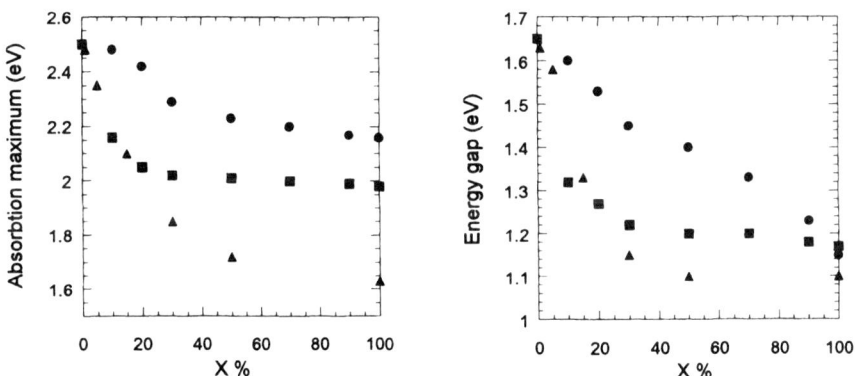

Figure 2. Plots of electronic absorption maximum (left) and of energy gap (right) of the copolymer series as a function of DTT1, DTT2 or DTT3 percentage (X%) in the starting co-monomers ratio; P(DTT0-DTT1) (●), P(DTT0-DTT2) (■) and P(DTT0-DTT3) (▲) series.

The addition into the PDTT0 backbone (the homopolymer with higher $\pi-\pi^*$ transition energy) of a small fraction of the other co-monomer provokes a significant decrease in the absorption maximum; this variation is less important when the narrow gap component is the prevalent moiety in the copolymer chains. The evolution of the energy gap with the chemical composition is similar, especially for P(DTT0-DTT2) and P(DTT0-DTT3). The same behaviour was predicted for random poly(p-phenylenevinylene) copolymers based on PPV and DMeOPPV [9], in which a small concentration of the narrow gap component causes a significant reduction in the HOMO-LUMO gap due to the formation of low energy gap copolymer segments.

Figure 3 shows the electronic spectra of the three copolymer series for different chemical composition of the polymer chains. The shift of the $\pi-\pi^*$ transition to low energy occurs without a significant band broadening, which confirms the copolymeric nature of these materials. The electronic absorption of the copolymers does not result simply from the weight average of the homopolymer spectra. By contrast, mechanical mixtures of different PDTT homopolymers, indeed, showed broad electronic bands and in some cases two absorption maxima.

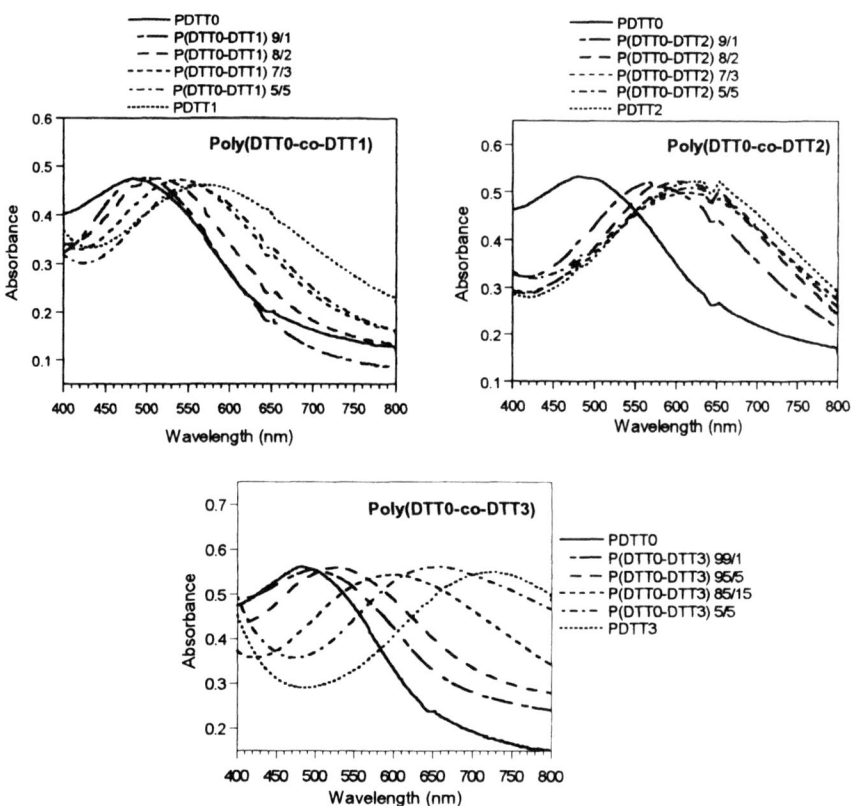

Figure 3. Electronic absorption spectra of the three copolymer series: P(DTT0-DTT1), P(DTT0-DTT2) and P(DTT0-DTT3) at different monomer ratios. Films grown on transparent electrodes.

The modulation in the energy gap reflects on the redox properties of the copolymers as the HOMO and LUMO energies are related to the polymer oxidation (p-doping) and reduction (n-doping) potentials, respectively. Figure 4 shows a comparison of the electrochemical performance of the P(DTT0-DTT3) copolymers and Table II lists the charge recovered (Q_d) in the reverse scan of the CVs, i.e. during the undoping processes, the coulombic efficiency (η%) and the doping level related to Q_d (y%) of the p- and n-doping processes of the P(DTT0-DTT3) copolymers.

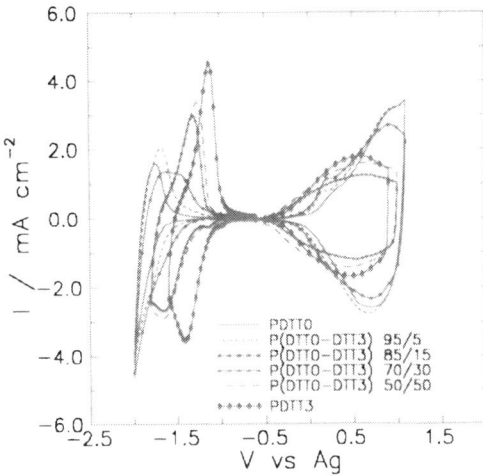

Figure 4. CVs in PC - 1 M Et₄NBF₄ at 50 mV s⁻¹ of P(DTT0-DTT3) copolymer series obtained from different co-monomer ratios.

Table II. Undoping charge Q_d, coulombic efficiency η% and doping level y% of p- and n-doping processes for P(DTT0-DTT3) copolymers with different co-monomer ratios

Copolymer	p-doping			n-doping		
	Q_d (mCcm⁻²)	η%	y%	Q_d (mCcm⁻²)	η%	y%
PDTT0	-38	97	37	8	48	8
P(DTT0-DTT3) 95/5	-42	98	41	13	58	12
P(DTT0-DTT3) 85/15	-39	98	38	14	51	14
P(DTT0-DTT3) 70/30	-25	98	23	18	85	16
P(DTT0-DTT3) 50/50	-31	98	28	22	90	20
PDTT3	-31	99	28	27	93	25

High coulombic efficiency is an important requirement for all electrochemical devices and we already knew that PDTT0 was useless for the application in symmetric devices because of the low n-doping coulombic efficiency. By contrast, the copolymers with 30-50% of DTT3 units show high coulombic efficiency, comparable to that of PDTT3. For the application in symmetric supercapacitors it is also important to attain high and comparable values for the capacitance of p- and n-doped electrodes. Figure 5 shows the capacitance data as a function of charging/discharging frequency of the fully n- and p-doped copolymers and homopolymers. The copolymer with a DTT0/DTT3 ratio of 50/50 exhibits comparable capacitance values for the p- and n-doped forms, which are not significantly lower than those of PDTT3, and this decrease in capacitance performance is balanced by a slight increase in the effective potential range.

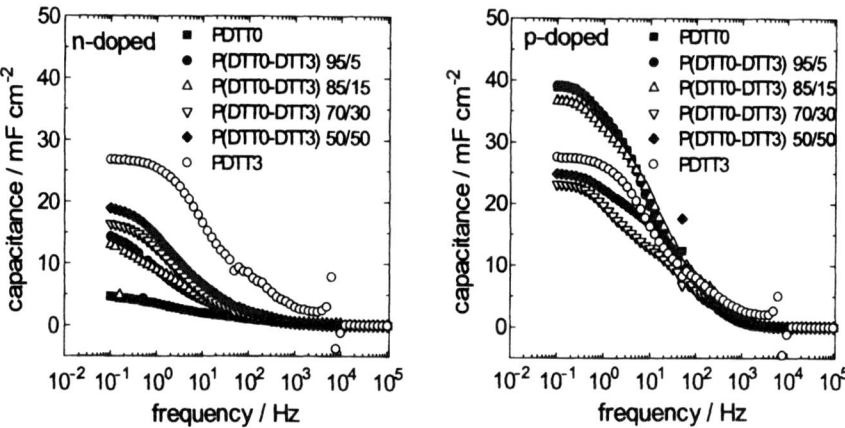

Figure 5. Capacitance values vs. frequency of PDTT0, PDTT3 and their copolymers in the fully n- and p-doped states.

CONCLUSIONS

The copolymerisation of two dithienothiophene isomers in well defined ratios leads to materials with different electronic properties depending on the chemical composition of the copolymer chains. The optical properties of the three copolymer series, P(DTT0-DTT1), P(DTT0-DTT2) and P(DTT0-DTT3), demonstrated that it is possible to modulate the band gap from 1.7 eV to 1.1 eV. This tuning of the energy gap reflects on the materials redox properties as the electrochemical tests on the P(DTT0-DTT3) copolymer series confirm .

ACKNOWLEDGEMENTS
Work supported by CNR, Progetto Strategico "Batterie leggere per auto elettriche. Supercapacitori redox e a doppio strato" and by University of Bologna (funds for selected research topics).

REFERENCES

1. J. Roncali, Chem. Rev. **97**, 173 (1997)
2. C.Arbizzani, M.Catellani, M.Mastragostino, M.G.Cerroni, J.Electroanal.Chem. **423**, 23 (1997)
3. C.Arbizzani, M.Catellani, M.Mastragostino, C.Mingazzini, Electrochim. Acta **40**,1871 (1995)
4. M. Mastragostino, C. Arbizzani, M.G. Cerroni, R. Paraventi, <u>Electrochemical Capacitors II</u>, F.M. Delnick, D. Ingersoll, X. Andrieu and K. Naoi (Eds.), The Electrochemical Society Proceeding Series, Pennington, **PV 96-25**, 109 (1997)
5. M. Catellani, S. Luzzati, R. Mendichi, A. Giacometti, P.C. Stein, Polymer **37**, 1059 (1996)
6. D. Lorcy, M.P. Cava, Adv. Mater., **4** , 562 (1992); S. Musmanni, J.P. Ferraris, J. Chem. Soc., Chem. Commun. 172 (1993)
7. Q. Pei, O. Inganas, J-E. Osterholm. J. Laakso, Polymer **34**, 247 (1993)
8. A.Bolognesi, M.Catellani, S.Destri, D.R.Ferro,W.Porzio, C.Taliani, R.Zamboni, P.Ostoja, Synth. Metals **28**, C 527 (1989)
9. D.A. dos Santos, C. Quattrocchi, R.H. Friend, J.L. Bredas, J. Chem. Phys. **100**, 3301 (1994)

Design and Synthesis of New Conjugated Porphyrin Copolymers for Optical-Electronic Applications

Biwang Jiang, Szu-Wei Yang, Phuong T. Lam, Wayne E. Jones Jr.*
Department of Chemistry and Institute for Material Research, State University of New York at Binghamton, Binghamton, NY 13902.

ABSTRACT

New linear conjugated porphyrin polymers were synthesized by a palladium-catalyzed cross-coupling reaction of [5,15-bis(ethynyl)-10,20-bis (mesityl)porphyrin]zinc and diiodobenzene derivatives. Enhanced solubility of the conjugated porphyrin polymers was achieved by attachment of long alkyl ether or dialkyl amide groups to the aryl moiety, resulting in unambiguous characterization by ^{1}H NMR, IR, GPC, UV-Vis and fluorescence spectroscopies. The introduction of alkyl ether (electron donor) or dialkyl amide (electron acceptor) results in significant modulation of the electronic properties of the conjugated porphyrin polymers due to strong electronic coupling. The spectroscopic and electronic characterization of these materials provides for comparison to earlier preparations of ethynyl bridged conjugated copolymers in which electronic coupling was substantially weaker.

INTRODUCTION

The preparation of soluble, linear conjugated porphyrin polymers is attractive not only for their central importance for elucidation of fundamental biological processes such as photosynthesis, but also as a new class of conjugated materials with potential applications in opto-electronic devices [1-7]. Previous work on porphyrin dimers and oligomers constructed through the building block approach have developed substantial information on basic electronic coupling in conjugated systems. Building on the previous dimeric work, conjugated polymers with porphyrins incorporated directly into the conjugation would allow for additional modification in both the ground and excited state electronic structures of these new materials. One of the major problems for synthesizing porphyrin polymers is their insolubility [2, 8-9], which seriously limits characterization and applications. However, a few soluble, conjugated porphyrin polymers have been studied [1].

Recently, we synthesized a series of soluble *meso*-phenylporphyrin-based polymers, 1, with tunable electronic and photophysical properties. Unfortunately, the electronic communication

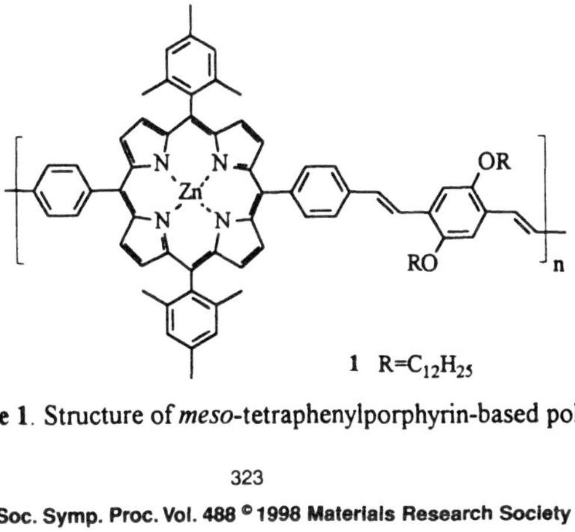

1   R=C$_{12}$H$_{25}$

Figure 1. Structure of *meso*-tetraphenylporphyrin-based polymer

along the polymer backbone was limited due to steric interactions between the phenyl *ortho* proton and the porphyrin β-hydrogen. The result of the steric interaction was a large dihedral angle between the porphyrin ring and the phenylenevinylene unit [6-7, 12] which limited electronic coupling through the π-conjugation. Through modifications of the synthetic chemistry we have now prepared more fully conjugated porphyrin polymers in which ethynyl linking groups are inserted between the porphyrin and the conjugated polymer main chain. The spectroscopic and electrochemical characterization of these new polymers demonstrate enhanced electronic and optical tunability relative to related ethenyl bridged materials.

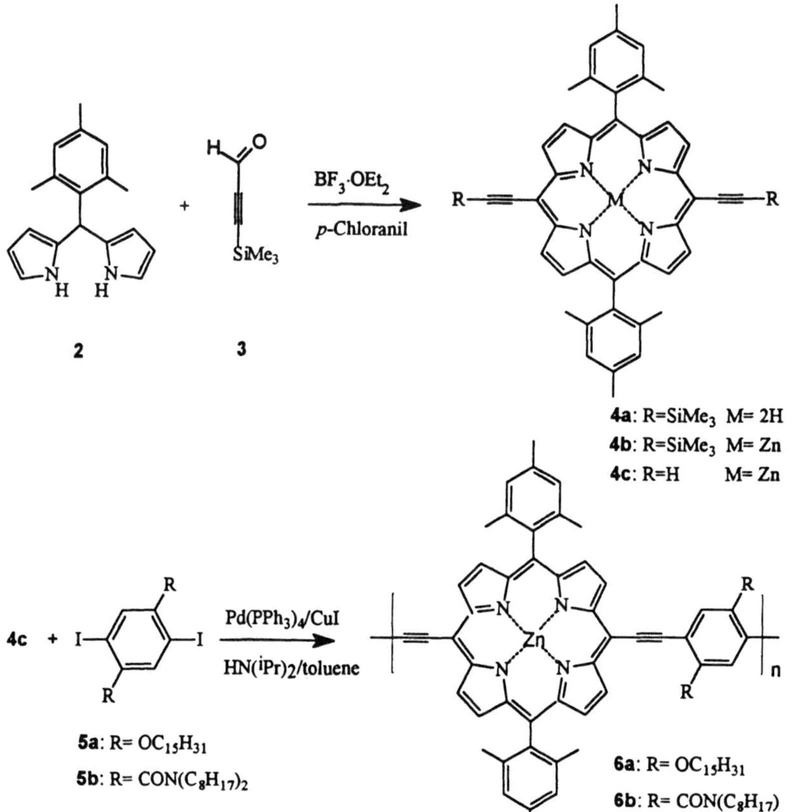
BF3·OEt2
p-Chloranil
SiMe3
2
3
4a: R=SiMe3 M= 2H
4b: R=SiMe3 M= Zn
4c: R=H M= Zn
4c
Pd(PPh3)4/CuI
HN(iPr)2/toluene
5a: R= OC15H31
5b: R= CON(C8H17)2
6a: R= OC15H31
6b: R= CON(C8H17)

Scheme 1. Synthesis of conjugated porphyrin polymers and their precursors.

EXPERIMENTAL

Meso-(mesityl)dipyrromethane **2** was synthesized from the condensation of excess pyrrole and mesitaldehyde in about 40% yield according to literature procedure[6]. Trimethylsilylpropynal **3** was prepared from trimethylsilyethyne in 80% yield by treatment with n-butyl lithium followed by dimethyl formamide. The condensation of dipyrromethane **2** with trimethylsilylpropynal **3** gave trimethylsilyl-protected porphyrin **4a** in 30% yield[10]. Metallation of porphyrin **4a** with

Zn(OAc)$_2$, followed by deprotection of the terminal acetylene with tetrabutylammonium fluoride leads to **4c**.

Preparation of Polymers (typical procedure). Under an atmosphere of argon, diisopropylamine (1.5 ml)/THF (3 ml) was added to a 15 ml Schlenk flask containing porphyrin **4c** (0.05 g, 0.076 mmol) diiodobenzene **5a** or **5b** (0.076 mmol), CuI (5 mg) and Pd(PPh$_3$)$_4$ (6.5 mg). This mixture was refluxed for 24 h and then subjected to a CHCl$_3$/H$_2$O. The combined organic phase was washed with NH$_4$OH (50%), H$_2$O and dried over MgSO$_4$. Most of the solvent was evaporated in under vacuum and the concentrated solutions poured into MeOH. The polymer precipitated as a green solid. (Yield:90%)

Similar synthetic methodologies can also be used to prepare oligophenylene bridged materials based on literature methods [10]. Examples of the oligophenylene materials prepared include polymer shown below in Figure 2.

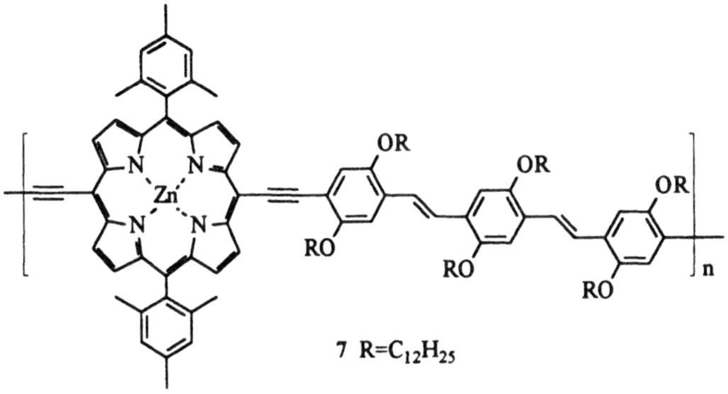

Figure 2. Ethynyl bridged oligophenylene porphyrin polymer.

RESULTS AND DISCUSSION

In previous preparations of soluble porphyrin polymers the electronic coupling was limited by steric interactions which forced the porphyrin conjugation to be perpendicular to the conjugation the polyphenylene bridge. The preparation of the ethyne bridge polymer, **6**, was designed to overcome this limitation by maximizing electronic coupling between the porphyrin and bridge. The polymers, **6a** and **6b**, were completely soluble in organic solvents such as THF. ^{1}H NMR, IR spectra demonstrated the expected polymer structure as shown in Scheme 1. The insertion of the triple bond was readily confirmed by both ^{1}H and ^{13}C NMR with no significant evidence of end groups in either the NMR or IR suggesting a high molecular weight.

Figure 3 shows the absorption spectra of polymers **6a** and **6b** compared with the monomer **4c**. In both cases, the porphyrin polymers exhibit dramatic red-shifts and broadening in both the B- and Q- type transition bands in comparison to porphyrin monomer **4c**. This would indicate strong electronic interactions along the conjugated porphyrin backbone.

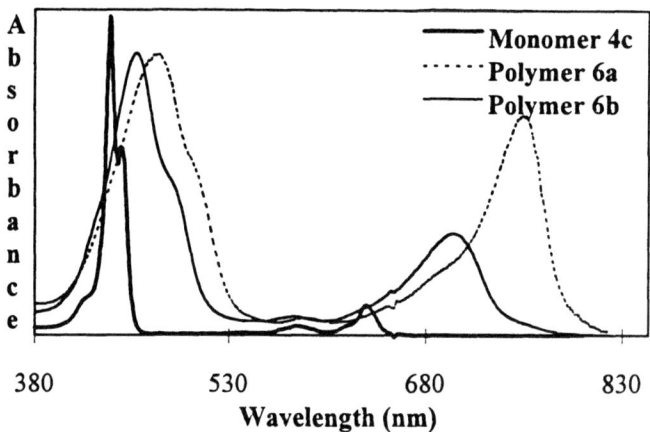

Figure 3. UV-Vis absorption spectra of monomer **4c** and polymers **6a** and **6b** in THF solution.

In the presence of strong electronic coupling, introduction of the aryl units into the porphyrin polymer main chain allows for facile modification of the electronic properties through the pendant aryl substituent. The observed decrease in absorption energy on going from polymer **6b** to **6a** is consistent with the electron donating character of the ether alkyl substituent and electron withdrawing character of the dialkyl amide substituent on the aryl units. The significant modulation of ground-state electronic properties of the polymer through aryl pendant groups and the broadening of the B-and Q-type transitions suggest high electron coupling along the polymer backbone (Figure 1). There was evidence for slight broadening of the Soret band in previous polymers where electronic coupling was found to be weak[10]. Cyclic Voltammetry measurements also showed evidence for coupling of the ground state reductions of the porphyrin groups in the ethyne bridged polymers[12].

The emission spectra of the polymers and the porphyrin monomer are shown in Figure **4**. Both polymers exhibit significant red-shifts in comparison to that of the monomer. Polymer **6a** displays a larger red shift than polymer **6b**, consistent with the electronic absorption spectra. The emission lifetimes of polymers **6a** and **6b** were 0.92 ns and 1.1 ns respectively, which is shorter than that of the porphyrin monomer (2.5 ns) measured in THF at room temperature, Table 1. The shifts in both the emission lifetime and the emission energy are consistent with *increased delocalization in the excited state*. The emission quantum yield is also shown to decrease from .071 to .022 upon polymerization, however these data are most consistent with a decrease in the rate of non-radiative decay that would be expected base on the observed lifetime and the increasing molecular weight of the polymers.

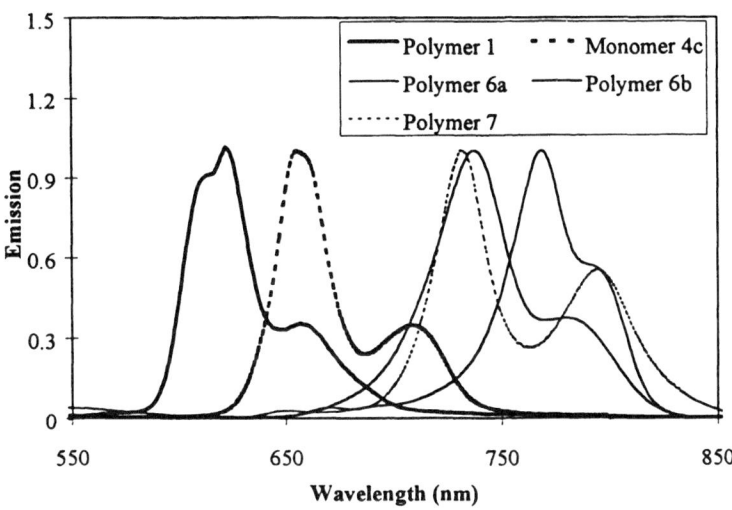

Figure 4. Emission spectra of monomer **4c** and polymers **1**, **6a**, **6b**, and **7** in THF solution.

Modifications in the synthetic chemistry allowed for extension of the oligophenylene bridge between the chromophores to form structures such as polymer **7**, Figure 2. This polymer and those with longer oligophenylene bridges represent direct analogues of the earlier work involving the ethenyl bridged materials[6]. Shown in Figure 4 are the emission spectra of the porphyrin monomer, the double bond bridged material **1**, and the triple bond bridged material **7**. There is an obvious shift in the emission energy, however the bandwidth of the observed emission does not change significantly when converted to a linear energy axis. The emission lifetime decreases from 7.2 to 1.38 ns upon insertion of the ethynyl linkage group between the porphyrin and the oligophenylene bridge.

	λ_{max}(Abs), nm	λ_{max}(Em), nm	Φ	τ, ns
Monomer **4c**	432	636	0.071	2.5
Polymer **6a**	460	756	0.022	0.916
Polymer **6b**	458	722	0.044	1.1
Polymer **1**	424	610	0.10	7.2
Polymer **7**	474	716	0.051	1.38

Table 1. Photophysical data of porphyrin polymers and monomer **4c**.

CONCLUSION

We have developed a new approach to synthesizing a family of fully conjugated porphyrin polymers in which the porphyrin units are connected through a triple bond linker group which significantly increases electron coupling along the polymer main chain. The porphyrin monomers are easily synthesized, and the high solubility of the polymers in common organic solvents allows routine manipulation, chemical characterization, and spectroscopic analysis. The photophysical and electronic properties can be fine-tuned through the structural modification of the more synthetically accessible aryl bridge units, rather than by direct modification of the porphyrin moiety. The unique coordinating ability of the central metal cation, combined with the high electron coupling of the polymers, may provide an opportunity to develop optical sensors based on this family of conjugated porphyrin polymers [13].

Acknowledgments

The authors thank Dr. Kirk Schanze, Dr. Janice Musfeldt, and Dr. Milissa Bolcar for helpful discussions. This research was funded by the Integrated Electronics and Engineering Center (IEEC) at the State University of New York at Binghamton. The IEEC receives funding from the NY State Science and Technology Foundation, the National Science Foundation and a consortium of industrial members.

Reference

1. (a) H. L. Anderson, S. J. Martin and D. D. C. Bradley, *Angew. Chem. Int. Ed. Engl.*, 1994, **33**, 655-657. (b) M. J. Therien, S. G. DiMagno, *U.S. Patent* 5,371,199.
2. H. L. Anderson, *Inorg. Chem.*, 1994, **33**, 972.
3. M. J. Crossley and P.L Burn, *J. Chem. Soc. Chem. Commun.*, 1987, 39-40.
4. M. J. Crossley and P. L. Burn, *J. Chem. Soc. Chem. Commun.*, 1991, 1569-1571.
5. V. S.-Y. Lin, S. DiMagno and M. J. Therien, *Science*, 1994, **264**, 1105.
6. B. Jiang, S. Yang, and W. E. Jones Jr., *Chem. Mater.*, **9**, 2031 (1997).
7. B. Jiang, S. Yang and W. E. Jones Jr., *Synthetic Metals.*, submitted.
8. (a) A. Osuka, N. Tanabe, R -P. Zhang, K. Maruyama, *Chem. Lett.* 1993, 1505-1508. (b) T. Nagata, A. Osuka, K. Maruyama, *J. Am. Chem. Soc.* 1990, **112**, 3054-3059.
9. H. L. Anderson, C. J. Walter, A. Vidal-Ferran, R. A. Hay, P. A. Lowden, J. K. M. Sanders, *J. Chem. Soc. Perkin. Trans. 1*. 1995, 2275-2279.
10. (a) P. J. Angiolilo, V. S.-Y. Lin J. M. Vanderkooi, and M. J. Therien, *J. Am. Chem. Soc.*, 1995, **117**, 12514. (b) M. Chachisvilis, V. Chirony, A. M. Shulga, B. Kallebring, S. Lesson, and V. Sundstrom, *J. Phys. Chem.* 1996, **100**, 13857. (c) V. S.-Y. Lin, M. J. Therien, *Chem. Eur. J.*, 1995, **I**, 645.
11. F. Bedioui and J. Devynck, *Acc. Chem. Res.*, **1995**, *28*, 30-36.
12. B. Jiang, S. Sahay, W. Jones, *J. Chem. Soc. Chem. Commun.*, submitted.
13. B. Jiang, S. Yang, D. Sarno, W. E. Jones Jr. manuscript in preparation.

STRUCTURAL APPROACH TO IMPROVE THE RESPONSE CHARACTERISTICS OF COPPER PHTHALOCYANINE THIN FILM-BASED NO$_2$ GAS SENSOR

Tadashi Nagasawa*, Kenji Murakami**, Kenzo Watanabe**
*Graduate School of Electronic Science and Technology, Shizuoka University, 3-5-1 Johoku, Hamamatsu 432, Shizuoka, JAPAN.
**Research Institute of Electronics, Shizuoka University, 3-5-1 Johoku, Hamamatsu 432, Shizuoka, JAPAN, k-murakami@rie.shizuoka.ac.jp

ABSTRACT

In order to realize a high-sensitivity, low temperature operable NO$_2$ gas sensor, thin films of α-form copper phthalocyanine (α-CuPc) have been deposited by vacuum sublimation. In this study, we have attempted to improve the gas-sensing characteristics through a modification of the film microstructure. Firstly, the gas sensitivity is remarkably increased by an insertion of higher-sensitive layer (vanadyl Pc film) between the α-CuPc film and the glass substrate in the low gas concentration range. Secondly, a reversibility in cycles of gas doping and dedoping is improved by film deposition on hydrofluoric acid-treated substrate. It is found from atomic force microscope analyses that this phenomenon may be closely related to a modification of the film microstructure.

INTRODUCTION

Due to increased concern about air pollution, monitoring of various noxious gases such as NO and NO$_2$ is of growing interest. One of the most simple and reliable methods to detect those gases is a measurement of the change in electric conductivity induced by an adsorption of gas molecules on the surface of inorganic oxide semiconductors or organic semiconductors. However, it is well known that gas sensors prepared from the inorganic oxide semiconductors show a poor sensing selectivity. Copper phthalocyanine (CuPc) is a p-type organic semiconductor which has higher thermal and chemical stability than similar organic semiconductors. The CuPc has been used in NO$_2$ gas sensors for long time because the electrical conductivity of CuPc increases upon adsorption of NO$_2$ gas [1]. However, there are still many problems for practical applications: the sensing mechanism is not yet clarified in detail and long-term stability has not been achieved. In particular, the response reversibility is insufficient for several cycles of alternated dopings and dedopings [2,3].

We have investigated a relationship between film structure and gas-sensing characteristics to improve the high-sensitive, low working-temperature NO$_2$ gas sensor characteristics prepared from the α-CuPc thin film. It is known that the structure and morphology of CuPc thin films strongly influence the gas-sensing characteristics [4-8]. Also in our case, the gas sensitivity and electrical conductivity were very sensitive to the film microstructure and morphology. Furthermore, this film microstructure was controlled by the deposition temperature, and higher gas sensitivity was obtained for α-CuPc thin films deposited at room temperature.

In this study, we have attempted to improve the gas-sensing characteristics of the α-CuPc film-based NO$_2$ gas sensor through a modification of the film microstructure. Firstly, an

329

improvement of the gas sensitivity has been attempted by an insertion of higher-sensitive layer between the α-CuPc thin film and the glass substrate. Secondly, we have attempted to improve the gas response characteristics of the α-CuPc thin films by means of a deposition on a hydrofluoric acid (HF)-treated glass substrate. Measured gas-sensing characteristics are discussed in relation to the detailed film microstructure analyzed by an atomic force microscopy (AFM).

EXPERIMENT

Since the preliminary studies suggest that vanadyl phthalocyanine (VOPc) shows a relatively higher sensitivity to the NO_2 gas than and a different crystal structure from the α-CuPc, the VOPc film was selected as an inserted layer between the α-CuPc film and the glass substrate. The α-CuPc and VOPc thin films were deposited by the vacuum sublimation method on the glass substrates at room temperature in a vacuum of 10^{-4} Pa. They were sublimed at between 450°C and 500°C. The film thickness was monitored by quartz oscillator to be 200 nm with a deposition rate of 0.1 nm/s. The HF treatment of the glass substrate was done in a 55% HF solution for one minute.

The 20 finger interdigitated gold electrodes were vacuum evaporated on the thin films to measure an electric conductivity in NO_2 gas diluted in N_2 gas, and in N_2 gas. The conducting current was monitored by using a Keithley 610C electrometer under the dc bias of 10 V. Details of the measurement are illustrated in Fig. 1.

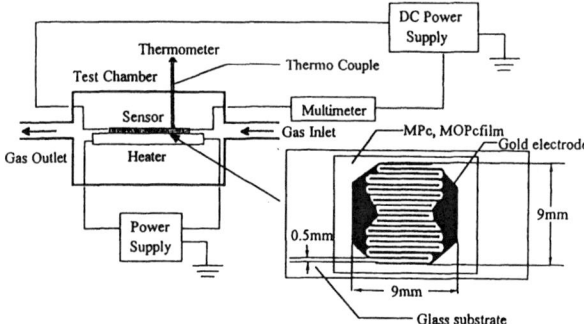

FIGURE 1   Schematic diagram of the measurement system for the α-CuPc thin film-based NO_2 gas sensors

The film crystallographic relation and microstructure were characterized by using the X-ray diffractometer (XRD, RIGAKU RAD-2A) with a Ni-filtered Cu-Kα radiation, and the AFM (JEOL JSTM-4200D) with a non-contact mode in air, respectively.

RESULTS

Structural Approach to Improve the Sensitivity

Figure 2 shows structures of the CuPc and VOPc molecules. It is expected from the difference in the structure (plane and pyramid) that microstructures of the deposited films must be quite different between these two sources. XRD profiles shown in Fig. 3 reveal that the α-CuPc thin film shows only one peak at 2θ=6.8° corresponding to an interplane distance of 13 Å [10], while no peak is detected from the VOPc films. Therefore, The α-CuPc film tends to orient uniaxially and the VOPc film seems to be amorphous like.

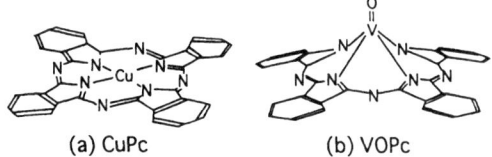

(a) CuPc   (b) VOPc

FIGURE 2   Molecule structure of (a) CuPc and (b) VOPc.

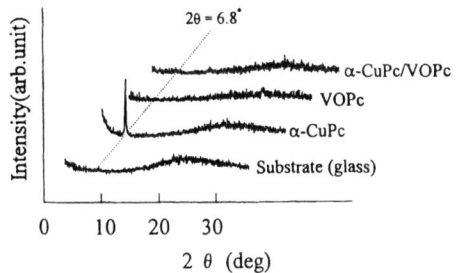

FIGURE 3   X-ray diffraction profiles of α-CuPc, VOPc and α-CuPc/VOPc thin films deposited on glass substrate at room temperature.

Gas sensitivity of the VOPc thin film is higher than that of the α-CuPc thin film in the gas concentration between 0 ppm and 100 ppm at a working temperature of 80°C, as shown in Fig. 4. The gas sensitivity is defined as a ratio of the measured current in NO_2 gas to the current in N_2. However, the conductive current of the α-CuPc thin film is higher than that of the VOPc films, as summarized in the figure. The actual flow current is very important for practical sensor devices, i.e. the higher conductive current of the sensor makes a design of peripheral circuits simpler.

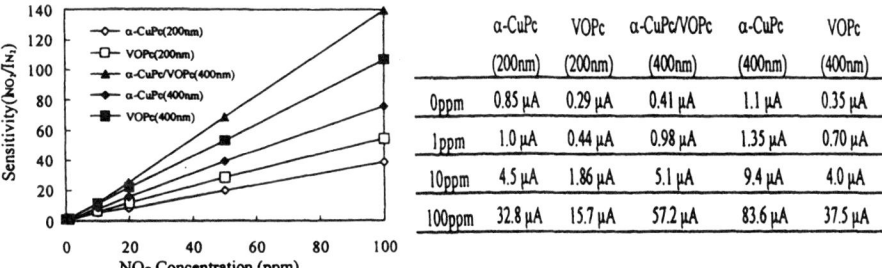

	α-CuPc (200nm)	VOPc (200nm)	α-CuPc/VOPc (400nm)	α-CuPc (400nm)	VOPc (400nm)
0ppm	0.85 μA	0.29 μA	0.41 μA	1.1 μA	0.35 μA
1ppm	1.0 μA	0.44 μA	0.98 μA	1.35 μA	0.70 μA
10ppm	4.5 μA	1.86 μA	5.1 μA	9.4 μA	4.0 μA
100ppm	32.8 μA	15.7 μA	57.2 μA	83.6 μA	37.5 μA

FIGURE 4   Dependence of the sensitivity and conductivity of the α-CuPc, VOPc and α-CuPc/VOPc thin films deposited on glass substrate at room temperature.

It has been already pointed out [9] that the gas sensitivity strongly depends on the film microstructure. Therefore, we tried to improve the gas sensitivity through a modification of the film microstructure. This attempt was realized by an insertion of the higher-sensitive VOPc layer between the higher-conductive α-CuPc film and the glass substrate. It is revealed from the XRD profiles that the peak has completely disappeared for the α-CuPc thin film with the VOPc layer (α-CuPc/VOPc) as also shown in Fig. 3. Figure 5(a), (b) and (c) show the AFM images of the α-CuPc, VOPc and α-CuPc/VOPc thin films, respectively. All the films are composed of small grains ranging from 50 nm to 100 nm in size. However, the shape and microstructure of the grains are different among the films. It is very interesting that the microstructure like a V-shape of the VOPc film is reflected in the α-CuPc/VOPc film with growing of the grain size.

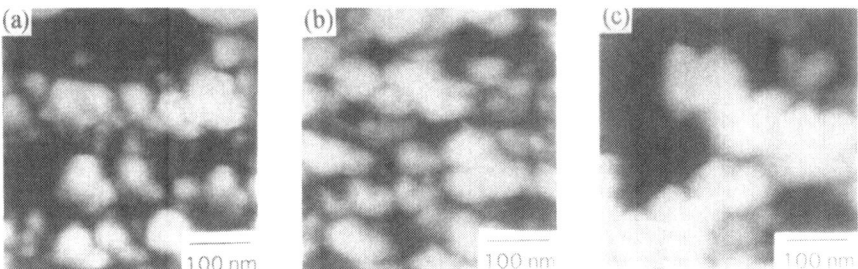

FIGURE 5   AFM images (500 nm x 500 nm) of (a) α-CuPc, (b) VOPc and (c) α-CuPc/VOPc thin films deposited on glass substrate at room temperature.

The gas sensitivity of the α-CuPc/VOPc thin film is shown in Fig. 4. The sensitivity of the α-CuPc film is drastically improved by the inserted layer. The actually measured current becomes close to that for the α-CuPc film. Furthermore, the response to 1 ppm NO$_2$ gas atmosphere is reproduced in cycles of NO$_2$ doping and N$_2$ dedoping at working temperature of 80°C, as shown in Fig. 6, however the recovery characteristics are still insufficient. The results suggest that the crystal structure of α-CuPc film becomes amorphous like which is considered to be higher sensitivity by the effect of underlying VOPc film. Furthermore, the drastic improvement of the sensitivity is attributed to the same V-shape microstructure as the VOPc film. The resulting conductivity is associated with the α-CuPc thin film.

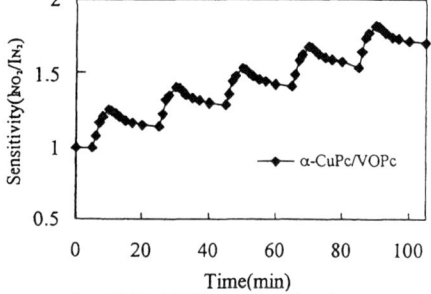

FIGURE 6   Time response of α-CuPc/ VOPc thin film deposited on the glass substrate to 1 ppm NO$_2$ gas at the working temperature of 80°C.

Structural Approach to Improve the Time Response Characteristics

Although the gas sensitivity of the α-CuPc film-based NO_2 gas sensor was improved by the insertion of the VOPc layer, its response characteristics are still insufficient, in particular, for several cycles of alternated dopings and dedopings. There have been several attempts to improve the characteristics [4-7]. In the present study, change of the grain structure of the α-CuPc film has been attempted in order to improve the response characteristics. If the grains have the simpler structure and their size is homogeneous resulting in closely packed grains in the film, the adsorption and desorption of the NO_2 gas might occur only in the vicinity of film surface. Therefore, we have attempted to create the undulation on a glass substrate surface by means of the HF treatment to reduce the interaction of the grains during the film growth.

Figure 7 shows the AFM image of the α-CuPc film deposited on the HF-treated glass substrate. Comparison with the image in Fig. 5(a) reveal that the shape of grains become simpler and the grain size homogeneous with about 90 nm in size for the film deposited on the HF-treated glass substrate. In addition, the film becomes more closely packed.

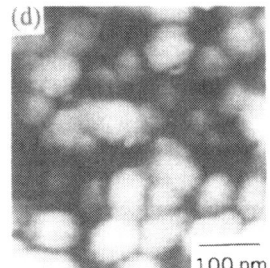

FIGURE 7   AFM images (500 nm x 500 nm) of α-CuPc thin film deposited on HF-treated glass substrate at room temperature.

Drastic improvement in the gas response characteristics is realized in Fig. 8 for the α-CuPc film deposited on the HF-treated glass substrate. The gas response characteristics were measured in cycles of 10 ppm NO_2 doping for 5 min and N_2 dedoping for 15 min for the films with and without the HF treatment at 150°C. Usually, the conductivity of the film does not recover to its initial value, and thus the film conductivity continues to increase with the NO_2 doping cycles, as shown in Fig. 8 (closed circles). However, for the film deposited on the HF-treated substrate, the film conductivity recovers completely to its initial value in every cycle. Furthermore, the initial response time of the α-CuPc film deposited on the HF-treated substrate glass substrate is shorter than that of the film deposited on the glass substrate. The results is very consistent with the model proposed from the difference in the microstructure of the AFM images.

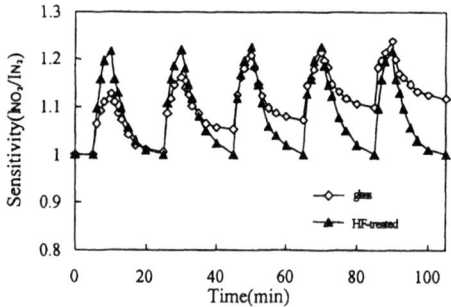

FIGURE 8   Time response of α-CuPc thin films deposited on the HF-treated glass substrate with 200 nm film thickness to 10 ppm NO_2 gas at working temperature of 150°C.

CONCLUSIONS

We have attempted to improve the sensing characteristics of α-CuPc thin film-based NO_2 gas sensor through a modification of the film microstructure.   The gas sensitivity is remarkably increased by the insertion of a higher-sensitivity layer (VOPc film) between the film and the substrate.   Furthermore, a reversibility in cycles of gas doping and dedoping is completely improved by the film deposition on the HF-treated substrate.   Although further intensive investigations are still necessary, the present study clearly shows possibility of improving the sensing characteristics of the α-CuPc film-based NO_2 gas sensors through structure control.

ACKNOWLEDGMENTS

The authors wish to thank Mr. Noboru Suzuki, Dainichiseika Color & Chemicals Mfg. Co., Ltd. for the supply of phthalocyanines.

REFERENCES

1. For example: Y. Sadaoka, N. Yamazoe and T. Seiyama, Denki Kagaku. **46**, p.597 (1978) (in Japanese).
2. S. Dogo, J.-P. Germain, C. Maleysson and A. Pauly, Thin solid films. **219**, p.251 (1992).
3. A. Cole, R.J. Mclloy, S.C. Thrope, M.J. Cook, J. McMurdo and A.K. Ray, Sensors and Actuators B. **13-14**, p.416 (1993).
4. B. Bott and T.A. Jones, Sensors and Actuators. **5**, p.43 (1984).
5. T.A. Jones, B. Bott and S.C. Thrope, Sensors and Actuators. **17**, p.467 (1989).
6. J. D. Wright, P. Roisin and G P. Rigby, Sensors and Actuators B. **13-14**, p.276 (1993).
7. Y. Sadaoka, M. Matsuguchi, Y. Sakai and Y. Mori. Sensors and Actuators B. **4**, p.495 (1991).
8. S. Pizzini, G.L. Timo, M. Beghi, N. Butta and C.M. Mari. Sensors and Actuators. **17**, p.481 (1989).
9. A.W. Snow and W.R. Barger, in C.C. Leznoff and A.B.P. Lever (eds)., <u>Phthalocyanines, properties and applications</u>, VCH, New York, 1989, pp. 362-367.

ELECTRICAL RECTIFICATION BY A MOLECULE OF HEXADECYLQUINOLINIUM TRICYANOQUINODIMETHANIDE

R.M. METZGER, B. CHEN, D. VUILLAUME, U. HÖPFNER , J.W. BALDWIN, T. KAWAI, H. TACHIBANA, H. SAKURAI, M. V. LAKSHMIKANTHAM, M.P. CAVA
Laboratory for Molecular Electronics, Chemistry Department, University of Alabama, Tuscaloosa, AL 35487-0336, USA, rmetzger@ua1vm.ua.edu

ABSTRACT

The asymmetrical forward versus reverse - bias DC electrical conductivity (macroscopic and also nanoscopic) through Langmuir-Blodgett multilayers and monolayers of γ-(n-hexadecyl)quinolinum tricyanoquinodimethanide, $C_{16}H_{33}Q$-3CNQ (1) is attributable to rectification of electrical current by a single molecule.

INTRODUCTION

Aviram and Ratner (AR) proposed in 1973 that a single, asymmetrical, molecule should be a rectifier of electrical current, because the electrical conductivity in one direction through the molecule is much larger than the conductivity in the opposite direction [1].

Rectification of electricity was discovered in vacuum diodes in 1904 [2], in pn junctions of semiconductors in 1949 [3], in macroscopic pn junctions of organic films in 1964 [4], in Langmuir-Blodgett (LB) multilayers in 1980 [5], in a LB monolayer photodiode in 1986 [6], and at a monolayer-modified electrode [7]. Asymmetric scanning tunneling spectroscopy (STS) currents have been seen in some molecules [8-10].

The Aviram-Ratner Ansatz proposes unimolecular rectification, i.e. asymmetrical electrical conduction through the molecular orbitals of a single D-σ-A molecule ("through-bond tunneling"), if D = "good" one-electron donor with low ionization potential, A = "good" one-electron acceptor with high electron affinity, and σ = insulating saturated covalent "bridge". The excited zwitterionic state D+-σ-A- should be reachable from the ground neutral state D-σ-A, while the opposite zwitterion D--σ-A+ should be several eV higher in energy [1].

Many D-σ-A molecules that form LB films were synthesized, but, using primitive techniques, rectification was not found [11-13]. Sambles found asymmetric currents in a sandwich "Pt I LB multilayer of dodecyloxyphenyl carbamate of bromohydroxyethoxytetracyanoquinodimethane I Mg I Ag" [14], but Schottky barrier could form in this system [15]. Ashwell introduced a new family of ground-state zwitterionic molecules of the type D+-π-A-, such as picolyltricyanoquinodimethane [16], and γ-(n-hexadecyl)quinolinum tricyanoquinodimethanide, $C_{16}H_{33}Q$-3CNQ (1) [17]. Z-type Langmuir-Blodgett multilayers of 1 have a large second-order nonlinear optical susceptibility $\chi^{(2)}{}_{zzz}$ = 180 pm V^{-1} [18]. Sambles found that the sandwich "Pt I LB multilayer of 1 I Mg I Ag " is a multilayer rectifier, but used dissimilar metal electrodes [19,20].

Following two preliminary studies of 1 [21,22] we recently found [23] a very convenient and efficient synthesis of 1, confirmed its ground-state zwitterionic structure, confirmed electrical rectification by both LB monolayers and multilayers by macroscopic and nanoscopic methods, and identified incontrovertibly the direction of higher conductivity relative to the orientation of the molecules in the multilayers: 1 is indubitably a unimolecular rectifier, and is probably the world's smallest active molecular device [23]. We review here these findings.

1 , $C_{16}H_{33}Q$-3CNQ

MOLECULAR PROPERTIES

1 is best made from N-cetyl-lepidinium tosylate and a two-fold excess of the lithium salt of TCNQ radical anion, in dry DMSO solution with added pyridine, in 59% yield; the NMR spectrum of 1 indicates a dipolar ground state [23]. The long-wavelength maximum in visible-ultraviolet spectrum of

335

1 is solvatochromic [23]. The cyclic voltammogram of **1** exhibits a one-electron quasi-reversible reduction ($E_{1/2}$ = -0.513 V vs SCE) [23].

The dielectric constants of solutions of **1** in CH_2Cl_2 were measured, as a function of temperature (-10°C < T < 30°C), at 1 kHz, using a coaxial capacitance cell and a Hewlett-Packard 4263B LCR meter. The extrapolation of the results to infinite dilution yields a dipole moment of 40 to 45 D (average = 43 ± 8 D) for **1** in CH_2Cl_2 [23].

MONOLAYER PROPERTIES

The Π-A isotherm of **1**, measured under a green safelight, shows a collapse point at 50 Å^2 molecule^{-1} and 34 mN m^{-1} (Fig. 1): this isotherm is moderately temperature-dependent [22,23]

Monolayers and multilayers of **1** were transferred to solid substrates at 15 mm/min at a film pressure of 20 mN/m: **1** first transfers on the downstroke onto hydrophobic HOPG, and on the upstroke onto hydrophilic Al. The hexadecyl group is thought to be the end of the

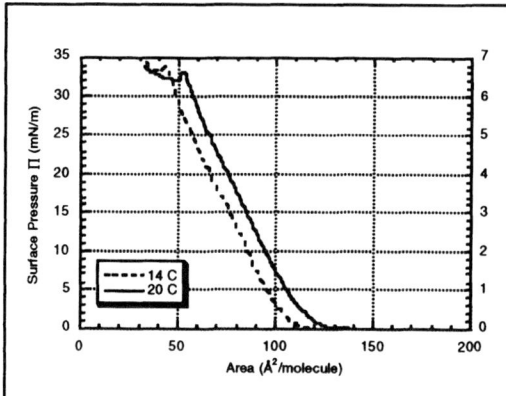

Fig. 1.
Pressure-Area isotherms for **1** at 14°C and 20°C [23].

molecule closest to HOPG, while the hydrophilic 3CNQ$^-$ (A$^-$) end should be closest to Al. At 15 mm/min the LB multilayers (past the first monolayer) are Z-type, while at 45 mm/min, the LB multilayers are Y-type (centro-symmetric). X-ray diffraction shows that the Y-type films apparently reorder to Z-type some time after transfer, with a thickness of 23 Å per monolayer: for multilayers of **1**, the Z-type ordering is more stable. The approximate molecular axis is thus tilted 48 ± 5° from the normal to the monolayer [23].

LB multilayers of **1** transferred onto HOPG at 15 mm/min have orientation:

HOPG | $\rightarrow \leftarrow \leftarrow \leftarrow \leftarrow \leftarrow \leftarrow$ etc. (1)

Here the arrow $\rightarrow$ indicates the rough molecular orientation; it points from the hexadecyl tail and the quinolinium ring D$^+$ (tail of arrow) towards the dicyanomethylene head of A$^-$ (head of arrow). The first monolayer adheres with the hexadecyl tail next to HOPG (X-type orientation), while the other layers have Z-type orientation (tails away from HOPG). The transfer ratios were 50 % for the first layer, and from 50 % to 100 % for the Z-type layers.

The film transfer to Al is Z-type for all layers:

Al | $\leftarrow \leftarrow \leftarrow \leftarrow \leftarrow \leftarrow$ etc.; (2)

the transfer ratio for the first layer is now over 90 %.

A sharp and narrow absorption band, with peak at 570 nm, is seen in LB films transferred onto quartz [22, 23]. This band disappears if the films are protonated, but returns when exposed to ammonia; it disappears irreversibly if the film, or a solution of **1** are exposed to bright light and air for some time [22, 23]. A grazing-angle Fourier transform infrared spectrum for a monolayer of **1** on Al shows absorbances at 2850 cm^{-1} (symmetric C-H), 2920 cm^{-1} (asymmetric -CH$_2$-), 2139 cm^{-1} (-C≡N) and 2175 cm^{-1} (-C≡N), indicating that the cyano groups and the methylene groups do not lie in the plane of monolayer [23].

MACROSCOPIC CONDUCTIVITY OF LB FILMS

Macroscopic "metal | organic | metal" sandwiches were studied. An Edwards 306A evaporator, with a liquid nitrogen-cooled substrate holder and a multiple-source evaporator, was used to form the sandwiches. The I-V curves were measured using both a PAR 270 potentiostat and also a Gateway 2000 microcomputer-controlled AC and DC conductivity measuring system (Hewlett-Packard Model 3245A universal source, and an Hewlett-Packard Model 3457A multimeter).

The glass or quartz or silicon substrates were covered by an Al layer, 25 mm × 75 mm × 100 nm thick, then LB monolayer or multilayer films were transferred under a green safelight, then the sample was dried for 2 days in a vacuum desiccator with P_2O_5. Twelve cylindrical "pads" of Al (4 each of areas 2.8, 4.5, and 6.6 mm^2, either 100 nm thick or 300 nm thick) were finally deposited per microscope slide, atop the organic layer, with the substrate cooled to 77 K, to form the sandwiches as shown for monolayers (Fig. 2a) and multilayers (Fig. 2b).

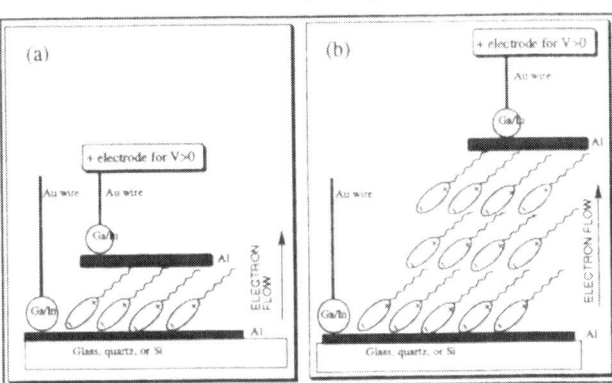

=================
Fig. 2.
Orientation of the LB monolayer (a) or multilayer (b) on the glass, quartz, or Si substrate; the electrode (+) for positive bias, and the direction of "easy" electron flow for V > 0 are marked. [23].
=================

Contact to the Al electrodes was made with a Ga/In eutectic. The electrodes (Al) are symmetrical, but probably covered by oxide layers. A very asymmetric I-V curve for the 4-monolayer film was found [23]; reversing the electrodes reversed the asymmetry. A similarly asymmetric I-V curve occurred when a Mg layer was deposited atop the organic layer below the Al pad [23].

The work with monolayers was more difficult. Of 39 pads checked (20 with areas of 2.8 mm^2, 15 with areas of 4.5 mm^2, 4 with areas of 6.6 mm^2) with two thicknesses of the Al pad (12 with thickness 300 nm, 27 with thickness 100 nm), 17 were electrical short circuits, either because of monolayer defects, or because the Ga/In or Ag paste created new defects during attachment. The 22 "good" junctions had DC resistances from 1 to 400 MΩ; of these, 4 exhibited rectifying behavior. Fig. 3 shows a typical rectification curve. Above a threshold voltage V_t, the monolayer shows larger currents for positive bias than for negative bias. This V_t varies from junction to junction in the range $V_t = 0.8$ to 1.3 V.

The rectification ratio RR is:

$$RR = (\text{current at } V_0) / (\text{current at } -V_0) \tag{3}$$

where V_0 is the highest positive bias used. RR ranges from 2.4 to 26.4 the first time the film is measured. As the measurement cycle is repeated, RR drops steadily, and disappears after 4 to 6 cycles. Applying a constant positive or negative bias, a slow decrease is seen, over 15 - 20 minutes, of the current for both polarities. It appears that, under the intense electric fields, the molecular dipoles tend to re-orient to minimize energy. For some junctions, short circuits form across the monolayer during the measurement [23].

Between -0.5 V and 0.5 V, Ohm's law is followed. At negative bias, $\ln_e(I) \propto V$ (through-space tunneling) from -0.5 V to the highest negative bias applied (-1.5 V or -1.8 V). At positive bias, $\ln_e(I) \propto V$ for V > 0.5, up to a threshold voltage $V_t = 0.8 - 1.3$ V, above which the current rises severalfold. This deviation from $\ln_e(I) \propto V$ may be due to an enhanced through-bond electron transfer [23], and resembles previous observations for multilayers of **1** [20].

The results on both monolayers and multilayers (not given here) show that the electrons preferentially flow by intervalence transfer (IVT) from 3CNQ (A$^-$) to quinolinium (D$^+$), in agreement with the AR Ansatz. For multilayers, electron transfers between successive monolayers from the D$^+$ of layer N to the A$^-$ of layer N-1 is possible (by tunneling through the alkyl chains), provided that the electric field is high enough to move the electron between the relevant molecular orbitals.

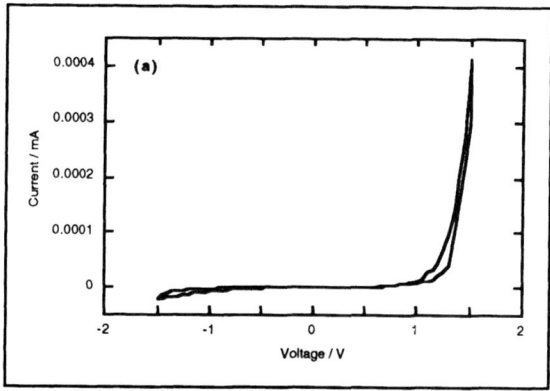

NANOSCOPIC STRUCTURE OF LB FILMS

In a low-current STM (Digital Instruments Nanoscope III) an LB monolayer of **1** showed a molecular image with low resolution and a repeat distance of 6 Å × 12 Å (Fig. 4) [23].

==========================

Fig. 3.
Rectification through a single monolayer of **1** sandwiched between Al electrodes (top Al pad area 4.5 mm^2, thickness 100 nm). DC voltage sweep rate=10 mV s^{-1} [23].

Fig. 4.
STM image of a LB monolayer of **1** on HOPG, with Pt/Ir tip (Nanoscope III). Scan size = 4.5 nm × 4.5 nm, Z-range = 2.3 pA, bias = -316 mV; setpoint current = 3.2 pA [23].

===========================

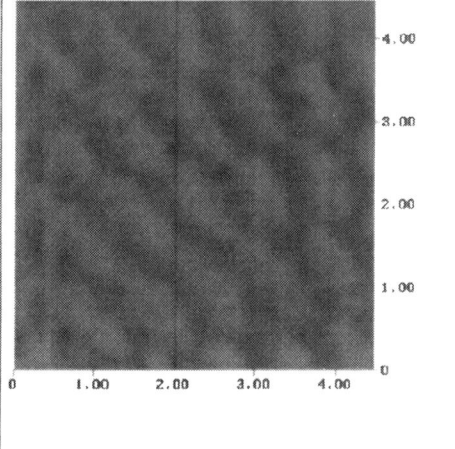

NANOSCOPIC CONDUCTIVITY OF LB FILMS

Multilayers of **1** on HOPG were studied using a Pt/Ir nanotip at room temperature in air by STM and STS [22]. Asymmetries were seen in the I-V plot in a 15-layer films (Fig. 5) [22], in agreement with the I-V characteristics of macroscopic films, with IVT between LB layers 2 through 15 (Fig. 6). The electrons must somehow jump between adjacent monolayers, and also get across the poorly transferred first monolayer (which is oriented the "wrong way", compared with the other 14 layers) [22].

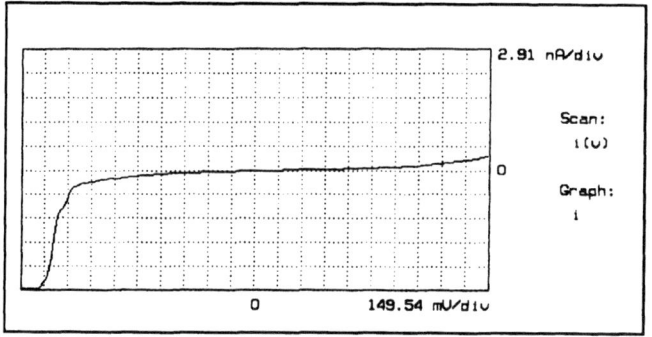

==============

Fig. 5.
STS I-V curve for an LB film of 15 Z-type monolayers of **1** on HOPG, with Pt/Ir tip. The higher current for V < -1.35 Volt corresponds to electron flow from HOPG through the molecules to the Pt/Ir tip [22].

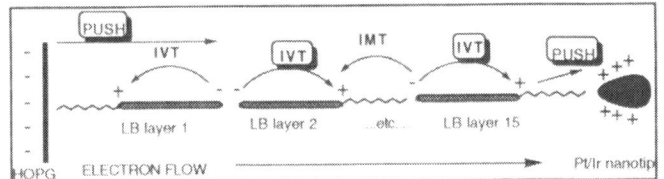

Fig. 6. Direction of STS electron current through 15-layer LB multilayer of **1** [22,23].

An STS spectrum for a partial monolayer of **1** adhering with its alkyl tail on HOPG shows a slightly higher current at positive bias, in the expected direction [23].

THE AVIRAM-RATNER MODEL ADAPTED TO A GROUND-STATE ZWITTERION

The AR model of unimolecular rectification [1] places a single molecule D-σ-A (D = good electron donor, A = good electron acceptor, σ = covalent linkage between D and A) between two macroscopic metal electrodes M_1 and M_2. Under "forward bias" an electron is transferred from M_2 to M_1 in two steps:

$$M_1 \mid D\text{-}\sigma\text{-}A \mid M_2 \quad \text{— (ET)} \rightarrow \quad M_1^- \mid D^+\text{-}\sigma\text{-}A^- \mid M_2^+ \quad (4)$$
$$M_1^- \mid D^+\text{-}\sigma\text{-}A^- \mid M_2^+ \quad \text{— (IVT)} \rightarrow \quad M_1^- \mid D\text{-}\sigma\text{-}A \mid M_2^+ \quad (5)$$

One electron hops by elastic tunneling (ET) from the HOMO of the neutral donor moiety D, onto the Fermi level of the metal electrode M_1 whose work function ϕ_1 is in resonance with the HOMO; an electron from the ϕ_2 of M_2 hops at resonance onto the LUMO of the neutral acceptor moiety A. This excited-state zwitterion D^+-σ-A^- decays inelastically to the ground state D-σ-A by IVT. The D^+-σ-A^- state may lie 1 to 2 eV above the D-σ-A ground state. The reverse process:

$$M_1 \mid D\text{-}\sigma\text{-}A \mid M_2 \quad \text{— (ET)} \text{—XX} \rightarrow \quad M_1^+ \mid D^-\text{-}\sigma\text{-}A^+ \mid M_2^- \quad (6)$$

can be neglected: the state D^--σ-A^+ is probably 4-5 eV higher in energy than D^+-σ-A^-.

The AR model can be adapted to a ground-state zwitterion D^+-π-A^-, whose first excited state is D^0-π-A^0. Under "forward bias" an electron is transferred from M_2 to M_1 in two steps:

$$M_1 \mid D^+\text{-}\pi\text{-}A^- \mid M_2 \quad \text{— (IVT)} \rightarrow \quad M_1 \mid D^0\text{-}\pi\text{-}A^0 \mid M_2 \quad (7)$$
$$M_1 \mid D^0\text{-}\pi\text{-}A^0 \mid M_2 \quad \text{— (ET)} \rightarrow \quad M_1^- \mid D^+\text{-}\pi\text{-}A^- \mid M_2^+ \quad (8)$$

The reverse bias process is:

$$M_1 \mid D^+\text{-}\pi\text{-}A^- \mid M_2 \quad \text{— (ET)} \rightarrow \quad M_1^+ \mid D^0\text{-}\pi\text{-}A^0 \mid M_2^- \quad (9)$$
$$M_1^+ \mid D^0\text{-}\pi\text{-}A^0 \mid M_2^- \quad \text{— (IVT)} \rightarrow \quad M_1^+ \mid D^+\text{-}\pi\text{-}A^- \mid M_2^- \quad (10)$$

The molecule rectifies, and electrons move from M_2 to M_1, if Eq. (9) is less likely than Eq. (7).

DISCUSSION

The synthesis of **1** was vastly improved. The molecule is zwitterionic: its solution dipole moment is $\mu = 43 \pm 8$ D (i.e. an electron-hole pair separated by 9 ± 1 Å (the estimated distance from the bridgehead C atom in the dicyanomethylene end of **1** to the quinolinium N atom is 10.5 Å). The monolayer thickness suggests molecules inclined by 45-55° from the surface normal. The first reduction of **1** is reversible, but **1** is not a powerful electron acceptor. The STM repeat areas of 6×12 Å (Fig. 4) are consistent with the calculated molecular cross-section (7.4×12.4 Å), and also roughly consistent with the observed collapse area of 50 Å^2 (Fig. 1) [23].

The STS rectification across an LB multilayer (Fig. 5) is very reproducible; the direction of higher electron flow at $V < -1.35$ V corresponds to IVT electron flow within layers 2 to 15 of a 15-layer film, as explained in Fig. 6 (boxed IVT), but the contrary flow across layer 1 (IVT) and the electron flow mechanism between layers (IMT) are not explained [22]. Rectification by a single molecule of **1** on HOPG is discernible, but not very dramatic [23].

Monolayer rectification between Al electrodes (Fig. 3) can be seen in 10% of the pads examined (one must find regions that are defect-free). Repeated cycling decreases the rectification ratio, as the molecules reorient in the intense electrical fields.

The current through a macroscopic film corrsponds to 0.33 electrons molecule^{-1} s^{-1} [23]. Recent preliminary work indicates that rectification persists down to 78 K in a monolayer [24].

CONCLUSIONS

A monolayer of molecule **1** rectifies by intramolecular tunneling; monolayers and multilayers of **1** rectify both as macroscopic films (using Al for both electrodes) and nanoscopically, by STS.

ACKNOWLEDGMENTS

Thanks are due to Profs. Geoffrey J. Ashwell, J. Roy Sambles, Michael P. Cava, and Ludwig Brehmer and Dr. M. V. Lakshmikantham for collaboration and advice, and to the National Science Foundation (grant DMR-94-20699), the Department of Energy (grant DE-FC02-91-ER-75678), and the Scientific Affairs Division of the North Atlantic Treaty Organization (grant CRG 960656) for their kind support.

REFERENCES
1. A. Aviram and M. A. Ratner, Chem. Phys. Lett. **29**: 277 (1974).
2. J. A. Fleming, U. K. Patent **24,850** (appl. 16 Nov. 1904).
3. W. Shockley, Bell System Techn. J. **28**: 435 (1949).
4. J. E. Meinhard, J. Appl. Phys **35**: 3059 (1964).
5. E. E. Polymeropoulos, D. Möbius and H. Kuhn, Thin Solid Films **68**: 173 (1980).
6. M. Fujihira, K. Nishiyama and H. Yamada, Thin Solid Films **132**: 77 (1986).
7. K. S. Alleman, K. Weber and S. E. Creager, J. Phys. Chem. **100**: 17050 (1996).
8. M. Pomerantz, A. Aviram, R. A. McCorkle, L. Li and A. G. Aschott, Science **255**: 1115 (1992).
9. A. Stabel, P. Herwig, K. Müllen and J. P. Rabe, Angew. Chem. Intl. Ed. Engl. **34**: 1609 (1994).
10. A. Dhirani, P.-H. Lin, P. Guyot-Sionnest, R. W. Zehner and L. R. Sita, J. Chem. Phys. **106**: 5249 (1997).
11. R. M. Metzger and C. A. Panetta, New J. Chem. **15**: 209-221 (1991).
12. R. M. Metzger, In Molecular and Biomolecular Electronics, R. R. Birge, Ed. ACS Adv. in Chem. Ser. **240** (American Chemical Society, Washington, DC 1994) p. 81.
13. R. M. Metzger, Mater. Sci. & Engrg. **C3**: 277-285 (1995).
14. N. J. Geddes, J. R. Sambles, D. J. Jarvis, W. G. Parker and D. J. Sandman, Appl. Phys. Lett. **56**: 1916 (1990).
15. N. J. Geddes, J. R. Sambles, D. J. Jarvis, W. G. Parker and D. J. Sandman, J. Appl. Phys. **71**: 756 (1992).
16. R. M. Metzger, N. E. Heimer and G. J. Ashwell, Mol. Cryst. Liq. Cryst. **107**: 133 (1984).
17. G. J. Ashwell, M. Szablewski and A. P. Kuczynski, in R. M. Metzger, P. Day, and G. C. Papavassiliou, Eds., Lower-Dimensional Systems and Molecular Electronics, NATO ASI Series (Plenum, New York, 1991) **B248**, p. 647.
18. G. J. Ashwell, in G. J. Ashwell and D. Bloor, Eds. Organic Materials for Nonlinear Optics, (Royal Soc. of Chem., Cambridge, 1993), p. 31.
19. G. J. Ashwell, J. R. Sambles, A. S. Martin, W. G. Parker and M. Szablewski, J. Chem. Soc. Chem. Commun. 1374 (1990).
20. A. S. Martin, J. R. Sambles and G. J. Ashwell, Phys. Rev. Lett. **70**: 218 (1993).
21. X.-L. Wu, M. Shamsuzzoha, R. M. Metzger and G. J. Ashwell, Synth. Metals **55-57**: 3836 (1993).
22. R. M. Metzger, H. Tachibana, X. Wu, U. Höpfner, B. Chen, M. V. Lakshmikantham and M. P. Cava, Synth. Metals **85**: 1359 (1997).
23. R. M. Metzger, B. Chen, U. Höpfner, M. V. Lakshmikantham, D. Vuillaume, T. Kawai, X. Wu, H. Tachibana, T. V. Hughes, H. Sakurai, J. W. Baldwin, C. Hosch, M. P. Cava, L. Brehmer and G. J. Ashwell, J. Am. Chem. Soc. **119**: 10455 (1997).
24. B. Chen and R. M. Metzger, unpublished results.

SOLVATO-CONTROLLED DOPING OF CONDUCTING POLYMERS: ENHANCED STABILTY IN SILVER-TRIFLATE DOPED FILMS

MICHAEL V. LAURITZEN, STEVEN HOLDCROFT[*]

[*]Department of Chemistry, Simon Fraser University, Burnaby, B.C., Canada,V5A 1S6,
steveh@bohr.chem.sfu.ca

ABSTRACT

Stable, conductive films of poly(3-alkylthiophenes) (P3ATs) can be obtained from a single chloroform based solution containing polymer, oxidant (silver triflate), and an oxidant-coordinating ligand (pyridine). Casting of this solution, followed by evaporation of the ligand, results in an electronically conductive film which is more stable and, notably, in the case of thick films, is free of the cracking which often results from swelling of the films during the doping process. This technique, coined solvato-controlled oxidative doping, results in films possessing enhanced stability compared to those oxidized by conventional dopants such as ferric chloride. Also, these silver-containing films are much more stable than those electrochemically doped with the triflate anion. Thus, the presence of metallic silver in the doped film has a profound effect on the stability of the conducting form.

INTRODUCTION

The incorporation of polythiophene films into technological applications has been retarded, in large part, by their poor processability and relative instability. Significant progress has been made in rendering their neutral insulating form soluble and processable[1]; however, the oxidized conducting form has seen little progress in this area. The stability of doped P3AT films has been found to be highly dependant on both the counter-ion incorporated into the doped film, as well as the nature of the substituents on the thiophene ring. Recent work has shown that P3AT films doped with the $AuCl_4^-$[2] or bis(trifluoromethanesulfonyl)imide[3] anions exhibit lifetimes much greater than their $FeCl_4^-$ doped counterparts; additionally, polythiophenes which are 3- or 3,4-substituted with electron releasing alkoxy groups have lower oxidation potentials and longer lifetimes.[4]

In this report, we describe (i) the stabilization of polymer/dopant solutions with dopant-coordinating ligands (ii) casting of the solutions described in (i) and evaporation of the coordinating ligand, yielding stable, conductive, silver-containing films doped with the triflate anion (iii) electrochemical oxidation of P3AT films (in the presence of the triflate anion) at various potentials and determination of their half-lifetimes (for comparison to silver triflate doped films). This work is part of a research program into the photolithography of electronically conducting polymers in which the stability and processability of the conducting polymer are important issues.[5]

EXPERIMENT

Regioregular poly(3-hexylthiophene) (P3HT) and poly(3-dodecylthiophene) (P3DDT) were prepared by a literature procedure.[6] Silver triflate (AgOTf) (Sigma Chemical Co.), acetonitrile (Caledon, HPLC grade), lithium triflate (LiOTf) and tetrabutylammonium tetrafluoroborate (Aldrich) were used without further purification. Pyridine (BDH) was dried over KOH and distilled prior to use. Chloroform (Mallinkrodt, reagent grade) was dried over calcium hydride and distilled before use. Diethyl ether (Caledon, reagent grade) was distilled from sodium. Tin-doped indium oxide (ITO) coated glass slides were obtained from Delta Technologies. UV/vis spectra were recorded on a Cary 3E spectrophotometer, NIR spectra were recorded on a Bomem MB-Series spectrometer. Electrochemical measurements were recorded using an EG&G Model 173 Potentiostat/Galvanostat, Fluke 8840A and 8000A digital multimeters, and a co-linear four-point probe. Film thickness measurements were recorded with a Tencor Alpha Step 100 profilometer.

In-situ conductivity measurements of electrochemically doped P3HT were recorded in 0.1M LiOTf/acetonitrile solutions using a platinum double-band electrode[7] and methods reported in literature.[8,9,10] Measurements of the electrochemical potentials of chemically doped films were performed in a 0.1M chloroform solution of tetrabutylammonium tetrafluoroborate, using an ITO slide as the working electrode, and Ag/AgCl as the reference.

For lifetime measurements of electrochemically doped P3HT films, ITO-coated glass slides were masked with resist and etched in a 6 M HCl solution (Anachemia, reagent grade) containing 0.2 M $FeCl_3$ (BDH)[11] for 20 minutes to give two-1 cm^2 ITO pads separated by a 0.7 mm gap. Wires were attached to each pad with Electrodag (Acheson Colloids Canada Ltd.) and insulated with 5 minute epoxy. P3HT was spin cast from chloroform. Each assembly was oxidized at a constant potential in a 0.1M LiOTf/acetonitrile solution until a constant current was attained, after which the film was dried and resistance measurements were recorded over time.

RESULTS

Solvato-controlled oxidative doping of conjugated polymer films is a new technique allowing intimate mixing of polymer and oxidant solutions while maintaining a neutral, processable solution. The basis of this technique is the ability of the oxidant to be shielded (protected) by a coordinating ligand which (in this case) decreases the reduction potential sufficiently to prevent oxidation of the polymer.[12] The dopant and polymer solutions can now be combined, and the polymer remains neutral until a film is cast and the shielding ligand is removed by heating or vacuum. This process (**Figure 1**) is demonstrated in this report using silver triflate as the oxidant and pyridine as the coordinating ligand.

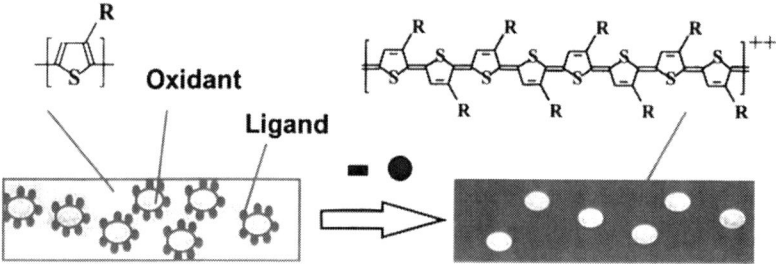

Figure 1: Schematic of the solvato-controlled oxidative doping process in which oxidant molecules are shielded by coordinating ligands. On removal of the ligands, the polymer film becomes doped and conductive.

The UV/vis absorption spectra shown in **Figure 2** shows the stability of a toluene/heptane/pyridine solution (volume ratio 1:1:0.1) of P3DDT following the addition of a silver triflate solution. Upon evaporation of the solvent and silver-coordinating ligand, the UV/vis (**Figure1: inset**) exhibits a decrease in the $\pi \rightarrow \pi^{*}$ transition of P3DDT, and the emergence of a new band at 844 nm indicating oxidative doping.

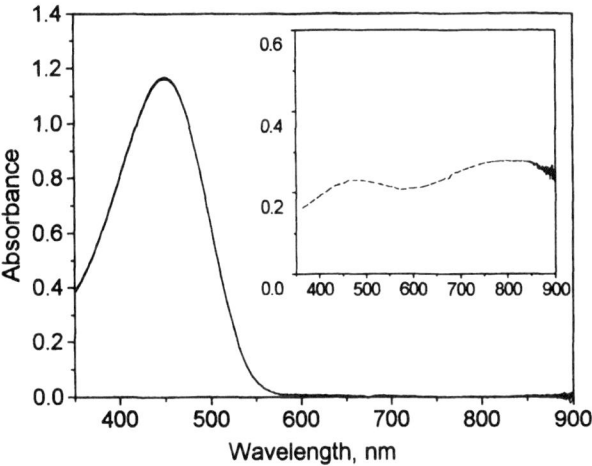

Figure 2: UV-vis spectra of a toluene/heptane/pyridine solution (volume ratio 1:1:0.1) of P3DDT (1mg/mL) in the presence of silver triflate (3.5 mM). **Inset:** Spectrum of the polymer film cast from the same solution.

The resulting films had similar conductivities, ca. 7-10 S/cm, as those oxidized with conventional dopants such as nitromethane solutions of ferric chloride. Similarly, their conductivity decreased in a first order manner so that a plot of $Log(\sigma/\sigma_o)$ vs. time yields a straight line (where σ is the conductivity). However, the half-lifetime for silver triflate doped polymers was found to be ca. 150 hours compared to that of 8 hours noted for $FeCl_3$ doped films of similar thickness. The other difference noticed was that films oxidized by AgOTf reproducibly exhibited lower potentials in dry chloroform (+0.80 ± 0.02 V vs. Ag/AgCl) compared to those oxidized by $FeCl_3$ (+1.00 ± 0.02 V), indicating a possible reason behind the enhanced stability found in silver triflate doped P3ATs.

In order to investigate the role of the standing electrochemical potential of the film on its stability, films of P3HT were oxidized at various potentials in LiOTf, and the stability measured. First, the conductivity-potential profile of the polymer was determined. **Figure 3** represents the in-situ conductivity of a P3HT film and shows the normalized drain current between the bands of a P3HT-coated dual-band platinum electrode as function of the applied potential. Each measurement was performed by setting both bands at the same potential against a Ag/AgCl reference (with a platinum gauze counter electrode) and allowing the steady state voltage drop between them (measured across a 150 kΩ resistor) to decrease to a constant value, thus indicating that the film was uniformly oxidized. The potential of one band was offset 20 mV in the negative direction and the resulting voltage drop was recorded after stabilization. The resulting graph shows that the conductivity of P3HT attains a maximum at ca. +0.6 V, and is insensitive to applied voltages up to +1.0 V. This potential region was used to determine whether the oxidized potential of a P3HT film strongly affects the half-lifetime of the conductivity.

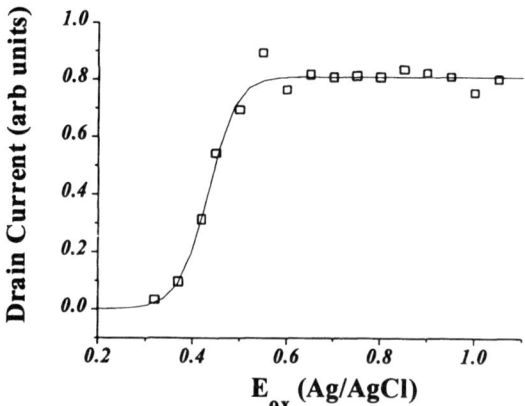

Figure 3: Normalized drain current (representing in-situ conductivity) between two 80 μm thick, P3HT-coated platinum band electrodes (separated by a 100 μm mylar film) as a function of the applied potential in an acetonitrile solution containing 0.1 M lithium triflate.

Figure 4 shows the half lifetimes of P3HT films (70 nm and 2.5 μm) oxidized at potentials ranging from +0.57 to +0.90 V (Ag/AgCl) after removal from the electrolyte and drying. Although absolute lifetimes differ, because of the dependence of the lifetime on thickness under ambient conditions, the lifetimes of electrochemically doped P3HT films exhibit no significant dependance on the standing potential to which they are oxidized.

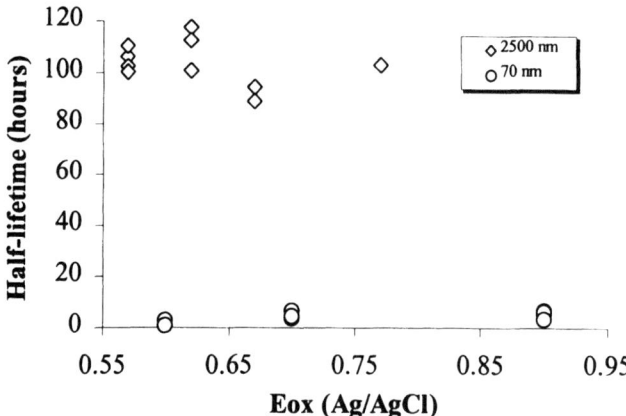

Figure 4: Half-lifetimes of P3HT films (2.5 μm and 70 nm thick) electrochemically doped with triflate anion from a 0.1 M LiOTf acetonitrile solution. Measurements were recorded after an initial 4 hour drying period.

Perhaps the most significant result of this study is the magnitude of the half-lifetimes of the electrochemically doped P3HT films (6-10 hours) compared to films of similar thickness (70 nm) doped with silver triflate (150 hours). Since both films possess the triflate anion as the counter-ion, and the lower doping potential does not appear to affect half-lifetimes, the enhancement of the stability may be due to the presence of metallic silver in the doped film. Recent transmission electron micrograph (TEM) studies[10] have shown that the silver is present in the form of small particles of roughly 12 nm (ca. 40 silver atoms) in diameter. Since metal particles of such size would span roughly 20 thiophene units, we postulate that they aid in delocalizing the positive charge of the charge carriers over a greater number of thiophene rings; furthermore, the counter-ions may be associated with bipolarons via coordination with silver particles, resulting in further delocalization of charge as shown below in **Figure 5**. As a result, the resulting polaron or bipolaron species would be less susceptible to nucleophilic attack, and should therefore exhibit greater stability.

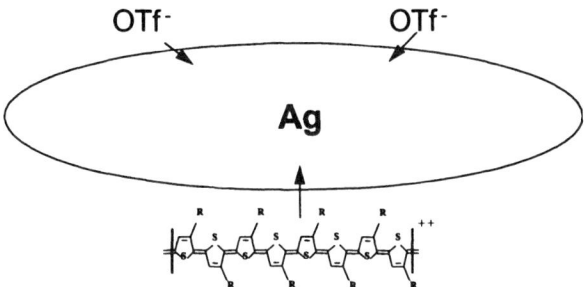

Figure 5: schematic diagram illustrating the delocalizing effect of Ag particles on charge carriers.

CONCLUSIONS

The potential to which P3HT films are oxidatively doped with the triflate anion has little to no affect on the half-lifetime of the conductive films. The observed stability of the doped films containing triflate ions, in the presence of silver particles, is much larger than exhibited by those lacking metallic silver. This is believed to be the result of enhanced charge delocalization due to the presence of conducting metallic particles.

ACKNOWLEDGMENTS

Financial support of this work by the Natural Science and Engineering Research Council of Canada, and the National Research Council, Canada is gratefully acknowledged.

REFERENCES

1. R.L.J. Elsenbaumer and K.Y.R. Oobodi, Synth Met. **15**, 169 (1986); J. Roncali, Chem. Rev., **92**,711 (1992).

2. M. Sandberg, S. Tanaka, and K. Kaeriyama, Synth. Met., **55**, 3587 (1993); M.S.A. Abdou and S. Holdcroft, Chem. Mater., **6**, 962 (1994).

3. H. Djellab, M. Armand, and D. Delabouglise, Synth. Met., **74**, 223 (1995).

4. S. Tanaka, M. Sato, and K. Kaeriyama, Synth. Met., **25**, 277 (1988); G. Dian and G. Barbey, Synth. Met., **13**, 281 (1986).

5. M.S.A. Abdou, Z.W. Xie, A.M. Leung, and S. Holdcroft, Synth. Met., **52**, 159 (1992); M.S.A. Abdou, G.A. Diaz-Quijada, M.I. Arroyo, and S. Holdcroft, Chem. Mater., **3**, 1003 (1991).

6. R.D. McCullough and R.D. Lowe, J. Chem. Soc. Chem. Comm., p.70 (1992).

7. G. Schivon, S. Sitran, and G. Zotti, Synth. Metals, **32**, 209 (1989).

8. E.W. Paul, A.I. Ricco, and M.S. Wrighton, J. Phys. Chem., **89**, 1441 (1985).

9. J.W. Thackeray, and M.S. Wrighton, J. Phys. Chem., **90**, 6674 (1986).

10. D. Ofer, R.M. Crooks, M.S. Wrighton, J. Amer. Chem. Soc., **112**, 7869 (1990).

11. J.E.A.M. van der Meerakker, P.C. Baarslag, and M. Scholten, J. Electrochem. Soc., **142**, 2321 (1995).

12. M.Y. Lebedev, M.V. Lauritzen, and S. Holdcroft, Chem. Mater., in press.

SYNTHESIS OF CONJUGATED POLYMERS BY VAPOR DEPOSITION POLYMERIZATION

CHAIN-SHU HSU AND TING-LI LIN
Department of Applied Chemistry, National Chiao Tung University,
Hsinchu, Taiwan, R.O.C.

ABSTRACT

The synthesis of poly(2,5-thienylene vinylene) and poly(4,7-thianaphthene) by vapor deposition polymerization is presented. 2,5-Di(bromomethyl)thiophene was subjected to vapor phase pyrolysis at 550 ℃ to form a reactive intermediate. Upon condensation, the reactive intermediate polymerized spontaneously at a temperature higher than -25 ℃ to produce poly(2,5-thienylene vinylene). The obtained poly(2,5-thienylene vinylene)s were fractionated into THF soluble and insoluble fractions. The number average molecular weights of the THF soluble fractions range from 1500 to 4000 as determined by GPC measurements. The obtained polymer shows no glass transition and melting point on the DSC scans. Doping of a poly(2,5-thienylene vinylene) film with I_2 vapor led to a conductivity of 1×10^{-4} S cm^{-1}. In the second part of this study, 2,3-diethynylthiophene was subjected to vapor phase pyrolysis at 300 ℃ to yield a reactive intermediate, 4,7-dehydrothianaphthene. Upon condensation, the obtained 4,7-dehydrothianaphthenene was polymerized spontaneously at a temperature higher than -25 ℃ to produce poly(4,7-thianaphthene). The obtained poly(4,7-thianaphthene) was insoluble in common laboratory solvents and shows also no melting point and glass transition on the DSC scans.

INTRODUCTION

Polymers with a fully conjugated structure such as poly(p-phenylene vinylene)s (PPVs) and poly(2,5-thienylene vinylene)s are currently attracting much interest due to their promising electrical, optical and most recently, electroluminescent properties.[1] However fully conjugated polymers are intractable due to their nonfusibility or insolubility in any solvents. In order to improve their processibility, a water soluble precursor route has been designed and a modification of the arylene unit, such as phenylene and thienylene groups, has been carried out by introducing long alkyl and alkoxy chain.[2-6] A varieties of structural derivatives of poly(phenylene vinylene)s and poly(2,5-thienylene vinylene)s have been prepared by the soluble precursor route and used for the preparation of light emitting diodes (LEDs).[7,8]

Vapor deposition polymerization was first reported by Gorham in 1996.[9] A strained [2,2]paracyclophane was converted by vacuum pyrolysis at 600 ℃ to form a reactive intermediate p-xylene, stable in the vapor phase, which spontaneously polymerized upon condensation to form a poly(p-xylene) film. This was a very useful method to obtain a polymer film free from contaminant because of using no solvent. This methodology was used recently by Staring et al.[10] and Shafer et al.[11] to generate PPV films by vapor phase pyrolysis of α, α'-dihalogenated p-xylenes.

Recently we reported the synthesis of poly(2,5-thienyleneethylene) and poly(2,5-furyleneethylene) by vapor phase pyrolysis of (5-methyl-2-thienyl)-methyl benzoate and (5-methyl-2-furyl)methyl benzoate.[12] In this study, the synthesis of poly(2,5-thienylene vinylene) and poly(4,7-thianaphthene) by vapor phase pyrolysis of 2,5-di(bromomethyl)thiophene and 2,3-diethynylthiophene is presented. The chemical structures and thermal properties of the obtained polymers are characterized by IR, NMR, differential scanning calorimetry and X-ray diffraction.

EXPERIMENT

Materials

2,5-Di(bromomethyl) thiophene (1M) was prepared by a procedure reported in the literatures.[13,14] 2,3-Dibromothiophene, bis(triphenylposphine)palladium chloride, 2-methyl-3-butyn-2-ol and all other reagents were obtained from Aldrich and used as received.

Techniques

^{1}H and ^{13}C NMR spectra (300 MHz) were recorded on a Varian VXR-300 spectrometer and infrared spectra were measured with a Nicolet 520 FT-IR spectrometer. Elemental analysis was determined using a Heraeus elementary analyzer. Thermal transitions and thermodynamic parameters were determined by using a Seiko SSC/5200 differential scanning calorimeter equipped with a liquid nitrogen cooling accessory. Heating and cooling rates were 10 ℃ /min. Thermogravimetric analysis was performed on a Seiko TG/DTA 200 thermal

analyzer at 10 ℃/min.   The thermal decomposition temperature (T_d) was read at a temperature at which 5% weight loss occurred.   Gel permeation chromatography (GPC) was run on an Applied Biosystems 400 LC instrument equipped with a differential refractometer, a UV detector, and a set of PL gel columns of 10^2, 5 × 10^2, 10^3, and 10^4 A.   The molecular weight calibration curve was obtained by using standard polystyrenes.   X-ray diffraction measurements were taken with nickel-filtered Cu Kα radiation with a Rigaku powder diffractometer.

Synthesis of monomers and polymers

Schemes 1 and 2 outline the synthetic routes used for the preparation of monomer 2M and poly(2,5-thienylene vinylene) and poly(4,7-thianaphthane)

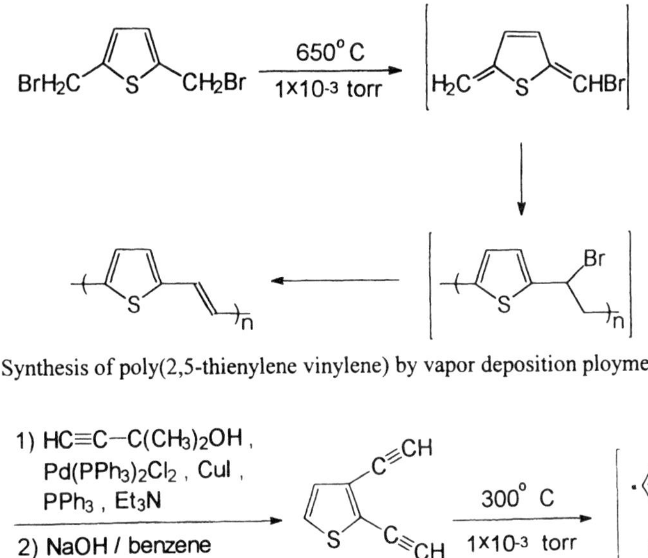

Scheme 1. Synthesis of poly(2,5-thienylene vinylene) by vapor deposition ploymerization.

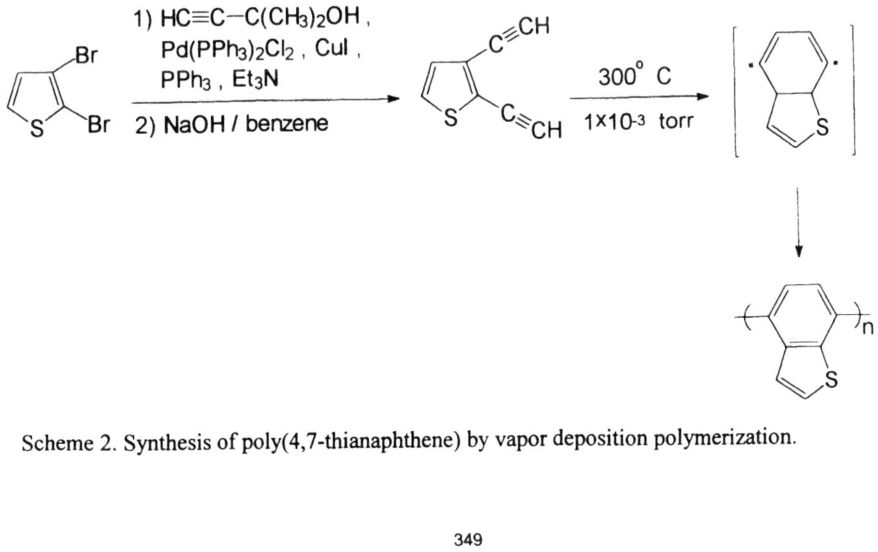

Scheme 2. Synthesis of poly(4,7-thianaphthene) by vapor deposition polymerization.

Synthesis of 2,3-diethynylthiophene (2M)

To a homogeneous solution of 2,3-dibromothiophene (5.0 g, 0.021 mol) and 2-methyl-3-butyn-2-ol (3.8 g, 0.046 mol) in triethylamine (150 mL) was added bis(triphenylphosphine)palladium (II) chloride (80 mg), cuprous iodide (80 mg) and triphenylphosphine (160 mg). The reaction mixture was heated to reflex temperature overnight, cooled to room temperature and treated with saturated ammonium chloride solution. The product was extracted into ethyl acetate, washed with water and dried over anhydrous $MgSO_4$. After removal of solvent, the remaining solid was purified by column chromatography (silica gel, ethyl acetate/n-hexane = 1/3 as eluent) to yield 4.0 g (76.8 %) of 2,3-bis(3-hydroxy-3-methylbutynyl)thiophene which was dissolved subsequently in toluene (40 mL), and aqueous NaOH (1.5 g, 3.84 mmol) was added. The reaction mixture was heated to reflex temperature for 8 hr, cooled and filtered. After the solvent in the filtrate had been removed, the crude product was purified by column chromatography (silica gel, n-hexane as eluent) to yield 1.2 g (50 %) of white crystals.

^{1}H NMR (CDCl3, TMS, ppm) δ 3.23 and 3.60 (two s, 2H, two - C $\equiv$ C-H̲), 7.01 and 7.15 (two d, two protons in thiophene ring)

Synthesis of poly(2,5-thienylene vinylene) and poly(4,7-thianaphthene) by vapor deposition polymerization.

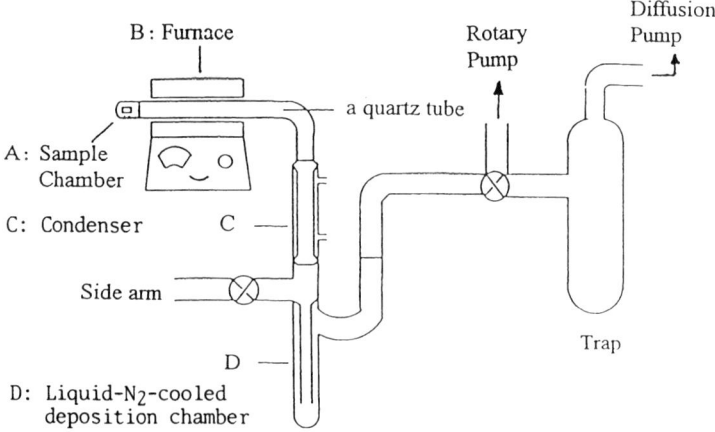

Figure 1. Apparatus for the vapor deposition polymerization.

The experimental set-up for the vapor deposition polymerization is depicted in Figure 1. The system was operated at low pressure, 0.001 Torr. Monomers 1 M and 2 M were placed in the sample chamber which was temperature controlled by a heater. Both monomers were converted to the reactive intermediates by pyrolysis in a quartz pipe heated to 100 - 600 ℃. The intermediates were trapped in the liquid-nitrogen-cooled deposition chamber. After pyrolysis was completed, the deposition chamber was warmed respectively to -78, -25, 0, 25 and 60 ℃. The polymeric films were subsequently formed in the deposition chamber. A large quanty of methanol was introduced into the deposition chamber through the sidearm to stop the polymerization. Finally the polymerization films were peeled from the wall, washed twice with methanol and dried at 100 ℃ under vacuum.

RESULTS AND DISCUSSION

Vacuum pyrolysis of monomer 1M was carried out under a pressure of 1×10^{-3} Torr at temperatures of 300, 400, 500, 550 and 600 ℃. The experimental results are summarized in Table 1. The optimal pyrolysis temperature for monomer 1 M was 550 ℃. Therefore we selected 550 ℃ as the pyrolysis temperature for further pyrolysis experiments. Monomer 1M was pyrolyzed at 550 ℃ to form an active intermediate, which was trapped in a liquid-nitrogen-cooled deposition chamber. After the pyrolysis was completed, the deposition chamber was warmed to -78, -25, 0, 25 and 60 ℃. Table 2 summarizes the polymerization results. When the active

Table 1   Preparation of poly(2,5-thienylene vinylene)by vapor deposition polymerization;   Runs 1 — 5

run	pyrolysis temp.(℃)	polymerization temp.(℃)	polymerization time(h).	yield(%)
1	300	25	6	6.3
2	400	25	6	19.2
3	500	25	6	41.7
4	550	25	6	54.4
5	600	25	6	47.3

intermediate was polymerized at -78 °C (run **6**), no polymer was obtained. When the polymerization temperature was raised to -25 °C (run **7**), a brown polymer film was obtained in the deposition chamber with a yield of 18.7 %. The total conversion to polymers increased basically as the polymerization temperature increased. However the maximum yield was obtained at 25 °C.

Table 2   Preparation of poly(2,5-thienylene vinylene) by vapor deposition
polymerization; Runs **6 - 10**.

run[a]	polym. temp. (°C)	yield (%)	wt.% of THF insoluble fraction	wt.% of THF soluble fraction	molecular weight of THF soluble fraction	
					$\overline{Mn}$	$\overline{Mw}$
6	-78	0	—	—	—	—
7	-25	18.7	0	100	1563	1664
8	0	42.3	0	100	2654	3236
9	25	53.3	29	71	3834	4578
10	60	49.4	37	63	3430	3977

a) Pyrolysis temperature was 550 °C and polymerization time was 6 hrs.

The obtained polymers were fractionated into THF soluble and insoluble fractions. The number average molecular weights of the THF soluble fractions range from 1500 - 4000 as determined by GPC measurements. Figures 2A and 2B present respectively the ^{1}H and ^{13}C NMR spectra of the polymer obtained from run **8**. Figure 2A exhibits a singlet peak at 4.67 ppm due to the vinylene protons and a singlet peak at 6.98 ppm due to the thiophene ring protons, while Figure 2B displays three carbon peaks at 121.45, 127.62 and 142.65 ppm. The NMR data demonstrate a perfect linear poly(2,5-thienylene vinylene) structure. Figure 3 shows the IR spectrum of the THF insoluble fraction obtained from run **10**. The characteristic absorption bands at 935 cm^{-1} and 1590 cm^{-1} are assigned to the C-H out-of-plane bending of the trans-vinylene and C = C stretching conjugated to the thienylene ring, respectively. The absorption at 935 cm^{-1} indicates the configuration is substantially trans. Absorption of the C-H bending of the thienylene ring appears at 802 cm^{-1}. These results indicate that the structure of poly(thienylene vinylene) is formed. The obtained polymers showed no glass transition and melting point before decomposition occurred on the DSC scans.

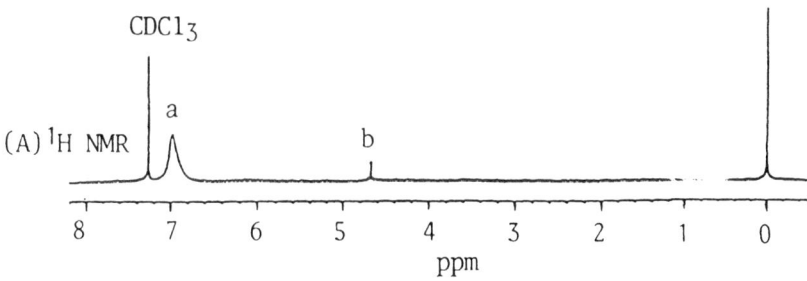

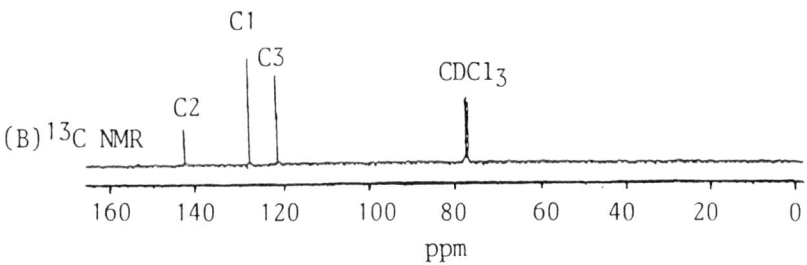

Figure 2. (A) ^{1}H NMR and (B) ^{13}C NMR of poly(thienylene vinylene)

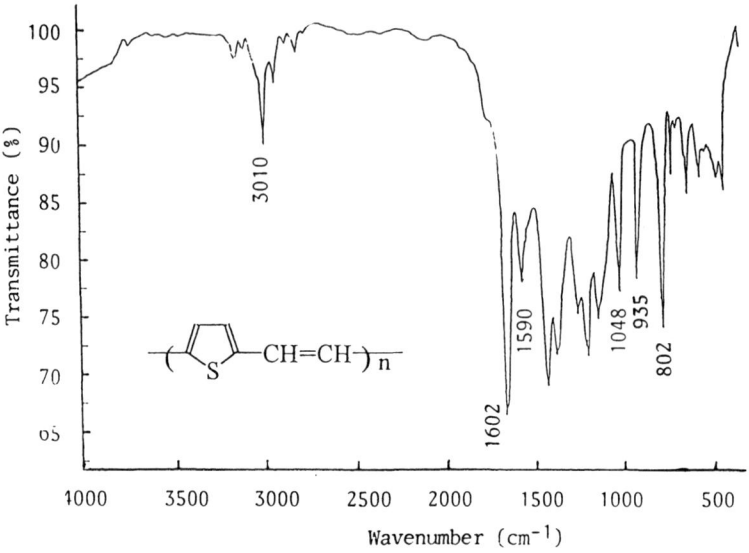

Figure 3.   FT-IR spectrum of poly(thienylene vinylene)

Vacuum pyrolysis of monomer 2 M was carried out under a pressure of 1×10^{-3} Torr at temperature of 100, 200, 250, 300 and 350 ℃.   The experimental results are summarized in Table 3. The optimal pyrolysis tempera⁻⁻⁻e for monomer 2 M was 300 ℃.   Therefore we selected 300 ℃ as the pyrolysis temperature for

Table 3   Vapor Deposition Polymerization of 4,7-dehydrothianaphthene ;
Runs **11 − 15**

Run	pyrolysis temp.(℃)	polymerization temp.(℃)	polymerization time(h).	yield(%)
11	100	25	6	26.5
12	200	25	6	38.4
13	250	25	6	46.3
14	300	25	6	48.6
15	350	25	6	45.8

further pyrolysis experiments. Monomer 2M was subjected to vapor phase pyrolysis at 300 ℃ to yield a reactive intermediate which was proven to be the 4,7-dehydrothianaphthene diradical by a free radical trapping experiment using 1,4-cyclohexadiene as a radical traping agent. (Scheme 3).[14-16] Benzene and thianaphthane were isolated as reaction products.

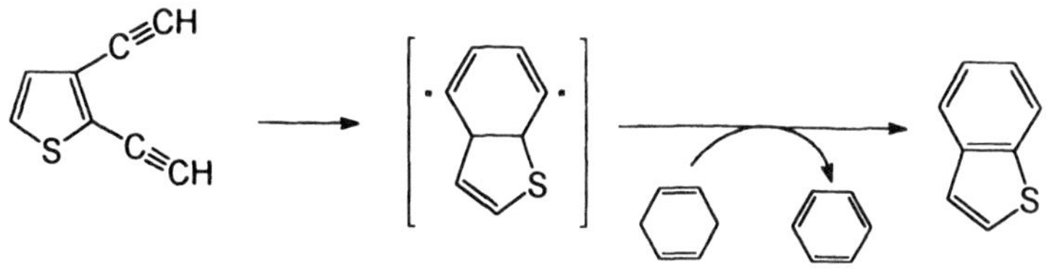

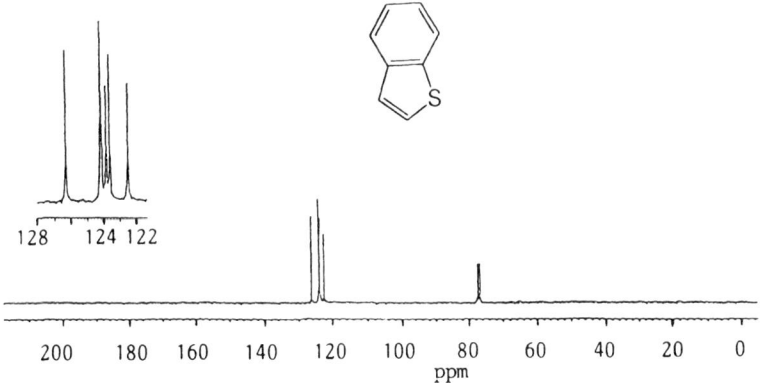

Scheme 3. The reaction mechanism of a radical trapping experiment.

Figure 4 show the ^{13}C NMR spectrum of thianaphthane. The obtained 4,7-dehydrothianaphthene diradical was polymerized spontaneously in bulk at a temperature higher than -25 ℃. Table 4 summarizes the polymerization results. The total conversion to polymer increased as the polymerization temperature increased. The tough films obtained in the deposition chamber were insoluble in common laboratory solvents. Figures 5 and 6 present the IR and solid state C[13] NMR spectra of the polymer obtained. Both spectra identify the chemical structure of ploy(4,7-thianaphthene). The polymers obtained by vapor deposition polymerization display no glass transition and melting point before decomposition occurred in the DSC scans.

Figure 4. ^{13}C NMR Spectrum of thianaphthalene

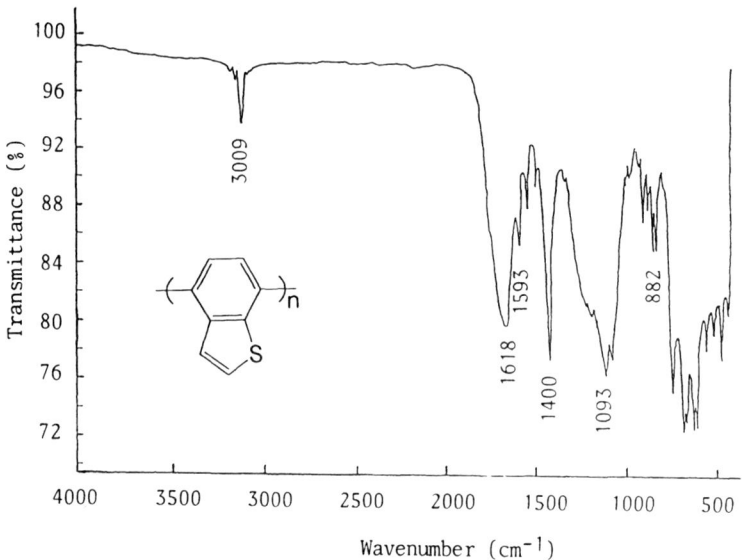

Figure 5. FT-IR Spectrum of poly(4,7-thianaphthene)

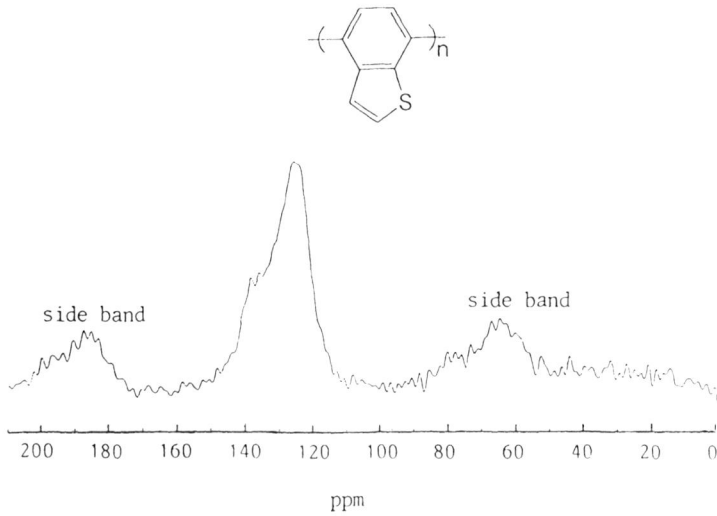

Figure 6.   Solid state ^{13}C NMR Spectrum of poly(4,7-thianaphthene).

ACKNOWLEDGEMENTS

The authors are grateful to the National Science Council of the Republic of China for financial support of this work (Grants NSC 87-2216-E-009-004)

REFERENCES

1. J. H. Burroughes, D. D. C. Bradley, A. R. Brown, R. N. Marks, K. Mackay, R. H. Friend, P. L. Burn and A. B. Holmes, Nature 347, 539 (1990).
2. T. Kaino, K. Kubodera, H. Kobayashi, T. Kurihara, S. Saito, T. Tsutsui, S. Tokito and H. Murata, Appl. Phys. Lett. 53, 2002 (1988).
3. K. J. Jen, H. Eckardt, T. R. Jow., L. W. Shacklette and R. L. Elsenbauner, J. Chem. Soc. Chem. Commun. 1988, 215.
4. M. L. Blohm, J. E. Pickett and P. C. Van Dort, Macromolecules 26, 2704 (1993).
5. I. Murase, T. Ohnishi, T. Noguchi and M. Hirooka, Synth. Met. 17, 639 (1987).
6. S. H. Agkari, S. D. D. V. Roughooputh and F. Wudl, Synth. Met. 17, 639 (1989).
7. D. Braun and A. J. Heeger, Appl. Phys. Lett. 58, 1982 (1992).
8. H. Westweber, A. Greiner, U. Lemmer, R. F. Mahrt, R. Richert, W. Heitz and H. Bassler, Adv. Mater. 4, 661 (1992).
9. W. F. Gorham, J. Polym. Sci. 4, 3027 (1966).
10. E. G. J. Staring, D. Braun, G. L. J. A. Rikken, R. J. C. E. Demandt, Y. A. R. R. Kessener, M. Bauwmans and D. Broer, Synth. Met. 67, 71 (1994).
11. O. Schafer, A Greiner, J. Pommerehne, W. Guss, H. Vestweber, H. Y. Tak, H. Bassler, C. Schmidt, G. Lussem, B. Schartel, V. Stumpflen, J. H. Wendorff, S. Spiegel, C. Moller and H. W. Spiess, Synth. Met. 82, 1 (1996).
12. Y. H. Lu, C. S. Hsu, C. H. Chou and T. H. Chou, Macromolecules 29, 5546 (1996).
13. K. Y. Jen, M. Maxfield, L. W. Shacklette and R. L. Elsenbaumer, J. Chem. Soc. Chem. Commun, 1987, 309.
14. M. Takeshita and M. Tashiro, J. Org. Chem. 56, 2837 (1991).

Double-strand Polyaniline as a Molecular Quasi-Memory to Chemical Stimuli

Gowri P. Kota, Linfeng Sun, Huaibing Liu, Sze C. Yang, University of Rhode Island, Dept of Chemistry, Kingston, RI 02881, U.S.A.

ABSTRACT

We report the apparent hysteresis cycles in the protonation/deprotonation induced conductor-insulator transition for a water-soluble double-strand polyaniline. Experimental evidences show that the polymeric complexes of polyaniline and poly(styrene sulfonic acid) are soluble in water as non-aggregated single molecules. Although the hysteresis loop becomes narrower within 40 hours and the bistability is not permanent, the rate of collapse of the hysteresis loop is extremely slow for a solvated polymer in water. The water soluble polyaniline was compared with solid state samples of PAN:HCl. It was found that, under thermodynamically identical conditions, the memory effect is stronger for the water-soluble polyaniline than the water-insoluble solid of PAN:HCl.

INTRODUCTION

Memory effects were mostly observed in solid-state materials. Examples include the magnetization/demagnetization hysteresis loop of ferromagnets, the intercalation/deintercalation hysteresis loop for inserting small ions or molecules in graphite or other inorganic solids with layers of two-dimensional networks. It is relatively rare to find memory effects exhibited by molecules dissolved in solutions, with the exception of bio-macromolecule such as DNA and proteins. If a water soluble polymer exhibits memory effect, it might be a result of the cooperative interactions within the polymer.

In this article we present experimental evidences for an apparent hysteresis loop for a polymer dissolved in solution as solvated independent molecules. The material is a water-soluble conducting polymer PAN:PSSA which is a molecular complex of polyaniline and poly(styrenesulfonic acid). The hysteresis loop involves the acid/base titration cycle of PAN:PSSA.

The aqueous solution of PAN:PSSA changes color from green to blue/purple when it is titrated by a base (NaOH). The green color of the solution is restored when PAN:PSSA is re-protonated (doped) by adding an acid (HCl) to the solution. The green color of the acidic form is due to the characteristic polaron absorption bands of the electrically conductive emeraldine salt with $\lambda = 420$ and 800 nm. The blue color deprotonated state has a $\lambda = 630$ nm band of the emeraldine base form. Although the color change is reminiscent to that of a pH indicator dye such as methylene blue, there is, however, an important difference between their titration curves when they are monitored spectroscopically. The titration curves of PAN:PSSA show a hysteresis loop. The forward and the backward titration curves of a common acid/base indicator are reversible at any point of the titration.

In this paper we report the time evolution of the hysteresis loop for dilute PAN:PSSA in pH buffered aqueous solutions. Although the hysteresis loop eventually becomes narrower in 40 hours and the bistability is not permanent, the rate of collapse of the hysteresis loop is extremely slow (days instead of seconds) considering the fact that the polymer is dissolved in an aqueous solution.

Conducting polymers in the solid insoluble state have been known to exhibit memory effects in both the electrochemical[1] and acid doping/dedoping cycles[2]. An apparent hysteresis reported in reference 2 for the acid doping/dedoping of polyaniline was explainable from considering the Donnan effect of the polymer network.[3,4] The previously proposed mechanisms for the memory effect in the solid state are not expected to be present if the molecules are not aggregated. In this context, the comparison between a solution of PAN:PSSA and the solid sample of PAN:HCl is surprising. The experimental data excludes a number of possible mechanisms for the hysteresis loop and sheds light on some possible mechanisms.

A WATER-SOLUBLE MOLECULAR COMPLEX OF POLYANILINE

The soluble PAN:PSSA is a molecular complex of two strands of polymers: (1) a polyaniline molecule, and (2) poly(styrenesulfonic acid). These two strands of polymers are bonded by non-covalent intermolecular interactions to form a stable molecular complex.[5] Experimental evidences suggest that the two strands are in side-by-side contact as depicted in the structure of the complex in Fig. 1.

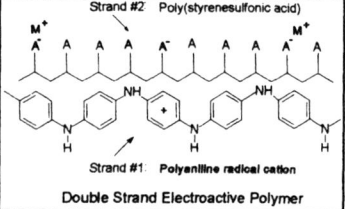

Figure 1 A molecular complex of polyaniline and poly(styrenesulfonic acid). A = -(C₆H₄SO₃H)

The "double-strand" PAN:PSSA was synthesized by a template-guided synthetic method: In the first step, the aniline monomers were adsorbed onto the poly(styrenesulfonic acid) chains in aqueous solution. The resulting adduct, PSSA:(Aniline)$_x$, has signatures that can be monitored and verified. In the second step, the attached aniline monomers were oxidatively polymerized to form the polymeric complex.[6]

An example of the synthesis of a water-soluble molecular complex of poly(styrenesulfonic acid) and polyaniline is described as the following: 4.05 ml of aniline monomer (from Aldrich, redistilled) is added into 30 gm of 30% poly(styrene-sulfonic acid) aqueous solution (Polysciences, M.W.= 70,000). After the solution was stirred for 1 hour, 30 ml of 3 M ferric chloride was added into the homogeneous solution. The reaction mixture soon became green-colored. While stirring, 80 ml of 0.5 M HCl and 1 ml of 30% hydrogen peroxide were added. Additional 4 ml of hydrogen peroxide was slowly added into the reaction mixture in successive portions. After rigorous stirring, the reaction mixture was poured through a filter paper to remove small particles. The filtrate is a dark green homogeneous aqueous solution. The reaction product was isolated from the solution by a series of purification procedure and was analyzed by elemental analysis to be a molecular complex between poly(styrenesulfonic acid) and polyaniline, or PAN:PSSA. The mole ratio between aniline and the styrene monomers in the complex was found to be 1:2.

A dried powder of PAN:PSSA is readily dissolved in water to form a clear green solution which shows no light scattering or turbidity. The solution passes through membrane filters (Poretics Corporation) with pore size of 200 nm without residues. The lack of aggregation in the solution is also supported by two other tests involving ultracentriгugation and light scattering. No precipitation was found upon ultracentrifugation at 1.6×10^5 g for 2 hours. To ensure that the lack of precipitation is not due to the accidental match of the specific gravities of the polymer and the solvent, the ultracentrifugation tests were repeated with mixed water/methanol solutions with different specific densities ranging from 0.79 to 1.0. Dynamic light scattering measurements show that the hydrodynamic diameter is less than 100 nm. Other properties also support that PAN:PSSA in water is a true solution without aggregation, as long as the number of aniline units in PAN is less than ½ that of the sulfonate units of PSSA.

We examined the transmission electron micrograph of a sub-monolayer film of PAN:PSSA prepared by evaporation from a dilute aqueous solution. The TEM images of single molecules of the polymer complexes were found to be flattened disks with about 10 nm diameter. This type of TEM image is similar to that of single molecules of polystyrene.[7] The size of the flat disk is consistent with that of a single molecule of PAN:PSSA. X-ray microanalysis of the single-molecule image shows that both the nitrogen and the sulfur elements are present as expected from the composition of the molecular complex. The TEM and the SEM images of this polymer dried from water solution were found to be very different from other polymeric complexes of PAN that form aggregates. From these observations, we expect that the molecular complex exists as single molecules in dilute solution, and the molecules are solvated random coils in water.

Solid film of PAN:HCl salt.

For the purpose of comparison, we prepared thin solid films of polyaniline:HCl salt (the chloride doped polyaniline, or PAN:HCl) coated on polycarbonate by an electrodeless chemical polymerization method[8]. A solution for coating PAN:HCl is prepared by dissolving 5 ml of aniline in 100 ml of 3 M HCl. To this solution is added 30 ml of 3% hydrogen peroxide and 1 ml of 0.1 M ferric chloride. A sheet (1/16" thick) of polycarbonate was dipped in the reaction mixture. The solution turns green colored indicating the formation of polyaniline. A green-colored thin film is adherent to the polycarbonate sheet. The sheet of polycarbonate is removed from the solution after a few minutes. The green solid film is uniform, electrically conductive, and shows the characteristic optical absorption spectra of polyaniline in both the emeraldine salt and the emeraldine base form. The film thickness is approximately 0.4 μm. Films were immersed in 0.1 M Hcl solution for 3 days before use. Strips of sample were cut from the same polycarbonate sheet as samples for study.

APPARENT HYSTERESIS LOOPS

Figure 2 shows the "forward titration curve" (marked by data points with the square symbols) and the "backward titration curve" (marked by the solid circle symbols) after the samples of PAN:PSSA were mixed with pH buffers for 5 minutes. In this experiment, the "titration" was carried out in a manner different from that of the conventional acid/base titration. In the more familiar conventional method of titration, a "forward titration" is started by adding a base (NaOH) to the acid form of a dye solution until the end point is reached and the color of the indicator is changed. The conventional "backward titration" is carried out by adding acid (HCl) to the basic solution until the color of the indicator changes back.

We do not use the conventional method of titration because the chemical environments are not thermodynamically equivalent for the conventional "forward" and the "backward" curves. In the conventional titration, when the "forward" and the "backward" solutions reach the same pH value, the ionic compositions and the ionic strengths are not the same. In this experiment, we obtain a pair of data points of a given pH on the titration curves by adding the protonated form (complex-A) and the deprotonated form (complex-B) to two separate but identical pH buffer solutions (100 ml). To speed up the mixing of the complex with the buffer, the complexes were first prepared as concentrated

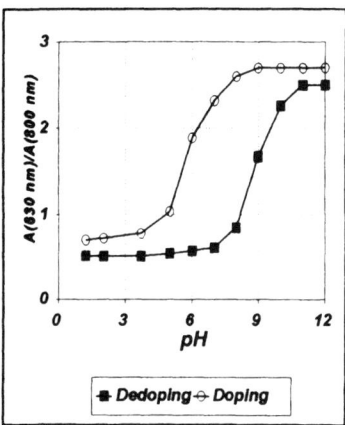

Figure 2 Hysteresis loop for water-soluble PAN:PSSA at a delay time of 5 minutes.

solutions before mixing with the buffer solution. The concentration of PAN:PSSA in the buffer is low (approximately 10^{-5} M in $-SO_3^-$) comparing with the total ionic strength of the buffers (0.1 M). The pH value of the solution was monitored periodically and was found to be stable.

A data point in Fig. 2 was obtained from a UV-visible absorption spectrum measured 5 minutes after a buffer solution of PAN:PSSA was prepared and mixed. The ratio of the absorbances at $\lambda = 630$ and $\lambda = 800$ nm, A(630 nm)/A(800 nm), was used as a relative measure of the population ratio of the deprotonated and the protonated forms. The mismatch between the forward and the backward titration curves leads to a hysteresis loop in Figure 2.

To illustrate the unusual hysteresis effect, we focus our attention for a moment on a pair of data points corresponding to two pH 8 buffer solutions of PAN:PSSA prepared by the "forward" and the "backward" titration. The values of A(630 nm)/A(800 nm) are widely different for these two samples with the same solution environment, except for the difference in the previous history of the PAN:PSSA complex. In fact, the color of the solution belonging to the forward leg of the titration curve is green, and the other solution is blue. Although the difference in color gradually diminishes over time, it changes very slowly. One can still visually distinguish the color of the two solutions after it was

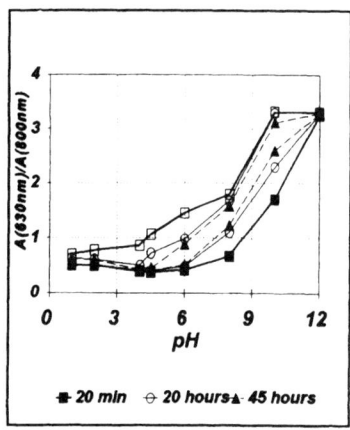

Figure 3 Hysteresis loop of an aqueous solution of PAN:PSSA at delay times of 20 min, 20 hours and 45 hours.

stored in room temperature for 2 to 3 months. This is an unusual memory effect for a molecule dissolved in water. It appears that the molecules "remember" their past history for a long period of time. If a water-soluble pH-sensitive indicator dye was used in a similar experiment, there would be no hysteresis loop. The common indicator dye in either the acid form or the basic form would have reached the same equilibrium instantaneously at the moment when it was added to a pH buffer.

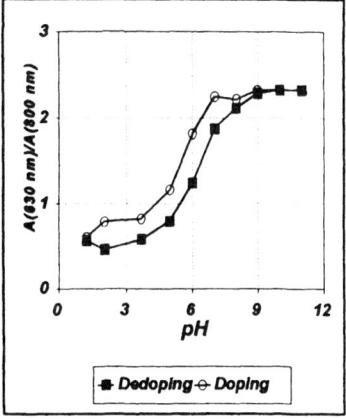

The pH buffer solutions of PAN:PSSA were stored in large-volume optically transparent cells in room temperature and the UV-visible spectra were monitored at regular time intervals. The time evolution of the hysteresis curves were generated from these measurements. The hysteresis curves at a time delay of 20 minutes, 20 hours and 45 hours are displayed in Figure 3. Figure 3 shows that the area within the hysteresis loop decreases with time. But the rate of decrease is surprisingly slow for molecules dissolved in water.

Figure 4 Hysteresis loop for solid film of PAN:HCl at a delay time of 5 minutes.

Considering that the collision frequency of the hydronium or hydroxyl ions with the macromolecule is on the order of 10^{12} per second, the observed slow evolution is not likely due to the inaccessibility of the protonation sites of polyaniline. Samples from forward and backward branches of the loop in the same pH buffer of 6, 7 and 8 were stored for an extended period of time. The difference in color can be recognized visually even when the delay time reaches 2 or 3 months, although the difference in optical spectrum appears to be less than the instrumental noise.

Comparison with a solid state sample of PAN:HCl

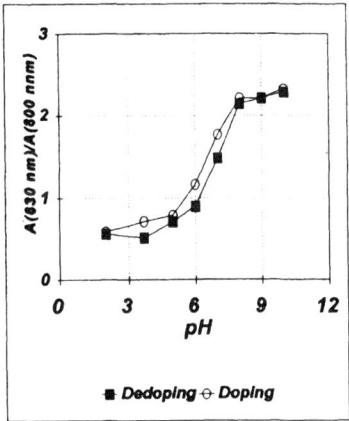

Figures 4 and 5 show the time evolution of the hysteresis curve for the solid films of PAN:HCl coated on transparent polycarbonate strips. The hysteresis loop is much smaller than that of PAN:PSSA initially, and the loop collapsed within 1 hour. The pH buffer solutions with ionic strength of 0.1 M is sufficient to reduce the Donnan effect that gave the apparent hysteresis for PAN:HCl reported in reference 2.

DISCUSSION

The hysteresis loop for PAN:PSSA is much larger than that of PAN:HCl samples. Since both samples contain polyaniline molecules, the slowness of the relaxation

Figure 5 Hysteresis loop for solid films of PAN:HCl at a delay time of 1 hour.

kinetics could not be attributed to the PAN part of the PAN:PSSA complex alone. This result suggests that the interaction between PAN and PSSA in the molecular complex plays a role in the macromolecular dynamics.

CONCLUSION

A water-soluble molecular complex of polyaniline PAN:PSSA shows hysteresis loop in an acid/base titration cycle described in this article. A hysteresis loop is unusual for a polymer dissolved as a dilute aqueous solution, considering that the water and hydronium ions should be readily accessible to the protonation sites of the solvated polymer. The memory effect was not permanent but the relaxation rate was found to be very slow. The apparent hysteresis effect in the solvated PAN:PSSA was found to be stronger and persist longer than that of a solid sample of single-strand polyaniline PAN:HCl. This result suggests that the interaction between the two strands of polymers in PAN:PSSA may be important for the observed apparent hysteresis.

ACKNOWLEDGMENT

This paper is dedicated in memory of the late Professor R. Ken Force who kindly provided helpful discussions in the early stage of this research. This research was partially supported by an Alcoa Foundation grant and by a URI Foundation grant.

REFERENCES

1.   A. F. Diaz and T. C. Clarke, J. Electroanal. Chem., 111 (1980) 115., A. F. Diaz, J. I. Castillo, J. A. Logan and W.-Y. Lee, J. Electroanal. Chem. 129 (1981) 115.], I. Rubinstein, J. Electrochem. Soc., 130 (1983) 1506. T. Kobayashi, H. Yoneyanma and H. Tamura, J. Electroanal. Chem., 177 (1984) 281. C. Odin and M. Nechtschein, Synth. Met. **41**, 2943 (1991).

2.   M. Nechtschein, F. Genoud, C. Menardo, K. Mizoguchi, J. P. Travers, B. Zilleret, Synth. Met. **29**, E211 (1989).

3.   A.G. MacDiarmid and A.J. Epstein, Mat. Res. Soc. Proc. **247**, 565 (1992).

4.   P. Chartier, B. Mattes and H. Reiss, J. Phys. Chem. **96**, 3556 (1992). S. Mate, J. A. Manzanares and H. Reiss, J. Chem. Phys. **98**, 2408 (1993)

5.   L. Sun, H. Liu, R. Clark, S. C. Yang, Syn Met **84**, 67 (1997).

6.   L. Sun, S. C. Yang, Mat. Res. Soc. Symp. Proc. **328**, 209 (1994).

7.   J. Kumaki, Macromolecules,, **19**, 2258 (1986)

8.   S. C. Yang and R. L. Clark, US Patent 4,842,383 (1989).

EFFECT OF MORPHOLOGY ON THE ELECTRICAL TRANSPORT PROPERTIES OF POLYANILINE FILMS FOR ELECTRONIC APPLICATIONS

S. S. Hardaker, K. Eaiprasertsak, J. Yon, R. V. Gregory, and G. X. Tessema*

School of Textiles, Fiber and Polymer Science, Clemson University, Clemson, SC 29634
*Department of Physics and Astronomy, Clemson University, Clemson, SC 29634

Abstract

Although it is well known that the oxidation state of polyaniline is an important characteristic, there are few reports of its influence on the development of morphology and electrical properties in fibers and films. In this work, differential scanning calorimentry is used in conjunction with measurements of temperature dependence of conductivity and thermoelectric power to elucidate the intimate relationship between structure and properties. By increasing the amount of chemical reduction of polyaniline solutions, films are prepared which exhibit a thermal transition between 300 and 385 °C, indicative of melting. Increasing the chemical reduction also increases the conductivity of iodine doped films. The most reduced film exhibited a semiconductor transport mechanism, while the other films could be modeled with a quasi-one dimensional variable range hopping mechanism. The temperature dependence of conductivity also showed increasing order for increasing reduction, consistent with the DSC results.

Introduction

The chemical structure of polyaniline in its base form is shown in Figure 1. When $y = 0$, the polymer is refered to as leucoemeraldine base. When $y = \frac{1}{2}$, it is called emeraldine base, and when $y = 1$, it is called pernigraniline base[1]. Although it is well known that the oxidation state of polyaniline is an important characteristic of this polymer, there are few reports of its influence on the development of morphology in fibers and films. Previous work has shown that both films and fibers produced from solutions of leucoemeraldine base in N,N'-dimethyl propylene urea (DMPU) exhibit crystallinity and a melting transition[2].

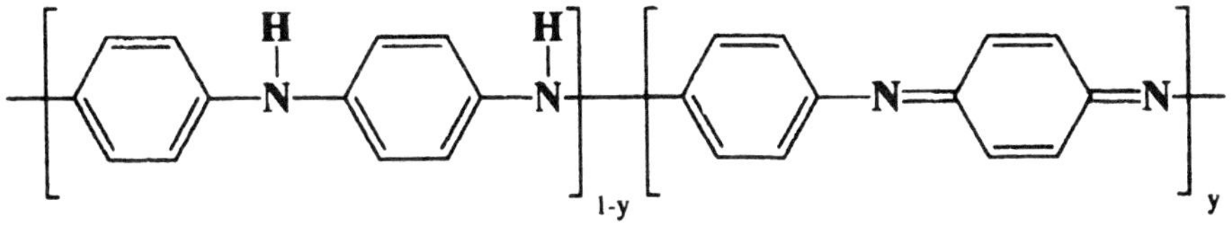

Figure 1. The chemical structure of polyaniline.

Electrical transport properties of chemically prepared polyaniline show a strong temperature dependence. In general, depending on the sample preparation and dopant species[3,4], the polymers may be metallic, insulating, or be in the critical region for a metal-insulator transition. It is also known that disorder plays a predominant role in the transport process. The disorder can be homogeneous on a molecular scale, mesoscopic heterogeneities with highly doped conducting clusters, or crystalline islands separated by insulating regions. The DC conductivity is thus influenced to a great deal by the extent of amorphous regions. However, because of the absence of a measuring current flow, the thermoelectric power is less dependent on these regions. Several models have been proposed for describing the temperature dependence and transport properties of charge carriers, including quasi-one dimensional variable range hopping (Q-1D VRH)[5], charging energy limited tunneling for granular metals[6], and three dimensional variable range hopping with a Coulomb gap[7].

The goal of the present work is to elucidate the interconnection between structure, both molecular (i.e. oxidation state) and morphological, and electrical transport properties. The techniques employed for this study are differential scanning calorimetry for characterization of thermal transitions, and temperature dependence of conductivity and thermoelectric power.

Experimental

Polyaniline was chemically synthesized by addition of an ammonium persulfate solution (in 1.0 M HCl) to a stirred aniline solution (in 1.0 M HCl) over two and a half hours. The temperature of the reaction mixture was controlled at -30 °C. LiCl (6.0 M) was present to prohibit freezing. For this reaction, the aniline:oxidant molar ratio was 2:1. The total reaction time was 24 hours after which the resultant emeraldine salt was washed with 6.0 L of deionized water. The emeraldine salt was deprotonated by stirring in 3 wt% NH_4OH for 25 hours and washed with 6.0 L deionized water. Oligomer was removed by stirring the emeraldine base in methanol for 45 minutes followed by rinsing with 6.0 L methanol. Gel permeation chromatography (polystyrene standards, N-methyl-2-pyrrolidinone/0.05 M LiBr as eluent, 80 °C) was employed to determine the molecular weight of the emeraldine base (M_n=32,000 and M_w=60,000).

Solutions (9 wt%) of the emeraldine base were prepared in N,N'-dimethyl propylene urea (DMPU) containing varying amounts of phenyl hydrazine. The molar ratios of phenyl hydrazine to emeraldine base (based on a four ring repeat unit) were R = 0.0, 0.36, 0.71, and 1.11. These solutions were spin coated onto silicon wafers at a spinning speed of 500 rpm and a spinning time of 30 seconds on a Headway Research Inc., PM101DT-R790 spin coater. These films were then placed on a heating plate for 10 minutes at ~75 °C. The films were removed from the substrate by submersion in deionized water and annealed at 200 °C for 20 minutes under a vacuum. I_2 vapor was used to dope the films.

Differential scanning calorimetry (DSC) was carried out on a TA Instruments, Model 2920 at a heating rate of 20 °C per minute under nitrogen, with indium as a calibration standard.

For the electrical measurements, samples were mounted across two copper blocks using carbon paint to make electrical contacts, with a silicon diode mounted near the sample to measure temperature. The resistance (using a standard four probe technique) was measured as a dip-stick containing the sample was slowly lowered into a liquid nitrogen dewar. After the temperature dependence of the resistance was measured, the sample was slowly withdrawn from the dewar and the thermoelectric power was measured. The two copper mounting blocks were controlled at two different temperatures and the gradient measured by a Au-Fe(0.07%)-Cu thermocouple anchored near the ends of the sample. Both the DC conductivity and thermoelectric power were measured in the temperature range between 77 and 300 K.

Results and Discussion

As stated earlier, films were spin coated from solutions containing molar ratios R=1.11, 0.71, 0.36, and 0.00 of phenyl hydrazine to emeraldine base (based on a four ring repeat unit). As one might expect, the appearance of the films depended strongly upon the level of oxidation. The fully reduced films were a grey color with a poorly reflecting surface. With increasing oxidation, the films were more reflective and had a dark grey color, while the unreduced films were quite reflective and had the characteristic coppery color of emeraldine base films.

This optical behavior is attributed to the presence of a crystalline phase (probably under ~15%) in the leucoemeraldine base form of the films which causes a rough surface to form. The surfaces of the films which were against the substrates are much more reflective As reported earlier[2], an endotherm was observed from differential scanning calorimetry consistent with melting of the crystallites.

For the present study, DSC was chosen to investigate the effect that oxidation state has on the morphological development of the prepared films. Figure 2 presents the DSC results on four annealed films. For the emeraldine base film (R = 0.00), as the temperature increases a broad exotherm centered at ~300 °C was observed, attributed to oxidation or crosslinking and characteristic of emeraldine base. With increasing R (i.e. more reduction) one can see the development of an endothermic peak between ~300 and ~385 °C, depending on oxidation state. As a greater proportion of the repeat units are reduced (as R increases), more of the chain segments can crystallize, which is accompanied by an increase in the magnitude of the endotherm. Another effect is that the endotherm becomes sharper and shifts to higher temperatures with increasing R, believed to be due to increasing crystallinity in conjunction with more perfect ordering in the crystallites.

When treated with I_2 vapor, the reduced segments are oxidatively doped to the conducting state[1]. Figure 3 presents the results of the temperature dependence of conductivity as a function of the oxidation level of films before doping. The room temperature conductivity of the R = 0.00 film doped with 1.0 M HCl was measured to be 0.22 S/cm. With increasing R, the conductivity of the I_2 doped films increases, and can be interpreted in terms of the quasi-1D variable range hopping mechanism (Equation 1),

$$\sigma_{dc} = \sigma_o \exp\left[-\left(\frac{T_o}{T}\right)^{\gamma}\right] \qquad (1)$$

where $\gamma = \frac{1}{2}$, and $T_o = 485$ K for the R = 0.36 film and $T_o = 137$ K for the R = 0.71 film. If fit to Equation 1, the R = 1.11 (most reduced) film exhibited a T_o of 70 K. A decreasing T_o can be attributed to more intra- and interchain ordering resulting in more delocalization, consistent with the morphological picture obtained from the DSC results. However, for the R = 1.11 film, a slightly better fit is obtained when plotted against 1/T, indicative of a semiconductor mechanism. This is in fact born out by thermoelectric power measurements on this film in which the thermoelectric power is proportional to 1/T. Figure 4 presents the thermoelectric power for the R = 0.36 film as a function of 1/T. A prominent feature of this figure is the transition at ~150 K, consistent with polaron scattering by phenyl ring vibrational modes[8]. This temperature corresponds to a β-transition observed by dynamic mechanical thermal analysis and temperature

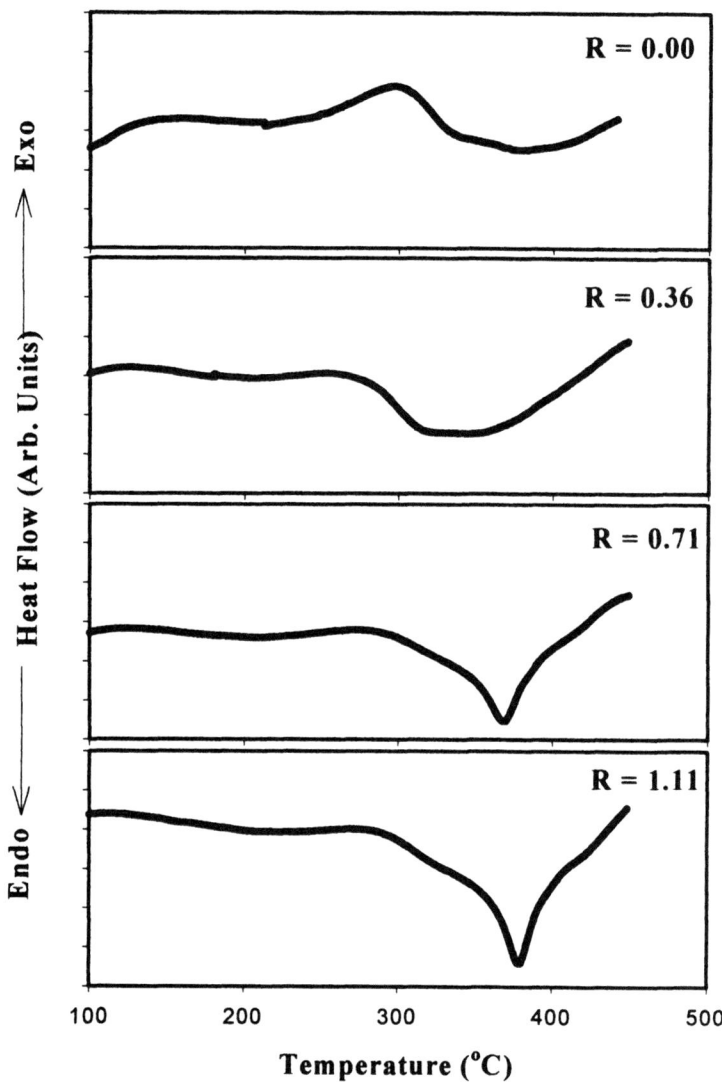

Figure 2. Differential scanning calorimetry of polyaniline spin coated films as a function of chemical reduction: R = 0.00: ~emeraldine base, R = 1.11: ~leucoemeraldine base.

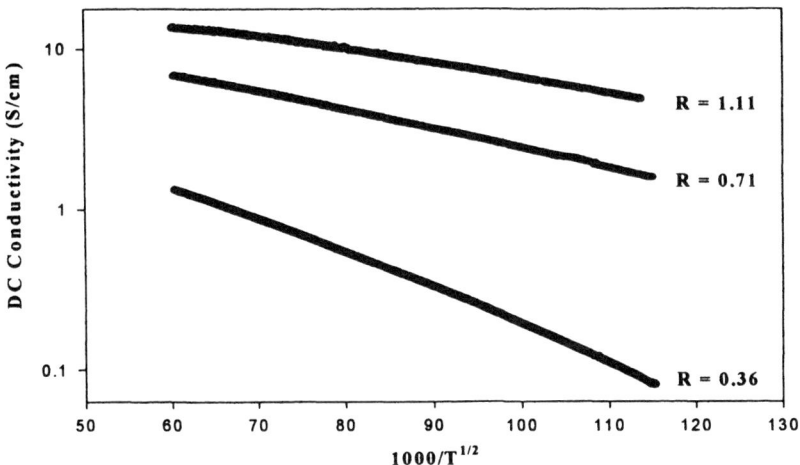

Figure 3. DC conductivity of iodine doped polyaniline films as a function of R, the molar ratio of reducing agent to emeraldine base (based on a four ring repeat unit) for the solution used to prepare films.

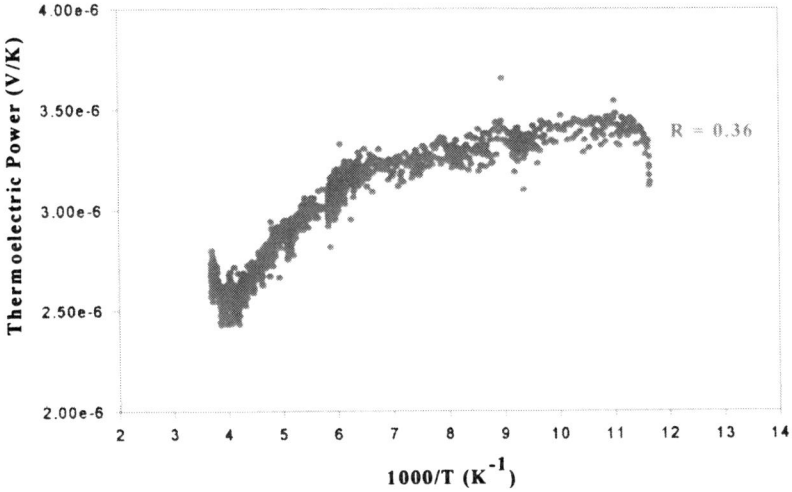

Figure 4. Thermoelectric power of a polyaniline spin coated film. R = 0.36. R is the molar ratio of reducing agent to emeraldine base (based on a four ring repeat unit) for the solution used to prepare films.

dependence of the anisotropy in σ_{dc} indicating a lack of phenyl ring rotation below 150 K[9]. Though still under investigation, there may be another transition in the thermoelectric power at about 250 K.

Conclusions

This work has shown that there is a clear connection between morphological structure as evidenced by the DSC results and electrical transport properties. With increasing chemical reduction, there is less hindrance to crystallization and sharper, higher temperature endotherms are obtained from the DSC, characteristic of larger, more perfect crystalline structures. This behavior is confirmed by the temperature dependence of conductivity in which T_o decreases with increasing chemical reduction. Unlike the other two samples (R = 0.36, 0.71), the R = 1.11 film exhibits the 1/T dependence of thermoelectric power indicative of semiconductor transport mechanism.

Acknowlegments

The authors wish to acknowledge K. Ivey and M. J. Drews for access to thermal analysis facilities. Financial support from the U. S. Department of Commerce of a grant administered by the National Textile Center is gratefully acknowledged.

References

1. A. G. MacDiarmid, Conjugated Polymers and Related Materials, edited by W. R. Salanek, I. Lundstrom, and B. Ranby (Oxford University Press, New York, 1993), pp. 73 - 98.
2. A. P. Chacko, S. S. Hardaker, R. V. Gregory, and T. W. Hanks, *Polymer Communications*, in press.
3. H. K. Chaudhari and D. S. Kelkar, *J. Appl. Poly. Sci.*, **62**, 15 (1996).
4. J. Joo, V. N. Prigodin, Y. G. Min, A. G. MacDiarmid, and A. J. Epstein, *Phys. Rev. B*, **50**, 12226 (1994)
5. E. P. Nakmedov, V. N. Prigodin, and A. N. Samukhin, *Sov. Phys. Solid. Stat.*, **31**, 368 (1989).
6. B. Abeles, P. Sheng, M. D. Coutts, and Y. Arie, *Adv. Phys.*, **24**, 407 (1975).
7. B. I. Shklovskii and A. L. Efros, Electronic Properties of Doped Semiconductors, (Springer-Verlag, New York), 1984.
8. F. L. Pratt, S. J. Blundell, W. Hayes, K. Nugamine, K. Ishida, and A. P. Monkman, *Phys. Rev. Lett.*, **79**, 2855 (1997).
9. A. P. Monkman, P. N. Adams, P. J. Laughlin and E. R. Holland, *Synth. Met.*, **69**, 183 (1995).

MELT PROCESSED ELECTRICALLY CONDUCTIVE BINARY AND TERNARY IMMISCIBLE POLYMER / POLYANILINE BLENDS

M. Zilberman [*], A. Siegmann [*], and M. Narkis [**]
Departments of [*]Materials and [**]Chemical Engineering,
Technion - Israel Institute of Technology, Haifa 32000, Israel.

ABSTRACT

In the present study, conductive binary and ternary blends of PANI with thermoplastic polymers were prepared by melt processing.

The binary blends' investigation focused on the morphology and on the resulting electrical conductivity. Generally, the level of interaction between the doped PANI and the matrix polymer determines the blend morphology, and thus, its electrical conductivity. The morphology of a conductive network is described by a primary structure of small dispersed polyaniline particles, interconnected by secondary short range fine fibrillar structure. In blends containing a semicrystalline matrix the doped PANI network locates within the amorphous regions, leading to a reduction of the percolation concentration.

The ternary blends' investigation focused on a system containing two co-continuous immiscible thermoplastic polymers and PANI. The PANI is preferably located in one of the matrix polymers. This concentration effect enables high electrical conductivities at low PANI contents.

INTRODUCTION

There is an increasing interest in processible polymeric materials, using conventional melt processing techniques, possessing electrical conductivity which can be tailored for a given application, and attractive physical and mechanical properties. Intrinsically conductive polymers (ICPs), recently making their first utilization appearance, are expected to yield such attractive combination of properties [1]. Polyaniline (PANI) is one of the most promising ICPs, due to its relatively high environmental and thermal stability, coupled with simple and economical production [2]. The main disadvantage of PANI, like other ICPs, is its limited thermal processability. One of the methods, still in development stage, to process PANI without altering its structure, is by proper blending with conventional polymers. These blends may combine the desired properties of the two components, i.e., electrical conductivity of PANI together with physical and mechanical properties of the matrix polymer. The morphology of such blends has a dominant effect on their properties.

Heeger and co-workers [2-4] reported the use of functionalized protonic acids to both, dope emeraldine base (EB, the non-conducting base form of polyaniline) and, simultaneously, render the resulting PANI complex solubility in common organic solvents. In most recent studies of electrically conductive polyaniline containing polymer blends, blending was performed in a solution. Only a few studies concerned with blends prepared by melt processing have recently been reported [1,5-7]. Shacklette et. al. [5] reported that Versicon[TM] (Allied-Signal Inc.), a conductive form of polyaniline, is dispersible in polar polymers such as polycaprolactone and poly(ethyleneterephthalate glycol). The percolation threshold for conductivity in these blends was observed in the range of 6-10 %v/v of Versicon. Ikkala et. al. [1] reported on conductive polymer blends, prepared by blending thermoplastic polymers with a Neste Complex, (Neste Oy), using conventional melt processing techniques. This research mainly addressed the electrical and mechanical properties of the blends.

In the present study, PANI was melt blended with thermoplastic polymers, and the resulting binary and ternary blends were investigated. This research focuses on the blends' morphology, which has a significant effect on the resulting electrical conductivity.

EXPERIMENTAL

Materials

Polyaniline: Versicon™, conductivity = 6 S/cm; Zipperling Kessler & Co, Germany.
Matrix Polymers: Polystyrene (PS), Galirene HH-102-E, Carmel Olefins, Israel, plasticized with di-octyl phthalate (DOP), PS:DOP = 85:15; Poly-ε-caprolactone (PCL), Capa 650, Solvy, U.K; Copolyamide 6/6.9 (CoPA), a random copolymer of 51% [HN - (CH$_2$)$_5$ - CO] and 49% [HN - (CH$_2$)$_6$ - NH - CO - (CH$_2$)$_7$ - CO], EMS, Switzerland; Linear low density polyethylene (LLDPE), Dowlex NG 5056, Dow Chemicals.

Blend Preparation

Binary and ternary polymer blends containing a single polymer or two immiscible polymers and a doped polyaniline were prepared by melt mixing in a Brabender plastograph, at 50 rpm for 12 minutes. The blending temperature varied with the matrix polymer as follows: blends containing LLDPE were blended at 180 °C, pasticized PS - 150°C, CoPA - 165°C, PCL - 70°C. Flat plaques, 3 mm thick, for conductivity measurements, were prepared by compression molding.

Conductivity Measurements

Electrical conductivity measurements were performed using the "four probe technique" (ASTM D 991-89) for samples (12x1.2x0.3 cm^3) of conductivity levels > 10^{-7} S/cm and the "two electrodes technique" (DIN-53596) for the less conductive blends. Samples for the latter technique were first coated with a silver paint, to reduce sample-electrode contact resistivity.

Morphological Characterization

Scanning Electron Microscopy (SEM) of cryogenically fractured surfaces was performed using a Jeol JSM-840, at an accelerating voltage of 10 kV. The SEM samples were gold sputtered prior to observation. High Resolution SEM of cryogenically fractured surfaces was performed using a Leo Gemini-982, at accelerating voltage of 1 kV. Transmission Electron Microscopy (TEM) of ultramicrotom samples was performed using a Jeol 2000 FX, at an accelerating voltage of 100 kV.

RESULTS

Binary Blends of PANI with Thermoplastic Polymers

The electrical conductivity as affected by PANI content for various melt blended matrix polymer / PANI blend series is presented in Fig. 1. The lowest percolation threshold, 5 wt% PANI, was observed for the PCL blend series. The PCL / PANI (90/10) blend is already conductive (~ 2x10^3 S/cm), due to formation of conducting paths of PANI within the PCL matrix. Beyond percolation, the conductivity level slowly increases with the increase in PANI content, due to the generation of a conductive network of an improved quality, attaining 0.1 S/cm for the 70/30 blend. Similar conductivity behavior was observed for the (PS/DOP) / PANI blend series. The percolation threshold of these blends was observed at a higher PANI content of 10 wt%, whereas higher conductivity levels were obtained for blends containing more than 15 wt% PANI. A different electrical conductivity behavior was observed for the CoPA / PANI blend series. The increase in electrical conductivity with the PANI content is more gradual than that observed for PCL / PANI and (PS/DOP) / PANI series, and the maximal conductivity level (7x10^{-4} S/cm for the CoPA / PANI 70/30) is lower. LLDPE / PANI blends are non-conductive even at PANI contents as high as 20 wt%.

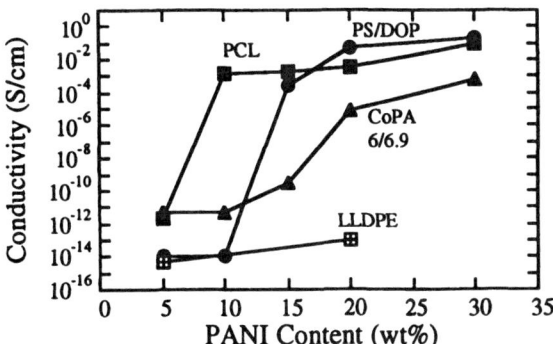

Fig. 1: Electrical conductivity vs. PANI content for various matrix polymer / PANI blend series.

The blends' morphology was studied to elucidate the structure responsible for the electrical conductivity behavior. SEM micrographs of cryogenically fractured surfaces of LLDPE / PANI and CoPA / PANI, containing 20 wt% PANI, are presented in Fig. 2, compared with the morphology of the neat matrix polymers. Large domains of PANI (1-5 μm) are observed within the LLDPE matrix (Fig. 2A), indicating that continuous networks of PANI have not been formed, resulting in a blend containing 20 wt% PANI which is practically insulative. In contrast, the SEM micrograph of the CoPA / PANI conductive blend (Fig. 2C) shows mainly the characteristic features of the blends' fracture surface and, in addition, very small, 0.1-0.2 μm, PANI particles. It is thus suggested, according to the blends' morphological behavior, that high conductivity at relatively low PANI contents is obtained mainly due to the good dispersability and structuring of the PANI particles within the matrix polymer. Smaller PANI particles were observed in the PCL and PS/DOP based blends.

The level of the PANI particle mechanical fracturing during melt mixing with a matrix polymer strongly depends on the specific matrix used. In polymer blending, both kinetic and thermodynamic considerations determine the degree of polymer dispersion within another polymer [8]. Thus, the interaction during melt blending between the matrix polymer and the PANI particles influences the level of PANI fracturing and dispersion within the matrix polymer. High interaction levels enable a higher level of PANI fracturing (at similar matrix viscosity and shear level), due to better interphase shear stress transferability [9]. High interactions and fine dispersions should occur in systems comprising components of similar solubility parameters, as shown by molecular simulations [10].

SEM observations of the blends enable us to estimate the PANI particles' size and their level of dispersion within the matrix polymer, but a continuous network structure, as expected in conductive blends, cannot be observed via SEM. Therefore, the blends' morphology was studied also by TEM. Micrographs of CoPA / PANI (80/20) and PCL / PANI (80/20) blends are presented in Fig. 3. The CoPA / PANI (80/20) blend (Fig. 3A) shows two types of regions : a region rich in PANI, in which continuous paths of PANI (in black) were developed within the CoPA matrix, and a region containing apparently discrete PANI particles. It is suggested that the ultramicrotom cutting in the former section is parallel to the network direction, while in the latter section it is perpendicular to the network's direction, showing, therefore, different micrographs of actually the same structure.

PCL is a semicrystalline polymer with a spherulitic morphology. As shown in Fig. 3B, the PANI in the PCL / PANI (80/20) blend is located around the spherulites, in the amorphous regions. Hence, The PANI effective content in these domains is higher than the nominal content. As a result, the percolation threshold of the PCL based blends is lower than that observed for amorphous matrix based blends, such as CoPA / PANI and (PS/DOP) / PANI (Fig. 1).

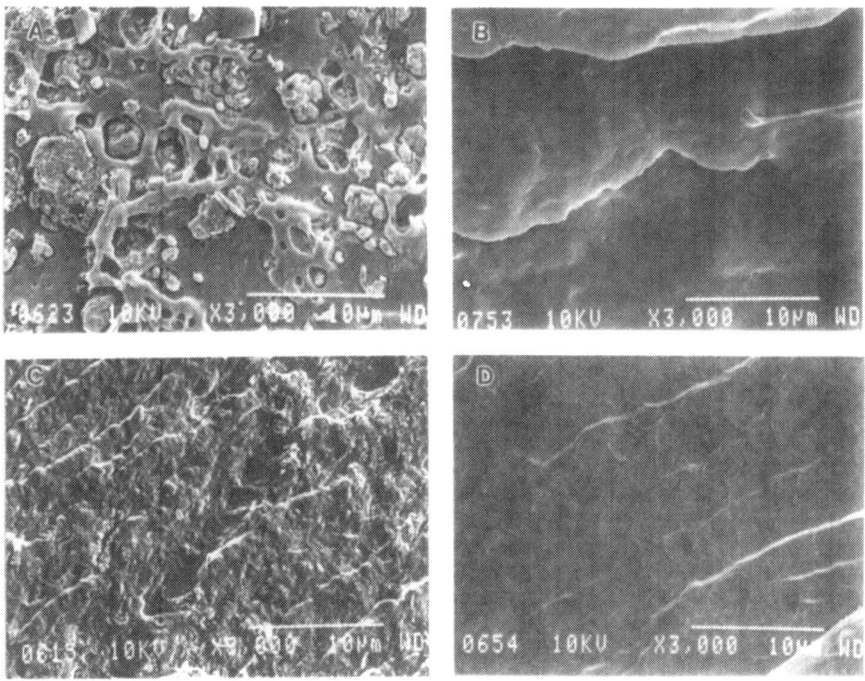

Fig. 2: SEM micrographs of: (A) LLDPE / PANI (80/20), (B) LLDPE. (C) CoPA / PANI (80/20), (D) CoPA.

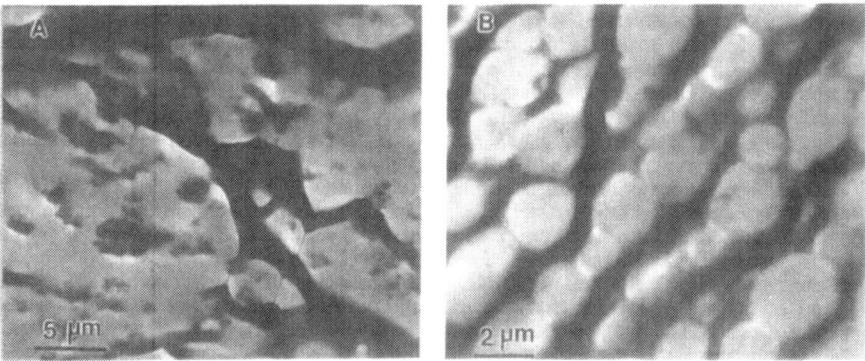

Fig. 3: TEM micrographs of: (A) CoPA / PANI (80/20), (B) PCL / PANI (80/20).

Figure 3A also indicates that the continuous network is actually composed of small particles (0.5-1 μm), consisting of even smaller particles, about 0.1 μm apart. These gaps, presumably, are too large to enable hopping of charge carriers between particles. It has thus been suggested, that in addition to the observed structural features by TEM, a very fine short-range PANI fibrillar-like morphology exists, presently unobservable by both, SEM and TEM, which interconnects the observable fine dispersed PANI particles. Since the solubility parameters of CoPA and PANI are quite similar [10], partial dissolution during blending of PANI in the molten CoPA matrix may occur. Upon cooling, at least partial precipitation of the PANI occurs, forming the secondary morphology, a short-range fibrillar-like structure, interconnecting the finely dispersed PANI particles (the primary structure), thus, generating continuous conducting paths. The secondary fine fibrillar structure resembles the structure obtained in solution cast polymer / PANI blends.

Ternary Blends of PANI with Two Immiscible Thermoplastic Polymers

Ternary blends of two immiscible polymers and the PANI, were investigated. The (PS/DOP) / CoPA / PANI system was chosen to demonstrate the behavior of a ternary blend. The electrical conductivity as a function of CoPA content, for blends containing 10 wt% and 20 wt% PANI is presented in Fig. 4. The conductivity levels of the blends containing 20 wt% PANI increases with a decrease in the CoPA content, and are higher than that of the CoPA / PANI (80/20) binary blend and lower than that of the (PS/DOP) / PANI (80/20) binary blend.

High resolution SEM (HRSEM) micrographs of cryogenically fractured surfaces of ternary blends containing 80 wt% matrix polymers and 20 wt% PANI are presented in Fig. 5. These micrographs show that the PANI (in white) is preferably located in the CoPA phase while the PS/DOP phase does not contain any PANI particles. Hence, the effective PANI content in the CoPA phase is higher than the nominal content. Since the two matrix polymers form a co-continuous blend structure in both 30/70 and 70/30 compositions, their ternary blends are both conductive. This concentration effect is even more impressive in ternary blends containing 10 wt% PANI (Fig. 5) ; while the (PS/DOP) / PANI (90/10) and CoPA / PANI (90/10) binary blends are practically insulative, the ternary blends containing 10 wt% PANi are highly conductive (at least 10^{-4} S/cm).

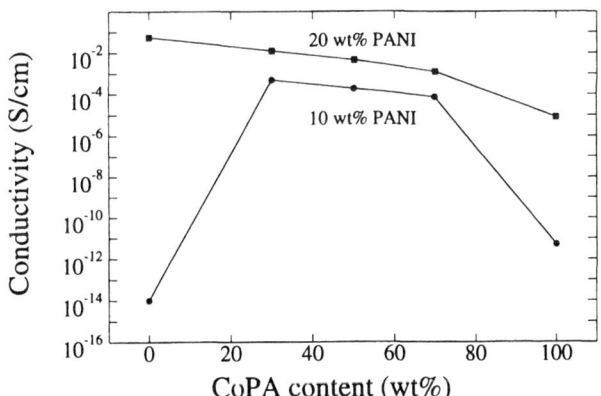

Fig. 4: Electrical conductivity vs. CoPA content for (PS/DOP) / CoPA / PANI ternary blends (CoPA content + PS/DOP content = 100% matrix polymers).

The electrical conductivity values of the (PS/DOP) / CoPA / PANI ternary blends (Fig. 4) are indeed similar to those of the corresponding CoPA / PANI binary blends (Fig. 1). For example, the conductivity level of the ternary (PS/DOP) / CoPA (30/70) based blend containing 10 wt% PANI is similar to that of the 73/27 CoPA / PANI blend. The CoPA phase of the former actually contains 27 wt% PANI, due to the preferential location of the PANI in the CoPA phase.

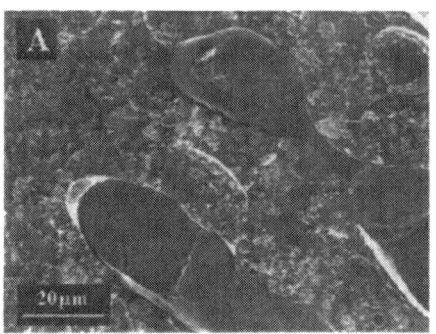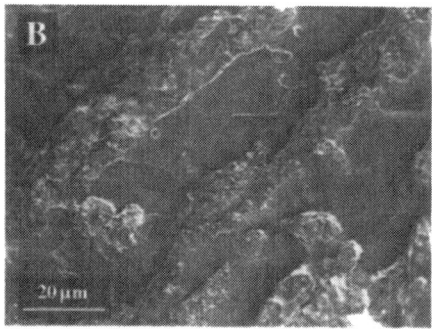

Fig. 5: High Resolution SEM micrographs of ternary blends containing 20 wt% PANI:
(A) (PS/DOP) / CoPA (30/70) / PANI, (B) (PS/DOP) / CoPA (70/30) / PANI.

CONCLUSIONS

Binary and ternary blends of a conductive PANI with various thermoplastic polymers were prepared by melt processing. Their investigation focused on the morphology and the resulting electrical conductivity.

Binary Blends: The level of interaction between the components determines the PANI particles' level of fracturing and mode of dispersion within the matrix polymer, and thus, the blends' morphology and electrical conductivity. The conductive network consists of two levels of structures: (a) a primary structure composed of small dispersed PANI particles, (b) a short-range very fine fibrillar structure, interconnecting the dispersed particles. The insulative blends show discrete large isolated particles within the matrix polymer. The conductive network is located within the amorphous regions of a semicrystalline matrix, which results in a low percolation threshold compared to systems of amorphous matrices.

Ternary Blends: The PANI is preferably located in one of the immiscible matrix phases. Therefore, ternary blends consisting of a matrix having a co-continuous structure exhibit high electrical conductivities at low PANI contents.

ACKNOWLEDGEMENTS

The authors acknowledge the support of the Israel Ministry of Science.

REFERENCES

1. O.T. Ikkala, J. Laakso, K. Vakiparta, E. Virtanen, H. Ruohnen, H. Jarvinen, T. Taka, P. Passiniemi and J.E. Osterholm., Synthetic metals, **69**, 97 (1995).
2. A.J. Heeger, Synthetic metals, **57**, 3471 (1993).
3. Y. Cao, P. Smith and A.J. Heeger, Synthetic metals, **57**, 3514 (1993).
4. C.Y. Yang, Y. Cao, P. Smith and A.J. Heeger, Synthetic metals, **53**, 293 (1993).
5. L.W. Shacklette, C.C. Han and M.H. Luly, Synthetic metals, **57**, 3532 (1993).
6. S.J. Davides, T.G. Ryan, C.J. Wilde and G. Beyer, Synthetic metals, **69**, 209 (1995).
7. P. Passiniemi, J. Laakso, H. Ruohnen and K. Vakiparta, Mat. Res. Soc. Symp. Proc., Vol. 413, P.577 (1996).
8. Y.J. Lee, I. Manas-Zloczower and D.L. Feke, Polym. Eng. Sci., **35**, 1037 (1995).
9. M. Narkis, M. Zilberman and A. Siegmann, Polym. Adv. Tech., **8**, 525 (1997).
10. M. Zilberman, A. Siegmann and M. Narkis, "Melt Processed Electrically Conductive Polymer / Polyaniline Blends", Journal of Macromolecular Science - Physics (B), in press.

THERMOREVERSIBLE GELS OF POLYANILINE:
VISCOELASTIC AND ELECTRICAL PROPERTIES

O.T. Ikkala*, T. Vikki*, J. Ruokolainen*, P. Hiekkataipale*, P. Passiniemi**,
T. Mäkelä*** and H. Isotalo***

* Helsinki University of Technology, Department of Engineering Physics and
  Mathematics, P.O. Box 2200, FIN-02015 HUT, Espoo, Finland
  Olli.Ikkala@hut.fi
** Neste Oy, P.O. Box 310, FIN-06101, Porvoo, Finland
*** VTT Electronics, Microelectronics, P.O. Box 1101, FIN-02044 VTT, Espoo, Finland

ABSTRACT

We demonstrate that polyaniline (PANI) dissolved in dodecyl benzene sulphonic acid (DBSA) shows *thermoreversible* gelation. The dissolution has been performed in formic acid which allows particle-free complexes according to optical microscopy. Below the gelation temperature the materials are rubber-elastic in compression experiments, the storage modulus G′ does not essentially depend on frequency, and the samples are electronically conductive. Above the gelation temperature, G′ indicates flow-like behavior and drastically lower ionic conductivity is observed. These results suggest reversible, i.e. fusible, network formation. The properties are compared with gels consisting of camphor sulphonic acid (CSA) doped PANI dissolved in *m*-cresol which are poorly thermoreversible.

INTRODUCTION

In the recent literature of electrically conducting polymers, considerable attention has been paid on finding methods to process them with controlled properties. In polyaniline, solvents have been identified [1, 2] allowing blends consisting of PANI-networks [3], and molecular level deposition processes have been described [4]. Other efforts describe plasticizers to allow fusible bulk PANI-complexes [5] and blends with reversible network formation in melt blending [6].

On the other hand, much progress has taken place in understanding (thermo)reversible gelation: In such systems polymers form solutions at certain temperatures, whereas *physically* connected networks are formed upon proper temperature change for particular solvents [7]. For example 9.5 %-wt of atactic polystyrene in carbon disulphide forms a solution for T > ca. -15 °C, whereas for T < -15°C a reversible network is formed. Many types of physical crosslinks have been reported, such as micellar crystallites, sidechain crystallinity, formation of triple helices, or glassy aggregates [7]. There are established methods to study the phase behavior of thermoreversible gels by, for example, dynamic rheology, "Tilted test tube method", and "Falling ball method".

Various gelling studies on emeraldine base form of PANI have been reported [8]. However, thermoreversible gels based on doped conductive polymers could combine viscoelasticity and electrical conductivity of networks and fusibility, on the other hand, in a challenging way. Astonishingly enough, only very recently such systems have been described: We showed that conducting *thermoreversible gel* is obtained if PANI emeraldine base is dissolved in DBSA using formic acid medium [9]. As a background, we describe viscoelastic

377

and conductivity properties of poorly reversible gels consisting of CSA-doped PANI dissolved in *m*-cresol.

EXPERIMENTAL

The emeraldine base form of PANI with a molecular weight $M_n = 25\ 000$ g/mol was prepared using conventional methods at Neste Oy (Finland). It contains 0.5 iminic nitrogens *vs* one aromatic ring where the iminic nitrogens can be protonated using strong acids such as DBSA or CSA to achieve doping. In the following definitions of compositions, 1 mol of PANI refers to 1 mol of PhN repeat units.

Preparation of PANI(CSA)$_{0.5}$/*m*-cresol samples

PANI emeraldine base and CSA (Aldrich) were dried at 60 °C in vacuum at least for 24 h. They (0.5 mol of CSA *vs* 1 mol of PANI) were added to formic acid (Fluka) at their total concentration of ca. 2 %-wt. After stirring for several hours, the formic acid was evaporated and the sample was dried in vacuum at 60°C for several days. The resulting material is denoted as PANI(CSA)$_{0.5}$. It was dissolved in *m*-cresol using a magnetic stirrer at different concentrations with a minimim exposure to atmospheric humidity.

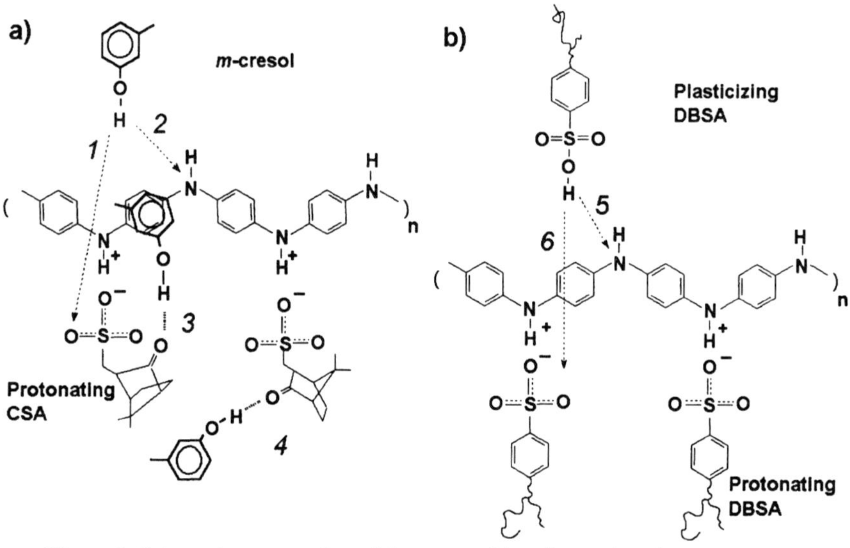

Figure 1. Schematic presentation of the expected bondings when the iminic nitrogens of PANI are nominally fully protonated (doping). a) PANI(CSA)$_{0.5}$ dissolved in *m*-cresol. The schemes 1-4 presents possible schemes how *m*-cresol molecules can interact with PANI(DBSA)$_{0.5}$. Note that only scheme 3 is unique to this system and may thus explain the specific properties [10]. b) PANI(DBSA)$_{0.5}$ dissolved in DBSA. The additional DBSA can further interact due to hydrogen bonding (schemes 5 and 6).

Preparation of PANI(DBSA)$_{0.5}$/DBSA samples

PANI-DBSA samples were prepared similarly, the only difference being that m-cresol was replaced by additional DBSA. PANI was dissolved in excess DBSA (Tokyo Kasei) in formic acid medium where the total concentration of PANI and DBSA in formic acid was 2.5 %-wt. The formic acid was evaporated and the sample was dried in vacuum at 60 °C for 1 week. Part of the DBSA molecules are consumed to protonate the iminic nitrogens of PANI. For example, taken PANI emeraldine base and DBSA at their ratio 10/90 w/w, nominally 0.5 mol of DBSA is first consumed for protonation, and the rest of DBSA serves as the background solvent that is able to form hydrogen bond with PANI(DBSA)$_{0.5}$. To emphasize the two roles of DBSA, and to help comparison with the corresponding PANI(CSA)$_{0.5}$/m-cresol gels, the resulting material is subsequently denoted as PANI(DBSA)$_{0.5}$/DBSA (with their ratio 28/72 w/w in the above example). Note that in practice slightly less than 0.5 mol of DBSA may actually be consumed for protonation vs 1 mol of PANI, a fact that is of no relevance in our studies. Note also that fusibility of PANI plasticized by excess amount of DBSA has been reported previously [11]. However, without formic acid treatment we did not achieve particle-free complexes even based on optical microscopy and thermoreversible gels.

GELLING STUDIES OF PANI(CSA)$_{0.5}$ DISSOLVED IN m-CRESOL

10 %-wt of PANI(CSA)$_{0.5}$ prepared using the above-mentioned method dissolves in 90 %-wt of m-cresol at room temperature during ca. 0.5 h to form particle-free solution according to optical microscopy. Immediately after preparation the sample behaves as a viscous fluid but changes later to a rubber-like solid. Quantitatively, the dynamic storage modulus has been determined as a function of time using Bohlin VOR dynamic rheometer (concentric cylinder geometry with the outer and the inner cylinders of diameters 15.4 and 14 mm, respectively. Before substantial gelation is observed, the storage modulus G' is strongly ω-dependent, typical for solutions (ideally $G' \approx \omega^2$ for solutions). For longer times times (> ca. 100h), $G' \approx \omega^0$ as characteristic for solids, i.e., gelation has taken place.

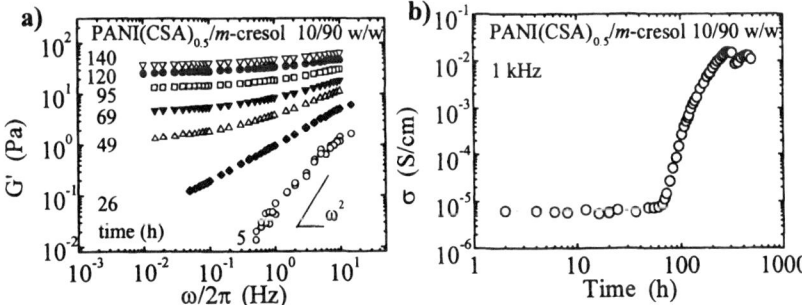

Figure 2. 10 %-m PANI(CSA)$_{0.5}$ dissolved in 90 %-m m-cresol a) Dynamic rheological storage modulus at various aging times. b) AC conductivity at 1 kHz as a function of aging time showing drastic increase of conductivity due to network formation when the gelation takes place.

AC impedance has been determined using HP 4192A LF Impedance Analyzer. The conductivity at 1 kHz is initially constant at ca. 6 x 10^{-6} S/cm. After 70 h has elapsed, a drastic change takes place where the conductivity increases 3-4 orders of magnitude. The rheology and the impedance measurements of Figure 2 are in perfect agreement: AC impedance suggests that the first conductive pathway extending over macroscopic distances is formed at the time = 70 h (leading to percolative conductive network), whereas already long before that time aggregation manifests in the viscoelastic behavior (due to networks that fill only part of the sample volume). Further results dealing also Bode-plots will be described elsewhere [12].

The above results illustrate how gelling manifests in viscoelasticity and impedance of conducting polymers. Note that $PANI(CSA)_{0.5}$/m-cresol gels are not clearly thermoreversible (at least at such high concentrations used) because heating does not cause melting. The crosslink sites are probably microcrystalline regions, and temperatures of the order of 200 °C may be required to achieve gel melting, as suggested by $PANI(CSA)_{0.5}$/resorcinol [5]. Such temperatures are far beyond the boiling point of m-cresol.

THERMOREVERSIBLE GELATION OF $PANI(DBSA)_{0.5}$ DISSOLVED IN DBSA

Dissolving 15.6 %-wt of PANI in 84.4 %-wt DBSA (i.e. $PANI(DSBA)_{0.5}$/DBSA 43.7/56.3 w/w) yields rubberlike material. Guenet has suggested that one of the critical issues to identify thermoreversible gels is to study rubber-elasticity upon compression [13]. Figure 3 shows that the above material recovers almost completely following a compression to 80% of the nominal initial thickness. Note that in the first compression the recovery is not complete, probably due to sample imperfections.

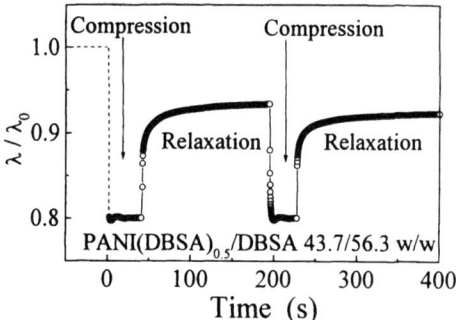

Figure 3. Sample thickness λ scaled by the original thickness λ_0 during isothermal compression/relaxation experiments at room temperature for a sample where 15.6 %-wt of PANI is dissolved in 84.4 %-wt DBSA [9].

The samples melt at elevated temperatures to yield liquid-like behavior. The easiest method to determine the gel melting is to suspend a steel ball on the surface of gel sample and to identify the gel melting at the temperature where the ball starts to penetrate the sample (Falling ball method [7]). We adopted a modification where the sample has been confined between two parallel plates with no constraining lateral walls. There the ball has been replaced by the upper plate of dynamic mechanical analyzer (DMA) oscillating at 1 Hz. Figure

4 describes gel melting at 180 °C. The strongly corrosive nature of the samples inhibited use of dynamic rheometer. However, the same parallel plate mode of DMA as a function of frequency could be used at different temperatures. The results show that below the gel melting essentially $G' \approx \omega^0$ whereas at the gel melting G' becomes strongly ω-dependent, in agreement with Figure 2a [9].

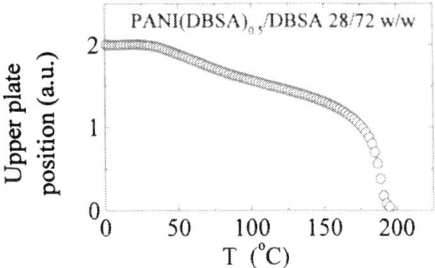

Figure 4. Gel melting at 180 °C demonstrated for a sample consisting 10 % PANI dissolved in 90 %-wt DBSA during heating at 3 °C/min [9]. The probe position denotes the position of the upper plate in parallel plate DMA measurement.

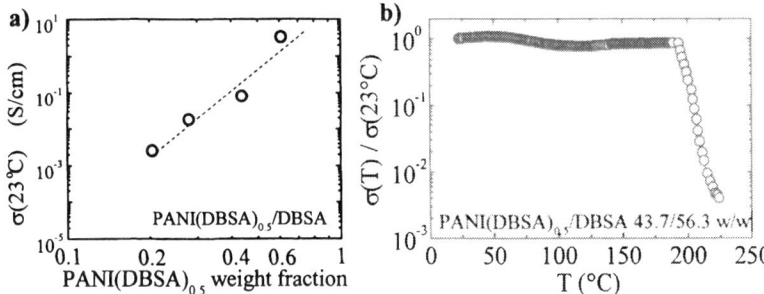

Figure 5. a) Room temperature conductivity of PANI(DBSA)$_{0.5}$ dissolved in DBSA. b) Conductivity as a function of temperature for a sample where 15.6 %-wt PANI is dissolved in 84.4 %-wt DBSA [9].

The AC-conductivity as a function of concentration using triangular current profile at 20 mHz is shown in Figure 5a. Note that the DBSA medium itself is ionically conducting. Therefore at sufficiently low PANI concentrations the conductivity due to the PANI(DBSA)$_{0.5}$ network is less than that of the background DBSA solvent and the network conductivity cannot thus be determined [9]. Figure 5b illustrates that the high conductivity due to percolating physical network drastically decreases upon gel melting. Although not shown in Figure 5b, cooling back to the room temperature allows recovery of the original conductivity [9]. That the conductivity is drastically higher in the case when the network is present is in agreement with Figure 2b.

CONCLUDING REMARKS

We have demonstrated that *thermoreversible* gelation takes place in PANI dissolved in DBSA when the complexation has been performed in formic acid. Formic acid probably helps homogeneous complex formation due to PANI swelling. Viscoelastic and conductivity data show that in the gel state the material is rubber-elastic and relatively highly conductive due to the electronically conductive network, whereas in the melt state it behaves as a viscous fluid and has low conductivity, probably due to poor chain-to-chain contacts. The state of the material can be reversibly switched between them by changing the temperature. On the other hand, a freshly prepared PANI(CSA)$_{0.5}$/m-cresol mixture is a viscous fluid but gel formation [14] takes place during time with changes in viscoelastic and conductivity properties. However, such gels cannot in general be denoted as thermoreversible because the solution state cannot easily be recovered once gelling has taken place, at least for higher concentrations.

ACKNOWLEDGEMENTS

We are grateful to Jan-Erik Österholm for discussions. Neste Chemicals is acknowledged for permission to publish the results. The work has been supported by Neste Chemicals, the Technology Development Centre (Finland) and the Neste Foundation.

REFERENCES

1. Y. Cao, P. Smith and A. J. Heeger, Synth. Met. **48**, p. 91 (1992).
2. A. Hopkins, P. G. Rasmussen and R. A. Basheer, Macromolecules **29**, p. 7838 (1996).
3. C. Y. Yang, Y. Cao, P. Smith and A. J. Heeger, Synth. Met. **53**, p. 293 (1993).
4. W. B. Stockton and M. F. Rubner, Macromolecules **30**, p. 2717 (1997).
5. T. Vikki, L.-O. Pietilä, H. Österholm, L. Ahjopalo, A. Takala, A. Toivo, K. Levon, P. Passiniemi and O. Ikkala, Macromolecules **29**, p. 2945 (1996).
6. J. Tanner, O. T. Ikkala, J. Laakso and P. Passiniemi, in Electrical, Optical and Magnetic properties of Organic Solid State Materials III, edited by A. K.-Y. Jen, C. Y.-C. Lee, L. R. Dalton, M. F. Rubner, G. E. Wnek, and L. Y. Chiang (Mater. Res. Soc. Proc. 413, Pittsburgh, PA 1996), p 565.
7. K. te Nijenhuis, Adv. in Polym. Sci. **130**, p. 1 (1997).
8. See, for example, A. G. MacDiarmid, Y. Min, J. M. Wiesinger, E. J. Oh, E. M. Scherr, and A. J. Epstein, Synth. Met. **55-57**, p. 753 (1993); and K. T. Tzou and R. V. Gregory, Synth. Met. **69**, p. 109 (1995).
9. T. Vikki, J. Ruokolainen, O. T. Ikkala, P. Passiniemi, H. Isotalo, M. Torkkeli and R. Serimaa, Macromolecules **30**, p. 4064 (1997).
10. O. T. Ikkala, L.-O. Pietilä, L. Ahjopalo, H. Österholm and P. J. Passiniemi, J. Chem. Phys. **103**, p. 9855 (1995).
11. K. Levon, K.-H. Ho, W.-Y. Zheng, J. Laakso, T. Kärnä, T. Taka and J.-E. Österholm, Polymer **36**, p. 2733 (1995).
12. To be published.
13. J. M. Guenet, Trends in Polym. Sci. **4**, p. 6 (1995).
14. Gels of PANI(CSA)$_{0.5}$/m-cresol have been mentioned in C. L. Gettinger, A. J. Heeger, D. J. Pine and Y. Cao, Synth. Met. **74**, p. 81 (1995).

Part IV

Molecular and Supramolecular Engineering

MOLECULAR SELF-ASSEMBLY ROUTES TO OPTICALLY FUNCTIONAL THIN FILMS: ELECTROLUMINESCENT MULTILAYER STRUCTURES

W. LI*, J. E. MALINSKY*, H. CHOU*, W. MA*, L. GENG*, T. J. MARKS*, G. E. JABBOUR[†], S. E. SHAHEEN[†], B. KIPPELEN[†], N. PEYGHAMBARIAN[†], J. ANDERSON[‡], P. LEE[‡], N. ARMSTRONG[‡]
*Department of Chemistry and the Materials Research Center, Northwestern University, Evanston, IL 60208-3113, tjmarks@casbah.acns.nwu.edu
[†]Optical Science Center, University of Arizona, Tucson, AZ 85721
[‡]Department of Chemistry, University of Arizona, Tucson, AZ 85721

ABSTRACT

This contribution describes the use of layer-by-layer self-limiting siloxane chemisorption processes to self-assemble structurally regular multilayer organic LED (OLED) devices. Topics discussed include: 1) the synthesis of silyl-functionalized precursor molecules for hole transport layer (HTL), emissive layer (EML), and electron transport layer (ETL) self-assembly, 2) the use of layer-by-layer self-assembly for ITO electrode modification/passivation/hole-electron balancing in a vapor-deposited device, 3) the microstructure/chemical characterization of HTL self-assembly using a prototype triarylamine precursor, 4) fabrication and properties of a hybrid self-assembled + vapor deposited two-layer LED, 4) fabrication and properties of a fully self-assembled two-layer OLED.

INTRODUCTION

Molecular self-assembly approaches involving self-limiting chemisorption represent an attractive new approach to the fabrication of functional thin films composed of molecular assemblies having "engineered" spatial relationships. A recent example involves the layer-by-layer construction of intrinsically acentric chromophoric superlattices for second-order NLO applications [1,2] using siloxane condensation methodology. Stepwise $\equiv$SiOH and ClSi$\equiv$ and stilbazole quaternization + chemisorption sequences are employed to produce robust, covalently interlocked multilayers in which high hyperpolarizability chromophores are "locked" into an acentric microstructure. An example of this approach is illustrated in Figure 1, where $\chi^{(2)}$ values as large as 320 pm/V have been achieved [3,4]. The chemisorption process has been characterized by a large battery of physical techniques [1,2]. In regard to structural regularity, both specular X-ray reflectivity [4-6] and standing wave X-ray [7] experiments indicate highly regular stacking architectures, while the linear dependence of $(I_{2\omega})^{1/2}$ on the number of assembled trilayers indicates that uniform microstructural acentricities are preserved as each trilayer is added [3, 4, 7]. AFM studies reveal that films of ~350 Å thickness have rms roughnesses of only ~12 Å with no evidence of pinholes [4,7]. Indeed, these thin films can be used in hybrid waveguide devices in which frequency doubling is achieved by evanescent coupling of light propagating through a proximate, spin-coated linear guiding layer. Hence, robust device-quality second-order NLO films can be constructed with Å -level control of layer dimensions and without electric field poling.

The above results raise the question of whether siloxane-based molecular self-assembly approaches can be employed for the fabrication of other optically functional thin film structures which might benefit from the ability to deposit conformal, covalently interlinked layers, composed of tailored molecular building blocks. In the present contribution, we report the first application of such techniques to multilayer electroluminescent device fabrication. Covalent self-assembly approaches are distinctly different from either conventional vapor deposition [8,9] or polymer spin-coating [8.9] techniques, and in principle, suggest means to address electrode passivation, balancing electron versus hole injection rates, the unavailability of suitably volatile building blocks, as well as failure modes such as atmospheric degradation, pinholes, layer cracking, layer interdiffusion, layer crystallization, and layer vaporization [8-15].

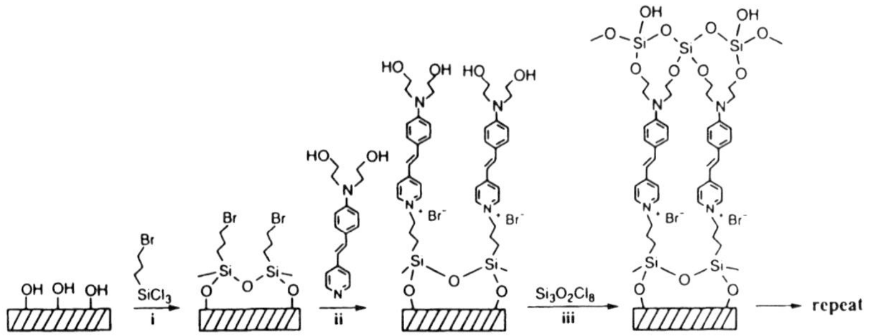

Figure 1. Approach to the layer-by-layer assembly of NLO-active chromophoric multilayers. The steps are: i, coupling layer introduction; ii, chromophore layer introduction; iii, capping layer introduction.

EXPERIMENTAL

All chlorosilane reagents were synthesized, stored, and manipulated with rigorous exclusion of oxygen and moisture using Schlenk and high vacuum line techniques as well as a Vacuum Atmospheres glove box with an efficient recirculator. All solvents were dried in an appropriate manner and distilled under inert atmosphere. The synthesis of the hole transport layer (HTL), emissive layer (EML), and electron transport layer (ETL) building blocks are outlined in Figures 2-4, respectively. All products were characterized as appropriate by ^{1}H and ^{13}C NMR, high resolution mass spectrometry, and elemental analysis. The basic self-assembly techniques and their general temporal characteristics are described elsewhere [4,5,7], as are the procedures for film thickness characterization by scanning ellipsometry [4,7] and specular X-ray reflectivity [4-6]. In the present case, specimens were protected from dust at all times, and specimen transfer from reaction containers was carried out in a class 1000 clean hood. ITO substrates were patterned and cleaned as described previously [16]. The instrumentation for electroluminescence studies is described elsewhere [16].

RESULTS AND DISCUSSION

This report addresses preliminary results in the following areas: 1) synthesis of hole transport layer (HTL), emissive layer (EML), and electron transport layer (ETL) building blocks for ultimate incorporation in three-layer devices, 2) self-assembling electrode modification/passivation layers, 3) microstructural/chemical respects of HTL self-assembly, 4) a hybrid self-assembled + vapor deposited device, 5) a totally self-assembled two-layer device.

Building Block Synthesis

Figures 2-4 outline synthetic approaches to reactive precursors for self-assembled layers having hole transport, emissive, and electron transport functions, respectively. The precursor core structures have established efficacies in these functions [8-15]. The chemistry makes heavy use of palladium-catalyzed Stille reactions for allyl group introduction and platinum-catalyzed hydrosilylation. Yields are acceptable, and purity has been established by standard spectroscopic and analytical techniques.

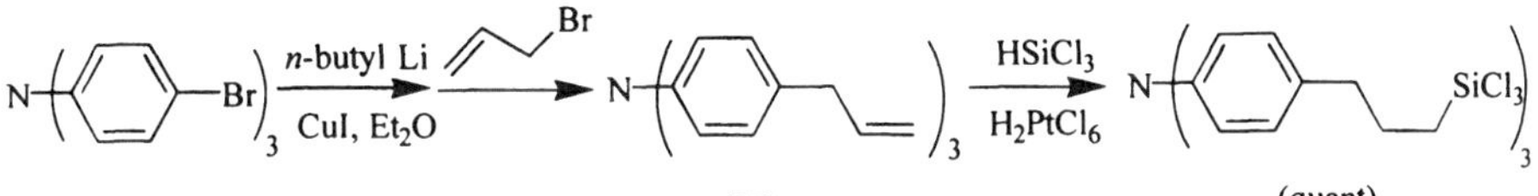

(97)   (quant)

Figure 2. Synthetic scheme for the hole transport layer precursor. Quantities in parentheses are isolated yields.

(15)   (98)

(65)

(10)   (quant)

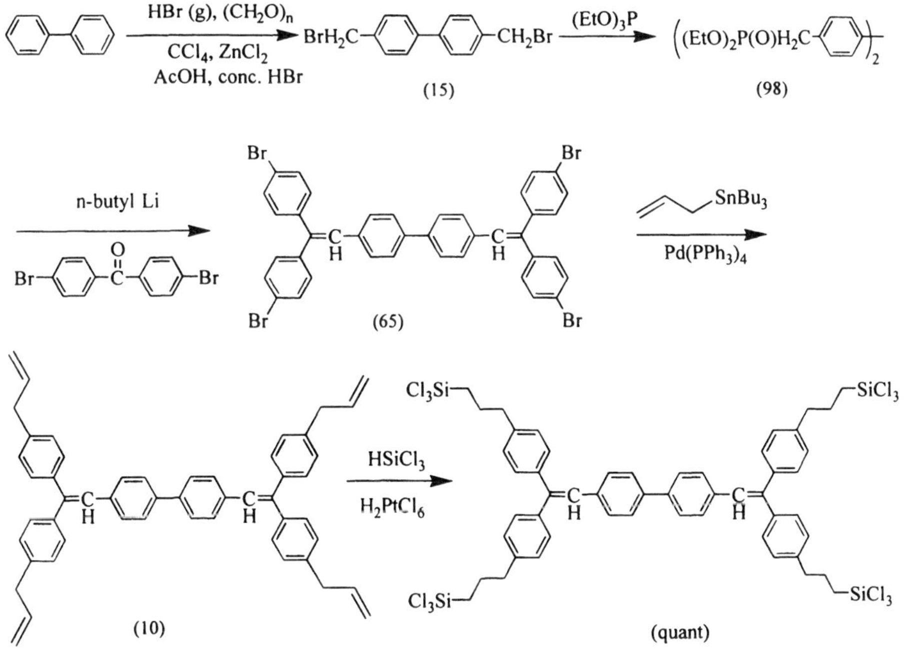

Figure 3. Synthetic scheme for the emissive layer precursor. Quantities in parentheses are isolated yields.

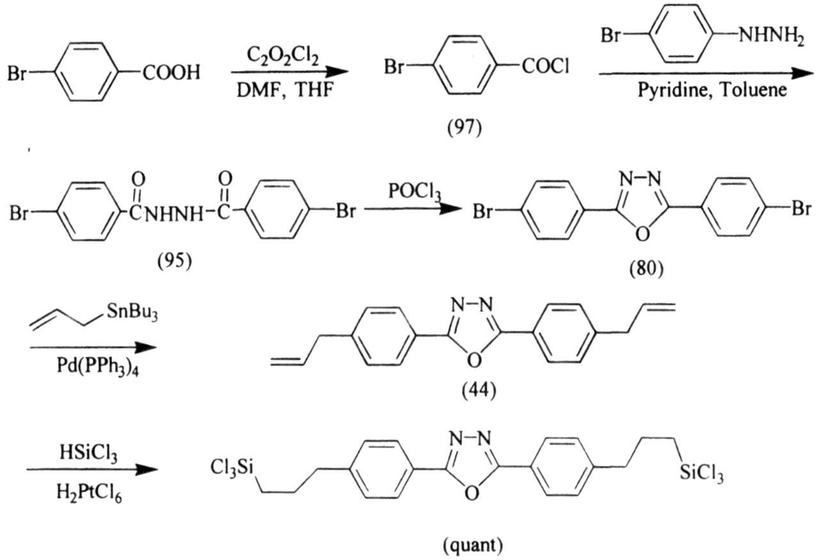

Figure 4. Synthetic scheme for the electron transport layer precursor. Quantities in parentheses are isolated yields.

Self-Assembling Electrode Derivatization

Layers of the "capping" reagent $Cl_3SiOSiCl_2OSiCl_3$ (Figure 1, step iii) were sequentially chemisorbed on clean ITO-coated glass in a sequence involving: i) anaerobic self-assembly, ii) rinsing, iii) curing/crosslinking in air. In a parallel set of experiments on single crystal Si substrates, X-ray reflectivity establishes that each layer is of ~8 Å thickness [17], in agreement with previous results on the NLO-active superlattices [2-7]. Conventional TPD (600 Å)/Alq (600 Å)/Mg (2000 Å) device structures were then vapor deposited on the derivatized ITO substrates. Figure 5 shows the characteristics of OLEDs fabricated with different numbers of self-assembled capping layers on the ITO surface. It can be seen in these preliminary results that the effect of two ITO derivatization layers is to significantly enhance device quantum efficiency at low voltages, while effects on the light output and current density are modest. Interestingly, the effects maximize at two 8 Å insulating layers, with a single layer exhibiting higher threshold voltages and depressed current densities, similar to behavior with n > 2. It appears that these effects reflect diminished current flow at the ITO surface (confirmed in electrochemical experiments, see below), resulting in better hole-electron injection balance and diminished oxidative/diffusional degradation [18,19] of the interface.

Self-Assembled Hole Transport Layers

The triarylamine precursor shown in Figure 2 was self-assembled in a layer-by-layer fashion on clean ITO and single crystal silicon substrates using the procedure outlined above. XPS studies reveal essentially instantaneous -$SiCl_3$ hydrolysis upon exposure to ambient. They also reveal that the ionization potential of the self-assembled molecular core is virtually identical

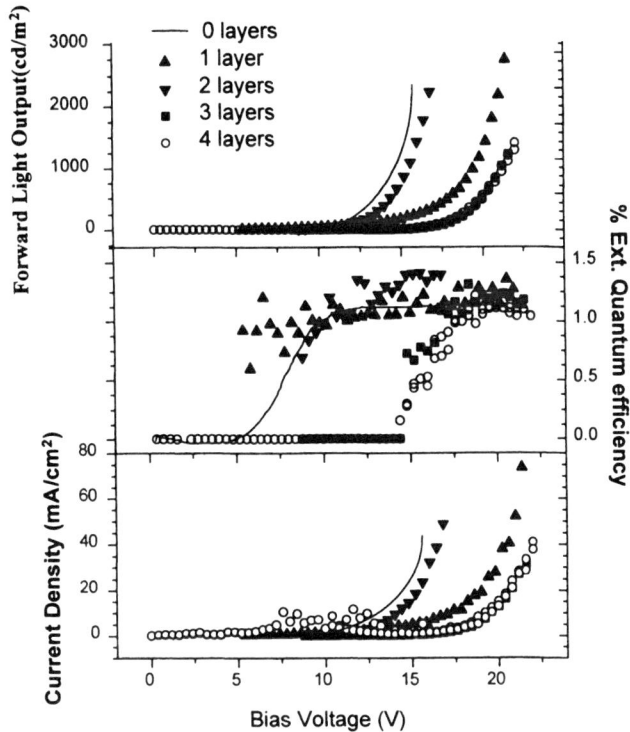

Figure 5. Performance characteristics of ITO/(CAP)n/TPD/Alq/Mg devices as a function of the number of self-assembled ITO-derivatizing "capping" layers.

to that of the vapor deposited parent triarylamine. HTL film thicknesses as assayed by scanning ellipsometry and X-ray reflectivity are in excellent agreement, and indicate that regular build-up of the multilayer architecture (e.g., Figure 6). Based on previous results [4-7], the layer thickness are in accord with the molecular dimensions. Atomic force microscopic examination of these films on ultra-smooth (Balzers) ITO indicates rms roughnesses on the order of ~5 Å for a 45 Å thick film, and ~11 Å for a 130 Å thick film, with no evidence of pinholes or other discontinuities. Cyclic voltammetry using these HTL functionalized specimens as electrodes yield complete blocking behavior for aqueous ferrocyanide, again indicating the absence of pinholes.

Hybrid Self-Assembled + Vapor Deposited OLED

LED devices were fabricated from the aforementioned ITO/HTL structures by vapor deposition of Alq (400 Å) followed by a Mg electrode (1500 Å). Extensive studies of response as a function of HTL layer thickness and fabrication/curing procedure are currently in progress. Figure 7 illustrates the characteristics of preliminary structure with a ~140 Å thick self-assembled HTL. Noteworthy is the low turn-on voltage (~4.5 V), high forward light output

(~2300 cd/m^2), and good external quantum efficiency (0.45%). Preliminary studies also indicate stabilities rivaling or exceeding those of comparable vapor-deposited devices.

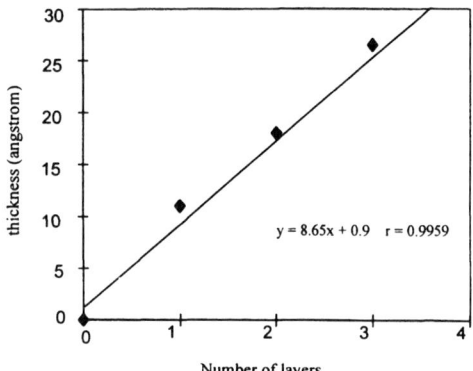

Figure 6. X-ray reflectivity-derived film thicknesses for a self-assembled HTL as a function of the number of layers.

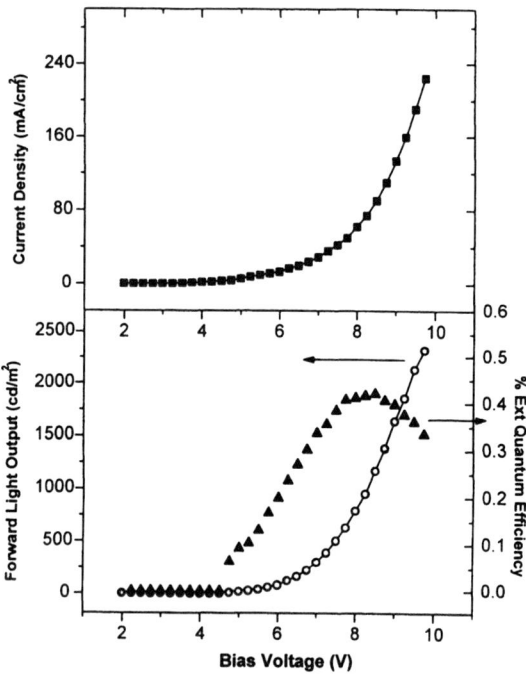

Figure 7. Device characteristics of a hybrid LED fabricated from ITO/140 Å self-assembled hole transport layer/400 Å Alq/1500 Å Mg.

Fully Self-Assembled Two-Layer OLED

LED devices were fabricated by sequential chemisorptive self-assembly of ten layers of the emissive layer precursor molecule shown in Figure 3 upon a self-assembled ITO/HTL substrate structure. A LiF buffer layer (15 Å) [16], and a Mg electrode were then vapor deposited. Detailed studies of electroluminescence and current voltage characteristics as a function of layer dimensions, fabrication/curing, passivation layers, ETL addition, and possible electrode quenching effects are in progress. Figure 8 illustrates the characteristics of one of the first prototype devices which emit *continuous blue light*. It can be seen that forward light outputs are as high as ~150 cd/m² and external quantum efficiencies as high as ~0.0023% with a turn-on voltage of ~6 V. Note also that current densities through this initial prototype structure are substantial.

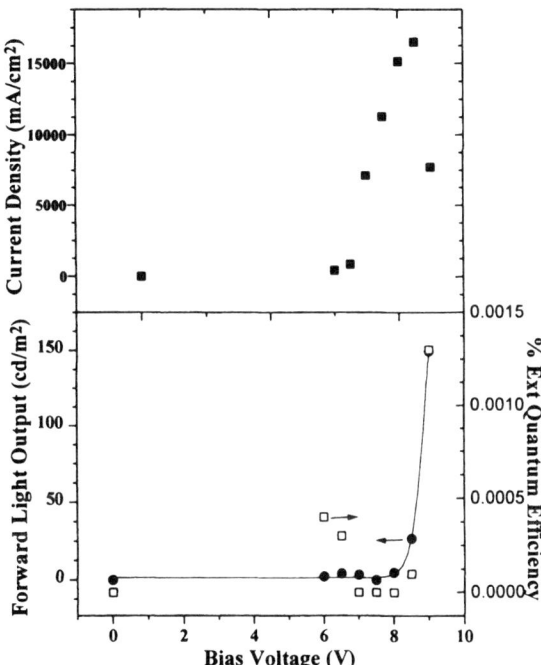

Figure 8. Preliminary performance data from a self-assembled OLED device consisting of ITO/self-assembled HTL/self-assembled EML/LiF/Mg. The line is drawn as a guide to the eye.

CONCLUSIONS

The results of this study indicate that self-limiting siloxane chemisorptive self-assembly approaches can be employed successfully for construction of multilayer OLED structures. Passivating multilayers deposited directly on the ITO surface lead to enhanced external quantum efficiencies at lower voltages without significant diminution of forward light output or current density. Studies of the deposition of a prototype triaryamine hole transport layer precursor on ITO using ellipsometry, X-ray reflectivity, AFM, XPS, and cyclic voltammetry reveal that smooth, contiguous, pinhole-free multilayer structures can be sequentially built up with high dimensional regularity. Vapor deposition of a conventional emissive/electron transport layer and a metal electrode on the self-assembled HTL structure yields stable OLED structures with low turn-on voltages, high light output, and good external quantum efficiencies. Finally, self-assembly of an emissive layer on the ITO/HTL structures, followed by a LiF buffer layer and metal electrode, yields the first totally self-assembled two-layer OLED devices which emit continuous blue light.

ACKNOWLEGMENTS

This research was supported by the ONR through the Center for Advanced Multifunctional Nonlinear Optical Polymers and Molecular Assemblies (CAMP) MURI (N00014-95-1-1319) and by the NSF-MRSEC program through the Northwestern Materials Research Center (DMR-9632472).

REFERENCES

1. T. J. Marks and M. A.Ratner, Angew. Chem. Int. Ed., Engl. **34**, p.155 (1995).
2. S. Yitzchaik, T. J. Marks, Acc. Chem. Res. **29**, p. 197 (1996).
3. J. E. Malinsky, W. Lin, T. J. Marks, W. Lin, and G. K. Wong, unpublished results.
4. W. Lin, W. Lin, J. K. Wong, and T. J. Marks, J. Am. Chem. Soc. **118**, p.8,034 (1996).
5. S. B., Roscoe, A. K. Kakkar, T. J. Marks, A. Malik, M. K. Durbin, W. Lin, G. K. Wong, P. Dutta, Langmuir, **12**, p. 4,218 (1996).
6. A. Malik, W. Lin, M. K. Durbin, T. J. Marks, P. Dutta, J. Chem. Phys. **107**, p. 645 (1997).
7. W. Lin, T.-J. Lee, P. J. Lyman, J. Lee, M. J. Bedzyzk, and T. J. Marks, J. Am. Chem. Soc. **119**, p. 2,205 (1997).
8. A. Dobabalapur and J. Kido, eds. IEEE Trans. Electron Dev. **44** (1997), and references therein.
9. T. Tetsui in A. J. Epstein and Y. Yang, eds. MRS Bulletin, **June**, p. 39-45 (1997), and references therein.
10. Y. Yang in A. J. Epstein and Y. Yang, eds. MRS Bulletin, **June**, p. 31-38 (1997), and references therein.
11. J. R. Sheats, H. Antoniadis, M. Hueschen, W. Leonard, J. Miller, R. Moon, D. Roitman, and A. Stocking, Science, **273**, p. 884 (1996).
12. C. Adachi, K. Nagai, and N. Tamoto, Appl. Phys. Lett. 66, p. 2,679 (1995).
13. P. E. Burrows, V. Bulovic, S. R. Forrest, L. S. Sampochak, D. M. McCarty, and M. E. Thompson, Appl. Phys. Lett. **65**, p. 2,922 (1994).
14. E. M. Han, L.-M. Do, N. Yamamoto, and M. Fujihara, Phys. Lett. 57 (1995).
15. L. M. Do, E. M. Han, Y. Nudorne, M. Fujihara, T. Kanno, S. Yashida, A. Masda, and A. J. Ikushima, J. Appl. Phys. **76**, p. 5,118 (1994).
16. J. E. Jabbour, Y. Kawabe, S. E. Shaheen, J. F. Wang, M. M. Maorrell, B.

Kippelen, and N. Peyghambarian, Appl. Phys. Lett. **71**, p. 1,762 (1997).

17. A. J. Richter, J. E. Malinsky, P. Dutta, and T. J. Marks, unpublished results.

18. C.-I. Chao, K.-R. Chuang, and S.-A. Chen, Appl. Phys. Lett. **69**, p. 2,894 (1996).

19. A. R. Schlatmann, D. Wilms Floet, A. Hilberer, F. Garten, P. J. M. Smulders, T. M. Klapwijk, and G. Hadziiannou, Appl. Phys. Lett. **69**, p. 1,764 (1996).

A SCANNING FORCE MICROSCOPY STUDY OF BLOCK COPOLYMERS CONTAINING A CONJUGATED SEGMENT

Ph. LECLÈRE[1], R. LAZZARONI[1], V. PARENTE[1], B. FRANÇOIS[2], J.L. BRÉDAS[1]
[1]Service de Chimie des Matériaux Nouveaux
Centre de Recherche en Electronique et Photonique Moléculaires
Université de Mons-Hainaut
Place du Parc 20, B-7000 Mons (Belgium)
[2]Laboratoire de Recherche sur les Matériaux Polymères, CNRS/UPPA
Avenue du Président Angot 2, F-64000 Pau (France)

ABSTRACT

Atomic Force Microscopy (AFM) and related techniques are used to investigate the morphology of diblock copolymers. We focus on compounds containing a conjugated segment, polyparaphenylene, associated to a polymethylmethacrylate or a polystyrene block. The influence of the presence of the conjugated segment on the microdomain morphology is analyzed as a function of chain composition. Separate microdomains are observed on the surface of thin films by means of phase-detection imaging tapping-mode AFM. Their shape and size are interpreted in terms of molecular aggregation, with the help of molecular dynamics calculations.

INTRODUCTION

Block copolymers are attractive compounds since they can combine the properties of their constituents, thereby leading to novel materials for technological applications. One well-known example is the case of thermoplastic elastomers, which consist of block copolymers associating one elastomeric segment, such as polybutadiene, and one or more thermoplastic segments, such as polystyrene or polymethylmethacrylate. Below their glass transition temperature, the thermoplastic blocks act as crosslinks in the elastomeric matrix; the material then behaves as a vulcanized rubber. When heated above the melting point of the thermoplastic segments, these compounds can be processed by classical techniques (injection molding, extrusion).

The properties of block copolymers are intimately related to their morphology on the microscopic scale, which in turn is governed by the phase separation between the components and the chemical composition of the chains, *i.e.*, the number and nature of the blocks and their mass ratio. In addition to scattering techniques, the morphology of block copolymers can be determined by Atomic Force Microscopy (AFM) and related techniques; such microscopic methods are especially useful to investigate the surface morphology of these materials, in the form of either thin films [1,2] or even thicker, relatively rough films [3]. Surface morphology can be best observed when the AFM apparatus is operated in Tapping-Mode [TM, 4] and the phase of the oscillation of the cantilever interacting with the surface is measured and compared to the phase of the exciting signal; this mode of operation is referred to as Phase Detection Imaging, PDI. It has been found that the difference in phase is very sensitive to the local mechanical properties of the surface [5,6].

In this contribution, we present preliminary results on the PDI-TMAFM investigation of the morphology of diblock copolymers containing a conjugated segment, typically polyparaphenylene (PPP); the PPP segment is associated to either a polymethylmethacrylate

(PMMA) or a polystyrene (PS) block (Figure 1). The initial motivation for the synthesis of such copolymers was to exploit the presence of the nonconjugated segment to provide solubility in common organic solvents (PPP being notoriously insoluble in any medium). In this work, we aim at using the intrinsic immiscibility of the blocks in the solid state to generate nanostructures of a conjugated material with well-defined shape and size, depending on chain length and mass ratio of the components. Thin films of the block copolymers are prepared from solution, analyzed with PDI-TMAFM, and the images are interpreted with the help of molecular dynamics calculations.

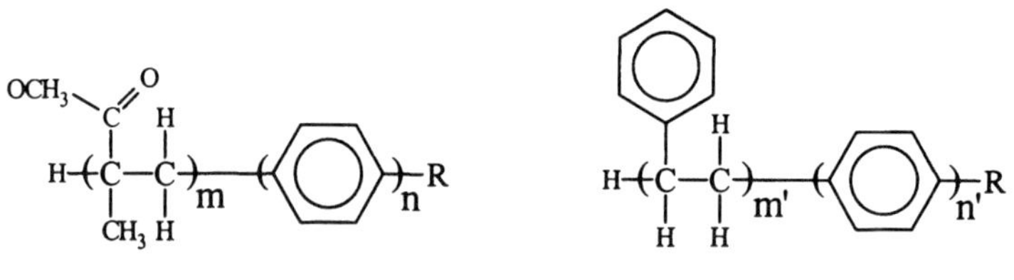

Figure 1: Chemical structure of PMMA-PPP (left) and PS-PPP (right) block copolymers.

EXPERIMENTAL

The PMMA-PPP and PS-PPP block copolymers were prepared from precursor copolymers, PMMA-Poly 1,3-cyclohexadiene (PMMA-PCHD) and PS-PCHD using previously described methods [7]. These precursors were synthetized by anionic polymerization with a lithium counter-ion in a non polar medium. This method allows a good control of the molecular weights and of the 1,4 binding of the resulting cyclohexene units. The PPP block is then obtained by dehydrogenation of the PCHD sequence. These soluble copolymers were characterized by size exclusion chromatography, light and neutron scattering. These techniques show a strong aggregation of copolymers in THF or CS_2 solutions and the formation of rather polydispersed micelles with a PPP hard core surrounded by a shell of PS. This polydispersity makes it difficult to determine the aggregate shapes. Thin films, typically 20 μm thick, were prepared on silicon or mica substrates by solvent casting from toluene solutions containing 1mg/ml of the compounds. After drying in air, the films were annealed at 150°C for 48 hours in vacuum (10^{-7} T).

All AFM images were recorded with a Nanoscope III microscope (Digital Instruments) operated at room temperature in air, using the microfabricated cantilevers provided by the manufacturer (spring constant of 30 Nm^{-1}). The system is equipped with the Extender electronics module to provide height and phase cartography. Images of each sample were taken at several locations, and the time for scanning was about 5 min. All images were obtained with the maximum number of pixels (512) in each direction. For image analysis, the Nanoscope image processing software was used. The images presented here were not filtered and are shown as captured. Repeated scans indicated that the observed structures were stable.

RESULTS AND DISCUSSION

As a first step into microscopic analysis, the surface smoothness of the area under study is checked using tapping-mode operation. For PDI measurements, it is best that the surface roughness be as small as possible, since topographic contrast usually makes the interpretation of phase images more complicated. Figure 2 (top) shows a typical 1x1 μm² topographic image for a PMMA-PPP copolymer, with molecular masses of the blocks on the order of 8000 and 2000 daltons, respectively (thus corresponding to a 80:20 mass ratio).

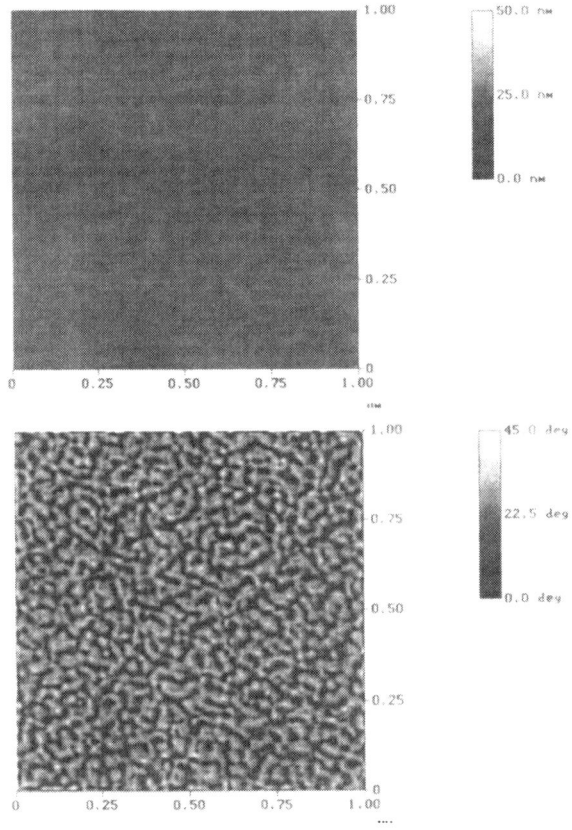

Figure 2: 1x1 μm² height image (top) and phase image (bottom)
of a 80:20 PMMA-PPP block copolymer.

The surface is smooth, with a root-mean square roughness of a few nm over a 1x1 μm² area. These topographic data, however, do not provide any information on the microdomain morphology. The information content of the phase-detection image (Fig. 2, bottom) is much higher; this image, obtained simultaneously to and on the same area as the topographic data of the top image in Figure 2, is made of a distribution of dark, elongated objects in a brighter matrix. The origin of the contrast in PDI-TMFAM is not yet fully understood; nevertheless the models proposed so far [5,6] all point to the local mechanical characteristics of the surface as the primary origin for the dephasing of the cantilever oscillation. Since PMMA (or PS) and PPP possess rather different mechanical properties and knowing the mass ratio of the two components, it is reasonable to assign the contrast in Figure 2 (bottom) to phase-separated microdomains of PMMA and PPP segments, the former constituting the bright matrix and the latter forming the dark, elongated structures. While the length and orientation of these dark domains appear to be rather variable, their width is nearly constant, around 12 nm (as measured directly on the images), over the whole surface.

The dimensions of the bright domains of the matrix are not as easy to determine since their shapes are not regular. We have thus performed a power spectral density analysis of the PDI images; it provides a typical correlation length, defined by the intersection between the plateau and the linear part of the curve, of $\approx$ 35 nm. In our opinion, this value corresponds to the average distance between the centers of domains of the same nature, *i.e.*, between the centers of adjacent PPP domains, and between the centers of adjacent PMMA domains. It thus represents the sum of typical sizes of PPP and PMMA domains (as two adjacent PPP domains are often separated by one PMMA domain, and vice versa). On this basis, we estimate a characteristic size of $\approx$ 23 (*i.e.*, 35 - 12) nm for the PMMA domains. Since the correlation between dark domains is expected to be strongest in areas where the elongated objects are parallel to each other, the 23 nm value likely represents the average width of the PMMA domains, *i.e.*, their shortest dimension.

In the 80:20 composition range, the morphology of block copolymers normally consists of cylinders of the minor component into a matrix of the major component. At first sight, the phase image of Figure 2 is consistent with such a morphology. However, the copolymers studied here are made of a very rigid segment (PPP) associated to a more flexible segment (PMMA or PS). In that sense, they are different from the "coil-coil" diblock copolymers, for which the "cylinders in a matrix" morphology is well known. They are more closely related to the "rod-coil" copolymers which have been described recently [8,9]. These compounds present particular morphologies as well as shifts in the compositions for which transitions between ordered phases (*i.e.*, spheres to cylinders to lamellae) occur [10]. Theoretical investigations indicate that the difference in stiffness between the components (the stiffness asymmetry) strongly influences the molecular aggregation and the phase behavior [11,12].

In our case, the PPP segments tend to closely pack in a parallel fashion. This kind of packing is most unlikely to produce elongated cylinders. Instead, it would rather result in lamellae or stripes. The width of the PPP domains measured on the AFM images should therefore rather correspond to the actual width of a lamella or a stripe than the diameter of a cylinder. At the present stage, we have no experimental indication favoring one or the other possibility; note, however, that: (i) the presence of stripes has been recently demonstrated in "rod-coil" copolymers by electron tomography [9]; and (ii) the composition of our sample (80:20) is quite far from the threshold for lamellae formation in classical block copolymers (64:36).

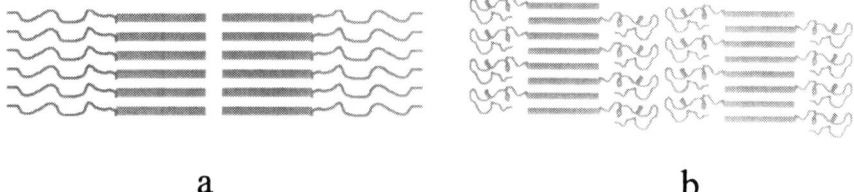

a   b

Figure 3: Schematic representation of the molecular packing in PMMA-PPP block copolymers: the PPP segments form a bilayer structure (a) or interdigitate (b).

A possible origin for the observed morphology, be it lamellae or stripes, would be the formation of a bilayer structure, as sketched in Figure 3a. However, in this case, the coil segments (here atactic PMMA) would have to pack as closely as the PPP chains; that would require full stretching of the PMMA chains, which is not favorable entropically. Moreover, such a morphology is in clear contradiction with the AFM data of Figure 2. According to the scheme of Figure 3a, the width of the PPP domains should be approximately twice the length of the PPP segments. The molecular weight of the PPP blocks corresponds to an average of 26 phenylene units per chain, hence a length of 11.2 nm. The measured width of the PPP domains ($\approx$12 nm) is very close to that value. Therefore, there is only one PPP chain, and not two, across the PPP domains. This then suggests that the PPP segments are actually interdigitated, as is illustrated in Fig. 3b, a packing configuration which has furthermore the advantage of leaving more space for the PMMA segments to relax into coils.

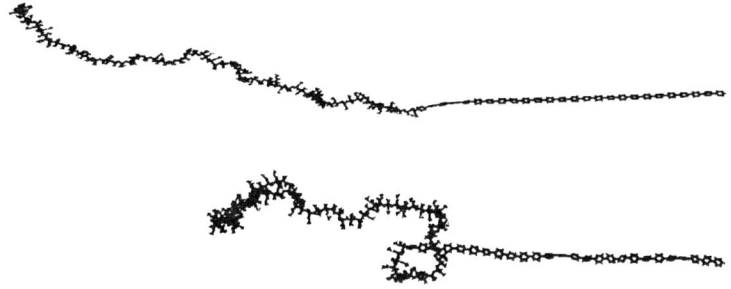

Figure 4: Conformation of a $(MMA)_{80}$-$(PP)_{26}$ chain, as obtained with molecular dynamics simulations; the PMMA segment, on the left part of the chain, is extended (top) or coiled (bottom) while the PPP segment, on the right part of the chain, is fully rigid.

In order to estimate the dimensions of the PMMA segments, we have performed canonical molecular dynamics simulations (Cerius2 program with Universal Force Field) on a single $(MMA)_{80}$-$(PP)_{26}$ chain, corresponding to the copolymer studied with AFM. Considering a single chain allows the PMMA segment to reach its most coiled conformation; the other extreme case is the fully extended chain. Both situations are depicted in Figure 4. In the fully extended conformation, the length of the PMMA segments is about 15 nm. Since the PMMA domains in Figure 3b are two-chain wide, such a length is inconsistent with the measured width; therefore, the PMMA chains cannot be fully extended. Neither can they be fully coiled, since the length of a coiled chain is only 6.7 nm. The PMMA segments are thus thought to be partially coiled to an average length of about 12 nm (23:2).

We have also investigated copolymers with lower PPP contents, for instance PS-PPP with molecular masses of 30000 and 3000, respectively, corresponding to a 91:9 mass ratio. In this composition range, classical block copolymers exhibit a "spheres in a matrix" morphology. However, as in the case of the 80:20 compounds, we find that the PDI-TMAFM image is mostly composed of dark, elongated objects, corresponding to PPP domains, in a lighter PS matrix. The absence of spheres is to be related to the compact parallel packing of the PPP segments, which cannot generate such a morphology. It has been shown [9,13] that the decrease in rod content in "rod-coil" copolymers leads to the breakup of stripes into shorter structures or even disk-shaped structures ("hockey puck" morphology). We therefore believe that the dark objects observed for the 91:9 PS-PPP copolymer consist of short stripes. Measurements of their width yield a value of 15 nm, which roughly corresponds to the length of a PPP segment of mass 3000 ($\approx$ 39 phenylene units = 16.7 nm). This suggests that the arrangement of the chains at this surface is similar to that of Figure 3b, *i.e.*, interdigitated PPP chains allowing the PS segments to coil. The results of the molecular dynamics simulations on this system are consistent with significant coiling of the PS chains.

CONCLUSIONS

We have used phase-detection imaging in tapping-mode AFM to investigate the morphology of block copolymers containing one conjugated PPP segment. The stiffness of the conjugated backbone induces parallel packing of the PPP chains, which in turn strongly influences the microdomain morphology. For compositions ranging from 10 to 20 mass % in PPP, the AFM data are consistent with the formation of short stripes or lamellae of PPP, one PPP chain in width, surrounded with partially coiled PMMA (or PS) segments.

Acknowledgments

We are grateful to Ph. Dubois and S.I. Stupp for stimulating discussions. Research on conjugated materials in Mons is supported by the Belgian Federal Government Office of Science Policy (SSTC) "Pôle d'Attraction Interuniversitaire en Chimie Supramoléculaire et Catalyse Supramoléculaire" (PAI 4/11), the Government of the Région Wallonne and the European Commission (project NOMAPOL - Objectif 1 - Hainaut), FNRS-FRFC, and an IBM Academic Joint Study. The Mons-Pau collaboration is supported by the Training and Mobility of Researchers programme of the European Commission (network SELOA). RL is Maître de Recherches du Fonds National de la Recherche Scientifique (FNRS).

REFERENCES
1. B. Collin, D. Chatenay, G. Coulon, D. Aussere, Y. Gallot, Macromolecules 25, 1621 (1992).
2. M.A. Van Dijck, R. Van den Berg, Macromolecules 28, 6773 (1995).
3. Ph. Leclère, R. Lazzaroni, J.L. Brédas, J.M. Yu, Ph. Dubois, R. Jérôme, Langmuir 12, 4317 (1996).
4. Q. Zhong, D. Inniss, K. Kjoller, V.B. Elings, Surf. Sci. 290, L688 (1993).
5. S.N. Magonov, V.B. Elings, M.H. Whangbo, Surf. Sci., 375, L385 (1997).
6. G. Bar, Y. Thomann, R. Brandsch, H.J. Cantow, M.H. Whangbo, Langmuir 13, 3807 (1997).
7. B. François, X.F. Zhong, Synth. Met. 41-43, 955 (1991). G. Widawski, M. Rawiso, B. François, J. Chim. Phys. 89, 1331 (1992). B. François, G. Widawski, M. Rawiso, B. César, Synth. Met. 69, 463 (1995).
8. R.S. Saunders, R.E. Cohen, R.R. Schrock, Macromolecules 24, 5599 (1991).
9. Radzilowski, S.I. Stupp, Macromolecules 27, 7747 (1994). L.H. Radzilowski, B.O. Carragher, S.I. Stupp, Macromolecules 30, 2110 (1997).
10. F.S. Bates, M.F. Schulz, J.H. Rosendale, Macromolecules 25, 5547 (1992).
11. A. Halperin, Macromolecules 23, 2724 (1990).
12. K.S. Schweizer, Macromolecules 26, 6050 (1993).
13. D.R.M. Williams, G.H. Fredriksson, Macromolecules 25, 3561 (1992).

MULTILAYER SELF-ASSEMBLIES AS ELECTRONIC AND OPTICAL MATERIALS

DeQuan Li, M. Lütt, Xiaobo Shi, and M. R. Fitzsimmons

Chemical Science and Technology Division (CST-4) and Manuel Lujan Jr. Neutron Scattering Center, Los Alamos National Laboratory, Los Alamos, NM 87545

ABSTRACT

The layer-by-layer growth of film structures consisting of sequential depositions of oppositely charged polymers and macrocycles (ring-shaped molecules) have been constructed using molecular self-assembly techniques. These self-assembled thin films were characterized with X-ray reflectometry, which yielded (1) the average electron density, (2) the average thicknesses, and (3) the roughness of the growth surface of the self-assembled multilayer of macrocycles and polymers. These observations suggest that inorganic-organic interactions play an important role during the initial stages of thin-film growth, but less so as the thin film becomes thicker. Optical absorption techniques were also used to characterize the self-assembled multilayers. Phorphyrin and phthalocyanine derivatives were chosen as the building blocks of the self-assembled multilayers because of their interesting optical properties.

INTRODUCTION

The sequential deposition of multilayers has attracted recent interest because it allows the fabrication of artificial materials with unusual properties[1]. Here, self-assembly techniques are used to produce material in a layer-by-layer manner, as opposed, for example, bulk crystal growth. However, the growth of so-called engineered or functional multilayers is extremely challenging because many parameters of the multilayer interfaces must be optimized in order to fabricate mechanically and chemically robust structures. For example, strains across components in the multilayer should be minimized, while chemical affinity across these layers should be maximized in order to promote structural integrity and adhesion. These issues are very complex, and often investigations focus on the growth of monolayers rather than multilayers.[2]

Recently, Decher[3] used charge interactions to assemble multilayers of polycations and polyanions. Others[4a,b] have adopted this method because it is an extremely simple and versatile technique to construct multilayer structures. Unlike covalently bound self-assembled mono- and multilayers,[1] charge interactions are very weak in water. In order to make reasonably robust films using charge interactions, previous work involved the fabrication of multilayer films using polymers containing many charged groups. More recently, Charged phthalocyanines and porphines were used to form multilayer assemblies alternatively with oppositely charged polymer ions.[5,6]

An important issue concerning the fabrication of multilayers using charge interactions is: what is the minimum number of charges needed for a molecule to self-assemble into a thin film? If this number, typically ≥ 200 for polymers, can be reduced to a very small number, e.g., less than six, then multilayers could be grown from a larger material base than presently possible using charged polymers.

In order to form a self-assembled film, the charge interactions which bind the film together must overcome forces that may dissociate the film. For example, the dissociation energy due to thermal motion of molecules is about 0.6 kcal/mole, while a single non-selective (random) secondary interaction in water, such as a charge-charge attraction or hydrogen bonding, are about 5 kcal/mole. Based on this consideration, a molecule with four charges which act in concert to hold the film together, should have enough bonding affinity to overcome non-selective interactions. In this paper, we describe a multilayer molecular self-assembly with alternating macrocycles containing four charges and positively charged polymers. The macrocycles used are α,β,γ,δ-tetrakis(5-sulfothienyl)-porphine (TSTP) and nickel phthalocyanine-tetrasulfonic acid (NiPc), tetrasodium salt; and the polymer is poly(diallyldimethylammonium chloride) (PDDA) (Fig. 1).

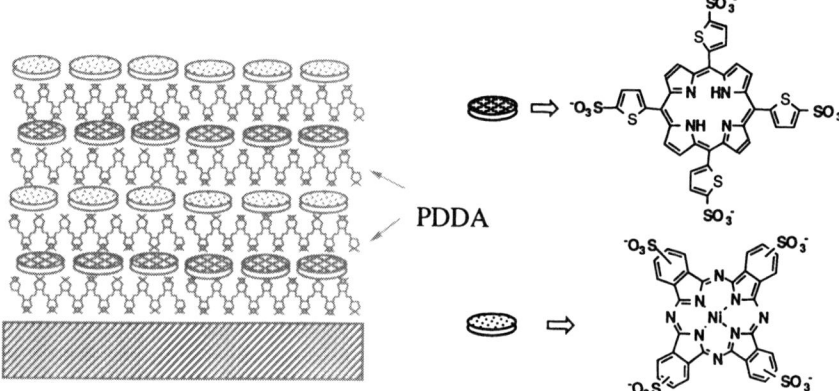

Figure 1. Structures of α,β,γ,δ-tetrakis(5-sulfothienyl)-porphine (TSTP) and nickel phthalocyanine-tetrasulfonate (NiPc), and their self-assembly of PDDA/NiPc/PDDA/TSTP.

The purpose of this study is to determine whether a macrocycle with four charges can be used as a building block to produce a self-assembled multilayer with optical properties. In order to gauge the success of thin-film formation, small-angle X-ray reflectometry is used to characterize the layered structure of this self-assembled system. In particular, X-ray reflectometry provides information about the electron density of the multilayer, its thickness, and the roughness of the growth interface.[7,8]

RESULTS AND DISCUSSION

In the process of fabricating multilayer assemblies through layer-by-layer growth pattern, PDDA (polycation) will attract NiPc or TSTP (anions) molecules, which results in the growth of alternating PDDA/NiPc or PDDA/TSTP multilayer. Multilayer growth is impossible with either component alone because polyions can not stack on top of themselves without multivalent counterions. The multilayer thin films are a result of sequential reactions between PDDA and NiPc or TSTP on the growth surface, which yields alternating layered structures of PDDA and NiPc or TSP. We do not expect that there is ordering or crystalline within each layer because of the nature of organic polymers. Considering long polymeric chains, a mixture formation through interpenetration and diffusion of the macrocycle molecules between layers is unfavorable while compared to layered structures.

Figure 2 shows the optical spectra of multilayer thin-film growth from alternating dipping from negatively charged macrocycles and positively charged polymer PDDA. For PDDA/NiPc system, we observed strong Q bands at 630 nm and 665 nm. For the PDDA/TSTP system, the strong phorphyrin Soret band at ~430 nm was observed. These characteristic bands are ideal for monitoring the growth of multilayer thin films. Complex systems such as PDDA/NiPc/PDDA/TSTP can be obtained while replacing one of the macrocycle with another in every other deposition sequence. This flexibility at nanoscale of thin-film growth will allow incorporation of molecules with unusual properties such as charge transfer and nonlinear optical response.

Figure 3 illustrates the successful growth of the multilayer in a layer-by-layer manner. Optical absorption spectrum shows that the amount of materials deposited is essentially the same for each layer. Moreover, X-ray reflectivity measurements confirm that the amount of materials deposited for each bilayer corresponds a certain increase of the film thickness. These prove the layered structure in the PDDA and NiPc system.

In order to obtain information about the structure of the multilayer, e.g., its electron density, ρ_e, thickness, Δ, and surface roughness, σ, reflectivity profiles of simple models were calculated and compared to the observed data. For the uncoated substrate (n=0), the model consisted of a Si substrate with electron density ρ_{Si}, terminated by an overlayer of SiO_2 with electron density ρ_{SiO_2} and thickness Δ_0. The roughness of the air-SiO_2 and SiO_2-Si interfaces were represented as error functions[9] whose derivatives— Gaussian peaks, have root-mean-square widths of σ_{Si} and σ_{SiO_2}, respectively. The X-ray reflectivity of the electron density depth profile, $\rho(z)$, was calculated using the Parratt formalism.[7] The structural parameters of the model, i.e., ρ, Δ, and σ, from which $\rho(z)$ is determined, were perturbed (adjusted) so as to minimize χ^2— the sum of the square differences between the calculated and fitted profiles weighted by the variance of the measurement.[9] The curvature of χ^2, as the adjustable parameters were independently perturbed about their optimal values, was used to deduce the degree of certainty (within 1-σ) to which a parameter was known.[9]

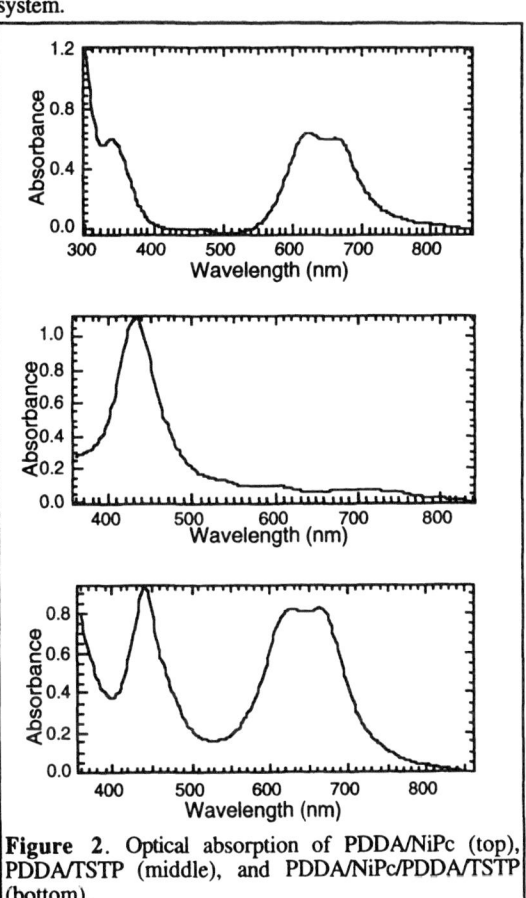

Figure 2. Optical absorption of PDDA/NiPc (top), PDDA/TSTP (middle), and PDDA/NiPc/PDDA/TSTP (bottom).

The solid curve in the upper profile of Fig. 4 is the profile calculated for the best fitting model. The electron density used for the Si substrate was that in the literature (0.71 e⁻'s/Å^3).[10] The electron density refined for the SiO_2 overlayer was 0.49±0.1 e⁻'s/Å^3. This value is in good agreement with that obtained from ref. 10. The thickness of the SiO_2 overlayer was determined to be 12±1Å, which is also in good agreement with pervious work.[10-12] The roughness of the SiO_2-

403

Si interface was determined to be 3±1Å, while the roughness of the air-SiO$_2$ interface was somewhat larger— 5±1Å. Since the deposition of PDDA onto the substrate is not expected to alter the roughness of the substrate, the electron density profile representing the uncoated substrate served as the foundation for later models used to fit the reflectivity profiles taken of the multilayer.

The model for the coated substrate consisted of a layer with uniform electron density, ρ_N, thickness Δ_N, and air-sample roughness σ_N, on top of the model for the sample substrate discussed earlier. By choosing to represent the multilayer as a material with uniform electron density, variations in the electron density across the PDDA-NiPc bilayer were ignored. This variation, which is expected to be on the order of 10%, occurs over a very small distance— a distance, $2\pi/Q_{max}$, much smaller than the thinnest film that can be detected when X-ray data are collected to $Q_{max} \sim 0.27$.[13] The

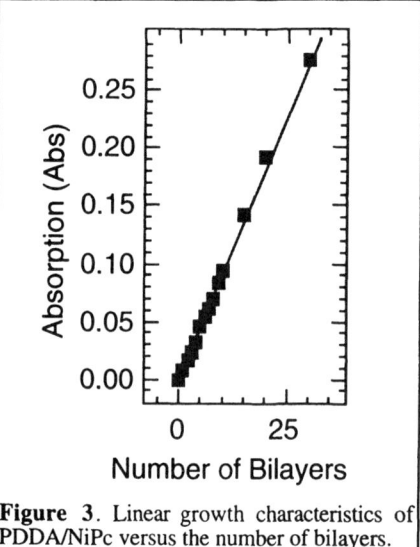

Figure 3. Linear growth characteristics of PDDA/NiPc versus the number of bilayers.

parameters, ρ_N, Δ_N, and σ_N, were adjusted to produce calculated profiles (solid curves in Fig. 4) that fitted the data very well.

To the extent the models adequately represent the structure of the multilayer, three conclusions can be drawn from the experiment. First, the electron densities of the bilayers did not change during the fabrication of the multilayer. In other words, the electron density of the first deposited bilayer was not significantly different than that of any other bilayer. Since the chemical composition of a bilayer is known on the basis of charge neutrality to be one unit NiPc, $\{Ni[C_8H_3N_2SO_3]_4\}^{4-}$, to four monomeric units of PDDA, $[C_8H_{16}N]^+$, and using the mean of the electron densities, the mass density of a bilayer is computed to be 1.16±0.06 g/cm^3.

The second conclusion drawn is that the average thickness of the material deposited during the first two complete dipping cycles was large (about 20Å thick), decreased sharply with subsequent depositions, and finally approached a value of about 8Å after the deposition of about 40 bilayers. The sharp decrease in the mean per layer thickness of multilayer occurring after two dippings suggests that the inorganic substrate may play a role in the initial growth of the multilayer. As the PDDA polymer and NiPc macrocycle were deposited on the silicon substrate, a new inorganic-organic interface was created. The polymer might reorganize at this interface to direct the charged moieties onto the inorganic surface and minimize exposure of the hydrophobic chain segments to the hydrophilic surface. Therefore, the first few layers could contain many polymers that were twisted and turned to produce thicker films. As the film grows thicker, the polymers might have taken on more relaxed and extended structures because of the absence of inorganic hydrophilic interactions with the substrate. This may have led to a dramatic decrease of the relative bilayer-thickness for the first few layers.

The third conclusion is that the roughness of the air-sample interface or growth surface increased with each dipping. The roughness after the first deposition ($\sigma = 14$Å) was larger than the roughness of the air-SiO$_2$ interface ($\sigma = 5$Å). The roughness of the growth surface generally increased with each deposition. Similar increases in the roughness of polymer-multilayers have been observed previously by Kellogg *et al.*[14] They attribute the increase to an accumulation of defects which may occur during deposition of the polymer film. In the PDDA-NiPc multilayer system, tilting of the NiPc macrocycle might produce defects that can lead to an increase in the roughness of the sample surface. To some extent, the PDDA polymer may "repair" or form bridges over defects, since the PDDA molecule is a long polymer chain compared to the NiPc macrocycle.

CONCLUSION

The growth of self-assembled multilayers from macrocycles and macromolecules was monitored with X-ray reflectometry and optical absorption. The X-ray measurements, were used to determine the electron density and thickness of the multilayer averaged over its lateral dimensions, and the roughness of the growth surface after different steps in the fabrication process. The electron density of a bilayer was found to be the same regardless of when the bilayer was deposited onto the sample. The average thickness of the first-deposited bilayers was large and decreased quickly to a value that did not change significantly after 40 bilayer-depositions. Furthermore, the roughness of the growth surface was found to increase with each deposition.

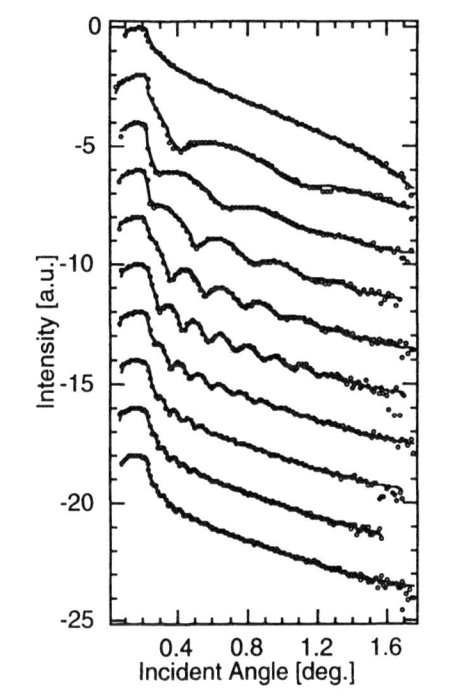

Figure 4. Reflectivity profiles (o) of the Si-substrate terminated by SiO_2 (n=0) and the multilayer sample after differing numbers of PDDA-NiPc bilayer depositions.

Optical absorption measurements show that there is an approximately linear relationship between the absorbance (the density of macrocycles) with the number of bilayers.

ACKNOWLEDGMENTS

We acknowledge the support of Los Alamos National Laboratory LDRD and U.S. Department of Energy, BES-DMS, under Contract No. W-7405-Eng-36

REFERENCES

1. Li, D. Q.; Ratner, M. A.; Marks, T. J.; Zhang, C. H.; Yang, J.; Wong, G. K. *J. Am. Chem. Soc.* **1990**, *112*, 7389.
2. Labinis, P. E.; Hickman, J. J.; Wrighton, M. S.; Whiteside, G. M. *Sci.* **1989**, *245*, 845.
3. Decher, G. Hong, J. D. Schmitt, J. *Thin Solid Films,* **1992**, *210*, 831-835.
4. (a) Cheung, J. H.; Fou, A. F.; Rubner, M. F. *Thin Solid Films* **1994**, *244*, 985.
   (b) Yoo, D.; Lee, J-K.; Rubner, M. F. *Mat. Res. Soc. Symo. Proc.*, **1996**, *413*, 395.
5. Lutt, M.; Fitzsimmons, M.; Li, D. Q. *J. Phys. Chem. B*, **1998**, *102* , 400 -405.
6. Ariga, K.; Lvov, Y.; Kunitake, T. *J. Am. Chem. Soc.* **1997**, *119,* 2224.
7. Parratt, L.G. *Phys Rev.* 1954, *95*, 359.
8. Lekner, J. *Theory of Reflection*, Nijhoff, Dordrecht, 1987.

9. Press, W.H.; Flannery, B.P.; Teukolsky S.A.; Vetterling, W.T. *Numerical Recipes*, Cambridge University Press, Cambridge, 1986.

10. Bruegemann, L.; Bloch, R.; Press, W.; Gerlach, P. *J Phys.* C **1990**, *2*, 8869.

11. Fuoss, P.H.; Norton, L.J.; Brennan, S.; Fischer-Colbrie, A. *Phys. Rev. Lett.* **1988**,*60*, 600.

12. Nitz, V.; Tolan, M.; Schlomka, J.-P.; Seeck, O. H.; Stettner, J.; Press, W.; Stelzle, M.; Sackmann, E. *Phys. Rev.* B **1996**, *54*, 5049.

13. Data covering a larger range of Q_z could be taken but only with the aid of a synchrotron source. Such measurements are planned for the future.

14. Kellogg, G. J. Mayes, A. M. Stockton, W. B. Ferreira, M. Rubner, M. F.; Satija, S. K. *Langmuir*, **1996**, *12* ,5109.

PERFORMANCES OF SEXITHIOPHENE BASED THIN-FILM TRANSISTOR USING SELF-ASSEMBLED MONOLAYERS

J. Collet*, O. Tharaud**, C. Legrand**, A. Chapoton**, D. Vuillaume*
* IEMN-CNRS, Dept of Physics (ISEN), BP69, F-59652 cedex, Villeneuve d'Ascq, France, vuillaume@isen.iemn.univ-lille1.fr
** IEMN-CNRS, University of Lille, Dept. Hyperfréquences et Semiconducteurs, BP69, F-59652 cedex, Villeneuve d'Ascq, France.

ABSTRACT

High performance thin-film transistors (TFT) made of conducting oligomers are obtained when the organic films are well ordered at a molecular level. Highly ordered films are obtained provided that oligomers have a sufficient mobility on the substrate surface during film formation. One possible way to fulfill such a condition is to evaporate oligomers on heated substrates [1,2]. In this work, we suggest that a high surface mobility is obtained by a chemical functionalization of the silicon dioxide surface, and the corresponding improvements of the TFT performances are evidenced. A self-assembled monolayer of octadecyltrichlorosilane (OTS) was deposited on the SiO_2 by chemisorption from solution before the evaporation of sexithiophene film. Room temperature current-voltage measurements indicate that the presence of the OTS monolayer improves TFT performances : threshold voltage is decreased, subthreshold slope is decreased, a high current ratio I_{on}/I_{off} is obtained for a reduced gate voltage excursion, the field-effect mobility is slightly increased. We have also fabricated and characterized a nanometer scale organic FET (gate length = 50 nm) made of 6T films and only with a self-assembled monolayer as the insulating film between the degenerated silicon substrate (gate) and the conducting channel (no thick SiO_2, we call it « oxide-free » organic FET). Performances of this nanometer size organic FETs are the following : subthreshold slope of 0.35V/dec, threshold voltage of -1.3V, effective mobility of 2×10^{-4} $cm^2/V.s$.

EXPERIMENT

Device fabrication

A self-assembled monolayer of octadecyltrichlorosilane (OTS, $SiCl_3(CH_2)_{17}CH_3$) was deposited on the SiO_2 (250 nm thermally grown oxide on a heavily doped silicon) by chemisorption from solution. This OTS monolayer reduces the surface tension (contact angle measurements) from 78 mN/m (untreated SiO_2) to 20.5-21 mN/m (silanized SiO_2 surface). Provided that simple thermodynamic rules are satisfied during the self-assembly of the monolayers [3], closely packed OTS monolayers are obtained (this was checked here by FTIR and ellipsometry). Then, a 50 nm thick layer of sexithiophene was deposited by thermal evaporation onto the treated substrate that was held at 80 °C. These devices were made to test the film-substrate interface effects, and no attempt was made to optimize the FET performances. The 6T was used as received (without purification that has been shown to influence the electrical performances [2,4]) and the temperature deposition was not varied to optimize the oligomer orientation into the film [1]. Such anneal treatment is not detrimental to the monolayer since OTS monolayers can sustain temperatures up to 350°C in an inert ambient or in vacuum [5].

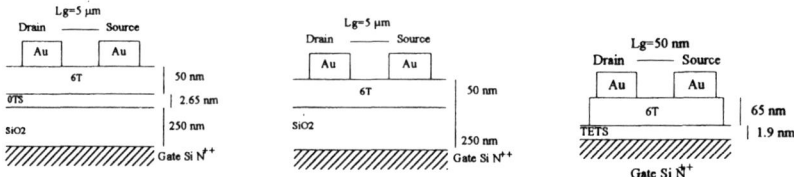

Figure 1 : Schematic structures of the three studied devices

Devices (Fig. 1) were completed by evaporating an 80 nm thick gold layer to form source and drain contacts (L=5μm, W=100μm). For comparison, similar devices without the OTS monolayers were made simultaneously (Fig. 1). Source and drain electrodes were patterned using optical lithography and lift-off techniques. This is made possible because 6T film is not soluble in current solvents (acetone) used to reveal the photoresist. We carefully checked that this source/drain process was not detrimental for the 6T films by comparing large devices (L=90 μm) made by the above process and using shadow masks to define the source and drain electrodes. Similar FET performances were obtained with both processes.

We have also fabricated and characterized a nanometer scale organic FET (Fig. 1, channel length = 50 nm) made of 6T films and only with a SAM as the insulating film between the degenerated silicon substrate (gate) and the conducting channel (no thick SiO_2). We call this device an « oxide-free » organic FET, neglecting the native oxide (ca. 1 nm thick) mandatory to self-assemble the tricholorosilane head groups onto the silicon substrate. It is well known that such a thin oxide is very leaky. We started by the self-assembly of a TETS (tetradecylenyltrichlorosilane, $SiCl_3(CH_2)_{12}CH=CH_2$) monolayer on the Si substrate. Then, we oxidized the monolayer surface according to chemical reaction described elsewhere [6] to transform the vinyl-terminated SAM in a COOH-terminated one. The 6T film (45 nm thick) was then evaporated on this functionalized substrate heated at 150 °C. Source and drain contacts (gold) were fabricated by e-beam lithography and lift-off techniques. This device was successfully fabricated and operated since we have previously demonstrated that a densely packed self-assembled monolayer of n-alkyltrichlorosilanes is a very efficient insulating barrier [7,8]. A high tunneling barrier (4.5 eV) at the Si/SAM interface reduces the tunneling current at a negligible level. For instance, a 1.9 nm thick SAM of dodecyltrichlorosilanes exhibits a reduced leakage current (10^{-8} A/cm^2 at 4 MV/cm) on a par with that of much thicker (> 4 nm) SiO_2 films [8]. The smallest channel length obtained by this process was 30 nm as shown on the scanning electron microscope (SEM) picture (Fig. 2). In this second device, we used COOH-terminated SAM's instead of OTS because the 6T film was fabricated by e-beam lithography and lift-off to pattern a reduced area of active organic semiconductor and to avoid excessive leakage current outside the current channel. It is not possible to spin coat photoresist on an OTS monolayer since its surface tension is too low (20.5 mN/m). COOH-terminated SAM exhibits higher surface tension (47mN/m, measured by contact angle techniques, Owens-Wendt method) due to the presence of carboxylic end groups, and thus allows PMMA to be spin coated.

We have demonstrated elsewhere [9] that TETS, COOH-terminated and OTS monolayers have similar molecular architecture and that they all exhibit very low leakage current mandatory to their use as molecularly thin insulating films. For instance, FTIR reveal that all SAM's have CH_2 stretching modes at 2850 ± 0.5 cm^{-1} (symmetric) and 2918 ± 0.5 cm^{-1} (asymmetric). These

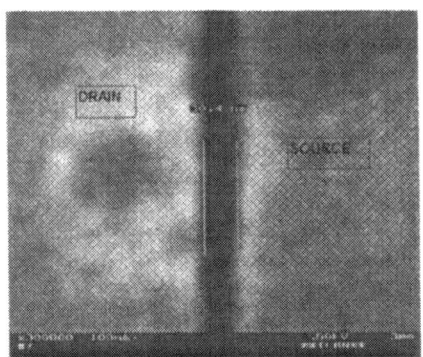

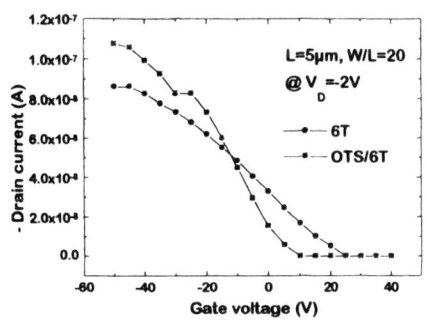

Figure 2 : SEM picture of a 30 nm channel « oxide-free » organic FET made of 6T and a COOH-terminated SAM as the gate insulator.

Figure 3: Comparison of the I_D-V_G curves of the SiO_2/OTS/6T and SiO_2/6T FET's.

values are recognized as the fingerprint of the presence of all-trans extended alkyl chains nearly perpendicular to the substrate as expected for a densely packed monolayers [3]. Thicknessmeasurements by ellipsometry and X-ray reflectivity are in agreement with such a conclusion [9]. Values (from ellipsometry/X-ray reflectivity/theory, respectively) of 2.65/x/2.62 nm, for OTS, 2.1/1.9/2.1 nm for TETS monolayers and 2.4/x/2.1 for COOH-terminated SAM's (x means not measured) well correspond to the length of the alkyl chain in its all-trans conformation, the chain being nearly perpendicular to the surface. This indicates a surface coverage by the monolayers nearly equal to 100%. For all monolayers, leakage currents at 4 MV/cm are in the 10^{-8}-10^{-7} A/cm^2 range and tunneling energy barrier at the SAM/silicon interface are 4-4.5 eV that reduces the tunneling current through the monolayers at a negligible level [7-9].

RESULTS

SiO_2/OTS/6T vs. SiO_2/6T transistors

Figures 3-5 show typical I_D-V_G and I_D-V_D characteristics of FET with and without the OTS monolayers. The main improvements with the OTS monolayer are summarized in table I. The threshold voltage decreases from 22 V to 8 V, the subthreshold slope decreases from 14 V/decade to 8 V/decade, a current ratio I_{on}/I_{off} of 10^6 is obtained for a gate voltage excursion of 75 V instead of 115 V and the field-effect mobility is slightly increased from 3.6×10^{-3} cm^2/V.s to 4.5×10^{-3} cm^2/V.s (values obtained from both I_D-V_G in linear regime and $I_D^{1/2}$-V_G in saturated regime).

Nanometer scale, « oxide-free » organic FET, SAM/6T

We carried out a study of the FET characteristics as a function of the channel length. Figures 6 and 7 show the typical I_D-V_D characteristics for a long channel (L=500 nm) and a short channel (L=50 nm) FET. It clearly appears that an usual field-effect behavior is not obtained for FET's with channel length below 200 nm. The typical performances obtained for FET's down to

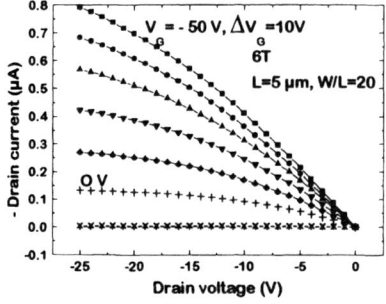

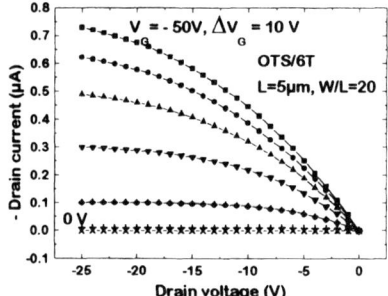

Figure 4 : Typical I_D-V_D characteristics of a SiO$_2$/6T transistor.

Figure 5 : Typical I_D-V_D characteristics of a SiO$_2$/OTS/6T transistor.

a channel length of 200 nm are : threshold voltage -1.3 V (normally off), subthreshold slope 0.35V/decade, $I_{on}/I_{off}=10^4$ (without correction of the ohmic current between source and drain, and of the leakage current through the total source and drain pads that are not isolated from the 6T film by a thick insulator), field-effect mobility 2×10^{-4} cm^2/V.s (for L=500 nm, V_D = -0.5 V).

DISCUSSION

SiO$_2$/OTS/6T vs. SiO$_2$/6T transistors

The presence of the OTS monolayer on the SiO$_2$ surface has increased the device performances by 25-63 % depending on the analyzed parameter (Table I). The same kind of improvements has been reported for pentacene based FET [2].We surmise that these improved performances are due to a better molecular order in the sexithiophene layer as elucidated by previous comparison between the performances of FET made of 6T and alkyl substituted α-6T [1]. Here, a low surface tension of the OTS treated SiO$_2$ substrate tends to increase the surface mobility of oligomers during the evaporation of the film. In previous approaches, this increased mobility during film evaporation has been obtained by heating the substrate up to 280 °C [1].

Our contact angle measurements with hexadecane (HD) and deionized water showed that the 6T films deposited on the OTS monolayers are slightly more hydrophobic than the 6T films directly evaporated in SiO$_2$. For instance, θ_{HD}=13.5 ± 1.5 ° and θ_{H20}=83±2 ° for the surface of 6T on OTS, while values were 7.7±1.5° and 78±2°, respectively, for the surface of 6T film on SiO$_2$. Preliminary results on an other organic semiconductor (polybutylcarbazole, PBuC) evaporated on SiO$_2$ and OTS treated SiO$_2$ showed a more highly marked difference. θ_{H20} was increased by more than about 30° when PBuC was deposited on an OTS monolayer. The higher hydrophobicity of the surface when 6T or PBuC are evaporated on an OTS monolayer may result from a more ordered surface, e.g., mainly made of the edges of the oligomers if they are mainly standing up-right nearly perpendicular to the substrate. An alternative explanation of the improved electrical performances may come from the passivation of the oxide surface traps by the deposited monolayer. More detailed studies are in progress.

Table I : Improvements of the main FET parameters with an OTS monolayer: threshold voltage (V_T), subthreshold slope (S), field-effect mobility (μ_{eff}).

This work	6T	OTS/6T	Improvement
V_T	22 V	8 V	63%
S	14V/decade	8V/decade	43%
ΔV_G for $I_{on}/I_{off}=10^6$	115 V	75 V	35%
μ_{eff}	3.6×10^{-3} cm^2/V.s	4.5×10^{-3} cm^2/V.s	25%
Ref. [2]	**pentacene**	**OTS/pentacene**	**Improvement**
V_T	90 V	60 V	33%
S	4.8V/decade	1.6V/decade	65%

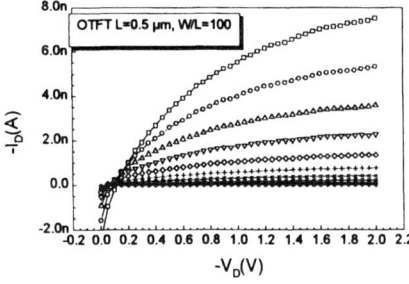

Figure 6 : Typical I_D-V_D characteristics of an « oxide-free » FET (L=500nm). V_G varies from -2 V (□) to - 0.7 V, step 0.1 V. Positive currents near V_D = 0 V are due to leakage currents through the large area of the S and D pads (compared to the area of the channel).

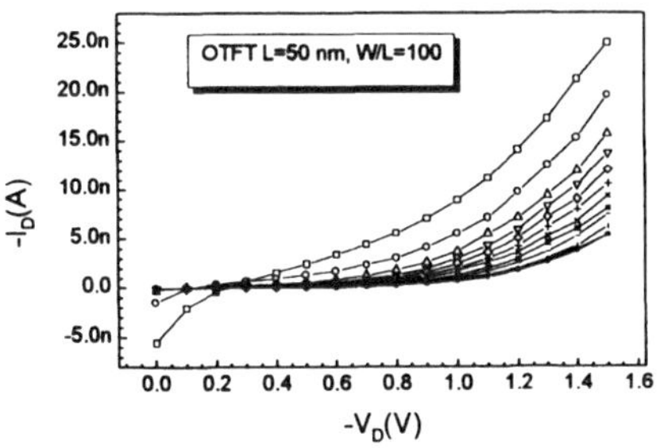

Figure 7 : Typical I_D-V_D characteristics of an « oxide-free » FET (L=50nm). V_G varies from -2.3 V to 1.5 V, step 0.4 V. Same as in Fig. 6 for positive currents near V_D = 0 V.

Nanometer scale, « oxide-free » organic FET, SAM/6T

     These devices were made as part of a program to test the ultimate limit in size reduction of organic FET's, i.e., the goal being to fabricate a FET on a single crystalline domain of the 6T film. Film deposition (temperature, evaporation rate,..) was not optimized here. This may explain the low mobility obtained here compared to the best results reported for 6T (ca. 3-5x10^{-2} cm^2/V.s) [1,4]. FET's with channel length down to 200 nm showed a low threshold voltage (normally off) and a remarkable low subthreshold slope (350 mV/decade) close to that of amorphous silicon (300 mV/decade). To our knowledge, this is the first time that such a low value is obtained with an organic FET.

     For channel length below 200 nm, the drain current is no longer well controlled by the gate moreover, the I_D-V_D curves (Fig. 6) follow a Fowler-Nordheim law. This means that the lateral electric field is too high and the energy barrier height for hole at the 6T/Au too low to allow a normal FET operation. From the Fowler-Nordheim plot, we deduced an energy barrier height of 0.17 eV, in close agreement with the known energy offset between the gold Fermi energy and the highest occupied molecular orbital (HOMO) of 6T (0.1-0.2 eV).

Finally, we note that our results extend the first attempt to fabricate an organic FET below 100 nm [10], and that according to the reduced channel length, our FET's exhibit a higher cut-off frequency (20 kHz at -3dB for L=200 nm) in agreement with the expected value deduced from the mobility using the equation $f_c = \mu V_D/2\pi L^2$ ($\mu=10^{-4}$ cm^2/V.s at V_D=-0.5V for L=200 nm).

CONCLUSION

These results show that the intercalation of an OTS self-assembled monolayer between the SiO$_2$ and the 6T film greatly improves the organic FET performances. Classical e-beam lithography and lift-off techniques have been used to fabricate metal contacts onto the 6T film, as well as to define isolated 6T FET. A metal electrode gap as small as 30 nm has been fabricated and has allowed the fabrication of a nanometer size organic FET using a self-assembled monolayer of alkyl chains as the gate insulator, instead of a thicker film of SiO$_2$ («oxide-free» organic FET). These devices exhibit long-channel type behavior down to channel length of 200 nm with a remarkable low threshold voltage and subthreshold slope.

ACKNOWLEDGEMENTS

We thank all the technical staff at the IEMN clean room facilities for advise and support. Research on self-assembled monolayers at IEMN has been financially supported by the CNRS research program ULTIMATECH.

REFERENCES

1. F. Garnier et al., J. Am. Chem. Soc. 115, 8716 (1993); B. Servet et al., Chem. Mat. 6, 1809 (1994); B. Servet et al., Adv. Mat. 5, 461 (1993).
2. D.J. Gundlach, Y.Y. Lin, T.N. Jackson, S.F. Nelson and D.G. Schlom, IEEE Electron Dev. Lett. 18, 87 (1997); Y.Y. Lin, D.J. Gundlach, S.F. Nelson and T.N. Jackson, IEEE Trans. Electron Dev. 44, 1325 (1997).
3. J.-B. Brozska, N. Shahidzadeh and F. Rondelez, Nature 360, 719 (1992); A.N. Parikh, D.L. Allara, I.B. Azouz and F. Rondelez, J. Phys. Chem. 98, 7577 (1994); J.-B. Brzoska, I. Ben Azouz and F. Rondelez, Langmuir 10, 4367 (1994); D.L. Allara, A.N. Parikh and F. Rondelez, Langmuir 11, 2357 (1995).
4. A. Dodabalapur, L. Torsi and H.E. Katz, Science 268, 270 (1995).
5. D. Vuillaume, in Amorphous and crystalline insulating thin films, edited by W.L. Warren, R.A.B. Devine, M. Matsumura, S. Cristoloveanu, Y. Homma, J. Kanicki (Mat. Res. Soc., Pittsburgh, 1997),. vol. 446, p. 79; P. Fontaine et al., Appl. Phys. Lett. 62, 2256 (1993).
6. S.R. Wasserman, Y-T. Tao and G.M. Whitesides, Langmuir 5, 1074 (1989).
7. C. Boulas, J.V. Davidovits, F. Rondelez and D. Vuillaume, Phys. Rev. Lett. 76, 4797 (1996).
8. D. Vuillaume, C. Boulas, J. Collet, J.V. Davidovits and F. Rondelez, Appl. Phys. Lett. 69, 1646 (1996).
9. J. Collet et al., this conference; J. Collet, M. Bonnier, O. Bouloussa, F. Rondelez and D. Vuillaume, Microelectronic Engineering 36, 119 (1997).
10. S. Franssila, J. Paloheimo and P. Kuivalainen, Electon. Lett. 29, 713 (1993).

DETERMINATION OF THE INTERNAL ELECTRIC FIELD IN THE ELECTROOPTIC ACTIVE LAYER OF MULTILAYER POLYMER STACKS

S. GROSSMANN, T. WEYRAUCH, AND W. HAASE
Darmstadt University of Technology, Institute of Physical Chemistry, Petersenstraße 20, 64287 Darmstadt, Germany, d54d@hrzpub.tu-darmstadt.de

ABSTRACT

We report on a method to investigate the inhomogeneous distribution of an electric dc field in multilayer polymer stacks. In situ electroabsorption (EA) measurements are applied in order to estimate the local electric fields in double layer polymer films. The observed time dependent behaviour is compared with a model equivalent circuit. The results indicate that besides the relation of ohmic resistivities and capacities of the different polymer layers in the investigated systems also the influence of the electric properties of polymer/electrode and polymer/polymer interfaces must be considered.

INTRODUCTION

The distribution ratio of an external electric dc field over a multilayer polymer stack is of basic interest for the fabrication as well as for the operation of a variety of integrated optic polymer devices. Particularly, the effective electric field during the poling process of polymeric multilayer structures which requires a large part of the applied voltage over the active layer [1,3] and the dc drift phenomenon in Mach-Zehnder modulators [4] should be stressed.

The electroabsorption spectroscopy (Stark spectroscopy) is used in order to measure the electric field strength in the active layer. In the experiment a superposition of an electric ac and dc field is applied to a polymer sample with a double layer structure. We distinguish between an active layer and an inactive layer. The active layer consists of polycarbonate which is doped with a electrooptical active dye. Only this layer contributes to the observable electroabsorption spectrum of the sample.

THEORETICAL BACKGROUND

Electroabsorption

Under the assumption of an orientationally fixed chromophore molecule characterized by two energy levels (the ground state and the lowest energy electronic excited state) the electroabsorption (i. e. the change of the absorbance band induced by an electric field) is due to

413

the Stark effect. The absorption band is shifted according to the differences of dipole moment $\Delta\mu$ and polarizability $\Delta\alpha$ in the ground and excited state. In the experiment one usually measures the effitive value of the absorbance change $\Delta A_{eff}(\omega)$ at fixed wavenumber $\tilde{\nu}$ under application of an ac electric field with frequency ω by lock-in technique. The quadratic effect (effective value of the absorbance change measured at the frequency 2ω) is given by [5,6]

$$\Delta A_{eff}(2\omega) = \frac{E_{eff}^2(\omega)}{\sqrt{2}}\left[\frac{\Delta\alpha}{hc}\tilde{\nu}\frac{\partial(A/\tilde{\nu})}{\partial\tilde{\nu}} + \frac{(\Delta\mu)^2}{10h^2c^2}\tilde{\nu}\frac{\partial^2(A/\tilde{\nu})}{\partial\tilde{\nu}^2}\right].$$

(1)

for the case of an isotropic or weak polar sample. The quadratic effect may also be measured at the frequency ω if a dc field E_0 is applied simultaneously. Thus

$$\Delta A_{eff}(\omega) = 2E_0E_{eff}(\omega)\left[\frac{\Delta\alpha}{hc}\tilde{\nu}\frac{\partial(A/\tilde{\nu})}{\partial\tilde{\nu}} + \frac{(\Delta\mu)^2}{10h^2c^2}\tilde{\nu}\frac{\partial^2(A/\tilde{\nu})}{\partial\tilde{\nu}^2}\right]$$

(2)

[5,6]. In the following this will be referred to as the quasilinear effect.

After the determination of the molecular parameters $\Delta\mu$ and $\Delta\alpha$ one can calculate the effective electric dc field in the active polymer layer. The measurement at a fixed wavenumber allows the determination of the temporal dependence of the electric field strength [7].

DC drift in multilayer films

Park et al. presented a model for the internal field distribution in multilayer films and its temporal dependence [4]. According to them an equivalent circuit as shown in Fig 1a can describe a double layer polymer film with one electrooptic active and one electrooptic inactive layer and corresponding capacities C_{al} an C_{il} and resistances R_{al} an R_{il}.

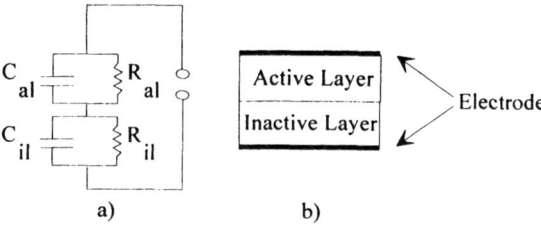

Fig. 1. Equivalent circuit of a double layer polymer stack (a) and its structural layout (b).

The time dependence of the dc voltage across the active layer V_{al} can be expressed by

$$V_{al} = V_{al}^{\infty} + (V_{al}^{0} - V_{al}^{\infty}) \times \exp(-t / \tau_{rel}) \tag{3}$$

were V_{al}^{0} and V_{al}^{∞} are the potential drop across the active layer at $t = 0$ and $t = \infty$, respectively. At $t = 0$ the voltage distribution ratio is defined by $V_{al}^{0} / V_{dc} = C_{al} / (C_{al} + C_{il})$, whereas at $t = \infty$ the distrubution ratio is defined by $V_{al}^{\infty} / V_{dc} = R_{al} / (R_{al} + R_{il})$. The relaxation time τ_{rel} is then given by

$$\frac{1}{\tau_{rel}} = \frac{1}{C_{al} + C_{il}} \left(\frac{1}{R_{al}} + \frac{1}{R_{il}} \right). \tag{4}$$

EXPERIMENT

Double layer polymer stacks were prepared by spin coating on ITO coated glass substrates. Top electrodes were provided by a second ITO substrate (in combination with epoxy resin) or prepared by sputtering a semitransparent gold electrode. The electrooptic active layer consist of the dye p-N,N-dimethylamino-p'-nitroazobenzene (DMANA) diluted in a polycarbonate bisphenol-A (PC) matrix (2% of weight). As inactive layers we used epoxy resin and polyimide (PI 2566). The thickness of the samples was determined using an interference microscope. Absorbance spectra were taken using a Cary 17 spectrometer.

Electroabsoprtion was measured using a spectrometer based on a monochromator (TOPAG LM-01) and a 100 W halogen lamp as light source. The ac and dc part of the intensity transmitted by the sample (ΔI and I, repectively) were measured using a photomultiplier and a lock-in amplifier (PAR 5210) and a digital multimeter (HP-34401A). The absorbance change was calculated from the measurements according to

$$\Delta A = - \Delta I / (I \ln 10) \tag{5}$$

RESULTS AND DISCUSSION

In Fig. 2 we present electroabsorption measurements at room temperature and $\lambda=488$ nm on a double layer film consisting of 3 μm PC/DMANA on the ITO/glass substrate and 13 μm epoxy resin. The filled circles represent the temporal dependence of the quasilinear effect. An ac voltage of 210 V was applied during the whole measurement. After switching on the dc voltage of 160 V at $t=250$ s a fast increasing signal was observed, which relaxed after the fast onset. After switching off the dc voltage at $t=1800$ s the signal changed the sign, but the absolute value was considerably lower. The quadratic effect (shown as open circles in fig. 1) was not effected by the external dc field and did not vary with time, i. e. according to eq. 1 $E(\omega)$ must remain constant over time. The quasilinear electroabsorption signal is additionaly proportional to the internal electric field in the active layer E_0 (cf. eq. 2), i. e. it is proportional to the part of the external dc voltage which appears across the active layer and especially the temporal variation of the electroabsorption must be attributed to this voltage.

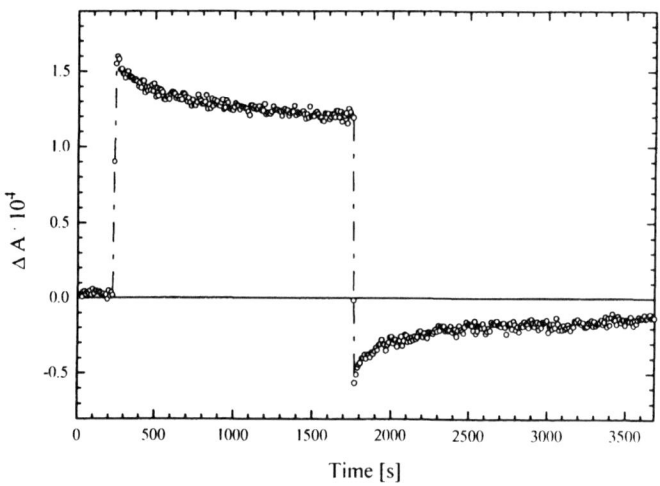

Fig. 2 Time dependency of the quasilinear electroabsorption of a double layer DMANA/PC - epoxy resin sample at room temperature.

With the model of Park et al. one can describe qualitatively the experimental results. Eq. 3 is in accordance with the behaviour which was observed after switching on the dc field and also the behaviour after switching off the dc field is in agreement with the equivalent circuit. Fitting eq. 3 to the temporal course of the signal between $t=250$ s and $t=1800$ s delivers the relaxation time $\tau_{rel} \approx 430$ s. The capacity of the sample was 2.5 nF. Using these data the term ($1/R_{al}$ + $1/R_{il}$) was estimated from eq. 4 to $6 \cdot 10^{-13}$ S. This mean value for the conductivity is in the order of magnitude which was estimated from polarization current measurements, but the measured values are somewhat to high to explain the relaxation time constant quantitatively. Thus an exact verification by careful measurements has still to be done to exclude experimental errors.

Fig. 3 shows the temporal dependence of $\Delta A(\omega)$ of a monolayer sample of DMANA in PC (thickness 3 µm) measured at room temperature. The ac voltage was 40 V. Between $t=0$ and $t=300$ s no dc field was applied and no signal was detectable. A dc voltage of 160 V was switched on at $t=300$ s and switched off at $t=3300$ s. As in the double layer sample the quadratic effect measured with the same ac voltage showed no temporal dependence. The magnitude is also shown in Fig. 3 (open circles). Surprisingly the temporal course of the quasilinear electroabsorption signal was qualitatively similiar to the measurements at the double layer sample. This behaviour can not be explained by the model of Park et al. Using only one part of the equivalent circuit for the double layer in fig. 1 (i. e. C_{al} and R_{al}) as equivalent circuit of the monolayer does not deliver a time dependence of V_{al} at all, it is always equal to the externally applied voltage V_{dc}. This signal must be attributed to an internal electric field, which has opposite direction in comparison to the external field applied before. This means the equivalent circuit of a monolayer (and consequently of multlayer films) film must be more complicated. Especially the behaviour of the interfaces polymer/electrodes should be considered carefully.

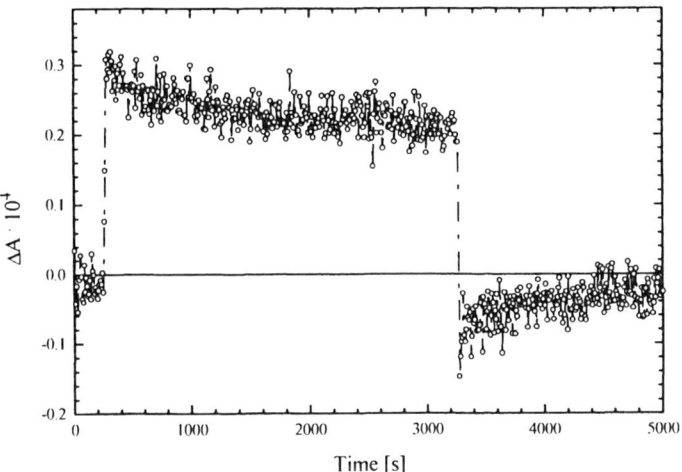

Fig. 3. Time dependency of the quasilinear electroabsorption of a monolayer DMANA/PC sample at room temperature.

In conclusion we have demonstrated the use of in situ electroabsorption measurements to investigate the distribution of internal dc electric fields in multi layer polymer films and its time dependence. The temporal course of the observed electroabsorption signal could be qualitatively described by the theoretical model from Park et al. However the time constant of the dc drift in the electrooptic active layer was higher than estimated from the conductivity and capacitance of the sample. Moreover, measurements on single layer films have shown similar behaviour as doublelayer films, which indicate that the model and the equivalent circuit must be improved. We assume that in this case the time dependent internal electric dc field strength is determined by nonuniformly distributed space charges, well known for polymer electrets [8]. Charge injection as well as redistribution of internal charge carriers has to be considered.

ACKNOWLEDGEMENT

The authors are grateful to S. Saal for stimulating discussions. Financial support by the Volkswagen-Stiftung is gratefully acknowledged.

REFERENCES

[1] K. D. Singer and J. H. Andrews, in Molecular Nonlinear Optics: Materials Physics and Devices, Ed. J. Zyss (Academic Press, Inc., San Diego, 1994).

[2] R. H. Page, M. C. Jurich, A. Sen, R. J. Twieg, J. D. Swalen, G. C. Bjorklund and C. G. Wilson, J. Opt. Soc. Am. **B 7**, pp 1239-1250, (1990).

[3] D. G. Girton, W. W. Anderson, J. A. Marley, T. E. Van Eck, S. Ermer, in Organic Thin Films for Photonics Applications, **Vol. 21**, (Optical Society of America, Washington DC, 1995), pp. 470-473.

[4] H. Park, W. - Y. Hwang and J. - J. Kim, Appl. Phys. Lett. **70** (21), pp. 2796-2798, (1997).

[5] L. M. Blinov, S. P. Palto, A. A. Tevosov, M. I. Barnik, T. Weyrauch, W. Haase, Mol. Mat., **5**, pp. 311-338, (1995).

[6] M. I. Barnik, L. M. Blinov, T. Weyrauch, S. Palto, A. A. Tevosov and W. Haase, in Polymers for Second-Order Nonlinear Optics, Ed. G. A. Lindsay, K. D. Singer, ACS Symposium Series 601, (Am. Chem. Soc., Washington DC, 1995), pp. 288-303.

[7] S. Großmann, T. Weyrauch, S. Saal and W. Haase, Optical Materials, in print.

[8] see e.g. J. van Turnhout, in Electrets, Ed. G. M. Sessler, Topics in Applied Physics, Vol. 33, (Springer, Berlin, 1987), pp. 81-215.

CONTROLLING MATERIALS ARCHITECTURE ON THE NANOMETER-SCALE: PPV NANOCOMPOSITES VIA POLYMERIZABLE LYOTROPIC LIQUID CRYSTALS

RYAN C. SMITH, HAI DENG, WALTER M. FISCHER, AND DOUGLAS L. GIN
Department of Chemistry, University of California, Berkeley, CA 94720-1460

ABSTRACT

We have developed a general strategy for the construction of ordered nanocomposites with hexagonal symmetry, using polymerizable lyotropic (i.e., amphiphilic) liquid crystals. In this approach, self-organizing lyotropic liquid-crystalline monomers are used to form an ordered template matrix in the presence of a reactive hydrophilic solution. Subsequent photopolymerization to lock-in the matrix architecture, followed by initiation of chemistry within the ordered hydrophilic domains to afford solid-state fillers, yields the anisotropic nanocomposites. Composites have been synthesized that have a regular hexagonal arrangement of extended poly(p-phenylenevinylene) (PPV) domains, with a regular interchannel spacing of 4 nm. The photoluminescence of these materials is significantly altered from that of bulk PPV. The dimensions of these nanocomposites can be tuned by varying the size of the hydrophobic tails and/or the nature of the counterion associated with the hydrophilic headgroup of the monomer.

INTRODUCTION

Nanometer-scale architectural control is the primary reason for the impressive properties of many biological materials such as bone, tendon, and nacre [1]. Nature is able to create truly incredible structural materials using components such as proteins and calcium salts by controlling their architecture on this size regime. Unfortunately, we lack the ability to assemble synthetic materials with this level of sophistication using conventional processing techniques [2]. Thus, a great deal of effort has recently been devoted to the development of new methods for assembling synthetic materials with nanometer control in the hopes of generating composites with unique or superior bulk properties [3–6].

We have developed a general approach to synthetic nanocomposites with hexagonal symmetry using polymerizable lyotropic (i.e., amphiphilic) liquid crystals (LLC's) [7,8]. Our basic strategy entails the following steps: (1) Design and synthesis of cross-linkable amphiphiles that self-assemble into the inverse hexagonal phase in the presence of hydrophilic reagents instead of pure water. (2) Photopolymerization of the matrix monomers into an ordered template network with retention of the original architecture. (3) In situ conversion of the reagents dissolved in the hydrophilic channels into solid "filler" materials in order to achieve the final composite. Both inorganic [7] and organic chemistry [8] can be performed inside the ordered hydrophilic domains, and the resulting nanocomposites can be shaped into films, fibers, etc. prior to cross-linking.

In order to investigate the effects of nanometer-scale engineering on the photophysical properties of encapsulated materials, we recently synthesized poly(p-phenylenevinylene) (PPV) inside the hydrophilic channels of the LLC phase formed by monomer **1** (Figure 1) [8]. PPV was chosen as the "filler" in this initial investigation because it is formed from a water-soluble precursor, and PPV and its derivatives have recently been used in a number of device applications [9,10]. The resulting nanocomposites have regular interchannel spacing of 4 nm and exhibit enhanced, blue-shifted fluorescence compared to that of pure PPV. In this paper, we present preliminary results on controlling the small-scale architecture of these composites through (1) choice of the metal ion on the headgroup of the monomer and (2) variation of the tail length. The incorporation of transition-metal and lanthanide cations also provides a simple

419

means of introducing a variety of interesting properties into these nanostructured materials.

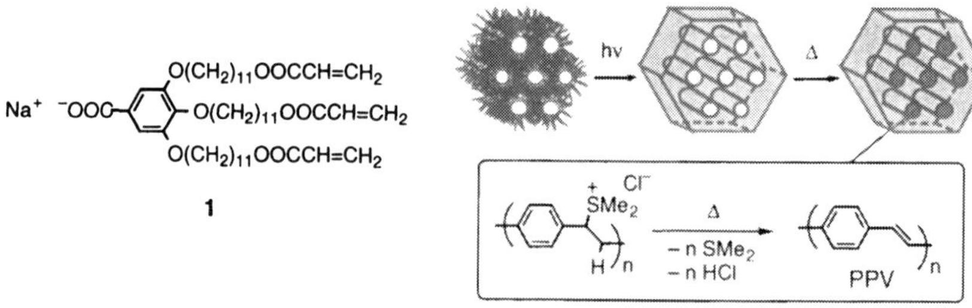

Fig. 1. Synthetic scheme for formation of PPV nanocomposites with hexagonal symmetry.

EXPERIMENTAL

General. All reagents and solvents were obtained from the Aldrich Chemical Co., Fisher Scientific, or Sigma Chemicals, and purified before use. Monomer **1** was prepared according to literature procedures (see Supporting Information of Ref. [8]). All syntheses were performed under inert atmosphere. Low angle X-ray diffraction studies were performed using an Inel CPS 120 powder diffraction system employing Cu K_{α} radiation. Fluorescence measurements were performed on a Spex Fluoromax 2. Elemental analyses were performed at Galbraith Laboratories, Knoxville, TN. Quantum yields were measured in the laboratory of Prof. G. Leising at the Technische Universität, Graz, Austria.

Typical preparation of transition-metal or lanthanide-containing LLC. Compound **1** (1.00 g, 1.16 mmol) was dissolved in a mixture of acetone (100 mL) and methanol (10 mL). The solution was titrated by aqueous NaOH solution until pH = 7. After 30 min of stirring, Co(II) nitrate hexahydrate (0.17 g, 0.58 mmol) was added slowly to the solution. The reaction mixture was stirred for an additional 2 h, and then the solvent was removed in vacuo. The residual solid was washed with distilled water (3 x 20 mL) and then redissolved in acetone (20 mL). The solution was then filtered, and the solvent removed in vacuo. The solid was suspended in an additional 20 mL of acetone and then pumped dry to remove any residual water. The resulting solid was dried in vacuo overnight to afford the Co(II) analog of **1** as a dark green solid in ca. 80% yield.

Preparation of analogs of 1 with different tail lengths. Analogs of **1** with different tail lengths were prepared according to the same procedure used to prepare **1** [8] but with 6-bromohexanol and 8-bromooctanol in place of 11-bromoundecanol.

Preparation and polymerization of the inverted hexagonal phase. Inverted hexagonal phases were all prepared with the same weight proportions. Polymerizable LLC (0.85 g), distilled water (0.10 g), and photoinitiator solution (0.05 g of a 20% w/w solution of 2-hydroxy-2-methylpropiophenone in degassed *p*-xylene) were combined in a 40-mL centrifuge tube. The mixture was centrifuged at 2800 rpm for 10 min, followed by hand-mixing with a spatula. The procedure was repeated three times until a homogeneous mixture formed. A small portion of the mixture (0.15 g) was pressed into a thin film and then exposed to 365 nm light (1800 μW/cm^2) overnight under a nitrogen atmosphere to ensure complete polymerization. The hexagonal architecture of the material was confirmed by polarized light microscopy and X-ray diffraction.

Preparation of a Eu(III)-PPV nanocomposite. An aqueous solution of poly(*p*-

xyxylenetetrahydrothiophenonium chloride) (~0.6 wt %) was used in place of distilled water to form the LLC phase with the Eu(III) salt of **1**, using the same mass proportions as described above. After photopolymerization, the film was placed in a tube furnace and heated at 220 °C under dynamic vacuum (10^{-4} torr) for 8 h to convert the PPV precursor in situ.

RESULTS AND DISCUSSION

Our PPV nanocomposites exhibit two intense emission bands at 504 and 534 nm when excited with 370 nm light. In comparison, pure PPV typically exhibits emission bands at 517 and 547 nm. In addition, the nanocomposites exhibit enhanced photoluminescence which is higher than that observed for pure PPV converted under the same conditions [8]. Fluorescence quantum yields of typically 27% (with some samples up to 85%) have recently been measured for these materials. This behavior is consistent with a short effective conjugation length for the PPV and dilution/isolation of emitting PPV segments to minimize self-quenching effects [11-13]. The PPV chains in our composites are most likely partially converted, consisting of conjugated segments separated by unconverted segments. This structure can be rationalized as a consequence of performing the PPV conversion inside a dimensionally constrained environment which limits the degree of conversion possible and hence the PPV conjugation length. This idea is supported by the fact that the emission wavelengths and intensities of the PPV composite reach a maximum after only 4 h of thermal conversion and remain unchanged even after heating for over 20 h at 220 °C in vacuo [8].

This apparent effect of nanostructure on the degree of conversion inspired us to develop methods for controlling the small-scale dimensions of our materials to further investigate this phenomenon. Israelachvili has developed a simple theory based on the size aspect ration between the hydrophobic tail section and hydrophilic headgroup of simple surfactants to describe their preference to form certain aggregates in dilute solution [14]. Percec has used similar shape considerations to design tapered thermotropic liquid crystals that adopt certain well-defined geometries in the pure form [15]. Packing considerations and molecular shape similarly dictate the preference of our polymerizable amphiphiles to form certain condensed LLC phases. For instance, monomer **1** adopts the inverted hexagonal phase because the amphiphile has a small hydrophilic headgroup and a broad, wedge-like tail section. By varying the nature of the ionic headgroup and the organic tails of our monomer, we thought that we might be able to modify the dimensions of our nanocomposites with retention of the overall microstructure.

In order to vary the effective size of the hydrophilic headgroup of monomer **1**, several transition-metal and lanthanide analogs were synthesized. Transition-metal and lanthanide cations were used in this initial study because these metal ions offer a range of ionic radii and charges (valencies) for potential modification of headgroup size. In addition, these d-and f-block ions exhibit a range of interesting properties ranging from optical effects to paramagnetism. Exchange of the sodium ion of **1** with transition-metals and lanthanides was accomplished by mixing a stoichiometric amount of the corresponding soluble transition-metal or lanthanide nitrate salt with **1** in acetone. The Co(II), Ni(II), Cd(II), Eu(III), and Ce(III) analogs of monomer **1** were obtained using this procedure. The structure and identity of the resulting LLC compounds were verified by ^{1}H NMR spectroscopy, FT-IR spectroscopy, and elemental analysis. All five heavy metal analogs of **1** readily absorb water, consistent with their amphiphilic nature. Low-angle X-ray diffraction analysis of the LLC phases at ambient temperature revealed that all the mixtures exhibit diffraction peaks with the ratio: 1, $1/\sqrt{3}$, $1/\sqrt{4}$ ..., corresponding to the d_{100}, d_{110}, and d_{200} planes indicative of a hexagonal assembly. These mixtures were then photocrosslinked with 365 nm light under nitrogen to lock-in their architecture. The extent of polymerization was determined by monitoring the loss of acrylate bands at 1630 and 810 cm^{-1} using FT-IR spectroscopy.

The X-ray diffraction data for the unpolymerized and polymerized inverted hexagonal phases of different analogs of **1** is summarized in Table I. The compositions of these mixtures were kept at the same weight percentages for comparison because the dominant component in the amphiphilic mixtures is the organic portion of the LLC. As can be seen from Table I, mixtures of LLC's with metal ions of the same charge exhibit very similar d_{100} spacings irrespective of the radius of the metal ion. For example, the Cd(II), Co(II) and Ni(II) LLC mixtures all exhibit d_{100} peaks of ca. 35–36 Å. The largest change in unit cell dimensions occurs when the charge on the metal ion is changed. The trivalent Ce(III) and Eu(III) salts both exhibit similar d_{100} values which are significantly smaller (ca. 30 Å) than that of the divalent transition-metal analogs. This trend of smaller unit cell size with increasing positive charge on the metal can be rationalized on the basis of stronger coordination or electrostatic interaction with the anionic carboxylate groups. Presumably, this stronger coordination or interaction decreases the effective size of the entire hydrophilic headgroup on the LLC, thereby decreasing the unit cell dimensions of the hexagonal assembly. However, this trend should not be over-generalized. The LLC mixture of 1, which has a monovalent sodium ion, exhibits a d_{100} spacing similar to that of analogs containing much smaller divalent transition-metal ions such as Ni(II) and Co(II). This difference may be due to stronger coordinative interactions between d- and f-block ions with carboxylates, in contrast to the purely ionic interactions between alkali metal ions and carboxylates [16]. Regardless, these results do demonstrate that the small-scale dimensions of the inverted hexagonal phase of 1 can be modulated by simply changing the nature of the cation on the LLC. This data also suggests that the strength of the interaction between the metal cation and the carboxylate anion is the dominant factor in determining the overall size of the LLC headgroup rather than simply the size contribution of the metal ion itself. We are currently examining the nature of these interactions using FT-IR spectroscopy, and the results will be presented in a forthcoming publication.

Table I. Effect of different metal ions on the dimensions of the inverted hexagonal phase of 1. System compositions are all 85/10/5 (w/w/w) LLC/water/photoinitiator solution.

Metal Ion (radius, Å)	d_{100}(Å)	d_{110}(Å)	d_{200}(Å)	interchannel spacing = d_{100} /cos 30°
Na(I) (0.97)	35.0	20.4	17.7	
after polymerization	34.5	20.2	17.5	
Ni(II) (0.69)	36.7	21.1	18.4	
After polymerization	35.8	20.8	18.2	
Co(II) (0.72)	36.7	21.4	18.1	
After polymerization	35.6	20.3	17.7	
Cd(II) (~0.97)	35.1	20.7	18.0	
After polymerization	35.0	20.6	18.0	
Eu(III) (0.98)	30.1	17.6	15.3	
After polymerization	30.2	17.7	15.7	
Ce(III) (1.07)	31.3	18.2	15.6	
After polymerization	30.9	17.8	15.5	

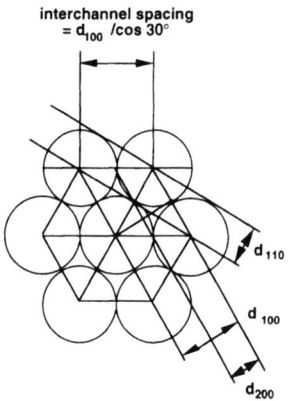

In order to determine the effect of the organic tail unit on the dimensions of the inverted hexagonal phase, two analogs of 1 with different tail lengths were initially synthesized. The X-ray diffraction data of these analogs are summarized in Table II. As can be seen in Table II, variation of the spacer length on the tails from n = 6 to n = 11 causes a change in d_{100} from 23 Å to 36 Å. This increase corresponds to a large change in interchannel spacing (d_{100}/cos 30°) from 26.6 Å to 41.5Å. Control of the organic tail length and hence the size of hydrophobic portion of the amphiphile is a viable means of controlling unit cell size in these systems. Tail length modifications appear to have a much more substantial effect on the dimensions of the inverted hexagonal phase than the nature of metal ion on the headgroup. The results of more detailed work on this subject will be presented in a separate publication.

Table II. X-ray diffraction data for analogs of **1** with different tail lengths

O(CH$_2$)$_n$OOCCH=CH$_2$

Na$^+$ $^-$OOC—⟨benzene ring⟩—O(CH$_2$)$_n$OOCCH=CH$_2$

O(CH$_2$)$_n$OOCCH=CH$_2$

System compositions:
85/10/5 (w/w/w) LC/H$_2$O/xylene

n	inv. hex. d$_{100}$ (Å)
6	23
8	26
11	36

The incorporation of transition-metals and lanthanides is also an excellent means of introducing new properties into the PPV nanocomposites. For example, the polymerized Eu(III) LLC network is extremely fluorescent under UV light. Strong emission bands at 592 and 616 nm characteristic of Eu(III) [17] were observed when the polymerized hexagonal phase of the Eu(III) salt of **1** was excited by 370 nm light. This phenomenon implies that the Eu(III) ions in the hydrophilic channels have very few water molecules in their primary coordination sphere since coordinated water molecules quench Eu(III) fluorescence [17]. In order to take advantage of the interesting optical properties of this material, a PPV nanocomposite was formed using the Eu(III) analog of **1**. An aqueous solution of poly(p-xyxylenetetrahydrothiophenonium chloride) (~0.6 wt %) was used to make the inverted hexagonal phase with the Eu salt of **1** instead of pure water. (This precursor is different from that shown in Figure 1.) The mixture was photopolymerized and then heated in vacuo to form the PPV in situ. As can be seen in Figure 2 below, the nanocomposite exhibits a new intense emission band at ca. 670 nm that is absent in the emission profiles of PPV nanocomposites made with monomer **1** and Eu(III) LLC networks containing only distilled water in the channels. This 670 nm emission band was also not observed in a mixture of PPV formed in the presence of Eu(III) nitrate. The presence of this new band suggests interaction between the Eu(III) cations and PPV chains in the periodic nanochannels and possibly energy transfer between the two components. The behavior of encapsulated PPV chains with this and other lanthanide and transition-metal LLC analogs are currently under investigation.

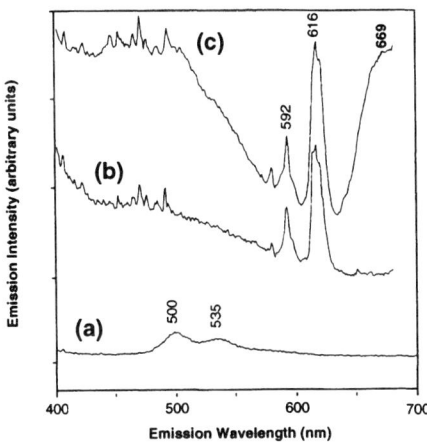

Fig. 2. Emission spectra of (a) PPV nanocomposite made with **1**, (b) the polymer of the Eu(III) salt of **1**, and (c) a PPV nanocomposite made with the Eu(III) salt of **1**. (Excitation: 370 nm).

CONCLUSIONS

PPV nanocomposites with hexagonal symmetry can be formed by using polymerizable LLC's that form the inverted hexagonal phase as a template. Using ion-exchange, a number of transition-metal and lanthanide cations can be easily incorporated into these amphiphilic monomers. This technique affords an avenue for modulating the small-scale dimensions of these polymerizable supramolecular assemblies and for introducing new properties into the systems. Additionally, variation of the tail length of these monomers offers a complementary approach to controlling the dimensions of these nanostructured materials.

ACKNOWLEDGMENTS

This research was supported primarily by the Office of Naval Research (N00014-97-1-0207). D.L. Gin gratefully acknowledges partial support from an NSF CAREER Award, a 3M Untenured Faculty Award and a research gift from the Raychem Corporation. R.C. Smith thanks Chevron for a graduate fellowship. We also thank Prof. G. Leising for providing detailed photophysical measurements on our initial systems.

REFERENCES

1. P.D. Calvert, MRS Bull. **17**, 37 (1992).
2. A.H. Heuer, D.J. Fink, V.J. Laraia, J.L. Arias, P.D. Calvert, K. Kendall, G.L. Messing, J. Blackwell, P.C. Rieke, D.H. Thompson, A.P. Wheller, A. Veis, and A.I. Caplan, Science **255**, 1098 (1992).
3. J.E. Mark and P.D. Calver, Mater. Sci. Eng. **C1**, 159 (1994) and references therein.
4. C.R. Martin, Chem. Mater. **8**, 1739 (1996).
5. J.H. Golden, F.J. DiSalvo, J.M.J. Fréchet, J. Silcox, M. Thomas, and J. Elman, Science, **273**, 5276 (1996).
6. P. Osenar, P.V. Braun, and S.I. Stupp, Adv. Mater. **8**, 1022 (1996).
7. D.H. Gray, S. Hu, E. Juang, and D.L. Gin, Adv. Mater. **9**, 731 (1997).
8. R.C. Smith, W.M. Fischer, and D.L. Gin, J. Am. Chem. Soc. **119**, 4092 (1997).
9. P.L. Burn, A.B. Holmes, A. Kraft, D.D.C. Bradley, A.R. Brown, R.H. Friend, R.W. Gymer, Nature **356**, 47 (1992).
10. F. Hide, B.J. Schwartz, M.A. Díaz-Gracía, A.J. Heeger, Chem. Phys. Lett. **256**, 424 (1996).
11. S.A. Jenekhe and J.A. Osaheni, Chem. Mater. **6**, 1906 (1994) and references therein.
12. W.P. Chang and W.T. Whang, Polymer **37**, 3493, (1996).
13. I.D.W. Samuel, B. Crystall, G. Rumbles, P.L. Burn, A.B. Holmes, and R.H. Friend, Chem. Phys. Lett. **213**, 472 (1993).
14. B.G. Deacon and J.R. Phillips, Coord. Chem. Rev. **33**, 227 (1980).
15. J.N. Israelachvili, *Intermolecular and Surface Forces*, Academic Press, New York, 1985, pp. 255–257.
16. V. Percec, J. Heck, G. Johansson, D. Tomazos, M. Kawasumi, P. Chu, and G. Ungar, J. Macromol. Sci. Pure Appl. Chem. **A31** (11), 1719 (1994).
17. D.M. Roundhill, *Photochemistry and Photophysics of Metal Complexes*, Plenum Press, New York, 1994, pp. 308–312.

HIGH MOBILITY AND LIQUID PHASE PROCESSABLE
ORGANIC SEMICONDUCTORS

H.E. KATZ, J. LAQUINDANUM, A.J. LOVINGER
Bell Laboratories-Lucent Technologies, 600 Mountain Avenue, Murray Hill, NJ 07974

ABSTRACT

New thiophene oligomers and fused ring compounds have been designed and synthesized with the aim of maximizing semiconductor mobility in thin film transistors (TFTs) while allowing for liquid phase processability. Thiophene hexamers with alkyl side chains of various lengths, some with an ether oxygen embedded in the chain, were synthesized, as were derivatives of the novel heterocycle anthradithiophene. Mobilities of the vapor-phase-deposited films ranged from 0.01 cm^2/Vs for the hexamers to 0.15 cm^2/Vs for dihexylanthradithiophene. The latter is the highest mobility yet reported for a polycrystalline film. Cast films of some of these compounds from solution gave mobilities within factors of 2-10 of the corresponding values from gas-phase-deposited films.

INTRODUCTION

TFTs where all of the materials utilized, including the semiconductor, are organic or polymeric are of interest for use in low-level memories and logic circuits, and as drivers for flexible displays.[1] A wide variety of materials have been screened as organic TFT semiconductors, primarily by groups at CNRS-Thiais, Philips-Eindhoven, IBM-New York, Penn State, and Bell Laboratories. The original reports of TFTs employing organic molecular solids as semiconductors were from the Garnier group at CNRS, where the linear hexamer of thiophene was viewed as promising[2] and the α,ω-dihexyl derivative was considered as a superior alternative.[3] Further structural improvements in these solids arose from the use of fused ring systems, culminating in the optimization of the mobility of pentacene to a value above 1 cm^2/Vs.[4]

It has become apparent that the principal advantages of organic TFTs will lie in the inexpensive options available for their processing, rather than in any performance advantage over silicon. To fully utilize the least expensive fabrication methods, such as printing and casting of features, the semiconductors must be deposited from solution to produce films that retain high mobilities as well as robustness and reproducibility. Regioregular polythiophene has been demonstrated to form a high-mobility spun film useful in TFTs.[5] Pentacene has been deposited via a precursor route as a solution, followed by thermal conversion in situ.[6] Besides a preliminary report of dihexylquaterthiophene forming high-mobility films from the condensed phase[7], there have been no accounts, particularly of high-melting compounds, so utilized in TFTs.

This paper describes new synthetic chemistry leading to high-melting organic semiconductors that may be deposited from solution as part of a TFT fabrication process. The value of the side chains in maximizing the solubility and preserving the mobility that could otherwise be obtained from these compounds by gas-phase deposition is discussed. Especially dramatic is the use of the hexyl substituent to solubilize the highly infusible anthradithiophene ring system, investigated for the first time as a part of this work.

SYNTHESES

Hexameric thiophenes were prepared by first synthesizing bithiophenes with the desired side chain at one 5-carbon and a tributyltin group at the other. For alkyl substituents, bithiophene was acylated and the monoketone purified, ensuring that the final product would not be contaminated by undersubstituted oligomers or oligomers of the wrong length. The ketones were easily reduced to alkylbithiophenes. For alkoxyalkyl substituents, bithiophene was alkylated with THP-protected iodopropanol. Again, the intermediate was purified, and subsequent chemistry effected the deprotection of the alcohol and O-alkylation to complete the ethereal group. The monosubstituted bithiophenes were coupled via their tin derivatives with dibromobithiophene using the Stille coupling.

Anthradithiophenes were synthesized by condensation of thiophene 2,3-dicarboxaldehydes with cyclohexanedione followed by reduction, analogously to the synthesis of pentacene. For dialkyl derivatives, the aldehyde groups were protected and the thiophene 5-lithiated and alkylated before deprotection and condensation. Because of the thiophene rings, the lithiation option provided a facile means of end-substitution that is not available for the pentacene ring system.

All of these syntheses are sketched in Scheme I. Details of the experimental conditions are described in two full papers, accepted for publication in 1998.[8]

TFT MOBILITIES

The behavior of these compounds as semiconducting films was assessed by incorporating the compounds in conventional TFTs, either in the "bottom" (source and drain electrodes prepatterned on the substrate before semiconductor deposition) or "top" (source and drain electrodes vacuum-deposited onto semiconductor films) contact geometry. Oligothiophenes were evaporated onto substrates at ambient temperature or cast from heated solutions in aromatic solvents. Anthradithiophenes were evaporated onto ambient or heated substrates, and the dihexyl derivative was also cast from solution. Solvents were removed in a vacuum oven held near the solution temperature, about 70-100 °C. Mobilities are listed in Table I.

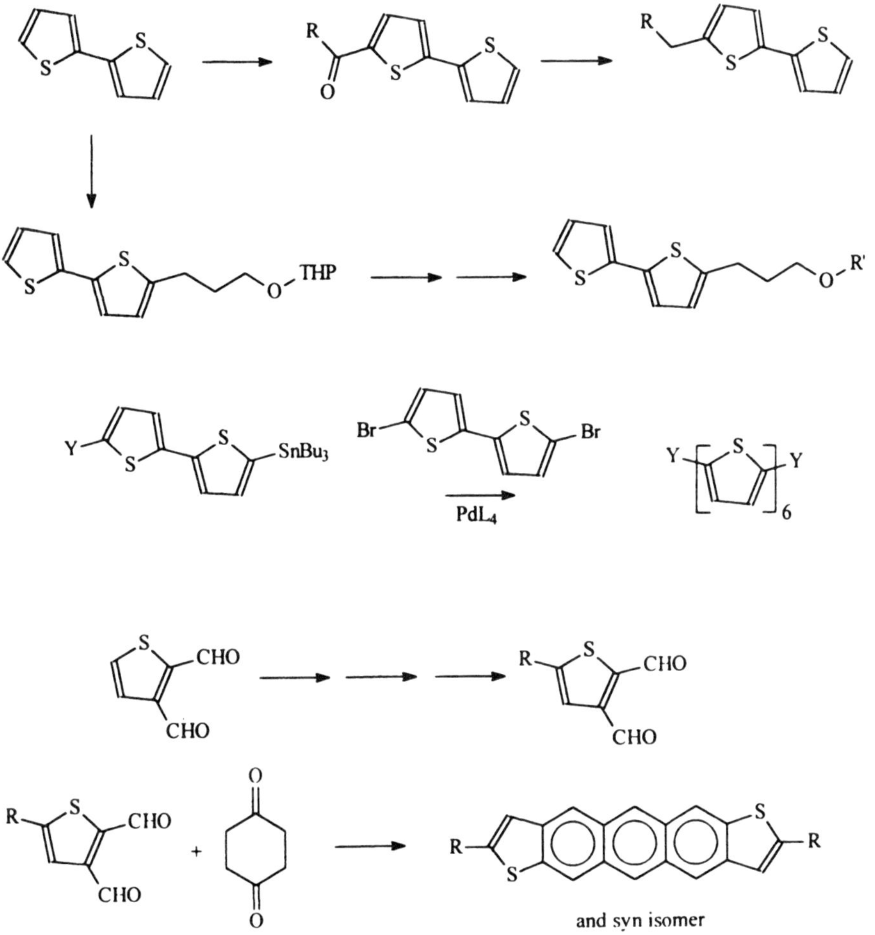

Scheme I

Table I. Representative Mobilities of Thiophene Hexamers and Anthradithiophenes

Compound (evaporated unless noted)	Mobility (cm^2/Vs)
sexithiophene: dihexyl	0.04
dihexyl, solution deposited	0.02
didodecyl	0.02
bis(butoxypropyl)	0.03
bis(octyloxypropyl)	0.01
anthradithiophene: unsubstituted	0.02-0.09
dihexyl	0.11-0.15
dihexyl, solution deposited	0.01-0.02
didodecyl	0.04-0.14
dioctadecyl	0.04-0.06

The incorporation of an oxygen atom in the side chain of a thiophene hexamer had little effect on the mobility of a sublimed film.   Lengthening the side chain much beyond six was slightly detrimental to the mobility, and did not aid solubility either since interchain interactions reinforce crystallinity when the chains are sufficiently long. The oxygen atom in a medium chain roughly doubled the solubility.

In the case of the anthradithiophenes, addition of side chains allowed the solubilization of a ring system that was otherwise insoluble for all practical purposes. Hexyl and dodecyl chains imparted significant increases in mobility as well, with the listed mobility ranges reflecting different substrate temperatures during deposition. The highest mobilities were obtained at about 80 °C. Even the octadecyl derivative had a substantial mobility, a tribute to the strict two-dimensional nature of the charge transport and the continuity of the layers, since the bulk dioctadecylanthradithiophene solid is predominantly insulating hydrocarbon.

The medium-chain hexamers and dihexylanthradithiophene formed polycrystalline films when cast from solution. Some of the films appeared rough even without magnification, but devices were operational over areas of many square millimeters, which were the areas of the films cast. The hexamer mobilities were only diminished by factors of two from the best reported values for sublimed films. The anthradithiophene was lowered by about an order of magnitude, but still gave a substantial mobility. The demonstration of solution casting of a compound whose melting point is nearly 400 °C is particularly noteworthy, and would not be possible for pentacene because of the difficulty of end substitution and the oxidative instability of the all-hydrocarbon semiconductor.

None of the processes described here could be considered optimized. There is considerable opportunity to refine the purification schemes for individual compounds and for the solvents from which films are cast. Concentrations, times, temperatures, and pressures can all be varied in the casting and evaporation procedures. The use of smoother, organic-based dielectric surfaces as substrates for the semiconductor films may result in more homogeneous films and will also be a further step toward the realization of all-organic and all-printed electronics.

REFERENCES

1. A.R. Brown, C.P. Jarett, D.M. de Leeuw, and M. Matters, Syth. Metals **88**, p.37 (1997); H.E. Katz, J. Mater. Chem. **7**, p. 369 (1997).

2. G. Horowitz, D. Fichou, X. Peng, and F. Garnier, Synth. Metals **41-43**, p. 1127 (1991), F. Garnier, G. Horowitz, X. Peng, and D. Fichou, Synth. Metals **45**, p. 163 (1991).

3. F. Garnier, A. Yassar, R. Hajlaoui, G. Horowitz, F. Deloffre, B. Servet, S. Ries, and P. Alnot, J. Am. Chem. Soc. **115**, p. 8716 (1993); H.E. Katz, A. Dodabalapur, L Torsi, and D. Elder, Chem. Mater. **7**, p. 2238 (1995).

4. Y.Y. Lin, D.J. Gundlach, S.F. Nelson, and T.J. Jackson, IEEE Transactions on Electronic Devices, **44**, p. 1325 (1997).

5. Z. Bao, Y. Feng, A. Dodabalapur, V.R. Raju, and A.J. Lovinger, Chem. Mater. **9**, p. 1299 (1997).

6. A.R. Brown, A. Pomp, C.M. Hart, and D.M. de Leeuw, Science **270**, p. 972 (1995).

7. F. Garnier, presented at the 1997 Materials Research Society Spring Meeting, April 1997, San Francisco, California.

8. H.E. Katz, J.G. Laquindanum, and A. J. Lovinger, Chem. Mater., accepted for publication; J.G. Laquindanum, H.E. Katz, and A.J. Lovinger, J. Amer. Chem. Soc., accepted for publication.

ELECTRICAL CHARACTERIZATION OF THIN SINGLE CRYSTALS OF SEXITHIOPHENE

C. Daniel Frisbie, Eric L. Granstrom, and Michael J. Loiacono
Department of Chemical Engineering and Materials Science, University of Minnesota,
Minneapolis, MN 55455

ABSTRACT

Extremely thin, single crystals of sexithiophene (6T), 2-14 nm thick and 2-5 μm in length and width, can be grown on flat gold substrates by thermal evaporation. The thickness dimension corresponds to 1-6 monolayers (ML) of 6T molecules arranged with their long axes nearly perpendicular to the substrate. We have measured the current-voltage (I-V) characteristics through the thickness of these crystallites, after doping them with iodine, using conducting probe atomic force microscopy (CPAFM). The I-V traces are linear in the ±50 mV regime. The conductance (I/V) of the doped 6T crystals does not decrease monotonically with increasing thickness as might be expected, but instead has a maximum at 3 ML thickness, and we discuss several possible explanations for this observation.

INTRODUCTION

Sexithiophene (6T) is a p-type organic semiconductor which has been studied extensively in thin film transistor (TFT) experiments because its hole mobility rivals that in amorphous silicon, making it a viable candidate material for all-plastic TFT technology.[1] In these TFT experiments, 6T is

6T

typically deposited as a polycrystalline thin film between metallic source and drain electrodes separated by at least 1 μm. The measured transport characteristics, for example transconductance and mobility, represent average film properties. Our research group is exploring strategies for probing transport over individual grains of organic semiconductors such as 6T with an ultimate aim to provide quantitative information on effects of grain boundaries and defects on transport properties in organic films.[2]

In this paper, we summarize results on the use of a new technique, conducting probe atomic force microscopy (CPAFM),[2,3] for measuring electrical conductance of extremely thin, doped 6T crystallites (Scheme I). CPAFM combines nanoscale electrical characterization with topographic imaging by employing a metal coated cantilever-tip assembly as both a scanning electrical contact and force sensor. The technique is well suited to electrical characterization of a variety of organic structures, having the advantage relative to scanning tunneling microscopy that the current-voltage (I-V) characteristic is decoupled from the sample position feedback mechanism.

B. CPAFM of 6T

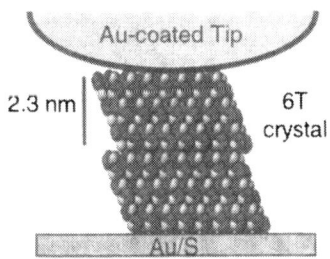

Scheme 1: Contact of a thin 6T crystal with a CPAFM probe.

The 6T crystallites examined in this study were grown on pretreated flat gold substrates by thermal evaporation. They ranged from 2-14 nm thick and 2-5 μm in length and width. AFM imaging demonstrated that these crystals consisted of consecutive layers of 6T molecules oriented with their long axes tilted slightly (~20°) with respect to the substrate normal. We have probed the conductance through the thickness of the crystallites, *i.e.*, in the direction perpendicular to the substrate, by contacting the top of the crystal with the CPAFM tip and using the gold substrate as the second electrical contact, as shown in Scheme I.

EXPERIMENTAL

Preparation of Gold Substrates. Flat gold substrates (mean roughness < 1 nm) were prepared using a previously reported templating procedure.[4] Briefly, 2000 Å of gold was deposited by thermal evaporation onto a heated mica sheet (300 °C) at 1×10^{-6} Torr. After cooling, the gold side of the gold coated mica sheet was glued to a glass cover slip (~1 cm diameter) with epoxy and left to cure at room temperature for 24 h. The mica was then cleaved off using a razor blade exposing a smooth Au surface. These gold substrates were immersed for 30 seconds in a 1 mM aqueous solution of NaSH (filtered). The purpose of this step was to coat the gold with a layer of S atoms. The presence of S on gold surfaces treated in this manner was verified by X-ray photoelectron spectroscopy (XPS).

Deposition of 6T. Thin crystallites of 6T were grown on the S-treated gold substrates by vacuum sublimation at a pressure of $\sim 10^{-4}$ Torr. The deposition was accomplished by ramping the temperature (~1 °C/min) of a sand bath surrounding the bottom part of an evacuated glass sublimation vessel containing a few grains of powdered 6T. When the sand temperature reached 260-280 °C, the sand bath was removed and the vessel allowed to cool. During this deposition process, the substrate temperature was maintained at ~170°C. We found that passivating the gold substrates with S atom was helpful in growing discrete crystals reproducibly on gold surfaces. The 6T crystals were doped by exposing them to I_2 vapor at room temperature for exactly 30 min. Doping resulted in no change in structure detectable by AFM.

Conducting Probe Atomic Force Microscopy. CPAFM was perfomed using a Digital Instruments Nanoscope III AFM as previously described.[2] A gold-coated AFM probe was mounted in a glass cell designed for AFM tapping mode imaging under liquids. The purpose of using the glass tapping mode cell (instead of the more common metal version) was to electrically isolate the tip from the rest of the head. The metal spring clip which holds the cantilever substrate in place was used to make electrical contact to the conductive AFM probe. Electrical contact was made to the gold substrate by connecting a fine gauge wire with silver paint. The leads to the AFM probe and the substrate were connected to the terminals of a Keithley model 236 source measure unit.

RESULTS AND DISCUSSION

Figure 1 shows an AFM image of a highly terraced 6T crystallite grown on a flat gold substrate. The height of the individual terraces is 2.3 nm which corresponds to half the c parameter of the 6T unit cell reported by Garnier.[1d] These terrace heights, in combination with molecular resolution imaging on individual terraces, demonstrate that the 6T crystallites consist of layers of 6T molecules oriented with their long axes approximately perpendicular to the substrate.

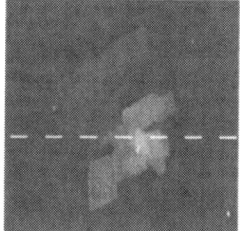

Figure 1: A 6T crystallite on gold.

Figure 2 shows AFM topographs of six individual 6T crystallites ranging in thickness from 1-6 6T monolayers (ML) (~2-14 nm). Figure 3a shows the I-V characteristics of each of these crystallites recorded using the same CPAFM probe (Scheme 1). Each of the I-V curves is linear over the ±50 window probed. The conductance (I/V) associated with each of these traces is plotted vs. crystallite thickness in Figure 3b. The surprising result is that the plot in Figure 3b shows a peak in conductance at 3 ML. If the transport were ohmic, one would expect conductance to be inversely proportional to thickness. That is, one expects the 1 ML crystal to be most conductive since it is thinnest. Indeed, for crystals 3 ML thick and thicker, the conductance appears to scale approximately inversely with thickness.

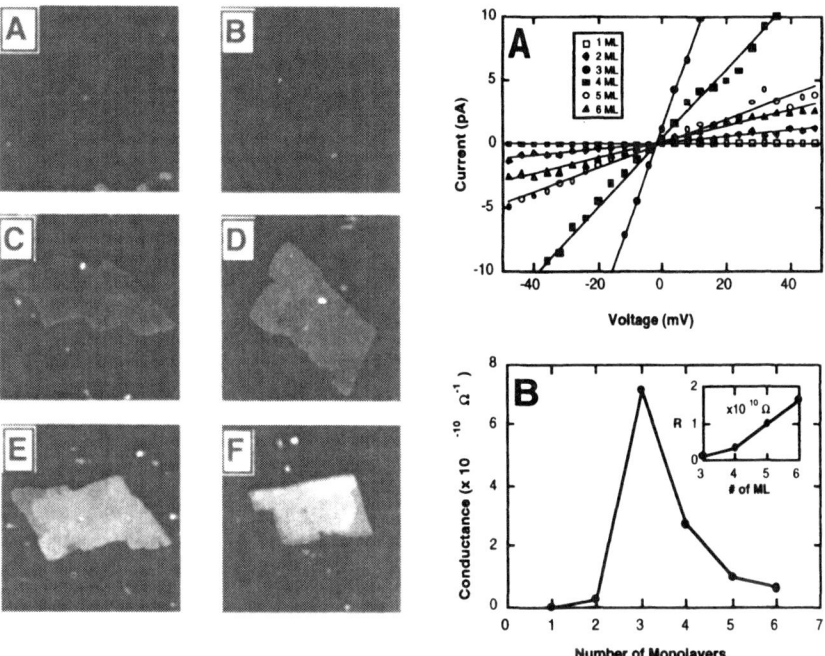

Figure 2: AFM topographs of 1-6 ML (A-F) 6T crystallites grown on gold. Crystallite thickness in (A) is 2.3 nm; in (F) it is 13.9 nm.

Figure 3: (A) I-V characteristics of 1-6 ML crystals shown in Figure 2. (B) Conductance vs. number of ML.

The challenge is to explain why the thinner crystallites have such low conductance. There are several possible explanations which we have discussed previously.[2] One is that the 1 and 2 ML crystallites are inherently less conductive because they have a different crystal structure or are less susceptible to doping. However, we do not have AFM evidence that the structures of 1 and 2 ML crystals are significantly different from the 3-6 ML crystals, and it is difficult to rationalize why both the 1 and 2 ML crystals would not be doped on exposure to iodine. A second possible explanation is that there is a charge injection barrier at the gold-6T interface. The data in Figure 3b are consistent with the idea that charge cannot be injected into any 6T layer in direct contact with gold.[5]

While we cannot as yet pinpoint the definitive reason for the peak in conductance in Figure 3b, we have shown here that CPAFM may be used to characterize transport in these delicate semiconductor crystallites. Future work will aim toward using CPAFM to measure resistance associated with individual grain boundaries in 6T and other materials and toward conductance measurements through single monolayers.

ACKNOWLEDGMENT

This work was supported by the National Science Foundation (DMR-9624154-001) and the University of Minnesota. ELG thanks the Department of Defense for a National Defense Science and Engineering Graduate Fellowship.

REFERENCES

1. (a) Torsi, L.; Dodabalapur, A.; Rothberg, L. J.; Fung, A. W. P.; Katz, H. E. *Science* **1996**, *272*, 1462. (b) Lovinger, A. J.; Rothberg, L. J. *J. Mater. Res.* **1996**, *11*, 1581. (c) Dodabalapur, A.; Katz, H. E.; Torsi, L.; Haddon, R. C. *Science* **1995**, *269*, 1560. (d) Dodabalapur, A.; Torsi, L.; Katz, H. E. *Science* **1995**, *268*, 270. (e) Lovinger, A. J.; Davis, D. D.; Ruel, R.; Torsi, L.; Dodabalapur, A.; Katz, H. E. *J. Mater. Res.* **1995**, *10*, 2958. (d) Horowitz, G.; Bachet, B.; Yassar, A.; Lang, P.; Demanze, F.; Fave, J.-L.; Garnier, F. *Chem. Mater.* **1995**, *7*, 1337. (e) Garnier, F.; Hajlaoui, R.; Yassar, A.; Srivastava, P. *Science* **1994**, *265*, 1684. (f) Garnier, F.; Yassar, A.; Hajlaoui, R.; Horowitz, G.; Deloffre, F. *Electrochimica Acta*, **1994**, *39*, 1339. (g) Garnier, F.; Yassar, A.; Hajlaoui, R.; Horowitz, G.; Deloffre, F.; Servet, B.; Ries, S.; Alnot, P. *J. Am. Chem. Soc.* **1993**, *115*, 8716. (h) Horowitz, G.; Peng, X.; Fichou, D.; Garnier, F. *J. Appl. Phys.* **1990**, *67*, 528. (i) Horowitz, G.; Fichou, D.; Peng, X.; Xu, Z.; Garnier, F. *Sol. State. Comm.* **1989**, *72*, 381.

2. Loiacono, M. J.; Granstrom, E. L; Frisbie, C. D. *J. Phys. Chem. B*, to be published (1998).

3. Dai, H.; Wong, E. W.; Lieber, C. M. *Science* **1996**, *272*, 523.

4. (a) Hegner, M.; Wagner, P.; Semenza, G. *Surf. Sci.* **1993**, *291*, 39. (b) Wagner, P.; Hegner, M.; Guntherodt, H. J.; Semenza, G. *Langmuir* **1995**, *11*, 3867.

5. (a) Garnier, F.; Kouki, F.; Hajlaoui, R.; Horowitz, G. *MRS Bulletin* June, **1997**, 52. (b) Parker, I. D. *J. Appl. Phys.* **1994**, *75*, 1656.

NOVEL LIQUID CRYSTAL DISPLAYS BASED ON HIGHLY POLARIZED PHOTOLUMINESCENT POLYMER FILMS

CHRISTIAN SARWA, ANDREA MONTALI, CEES BASTIAANSEN, CHRISTOPH WEDER, PAUL SMITH
Department of Materials, Institute of Polymers, ETH Zürich, CH-8092 Zürich, Switzerland.

ABSTRACT

Since the early 90's, much research has focused on the photoluminescence (PL) and electroluminescence (EL) properties of conjugated polymers, because of their potential application as emitting layer in EL devices. The introduction of uniaxial molecular orientation into films of luminescent polymers was naturally found to yield structures that emit *polarized* light. Rather surprisingly, the photoluminescence properties of oriented, conjugated polymers have attracted substantially less attention, especially from an application point of view. In this paper we report the fabrication of highly-polarized photoluminescent polymer films based on poly(2,5-dialkoxy-*p*-phenyleneethynylene)s (PPE), and their use in a new family of liquid crystal displays (LCDs). As one relevant example, a back-lit twisted-nematic configuration of an LCD was built, in which one of the absorbing polarizers was replaced by a polarized PL film, characterized by a dichroic ratio in excess of 70. Such devices can exhibit a substantial improvement in brightness, contrast and viewing angle, since the polarized photoluminescent films can combine two separate features, i.e. the functions of a polarizer and an efficient color filter.

INTRODUCTION

Liquid crystal displays, which currently represent the dominant flat panel display technology, exhibit severe limitations in brightness and energy efficiency caused by the use of absorbing polarizers and color filters which convert a major amount of the incident light into thermal energy.[1] The use of photoluminescent materials which act as 'active' color filters and, therefore, might enhance the visual performance of LCDs, was previously suggested.[2-4] Several principal possibilities exist to incorporate PL materials into LCDs, including the use of fluorescent liquid crystals or the dispersion of luminescent molecules in a conventional LC layer,[2] the application of PL plates[3] or front-face screens.[4] However, the proposed devices suffer from a number of drawbacks, related to limited stability, the difficulty to produce pixilated devices, depolarization effects, or the required thickness and large area of the luminescent layer.[3]

Here, we introduce a new concept for the design of photoluminescent LCDs which are based on photoluminescent polarizers. The latter were recently demonstrated to combine two separate features: the linear polarization of light and the efficient generation of bright color.[5] Photoluminescent polarizers rely on the fact that macroscopically uniaxially oriented photoluminescent matter usually also exhibits anisotropic, i.e. *linearly polarized*, absorption and emission. This phenomenon has been known for inorganic crystals for more than a century[6] and recently was also reported for uniaxially oriented films of conjugated, luminescent polymers.[7] We suggest that the use of such polarized PL films in liquid-crystal-based photoluminescent display devices may lead not only to a simplification in device design, but also to a substantial increase in device brightness and efficiency, and, in addition, in specific device configurations to a significant improvement of the viewing angle performance.

RESULTS AND DISCUSSION

The PL polarizers used in this work are based on uniaxially oriented blend films of EHO-OPPE[8,9] (2 % w/w), a poly(2,5-dialkoxy-p-phenyleneethynylene) derivative substituted with linear and sterically hindered alkyloxy groups in an alternating pattern (Figure 1), and ultra-high molecular weight polyethylene (UHMW PE). Uniaxially oriented films were prepared by solution casting and subsequent tensile drawing in the solid state. This process is extremely advantageous, because it not only leads to highly oriented, strongly luminescent and extraordinarily stable photoluminescent films, but also allows to produce mechanically coherent, PL polarizers of minute thickness, a feature that is essential for effective use in PL LCDs.[5] The prepared, yellow-green light-emitting PPE-based PL polarizers were of a thickness of about 2 μm (i.e. a fraction of the typical thickness of the LC layer in standard devices), and were characterized in a high degree of polarization in both, absorption and emission, as is evidenced by the dichroic ratios (defined as the ratios of the *integrals* of the respective transitions) of 15 in absorption and 22 in emission (Figure 2).

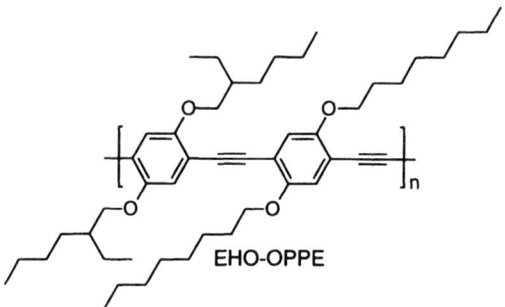

Figure 1: Molecular structure of EHO-OPPE.

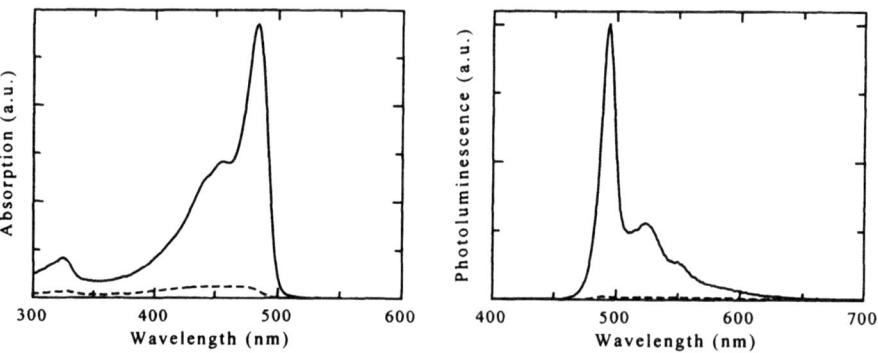

Figure 2: Polarized absorption (left) and photoluminescence (right) spectra of an oriented film (draw ratio = 80) of a 2 % w/w EHO-OPPE / UHMW PE blend. Spectra were recorded with polarizers oriented parallel (solid line) and perpendicular, respectively, (dashed line) to the orientation direction of the polymer blend film.

It is obvious, that in photoluminescent display devices, principally both radiation from the light source employed for the photoexcitation of the PL polarizer or, alternatively, light emitted from the exited polarized PL film may be switched by the LC electrooptical light valve. Depending on the selected configuration, the predominantly relevant polarization characteristic of the polarized PL film is either a high degree of anisotropy in absorption or PL emission. Of course, a variety of device configurations can be envisioned.[10]

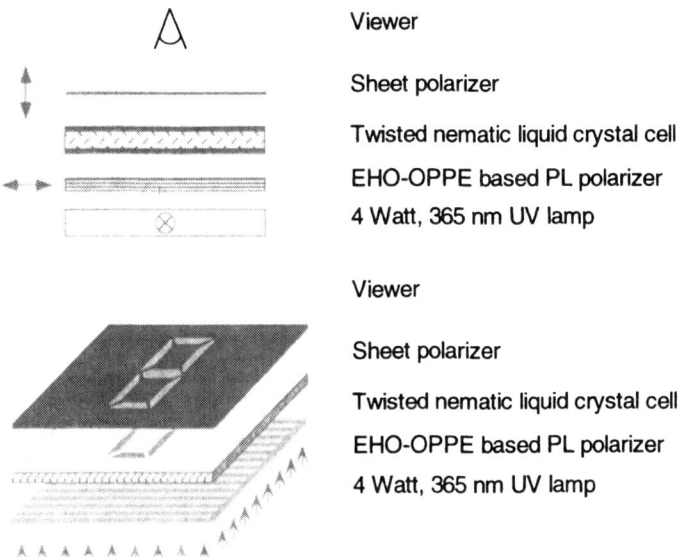

	Viewer
	Sheet polarizer
	Twisted nematic liquid crystal cell
	EHO-OPPE based PL polarizer
	4 Watt, 365 nm UV lamp
	Viewer
	Sheet polarizer
	Twisted nematic liquid crystal cell
	EHO-OPPE based PL polarizer
	4 Watt, 365 nm UV lamp

Figure 3: Schematic structure of a PL liquid crystal display device.

As one relevant example, we here describe a back-lit twisted nematic (TN) configuration, in which light emitted from the polarized PL film is switched (Figure 3). The arrangement consists of a linear polarizer (here used as analyzer), an electrooptical light valve, the polarized PL film and a light source. In this device configuration the light emerging from the light source is at least partially absorbed by the PL polarizer, from where subsequently *polarized* light is emitted. The light emitted in the direction of the viewer either passes the combination of electrooptical light valve and polarizer (switching state "bright") or is blocked (switching state "dark"). Maximum contrast is obtained when (i) the PL polarizer is characterized by a high dichroic ratio for PL emission, and (ii) the portion of light visible to the human eye that is emitted by the light source, but not absorbed by the polarized PL film - and thus exits the device in the direction of the viewer - is minimal. This can be achieved by different means; for example, by using an additional cut-off-filter or employing a UV light source. In the latter case the portion of UV light that is not absorbed by the PL polarizer, may be absorbed by the electrooptical light valve and the polarizer, since both these elements are usually absorbing in the UV regime. The device investigated in this work comprised a PPE-based PL polarizer and a UV lamp emitting at 365 nm as the light source. Of course, the emission wavelength of the light source is by no means optimized for the absorption of the PPE-based polarized PL layer, as is evident from the PPE's absorption spectrum shown in Figure 2; however, this light source was used because of its wide availability. The switching of the device yielded a significant change in brightness, that was perceived by the human eye as a change from a very bright yellow-green to almost completely dark. The outstanding contrast and the absolute brightness of the "on" and the "off" state were quantified with a standard luminance meter: a brightness of 30 ("bright") and 4 cd/m^2 ("dark"), respectively, was measured for the respective switching states. These values for brightness and contrast compare favourably with those of a similar, commercial, direct-view LCD (17 vs. 7 cd/m^2) that was analyzed under identical ambient conditions.

Of course, the absolute brightness of these PL-based LCDs can easily be further enhanced by increasing the absorption of the PL polarizer at the excitation wavelength. This, obviously, can be easily achieved by increasing the thickness of the PL polarizer or the concentration of the luminophor. Preliminary experiments have indeed resulted in devices according to Figure 3 with a brightness in excess of 60 cd/m^2.

CONCLUSIONS

In summary, we have introduced a new concept for the design of LCDs that comprise one or multiple, polarized photoluminescent layers that are characterized in a high degree of polarization in absorption or emission. These thin, anisotropic photoluminescent films efficiently combine two separate features, i.e. the polarization of light, and the generation of bright color. Hence, the new photoluminescent display devices can offer a substantial increase in device brightness and efficiency.

EXPERIMENTS

Materials

EHO-OPPE (M$_n$ ~ 1x10^4 gmol^{-1}) was prepared according to the procedures given elsewhere[8], and ultra-high molecular weight polyethylene (Hostalen Gur 412, M$_w$ ~ 4x10^6 gmol^{-1}) was obtained from Hoechst AG.

Preparation of PL Polarizers[5]

Blend films were prepared by casting a solution of EHO-OPPE (10 mg) and UHMW PE (500 mg) in xylene (50 g) (dissolution at 130 °C, after degassing the mixture in vacuum at 25 °C for 15 min) into a petri-dish. The resulting gels were dried under ambient conditions for 24 h and the resulting blend films were drawn at temperatures of 90 - 120 °C to a draw ratio (final length / initial length) of about 80, to yield oriented films of a final thickness of about 1 - 2 μm.

Devices

The display devices under investigation (Figure 3) comprised a commercially available, standard TN cell (Texas Instruments) fitted with one linear sheet-polarizer. The PL polymer films were sandwiched between two slides of fused silica and silicon oil was applied to minimize light scattering at the film surfaces. A UV lamp (Bioblock, VL-4LC, 4 Watts), operated at 365 nm, was employed as the light source.

Photophysical Characterization

Polarized UV-vis spectra were recorded with a Perkin Elmer Lambda 900 instrument, fitted with motor-driven Glan-Thomson polarizers. Corrected PL spectra were recorded on a SPEX Fluorolog 2 (Model F212 I). All emission dichroic ratios were obtained under isotropic excitation at 365 nm. The absolute brightness of the devices was measured with a Minolta LS 100 luminance meter, which was fitted with a No. 110 and a No. 122 close-up lens.

REFERENCES

1. T.J. Nelson and J.R. Wullert II, Electronic Information Display Technologies (World Scientific Publishing, Singapore, 1997).
2. H.J. Coles, Liq. Cryst. **14**, 1039 (1993).
3. G. Baur and W. Greubel, Appl. Phys. Lett. **31**, 4 (1977).
4. W.A. Crossland, I.D. Springle and A.B. Davey, Proc. SID Symp. Digest of Technical Papers, *27*, (Society for Information Display, Santa Ana, 1997), p. 837.
5. C. Weder, C. Sarwa, C. Bastiaansen, and P. Smith, Adv. Mater. **9**, 1035 (1997).
6. E. Lommel, Ann. d. Phys. und Chem. **244**, 534 (1879).
7. T.W. Hagler, K. Pakbaz, J. Moulton, F. Wudl, P. Smith and A.J. Heeger, Polymer Comm. **32**, 339 (1991).
8. C. Weder, and M.S. Wrighton, Macromolecules **29**, 5157 (1996).
9. D. Steiger, P. Smith and C. Weder, Macromol. Chem. Rapid. Commun. **18**, 643 (1997).
10. C. Weder, C. Bastiaansen, C. Sarwa and P. Smith, EP 97111228.9.

ELECTRO-OPTIC EFFECTS IN NANOPHASE POLYMER DISPERSED LIQUID-CRYSTAL SYSTEMS

R. S. Blacker, K. L. Lewis, I. Mason, I. Sage, K. Webb
DERA Malvern, St Andrews Road, Malvern, Worcs, WR14 3PS, ENGLAND.

ABSTRACT

Research into electro-optic effects in nanophase polymer dispersed liquid crystal (PDLC) materials has highlighted their potential as materials for a new class of tuneable filters. The structures, based on UV cured phase separated composites, contain liquid crystal both as discrete nano-scale droplets, and as material dissolved in the polymeric host. The essential difference between these materials and more conventional PDLC's is the scale of the refractive index inhomogeneity which is considerably smaller than the wavelength of visible light. Based upon effective medium approximations, the composite thus acts as a single isotropic medium, whose average refractive index is dependant on the level of applied electric field. Tuneable filters have been fabricated using the composite material for use in the visible spectral band.

INTRODUCTION

Electrooptical systems based on composites of liquid crystals (LCs) and polymeric hosts have been widely investigated by many workers [1-4]. Such systems are commonly known as polymer dispersed liquid crystals (PDLCs). The driving force for much of this work is the possibility of cheap and effective large area visible light shutters [5]. Today such systems are readily available, and, indeed, large electrooptic PDLC privacy windows are available that can be switched from opacity to a transparent state by applying a small electric field across the PDLC.

PDLCs are formed when an intimate mixture or solution of liquid crystal and [suitable] polymeric host are made incompatible [6]. This occurs as the interaction energy between the polymer (monomer or oligomer) and the LC is increased, or as the temperature or pressure applied to the system is reduced. This effect is described by the Flory-Huggins Model which derives the unitless interaction parameter χ. As the interaction parameter is increased the solubility of the LC in the polymer phase is reduced thereby driving the LC out of solution and allowing it to form the discrete areas of LC that give the PDLC its unique dispersed phase properties.

It is thus apparent that the final properties of a given PDLC system will depend strongly on the initial solubility of the LC in the polymer precursor, the degree of change of solubility and the rate at which the change occurs (i.e. the rate of polymerisation or freezing). Current systems are generally based on phase separation processes which result in LC droplet sizes of the order of a few microns in dimension. This paper is concerned with material in which the droplet size is reduced to the order of a few tens of nanometers, by control of the kinetics of the phase separation process.

MATERIALS AND CELL MANUFACTURE

The composite system investigated is based on the photocurable monomer di-penta-erythritol hexa acrylate (DPEHA) and up to 40% of fluorinated liquid crystal (Merck BL036). The system undergoes the photocuring reaction illustrated in figure 1.

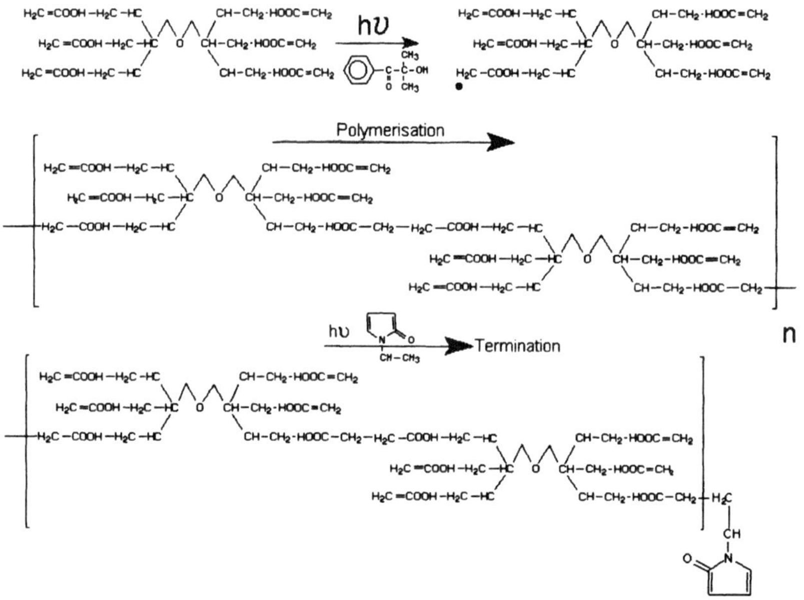

Figure 1. Polymerisation of PDLC Host.

In this reaction, the photoinitiator is phenyl-2-hydroxy-2-propyl ketone, whilst the termination step is provided by the photonically activated 1-vinyl-2-pyrollidone (1V2P) radical. Different degrees of system crosslinking can be achieved by using fewer functional groups on the base monomer or increased levels of the reaction terminator (1V2P). A typical composition would consist of 30% DPEHA, 30% 1V2P, 35% liquid crystal and 5% photoinitiator.

In the uncured and fluid state the system is suitable for cell manufacture using several techniques. The present work has concentrated on two techniques: draw bar methods, and vacuum filling of preassembled cells. The results presented here were obtained from cells manufactured using the latter method. Transparent cells were assembled with a 12.5µm air spacing (using bead spacers) and opposing, overlapping indium tin oxide (ITO) electrodes. This allowed an electric field to be applied to the PDLC film in a longitudinal direction.

Enhanced optical characteristics were achieved by exploiting Fabry Perot cavities as a feature of the cell design. To achieve this, the ITO glass substrates were vacuum coated with a simple quarter wave stack. Designs using SiO_2 and TiO_2 (or Ta_2O_5) as the low and high refractive index layers were used to obtain transmission maxima close to 60% at 550nm. An appropriate mask was used to allow electrical contact to be maintained with the transparent ITO electrode underlayer.

After vacuum filling the cells, curing was carried out using a low power, broad band UV light source. This provided $32\mu Wcm^{-1}$ of radiant energy at the surface of the cell being cured. Under standard conditions (in the absence of the mirror stack) the system described achieved an initial 'gel' cure in under 1 second, whilst full cure was achieved in less than 5 seconds. However, the transmission characteristics of the mirror stack led to significant attenuation of the UV component of the irradiating light with a consequent effect on cure rate. This had to be compensated by increase in source intensity. A typical cell configuration is illustrated in figure 2.

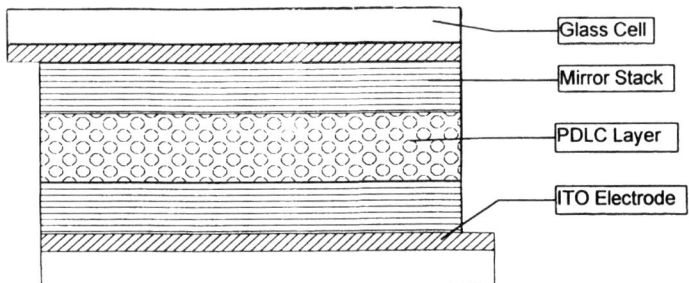

Figure 2. Standard 12.5mm Fabry Perot Cell Configuration.

EXPERIMENTAL PROCEDURE

Fabry-Perot etalons containing the nano-phase PDLC material were assessed using optical spectroscopy techniques under both dc and ac fields. Measurements were made initially using a Lambda-9 spectrometer allowing one minute settling time at each voltage step. Dynamic measurements of electrooptic effect were made using an EG&G optical multichannel analyser capable of capturing one full spectrum every 30ms. These were supplemented by measurements at a single wavelength using a diode laser in conjunction with a silicon photodiode to measure the transmitted beam intensity.

The measured response was compared with data calculated for the multilayer system using standard matrix inversion techniques. This allowed the determination of the optical constants of the PDLC system, and a good estimation of the effective quadratic electrooptic coefficient using equation 1.

$$\overline{\Delta n} = -\frac{1}{2} n_0^3 R E_3^2 \qquad (1)$$

The morphology of the PDLC system was investigated by scanning electron microscopy (SEM) using cleaved samples formed using the draw bar technique. The LC phase was extracted using isopropanol before gold coating to reduce charging effects. The SEM examinations were carried out using a JEOL JSM 6400F field emission system at 20kV.

RESULTS and DISCUSSION

Figure 3 shows the effect on morphology of adding liquid crystal to the polymer system. The difference in morphologies is significant, with the pure host polymer exhibiting very little contrast. Indeed the features shown in figure 3(a) could only made visible by imaging at 45°. In comparison, the LC loaded material exhibited significant texture and in the example presented in figure 3(b) clearly highlights the presence of spherulites of ball-polymer morphology. Iso-propanol (IPA) extraction removes the LC phase which in this particular example resides largely at the interstices between the spherulites. The size of the interstices varies considerably, however at between 20 to 50nm they remain non-scattering in the visible spectrum, thus enabling the rapidly cured PDLC system to remain highly transparent across the visible band under all conditions of applied field. Studies of the electrooptic characteristics of the PDLC system were carried out at wavelengths near 550nm.

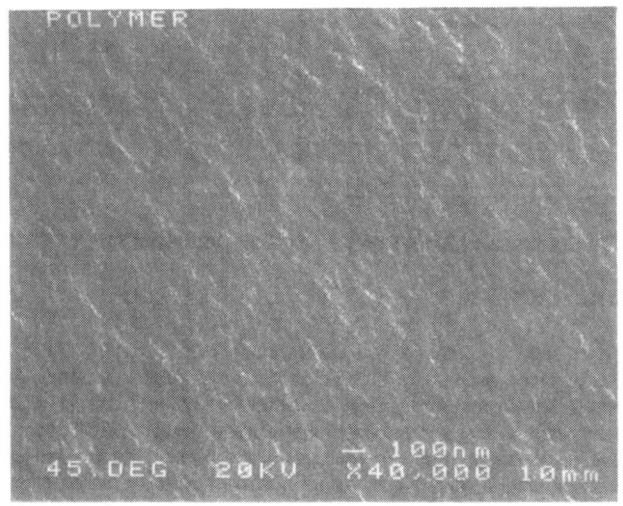

Figure 3(a) SEM Micrograph Of Pure Polymer Host

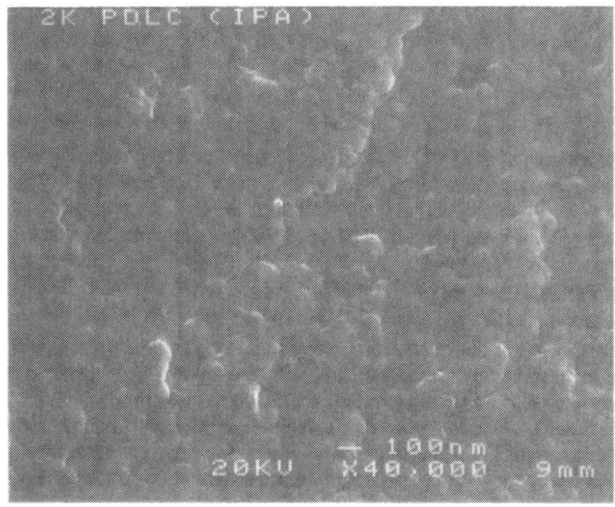

Figure 3(b) SEM Micrograph Of Nano-Phase PDLC Material, Highlighting The
Formation Of Ball Polymer Morphologies.

Data collected using the optical multichannel analyser is shown in figure 4. This waterfall plot has a time interval of 30ms between each collected spectrum. At 150ms into the baseline data collection, a dc voltage of 300V was applied to the ITO electrodes. This equates to a field level across the entire device of 24Vμm^{-1}. The apparently high value of the required field is largely a result of the effect of the mirror stacks and the permittivity of the polymer host.

For a model based on discrete droplets of LC in the host, with 50% of the LC remaining

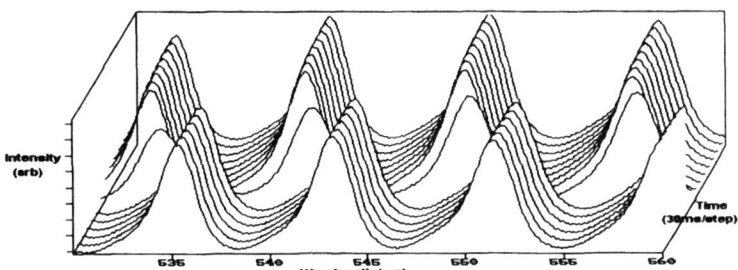

Figure 4. Frequency Tuning of 11.6μm Fabry Perot PDLC Device.

in the polymer after curing, it can be calculated that for a host of natural dielectric permittivity 3.5, and an LC phase of permittivity 22.0, the host permittivity is proportionately increased to 12.75. Thus, by applying the Laplace solution [8] to dielectric (LC) spheres in a lower permittivity dielectric host, the effective electric field seen by the active LC droplets is calculated to be as low as 80% of the applied field.

This reduction in field intensity is further attenuated by the use of the dielectric layers in the Fabry Perot mirror structure. These layers attenuate the field level by a further 80%. Consequently, for an applied field of 24Vμm^{-1}, the liquid crystal phase has an apparent field of 3.8Vμm^{-1} applied across it.

Optical modelling (figure 5) of the measured data indicates that the initial PDLC system

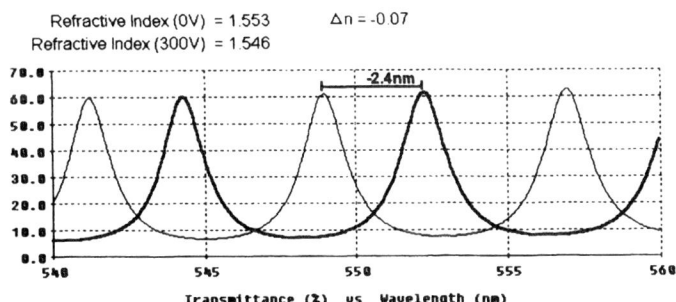

Figure 5. Modelled Frequency Tuning Of A PDLC Device.

refractive index of 1.553 is reduced to 1.546 by the electrooptic effect.

Without any modifications to the applied electric field intensity, the calculated quadratic electrooptic effect is $6.5 \times 10^{-18} V^2 m^{-2}$. However, reducing the apparent electric field for the reasons stated previously yields an apparent increase in the calculated electrooptic Kerr effect to $5.05 \times 10^{-16} V^2 m^{-2}$.

Such levels of applied electric field are still considerably greater that that found necessary by other workers [7]. The LC phase dimensions of 50nm are an order of magnitude smaller than those typically reported for conventional PDLC systems that rely on scatter modes of operation. At this scale it is possible that the olefinic tails of the LC phase are being confined to the polymeric host phase which itself has a largely olefinic character. The mesogenic units of the LC are thought to be entrained within the interstices between the ball like polymer host morphology in concentrated regions of LC. Thus, when an electric field is applied, the field levels required to align the constrained mesogens are greater than in the case where the olefinic tail is not constrained by entanglements with the host polymer. Evidence for this reasoning is found in the low time constants measured for this class of nanophase PDLC.

Temporal response measurements taken by measuring the attenuation of a diode laser beam at 670nm show that significant modulation can be detected at up to 16kHz. This indicates a response time of <65μs for this particular PDLC system. The modulation rate achieved by conventional scatter mode PDLC devices is often an order of magnitude greater than this value, thus indicating that such composites rely on the alignment of the LC phase which then takes several milliseconds to relax into the disordered state. In the nanoscale structures discussed in this paper such relaxation is 'forced' by the constrained olefinic LC tails to realign with the preferred low energy conformation. The probable alignment of the mesogenic units perpendicular to the host phase domain walls acts as a 'seed' to the non-attached LC units, thus allowing faster switch times to be obtained, with little hysteresis.

CONCLUSION

A new class of polymer dispersed liquid crystal devices have been investigated. It has been found that rapidly cured polymer host / guest systems will form nanoscale inclusions of electrooptically active liquid crystal phases. Such inclusions are below the size at which light scattering takes place, and thereby allow the formation of optically transparent PDLC structures. Optical spectroscopic measurements have led to the determination of the quadratic electrooptic coefficient of this system. Values in the region of $6.5 \times 10^{-18} V^2 m^{-2}$ are attractive for exploitation in several device applications.

Fabry Perot type cells have been utilised to enhance optical contrast and allow signal modulations of 50% to be recorded over tuning ranges of greater than 2π. Fast laser modulation techniques have allowed response times of less than 65μs to be recorded.

REFERENCES

1.   P S Drzaic *Liquid Crystal Dispersions*, World Scientific 1995
2.   J W Doane: *Liquid Crystals, Applications and Uses* (Ed B Bahadur), 361-395 (World Scientific 1990.
3.   H S Kitzerow: Liq Cryst **16** 1 (1994)
4.   H.Nomura, S.Suzuki, Y.Atarashi, Jap. J. Appl. Phys. 30 327-330 (1991)
5.   Y.Hirai, S.Niiyama, Y.Ooi, M.Kunigita, H.Kumai, M.Yuki, T.Gunjima, SID Dig. 594-597 (1991)
6   J.L.West, Mol. Cryst. Liq. Cryst., Nonlin. Opt. 157 427 1988
7   S. Tahata, M. Mizunuma, A. Tsumura, T. Fujimoto, T. Ando, T. Masumi. Proc. SPIE, 2651 101 1996
8   B I Bleaney *Electricity and Magnetism*, Oxford Press 1957

STRUCTURAL AND ELECTRONIC PROPERTIES OF SELF-ASSEMBLED SUPRAMOLECULAR GRID STRUCTURES: DOPING OF SUPRAMOLECULAR THIN FILMS

J. HASSMANN*, C.Y. HAHN*, O. WALDMANN*, E. VOLZ*, H-J. SCHLEEMILCH*, N. HALLSCHMID*, P. MÜLLER*, G.S. HANAN**, D. VOLKMER**, U.S. SCHUBERT**, J-M. LEHN**, H. MAUSER***, A. HIRSCH***, T. CLARK***

*Physikalisches Institut III, Universität Erlangen-Nürnberg, 91058 Erlangen, Germany
hassmann@physik.uni-erlangen.de
**Université Louis Pasteur, 4 rue Blaise Pascal, 67000 Strasbourg, France
***Institut für Organische Chemie, Universität Erlangen-Nürnberg, Henkestr. 42, 91058 Erlangen, Germany

ABSTRACT

Thin films of self-assembled supramolecular grid structures were prepared by electrochemical deposition. The surface structure of the films was determined by electrochemical in-situ scanning tunneling microscopy (STM), revealing ordered monolayers of the grid molecules on a Au(111) surface. The electronic structure of the films was studied by photoelectron spectroscopy. The thin films show semiconducting behavior with an insulating gap of ca. 2.5 eV. The size of insulating gap can be reduced to 0.5 eV by doping the films with additional Cd^{2+} ions, which is in agreement with molecular orbital calculations. Additionally, conductivity measurements were performed. By doping of the thin films, their conductivity could be increased by several orders of magnitude. This is due to additional electronic states in the insulating HOMO-LUMO gap, which is confirmed by UPS measurements.

INTRODUCTION

One major goal in current metallo-supramolecular chemistry is the controlled synthesis of specific arrays of metal ions in a designed organic ligand system. In view of the possible analogy of the metal centers in an insulating ligand system to quantum dots, these inorganic architectures are interesting model systems for functionalizations [1]. One important prerequisite is the preparation of ordered mono- and multilayer assemblies on solid substrates. Self assembled monolayers (SAMs) of organometallic nanoclusters have been reported, e.g. consisting of Li^+TCNQ^- [2]. Another promising approach is the electrochemical deposition of thin films. Such films have been prepared using pure organic structures like DNA bases [3] or pure metallic adsorbates like Ag on Au [4]. Nevertheless, there are few reports about the transfer and electronic functionalization of ordered monolayers of complex organometallic architectures. A striking result was the observation of a Coulomb staircase on a Langmuir-Blodgett film consisting of stearic acid and organometallic carborane clusters with a STM, proving single electron transport [3].

The molecules used in this work are novel supramolecular grid structures consisting of four Co^{2+} or Cd^{2+} ions located at the crossing points of four bis(bipyridyl)pyrimidine ligands. These grids are denoted as Co-[2×2] and Cd-[2×2], respectively. As a first important result, we have shown that the four Co^{2+} electron spins in the Co-[2×2] grid form an intramolecular antiferromagnetic domain [6]. Here, we present the structural and electronic properties of these supramolecular grid architectures. Thin films on solid substrates were prepared by electrochemical underpotential deposition. The electronic valence band structure was determined by photoelectron spectroscopy with vacuum-UV-radiation (UPS) and molecular orbital calculations. The grids reveal semiconducting behavior with an insulating gap of approximately 2.5 eV. Nevertheless, the conductivity of the films was too low for reliable electronic functionalizations. Therefore, the conducting properties had to be "tuned" by the incorporation of additional metal ions in the films. The size of the insulating gap could be reduced by doping of the thin films by addition of Cd^{2+} ions, which results in a strongly increased conductivity.

EXPERIMENT

The molecules

The structure of the 4,6-bis(6-(2,2'-bipyridyl)-pyrimidine) ligands and the crystal structure [7] of the supramolecular Co-[2×2] grid are depicted in Fig. 1. The ligands were synthesized as published [8]. The grids form by self-assembly in methanol and were investigated in acetonitrile solution. They were characterized by X-ray crystallography, UV spectroscopy, mass spectrometry, elemental analysis and electrochemical methods [9]. Characterization on solid substrates was performed with XPS and mass spectrometry. The length of the ligand is approximately 16 Å, the distance between the metal centers is 6.4 Å. The positive charges are countered by PF_6^- ions.

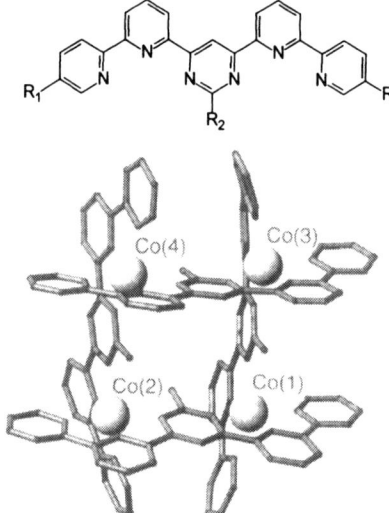

Figure 1: Crystal structure of the supramolecular Co-[2×2] grid (bottom), consisting of four Co^{2+} ions and four 4,6-bis(6-(2,2'-bipyridyl)-pyrimidine) ligands (top). The grids form by self-assembly in methanol. The positive charges are countered by PF_6^- ions R_1= -H, R_2= -CH_3.

Preparation and characterization of the thin films; molecular orbital calculations

Electrochemical preparation was performed using a EG&G model 362 potentiostat. Multilayer systems like 50 monolayers for UPS were also produced by depositing a defined volume on the substrate with a microliter syringe and slow evaporation of the solvent in a saturated solvent atmosphere. The same method was used for the doped films. The dopant was dissolved with the grids in acetonitrile in a defined stoichiometry.

Electrochemical in-situ STM measurements were performed using a commercial Nanoscope ECM with a Pt wire as counter-, and a Ag wire as reference electrode. UPS measurements were carried out at the HASYLAB,DESY/Hamburg using vacuum UV synchrotron radiation in an energy range between 40 eV and 110 eV. For detection of the photoelectrons, a cylindrical mirror analyzer was used. Base pressure was $5\,10^{-11}$ mbar. Semiempirical molecular orbital calculations with a VAMP 6.5 program package [10] were performed using the PM3 method [11]. The positions of the dopants were optimized using the constant crystal structure of the grids. Conductivity measurements were carried out using a Keithley 238 Source measure unit in a four point geometry.

RESULTS

Electrochemical in-situ STM

One important condition for physical functionalizations of the supramolecular architectures on solid substrates is the preparation of ordered structures. A promising approach is the electrochemical deposition, but the surface structure depends significantly on the adsorption voltage. Using an electrochemical in-situ STM setup, a layer of the Co-[2×2] grids was deposited by electrochemical underpotential deposition and imaged at the same time. In Fig. 2, we show a layer deposited at an adsorption voltage of -1V vs. Ag.

Figure 2: STM image of an ordered layer of Co-[2×2] grids on a Au(111) surface in an electrochemical cell. The adsorption potential was -1V vs. Ag. Measurement in nitrobenzene solution. The film is deposited over a monoatomic step on the gold surface.

At this voltage, an ordered surface structure of the Co-[2×2] grids on a Au(111) substrate is clearly visible. Details of the grid structure can not be resolved at room temperature.

UPS and molecular orbital calculations

The electronic structure of the grid films was studied by UPS. A thin film of pure Cd-[2×2] grids reveals an insulating gap of approximately 2.5 eV, independent of the preparation. This results in conductivities down to 10^{-7} S/cm. In any case, this is too small for functionalizations of the films as electronic devices or for detection of specific electronic phenomena like single electron transport or hopping. Therefore, the thin films were doped by the addition of $Cd(CF_3SO_3)_2$ salt. In Fig. 3, we show a UPS spectrum of 50 monolayers Cd-[2×2] film with two $Cd(CF_3SO_3)_2$ entities per grid. The UPS spectrum is in good agreement with the molecular orbital calculation. The most important feature is an additional state in the insulating gap, which is visible as a clear foot in the UPS spectrum, starting at a binding energy of ca. 0.5 eV. Calculations show that the $CF_3SO_3^-$ anions do not have any effect on the insulating gap; the state in the HOMO-LUMO gap is caused by adding Cd^{2+} ions to the film. In the measurement, this effect is weaker due to shielding of the positive charges of the additional Cd^{2+} ion by $CF_3SO_3^-$ or PF_6^- anions, which is not accounted for in the calculations. But this might have a strong influence on the size of the insulating gap. This additional state is not visible at all in UPS spectra or molecular orbital calculations of undoped films. The additional state in the gap results in a significant increase of the conductivity.

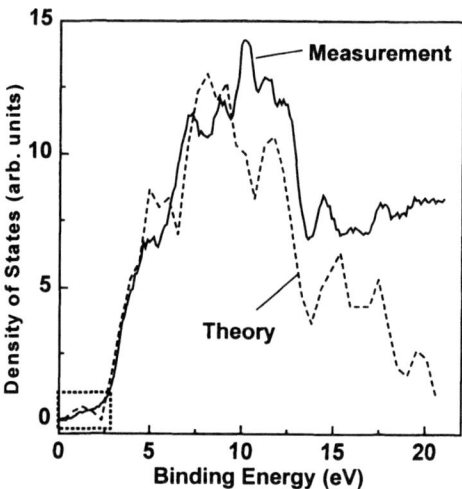

Figure 3: UPS spectrum and molecular orbital calculation of a doped Cd-[2×2] film: Theory: one Cd-[2×2] molecule with an additional Cd^{2+} ion. The density of states is depicted with an approximate experimental resolution of 0.5 eV. Measurement: 50 monolayers Cd-[2×2] with two $Cd(CF_3SO_3)_2$ per grid. Photon energy was 60 eV. Dotted box: Additional state in the HOMO-LUMO gap.

The width of the valence band is well described by the calculations. The states between approx. 2.5 eV and 8 eV are predominantly caused by the metal ions and the right half of the valence band is mainly made up of organic contributions. In the measurement, the metallic contributions are less pronounced because of their lower photon cross section compared to the lighter atoms in the organic parts at a primary photon energy of 60 eV. Nevertheless, the rather low energy of 60 eV was chosen due to the high resolution and low background of secondary electrons. The highest peak at approximately 9 eV binding energy mainly consists of contributions of the N donor atoms and Cd^{2+} metal centers in the grid. In the raw data material of the calculations, a sharp peak is visible next to a minimum, which is smeared out due to the averaging with the experimental resolution. Additionally, we find the well-known states of the carbon π and carbon σ bonds in the ring structures at 14.3 and 17.5 eV, respectively [12].

Conductivity measurements

Clear evidence for the doping of the supramolecular thin films is provided by conductivity measurements at various dopant concentrations. In Fig. 4, we show the conductivity of doped thin films of the Co-[2×2] grid as a function of the concentration of Cd^{2+} dopants. The concentrations were denoted in units of doping ions per grid. By the addition of one Cd^{2+} ion per grid, the conductivity is increased by a factor of four in spite of shielding of the ions by $CF_3SO_3^-$ or PF_6^- anions. The incorporation of four Cd^{2+} ions per grid leads to an increase of the conductivity almost by a factor of 200. The conductivity saturates at an increase of four orders of magnitude by incorporation of 25 Cd^{2+} ions per grid.

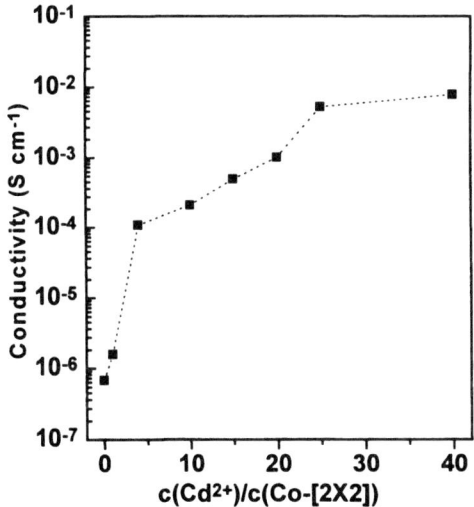

Figure 4: Conductivity measurements on a thin film (50 monolayers) of Co-[2×2] with various concentrations of $Cd(CF_3SO_3)_2$. We show the conductivity at 300 K vs. the number of Cd^{2+} ions per grid. The dotted line is guide to the eye.

In order to check if the increase in conductivity is caused by ionic conductance, additional K^+ ions were incorporated into the film in the typical doping concentrations. No increase in conductivity could be observed, which implies that the enhanced conductivity is most likely caused by the additional electronic states in the gap. This will be further verified by Hall effect measurements.

CONCLUSIONS

Ordered superstructures of organometallic grid structures can be produced on a Au(111) surface. By incorporation of additional metal centers, the electronic properties of the films can be changed considerably. The incorporation of one additional Cd(II) metal ion leads to an additional electronic state in the insulating HOMO-LUMO gap. This was shown by UPS measurements, which are in good agreement with molecular orbital calculations. The doping leads to a strong increase of the conductivity of the films of up to four orders of magnitude, depending on the concentration of the additional metal ions.

ACKNOLEDGMENTS

The authors thank the Bayerische Forschungsstiftung FORSUPRA and the Deutsche Forschungsgemeinschaft for partial financial support. We thank J. Schülein and S. Schromm for valuable discussions and R.L. Johnson, Universität Hamburg/DESY, for providing the photoelectron spectrometer.

REFERENCES

1. J.-M. Lehn, Supramolecular Chemistry- Concepts and Perspectives, VCH, Weinheim, 1995, pp. 89-139.

2. J.H. Scott, C.M. Yip, M.D. Ward, Langmuir **11**, 177 (1995).

3. N.J. Tao, J.A. DeRose, S.M. Lindsay, J. Phys. Chem **97**, 910 (1993).

4. C. Chen, S.M. Versecky, A.A. Gewirth, J. Am. Chem. Soc. **114**, 451 (1992).

5. E.S. Soldatov, V.V. Khanin, A.S. Trifonov, D.E. Presnov, S.A. Yakovenko, G.B. Khomutov, C.P. Gubin, V.V. Kholesov, JETP Lett. **64**, 556 (1996).

6. O. Waldmann, J. Hassmann, P. Müller, G.S. Hanan, D. Volkmer, U.S. Schubert, J.-M. Lehn, Phys. Rev. Lett. **78**, 3390 (1997).

7. G.S, Hanan, U.S. Schubert, D. Volkmer, J.-M. Lehn, J. Hassmann, C.Y. Hahn, O. Waldmann, P. Müller, G. Baum, D. Fenske in 9th Internationl Symposium on Molecular Recognition and Inculusion, edited by A. Colman, Kluwer Academic Press, in press.

8. G.S. Hanan, C.R. Arana, J.-M. Lehn, D. Fenske, Angew. Chem. **107**, 1191 (1995).

9. G.S, Hanan, U.S. Schubert, D. Volkmer, J.-M. Lehn, G. Baum, D. Fenske, Angew. Chem. **109**, 1929 (1997).

10. G. Rauhut, A. Alex, J. Chandrasekhar, T. Steinke, W. Sauer, B. Beck, M. Hutter, P. Gedeck, T. Clark, VAMP 6.1, Oxford Molecular Ltd., Madwar Centre, Oxford Science Park, Standford-on-Thames, England, 1996.

11. J.J. Stewart, J. Comput. Chem. **10**, 209 (1989).

12. J. Hassmann, E. Volz, H.-J. Schleemilch, P. Müller, D. Volkmer, G.S. Hanan, J.-M. Lehn, H. Mauser, A. Hirsch, R.L. Johnson, to be published.

PHOTOCHROMIC LIQUID HYDROGELS
AS HOSTS FOR HOLOGRAPHIC MATERIALS

R.L. WHITE[*], Y.Y. HSU[*], T.M. COOPER[**], J.D. GRESSER[*], D.L. WISE[***], and
D.J. TRANTOLO[*]
[*]Cambridge Scientific, Inc., 195 Common Street, Belmont, MA   02178
[**]Materials Directorate, Wright-Patterson AFB, OH   45433
[***]Center for Biotechnology Engineering, Northeastern University, Boston, MA   02115

ABSTRACT

The goal of this project is to develop, fabricate, and test advanced optical materials for potential applications to real-time holography based on liquid crystalline polymer hydrogels. In this project, we are investigating the feasibility of increasing holographic capacity and lifetime by coupling a photochromic spyropyran dye to in a liquid crystalline polymer in which cholesteric order has been 'captured.' 'Capture' is being approached using a unique in-plane poling process with the helical polypeptide poly(α-benzyl-L-glutamate), PBLG, a biopolymer which is capable of maintaining cholesteric order in a liquid crystalline state. Subsequent *in situ* crosslinking of this aligned biopolymer is projected to offer increased birefringence of the host in the writing of a hologram. Given that a key issue is the magnitude of the real component of the refractive index, increasing the birefringence may be a useful approach. In writing the hologram, the liquid crystals (LC's) go from isotropic to an ordered dispersion, a property which can be captured via crosslinking to improve holographic lifetime. In the following, the characterization of an aligned host LC system based on the biopolymer poly(α-benzyl-L-glutamate), PBLG, is presented. In-plane alignment is shown to depend on a number of variables, most notably the choice of solvent, polymer molecular weight, and field strength.   The results show that optimal alignment of the PBLG LC is achieved with a 2.5% (w/w) concentration of a 118kD biopolymer in methylene chloride in an applied field of 10 kV/cm. Subsequent work will exploit this system as a host for a spiropyran dye for improved holographics.

INTRODUCTION

The basic requirements of holographic memory design were established in the 1960's. Holographic optical techniques allow for the storage of high information densities. Both planar and volume holographic memories can be achieved in a page oriented format. In the most typical configuration, the subhologram is the recording of the interference pattern between the Fourier transform of two dimensional binary data in each page and the corresponding plane wave reference beam. In order to achieve angular multiplexing of multiple images, thick holograms are needed to achieve high angular selectivity, and, furthermore, phase holograms are required for have high diffraction efficiency. The overall goal of using arrays of such holograms in page oriented holographic memories is to decrease the dimensions of the memory system.

Many different types of materials have been investigated for holographic storage. Among them are silver halides, dichromated gelatin (DCG), ferroelectric crystals, photochomics, photodichroics, photorefractives, azoaromatic polymers, and photodynamic proteins (such as a PBLG–like biopolymer appropriately derivatized with a photochromic dye). Biopolymers such as derivatized polypeptides are very promising candidates for holographic applications.[1] Biopolymers effectively compete with inorganic, organic and polymeric optical materials particularly with respect to cost, environmental stability, and performance. However, among the drawbacks is the extremely short lifetime of these holograms. If the advantages of a polymer system could be coupled to longer holographic lifetimes and capacities, then technology for large data storage and rapid information retrieval could be realized in a system more compact than current CD-ROM technology.

In this project, we are investigating the feasibility of increasing holographic capability by coupling a spiropyran dye to a liquid crystal (LC) system which offers increased birefringence in the writing of the hologram. Given that a key issue is the magnitude of the real component of the refractive index, increasing the birefringence may be a useful approach. In writing the hologram, the LC's go from isotropic to an ordered dispersion, a property which can be linked to those of bR to improve holographic lifetime.

An attractive aspect of biopolymers is their liquid crystalline properties. This work exploits a model LC biopolymer with which we have had extensive experience, namely poly(γ –benzyl–L–glutamate), "PBLG"[2,3]. Polymer gels with LC order have been prepared by the crosslinking of a LC solution of PBLG. These gels with cholesteric LC order have been prepared, and they show cholesteric-isotropic reversible transition accompanied by the helix-coil transition of PBLG molecules[4]. These gels have been used to orient dye molecules introduced either by doping or by covalent linkages[5]. This overall project will investigate the feasibility of a PBLG 'host' to increase the birefringence of the written hologram and thusly refine its holographic potential (vis a vis its lifetime). The experimental work, reported herein, has first focused on characterizing the ordering behaviors of the LC host prior to introduction of the photochromic dye.

EXPERIMENTAL

Materials

All PBLG's were obtained from Sigma Chemical Co., St. Louis, MO and used as received. Three molecular weight ranges with nominal weight averages of 26kD, 118kD, and 236kD were investigated in this study. All solvents were obtained from Fisher Scientific, Inc., and used as received.

In-Plane Electric Field Alignment of PBLG

The cell design for in-plane poling uses teflon blocks machined to include rectangular reservoirs with electrodes imbedded at either end and providing electrode gaps of 1.2, 2.8, and 6.2 cm. A power supply (Del Electronics Corp., Valhalla, NY, Model RHVS60–300P) capable of delivering 60kV DC enables preparation of large films at high field strengths. Electrodes were placed so that their upper and lower edges were above and below the solution layer. The entire apparatus is contained within a glass housing. This cell design provides for an optically clearer film of more uniform thickness by reducing the problem of electroconvection via control over the solvent atmosphere within the cell and the rate of solvent evaporation. In addition, the teflon material exhibits good chemical resistance.

Characterization of the Aligned PBLG Hosts by Infrared Spectroscopy

Films were characterized for alignment using FTIR using the method of Marcher et al.[6]. Infrared spectra were taken on a Perkin-Elmer (Norwalk, CT) 1600 Series FTIR equipped with gold wire grid polarizer (gold vapor deposited on a silver bromide grid, Perkin-Elmer Part 186-0243). IR spectra were taken in the absorbance mode with the polarizer set to obtain minimum and maximum values of the amide A, I, and II bands. With the film mounted in the spectrometer so that the field direction was vertical, the polarizer settings giving maximum/minimum absorbances were 45° and 135°. The bands of interest are as follows:

Wave number	Band ID	Polarization	Assignment
3292	Amide A	parallel	>N-H stretch
1650	Amide I	parallel	>C=O stretch
1550	Amide II	perpendicular	>N-H deformation

Representative spectra of aligned PBLG films are shown in Figures 1a through 1d (a film of molecular weight 20,000, cast from a 2.5 mg/ml solution in a field of 0.8 kV/cm) showing spectra at polarizer settings of 0, 45, 90, and 135°. Figure 2 shows the change of relative absorbances of the Amide A and II bands of this film with the angular setting of the polarizer.

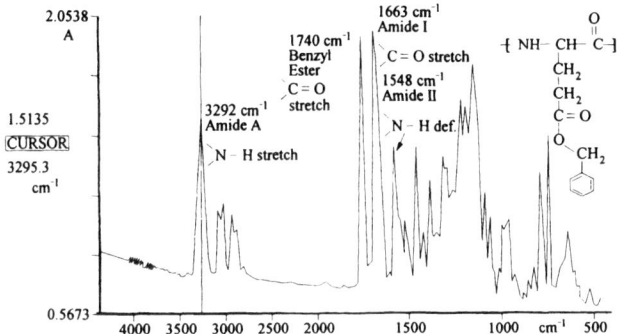

Figure 1a. IR Spectrum of PBLG Film; Aligned at 0.8kv/cm; Polarizer 0°

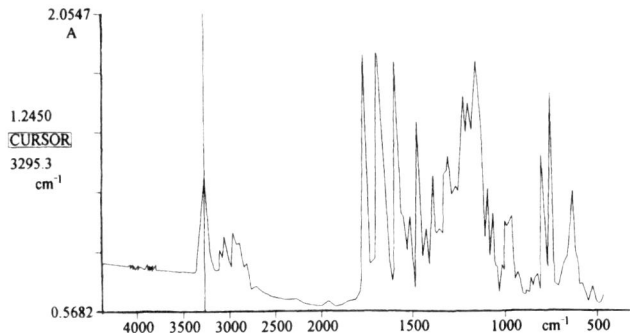

Figure 1b. IR Spectrum of PBLG Film; Aligned at 0.8kv/cm; Polarizer 45°

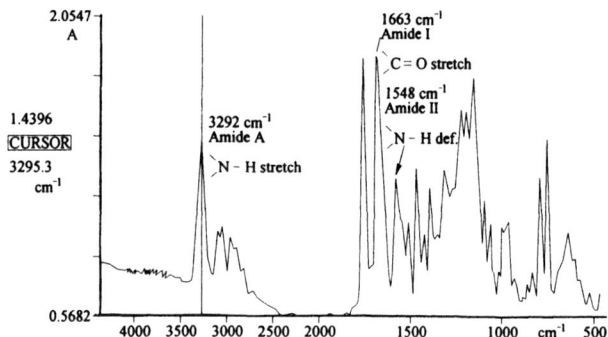

Figure 1c. IR Spectrum of PBLG Film; Aligned at 0.8kv/cm; Polarizer 90°

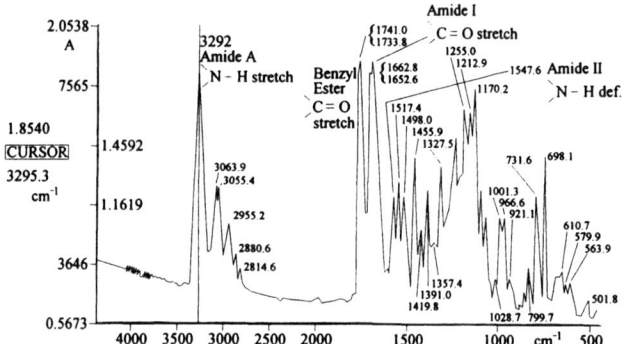

Figure 1d. IR Spectrum of PBLG Film; Aligned at 0.8kv/cm; Polarizer 135°

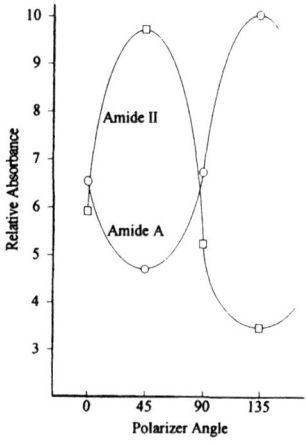

Figure 2. Relative Absorbances of Amide A and Amide II Bands of a PBLG Film at Various Polarizer Settings

RESULTS

Order parameters for electric field processed PBLG films

Application of electric fields of increasing strength to PBLG solutions results in films with increasing alignment and therefore in higher bulk dipole moments. The dependence of the mean value of the order parameter, S_θ, on molecular weight and solution concentration illustrated in Figure 3 for PBLG's of varying molecular weights prepared from solutions containing 1.0–7.0% (w/w) of polymer. This Figure, illustrating the data for PBLG films cast in a 4.9kV/cm field, demonstrates that optimum conditions for alignment appear to be with a molecular weight of 118,000 and a solution concentration of 2.5% (w/w). Additional data collected at field strengths of 2.2 and 10.4 kV/cm (not illustrated here), show a similar MW and concentration optimum.

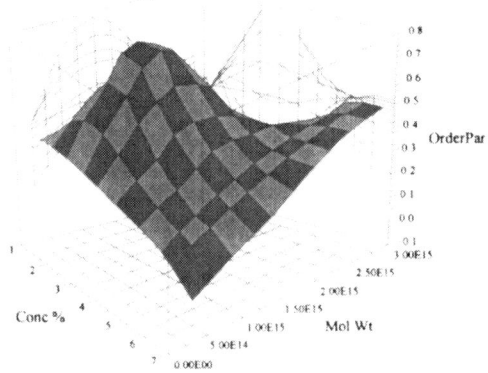

Figure 3. Variation of the Order Parameter, S_θ, with Molecular Weight and Solution Concentration at a Field Strength of 4.1 kV/cm.

Data for the 118 kD PBLG strongly suggest that a minimum field strength of about 2.0 kV/cm is required before alignment as measured by the mean value of the order parameter can be detected. The order parameter, S_θ, increases rapidly with field strengths to 4.0 kV/cm and more slowly above 4.0 kV/cm, appearing to approach asymptotically the maximum order value of 1.0 (Figure 4).

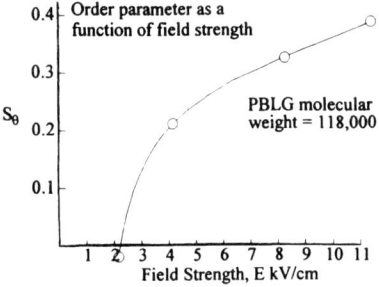

Figure 4. Electric field alignment of Poly(γ-Benzyl-L-Glutamate)

CONCLUSIONS

The magnitude of the order parameter is dependent on not only the field strength, but also on the concentration of the solution from which the film is cast and the polymer molecular weight. Identifying the conditions which yields films with precise and reproducible geometries will potentially enhance the application of aligned host LC polymers to holographics, as well as other electrooptical, applications. Subsequent work will exploit this understanding to 'capture' crosslinked LC hosts for support of photochromic dyes for holographic applications.

ACKNOWLEDGMENTS

The authors gratefully acknowledge the advice, guidance and support of the Materials Directorate of Wright-Patterson AFB under Subcontract No. TMC-96-5835-0032-01.

REFERENCES

1. T.M. Cooper, and L.V. Natajarajan, in Photonic Polymers: Fundamentals, Methods and Applications, edited by D.L. Wise, G.E. Wnek, D.J. Trantolo, J.D. Gresser, and T.M. Cooper, Marcel Dekker, Inc., NY, in press.

2. D.J. Trantolo, M.G. Mogul, D.L. Wise, G.E. Wnek, D.O. Frazier, and J.D. Gresser, Space Processing of Materials, SPIE, **2809**: 106, 1996.

3. M.G. Mogul, J.D. Gresser, D.L. Wise, G.E. Wnek, and D.J. Trantolo, in Photonic Polymers: Fundamentals, Methods and Applications, edited by D.L. Wise, G.E. Wnek, D.J. Trantolo, J.D. Gresser, and T.M. Cooper, Marcel Dekker, Inc., NY, in press.

4. R. Kishi, M. Sisido, and S. Tazuke, Macromolecules, **23**: 3779, 1990.

5. M. Sisido, and R. Kishi, Macromolecules, **24**: 4110, 1991.

6. L.L. Marcher, L.L. Chapoy, and D.L. Christensen, Macromolecules, **21**: 677, 1988.

Part V

Organic Metals and Magnetic Materials

ORGANIC/INORGANIC LANGMUIR-BLODGETT FILMS
BASED ON METAL PHOSPHONATES

DANIEL R. TALHAM, GAIL E. FANUCCI, MELISSA A. PETRUSKA AND
CANDACE T. SEIP
Department of Chemistry, University of Florida, Gainesville, FL 32611-7200

ABSTRACT

   Langmuir-Blodgett (LB) bilayers of organophosphonic acids can be prepared
where the phosphonic acid headgroups bind metal ions to form the same layered
extended-solid structures present in solid-state metal phosphonates. The inorganic
extended-solid network enhances the stability of the LB films, but can also be
designed to introduce physical properties, such as magnetism, that are typical of the
inorganic solid-state. By preparing films based on functionalized organophosphonic acids, the metal phosphonate approach can be used to produce "dual-
network" LB films, where both the organic and inorganic networks add function to
the thin film assembly. To begin to understand the design constraints associated
with dual-network metal phosphonate films, LB bilayers of a phosphonic acid-
derivatized azobenzene amphiphile are formed with Cd^{2+} and La^{3+} and the
structures are compared to octadecylphosphonate LB films prepared with the same
metals.

INTRODUCTION

   The Langmuir-Blodgett film method provides an elegant approach for
organizing organic molecules into functional thin film assemblies, many examples
of which have been demonstrated as active components in a variety of electrical,
optical and magnetic materials applications [1]. Molecular LB films are most often
fabricated from amphiphilic derivatives of the molecules of interest, and the
traditional approach is to prepare a functionalized fatty acid. Amphiphilic
carboxylic acids commonly form stable monolayers at the air/water interface and
are made more processable by the presence of metal ions in the aqueous subphase.
LB films prepared as salts tend to be more stable than films of neutral molecules
because the ionic metal-headgroup interactions help bind the amphiphiles
together. But even layered films of the metal carboxylate salts are often metastable
structures, and despite the elegance of the supramolecular assembly, the long term
stability of most LB films is not sufficient for materials applications.
   It is possible, however, to consider other metal-ligand interactions when
constructing LB films, and we have recently been exploring the formation of LB
films based on known organic/inorganic layered solids [2-6] such as the layered
metal phosphonates shown in Figure 1. In the solid-state, organophosphonates
with divalent, trivalent and tetravalent metal ions form layered structures where
metal/phosphonate extended-solid layers are separated by layers of the organic
group [7-9]. Figure 1 shows crystal structures of two examples of solid-state metal
phosphonates, $Mn(O_3PC_6H_5)\cdot H_2O$ and $La(O_3PC_6H_5)(HO_3PC_6H_5)$ [8, 10]. Several

divalent metals form layered phosphonates that are isostructural with $Mn(O_3PC_6H_5) \cdot H_2O$, and similarly, $La(O_3PC_6H_5)(HO_3PC_6H_5)$ is a prototype for a larger family of trivalent metal phosphonates. Also, layered structures are not restricted to phenylphosphonates. As the organic group is changed from a phenylphosphonate to any of a number of alkylphosphonates, the interlayer distances vary depending on the length of the alkyl chain, but the in-plane lattice constants remain virtually unchanged [8]. Since the solid-state lattice energy of the ionic/covalent metal phosphonate sheets favors a layered structure, LB films based on metal phosphonates should no longer be metastable.

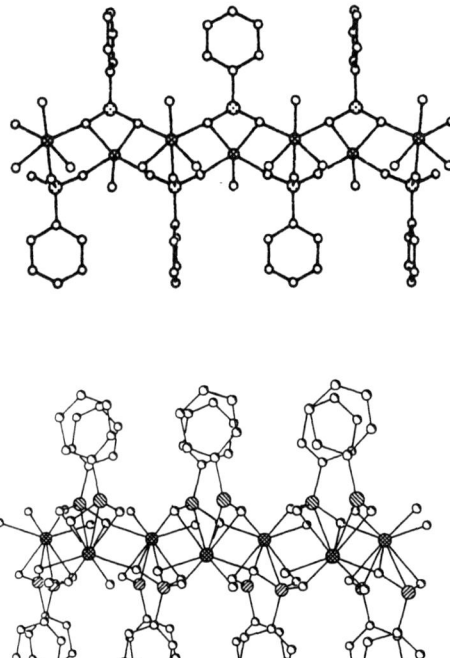

Figure 1. Top. A single layer of the solid-state manganese phenylphosphoponate, $Mn(O_3PC_6H_5) \cdot H_2O$, viewed parallel to the manganese ion plane. Crystallographic data were taken from reference [8]. Bottom. A single layer of the solid-state lanthanum phenylphosphonate, also viewed parallel to the metal ion plane. Crystallographic data were taken from reference [10].

Another advantage of modeling LB films after organic/inorganic layered solids is the possibility of building into the films physical properties that are typical of the inorganic solid-state. We recently demonstrated this idea by characterizing a magnetic LB film [6]. LB films of manganese octadecylphosphonate undergo a magnetic ordering transition at 13.5 K to a canted antiferromagnetic state. The same magnetic ordering is seen in the analogous solid-state materials [11]. The spontaneous magnetization in the LB film is attributed to the inorganic metal phosphonate layer and is not affected by changes in the interlayer spacing that result from changing the organic layer [6, 11].

The metal phosphonate LB framework provides an opportunity to form mixed organic/inorganic LB films where both components add function to the LB assembly. Solid-state layered metal organophosphonates are not limited to alkylphosphonates, and metal phosphonate LB films with functional organic groups can also be formed. Ultimately, structures should be possible where both the organic network and inorganic network contribute useful properties to the assembly to form a composite material, but there is also the possibility of the two networks interacting to produce new phenomena.

In this report, we demonstrate how the metal phosphonate method for assembling LB films can be extended beyond simple alkylphosphonates to phosphonic acid functionalized azobenzene chromophores. We compare octadecylphosphonate LB films formed with two metals, Cd^{2+} and La^{3+}, to the azobenzene films prepared with the same metals. We also show that there are limitations to the approach when there are significant differences between the preferred lattice spacings of the separate organic and inorganic networks.

EXPERIMENT

Octadecyltrichlorosilane (OTS), $CdCl_2 \cdot 2.5H_2O$ and $La(NO_3)_3$ were purchased from Aldrich (Milwaukee, WI) and octadecylphosphonic acid (OPA), $C_{18}H_{39}O_3P$, was obtained from Alfa Aesar (Ward Hill, MA). Commercial reagents were used as supplied. The preparation of 4-(4'-tetradecyloxyphenyldiazenyl)phenylbutyl-phosphonic acid (A4) will be described elsewhere [12]. A Barnstead NANOpure purification system produced water with an average resistivity of 18 MΩ cm for all experiments. The subphase pH was adjusted with either a 0.012M HCl solution or 0.10 g/L KOH solution. The LB experiments were performed using KSV 3000 and 5000 Instruments (Stratford, CT) modified to operate with hydrophobic double barriers on home-made Teflon troughs with surface area of 794 cm^2 (12.8 cm x 67 cm) and 960 cm^2 (12.8 cm x 75 cm), respectively. Solid supports for film deposition, glass or quartz slides, silicon wafers, and silicon or germanium attenuated total reflectance crystals, were made hydrophobic with OTS. All methods for characterizing transferred films have been described in detail elsewhere [3, 6].

RESULTS

Transition Metal Octadecylphosphonate LB Films

Metal phosphonate LB films are formed on hydrophobic substrates using standard LB vertical deposition procedures. The hydrophobic surface allows one organophosphonate layer to be transferred on the deposition downstroke and a second layer on the upstroke forming a head-to-head bilayer. Metal ions from the aqueous subphase are bound by the phosphonate headgroups, so each "bilayer" is made up of one metal layer sandwiched between two organophosphonate layers, the same arrangement seen in the solid-state metal phosphonates (Figure 1). The deposition can be thought of as a "crystallization." As water drains from the film on the upstroke, the metal phosphonate lattice "crystallizes" in the bilayer [6]. The subphase pH required to achieve the crystalline layers varies from metal to metal, reflecting the affinity of the different metal ions for the phosphonate headgroup. The cadmium octadecylphosphonate films are deposited within a pH range 5.2-5.6 and the lanthanum films within the range 2.6-2.9. At lower pH values, not enough metal ions bind the film to form the solid-state structures, and at higher pH, the metal ions bind the Langmuir monolayers so tightly that the film becomes too rigid to transfer. Within the appropriate pH range, transfer ratios of 1.0 ± 0.1 are obtained for each deposition.

The expected layered nature of the LB films is confirmed with X-ray diffraction. Figure 2 includes the X-ray diffraction pattern from a ten-bilayer lanthanum octadecylphosphonate film. Several orders of the (00l) reflection are seen corresponding to an interlayer spacing of 51 ± 0.5 Å, which is reasonable for an octadecylphosphonate bilayer. The cadmium octadecylphosphonate film gives a similar pattern, corresponding to an interlayer spacing of 48 ± 0.5 Å. The metal ion content in the transferred films can be quantified with XPS. For the lanthanum film, the La:P ratio determined from XPS is 1:2, consistent with the stoichiometry of the lanthanum phosphonate solids, like $La(O_3PC_6H_5)(HO_3PC_6H_5)$ whose structure is shown in Figure 1. For the cadmium octadecylphosphonate film, the Cd:P ratio is 1:1, again consistent with the solid-state analogs.

While X-ray diffraction and XPS show that the LB films have their stoichiometry and layered structure in common with the solid-state phosphonates, neither method gives information about the in-plane structure of the films. However, careful comparison of IR vibrational modes in the films to those of the analogous solids shows that the metal/phosphonate binding is similar. Phosphorus-oxygen stretching modes of the phosphonates have been shown to be quite sensitive to the mode of binding in the layered metal phosphonates [6]. The infrared P-O stretches in the cadmium octadecylphosphonate LB film are compared to those from a KBr pellet of the solid-state cadmium ethylphosphonate in Figure 3A. The asymmetric P-O stretch at 1089 cm^{-1} and symmetric P-O stretch at 956 cm^{-1} of the solid-state sample are seen at 1092 cm^{-1} and 959 cm^{-1}, respectively in the LB film. In both cases, the bands are split due to the low symmetry of the phosphonate group in the lattice. The same infrared region is shown in Figure 3B where the lanthanum octadecylphosphonate LB film is compared to the solid-state lanthanum butylphosphonate. The phosphonate P-O stretches are less well resolved for these materials because of the presence of both a monobasic and a

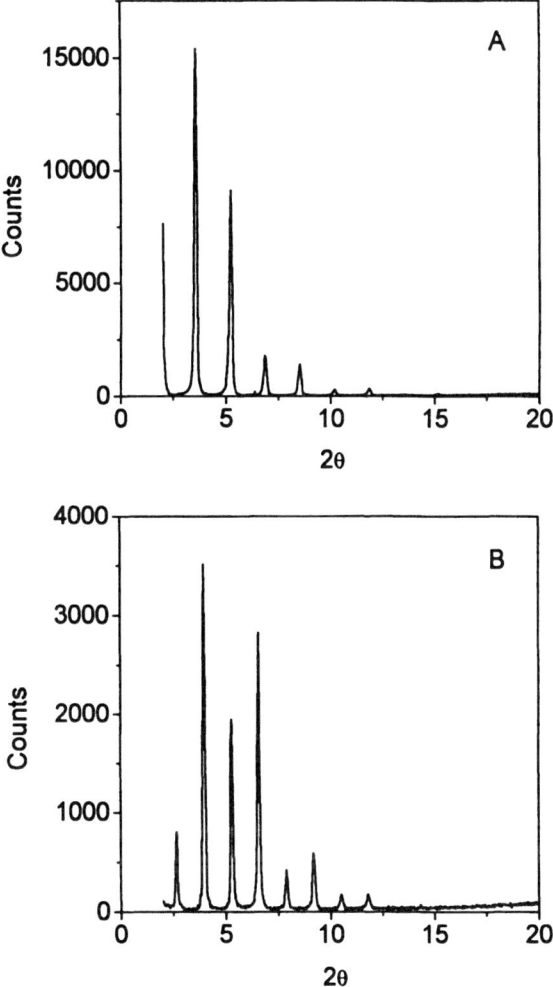

Figure 2. A) X-ray diffraction from a ten bilayer LB film of lanthanum octadecylphosphonate. The Bragg progression corresponds to an interlayer spacing of 51.0 ± 0.5Å. **B)** X-ray diffraction from a ten bilayer LB film of lanthanum A4, corresponding to an interlayer spacing of 67 ± 1Å.

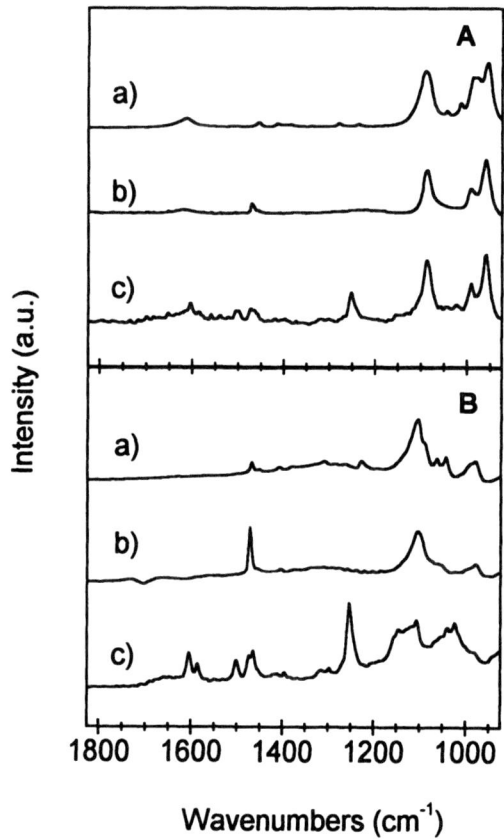

Figure 3. A) FTIR spectra of cadmium phosphonates: a) KBr pellet of cadmium ethylphosphonate, b) cadmium octadecylphosphonate LB film and c) cadmium A4 LB film. **B)** FTIR of lanthanum phosphonates: a) KBr pellet of lanthanum butylphosphonate, b) lanthanum octadecylphosphonate LB film and c) lanthanum A4 LB film. The intensity scale is arbitrary in order to fit all spectra on the same plot.

dibasic phosphonate in the La(HO₃PR)(O₃PR) structure. Nevertheless, the fingerprinting of the two spectra makes it clear that the phosphonate binding in the LB film is the same as seen in the solid-state.

The analyses described above show that metal phosphonate extended solids can be formed in LB films. Additional studies have shown the concept is general, extending the idea to other divalent and trivalent metals [6]. The most extensively characterized example has been manganese octadecylphosphonate, which is isostructural with the cadmium film described above, where magnetic studies confirm that the LB films have the same structure as the solids [5, 6]. In these materials, where antiferromagnetic exchange is mediated by the phosphonate ligands, differences in the mode of binding within the metal phosphonate layer would lead to different values of magnetic exchange. However, electron spin resonance studies show that magnetic exchange in the films and the solids is the same [5]. Furthermore, SQUID magnetometry shows that the LB films undergo a magnetic ordering transition to a canted antiferromagnetic state at 13.5 K [6], the same temperature at which the solid-state alkylphosphonates order [11]. Canted antiferromagnets are sometimes called "weak ferromagnets" because of the spontaneous magnetization of the ordered state. Manganese octadecylphosphonate is the first example of a magnetic LB film.

Azobenzene Functionalized Metal Phosphonate Films

An important question to ask is whether metal phosphonate based LB films are limited to long chain alkylphosphonates, or can they be extended to functional organic groups. To begin to probe both the utility and limitations of the method, we have prepared an amphiphilic azobenzene functionalized phosphonic acid, A4, and investigated metal phosphonate LB film formation using the same divalent, Cd²⁺, and trivalent, La³⁺, metal ions used in the octadecylphosphonate LB films discussed above. We observe different results for the two metals. In the La³⁺ case, we find that LB bilayers with A4 can be prepared where both the azobenzene groups in the organic network are close packed and the lanthanum phosphonate extended-layer forms with the same structure seen in the solids. On the other hand, we are not able to form continuous LB films of A4 with Cd²⁺ in which both the organic groups are organized and the cadmium phosphonate lattice crystallizes.

$$H_{29}C_{14}O-\!\!\!\!\bigcirc\!\!\!\!-N\!\!=\!\!N-\!\!\!\!\bigcirc\!\!\!\!-C_4H_8PO_3H_2$$

A4

X-ray diffraction from 10 bilayers of the A4/La/A4 film, Figure 2, yields an interlayer spacing of 67±1 Å that reflects the presence of the azobenzene group in the organic layer. XPS analysis gives a La:P ratio of 1:2, which is the same ratio seen for the octadecylphosphonate film and is consistent with the stoichiometry of the layered solids with formula $La(O_3PR)(HO_3PR)$. The binding of the phosphonate groups can again be probed by analyzing the infrared P-O stretches shown in Figure 3B. This spectral region is now complicated by several overlapping azobenzene related bands, but the significant P-O stretch at 1107 cm^{-1} appears at the same frequency in the A4 film as it does in the lanthanum octadecylphosphonate LB film and lanthanum butylphosphonate powdered solid, suggesting that the lanthanum phosphonate layers have the same in-plane extended structure.

Organization within the organic network can be inferred from UV-vis spectra where the azobenzene chromophore gives rise to a n-$\pi^*$and an intense π-π^* transition (Figure 4). The π-π^* band is especially sensitive to molecular aggregation [13]. In a chloroform solution, the π-π^* band appears at 354 nm, but shifts to 298 nm in the lanthanum phosphonate LB film. The blue-shift of this band indicates that the azobenzene groups form H-aggregates, and the large magnitude of the shift suggests that there is long-range organization of the azobenzene packing [14].

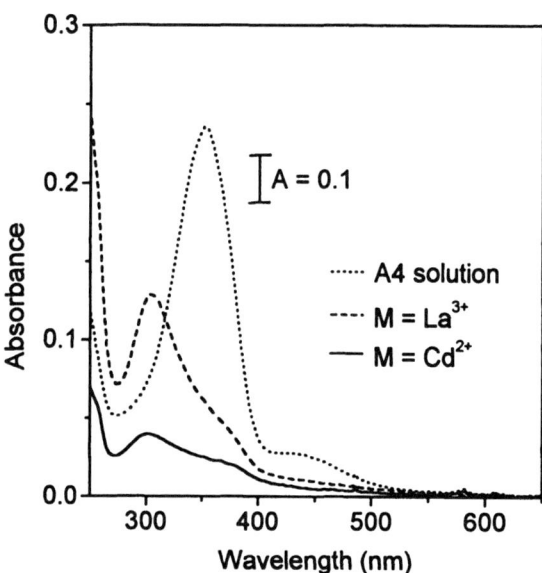

Figure 4. UV-visible spectra of five LB bilayers of A4 formed with La^{3+} and Cd^{2+}. The π-π^* transition is seen at 298 nm in the La^{3+} film and 302 nm in the Cd^{2+} film. This band is blue shifted with respect to the spectrum of A4 in chloroform solution (3.9 x 10^{-5} M), indicating that the azobenzene chromophores are H-aggregated. The absorbance scale at left refers to the LB films.

Further evidence for the close packing arrangement of the azobenzene groups comes from the inability to observe trans-cis isomerization in the lanthanum A4 LB film. Normally, azobenzene chromophores undergo a photochemically induced isomerization when excited in the $\pi-\pi^*$ band [15]. However, if the azobenzene groups are tightly aggregated, it will not be observed. Such is the case for the lanthanum A4 film. The lack of any trans-cis isomerization indicates that there is tight aggregation of the azobenzene groups and a small percentage of defect areas. The characterization of the lanthanum film can be contrasted to the cadmium A4 film. We have not been able to find conditions where the A4 film both transfers effectively and binds Cd^{2+} ions in the expected Cd:P ratio of 1:1. When working at a subphase pH lower than 4.3 ± 0.1, the A4 film transfers completely, but does not bind cadmium. At subphase pH values at or above 4.3 ± 0.1 only partial transfer of the film is achieved. In the partially transferred films formed near pH 4.3, the cadmium phosphonate lattice appears to be similar to the octadecylphosphonate film as evidenced by the presence of the P-O symmetric and asymmetric stretches at 959 cm^{-1} and 1088 cm^{-1}, respectively, shown in Figure 3A. However, the LB film transfer ratios are very low. This is seen by comparing the intensity of the $\pi-\pi^*$ band in the cadmium A4 film to that of the lanthanum A4 film in Figure 4. For the same number of bilayers, the intensity in the cadmium A4 film is about 1/3 that of the lanthanum phosphonate film. Interestingly, the $\pi-\pi^*$ band is blue shifted to nearly the same extent as it is in the lanthanum A4 film, indicating that despite the low transfer ratios, the azobenzene groups are still tightly aggregated.

The behavior observed for the cadmium A4 film might be explained by differences in the preferred spacing of the azobenzene groups in the organic network and the phosphonate groups of the cadmium phosphonate lattice in the inorganic network. If there is a mismatch in the preferred area-per-molecule in the two components of the assembly, then strain will develop limiting the size of well organized domains. In the cadmium A4 film, this strain appears to hinder the successful transfer of a continuous film. Azobenzene groups are known to aggregate on the water surface, and this aggregation provides a barrier to any reorganization that might be required to achieve the phosphonate spacing needed to form the cadmium phosphonate extended-structure. The anticipated area-per-molecule of the azobenzene group in a layered assembly of H-aggregates is approximately 22 $Å^2$ [14], while the area-per-molecule of the organophosphonates in the solid-state cadmium alkylphosphonates is close to 28 $Å^2$ [8]. So, although the small amount of film that transfers has both well organized organic and inorganic networks, this lattice mismatch creates a barrier to binding the inorganic ions, reducing the film transfer. In contrast, there is a much better lattice match in the lanthanum A4 film. Here, the expected area-per-molecule of the phosphonate in the solid-state lanthanum alkylphosphonates is 21 $Å^2$ [10], much closer to the expected packing of the azobenzene groups. This results in easy transfer, leading to LB films where both the organic and inorganic networks can achieve their preferred organization.

CONCLUSIONS

LB films of amphiphilic organophosphonates can be prepared where the phosphonate headgroup binds metal ions to form extended-solid structures that are isostructural with the known solid-state layered metal phosphonates. The method can be used to add inorganic solid-state properties to the LB assemblies. The approach is demonstrated here for both divalent Cd^{2+} and trivalent La^{3+} with octadecylphosphonic acid. Metal phosphonate LB films can also be prepared when the organophosphonate contains an azobenzene chromophore, such as the molecule A4. However, in that case, the quality of the transferred film depends on the choice of metal ion. The strong aggregation of the azobenzene moieties on the water surface restricts the ability of the phosphonate headgroups to optimize their spacing to match those in the solid-state layered phosphonates. As a result, the A4 metal phosphonate bilayers form easily with La^{3+} because the preferred phosphonate spacing nicely matches the azobenzene spacing. In contrast, a mismatch in the preferred spacings of the organic and inorganic components of the cadmium A4 film results in poor quality films.

ACKNOWLEDGMENTS

We are grateful to the National Science Foundation for financial support. M.A.P. thanks the National Science Foundation for a predoctoral fellowship.

REFERENCES

1. G. G. Roberts, *Langmuir-Blodgett Films* (Plenum Press, New York, 1990).
2. H. Byrd, J. K. Pike, D. R. Talham, *Chem. Mater.* **5**, 709-715 (1993).
3. H. Byrd, S. Whipps, J. K. Pike, J. Ma, S. E. Nagler, D. R. Talham, *J. Am. Chem. Soc.* **116**, 295-301 (1994).
4. H. Byrd, J. K. Pike, D. R. Talham, *Thin Solid Films* **242**, 100-105 (1994).
5. C. T. Seip, H. Byrd, D. R. Talham, *Inorg. Chem.* **35**, 3479-3483 (1996).
6. C. T. Seip, G. E. Granroth, M. W. Meisel, D. R. Talham, *J. Am. Chem. Soc.* **119**, 7084-7094 (1997).
7. D. Cunningham, P. J. D. Hennelly, *Inorg. Chim. Acta* **37**, 95-102 (1979).
8. G. Cao, H. Lee, V. M. Lynch, T. E. Mallouk, *Inorg. Chem.* **27**, 2781-2785 (1988).
9. Y. Ortiz-Avila, P. R. Rudolf, A. Clearfield, *Inorg. Chem.* **28**, 2137-2141 (1989).
10. G. Cao, V. M. Lynch, J. S. Swinnea, T. E. Mallouk, *Inorg. Chem.* **29**, 2112-2117 (1990).
11. S. G. Carling, P. Day, D. Visser, R. K. Kremer, *J. Solid State Chem.* **106**, 111-119 (1993).
12. M. A. Petruska, D. R. Talham, *manuscript in preparation*.
13. E. G. McRae, M. Kasha, in *Physical Processes in Radiation Biology* L. Augenstein, R. Mason, B. Rosenberg, Eds. (Academic Press, New York, 1964) pp. 23-42.
14. X. Song, J. Perlstein, D. G. Whitten, *J. Am. Chem. Soc.* **119**, 9144-9159 (1997).
15. J. Griffiths, *Chem. Soc. Rev.* **1**, 481-493 (1972).

AC SUSCEPTIBILITY STUDIES OF NEW AND FAMILIAR MAGNETIC MOLECULAR SOLIDS

Roger D. Sommer, Brenda J. Korte, Scott P. Sellers and Gordon T. Yee*
Department of Chemistry and Biochemistry, University of Colorado, Boulder CO 80309
yeeg@colorado.edu

ABSTRACT

ac Susceptometry has been used to study a number of magnetic molecular solids including a new compound, Mn(II) octaethyltetraazaporphyrin, α-MnOETAP, and decamethylmanganocenium tetracyanoethenide, Mn(Cp*)$_2$•TCNE, a previously reported molecule-based ferromagnet. Both of these compounds exhibit signatures of ferromagnetism including significant hysteresis below 2 K and rapidly increasing χT (where χ is the molar susceptibility) with decreasing temperature. However, their ac susceptibility data show relatively strong dependence of χ' and χ'' on the frequency of the applied field, indicating a spin-glass state. Other molecular ferromagnetic solids examined show much less sensitivity. These studies indicate that the standard practice of characterization by dc and ac susceptometry at a single frequency are clearly insufficient for identifying the magnetic state of a molecular solid.

INTRODUCTION

Temperature and field dependent dc susceptometry has historically been used to study the magnetic properties of molecular solids. In particular, the discovery of a rapid increase in χT vs. T as temperature decreases (where χ is the molar susceptibility), along with evidence for spontaneous magnetization (from low field measurements) and for hysteresis have been taken to identify, definitively, a transition to a ferromagnetic or ferrimagnetic state. Occasionally, ac susceptometry at a single frequency is reported where a peak in the in-phase signal (χ' or χ_{real}) together with a non-zero out-of-phase component (χ'' or $\chi_{imaginary}$) is taken as further proof of a phase transition.

In this contribution, we demonstrate that these techniques, while valuable, are insufficient to unambiguously identify the magnetic ground state of a compound. In particular, we have recently discovered a solid that shows all of the features above, most notably hysteresis at 1.8 K with a coercive field of 2.5 kemu-G/mole and remanence of 4 kG.[1] This compound, manganese(II) octaethyltetraazaporphyrin, α-MnOETAP, (Figure 1b) is structurally and electronically similar to the known molecular ferromagnet, manganese(II) phthalocyanine, β-MnPc (Figure 1a).[2] However, unlike for β-MnPc, ac susceptometry on α-MnOETAP reveals a strong dependence of the susceptibility on the frequency of the applied field suggesting the compound is a spin-glass. For comparison, we show ac susceptibility data for two molecular ferromagnets, α-FeOETAP, which is isomorphous with its manganese analog and β-MnPc.

To explore the ubiquity of the frequency-dependent ac susceptometry result, we have examined the molecule-based solid, decamethylmanganocenium tetracyanoethenide, Mn(Cp*)$_2$•TCNE, (Figure 1 c and d). This compound had been previously characterized by dc techniques to be a hard molecular ferromagnet with T_c = 8.8 K.[3] Our results show that it too is a spin glass.

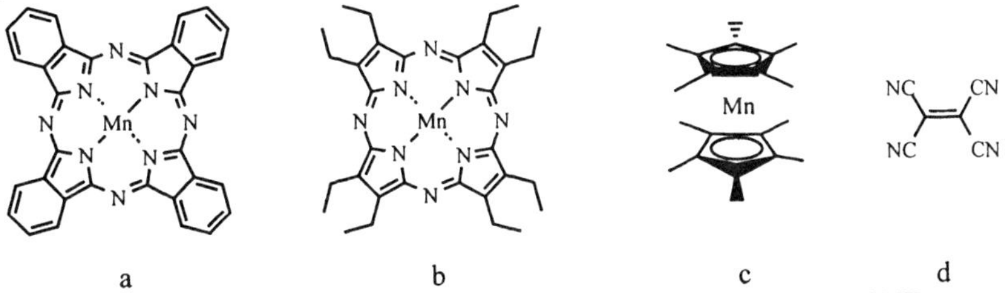

| a | b | c | d |

Figure 1. Structures of a) MnPc b) MnOETAP c) Mn(Cp*)$_2$ d) TCNE

Mat. Res. Soc. Symp. Proc. Vol. 488 © 1998 Materials Research Society

Palacio, et al. have previously made the point that dc measurements were unsufficient,[4] but an examination of the literature suggests that this warning has been unheeded. For the most part, variable frequency ac measurements have only been utilized when magnetic frustration was built into the system by design and hence spin-glass properties could be anticipated. For instance, Wynn, et al. have prepared an intentionally frustrated molecule-based magnet on a tridentate chelate which shows spin glass properties and frequency-dependent ac susceptibility.[5] We are unaware of any studies that suggest that the phenomenon may be more widespread among molecule-based solids.

EXPERIMENT

All magnetic measurements were carried out on either a 5 Tesla or 7 Tesla Quantum Design SQUID magnetometer. Full temperature scan dc susceptibility measurements were performed in 100 G applied field from 1.7 K to 350 K. Low field dc measurements were performed in 0 - 10 G or 0 - 3 G. The amplitude of the applied ac magnetic field for ac measurements was 5 G with no additional dc bias field. A flame-sealable air-tight sample holder was constructed from segments of Wilmad 507 5 mm NMR tube.[6] In the glove box, the sample ($\sim$ 15 mg) was placed into the holder, packed with a small piece of glass wool ($\sim$10 mg) and attached to a Teflon stopcock by means of an Ultratorr o-ring connector. The tube was evacuated in glove box by mechanical pump, sealed, brought out of the glove box, evacuated to 10^{-5} torr by oil diffusion pump and flame sealed. The diamagnetic correction for the complexes was estimated from Pascal's constants and from the measured magnetic susceptibility of diamagnetic analogs. The diamagnetic susceptibility of the glass wool was calculated from its measured average gram susceptibility as determined from several independent samples. The diamagnetic susceptibility of the sample holder was taken to be the average value of the measurements on several identical sample holders.

RESULTS AND DISCUSSION

MnOETAP. Manganese(II) octaethyltetraazaporphyrin, α-MnOETAP, is a square planar tetrapyrrolic complex possessing three unpaired electrons formally associated with the Mn(II) center. It is closely related to manganese(II) phthalocyanine, β-MnPc, a widely studied canted molecular ferromagnet with a Curie temperature of 8.6 K.[2] Canting of the moment occurs because there is large local magnetic anisotropy which causes the local moments to align in a direction other than the stacking direction. Both of these compounds, as well as α-FeOETAP, (vide infra) are somewhat unusual in that they are *true molecular solids* in contrast to simply "molecule-based". As such, only van der Waals interactions hold the solid together.

The temperature dependence of the χT product for α-Mn(OETAP) is shown in Figure 2 (where χ is the molar susceptibility). A dramatic rise in χT (from 2.3 up to a maximum of $\sim$55 emu-K/mole) is evident. The room temperature value corresponds to intermediate spin S = 3/2 with a g value of 2.2. However, the plot of χ^{-1} vs. T (not shown) is not linear over any meaningful temperature range and possesses an inflection point at $\sim$200 K. The low field data (Figure 3) appear to indicate a ferromagnetic transition with a Curie temperature of $\sim$14 K (based on extrapolation of the field-cooled magnetization data). A hysteresis plot at 1.8 K (Figure 4) indicates substantial remanence (4 kemu-G/mole) and coercive field (2.5 kG) and evolution to a soft ferromagnetic state as the temperature is raised to 5.0 K. The low value of the saturation magnetization is indicative of canting of the magnetic moments in the ordered phase. All these data seem to indicate a transition to a canted ferromagnetic state. However, more careful examination of the hysteresis loop reveals that it is displaced to the left along the applied field axis. This behavior is rather unusual and has been observed previously in spin-glasses such as 0.5 at.%CuMn.[7]

The ac magnetic susceptibility measurements on this compound (Figure 5) also reveal a situation that is not indicative of a true ferromagnet. Although the in-phase component of the susceptibility, χ', does peak ($\sim$5 K at 1 kHz) and the out-of-phase component, χ'', becomes non-zero (5.5 K at 1kHz), the data show a pronounced dependence on the frequency of the applied field and the transition is considerably broader and less sharply peaked than for α-Fe(OETAP) (vide infra). The peak shift is over 1 K between 10 Hz and 1 kHz. In Figure 3, the field cooled magnetization and zero-field cooled magnetization do not indicate a sharp transition, only falling to zero above 30 K in 10 G applied field. These features are not expected of a true ferromagnet and are evidence for a spin-glass state.[7]

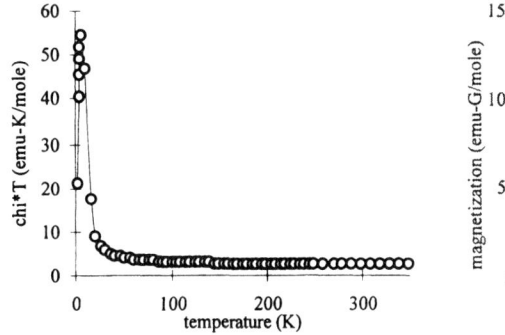

Figure 2. χT vs. T for MnOETAP

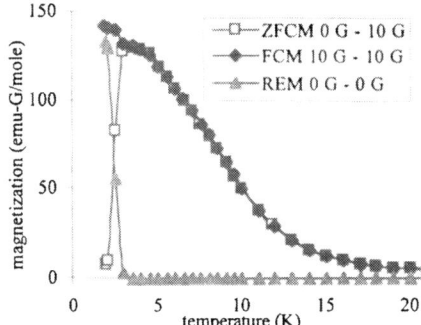

Figure 3. Low field data for MnOETAP

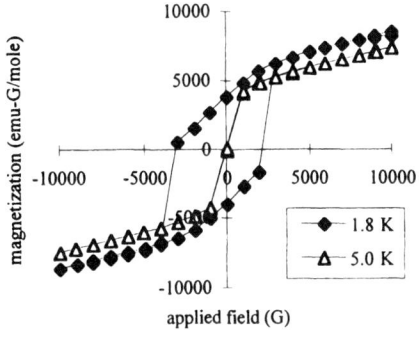

Figure 4. M vs. H data for α-MnOETAP

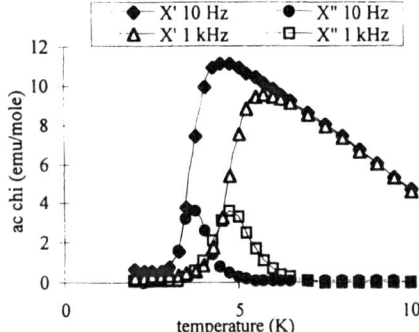

Figure 5. ac Susceptibility for α-MnOETAP

A spin glass is a metastable state of frozen spins that does not possess true long-range order but may possess net magnetization in the absence of an applied magnetic field. It is characterized by a ferromagnetic or antiferromagnetic local interactions modified by elements of randomness and disorder which prevent onset of a true phase transition. Instead, it may possess a large number of nearly degenerate ground states and the property of magnetic viscosity. In this case, the peak in χ' is no longer referred to as the Curie temperature. Instead, one identifies it with a temperature, T_f, below which the spins are frozen. It is unclear if the freezing process represents a true phase transition.[7]

FeOETAP. In contrast, we believe that both ac and dc susceptiblity measurements show that iron(II) octaethyltetraazaporphyrin, α-Fe(OETAP), undergoes a true phase transition to a canted ferromagnetic state. The χT product at room temperature is most consistent with intermediate spin $S = 1$, with a significant orbital contribution such that g = 2.7 (Figure 6). The dramatic increase in the χT product (from approximately 1.8 to over 100 emu-K/mole) and the linear χ^{-1} vs. T plot (not shown) indicate the presence of ferromagnetic coupling in α-Fe(OETAP). These data may be fitted to a Curie-Weiss Law giving g = 2.7 and θ = 8 K.

The hysteresis plot at the lowest temperature attainable on the SQUID, 1.8 K, (Figure 7) shows that the transition is to a soft ferromagnetic state. Canting of the moments is inferred from the low value of the saturation magnetization below T_c (8000 emu-G/mole (experimental) vs. ~15000 (theoretical based on g = 2.7)). Given the evidence for canting in α-Fe(OETAP) and its similar molecular structure, it seems reasonable to assume that the stacking and mechanism for intrastack magnetic coupling (canted stacks and superexchange mediated by the *meso*- nitrogen atoms) are similar to that in β-MnPc.[2]

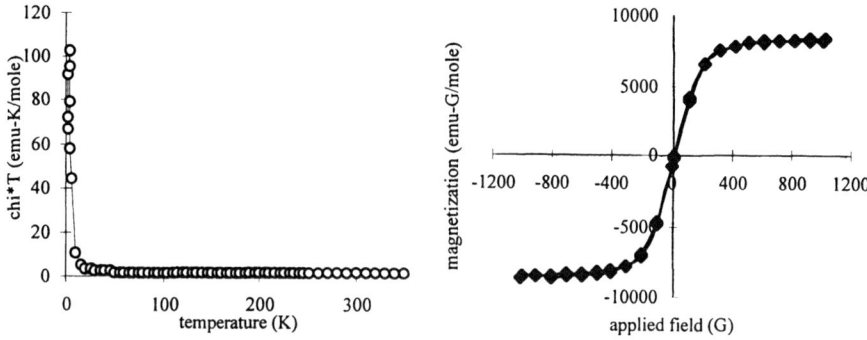

Figure 6. χT vs. T for α–FeOETAP   Figure 7. M vs. H data for α-FeOETAP

The peak in the in-phase component (χ') of the ac susceptibility at 2.8 K (Figure 8) indicates a magnetic phase transition to an ordered state. The non-zero out-of-phase component (χ'') and, importantly, its near frequency-independence supports that the transition is ferromagnetic in nature. At the present time, we cannot explain the very weak frequency-dependence affecting mostly the magnitude of the peak and not its position. The ac susceptibility of ferromagnetic β-MnPc shows a similar, weak, dependence on the frequency of the applied field (Figure 9). It is possible that α-Fe(OETAP)and β-MnPc are both spin-glasses as well, although the differences between them, and α-Mn(OETAP) are quite dramatic. These differences include the height and width of their respective peaks in χ' and the fact that the former compounds obey the Curie-Weiss law.

While the observed spin-glass behavior in α-Mn(OETAP) could have extrinsic reasons associated with small particle sizes, we presently favor an intrinsic explanation which assumes that this compound is magnetically frustrated. The source of this frustration could be competing ferromagnetic and antiferromagnetic interstack coupling. The fact that this compound does not obey the Curie-Weiss law supports this position.

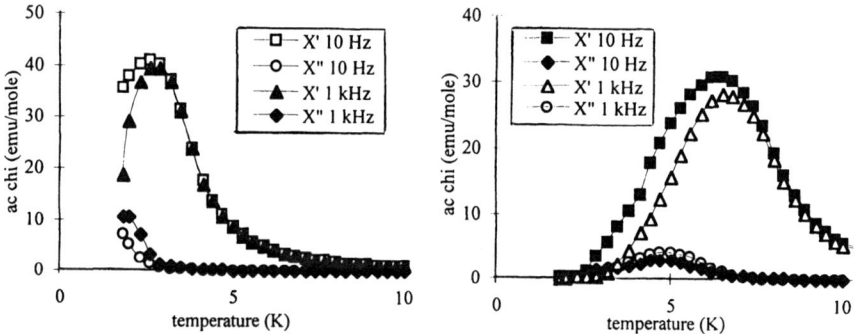

Figure 8. ac Susceptibility for α-FeOETAP   Figure 9. ac Susceptibility for β-MnPc

Epstein et al have recently postulated that in similar stacked compounds, two coupling mechanisms with opposite signs may be present.[8] The ferromagnetic pathway is dipolar and though-space, whereas the antiferromagnetic pathway is based on orbital overlap and is through bond. As applied to α-Mn(OETAP), we imagine that these two interactions are more nearly equal than for either α-Fe(OETAP)and β-MnPc, resulting in frustration and spin-glass behavior.

Mn(Cp*)$_2$•TCNE. Having observed the phenomenon of frequency-dependent ac susceptibility in a true molecular solid system, we turned to an examination of a well-known charge-transfer salt magnet, decamethylmanganocenium tetracyanoethenide. This compound was previously reported by Miller, Yee et al. to have a Curie temperature of 8.8 K as determined by dc techniques, although we are unaware of any ac susceptibility studies.

Decamethylmanganocenium tetracyanoethenide (Figure 1) is a copper-brown, microcrystalline, solid consisting of $S = 1$ donor cations and $S = 1/2$ acceptor anions in a mixed stack arrangement. Although we have prepared it by exactly the same procedure as previously described, we have been unable to reproduce, quantitatively, all of its reported dc magnetic properties. As may be seen in Table I, most of the data are comparable, though the plot of χ^{-1} vs. T extrapolates to a somewhat lower value of θ and the coercive field we measure is considerably higher. All the new results shown in Table I were averaged over three experiments from separate preparations. We believe that we are examining the same compound as reported before and offer the following explanations for the discrepancy in these results.

Table I. Comparison of magnetic properties for Mn(Cp*)$_2$•TCNE

	previous report	this work
room temperature moment[a]	$3.78 \pm 0.2\mu_B$	$3.8 \pm 0.2\mu_B$
Curie-Weiss θ[a]	22.6 ± 2 K	10 ± 2 K
Curie temperature[b]	8.8 K	8.8 ± 0.1 K
coercive field[c] (G)	1200 ± 200	5400 ± 200
remanence[c] (emu-G/mole)	6300	5500 ± 800

a) estimated ranges based on published values

b) as determined by extrapolation of the steepest part of the 3 G field cooled magnetization data

c) at 2 K

The susceptibility data reported here were obtained in 100 G applied field, whereas previous measurements were made at 19.5 kG.[9] The lower field is expected to give more reliable results because it minimizes the risk of significant reorientation of the crystals. This can occur because the interaction of a magnetic field with a crystal possessing anisotropic magnetic properties may produce a torque on the crystal large enough to align or partially align it in a preferred direction.[3] This would result in data which are not representative of the properties randomly averaged over all possible orientations as is presumed for a powder sample. Reorientation is particularly a problem with samples that are unrestrained in an open quartz bucket, as was the case in the original report on this compound. As expected, the low field determination of T_c is not affected by reorientation effects.

We also observe a hysteresis loop feature at 2 K which was present in the original report (but not commented on): a discontinuity around H = 0 G. A possible explanation is that it represents the presence of a contaminating phase which is a soft ferromagnet. The largest difference between the two data sets is in the coercive field value. Our data consistently place it at around 5000 G which is quite large among molecule-based magnets. We have found that the magnitude of the coercive field may depend on the age of the sample which might explain the higher value reported here. Alternatively, the unrestrained geometry utilized in previous work would be expected to decrease the observed coercive field by allowing the crystals to realign with the field.

The plot of ac susceptiblity vs. temperature at 10 Hz and 1 kHz is shown in Figure 11. The marked frequency dependence again suggests spin-glass behavior and not true ferromagnetism. The χ' peak shift amounts to 0.6 K between 10 Hz and 1 kHz. However, in this case, the compound does obey the Curie-Weiss law and the low field onset of spontaneous ordering is sharp, making the case for spin-glass behavior somewhat more ambiguous. More detailed study of this compound is required including low temperature heat capacity measurements.

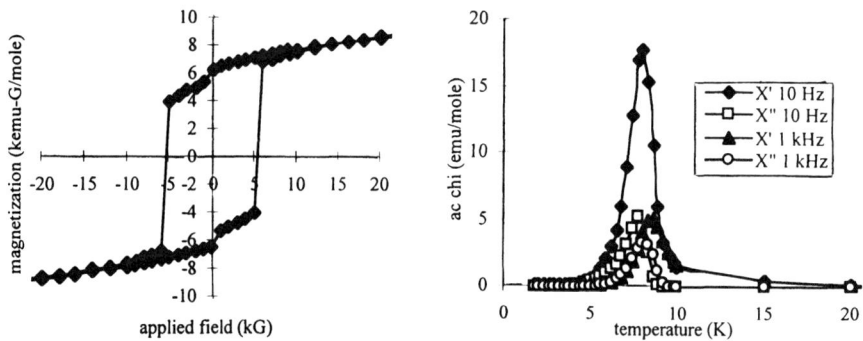

Figure 10. M vs. H data for Mn(Cp*)$_2$•TCNE   Figure 11. ac Susceptibility for Mn(Cp*)$_2$•TCNE

CONCLUSION

We have demonstrated that dc measurements are not adequate for characterizing at least two subclasses of molecular and molecule-based magnets. These results suggest that variable frequency ac susceptometry is an indispensable tool and that identification of ferromagnetism should not be made without the data from this technique. Future efforts with the compounds described here will focus on discovering why they are spin glasses, in particular, ruling out particle size as a factor.

ACKNOWLEDGMENTS

We thank Dr. Ron Goldfarb for many helpful discussions and the National Institute of Standards and Technology for use of the SQUID magnetometers. We acknowledge the Donors of the Petroleum Research Fund administered by the American Chemical Society for partial support of this work.

REFERENCES

1.   S. P. Sellers, B. J. Korte, J. P. Fitzgerald, W. M. Reiff and G. T. Yee, submitted to J. Am. Chem. Soc.
2.   J. M. Robertson, J. Chem. Soc. 615 (1935) and 219 (1937) b) B. N. Figgis, R. Mason, and G. A. Williams, Acta Crystallogr., Sect. B **B36**, 2963 (1980) c) C. G. Barraclough, R. L. Martin and S. Mitra, J. Chem. Phys. **53** 1638-1642 (1970) d) S. Mitra, A. K. Gregson, W. E. Hatfield and R. R. Weller, Inorg. Chem. **22**, 1729-1732 (1983).
3.   G. T. Yee, J. M. Manriquez, D. A. Dixon, R. S. McLean, D. M. Groski, R. B. Flippen, K. S. Narayan, A. J. Epstein and J. S. Miller, Adv. Mater. **3**, 309-311 (1991).
4.   F. Palacio, F. J. Lazaro and A. J. van Duyneveldt, Mol. Cryst. Liq. Cryst. **176**, 289-306 (1989).
5.   C. M. Wynn, A. S. Albrecht, C. P. Landee, C. Navas and M. M. Turnbull, Mol. Cryst. Liq. Cryst. **274**, 1-10 (1995).
6·   Quantum Design Application Note #1, Sample Mounting.
7.   J. A. Mydosh, *Spin Glasses, An Experimental Introduction* (Taylor and Francis, London, 1993).
8.   C. M. Wynn, M. A. Girtu, W. B. Brinckerhoff, K.-I. Suguira, J. S. Miller and A. J. Epstein, Chem. Mater. **9**, 2156-2163 (1997).
9.   R. S. McLean, private communication. A Faraday balance was used to obtain these data. A large applied field improves signal to noise with this technique.

MAGNETIC PROPERTIES OF ORGANIC FERROMAGNETIC TDAE-C_{60} SINGLE CRYSTALS AND THIN FILMS

M. TOKUMOTO*, Y. TSUBAKI**, K. POKHODNYA***, Y. SAKAKIBARA*,
A. OMERZU***, D. MIHAILOVIC***, T. UCHIDA**
*Electrotechnical Laboratory, Tsukuba, Ibaraki 305, Japan
**Science University of Tokyo, Noda, Chiba 278, Japan
***Jozef Stefan Institute, Jamova 39, Ljubljana, Slovenia

ABSTRACT

Recent progress in the study on magnetic properties of molecular soft ferromagnet TDAE-C_{60} (where TDAE is tetrakis(dimethylamino) ethylene) is reported. Both crystalline and thin film samples are prepared by the diffusion method and the temperature and field dependence of magnetization are studied by SQUID magnetometer. The magnetic behaviors of single crystals were found to vary from sample to sample. In best single crystals, the **saturation magnetization** as high as **1.0 μ_B/C_{60}** has been observed. **Spontaneous magnetization** and its temperature dependence was evaluated from the Arrott plot. The ferromagnetic phase transition was also observed in TDAE-C_{60} thin films.

INTRODUCTION

TDAE-C_{60} was first prepared in powder sample and reported as an organic molecular soft ferromagnet with $T_c=16K$[1]. Since then, many experimental studies have been reported. However, most of them were on powder samples until recently. Preparation of single crystals have been reported by Suzuki et al.[2], but they did not observe the ferromagnetic behavior. On the other hand, Blinc et al.[3] reported observation of antiferromagnetic correlation and weak ferromagnetism in single crystals. The static magnetic properties of single crystals are reported to be quite different from those for powder samples.[4-6]   In spite of intensive studies both experimental and theoretical, the origin and the true nature of the low temperature magnetic phase of TDAE-C_{60} is still not well understood. So, the preparation of good single crystals and further extended study of its magnetic properties are necessary to elucidate the nature of the electronic ground state of this compound. On the other hand, the preparation of TDAE-C_{60} thin films and optical and electrical properties were reported by Mihailovic et al.[7-9], but magnetic properties have not been reported so far.

In this paper, we report on the preparation and characterization of ferromagnetic properties of TDAE-C_{60} single crystals and thin films.

EXPERIMENTAL

Single crystals are prepared by the diffusion method with dry toluene as solvent at temperatures 15°C ,20°C, 25°C and 30°C. Pure C_{60} and TDAE were placed in opposite sides of the H-shaped cell separated by a glass filter, and placed in an inert atmosphere. The crystals were observed to grow on the C_{60} side of the filter after about one month. Single crystals used for magnetic measurements were oriented by the X-ray diffraction method. Typical dimension of plate shaped crystals was $1.0 \times 0.5 \times 0.2$ mm^3 with c-axis perpendicular to the extended a-b plane. Oriented single crystal was mounted in a quartz tube under inert atmosphere and sealed with He gas. The DC magnetization was measured by Quantum Design SQUID magnetometer. Thin films were prepared in two steps. First, pristine C_{60} thin films were prepared by vacuum deposition of pure C_{60} onto quartz substrates. Then, the pristine C_{60} film was immersed into liquid TDAE in inert atmosphere to obtain a film of the charge transfer complex, TDAE-C_{60}.

RESULTS AND DISCUSSION

Temperature and Field Dependence of Magnetization

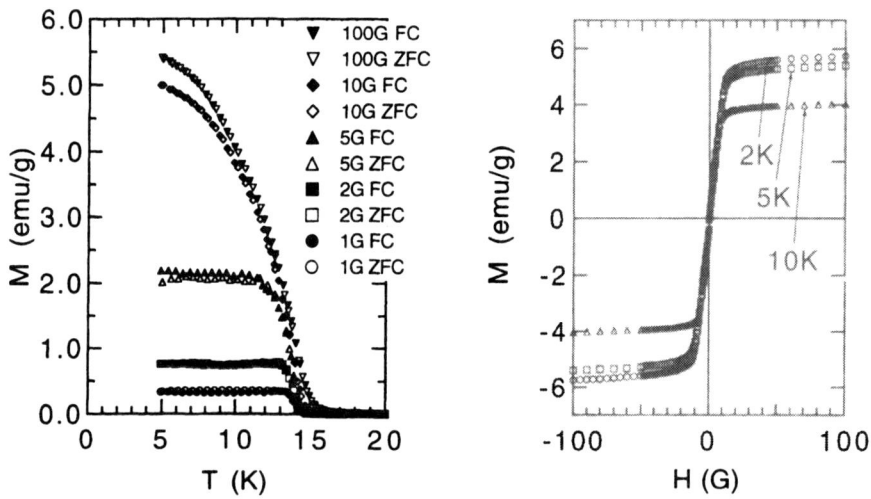

Fig. 1 Temperature dependence M-T (a) and field dependence M-H (b) of magnetization of TDAE-C_{60} single crystal #8 (H⊥c).

478

Fig.1(a) shows the temperature dependence of d.c. magnetization M of our single crystal sample #8. The sample was first cooled at zero-field, then the magnetic field perpendicular to the c-axis was applied. Although both zero-field-cooled (ZFC) and field-cooled (FC) magnetization are shown in the figure, the difference between the two is not clear. Note that the magnetization is almost independent of temperature at low fields.

Fig.1(b) shows the magnetization curves of the same sample at different temperatures. The magnetic field was applied along the a-b plane, i.e. H⊥c, and correction due to demagnetization field is not taken into account. It is to be noted that, at low fields below about 10 Gauss, magnetization is proportional to the applied field, and the slope i.e. magnetic susceptibility is independent of temperature as shown in Fig.1(b). This temperature independent magnetic susceptibility, commonly observed in ferromagnet, is due to the demagnetization effect. This also explains why the magnetization at low fields is temperature independent as shown in Fig.1(a). However, the correction of demagnetization effect is not straightforward because our single crystals are not well shaped. Almost no hysteresis and zero coercivity were observed in Fig. 1(b). The **saturation magnetization** at low temperature amounts to almost **6 emu/g** corresponding to **1.0 μB/C60** in this sample. Other samples showed smaller values. In the temperature dependence, the remanence or the irreversibility, i.e. the difference between the zero-field-cooled (ZFC) and field-cooled (FC) magnetization was not clear in most single crystals,

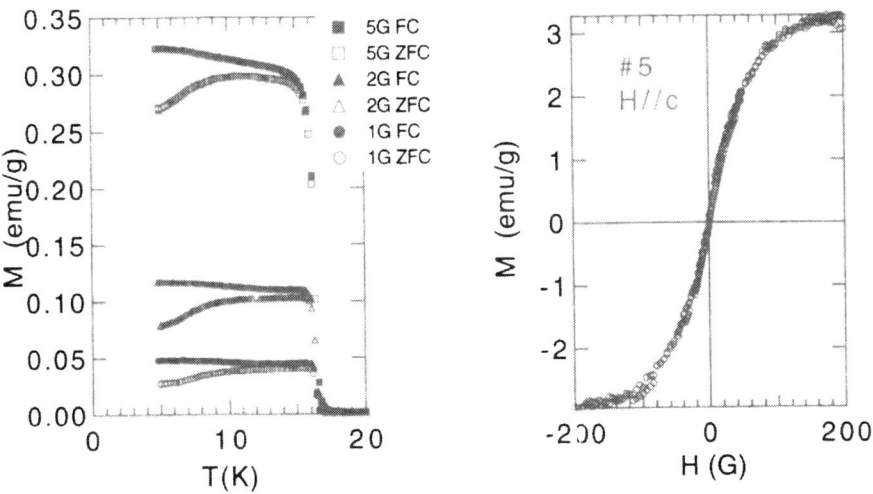

Fig. 2 Temperature dependence M-T (a) and field dependence M-H (b) of magnetization of TDAE-C60 single crystal #5 (H//c).

in contrast to the behavior observed in powder samples.[10]   However, a clear difference between ZFC and FC was observed at low fields in some single crystal samples with smaller saturation magnetization and slightly higher T_C as shown in Fig. 2(a).   The sample was first cooled at zero-field, then the magnetic field parallel to the c-axis was applied.   The difference becomes negligibly small at field higher than 20 G.   Fig. 2(b) shows the the magnetization curves of the same sample at 5 K without correction due to demagnetization field.   The magnetic field was applied perpendicular to the a-b plane, i.e. H//c.

Estimate of Spontaneous Magnetization

By definition, a material is ferromagnetic if there can exist regions within the material in which a spontaneous magnetization exists.[11]   The magnetization curves of TDAE-C_{60} show almost no hysteresis and no remanent magnetization, suggesting absence of macroscopic spontaneous magnetization at zero field.   However, the equivalent spontaneous magnetization and its temperature dependence can be evaluated from the analysis of field dependence of magnetization at higher fields.[11]   Fig. 3-a shows the **Arrott plot**, i.e. M^2 vs. H/M, of magnetization data for another single crystal (#K2) at various temperatures.   At high fields above H/M=30 G·g/emu in Fig. 2-a, a linear relation between M^2 and H/M is observed as shown by dashed lines.   If the extrapolation of each line to H/M=0 gives a

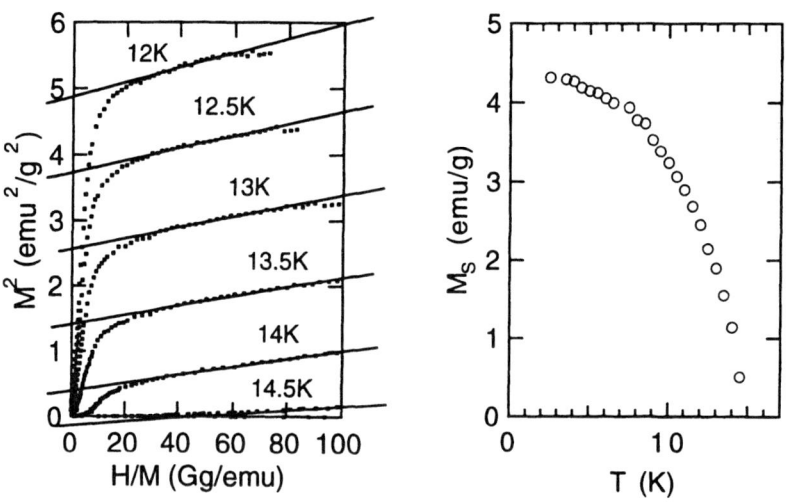

Fig. 3 Arrott plot, i.e. M^2-H/M (a), and temperature dependence of spontaneous magnetization M_S-T (b) of TDAE-C_{60} single crystal #K2.

positive intercept, it gives us an estimated value of M_S^2 , where M_S represents the **spontaneous magnetization** without external field. The intercept, corresponding to M_S^2, decreases with increasing temperature, crosses zero at T_C, and becomes negative above T_C. Fig. 3(b) shows the derived M_S as a function of temperature. M_S becomes zero at about $T_C = 15$ K in this case.

Thin films

Thin films of TDAE-C_{60} were prepared by dipping pristine C_{60} thin film (thickness of $1\mu m$) into liquid TDAE in inert atmosphere. After doping, the color of thin films changed from brown to dark green, suggesting charge transfer occurred by diffusion of TDAE into solid C_{60} thin film. Fig. 4 shows the temperature dependence of magnetization of TDAE-C_{60} thin film on the quartz substrate ($3 \times 10 mm^2$). This measurement was performed under the magnetic field of 1 G. Thus, we have succeeded in preparation of ferromagnetic thin film of TDAE-C_{60}, which will enable us to study its magneto-optical property. In this measurement magnetic transition was observed at $T_C = 16$ K.

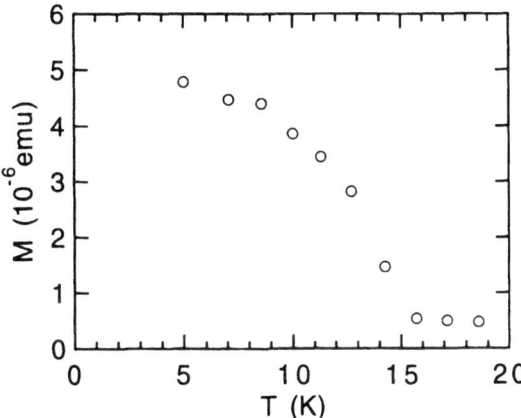

Fig. 4   Temperature dependence of magnetization of TDAE-C_{60} thin film.

SUMMARY

Both crystalline and thin film samples are prepared by the diffusion method and the temperature and field dependence of magnetization are studied by SQUID magnetometer. The magnetic behaviors of single crystals were found to vary from sample to sample. In best single crystals, the **saturation magnetization** as high as **1.0 μ_B/C_{60}** has been observed. In the temperature dependence, the difference between the zero-field-cooled (ZFC) and field-cooled (FC) magnetization was not clear in most single crystals. However, a clear difference between ZFC and FC was observed at very low fields in some single crystal samples. **Spontaneous magnetization** and its temperature dependence was evaluated from the Arrott plot. The ferromagnetic phase transition was also observed in TDAE-C_{60} thin films.

ACKNOWLEDGMENT

We wish to thank Dr. M. Mizuno at National Institute of Materials and Chemical Research for his valuable advice in the X-ray diffraction analysis.

REFERENCES

1. P.M. Allemand, K. Khemani, A. Koch, F. Wudl, K. Holczer, S. Donovan, G. Gruner, J.D. Tompson: Science **253** (1991) 301.
2. A. Suzuki, T. Suzuki and Y. Maruyama: Solid State Commun. **96** (1995) 253.
3. R. Blinc, K. Pokhodnia, P. Cevc, D. Arcon, A. Omerzu, D.Mihailovic, P. Venturini, L. Golic, Z. Trontelj, J. Luznik, Z. Jeglicic and J. Pinat: Phys. Rev. Lett. **76** (1996) 523.
4. M. Tokumoto, K.Tanaka, T. Sato, T. Yamabe, A. Omerzu, K. Pokhodnia and D. Mihailovic in Proc. Int. Winterschool on Electronic Properties of Novel Materials "Fullerenes and Fullerene Nanostructures", Kirchberg, Austria, 1996, ed. H. Kuzmany et al. (World Scientific, 1996) p.535.
5. M. Tokumoto, Y.S. Song, K. Tanaka, T. Sato, T. Yamabe, A. Omerzu, K. Pokhodnia and D. Mihailovic: Synthetic Metals **86** (1997) 2315.
6. Z. Jaglicic, Z. Trontelj, J. Luznik, J. Pirnat, R. Blinc: Solid State Commun. **101** (1997) 591.
7. D. Mihailovic, P. Venturini, A. Hassanien, J. Gasperic, K. Lutar, S. Milicev, V. Srdanov in Proc. Euroconference on Fullerides and Fulleroides, Kirchberg, Austria, 1994, ed. H. Kuzmany et al. (Springer-Verlag, 1994) p.275.
8. F. Bommeli, L. Degiorgi, P. Wachter, D. Mohailovic, A. Hassanien, P. Venturini, M. Schreiber and F. Diederich: Phys. Rev. B **51** (1995) 1366.
9. A. Hassanien, I. Musevic, D. Mihailovic, A. Omerzu and S. Tomic in Proc. Int. Winterschool on Electronic Properties of Novel Materials "Physics and Chemistry of Fullerenes and Derivatives", Kirchberg, Austria, 1995, ed. H. Kuzmany et al. (World Scientific, 1995) p.489.
10. M. Tokumoto, Y.S. Song, K. Tanaka, T. Sato, T. Yamabe: Solid State Commun. **97** (1996) 349.
11. A. Arrott: Phys. Rev. **108** (1957) 1394.

FUNCTIONALIZED TETRATHIAFULVALENE DERIVATIVES AND THEIR RADICAL CATION SALTS: SYNTHESIS AND X-RAY CRYSTAL STRUCTURES

MARTIN R. BRYCE, ADRIAN J. MOORE, ANDREI S. BATSANOV,
ANTONY CHESNEY, CLARE L. WOOD, JUDITH A.K. HOWARD
Department of Chemistry, University of Durham, Durham, U.K. DH1 3LE.
E. mail: m.r.bryce@durham.ac.uk

ABSTRACT

We report our investigations into tetrathiafulvalene (TTF) derivatives bearing N-methylthiocarbamoyl and halogen substituents which engage in intermolecular interactions in the solid state. The synthesis of new donors is presented, along with the X-ray crystal structures of some of these derivatives and their radical ion salts.

INTRODUCTION

Within the field of molecular conductors [1] the control of intermolecular interactions in the solid state by chemical modification remains a very challenging topic, and the key role played by chalcogen atoms in salts of tetrachalcogenofulvalenes is well known. There is continuing interest in the synthesis of new TTF derivatives which bear substituents (*e.g.* halogens [2], thiocarbamoyl [3] and carbamoyl groups [4]) which can engage in intermolecular interactions, leading to radical cation salts and charge-transfer complexes which possess increased dimensionality [5] thereby providing a means of 'crystal engineering' of molecular conductors. Herein we report our recent results on this topic, focussing on TTF derivatives which possess N-methylthiocarbamoyl, iodine and bromine substituents.

EXPERIMENTAL

Synthesis

TTF derivatives **1** [3a] and **2** [2c] were prepared as described in the literature by trapping TTF-Li (generated from TTF [6]) with N-methylisothiocyanate and perfluorohexyl iodide, respectively. TTF derivatives **4a-c** and **9** were prepared as shown in Schemes 1 and 2. Lithiation of **3a** [7], **3b** [7,8] and **3c** [9], respectively, using lithium diisopropylamide-tetrahydrofuran (LDA-THF) complex (from Aldrich) at -78 °C, followed by addition of N-methylisothiocyanate, gave **4a-c** which were isolated by column chromatography on silica, eluent hexane / toluene (1:1 v/v): **4a**, 66% yield, black plates, mp 171-172 °C (from toluene); **4b**, 70% yield, dark red needles, mp 174-175 °C (from toluene); **4c**, 63% yield, black needles, mp 134-135 °C (from acetonitrile). Compound **9** was prepared as follows: vinylene trithiocarbonate **5** was lithiated using LDA in THF at -78 °C; addition of 1,2-dibromotetrafluoroethane gave the dibromo derivative **6** in *ca.* 50% yield, mp 90-92 °C. Cross-coupling of **6** (1 equivalent) with ketone **7** [10] (4 equivalents) in refluxing toluene in the presence of triethylphosphite gave TTF derivative **8**, 45-50% yield, yellow crystals, mp 131-133 °C. Treatment of **8** with cesium hydroxide in THF-MeOH at 20 °C, followed by addition of methyl iodide afforded compound **9**, 95% yield, red crystals, mp 119-120 °C. This route to **9** is notionally different from that reported by Iyoda *et. al.* [2g], and offers the added advantage that **8** can serve as a versatile precursor to a range of alkylsulfanyl derivatives.

Mat. Res. Soc. Symp. Proc. Vol. 488 © 1998 Materials Research Society

Salt $1^{+\cdot}$ B r⁻ was prepared by cooling a solution of compound **1** and Bu_4N^+ B r₃⁻ in acetonitrile. Salt $2^{+\cdot}$ H S O₄⁻ was prepared by electrocrystallization of **2** in acetonitrile containing concentrated H_2SO_4 at a constant current of 1μA [2f]. Complex **4c** · TCNQ was obtained by cooling a solution of compound **4c** and 7,7,8,8-tetracyano-*p*-quinodimethane (TCNQ) in acetonitrile. Salt $9^{+\cdot}$ I₃⁻ was prepared by diffusion of iodine vapor into an acetonitrile solution of **9** in a sealed container.

X-Ray Crystallography

Single-crystal X-ray diffraction experiments were carried out on a Rigaku AFC6S 4-circle diffractometer or a Siemens 3-circle SMART diffractometer with a CCD area detector. Graphite-monochromated Mo-K_α radiation ($\overline{\lambda}$ =0.71073 Å) was used. The structures were solved by direct methods and refined by full-matrix least squares against F^2 of all data, using SHELXTL software [11].

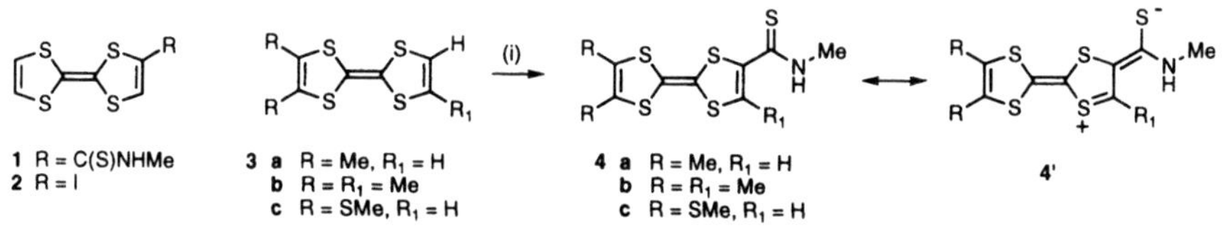

Scheme 1. (i) LDA (1 equiv.), Et₂O, -78 °C, then MeNCS, -78 °C to 20 °C, then H₂O

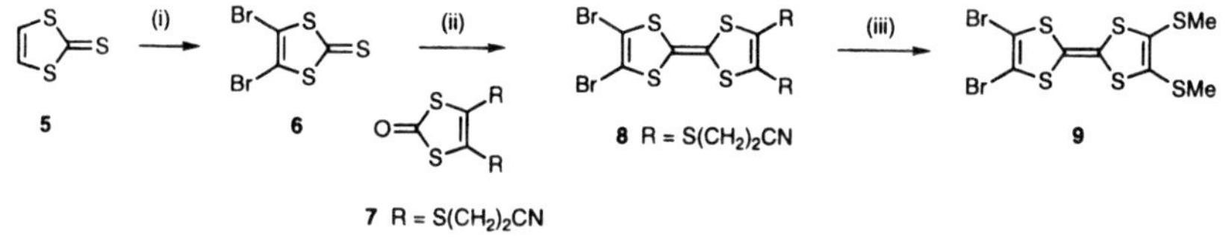

1 R = C(S)NHMe
2 R = I

3 a R = Me, R₁ = H
  b R = R₁ = Me
  c R = SMe, R₁ = H

4 a R = Me, R₁ = H
  b R = R₁ = Me
  c R = SMe, R₁ = H

4'

5

6

7 R = S(CH₂)₂CN

8 R = S(CH₂)₂CN

9

Scheme 2.   (i) LDA (3 equiv.), THF, -78 °C, then (CF₂Br)₂, -78 to 20 °C; (ii) compound **7** (4 equiv.), P(OEt)₃, PhMe, reflux; (iii) CsOH.H₂O, MeOH-THF, 20 °C, then MeI (excess).

RESULTS AND DISCUSSION

N-Methylthiocarbamoyl-TTF Derivatives

Compounds **1** and **4a-c** were designed to explore intermolecular hydrogen bonding interactions involving the thiocarbamoyl group, in conjunction with stacking of the TTF system. The attachment of this electron-withdrawing substituent is known to raise the first oxidation potential of TTF by 90 mV [3a]. We have previously shown that compound **1** packs in a *kappa* fashion in the crystal structure [3a,b] (that is, the π-donor forms orthogonal dimers which are coupled through weak intermolecular forces); such an arrangement in the neutral state is very rare [4, 12]. This is an important packing motif to explore, as all the TTF-based superconductors with T_c > 10 K possess *kappa*-phase structures [13]. We, therefore, synthesized analogues **4a-c**, and we now report the molecular structures of **4a** and **4b**, the 1:1 salt **1⁺·** Br⁻ and the 1:1 complex **6c ·** TCNQ.

For both compounds **4a** (Fig. 1a) and **4b** (Fig. 1b) it is notable that the S(2)-C(3) bond is significantly shorter than the other C-S bonds, by 0.036 Å in **4a** and 0.013 Å in **4b** (*cf.* 0.03 Å in **1** [3a]); this is in agreement with our suggestion [14] that the mesomeric effect of the C=S group gives rise to a significant contribution from the dipolar canonical form **4'**. The S(2)-C(3) bond contraction is smaller [and the C(2)-C(7) bond marginally longer] in **4b**, wherein the methyl substituent at C(3) interferes with the conjugation.

Compound **4a** exhibits a highly-distorted *kappa*-packing of molecular dimers, with the TTF moieties in a boat conformation, with 3.4 Å separation between their central C_2S_4 planes, and rings A and B folding along S···S vectors by 17.5 and 4.9°, respectively (Fig. 2). Compound **4b** exhibits a similar packing of individual molecules (not dimers) which are closer to a planar conformation (both rings fold by 4°). The NH group forms a bifurcated intermolecular hydrogen bond to S(1) and S(5) with the respective N···S distances of 3.783(3) and 3.425(3) Å in **4a**, and 3.456(2) and 3.697(2) Å in **4b** (Figs. 1a and 1b). The hydrogen bonding in both structures, is, therefore, rather weak.

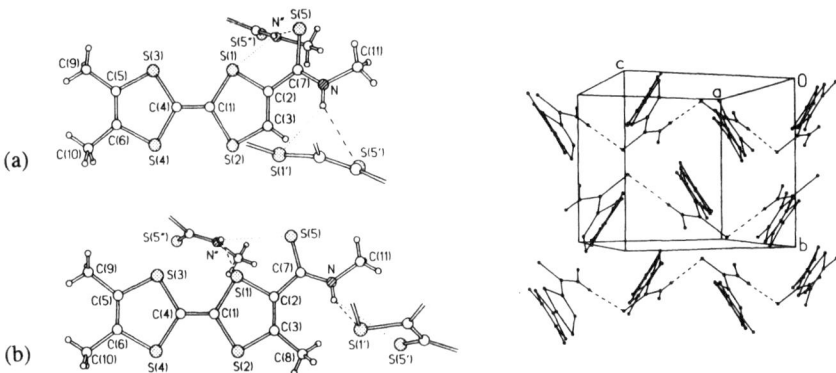

Fig. 1. Molecular structures of **4a** (a) and **4b** (b) showing intermolecular hydrogen bonds.

Fig. 2. Crystal packing of **4a**.

The motif of **1⁺·** Br⁻ is entirely different from that of neutral **1** [3a,b]. Here, **1⁺·** radical cations have almost planar TTF moieties and form an infinite stair-like stack of dimers (Fig. 3). Adjacent cations in a stack are inversion-related. Within a dimer, TTF moieties are nearly eclipsed with the interplanar separation d = 3.34 Å and S···S contacts 3.35 - 3.39 Å. Between these dimers, d = 3.46 Å (shortest S···S distances 3.77 - 3.94 Å) and the overlap of TTF moieties is only partial. However, this arrangement corresponds to the best possible overlap of the *entire* cations of **1** (*i.e.* including the *N*-methylthiocarbamoyl substituent). All TTF planes within a single stack of **1⁺·** Br⁻ are parallel but inclined by 46° to those of adjacent stacks. The anion is situated between stacks and is linked with the cation by a unique strong N-H···Br hydrogen bond [H···Br 2.37(4) Å] into an ionic pair. Even the N···Br distance of 3.210(2) Å is shorter than the sum of their Van der Waals radii (3.47 Å). The contacts Br···S(4) of 3.38 and Br···S(3) of 3.57 Å with other cations (*cf.* the sum of their Van der Waals radii is 3.81 Å) are obviously affected by electrostatic attraction.

The structure of **4c** · TCNQ comprises mixed stacks of a ...DDAADD... type (Fig. 4) with interplanar separations of TTF···TTF *ca.* 3.5 Å, TTF···TCNQ *ca.* 3.2 Å, and TCNQ···TCNQ *ca.* 3.25 Å. The TTF moiety shows only insignificant folding of the dithiole rings and a small (3.8°) twist around the central C=C bond. The TCNQ radical anion is planar to within ±0.04 Å. The TTF moieties are laterally slipped and the TCNQ moieties overlap in the common ring over C=C bond manner. The NH group of **4c** is engaged in an N-H···N interstack hydrogen bond with an anion [H···N 2.24(3) Å].

Charge transfer from the TTF moiety in both **1**$^{+\cdot}$ Br$^-$ and **4c** · TCNQ has the usual effect upon bond distances [15]. The C=C bonds of the donor unit lengthen (especially the central C=C bond, by 0.04 - 0.05 Å) whereas the C-S bonds contract (by 0.02 - 0.04 Å), with the exception of the S(2)-C(3) bond which shortens insignificantly in **4c**$^{+\cdot}$ and not at all in **1**$^{+\cdot}$. Thus, the non-equivalence of the C-S bonds diminishes but persists. This can be understood with reference to the resonance form **4'**: electron density is withdrawn only from the TTF ring system, while the thiocarbamoyl group is practically unaffected. Consideration of the TTF and TCNQ bond lengths [16] in **4c** · TCNQ suggests that the degree of charge transfer (ρ) is between 0.5 and 0.7 electrons per molecule. The conductivity values of **1**$^{+\cdot}$ Br$^-$ and **4c** · TCNQ are σ_{rt} = 5 x 10^{-5} and 5 x 10^{-6} Scm^{-1}, respectively (two-probe, compressed pellet data).

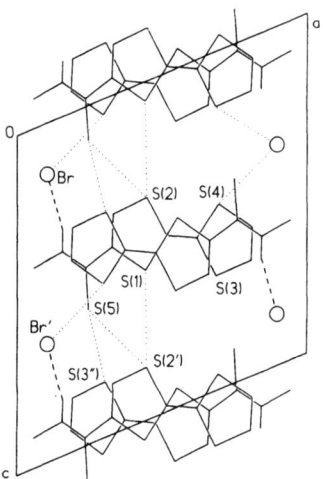

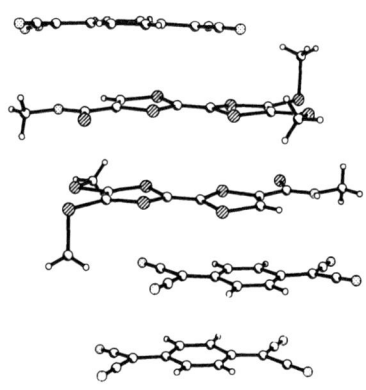

Fig. 3. Crystal structure of **1**$^{+\cdot}$ Br$^-$.   Fig. 4. Crystal structure of **4c** · TCNQ.

Halogenated TTF Derivatives

It is known in solid state chemistry that halogen atoms can participate in relatively strong and directional intermolecular interactions [17]. Halogenation of TTF raises the oxidation potential [2d]; therefore, we have studied derivatives bearing only one or two halogens atoms.

We have obtained the first radical cation salts of 4-iodo-TTF **2** [2f]. In the structure of the salt **2**$^{+\cdot}$ HSO$_4^-$ the radical cations, which are almost planar, form a stair-like stack with two alternating kinds of overlap: (i) between two TTF moieties with a lateral shift of *ca.* 0.5 Å, and (ii) between substituted dithiole rings only; the interplanar separations being 3.33 and 3.40 Å, respectively. Parallel to this stack, runs a chain of hydrogensulphate anions, linked by strong hydrogen bonds O(2)···O(2') 2.555(7) and O(3)···O(3') 2.627(7) Å. The anion-cation contacts I···O(4) 2.92 (C-I-O 168°) and S(3)···O(2) 2.82 Å are substantially shorter than the sums of the Van der Waals radii (3.6 and 3.4 Å, or 3.16 and 3.00 Å, respectively, correcting for the ellipsoidal shape of the I and S atoms [18]) which implies significant polarization of the 'soft' I or S atoms. On the other hand, the H(3)···O(1) and H(6)···O(1) contacts of 2.19(5) and 2.24(6) Å can be regarded as hydrogen bonds. Thus, the anionic chain contributes significantly to the close packing of cations. The conductivity of salt $2^{+\cdot}$ HSO$_4^-$ is $\sigma_{rt} = 5 \times 10^{-7}$ Scm^{-1} (four-probe, compressed pellet data).

While our work was in progress, a triiodide salt of donor **9** (1:1 stoichiometry) was prepared electrochemically, and its structure determined by Iyoda *et. al.* [2f]. We have characterized the same salt $9^{+\cdot}$ I$_3^-$ obtained by simpler diffusion methods (see Experimental section). The crystal structure (Fig. 6) shows cations and anions in strictly planar layers with short Br···I (3.604-3.872 Å) and S···I (3.719-3.774 Å) contacts, while mixed stacks (interlayer separation of 3.55Å) exist in the perpendicular direction.

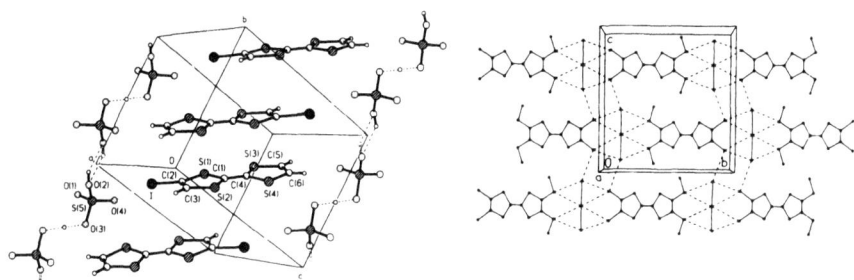

Fig. 5. Crystal structure of $2^{+\cdot}$ HSO$_4^-$.   Fig. 6. Crystal structure of $9^{+\cdot}$ I$_3^-$.

CONCLUSIONS

Recent advances in synthetic methodology in TTF chemistry have enabled the synthesis of a range of new functionalized derivatives. X-ray crystallographic studies have established that intermolecular interactions involving *N*-methylthiocarbamoyl or halogen substituents play an important role in regulating the solid state structures in the neutral and the radical cation species. Future work will concern crystal engineering of mixed-valence salts of these, and related, TTF derivatives.

ACKNOWLEDGMENTS

We thank EPSRC for funding, the Royal Society for a Leverhulme Senior Research Fellowship (to J.A.K.H.) and a Leverhulme Visiting Fellowship (to A.S.B.), Durham University for financial support (to A.S.B.) and Dr N. Robertson for crystal growth of $1^{+\cdot}$ Br$^-$ and $2^{+\cdot}$ HSO$_4^-$.

REFERENCES

1. J. Mater. Chem. **5**, 1469-1760 (1995) (Special Issue on Molecular Conductors).
2. a) M. Jørgensen and K. Bechgaard, Synthesis 207 (1989); b) M. R. Bryce and G. Cooke, Synthesis 263 (1991); c) C. Wang, A. Ellern, V. Khodorkovsky, J. Bernstein and J. Y. Becker, J. Chem. Soc., Chem. Commun. 983 (1994); d) C. Wang, J. Y. Becker, J. Bernstein, A. Ellern and V. Khodorkovsky, J. Mater. Chem. **5**, 1559 (1995); e) R. Gompper, J. Hock, K. Polborne, E. Dormann and H. Winter, Adv. Mater. **7**, 41 (1995); f) A. S. Batsanov, A. J. Moore, N. Robertson, A. Green, M. R. Bryce, J. A. K. Howard and A. E. Underhill, J. Mater. Chem. **7**, 387 (1997); g) M. Iyoda, Y. Kuwatani, K. Hara, E. Ogura, H. Suzuki, H. Ito and T. Mori, Chem. Lett. 599 (1997).
3. a) A. S. Batsanov, M. R. Bryce, G. Cooke, J. N. Heaton and J. A. K. Howard, J. Chem. Soc., Chem. Commun. 1701 (1993); b) A. S. Batsanov, M. R. Bryce, G. Cooke, A. S. Dhindsa, J. N. Heaton, J. A. K. Howard, A. J. Moore and M. C. Petty, Chem. Mater. **6**, 1419 (1994); c) A. J. Moore, M. R. Bryce, A. S. Batsanov, C. W. Lehmann and J. A. K. Howard, Synth. Met. **86**, 1901 (1997).
4. A. S. Batsanov, M. R. Bryce, J. N. Heaton, A. J. Moore, P. J. Skabara, J. A. K. Howard, E. Ortí, P. M. Viruela and R. Viruela, J. Mater. Chem. **5**, 1689 (1995).
5. Review: M. R. Bryce, J. Mater. Chem. **5**, 1481 (1995).
6. A. J. Moore and M. R. Bryce, Synthesis 407 (1997).
7. K. Lerstrup, I. Johannsen and M. Jørgensen, Synth. Met. **27**, B9 (1989).
8. A. J. Moore, M. R. Bryce, A. S. Batsanov, J. C. Cole and J. A. K. Howard, Synthesis 675 (1995).
9. M. Fourmigué, F. C. Krebs and J. Larsen, Synthesis 509 (1993).
10. N. Svenstrup, K. M. Rasmussen, T. K. Hansen and J. Becher, Synthesis 809 (1994).
11. G. M. Sheldrick, SHELXTL, Version 5, Siemens Analytical X-ray Instruments Inc. Madison, WI, USA (1995).
12. Y. Misaki, T. Ohta, N. Higuchi, H. Fujiwara, T. Yamabe, T. Mori, H. Mori and S. Tanaka, J. Mater. Chem. **5**, 1571 (1995).
13. J. M. Williams, J. R. Ferraro, R. J. Thorn, K. D. Carlson, U. Geiser, H. H. Wang, A. M. Kini and M-H. Whangbo, Organic Superconductors (Including Fullerenes), Prentice Hall, Englewood Cliffs, New Jersey, 1992.
14. A. S. Dhindsa, Y. P. Song, J. P. Badyal, M. R. Bryce, Y. M. Lvov, M. C. Petty and J. Yarwood, Chem. Mater. **4**, 724 (1992).
15. a) C. Katayama, M. Honda, H. Kumagai, J. Tanaka, G. Saito and H. Inokuchi, Bull. Chem. Soc. Jpn. **58**, 2772 (1985); b) D. A. Clemente and A. Marzotto, J. Mater. Chem. **6**, 941 (1996).
16. S. Flandrois and D. Chasseau, Acta Crystallogr. Sect. B **33**, 2744 (1977).
17. G. R. Desiraju, Crystal Engineering, The Design of Organic Solids, Elsevier, Amsterdam, 1989, ch. 6.
18. a) A. Gavezzotti, J. Amer. Chem. Soc, **105**, 5220 (1983); b) S. C. Nyburg and H. C. Faerman, Acta Crystallogr. Sect. B **41**, 274 (1985).

ELECTRONIC STRUCTURE OF THE ORGANIC METALS κ-ET₂Cu(SCN)₂ AND κ-ET₂Cu[N(CN)₂]Br MEASURED BY SOFT X-RAY EMISSION AND SOFT X-RAY ABSORPTION

CRISTIAN B. STAGARESCU [1†], L.-C. DUDA [1], KEVIN E. SMITH [1‡], J. H. GUO [2], J. NORDGREN [2], R. HADDON [3], J.S. BROOKS [4], and D. JEROMÉ [5]
[1] Department of Physics, Boston University, Boston, MA, 02215
[2] Department of Physics, Uppsala University, Box 530, 75121 Uppsala, Sweden
[3] Bell Laboratories, Lucent Technologies
[4] Department of Physics, Florida State University, Tallahassee, FL
[5] Laboratoire de Physique des Solides, Université de Paris-Sud

ABSTRACT

The electronic structure of the two organic, layered metals κ-ET₂Cu(SCN)₂ and κ-ET₂Cu[N(CN)₂]Br has been studied using a combination of soft X-ray emission (SXE) and soft X-ray absorption (SXA) spectroscopy. These techniques are powerful probes of the site and angular-momentum resolved partial density of states (PDOS) for both occupied and unoccupied states. Therefore these spectroscopies are particularly suited for an analysis of the density of states of multi-atomic, complex materials as the ET-based organic conductors, allowing site-specific electronic structure to be measured. Furthermore, for certain materials, the electronic structure of specific layers can be measured. We present a preliminary picture of the electronic structure of κ-ET₂Cu(SCN)₂ and κ-ET₂Cu[N(CN)₂]Br as measured by SXE and SXA performed at the C-1s and N-1s core levels.

INTRODUCTION

Among the large number of organic molecular conductors exhibiting superconducting properties at ambient pressure κ-ET₂Cu(SCN)₂ [1] and κ-ET₂Cu[N(CN)₂]Br [2] are some of the most remarkable compounds, having the highest superconducting transition temperatures, of 10.4 K and 11.6K. Most of the up to date experimental knowledge on the electronic structure of these materials resulted from magnetic and/or transport measurements. These probes are limited to the part of the electronic structure directly responsible for conduction, i.e., the electrons at the Fermi surfaces. Hence, they can afford only a restricted comparison with results of theoretical calculations through physical quantities directly involved in conductive properties, such as density of charge carriers at the Fermi surface or their effective masses. It is therefore highly desirable to have a more comprehensive picture of the electronic structure in these molecular conductors, involving both occupied and unoccupied electronic states over a larger range of energies. Recently, the electronic structure of κ-ET₂Cu(SCN)₂ and κ-ET₂Cu[N(CN)₂]Br [3] has been studied using photoemission spectroscopy. Spectral features representing occupied electronic states have been identified and, for some of them, their orbital and atomic character could be inferred from the dependence of their intensities with the excitation energy. Similar studies were performed on members of other ET-derived structural phases, like α-ET₂I₃ [4]. Nevertheless, a proper interpretation of the photoemission measurements remains highly dependent on the quality of the surfaces measured. We have used a combination of bulk probes, high-resolution SXA and SXE to investigate the C-2p and N-2p contributions to the electronic structure in κ-ET₂Cu(SCN)₂ and κ-ET₂Cu[N(CN)₂]Br. For such materials, their electronic structure is a complex "sum" of contributions given by molecular orbitals of well defined angular momentum character, located at various distinct chemical sites. SXA and

SXE are particularly suited for this study. The absorption and emission of a soft X-ray photon involve participation of a core-level atomic orbital, in our case C-1s and N-1s. At the spatial scale of the typical crystalline structures, the localized nature of the core-level orbital leads to the predominance of the intra-atomic processes, and thus to a chemical or atomic-site sensitivity. In addition, the dominant contribution to absorption and emission of a photon comes from the dipolar term of the expansion of its electromagnetic field. Therefore, the subsequent dipole-selection rules will restrict the orbital character of the probed occupied or unoccupied states. This leads to the interpretation of the measured X-ray absorption and X-ray emission spectra as representing the site-projected and orbital symmetry restricted partial density of unoccupied, respectively occupied states (PDOS).

MATERIALS

κ-ET$_2$Cu(SCN)$_2$ and κ-ET$_2$Cu[N(CN)$_2$]Br belong to the κ-phase structural family of charge-transfer salts of the type κ-ET$_2$X, where ET stands for the organic donor molecule of bis(ethylenedithio)-tetrathiafulvalene (BEDT-TTF) and X for an inorganic acceptor radical [5]. Generally, they can be defined as molecular solids in the sense that they form well-defined crystals, with long range order, in which the bases of the crystalline lattices are composed of molecules rather than atoms. More specifically, their typical molecular arrangement (Fig. 1) consists of a stacking of two alternating layers: an inert, inorganic layer, in which the inorganic acceptor radicals (Cu(SCN)$_2^-$ or Cu[N(CN)$_2$]Br$^-$) build a zigzag, planar network, and a conducting, organic layer containing face-to-face ET-dimers.

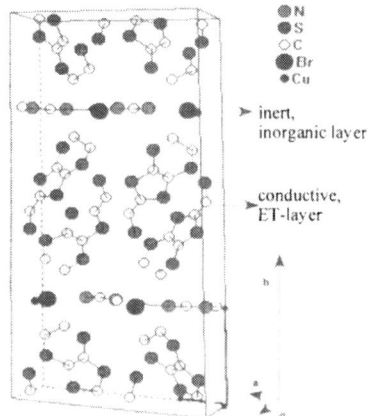

Fig.1 Molecular arrangement in the crystalline structure of κ-ET$_2$Cu[N(CN)$_2$]Br

The overlaps of the ET-derived π-molecular orbitals, provide a two-dimensional conductive network in the ET-layers in the form of narrow, quasi-planar bands and in turn, a strong anisotropy of the electrical conductivities along and perpendicular to the layers.

SPECTROSCOPIC TECHNIQUES

In the SXA experiment, the sample is illuminated with monochromatic radiation in the 100 - 1000 eV range. The photon energy is swept through an X-ray absorption edge and the current through the sample is measured. The photons excite core electrons into the unoccupied bands, and assuming a non-dispersive core state, the current is proportional [6] to the PDOS of the empty bands. The resolution of a SXA measurement is determined by that of the synchrotron radiation monochromator. In the SXE experiment the sample is again illuminated with monochromatic synchrotron radiation in the 100 - 1000 eV range. For a fixed photon energy (excitation energy), core holes are selectively created. The de-excitation of the core holes takes place predominantly by non-radiative Auger type transitions. Nevertheless, two weak, distinct radiative processes follow the creation of the core holes (Fig. 2). Firstly, some of these de-excite by the radiative decay of an electron from a normally filled state into the hole (normal fluorescence). By use of a suitable high resolution spectrometer, the spectrum of emitted photons can be measured [7], with a resolution primarily determined by the resolution of the emission spectrometer. The transition is governed by dipole selection rules, and assuming a non dispersive core hole, the resulting spectrum of emitted photons reflects the occupied valence band PDOS [7]. The sampling depth for SXE in this energy range is on the order of 1000 Å, and so it is the bulk electronic structure that is being probed.

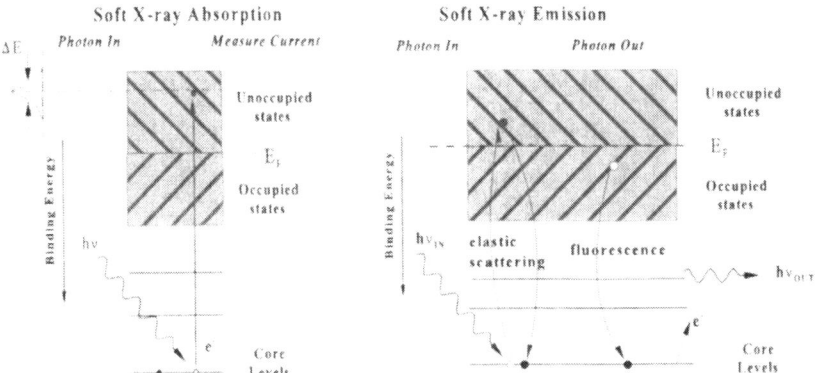

Fig. 2: Schematic illustration of the X-ray absorption and X-ray emission processes

Secondly, a direct radiative recombination involving the electron excited into the previously empty state can take place. This process, particularly strong for excitations in the close vicinity of the absorption edge and dependent on the localized nature of the empty state involved, appears in the SXE spectrum as a sharp peak (elastic peak) at the same energy as the excitation energy. When the excitation energy is tuned to that of a sharp absorption resonance, the presence of the elastic peak can be used for the alignment of the SXA and SXE spectra on a common energy scale.

EXPERIMENTAL DETAILS

The C-1s edge measurements were carried out at the beam line 7 of the Advanced Light Source [8]. The beamline is equipped with a 5 cm period undulator, a spherical grating monochromator and adequate soft x-ray optics, providing a small spot, high flux soft X-rays beam in the energy range from 60 to 1000 eV at a typical resolving power of 2000. The N-1s edge SXA/SXE measurements were performed at the HASYLAB synchrotron, DESY, Hamburg, on undulator beamline BW3, which is equipped with a modified SX-700 monochrometer. The SXA spectra were taken in the sample drain current mode, with energy resolutions of 0.200 eV at the C-1s edge (approx. 290 eV) and 0.250 eV at the N-1s edge (approx. 400 eV). The SXE spectra were recorded with a grazing-incidence grating spectrometer [9]. For the X-ray emission measurements, the resolutions were set to 0.250 eV for the beamline's monochromator and to 0.50 eV for the spectrometer. At these resolutions, the typical amount of time to record an SXE spectrum with a reasonable signal to noise ratio was of the order of 1 hour. The samples used were single crystals having typical dimensions of 2 x 2 x 0.5 mm, mounted with a vacuum compatible conductive epoxy on the sample holder. The experiments were carried in a vacuum base pressure of 1×10^{-8} torr. This vacuum is sufficient for these bulk probes.

RESULTS AND DISCUSSION

C-2p PDOS

As shown in Fig. 3, the X-ray absorption spectra of the two organic metals are highly similar. Above the C-1s absorption edge (approx. 288.0 eV), three spectral features (a, b, c) at energies of 289.6, 291.8 and 297.5 eV are common to both spectra. By comparing to the general characteristics of the NEXAFS spectra of large organic compounds [10], (a) and (c) are easily identified with the π*(C=C) resonance and the σ* shape resonance of the ET molecule. In this picture (a) results when electrons from the C-1s core level of the double bonded carbon atoms are excited into the previously empty lowest unoccupied molecular orbital (LUMO) . By reference to a previous XPS study of κ-ET$_2$Cu(SCN)$_2$ [11], the (b) spectral feature was assigned to the X-ray absorption process from the inequivalent

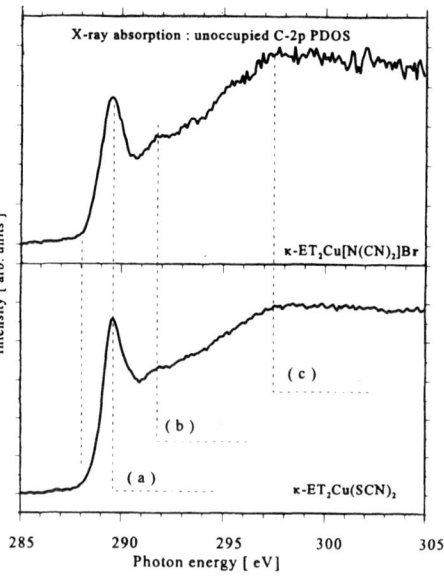

Fig.3 X-ray absorption spectra of κ-ET$_2$Cu[N(CN)$_2$]Br and κ-ET$_2$Cu(SCN)$_2$

C-1s core of the single bonded ET carbon atoms into the LUMO. Our results, showing a much smaller intensity of the absorption from the (C-C)-1s cores into the LUMO compared to the absorption from (C=C)-1s cores, confirms theoretical studies [12] that have found most of the charge density of the LUMO to be centered over the central part of the ET molecule, containing double bonded carbon atom, with very little weight present over the outer, single bonded carbon atoms. Since we have found the X-ray absorption of κ-ET$_2$Cu[N(CN)$_2$]Br and κ-ET$_2$Cu(SCN)$_2$ to be highly similar, in Fig.4 we compare SXE spectra taken from the two compounds with the same excitation energy. For the two spectra presented the excitation energy was tuned over the prominent π*(C=C) feature of the absorption spectra, at 289.6 eV. The strong, sharp peak centered at 289.6 eV (scaled down in the figure) is the elastic peak. The remaining SXE weight visible in the spectra, can be asigned to contributions of the π and σ-bands. As expected for the electronic structure of molecular crystals formed by organic molecules [13] the π-bands fill the region of binding energies from the Fermi level down to approx. 4.5 eV, while the σ-bands form the broad, dominant feature centered at approx. 8.5 eV binding energy. The overall width of the occupied electronic structure having a C-2p character is found to be of approx. 18.0 eV. As a final remark, the SXE spectra of κ-ET$_2$Cu(SCN)$_2$ and κ-ET$_2$Cu[N(CN)$_2$]Br are very similar, as were the SXA spectra (Fig. 3). This indicates that the information contained in the X-ray emission spectra is coming predominantly from the ET conductive layers of both materials and we can not discern the small, possibly different, contribution to the C-2p PDOS of the carbon atoms residing in the inorganic, inert layers.

N-2p PDOS : a layer specific measurement

A particular advantage of SXA and SXE is their chemical selectivity. For both κ-ET₂Cu(SCN)₂ and κ-ET₂Cu[N(CN)₂]Br nitrogen atoms are present only as components of the inorganic acceptor radicals Cu(SCN)₂⁻ and Cu[N(CN)₂]Br⁻. Therefore an atomic site specific measurement of the N contributions to the electronic structure, as provided by SXA and SXE, is at the same time layer specific. In Fig. 5 we present the recently measured *layer-resolved* N-2p occupied and unoccupied PDOS of the two organic compounds. In the vicinity of the Fermi level, the absorption spectra are dominated by the presence of three absorption resonances situated at correspondingly similar energies. The absorption resonance closest to E_F is stronger in κ-ET₂Cu[N(CN)₂]Br, a possible signature of the different nature of the two inorganic radicals (unlike Cu(SCN)₂⁻, Cu[N(CN)₂]Br⁻ has two inequivalent nitrogen sites). The SXE spectra were taken with the excitation energy tuned over the most prominent feature of the absorption spectra, as seen from the position of the elastic peaks. The SXE spectra are somewhat similar, showing three regions of spectral weight spanning a width of approx. 12 eV, with the main spectral feature situated at approx. 6.5 eV binding energy. Finally we mention that in a series of measurements recorded with photon energies swept from the Fermi level along the main absorption resonances we have found a *strong dependence of the shape of the SXE spectra on the excitation energy.* An analysis (partly in terms of different contributions from distinct sites) and a comparison with recent extended Hückel tight-binding calculations is in progress [14].

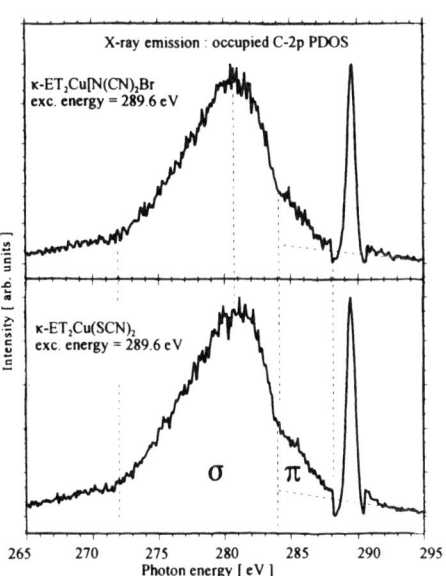

Fig.4 Occupied C-2p PDOS of κ-ET₂Cu[N(CN)₂]Br and κ-ET₂Cu(SCN)₂ measured by soft X-ray emission

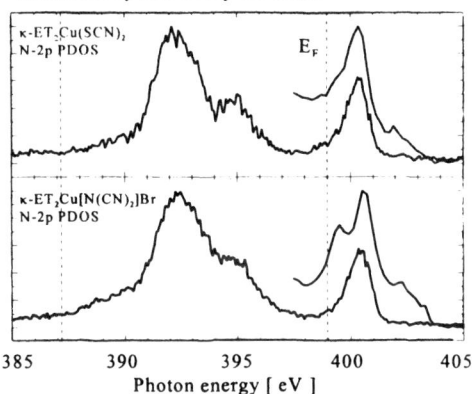

Fig. 5: Layer specific N-2p PDOS of κ- ET₂Cu(SCN)₂ and κ-ET₂Cu[NCN)₂]Br

SUMMARY

We have used for the first time a combination of high-resolution bulk spectroscopies, soft X-ray emission and soft X-ray absorption, to investigate the electronic structure of two layered, organic metals: κ-ET₂ Cu(SCN)₂ and κ-ET₂Cu[N(CN)₂]Br. The main contribution to the C-2p PDOS is given by the ET molecular orbitals, comprising the conductive layers. In the proximity of the Fermi level, two absorption resonances are found to correspond to transitions from the inequivalent double / single bonded C 1s cores into the ET LUMO; their relative intensity is in general agreement to theoretical studies [12] that have found most of the charge density of the LUMO to be centered over the central part of the ET molecule. The C-2p X-ray emission spectra are highly similar for the two compounds. The equivalent measurements performed at the N-2p edge express, given the structure of κ-ET$_2$ Cu(SCN)$_2$ and κ-ET$_2$Cu[N(CN)$_2$]Br, a *layer resolved PDOS*. Strong spectral changes of the N-2p X-ray emission spectra with the excitation energy are noted. Finally, a detailed analysis of each of the two studies (C-2p and N-2p) and comparisons with recent extended Hückel tight-binding calculations is in progress [14].

ACKNOWLEDGMENTS

This work was supported in part by the National Science Foundation under DMR-9501174 and INT-9515370. The ALS is operated by the University of California as part of the U.S. Department of Energy, Office of Basic Energy Sciences. L.C.D gratefully acknowledges the Swedish Natural Research Council for a postdoctoral fellowship at Boston University.

REFERENCES

†: *On leave from the Institute of Microtechnology, Bucharest, Romania.*
‡: *author to whom correspondence should be addressed. Electronic mail: ksmith@bu.edu*

1. H. Urayama, H. Yamochi, G. Saito *et. al.*, Chem. Lett. **463**, 1988 (1988).
2. A. M. Kini, U. Geiser, H. H. Wang, K. D. Carlson, J. D. Williams *et. al.*, Inorg. Chem. **29**, 2555 (1990).
3. R. Liu, H. Ding, J. C. Campuzano *et. al.*, Phys. Rev. B **51**, 6155 (1995); *ibid.* **51**, 13000 (1995).
4. S. Söderholm, B. Loppinet, D. Schweitzer, Synth. Met. **62**, 187 (1994); S. Söderholm, R. T. Girard, D. Schweitzer, Phys. Rev. B **52**, 4267 (1995); S. Söderholm, P. Varekamp, D. Schweitzer, *ibid.*, **55**, 9629 (1995).
5. T. Ishiguro and K. Yamaji, Organic Superconductors, (Springer-Verlag, New York, 1990).
6. See for example X-Ray Absorption: Principles, Applications, Techniques of EXAFS, SEXAFS, and XANES, Ed. by D.C. Koningsberger, R. Prins (Wiley, New York, 1988).
7. J. Nordgren and N. Wassdahl, Phys. Scr. **T31** 103 (1989); J. Nordgren, J. Physique **C9,** 693 (1987). T.A. Callcott, C.H. Zhang, D.L. Ederer, D.R. Mueller, J.E. Rubensson and E.T. Arakawa, Nucl. Inst. Methods **A291**, 13 (1990).
8. T. Warwick, P. Heinemann, D. Mossessain *et. al.*, Rev. Sci. Instrum. **66**, 2037 (1995).
9. J. Nordgren, G. Bray, S. Cramm, R. Nyholm, J-E. Rubensson, and N. Wassdahl, Rev. Sci. Instrum. **60**, 1690 (1989); J. Nordgren and R. Nyholm, Nucl. Inst. Methods, **A246**, 242 (1986).
10. J. Stöhr, NEXAFS Spectroscopy, (Springer-Verlag, New York, 1996).
11. R. Itti, H. Mori, K. Ikeda *et. al.*, Physica C **185-189**, 2673 (1991).
12. E. Demilrap and W. A. Goddard III, J. Phys. Chem. **98**, 9781 (1994).
13. W. D. Grobman and E. E. Koch, in Photoemission in Solids, edited by L. Ley and M. Cardona (Springer-Verlag, New York, 1979), vol. 2.
14. Cristian B. Stagarescu, L.-C. Duda, Kevin E. Smith, J. H. Guo, J. Nordgren, D.-K. Seo, M.-H. Whangbo, R. Haddon, J.S. Brooks and D. Jeromé (unpublished); (unpublished).

PREPARATION OF C_{60} CHARGE TRANSFER COMPLEXES WITH ORGANIC DONOR MOLECULES AND ALKALI DOPING

A. OTSUKA, G. SAITO, S. HIRATE, S. PAC, T. ISHIDA, A. A. ZAKHIDOV*, AND K. YAKUSHI*
Division of Chemistry, Graduate School of Science, Kyoto University, Kyoto 606-01, JAPAN
*Institute for Molecular Science, Okazaki 444, JAPAN

ABSTRACT

Solid charge transfer (CT) complexes of C_{60} with $TSeC_1$-TTF, EDT-TTF, EOET-TTF, and TDAP (1, 3, 6, 8-tetrakis(dimethylamino)pyrene) were newly prepared. All the obtained black crystals were proved to be neutral despite of their rather strong electron donor ability. Lattice parameters of them except for EOET-TTF complex were determined together with those of HMTTeF·C_{60}, which had been reported with different values. Rubidium doping under a mild condition was examined on the complexes of TDAP, EOET-TTF, HMTTeF, BEDT-TTF, hydroquinone and ferrocene to search for the superconductors of new crystal and electronic structures. Among them, the rubidium-doped ferrocene complex easily showed an apparent superconducting signal in SQUID magnetization measurements. The doping effect on these CT complexes is compared to that on OMTTF complex.

INTRODUCTION

Superconductivity so far observed in the C_{60} compound is mainly restricted to the -3 charged C_{60} substances generated by alkali, alkaline ammonia, or alkaline earth doping. In such compounds, the crystal and electronic structures have also been restricted to three-dimensional ones because of the cubic-like arrangement of C_{60} molecules in the lattice. An extensive study on development of new C_{60} compounds has been achieved by several groups [1-6]. In order to explore new superconductors based on C_{60} which possess different characters in dimensionality of crystal and electronic structures, we have been investigating the alkali doping to CT complexes of C_{60} with organic donor molecules. We have reported that, after potassium or rubidium doping under a mild condition to the 1:1:1 CT complex of OMTTF·C_{60}·benzene, it showed superconductivity at 17-18.8 K or 23-26 K, respectively, with its pristine crystal structure constructed by two-dimensional arrangement of C_{60} preserved [7]. As the starting material for alkali doping, new C_{60} complexes as well as the known ones have been prepared in this work. The donor molecules used are shown in Figure 1 [8]. The results of rubidium doping to the C_{60}

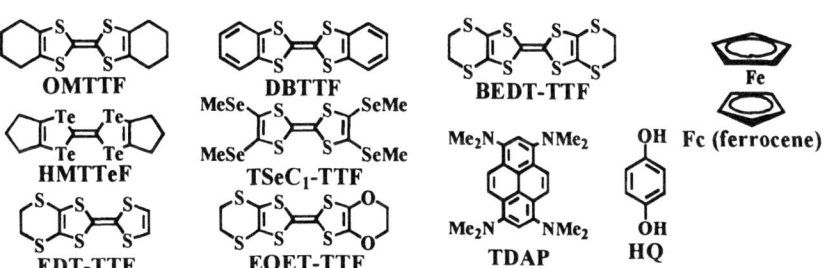

Figure 1. The donor molecules used for the C_{60} complex preparation.

complexes of TDAP, EOET-TTF, HMTTeF, BEDT-TTF, hydroquinone (HQ), ferrocene (Fc) are presented, and compared with those of the OMTTF complex.

EXPERIMENT

The CT complexes of C_{60} with TDAP [9] and EOET-TTF were prepared by direct mixing of benzene solutions of appropriate components followed by slow evaporation of benzene in the ambient atmosphere. The C_{60} CT complexes with HMTTeF [3], BEDT-TTF [2], HQ [1], ferrocene [4] were also prepared approximately according to the literature methods. Similar to the preparation of OMTTF·C_{60}·benzene, the excess amount of donor in the mixing ratio was preferred. The use of teflon (PFA) beaker was occasionally successful. The composition of the complexes were examined by elemental analyses. The CT bands in the solid state were measured in KBr disks.

The rubidium doping to the CT complexes was carried out by the previously reported method. In the seal-off procedure of the sample tubes, the base pressure near the oil diffusion pump was 1.5-5×10^{-6} torr. The sealed tubes were warmed in a horizontal furnace in which the source CT complex side was kept at 67°C and the rubidium side at 64°C. The sample tubes were kept in the furnace for from 24 hours to 53 days. A reference sample tube was prepared and warmed at the same time in which only C_{60} and rubidium were loaded. At some stages of doping period of each sample, the occurrence of superconductivity was checked by the magnetization measurements with a SQUID magnetometer (Quantum Design, MPMS-2). To minimize the contribution of the quartz sample holder, a piece of blank quartz tube was attached to the bottom of the sample tube. First the sample was cooled to 5 K (or 2 K) in the zero field (remanence field was less than ±2 mG), the field dependence of the magnetization was measured in the course of increasing the field from 0 G to 10 G. After the field of 10 G was applied at that temperature, the magnetization under 10 G was measured on heating to more than 30 K (ZFC). Then the magnetization was measured under the same magnetic field on cooling (FC). As we did not introduce any thermal exchange gas into the sample tube, a 30 seconds of pause was inserted before every measurement at intervals of 0.5 K to get thermal equilibrium. The volume fraction of superconductivity (relative to $-1/(4\pi)$) was calculated from the value of ZFC at 5 K under 10 G followed by its normalization to the volume susceptibility by use of the mass and the specific gravity of the sample.

RESULTS AND DISCUSSION

CT complexes

Some characters of the newly obtained CT complexes were summarized in Table I. Among them, the optical spectra of TDAP·$(C_{60})_2$·benzene and (EOET-TTF)·(C_{60}) are shown in Figure 2.

The first redox potential of TDAP is +0.02 V vs. SCE and it is a stronger electron donor than OMTTF (+0.29 V) or TMPD (+0.15 V). Its CT complex with C_{60} was obtained from benzene solution as a 1:2:1 composition including benzene, TDAP·$(C_{60})_2$·benzene. The TDAP complex showed the CT band at around 11.4×10^3 cm^{-1}. The IR vibration spectra were explained by superposition of the respective component molecules of neutral state. It was proved to be a neutral CT complex despite of the strong electron donor ability of TDAP (Figure 2). The lattice parameters of TDAP·$(C_{60})_2$·benzene were obtained by a four-circle X-ray diffractometer (Table I), but the structure analysis has not succeeded due to an insufficient reflection intensity.

We obtained HMTTeF~C_{60} CT complex both in powder and crystals. When the

Table I.   Some characters of the solid C_{60} CT complexes.

Donor lattice data	solvent used	Donor:C_{60}:solvent	shape	$h\nu_{CT}$ / 10^3 cm^{-1}
TDAP	benzene	1:2:1	rod	11.4
triclinic, a=10.037(6), b=14.401(5), c=15.475(4) Å, α=93.57(2), β=91.37(4), γ=104.58(4)°, V=2158.8(15) Å^3				
EOET-TTF	benzene	1:1	plate	11.8
EDT-TTF	benzene	1:1:1	plate	12.0
monoclinic, a=10.084(4), b=17.730(3) c=10.917(3) Å, β=101.38(3)°, V=1914(1) Å^3				
TSeC$_1$-TTF	benzene	1:1	plate	11.8
monoclinic, a=13.516(8), b=19.761(6), c=16.647(7) Å, β=98.05(4)°, V=4402(3) Å^3				
HMTTeF	toluene	1:1	rhombus	11.7
triclinic, a=9.950(7), b=13.159(5), c=9.935(5) Å, α=95.91(3), β=118.36(4), γ=106.96(4)°, V=1048.9(9) Å^3				

concentration of mixing in boiling toluene was relatively high (54.1-63.4 mg of HMTTeF and 64.4-68.5 mg of C_{60} in 100 cm^3 of toluene), the powdered sample of 1:1 composition in which the CT band was observed at 11.7×10^3 cm^{-1} (may be the same material as described in ref. [3]) was obtained.   A relatively low concentration in a small scale (2.7-3.5 mg of HMTTeF and 3.0-4.4 mg of C_{60} in 10 cm^3 of toluene) gave the rhombus crystals.   Its lattice constant was determined by a four-circle diffractometer (Table I), but the structure analysis has not succeeded due to a poor quality of the crystal.   The data of all new complexes in this work are summarized in Table I.

HQ is known to give (HQ)$_3$C$_{60}$ by mixing HQ and C_{60} in 3:1 ratio in benzene followed by slow evaporation of benzene [1].   Even though we used excess amount of HQ (HQ:C_{60}≅6:1) and boiled the benzene solution mixture for about 1 hour before the slow evaporation of benzene, the 3:1 compound was obtained.   The solution mixture resulted in completely colorless liquid phase in addition to black powder in the mixing ration of 6:1 whereas the solvent remained slightly purple in the mixing ratio of 3:1.   In the 6:1 mixing, the obtained black powder was contaminated by the colorless crystals of HQ which was easily washed out by benzene.   The absorption

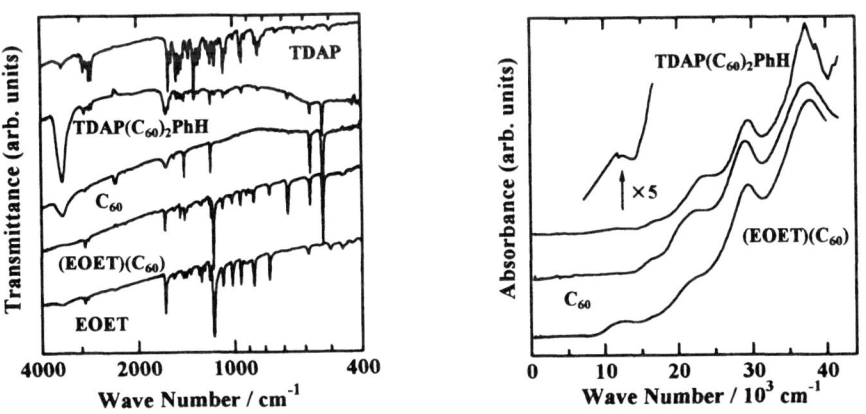

Figure 2.   Optical spectra of TDAP·(C_{60})$_2$·benzene and (EOET-TTF)·(C_{60}) together with those of the respective components, indicating that both the complexes are neutral.

observed at 12×10^3 cm^{-1} in our sample is still controversial since the donor strength of HQ should be too weak to show CT band in that region with C_{60}.

The elemental analyses on the Fc~C_{60} complex we obtained did not show the 2:1 composition but indicated a slightly larger content of C_{60} ((Fc)$_2$(C_{60})$_{1.45}$ or (Fc)$_2$(C_{60})$_{1.39}$(benzene)$_{0.55}$). It is not clear at present whether we have obtained a new composition or a part of Fc was removed in our washing procedure by acetonitrile.

Rubidium doping

Table II shows the superconducting volume fractions and T_c's of the TDAP and Fc complexes at various doping periods together with those of pristine C_{60} and the OMTTF complex doped under our mild condition (64/67°C). It is remarkable that the EOET-TTF, HMTTeF, BEDT-TTF, and HQ complexes have not shown any detectable superconductivity under the same condition for a period of 34 hours-53 days.

(TDAP)·(C_{60})$_2$·(benzene) did not show superconductivity after 1 day warming, but after 36 days it showed diamagnetic onset below 20 K. The magnitude of the diamagnetic susceptibility corresponds to only 0.004% of $-1/(4\pi)$. Although (TDAP)·(C_{60})$_2$·(benzene) has a moderate thermal stability (dec.>300°C), we should be aware of that C_{60} itself shows 0.03% volume fraction of onset below 27 K after 53 days in the same doping condition.

Fc~C_{60} showed a distinguishable behavior on rubidium doping. It started showing apparent superconductivity even after 1 day warming. The volume fraction increased as the doping period elongated and amounted to 3% (onset of about 23 K) after 34 days warming, assuming the specific gravity of 1.78 g/cm^3 [4] (Figure 3). Paramagnetic contributions were also seen to accumulate. An important observation is that some material has migrated from the 67°C side to the 64°C side. After 50 hours warming, the inside wall of the sample tube of the 64°C side (most of rubidium is condensed around this position) started being colored pale brown. The color changed deeper as the doping period prolonged. After 34 days warming, the half of the sample tube kept at lower temperature was coated with black material. It was confirmed by observation under microscope that the surface of the source crystals was roughened during the doping. In order to know whether the sublimation of Fc from the Fc~C_{60} CT complex takes place under the same vacuum and thermal condition, we warmed the Fc~C_{60} CT complex under the same procedure in the

Table II. Rubidium doping period at 64/67°C, onset T_c in ZFC measurements, and superconducting volume fraction at 5 K under 10 G of C_{60} complexes.

Donor	Period	onset T_c / K	Volume fraction / %	
TDAP	24 h	no	no	
	36 d	20	0.004	
ferrocene	24 h	17	0.02	
	50 h	19	0.06	
	82 h	21	0.2	
	34 d	23	3	
OMTTF	19 d	26	ca. 10	[7]
no (C_{60} only)	24 h	10	<0.001	
	13 d	22	0.01	
	41 d	25	0.02	
	53 d	27	0.03	

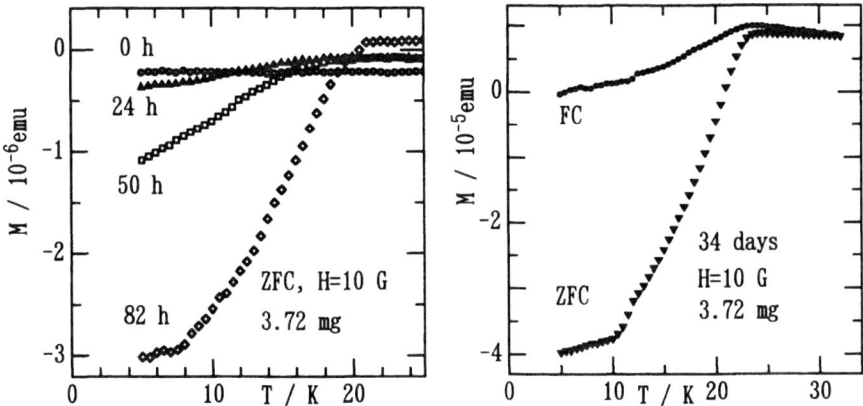

Figure 3. Temperature dependence of magnetization of the Fc~C_{60} CT complex at the various doping periods. The left panel shows the data of 0-82 hours doping, while the right one shows the data of 34 days doping. Any corrections to the ordinate were not made in the both panels.

absence of rubidium. Contrary to the case in which rubidium existed in the tube, the CT complex alone did not show any material migration even after 52 hours. It is also known that $(Fc)_2 \cdot C_{60}$ is stable for heating up to 60-70°C. The origin of the observed superconductivity is not clear at present, but it is suggested that rubidium pushes out a part of Fc from the pristine complex, leaving voids and creates a doped phase.

It is worthwhile to speculate the difference between the CT complexes in which superconductivity is easily observed by alkali doping and ones in which such a phenomenon is hardly realized in the same procedure.

The former complexes are OMTTF·C_{60}·benzene, DBTTF·C_{60}·benzene, and Fc~C_{60} at present. The common feature among them might be a relatively loose molecular packing. In OMTTF·C_{60}·benzene and DBTTF·C_{60}·benzene [6], C_{60} molecules are arranged two- and three-dimensionally, respectively, but the interaction between C_{60} molecules is not so strong. In $(Fc)_2 \cdot C_{60}$, two-dimensional arrangement of C_{60} molecules is packed more densely compared with that in OMTTF·C_{60}·benzene. On the other hand, the interaction between Fc and C_{60} in the crystal is expected to be weak.

The latter complexes are (EOET-TTF)·C_{60}, HMTTeF~C_{60}, (BEDT-TTF)$_2$·C_{60}, and (HQ)$_3$·C_{60}. Though the crystal structures are known only for (BEDT-TTF)$_2$·C_{60} and (HQ)$_3$·C_{60}, these two complexes are characterized by a strong interaction between donor and C_{60} [1, 2]. It is a rational expectation that those CT complexes are tough against the alkali doping.

To know the valence change of C_{60}, Raman scattering experiments were performed on the pristine and doped crystals of OMTTF·C_{60}·benzene. The signal ascribable to the C=C Ag mode of C_{60} appeared at 1467, 1453, and 1446 cm^{-1} for the pristine, rubidium-doped, and potassium-doped complexes, respectively. Although the frequency observed in the potassium-doped complex was close to the value of K_3C_{60}, the rubidium-doped complex showed a much higher frequency suggesting a lower reduction state of C_{60}. If the ionization shift of the C=C Ag mode from the pristine to potassium-doped complex (Δ=21 cm^{-1}) is due to the transfer of three electrons to C_{60}, the shift for the rubidium-doped complex (Δ=14 cm^{-1}) may be ascribed to dianion state of

C_{60} provided that the ionization shift is linear to the charge on the C_{60} molecule.   A preliminary measurement of reflectance spectra on rubidium-doped OMTTF·C_{60}·benzene crystal were made at room temperature.   A weak enhancement of reflectance was observed below 1000 cm^{-1} which disappeared by exposure to the air.

CONCLUSIONS

(TDAP)·$(C_{60})_2$·benzene, (EOET-TTF)·C_{60}, (HMTTeF) ·C_{60}, (BEDT-TTF)$_2$·C_{60}, (HQ)$_3$·C_{60} Fc~C_{60} were examined for rubidium doping at 64/67°C, of which the former two complexes were newly prepared.   The volume fraction of superconductivity in rubidium-doped complexes was apparent (3%) in Fc~C_{60}, very small in (TDAP)·$(C_{60})_2$·benzene, and hardly observed in (EOET-TTF)·C_{60}, (HMTTeF) ·C_{60}, (BEDT-TTF)$_2$·C_{60}, and (HQ)$_3$·C_{60}.

ACKNOWLEDGMENTS

The authors are greatly indebted to Prof. S. Sugai at Nagoya University for the Raman investigations.   This work was supported by the Grant in Aid for Scientific Research, Ministry of Education, Science, Culture, and Sports, Japan, the Joint Studies Program (1997) of the Institute for Molecular Science, and a fund for "Research for Future", from Japan Society for Promotion of Science.

REFERENCES

1.   O. Elmer, Helvetica Chimica Acta, **74**, 1339 (1991).

2.   A. Izuoka, T. Tachikawa, T. Sugawara, Y. Suzuki, M. Konno, Y. Saito, and H. Shinohara, J. Chem. Soc., Chem. Commun., 1992, 1472.

3.   T. Pradeep, K. K. Singh, A. P. B. Sinha, and D. E. Morris, J. Chem. Soc., Chem. Commun., 1992, 1747.

4.   J. D. Crane, P. B. Hitchcock, H. W. Kroto, R. Taylor, and D. R. M. Walton, J. Chem. Soc., Chem. Commun., 1992, 1764.

5.   G. Saito, T. Teramoto, A. Otsuka, Y. Sugita, T. Ban, M. Kusunoki, and K. Sakaguchi, Synthetic Metals, **64**, 359 (1994); A. Otsuka, T. Teramoto, Y. Sugita, T. Ban, and G. Saito, ibid., **70**, 1423 (1995).

6.   D. V. Konarev, Y. V. Zubavichus, Y. L. Slovokhotov, Y. M. Shulga, V. N. Semkin, N. V. Drichko, and R. N. Lyubovskaya, Synthetic Metals, **92** (1), (1998), in press.

7.   A. Otsuka, G. Saito, T. Teramoto, Y. Sugita, T. Ban, A. A. Zakhidov, and K. Yakushi, Mol. Cryst. Liq. Cryst., **284**, 345 (1996); A. Otsuka, G. Saito, A. A. Zakhidov, K. Yakushi, M. Kusunoki, and K. Sakaguchi, ibid., **285**, 187 (1996); A. Otsuka, G. Saito, A. A. Zakhidov, and K. Yakushi, Synthetic Metals, **85**, 1459 (1997).

8.   TSeC$_1$-TTF, tetrakis(methylseleno)-tetrathiafulvalene; EDT-, ethylenedithio-; EOET-, ethylenedioxyethylenedithio-; HMTTeF, hexamethylenetetratellurafulvalene; BEDT-, bis(ethylenedithio)-; OMTTF, octamethylenetetrathiafulvalene; DBTTF, dibenzotetrathiafulvalene.

9.   N. Ueda, Y. Sakata, and S. Misumi, Bull. Chem. Soc. Jpn., **59**, 3289 (1986).

Part VI

Poster Presentations

METAL COMPLEX POLYMERS FOR ELECTROLUMINESCENT APPLICATIONS

X. T. Tao [†], H. Suzuki[‡], Y. D. Zhang[†], T. Watanabe[‡], S. Miyata [‡], T. Wada[†], and H. Sasabe[†]

‡ Graduate School of Bio-Applications and Systems, Tokyo University of Agriculture and Technology, 2-24-16 Nakamachi, Koganei-shi, Tokyo 184,
† Bio-polymer physics laboratory, The Institute of Physical and Chemical Research (RIKEN), 2-1 Hirosawa, Wako, Saitama 351-01, Japan
e-mail: txt@postman.riken.go.jp

ABSTRACT

We report the synthesis and characterization of a soluble metal complex polymer for electroluminescent (EL) applications. The polymer was prepared by the reaction of a zinc Schiff base with 4,4'-diphenylmethane-diisocyanate. The polymer is amorphous and with glass transition temperature of 156 °C and is soluble in common organic solvents such as chloroform, tetrahydrofuran (THF), and N-methylpyrrolidinone (NMP). High optical quality films were obtained by spin-coating the polymer solution of THF. The zinc Schiff base, and the polyurethane (PU) shows strong photoluminescence under a UV-lamp illumination. Single and double layer EL devices consisting ITO/hole transfer layer (HTL)/PU/Al have been fabricated and characterized. The results indicated that the complex polymer could act as both electron transport and emissive layers for EL devices.

INTRODUCTION

There has been considerable recent interest in the use of organic materials, including vapor evaporated low-mass molecules and conjugated polymers, as electroluminescent layers in light emitting devices (LEDs) and large area displays[1-7]. The merits of organic materials for electroluminescent (EL) applications include high luminance and low drive voltage. Since the initial report of a multilayer LED based on vacuum-deposited low-mass metal complexes thin films of tris(8-hydroxyquinoline)-aluminum (AlQ3) [8,9], low-mass metal complexes have been widely used as emitter materials in EL devices. Compared to corresponding organic materials, metallic complexes have better mechanical strength and thermal stability, and a wide variety of central metal atoms, their size and oxidation states, as well as the size and nature of the ligands, provide architectural flexibility to tailor EL up to a maximum level. To date, however, metal complex polymers have not been extensively investigated for EL applications [4(a), 10].

We report here the synthesis and characterization of a metallic complex polyurethane (PU) which containing neutral metal complex of bis (N-salicylidene-p-2-hydroxyethylphenylamino)zinc in the main chain, for EL applications.

EXPERIMENT

The synthesis of the monomers and polymer is shown in the following scheme. N-salicylidene-p-2-hydroxyethylphenylamine (3) was obtained by the reaction of salicylaldehyde (1) with 4-aminophenethyl alcohol (2) in methanol at room temperature (yield 95%). The zinc coordination compound was obtained by the reaction of (3) with zinc acetate dihydrate Zn(COO)$_2$. 2H$_2$O in hot methanol using piperidine as catalyst (yield 86%). Preparation of polyurethane is carried out by treatment of zinc Schiff base with 4,4'-diphenylmethane-diisocyanate in dimethyl-sulfoxide (DMSO) at 100 °C for 12 h, the metal complex

503

polyurethane is precipitated in methanol and purified by washing with methanol.

4-aminophenethyl alcohol   2   3
        1   (Salicylaldehyde)   N-Salicylidene-p-2-hydroxyethyl phenylamine
                                        Yield 95%

4 (Yield 86%)
Bis (N-Salicylidene-p-2-hydroxyethyl phenylamino)zinc

4,4'-methylenebis(phenyl isocyanate)

Polymer
(Yield 82%)

The T_g of the polymer and melting points of N-salicylidene-p-2-hydroxyethylphenylamine (3) and neutral metal complex of bis (N-salicylidene-p-2-hydroxyethylphenylamino)zinc (4) was obtained by using a Nigaku 220 DSC-TG-DTA instrument at a heating rate of 10 °C/min in nitrogen atmosphere. The T_g was taken from the midpoint of the change in slope of the baseline, while melting temperatures were taken from the onset of the change in slope to the minimum of the endothermic peak. The weight loss data were obtained from a Nigaku 220 TG-DTA instrument at a heating rate of 10 °C/min in nitrogen. The UV/VIS spectrum was studied for thin films spin coated from THF solution onto quartz substrates using a SHIMADZU UV-3100PC UV/VIS/NIR scanning spectrophotometer.

The LEDs were fabricated as follows: Before deposition of the polymeric films, indium tin oxide (ITO) coated glasses were cleaned by sequential ultrasonic treatment in ionized water, acetone and methanol. Between each cleaning step, the substrates were dried in high purity nitrogen. The device structures were double-layer. After spin-coating the hole transfer layer (HTL): Poly(9-tetradecanyl-3,6-dibutadiynyl-carbazole) (PTDDBC) at 12 mg/ml concentration in chloroform at 4000 rpm for 30 s. The electron transfer layer (ETL) of metal complex polyurethane was spin cast from a THF solution (10 mg per ml of THF) at 4000 rpm for 30 s. The solutions were filtered through a 0.45 μm Teflon filter before spin-coating. The thicknesses of the films were measured with a Sloan Dektak IIA stylus profilometer. The film thicknesses of HTL and ETL were ~100 nm. The metallic cathodes were evaporated from tungsten wire baskets in vacuum at pressures below 4×10^{-6} Torr., using a mask capable of patterning contacts for nine electroluminescence cells. Active areas for these devices were about 12 mm^2.

RESULTS AND DISCUSSIONS

The melting point of (3) was 93 °C. After the formation of complex with a central zinc atom, the melting point of the zinc Schiff base was increased to 223 °C. This is indicating that the metal complex have better thermal stability than the corresponding organic compounds. The glass transition temperature of the PU is 156 °C and the weight loss temperature is higher than 300 °C. The polymer is soluble in common polar organic solvents such as chloroform, THF, and N-methyl-pyrrolidone.

The complex monomer and polymer show strong green-blue photoluminescence under the UV-lamp irradiation. Shown in Figure 1 is the absorption spectrum of the polymer.

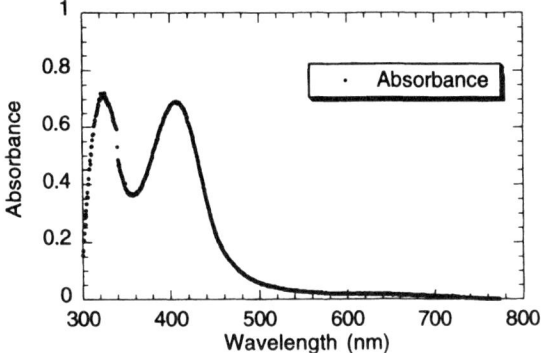

Fig. 1. Absorption spectrum of the metal complex polymer films spin coated on ITO glasses.

The absorbence maximum for PU is at 408 nm.

The ionization potentials of ≈5.8 and ≈5.4 eV for the metal complex PU and a newly synthesized main-chain polymer: Poly(9-tetradecanyl-3,6-dibutadiynyl-carbazole) (PTDDBC) used as hole transfer layer were determined by a photoemission apparatus (Riken keiki AC1), respectively. The technique involves illuminating a thin film of the polymer over a range of energies using a monochromated light source and measuring the photocurrent produced from the sample at each energy. The electron affinities of ≈3.4 eV was estimated by subtraction of optical gap energies determined from absorption spectra (see Fig.1). The energy levels of a double layer device ITO/PTDDBC/PU/Al are shown schematically in Fig. 2. The results indicate PU can be used as electron transfer layer (ETL) and emitting layer in this double layer device. The injection barriers for electrons and holes are about 0.8 and 0.6 eV, respectively.

The photoluminescence (PL) spectrum of the PU was measured by using a JASCO FP-777 spectrofluorometer. The electroluminescent (EL) spectrum was determined with the same instrument as used in PL measurement, by blocking the xenon excitation lamp and scanning the emission from forward-biased cells. The bias voltage was 12 V. The PL spectrum of the PU as well as the EL spectrum of the double layer device, are shown in Fig. 3. The PL spectrum was obtained with an excitation wavelength of 350 nm. The EL spectrum resembled the PL spectrum of the metal complex polymer in the short-wavelength parts, indicating that emission is mainly due to come from the radiative deactivation of ETL singlet excited states. The enhanced low energy tail emission in the EL spectrum relative to the PL spectrum may be due to the structural defects of the metal complex polymer which form a manifold of emitting trapped states act as recombination centers leading to a long-wavelength broadening of the emission spectrum. Similar phenomenon has also been observed in low tris-(-8-hydroxyquinoline) aluminum (Alq3) doped N,N'-diphenyl-N,N'-bis(3-methyl)-1,1'-biphenyl-4,4'-diamine (TPD) LEDs and many other systems [11]. The total absence of any emission similar to that of HTL suggests that recombinations of holes and electrons occur only in the ETL, and that in these devices the PTDDBC is a hole transporting layer only.

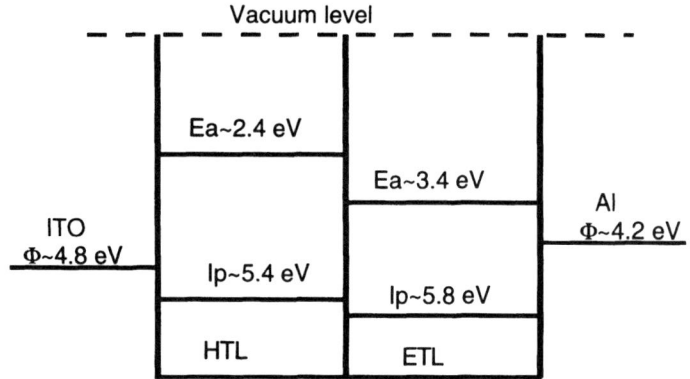

Fig. 2. A schematic energy diagram for the double-layer device (ITO/HTL/ETL/Al).

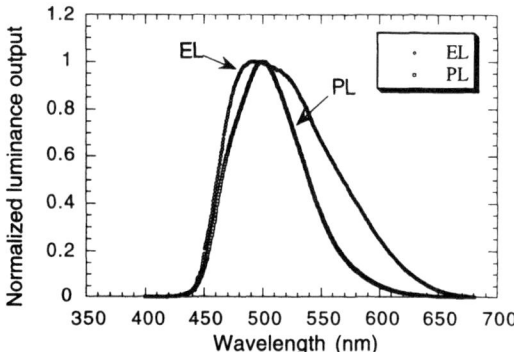

Fig. 3. Normalized PL of polyurethane and EL of a two layer device (ITO/HTL/PU/Al).

The cell is biased via a IWATSU SP 6005P dc power supply and a Advantest R5451A auto ranging multimeter is used for current measurements. Cell luminances were determined under dark conditions through the front end of the ITO glasses with a Minolta LS-100 luminescence meter. The luminance-voltage curve of the double layer devices ITO/HTL /PU /Al are presented in Fig. 4.

The LEDs show turn-on voltage of about 6 volts for luminance. The external quantum efficiency (photons/electron) of the device is 0.06%, which can be improved by using low work function metals as cathode and using other polymers as hole transfer layers. Studies on these aspects are underway. In the experiments, we found that the external quantum efficiency has no apparent variation when the ETL film thickness changed from 100 nm to 130 nm, although the turn on voltage is film thickness dependent. For the single layer devices, visible emission and apparent current can also be observed in the experiment.

CONCLUSIONS

A metal complex polymer soluble in common organic solvents was prepared and characterized. The results indicated that metal complex polymers can be used as electron transfer and luminescent layer in EL devices. The good thermal stability and structural variability indicated that metal complex polymers are attractive candidates as potential EL

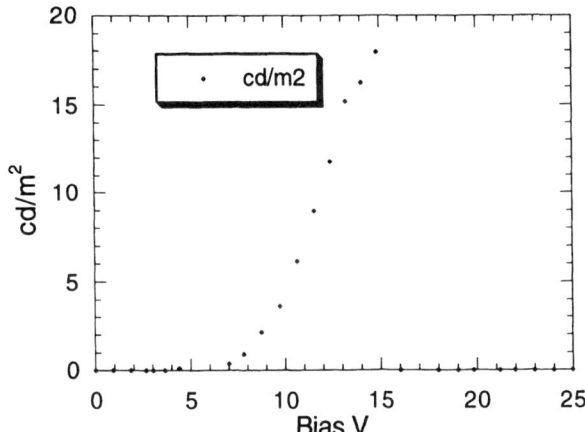

Fig. 4. Luminescent-voltage curve of a two layer EL device, ITO/HTL/PU/Al.

materials. Metal complex polymers with good EL characteristics as well as thermal stability will be obtained by appropriate selection of central metals or/and ligands.

ACKNOWLEDGEMENTS: This research was partially supported by the Grant-in Aid for Scientific research, Nos. 07246105 and 07750049, from the Ministry of Education, Science, and Culture, Japan.

REFRENCES

1. J. H. Burroughes, D. D. C. Bradley, A. R. Brown, R. N. Marks, K. Mackay, R. H. Friend, P. L. Burns, A. B. Holmes, Nature **347**, 539 (1990).
2. G. Gustafsson, Y. Cao, G. M. Treacy, F. Klavetter, N. Colaneri, A. J. Heeger, Nature **357**, 477-479 (1992).
3. E. Buchwald, M. Meier, S. Karg, W. Rieβ, M. Schiwoerer, P. Pösch, H. W. Schmidt, S. Strohriegl, Adv. mater. **7**, 839 (1995).
4. (a) X. T.Tao, H. Suzuki, T. Watanabe, S. H. Lee, S. Miyata, H. Sasabe, Appl. Phys. Lett. **70**, 1503 (1997). (b) X. T.Tao, Y. D. Zhang, T. Wada, H. Sasabe, H. Suzuki, T. Watanabe, S. Miyata, Appl. Phys. Lett. **71**, 1921(1997).
5. X. C. Li, F. Cacialli, M. Giles, J. Gruner, R. H. Friend, A. B. Holmes, S. C. Moratti, T. M. Yong, Adv. Mater. 7, 898 (1995).
6. J. Kido, K. Nagai, Y. Okamoto, T. Skotheim, Chem. Lett. 1267 (1991).
7. D. O'Brien, A. Bleyer, D. G. Lidzey, D. D. C. Bradley, T. Tsutsui, J. Appl. Phys. 82, 2662 (1997).

8.   C. W. Tang and S. A. Van Slyke, Appl. Phys. Lett. **51**, 913 (1987).
9.   C. Adachi, S. Tokito, T. Tsutsui, S. Saito, Jpn. J. Appl. Phys. **27**, L269 (1988).
10. J. K. Lee, D. Yoo, M. F. Rubner, Chem. Mater. **9**, 1710 (1997).
11. J. Kalinowski, in Organic electroluminescent materials and devices S. Miyata, H. S. Nalwa, Eds.; OPA, Amsterdam, 1997, pp. 1-65.

INVESTIGATIONS ON THE ALUMINUM/PARA-HEXAPHENYL INTERFACE IN LIGHT EMITTING DEVICES

N.Koch[a], L.-M.Yu[b], J.-L.Guyaux[b], Y.Morciaux[c], G.Leising[a], J.-J.Pireaux[b], G.Demoustier[c]
[a] Institut für Festkörperphysik, Technische Universität Graz, A-8010 Graz, Austria
[b] L.I.S.E., University of Namur, B-5000 Namur, Belgium
[c] L.A.R.N., University of Namur, B-5000 Namur, Belgium

ABSTRACT

Blue light emitting devices (LED) with *para*-hexaphenyl (PHP) as the active material and aluminum as cathode exhibit very high quantum efficiencies. To further optimize device performance it is crucial to understand the physical properties of the involved interfaces. We have performed Rutherford-Backscattering experiments on actual devices to show the importance of oxygen in the interface formation at the cathode as this leads to the formation of a layer of Al_xO_y between PHP and aluminum. In devices, where the organic film is exposed to air before the metal electrode is evaporated, an insulating layer on the metal-side therefore is inherent. It has been shown that the introduction of an intermediate layer between active material and electrodes results in a higher quantum efficiency of the LED, the most common concepts being charge-transport-layers, or insulators on the other hand. Our results underline the need for a better control of the LED processing. Ultraviolet- and X-ray photoelectron spectroscopy *in situ* growth studies of thin aluminum films on PHP have been made to reveal the change in the electronic structure of the active medium in a LED in the absence of oxygen. Also the direct interaction of oxygen with this organic material is investigated by photoelectron spectroscopy.

INTRODUCTION

Since it has been demonstrated that it is possible to build highly efficient light emitting devices with conjugated polymers and oligomers sandwiched between two electrodes [1-7], many attempts have been made to increase the device performance by introducing additional layers in the device structure [8-14]. Before being able to tune these layers to an optimum one has to understand the principal processes involved in the interface formation in the ordinary triple systems. The general way to produce a LED with a conjugated organic material as active medium is to evaporate or spin-coat it onto the transparent, high work function electrode (most commonly indium-tin-oxide, ITO), and to evaporate the other, low work function, metal electrode (Ca, Mg, Al, alloys...) on top of it [4]. If ultra-clean conditions are not maintained throughout the production process, or the device is not kept in protective environments afterwards, the simple triple system may be altered into a more complicated one. Interaction of water or oxygen with the organic material may change its electronic structure [15]. Not affected by the environment are the diffusion and the possible doping processes of the metal atoms from the top electrode, which nevertheless may take place [16]. This will result in regions of virtually unknown electronic properties in the LED, most likely confined to the interface of the active material and the low work function electrode, which will strongly modify the characteristics of charge carrier injection.

In the present investigation, LEDs with *para*-hexaphenyl (PHP) as active material, ITO and aluminum as electrodes are considered. As a first step the influence of oxygen on the electronic structure of PHP is examined by *in situ* photoelectron spectroscopy. The same method is applied to gain insights into the formation of the interface between PHP and aluminum. Finally, LEDs were produced in a conventional way, with the PHP film being exposed to air before the aluminum electrode was evaporated onto it, and no sealing of the device. On these devices Rutherford-backscattering experiments (RBS) were performed after storage in air for several days, to test if a LED like this can be treated as the simple triple system mentioned above.

EXPERIMENT

Photoemission experiments were performed with a SCIENTA ESCA 300 photoelectron spectrometer for X-ray photoelectron spectroscopy (XPS), and at the FLIPPER II experiment at Hasylab, Hamburg for ultraviolet photoelectron spectroscopy (UPS) [17]. There all depositions of PHP and aluminum were made in situ, the residual pressure in the recipients never exceeding 5×10^{-9} mbar. The LEDs were prepared by vacuum sublimation (residual pressure 2×10^{-6} mbar) of the organic material onto an ITO covered glass substrate, cleaned following a specially developed procedure, described elsewhere [18]. Afterwards, the sample was transferred under air to the evaporation chamber for applying the aluminum electrode. RBS experiments on devices produced by this way have been performed at LARN at the University of Namur/Belgium, so devices prepared in Graz were exposed to air for several days before the measurements. As probing ions, α-particles have been used, which were accelerated to a kinetic energy of 1283 keV. In the given setup, the backscattering angle was 175°; the Si p-n junction type detector gave a resolution of 12 keV.

RESULTS

A thick PHP film (approximately 500 Å as indicated by a quartz microbalance) was evaporated in UHV onto an Ar^+-sputtered ITO/glass substrate. Without breaking the vacuum the sample was transferred from the preparation chamber of the SCIENTA ESCA 300 photoelectron spectrometer to the analysis chamber. An XPS survey scan showed nothing but carbon to be present on the surface. Then the sample was exposed to pure O_2 at atmospheric pressure for 10 minutes and reintroduced into the analysis chamber. Still the only signal in XPS came from carbon, no trace of oxygen could be detected. Nevertheless, the C1s peaks was shifted towards a lower binding energy by 0.3 eV, from 284.56 eV to 284.26 eV as shown in Fig.1a) and b). The lineshape of the curves is practically identical, no broadening or an asymmetric tail is observable. This indicates that no new chemical species are formed by the exposure of PHP to oxygen, yet some kind of interaction does take place.

Principally the same experiment has been repeated at the FLIPPER II experiment at Hasylab for UPS, the only difference being that a stainless steel plate was used as a substrate for PHP evaporation instead of ITO, which does not influence the spectra, as tested in a separate experiment. The valence band region was recorded at a photon energy of 32 eV, because of the high crossection for the highest occupied molecular orbital (HOMO) of PHP at this photon energy [19]. For curve c) in Fig.1, which was recorded for the pristine PHP, the HOMO onset is located at 1.40 eV binding energy (E_b) relative to the Fermi energy. After exposure to oxygen for

330 ML, the binding energy of the onset decreased to 0.90 eV as can be seen in Fig.1 d). But not only the onset of the HOMO exhibits this shift in binding energy, also the whole valence band region shifts rigidly by the same amount. Principally the lineshape of the recorded energy region remains the same, except for a dramatic change in the peak intensity of the first predominant peak (at E_b = 3.5 eV for pristine PHP) with respect to the following ones. Simultaneously the ionization potential (IP) of the sample was measured after each step. The IP of pristine PHP was determined to be 6.15 eV in this experiment, and no change after oxygen exposure was observed within the experimental error of 0.15 eV.

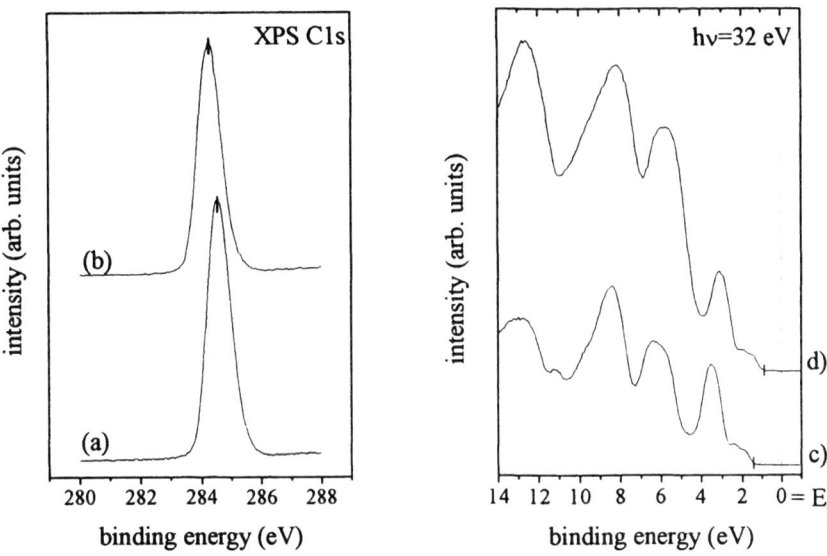

Fig.1. XPS spectra of pristine PHP a) and after exposure to oxygen b); and UPS spectra of pristine PHP c) and after exposure to 330 ML oxygen d); binding energies are given with respect to the Fermi-level

Because of the high electron affinity of oxygen, one could expect an electron transfer *from* PHP *to* the oxygen, corresponding to p-doping of the organic material. As the mean electron density on the molecule is reduced by p-doping, the shift, in particular for the core levels, should be towards higher binding energy. But in all experiments presented here, the binding energies (referred to the Fermi-level) are lowered by exposure to oxygen. But the energy reference of chemical significance is not the Fermi-level E_F, instead the vacuum level must be taken. This reference level is accounted for in the IP measurements. There the position of the HOMO-onset relative to the vacuum level does not change, only E_F moves towards the HOMO-onset by 0.5 eV. Together with the shift of the C1s peak of 0.3 eV towards E_F, a resulting shift of the C1s level away from the vacuum level by 0.2 eV results. The increased binding energy (vs. vacuum) of the carbon core-level and the movement of E_F towards the HOMO-onset resembles the situation expected for p-doping a semiconductor [20,21]. The conclusion derived from these experiments is that we observe a phenomenon related to a weak p-doping of PHP by minute amounts of oxygen.

The formation of the aluminum/PHP interface was studied by means of *in situ* XPS and UPS growth studies. Deposition of Al on a PHP film results in no shift or shape-change of the C1s core-level. With UPS, also substantially no changes in the valence region of PHP was observed upon increasing Al coverage. This gives clear evidence for simple physisorption of aluminum on PHP, details being described in [22].

Now, to reveal the conditions in an actual LED, which is not sealed in a protective environment, devices produced in the way outlined above were examined by RBS. Fig.2b) shows the as recorded RBS spectrum taken on a spot of a LED stored in air for several days. For interpretation of RBS data it is necessary to compare the experimental results with numerically simulated spectra. With certain assumptions on the geometrical and compositional properties of the device, in our case a stacking as shown in Fig.3, it is possible to obtain simulated spectra fitting the experimental data, shown in Fig.2a).

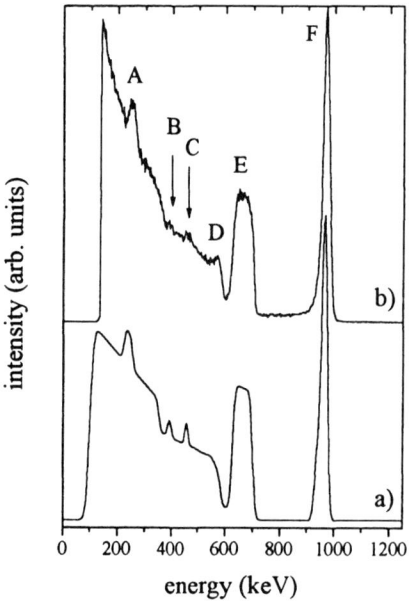

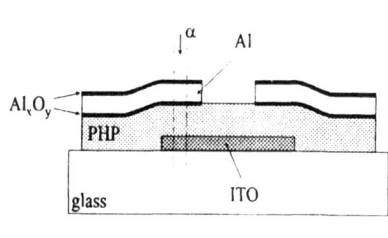

Fig.2. Simulated a) and as measured b) RBS spectra used of a conventional ITO/PHP/Al LED; the peak assignment is explained in the text

Fig.3. Side view of the LED-model to calculate the simulated RBS spectrum

Within this fitting procedure, peak F corresponds to alpha particles backscattered by the In and Sn atoms in the ITO electrode. The 12 keV resolution of the detector is not able to resolve the two structures. The square-like E is assigned to the particles backscattered by the Al film on top of the PHP. The sharp step labeled D is the fingerprint of the particles backscattered by Si atoms in the glass substrate, while the extended edge around 360 keV comes from the oxygen atoms in both, ITO and glass substrate. Finally, peak A originates from the backscattered particles by the carbon in the PHP region. Feature C corresponds to aluminum oxide on top of the Al contact. Within this fitting procedure, the feature appearing around 400 keV (labeled B) corresponds to oxygen localized at the interface between the aluminum and the PHP layer, most probably forming aluminum oxide (Al_xO_y).

This gives clear evidence that a layer of aluminum oxide of undefined thickness is formed between PHP and the aluminum contact when exposing the PHP film to air before evaporation of the electrode, and/or by diffusion of oxygen to the aluminum through the PHP film afterwards [23]. Due to the limited spectra quality, the thickness for the top aluminum oxide layer (peak C) only can be estimated roughly to be between 150 Å to 250 Å. But it can be seen that peak B is within the same order of magnitude as peak C, suggesting a layer of Al_xO_y between PHP and Al, which is much thicker than suggested for an intermediate layer for increased electron injection efficiency [8,12]. Since the two layers of aluminum oxide are rather thick, a porous Al film seems to be formed by thermal evaporation, plus diffusion of oxygen through the PHP layer is favored, because the interfacial Al_xO_y layer B nearly equals the oxide layer at the Al/air side. This is a very important fact one has to be aware of, when discussing the processes of charge carrier injection in such a device, since this introduction of a dielectric layer (i.e. an insulator in this case) will change the interface properties dramatically. Also the simple model of a triple system cannot be applied in this case.

CONCLUSIONS

With photoelectron spectroscopy (XPS and UPS) it has been shown that there is a weak, but significant interaction of PHP and oxygen. The experimental data suggest this interaction to be related to a p-type doping phenomenon of the organic material. As expressed in detail elsewhere [22], *in situ* photoemission experiments yield evidence for physisorption of aluminum atoms on a PHP film. But the situation of a pure Al/PHP interface in a LED seems not to be present in devices, which are exposed to air after production. RBS measurements on a conventional device revealed that there is a layer of aluminum oxide between the organic film and the aluminum electrode. Due to oxygen diffusion through the PHP film towards the aluminum electrode, over time an oxide layer practically as thick as the one on the Al/air contact is formed. This will change the electron injection efficiency dramatically, since insulation layers introduced at that interface on purpose are reported to be much thinner. A way to prohibit oxide formation would be encapsulation of the device right after production.

ACKNOWLEDGEMENT

The financial support of the Spezialforschungsbereich (SFB) *Elektroaktive Stoffe* is gratefully acknowleged.

REFERENCES

[1]   J.H.Burroughes, D.D.C.Bradley, A.R.Brown, R.N.Marks, K.Mackay,
        R.H.Friend, P.L.Burns and A.B.Holmes, Nature 347, 539 (1990)
[2]   D.Braun, A.J.Heeger, Appl.Phys.Lett. 58, 1982 (1991)
[3]   Y.Ohmori, M.Ochida, K.Muro, K.Yoshino, Japan.J.Appl.Phys. 30, L1938 (1991)
[4]   G.Grem, G.Leditzky, B.Ullrich, G.Leising, Adv.Mat. 4, 36 (1992)
[5]   W.Graupner, G.Grem, F.Meghdadi, Ch.Paar, G.Leising, U.Scherf,
        K.Müllen, W.Fischer, F.Stelzer, Mol.Cryst.Liqu.Cryst. 256, 549 (1994)
[6]   F.Meghdadi, G.Leising, W.Fischer, F.Stelzer, Synth.Met. 76, 113 (1996)
[7]   U.Scherf, K.Müllen, Makromol.Chem., Rapid Commum. 12, 489 (1991)
[8]   F.Li, H.Tang, J.Anderegg, J.Shinar, Appl.Phys.Lett. 70, 1233 (1997)
[9]   R.N.Marks, F.Biscarini, T.Virgili, M.Muccini, R.Zamboni, C.Taliani,
        Phil.Trans.R.Soc.Lond.A (1996) submitted
[10]   Y.E.Kim, H.Park, J.J.Kim, Appl.Phys.Lett. 69, 1 (1996)
[11]   M.Era, T.Tsutsui, S.Saito, Appl.Phys.Lett. 67, 2436 (1995)
[12]   H.H.Kim, T.M.Miller, E.H.Westerwick, Y.O.Kim, E.W.Kwock, M.D.Morris,
        M.Cerullo, J.Lightwave Technol. 12, 2107 (1994)
[13]   A.Schmidt, M.L.Anderson, N.R.Armstrong, J.Appl.Phys. 78, 5619 (1995)
[14]   G.Leising, S.Tasch, W.Graupner, Fundamentals of Electroluminescence in
        Paraphenylene-Type Conjugated Polymers and Oligomers, to be published
        in Handbook of Conducting Polymers II
[15]   K.Xing, M.Fahlman, M.Lögdlund, D.A.Santos, V.Parentè, R.Lazzaroni,
        J.L.Brédas, R.W.Gymer, W.R.Salaneck, Adv.Mater. 8, 971 (1996)
[16]   Y.Hirose, A.Kahn, V.Aristov, P.Soukiassian, Appl.Phys.Lett. 68, 217 (1996)
[17]   R.L.Johnson, J.Reichardt, "FLIPPER II - a new photoemission system in
        Hasylab", MPI f. Festkörperforschung, Stuttgart, Germany
[18]   C.Hochfilzer, Diploma Thesis, TU-Graz, 1997
[19]   S.Narioka, H.Ishii, K.Edamatsu, K.Kamiya, S.Hasegawa, T.Ohta, N.Ueno,
        K.Seki, Phys.Rev. B52, 2362 (1995)
[20]   R.H.Williams, Phys.Bl. 45, 219 (1989)
[21]   W.R.Salaneck, Rep.Prog.Phys. 54, 1215 (1991)
[22]   N.Koch, L.M.Yu, V.Parenté, R.Lazzaroni, G.Leising, J.J.Pireaux, J.L.Brédas, to be
        submitted
[23]   T.P.Nguyen, G.Leising, Interface effects in metal-polyparaphenylene-metal structures, in
        Polymer-Solid Interfaces, Eds. Pireaux, Bertrand and Brédas, IOP Publ., 1992

CHARACTERIZATION OF POLYPARAPHENYLENE SUBJECTED TO DIFFERENT HEAT TREATMENT TEMPERATURES

S. D. M. Brown[a], M. J. Matthews[a], A. Marucci[a], M. A. Pimenta[a], M. S. Dresselhaus[a,b],
M. Endo[c], T. Hiraoka[c]
[a]Department of Physics, Massachusetts Institute of Technology, Cambridge, MA 02139
[b]Department of Electrical Engineering and Computer Science, Massachusetts Institute of
Technology, Cambridge, MA 02139
[c]School of Engineering, Shinshu University, Nagano, 380 Japan

ABSTRACT

We investigated the structural and electronic properties of samples of polyparapheny-
lene (PPP), derived from two synthesis methods (the Kovacic and Yamamoto methods).
These samples have been subjected to different heat-treatment temperatures ($650°C \leq T_{HT} \leq 2000°C$) and their properties are compared to the polymer prior to heat-treatment
($T_{HT}=0°C$). The photoluminescence (PL) spectra of heat-treated PPP based on the two
synthesis methods reflects the differences in electronic structure of the starting polymers.
The PL emission from the heat-treated Yamamoto polymer is quenched at much lower T_{HT}
than from the Kovacic material. However, Raman spectra taken of the material resulting
from heat-treatment of the polymer (using both preparation methods) indicate the presence
of phonon modes for PPP in samples at T_{HT} up to $650°C$.

INTRODUCTION

Research and development on lithium ion batteries, having a carbon host doped with
lithium as the anode, is currently a very active research field because of the general need for
light weight, safe, rechargeable battery cells for portable electronic devices.[1, 2] Previous
studies on PPP prepared by the Kovacic method, and heat treated up to temperatures
$T_{HT} \sim 3000°C$ have examined the electronic properties[3, 4] and structural characteristics[4,
5] of these PPP-based carbons. It has been found that materials derived from a Kovacic-
PPP precursor (eg. after heat-treatment at $700°C$) have a large lithium uptake capacity,
while PPP-based materials derived from a Yamamoto-PPP precursor do not exhibit this
high uptake capacity. In this work we examine the optical properties of PPP-based carbons,
derived from PPP from both the Kovacic and Yamamoto synthesis methods, which are
then heat-treated to various temperatures (T_{HT}). We use Raman scattering as a probe of
the vibrational spectra, and PL to study the electronic structure related to the complicated
defect structure within the carbon-polymer matrix. Our goal is to gain further understanding
of the carbonization stages of the PPP polymer (from both preparation methods), which is
studied as a function of T_{HT}, and thus we expect to gain further insight into why Kovacic-
PPP heat treated to $700°C$ has such a high lithium uptake.[6]

EXPERIMENTAL DETAILS

The granular PPP samples in this work were synthesized by the Kovacic method[7], and
the Yamamoto method[8]. The Kovacic PPP samples were pressed into $\sim$ 15mm diameter discs to simulate an electrode in a rechargeable battery, and heat treated to various

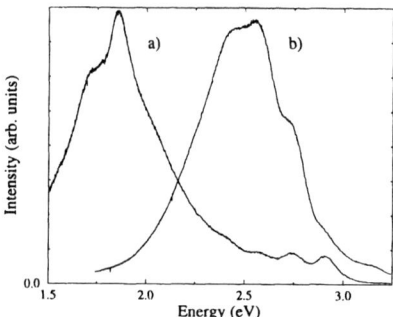

Figure 1: Plot of PL spectra from PPP (T_{HT}=0°C) derived from the (a) Kovacic and (b) Yamamoto synthesis methods. The laser excitation was at 3.54 eV.

temperatures (T_{HT}) between 600 and 2000°C for 1 hour in high purity argon gas, using a conventional resistance furnace. The Yamamoto samples were subjected to the same heat-treatment conditions but due to limited quantities could not be pressed into pellets. Raman scattering experiments were performed under ambient conditions using a back-scattering geometry with 488.0 nm as the laser excitation wavelength (λ_{exc}). A SPEX FluoroLog Model FL212 Spectrofluorimeter was used to measure the PL spectra. Energy dispersive X-ray analysis was performed on the polymer prior to heat-treatment from both synthesis methods.

RESULTS AND DISCUSSION

Energy dispersive X-ray analysis of Kovacic-PPP-000 showed the presence of carbon, some silicon (as impurities), copper, aluminum and chlorine. The presence of copper, aluminum and chlorine are indicative of doping of the polymer from residual catalyst after synthesis. Similar analysis of Yamamoto-PPP-000 showed the presence of silicon impurities, magnesium, bromine as residues from the catalyst.

PL spectra, with λ_{exc}= 3.54 eV, were collected from the non-heat-treated samples of the polymer from both synthesis methods, and the results are shown in Figure 1. All PL measurements were performed under ambient conditions. The dominant feature in the PL spectra from the Kovacic-PPP-000 sample, i.e. before heat-treatment (T_{HT}=0°C), is the broad red photoluminescence centered at 1.86 eV, with a shoulder at 1.71 eV. The broad red PL emissions can be attributed to defects in the samples (such as macrocycles)[9], arising from the Kovacic preparation method. However, the presence of chlorine as a chromophore can also contribute to this red PL emission.[10] The PL from the Kovacic-PPP-000 sample also shows well-defined spectral features in the blue-green located at 2.92 eV, 2.75 eV, 2.59 eV, and 2.41 eV, associated with vibronic transitions in the polymer. The separation of the peaks (~0.17 eV) corresponds to a maximum in the density of states of phonon modes for PPP. After heat-treatment at 600°C, the intensity of the red PL emission is significantly reduced; the ratio of the total blue-green emission to that of the red is 3% for Kovacic-PPP-000, but is 14% for Kovacic-PPP-600.

Under similar conditions, the PL emission from the Yamamoto-PPP-000 polymer has a

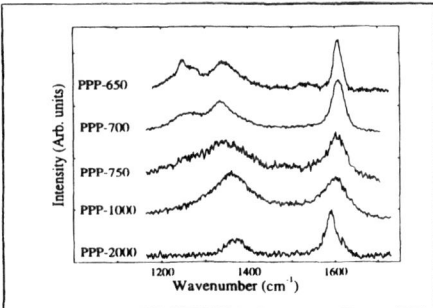 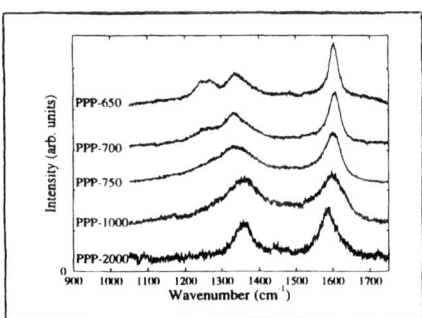

Figure 2: Raman spectra for λ_{exc}= 488.0 nm of heat-treated polyparaphenylene (PPP) derived from the Kovacic (a) and Yamamoto (b) synthesis methods, then heat-treated to 650°C, 700°C, 750°C, 1000°C and 2000°C.

higher intensity than that from the Kovacic-PPP-000 polymer; however, the spectra from the two polymers have been scaled in order to compare the spectral features (Fig.1). The PL spectra from the Yamamoto-PPP samples consist of features in the blue-green at 2.75 eV, 2.59 eV, 2.41 eV, and 2.24 eV, with very little background emission in the red. The intensity of the PL emission decreases as T_{HT} increases for both types of PPP, but it is quenched above T_{HT}=650°C for Kovacic-PPP as compared to T_{HT}=500°C for Yamamoto-PPP.

PL provides an effective probe of the presence of the polymer in the PPP samples after the heat-treatments, since the polymer fragments exhibit strong photoluminescent emissions, but the disordered carbon regions formed from heat-treatment of the polymer precursor generally will not luminesce. Our experimental PL results clearly indicate the presence of PPP polymer segments in the Kovacic-PPP samples heat-treated up to 650°C. Our results further show that the PL disappears at a much lower heat-treatment temperature for the Yamamoto-PPP sample, indicating that disruption of the phonon-assisted PL mechanism occurs at a much lower carbonization temperature for Yamamoto-PPP than for PPP-Kovacic. The phonon-assisted PL mechanism will be discussed in detail in a future publication.

The vibrational spectra of PPP (before heat-treatment) has been studied extensively.[11, 12, 13, 14] There are 3 A_g modes which produce strong Raman lines in the range 1100 cm^{-1} to 1800 cm^{-1}, namely at $\sim$1220 cm^{-1}, $\sim$1280 cm^{-1} and $\sim$1600 cm^{-1}, characteristic of the benzenoid configuration. It is also well known that a (higher energy) quinoid state exists for the molecule,[15, 16] producing Raman lines near 1240 cm^{-1} and 1330 cm^{-1}, slightly higher in frequency from those for the benzenoid configuration, due to a redistribution of C-C double-bonds. Such an excited state of the molecule (induced here by heat-treatment) has been observed previously through doping.[15]

Figure 2 shows Raman spectra of the heat-treated samples from both synthesis methods, collected using an excitation wavelength, λ_{exc}, equal to 488.0 nm. For all samples heat-treated between 650°C and 750°C, a single well-defined band at 1605 cm^{-1} is observed, along with weaker bands in the range 1200–1500 cm^{-1}. The high-frequency band at 1605 cm^{-1} broadens with increasing T_{HT}, with a full-width-half-maximum (FWHM) ranging from $\sim$25 to $\sim$50 cm^{-1} for samples PPP-650 through PPP-750, respectively. After heat-treatment at 750°C, the low-frequency structure in the Raman spectra of both types of PPP become less defined, leading to a single, broad peak near $\sim$1330 cm^{-1}. Since the structures at

1245, 1270 and 1335 cm^{-1} for the two types of PPP-650 samples are well resolved, we have therefore chosen to focus on the T_{HT}=650°C data for detailed analysis and comparison.

A detailed least-squares fit to the Raman data collected from both types of PPP indicates the presence of several low frequency peaks in samples PPP-650 and PPP-700. Figure 3 shows the results of a lineshape analysis for PPP-650 for the Yamamoto and Kovacic samples. All peaks are well described by Lorentzian lineshapes, after subtraction of a linear background term in the fitting routine. In the case of heat-treated Kovacic-PPP, the aforementioned bands ($\sim$1245, $\sim$1270 and $\sim$1335 cm^{-1}) were well accounted for by the fit, but a better fit is obtained if small Lorentzians centered at 1216 and 1360 cm^{-1} are also used to fit the PPP-650 spectra. In the case of heat-treated Yamamoto-PPP, only a line at $\sim$1365 cm^{-1} is needed to improve the fit. The ratio of the intensities of the two benzenoid modes (I_{1220}/I_{1280}) increases with the number of phenyl rings [15]. Although the benzenoid phonon mode at 1270 cm^{-1} is evident in the spectra from Yamamoto-PPP-650, no 1220 cm^{-1} peak is found in the fit, and thus we cannot use this ratio for quantitative comparisons. A similar analysis of the ratio of the intensities of the quinoid phonon modes gives I_{1245}/I_{1335}=0.59 for Yamamoto-PPP-650, and I_{1245}/I_{1335}=0.74 for Kovacic-PPP-650, indicates that there are longer quinoid chains in Kovacic-PPP-650 than in Yamamoto-PPP-650.

Due to the large PL background, the benzenoid modes of Kovacic-PPP-000 cannot be observed with excitation in the visible (λ_{exc}=488 nm). The reduction in PL emission with heat-treatment allows us to observe these benzenoid modes. We also observe a stabilization of the quinoid modes of PPP by the heat-treatment-induced disorder, and these quinoid modes has previously been observed only in the doped polymer. We see evidence of an earlier onset of carbonization in Yamamoto-PPP, through quenching of its PL spectra at lower T_{HT}, and the quenching of the benzenoid mode at 1216 cm^{-1} in the PPP-650 sample.

After heat-treatment at much higher T_{HT} (2000°C), where the samples are thoroughly carbonized, the Raman spectra of both types of PPP contained the disorder-induced peak at $\sim$1360 cm^{-1} and a graphite line at $\sim$1590cm^{-1}. The crystallite sizes[17] were 110 Å for the Kovacic-PPP-2000 based carbon, but only 58 Å for the Yamamoto-PPP-2000 based carbon. Thus, at higher T_{HT}, the Kovacic-PPP samples yield carbons with greater in-plane order than Yamamoto-PPP. However, x-ray measurements show that the c-axis stacking is more ordered for Yamamoto-PPP based carbons, and for this reason PPP-derived carbon is an unusual carbon.

We observed that the Raman spectra for PPP (both types) heat-treated near 650°C is the result of a superposition of peaks derived from both the quinoid and benzenoid segments, the former being induced by disorder.[5] Ultimately, a planar configuration of phenyl groups is energetically favored as the heat treated solid loses increasing amounts of hydrogen and eventually forms small graphene ribbons or sheets. The difference in carbonization of the two types of PPP is seen in the chain length of the quinoid-like polymer remaining in the samples.

CONCLUSION

The difference in the PL spectra from the two types of PPP polymer results from the synthesis method. Evidence of the phonon-assisted PL process is common to both polymers, while the large red PL emission (due to defects) is seen only in the Kovacic-PPP. The quenching of the PL that occurred with heat-treatment allowed the observation of the Raman spectra with visible excitation for both types of PPP.

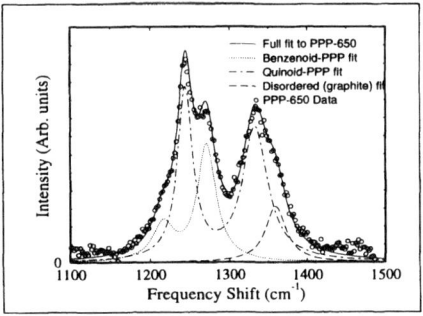 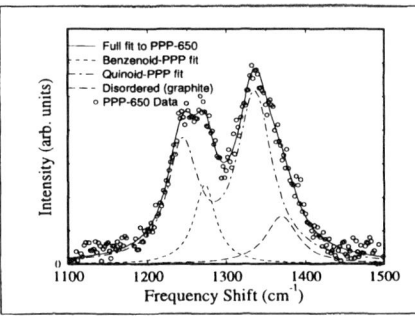

Figure 3: Low frequency Raman spectra of PPP-650 from the Kovacic (a) and Yamamoto (b) synthesis methods, showing the superposition of several peaks due to the coexistence of benzenoid and quinoid segments.

Raman studies reveal that, for both types of PPP-650 samples, features in the Raman spectrum between 1200 cm^{-1} and 1500 cm^{-1} may be identified with phonon modes for both benzenoid and quinoid configurations. Thus, at low T_{HT}, we find the co-existence of a mostly quinoid-like disordered PPP material with smaller amounts of benzenoid-PPP and turbostratic graphene constituents. However, the Yamamoto-PPP-650 sample has shorter lengths of quinoid-like disordered PPP material than Kovacic-PPP-650. The crosslinking of polymer chains that occurs with heat-treatment results in pockets of polymer surrounded by regions of turbostratic graphene constituents. The larger crystallite size of Kovacic-PPP (heat-treated to 2000°C), as compared to Yamamoto-PPP, indicates that greater in-plane order is achieved through the heat-treatment in Kovacic-PPP.

Acknowledgments

The authors are thankful to Dr. G. Dresselhaus for stimulating discussions and to Dr. Don Heiman for the generous use of optical equipment and for useful discussions. This work made use of MRSEC Shared Facilities supported by the National Science Foundation under Award Number DMR 94-00334. We gratefully acknowledge support for this work under NSF grant DMR 95-10093.

REFERENCES

[1] M. Mohri, J. Yanagisawa, Y. Tajima, H. Tanaka, T. Mitate, S. Nakajima, M. Yoshida, Y. Yoshimoto, T. Suzuki, and H. Wada, J. Power Sources 26, 545 (1989).

[2] I. Kuribayashi, M. Yokoyama, and M. Yamashita, J. Power Sources 54, 1 (1995).

[3] M. J. Matthews, M. S. Dresselhaus, N. Kobayashi, T. Enoki, M. Endo, and T. Takahashi, Appl. Phys. Lett. 69, 2042 (1996).

[4] M. J. Matthews, M. S. Dresselhaus, M. Endo, Y. Sasabe, T. Takahashi, and K. Takeuchi, J. Mat. Res. 11, 3099 (1996).

[5] M. J. Matthews, X. X. Bi, M. S. Dresselhaus, M. Endo, and T. Takahashi, Appl. Phys. Let. **68**, 1078 (1996).

[6] K. Sato, M. Noguchi, A. Demachi, N. Oki, and M. Endo, Science **264**, 556 (1994).

[7] P. Kovacic and A. Kyriakis, J. Am. Chem. Soc. **85**, 454 (1963).

[8] T. Yamamoto, Y. Hayashi, and A. Yamamoto, Bull. Chem. Soc. Japan **51**, 2091 (1978).

[9] G. Froyer, F. Maurice, P. Bernier, and P. McAndrew, Polymer **23**, 1103 (1982).

[10] P. Kovacic and L.-C. Hsu, J. of Polymer Science. Part A-1 **4**, 5 (1966).

[11] S. Krichene, S. Lefrant, G. Froyer, F. Maurice, and Y. Pelous, J. Phys. (Paris) Colloq. **44**, C3–733 (1983).

[12] Z. Iqbal, H. Bill, and R.H. Baughman, J. Phys. (Paris) Colloq. **44**, C3–761 (1983).

[13] J. P. Buisson, S. Krichene, and S. Lefrant, Synth. Met. **21**, 229 (1987).

[14] G. Zannoni and G. Zerbi, J. Chem. Phys. **82**, 31 (1985).

[15] S. Krichene, J.P. Buisson, and S. Lefrant, Synth. Met. **17**, 589 (1987).

[16] Y. Furukawa, H. Ohtsuka, and M. Tasumi, Synth. Met. **55-57**, 516 (1993).

[17] D. S. Knight and W. B. White, J. Mater. Res. **4**, 385 (1989).

MORPHOLOGICAL EFFECTS IN THE
CHEMICAL AND PHOTOLUMINESCENT BEHAVIOR OF
ALUMINUM TRIS(8-HYDROXYQUINOLINE) (ALQ3)

K. A. HIGGINSON, X.-M. ZHANG[†], F. PAPADIMITRAKOPOULOS[*]
Department of Chemistry, Polymer Science Program, Nanomaterials Optoelectronics Laboratory,
Institute of Materials Science, University of Connecticut, Storrs, CT 06269-3136

ABSTRACT

Aluminum tris(8-hydroxyquinoline) (Alq3) is presently considered one of the most reliable electron transporting and emitting materials for molecular-based organic light emitting diodes (OLEDs). This paper discusses the effects of sample morphology on the hydrolysis and photoluminescence response of Alq3. The evolution of 8-hydroxyquinoline (8-Hq), a volatile byproduct of the hydrolysis of Alq3, was quantified using gas chromatography/mass spectroscopy (GC/MS) analysis on samples of different morphology. Annealed (more crystalline) samples exhibited greater chemical stability than freshly sublimed films, at the expense of photoluminescence efficiency. These phenomena are discussed with respect to possible failure mechanisms that Alq3-based OLEDs might undergo during prolonged operation.

INTRODUCTION

Light emitting diodes (LEDs) based either on sublimable organics or polymeric materials have attracted considerable attention from both the scientific and technological fields.[1-3] The promise of high brilliance, cost-effective, flexible displays and illuminators has intensified research and development in both academic and industrial laboratories on a global basis, yielding many advancements.[1-5, and references therein]

One of the key issues in the development of organic LEDs (OLEDs) is lifetime enhancement. Morphological changes and electrode deterioration are believed to contribute to the rapid decay of unoptimized devices.[2,6] At film pinholes and substrate imperfections, localized heating due to uneven current distribution through the irregular film facilitates a number of failure processes.[5-9] Rapid physical changes which limit OLED performance are slowly being controlled (for example, through the use of morphologically stable glasses[10-12], substrate smoothing, or composite films[13]), particularly for the case of hole transporting materials.[14-16] Effects noticed at longer times, however, such as the gradual electroluminescence (EL) decay and accompanying increase in device resistance,[5,6] warrants investigation of typically slower physical aging[17] and chemical degradation[18,19] processes.

Aluminum tris(8-hydroxyquinoline) (Alq3) has risen to a prominent position in the development of robust OLEDs due to its relative stability as an electron transporting and emitting material.[4,5] The glass transition temperature (T_g) of Alq3 is relatively high (at 172 °C), but crystallization may still occur below T_g at greater time scales. These processes are closely related to device stability since photophysical and electrical properties are affected by the molecular or supramolecular order. Furthermore, Alq3 absorbs moisture, and can hydrolyze significantly at elevated temperatures to form free 8-hydroxyquinoline (8-Hq) species.[20-22] The freed 8-Hq may then undergo further reactions to produce nonemissive species which can act as luminescence quenchers.[21] The observation of degradation reactions and byproducts in an actual device configuration (with thicknesses of 600-1200 Å), has proven a formidable task,[18] since the amount of material that can be generated in an operating OLED is undetectable by most instruments. The volatile nature of 8-Hq, however, provides a measurable quantity: gas chromatography and mass spectroscopy (GC/MS) were utilized to investigate the degradation of Alq3.[22,23].

---

[*] To whom correspondance should be addressed.
[†] Department of Chemistry, Wesleyan University, Middletown, CT 06459

EXPERIMENTAL

Alq$_3$ was synthesized according to the literature method and purified by sublimation[21].

Differential scanning calorimetric (DSC) measurements were conducted with a Perkin Elmer DSC-7, employing a 20 mL/min. flow of dry nitrogen as a purge gas for the sample and reference cells. The temperature and power ordinates of the DSC were calibrated with respect to the known melting point and heat of fusion of high purity indium. In a nitrogen dry box, about 10 mg of sublimed or annealed Alq$_3$ was scraped from its substrate, packed, and compression sealed with a Perkin Elmer crimper press. The scan rate was 10 °C/min. For cooling, a two stage intercooler based on freon 502 and ethane 170 was used.

X-ray diffraction (XRD) was performed on a Norelco/Phillips diffractometer using Cu K_α radiation (λ=1.5418 Å). An Alq$_3$ film of several μm was evaporated at 7-10 Å/s on a glass slide for measurement of its XRD spectrum before and after annealing, followed by subtraction of the substrate background. The film was annealed at 200 °C in a vacuum oven (1×10^{-5} Torr) equipped with a liquid nitrogen trap to prevent contamination by diffusion pump oil.

A Hewlett/Packard 5985 GC/MS equipped with a direct dynamic headspace injector (DDHI, a specialized accessory used for the analysis of volatile materials formed from non-volatile solids[22,24]) was used to study the degradation reaction. The DDHI allows automatic transfer to and from the GC injection port. 10 to 20 μg Alq$_3$ samples (powder or evaporated films, ca. 1000 Å on aluminum foil) were placed in small glass tubes, exposed to moisture, and placed in the DDHI, using a new sample each time. During the entire course of the measurement, the sample remains under high purity helium. The column used was OV-1 fused silica capillary column 0.32 mm × 50 m, 1.2 μm in size. The injection port temperature was varied to probe the temperature dependence of the reaction, and the samples were normalized with respect to their initial masses. The chromatography was carried out by temperature programming at 15 °C/min. rate, from 35 to 300 °C. The mass spectroscopy had a range of 10 to 1000 atomic mass units. Electron impact ionization was used at 70 electron volts and the scanning rate was 400 atomic amu/s.

A Perkin Elmer TGA-7 was used for the thermogravimetric analysis (TGA) at a scan rate of 10 °C/min. A 10 mL/min. flow of dry nitrogen was used to purge the sample at all times.

Photoluminescence (PL) spectra were obtained with a Perkin Elmer LS50B Luminescence Spectrometer. A special cuvette was fabricated which could be evacuated (to approximately 8×10^6 Torr) and flame sealed. Alq$_3$ was evaporated on a glass slide (1000 Å), followed by 1000 Å of magnesium, and secured within the cuvette. The films were annealed at 175 °C by placing the entire cuvette in a constant temperature bath. Reflection UV-VIS spectra were collected on a Perkin-Elmer Lambda 3840 array spectrophotometer after each annealing period to verify a constant optical density of the sample.

RESULTS AND DISCUSSION

Figure 1A illustrates the DSC measurements of freshly sublimed Alq$_3$ samples. Sharp exotherms are observed above 125 °C and their magnitude increases dramatically above 165 °C, the glass transition temperature(T_g). The rather "noise-reminiscent" exotherms below T_g are not artifacts due to sample preparation or chemical reactions (GC/MS and TGA show that these dry films are not undergoing hydrolysis in this range), and are quite reproducible, although the specific temperatures at which they occur vary from sample to sample. Subsequent scans of this or scans of annealed samples (held 2 hr at 200 °C under vacuum) show the absence of these exotherms and a broad, featureless T_g (at about 172 °C), typical of a semicrystalline glass with a broad spectrum of relaxation times. The intensity increase and narrowing of the X-ray diffraction peaks during the annealing of Alq$_3$ films (at 200 °C) indicates an increase in both crystal size and overall degree of crystallinity.[20] Parameters such as the rate of deposition, substrate temperature and surface treatment, however, are expected to influence both crystallite size and the extent of amorphousness in such films. For this study, the films were deposited onto room temperature glass slides, cleaned with a 7:3 H$_2$SO$_4$:H$_2$O$_2$ solution and deionized water, at a rate of about 7-10 Å/s.

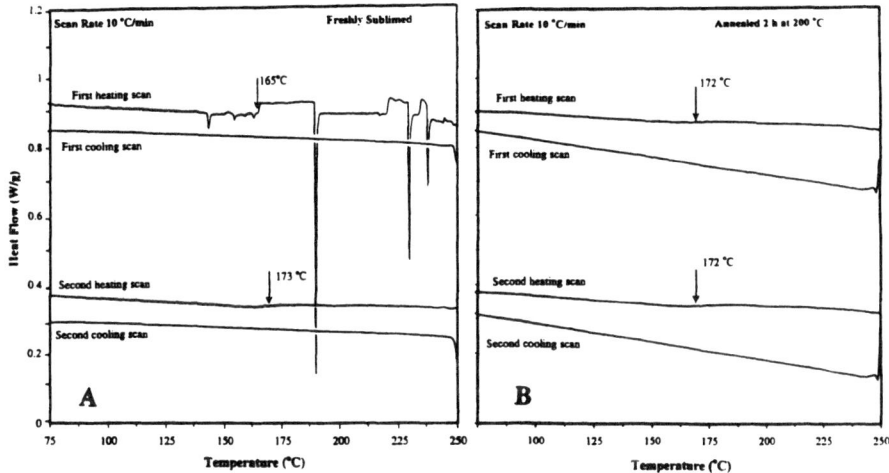

Figure 1. DSC traces for (A) freshly sublimed and (B) 200 °C, 2hr. annealed Alq3 samples.

The decomplexation of Alq3 in the presence of water and/or weak acids has been observed with the help of spectroscopic (NMR, IR, etc.) and electrochemical methods.[25] On the other hand, the reaction environment within the device is markedly different since it involves sorbed water within the semicrystalline layer of Alq3. Freshly sublimed Alq3 powders or thin films were exposed to ambient moisture before the chemical reaction was probed with the GC. Significant amounts of 8-Hq were detected when the temperature of the injection port was set above 90 °C. Water was virtually absent from the GC trace, leaving 8-Hq as the only volatile compound detected in this temperature range. The area of the 8-Hq elution peak (of roughly 15 seconds width) was normalized with respect to the initial sample mass and used to quantify the extent of reaction. Qualitative sorption kinetics were obtained using this GC/MS technique by observing the elution of 8-Hq as a function of exposure time in air. The Alq3-containing sample tube was placed into the instrument after exposure (at ~40% humidity) for different time intervals. The injection time was sufficiently long for the completion of the reaction and 8-Hq desorption (about four minutes). It was found that the amount of freed 8-Hq saturates in less than 2 minutes.[22] This rapid sorption of moisture on fresh Alq3 indicates that considerable care must be exerted to avoid water contamination of the finished devices.

Figure 2 shows the (normalized) peak area of 8-Hq plotted as a function of temperature. Annealed Alq3 showed a greater resistance to hydrolysis than the unannealed samples.[20] At fixed injection times, the amount of 8-Hq was monitored in the GC as a function of reaction temperature. These data exhibited an Arrhenius dependence, and the activation energy was used to compare the thermal stability of both samples (see inset). The difference in activation energy is probably a result of the difference in free volume between the two sample morphologies. This would be reflected in water diffusivity as well as in the amount of amorphous Alq3 participating in the reaction, since the material within the crystallites is comparatively inaccessible to moisture.[26] Care was taken to conduct these studies below the measured glass transitions and to limit the injection time so that only a fraction of the water is consumed.

At the end of all GC/MS tests, the samples continue to exhibit characteristic bright greenish-yellow photoluminescence. Although the sample form and geometry hinder quantitative comparison in relative photoluminescence (PL) intensity, PL quenching by formation of 8-Hq degradation products has not been observed[21]. This is attributed to the efficient removal of the volatile 8-Hq, as well as to the absence of oxygen, which we believe an important step in the degradation of 8-Hq. Even in LED configurations, the further oxidation of 8-Hq is probably important only at the initial flare of a "hot spot," while the production of volatile 8-Hq is consistent with some of the more dramatic failures observed as the electrodes delaminate.[8,15]

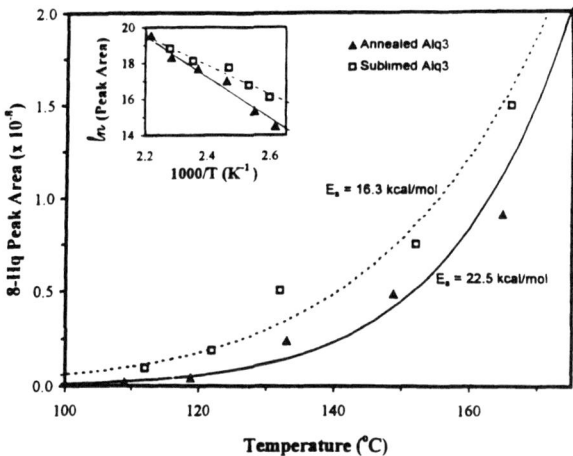

Figure 2. Evolution of 8-Hq as a function of temperature for freshly sublimed (unfilled squares) and 2 hours annealed at 200 °C (filled triangles) Alq3, as measured by GC/MS. Dashed and solid lines are Arrhenius fits to the data.

Figure 3 elucidates the effect of sample morphology on the photoluminescence (PL) of 1,000 Å Alq3 films.[20] Annealing these films at 175 °C, slightly above their T_g, resulted in a decrease of PL intensity. This was also accompanied by a slight red shift of the emission peak (507 nm at 0 min. to 509 nm at 13.5 min.). Both the PL decrease and red shift may be explained by the formation of aggregates or crystals, as has been observed by XRD,[20] which promote formation of excimers or ground state complexes. As the crystals get large, however, an apparent increase in intensity and spectral blue shift is observed. This is not due to change of the emission characteristics but rather to Rayleigh scattering for shorter wavelengths (see Fig. 3f). After long periods of annealing in this fashion, crystallinity is visible to the naked eye.

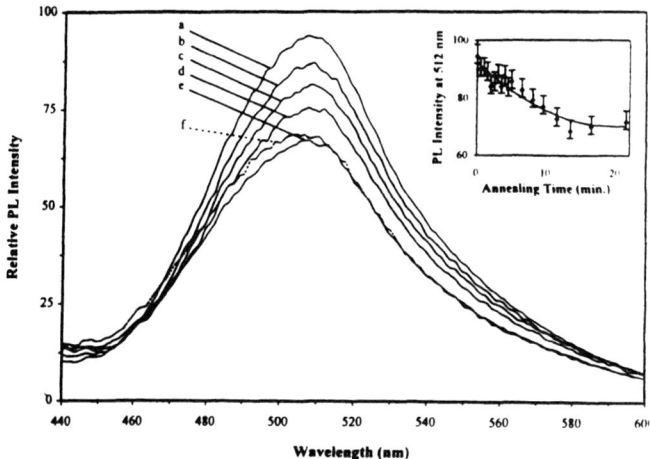

Figure 3. Photoluminescence spectra of a 1000 Å Alq3 film at 175 °C under 8×10^{-6} Torr at (a) 0 min., (b) 3 min., (c) 6.5 min., (d) 9.5 min., (e) 16.5 min., and (f) 21.5 min. annealing times. Inset plots the intensity at 512 nm as a function of annealing time.

The morphological characteristics of Alq3 are expected to contribute to three aspects of possible failure of its OLEDs: dimensional instability, luminance decay and uneven charge transport. These elaborate on the present understanding of device failure and may add upon other existing problems.

The substantial volume shrinkage observed as a result of crystallization could create drastic thickness variations in molecular sublimed thin films,[10,14,15] resulting in variations of electric field, if the cathode conforms to the altered geometry, assisting in the formation of high-current hot spots, and failures similar to those observed from pinholes and ITO asperities.[7-9,15,27,28] Alternatively, if the electrode does not adhere, it separates from the receding organic layer, a process which may be facilitated by water-assisted cathode oxidation[7,8,28] or production of volatile products upon chemical decomposition.[22,27]

Although an OLED is far more complex than the simple emission decay observed in Figure 3, the fact that the decay rate observed in long-lived devices is proportional to the current density (which to the first approximation can be related to resistive heating)[6-8], is in good agreement with Alq3 crystallization. Furthermore, the observed luminance decay of Alq3 devices can be slowed with molecular doping,[4,5,29] which has been proposed to act beneficially in suppressing crystallinity[13] as well as containing the emission within localized, high-efficiency chromophores.

The altered charge mobility in the crystalline domains may also lead to uneven charge transport within a single microcrystalline layer of constant thickness. The gradual luminance decay and resistance increase observed in long-lived devices may hinge on the formation and growth of crystalline domains within the Alq3 layer. At long times, the continued growth of crystals at the expense of the amorphous region may lead to preferential conduction pathways, and consequently, localized damage. Still, competing processes, such as the intrinsic degradation of the Alq3 chromophore, layer interdiffusion, contamination from ITO, and crystallization of hole transport materials will probably cause similar effects. Significant work is still needed to distinguish and prioritize the different mechanisms, leading to a more thorough understanding of device failure.

CONCLUSIONS

The hydrolytic stability of Alq3 was discussed in light of sample morphology. Samples of higher crystallinity were found to be less susceptible to chemical attack by moisture, although they exhibited lower photoluminescence efficiency. The ramifications of these findings were discussed with respect to the various modes of device failure. The witnessed chemical and morphological transformations below the glass transition temperature of Alq3 (ca. 172 °C) raise the importance of these effects during prolonged, low-temperature device operation.

ACKNOWLEDGMENTS

This research was financially supported by the Institute of Materials Science, Critical Technology Program, and by a Hewlett Packard Fellowship to K. A. H. The authors wish to thank Mr. G. Lavigne and Dr. Dorab E. Bhagwagar for their expert technical assistance and Dr. H. Antoniadis for helpful discussions.

REFERENCES

1. Burrows, P. E.; Forrest, S. R.; Thompson, M. E. *Current Opinion in Solid State & Materials Science* **1997**, *2*, 236.
2. Sheats, J. R.; Antoniadis, H.; Hueschen, M.; leonard, W.; Miller, J.; Moon, R.; Roitman, D.; Stocking, A. *Science* **1996**, *273*, 884.
3. Rothberg, L. J.; Lovinger, A. J. *J. Mater. Res.* **1996**, *11*, 3174.
4. Tang, C. W.; VanSlyke, S. A.; Chen, C. H. *J. Appl. Phys.* **1989**, *65*, 3610.
5. Shi, J.; Tang, C. W. *J. Appl. Phys.* **1996**, *70*, 1665.
6. VanSlyke, S. A.; Chen, C. H.; Tang, C. W. *Appl. Phys. Lett.* **1996**, *69*, 2160.

7. McElvain, J.; Andoniadis, H.; Hueschen, M.; Miller, J.; Roitman, D.; Sheats, J.; Moon, R. *J. Appl. Phys.* **1996**, *80*, 6002.
8. Antoniadis, H.; Hueschen, M. R.; McElvain, J.; Miller, J. N.; Moon, R. L.; Roitman, D. B.; Sheats, J. R. *ACS Polymer Preprints* **1997**, *38*, 382.
9. Burrows, P. E.; Bulovic, V.; Forrest, S. R.; Sapochak, L. S.; McCarthy, D. M.; Thompson, M. E. *Appl. Phys. Lett.* **1994**, *65*, 2922.
10. Shirota, Y.; Kuwabara, Y.; Inada, H.; Wakimoto, T.; Nakada, H.; Yonemoto, Y.; Kawami, S.; Imai, K. *Appl. Phys. Lett.* **1994**, *65*, 807.
11. Tanaka, H.; Tokito, S.; Taga, Y.; Okada, A. *Chem. Comm.* **1996**, 2175.
12. Adachi, C.; Nagai, K.; Tamoto, N. *Appl. Phys. Lett.* **1995**, *68*, 2697.
13. Sano, T. *Proceedings of the Electroluminescence Workshop*; Wissenschaft and Technic: Berlin, 1996, p. 196.
14. Han, E.-M.; Do, L.-M.; Yamamoto, N.; Fujihira, M. *Thin Solid Films* **1996**, *273*, 202.
15. Do, L.-M.; Han, E. M.; Niidome, Y.; Fujihira, M.; Kanno, T.; Yoshida, S.; Maeda, A.; Ikushima, A. J. *J. Appl. Phys.* **1994**, *76*, 5118.
16. Joswick, M. D.; Campbell, I. H.; Barashkov, N. N.; Ferraris, J. P. *J. Appl. Phys.* **1996**, *80*, 2883.
17. Struik, L. C. E. *Physical Aging in Amorphous Polymers and Other Materials*; Elsevier: 1978.
18. Papadimitrakopoulos, F.; Yang, M.; Rothberg, L. J.; Katz, H. E.; Chandross, E. A.; Galvin, M. E. *Mol. Cryst. Liq. Cryst.* **1994**, *256*, 663.
19. Chadha, S. S. In *Solid State Luminescence*; A. H. Kitai, Ed.; Chapman & Hall: New York, 1993; pp 159.
20. Higginson, K. A.; Zhang, X.-M.; Papadimitrakopoulos, F. *Chem. Mater.* **1997**, Accepted for Publication.
21. Papadimitrakopoulos, F.; Zhang, X.-M.; Thomsen, D. L.; Higginson, K. A. *Chem. Mater.* **1996**, *8*, 1363.
22. Papadimitrakopoulos, F.; Zhang, X.-M. *Synthetic Metals* **1997**, *85*, 1221.
23. Ohring, M. *The Materials Science of Thin Films*; Academic Press: NY, 1991.
24. Ezrin, M.; Lavigne, G. *SPE-ANTEC Proceedings 1991*, 2230.
25. Phillips, J. P. *Chem. Rev.* **1956**, *56*, 271.
26. Yeh, Y.-S. *J. Polym. Sci.* **1990**, *28*, 545.
27. Fujihira, M.; Do, L.-M.; Koike, A.; Han, E.-M. *Appl. Phys. Lett.* **1996**, *68*, 1787.
28. Do, L.-M.; Oyamada, M.; Koike, A.; Han, E.-M.; Yamamoto, N.; Fujihira, M. *Thin Solid Films* **1996**, *273*, 209.
29. VanSlyke, S. A.; Bryan, P. S.; Tang, C. W. *Proceedings of the Electroluminescence Workshop*; Wissenschaft and Technic: Berlin, 1996, p. 195.

FABRICATION OF POLYMER LIGHT EMITTING DIODES
BY LAYER-BY-LAYER COMPLEXATION TECHNIQUE

Jaehyun Kim*, Kethinni G. Chittibabu**, Mario J. Cazeca**, Woohong Kim***, Jayant Kumar*, and Sukant K. Tripathy*

*Center for Advanced Materials, Department of Chemistry and Physics, University of Massachusetts, Lowell, MA, 01854, U.S.A.

** Molecular Technologies, Inc., Westford, MA, 01886, U.S.A.

*** Samsung Central Research Institute of Chemical Technology, Taejeon, 305-380, Korea

ABSTRACT

Multilayer films were fabricated employing a new layer-by-layer complexation technique by alternatively dipping substrates in solutions of macromolecular ligands and Eu^{3+} ions. A luminescent poly[2-(3-thienyl)ethanol hydroxycarbonyl-methyl urethane] (H-PURET) was prepared by hydrolysis of thiophene polymer, poly[2-(3-thienyl)ethanol butoxycarbonyl-methyl urethane] (PURET), which was developed in our laboratory. H-PURET was used as macromolecular ligands. The multilayer deposition was monitored using UV-visible spectroscopy. As the number of bilayers increases, the absorption due to the polymer increases. The multilayer films were characterized by infrared and fluorescence spectroscopic techniques. Electroluminescence brightness easily obtained from the multilayer film of conjugated polymer-Eu^{3+} complex was measured to be about 40nW.

INTRODUCTION

Electroluminescent devices from organic materials have attracted considerable interest, since Tang and VanSlyke[1,2] introduced injection-type electroluminescent devices based on sublimed organic materials, due to their potential application as future flat panel displays.[3-5] The devices were fabricated with charge transport layers and an emissive layer, which provide high brightness outputs under low driving voltage. Lanthanide metal complexes are known to exhibit photoluminescence with very high quantum efficiencies[6] and are expected to show efficient electroluminescence.[7] However, long-term stability is a serious concern for those kinds of devices. The morphological rearrangement of the sublimed luminescent and charge transport molecules, as well as the diffusion of the electrode metals into the emitting layer[8] during operation, results in the degradation of these devices.

Recent discovery of using conjugated polymers such as poly(phenylene vinylene) for electroluminescent devices has provided an interesting approach to realize the potential applications of the devices, leading to intensive investigations to identify new materials with enhanced light emitting properties.[9-11] However, the conjugated polymers still suffer from relatively low quantum efficiency.

It is possible to make self-assembled multilayer thin films from the combinations of conjugated polymers and highly luminescent lanthanide metal ions through layer-by-layer

527

complexation technique. This technique may be a promising process for improving the stability of present organic devices without losing their electroluminescence. This improvement can be achieved by incorporation of highly efficient metal complexes into relatively stable polymers.

In this paper, we report on the electroluminescence characteristics of multilayer thin films, which were prepared by employing a layer-by-layer complexation technique. A highly luminescent thiophene-based conjugated polymer, poly[2-(3-thienyl)ethanol butoxycarbonyl-methyl urethane] (PURET), developed in our laboratory was hydrolyzed to water soluble, luminescent poly[2-(3-thienyl)ethanol hydroxycarbonyl-methyl urethane] (H-PURET). Multilayer films of H-PURET-Eu(III) complexes were successfully fabricated by alternatively dipping the substrates in aqueous H-PURET and EuCl3 solutions.

EXPERIMENTAL

Materials:

PURET was prepared according to the literature.[12] Europium chloride(III) hexahydrate and other chemicals and organic solvents used in this study were obtained from Aldrich Chemical Co. Inc., and were used without further purification.

Preparation of poly[2-(3-thienyl)ethanol hydroxycarbonyl-methyl urethane] (H-PURET):

PURET polymer was dissolved in tetrahydrofuran (THF) (1.0 g in 100 mL) and was transferred into a round-bottomed flask equipped with a reflux condenser. The temperature of the bath was raised to about 40 °C. Saturated NaOH solution in methanol (about 10 drops) was added to the PURET solution with vigorous agitation. A bright red precipitate appears upon adding sodium hydroxide solution. The precipitate was filtered and was washed with THF. The precipitate is the sodium salt of hydrolyzed PURET (H-PURET) and is soluble in water. H-PURET was precipitated from water by reducing the pH of the solution and filtered. The precipitate was washed with water and dried under vacuum. H-PURET is soluble in DMF, NMP and water at high pH. However, once the polymer is dissolved in water, the pH can be reduced to about 3.0 without precipitation. The number- and weight- average molecular weights were calculated as 42,800 and 276,000 g/mol respectively, using Gel Permeation Chromatography (GPC-Waters Model 510 pump, Model 410 refractive index detector, and model 730 module with 500-10^5 Å Ultrastyragel columns in series), relative to polystyrene standards. The column was injected with 100 μL of the polymer solution in DMF (1 mg/mL) and was eluted with DMF. The hydrolysis of PURET is shown in Figure 1.

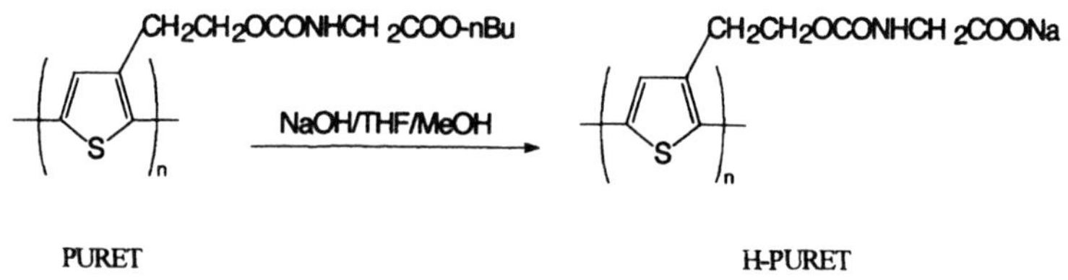

Figure 1   Preparation of H-PURET

Layer-by-Layer Complex Deposition of Polymer-Metal complexes:

The concentrations of all polymer solutions were calculated based on their repeating units. A transparent indium-tin-oxide coated glass slide surface rendered to a hydrophilic and a molecular layer of polycation was adsorbed onto the substrate by electrostatic interactions.[13] The slide was dipped into an aqueous solution of a bidentate ligand functionalized thiophene polymer and a molecular layer of it was adsorbed onto the slide by well-established electro-static interaction forces. This type of multilayer assembly using electrostatic interactions has been well studied.[13] The exposed bidentate ligands on the slide was treated with Eu^{3+} metal ion solution followed by alternating treatment of the slide with polymer and metal ion solutions at appropriate conditions to form stable complex multilayers.

Characterization of multilayer films:

The infrared (IR) spectra were recorded on a Perkin-Elmer 1760X FT-IR spectrometer. The UV-Vis spectra were recorded using a GBC UV/VIS 916 spectrophotometer. The emission spectra were measured with SLM-AMINCO Model 8100 spectrofluometer from 500 nm to 700 nm.

EL device fabrication:

Polyaniline (PANI) was incorporated as a hole-transporting layer into the device. PANI was dissolved in trifluoro acetic acid (TFA, 2% wt/vol), filtered through 0.2 μm Gelman filters and was spin-coated on ITO glass slides (1000 Å thickness; surface resistivity of 10 Ωcm^2). H-PURET (5 mM solution in water) was deposited onto this PANI coated ITO slide by spin coating (350 Å thickness). The ITO-coated glass slides were patterned before coating PANI and H-PURET, as shown in Figure 2. Finally, a 500 Å thick Al film was deposited on top of the PURET by thermal evaporation to form the LED device as shown in Figure 2. The ITO film served as anode (+) and the aluminum electrode was used as cathode (-), during EL measurements.

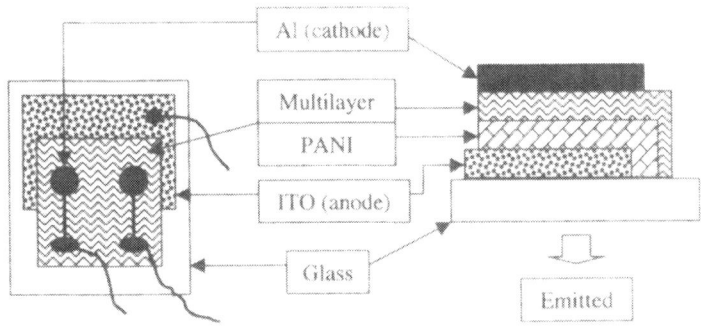

Figure 2 Polymer light emitting diode (LED) configuration.

RESULTS AND DISCUSSION

Visible Absorption Studies of Multilayer films:

Layer-by-layer conjugate polymer-metal ion complexation process can be used to build very thick layers by alternatively dipping the desired substrates in metal ion and polymer solutions. We have fabricated multilayer films using H-PURET solution in water (5 mM; at pH 6.5) and europium chloride solution (5 mM; pH 5.5) at 50°C. The absorbance increased monotonically up to 200 bilayers (Figure 3). Much thicker films can be obtained if the experiments were carried out at higher temperatures and at optimized pH.

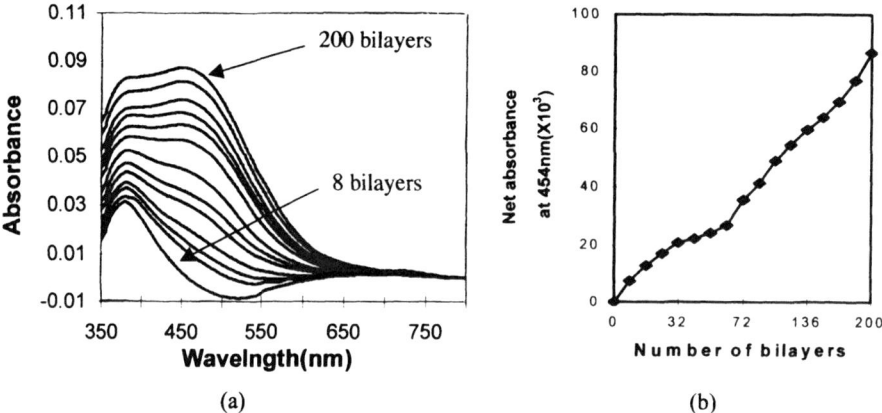

(a)   (b)

Figure 3. (a) Absorption of multilayer films deposited on ITO. (from 8 to 200 bilayers)
(b) Net absorbance (absorbance of multilayer – absorbance of priming layer) vs. number of bilayers

Infrared spectrometry (IR) was used to characterize the complex formation between H-PURET and Eu^{3+} ions The strong absorption bands of 1580 cm^{-1} and 1420 cm^{-1} in IR spectrum of multilayer film shows asymmetric and symmetric stretching of coordinated carboxylate, which are not found in H-PURET itself. (Figure 4)

Photoluminescence of multilayer films:

Fluorescence behavior of a multilayer film was measured as shown in Figure 5. Fluorescence spectra of spin-coated and multilayer films show that there is no contribution of europium or europium complex to the fluorescence of multilayer film. Although carboxyl groups are well-known bidentate ligands[14], their complexes with metal ions are not highly luminescent. High luminescence in metal ion complexes is achieved only, when energy transfer from excited ligand molecules to the metal ions takes place [ligand-to metal energy transfer (LMET)]. Emission maxima and peak shapes of two solid films are almost identical, but the emission spectrum of solution is quite different presumably due to shorter conjugation length in solution.

Electroluminescence of multilayer films:

EL of multilayer films was measured in a black painted box to prevent background light and under nitrogen environment to avoid oxidation (Figure 6). The observed EL signal was not very high, which can be explained based on the poor energy transfer between carboxylate ligands and europium metal ions in the fabricated devices. Enhanced electroluminescence signal can be achieved if we use appropriate ligands that help in ligand-to-metal energy transfer (LMET)].

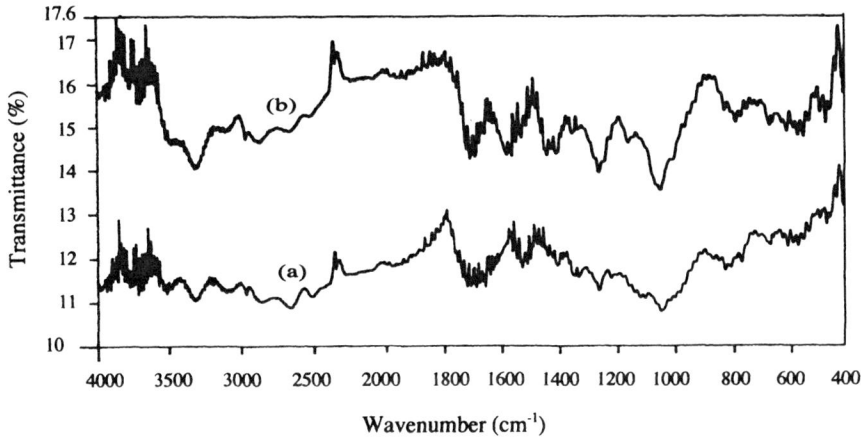

Figure 4 Infrared spectrum of (a) H-PURET and (b) multilayer films of H-PURET and Eu³⁺.

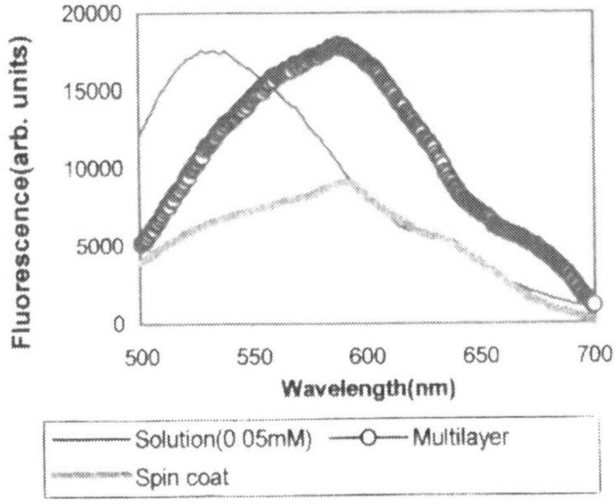

Figure 5   Fluorescence of solution, multilayer, and spin-coated film of H-PURET.

Solution: 0.05mM, pH 7.5
   λ_ex : 384nm, λ_em, max : 528nm
Multilayer film: 20 bilayers on slide glass
   λ_ex : 387nm, λ_em, max : 591nm
Spin-coated film: 680Å on slide glass
   λ_ex : 389nm, λ_em, max : 590nm
(Each spectrum was manipulated to be compared, therefore, fluorescence intensity of each has no meaning)

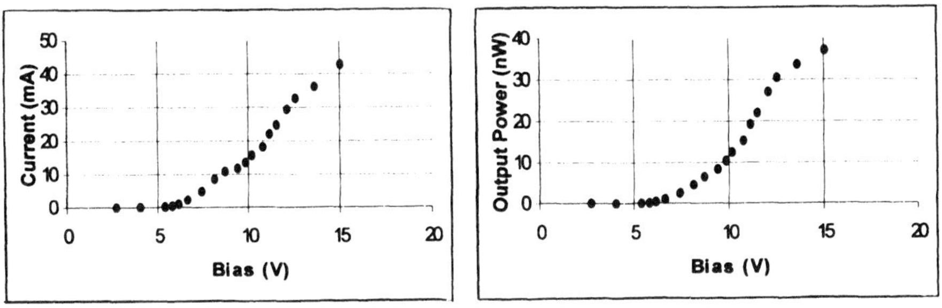

Figure 6   Electroluminescence properties of multilayer (12 bilayers) under forward bias condition.

CONCLUSION

In this research, we have successfully developed a technique to fabricate multi-layers of H-PURET-Eu(III) complexes employing layer-by-layer complexation technique. It was accomplished by alternatively dipping the substrates in aqueous H-PURET and EuCl$_3$ solutions. FT-IR spectra showed the presence of coordination bonding between ligands and Eu^{3+} ions. Light emitting diodes fabricated using multilayer films showed low luminescence due to low energy transfer between ligands and metal. The electroluminescence observed is primarily from the polymer. We believe the luminescence property can be improved through the selection of appropriate ligands, for example, β-diketone and hydroxyquinoline, which aid in ligand-to-metal energy transfer process (LMET). Various lanthanide and transition metal ions can be used with macromolecular bidentate ligands to achieve blue, green and red emission.

Acknowledgments   We thank Steve Kumiega and Arvind Viswanthan for GPC measurements. K.G.C. and M.J.C. acknowledge financial support from NSF (SBIR). Support from Samsung Central Research Institute of Chemical Technology is also gratefully acknowledged.

REFERENCES

1.   C. W. Tang and S. A. VanSlyke, Appl. Phys. Lett. **51**, 913 (1987).

2.   C. W. Tang, S. A. VanSlyke, and C. H. Chen, J. Appl. Phys. **65**, 3610 (1989).

3.   C. Adachi, S. Tokito, T. Tsutsui, and S. Saito, Jpn. J. Appl. Phys. **27**, L269 (1988).

4.   Y. Hamada, C. Adachi, T. Tsutsui, and S. Saito, Jpn. J. Appl. Phys. **31**, 1812 (1992).

5.   J. Kido, C. Ohtaki, K. Hongawa, K. Okuyama, and K. Nagai, Jpn. J. Appl. Phys. **32**, L917 (1993).

6.   L. R. Melby, N. J. Rose, E. Abramson, and J. C. Caris, J. Am. Chem. Soc., **86**, 5117 (1964).

7.   J. Dresner, RCA Rev., **30**, 332 (1969).

8.   V.-E. Choong, Y. Park, B. R. Hsieh, C. W. Tang and Y. Gao, Polymer Preprints, **38**, 392 (1997).

9.   J. H. Burroughes, D. D. C. Bradley, A. R. Brown, R. N. Marks, K. Mackay, R. H. Friend, P. L. Burn, and A. B. Holmes, Nature, **347**, 539 (1990).

10.   D. Braun and A. J. Heeger, Appl. Phys. Lett. **58**, 1982 (1991).

11.   P. L. Burn, A. B. Holmes, A. Kraft, D. D. C. Bradley, A. R. Brown, R. H. Friend, and R. W. Gymer, Nature, **356**, 47 (1992).

12.   K. G. Chittibabu, S. Balasubramanian, W. Kim, A. L. Cholli, J. Kumar, and S. K. Tripathy, J. Macro. Sci. A. Pure & Appl. Chem., **A33(9)**, 1283 (1996).

13.   M. Ferreira, O. Onitsuka, A. C. Fou, B. Hsieh, and M. F. Rubner, Mater. Res. Soc. Symp. Proc. **413**, 49 (1996).

14.   A. D. Pomogailo and D. Wöhrle, in Macromolecule-Metal Complexes, edited by F. Ciardelli, E. Tsuchida, and D. Wöhrle (Springer, New York, 1996) p. 60.

OPTICAL PROPERTIES OF DISTYRYLBENZENE CHROMOPHORES AND THEIR SEGMENTED COPOLYMERS

Nicholas Benfaremo[*], Daniel J. Sandman[*], Sukant Tripathy[*], Jayant Kumar[**], Ke Yang [**], Michael F. Rubner[***], Cormac Lyons[***]

[*]Center for Advanced Materials, Department of Chemistry, University of Massachusetts Lowell, Lowell, MA 01854
[**]Center for Advanced Materials, Department of Physics, University of Massachusetts Lowell, Lowell, MA 01854
[***]Department of Materials Science and Engineering, Massachusetts Institute of Technology, Cambridge, MA 02139.

Abstract

A new segmented polymer (**3**) consisting of a distyrylbenzene chromophore separated by polyethylene glycol segments has been prepared by two independent methods: a novel, polymer analogous Mitsunobu reaction and conventional double displacement reaction. The polymer is soluble in a variety of organic solvents, forms excellent, optically clear films and exhibits strong fluorescence. The properties of the chromophore and the polymer, as well as the scope and limitations of the novel Mitsonobu polymerization are presented. Attempts to use polymer (**3**) in electroluminescent devices are also discussed.

Introduction

In recent years, poly(p-phenylenevinylene) (PPV) has been extensively investigated because it exhibits efficient photoluminescence and potentially useful electroluminescent properties.[1] Numerous phenylene vinylene oligomers have been synthesized in order to improve the solubility and stability of these materials as well as to tune their emission frequencies.[2,3,4,5,6,7,8,9] In a light-emitting electrochemical cell (LEC), emission occurs because of hole-electron combination, from electrochemically generated p-type and n-type materials. In this case, good counterion mobility is required. This concept has been used to fabricate numerous devices using poly(ethylene glycol), as the ion transporting medium blended with a photoluminescent material.[10] A potential problem with this approach has been phase separation of a hydrophobic chromophore and the oligo(ethylene glycol).[11] Only one other electroluminescent polymer, in which diethylene glycol units are attached as side chains on a oligo(p-phenlyene) backbone, has so far been reported.[12]

Here we report the first synthesis of a polymer containing a photoluminescent distyrlbenzene chromophore separated by poly(ethylene

533

glycol) spacers. This polymer exhibits efficient solution fluorescence and was characterized by ^{1}H NMR, ^{13}C NMR, IR, elemental analysis, X-ray powder diffraction, TGA and DSC.

EXPERIMENTAL SECTION

Materials. All solvents and materials were used as received unless otherwise noted. α, α'-Para xylylene dichloride and polyethylene glycol were purchased from Aldrich. Poly(ethylene glycols) were dried at 45°C under vacuum over P_2O_5 for a minimum of three days. Tetrahydrofuran (THF) was distilled from sodium under inert atmosphere. All reactions were performed under a dry argon atmosphere.

Chromophore. The distyrylbenzene chromophore (**2**) was synthesized by direct condensation of the bis phosphonate (**1**) with p-hydroxybenzaldehyde using excess sodium hydride in DMF (Scheme 1).[13] The expected all trans configuration was confirmed by ^{1}H NMR and IR spectroscopy.[14] The chromophore itself is a yellow, crystalline material, very sparingly soluble in most organic solvents at room temperature. It exhibits a fluorescence in the solid state and even very dilute solutions exhibit an extremely strong blue fluorescence (0.188 g/L, λ_{max} = 395 nm).

Scheme 1

Synthesis of the Distyrylbenzene Polymer

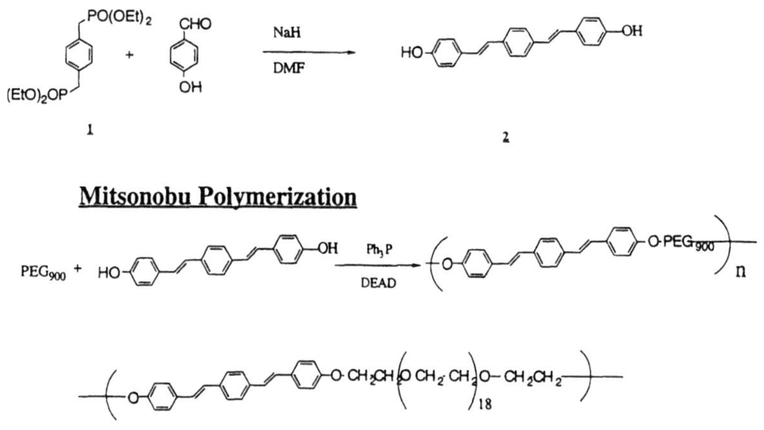

Mitsonobu Polymerization

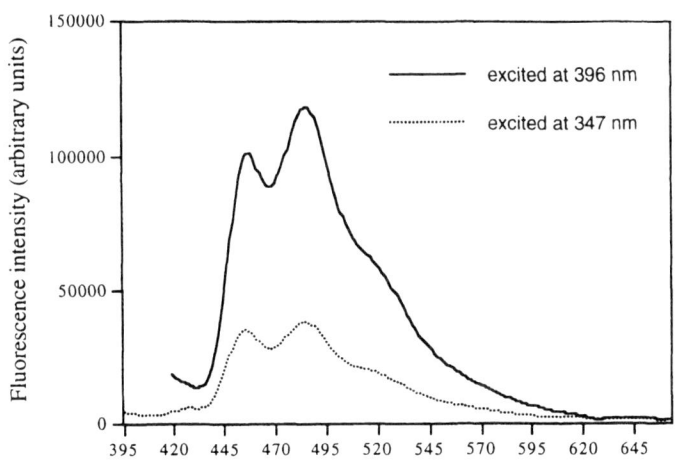

Figure 1. Emission spectra. Both spectra have peaks at 455 nm and 484 nm.

Polymer. The commercially available PEG that was used in this study (Mn 900, Pd 1.1) corresponds to approximately twenty ethylene glycol units. Polycondensation of the distyrlbenzene segment (**2**) with poly(ethylene glycol) was initially carried out via a Mitsunobu reaction and are the first examples of what seems to be a new and useful polymerization method [15,16]. Upon addition of diethyl azodicarboxylate (DEAD) to a suspension of the chromophore to a solution of poly(ethylene glycol) and triphenyl phosphine in THF at 45°C, the chromophore slowly dissolves. Extended heating over 2 days gave a viscous solution. Precipitation into ethanol and drying under vacuum over P_2O_5 gives the desired polymer as a yellow rubbery powder in 90% yield.

RESULTS AND DISCUSSION

To our knowledge, this is the first example of the use of the Mitsunobu reaction to carry out a condensation polymerization. We have used this reaction to also prepare a number of end functionalized poly(ethylene glycols) and poly(ethylene glycol) copolymers that we will report in subsequent publications. It should be noted, however, that large significant IR absorptions at 1740 and 1700 cm^{-1} are indicative of carbonyl groups. Together with elemental analyses and comparison to the IR spectrum of diethyhydrazine dicarboxylate in the 1700 to 1800 cm^{-1} region, the incorporation of hydrazino carboxylate residues in the Mitsunobu prepared polymer was indicated. Preparation of a sample by

535

an alternate route (Scheme 2) from poly(ethylene glycol) bis(mesylate)[17] gave a product with no absorptions in the carbonyl region.

Scheme 2
Alternate Polycondensation

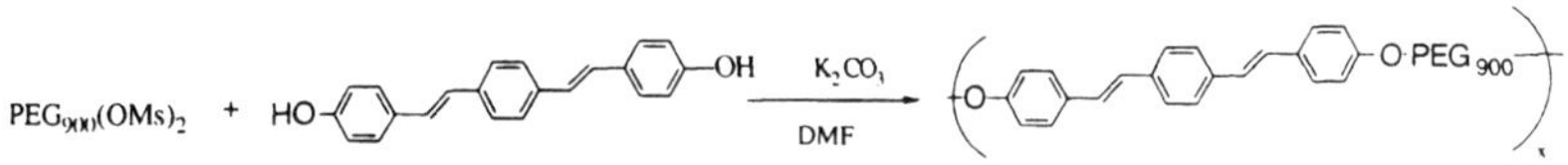

The polymer is soluble in THF, DMF, chloroform and acetonitrile. In addition, the lack of any detectable peak at 4.56 ppm indicated <1.0% of ethylene glycol end groups.[18] The ratio of peaks from the chromophore to those from the ethylene glycol segment is also in good agreement with a 1:1 copolymer. The known insolubility of the chromophore suggested that we have a reasonably high molecular weight product. Size Exclusion Chromatography (SEC) analysis of the polymer in chloroform gave good agreement with commercially available PEG samples and indicates a clean monomodal distribution with M_n 7400, M_w 18,700, P_d 2.5. The peak molecular weight is 10,800 and corresponds to a DP of 9.

The polymers synthesized were investigated for their fluorescence properties. In DMF solution, when excited in the wavelength range 320-400 nm, the polymer exhibited strong fluorescence in the range 410-470 nm. When pumped with a 337 nm nitrogen 3 nsec laser pulse, a solution of the polymer exhibited laser action due to the amplified spontaneous emission. The intensity of this laser action is comparable when only the chromophore at comparable number densities was used. This clearly establishes that the chromophore does not have any significant loss in quantum efficiency when incorporated into polyethylene oxide segments. The absorption spectrum of a spin coated film (1129 Å) has a maximum at 318 nm. The fluorescence spectra of a cast film on a quartz substrate is shown in Figure 1. The structure observed in both absorption and fluorescence emission spectra is attributed to vibronic effects and has a frequency separation of approximately 1300 cm^{-1}. A thin film of the polymer also showed very strong fluorescence. Stimulated emission was observed when a DMF solution was pumped with a nitrogen pulse laser (337 nm). The observed emission (435 nm) showed a narrow bandwith of 3 nm (FWHM). Efforts are underway to achieve the same effect with a film.

Thermal Properties.

The thermogravimetric analyses (TGAs) of the polymer in air and under N2 show a gradual weight loss above 50°C, 98.18% of the material is still present at 258°C and nearly 95% at 336°C (under N_2). Mates and

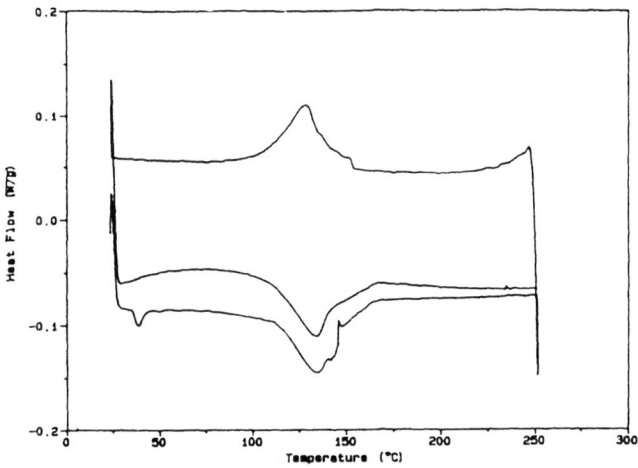

Figure 2 DSC of the Polymer (**3**)

Ober[19] prepared a number of alkyl derivatives of the same chromophore and showed that several of them demonstrated liquid crystalline behavior. The DSC trace (Figure 2) shows a transition at approximately 55°C which is not observed in subsequent traces. This indicates a possible liquid crystalline transition.

Device Experiments

We have made several attempts to incorporate (**3**) into a light-emiting device. Attempts to use (**3**) in a LEC have not been successful to date. On top of a spin coated film of (**3**) on an indium tin oxide (ITO) slide was coated a layer (300-400 Å) of a sulfonated tris-phenanthroline Ru^{+2} complex that has been previously used in light emitting devices.[20] Red orange light similar to that previously reported is obtained in these devices at voltages up to +6 volts with a luminance of 40 candelas/m^2. The current density at this voltage is 50 ma/cm^2. Use of higher voltages led to irreversible malfunctioning of the devices.

CONCLUSIONS

We have prepared a novel PPV analog containing a known ion transport moiety (PEG) by both a novel Mitsunobu reaction and by a convention double displacement reaction. The polymer was fully characterized and is easily processible yielding optically clear, highly fluorescent films. Laser action, with an emission bandwidth of 3 nm, has been observed in solution. Studies of light emitting devices using (**3**) have been initiated.

ACKNOWLEDGEMENTS

The authors thank Milton J. Downey for the X-ray powder diffraction data of the polymer. This work was supported, in part, by a Multidisciplinary University Research Initiative (MURI) grant from the Office of Naval Research and by the Air Force Office of Scientific Research (GRANT 49620-95-1-0179).

REFERENCES

1   J. Travis, Science, 263, 1700 (1994) and references therein.
2   U. Stalmach, H. Kolshorn, I. Brehm, H. Meier, Liebigs Ann. 1996, 1449.
3   R. Cumming, R.A. Gaudiana, K. Hutchinson, E. Kolbe, R. Ingwall, P. Mehta, R.A. Minns, C.P. Petersen, D. Waldman, J. Mater. Sci.-Pure Appl. Chem. A33(9), 1301 (1996).
4   M. Hay, F.L. Klavetter, J. Am. Chem. Soc. 117, 7112 (1995).
5   Z. Yang, I. Sokolik, F.E. Karasz, Macromolecules 26, 1188 (1993).
6   A. Greiner, B. Bolle, P. Hesemann, J.M. Oberski, R. Sander, Macromol. Phy. and Chem. 197, 113 (1996).
7   R.E. Gill, P.F.van Hutten, A. Meetsama, G. Hadziioannou, Chem. Mater. 8, 1341 (1996).
8   V. Stalmach, H. Kolshorn, I. Brehn, H. Meier, Liebigs, Ann. 1996, 1449.
9   F. Hide, M.A. Díaz-García, B.J. Schwartz, A.J. Heeger, Acc. Chem. Res. 30, 430 (1997).
10 Q. Pei, Y. Yang, G. Yu, C. Zhang, A.J. Heeger, J. Am. Chem. Soc. 118, 3922 (1996).
11 M.F. Rubner, C. Lyons, Manuscript in preparation.
12 Q. Pei, Y. Yang, J. Am. Chem. Soc. 118, 7416 (1996).
13 R.S. Tewari, N. Kumari, P.S. Kendurkar, Ind. J. Chem. 15B, 753 (1977).
14 Z. Yang, H.J. Geise, M. Mehbod, G. Debrue, J.W. Visser, E.J. Sonneveld, L. Van't dack, R. Gijbels, Syn. Metals 39, 137 (1990).
15 O. Mitsunobu, Synthesis 1981, 1.
16 D.L. Hughes, Organic Reactions, edited by Leo A. Paquett (John Wiley and Sons, New York, 1992), vol. 42, p. 335-656.
17 J.M. Harris, E.C. Struck, M.G. Case, M.S. Paley, M. Yalpani, J.M. Van Alstine, D.E. Brooks, J. Poly. Sci.: Poly. Ed. 22, 341 (1984).
18 J.M. Dust, Z-h. Fang, J.M. Harris, Macromolecules 23(16), 3742 (1990).
19 T.E. Mates and C.K. Ober, J. Poly. Sci.: Part C: Poly. Lett. 28, 331 (1990).
20 J-K. Lee, D.S. Yoo, E.S. Handy, M.F. Rubner, Appl. Phys. Lett. 69, 1686 (1996).

LIGHT-EMITTING DIODES WITH VOLTAGE-SWITCHABLE COLORS FROM SEMICONDUCTING POLYMER/POLYMER HETEROJUNCTIONS

XUEJUN ZHANG, SAMSON A. JENEKHE
Departments of Chemical Engineering and Chemistry, University of Rochester,
Rochester, New York 14627-0166

ABSTRACT

Reversible electroluminescence color changes with applied voltage have been observed in light-emitting diodes fabricated from semiconducting polymer heterojunctions consisting of an electron transporting polybenzobisthiazole and hole transporting poly(p-phenylene vinylene) when layer thicknesses are less than 60–100 nm. Enhanced device performances such as lower turn-on voltage and higher efficiency and luminance were also obtained compared to single-layer devices. The observed voltage-switchable emission colors in these nanoscale heterojunction light sources can be understood in terms of spatial confinement effects which are related to field-dependent charge transport and trapping processes in the materials. These results also demonstrate the use of new high temperature rigid-rod polymers as electron transport and emissive layers in electroluminescent devices.

INTRODUCTION

Semiconducting polymer/polymer heterojunction structures [1-4] have been used to improve the performances of polymer light-emitting diodes (LEDs) since the first report on conjugated polymer LED [5]. Most of the reported electroluminescent (EL) conjugated polymers are p-type (hole-transport) materials, such as poly(p-phenylene vinylenes) (PPV), polythiophenes and poly(phenylenes). Recently, n-type (electron-transport) polymers were reported as the emissive layer or the electron transporting layer in heterojunction LEDs [1-4]. Polymer/polymer heterojunction light-emitting diodes showed balanced electron and hole injection and transport, and hence improved performance of the devices. Yamamoto *et al.* [2] reported the use of an n-type polymer, poly(2,3-diphenylquinoxaline-5,8-diyl), as the emissive layer, and by introducing a p-type hole transporting layer, the electroluminescence efficiency increased about two orders of magnitude. Greenham *et al.* [1] showed the use of a high electron affinity cyano-PPV derivative to form a heterostructure with PPV, resulting in up to 4% EL quantum efficiency. O'Brien *et al.* [4] demonstrated the use of poly(phenyl quinoxaline) as the electron transport layer to give a ten-fold enhanced quantum efficiency compared with the corresponding single layer devices. All these polymer/polymer heterojunction LEDs showed EL emission only from either the n-type layer [1,2] or the p-type layer [4], and no voltage-tunable EL color was observed as reported for polymer blends [6] and organic multilayer devices [7]. Recently we reported [3] finite size effects on electroluminescence of p-type/n-type semiconducting polymer heterojunction LEDs, in which PPV was used as the p-type layer, and a series of polyquinolines were used as the n-type layer. It was found that the PPV/polyquinoline heterojunction light-emitting diodes switch colors reversibly (orange/red ↔ green) with applied voltage and showed enhanced performance compared to either single layer devices or bilayer devices with only single-layer emission.

Organic hole-transporting and electron-transporting molecules, such as N,N'-diphenyl-N,N'-bis(3-methylphenyl)-1,1'-biphenyl-4,4'-diamine (TPD) and 2-(4-biphenyl)-5-(4-*tert*-butylphenyl)-1,3,4-oxadiazole (PBD), respectively, which have been widely used as charge transport materials in LEDs suffer from their low glass transition temperature or propensity to crystallize. Conjugated rigid-rod polymers, however, have the advantages of high temperature stability, ease of fabrication into large area thin film devices by solution spin-coating, and morphological stability. In this paper, we report semiconducting polymer heterojunction light-emitting diodes with voltage tunable colors using PPV as the p-type layer and polybenzobisthiazoles (PBTPV and PPyBT) as the n-type layer. Poly(benzobisthiazole-1,4-phenylenebisvinylene) (PBTPV) and poly(2,5-pyridylene benzobisthiazole) (PPyBT) are two examples of n-type semiconducting polymers synthesized in our laboratory. Their synthesis, characterization, optical, electrochemical, and charge photogeneration properties have been reported elsewhere [8]. Their n-type characteristics were revealed by electrochemical studies (cyclic voltammetry) [8]. The chemical structures of PPV, PBTPV, and PPyBT are shown in Figure 1.

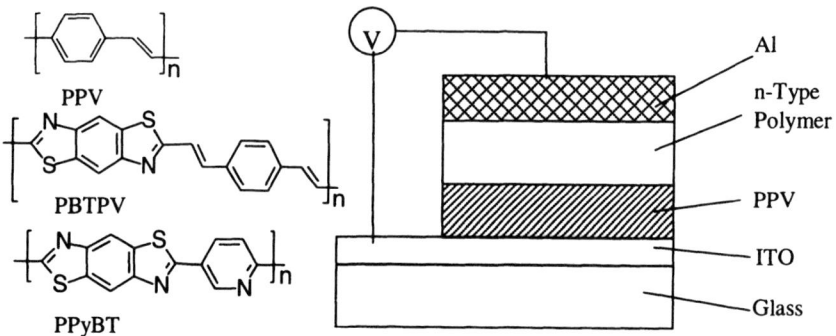

Figure 1. Chemical structures of PPV, PBTPV, and PpyBT, and
the schematic of the heterojunction light-emitting diode.

EXPERIMENTS

The semiconducting polymer heterojunction LEDs were prepared and investigated as sandwich structures between aluminum and indium-tin-oxide (ITO) electrodes as shown in Figure 1. The PPV thin films were deposited onto ITO coated glass by spin coating of the sulfonium precursor from methanol solution followed by thermal conversion in vacuum (250 °C for 1.5 hr). Thin films of PPyBT and PBTPV were spin coated on the PPV layer from their reversibly soluble Lewis acid (GaCl₃) coordination complexes in nitromethane[9]. The film thickness was measured by an Alpha-step profilometer (Tencor Northern) with an accuracy of ±1 nm and confirmed by an optical absorption coefficient technique. Finally, 100–130 nm aluminum electrodes were vacuum ($< 10^{-5}$ torr) evaporated onto the resulting polymer bilayers. Electroluminescence spectra were obtained by using a calibrated Photo Research Model PR-60 photo-colorimeter or using a Spex Fluorolog-2 spectrofluorimeter. Current-voltage-luminance curves were recorded simultaneously by hooking up an HP4155A semiconductor parameter analyzer together with a Grasby S370 optometer equipped with a calibrated luminance sensor head. The EL quantum efficiency of the LEDs was estimated by using procedures similar to that previously reported [10]. All the fabrication and measurements were done under ambient laboratory conditions.

RESULTS

Figure 2 shows the optical absorption and photoluminescence spectra of PBTPV and PPyBT and the absorption spectrum of PPV. PBTPV has absorption peaks at 475 nm and 508 nm and absorption band edge at 2.1 eV. The emission peak of PBTPV is 630 nm. PPyBT has absorption peaks at 440 and 470 nm and absorption band edge at 2.48 eV. The emission peak of PPyBT is 560 nm. The absorption band edge of PPV is well known to be 2.4 eV [1]. Clearly, there is little overlap between the absorption spectrum of PPV and the emission spectra of PBTPV and PPyBT. Therefore, the emission from the n-type layer will not be absorbed by the p-type layer according to the heterojunction configuration illustrated in Figure 1.

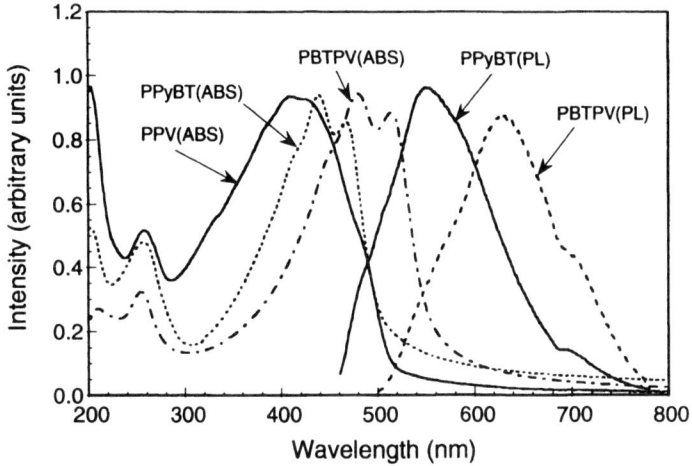

Figure 2. Optical absorption and photoluminescence spectra of PPV, PBTPV and PPyBT.

Figure 3(a) shows representative voltage-switchable electroluminescence spectra for the ITO/PPV(25 nm)/PBTPV(35 nm)/Al device which exhibits a turn-on voltage of 3.5 V. At lower voltages (3.5–17 V), the emission is from the PPV layer, showing its characteristic green EL color; at higher voltages (> 17 V), the emission from the PBTPV layer becomes dominant, showing orange/red EL color. This novel voltage-switchable electroluminescence color is reversible. We also observed that, when the relative thicknesses of the n-type layer and the p-type layer were changed, the emission zone can be within either the p-type layer or the n-type layer. For example, we found that the device ITO/PPV(25nm)/PBTPV(50nm)/Al shows EL emission from only the PBTPV layer at all forward bias voltages as shown in Figure 3(b). Similar results were observed in PPyBT/PPV heterojunction devices. For example, a green EL color was observed in the PPyBT(30 nm)/PPV(25 nm) device; in the PPyBT(65 nm)/PPV(25 nm) device, the EL emission is orange; whereas the PPyBT(40 nm)/PPV(25 nm) device showed strong voltage dependence of EL color going from orange (< 13 V) to green (> 15 V). The observed novel voltage-switchable electroluminescence color

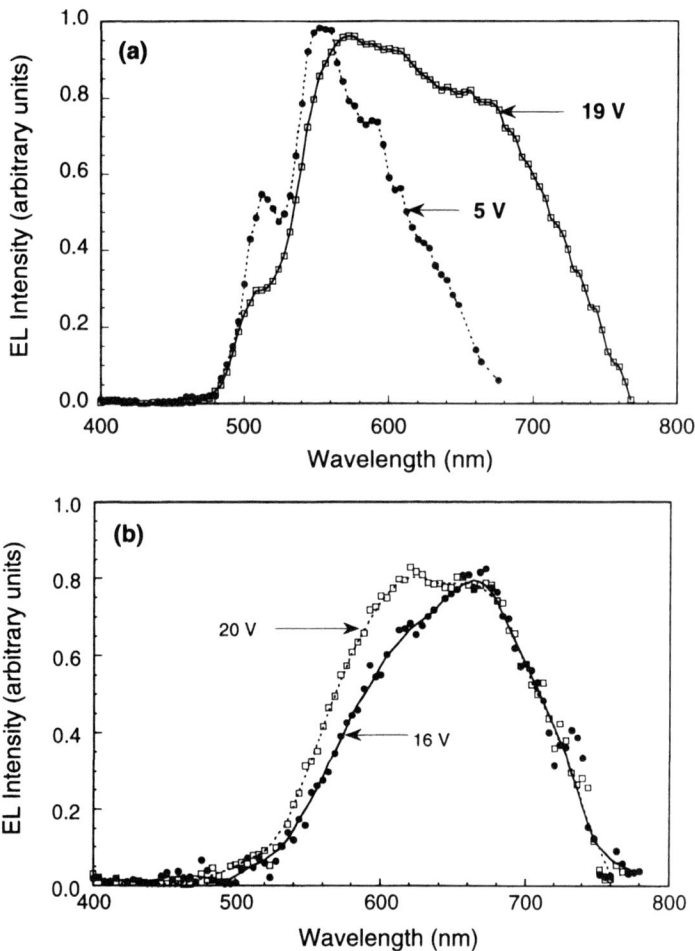

Figure 3. Electroluminescence spectra for: (a)the device ITO/PPV(25 nm)/PBTPV(35 nm)/Al, showing reversible voltage-tunable color emission; (b) the device ITO/PPV(25nm)/PBTPV(50nm)/Al, showing the emission only from the PBTPV layer.

is related to the field-dependent charge carriers transport and trapping processes in the polymers as well as the electric-field-dependent bipolar charge transfer characteristic of the polymer/polymer interface [3]. The ranges of charge carriers (electrons and holes) can be expressed as $x = \mu\tau E$, where x is the range of charge carriers, μ the mobility, τ the lifetime, and E the electric field. It can be seen that the ranges of charge carriers can be regulated by the electric field through the applied voltage. If the range of electrons is smaller than the thickness of the n-type layer, the polymer/polymer interface is unipolar (only holes can cross the interface) and the emission from only the n-type layer is obtained. Similar arguments apply to the p-type layer. If the thicknesses of n-type and p-type layers are appropriate, the polymer/polymer interface is unipolar at low field and bipolar (both electrons and holes can cross the interface) at high field and voltage-switchable electroluminescence color can be achieved.

The best results in terms of luminance were obtained by applying a thin layer (15 nm) of PBTPV or PPyBT as the electron transport layer and PPV (60 nm) as the emissive layer. A turn-on voltage of ~5V, luminance of ~100 cd/m^2, and quantum efficiency of ~0.04% were obtained for the ITO/PPV(60nm)/PPyBT(15nm)/Al device compared to the single layer device ITO/PPV(60 nm)/Al with turn-on voltage of ~10 V, luminance of ~10 cd/m^2 and quantum efficiency of ~0.006%. Figure 4 shows the luminance-electric field characteristics of various devices. It can be seen that all the heterojunction devices have lower turn-on voltages than the PPV single layer devices even though the bilayer devices have a larger total thickness. For example, the ITO/PPV(25nm)/PBTPV(35nm)/Al turns on at 2.4×10^5 V/cm which corresponds to ~3.5 V; the ITO/PPV(60nm)/PPyBT(15nm)/Al turns on at ~6.7×10^5 V/cm which corresponds to ~5 V; and the single layer device ITO/PPV(60nm)/Al turns on at ~1.7×10^6 V/cm which corresponds to ~10 V.

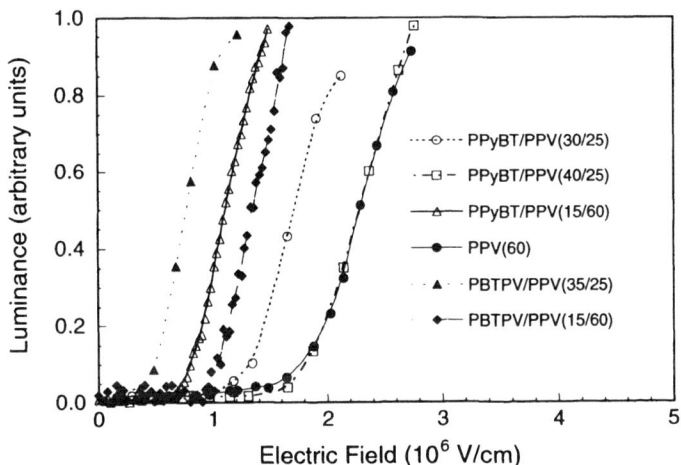

Figure 4. Luminance-electric field characteristics for various devices, the numbers in the parentheses are film thicknesses in nanometers.

Voltage-tunable polymer LEDs have previously been reported for blends of polythiophene derivatives [6]. The voltage-tunable behavior arises from the phase-separated morphology of the blends coupled with radiative and nonradiative energy transfer processes [6]. Such energy transfer processes can prevent EL emission from the higher energy polymer. The morphology of polymer blends strongly depends on the compositions and processing conditions. In polymer/polymer heterojunction devices, we only need to control the thicknesses of the two layers in order to get voltage-switchable EL colors. Because PPV is insoluble in the solvent of the n-type polymer, the polymer/polymer interfaces in these heterojunctions are well-defined. In polymer/polymer heterojunction LEDs, there is little or no energy transfer because of negligible overlap between the absorption spectrum of PPV and the emission spectra of the n-type polymers and the small interfacial area. Light sources with voltage-tunable colors offered by polymer heterojunctions may have applications in multicolor and full-color displays as well as signaling/control systems analogous to traffic control lighting.

CONCLUSIONS

We have demonstrated the use of new high temperature, conjugated rigid-rod polymers as electron transport and emissive layers in semiconducting polymer heterojunction LEDs. It was found that reversible voltage-switchable LED colors can be achieved from n-type/p-type polymer heterojunctions by controlling the thicknesses of both layers. Furthermore, the EL emission can be from either the n-type or the p-type layer depending on the thicknesses of those layers. The observed voltage-switchable EL colors in these nanoscale heterojunction LEDs are related to the field-dependent charge transport and trapping processes in semiconducting polymers and result from the small and field-dependent ranges of charge carriers. The performances of the heterojunction LEDs were greatly improved compared to single-layer devices. One order of magnitude enhancement in luminance and a turn-on voltage as low as 3.5 V were observed in heterojunction devices compared to PPV single-layer devices.

ACKNOWLEDGMENTS

This research was supported by the Office of Naval Research and National Science Foundation (CTS-9311741, CHE-9120001). We thank J. A. Osaheni, A. K. Alanko, and B. R. Hsieh for providing materials, and V.-E. Choong and Y. Gao for assistance in device fabrication.

REFERENCES

1. N. C. Greenham, S. C. Moratti, D. D. C. Bradley, R. H. Friend, and A. B. Holmes, Nature **365**, p.628 (1993).
2. T. Yamamoto, T. Inoue, and T. Kanbara, Jpn. J. Appl. Phys. **33**, p.L250 (1994).
3. S. A. Jenekhe, X. Zhang, X.L. Chen, V.-E. Choong, Y. Gao, and B. R. Hsieh, Chem. Mater. **9**, p.409, (1997).
4. D. O'Brien, M. S. Weaver, D. G. Lidzey, and D. D. C. Bradley, Appl. Phys. Lett. **69**, p.881 (1996).
5. J. H. Burroughes, D. D. C. Bradley, A. R. Brown, R. N. Marks, K. Marckay, R. H. Friend, P. L. Burns, and A. B. Holmes, Nature **347**, p.539 (1990).
6. M. Berggren, O. Inganas, G. Gustafsson, J. Rasmusson, M. R. Andersson, T. Hjertberg, and O. Wennerstrom, Nature **372**, p.444(1994).
7. J. Kalinowski, P. Di Macro, M. Cocchi, V. Fattori, N. Camaioni, and J. Duff, Appl. Phys. Lett. **68**, p.2317 (1996).
8. (a) J. A. Osaheni and S. A. Jenekhe, Macromolecules **26**, p.4726 (1993); (b) J. A. Osaheni and S. A. Jenekhe, Chem. Mater. **7**, p.672 (1995); (c) A. K. Alanko and S. A. Jenekhe, manuscript in preparation.
9. S. A. Jenekhe and P. O. Johnson, Macromolecules **23**, p.4419 (1990).
10. N. C. Greenham, R. H. Friend, and D. C. C. Bradley, Adv. Mater. **6**, p.491 (1994).

TOWARD ELECTRICALLY PUMPED ORGANIC DIODE LASERS: ELECTROLUMINESCENCE OF PROTON TRANSFER POLYMERS

XUEJUN ZHANG[1], RICHARD M. TARKKA[1], SAMSON A. JENEKHE[1],
JOSEPH B. SCHLENOFF[2]
[1]Departments of Chemical Engineering and Chemistry, University of Rochester,
Rochester, NY 14627-0166.
[2]Department of Chemistry, The Florida State University, Tallahassee, FL 32306-3006.

ABSTRACT

Electroluminescence from a series of polymers and copolymers which exhibit photoinduced excited-state intramolecular proton transfer (ESIPT) is reported. Among these polymers containing ESIPT moieties in the main chain or side chain are poly(1,4-(2-hydroxy)phenylene benzobisthiazole-co-decamethylene benzobisthiazole), poly(styrene-co-3-hydroxy-4'-ethenylflavone) and poly(2,5-benzoxazolediyl(2,2,2,-trifluoro-1-(trifluoromethyl) ethylidene)-5,2-benzoxazolediyl-4-hydroxy-1,3-phenylene). The electroluminescence spectra of these polymers are identical to their photoluminescence spectra, indicating that electroluminecence originates from electrically-generated intramolecular proton transfer (EGIPT). Because electroluminescence from EGIPT implies electrically-induced population inversion, these results demonstrate the possibility of electrically-pumped organic solid-state laser diodes.

INTRODUCTION

Photoinduced excited-state intramolecular proton transfer (ESIPT) in heterocyclic aromatic molecules has been widely studied [1]. Molecules exhibiting ESIPT, such as 2-(2-hydroxyphenyl)benzimidazole (HBI) [2], 2-(2-hydroxyphenyl)benzothiazole (HBT) [3], 2-(2-hydroxyphenyl)benzoxazole (HBO) [4], and 2-(2-hydroxyphenyl)benzotriazole (HPB) [5], show a broad and structureless emission with a large Stokes shift that is characteristic of the keto tautomer. In most cases, the fluorescence from the primary excited state (enol form) is quenched due to the very rapid (< 1 ps) photoinduced proton transfer. Molecules exhibiting ESIPT have been used as photostabilizing additives to protect polymers from photochemical degradation [6,7].

The inherent population inversion that ocurs in molecules exhibiting ESIPT was early reconized and proposed as the basis of new laser materials [8]. Proton transfer dye lasers have been demonstrated using salicylamide [9], sodium salicylate [10], and HBI [10] in nonpolar solvents. Optically-pumped solid state dye lasers which show significantly increased lasing efficiency and dye photostability have been realized by covalently attaching HBI derivatives onto poly(methyl methacrylate) (PMMA) chains compared to physical blends of the HBI with PMMA [11].

There is a growing interest in applications of conjugated polymers in optoelectronic devices such as light-emitting diodes [12]. Incorporation of proton transfer moieties into conjugated polymers might lead to new electroluminescent materials with the thermal stability of conjugated polymers and the photochemical stability of proton transfer molecules. Systematic studies on the effects of molecular size, conjugation length, and competition with excimer formation suggested that extended conjugation is the most important factor that can inhibit the proton transfer in such polymers [13,14]. The results also showed that molecular size does not inhibit ESIPT and that the barrier to ESIPT can be lower than the barrier to excimer formation in polymers exhibiting ESIPT [13,14].

The successful development of semiconducting polymer light emitting diodes (LED) [12] and recent demonstration of optically-pumped lasers [15] in the same class of materials have now focused interest in the prospects of making electrically-pumped laser diodes [15-17]. The question of whether electrical pumping can produce the necessary population inversion in organic semiconductors has become of central importance to this quest. The recent report of electrically-generated intramolecular proton transfer (EGIPT) and stimulated emission from proton transfer polymers has provided the first affirmative answer to this question [17]. In this paper we report investigation of the EGIPT process in several proton transfer polymers with different structures shown in Figure 1. The investigated polymers include poly(1,4-(2-hydroxy)phenylene benzobisthiazole-co-decamethylene benzobisthiazole) (HPBT-co-PBTC10), poly(styrene-co-3-hydroxy-4'-ethenylflavone) (PS-co-V3HF), and poly(2,5-benzoxalediyl(2,2,2-trifluoro-1-(trifluoromethyl)ethylidene)-5,2-benzoxazodiyl-4-hydroxy-1,3-phenylene) (mH6FPBO).

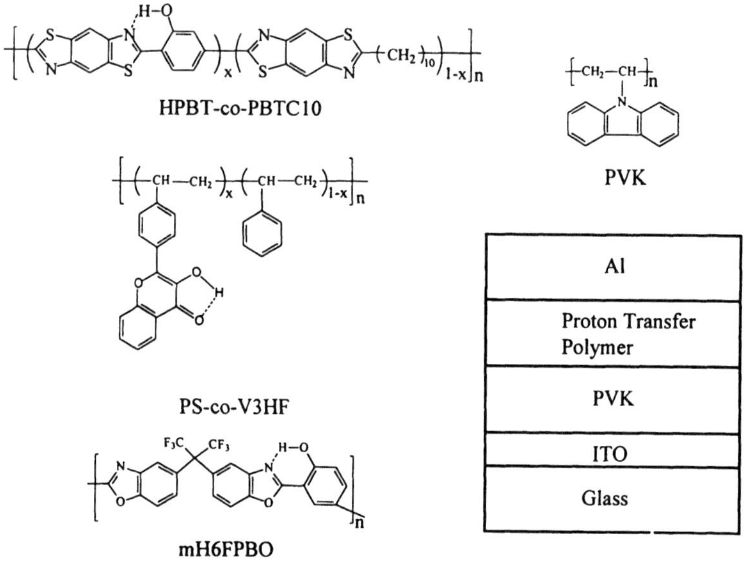

HPBT-co-PBTC10

PVK

PS-co-V3HF

mH6FPBO

| Al |
| Proton Transfer Polymer |
| PVK |
| ITO |
| Glass |

Figure 1. Molecular structures of proton transfer polymers and a schematic the LED.

EXPERIMENTS

The synthesis and characterization of the proton transfer polmers shown in Figure 1 have previously been reported [14,17,18]. The structure of the electroluminescence (EL) devices is also shown in Figure 1. First, a layer of 50 nm poly(vinyl carbazole) (PVK) was deposited onto the ITO (indium tin oxide) coated glass substrate by spin coating from dichloroethane solution. The PVK layer functions as the hole transport layer and electron blocking barrier to confine electrons within the emissive layer. Next, a 50 nm layer of a proton transfer polymer, the emissive layer, was spin coated from formic acid (HPBT-co-PBTC10 and mH6FPBO) or toluene (PS-co-V3HF) solutions. After drying the films in vacuum oven at 80 °C for 12 hours, an aluminum electrode (electron injecting electrode) of 50–100 nm thick was thermally evaporated at high vacuum (below 10^{-5} torr). Optical absorption measurements were done using a Perkin-Elmer Lambda-9 UV/vis/near-IR spectrophotometer. Photoluminescence (PL) and electroluminescence spectra were taken by using a Spex Fluorolog-2 spectrofluorimeter. Current-luminance-voltage curves were recorded simultaneously by hooking up an HP4155A semiconductor parameter analyzer together with a Grasby S370 optometer equipped with a calibrated luminance sensor head. All measurements were performed under ambient conditions.

RESULTS

Figure 2 shows the optical absorption and PL spectra of 5 and 15 % HPTB-co-PBTC10 and the corresponding conjugated homopolymer, poly(1,4-(hydroxy)phenylene benzobisthiazole) (HPBT). The copolymers have an absorption peak at 396 nm whereas the corresponding homopolymer HPBT has an absorption peak at 448 nm. It can be seen that HPBT homopolymer shows only excimer emission [19] with the emission peak at 590 nm. But the emission peak of 540 nm in HPBT-co-PBTC10 copolymers is independent of compositions (x=1–40 %). Therefore, this peak originates from the ESIPT emission [13,14]. The blue shoulder in the PL spectra of HPBT-co-PBTC10 copolymers depends on the rigid-rod segment composition. Below 5%, the intensity of the blue shoulder is negligibly small compared to the ESIPT emission band. In the 10–40 % copolymers, the relative intensity of the blue shoulder to the ESIPT emission band increases with increasing HPBT segment composition. The blue shoulder emission can be assigned to the enol tautomer [13,14].

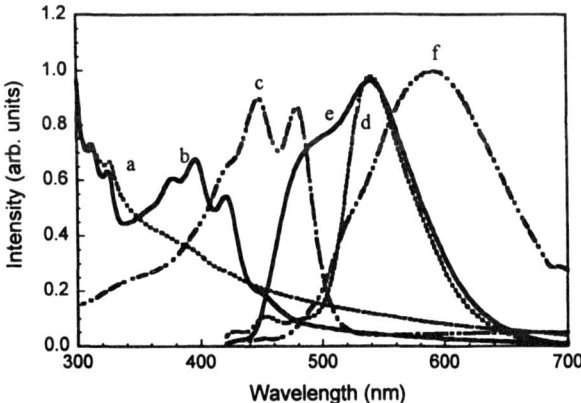

Figure 2. Optical absorption spectra of HPBT-co-PBTC10 copolymer system: (a)5 %, (b)10 % and (c)100 % HPBT. The PL spectra are for 5 % (d), 10 % (e) and 100 % HPBT (f), all with excitation wavelength of 400 nm.

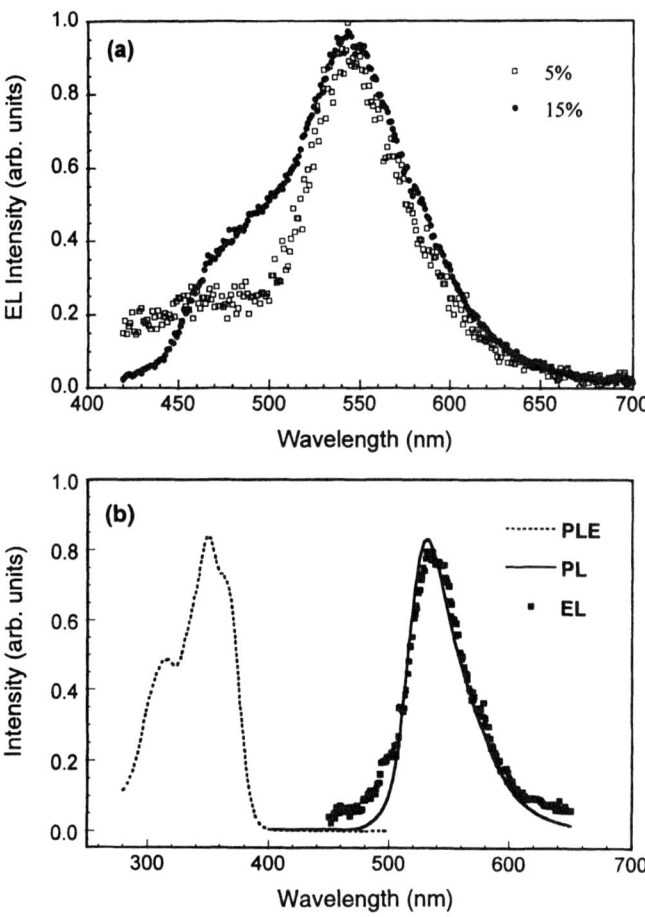

Figure 3. (a) Electroluminescence spectra of 5% and 15% HPBT-co-PBTC10.
(b) PL excitation, PL, and EL spectra of 10% PS-co-V3HF.

All the HPBT-co-PBTC10 copolymers (x=1, 5, 10, 15, and 40 %) exhibited
electroluminescence. Figure 3a shows representative EL spectra for the 5 % and 15 % HPBT-
co-PBTC10 copolymers. The EL emission peaks in both the 5 % and 15 % copolymers are the
same as those in PL spectra (Figure 2). The 5 % HPBT-co-PBTC10 device shows an ESIPT
emission band at 540 nm with only a very weak band at ~460 nm. The 15 % copolymer EL
device shows a blue shoulder (~475 nm) as well as the 540 nm ESIPT emission band. Figure
3b shows the PL excitation, PL, and EL spectra of the 10 % PS-co-V3HF (The 2% copolymer
showed the same results). An excitation peak at 349 nm was observed whereas both PL and
EL spectra have the same peak at 532 nm. Similar results were obtained for mH6FPBO proton
transfer polymer which showed an emission peak at 505 nm for both PL and EL. The fact that

these proton transfer polymers have identical PL and EL spectra shows that the EL emission of these proton transfer polymers originates from electrically-generated proton transfer (EGIPT) reaction. The excited keto tautomer (K*) formation in the photoinduced ESIPT process is well documented to occur by proton transfer from the primary excited enol tautomer state (E*). However, the detail mechanism of formation of K* by electrical pumping is yet to be elucidated but the overall EGIPT reaction can be expressed by [17]:

$$E^- + E^+ \rightarrow K^* + E \rightarrow E + E + h\nu \qquad (1)$$

After formation of enol radical anions E^- and radical cations $E^{\cdot+}$ by electrical injection, it is not clear whether K* forms from E* or from K^- and $K^{\cdot+}$ recombination. Nevertheless, electrical generation and EL emission from K* which does not exist in the ground state demonstrate electrically induced population inversion in an organic LED. To achieve laser emission by electrical pumping requires that the excitation density be sufficiently large, just as in the case of optical pumping. This means that the concentration of K* produced by EGIPT needs to be much higher than exists in current devices

Figure 4 shows the current-voltage (I-V) and luminance-voltage curves of representative ITO/PVK/15 % HPBT-co-PBTC10/Al device under forward bias (positive polarity at the ITO electrode). It can be seen that the current starts to rise at 10 V and luminance starts at about 13 V. All proton transfer polymers studied here showed similar LED turn-on voltages of 12–13 V. The devices presented here are not optimized yet. The luminance of these devices is relatively low, 2–5 cd/m². Clearly, significant improvement in the LED performance will be needed bfore possible laser emission can be expected.

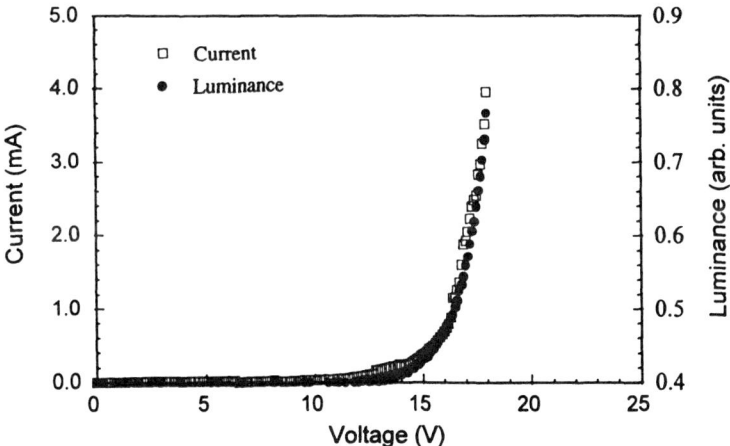

Figure 4. Current-voltage and luminance-voltage curves of the
15% HPBT-co-PBTC10 LED under forward bias.

CONCLUSIONS

Our results demonstrate that proton transfer can occur by current injection into polymers with ESIPT moieties. The electrically generated intramolecular proton transfer process and the resulting electroluminescence demonstrate electrically induced population inversion in an organic LED, a necessary first step toward electrically pumped laser diodes. Together with the prior observation of stimulated emission in the same EGIPT polymers [17], our results suggest that proton transfer polymers exhibiting EGIPT are promising candidates for electrically-pumped organic laser diodes.

ACKNOWLEDGEMENTS

This work was supported by the Office of Naval Research and the National Science Foundation (CHE-9120001, CTS-9311741).

REFERENCES

1. (a) See special issues: Chem. Phys. **136**, 153-360 (1989); J. Phys. Chem. **95**, 10215-10524 (1991). (b) S. J. Formosinho and L. G. Arnaut, Photochem. Photobiol. A: Chem. **75**, 21 (1993).
2. H. K. Sinha and S. K. Dogra, Chem. Phys. **102**, 337 (1986).
3. R. S. Becker, C. Lenoble and A. Zein, J. Phys. Chem. **91**, 3509 (1987).
4. M. F. Rodriguez-Preito, B. Nickel, K. H. Grellman and A. Mordzinski, Chem. Phys. Lett. **146**, 387 (1988).
5. M. Wiechman, H. Port, F. Laemer, W. Frey and T. Elsaesser, Chem. Phys. Lett. **165**, 28 (1990).
6. P. F. Barbara, P. K. Walsh and L. E. Brus, J. Phys. Chem. **93**, 29 (1989).
7. Z. Nir and O. Vogl, J. Polym. Sci. Polym. Chem. Ed. **20**, 2735 (1982).
8. A. U. Khan and M. Kasha, Proc. Natl. Acad. Sci. USA, **80**, 1767 (1983).
9. A. U. Acuna, A. Costela and J. M. Munoz, J. Phys. Chem. **90**, 2807 (1986).
10. A. U. Acuna, F. Amat, J. Catalan, , A. Costela, J. M. Figurera and J. M. Munoz, Chem. Phys. Lett. **132**, 567 (1986).
11. M. L. Ferrer, A. U. Acuna, F. Amat-Guerri, A. Costela, J. M. Figurera, F. Florido and R. Sastre, Appl. Opt. **33**, 2266 (1994).
12. J. H. Burroughes, D. D. C. Bradley, A. R. Brown, R. N. Marks, K. Mackay, R. H. Friend, P. L. Burn and A. B. Holmes, Nature, **347**, 539 (1990).
13. R. M. Tarkka and S. A. Jenekhe, Mater. Res. Soc. Symp. Proc. **413**, 97 (1996).
14. R. M. Tarkka and S. A. Jenekhe, Chem. Phys. Lett. **260**, 533 (1996).
15. N. Tessler, G. J. Denton, and R. H. Friend, Nature **382**, 695 (1996).
16. F. Hide, M. A. Diaz-Garcia, B. J. Schwartz, M. R. Andersson, Q. Pei, and A. J. Heeger, Science **273**, 1833 (1996).
17. R. M. Tarkka, X. Zhang and S. A. Jenekhe, J. Am. Chem. Soc. **118**, 9438 (1996).
18. J. R. Dharia, K. F. Johnson, and J. B. Schlenoff, Macromolecules **27**, 5167 (1994).
19. S. A. Jenekhe and J. A. Osaheni, Science, **265**, 765 (1994).

ELECTROACTIVE AND PHOTOACTIVE ROD-COIL BLOCK COPOLYMERS: SELF-ORGANIZATION AND PHOTOPHYSICAL PROPERTIES

X. LINDA CHEN AND SAMSON A. JENEKHE

DEPARTMENTS OF CHEMICAL ENGINEERING AND CHEMISTRY, UNIVERSITY OF ROCHESTER, ROCHESTER, NEW YORK 14627-0166

ABSTRACT

  Two series of new electroactive and photoactive coil-rod-coil and rod-coil-rod triblock copolymers, poly (pentadecamethylene carboxester)-*block* - poly(p-phenylene benzobisthiazole) - *block* - poly (pentadecamethylene carboxester) (1), and poly(2,6-benzoxazole)-*block*-poly(benzobisthiazole decamethylene)-*block*-poly(2,6-benzoxazole) (2), were synthesized, characterized, and used to investigate the self-assembly properties of rod-coil block copolymers. The progressive band narrowing of the absorption spectrum of thin films of 1 confirmed the effects of spatial confinement with increasing coil block size. Photoluminescence studies of thin films of 1 and 2 showed the effects of self-organization, annealing at 110 °C, block lengths, and composition on photophysical properties. Bilayer photoreceptors consisting of a layer of block copolymer as the charge generation layer and a layer of tris(p-tolyl)amine dispersed in polycarbonate as a trap-free hole transport layer were oberved to have high quantum efficiency, good photosensitivity and good dark decay.

INTRODUCTION

  Rod-coil block copolymers are currently of theoretical [1-4] and experimental [5-7] interest because of their complex morphologies and phase behavior. Because of the thermodynamic incompatibility of the two vastly different conformations of rigid-rod and coillike blocks and the constraint imposed by the chemical bonding between the blocks, rod-coil block copolymers exhibit a wide range of microphase-separated morphologies. The rich variety of ordered morphologies in this class of block copolymers are being explored and the factors determining their phase behavior are being defined [1-7]. However, rod-coil block copolymers which have been studied so far are non-electroactive and non-photoactive[5-7], thus limiting both the scope of the techniques suitable for investigating the morphologies and applications of the resulting nanophase materials.

  In this paper, two series of new electroactive and photoactive rod-coil-rod and coil-rod-coil triblock copolymers: poly (pentadecamethylene carboxester)-*block* - poly(p-phenylene benzobisthiazole) - *block* - poly (pentadecamethylene carboxester) (1, CRCA), and poly(2,6-benzoxazole)-*block*-poly(benzobisthiazole decamethylene)-*block*-poly(2,6-benzoxazole) (2, RCRA-1) have been synthesized, characterized, and used to explore the self-assembly properties of block copolymers and the effects of the morphologies of the resulting self-organized nanostructured materials on their solid-state photophysical properties. The chemical structures of the triblock copolymers we investigated are shown below. The poly(p-phenylene benzobisthiazole) (3, PBZT) homopolymer, is a well-known conjugated rigid-rod polymer which has interesting photoconductive [8] and light emitting [9] properties. The coillike blocks of 1 consist of blocks of various lengths of the nonphotoactive and nonelectroactive polyester 4. The poly(2,6-benzoxazole) (5, 2,6-PBO) homopolymer is a conjugated polymer with high modulus and thermal stability [10], that also exhibits liquid crystalline ordered phases in solution [11]. However, its electroactive and photoactive properties have not been reported before. The poly(benzobisthiazole decamethylene) (6, PBTC10) is a non-photoactive and non-electroactive polymer, soluble in common organic solvent such as THF, chloroform. Various techniques, such as differential scanning calorimetry, polarized optical microscopy, optical absorption and photoluminescence spectroscopies, and cyclic voltammetry, were used to probe and elucidate the morphologies and properties of the block copolymers. Absorption and photoluminescence measurements on the thin films were performed, and confirmed the self-organizaton of the rigid-coil block copolymers.

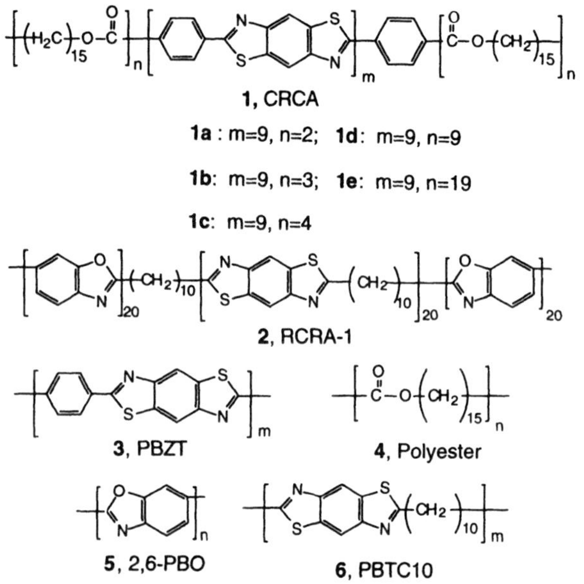

1, CRCA

1a : m=9, n=2;   **1d**: m=9, n=9

1b: m=9, n=3;   **1e**: m=9, n=19

1c: m=9, n=4

2, RCRA-1

3, PBZT   **4, Polyester**

5, 2,6-PBO   **6, PBTC10**

Figure 1. Chemical structures of **1** and **2** copolymers and their related homopolymers.

EXPERIMENTS

The coil-rod-coil triblock copolymers **1a-1e** were synthesized by copolymerization of a carboxylic acid-terminated PBZT (HOOC-PBZT-COOH) [11] with the AB-type monomer 16-hydroxyhexadecanoic acid (16-HA). The resulting products of copolymerization were purified by extensive extraction with refluxing acetone, which is a selective solvent for the polyester. Triblock copolymer **2**, was synthesized according to modified literature method [11]. In short, carboxylic acid-terminated flexible poly(benzobisthiazole decamethylene) (PBTC10) block (HOOC-B$_m$-COOH) was synthesized by reacting 2,5-diamino-1,4-benzenedithiol dihydrochloride (DABDT) with excess 1,10-decanedicarboxylic acid in polyphosphoric acid (PPA). Then, 4-amino-3-hydroxybenzoic acid (A) was added to generate the rigid-rod blocks. The rod-coil-rod triblock sample we investigated here is A$_{20}$B$_{20}$A$_{20}$, with average numbers of repeat units of 20 for 2,6-PBO and PBTC10 repeat unit respectively.

Optical quality thin films of **1**, **2**, and their blends with PMMA were obtained by spin coating onto silica substrates from nitrobenzene(NB)/GaCl$_3$ solution. The films were then regenerated in methanol and dried in a vacuum oven. The bilayer receptors for the photodischarge measurements were prepared by spin coating the solution of a copolymer onto poly(ethylene terephthalate) (PET) substrates, which were pre-coated with nickel. The copolymer films were then coated with a 20-mm tris(p-tolyl)amine/polycarbonate (TTA:PC 40/60, w/w) layer. Photophysical measurements, such as, UV-Vis, steady-state photoluminescence(PL), and charge photogeneration, were all performed on thin films at room temperature. Details of instruments and methods have been previously reported [8,9].

RESULTS AND DISCUSSION

The structures and compositions of the triblock copolymers **1a-1e** and **2** were established by various techniques, including ^{1}H NMR, ^{13}C NMR, FTIR, DSC, TGA, UV-Vis, and photoluminescence spectroscopies. The expected microphase separation or self-assembly processes of rod-coil triblock copolymers are illustrated in Figure 2. In isotropic solution, the triblock copolymers are in a disorder state. In the solid state, however, phase-separated ordered structures in which the rodlike blocks aggregate into anisotropic domains are the thermodynamically stable structures [1-7]. Thus, the self-assembly process is a kinetic process that occurs during the coagulation of solutions into solids as well as during subsequent processing (e.g. annealing) of the solids. Because the rodlike blocks are electroactive and photoactive, optical techniques can be used as powerful probes of the self-assembly process and the resulting nanostructures.

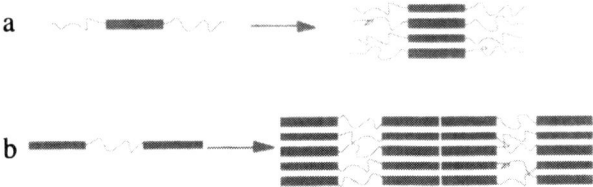

Figure 2. Schematic illustration of self-assembled structures from coil-rod-coil and rod-coil-rod triblock copolymers.

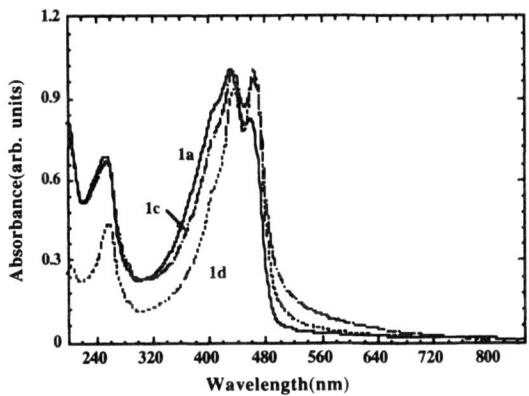

Figure 3. Absorption spectra of thin films of **1a, 1c** and **1d**.

Figure 3 shows the optical absorption spectra of **1a, 1c,** and **1d** films cast from nitrobenzene/AlCl$_3$. The absorption bands of the triblock copolymers **1a-1e** have identical maxima at about 438 and 468 nm and a π-π^* absorption edge of 500 nm (2.48 eV). These triblocks with average 9 repeat units of PBZT have essentially identical electronic absorption spectra as the high molecular weight homopolymer(DP$_n$ ~ 150) [9]. However, a novel feature of the UV-Vis spectra of the triblocks is the narrowing of the bandwidth of the main absorption band as the polyester block length increases (**1a** to **1d** in Figure 3). The origin of these effects is the increasing chromophore confinement as the coillike block length increases.

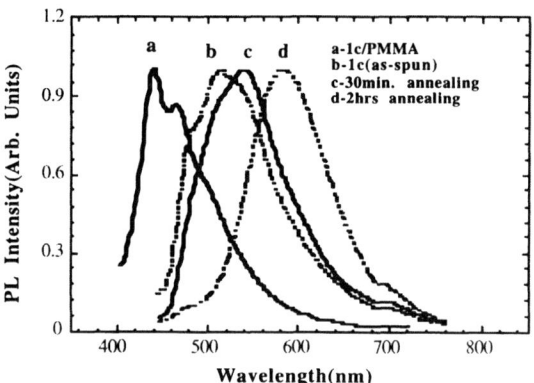

Figure 4. PL spectra of thin films of **1c**: (a) **1c** in PMMA (0.1 wt%); (b) as-spun; (c) annealed at 110 °C for 30 min.; (d) annealed at 110 °C for 2 hours.

Figure 4 show the PL spectra of thin films of **1c**. The PL spectrum of the as-spun film shows a slightly structured emission band with peak at 513 nm and a shoulder peak at 480 nm (Fig. 4b). When the film was annealed at 110 °C for 30 min, the emission band is red-shifted and the PL λ_{max} increase to 540 nm (Fig. 4c). A longer annealing time (2 hours) gave rise to a broad, featureless emission band centered at 583nm and a large Stokes shift (115 nm) (Fig. 4d). For comparison, the emission spectrum of **1c** dispersed in inert PMMA matrix (0.1 wt%) is also shown (Fig. 4a). The PL spectrum of 1c/PMMA blend shows a well-resolved vibronic structure with peak at 440 nm and a shoulder peak at 465 nm, which is assigned to the single-chain emission of isolated PBZT blocks. Because the broad and structureless emission of conjugated polymers originate chromophore aggregation and excimer formation [9], the PL results of copolymer **1c** not only reflect and evidence three degrees of chromophore aggregation (b,c,d) during the self-assembly process of the same triblock, but also suggest that the self-assembly process is a kinetic one.

The UV absorption spectrum of a thin film of **2** is shown in Figure 5 (curve a). The absorption spectrum of **2** is simply the superposition of those of the two corresponding homopolymers, showing peaks at 364, 380, 404 and 431 nm, and a π-π^* transition band edge of 2.73 eV. Figure 5 also shows the PL emission spectra of thin films of 2,6-PBO homopolymer (Fig. 5d), copolymer **2** as a pure film (Fig. 5c) and as a blend in PMMA (0.1wt%, curve 5b). The thin films were all excited at 360 nm, where the absorption band of the 2,6-PBO is located. The emission spectrum of **2** shows a broad, featureless emission band with λ_{max} at 510 nm, and a significant Stokes shift. By dispersing the triblock in PMMA, the emission showed a well-resolved vibronic structure with peaks at 433, 460 and 490 nm and a very small Stokes shift (2 nm). The spectrum of 2,6-PBO shows a broad, featureless emission band centered at 560 nm (Fig. 5d), which is characteristic of the excimer aggregate emission [9]. The PL spectra evolution was also accompanied by quantum efficiency enhancement. More than 15-fold enhancement of PL quantum efficiency was observed, being 1% for 2,6-PBO, 3% for **2**, and 15% for 2/PMMA blend (0.1wt%). The large blue shift of the PL spectrum of 2/PMMA (Fig. 5b), together with the dependence of PL quantum efficiency on concentration suggest that the broad and featureless emission of **2** (Fig. 5c) is from the aggregation of the 2,6-PBO blocks, which leads to the formation of excimer, a phenomenon which severely quenches the PL quantum efficiency [9].

554

Because the emission peak of the copolymer is 50 nm blue-shifted compared to that of the homopolymer, one can infer that the domains of 2,6-PBO blocks are much smaller than those of the homopolymer. The PL studies indicate that for **2**, microphase separation does occur in the solid state and that the resulting aggregates have strong effects on photophysical properties.

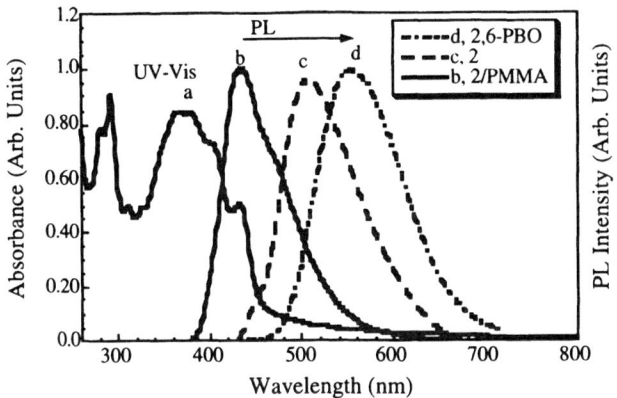

Figure 5. Optical absorption spectrum of **2** (a) and PL emission spectra of thin films excited at 360 nm: (b) **2**/PMMA(0.1wt%), (c) **2**, and (d) 2,6-PBO.

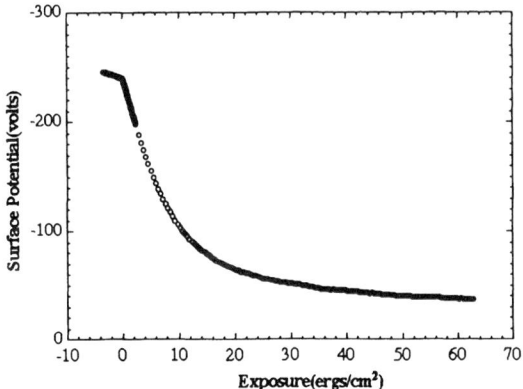

Figure 6. Photoinduced discharge curve for **1d** /TTA bilayer device. The excitation wavelength is 470 nm.

Figure 6 shows the photoinduced discharge curve for the **1d** /TTA bilayer device (80 nm thickness of **1d**) which was initially charged to a surface potential of ~250 V. Three important device parameters: photosensitivity, dark decay and residual potential can be obtained from Figure 6. For the **1d**/TTA bilayer device, photosensitivity of ~8 ergs/cm^2, dark decay of ~4 V/cm, and residual potential of ~30 V were obtained. At high electric field (~8x10^5 V/cm), photogeneration quantum efficiency as high as 28% was achieved. Compared to PBZT/TTA bilayer device which has a photosensitivity of 12 ergs/cm^2 and a quantum efficiency of ~25% at a field of 8x10^5

V/cm [8], **1d** /TTA device has a similar quantum efficiency but a better photosensitivity. The superior performance of block copolymers as charge photogeneration layers is not currently understood but will be further investigated.

CONCLUSION

Two new series of electroactive and photoactive rod-coil triblock copolymers have been synthesized, characterized and their self-assembly properties exploited to construct nanostructured materials. Our photophysical studies demonstrate the aggregation of the rigid-rod blocks and the effects of self-organization, annealing at 110 °C, block lengths, and composition on photophysical properties. Bilayer photoreceptors using **1d** as the charge generation layer exhibit high quantum efficiency (> 28 % at $\sim 10^6$ V/cm), good photosensitivity (8 ergs/cm^2) and good dark decay (< 5 V/s). These results suggest that rod-coil block copolymers with conjugated rodlike blocks are promising building blocks for preparing functional, electroactive and photoactive nanostructured polymer systems.

AKNOWLEDGMENTS

This research was supported by the Office of Naval Research and in part by the National Science Foundation (CTS-9311741, CHE-9120001).

REFERENCES
1.   (a) A. N. Semenov and S. V. Vasilenko, Sov. Phys. **63**, 70 (1986).
        (b) A. N. Semenov, Mol. Cryst. Liq. Cryst. **209**, 191 (1991).
2.   A. Halperin, Macromolecules **23**, 2724 (1990).
3.   D. R. M. Williams and G. H. Fredrickson, Macromolecules **25**, 3561 (1992).
4.   (a) E. Raphael and P. G. de Gennes, Physica A **177**, 294 (1991).
        (b) E. Raphael and P. G. de Gennes, Makromol. Chem.; Macromol. Symp. **62**, 1 (1992).
5.   D. Vernino, D. Tirrell and M. Tirrell, Polym. Mater. Sci. Eng. **71**(2), 496 (1994).
6.   (a) J. T. Chen, E. L. Thomas, C. K. Ober and G. -P. Mao, Science **273**, 343 (1996).
        (b)J. T. Chen, E. L. Thomas, C. K. Ober and S. S. Hwang, Macromolecules **28**, 1688 (1995).
7.   (a) L. H. Radzilowski, J. L. Wu and S. I. Stupp, Macromolecules **26**, 879 (1993).
        (b) L. H. Radzilowski and S. I. Stupp, Macromolecules **27**, 7747 (1994).
8.   J. A. Osaheni, S. A. Jenekhe and J. Perlstein, J. Phys. Chem. **98**,12727 (1994).
9.   (a) S. A. Jenekhe and J. A. Osaheni, Science **265**, 765 (1994).
        (b) J. A. Osaheni and S. A. Jenekhe, J. Am. Chem. Soc. **117**, 7389 (1995).
10.   (a) S. J. Krause, T. B. Haddock, D. L. Vezie, P. G. Lenhert, W. -F. Hwang, G. E. Price, T. E. Helminiak, J. F. O'Brien and W. W. Adams, Polymer **29**, 1354 (1988).
        (b) R. J. Young and P. P. Ang, Polymer **33**, 975 (1992).
11.   J. F. Wolfe, In Encyclopedia of Polymer Science and Engineering, Vol 11, 2nd ed. Wiley: New York, 1988, pp601-635.

CONCENTRATION DEPENDENCE OF PHOTOLUMINESCENCE AND ELECTROLUMINESCENCE IN 2,5 BIS[2'-(4"-HYDROXYHEXANOL BENZYL)-1'-ETHYL]3,4-DIBUTYL THIOPHENES

S. BLUMSTENGEL[a], I. SOKOLIK[*] AND R. DORSINVILLE[a,b]
[a]Department of Electrical Engineering, City College of the City University of New York, 138th Street and Convent Ave., New York, NY 10031
[b]New York State Center for Advanced Technology for Ultrafast Materials and Application at CUNY and Center for Analysis of Structure and Interfaces, The City College and Graduate Center of CUNY
D. VOLOSCHENKO, M. HE[**], L.-C. CHIEN, O. LAVRENTOVICH
Kent State University, Liquid Crystal Institute and Chemical Physics Program, Kent, OH

ABSTRACT

A new thiophene-based compound 2,5 bis[2'-(4"-hydroxyhexanol benzyl)-1'-ethyl]3,4-dibutyl thiophene (HBDT) as well as a copolymer (HBDT-PU) consisting of alternating HBDT and urethane spacer units were synthesized. Absorption and luminescence properties of both compounds were studied in solution and in different polymer matrices. Absorption and luminescence spectra of HBDT and HBDT-PU coincide indicating that emission in HBDT-PU occurs from the thiopene containing unit. Photoluminescence (PL) is emitted in the blue-green region of the visible spectrum with a maximum at 460 nm. The PL efficiency of both compounds is strongly enhanced when dispersed in a PVK or PMMA matrix indicating that concentration quenching occurs in the pure material. Light emitting devices were fabricated utilizing a PVK/PBD blend doped with HBDT and HBDT-PU at different concentrations as emitter material. The electroluminescence (EL) spectra coincided with the PL spectra of HBDT indicating that EL emission originates from the dopant molecules. The dependence of the EL efficiency on the doping concentration was measured and found to be close to the concentration dependence of the PL quantum yield.

INTRODUCTION

Thiophene containing conjugated polymers are among the most extensively studied materials for use in organic light-emitting diodes (OLED). It has been shown that the optical as well as charge transport properties of polythiophenes (PT) can be changed by attaching different side chains [1-3]. By selecting the proper sidegroup the band gap of PTs can be tailored. PTs with emission ranging from blue to red were synthesized [4-6]. Blending of PTs with different side chains yields white emitting LEDs [7].

In this work we follow another approach to tailor the optical properties which consists in the application of oligomers [8] and copolymers [9]. Here the number of monomer units determines the band gap and thus the emission wavelength. In this paper we report about the optical properties of a new thiophene-based oligomeric compound of 2,5 bis[2'-(4"-hydroxyhexanol benzyl)-1'-ethyl]3,4-dibutyl thiophenes (HBDT) which consists of thiopene and phenyl rings as well as a copolymeric compound (HBDT-PU) where the HBDT units are separated by urethane

[*] present address: FED Corp., 1580 Route 52, Hopewell Junction, NY 12533
[**] present address: University of Southern California, Loker Hydrocarbon Research Institute, Los Angeles, CA

spacer units. Since concentration quenching is an important issue in both low molecular weight organic [10] and polymeric [11] LEDs we have studied the dependence of PL quantum yield (φ_{PL}) on the HBDT as well as HBDT-PU doping concentration in a polymer matrix and show that φ_{PL} can be dramatically enhanced by diluting the emitting molecules in a host material. It has been shown before that the luminescence properties of PTs, i.e. φ_{PL} and EL quantum yield (φ_{EL}) can be improved by application of polymer blends [11,12] or attaching bulkier sidegroups to the main chain[13-15].

Single-layer LEDs were fabricated utilizing blends of poly(N-vinylcarbazole) (PVK) and 2-(4-biphenylyl)-5-(4-tert-butylphenyl)-1,3,4-oxadiazole (PBD) doped with HBDT. PVK serves as the hole transporting host matrix. PBD was added to improve the electron transport properties. To elucidate the underlying physics of the EL emission, EL spectra and wavelength selective photoexcited fluorescence spectra were measured and the dependence of φ_{EL} and current-voltage characteristics on the doping concentration studied.

EXPERIMENTAL

The chemical structure of HBDT (a) and HBDT-PU (M_w = 2419) (b) are shown in Fig. 1. The synthesis will be described elsewhere.

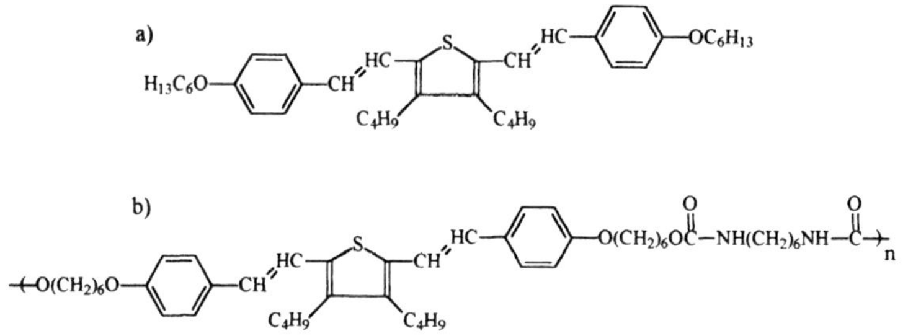

Figure 1: Chemical structure of HBDT (a) and HBDT-PU (b)

Absorption, photoluminescence (PL) and photoluminescence excitation (PLE) spectra were studied in solution and in thin film. Thin films were spincast from chloroform solutions of the pure materials and polymer blends onto quartz substrates. PMMA and PVK/PBD blends were used as polymer matrices, the concentration of HBDT and HBDT-PU in the host material ranged from 50% to 0.2%.

LEDs were fabricated using blends of PVK with PBD (2:1 by weight) doped with HBDT and the dopant concentration varied between 15% and 1%. An ITO covered glass substrate was used as anode material and vacuum evaporated calcium as cathode protected by an aluminum layer to retard oxidation. The organic layer (thickness ca. 100 nm) was spincast from chloroform.

The absorption and PL spectra were taken with Perkin Elmer Lambda 9 UV/VIS/NIR spectrometer and a Perkin Elmer LS50 fluorescence spectrometer respectively. EL spectra were recorded using an Acton Research Corporation SpectraPro 150 spectrograph connected to a Princeton Applied Research Corporation 1420 diode array. Current-voltage characteristics were measured with a Keithley 6517 electrometer and brightness-voltage with an International Light

Corporation IL1700 radiometer. Fabrication of the LEDs and the EL-measurements were done in nitrogen atmosphere to avoid oxidation. All measurements were carried out at room temperature.

RESULTS

Photoluminescence

Absorption, PL and PLE spectra of HBDT and HBDT-PU were measured in dilute chloroform solution. The spectra of HBDT and HBDT-PU coincide indicating that the thiophene containing group is solely responsible for the optical properties of HBDT-PU and the urethane unit acts only as a spacer. The HBDT spectra are depicted in Fig. 2. The absorption spectrum shows a clearly resolved vibronic structure. PL lies in the blue-green region of the visible spectrum with vibronic peaks at 460 nm and 491 nm. φ_{PL} in dilute chloroform solution was determined to be (20 ± 2) % for HBDT and HBDT-PU.

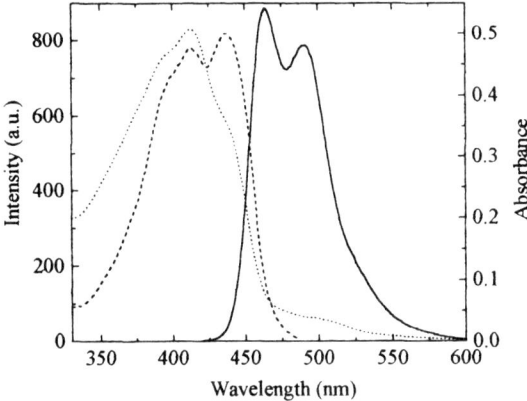

Figure 2: Absorption (dotted line), PL (solid line) and PLE (dashed line) spectra of HBDT (0.01mg/ml) in chloroform solution, $\lambda_{ex} = 412$ nm.

Thin films of pure HBDT do not fluoresce. Surprisingly, thin films of the copolymer HBDT-PU where the emitting units are separated by inert spacers do not show any emission either. When either compound is dispersed in a polymer matrix like PMMA strong blue-green PL emission arises. The spectral features of the absorption, PLE and PL spectra of HBDT and HBDT-PU in PMMA coincide with the spectra taken in solution. φ_{PL} strongly depends on the concentration of the chromophor in the polymer matrix whereas the shape of neither the absorption nor the emission spectra is affected. Fig. 3 shows that the concentration dependence of φ_{PL} of HBDT and HBDT-PU in PMMA coincide. A strong increase of φ_{PL} is observed when the chromophor concentration decreases. Saturation of the effect occurs at concentrations about 2 wt% of the chromophor in PMMA. Since concentration quenching occurs in HBDT-PU too it can be concluded that interchain interactions are responsible for the quenching of the PL. Spatial separation of the molecules suppresses the formation of non-emissive [16] interchain excitons, which compete with the generation of emissive intrachain excitons and leads therefore to an increase of φ_{PL}.

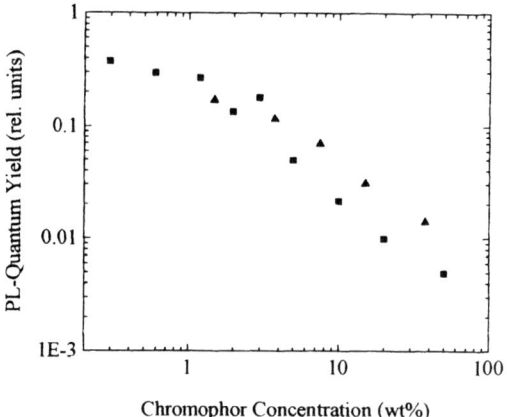

Figure 3: Dependence of φ_{PL} of HBDT (■) and HBDT-PU (▲) dispersed in PMMA on the concentration of HBDT units, $\lambda_{ex} = 412$ nm, ($M_{W(Chromophor)} : M_{W(Spacer)} = 3:1$)

Electroluminescence

Since the emission properties of HBDT and HBDT-PU resemble each other only the results obtained for HBDT will be presented here. To optimize the efficiency of HBDT based LEDs the concentration dependence of φ_{PL} and φ_{EL} of HBDT doped PVK+PBD (2:1) thin films was studied. The PL spectra of PVK and PBD overlap with the absorption spectrum of HBDT, therefore excitation energy transfer to HBDT molecules is very efficient. The concentration dependence of φ_{PL} was measured when HBDT molecules were excited directly ($\lambda_{ex} = 412$ nm) or

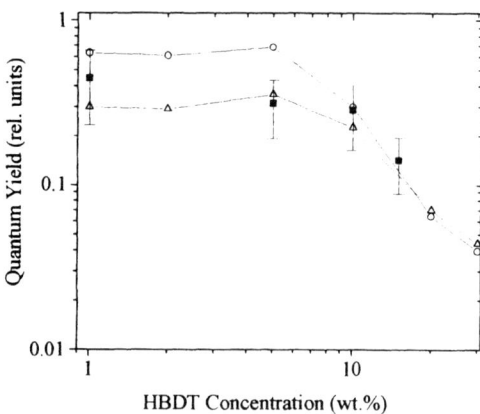

Figure 4: Dependencies of φ_{PL} (Δ: $\lambda_{ex} = 412$ nm, o: $\lambda_{ex} = 344$ nm) and φ_{EL} (■) on HBDT concentration in PVK+PBD+HBDT thin film and respective LED.

via energy transfer from PVK or PBD (λ_{ex} = 344 nm). The results are depicted in Fig. 4. φ_{PL} and φ_{EL} increase with decreasing dopant concentration although φ_{PL} (λ_{ex} = 412 nm) shows a weaker dependence then φ_{PL} (λ_{ex} = 344 nm). At HBDT concentrations lower than 5% the PL spectrum of the composite films demonstrate contributions from PVK and/ or PBD at wavelengths shorter than 450 nm, see Fig. 5 since excitation energy transfer becomes less sufficient. Within the experimental error the concentration dependence of φ_{PL} and φ_{EL} coincide. I-V curves where recorded varying the HBDT concentration between 1% and 15% but no difference was found. Therefore we conclude that HBDT doping does not alter charge injection or the transport properties of the LED and the increase of φ_{EL} is due to the increase of the emission efficiency.

Fig. 5 shows the EL spectrum of a LED with PVK+PBD+HBDT (1 wt.%) as emitter layer in comparison with the PL spectrum of the respective film excited at 344 nm. EL lies in the blue-green region of the visible spectrum with a maximum at 460 nm. The brightness of the diodes was 500 cd/m^2 and the onset voltage 15 V for a 100 nm thick organic layer.

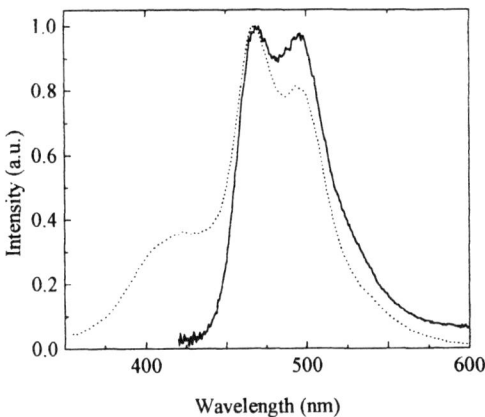

Figure 5: PL spectrum (dotted line) of PVK+PBD (2:1) doped with 1wt. % HBDT (λ_{ex} = 344 nm) and EL spectrum (solid line) of respective LED

The EL spectrum coincides with the PL spectrum of HBDT. Whereas the depicted PL spectrum is a superposition of the PL spectra of PVK and HBDT no emission from PVK was observed under the condition of electroexcitation leading to the conclusion that recombination of electrons and holes and subsequent formation of radiative singlet excitons occurs also at HBDT molecules.

CONCLUSIONS

In this paper we have reported about the luminescence properties of a new thiophene based compound HBDT and the related copolymer HBDT-PU. We have shown that the luminescence efficiencies of both substances can be dramatically enhanced when dispersed in a polymer matrix. Decreasing the HBDT concentration in a PMMA film from 50% to 2% enhances φ_{PL} of almost two orders of magnitude. The effect might be explained by the suppression of the formation of nonradiative interchain excitons due to the spatial separation of the HBDT molecules in a polymer matrix.

LEDs utilizing a blend of PVK+PBD doped with HBDT were fabricated. It was shown that the HBDT doping concentration has also strong influence on the efficiency of the LEDs. Decreasing the HBDT concentration from 15% to 1% enhances φ_{EL} almost 3 times. Since HBDT doping (in the studied concentration range) does not affect the transport properties of the diode the increase of the EL efficiency is due to the increase of the luminescence quantum yield.

The results of our investigations demostrate the significance of concentration quenching of φ_{PL} and φ_{EL}. When designing LEDs this fact should be considered.

REFERENCES

[1]   T.J. Kang, J.Y. Kim, K.J. Kim, C. Lee and S.B. Rhee, Synth. Met. **69**, 377 (1995).

[2]   M. Catellani, C. Botta, P.C. Stein, S. Luzzati and R. Consonni, Synth. Met. **69**, 375 (1995).

[3]   C. Lee, K.J. Kim and S.B. Rhee, Synth. Met. **69**, 295 (1995).

[4]   M.R. Andersson, M. Berggren, G. Gutafsson, T. Hjertberg, O. Inganäs and Wennerström, Synth. Met. **71**, 2183 (1995).

[5]   M. Berggren, O. Inganäs, G. Gustafsson, J.C. Carlberg, J. Rasmusson, M. Andersson, T. Hjertberg and O. Wannerström, Nature **372**, 444 (1994).

[6]   M. Berggren, O. Inganäs, G. Gustafsson, M.R. Andersson, T. Hjertberg and O. Wennerström, Synth. Met. **71**, 2185 (1995).

[7]   M. Granström and O. Inganäs, Appl. Phys. Lett **68**, 147 (1996).

[8]   Y. Kanemitsu, K. Suzuki, Y. Tomiuchi, Y. Shiraishi, M. Kuroda and O. Nabeta, Synth. Met. **71**, 2209 (1995).

[9]   I. Sokolik, Zhou Yang, F.E. Karasz and D.C. Morton, J. Appl. Phys. **74**, 3584 (1993).

[10]   S.A. VanSlyke and C.W. Tang, 1995 Digest of LEOS Summer Topical Meetings, p. 3.

[11]   M. Yan, L.J. Rothberg, E.W. Kwock and T.M. Miller, Phys.Rev.Lett. **75**, 1992 (1995).

[12]   G. Yu, H. Nishino, A.J. Heeger, T.-A. Chen and R.D. Rieke, Synth.Met. 72, 249 (1995).

[13]   K. Yoshino, Y. Manda, K. Sawada, M. Onoda and R.Sugimoto, Solid State Commun. **69**, 143 (1989) 143.

[14]   Y. Ohmori, M. Uchida, K. Muro and K. Yoshino, Jpn. J. Appl. Phys. **30**, L1938 (1991).

[15]   Y. Ohmori, M. Uchida, K. Muro and K. Yoshino, Solod State Commun. **80**, 605 (1991).

[16]   M. Yan, L.J. Rothberg, F. Papadimitrakopoulos, M.E. Galvin and T.M. Miller, Phys. Rev. Lett. **72**, 1104 (1994).

IONIC COMPLEXES OF CONJUGATED OLIGOQUINOLINES

ASHOK S. SHETTY AND SAMSON A. JENEKHE

Departments of Chemical Engineering and Chemistry, University of Rochester, Rochester, New York 14627-0166.

ABSTRACT

Single crystal X-ray structures of salt complexes of oligoquinoline, 2,2'-bis(4-phenylquinoline)-1,4-phenylene (**4**), with *p*-toluenesulfonic acid, trifluoromethanesulfonic acid and a laser dye, Sulforhodamine 101 hydrate have been determined and used to study the effects of molecular packing on physical and photophysical properties of conjugated oligomer and polymers. UV-vis and emission spectra of thin films of the oligoquinoline-*p*-toluenesulfonate complex were also performed. These results indicate that they form an intermolecular charge-transfer complex. These complexes serve as models of quarternized polyquinolines such as the self-assembled bilayer films of polyquinoline and polystyrenesulfonic acid.

INTRODUCTION

Conjugated polymers are the subject of much research effort because of their potential applications in electronic, optoelectronic, and optical devices [1]. One class of conjugated polymers that we have extensively investigated are the conjugated rigid-rod polyquinolines, with the general structure **1**, because of their excellent thermal stability and high mechanical strength. They are also intrinsic n-type (electron transport) semiconducting polymers [2] with interesting electronic [3], photoconductive [4], and nonlinear optical [5] properties. Recently, they were also used as both the electron-transport layer and the emission layer in polyquinoline/poly(*p*-phenylenevinylene) heterojunction light-emitting diodes (LEDs) [6].

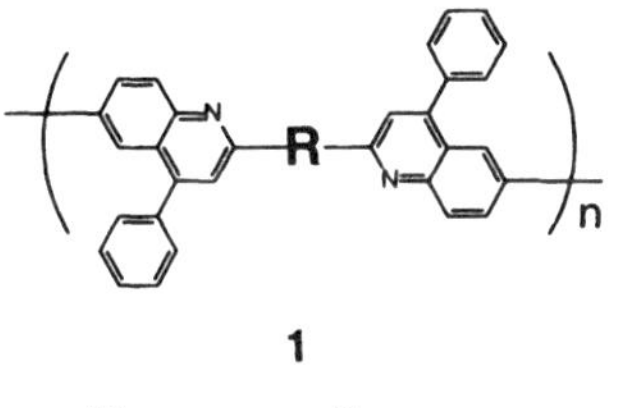

1

R = aromatic groups

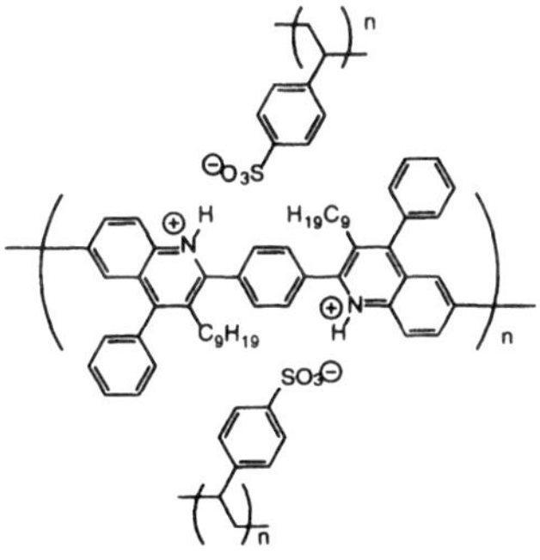

BuN-PPQ (**2**).PSSA (**3**) complex

Precise control over the molecular and supramolecular organizations of these polymeric materials could enhance the efficiency and overall performance of the devices they are incorporated in [7]. One of the methods that exemplifies this approach is the construction of layer-by-layer thin films by consecutively alternating adsorption of oppositely charged polyelectrolytes from dilute solutions [8,9]. The advantage of this

Mat. Res. Soc. Symp. Proc. Vol. 488 © 1998 Materials Research Society

method is that the thickness and supramolecular architecture of thin films could be controlled very precisely. Multilayer thin film materials have been successfully fabricated in this way with a large variety of polymers, such as conjugated [10] and non-conjugated [8,9] polyions, conjugated polymer precursors [11,12], and p-type doped conducting polymers [7].

We have recently used this method for the construction of bilayer films of polyquinoline, poly(2,2'-(p-phenylene)-6,6'-bis(3-nonyl-4-(p-tert-butyl phenyl)quinoline)) (BuN-PPQ, **2**), and polystyrenesulfonic acid (PSSA, **3**) [13]. This was accomplished by dipping thin films of the polyquinoline into a dilute solution of PSSA in methanol. The sulfonic acid groups of **3** formed an acid-base complex with the imine nitrogens of the polyquinoline leading to the adsorption of PSSA onto the polyquinoline layer. This adsorption was monitored using optical absorption and photoluminescence spectroscopies. The photoluminescence emission efficiency of the polyquinoline/PSSA complex was enhanced by over one order of magnitude, suggesting that such complexes are promising candidates for incorporation into optoelectronic devices.

To gain a better understanding of the polyquinoline/PSSA complex and quarternized polyquinolines in general, on a molecular and supramolecular level, we initiated investigation of complexes of oligoquinolines with various sulfonic acids. Results for oligoquinoline 2,2'-bis(4-phenylquinoline)-1,4-phenylene (**4**) which is the model for polyquinoline and its complexes are reported here. Photophysical studies and X-ray structures of single crystals of these complexes together with their photophysical properties provide a good basis for understanding the molecular structure and properties of the polyquinoline complexes in particular and polyelectrolyte complexes in general. To our knowledge, this comprises the first such study.

4

p-toluenesulfonic acid /methanol → *p*-toluenesulfonate salt **5**

trifluorosulfonic acid /methanol → trifluorosulfonate salt **6**

Sulforhodamine 101 hydrate/methanol → sulforhodamine salt **7**

EXPERIMENTAL SECTION

Materials. All the starting materials were obtained from Aldrich and used as received. The oligoquinoline **4** was synthesized by condensing 2-aminobenzophenone and 1,4-diacetylbenzene in a *m*-cresol:diphenylphosphate mixture [14]. It was recrystallized from a methanol/THF mixture prior to use. Sulfonate salts were prepared by dissolving stoichiometric amounts of oligoquinoline **4** and the corresponding sulfonic acid in methanol. Single crystals of the complexes were obtained by the slow evaporation of the methanol.

Characterization. Crystals of **5, 6** and **7** were each cut and mounted under Paratone-8277 on glass fibers, and immediately placed on the X-ray diffractometer in a cold nitrogen stream supplied by a Siemens LT-2A low temperature device. The X-ray intensity data were collected on a standard Siemens SMART CCD Area Detector System equipped with a normal focus molybdenum-target X-ray tube operated at 2.0 kW (50 kV, 40 mA). Thin films of the complexes were prepared by spin coating of the methanol solutions. Optical absorption spectra were recorded with a Perkin-Elmer Model Lambda 9 UV-Vis-Nir spectrophotometer. Steady state photoluminescence spectra were recorded by using a Spex Fluorolog-2 spectrofluorometer equipped with a Spex DM3000f spectroscopy computer.

RESULTS AND DISCUSSION

The molecular and single-crystal X-ray structures of the complex **5** are shown in Figure 1. In the crystal structure, the complex is located on a crystallographic center of symmetry at the midpoint of the central 1,4-biphenylene ring of the oligoquinoline as would be expected. The quinoline moieties are in a crystallographically imposed *anti*-orientation with respect to each other. The p-phenylene ring is twisted 21° from the mean plane of the quinoline moieties. The phenyl groups appended onto the quinoline moieties in the oligomer, are twisted at *ca.* 50° from each of the quinolines. The *p*-toluenesulfonate moieties are oriented such that the tolyl groups form an "edge-to-face" π-stacking interaction [15] with the central *p*-phenylene ring of the oligoquinoline. Hence they are predisposed to engage in an intermolecular charge-transfer interaction between the sulfonates and the oligoquinoline moiety. This is reflected in the photophysical properties of thin films of the complex (Figure 2).

The UV-vis absorption maximum of the complex (370 nm) is red shifted by 20 nm compared to the pure oligoquinoline **4** (350 nm). The emission spectra (Figure 2), display a red shift of 50 nm between the peak maxima of **4** (420 nm) and **5** (470 nm). As the peaks in the absorption and the emission spectra cannot be assigned simply to a π–π* transition, the above observations indicate that there is a intermolecular charge transfer component between the negatively charged sulfonates and the positively charged oligoquinoline. UV-vis and emission studies of the solution of the complex could not be accomplished as they dissociate in solution. The photophysical and X-ray structural studies of the oligoquinoline:*p*-toluenesulfonic acid complex (**5**), which may be considered to be the model compounds of the polyquinoline BuN-PPQ (**2**) :PSSA (**3**) complex, indicate that there could be similar charge-transfer between the two polymers which could account for the enhanced quantum efficiency and the bathochromic shift in the emission spectra of the complex.

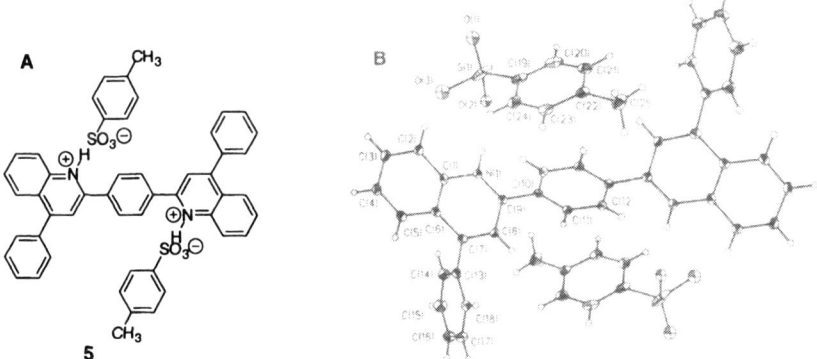

Figure 1. Molecular structure (A) and 30% ORTEP diagram (B) of **5** with the atom numbering. The dotted line represents the hydrogen-bonding between the oxygen atom of the *p*-toluenesulfonate molecules and the protonated nitrogens of the oligoquinoline.

To further study the effects of the anion structure on intermolecular charge-transfer, we studied complexes of the **4** with trifluoromethyl sulfonic acid (**6**) and with Sulforhodamine 101 Hydrate (**7**). In the former salt, we expected a diminished intermolecular charge transfer because of the absence of aromatic groups on the sulfonate. In the case of **7** we expected enhanced charge transfer interaction with **4** because of the highly conjugated nature of the sulforhodamine. The molecular structure of the complexes are shown in Figures 3 and 4. In the crystal structure of **6**, the complex is located on a crystallographic center of symmetry at the midpoint of the central 1,4-phenylene ring of the oligoquinoline. The quinoline moieties are in a crystallographically imposed anti-orientation with respect to each other. The *p*-phenylene ring is twisted 25° to the mean plane of the quinoline moieties. The phenyl groups appended onto the quinoline moieties in the oligomer, are twisted *ca.* 65° to each of the quinolines. However, in this case, two methanol molecules participate in the hydrogen bonding as is shown in Figure 3. Thus the methanols play a key role in stabilizing the crystals. On exposure to air, the crystals crumble on the loss of the methanol molecules. The trifluoromethane sulfonates are oriented such that the oxygen atoms are at *ca.* 3 Å from the *p*-phenylene ring indicating there is an interaction between the electron deficient *p*-phenylene ring and the sulfonates

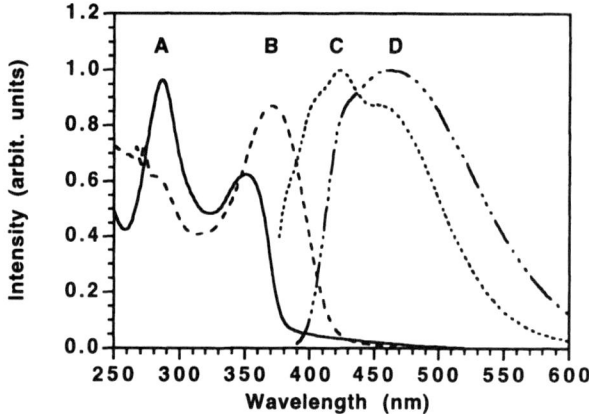

Figure 2 : UV-vis fluorescence spectra of thin films of oligoquinoline **4** and its *p*-toluenesulfonate salt **5**.

UV-vis	Emission
oligoquinoline **4** (curve A)	oligoquinoline **4** (curve C)
p-toluenesulfonate salt **5** (curve B)	*p*-toluenesulfonate salt **5** (curve D)

In the crystal structure of **7**, the salt molecule is not located on a crystallographic center of symmetry at the midpoint of the central 1,4-phenylene ring of the oligoquinoline as would be expected. The quinoline moieties are in a crystallographically imposed anti-orientation with respect to each other. The *p*-phenylene ring is twisted at 25° to the mean plane of the quinoline moieties. The phenyl groups appended onto the quinoline moieties in the oligomer, are twisted at *ca* 45° to each of the quinolines. The sulforhodamine is π-stacked with the quinoline moieties of the oligoquinoline. Eight methanol molecules are in the unit cell and hold the integrity of the crystal. The crystals crumble on exposure to air on losing the methanols. Photophysical studies these complexes are in progress.

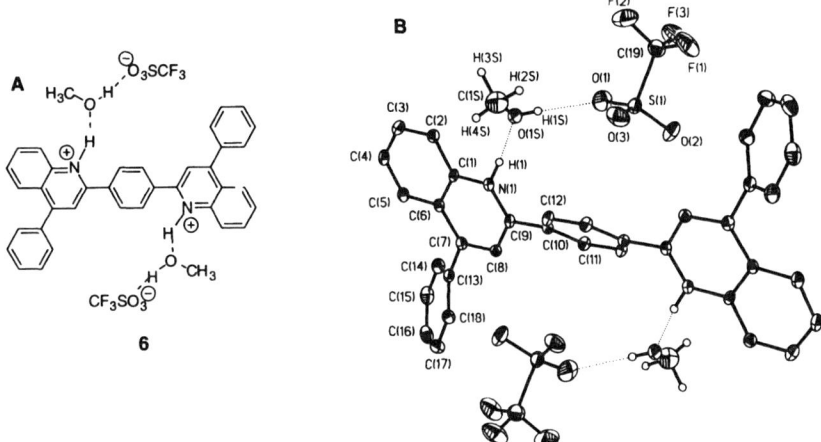

Figure 3. (B) 30% ORTEP diagram of **6** with the atom numbering. The dotted line represents the hydrogen-bonding between the oxygen atom of the trifluoromethane sulfonate molecules, the methanols, and the protonated nitrogens of the oligoquinoline.

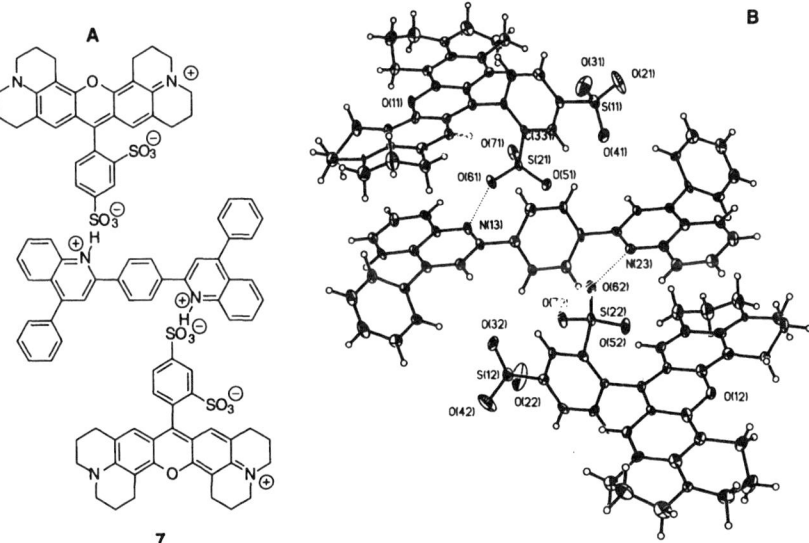

Figure 4. (B) 30% ORTEP diagram of **7** with the numbering of the selected atoms. The dotted line represents the hydrogen-bonding between the oxygen atom of the sulforhodamine molecules and the protonated nitrogens of the oligoquinoline. The eight methanol atoms have been omitted for clarity.

CONCLUSIONS

We have synthesized new complexes of oligoquinolines with various sulfonic acids and determined their X-ray crystal structures. We have also performed photophysical studies of one of these complexes which indicate there could be a significant intermolecular charge transfer between the negatively charged sulfonates and the positively charged oligoquinoline. These complexes serve as electronic models of the self-assembled bilayer films of polyquinoline, BuN-PPQ 2 and PSSA 3 and the results from studies on the model complexes indicate that there could be charge-transfer between the two polymers in particular and polyelectrolyte complexes in general.

ACKNOWLEDGMENTS

We thank Dr. Rene J. Lachicotte for the single crystal X-ray structures. This research was supported by the Office of Naval Research.

REFERENCES

1.   Bredas, J. L., Chance, R. R., Eds. Conjugated Polymeric Materials: Opportunities in Electronics, Optoelectronics, and Molecular Electronics; Kluwer Academic Publishers; Dordrecht, the Netherlands, 1990.
2.   A. K. Agrawal, S. A. Jenekhe,Chem. Mater., 8, 579 (1996).
3.   A. K. Agrawal, S. A. Jenekhe, Macromolecules, 26, 895 (1993).
4.   M. A. Abkowitz, M. Stolka, H. Antoniadis, A. K. Agrawal, S. A. Jenekhe, Solid State Commun., 83, 937 (1992).
5.   A. K. Agrawal, S. A. Jenekhe, H. Vanherzeele, J. S. Meth, J. Phys. Chem., 96, 2837 (1992).
6.   S. A. Jenekhe, X. Zhang, X. L. Chen, V.-E. Choong, Y. Gao, B. R. Hsieh, Chem. Mater., 9, 409 (1997).
7.   M. Ferreira and M. F. Rubner, Macromolecules, 28, 7107 (1995)
8.   Decher, G.; Hong, J. D.; Schmitt, J.; Thin Solid Films, 210, 831 (1992)
9.   Lvov, Y; Decher, G.; Mohwald, H; Langmuir, 9, 520 (1993)
10.   J. Tian, C.-C. Wu, M.E. Thomson, J. C. Sturn, R. A. Register, M. J. Marsella, and T. M. Swager, Adv. Mater., 7, 395 (1995)
11.   A. C. Fou, O.Onitsuka, M. Ferrera, and M. F. Rubner, J. Appl. Phys., 79, 7501 (1996)
12.   G. Mao, Y. Tsao. M. Tirrell, and H. T. Davis, Langmuir, 9, 3461 (1993)
13.   S. Yi, A.S. Shetty and S.A. Jenekhe, manuscript in preparation.
14.   A.S. Shetty, E.B.Liu, R.J. Lachicotte and S.A. Jenekhe, manuscript in preparation.
15   (a) A.S. Shetty, J. Zhang, J. S. Moore, J.Am.Chem. Soc. 118, 1019 (1996) (b) J.H. Williams, Acc. Chem. Res. 26, 593 (1993) and references therein.

CASCADE CONNECTED ORGANIC EL DIODES FOR MULTI-COLOR EMISSION

Yutaka OHMORI, Norio TADA, Yoshitaka KUROSAKA, Akihiko FUJII and Katsumi YOSHINO
Department of Electronic Engineering, Osaka University, Yamada-oka, Suita, Osaka 565 Japan
ohmori@ele.eng.osaka-u.ac.jp

ABSTRACT

Multicolor electroluminescent (EL) device which emits red (R), green (G) and blue (B) light has been realized by stacking a two-color emission part and a single-color emission part. The former part consists of two emissive layers of red and blue light, which can be selected by changing the polarity of applied field. The latter part consists of a single emissive layer of green light. The emission color from the R-G-B emission device can be modulated by the combination of applying various voltages to the two-color and to the single-color emission parts, separately.

INTRODUCTION

Organic electroluminescent (EL) devices have attracted great interest because of their potential applicability to full-color flat panel displays, for example, due to wide spectral range and low drive voltage. Since highly efficient electroluminescence was achieved by Tang and VanSlyke [1], various EL devices were demonstrated in order to obtain high brightness, long life time [2], white color [3, 4] and multicolor emissions [5-13] for display applications. Multicolor organic EL devices are classified into four types; (1) dye dispersed polymer emissive device [5], (2) single quantum-well layer inserted in the emissive layer device [6], (3) two emissive layers separated by carrier blocking layer device [7-9], (4) two emissive parts stacked on a transparent electrode device [10, 11].

We have already realized the multicolor organic EL devices utilizing multilayer of fluorescent dyes [7, 9], in which, emission colors have been changed by applying opposite polarity of applied field. In this paper, we report the realization of EL device which can emit completely different colors of R-G-B emission, and the emission characteristics are discussed by the combination of various aplied voltage to the device for realizing the full color emision.

EXPERIMENTAL

Figure 1 shows molecular structure of fluorescent dye materials used for multicolor EL diode. In Fig. 1, (a) 1,2,3,4,5-pentaphenyl-1,3-cyclopentadiene (PPCP), (b) 8-hydroxyquinoline aluminum (Alq_3), (c) N,N'-bis(2,5-di-tert-butylphenyl)-3,4,9,10-perylenedicarboximide (BPPC) and (d) N,N'-diphenyl-N,N'-(3-methyl phenyl)-1,1'-biphenyl-4,4'-diamine (TPD) are for blue, green, red emissive layers and carrier transport/blocking layers, respectively.

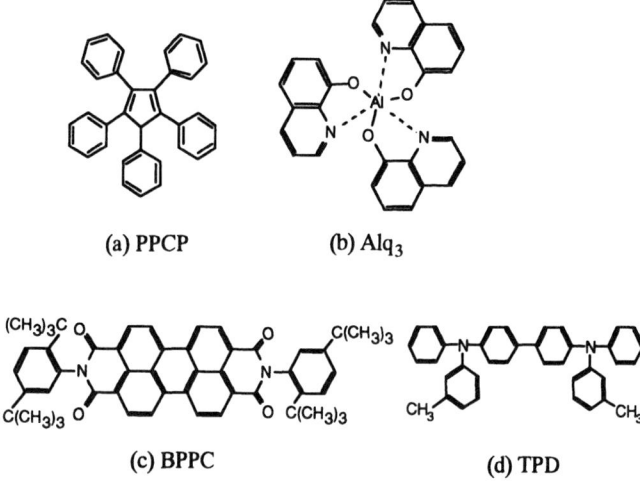

(a) PPCP   (b) Alq$_3$

(c) BPPC   (d) TPD

Fig.1. Molecular structure of fluorescent dyes used for fabricating multi-color emission device.
(a) 1,2,3,4,5-pentaphenyl-1,3-cyclopentadiene (PPCP), (b) 8-hydroxyquinoline aluminum (Alq$_3$),
(c) N,N'-bis(2,5-di-*tert*-butylphenyl)-3,4,9,10-perylenedicarboximide (BPPC), (d) N,N'-diphenyl-
N,N'-(3-methylphenyl)-1,1'-biphenyl-4,4'-diamine (TPD)

In Fig. 2, device structure which has been realized R-G-B emission is schematicallly shown.
The device consists of two parts, *i.e.*, two-color emission part and single-color emission part. It
consists of an indium-tin-oxide (ITO) coated glass substrate, fluorescent dye layers, a half-transparent
aluminum (Al) electrode, fluorescent dye layers, and an indium-containing magnesium (Mg:In) as
shown in Fig. 2. The layer structure was fabricated by organic molecular beam deposition (OMBD)
on an ITO-coated glass substrate at a background pressure of about 10^{-5} Pa for EL devices.

For optical measurement, the layers were deposited on a quartz substrate. The powders of
fluorescent dyes were loaded into separate Knudsen cells (K-cells), then the cells were subsequently
heated upto their sublimation temperatures, and deposited onto a substrate. The typical deposition
rates were 2-3 nm/min in this experiment. The layer thickness of the deposited materials was
determined *in-situ* using an oscillating quartz thickness monitor. Al and Mg:In electrodes were
vapor deposited at a background pressure of about 10^{-5} Pa.

For two-color emission part, 25-nm-thick BPPC as red emissive layer, 40-nm-thick TPD as
electron blocking layer, 30-nm-thick PPCP as blue emissive layer, 5-nm-thick Alq$_3$ as electron
transporting layer, and a half-transparent Al electrode were deposited on an ITO electrode. The
Alq$_3$ layer, which is inserted between the Al electrode and the PPCP layer, act as to increase electron
injection from Al electrode to the PPCP layer. Since the lowest unoccupied molecular

orbital (LUMO) of Alq₃ is lower than that of PPCP, insertion of Alq₃ layer helps electron injection into PPCP layer. For single-color emission part, 40-nm-thick TPD as hole transporting layer, 40-nm-thick Alq₃ as green emissive layer, and a Mg:In cathode were deposited on the half-transparent Al electrode. The Al electrode, whose transmittance is about 17% at 515nm of the green emissive wavelength, plays roles as cathode to the two-color emission part and, at the same time, as anode to the single-color emission part, simultaneously. The active electrode area of the device was 4mm². Electrical properties such as current-voltage (I-V) ,current-emission intensity (I-L) characteristics, and the emission spectra were measured by conventional methods, under DC, pulsed or AC bias conditions, which was reported by our previous paper [5]. The measurements were carried out at liquid nitrogen temperature (77K) or at room temperature (RT). The devices showed similar emission characteristics to those measured at 77K, but the emission intensity is not so high as those at 77K, compared with those at RT.

RESULTS AND DISCUSSION

Forward bias conditions are defined as the case in which the ITO electrode is positively biased against Al electrode in the two-color emission part (red and blue light) and as that in which the Al electrode is biased against Mg:In electrode in the single-emission part, whereas reverse bias conditions as the opposite case. The current increases superlinearly with increasing applied voltage in both the forward and reverse bias conditions in the two-color emission part, which is obtained by current-voltage (I-V) and EL intensity-voltage (L-V) characteristics measurements of the two-color emission part. Emission starts to increase at 15V, and EL intensity increases proportionally

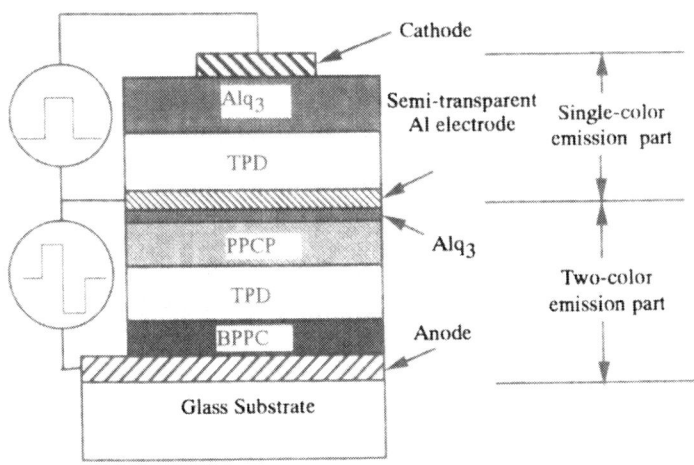

Fig. 2 Schematic of R-G-B emission EL device.

with increasing injection current in both bias conditions. The L-V characteristics of the two-color emission part show nearly symmetrical behavior with respect to the bias fields, however, the emission in the forward bias condition shows stronger than that in the reverse one. Blue light is rather stronger than red one under the same injection current in the two-color part.

The current and emission intensity increased only in forward bias condition with increasing applied voltage for the single-emission part (green light). By measuring I-V, L-V, and L-I characteristics, we can see that the emission starts to increase at 10V in the forward bias case. The emission from the green emission part is stronger than that of two-color emission part due to high emission efficiency of Alq$_3$.

The EL spectra from the stacked R-G-B emission device are shown in Fig. 3, in which the peak emission intensities are normalized. In case of forward bias condition in the two-color emission part, the device emits blue light originating from PPCP which has a peak emission at 450nm. On the other hand, in the reverse bias case, it emits red light from BPPC, whose peak was at 640nm. The emission color changes depending on the polarity of applied voltage, because the emission region is restricted only to the layer near the electron injecting electrode by electron blocking layer of TPD.

In the case of forward bias condition for the single-emission part, the device emits green light originating from Alq$_3$ with an emission peak at 515nm and the dips of emission at 490nm and 530nm correspond to the absorption by BPPC layer which exists between Alq$_3$ layer and the window of ITO electrode. The device can emit three colors of red, green, and blue lights by switching the polarity and by changing the biasing part. The EL intensities of R-G-B spectrum are

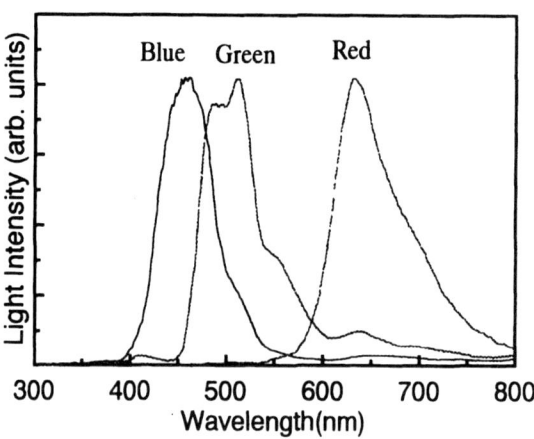

Fig. 3. EL spectra from the stacked R-G-B emission device.

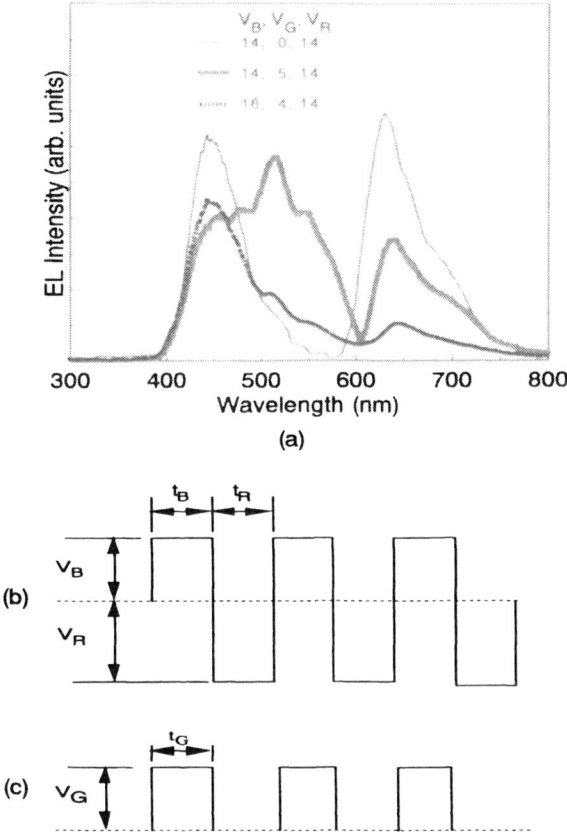

Fig.4. The EL spectra and the waveform of applied voltage to the two-color and single-color emission part. (a) R-G-B EL spectra from the device, (b) applied waveform to the two-color part, (c) that to the single-color emision part.

comparable with respect to the bias field. The green emission is the strongest compared with other blue and red emissions, however, we can control the total intensity of the color controlling the applied voltage and the pulse width of injection current.

Figure 4 shows the EL spectra from the device driven by pulsed current of various waveform. The applied voltage waveform is a bipolar curent to the two-color emission part with pulse height of V_B and V_R, at the repetition rate of 1KHz of 50% duty cycle as shown in Fig. 4(b). To the single-emission part, simple pulsed current of aplied voltage of V_G. The emission color is the superposition of the emission from PPCP, BPPC and Alq$_3$,and can be continuously modulated from blue to red by changing the pulse height of V_B, V_R and V_G as shown in Fig. 4. The device can emit multicolor based on the superposition of red, green and blue components by modulating the combination of the pulse height of V_R, V_B, and V_G, and the emission time of t_R, t_B, and t_G, respectively.

CONCLUSIONS

Multicolor EL device which emits red, green, and blue lights has been realized by stacking a green single-color emission part on a red and blue two-color emission part. The emission color can be controlled by changing the polarity of the applied field and controlling applied voltage, separately. The color can be continuously selected by modulating the applied voltage with changing the polarity and by controlling the pulse width of the injection current to the emission part.

ACKNOWLDGEMENTS

A part of this work was supported by a Grant-in Aid for Scientific Research from the Ministry of Education, Science, Sports and Culture.

REFERENCES

1. C. W. Tang and S. A. VanSlyke, Appl. Phys. Lett. **51**, pp. 913-915 (1987).
2. S. A. VanSlyke, C. H. Chen and C. W. Tang, Appl. Phys. Lett. **69**, pp. 2160-2162 (1996).
3. J. Kido, M. Kimura and K. Nagai, Science **267**, pp. 1332-1334 (1995).
4. J. Kido, H. Shionoya and K. Nagai, Appl. Phys. Lett. **67**, pp. 2281-2283 (1995).
5. M. Uchida, Y. Ohmori, T. Noguchi, T. Ohnishi and K. Yoshino, Jpn. J. Appl. Phys. **32**, pp. L921-L924 (1993).
6. A. Fujii, M. Yoshida, Y. Ohmori and K. Yoshino, Jpn. J. Appl. Phys. **34**, pp. L499-L502, (1995).
7. M. Yoshida, A. Fujii, Y. Ohmori and K. Yoshino, Appl. Phys. Lett. **69**, pp. 734-736 (1996).
8. M. Hamaguchi and K. Yoshino, Appl. Phys. Lett. **69**, pp. 143-145 (1996).
9. N. Tada, A. Fujii, Y. Ohmori and K. Yoshino, IEEE Trans. Electron Devices vol. **44**, pp. 1234-1239 (1997).
10. V. Bulovic, G. Gu, P. E. Burrows, S. R. Forrest and M. E. Thompson, Nature **380**, p. 29, (1996).
11. P. E. Burrows, S. R. Forrest, S. P. Sibley and M. E. Thompson, Appl. Phys. Lett. **69**, pp. 2959-2961 (1996).
12. Y. Yang and Q. Pei, Appl. Phys. Lett. **68**, pp. 2708-2710 (1996).
13. Y. Z. Wang, D. D. Gebler, D. K. Fu, T. M. Swanger, and A. J. Epstein, Appl. Phys. Lett. **70**, pp. 3215-3217 (1997).

CONTROL OF EMISSIVE LAYER INTERFACES WITH INORGANIC THIN LAYER BETWEEN 8-HYDROXYQUINOLINE ALUMINUM AND DIAMINE LAYERS IN ORGANIC EL DIODE

Yutaka OHMORI, Yoshitaka KUROSAKA, Norio TADA, Akihiko FUJII and Katsumi YOSHINO
Osaka University, Department of Electronic Engineering, Yamada-oka, Suita, Osaka 565, Japan,
ohmori@ele.eng.osaka-u.ac.jp

ABSTRACT

Control of organic interfaces by insertion of a thin inorganic film (SiO) has been investigated for an organic electroluminescent (EL) diode which consists of 8-hydroxyquinoline aluminum (Alq_3) and diamine derivative (TPD). In order to evaluate optical quality of the emissive layer at the interface, thin film of emissive marker layer was inserted into the emissive layer at the interface between the emissive layer and the carrier transport layer. The EL emission spectrum and the optical characteristics have been discussed for the EL diode with and without inorganic film.

INTRODUCTION

Recently, organic electroluminescent (EL) diodes [1-14] have attracted great attention because of their capability of an emission in a wide visible spectral range and their potential use in large-area flat panel display devices driven at low voltages. Since Tang and VanSlyke have demonstrated a high efficient EL diode [1, 2], consisting of light-emitting layers and carrier transport layers, many kinds of conducting polymers [4, 8] and fluorescent dyes [3, 9] have been developed in order to obtain high efficiency, long lifetime, multi-color emissions. [10-14]

In this paper, we report the control of the emission from 8-hydroxyquinoline aluminum (Alq_3) in an organic EL diode which consists of Alq_3 and diamine derivative (TPD) by the insertion of a layer of thin film silicon monoxide (SiO) at the Alq_3 and TPD interface. The mechanism of the emission enhancement has also been discussed.

EXPERIMENTAL

Figure 1 shows the molecular structure of the organic materials used in this study. 8-hydroxy-quinoline aluminum (Alq_3) and N,N'-diphenyl-N,N'-(3-methylphenyl)-1,1'-biphenyl-4,4'-diamine (TPD) were used as an emissive and a hole transporting materials, respectively. 4-(dicyanomethylene)-2-methyl-6-(p-dimethylaminostyryl)-4H-pyran (DCM) was used as a marker layer which was inserted in the emissive layer.

Figure 2 shows the schematic description of the device structure for investigating an emissive layer interface at the carrier transport layer. An organic EL diode consists of an Alq_3 emissive layer

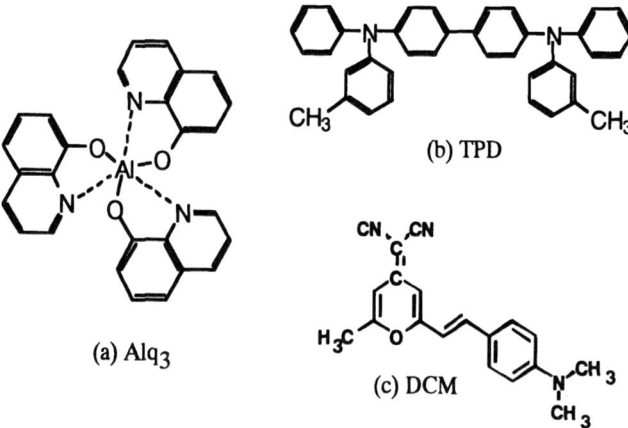

(a) Alq₃

(b) TPD

(c) DCM

Fig. 1 Organic materials used in this study.

and a TPD carrier transport layer was commonly used for organic EL diode, which is shown in Fig. 2(a). As shown in Fig. 2(b), thin film of DCM layer was inserted into the emissive layer of Alq_3. The emission characteristics of the device has been investigated by comparing the emission spectrum from the Alq_3 and DCM layers, since the emission from Alq_3 and DCM appears at 510nm and 560nm, respectively.

ITO coated glass substrate was used as a substrate, which acts as anode. Vapor deposited indium containing magnesium (Mg:In) was used as a cathode. Organic thin films and the SiO layer were

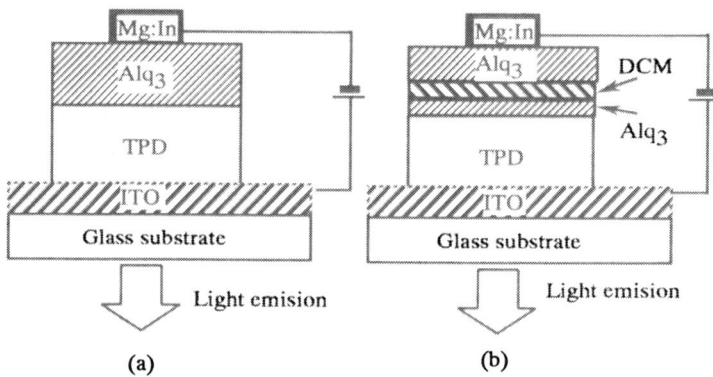

(a)

(b)

Fig. 2 Schematic description of device structure to determine
the emissive layer interface using marker layer.

fabricated using the organic molecular beam deposition (OMBD) system. The deposition was carried out at a background pressure of under 10^{-5} Pa. Organic materials were loaded into separate Knudsen cells (K-cells) and subsequently heated to their sublimation temperatures, and then deposited onto the substrates. The layer thickness of the deposited material was monitored in situ using an oscillating quartz thickness monitor. The active device area was 4 mm^2. All measurements were carried out at room temperature, and the electrical and optical characteristics were measured using conventional method as shown in our previous papers.

RESULTS AND DISCUSSION

The effect of the insertion of the SiO layer was investigated from the emission spectra from the diode shown in Fig. 3. In order to determine the emission from the Alq$_3$ at its interface with TPD, a thin film of DCM was inserted into the Alq$_3$ layer near the interface of the TPD and Alq$_3$ layers as shown in Fig. 3. The device consists of an ITO-coated glass substrate, 60-nm-thick-TPD as a hole transporting layer, a 9-nm-thick SiO layer as an interface modulation layer, 10-nm-thick Alq$_3$ emissive layer, 10-nm-thick DCM marker layer, 40-nm-thick Alq$_3$ as an electron transporting layer and a Mg:In cathode electrode. In order to determine the emission from the Alq$_3$ layer at the interface of Alq$_3$ and TPD, a thin film of DCM was inserted in the Alq$_3$ layer near the Alq$_3$ and TPD interface. The device is sealed with epoxy resin in order to prevent oxidation from air. The layer thickness of the SiO thin film inserted between Alq$_3$ and TPD interface was changed as a parameter

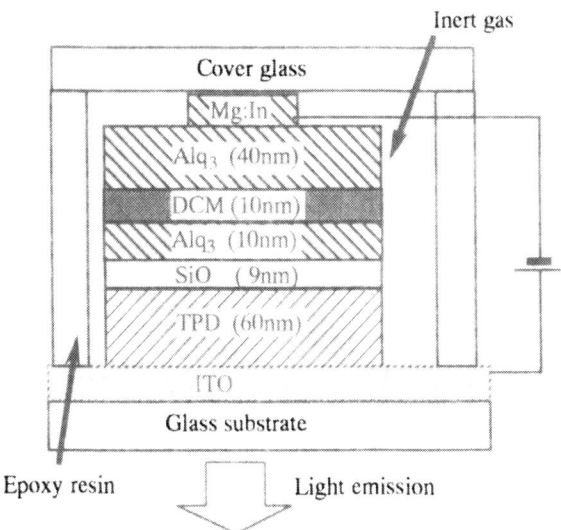

Fig. 3 Schematic description of device structure used for determining the effect of insertion of inorganic layer.

to realize efficient emission from the Alq$_3$ layer. The diode shows the emission from both the Alq$_3$ and DCM layers, which emit light at 510 nm and 650 nm, respectively. That means that carriers are injected both the Alq$_3$ and the DCM layers and recombine in both the layers. The emission spectrum can be used as a measure of the emission from the Alq$_3$ layer at the interface of Alq$_3$ and TPD and as that from the DCM layers.

The emission spectra from the diodes with and without the SiO layer are shown in Fig. 4(a) and (b), respectively. The emission of the diodes with the SiO layers is about 3 times stronger at the emission wavelength of Alq$_3$ layer, compared with that from the diodes without SiO layers under the same driving voltage. The emission intensity from the DCM layer in the diodes with the SiO layers is similar to that without the SiO layers, indicating that the recombination of carriers in the DCM layers are the same in the diodes with and without the SiO layers under the same driving voltage and current.

The results are explained using the energy band diagram as shown in Fig. 5. The energy band diagrams are schematically shown in the forward bias conditions. Holes are injected from the ITO electrode into the highest occupied molecular orbital (HOMO) state of the TPD layer, and electrons from the Mg:In electrode into the lowest unoccupied molecular orbital (LUMO) state of the Alq$_3$ layer. The Alq$_3$ layer next to the Mg:In electrode acts as an electron transporting layer to DCM and the Alq$_3$ emissive layer next to the SiO layer. Hole injection from TPD into the Alq$_3$ layer should be performed by the tunneling process, since a high barrier exists at the SiO layer which was inserted at the TPD and Alq$_3$ interface. However, the inserted SiO film is thin enough for holes to tunnel through to reach Alq$_3$ emissive layer.

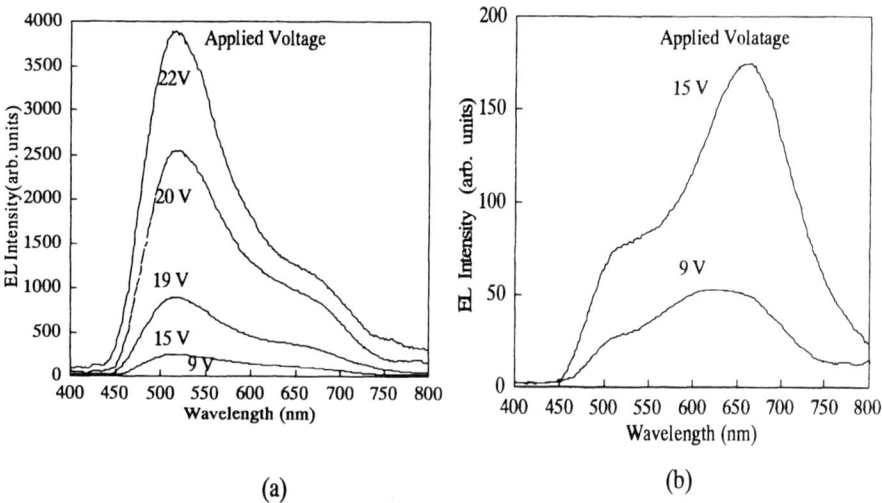

(a)   (b)

Fig. 4 EL emission spectra from the diode with Alq$_3$ and TPD inserted with SiO (a) and without SiO (b) layers between Alq$_3$ and TPD layers.

Electrons injected from the Mg:In cathode migrate through the electron-transporting Alq_3 and DCM layers to the emissive Alq_3 layer and then finally recombine with the holes injected through the SiO layer. Carrier recombination also occurs in the DCM layer. Emissions from both the Alq_3 and DCM layers are obtained in the devices with and without SiO layers as shown in Figs. 4(a) and 4(b), respectively.

In the device fabrication process, TPD was deposited on the ITO electrodes. The surface morphology [15] of the diodes, i.e., the ITO electrode, is not as perfect as the TPD layer to establish a clear interface between the layers. It is expected that the Alq_3 and TPD interface is not smooth enough to enable all the injected carriers to recombine for emission. There are some defects or non-radiative center at the interfaces between the TPD and Alq_3 layers. It was pointed out that the inter-diffusion of Alq_3 and TPD atoms at the hetero interface results in the quenching of the emission from Alq_3 layer.[16] As a result, the insertion of the SiO layer in the interface separates the interface completely to improve the carrier injection process and to decrease the number of non-radiative centers in the emissive layer at the interface.

The current vs. applied voltage (I-V) characteristics have been investigated as a function of the thickness of SiO layer inserted between TPD and Alq_3 layers, when the DCM layer was not inserted in the Alq_3 layer. From the results of the I--V characteristics of the device with SiO layers, the inserted thin SiO layer does not act as carrier blocking layer. It also indicates that the SiO layer of an appropriate thickness does not prevent the current flow through the interface between TPD and Alq_3 layers. The result that the emission from the Alq_3 layer is stronger in the diodes with the SiO layer indicates that the insertion of the SiO layer decreases the numbers of non-radiative recombination centers in the emissive layer at the interface of the TPD and Alq_3 layers.

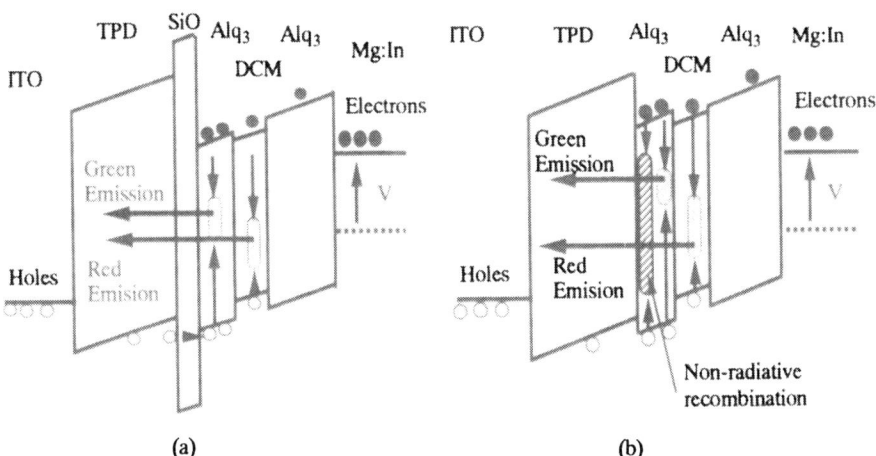

(a)   (b)

Fig. 5 Schematic description of energy band diagram of EL diode with SiO (a) and without SiO (b) layers.

579

Emission intensity vs. injection current density characteristics were investigated for different values of SiO layer thickness. Diodes with 12-nm and 9-nm-thick SiO layers exhibit the maximum emission intensity compared with other layer thicknesses of SiO. The emission intensity of the diode with the 12-nm-thick SiO layer is about 1.5 times stronger than that of the diode without the SiO layer, and in addition, the diodes with the SiO layer can be operated under high injection currents or applied voltages. The results show that the insertion of an SiO layers having an appropriate layer thickness at the interface of the TPD and the Alq_3 layers improves the emission intensity in the EL devices.

In this experiment, we use the SiO layer as an inorganic barrier layer, as we can easily control the layer thickness which can be made by the evaporation technique. The insertion of other insulating layers such as fluoride, nitride or oxide layers in the interface of the emissive and the carrier transporting layers may improve the emission efficiency in organic EL devices. Long running operations and operations at high temperatures are under study.

CONCLUSIONS

In conclusion, the insertion of an SiO layer having an appropriate value of layer thickness at the interface of emissive and carrier transport layers improves the emission intensity in EL devices. The improvement of the interface between the carrier transporting layer and the emissive layer enhances the emission efficiency.

ACKNOWLEDGEMENTS

Parts of this work was supported by a Grant-in-Aid for Scientific Research from the Ministry of Education, Science, Sports and Culture.

REFERENCES

1. C. W. Tang and S. A. VanSlyke, Appl. Phys. Lett. **51**, p.913(1987).
2. C. W. Tang, S. A. VanSlyke and C. H. Chen, J. Appl. Phys. **65**, p. 3610 (1989).
3. C. Adachi, T. Tsutsui and S. Saito, Appl. Phys. Lett. **56**, p. 799 (1990).
4. J. H. Burroughes, D. D. C. Bradley, A. R. Brown, R. N. Marks, K. Mackay, R. H. Friend, P. L. Burns and A. B. Holmes: Nature **347**, p. 539 (1990).
5. Y. Ohmori, M. Uchida, K. Muro and K. Yoshino, Jpn. J. Appl. Phys. **30**, p. L1938 (1991).
6. Y. Ohmori, M. Uchida, K. Muro and K. Yoshino, Jpn. J. Appl. Phys. **30**, p. L1941 (1991).
7. D. Braun and A. J. Heeger, Appl. Phys. Lett. **58**, p. 1982(1991).
8. G. Grem, G. Leditzky, B. Ullrich and G. Leising, Adv. Mater. **4**, p. 36 (1992).
9. C. C. Wu, J. K. M. Chun, P. E. Burrows, J. C. Sturm, M. E. Thompson, S. R. Forrest and R. A. Register, Appl. Phys. Lett. **66**, p. 653(1995).
10. J. Kido, H. Shionoya and K. Nagai, Appl. Phys. Lett. **67**, p. 2281(1995).
11. M. Uchida, Y. Ohmori, T. Noguchi, T. Ohnishi and K. Yoshino, Jpn. J. Appl. Phys. **32**, p. L921(1993).
12. A. Fujii, M. Yoshida, Y. Ohmori and K. Yoshino, Jpn. J. Appl. Phys. **34**, p. L499(1995).
13. M. Yoshida, A. Fujii, Y. Ohmori and K. Yoshino, Jpn. J. Appl. Phys. **35**, p. L397(1996).
14. Y. Ohmori, M. Uchida, C. Morishima, A. Fujii and K. Yoshino, Jpn. J. Appl. Phys. **32**, p. L1663 (1993).
15. E. M. Han, L.M. Do, N. Yamamoto and M. Fujihira, Thin Solid Films **273**, p. 202(1996).
16. E. M. Han, L. M. Do, Y. Niidome and M. Fujihira, Chem. Lett. p. 969(1995).

SYNTHESIS AND PHOTOPHYSICAL PROPERTIES OF SOME CONJUGATED POLYMERS FUNCTIONALIZED WITH RUTHENIUM POLYPYRIDINE COMPLEXES

PO KING NG, WAI YUE NG, XIONG GONG, WAI KIN CHAN*
Department of Chemistry, University of Hong Kong, Pokfulam Road, Hong Kong

ABSTRACT

A series of conjugated polymers functionalized with different ruthenium polypyridine metal complexes were synthesized by the palladium catalyzed reaction. Two conjugated polymer systems have been studied: 1. poly(phenylenevinylene) with bis(2,2':6',2"-terpyridine) ruthenium (II) on the mainchain and 2. quinoxaline based polymers with tris(2,2'-bipyridine) ruthenium (II). The ruthenium polypyridine complexes exhibit a long-lived metal to ligand charge transfer excited state which can enhance the photosensitivity of the resulting polymers. Different physical properties such as the photoconductivity and charge mobility in these polymers are also studied.

INTRODUCTION

Organic conjugated polymers and oligomers have been used as the active semiconducting layer in various organic-based thin film devices such as solid state laser materials,[1] organic transistors,[2] light emitting diodes,[3] photodiodes,[4] and electrode active materials.[5] They combine the electronic and photonic properties of traditional inorganic semiconductors and the synthetic and processing advantages of organic polymers. In addition, organic materials also enjoy the advantages of ease of structural design and modification. The physical properties, solubility, and processibility can be easily "tuned" by attaching different functionalities to the polymer backbone or side chain. When conjugated polymers are coupled with transition metal complexes, they are expected to exhibit new properties because many transition metal complexes have specific catalytic, redox, and photophysical properties. A polymer-metal complex is a metal complex containing a polymer ligand which presents a remarkably specific structure in which central metal ions are surrounded by an enormous polymer chain. Based on this polymeric ligand, the polymer-metal complex shows interesting and important characteristics. Most of the studies on polymer-metal complexes are directed toward their catalytic activities not only because they are excellent models for the metalloenzymes, but also lead to the development of highly efficient catalysts.[6]

Here, we report the synthesis and characterization of two series of conjugated polymers which contain polypyridine complexes. The first system is a poly(p-phenylenevinylene) (PPV) incorporating ruthenium terpyridine complexes $[Ru(tpy)_2]^{2+}$ in the polymer mainchain, while the second system is quinoxaline-based poly(p-phenylene) (PPP), which contains a ruthenium bipyridine $[Ru(bpy)_3]^{2+}$ metal complex. The chemistry and photophysics of ruthenium polypyridine complexes have been studied extensively.[7] They are well known photosensitizers because of the relatively long-lived metal-ligand charge transfer (MLCT) excited states. When incorporated into conjugated polymers, the ruthenium polypyridine complexes may act as

581

photosensitizers and thus enhance the photoconductivity. Moreover, the ruthenium complexes are also electrochemically active exhibiting a reversible Ru[II,III] redox process and a number of reductive ligand-centered processes.[8] In both polymer systems, the conjugated backbones were synthesized by the palladium catalyzed coupling reaction. This is a new approach to the design and synthesis of novel organic polymers for optoelectronic applications.

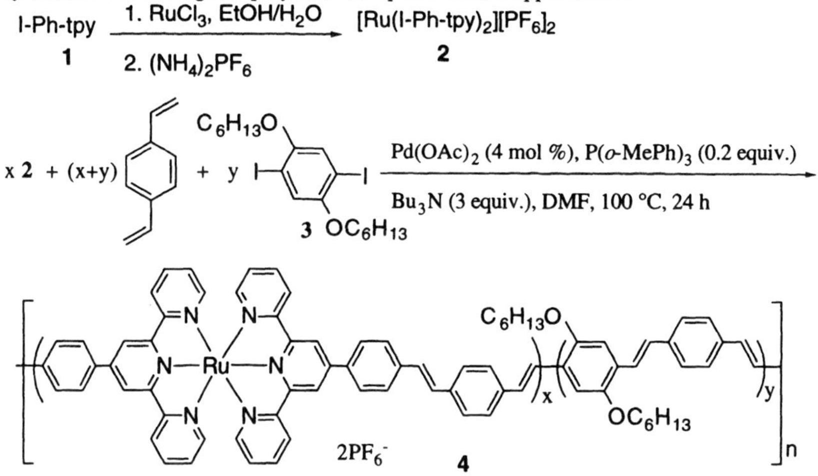

Scheme 1. Synthesis of [Ru(tpy)$_2$]$^{2+}$ containing PPVs.

EXPERIMENT

The synthesis of the monomers and polymers for the [Ru(tpy)$_2$]$^{2+}$ containing PPVs are shown in scheme 1.[9] Monomer **2** was synthesized by refluxing two equivalents of **1** with RuCl$_3$ and was isolated as the hexafluorophosphate salt. Polymers **4a-4e** were obtained in good yield by reacting 1,4-divinylbenzene with different ratio of **2** and **3** under standard Heck reaction conditions.[10] Dimethylformamide (DMF) was used as the solvent and the catalyst system was composed of palladium (II) acetate, tributylamine, and tri-o-tolylphosphine.

The synthesis of the quinoxaline-based monomers and polymers are shown in scheme 2. 5,8-Dibromo-2,3-dihexylqunioxaline **6** and 6,9-dibromodipyrido[3,2:a-2'3':c]phenazine **7** were synthesized by condensation reaction of **6** with tetradecane-7,8-dione and 1,10-phenanthroline-5,6-quinone, respectively. The direct complexation of **7** with cis-dichlorobis(2,2'-bipyridine)ruthenium dihydrate [cis-Ru(bpy)$_2$Cl$_2$•H$_2$O] failed to give monomer **8**. Instead, monomer **8** was synthesized by refluxing an ethanol solution of Ru(bpy)$_2$(acetone)$_2$(OTf)$_2$ with **7**.[11] The quinoxaline based PPPs **9a-9g** were synthesized by the Suzuki coupling reaction using benzene-1,4-diboronic acid as comonomer.[12]

The photoconductivity of the polymers was measured according to the literature method with a lock-in-amplifier.[13] A 150 W xenon lamp with band pass filter was used as the light source. The charge mobility was determined by the conventional time-of-flight technique using a nitrogen laser (pulse energy = 120 μJ, pulse width = 3 ns) as the light source. The polymer film for measurement was prepared by casting the polymer solution on an indium-tin-oxide (ITO) glass. A thin layer of gold electrode (120Å) was coated on the polymer film by sputtering.

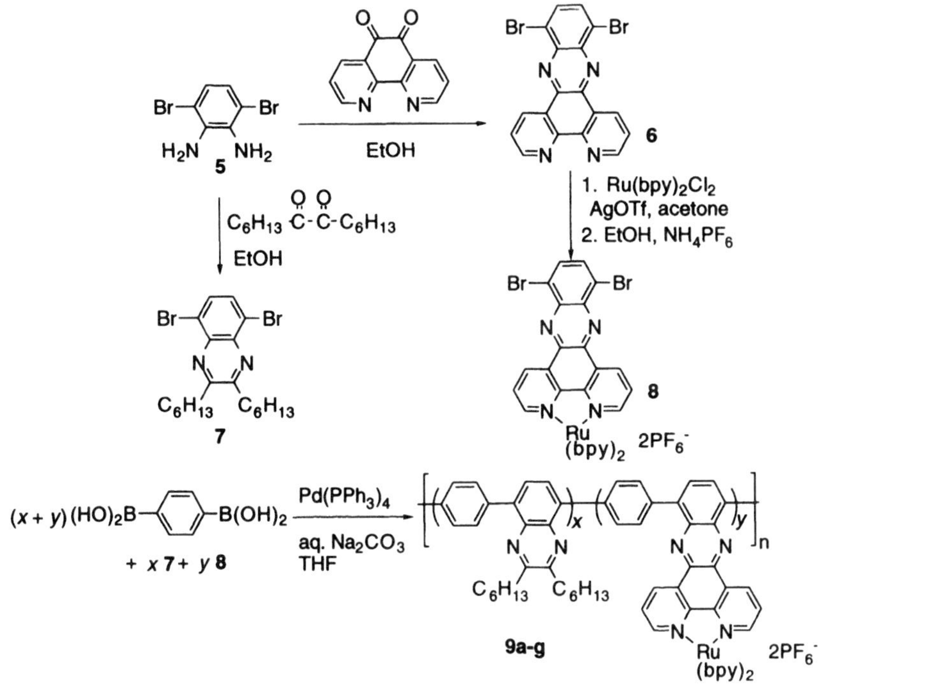

Scheme 2. Synthesis of quinoxaline-based PPPs.

RESULTS AND DISCUSSION

Some properties of the metal-containing polymers are summarized in Table I.

Table I. Properties of the metal containing polymers

Polymer	x	y	Reaction yield (%)	Inherent viscosity (dL/g)^a	$M_n{}^b$	Decomp. Temp. (°C)
4a	0.05	0.95	94	0.16	--	380
4b	0.1	0.9	98	0.38	--	375
4c	0.2	0.8	90	0.55	--	371
4d	0.3	0.7	94	0.41	--	386
4e	1	0	95	0.35	--	440
9a	1	0	87	--	26300	390
9b	0.9	0.1	76	--	9900	382
9c	0.8	0.2	69	--	11000	385
9d	0.7	0.3	51	--	8700	405
9e	0.6	0.4	48	--	6370	410
9f	0.5	0.5	49	--	5800	408
9g	03	0.7	41	--	15500	420

^{a}Measured in DMF solution at 30 °C with concentration $c = 0.5$ g/dL.
^{b}Number averaged molecular weight measured by GPC.

The metal containing PPVs **4a-4e** were synthesized by the Heck reaction in good yield. The reaction yield was not affected by the amount of the metal complex **2** present in the polymers. The quinoxaline-based PPPs **9a-9g** were synthesized by the Suzuki coupling reaction using Pd(PPh₃)₄ as the catalyst. The polymerization was carried out via a two-phase system with THF as the solvent and aqueous sodium carbonate solution as the base. In this system, the polymers exhibit lower yield and molecular weight when the proportion of the Ru(bpy)₂(dppz)²⁺ (dppz = dipyridophenazine-6,9-diyl) unit is increased. One possible explanation is that the bulkiness of monomer **8** gives rise to its low reactivity. The low reactivity of monomer **8** was further confirmed from the synthesis of model compounds.

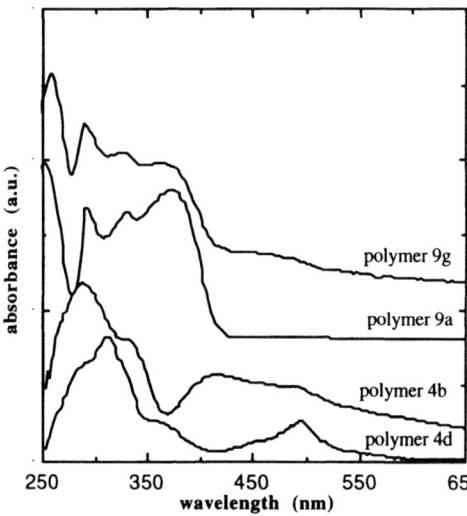

Fig. 1. UV/vis spectra of some
metal containing polymers.

The UV-vis spectra of some polymers are shown in Fig. 1. Polymers **4b** and **9a** show absorption peaks at 420 and 380 nm, which are attributed to the electronic absorption of the conjugated PPV and PPP backbone, respectively. In polymers **4b** and **4d**, another peak was observed at ca. 500 nm due to the MLCT transition in the $[Ru(Ph-tpy)_2]^{2+}$ moiety. When the $[Ru(Ph-tpy)_2]^{2+}$ content in the polymer is increased, the intensities of the peak at 420 nm become lower because the extent of conjugation is decreased. For the PPP system, incorporation of monomer **8** into the polymer mainchain causes a small absorption peak to appear at 450 nm, which is due to the $Ru(bpy)_2(dppz)^{2+}$ unit. The intensity of this MLCT band increases further when the content of ruthenium complex in the polymer is increased.

The photoconductivity and charge transporting properties of the PPVs were also studied in more detail. Under the same applied electric field, the photoconductivity of the polymers increases with the $[Ru(Ph-tpy)_2]^{2+}$ content. Moreover, the photoconductivity of the polymers at different wavelength resemble their absorption spectra. This clearly shows the ruthenium terpyridine complexes increases the photoconductivity by enhancing the sensitivity.

The charge carrier mobility of the polymers was determined by the time-of-flight experiment. The transient photocurrent signal shows a featureless decay and the transient time cannot be determined directly from the photocurrent trace. The signal indicates that the charge transport is dispersive with non-Gaussian carrier distribution.[14] At room temperature, the hole mobilities of polymers **4b** and **4d** were determined to be 6.7×10^{-5} and 7.5×10^{-5} cm²V⁻¹s⁻¹, respectively, which are both electric field and temperature dependent. These hole mobilities are three orders of magnitude higher than those of some phenyl-substituted PPVs.[15] These results clearly show the participation of the ruthenium complex in charge transport.

An Arrhenius plot of the hole mobility for polymer **4d** under different electric fields is shown in Fig. 2a. The graph indicates a thermally activated charge migration process with an activation energy of 0.19 eV at E = 160 kV/cm. The activation energy of polymer **4b** is slightly higher (0.22 eV) under the same applied field. The difference in activation energy may be related to the composition of the polymers. However, the exact reason for this is not clear. It was also found that the charge carriers are mainly holes as the electron mobilities are approximately one-tenth of the hole mobilities.

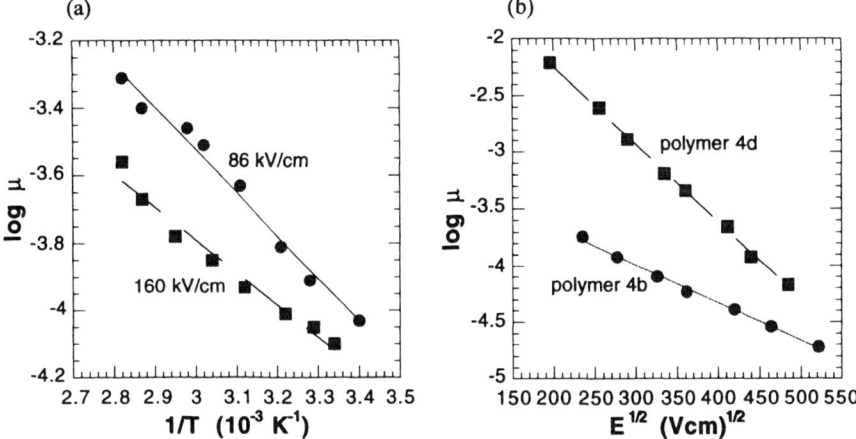

Fig. 2. (a) Temperature dependence of the hole mobility for polymer **4d** at different field strength. (b) Field dependence of the hole mobility at T = 298 K for polymer **4b** and **4d**.

Fig. 2b shows the hole mobilities of polymers **4b** and **4d** at room temperature. It can be seen that the log μ vs. $E^{1/2}$ plot shows a linear relationship with negative slopes. This interesting phenomenon can be explained by the presence of off-diagonal disorder in the hopping sites. The charge carriers have to jump against the field direction in order to open a faster route.[16] Other PPV derivatives[17] and triarylamine doped polycarbonate systems[18] reported in the literature also exhibit the same behavior.

CONCLUSIONS

We have synthesized two types of conjugated polymers functionalized with ruthenium polypyridine complexes using the palladium catalyzed reactions. Some photophysical properties of these polymers were studied by various characterization methods. It was found that the metal complexes can act as photosensitizers, as can be seen from the enhancement in photoconductivity. These polymers also exhibit modest hole carrier mobilities, whose migration is dispersive with non-Gaussian behavior. The charge mobilities are also affected by the amount of metal complex present in the polymer, indicating that the metal complex may also contribute to charge transport.

ACKNOWLEDGMENTS

This work was supported by the Research Grant Council of Hong Kong and by the Committee on Research and Conference Grants (University of Hong Kong). Partial financial support from Hung Hing Ying Physical Science Research Fund and The Run Run Shaw/Leung Kau Kui Research and Teaching Endowment Fund is also gratefully acknowledged.

REFERENCES

1   N. Tessler, G. J. Denton, and R. H. Friend, *Nature* **382**, 695(1996); F. Hide, M. A. Diaz-Garcia, B. J. Schwartz, M. R. Anderson, Q. Pei, and A. J. Heeger, *Science* **273**, 1833(1996).

2   Y. Yang, A. J. Heeger, *Nature* **372**, 344(1994).

3   For example: J. H. Burroughes, D. D. C. Bradley, A. R. Brown, R. N. Marks, K. Mackay, R. H. Friend, P. L. Burns, A. B. Holmes, *Nature* **347**, 539(1990); G. Gustafsson, Y. Cao, G. M. Treacy, F. Klavetter, N. Colaneri, A. J. Heeger, *Nature* **357**, 477(1992); N. C. Greenham, S. C. Moratti, D. D. C. Bradley, R. H. Friend, A. B. Holmes, *Nature* **365**, 628(1995); B. R. Hsieh, H. Antoniadis, D. C. Bland, W. A. Feld, *Adv. Mater.* **7**, 36(1995).

4   J. J. M. Halls, C. A. Walsh, N. C. Greenham, E. A. Marseglia, R. H. Friend, S. C. Moratti, A. B. Holmes, *Nature* **376**, 498(1995); G. Yu, C. Zhang, A. J. Heeger, *Appl. Phys. Lett.* **64**, 1540(1994).

5   T. Kawai, T. Iwasa, T. Kuwabara, M. Onoda, K. Yoshino, *Jpn. J. Appl. Phys.* **25**, 1833(1990).

6   *Catalysis by Polymers*, edited by E. A. Bekturov and S. E. Kudaibergenov (Huthig & Wepf Verlag, Heidelberg, 1996); M. Kaneko, *J. Polym. Sci., Macromol. Rev.* **16**, 397(1981).

7   E. C. Constable, *Adv. Inorg. Chem. Radiochem.* **30**, 69(1986); E. C. Constable, J. Lewis, M. C. Liptrot, P. R. Raithby, *Inorg. Chim. Acta.* **178**, 47(1990); J.-P. Sauvage, J.-P. Collin, J.-C. Chambron, S. Guillerez, C. Coudret, V. Balzani, F. Barigelletti, L. De Cola, L. Flamigni, *Chem. Rev.* **94**, 993(1994).

8   M. Beley, J.-P. Collin, J.-P. Sauvage, H. Sugihara, F. Heisel and A. Miehé, *J. Chem. Soc. Dalton Trans.* 3157(1991); E. C. Constable, A. M. W. Cargill Thompson, N. Armaroli, V. Balzani and M. Maestri, *Polyhedron* **20**, 2702(1992).

9   W. Y. Ng and W. K. Chan, *Adv. Mater.* **9**, 716(1997).

10 R. F. Heck, *Organic Reactions* **27**, 345(1982).

11 B. P. Sullivan, D. J. Salmon, and T. J. Meyer, *Inorg. Chem.* **17**, 3334(1978).

12 N. Miyaura, T. Yanagi, and A. Suzuki, *Synth. Commun.* **11**, 513(1981).

13 L. Li, J. Y. Lee, Y. Yang, J. Kumar,and S. K. Tripathy, *Appl. Phys. B* **53**, 279(1991).

14 H. Scher and E. W. Montroll, *Phys. Rev. B* **12**, 2455(1975); G. Pfister and H. Scher, *Adv. Phys.* **27**, 747(1978).

15 A. Y. Kryukov, A. C. Saidov, and A. V. Vannikov, *Thin Solid Films* **209**, 84(1992).

16 M. Van der Auweraer, F. C. De Schryver, P. M. Borsenberger, and H. Bässler, *Adv. Mater.* **6**, 199(1994).

17 T. Takiguchi, D. H. Park, H. Ueno, and K. Yoshino, *Synth. Met.* **17**, 657(1987); M. Gailberger and H. Bässler, *Phys. Rev. B* **44**, 8643(1991).

18 P. M. Borsenberger, L. Pautmeier, and H. Bassler, *J. Chem. Phys.* **94**, 5447(1991).

Low Pressure Organic Vapor Phase Deposition of Small Molecular Weight Organic Device Structures

M.A. Baldo[*], V.G. Kozlov[*], P.E. Burrows[*], S.R. Forrest[*], V.S. Ban[**], B. Koene[†], and M.E. Thompson[†]
[*]Department of Electrical Engineering, Center for Photonics and Optoelectronic Materials, and Princeton Materials Institute, Princeton University, Princeton, New Jersey
[**]PD-LD Inc., 209 Wall Street, Princeton, New Jersey 08544
[†]Department of Chemistry, University of Southern California, Los Angeles, CA 90089-0744

ABSTRACT

A new technique for the deposition of amorphous organic thin films, low pressure organic vapor phase deposition (LP-OVPD), was used to fabricate organic light emitting devices (OLEDs) and optically pumped organic lasers. The OLED consisted of a film of aluminum tris-(8 hydroxyquinoline) (Alq_3) grown on the surface of a film of N'-diphenyl-N,N'-bis(3-methylphenyl)1-1'biphenyl-4-4'diamine (TPD). Growth on both glass and polyester substrates was accomplished and the resulting heterojunction devices were found to have a performance similar to conventional, small molecular weight OLEDs grown using thermal evaporation in vacuum. The LP-OVPD grown OLED has an external quantum efficiency of $0.40 \pm 0.05\%$ and a turn-on voltage of approximately 6V. The optically pumped organic laser consisted of a film of Alq_3 doped with the laser dye, benzoic acid, 2-[6-(ethylamino)-3-(ethylimino)-2,7-dimethyl-3H-xanthen-9-yl]-ethyl ester, monohydrochloride (Rhodamine 6G). The laser output was centered at approximately 610nm and the lasing threshold was $30\mu Jcm^{-2}$. The rapid throughput of LP-OVPD and its use of low vacuum in a horizontal reactor demonstrate its potential to facilitate low cost, roll-to-roll deposition of organic films for many photonic device applications.

INTRODUCTION

The field of organic electroluminescence is attracting rapidly growing commercial interest, spurred by possible applications in lucrative markets such as video displays. Both small molecule and polymer-based organic devices have been investigated, but polymer-based devices have the advantage of simple and perhaps inexpensive fabrication by spin-on deposition techniques. In contrast, small molecular weight devices are usually fabricated by thermal evaporation in vacuum. This is often considered to be a more expensive process. However, a new organic film growth technique, organic vapor phase deposition (OVPD) [1,2,3] may eliminate this apparent disadvantage, leading to very low cost, large scale deposition of small molecular weight organic layers for numerous photonic device applications.

Similar to the process of vapor phase epitaxy commonly used in the growth of III-V semiconductors, OVPD uses carrier gases to transport vapor from the source to the substrate where it condenses to form the desired thin film. However, films grown using this technique at atmospheric pressure are often found to have non-uniform surface morphologies due to gas phase nucleation and a diffusion-limited growth process [2]. These effects were also observed by us when atmospheric pressure OVPD was used to deposit small molecular weight organic materials for use in OLEDs. Following the example of the growth of III-V films by organometallic vapor phase epitaxy, more uniform organic film growth should be possible at reduced pressures. This is due in part to lower concentrations of organic materials in the deposition chamber, minimizing gas phase nucleation. Also, low pressures can result in kinetically-controlled growth, leading to considerably smoother and more uniform films.

EXPERIMENT

The LP-OVPD system is shown schematically in Fig. 1. A mechanical vacuum pump is attached to the exhaust port of a 14cm diameter by 70cm long pyrex reactor tube, which lies horizontally within a three zone furnace. Pressure within the reactor tube is maintained by a combination of a Barotron [4] pressure gauge and an electronically controlled throttle valve positioned before the input of the vacuum pump. The organic source materials are contained in glass boats located in four, 5cm diameter feeder tubes connected at the upstream, or input end, of the reactor tube. The nitrogen carrier gas is also introduced via the feeder tubes, with the gas flow streams adjusted using three mass flow controllers. Each feeder tube contains a glass push rod valve used to regulate transport of organic material down the tube. The fourth feeder tube is used to flush the system with nitrogen, and to independently adjust the reactor pressure. The source boats are moved within the feeder tubes to control the temperature and sublimation rates of the source materials. Using this arrangement, up to three different organic materials can be co-deposited onto a cooled substrate located in the condensation zone near the exhaust port. Excess organic materials are condensed in a cold trap attached to the reactor exhaust port.

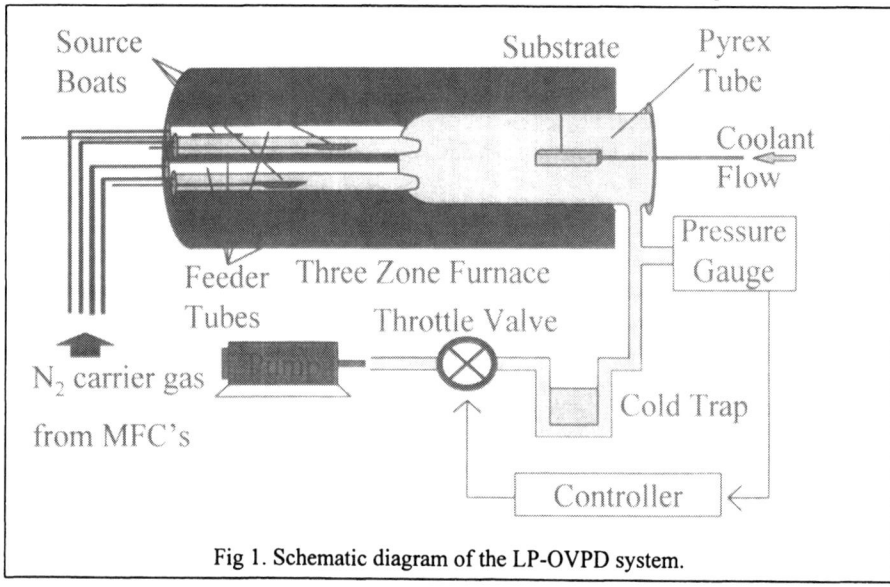

Fig 1. Schematic diagram of the LP-OVPD system.

The OLEDs were grown using both glass and flexible polyester substrates [5]. The flexible substrates are similar to those which might be used for high volume roll-to-roll production of OLEDs, using the LP-OVPD process. Both substrates were precoated with transparent layers of indium tin oxide (ITO). The ITO forms the anode of the OLED with a thickness of 1700Å and 1200Å [6] for the glass and polyester substrates, respectively. Glass substrates were cleaned by rinsing in a solution of detergent and finally rinsing in 2-propanol. To avoid damage due to exposure to organic solvents, the flexible substrates were cleaned by rinsing only in the detergent and 2-propanol.

During growth, the temperature in the center zone of the furnace was maintained at 300°C, decreasing towards the exhaust port with a thermal gradient of ~ -1°C/cm. The glass substrate was placed within the condensation zone where the temperature was ~ 220°C. The first layer deposited on the ITO surface was a hole transporting material, N'-diphenyl-N,N'-bis(3-methylphenyl)1-1'biphenyl-4-4'diamine (TPD). The TPD growth was accomplished using a source temperature of 200 ± 5 °C, a nitrogen carrier gas flow rate of 100 sccm, a reactor pressure of 0.50 Torr and a growth time of 20 minutes. The pressure was not controlled directly, rather it was a function of flow rate, pumping rate, and the diameter of the injection nozzle between each feeder tube and the reactor tube. At a flow rate of 100 sccm, the flow was laminar with a Reynolds number of ~ 500.

After growth, the TPD boat was retracted from the feeder tube hot zone, and the nitrogen flow was shut off. Next, an electron transporting layer of aluminum tris-(8 hydroxyquinoline) (Alq_3) was grown under the following conditions: a source temperature of 247 ± 8 °C, a nitrogen flow rate of 50 sccm, a pressure of 0.65 Torr and a growth time of 10 minutes. During both the TPD and Alq_3 growths, the substrate was maintained at 15 °C using a water cooled stainless steel substrate holder. The substrate was removed from the reactor by opening a flange positioned behind the exhaust port. Organic layers were also deposited on flexible substrates that can possibly be deployed from reels of material translated through the condensation zone during growth. The polyester substrates are more easily damaged by heat, making it necessary to reduce the furnace temperature to 280°C and to deposit thicker organic layers to compensate for possible interface nonuniformities.

For comparison purposes, an OLED with a similar structure to that grown by LP-OVPD was grown by thermal evaporation onto a glass substrate in vacuum. The glass/ITO substrate and top contact of the evaporated device were identical to those used in the LP-OVPD grown device. The evaporated device consisted of 350Å of TPD and 500Å of Alq_3. Finally, the top contacts for all devices were deposited by thermal evaporation in vacuum with a base pressure of 10^{-6} Torr, using a shadow masks to define arrays of 1mm diameter, 500Å thick, 25:1 Mg:Ag alloy cathodes. The contacts were completed with the evaporation of a 1000Å thick protective Ag layer. Since the operating pressure of LP-OVPD is compatible with cathode deposition by either sputtering or organometallic VPE, the thermal evaporation step could be eliminated in a commercial LP-OVPD system.

To examine LP-OVPD film structure and thickness, layers of TPD and Alq_3 were grown on silicon substrates. The RMS roughness of the LP-OVPD grown TPD film was measured using atomic force microscopy (AFM) to be from 6 to 8 Å, and that of the Alq_3 film was from 9 to 11 Å. Previously, it was found that Alq_3 and TPD films grown using thermal evaporation in vacuum had slightly smoother RMS roughnesses of ≤ 4.5 Å [7].

Ellipsometry was used to measure the thickness and uniformity of the films grown on silicon. The films grown with LP-OVPD were found to have a significant thickness gradient due to the temperature gradient within the condensation zone, and also to kinetic gas flow effects over the substrate surface. The Alq_3 and TPD films both had growth rate gradients of ~1.2 $Ås^{-1}cm^{-1}$ across the surface. Accounting for these gradients and also run-to-run variations, the TPD layer thickness in the OVPD-grown OLED was estimated to be between 100 Å and 300 Å, and the Alq_3 layer thickness was between 700 Å and 1100 Å.

The current-voltage (I-V) characteristics measured for all devices grown in this study are shown in Fig. 2. It is evident that the performance of the LP-OVPD device on a glass substrate is similar to the evaporated device, indicating that they both have similar organic layer thicknesses. The voltage at which the power law dependence [8] of I on V changes is the turn-on voltage, V_T.

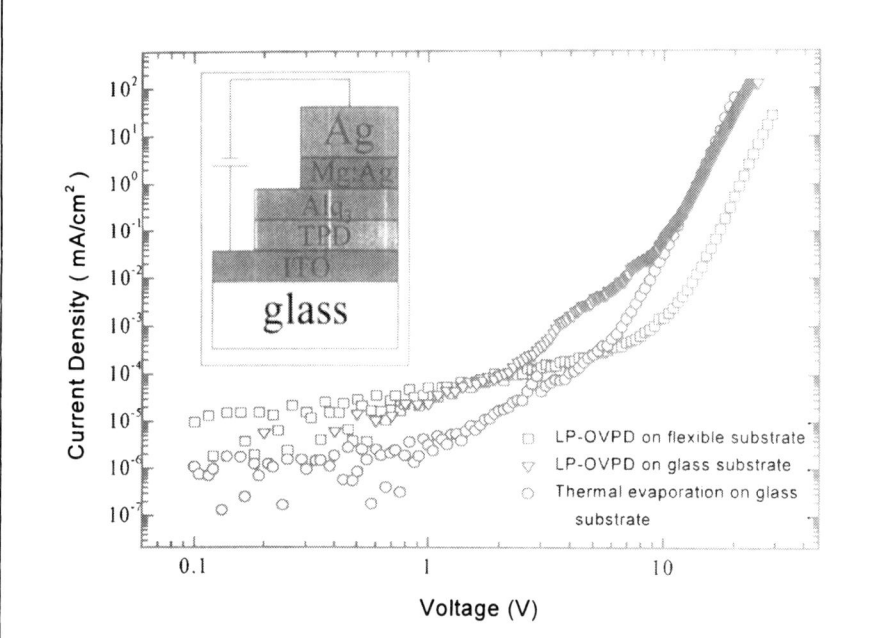

Fig 2. Current vs. voltage characteristic of 1mm diameter OLEDs grown by LP-OVPD on an ITO/glass substrate (inverted triangles), compared to a similar device grown by thermal evaporation in vacuum (circles). Also shown is the current vs. voltage (squares).

The LP-OVPD device on a glass substrate has $V_T = 6V$, compared to $V_T = 4V$ for the evaporated device. The thicker organic layers of the LP-OVPD device grown on a flexible substrate resulted in a somewhat increased V_T of 8V. The external quantum efficiencies of the devices were also both measured to be 0.40 ± 0.05%. The flexible device required 20V at 1mA/cm^2, and the external quantum efficiency was reduced to only 0.040 ± 0.005%, due to increased non-radiative recombination in the thicker layers. The electroluminescent spectrum of the LP-OVPD grown devices were consistent with that typical of Alq$_3$-based OLEDs, peaking at 510nm with a full width at half maximum of 100nm.

Since LP-OVPD uses individual carrier gases for each organic deposition source, independent control can be exercised over the deposition rate of multiple organic thin film materials. Assuming that the system can be modeled using boundary layer analysis, it can be shown that mass transport is proportional both to the partial vapor pressure of the source and to the square root of the carrier gas flow velocity [9]. Near the source solid to gas phase transition, vapor pressure varies rapidly with temperature, making accurate control of transport rates by varying source temperature difficult. However, if temperature can be precisely maintained, doping concentrations can be accurately controlled by varying carrier gas flow rates. Thus, to our knowledge, OVPD is the only method currently available for the highly accurate addition of small concentrations of dopants. Hence, LP-OVPD is ideally suited to the fabrication of OLEDs and organic lasers consisting of organic dye molecules doped into a host material [10]. Also, many common laser dyes are organic salts. OVPD may be the preferred method for co-deposition of such salts with host materials such as Alq$_3$ due to it demonstrated capability for controlled co-deposition of multiple materials with radically different vapor pressures [3].

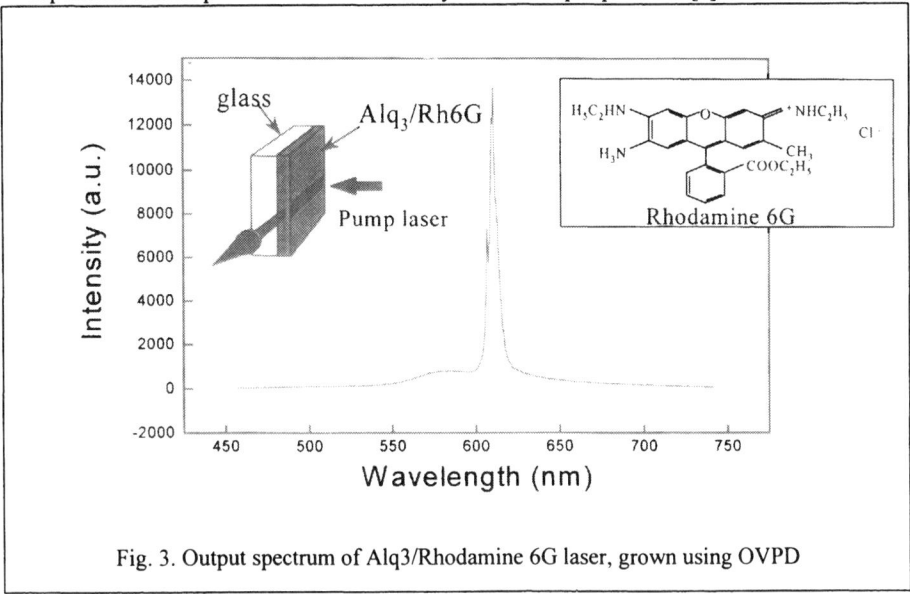

Fig. 3. Output spectrum of Alq3/Rhodamine 6G laser, grown using OVPD

To test these concepts, LP-OVPD was used to grow an optically pumped organic laser using Alq$_3$ as the host material and Rhodamine 6G [11], a typical organic salt laser dye. The laser was grown on a cleaned, 2 cm by 5 cm, glass substrate. The Alq$_3$ was co-deposited with the Rhodamine for 70 minutes at a pressure of 0.84 Torr, with nitrogen flow rates of 50 sccm for both materials. The Alq$_3$ source temperature was 251°C and the Rhodamine 6G temperature was 102°C. The laser thickness was approximately 3000Å, with facets defined by the edges of the glass slide. The device was optically pumped using a nitrogen laser (λ=337nm), which generates 500 ps pulses at a 50Hz repetition rate. Lateral confinement was achieved by focusing the pump beam into a 100 μm wide by 2cm long strip on the film surface, oriented orthogonal to the edges of the film. The threshold for lasing action was found to be 30μJcm^{-2} and the output spectrum above threshold is shown in Fig. 3. Host materials which possess a large Stokes shift between absorption and emission, such as Alq$_3$, can achieve low threshold lasing because they are essentially transparent to their own radiation. This effect can be enhanced by using multiple dopants, resulting in multiple transfers of energy between molecular species, such that the luminescence is far removed from the absorption band of the host [12]. LP-OVPD offers the opportunity to exploit this by the accurate control of concurrent doping of multiple small molecule materials.

The lasing frequency of ~ 610 nm depends on doping concentration but the LP-OVPD chamber used in these experiments lacks *in situ* thickness monitoring, making measurement of doping concentrations difficult. However, from comparison with devices grown using thermal evaporation in vacuum where the film was characterized using a crystal thickness monitor, the doping concentration of the LP-OVPD Rhodamine/Alq$_3$ laser was estimated to be 1%.

CONCLUSIONS

The new deposition technique of low pressure organic vapor phase deposition has been used to successfully grow single heterostructure OLEDs and optically pumped organic lasers based on small molecular weight organic thin films. Due to its high throughput rate and accurate control over doping concentration, LP-OVPD has advantages in the large scale production of small molecule organic thin film devices. In addition, LP-OVPD has the advantage of supporting metal contact deposition using metal organic sources used in OMVPE, forming a single system in which to deposit both the metal cathodes and organic thin films. Such a system can provide an integrated, inexpensive, high throughput solution for organic device fabrication.

ACKNOWLEDGEMENTS

The authors thank DARPA / Wright Laboratories, Universal Display Corporation and AFOSR for their financial support of this work. Also, thanks are due to Mike Souza for his invaluable glass blowing skills.

REFERENCES

1. M.A. Baldo, V.G. Kozlov, P.E. Burrows, S.R. Forrest, V.S. Ban, B. Koene and M.E. Thompson, Appl. Phys. Lett. **71**, 3033 (1997)

2. S.R. Forrest, P.E. Burrows, A. Stroustrup, D. Strickland and V.S. Ban, Appl. Phys. Lett. **68**, 1326 (1996)

3. P.E. Burrows, S.R. Forrest, L.S. Sapochak, J. Schwartz, P. Fenter, T. Buma, V.S. Ban and J.L. Forrest, J. Cryst. Growth **156**, 91 (1995)

4. MKS Barotron pressure gauge type 107, MKS instruments, Andover, MA, 01810-2449

5. G.Gu, P.E. Burrows, S.Venkatesh, S.R. Forrest and M.E. Thompson, Opt. Lett. **22**, 172 (1997)

6. Southwall Technologies Inc., 1029 Corporation Way, Palo Alto, CA, 94303, Part # 903-6011

7. P. Fenter, F. Schreiber, V. Bulovic and S.R. Forrest, Chem. Phys. Lett. **277**, 521 (1997)

8. P.E. Burrows, Z. Shen, V. Bulovic, D.M. McCarty, S.R. Forrest, J.A. Cronin and M.E. Thompson, J. Appl. Phys. **79**, 7991 (1996)

9. G.B. Stringfellow. Organometallic Vapor-Phase Epitaxy. Academic, London. 1989.

10. V.G. Kozlov, V. Bulovic, P.E. Burrows and S.R. Forrest, Nature **389**, 362 (1997)

11. U. Brackmann., Lambdachrome Laser Dyes. Lambda Physik Gottingen. 1997.

12. M. Berggren, A. Dodabalapur, R.E. Slusher and Z. Bao, Nature **389**, 466 (1997)

Photolithographic Patterning of Vacuum-Deposited Organic Light Emitting Devices

P. F. Tian, P. E. Burrows, V. Bulovic, and S. R. Forrest,
Department of Electrical Engineering, Center for Photonics and Optoelectronic Materials,
Princeton University, Princeton, New Jersey 08544

ABSTRACT

A photolithographic technique to fabricate vacuum-deposited <u>organic light emitting devices</u> has been demonstrated. The photolithographically patterned, yet unpackaged devices show no sign of deterioration in room ambient when compared with devices fabricated using shadow masks. Furthermore, the environmental robustness of the encapsulated devices makes them particularly useful in fabricating high resolution, full color displays.

INTRODUCTION

Organic light emitting devices (OLEDs) are of great interest for potential use in <u>flat panel displays</u> (FPDs) because of their high efficiency[1], long lifetime[2,3], low driving voltage, wide viewing angle, light weight and potential low cost[4-6]. High <u>resolution</u>, one of the key manufacturing challenges[4], has to be demonstrated in order to commercialize full color FPDs based on OLEDs using either polymers or small molecular weight vacuum-deposited thin films.

A typical small molecular weight OLED consists of organic multi-layers sandwiched between a transparent indium tin oxide (ITO) anode coated on a glass substrate, and a metal cathode such as an alloy of Mg and Ag[7].Both top and bottom electrodes need to be patterned to form OLEDs less than 100 μm in diameter to achieve high resolution and full color FPDs. While the bottom electrode is readily patterned using standard <u>photolithography</u> combined with wet etching, the top electrode is typically defined by a shadow mask. Because of 1) the low resolution of shadow masks and 2) the degradation or complete failure of OLEDs after exposure to water vapor, O_2, or solvents and developers used in the removal and patterning of photoresists[8-12], a photolithographic method using device encapsulation and/or dry processes must be established to microfabricate the top cathode for practical manufacturing purposes. Thick insulator walls combined with the evaporation of cathode materials at a suitable angle have previously been suggested as a means to micro-pattern cathode contacts[13]. Recently, micro-patterned cathodes were demonstrated by using thick photoresist strips in a multicolor OLED display[14]. In both cases, the exposed edges of active organic materials were subjected to chemical attack in subsequent processing. Thus, we have developed a photolithographic method to pattern top cathodes at high resolution and device yield without incurring a deterioration in the device performance. Furthermore, the completed devices are encapsulated by metal caps, making them resistant to attack from subsequent processing.

EXPERIMENT

Figure 1 shows the steps used in the fabrication process. Prior to thin film deposition, glass substrates pre-coated with ITO (10 Ω/□) were cleaned by steps described in Ref. 15. Next, a 2000 Å thick polyimide layer was spun on the substrate and cured in atmosphere. The OLED active regions, with dimensions of 300 μm × 1.5 mm spaced on 500 μm centers, were formed by photolithography and exposure to O_2 plasma reactive ion etching. This was followed by depositing a 2 μm thick SiO_2 layer using plasma-enhanced chemical vapor deposition. Next, a 2 μm thick photoresist layer was spun on and patterned using a second mask consisting of 400 μm × 1 cm strips centered over the holes in the

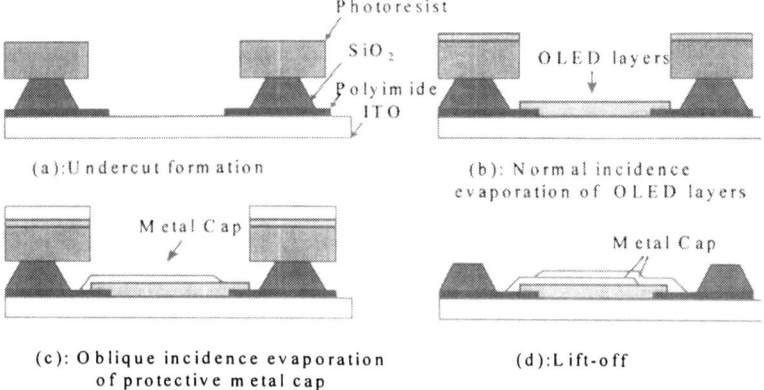

(a):Undercut formation

(b): Normal incidence
evaporation of OLED layers

(c): Oblique incidence evaporation
of protective metal cap

(d):Lift-off

Figure 1. Schematic diagram of steps used in the photolithographic patterning of vacuum-deposited, small molecular weight OLEDs

polyimide layer. Undercuts were formed by wet etching the SiO_2 layer (Fig. 1a). Next, the OLED layers were evaporated in vacuum under a base pressure of 3×10^{-6} Torr through the photoresist patterns in a direction normal to the substrate plane (Fig. 1b). The growth sequence was: 1) a 400Å thick α-NPD hole transport layer; 2) a 600 Å thick Alq_3 layer serving as both the light emitting and electron transport material; 3) a 250 Å thick Mg:Ag cathode co-evaporated with an atomic ratio of 24:1; 4) a 200 Å thick Ag protection layer. Next, a 0.5 μm thick Ag layer was obliquely deposited onto the substrate to cover one side of the devices (Fig. 1c), followed by rotating the substrate 180° and evaporating a second 0.5 μm thick Ag layer to cover the opposite side of the devices. This process can be simplified by installing a planetary mount into the vacuum chamber which allows for substrate rotation during deposition. Finally, lift-off of the photoresist mask was done by soaking in acetone for several minutes, followed by drying in pure N_2 (Fig. 1d). A more detailed description of the process can be found in Ref. 15.

RESULTS

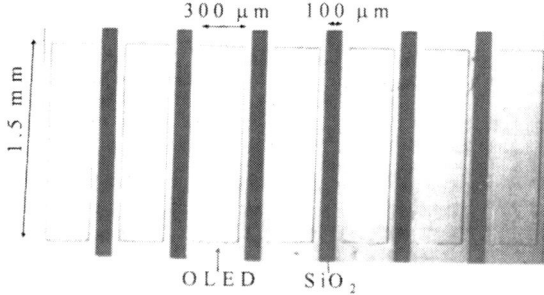

Figure 2. Photograph of seven completed, photolithographically patterned devices.

The overall smooth appearance of the devices shown in Fig. 2 indicates no obvious solvent penetration into the organic materials due to the encapsulation by the Ag caps. A few defects, possibly caused by particles preexisting on the substrate, or introduced during subsequent film deposition, can be eliminated by proper handling in a suitably clean environment.

Figure 3 shows the forward biased current-voltage (I-V) characteristics of the devices before and after lift-off. The inset shows the electroluminescence (EL) spectrum of a device after lift-off. Both the I-V characteristics and EL spectrum are similar to those previously reported for conventional devices

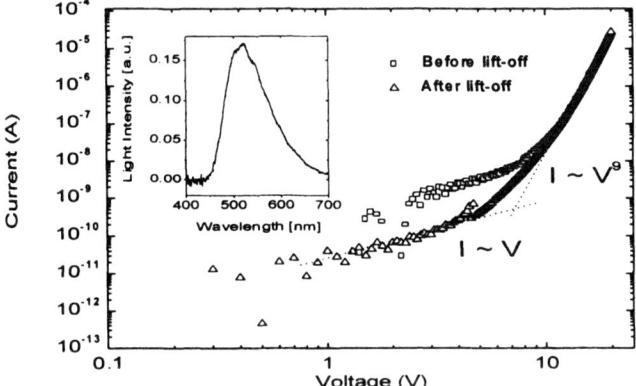

Figure 3. Current vs. voltage characteristics of a 0.45mm^2 device obtained prior to and after photoresist lift-off in acetone.
Inset: Electroluminescence spectrum of the patterned device.

fabricated by shadow mask techniques[16]. We note that a few devices were shorted before lift-off, possibly due to metal bridges causing shorts between the metal on the photoresist surfaces and the device caps. After lift-off, however, all devices were electrically isolated. Figure 4 shows the light output vs.

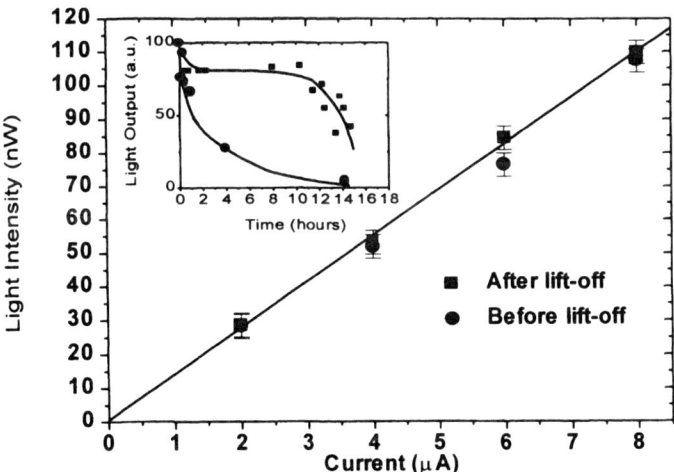

Figure 4. Light output vs. current characteristics of a 0.45mm^2 device obtained prior to and after lift-off.
Inset: Time evolution of the light output of the cathode patterned device at a constant current density of 6.4 mA/cm^2 (squares), and of the conventional unpackaged 0.1cm^2 devices fabricated using shadow mask technique at a constant current density of 10mA/cm^2 (circles).

The initial luminance was 130 cd/m² for patterned device and 30 cd/m² for the conventional devices.

current (L-I) characteristics for devices prior to and after lift-off. The external quantum efficiencies measured only in the forward scattered direction are ~ 0.6%, typical of undoped Alq₃-based OLEDs. The similarity between devices before and after lift-off, as shown in Figs. 3 and 4, implies that the thick metal cap can prevent solvent from penetrating into the organic layers during lift-off, the only wet process after the deposition of organic materials. About 70% of the devices after lift-off were defect-free, exhibiting I-V and L-I characteristics as shown in Figs. 3 and 4. The remaining 30% either had a low quantum efficiency or did not emit light due to defects. The yield should approach 100% if the ITO surface is properly prepared to remove defects and particles prior to device fabrication.

The inset of Fig. 4 shows the time evolution of light output from a typical, unpackaged patterned device (squares) obtained at a constant current density of 6.4mA/cm². Prior to the test, the devices were exposed to air (humidity ~ 60%) at room temperature for approximately 100 hours. The lifetime, at which the luminance dropped to half of its initial value, was ~ 14 hours. Failure of the device was due to a 300 μm diameter dark spot, attributed again to particles on the ITO surface. The time evolution of the average light output of unpackaged devices fabricated using conventional shadow mask methods (circles)[17] is shown for comparison. The lifetime of the devices, tested immediately after fabrication, was ~ 2 hours. Both tests were performed at room ambient and temperature. In contrast to the patterned OLEDs, prolonged exposure of unpackaged, conventional devices leads to rapid degradation and a "shelf-life" of only several hours. Though more systematic studies are required, it is very likely that the unpackaged patterned devices have significantly longer shelf and operational lifetimes than unpackaged, conventional OLEDs. The environmental robustness of the encapsulated devices allows for prolonged exposure to the environment should subsequent device processing or handling be required prior to packaging.

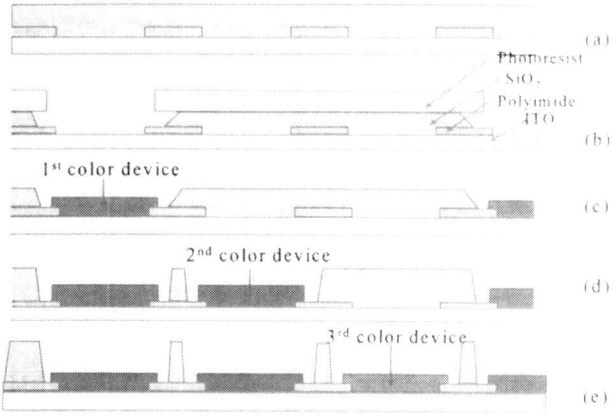

Figure. 5 Schematic diagram of photolithographically patterning side by side 3 color devices

A side-by-side 3 color display can be fabricated using this method as follows: 1) Form a polyimide pattern on ITO pre-coated glass substrate, followed by deposition of a 2 μm thick SiO₂ (Fig. 5a); 2) Spin on a 2 μm thick photoresist and pattern it to open windows for subsequent OLED layer depositions. This is followed by wet etching the SiO₂ to form the undercuts for the first color devices (Fig. 5b); 3) Follow the same procedures shown by Fig. 1b to 1d to obtain the first color devices (Fig. 5c). 4) Repeat steps 2) and 3) for the second color devices (Fig. 5d) and then the third (Fig. 5e). The aperture ratio of a 3 color pixel equals active region diameter/(active region diameter + 3 × SiO₂ width + 6 × insulator pattern tolerance + 6 × undercut gap), which is 80% for a 150 μm active region diameter, a 5 μm SiO₂ width, a 2 μm insulator pattern tolerance and a 2 μm undercut gap.

This patterning technique can also be applied to accessing the leads of a recently demonstrated three color stacked OLED (SOLED) which offers a three-fold improvement in resolution and display fill-factor as compared to the side-by-side strategies[18]. Fig. 6a shows the schematic cross section of a

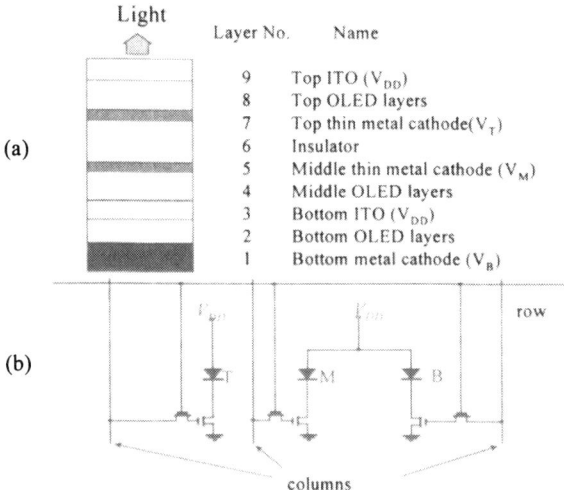

Figure 6. a) Schematic cross section of a SOLED with an insulating layer between the top and middle devices. b) A full color pixel driven by active matrix.

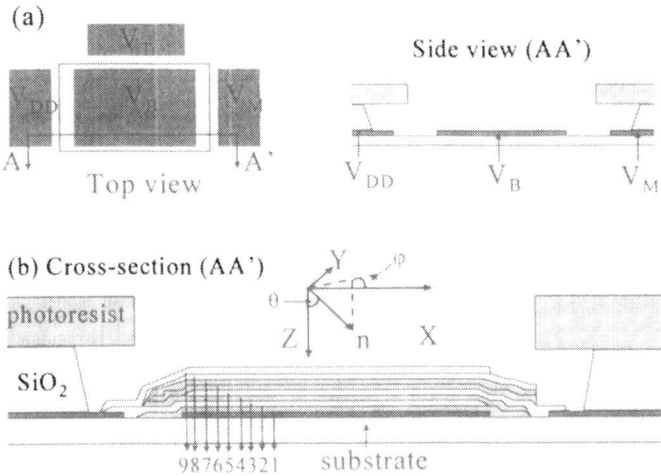

Figure 7. The micro-fabrication steps used to pattern the SOLED as shown in Fig. 6a.

SOLED with an insulating layer inserted between the top and middle devices so that only six transistors are needed in the active matrix driving scheme as shown by Fig. 6b[19]. The devices share the same anode: V_{DD}, while each has its own cathode: V_B, V_M and V_T. A SOLED can be micro-patterned by 1) forming four contact pads for V_{DD}, V_B, V_M and V_T on the substrate, followed by fabricating the undercut pattern using SiO_2 and photoresist(Fig. 7a); 2) evaporating layers 2 to 9 at the following angles: for layers 2, 4, 6 and 8, $\theta = 0°$; for all the other layers, $\theta = 45°$, and $\varphi = 180°$ for layer 3 and 9, $0°$ for layer 5 and $90°$ for layer 7. θ and φ are defined in Fig. 7b. Thus, proper contacts are made between SOLED and the pre-patterned contact pads on the substrate. The contact between the top device ITO and V_T is not shown in Fig. 7b.

CONCLUSIONS

We have demonstrated a simple photolithographic method capable of microfabricating vacuum-deposited, small molecular weight organic devices such as OLEDs, without apparent damage to the devices. The OLEDs are encapsulated by a thick metal layer, significantly increasing shelf and operational lifetimes. The extended shelf-life is important if the devices require subsequent processing steps after patterning, as is the case in fabricating multicolor displays. This simple patterning technique illustrates a potential foundation for the volume production of very high resolution, multicolor flat-panel displays or other active optoelectronic devices based on small molecular weight vacuum-deposited organic thin films.

ACKNOWLEDGEMENTS

The authors thank Universal Display Corp. and DARPA/Wright Labs for their support.

REFERNCES

1. S. A. VanSlyke, and C. W. Tang, 1995 Digest of LEOS Summer Topical Meetings, p. 3 (1995).
2. J. Shi, and C. W. Tang, Appl. Phys. Lett., 70, 1665, March (1997).
3. S. A. VanSlyke, C. H. Chen, and C. W. Tang, Appl. Phys. Lett., 69, 2160 (1996).
4. L. J. Rothberg, and A. J. Lovinger, J. Mater. Res., 11, 3174 (1996).
5. P. E. Burrows, S. R. Forrest, and M. E. Thompson, Current Opinion in Solid State & Materials Science, Vol 2, No 2, 236 (1997).
6. S. R. Forrest, P. E. Burrows, and M. E. Thompson, Laser Focus World, 99, Feb. (1995).
7. C. W. Tang, and S. A. VanSlyke, Appl. Phys. Lett., 51, 913 (1987).
8. D. Yap, P. E. Burrows, and S. R. Forrest, Organic Thin Films for Photonics Applications, Technical Digest, 302 (1995).
9. R. B. Taylor, P. E. Burrows, and S. R. Forrest, IEEE Photonics Technology Letters, Vol. 9, No. 3, 365, March (1997).
10. C. C. Wu, J. C. Sturm, R. A. Register, and M. E. Thompson, Appl. Phys. Lett., 69, 3117 (1996).
11. D. G. Lidzey, M. A. Pate, M. S. Weaver, T. A. Fisher, and D. D. C. Bradley, Synthetic Metals, 82, 141 (1996).
12. F. Papadimitrakopoulos, X. Zhang, D. L. Thomsen, and K. A. Higginson, Chem, Mater, 8, 1363 (1996).
13. C. W. Tang, U.S. Patent No. 5 276 380; C. W. Tang, and J. E. Littman, U. S. Patent No. 5 294 869; C. W. Tang, D. J. Williams, and J. C. Chang, U. S. Patent No. 5 294 870.
14. C. Hosokawa, M. Eida, M. Matsuura, K. Fukuoka, H. Nakamura, and T. Kusumoto, SID 97 Digest, 1073 (1997).
15. P. F. Tian, P. E. Burrows, and S. R. Forrest, Appl. Phys. Lett., 71, 3197, December (1997).
16. P. E. Burrows, Z. Shen, V. Bulovic, D. M. McCarty, S. R. Forrest, J. A. Cronin, and M. E. Thompson, J. Appl. Phys. 10, 79, May (1996).
17. P. E. Burrows, V. Bulovic, S. R. Forrest, L. S. Sapochak, D. M. McCarty, and M. E. Thompson, Appl. Phys. Lett., 65, 2922 (1994).
18. Z. Shen, P. E. Burrows, V. Bulovic, S. R. Forrest, and M. E. Thompson, Science, 276, 2009, 1997.
19. G. Gu, P. E. Burrows, and S. R. Forrest, IEEE J. Selected Topics in Quantum Electronics, submitted.

Moisture Effect on Polythiophene Electroluminescence

TSU-CHIEN WENG[a1], BIH-YAW JIN[a*], DING-KUO DING[b], MITSUHIKO SAKAKIBARA[c]

[a]Department of Chemistry, National Taiwan University, Taipei, Taiwan 106 ROC
[b]TECO Electric & Machinery Co., LTD., Taipei, Taiwan 102 ROC &International Center for Materials Research, Kawasaki, Kanagawa 213 Japan
[c] Japan Synthetic Rubber Co., LTD. Tsukuba, Ibaraki 305, Japan

ABSTRACT

The moisture effect on the neutral and charged thiophene oligomers was investigated using *ab initio* methods. The influence of water molecules on the geometric relaxation of the neutral and charged tert-thiophenes is found to be important. The torsional distortion in neutral water/tert-thiophene complexes is larger than that of undoped or charged complexes. The relative orientation of the water molecule to thiophene changes dramatically upon the hole injection. The existence of water molecules in the polythiophene suppresses the hole mobility and, therefore, diminishes the efficiency of electroluminescence.

INTRODUCTION

The electroluminescence from the conjugated polymeric materials has been under extensive experimental and theoretical studies in the last few years. It is now possible to make light-emitting diodes (LED) by using many conjugated polymers such as polyphenylenevinylene (PPV) and polythiophene (PT) as the active materials[1]. To improve the efficiency of LED, it is important to understand the intrinsic and extrinsic factors that influence the radiative and nonradiative relaxation channels. There are studies on the extrinsic factor of exposing the LED devices based on PPV and PT to the air. The performance of the devices is considerably affected due to the interaction between polymers and the water molecules in the air[2].

In order to understand the underlying mechanism of the interaction between water molecules and conjugated polymers in the LED devices, we first need to identify the major species in the conjugated polymers[3]. In a typical electroluminescence experiment, electrons and holes are injected into the polymeric material from the electrodes. Unlike the traditional semiconductors, the electrons and holes interact strongly with the polymer chains. Therefore, strongly bound electron-phonon quasi-particles such as electron- and hole-polarons will form quickly. These quasi-particles move around until they meet together and form singlet excitons, and then radiate through electron-hole recombinations.

The mobilities of these quasi-particles are quite different. The hole mobility is usually higher than the electron mobility. The reason for the difference in the mobilities of electrons and holes is due to the different strength of electron-phonon coupling. When an electron is taken away from a conjugated molecule, a hole polaron, which is just a planar positive carbocation consisting of an empty p_z orbital and three occupied sp^2 hybrid orbitals, forms. The local distortion induced by the hole is small in this situation. However, if the conjugated

---

[1]Current address: Department of Chemistry, University of Michigan

Mat. Res. Soc. Symp. Proc. Vol. 488 © 1998 Materials Research Society

polymer is reduced by adding an electron into the chain, a carboanion will form. This extra electron tends to combine one of the π-electrons in the backbone and forms a lone electron-pair which occupies an sp^3 orbital. The stronger interaction between the lone pair and the saturated σ-bond produces a much larger out-of-plane electron-phonon interaction for the electron-polarons. The electron mobility is suppressed strongly. One can assume that the majority of charge carriers in the device are hole polarons. Therefore, it is reasonable to consider only two forms of the polythiophenes, neutral and positively charged, in the studies of the interaction with water molecules.

METHOD AND EXPERIMENT

We studied the effect of moisture on the PT electroluminescence by carrying out quantum-chemical studies on model systems composed of neutral (or positively charged) thiophene oligomers and water molecules. To minimize the chain-end effect, we chose the tert-thiophene which contains only three thiophene rings in the present study. The quantum-chemical calculations were carried out by several steps in order to minimize computational cost. The geometry of a water/tert-thiophene complex was first optimized using AM1 semiempirical method in order to get a preliminary molecular conformation. Thereafter, a precise geometry optimization using Hartree-Fock (HF) and density functional theory methods with 6-31G* basis set built in the Gaussian94 program package[4] was carried out. The functionals we used were Becke's three parameter hybrid method with the Lee-Yang-Parr correlation functional. All geometries were optimized without using any constraint.

The LED devices for the electroluminescence experiment consist of an ITO glass electrode, a subsequently spin coated polyhexylthiophene film, and an Al vapor evaporated cathode. These samples were kept inside a closed box under ambient atmosphere with chemicals as drying agents, which are corresponding to different relative humidity. The relative humidities inside closed box for P2O5, KOH, and NaOH are 0.0001%, 0.1%, and 3%, respectively. The electroluminescence spectra were measured by Shimadsu RF-5000 fluorophotometer at room temperature (see Figure 1). The luminance were measured by Minolta LS-100 luminance meter. The measured luminances of PT LED devices at 4.5 voltage are 0.08, 0.047, and 0.006 (cd/m^2).

RESULTS AND DISCUSSION

The bond lengths of the optimized geometries of neutral and positively charged water/tert-thiophene complexes found in these calculations are listed in the table 1. The backbone structure of an undoped tert-thiophene is actually a cis-transnoid polyene with linked sulfur atoms. The bond length alternation pattern is similar to that of polyenes but with a smaller dimerization amplitude. The bond lengths of carbon-carbon single bonds and double bonds are 1.42 and 1.38 Å, respectively. The inter-ring carbon-carbon single bonds (~ 1.45 Å) are longer than single bonds in the thiophene rings. Since the lengths of carbon-sulfur bonds are longer than 1.7Å, the major effect of heteroatom is the inductive effect and the mesomeric effect is not important.

From our previous studies based on the semi-empirical QCFF/PI model[5], the lattice distortion of a polythiophene is similar to that of an all-trans polyene[6]. When an electron is removed from the backbone of a chain with infinite length, a localized charged polaron

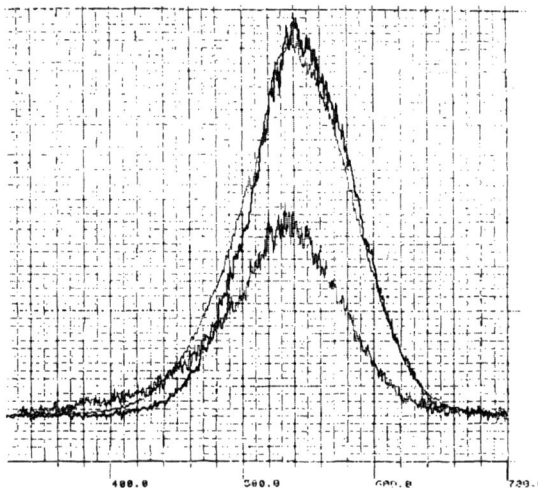

Figure 1: The electroluminescence spectra of polyhexylthiophene measured by Shimadsu RF-5000 fluorophotometer at room temperature.

	Bond	$Th_3(\text{Å})$	I	II	III	Th_3^{+1}	I^{+1}	II^{+1}	III^{+1}
Ring 1	a	1.74	1.74	1.73	1.74	1.72	1.72	1.72	1.72
	b	1.37	1.37	1.37	1.37	1.39	1.38	1.38	1.39
	c	1.42	1.42	1.42	1.42	1.40	1.40	1.40	1.40
	b'	1.38	1.38	1.38	1.38	1.40	1.40	1.40	1.40
	a'	1.76	1.75	1.76	1.76	1.77	1.76	1.76	1.76
Inter-Ring	d_{12}	1.45	1.45	1.45	1.45	1.41	1.41	1.41	1.41
Ring 2	a	1.76	1.75	1.76	1.76	1.76	1.75	1.76	1.76
	b	1.38	1.38	1.38	1.38	1.41	1.42	1.41	1.41
	c	1.42	1.42	1.42	1.42	1.38	1.38	1.38	1.39
	b'			1.38				1.41	
	a'			1.76				1.75	
Inter-Ring	d_{23}			1.45				1.41	
Ring 3	a			1.76				1.76	
	b			1.38				1.40	
	c			1.43				1.40	
	b'			1.37				1.39	
	a'			1.74				1.42	

Table 1: Optimized carbon-carbon bond lengths of the neutral tert-thiophene (Th_3), three neutral water/tert-thiophene complexes (I, II, III), charged tert-thiophene(Th_3^+) and three charged complexes (I^+, II^+, III^+). See Fig. 2 for the definition of symbols.

forms and extends around ten unit cells. However, for a tert-thiophene with only three unit cells, the hole polaron (Th_3^+) is strongly confined and the corresponding bond length alternation (BLA) becomes smaller with the carbon-carbon bond lengths spread over 1.38-1.41Å(Table 1). The bond lengths of single and double bonds in the central thiophene ring of a charged tert-thiophene are reversed in magnitude. This is reminiscent of the soliton in polyacetylene, albeit the ground state of polythiophene is not degenerate[3].

There are three stable conformations of the neutral water/tert-thiophene complexes (Fig. 3), which are obtained by taking the optimized structures of the HF level as the initial guesses for the geometry optimization. Two of these conformations, I and II, are considerably distorted from the molecular plane. The water molecule is located on the side of the molecular plane and close to the sulfur atom in each of the three cases. The strong interaction between the hydrogen atom of water molecule and the sulfur atom confines the water molecule to the sulfur side of the thiophene. There were several attempts of initial configuration other than the HF geometries were made, putting the water molecule with different O-H bond orientations on top and between the thiophene rings, and they all optimized to the same minimum conformation, complex II. This might suggest that complex II is more stable than the other two conformation. This result is consistent with the previous work of the water/PPV complex[2], where there are also three stable structures.

Using a similar optimization procedure, we have obtained three stable conformations for positively charged water/tert-thiophene molecule (Fig. 3). The planar feature of the tert-thiophene molecule is preserved in all three structures. The water molecule is always found close to the sulfur atom of the center thiophene ring. However, the distance between the hydrogen atom of the water molecule and the sulfur atom of the tert-thiophene complex varies from 2.7 Å (III^+) to 3.8 Å (II^+), while it is almost constant, around 2.7 Å, for the neutral complex structures.

Comparing the corresponding bond lengths of neutral and charged species, it is obvious that the major perturbation of tert-thiophene BLA comes from charging of the thiophene molecule. The Mulliken population analysis indicates that there is some positive charge on the water molecule, however, the effect of partial charge transfer to the water molecule on BLA seems to be minute. The existence of water molecule only slightly perturbs thiophene BLA for both neutral and charged complexes.

On the other hand, the existence of water molecules does cause notable changes in the inter-ring angles for neutral complexes. The torsional angles between two thiophene rings are more than 25° for complexes I and II. When the water molecule was introduced to the planar neutral tert-thiophene, the geometry of the tert-thiophene began to relax via an inter-ring torsional channel. The position and the orientation of the water molecule are crucial to the relaxation. As in the case of complex III, the torsional relaxation is negligible due to the symmetry of the system. In the case of the charged complex, the tert-thiophene molecule is almost planar with no torsional angle between rings in each of the three conformations. The influence of the water molecule on the torsional distortion is small in these cases. The preservation of molecular planarity of charged thiophene might be due to the subtle balance between the effect of positive charge and the perturbation of the water molecule.

The binding energies of water molecule to both neutral and charged thiophene are listed in Table 2. These energies are all small, which is consistent with the experimen-

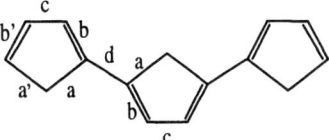

Figure 2: The definition of C-C bonds used in Table 1.

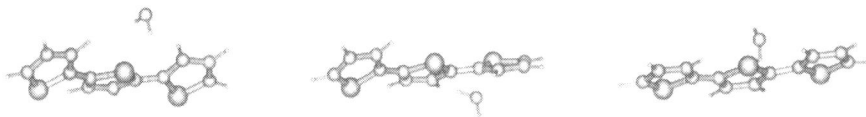

Figure 3: Three optimized geometries of neutral water/tert-thiophene complexes. (Left: I, center: II, and right: III)

I	II	III	I$^+$	II$^+$	III$^+$
-3.52	-4.73	-3.26	-7.60	-5.08	0.05

Table 2: The binding energies (in kcal/mol) of water to the neutral and positively charged complexes.

tal observation that the moisture effect is reversible. The three conformations of neutral water/tert-thiophene appeared to have similar water binding energies, the difference is less than $1.5 kcal/mol$, therefore, it is very likely that they all exist with roughly the same populations. For charged system, the difference of binding energies is more significant. Structure II$^+$ becomes energetically unfavorable, and I$^+$ the most stable structure.

The energy differences for the charged structures can be attributed to the orientation of the water molecules. In Figure 3, the neutral structures all have the hydrogen atoms pointing toward the sulfur atoms, as expected. Figure 4, however, shows that the water molecules orient differently in the positively charged complex structures. Structures I$^+$ and II$^+$, with different water to thiophene distances, both have the oxygen atom pointing toward the sulfur atom of the middle thiophene ring. Since the oxygen of the water molecule has the ability to donate electronic density, it is energetically favorable to directs itself toward the positively charged tert-thiophene. The molecular structures of III and III$^+$ are almost the same, both of them have an approximate C_{2v} symmetry with two hydrogen atoms pointing to the sulfur atom of the middle thiophene ring. This makes the complex III$^+$ less stable. After the injection of electrons and holes, the amount of the charged conformation III$^+$ will decrease gradually, although all three neutral species co-exist almost equally before the electron-hole injection.

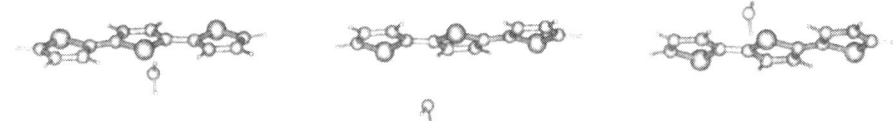

Figure 4: Three Optimized geometries of charged water/tert-thiophene complexe. (Left: I^+, center: II^+, and right: III^+)

The blue shift in the electroluminescence spectra is due to the extra luminescent centers created by the existence of water molecule in the neutral complex. The torsional distortion of the neutral water/tert-thiophene complex interrupts the effective conjugation length and, thus, increases the optical gap. Using the SCI calculations at AM1 level[7], we found that the vertical gap is 2.55 eV for the planar thiophene and 2.74 eV for a distorted tert-thiophene with a torsional angle of 30°. The electroluminescence spectra in Figure 1 also show some broadening effect, which is due to the inhomogeneous distribution of luminescent centers with different effective conjugation lengths induced by water molecules.

CONCLUSIONS

From the results of quantum-chemical calculation, we observed that the major effect of moisture is to produce considerable torsional distortions in the neutral water/tert-thiophene complex and to change the relative stabilities of different conformations. The torsional deformation in the neutral complex produces a large Franck-Condon factor for the intermolecular hole transfer from a charged to a neutral water/tert-thiophene complex. Besides the large Franck-Condon factor, the extra stability of the charged complex can also trap or slow down the motion of the holes.

Since all extra paths for the hole migration in the presence of moisture are unfavorable, the electron and hole recombination rate is reduced and the electroluminescence consequently quenches. Through the quantum-chemical calculation on the stabilities and geometries of neutral and charged polythiophene complexes, we have clarified the major effect of water molecules on the electroluminescence of LED based on polythiophene materials.

ACKNOWLEDGEMENTS

B.-Y. Jin and T.-C. Weng acknowledge the National Science Council for the financial support for this work.

References

[1] J. H. Burroughes, D. D. C. Bradley, A. R. Brown, R. N. Marks, K. Mackey, R. H. Friend, P. L. Burns, and A. B. Holmes. *Nature*, 347:539, 1990.

[2] K. Xing et al. *Adv. Mater.*, 8(12):971, 1996.

[3] A. J. Heeger, S. Kivelson, J. R. Schrieffer, and W. P. Su. *Rev. Mod. Phys.*, 60:781, 1988.

[4] Gaussian (Revision), M. J. Frisch, G. W. Trucks, H. B. Schlegel, P. M. W. Gill, *et al.*, Gaussian, Inc. Pittsburgh, PA, 1995.

[5] A. Warshel and M. Karplus. *J. Am. Chem. Soc.*, 94:5612–5625, 1972.

[6] Tsu Chien Weng and Bih-Yaw Jin. unpublished.

[7] M.J.S. Dewar, E. G. Zoebisch, E. F. Healy, and J. J. P. Steward. *J. Am. Chem. Soc.*, 107:3902–3909, 1985.

DESIGN OF FLAT PANEL DISPLAYS
BASED ON ORGANIC LIGHT EMITTING DEVICES

G. Gu, P. E. Burrows, and S. R. Forrest
Center for Photonics and Optoelectronic Materials (POEM), Department of Electrical Engineering and
Princeton Materials Institute, Princeton University, Princeton, NJ 08540

ABSTRACT

We provide a systematic and quantitative analysis of the design of flat panel displays (FPDs) based on organic light emitting devices (OLEDs). Key performance parameters are estimated for OLED-based displays; system issues, including addressing schemes and strategies required to achieve full color displays, are treated quantitatively. Furthermore, addressing schemes for recently demonstrated, full color, high resolution stacked OLEDs are discussed. Our analysis shows that OLED technology is suitable for many FPD applications, and can provide performance superior to that achieved using alternative display technologies.

INTRODUCTION

The rapid progress in organic light emitting devices (OLEDs) creates great potential for developing lightweight, high resolution, full color flat panel displays (FPDs).[1-3] In this paper we report a systematic study of the design of OLED-based FPDs. We focus on high performance small-molecule OLEDs.

ANALYSIS AND RESULTS

Photometry and Efficiency

The structure of a conventional small-molecule OLED remains essentially unchanged since their first demonstration.[4] This study also considers a transparent OLED (or TOLED),[5] where a very thin metal film (~100 Å) combined with an ITO cap is used as the transparent cathode. The TOLEDs are 60 ~ 80% transparent in the visible spectral region.

Consider an OLED with an emissive area A, driven by a voltage V and a current density J, emitting at a quantum efficiency η. The angular distribution of luminous intensity is near Lambertian.[6] The luminance, $L(\theta)$, in direction θ from the normal increases monotonically from $L(0)$ to $L(90°) = 1.24L(0)$, for a refractive index of the light emitting material of $n_a = 1.7$ which is typical for organic EL materials such as tris-(8-hydroxyquinoline) aluminum (Alq$_3$).[7] Hence, the viewing cone of an OLED extends over the entire forward hemisphere.

The current density required to achieve a given luminance, $L(0)$, is: $J = C_n C_E \pi L(0) / (\eta C_V)$. Here, C_n, C_V, and C_E are constants determined by the EL material: $C_n \equiv 2n_a^2(1 - \sqrt{1 - 1/n_a^2})$; $C_V \equiv 683 \int_0^\infty V(\lambda)\phi(\lambda)d\lambda$, where $\phi(\lambda)$ is the OLED emission spectrum normalized to $\int_0^\infty \phi(\lambda)d\lambda$ $=1$, $V(\lambda)$ is the photopic vision function;[8] and $C_E \equiv q \int_0^\infty (\lambda / hc)\phi(\lambda)d\lambda$, where h is Planck's constant, c is the speed of light, and q is the electron charge. For undoped Alq$_3$-based OLEDs, C_n=1.1, C_V = 427 lm/W, and C_E = 0.44 V^{-1}. A luminance of 100 cd/m^2, usually considered sufficient for video displays, requires a current density of only 3.6 mA/cm^2 for $\eta = 1\%$, typical of undoped Alq$_3$-based devices. The energy conversion efficiency is $\eta_E = \eta / C_E V$, and the luminous efficiency is $\eta_L = C_V \eta_E$. Figure 1 shows the energy conversion and luminous efficiencies versus operating voltage for Alq$_3$-based OLEDs with η as a parameter. The OLED luminous efficiency under typical operation conditions is higher than that achieved with plasma displays, and superior to or comparable with that of LCDs, as shown in Fig. 1.

The high reflectance of the OLED alloy cathode leads to low contrast. Here we quantify the contrast by defining the discrimination ratio as the ratio of the brightness of an "on" pixel to that of an "off" pixel. Assuming that the brightness of an "off" OLED is solely due to the reflection of the uniform ambient illuminance, E_a, the discrimination ratio is: $D = 1 + 2\pi L_d / (RE_a)$. Here, L_d is the display brightness which may differ from the OLED instantaneous luminance, $L(0)$, due to addressing, and R is the reflectance of the device. For $L_d = 100$ cd/m^2, $R = 80\%$, and $E_a = 500$ lux (ANSI standard), D is only 2.6. A circular polarizer (CP) placed in front of the display is found to increase D to 113 at a ~3dB power penalty (ΔP). ΔP is defined as the increase in *electrical* power resulting from a particular contrast-enhancing, addressing, or full color scheme to achieve the same brightness which would be achieved without employing the scheme. To compare this and the following schemes, we assume that the front surface of the display is anti-reflection coated. Polarizers of any type reduce the useful viewing angle of a display to 45° from the normal.[9] Another problem is cost and complexity. An alternative scheme employs an absorber with an optical transmittance, T_a, placed in front of the display. In this case, $D = 1 + 2\pi L_d / (T_a RE_a)$. If $T_a = 0.5$, then D increases to 7.3 for a 100 cd/m^2 perceived brightness and $\Delta P \approx 3$ dB. Finally, a high contrast display can also be realized using TOLEDs[5] with an absorbing background. The discrimination ratio increases to 13.6 at a ~3 dB power penalty for a 100 cd/m^2 perceived brightness, assuming a 10% residual reflectance for the semitransparent cathode.[5] This simple scheme provides sufficient improvement in D at the same power penalty as the other schemes, without encountering problems of reduced viewing angle and extra cost arising from expensive optics and filters.

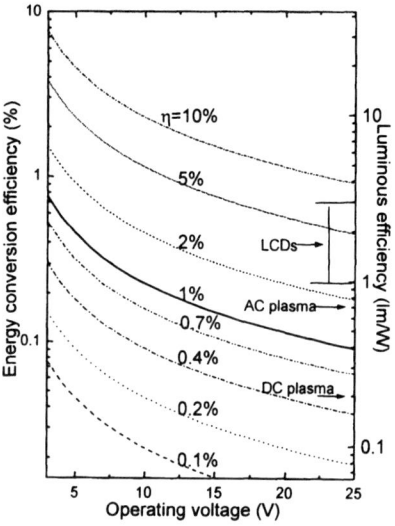

Fig. 1. Current density versus luminance curves for non-doped Alq$_3$-based OLEDs.

Addressing Schemes for Monochromatic Displays

A passive matrix OLED (PM OLED) array consists of electrically isolated rows and columns arranged orthogonally with an OLED at each intersection, as has been demonstrated.[10] To achieve gray levels, the column drivers must be current sources since the pixel luminance is proportional to the drive current, and the OLED I-V characteristics are highly nonlinear.[11] Here, we consider an N (columns) × M (rows) PM array of identical OLEDs, with W_r wide rows separated by gaps of width D_r, and W_c wide columns separated by gaps of width D_c. An addressed pixel must be driven with a pulsed current at a duty cycle, d, to avoid catastrophic, irreversible breakdown (or shorting) caused by long term, unintentional reverse bias on the non-selected pixels. For a display addressed in the row-at-a-time mode to achieve an average display luminance of L_d, the pixels must be driven to an instantaneous luminance of: $L(0) = ML_d/(r_d d)$, where r_a is the aperture ratio. This requirement limits the number of rows in a display. Hosokawa *et al* reported[12] a pulse emission of 5×10^4 cd/m^2, enabling 250-line PM addressing for $d = 0.5$, $r_a \sim 1$, and $L(0) = 100$ cd/m^2. Even a $\sim10^5$ cd/m^2 instantaneous luminance will limit row number to 500 for $d = 50\%$ or ~1000 for $d \approx 100\%$.

Leakage currents flowing through non-selected, reverse-biased pixels limit the display size. For a display with N_g gray levels based on high rectification ratio OLEDs, the condition, $f(V)/f_{rev}(V-2V_{on}) \geq MNN_g$, must be satisfied to resolve adjacent gray levels, where V_{on} is the turn-on voltage (defined as the voltage at which the currents due to ohmic and trap-limited conduction are equal),[11] and $I = f(V)$ and $I_{rev} = f_{rev}(V_{rev})$ are the forward and reverse I-V dependence, respectively. For OLEDs with $V_{on} = 3.5$ V operating at $V = 10$ V, the ratio $f(V)/f_{rev}(V - 2V_{on})$ is typically $\sim10^5$, limiting the pixel number, $N \times M$, to $\sim(84 \times 72) = 6048$ for a display with 16 gray levels.

The voltage drop on a row line is $V_r = \dfrac{C_n C_E \pi}{C_V \eta} \cdot \dfrac{L_d}{r_a d} \cdot MW^2 R_{sqr}/2$, where R_{sqr} is the row line sheet resistance, and W is the width of the array. The voltage drop along a column line is $V_c = \dfrac{C_n C_E \pi}{C_V \eta} \cdot \dfrac{L_d}{r_a d} \cdot \dfrac{W_r}{W_r + D_r} \cdot H^2 R_{sqc}$, where H and R_{sqc} are height of the array and sheet resistance of the column lines, respectively. If the row and column lines are connected to the drive circuit at both ends, V_r and V_c are halved. Since $V_r \propto MW^2$ while $V_c \propto H^2$, metal cathode lines are used as row lines due to their low resistance. Figure 2 shows V_r versus W for $L_d = 100$ cd/m^2, $r_a = 80\%$ and $d = 50\%$, when both ends of the row line are connected to the drive circuit. Here, $R_{sqr} = 0.03$ $\Omega/\square$ is typical for top metal contacts. Figure 3 shows V_c versus H for $r_a = 80\%$, $d = 50\%$, $W_r/(W_r+D_r) = 0.9$, and $R_{sqc} = 20$ $\Omega/\square$, assuming both column line ends are connected to the drivers. The resulted power penalty, $\Delta P = 1+V_c/(2V)$, is shown for an operating voltage $V = 10$ V. Also shown in Fig. 3 is V_c in the case where a 150 Ω/cm metal strip (10 μm wide, 1000 Å thick Ag) is used to improve the column line conductance, assuming $W_r = W_c = 100$ μm.

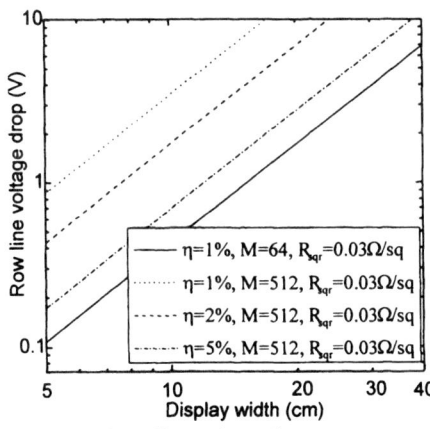

Fig. 2. Row line voltage drop versus array width for a PM OLED display.

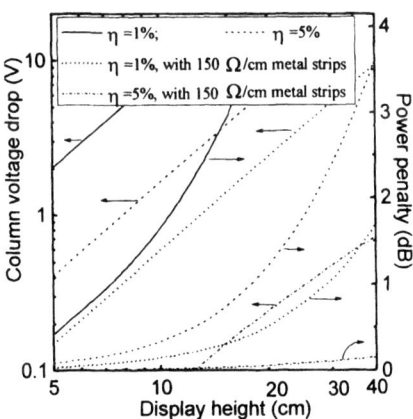

Fig. 3. Column line voltage drop versus array height for a PM OLED display.

Due to the limits on PM OLED array size, AM addressing is necessary for high information content, large area FPDs. A pixel design for an AM OLED display using n-channel FETs and top emitting, inverted OLEDs[13] is shown in the Fig. 4. The field effect transistors (FETs) may be crystalline Si MOS FETs or thin film transistors (TFTs) made from Si of various crystallinity ranging from polycrystalline Si (poly-Si) to amorphous Si (a-Si). Organic TFTs, with perform-ance[14] superior to that of a-Si TFTs, may also be used . The supply voltage, V_{DD}, must satisfy:

$V_{DD} \geq V + \sqrt{2JA/[\mu_{FET}C_{OX}(w_2/l_2)}$, where μ_{FET}, C_{OX}, w_2, and l_2 are the field effect mobility, insulator layer capacitance per unit area, channel width, and channel length of T_2, respectively. The voltage drop on T_2, $V_{DD} - V$, and power penalty, ΔP, are plotted versus $\mu_{FET}(w_2/l_2)$ in Fig. 5. By analyzing T_1 in a similar manner to that done with the accessing transistor in AMLCDs,[15] we point out the following: (1) A separate storage capacitance, $C_S \gg C_{GS1}$, is necessary to avoid charge redistrbution.[15] (2) The on-off ratio and field effect mobility of a-Si, poly-Si, or organic TFTs are sufficient for use as T_1 in AM addressing. An array of OLEDs grown in the conventional sequence can be addressed by p-channel TFTs, and the analysis is similar and straightforward.

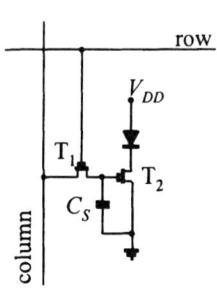

Fig. 4. A pixel circuit design for
AM OLED displays.

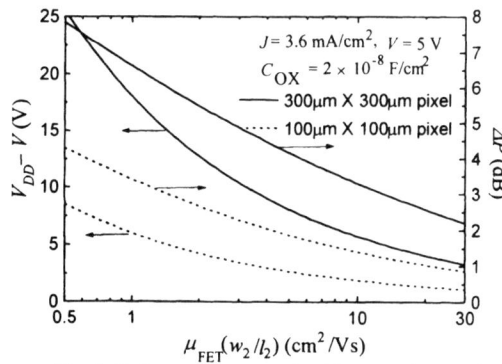

Fig. 5. Voltage drop across the drive transistor and
power penalty of an AM OLED display.

Assuming the edges of the top, continuous contact are at the same potential, the top contact voltage drop from the edges to the center of the display is:

$$|V_0| = JR_{sq}W^2\left[\frac{1}{8} - \frac{4}{\pi^3}\sum_{n=1}^{\infty}\frac{(-1)^{n+1}}{(2n-1)^3\cosh(2n-1)(\pi H/W)}\right] \equiv JR_{sq}W^2 \cdot G\left(\frac{H}{W}\right),$$ where R_{sq} is the

sheet resistance of the top contact. For an AM OLED array using p-channel FETs as switches and a metal cathode for the top contact, $|V_0|$ is negligibly small ($|V_0| = 57$ mV for $J = 3.6$ mA/cm^2, $R_{sq} = 0.03$ $\Omega/\square$, $W = 100$ cm, and $H/W = 3/4$). For arrays using inverted OLEDs with a continuous ITO anode, however, $|V_0|$ is high for large arrays. If $J = 3.6$ mA/cm^2 and $R_{sq} = 300$ $\Omega/\square$ (typical for ITO sputter deposited at low power and room temperature),[5] then $|V_0| = 90$ V for $W = 40$ cm and $H/W = 3/4$. $|V_0|$ can be decreased by depositing a thin (~ 100 Å) metal cap or a metal grid on the ITO anode without appreciably lowering the its transparency. A 3 $\Omega/\square$ top contact leads to a voltage drop of only 0.9 V for the same conditions as above.

Full Color Displays

There are five potential strategies[2] for making OLED-based, full color displays: (1) side-by-side patterning of R, G, and B OLEDs, (2) absorptive filtering of white OLEDs, (3) microcavity filtered OLEDs, (4) fluorescent down-conversion of blue OLEDs, and (5) color-tunable stacked

OLEDs (or SOLEDs).[16] Our quantitative comparison shows the following: Strategy (1), with no extra optical loss, is a good candidate if a high definition, low cost patterning process for organic thin films is established. The high optical loss (> 70%) makes strategy (2) less attractive than the others. However, it may be used before other techniques mature, since filter arrays widely used in LCDs can be readily transplanted. Strategy (3) is limited by the small viewing angle, although extra loss may be avoidable. Using high quantum efficiency (~70%) photoluminescent materials, strategy (4) is also a good candidate with moderate losses (40 ~ 50% for R and G sub-pixels). With optical loss better than or comparable with strategy (4), and the unique three-fold improvement in resolution, strategy (5) represents a promising approach to achieving high resolution, full color FPDs.

The addressing schemes for an $N \times M$ full color display implemented with strategies (1) though (4) are identical to those used for a $3N \times M$ monochromatic display. The following discussion will therefore focus on the addressing of SOLED arrays. A PM SOLED array can be constructed as shown in Fig. 6. There are two drive lines for each column: the bottom anode, c_1, and the common anode of the middle and top devices, c_2. Further, two addressing lines are used for each row: the common cathode of the bottom and middle devices, r_1, and the top cathode, r_2. Hence, the addressing time of each row is divided between r_1 and r_2. To avoid decreasing the addressing time, an alternative PM addressing scheme employs an insulating layer inserted between the top and middle devices to electrically isolate these two elements. This design is again equivalent to a $3N \times M$ pixel PM array. An AM SOLED pixel design making use of such an insulating layer is shown in Fig. 7, which is similar to that of a $3N \times M$ monochromatic display.

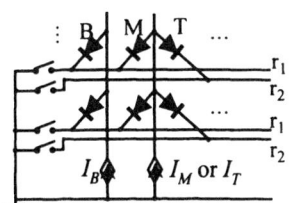

Fig. 6. A PM OLED array circuit design.

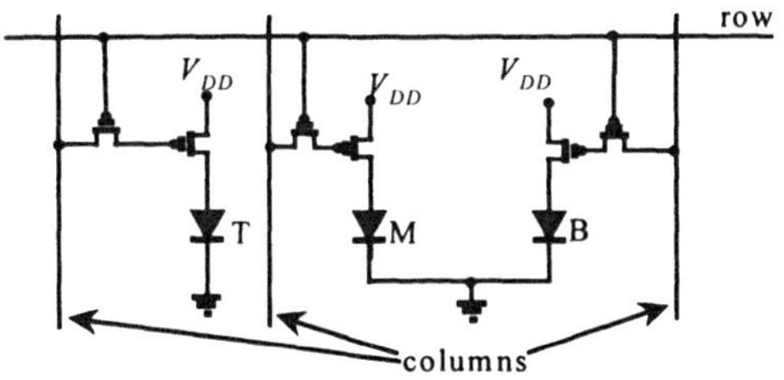

Fig. 7. A full color, AM OLED pixel circuit design using a SOLED with a insulating layer.

The above described addressing schemes are compatible with the recently demonstrated SOLED structure, possibly with an extra transparent, insulating layer to separate the shared contact. Using a simplified model employed in previous work,[2] we obtain the optical loss of the bottom, middle, and top devices: $L_B = 1 - (1 + T_C^2 T_P^2)/2$, $L_M = 1 - (1 + T_P^2)T_C/2$, and $L_T = 1 - T_C T_P$, where T_C and T_P are the transmission of the thin metal cathode[16] and the protection layer.[13, 16] For typical values of $T_P = 80\%$ and $T_C = 70\%$, the losses are: $L_B = 33\%$, $L_M = 43\%$, $L_T = 44\%$. To lower the optical loss, a highly transparent protection layer is desired. If this layer is fully transparent, then $L_B = 25\%$, $L_M = 30\%$, $L_T = 30\%$ for $T_C = 70\%$. Active matrix addressing can also be implemented with n-channel TFTs using a modified, top emitting SOLED structure where the bottom and top devices are inverted OLEDs[13] and the middle element shares the ITO anode with the bottom one. This modified device configuration, along with the n-channel transistor drive scheme, allows for the use of low cost a-Si TFTs in the AM addressing of SOLED arrays.

SUMMARY

Current OLED technology can provide sufficient luminance and contrast required by video displays. OLEDs are superior to LCDs in terms of viewing angle and brightness, and their luminous efficiency is comparable with, or better than those of other FPD technologies. Both AM and PM addressing can be used for OLED displays. AM addressing is needed for large size and high information content. Current a-Si and poly-Si TFT performance satisfies the requirements of AM OLED displays. High performance organic TFTs create the potential for developing "all-organic" AM displays. A novel approach toward full color displays, the SOLED, has moderate loss and the potential to achieve the highest possible resolution. Vacuum-deposited OLEDs have potential applications ranging from monochromatic alphanumeric displays to high resolution, full color video displays if major difficulties such as relatively short operational lifetime are overcome.

ACKNOWLEGEMENT

We gratefully acknowledge the generous support from Universal Display Corp. and DARPA.

REFERENCES

1. S. R. Forrest, P. E. Burrows, and M. E. Tompson, Laser Focus World, **31**, 99 (Feb. 1995).

2. P. E. Burrows, G. Gu, V. Bulović, S. R. Forrest, and M. E. Thompson, IEEE Transactions on Electron Dev. **44**, 1188 (1997).

3. J. R. Sheats, H. Antoniadis, M. Hueschen, W. Leonard, J. Miller, R. Moon, D. Roitman, and A. Stocking, Science, **273**, 884, (1996).

4. C. W. Tang and S. A. VanSlyke, Appl. Phys. Lett., **51**, 913 (1987).

5. G. Gu, V. Bulović, P. E. Burrows, S. R. Forrest, and M. E. Thompson, Appl. Phys. Lett. **68**, 2606, (1996).

6. G. Gu, D. Z. Garbuzov, P. E. Burrows, S. R. Venkatesh, S. R. Forrest, and M. E. Thompson, Opt. Lett. **22**, 396, (1997).

7. D. Z. Garbuzov, S. R. Forrest, A. G. Tsekoun, V. Bulovic, and M. E. Thompson, J. Appl. Phys. **80**, 4644 (1996).

8. D. A. Roberts, Photonics Spectra, pp.59, (Apr. 1987).

9. Lawrence E. Tannas, Jr., Flat Panel Displays and CRTs, (New York, 1985), p. 46.

10. H. Nakada and T. Tohma, Inorganic and Organic Electroluminescence/EL 96 Berlin, R. H. Mauch and H. -E. Gumlich, Eds. Berlin: Wissenschaft und Technik Verl., 1996, pp. 385-390.

11. P. E. Burrows, Z. Shen, V. Bulović, D. M. McCarty, S. R. Forrest, J. A. Cronin, and M. E. Tompson, J. Appl. Phys. **79**, 7991, (1996).

12. C. Hosokawa, E. Eida, M. Matsuura, K. Fukuoka, H. Nakamura, and T. Kusumoto, *Soc. for Information Display International Symp. Dig. of Tech. Papers*, **28**, 1073, (1997).

13. V. Bulović, P. Tian, P. E. Burrows, M. R. Gokhale, S. R. Forrest, and M. E. Thompson, Appl. Phys. Lett. **70**, 2954, (1997).

14. Y. Y. Lin, D. J. Gundlach, S. F. Nelson, T. N. Jackson, IEEE 55th Annual Dev. Res. Conf. Dig., pp. 60, (1997).

15. M. S. Shur, M. D. Jacunsky, H. D. Slade, and M. Hack, J. of the SID, **3-4**, 223, (1995)

16. Z. Shen, P. E. Burrows, V. Bulović, S. R. Forrest, and M. E. Thompson, Science, **276**, 2009, (1997).

EFFECT OF IMPURITY IONS AND PERMANENT DIPOLES FOR DEVICE PERFORMANCE OF THIN-FILM ELECTROLUMINESCENT DIODES

TETSUO TSUTSUI[*, **], MASAYUKI YAHIRO[*], DECHUN ZOU[**]
*Graduate School of Engineering Sciences, Kyushu University, Kasuga, Fukuoka 816 Japan,
**Tsutsui Team, CREST/JST, Kasuga, Fukuoka 816 Japan

ABSTRACT

Double-layer EL diodes composed of a spin-coated Polyvinylcarbazole (PVCz) layer and a vacuum-sublimed tris-(8-hydroxyquinoline)aluminum (Alq) layer were prepared. The diodes with the same device structure but with PVCz layer with added ionic impurities were also prepared. The diodes were driven at constant voltage and allowed to stand under short-circuit or reverse bias conditions. Observations of luminance-current density-voltage relations at constant voltage driving were repeated. The decrease of both luminance and current density during constant voltage driving were observed. Both spontaneous and reverse-bias assisted recovery of device performances were observed and these degradation and recovery phenomena were discussed in terms of the movement of ionic impurities in organic layers.

INTRODUCTION

Today, basic working mechanism for charge-injection type electroluminescent (EL) diodes, which consists of double-injection of positive and negative charges, their transport, recombination and production of neutral excited states, and emissive transition to ground states, has been well understood. Molecular materials useful for EL diodes, whether small molecules or polymers, are classified into insulators with high resistivity, even though high charge-carrier mobility and high quantum efficiency of fluorescence are important requisites[1,2]. In other words, very little amount of electronic charges are present in molecular films without charge injection from outer electrodes. Once electric field is applied, large amount of positive and negative charges pass through molecular solid films. One should note that fundamental electric properties of molecular materials for EL diodes are governed by insulating characteristics and dielectric properties. Little attention has been paid for this view points, because all the attentions have been focused on the design of specific small molecules and polymers with high hole and electron transporting capabilities. It is well known that purities of molecular materials are not so high enough compared with the cases of inorganic semiconductors. Thus it is always assumed that organic thin films contain small amounts of impurities, even when they are fabricated from high purity materials under careful processing technique. Impurities, especially ionic species in insulating films move slowly under applied electric field and form an internal electric field to the opposite direction to the applied field, even though their mobilities are extremely low. Very recently, we found that ionic impurities give significant effects on device performances of EL diodes.

Possible origins of long-term irreversible degradation processes in EL diodes, such as degradation of organic layers due to spontaneous crystallization of dye molecules, corrosion of metal electrodes and exfoliation of organic layers from electrodes, have been extensively studied [3-6]. Thus the origins of irreversible degradation have been reasonably ascribed to chemical or physical changes of organic layers or electrode materials. In contrast, little argument has been provided on the short-term decrease in device performances although both absolute luminance and EL efficiency decrease as large as several tens of percent of the initial values at continuous driving time scales spanning from a few second to several tens of hours [7,8].

Recovery phenomena in device performances have been observed related with initial drops of luminance and efficiency. Hamada and his coworkers reported the recovery of luminance when the diodes were allowed to stand for several hours after the cessation of constant current driving in mutilayer vacuum-sublimed dye diodes [9]. Riess observed a similar recovery phenomenon in a single layer PPV diode [10]. Moreover, it has been known that pulse-mode driving combined with reverse-bias application is quite effective for extending lifetimes at continuous driving [11]. Few interpretations which explain these recovery phenomena have been given so far. The argument that the performance recovery is related with the spontaneous decrease of large amounts of accumulated trapped-charges within organic layers or organic layer/electrode interfaces, which cause suppression of device performances during continuous driving, has been given, although no experimental evidence for this interpretation has been given.

The existence of ionic impurities are assumed to considerably influence the durability of EL diodes. Even though the amount of ionic species present in organic layers assumed to be very small, the effect of internal local field can be considerably large and never be ignored. The formation of internal field leads to the decrease in effective applied electric field. In particular, the initial degradation mechanisms occurring within the first few minutes after a bias voltage is applied is assumed to be closely related to the movement of ionic impurities. The reason why EL diodes prepared via solution-processing generally have poor durability as compared to those prepared via dry processing techniques is easily understood. When external electric field is removed, the formed internal local field relax spontaneously due to diffusion of ions. In addition, the internal field can be removed much faster when external field reverse to the drive voltage is applied. In this report, we will propose our new assumption that movement of ionic impurities is one of the key factors to cause short-term degradation and its recovery in organic EL diodes. We will show our experimental results which support our proposed mechanism due to ionic impurities.

Dielectric properties sometimes are closely related with device performances of EL diodes. Under the influence of strong external electric field, typically in the order of 10^6 V/cm, permanent dipoles within dielectric films may find a chance to align along the external electric field. The alignment of permanent dipoles may also contribute to form an internal electric field. At present we have no experimental evidence that shows dipole-orientation effect surely contribute to recoverable degradations. Thus this aspect will only be mentioned briefly and will be discussed in future publications.

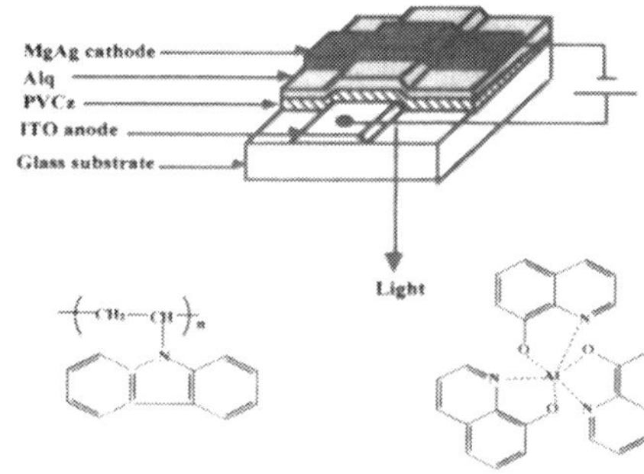

Fig.1. Sturctures of ITO/PVCz/Alq/MgAg device.

EXPERIMENT

Double-layer EL diodes composed of a spin-coated polyvinylcarbazole (PVCz) film as a hole transport layer and vacuum-sublimed tris-(8-hydroxyquinoline)aluminum (Alq) film as an electron transport layer were prepared (Fig. 1). On a patterned indium-tin-oxide (ITO) substrate, a PVCz layer (70 nm) was spin-coated from a dichloromethane solution . Alq layer (50 nm) and a MgAg alloy cathode were formed using a standard vacuum-vapor deposition procedure. The size of emitting area was 2x2 mm and 8 devices were formed on the same glass substrate at the same time. The ITO/PVCz/Alq/MgAg diodes on a substrate were transferred into a vacuum cryostat. The EL diodes were kept under vacuum and measurements on every diode on the same substrate were conducted under vacuum without breaking a vacuum. This ensured reproducibility of our luminance-current density-voltage (L-J-V) data. A source-measure unit (Keithley 238) and luminance meter (Topcon BM-5A) were used for L-J-V measurements.

We adopted a combination of wet-processed PVCz and dry-processed Alq films. It is assumed that the PVCz layer contains larger amounts of impurities than the Alq layer does. Thus impurity effects are supposed to be originated from the PVCz layer. For comparison, EL devices which contain larger amounts of impurity ions were also prepared. By using organic layers with larger amount of known ionic impurities, the effects of impurities are expected to be much clearer. To the PVCz solutions, 0.5 wt% $LiClO_4$ was added, and PVCz layers were spun on ITO electrodes. Thus ion-doped diodes, ITO/PVCz(ion)/Alq/MgAg, were also prepared.

RESULTS AND DISCUSSIONS

Recoverable Degradation in EL Diodes without Added Ionic Impurities

The emission spectra from the both ITO/PVCz/Alq/MgAg and ITO/PVCz(ion)/Alq /MgAg diodes were identical to the PL spectrum of an Alq thin film and thus it was confirmed that the PVCz layers behave as a simple hole-transport layer and the Alq layer takes the roles of electron transport and emission. First, one of the ITO/PVCz/Alq/MgAg diodes was driven at a constant bias voltage of 9 V for 10 minuets. Then the diode was allowed to stand for 10 minutes under short-circuit and was driven again for another 10 minutes. The diode was allowed to stand for 10 minutes under reverse bias of -10 V, and again was driven under forward bias of 9 V. Figure 2 shows temporal traces of current density and luminance during above procedure. During the first ten-minute driving, both current density and luminance decreased very rapidly. After 10-minute storage under short-circuit condition, both current density and luminance showed spontaneous recovery, although initial values were not restored. Extents of recovery were larger when reverse bias was applied to the degraded device. These degradation and recovery processes were repeatedly observed on the

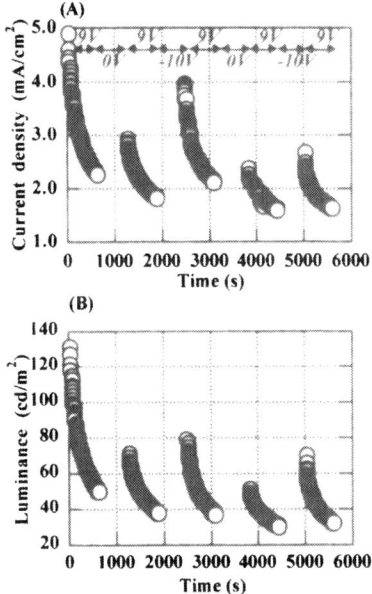

Fig.2.Temporal profiles of current density (A) and luminance (B) in repeated constant voltage driving.

same diode, although average device performances showed gradual degradation during repeated cycles. Evidently the extent of forced recovery with reverse bias is much larger than that of spontaneous recovery. These observations support our assumption that built-up of internal reverse electric field causes the decrease in both current density and luminance. When a forward bias is applied to the diodes, ionic impurities diffuse towards electrodes, cations towards a metal cathode and anions towards an ITO anode. Consequently, an internal electric field in opposite direction to the external field is formed. This internal field reduces the effective electric field for charge injection and transport. Thus the current density decrease under a constant bias voltage condition, followed by the decrease in luminance.

Figure 3 plots the time trace of external quantum efficiencies for the same diode. If the decrease of luminance is simply caused by the decrease in current density under constant voltage driving, quantum efficiency should remain constant, even when degradation proceeds. The quantum efficiency values, however, showed significant amounts of decrease and recovery. This observation tell us that not all the luminance degradation at constant voltage driving can be ascribed to the decrease in current density due to reverse internal field. It seems that the charge injection balance may also be affected by the formation of internal field.

Recoverable Degradation in EL Diodes with Added Ionic Impurities

Figure 4 shows temporal traces of luminance and current density when the diode with ionic impurities was driven with the same time sequences used for the previous measurement. Very similar behaviors with the case of the diodes without ionic impurities were observed, except for the fact that the absolute values of both current density and luminance at the

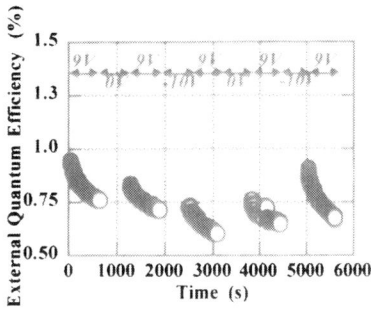

Fig.3.Temporal profiles of external quantum efficiency in repeated constant voltage driving.

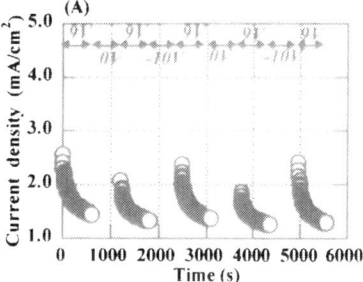

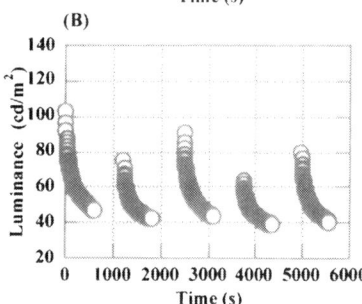

Fig.4. Temporal profiles of current density (A) and luminance (B) in repeated constant voltage driving for the diode with impurity doped PVCz layer.

initial stage of driving were a little smaller than the case of the device without impurity ions. Contrary to our expectation that the addition of ionic impurities would enhance both degradation and its recovery, no tendency of accelerated degradation due to added ionic impurities was observed. Extent of ionization of $LiClO_4$ in low-dielectric organic solids is small and the movement of free Li ions might not be so effective to overcome the effects due to unknown impurities in PVCz layers.

Spontaneous and Field-Assisted Recovery

In order to examine the mechanism of degradation recovery, the effect of duration of application of reverse bias on the extent of recovery was measured. In every measurement, a fresh device which was fixed in the same cryostat was used. First, a diode was driven for 10 minutes at 9 V, and then the diode was once shorted and allowed stand under reverse 10 V bias for fixed durations. The current density and luminance values at just before cessation of driving, J_1 and L_1 were recorded. Again, the diode was driven at 9 V. The initial current density and luminance values of this second driving, J_2 and L_2 were recorded. The extents of current density and luminance recovery are defined as $(J_2-J_1)/J_1$ and $(L_2-L_1)/L_1$, respectively. Figure 5 plots the current density and luminance recovery against the duration of applied reverse bias. It is evident that the

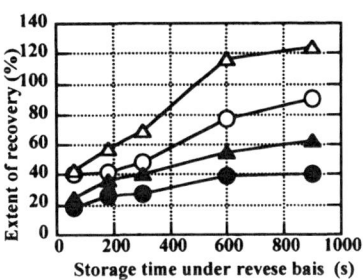

Fig.5. ($\bullet$, $\circ$)Recovery of luminance and ($\blacktriangle$, $\triangle$) current density as a function of the period of applied reverse bias. Filled and open marks correspond for the diode without and with added impurity ions, respectively.

extents of both current and luminance recovery linearly increase with the increase in the duration of electric field-assisted recovery. In addition, the extents of recovery are larger in the diode with added impurity than in the diode without added impurity. Application of reverse bias for a long period contributes to blow out the formed internal electric field and may further help for forming internal forward field.

CONCLUSIONS

We have observed device degradation and its spontaneous and field-induced recovery in double-layer EL diodes. Our observations were interpreted with the assumption that impurities incorporated in organic layers caused recoverable degradation by forming internal field opposite to the applied voltage. When a standard dry-processed TPD layer is used in place of a wet-processed PVCz layer, features in degradation may be somewhat different and degradation and its recovery may be much sensitive to a trace amount of water. Thus we believe the PVCz-Alq double layer system used in this study was just suitable for a model system for understanding recoverable degradation. The careful experimental work for the corresponding cases for standard double-layer TPD/Alq diodes is in progress.

ACKNOWLEDGMENT

This work has been partly supported by the Core Research for Evolutional Science and Technology, Japan Science and Technology Cooperation (CREST/JST).

REFERENCES

1. T. Tsutsui and S. Saito, in Intrinsically Conducting Polymer: An Emerging Technology, edited by M. Aldissi (Kluwer Academic Pub., 1993) p.123-134.
2. T. Tsutsui, MRS Bulletin, 22(6), p.39 (1997).
3. C. Adachi, T. Tsutsui and S. Saito, Appl. Phys. Lett. 56, p.799 (1990).
4. L. M. Do, E. M. Han, Y. Niidome, M. Fujihira, T. Kanno, S. Yoshida, A. Maeda and A. J. Ikushima, J. Appl. Phys., 76, p. 5118 (1994).
5. P. E. Burrows, v. Bulvic S. R. Forrest, L. S. Sapochak D. M. McCarty and M. E. Thompson, Appl. Phys. Lett., 65, p.2992 (1994).
6. J. R. Seats, H. Antoniadis, M. Hueschen, W. Leonard, J. Miller, R. Moon, D. Roitman and A. Stocking, Science, 273, p.884 (1996).
7. Y. Shirota, Y. Kuwabara, H. Inada, T. Wakimoto, H. Nakada, Y. Yonemoto, S. Kawami and K. Imai, Appl. Phys. Lett., 65, p.807 (1994).
8. Y. Hamada, T. Sano, K. Shibata and K. Kuroki, Jpn. J. Appl. Phys. 34, p.L824 (1995).
9. Y. Hamada, C. Adachi, T. Tsutsui and S. Saito, Jpn. J. Appl. Phys., 31, p.1812 (1992).
10. W. Riess, in Organic Electroluminescence Materials and Devices, edited by S. Miyata and H. S. Nalwa, (Gorden Breach Sci. Pub., 1997) p.73-146.
11. J. Shi and C. W. Tang, Appl. Phys. Lett., 70, p.1665 (1997).

LIGHT EMITTING DEVICES WITH MOLECULARLY DOPED POLYMER LAYERS. A NEW DIAMINE AS A HOLE TRANSPORTING MOLECULE

V. FATTORI[a], M. COCCHI[a], P. DI MARCO[a], G. GIRO[a], J. KALINOWSKI[a,b],
P. DEMBECH[c], G. FABBRI[c], G. SECONI[c]

[a] CNR-FRAE, Area della Ricerca, via P. Gobetti 101, 40129 Bologna, Italy
[b] Department of Molecular Physics, Technical University of Gdansk, ul. G. Narutowicza 11/12, 80-952 Gdansk, Poland
[c] CNR-ICoCEA, Area della Ricerca, via Gobetti 101, 40129 Bologna, Italy

ABSTRACT

In the present contribution a new hole transporting molecule belonging to the triphenyldiamine family (N,N'-diphenyl-N,N'-bis(4-metylphenyl)-1,1'-biphenyl-4,4'-diamine) in a polycarbonate matrix is studied as the hole transport layer in a two layer device with Alq_3 as emitting layer. The dependence of the electrical and emitting behaviour on the doping level is studied and compared with the vacuum deposited diamine layer also interfaced with Alq_3.

INTRODUCTION

Molecularly doped polymers (MDP) are widely studied as active materials for organic large area electroluminescent displays because they combine the easiness of the spreading process on large areas and good mechanical properties of a polymer with the big variety of molecules which can be mixed into the polymer matrix to play the active roles in hole transport, electron transport and light emission.

Since Tang and Van Slike's report on a double layer electroluminescent structure [1], a great deal of work has been done in improving both efficiency and durability of organic LEDs.

The electroluminescence (EL) in organic based devices is a process involving charge carrier injection in the organic layers by the electrodes, their migration through the layers and their recombination to give excited species which loose their excess electronic energy by light emission [2]. The optimization of anyone of these processes is the way to get higher efficiency of the device. Another important parameter to be improved in order to achieve a real commercial interest for organic based devices is their durability.

In multilayer devices, the hole transport layer, besides having good hole mobility, plays the important role of confining the charge recombination region far from injecting electrodes thus preventing quenching of the excited states formed by charge recombination.

Aromatic diamine derivatives with bulky substituents on the aromatic rings are usually employed as hole transport molecules as they can be layered, by vacuum sublimation, in good quality amorphous films with high hole mobility. In particular a methyl-substituted diamine (TPD) has been largely used as the active molecule in the hole transporting layer of multilayer devices [3-7].

Mat. Res. Soc. Symp. Proc. Vol. 488 © 1998 Materials Research Society

A drawback associated with evaporated layers is a poor durability due to the crystallization of the initially amorphous film [8] with a fall down of the electrical properties and in particular of the electrode/organic-layer interface [9].

Dispersion of the active molecule in a polymer matrix is expected to prevent crystallization of the hole transporting layer and increase the device durability. Such molecularly doped polymers (MDP) allow also high weight or thermally labile molecules, which wouldn't sublime without being destroyed, to be used as the electrically active species in electroluminescent device.

MDP have been widely studied as photoconductive layers for electrophotography [10-12] and have also been used in light emitting devices [13-16].

The present paper deals with a recently synthesized metyl substituted diamine derivative (N,N'-diphenyl-N,N'-bis(4-metyl phenyl)-1,1'-biphenyl-4,4'-diamine) [17] which acts as the hole transporting molecule in a double layer electroluminescent device. The diamine is studied both as dispersed in a polycarbonate matrix at different weight ratios and as a pure evaporated film as the hole transport layer interfaced with an Alq_3 evaporated film as the emitting layer.

EXPERIMENT

Figure 1 shows the schematic device configuration and the molecular structures of the materials used. The PC:pTPD layer was prepared by spinning a dichloromethane

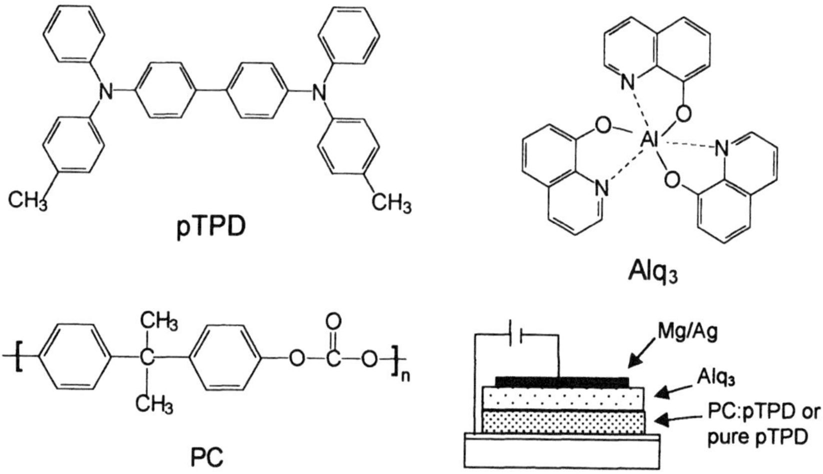

Figure 1. Molecular structures and device configuration.

solution, containing the appropriate amount of PC and pTPD, on an ITO coated glass having a sheet resistance of 20 Ω/cm^2. Three different solutions have been cast in order to achieve the concentrations of pTPD in the PC matrix of 75%, 50% and 25% by weight. The pure pTPD layer was vacuum deposited, at 3-4 Å/s deposition rate. The film thickness was 70 nm for all the samples.

On top of the MPD !ayer, a 60 nm Alq₃ film was deposited by vacuum sublimation (deposition rate 3-4 Å/s), and finally a Mg/Ag metal film was deposited with the same technique on the Alq₃ surface as the top electrode.

Electroluminescence was measured in dc mode by applying a positive polarity to the ITO electrode by means of a calibrated Si photodetector.

RESULTS

The EL spectra (not reported) show that in each of the double layer devices here analyzed the electroluminescence is due to Alq₃, while no light is generated in the pTPD containing layer which only acts as hole transport medium. This behaviour is analogous to many other devices reported in the literature where a diamine derivative is interfaced to Alq₃.

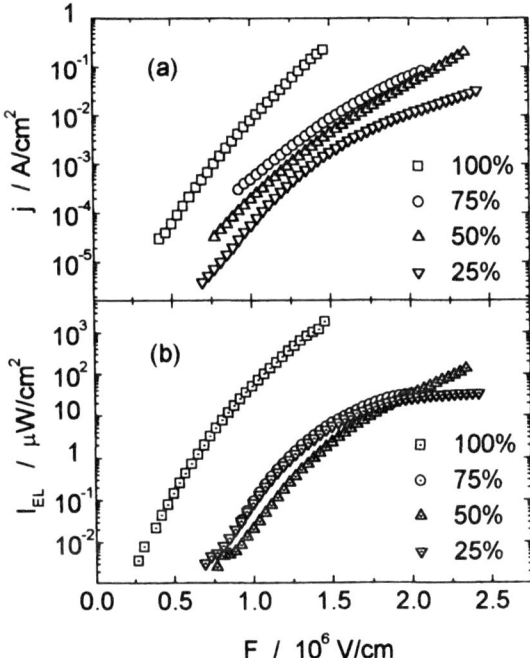

Figure 2. Current density vs. field (a) and EL intensity vs. field (b) characteristics of the cells ITO/PC:pTPD(w%)/Alq₃/Mg/Ag with pTPD weight percentage (w%) in the polymer matrix varying from 25% to 75%. The 100% curves refer to the cell ITO/pTPD/Alq₃/Mg/Ag with vacuum deposited pTPD layer. All cells were operated in dc mode with positive polarity on the ITO electrode.

Figures 2a,b report current density and EL intensity as a function of electric field for the ITO/PC:pTPD(w%)/Alq3/Mg/Ag cells with the pTPD content in the polymer matrix (w%) varying from 75% to 25% together with the ITO/pTPD/Alq3/Mg/Ag cell with a vacuum deposited pTPD layer (w%=100%). The current vs. field curves show the current to be an increasing function of the pTPD concentration in the hole transport layer (HTL), practically, no variation in EL intensity is observed within the 25%-75% concentration range.

The different behaviour of current and brightness with HTL composition, as put in evidence in figure 3, is an indication of a complex correlation between EL and current flowing in the cell.

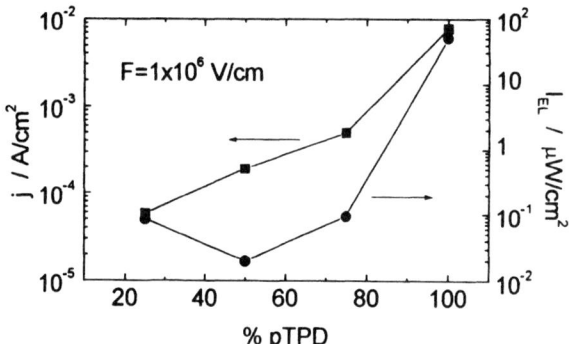

Figure 3. Current density and EL intensity against pTPD weight percentage in HTL. The data are derived from Fig. 2 a,b at the field value of 1×10^6 V/cm.

The EL quantum yield (φ_{EL}) is determined by the quantum yield of luminescence (φ_{PL}) of the emissive material and charge recombination probability (P_{rec}), the latter being a function of the recombination (τ_{rec}) to the charge transit (τ_T) time ratio[18] in the recombination zone:

$$\varphi_{EL} = \varphi_{PL}\, P_{rec} = \varphi_{PL} / (1 + \tau_{rec}/\tau_T) \qquad (1)$$

As no recombination occurs in the HTL and in Alq3 $\mu_e \gg \mu_h$, it follows that $\tau_{rec} = \varepsilon_0\varepsilon/en_e\mu_e$ and $\tau_T = d/\mu_h F$, hence:

$$\varphi_{EL} = \varphi_{PL} / (1 + \varepsilon_0\varepsilon\mu_h F / e\mu_e n_e d) \qquad (2)$$

where μ_h and μ_e are the hole and the electron mobility respectively and n_e the concentration of electrons in the Alq3 layer close to its interface with HTL.

Figure 4 shows the experimental electroluminescence quantum yield $\varphi_{EL} = (I_{EL}/h\nu)/(j/e)$ as a function of the electric field for all the cells studied in the present work.

EL yield varies with field showing a maximum value at a field strength which depends on pTPD concentration. As comes from relation (2), the φ_{EL} increase with the electric field (lower field section of the curves in figure 4) is due to a stronger increase of the electron concentration (n_e) with respect to the term $\mu_h F/\mu_e$; the opposite behaviour

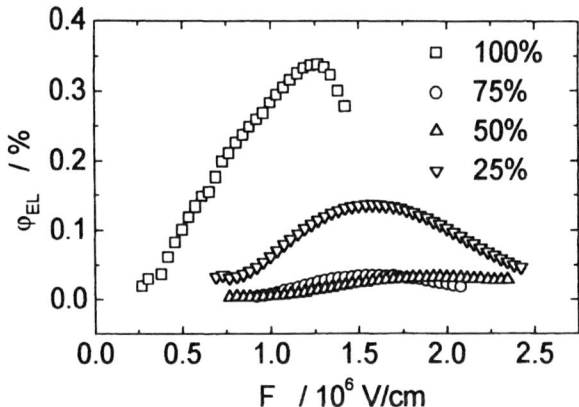

Figure 4. External EL quantum yield against applied electric field for the different pTPD percentages reported in legend (see caption of Fig. 2).

can be observed in the decreasing section of the curves. This suggests that charge injection mechanism or field distribution among the two LED layers changes when the applied field gets higher.

The EL quantum yield El (φ_{EL}) and the energy conversion efficiency ($\eta_{EL}=I_{EL}/Uj$)

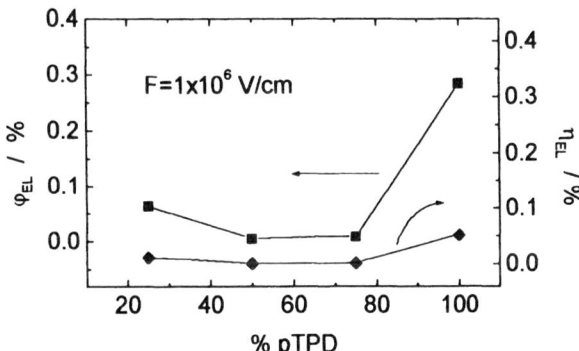

Figure 5. EL quantum yield (φ_{EL}) and power conversion efficiency (η_{EL}) against pTPD percentage in HTL. The data are derived from Fig. 4 at the field of 1×10^6 V/cm.

are not strongly depending on pTPD concentration in the PC matrix, as shown in figure 5, except for the highest value of pTPD content. This behaviour is consistent with relation (2) where φ_{EL} is determined by what is occurring in Alq$_3$. The highest quantum yield value, corresponding to the evaporated p-TPD layer, can be due to the electrode/organic layer interface which is not easily controllable and reproducible even when prepared under same conditions [19].

CONCLUSIONS

The present results demonstrate that, in contrast to the injection current, the EL emission is only a weakly varying function of the concentration of hole transporting molecules incorporated into the inert matrix of polycarbonate forming a hole transport layer in the double-layer structure of (PC:pTPD)/Alq$_3$. On the other hand, the EL quantum yield shows strong electric field dependence with a maximum which suggests a possibility of change in the mechanism of the charge carrier injection or field distribution among the two LED layers.

REFERENCES

1. C.W. Tang and S.A. Van Slyke, Appl. Phys. Lett. **51**, 913 (1987).
2. J. Kalinowski in <u>Organic Electroluminescent Materials</u>, edited by S. Miyata and H.S. Nalwa (Gordon and Breach Science Publisher, Amsterdam, 1997), p. 10.
3. C. Adachi, K. Nagai and M. Tamoto, Appl. Phys. Lett. **66**, 2679 (1995).
4. C. Adachi, T. Tsutsui and S. Saito, Appl. Phys. Lett. **57**, 531 (1990).
5. Y. Omori, A. Fujii, M. Uchida, C. Morishima and K. Yoshino, Appl. Phys. Lett. **62**, 3250 (1993).
6. P.E. Burrows, Z. Shen, V. Bulovic, D.M. McCarty, J.A. Cronin, M.E. Thompson and S.R. Forrest, J. Appl. Phys. **79**, 7991 (1996).
7. J. Kalinowski, N. Camaioni, P. Di Marco, V. Fattori and G. Giro, Int. J. Electronics **81**, 377 (1996)
8. E.M. Han, L.M. Do, N. Yamamoto, M. Fujihira, Chem. Lett. <u>1995</u>, 57.
9. P.E. Burrows, V. Bulovic, S.R. forrest, L.S. Sapochak, D.M. McCarty and M.E. Thompson, Appl. Phys. Lett. **65**, 2922 (1994).
10. M. Abkowitz and D.M. Pai, Phil. Mag. B **53**, 193 (1986).
11. G. Giro, P. Di Marco, V. Fattori and R. Degli Esposti, Mol. Cryst. Liq. Cryst. **186**, 107 (1990)
12. P.M. Borsenberger, E.H. Magin and S. Heun, Macromol. Symp. **116**, 51 (1997).
13. J. Kido, K. Hongawa, M. Kohnda, K.Nagai and K. Okuyama, Jpn. J. Appl. Phys. **31**, L960 (1992).
14. G. Giro, N. Camaioni, P. Di Marco, V. Fattori and J. Kalinowski, Synth. Metals **84**, 379 (1997).
15. T. Uemura, N. Okuda, H. Kimura, Y. Okuda, Y. Ueba and T. Shirakawa, Polymers for Adv. Technol. **8**, 437 (1997).
16. J. Kalinowski, G. Giro, P. Di Marco, V. Fattori and E. Di Nicolò, (to be published)
17. P. Dembech, G. Fabbri et al. Angew. Chem. (submitted)
18. J. Kalinowski in <u>Electrical and Related properties of Organic Solids</u>, edited by R.W. Munn, A. Miniewicz and B. Kuchta, NATO ASI Series vol. 24, (Kluwer, Dordrecht, 1997), p. 167.
19. M.A. Abkowitz, H. Antoniadis, J.S. Facci, B.R. Hsieh and M. Stolka, Synth. Metals **67**, 187 (1994)

NANOSTRUCTURE OF ELECTRICALLY CONDUCTING POLYANILINE PREPARED BY A NOVEL EMULSION POLYMERIZATION PROCESS

J. LIU*, P. J. KINLEN** AND C. R. GRAHAM**
* Monsanto Corporate Research, 800 N. Lindbergh Blvd., St. Louis, MO 63167
** Growth Enterprise Dvision, Monsanto, 800 N. Lindbergh Blvd., St. Louis, MO 63167

ABSTRACT

A soluble polyaniline (PANI) salt with moderate conductivity was synthesized by a novel emulsion polymerization process. The conductivity of the processed PANI films can be substantially increased by treating the polymer films with surfactants or with low molecular weight alcohols. Transmission electron microscopy (TEM) images of thin polymer films revealed the existence of small islands of conducting PANI embedded in a non-conducting, dopant matrix. The conductivity of the PANI films is affected by the spatial distribution and the connectivity of these small islands. The conductivity enhancement observed upon treatment with surfactants is due to self-assembly of conducting PANI molecules into an interconnected network morphology. In the case of alcohol treatment the film conductivity is enhanced due to extraction of excess dopant phase and the subsequent densification of PANI islands to form highly conducting pathways.

INTRODUCTION

Polyaniline (PANI) is one of the most intensively investigated inherently conductive polymers (ICP) because of its straightforward polymerization and excellent chemical stability combined with relatively high levels of conductivity [1-3]. PANI is unique among conducting polymers in that its electrical properties can be reversibly controlled either electrochemically (by oxidation-reduction process) or chemically (by protonation/deprotonation process). One of the main problems for industrial applications of PANI is, however, its solubility and processability. PANI can be regarded as a semi-rigid polymer, easily resulting in intractability. Recently, the processability of PANI has been improved by a post processing step [4, 5]

PANI is considered as a structurally inhomogeneous material because of the existence of both crystalline and amorphous regions in PANI polymer films. The crystalline regions are thought to be more conductive because the charge delocalization will be better achieved in the crystalline regions [2]. The electrical and optical properties of PANI polymers and polymer blends are affected by the local structure of the PANI films. The microstructure of conductive PANI polymers, however, is not well understood because the local structure of PANI depends critically on the specific preparation methods.

We recently discovered a novel emulsion process for direct synthesis of soluble PANI emeraldine salt [6]. This unique PANI salt exhibits high solubility in low dielectric constant solvents, has a high molecular weight (>22,000) and moderate conductivity (10^{-5} S/cm), and is miscible and compatible with a host of other polymer systems. Treatment of the PANI films with quaternary ammonium salts or solvents such as methanol or acetone increases the conductivity 2-3 orders of magnitude [6].

This work deals mainly with the characterization of the nanostructure of PANI polymer films. Transmission electron microscopy techniques were employed to extract information about the structure and morphology of conducting PANI films and their structural modifications by surfactants or solvents.

EXPERIMENT METHODS

A detailed discussion of the synthesis of the PANI salt and the emulsion polymerization process is reported elsewhere in these proceedings [6]. Briefly, the emulsion process entails formation of emulsion particles with a mean hydrodynamic diameter of 210 nm and consists of a water-soluble organic solvent (e.g., 2-butoxyethanol), a water-insoluble organic acid (dinonylnaphthalene sulfonic acid (DNNSA)), aniline and water. Aniline is protonated by the organic acid to form a salt that partitions into the organic phase. As oxidant (ammonium peroxydisulfate) is added to the reaction mixture, PANI intermediates are formed in the organic phase. As the reaction proceeds, the emulsion flocculates to form a two-phase system with soluble PANI remaining in the organic phase. The concentrated PANI-DNNSA can be readily dissolved in xylene or other solvents [6].

A 5% solution of the PANI-DNNSA complex in pure xylene was produced. Electron beam transparent films were prepared by dip coating a gold grid at various speeds. These ultra thin films were then directly examined in a JEOL 2000FX TEM microscope equipped with a Gatan slow-scan CCD camera. It was found that the PANI-DNNSA films were sensitive to electron beam irradiation. At normal electron beam doses, the crystalline nature of the PANI-DNNSA was destroyed. The radiation damage necessitates the use of low-dose procedures to record the TEM images of PANI-DNNSA polymer films. All TEM images reported here were acquired digitally with the slow-scan CCD camera at low electron doses.

After TEM images were recorded, the PANI films were taken out of the microscope for treatment with surfactants such as benzyltriethylammonium chloride (BTEAC) or with low molecular alcohols and ketones such as methanol and acetone. These treated films were then re-examined in the microscope to observe the structural and/or morphological modifications of the PANI films due to the chemical treatment.

RESULTS AND DISCUSSIONS

Nanostructure of PANI-DNNSA Films

Figure 1 shows a typical bright field TEM image of the PANI-DNNSA films. A large number of small dark spots with diameters ranging from 10 to 30 nm were incorporated into a less dense matrix (white regions). The darker spots or domains represent conducting PANI-DNNSA islands. The brighter regions represent non-conducting dopant phase. The distribution of the PANI-DNNSA islands was not uniform within the polymer film. In some regions, these small islands aggregated to form domains with a high density of PANI-DNNSA islands. The conducting PANI-DNNSA regions were clearly not very well inter-connected in the polymer film. These small dark spots can be viewed as colloidal particles of PANI-DNNSA complexes in the diluted xylene/PANI-DNNSA solution. The observation of the existence of conducting islands in thin PANI-DNNSA films supports the proposed emulsion

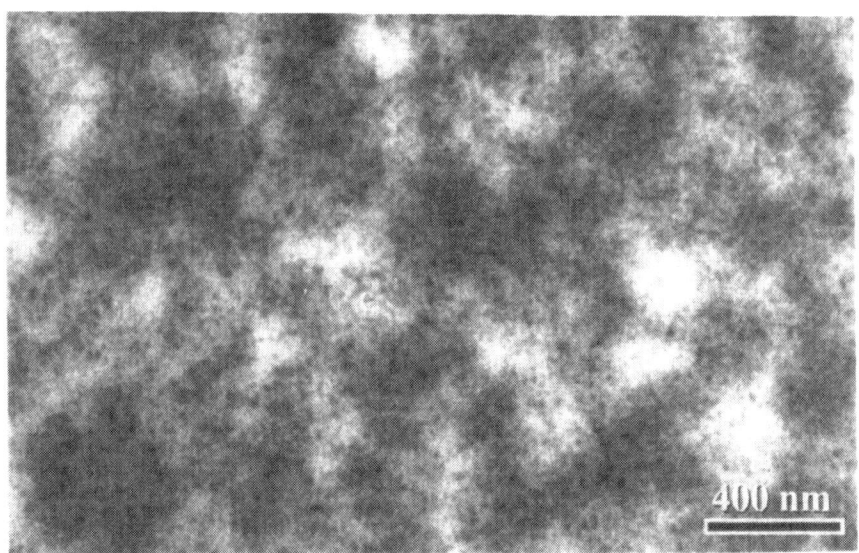

Figure 1. Bright-field TEM image of a thin PANI-DNNSA conducting film prepared by a novel emulsion process. The small dark spots represent PANI-DNNSA conducting islands. The bright regions represent the non-conducting, dopant phase.

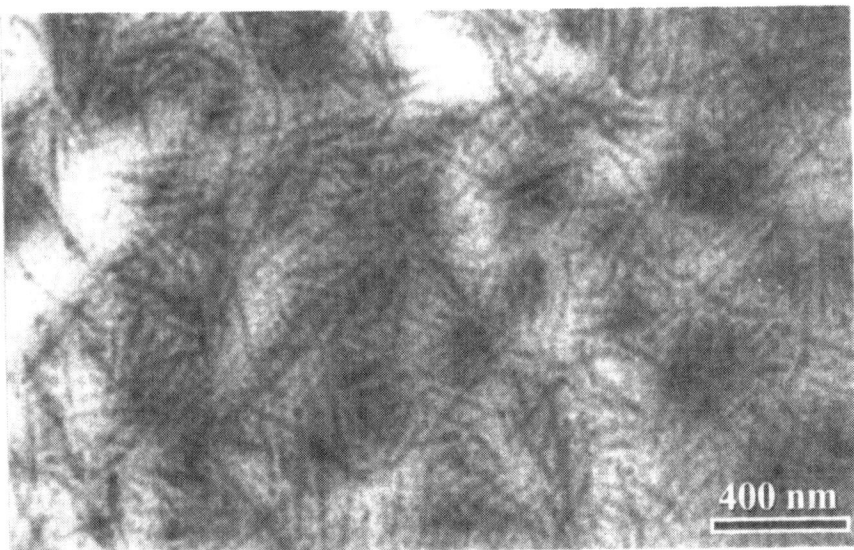

Figure 2. Bright-field TEM image of a thin PANI-DNNSA film showing long fibrils of conducting PANI. The film was formed after sonicating a xylene/PANI-DNNSA solution for 5 minutes prior to dip coating a gold TEM grid.

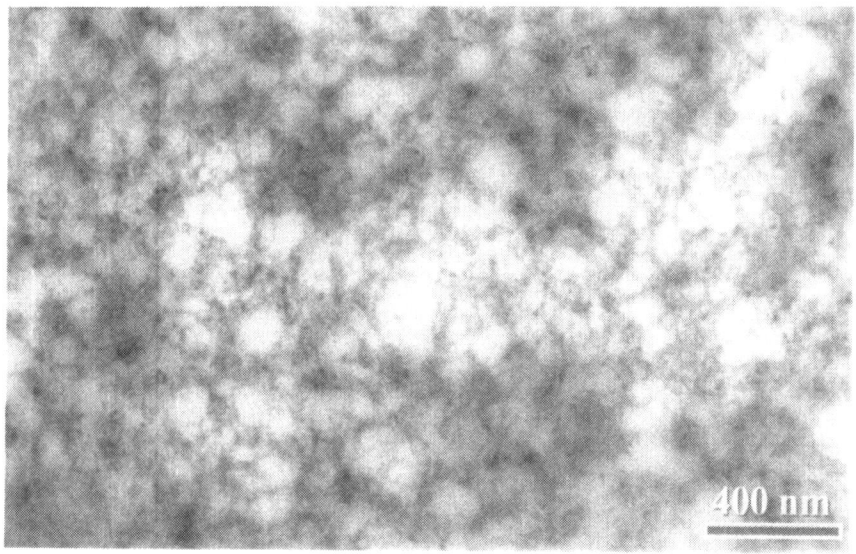

400 nm

Figure 3. Bright-field TEM image of a thin PANI-DNNSA film after approximately 2 minutes treatment with BTEAC solution. The BTEAC solution treatment facilitated the movement of the conducting islands and the self-assembly of PANI molecules to form multiply connected conducting pathways.

polymerization process [6] and is in agreement with the measurement of the hydrodynamic diameter of emulsion particles [6]. The conductivity of PANI-DNNSA films is affected by the spatial distribution and the connectivity of these conducting islands.

Depending on the specific preparation conditions, these small islands may 'aggregate' to form conducting domains or to form long fibrils of PANI-DNNSA. Figure 2 shows a typical bright field TEM image revealing fibrils of PANI-DNNSA up to 500 nm in length. The thin film was formed after sonicating a xylene/PANI-DNNSA solution for 5 minutes prior to dip coating the gold TEM grid. There were not many isolated conducting islands shown in figure 2 as compared to those in figure 1. Apparently, the sonication process facilitated the formation of the fibrils. These fibrils may be formed due to the enhanced interaction of the PANI-DNNSA complexes in the xylene solution during the sonication process. The conductivity of these films may significantly increase because of the increased connectivity of the conducting pathways. Further experiments are under way to investigate the crystal structure of these fibrils and their effect on the conductivity enhancement of the PANI-DNNSA polymers.

<u>Structure of PANI-DNNSA Films Treated with Surfactants</u>

Treatment of moderately conducting PANI-DNNSA films by surfactants such as quaternary ammonium salt solutions increased the film conductivity 2-3 orders of magnitude [6]. Figure 3 shows a typical TEM image of PANI-DNNSA films after treatment with BTEAC solution for approximately 2 minutes. The image clearly shows that the morphology

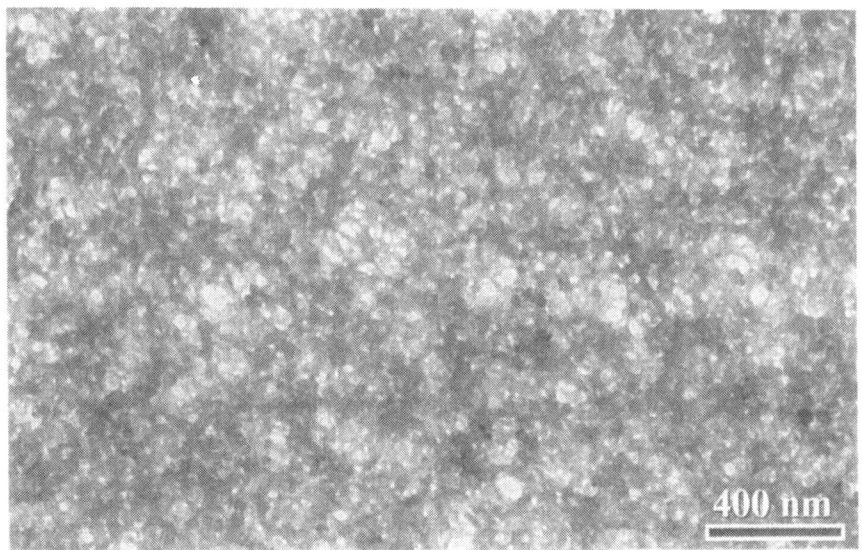

Figure 4. Bright-field TEM image of a thin PANI-DNNSA film after approximately 2 minutes treatment with methanol revealing a 'Swiss cheese' like morphology. Methanol extracted away the excess non-conducting dopant phase. Thus, the conducting PANI islands collapsed to form stringy networks consisting of links, nodes and large agglomerations.

of the PANI-DNNSA films changed significantly after treatment in the quaternary ammonium ion solution. In the non-treated films (see figure 1), small islands of conducting PANI-DNNSA are "embedded" in the non-conducting dopant matrix. Some of these islands may aggregate to form loosely connected conducting domains. For the BTEAC treated PANI-DNNSA, highly inter-connected networks of conducting PANI-DNNSA were observed. The BTEAC solution apparently facilitated the movement of the conducting islands and the self-assembly of PANI molecules to form multiply connected conducting pathways. Thus, the significant increase in conductivity for the BTEAC treated PANI-DNNSA films can be explained by the change of the microstructure of the polymer films.

Structure of PANI-DNNSA Films Treated with Alcohols

Treatment of moderately conducting PANI-DNNSA films with methanol or acetone for a few seconds increased the film conductivity to 5 orders of magnitude [6]. This dramatic change of conductivity was accompanied by a significant change of the physical properties of the polymer films. To understand the effect of alcohol treatment on the physical properties of PANI-DNNSA polymer films, TEM images of both non-treated and methanol treated PANI-DNNSA films were obtained. For non-treated films, all the TEM images were similar to that shown in figure 1. Figure 4 shows a typical TEM image of PANI-DNNSA films treated with methanol for approximately 2 minutes, revealing a 'Swiss cheese' like morphology.

After detailed analysis of a large number of TEM images, a model for the structural modification of the PANI-DNNSA polymer films was proposed as the following. First, methanol apparently dissolved away the excess non-conducting dopant phase. Then, The conductive PANI-DNNSA islands/domains collapsed or 'condensed' to form stringy PANI-DNNSA networks consisting of links, nodes and large agglomerations. The final result of the methanol treatment was to form a highly conducting PANI film with a large number of holes or voids. The existence of these holes and voids in the treated polymer films certainly changed their physical properties. The significant increase in conductivity originated from the formation of continuous conducting films of PANI-DNNSA.

CONCLUSIONS

A soluble PANI salt with moderate conductivity was synthesized by a novel emulsion polymerization process. The conductivity of the PANI films was further increased substantially by treating the films with surfactants or with low molecular weight alcohols. Electron microscopy images revealed the existence of small islands of conducting PANI-DNNSA embedded in a non-conducting, amorphous, dopant matrix. The conductivity of the PANI-DNNSA films is affected by the spatial distribution and the connectivity of these small islands. The conductivity enhancement observed upon treatment with surfactants is due to self-assembly of conducting PANI-DNNSA molecules into an interconnected network morphology. In the case of alcohol treatment the film conductivity is enhanced due to extraction of excess dopant and the subsequent densification of PANI-DNNSA islands to form highly conducting pathways.

REFERENCES

1. F. Lux, Polymer **35**, p2,915 (1994).

2. J. P. Pouget, Z. Oblakowski, Y. Nogami, P. A. Albouy, M. Laridjani, E. J. Oh, Y. Min, A. G. MacDiarmid, J. Tsukamoto, T. Ishiguro and A. J. Epstein, Synthetic Metals **65**, p131 (1994).

3. T. Vikki, L-O Pietila, H. Osterholm, L. Ahjopalo, A. Takala, A. Toivo, K. Levon, P. Passiniem and O. Ikkala, Macromolecules **29**, p2,945 (1996).

4. Y. Cao, P. Smith and A. J. Heeger, Synthetic Metals **48**, p91 (1992).

5. K. Tzou and R. V. Gregory, Synthetic Metals **53**, p365 (1993).

6. P. J. Kinlen, Y. Ding, C. R. Graham, J. Liu and E. E. Remsen, in these proceedings.

INJECTION EFFICIENCY FROM VARIOUS METALS INTO TRAP FREE MOLECULARLY DOPED POLYMERS EVALUATED FROM COMBINED ANALYSIS OF CURRENT-VOLTAGE AND TIME-OF-FLIGHT DRIFT MOBILITY DATA

M. ABKOWITZ[*†], J. S. FACCI[*†], J. REHM[*]
* Center for Photoinduced Charge Transfer, University of Rochester, Rochester, N.Y.
† Wilson center for Research and Technology, Xerox Corporation, 114-39D, Webster, N.Y.

ABSTRACT AND SUMMARY

An experimental procedure which can provide a direct quantitative measure of contact injection efficiency into a trap free transport polymer is described. This technique combines coupled analyses of bias dependent injection current supplied by the contact under test with time-of-flight drift mobility measurements carried out on the same specimen film. This general procedure has been used in a broad range of experiments to evaluate the injection efficiencies from a series of hole injecting contacts of differing work functions into the molecularly doped polymer, TPD/Polycarbonate Two key results are emphasized: (1) Injection efficiency does scale on average with the interfacial barrier, the energy step estimated from substrate workfunction and MDP electrochemical data. In individual cases however, large variations about the norm are clearly identified and can be attributed to variations in the contact surface prior to coating, for example the relative degree of oxidation or, as TOFSIMS reveals, the interdiffusion of metal atoms from adhesive layers which secure the metal coating to a Si wafer substrate. (2) For the case of Au, the injection efficiency of a Au substrate was compared to the behavior of Au vacuum deposited onto the already formed coating under a variety of conditions. The injection efficiency from Au evaporated onto preformed coating initially varies with deposition conditions but relaxes with time under ambient conditions, eventually becoming ohmic. Significant details of this relaxation behavior are described and analyzed in an accompanying paper.

INTRODUCTION

Hole transport in molecularly doped polymers, (MDPs), can be regarded as a field driven chain of redox reactions in which the carrier hops from a neutral molecule to an adjacent radical cation derivative[1] The hole transport process is triggered by an initial energetically favorable charge transfer in which a neutral molecule donates an electron either to an adjacent metal contact or photoexcited layer[2]. If the neutral transport molecule is designed to have an ionization potential lower than any competing species present, then the latter, independent of their concentration, cannot act as traps; thus they are invisible to transiting holes. A complementary situation can be described in terms of relative electron affinities for electron transporting systems. Such molecularly engineered transport layers are trap-free insulators with finite unipolar mobility which means that while there are no intrinsic free carriers, an extrinsic carrier once injected will, under the influence of an electric field, traverse the sample bulk without being locally neutralized or immobilized. An ohmic contact is defined as an infinite reservoir of charge that is able to satisfy the demands of the bulk for injected charge up to and including a supply sufficient to sustain steady state space charge limited current[3]. The bulk demands a space charge limited carrier supply when the transit time of any excess injected carrier is less than the bulk relaxation time $\rho\varepsilon$. These conditions always prevail for an ohmic contact on a finite mobility insulator. The trap free space charge limited current (J_{TFSCLC}), which only an ohmic contact can supply, takes a particularly simple form namely[3]:

$$J_{TFSCLC} = 9/8\ \varepsilon\ \varepsilon_0\mu\ E^2/\ L \qquad (1)$$

Here ε is the relative dielectric constant, μ is the drift mobility, E is the field defined as V/L where L is the film thickness. At a given field for fixed specimen geometry the current supplied by an ohmic contact is uniquely specified by the drift mobility of the injected carrier at that field. If we

629

measure the drift mobility at a given field and calculate J_{TFSCLC} then compare this calculated current density to the current injected in the dark J_{DC} by a contact under test on the same specimen we can define an injection efficiency figure of merit F, under fully self-consistent conditions namely:

$$F = J_{DC}/J_{TFSCLC} \qquad\qquad (2)$$

EXPERIMENTAL

Figure 1 is a schematic of the experimental apparatus[4] Unipolar hole transport specimen films were configured as thin film parallel plate capacitors with the contact under test on one face, and the opposite contact blocking and transparent (typically a 300 Å semi-transparent evaporated Al film) as required for the execution of small signal current mode TOF experiments

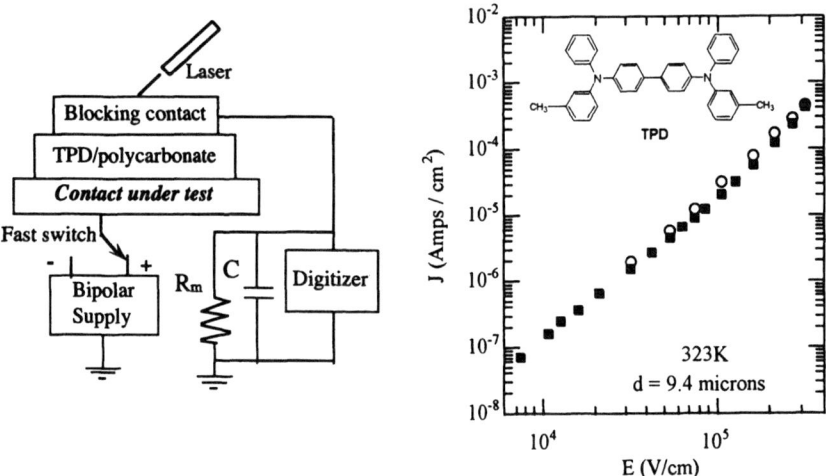

Figure 1. Experimental apparatus for analyzing injection efficiency of a test contact.

Figure 2. Measured J vs. E (filled squares) compared with J computed from drift mobility data using equation 1.

The TOF technique was used to make measurements of hole drift mobility. With positive bias applied to the Al semi-transparent blocking contact the specimen film was photoexcited with a highly attenuated 337 nm pulse from a nitrogen gas laser. Transit times t_{tr} for photoinjected holes at the applied bias could then be determined from the current transient and the mobility calculated from $m = L^2/(t_{tr}V)$. In the present case, e.g. evaluation of hole injecting contacts, steady state J-E measurements were done by application of a positive step voltage to the contact under test, capturing the transient for additional analysis, and then waiting for the current to achieve a steady state value.

RESULTS

Figure 2 is a plot of current density J versus electric field E for a thin film of a MDP composed of a 40 wt% solid solution of TPD in a polycarbonate binder. The molecular structure of the tertiary arylamine TPD is shown in the figure. The polymer film was solvent coated onto a carbon loaded polymer film substrate (Myst-R®) and then overcoated with an evaporated semi-transparent Al top contact. The hole drift mobilities, $\mu(E)$, were measured over a wide range of applied field and these values used to calculate $J_{TFSCLC}(E)$ according to eq. 1. These data are represented by the open circles in the figure. The filled squares represent the measured steady state hole injection current density J_{DC} supplied by the carbon filled polymer substrate contact. The measured values of the steady state dark currents are coincident within experimental error to the calculated J_{TFSCLC}. Thus taken together these results constitute *prima facie* evidence that the carbon filled contact is ohmic for hole injection. The comparisons presented in Figure 2 are taken as a benchmark representation of ohmic contact behavior on a molecularly doped polymer and illustrate application of the experimental technique to be used in the following study for evaluation the injection efficiency of each of the contacts under test.

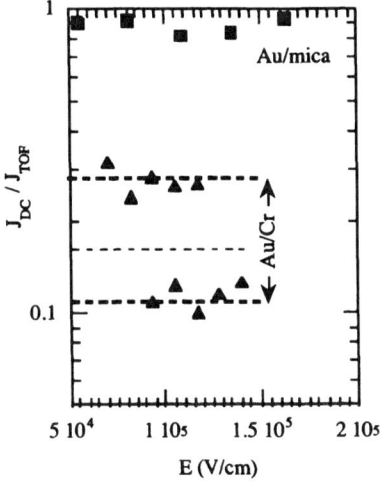 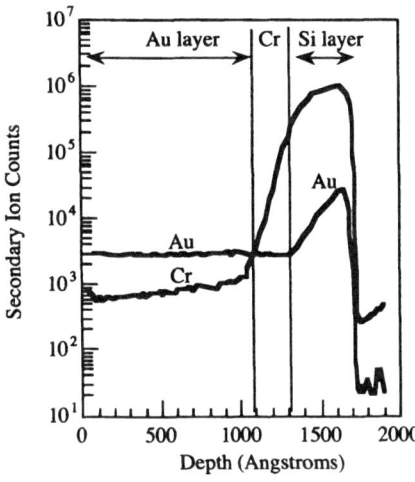

Figure 3. Injection efficiency figure of merit vs. E for Au/Cr and Au/mica contacts.

Figure 4. TOFSIMS elemental depth profiles for Au/Cr substrate. (See text)

The injection behavior illustrated in Figure 2 was also observed for injection into TPD/polycarbonate from a substrate contact consisting of Au evaporated onto mica. The injection efficiency is defined by the quotient J_{DC} / J_{TOF} where J_{DC} is the measured steady state current density supplied at a field E and J_{TOF} is the trap free space charge limited current calculated from injected hole transit times measured at the same field in the same specimen as illustrated in Figure 1. Hole injection efficiencies J_{DC} / J_{TOF} as a function of field for the Au/mica contact (filled squares) are presented in Figure 3. Note that the latter values are very nearly unity clearly demonstrating the ohmic nature of Au/mica substrate for hole injection into TPD/polycarbonate. On the other hand, when Au is evaporated onto Si using a thin (5 nm) Cr adhesive layer, the hole injection efficiency is significantly degraded The latter is illustrated in Figure 3. The two sets of triangles are the hole injection efficiencies for Au/Cr/Si substrate contacts made to two different specimens. The dashed curve represents an average value of hole injection efficiency. Variability in J_{DC}/J_{TOF} is observed despite the care taken to cast TPD films from the same solution onto different Au/Cr/Si substrates diced from the same wafer. The error bar is presented as an estimated representation of the variability since sufficient replicate runs for a complete statistical analysis was not undertaken. The severe depression, with respect to Au/mica, of the hole injection efficiency from the Au/Cr substrates appear to be associated with known contamination of the Au surface by migration of the "glue" metal Cr to the polymer/metal interface. We have obtained independent spectroscopic evidence that this occurs in our Au/Cr/Si substrates. Measurements of the surface composition and elemental depth profile analyses of the Au/Cr substrate were obtained by the

TOFSIMS technique and the salient results are presented in Figure 4. Figure 4 presents an elemental depth profile of 1000 Å of evaporated Au on highly polished Si(100) with a thin (5 nm) intermediate layer of Cr used to enhance adhesion. One striking feature is the presence of Cr atoms in the Au layer all the way to the Au surface. The mechanism of Cr migration to the Au surface probably involves Cr atom diffusion along the grain boundaries[5] of polycrystalline Au. Subsequent contamination of the Au surface with Cr suggests that hole injection occurs from a lower work function surface consistent in the present case with an apparent increase in the estimated barrier to hole injection.

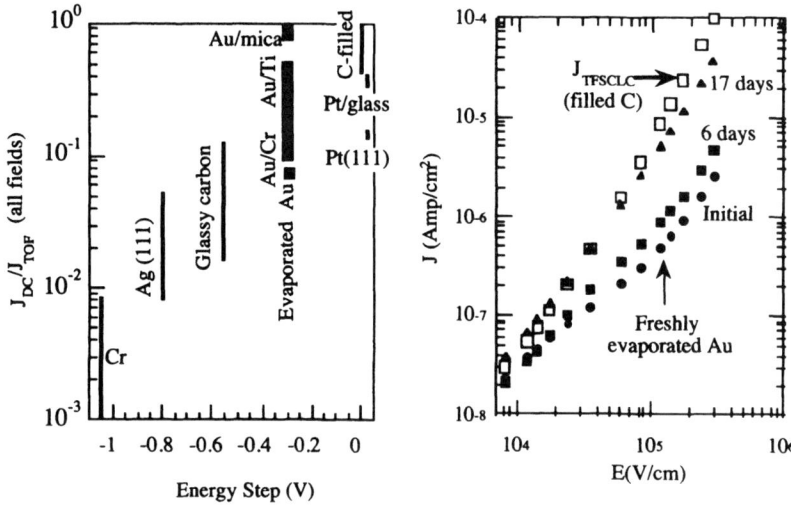

Figure 5. Injection efficiency figure of merit for contacts of varying work function across the contact-polymer interface.

Figure 6. J vs. E plots for hole injection from a Au contact as a function of room temperature annealing time following rapid evaporation on the free surface of a specimen film.

Figure 5 presents for various contact systems a compilation of hole injection efficiency data expressed as J_{DC}/J_{TOF} versus estimated interfacial barrier heights. The latter barrier heights are estimated from published values of substrate contact work functions[6] together with our estimate of the transport molecule ionization potential (5.5eV) based on the measured value of the electrochemical half wave potential of neutral TPD and its analogues in solution[7]. Such estimates have an uncertainty associated with them because of the difference in the solvent and MDP polarization energies. A value of -4.7 V was assumed for the electrochemical potential of the SCE (saturated calomel electrode) in the conversion of the electrochemical potential scale to the vacuum scale[8]. While the absolute value of the barrier height are associated with some uncertainty the relative differences in injection barriers are more certain. The vertical bars in Figure 5 encompass in each case the experimental variation observed in specimens prepared under ostensibly identical conditions. For the cases of Cr and Ag(111) it is not unreasonable to ascribe the large variation observed to uncontrolled air oxidation of the surface of the substrate contact prior to coating with TPD/polycarbonate. In the case of Cr, one of the substrate contacts was solvent coated immediately after evaporation while the other was exposed to ambient for several months prior to coating. The injection efficiency of these two Cr contacts with TPD/polycarbonate are, respectively, 8×10^{-3} and 1×10^{-3}.

Glassy carbon is hard and impervious to liquids or gases. The structure of GC is believed to be graphitized ribbons with a coherence length on the order of tens of angstroms and the electrical conductivity is about one-fourth that of randomly ordered graphite[9]. Commercially obtained glassy carbon (GC) electrode disk substrates were extensively pretreated prior to solvent coating. Their surfaces were first brought to a mirror polish with a graded series of abrasives ending in either (0.05 micron) aluminum oxide dispersed in highly purified water or 1 micron diamond polish (pasted in a high molecular weight hydrocarbon). The aluminum oxide polishing was done against a clean substrate. As exhibited in Figure 5 the aluminum oxide polishing pretreatment systematically yielded an interface with higher hole injection efficiency than similarly prepared but diamond paste polished GC. It appears that polishing preparation of GC electrode surfaces with diamond paste polishing leads to a surface that contains irreversibly adsorbed organic species likely derived from the pasting vehicle. These sites may be blocking for hole injection. In contrast the aluminum oxide abrasive was a highly purified sample that had no history of contact with organic impurities. Extensive analogous electrochemical studies of interfacial electron transfer rate constants k_s, for the oxidation-reduction of aqueous solutions of ferro/ferricyanide at glassy carbon electrodes subject to both aluminum oxide and diamond polish surface pre-treatments have been carried out[9]. The electrochemical studies pointed out a parallel enhancement of k_s in the case of GC electrodes treated with aluminum oxide relative to diamond polished surfaces. A striking enhancement in hole injection efficiency for another type of carbon surface is illustrated in the upper right hand portion of Figure 5. In this case the substrate contact consists of highly graphitized carbon particles dispersed in a polymer binder at concentrations exceeding the percolation threshold. Carbon filled polymers are complex percolative systems whose electrical behavior is a function of the filler, the filler interparticle contacts, the host matrix and the resulting internal network topology[10]. Such contact systems were in certain cases found, using the present techniques, to be ohmic for hole injection into TPD/polycarbonate. One such case was illustrated in Figure 2. Although we generally have no independent information regarding the work function of these carbon filled media *a priori*, it is evident from a rudimentary analysis based on thermionic emission theory that the barrier to hole injection must, in the illustrated cases, be very small.

The behavior of platinum electrodes is known to be especially complex. The two cases described for platinum substrates are, respectively, Pt thermally evaporated onto glass under high vacuum conditions and a Pt(111) single crystal precleaned by a brief immersion in acidic peroxide and copious rinsing in distilled water. It is well known from electrochemical studies that the electrochemistry of the Pt surfaces and the electron transfer reactions at Pt are highly sensitive to mono- and submonolayer levels of contaminants and adsorbates as well as to the electrode surface texture and morphology[11]. Moreover, it is known from controlled ultrahigh vacuum studies that the work function of platinum varies by approximately 1 eV when a pristine Pt(111) surface is dosed with a single monolayer of water. Thus the sensitivity of the Pt work function and associated interfacial charge transfer reactions to relatively minor variations in surface condition likely underlies the apparent misfit of the Pt data in Figure 5 which is based on published values of the Pt workfunction obtained by photoemission under high vacuum conditions.

With the possible exception of Pt shown on the extreme right hand side of Figure 5 the average behavior of the injection efficiency nominally scales with the estimated interfacial barrier height. Variations about the average behavior are, however, significant and it is suggested that they reflect significant perturbations of the effective work functions of the contacts. These perturbations involve the influence of factors extrinsic to the two phases involved in the formation of the electrical interface but likely to be encountered in realistic manufacturing environments which exploit the advantages inherent in devices based on solution coatable organic layers.

In Figure 6 we augment the hole injection efficiency from various substrates with data obtained when Au is vacuum deposited (3-10 Å/sec) on the *top surface* of a TPD/polycarbonate film which is in turn coated onto a substrate which is ohmic for hole injection, e.g., when the positive bias is applied to the substrate contact $J_{DC} / J_{TOF} = 1$. The evaporated Au data in the figure corresponds to hole injection when positive bias is applied to the top contact. It is in fact found that the hole injection efficiency from the evaporated Au contact depends on the details of deposition, and more interestingly the extent to which the system has been allowed to relax at a given temperature following the initial deposition. To our knowledge this is the first observation of time dependent contact forming on an organic film[12]. We note that the reported behavior while systematic for the case of evaporated Au is not observed when Al is similarly evaporated on a preformed polymer film. On the other hand contact forming phenomena were never observed when TPD/polycarbonate was coated on any of preformed metal substrates shown in Figure 5. Typical results for an evaporated Au top contact are shown in Figure 6. The figure plots the current density versus field for hole injection from an evaporated Au contact within an hour of solvent coating (solid circles), after 6 days under ambient conditions (solid squares), and after 17 days under ambient conditions (solid triangles). Injection from the ohmic substrate contact Myst-R® in the same specimen film is represented by the open squares and provides an internally consistent comparison. Extended aging studies under ambient conditions demonstrate the eventual convergence of the hole injection currents from the evaporated contact to the ohmic case. The relaxation rate(s) is/are temperature dependent, increasing with increasing temperature. This relaxation behavior which is reminiscent of contact forming phenomena in inorganic semiconductors has been studied in considerable detail. The results of these latter studies are discussed in an accompanying publication.

REFERENCES

[1] P. M. Borsenberger and D. S. Weiss, *Organic Photoreceptors for Imaging Systems* (Marcel-Dekker, New York, 1993).

[2] M. A. Abkowitz and D. M. Pai, Philos. Mag. B **53**, 193-216 (1986).

[3] M. A. Lampert and P. Mark, *Current Injection in Solids* (Academic Press, New York, 1970).

[4] M. A. Abkowitz, J. S. Facci, and M. Stolka, Applied Physics Letters **63**, 1892 (1993).

[5] M. Josowicz, J. Janata, and M. Levy, Journal of the Electrochemical Society **135**, 112 (1988).

[6] H. B. Michaelson, Journal of Applied Physics **48**, 4729 (1977).

[7] J. S. Facci, M. Abkowitz, W. Limburg, F. Knier, J. Yanus, and D. Renfer, Journal of Physical Chemistry **95**, 7908 (1991).

[8] S. U. M. Khan, R. C. Kainthla, and J. O. M. Bockris, Journal of Physical Chemistry **91**, 5974 (1987).

[9] R. L. McCreery, *Carbon Electrodes: Structural Effects on Electron Transfer Kinetics*, Vol. 17 (Marcel-Dekker, New York, 1991).

[10] M. B. Heaney, Applied Physics Letters **69**, 171 (1996).

[11] H. Angerstein-Kozlowska, *Surfaces, Cells, and Solutions for Kinetic Studies* (Plenum Press, New York, 1984).

12. M.A.Abkowitz, Journal of Imaging Science and Technology, 40, 319 (1996).

ACKNOWLEDGEMENTS
The authors gratefully acknowledge a Science and Technology Center Grant (CHE-9120001). The authors also wish to thank Dr. Dominic Salamida for TOFSIMS experimental assistance.

ELECTRICAL PROPERTIES OF Au/POLYMER INTERFACES

A. Ioannidis, J. S. Facci, M. Abkowitz
Center for Photoinduced Charge Transfer, University of Rochester, Rochester NY 14627.
Wilson Center for Research and Technology, Xerox Corp.,800 Phillips, 114-39D,Webster, NY 14580.

ABSTRACT

A trap-free Molecularly Doped hole-transport Polymer (MDP) provides a model system for the study of metal/polymer interfaces by enabling the use of a recently developed technique that supplies a quantitative measure of contact injection efficiency. The technique combines field-dependent injection current measurements from a contact under test with time-of-flight (TOF) mobility measurements on the same sample. In the present case, MDP films were prepared with two top vapor-deposited contacts, one of Au (test contact) and one of Al (for TOF), and a bottom carbon-loaded polyimide electrode. An unusual phenomenon is investigated whereby injection from Au is initially blocking, evolving to ohmic over time. This contrasts with the behavior expected according to the relative workfunctions of Au and of the polymer whereby Au should inject holes efficiently. The samples were aged at various temperatures below the glass transition of the polymer (85°C) and the evolution of current-field measurements and of capacitance is followed in detail over time and analyzed. All Au contacts eventually achieved ohmic injection. Rapid sequence data acquisition enabled the separation of two main processes in the injection evolution. Control measurements ensure that the evolution of the electrical properties is due to the Au/polymer interface behavior and not the bulk. The behavior of Au contacts evaporated under various deposition conditions was compared. Transmission electron microscopy results at the Au/MDP interface were obtained as both a function of time and of deposition conditions. Mechanisms operating at the interface of the evaporated Au films on the polymer are discussed.

INTRODUCTION

Fundamental questions concerning metal/organic interfaces are receiving much attention, stimulated by wide application of organic materials in electrophotography[1] and the rapidly expanding field of organic electronic devices. In any such applications, how electrical contact is made to the molecular material is crucial. Contacts of metals to classical semiconductors and insulators are understood within the framework of band theory. However, it is unclear whether or how band theory applies to the case of organic (molecular) materials, eg. polymers. Controversy surrounds the issue of whether charge injection in organic materials is directly a function of relative workfunctions, that is, of interfacial energy barriers described in classical treatments such as the Richardson-Schottky theory of thermionic emission or the Fowler-Nordheim tunnelling approach[2]. Direct correlation to energy barriers is difficult to rationalize in most molecular materials where a distribution of energy states exists, carriers are localized and transport is by hopping. Recently, models of injection and charge transport geared to hopping systems have been proposed, notably, by E. Conwell and coworkers[3] in the U.S. and by H. Bässler and coworkers[4] in Europe. Implications of the theoretical treatments are far-reaching, both fundamentally and for practical applications. An important limitation to the study of the nature of organic/metal interfaces is the dearth of experimental results that directly measure the injection efficiency of contacts, as measured injection currents are also dependent on the charge transport properties of the material. The problem of separating charge transport and interface effects on the injection current is solved in the case of trap-free molecularly doped materials. The cornerstone of the present study is the method, pioneered by Abkowitz and coworkers[5a,b] for obtaining the injection efficiency of a contact on a trap-free molecularly doped transport material. Using time-of-flight (TOF) mobility and field dependent current measurements, we directly probe the electrical behavior of an evaporated Au/MDP interface. The trap-free material is a model molecular system enabling a quantitative study of contact injection efficiency. The MDP consists of an electroactive triarylamine in an electrically inert polycarbonate matrix. We describe an unusual phenomenon where the efficiency for hole injection from the evaporated Au contact into the MDP increases over time, stabilizing at the maximum, ohmic contact level. An initially blocking nature of the Au contacts was contrary to the expectation from relative workfunctions and apparently contrary to the case where the MDP is coated onto a preformed Au film [6]. We therefore investigated how much of the Au contact behavior may be due to the method of contact fabrication and what mechanisms could account for the evolution in injection efficiency.

EXPERIMENTAL

The sample configurations used for the measurements are shown in Fig.1a and an illustration of graphical mehods in Fig.1b. The TOF measurement is performed using a semitransparent Al top contact [5a-b,6] enabling the calculation of the trap-free space charge limited current, which is the maximum current the bulk will support, according to [2]

$$J_{TFSCLC} = 9/8 \ \mu\varepsilon\varepsilon_o \ V^2/d^3 \qquad (1)$$

Where d=MDP thickness (typically 20μm), V= potential, μ=drift mobility, ε=dielectric constant and ε_o=vacuum permitivity. Any injection current measured can then be compared against the maximum value calculated so that injection efficiency of that contact is simply the ratio of measured to calculated value. Details of the rationale for this method are described in an accompanying paper as well as in [5,6]. At the same time, monitoring the mobility during the evolution and during the sample manipulations to be described ensures that any effects noted are not due to changes in the bulk transport properties. The injection current is measured for the contact under test, evaporated Au, as well as the bottom contact which then acts as a control contact as well. The bottom contact is an ohmic contact for hole injection into this MDP, as seen in Fig. 1b where the TFSCLC current density calculated from the mobility data coincides with the injection measured from the bottom contact. Equivalently therefore, the injection from the Au contact can be compared to that from the bottom contact to obtain the injection efficiency. Au contacts were evaporated by resitive heating of a Au source, producing films of 220-250Å at a rate of 10Å/s, unless otherwise indicated.

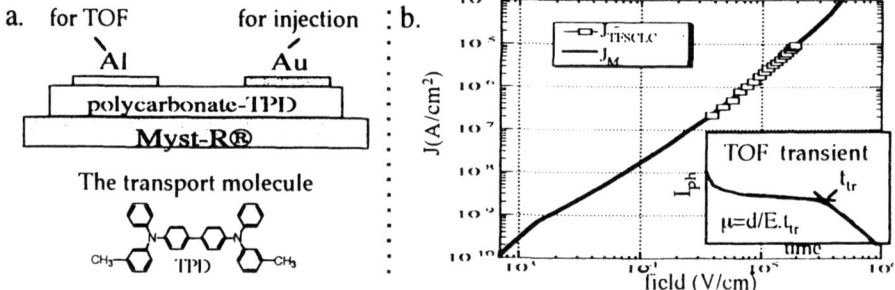

a. for TOF | for injection : b.
The transport molecule
TOF transient
μ=d/E.t_{tr}

Fig.1. a. Sample configuration; b. Illustration of graphical methods

RESULTS AND DISCUSSION

<u>Evolution parameters - time constants and thermal activation</u>

The hole injection current from evaporated Au contacts was studied as a function of applied field, E, over a one month period for three samples aged at 23, 40 and 65°C. Results were reproducible for two subsequent sample sets. The hole injection current from the control contact, which was the commercially available carbon-filled polyimide Myst-R®, was monitored during this time, as was the TOF drift mobility. Current density from Myst-R®, J_m, remained constant throughout the experimental time scale, as did the drift mobility, yielding constant J_{TFSCLC} results, such that $J_m=J_{TFSCLC}$ throughout this study. Beginning at 5 minutes after Au deposition, J-E curves were obtained ever ~15 minutes for the first few hours, tapering off to every few hours and finally every 2 days as the rate of change in injection current decreased. Indicative results are illustrated in Fig.2 for a sample aged at 40°C. The initial behavior of the Au contact is fairly blocking, with the injection current J_{Au} being ~3 orders of magnitude below the maximum sustainable by the bulk (exemplified by J_m). This is especially evident in the mid-to-upper field region 10^4-10^6V/cm, where the demand of the bulk for charge is greatest. J_{Au} increases continually over time, reaching J_m first at the lower fields and finally at the highest fields as well, showing that Au eventually becomes an ohmic contact, ie.infinite reservoir of charge available to the MDP. The latter is the expected behavior of Au regarding injection of holes into this MDP, since the ionization potential of the MDP is ~5.5 eV and the Au workfunction is between 5.2 and 5.4eV, depending on Au

636

crystallinity[7]. Furthermore, in the inverse sample geometry, when the MDP is cast onto a preformed Au film, the Au contact behavior has been shown to be ohmic for hole injection. Fabricating the Au/MDP interface by evaporating a Au contact has a substantial and unexpected effect on the Au injection behavior.

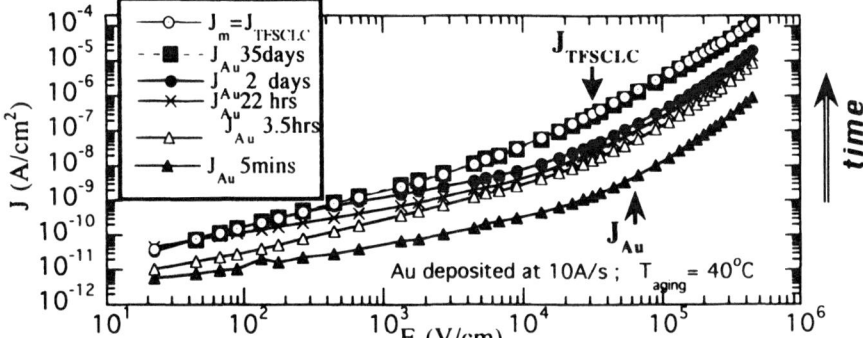

Fig.2. Typical evolution in hole injection current density from evaporated Au contacts onto MDP

In order to quantify the kinetics involved in the evaporated-Au/MDP interface formation, the injection efficiency was calculated by the ratio of J_{Au} to J_m at a given field (10^5V/cm) and plotted as a function of time. The results are shown in Fig.3 for the three samples. A non-linear behavior is observed, composed of a rapid increase in injection efficiency at early times which accounts for approximately half the increase in injection efficiency overall, followed by a much slower increase over a period of weeks.

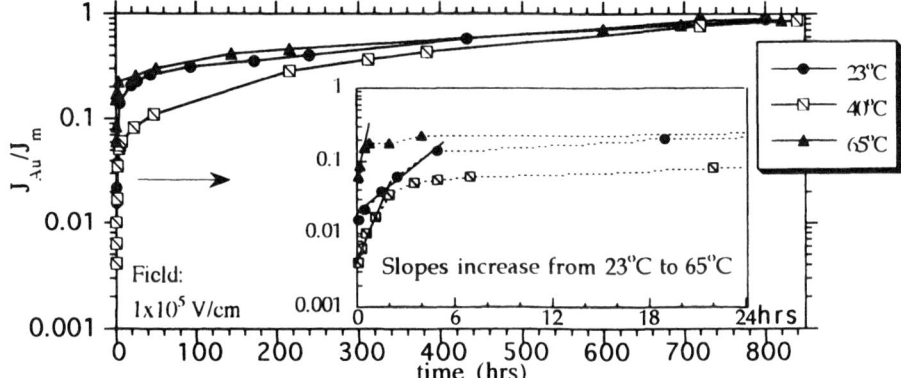

Fig.3. Two time domains in evolution of injection efficiency of evaporated Au on MDP

The inset to Fig.3 truncates the time scale in order to show the evolution in injection efficiency at early times. The results were very well fit by two exponentials, yielding two distinct time constants. It was possible therefore to distinguish two main processes governing the injection evolution. The rate (slope) of the evolution at early times was temperature dependent and an Arrhenius plot of the slopes of the three samples gave an activation energy of ~0.3eV. However, with only three points in the Arrhenius plot, the value of 0.3eV may only be considered as an estimate. It is noteworthy that the J_{Au}/J_m ratio itself shows no dependence on temperature.

Mechanisms

In order to investigate what mechanisms may be responsible for the two evolution processes,a sample is identified as a three-component system, comprised of the MDP, the Au and their interface and effects on each of these were considered. A major mechanism postulated to operate during vapor deposition of various metal contacts on organic thin films is diffusion of metal atoms into the films,as was recently reported for PTCDA thin films and a series of metal contacts[8]. In that case, no diffusion was observed for the least reactive metals, Ag and Au and the diffusion of more reactive metals was driven by metal-organic complex formation. Metal-organic complexes were also found for thiophene containing thin films and metals[9]. For the MDP, the possibility of Au penetration and diffusion was investigated by Transmission Electron Microscopy(TEM). TEM images of the evaporated Au/MDP interface were obtained with Au films prepared under the same deposition conditions of 10Å/s and the interface features were monitored over a period of 2 months in any given sample. Time-lapsed results at the highest obtained resolution of 1nm are shown in Fig.4. The interface is seen to be abrupt to within 1nm (10Å or ~2 monolayers).

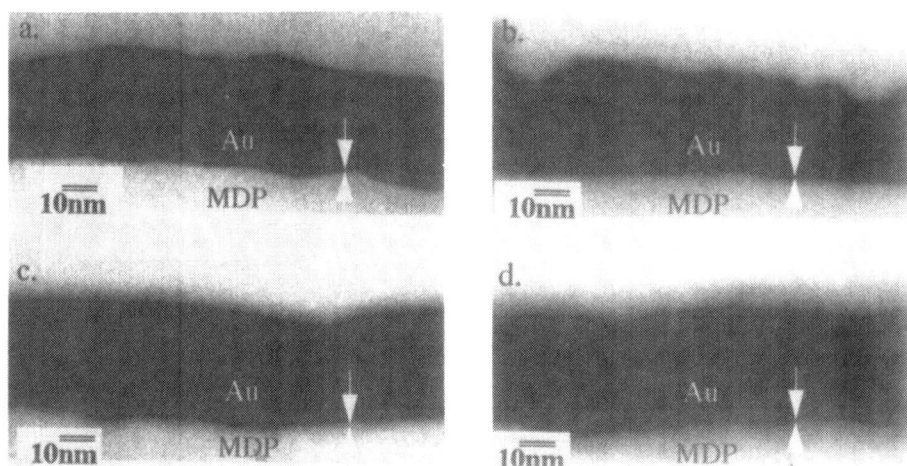

Fig.4. TEM of evaporated Au/MDP interface at a. one hour, b. 4 hours, c.24 hours and d. 2months after Au deposition.

More strikingly, the roughness (smoothness) is invariant over the entire period of injection evolution. Careful handling allowed injection measurements to be obtained periodically for the same sample from which sections were being cut for the TEM to verify the usual evolution in injection efficiency. Au/MDP interpenetration is not observed on this scale and seems unlikely to account for the evolution in the injection from the Au contact.

The lack of interdiffusion does not however affect the possibility that the surface of the polymer may be thermally damaged during the evaporation of Au, either directly through exposure to the heating source or by incoming energetic Au atoms. The fact that Au is perfectly injecting when the MDP is overcoated onto it as opposed to when Au evaporated onto the MDP points to a surface damaging effect of the evaporation. Indeed, the slow process occurs on a time scale consistent with various polymer chain motions which would act to repair the surface[10]. Accordingly, the evaporation conditions of the Au were systematically varied, to affect the rate and therefore energy of evaporating Au atoms and the exposure of the MDP surface to heat. An indicative comparison of results from four variations in deposition conditions is presented in Fig.5. The slow, long term process is dramatically curtailed by any type of layer-by-layer Au deposition (Fig.5a-c) and ohmic injection level is achieved after several hours, compared to several weeks for the usual one-shot depositions of Au at 10Å/s. On the other hand, an approximately one order of magnitude change in the rate of the deposition, the rate of incoming Au atoms to the surface, has no effect whatsoever on the kinetics of the phenomenon, as seen is Fig.5d. The effect of the thermal evaporation of Au onto the MDP seems to be one of thermal damage to the MDP surface, either due to its exposure to the hot source (radiative damage) or due to the heat cumulatively applied to the surface by the hot 220Å Au layer, or both. Evaporating a thin Au layer first has a protective effect. Even more significantly however, no deposition variation had any effect on the initially blocking nature of the contact and no consistent effect on the early, fast process.

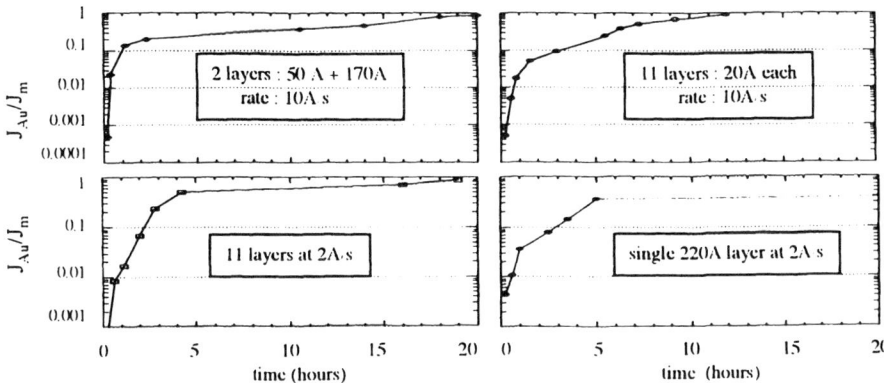

Fig.5. Effect of 4 changes in Au evaporation conditions on the Au injection efficiency to the MDP

The initial blocking behavior and subsequent rapid evolution to ohmic injection are independent of Au evaporation conditions and may be intrinsic to the Au/MDP interface. This is not contradicted by the "initially" injecting behavior noted for the inverse sample geometry, where MDP is solution coated onto a preformed Au film, since the initial measurement of injection current in the latter case is actually several hours after MDP and Au come in contact due to the need to wait for the drying and curing of the MDP. Furthermore, the state of the MDP when it comes into contact with the Au is evidently different in that case. Accordingly, the state of the MDP onto which Au is evaporated was slightly altered in an attempt to approach the conditions of the inverse geometry. Films of MDP that had been coated as usual on Myst-R and cured were subsequently exposed to a saturated vapor atmosphere, the vapor being that of the MDP casting solvent methylene chloride. Au contacts were then evaporated onto such films at the usual 10Å/s and ~250Å thickness, to compare with the series of samples represented in Figs.2 and 3. The results are seen in Fig.6a, where the initial measurement of injection current, taken at the same time as for Fig.2, is seen to be only ~one order of magnitude below ohmic at a field of 10^5 V/cm. This compares very favorably with the ~three orders of magnitude depression of injection current observed for the films in Figs.2 and 3 at the same field. Furthermore, the injection efficiency becomes ohmic in the shortest time period of four hours. Similarly, if vapor doping is performed some time after Au is evaporated onto cured MDP, abrupt improvements in injection efficiency are observed during the usual evolution, as shown in Fig.6b.

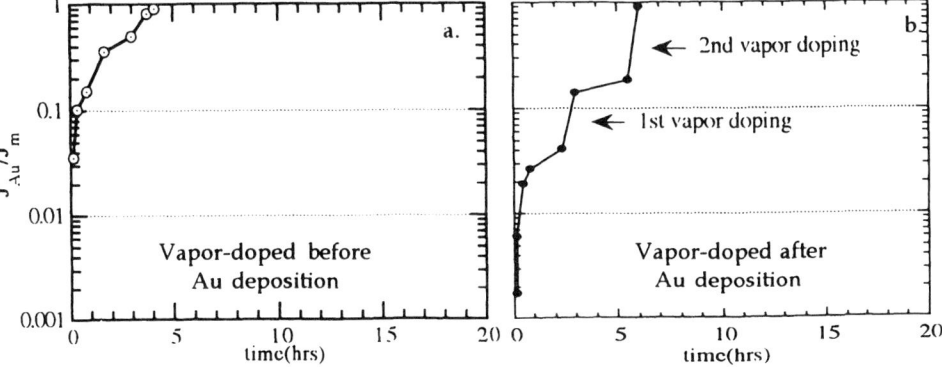

Fig.6 Effect of doping MDP with methylene chloride vapor a.before ; b.after Au contact deposition

The increased injection current cannot be ascribed to an increase in mobility of charge carriers as both TOF measurements and injection current from the control contact, Myst-R, remained invariant at the vapor doping levels used. While not a quantitave study, the vapor doping demonstrates clearly a dramatic effect on injection efficiency from the metal contact. The results in Fig.6, taken together with the above investigations of the interface abruptness to within 10Å and the independence of the initial blocking behavior on Au fabrication conditions, may have fundamental implications for MDP/metal interface formation: injection from a metal into the hopping transport molecular material is not predictable solely based on the relative workfunctions, as it is for band type materials. Bringing the MDP and the metal into direct contact did not immediately cause injection of charge carriers. For injection to occur across a metal/organic interface, carriers must transfer from the Fermi sea of the metal into charge transport states in the molecular material. This charge transfer requires electronic wavefunction overlap between the transport molecules and the metal in order to occur. The wavefunction overlap may be enhanced in the case of the evaporated Au/MDP interface by configurational reorganization of the molecular surface in contact with the metal. This is entirely consistent with the results on vapor doping, since the vapor doping of the MDP allows a greater motility of the TPD charge transport molecules, facilitating the reorganization of the MDP surface. At the same time, the vapor doping raises the possibilty of chemical changes at the interface caused or mediated by the solvent and this will form the subject of future studies.

Finally, for consistency, the surface morphology and the crystallinity of Au films evaporated onto the MDP were examined for any changes with time by TEM in transmission and diffraction mode. No change in the Au film is observed over time.

CONCLUSION

The unusual phenomenon of evolution in injection efficiency of evaporated Au contacts into a molecularly doped polymer was quantitatively studied by using a model, trap-free hole transport material. The kinetics of the evolution were characterized, enabling the detection of two main processes operating during the evolution. A slow process, on the time scale of one month, has been assigned to thermal damage of the MDP surface during the evaporation of the 220-250Åthick Au contacts at 10Å/s. Evaporation of a thin, 20-50Å, Au layer at rates from 2 to 15 Å/s onto the MDP first has a protective effect, drastically shortening the time scale of the evolution. No Au penetration or diffusion is detected by TEM of the interface, obtained sequentially during the injection evolution. A fast process, occuring early in the evolution over a period of several hours, and the initial blocking behavior of the contact are consistent with the operation of a mechanism whereby the transport molecules in the MDP reorganize to enable interfacial charge transfer from the Au to the molecular material. An immediate consequence of the proposed mechanism is that, dependent as it is on the MDP, it should occur for other metals, barring any extraneous effects from particular metals. Indeed, preliminary results using the similarly non-reactive Ag as the evaporated contact also show an evolution in injection efficiency. Measurements are under way to characterize the kinetics.

REFERENCES
1. P.M.Borsenberger, D.S.Weiss, Organic Photoreceptors for Imaging Systems, Marcel Dekker(1993)
2. M.A. Lampert, R.Mark, Current Injection in Solids, Academic Press, NY (1970)
3. E.M. Conwell, in Handbook of Organic Conductive Molecules and Polymer, Vol.4, p.1. Ed. H.S.Nalwa, John Wiley & Sons (1997)
4. H.Bässler, Phys.Stat. Sol.(B) 175, 15 (1993)
5.a. M. Abkowitz and D.M. Pai, Phil.Mag. B., 53, 193 (1986).
  b. M. Abkowitz, J.S. Facci; M. Stolka, Appl.Phys.Lett., 63, 1892 (1993).
6. M.Abkowitz, J.S. Facci and J.Rehm, J.Appl.Phys., March 1998 in press.
7. H.B. Michaelson, J.App.Phys. 48, 4729 (1977)
8. Y. Hirose, A. Kahn,V.Aristov, P.Soukiassian, V.Bulovic, S.R.Forrest, Phys.Rev.B, 54,13748(1996)
9. A.Lachkar, A.Selmani, E.Sacher, M.Leclerc, R.Mohkliss, Synthetic Metals, 66, 209 (1994)
10. R.P. Wool, Polymer Interfaces,chs. 3 and 4, Hanser Verlag, Munich (1995)

ACKNOWLEDGEMENTS
The authors gratefully acknowledge the technical assistance of Harry Freitas and Jack Czerniawski for TEM measurements. The work was supported by a Science and Technology Centre grant CHE-91-20001.

HEXACYANODIAZAHEXADIENE (HCDAH) DIANION
AS A COMPONENT OF CONDUCTING COMPLEXES

Hideki Yamochi* **, Kiyohiko Tsutsumi*, Takeshi Kawasaki*, Gunzi Saito*
* Division of Chemistry, Graduate School of Science, Kyoto University,
 Sakyo-ku, Kyoto 606-01, JAPAN,   yamochi@kuchem.kyoto-u.ac.jp
** CREST, Japan Science and Technology Corporation (JST)

ABSTRACT

In the course of the preparation of conductive molecular crystals using 'percyano anions', the molecular structure of $HCDAH^{2-}$ was revealed to be a planar geometry having a long diagonal line ($\approx$ 10.4 Å). In $(TTF)_2(HCDAH)(CH_3CN)$ [block-shaped TTF complex] (triclinic, $P\bar{1}$, a = 8.258(5), b = 9.572(6), c= 10.070(4) Å, α = 89.89(5), β = 70.76(4), γ = 80.76(5)°, V = 740.7(7) Å^3, Z = 2, final R-value = 0.074), the face-to-face type overlap between the side-by-side arranged HCDAH dianions was observed. The donor packing pattern in (BEDO-TTF)$_6$(HCDAH) (monoclinic, P2$_1$/a, a = 20.243(3), b = 4.086(1), c = 16.820(4) Å, b = 99.45(1)°, V = 1372.4(5) Å^3, Z = 2/3, final R-value = 0.048) was determined to be HCP-type which is the second example in BEDO-TTF complexes.

INTRODUCTION

All of the molecular crystals which show metallic conducting properties belong to the charge transfer (CT) complexes in broad sense. CT complexes can be classified into two categories. One is the donor - acceptor complexes of which constituent molecules are isolated as stable neutral molecules. Another is the radical ion salts, at least one of the components of which can not be isolated as a stable neutral species. TTF-TCNQ and TTF-Cl are the examples which belong to the former and latter categories, respectively. In both cases, partially charged organic molecules dominate the conducting properties of the complexes. The counter ions in radical ion salts contribute to determine the total crystal structures and only perturb the electronic structure of the salts. Generally, inorganic ions have been utilized as the counter components. Although the variety of inorganic ions has provided the numerous organic conductors and superconductors, the control of the ion size and shape has been realized within the limited range.[1] Contrary, the wide variety of stable organic ions will enlarge the degree of freedom to control the size, shape, and formal charge of the counter ion.

For the conducting radical ion salts, stable organic ions have been studied in the limited range. Representatively, R_4X^+ (R = alkyl, aryl, X = N, P, As, Sb)[2] and sulfonic acid derivatives[3] have been studied as cations and anions, respectively. However, especially for the anions, the studies to expand the variety of the chemical species are rare.

Concerning the theoretical prediction[4] that a counter component having big polarizability can stabilize the metallic state of the neighboring conducting path in a charge transfer complex, π-conjugated anions are adequate for conducting CT complexes. We focused on 'percyano anions' which are defined as the stable organic anions consist of fully cyano substituted π-systems. This class of anions form a subgroup of so called 'cyanocarbons'[5]. Although the investigations to apply these anions to the conductive CT complexes have rarely carried out[6], a systematic study of these anions will reveal the relationship between the structures of anions and CT complexes and will aid the expectation of the structures and hence the properties of new complexes.

As a subsequent study to our previous investigation on small sized anions of CF[-], PCA[-], HCTMM[2-] [7], here we report our recent results on the complexes of HCDAH[2-] for which a planar geometry having a big size is expected.

CF[-]   PCA[-]   HCTMM[2-]   HCDAH[2-]

EXPERIMENTAL

Electrooxidation experiments in CH_3CN, Cl_2CH-$CHCl_2$+MeOH, THF, Ph-CN of tetrathiafulvalene (TTF) derivatives were carried out in constant current conditions (0.1 - 2.0 μA) in the presence of bistetrabutylammonium salt of HCDAH $((Bu_4N)_2(HCDAH))$ which was prepared according to the synthetic procedure for the bistetraethylammonium salt[8]. The results were summarized in Table 1 along with the first redox potentials of the donor molecules used. Metathesis of $(TTF)_3(BF_4)_2$ and $(Bu_4N)_2(HCDAH)$ was carried out in an acetonitrile solution to give black needles and black block-shaped crystals as major and minor product, respectively. Among these two modifications, only the latter one gave crystals which were suitable for the crystal structure analysis though the dimension and total amount obtained were too small to measure the resistivity and carry out the elemental analysis.

The reflection data for X-ray single crystal structure analyses were collected on a four-circle diffractometer irradiating monochromated $MoK\alpha$ radiation at room temperature.

RESULTS and DISCUSSION

Formation of Solid Charge Transfer Complexes

The electrocrystallization of TTF derivatives in the presence of $(Bu_4N)_2(HCDAH)$ afforded no solid CT complexes with weak donor molecules so far. Including the results of metathesis reaction with $(TTF)_3(BF_4)_2$, it seems that the lower first oxidation potential $(E_{1/2})$ of the donor molecule than approximately that of HCDAH[2-] is needed to give solid CT complexes of HCDAH. The exception is the case of (EDO)2DBTTF which is known to have a strong tendency to give solid CT complexes regardless the properties and structures of counter component molecules.[9]

Among the obtained CT complexes, the composition of the black block-shaped TTF complex and BEDO-TTF complex were determined as $(TTF)_2(HCDAH)(CH_3CN)$ and (BEDO-TTF)$_6$(HCDAH) by the single crystal X-ray structure analysis and the elemental analysis (C, H, N, O, S), respectively.

Block-shaped TTF complex [(TTF)$_2$(HCDAH)(CH$_3$CN)]

Figure 1 shows the b-axis projection of the crystal structure in which the layers consist of $(TTF)_2(CH_3CN)$ and of HCDAH pile up alternately along the c-axis. Crystallographically one and half molecule of TTF and HCDAH are unique, respectively. The bond lengths in the donor molecule indicate that the charge on TTF is unity or at least very close to +1. As a result, HCDAH anion is in the dianion state in this complex. In the $(TTF)_2(CH_3CN)$ layer, face-to-face

Table 1. Summary of the electrocrystallization*[1] of TTF derivatives in the presence of $(Bu_4N)_2(HCDAH)$.

Donor*[2]	R, R	$E_{1/2}$*[3]	Result
TMTTF	CH_3, CH_3	0.29	Powder
OMTTF	$-(CH_2)_4-$	0.29	Few amount of powder
HMTTF	$-(CH_2)_3-$	0.29	Few amount of powder
BEDO-TTF	$-OCH_2CH_2O-$	0.43	Black Plates
TMTSF	*	0.45	No solid complexes
TTC1-TTF	SMe, SMe	0.52	No solid complexes
(EDO)2DBTTF	*	0.51	Powder
BEDT-TTF	$-SCH_2CH_2S-$	0.53	No solid complexes
DBTTF	benzo	0.63	No solid complexes

*1: Solvents used were $Cl_2CH-CHCl_2$+MeOH, THF, and Ph-CN.
*2: Chemical structures of donor molecules are as following;

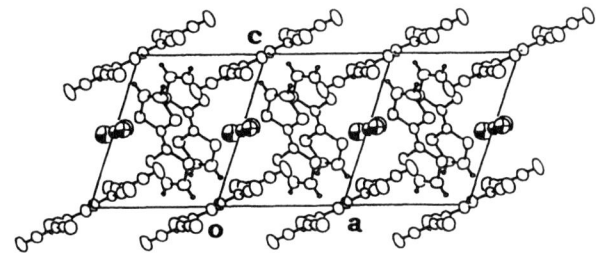

TTF derivatives   TMTSF   (EDO)2DBTTF: R, R =

*3: in 0.1M $(Bu_4N)BF_4/CH_3CN$ vs. S.C.E. Under the same condition, the redox processes of $HCDAH^{2-} \leftrightarrows HCDAH^-$ and $TTF \leftrightarrows TTF^+$ were observed at $E_{1/2} = 0.40$ and 0.37 V, respectively.

type dimers of TTF molecules are isolated by the disordered CH_3CN. The best planes of TTF and the dianion are almost perpendicular to each other (dihedral angle = 82.5°). This relative arrangement is similar to those observed in the complexes of TTF derivatives and planar inorganic anions.[1, 3]

$HCDAH^{2-}$ showed a planar and well spread molecular geometry as was expected. This configuration in which cyano groups were aligned in a zig-zag arrangement and the intramolecular bond lengths and angles were very similar to those in $(Bu_4N)_2(HCDAH)$. The longest intramolecular atomic distance was 10.4 Å which is rather big comparing to those of the usual discrete anions. It should be noted that $HCDAH^{2-}$ showed face-to-face type geometrical overlap in the anion layer, though the packing motif in a layer is a side-by-side

Figure 1. Crystal structure of $(TTF)_2(HCDAH)(CH_3CN)$ [block-shaped TTF complex]. Projected along the b-axis. Non hydrogen atoms of CH_3CN were shaded all of which showed disorder around the center of inversion.

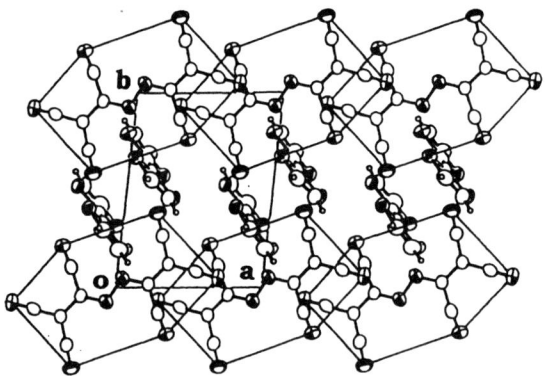

Figure 2. Crystal structure of $(TTF)_2(HCDAH)(CH_3CN)$ [block-shaped TTF complex]. Anion best plane projection. Sulfur and nitrogen atoms were shaded. Only TTF and HCDAH at z ≈ 0.5 and 1.0, respectively, were drawn. To distinguish the anion plane, the terminal nitrogen atoms in an HCDAH were linked by solid lines. Attention should be paid to that the parallelogram indicates only an ab-plane and not a unit cell.

type along the a-axis in principle (figure 2). This peculiar packing pattern may be originated from the big dimension of the dianion.

$(BEDO-TTF)_6(HCDAH)$

The electrocrystallization of BEDO-TTF(BO) in the presence of $(Bu_4N)_2(HCDAH)$ gave black prisms as the sole product. Although the exact estimation of the charges on the donor and HCDAH in this complex could not be done yet, this material is the most conductive HCDAH complex so far obtained ($\rho_{rt} \approx 0.06 - 9.6 \, \Omega cm$). The crystal structure analysis revealed the layered structure of this complex in which the donor and anion layers pile up along the c-axis alternately. The atomic parameters of BO molecules were refined according to the usual procedure and it was found that BO formed face-to-face type columnar stacks along the b-axis. The direction of the BO molecular plane was changed for every two columns along the transverse (a-axis) direction (Figure 3a).

In the anion layer, the crystallographic analytical procedure of Fourier and differential synthesis gave the peak positions running parallel to the b-axis, from the distance between which (a-axis direction) no chemical bonds could be considered (Figure 3b). It was obvious that HCDAH could not be placed in a unit cell aligning the longitudinal axis (≈ 10.4 Å) parallel to the shortest unit cell axis (b = 4.086(1) Å). To account the peak positions and the composition determined by the elemental analysis, we present a model in which one HCDAH occupies three unit cells along the b-axis. In this model, for example, an HCDAH molecule is packed as indicated in Figure 3b occupying 18 sites out of the 24 peak positions of $0 < x < 0.5$ in three unit cells. One can derive further five types of packing patterns concerning the one-unit-cell translation along the b-axis and screw axis at $x = 0.25$, $z = 0$. The six types of occupation patterns are regarded to occur randomly in the crystal. As a result, HCDAH is crystallographically disordered.

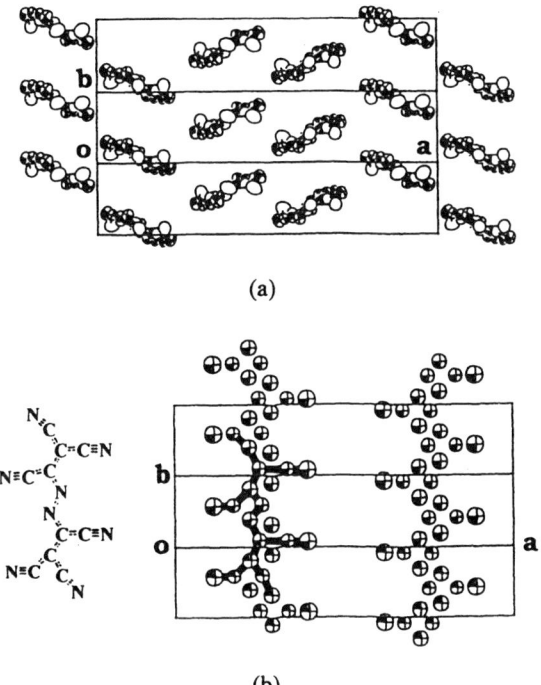

(a)

(b)

Figure 3. (a) Donor layer structure in (BEDO-TTF)₆(HCDAH) projected along the c-axis. Oxygen and sulfur atoms were shaded. Hydrogen atoms were omitted for simplicity. (b) Shaded circles indicate the peak positions obtained from Fourier and differential synthesis in the anion layer of (BEDO-TTF)₆ (HCDAH). The thick lines indicate an example to assign them to an HCDAH molecule. The configuration in this case is indicated schematically in left side.

BO is known to have the strong ability to form metallic complexes with variety of acceptor molecules and anions. The crystal structure of such conductive complex contains the donor layer in which BO molecules form face-to-face type stacking columns. The packing patterns in the donor layer had been classified into three types, namely I₃-, Cl-, and HCP-type.[10] The classification corresponds to the periodicity of the inter-column structure. In Cl-, and HCP-typed ones, the direction of BO molecular plane changes every and every two columns along the perpendicular direction to the stacking axis, respectively. In I₃-type complexes, there occurs no alternations of the BO molecular plane (i.e. all the BO molecules are (almost) parallel). I₃-type is the most frequently appearing packing motif in which the intermolecular weak hydrogen bonds and heteroatom contacts between BO molecules are regarded to effect the stabilization of the crystal structure. Only one Cl-type and one HCP-type complexes have been reported so far.

The packing motif of BO molecules in (BO)₆(HCDAH) is HCP-type. Although the further detailed analysis is needed, the size and the unique molecular shape of HCDAH should be the origin of this peculiar packing pattern of the donor molecules. In fact, the alternation of the incline of BO molecular plane occurs at x = 0.25 and 0.75 at where HCDAH lies.

CONCLUSION

The crystal structure analyses of (TTF)$_2$(HCDAH)(CH$_3$CN) clarified that HCDAH^{2-} tends to show an elongated planar molecular structure in the complex. The big dimension and probably the characteristic shape of this organic anion afforded the peculiar crystal structures of the CT complexes. In the block-shaped TTF complex, the face-to-face type overlap between the side-by-side arrangement of HCDAH^{2-} was observed (see Figure 2). In the BO complex, a very rare packing pattern of the donor molecules was constructed.

ACKNOWLEDGIMENTS

Financial supports were provided by Toyota Physical and Chemical Research Institute, a Grant-in-Aid for Science Research from the Ministry of Education, Science, Sports, and Culture, Japan, and the fund from Japan Society for the Promotion of Science for 'Research for the Future'.

REFERENCES

[1] G.Saito, in Organic Molecular Solids - Properties and Applications, edited by W.Jones (CRC Press, New York, 1997) p.309.

[2] A.Kobayashi and H.Kobayashi, in Handbook of Organic Conductive Molecules and Polymers, vol. 1, edited by H.S.Nalwa (John Willey & Sons Ltd, New York, 1997) p.249.

[3] For example, a) J.M.Fabre, D.Serhani, K.Saoud, S.Chakroune, and M.Hoch, Synthetic Metals, 60, 295(1993). b) U.Geiser, J.A.Schlueter, H.H.Wang, A.M.Kini, J.M.Williams, P.P.Sche, H.I.Zakowicz, M.L.VanZile, J.D.Dudek, P.G.Nixson, R.W.Winter, G.L.Gard, J.Ren, and M.-H.Whangbo, J. Am. Chem. Soc., 118, 9996(1996).

[4] O.H.LeBlanc,Jr., J.Chem.Phys., 42, 4307(1965).

[5] a) W.R.Hertler, W.Mahler, L.R.Melby, J.S.Miller, R.E.Putscher, and O.W.Webster, Mol. Cryst. Liq. Ceyst., 171, 205(1989). b) E.Ciganek, W.J.Linn, and O.W.Webster, in The Chemistry of the Cyano Group, edited by Z.Rappoport (Interscience Publishers, London, 1970) p.423.

[6] a) T.Fukunaga, M.D.Gordon, and P.J.Krusic, J. Am. Chem. Soc., 98, 611(1976). b) M.A.Beno, H.H.Wang, L.Soderholm, K.D.Carlson, L.N.Hall, L.Nuñez, H.Rummens, B.Anderson, J.A.Schlueter, J.M.Williams, M-.H.Whangbo, and M.Evain, Inorg.Chem., 28, 150(1989). c) W.H.Watson, A.M.Kini, M.A.Beno, L.K.Montgomery, H.H.Wang, K.D.Carlson, B.D.Gates, S.F.Tytko, J.Derose, C.Cariss, C.A.Rohl, and J.M.Williams, Synthetic Metals, 33, 1(1989).

[7] H.Yamochi, T.Tsuji, G.Saito, T.Suzuki, T.Miyashi, and C.Kabuto, Synthetic Metals, 27, A479(1988).

[8] W.J.Middleton, E.L.Little, D.D.Coffman, and V.A.Engelhardt, J. Am. Chem. Soc., 80, 2795(1958).

[9] T.Senga, K.Kamoshida, L.A.Kushch, G.Saito, T.Inayoshi, and I.Ono, Mol. Cryst. Liq. Cryst., 296, 97(1997).

[10] S.Horiuchi, H.Yamochi, G.Saito, K.Sakaguchi, and M.Kusunoki, J. Am. Chem. Soc., 118, 8604(1996).

SYNTHESIS AND INCORPORATION OF THIENYLENE VINYLENE OLIGOMERS IN MAIN-CHAIN COPOLYMERS

L. G. MADRIGAL, E. H. ELANDALOUSSI, C. W. SPANGLER
Department of Chemistry and Biochemistry, Montana State University, Bozeman, MT 59717,
uchcs@earth.oscs.montana.edu

ABSTRACT

Poly [2,5-thienylene vinylene] (PTV) has been studied extensively over the past decade for both its metallic conductivity behavior upon chemical doping, as well as its interesting third order nonlinear optical properties. PTV oligomers have been synthesized by our group[1], as well as others[2], and the formation of polaron-like radical-cations or bipolaron-like dications by oxidative doping has been demonstrated. In this paper we describe a general synthetic approach to PTV oligomers functionalized for copolymer formation by step-growth reaction.

INTRODUCTION

We have recently reviewed the synthesis of thienylene polyenylene oligomers and polymers, with particular emphasis on charge-state incorporation in these materials[3]. When examining the oxidative doping of these oligomers in solution ($SbCl_5$ / CH_2Cl_2), it is clear that both the neutral π - π^* transition and the shift in oscillator strength upon doping is more well-defined than that observed in the fully conjugated polymer. Doping studies of the oligomers in solution, however, are complicated by the insolubility of oligomers longer than the pentaene[1]. Over the past few years, it has been shown that incorporation of alkyl groups (e.g. butyl) in the 3 and 4 positions of each thiophene ring greatly enhances the solubility of these compounds to the point that they can now be derivatized in the terminal α positions with the expectation that they can be incorporated as formal repeat units in well-described copolymers. For example, we have described the incorporation of dithienyl polyene repeat units with up to ten ene units in various copolymer formulations utilizing this approach[4], exemplified by the dithienylhexatriene incorporated polyurethane shown below:

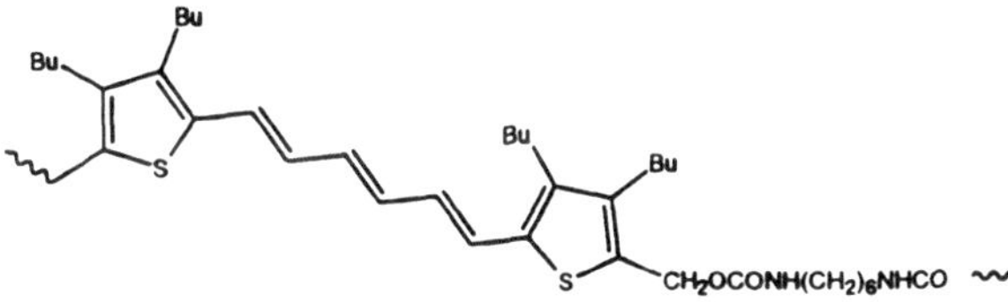

SYNTHESIS OF PTV-OLIGOMERS CONTAINING SOLUBILIZING GROUPS

During the past two years we have synthesized PTV oligomers solubilized with butyl substituents from dimer through pentamer. The synthetic approach is outlined in Schemes 1 and 2. Roncali and coworkers have recently described an almost identical approach to these materials[5]. The alkyl group substitution pattern renders these materials soluble in most organic solvents. The α positions on the terminal rings of these oligomers remain reactive under Vilsmeier conditions, yielding reactive CHO groups. Reduction to terminal CH_2OH groups thus provides functionalities that can be polymerized in a variety of step-growth condensations.

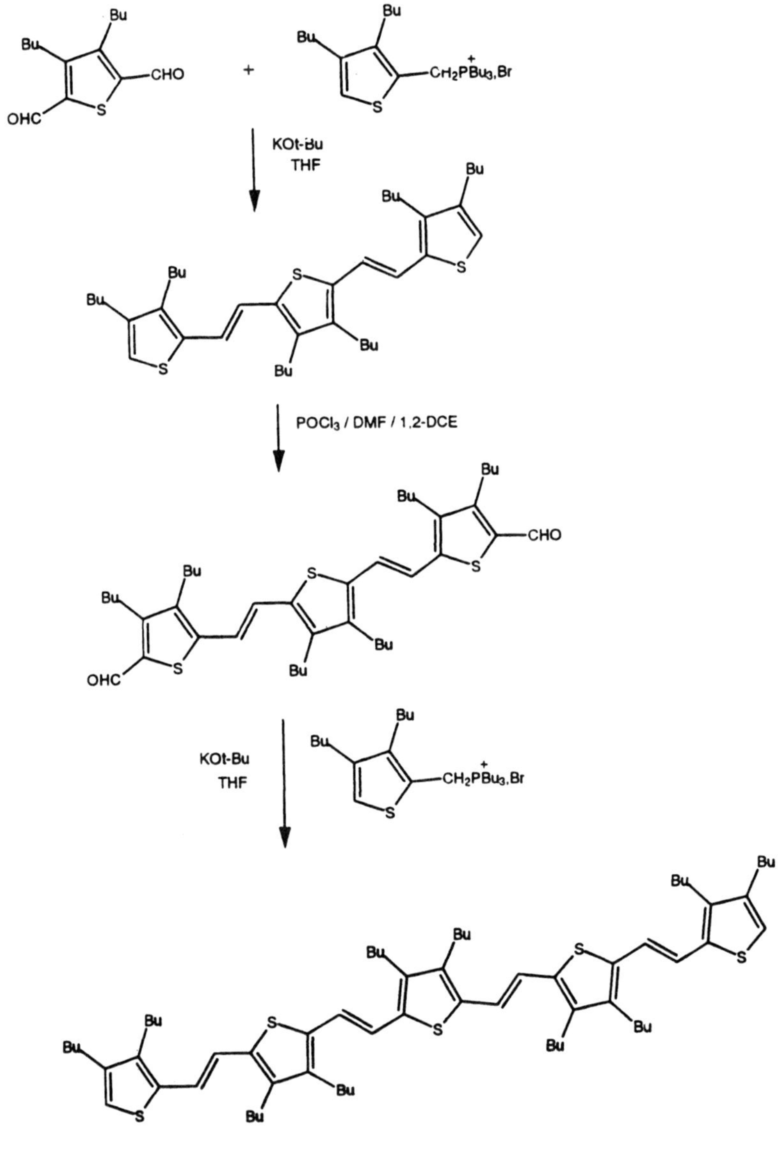

Scheme 1. Synthesis of PTV-dimer and tetramer

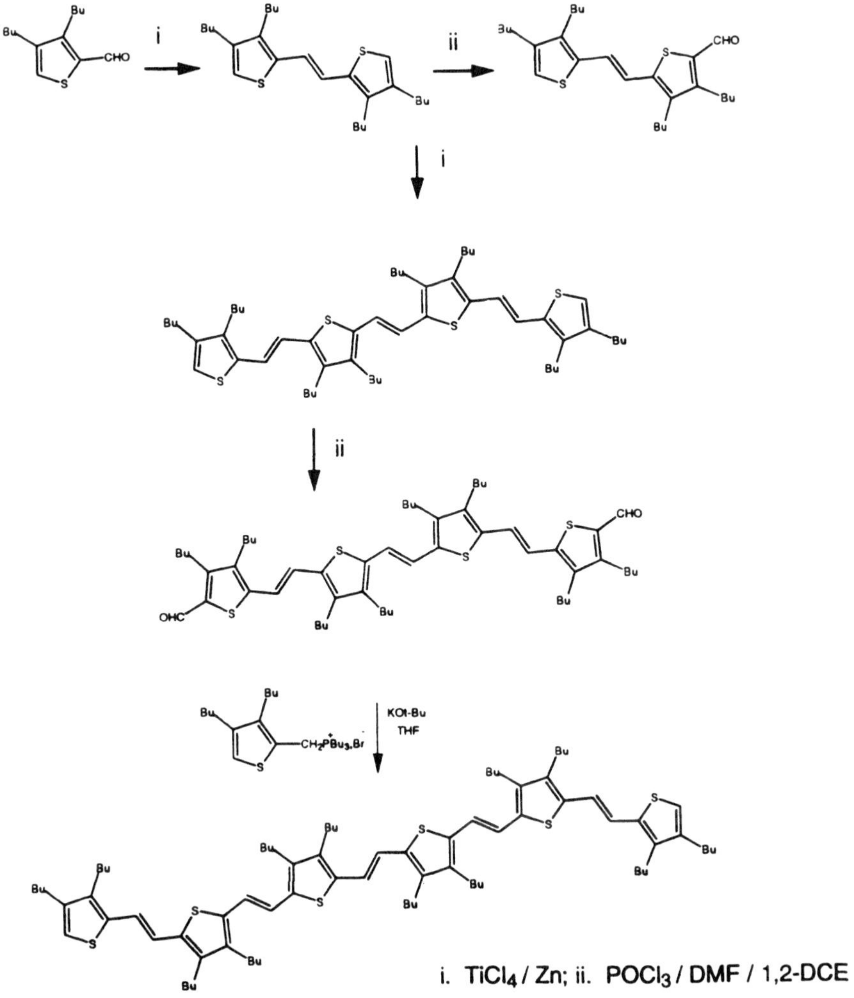

Scheme 2. Synthesis of PTV-trimer and pentamer

i. TiCl₄ / Zn; ii. POCl₃ / DMF / 1,2-DCE

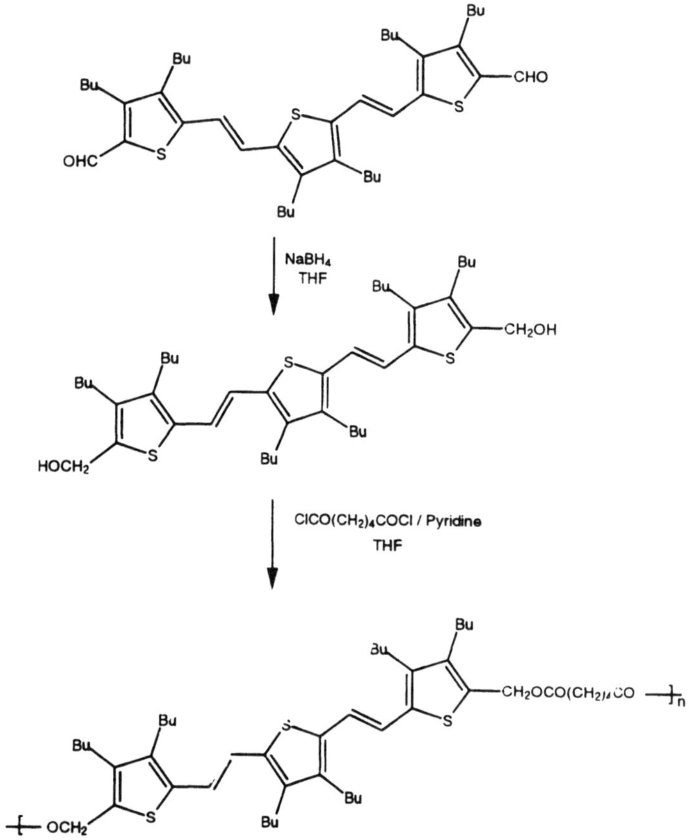

Scheme 3. Incorporation of PTV-dimer into a polyester copolymer

INCORPORATION OF PTV-DIMERDIOL IN POLYESTERS

The synthesis and incorporation of PTV-dimerdiol into a polyester copolymer is illustrated in Scheme 3. The diol, obtained in good yield by reduction of the dimerdialdehyde was polymerized in THF with both terephthaloyl chloride and adipoyl chloride. The PTV-dimer terephthalate was not particularly soluble in common organic solvents, but the adipate had good solubility in DMF. In a typical run, 0.26g (0.38 mmol) of the diol was dissolved in 25 mL of dry THF. Adipoyl chloride (0.07g, 0.38 mmol) was added all at once along with 1-2 drops of pyridine. The reaction was stirred at 85° for 24 h. Water (25 mL) was then added and the resulting mixture neutralized with sodium carbonate. The THF was removed under vacuum and MeOH (25 mL) was added. The resulting mixture was refluxed for 1 h, and the polymer obtained by filtration. After heating with THF to remove low MW material and washing with MeOH, the final polymer was obtained (0.13g, 49%).

POTENTIAL APPLICATIONS

PTV has previously been shown to have excellent third order nonlinear optical (NLO) properties[6]. It has also been demonstrated that under laser irradiation, stable bipolarons can be photogenerated, with lifetimes that can apprach 0.1 sec.[7]. Our previous chemical doping studies[1] showed that either polaron or bipolaron states can be highly absorbing. Thus we have proposed that these, and other materials that yield stable, highly absorbing excited states, may be excellent optical power limiting (OPL) polymers *via* reverse saturable absorption (RSA) from these photogenerated charge states. We have recently shown, in collaboration with coworkers at Laser Photonics Technology, that polyene-type molecules or polymers may function as OPLs by this mechanism[8]. Using PTV oligomer repeat units rather than the fully conjugated polymer offers the advantage that both the π - π^* transition of the copolymer and the nonlinear absorption of the excited states can be predicted and synthetically controlled. The PTV-dimer adipate discussed above is the first such polymer specifically designed to take advantage of the potential RSA characteristics of these main-chain copolymer repeat units. These polymers are currently being evaluated as OPL materials as part of our continuing collaboration with Laser Photonics Technology, and their merits will be presented at a future MRS meeting.

CONCLUSIONS

PTV oligomers can be solubilized and functionalized to allow their incorporation in a variety of copolymer formats. These materials form exceptionally stable polarons or bipolarons when chemically doped, and it is proposed that the absorption characteristics of the chemically produced species will model or approximate the absorption profile of the corresponding photogenerated charge states when these copolymers are employed as passive smart OPLs by RSA.

ACKNOWLEDGMENTS

The authors gratefully acknowledge financial support for this research by the Air Force Office of Scientific Research under Grant # F49620-96-1-0440, and also acknowledge helpful advice and discussioin with the following: Martin Casstevens and Ryszard Burzynski (Laser Photonics Technology), Paras Prasad (Photonics Research Center, SUNY-Buffalo) and Tom Cooper (Wright Labs-MLPJ).

REFERENCES

1. C. W. Spangler and P.-K. Liu, Synth. Metals 44, 259 (1991).

2. E. H. Elandaloussi, P. Frère and J. Roncali, J. Chem. Soc. Chem. Commun. 1997, 301.

3. C. W. Spangler and M. He in <u>Handbook of Organic Conductive Molecules and Polymers</u>: Vol. 2, edited by H. S. Nalwa (John Wiley & Sons Ltd., Chichester, 1997) pp. 389-414.

4. C. W. Spangler and M. He in <u>Electrical, Optical and Magnetic Properties of Organic Solid State Materials III</u>, edited by A. K.-Y. Jen, C. Y.-C. Lee, L. R. Dalton, M. F. Rubner, G. E. Wnek and L. Y. Chiang (Mater. Res. Soc. Proc. <u>413.</u> Pittsburgh, PA, 1996) pp. 221-229.

5. E. H. Elandaloussi, P. Frère, P. Richomme, J. Orduna, J. Garin and J. Roncali, J. Amer. Chem. Soc. <u>119</u>, 10774 (1997).

6. T. Kaino, K. Kubodera, H. Kobayashi, T. Kurihara, S. Saito, T. Tsutsuni, S. Tokito and H. Murata, Appl. Phys. Lett. <u>53(21)</u>, 2002 (1988).

7. S. Luzzati, C. Botta, M. Romanoni, D. Comeretto, R. Tubino, M. Nisoli, A. Cybo-Ottone and S. De Silvestri in <u>Nonlinear Properties of Organic Materials VI.</u> edited by G. Möhlmann (Proc. SPIE, <u>2025</u>, Bellingham, WA, 1993) pp. 450-458.

8. M. K. Casstevens, R. Burzynski, J. F. Weibel, C. Spangler, G. He and P. N. Prasad in <u>Nonlinear Optical Liquids and Power Limiters</u> (Proc. SPIE, Bellingham, WA, 1997) pp. 152-159.

CORROSION PROTECTION OF METALS BY CONDUCTIVE POLYMERS
III. IMPROVED PERFORMANCE AND INHIBITION IN NaCl

WEI-KANG LU *, RONALD L. ELSENBAUMER*, T. CHEN** AND V. G. KULKARNI**
*Materials Science and Engineering Program, The University of Texas at Arlington, Arlington, TX 76019, elsenbaumer@uta.edu
**Americhem Inc., 225 Broadway East, Cuyahoga Falls, OH 44221

ABSTRACT

The use of conducting polymers for corrosion prevention is an area which has gained increasing attention during the last decade [1]. This study explores the use of polyaniline based polymer coatings for corrosion prevention on mild steel. Data on coating degradation and passivation on electrochemically polarized painted metal specimens exposed to acid chloride solutions and artificial seawater at an ambient temperature are presented. A Systematic comparison between controls and designated coated sample sets has been made to demonstrate good corrosion protection efficiency with synergistic effects between conductive polymers and metals by classical DC monitoring techniques. Brief comparisons are made with data from simulated marine exposure. Meanwhile, in separate experiments, electrochemical data were obtained for conductive polymer primer coatings with epoxy top-coat under fully immersed conditions by using electrochemical noise (ECN) monitoring and scanning electrochemical microscopy (SECM) techniques to discover the initial localized corrosion phenomena in order to achieve further understanding of the protection mechanism. Additionally, electrochemical impedance (EIS) spectra were utilized for the assessment of anti-corrosion performance provided by conducting polymers to mild steel.

INTRODUCTION

Organic maintenance coatings are widely used to protect ferrous industrial structures, tools and storage utilities from corrosive detriment caused by exposure in natural and artificial environments. The studies of corrosion inhibition of metals with the exploitation of intrinsically semiconductive polymers is getting an upsurge of attention lately [2]. More recent investigations indicated polyaniline and polypyrrole can provide some corrosion protection to steel in acidic media [3]. Primarily, the corrosion protection by conductive polymers in low pH protonic acids is much better than that provided in neutral NaCl solutions [4]. Formulation efforts are being taken into account in order to improve anticorrosion performance on steel in dilute NaCl environments. Increased inhibition efficiency has now been obtained for controlled scratch areas in both atmospheric and immersed conditions.

EXPERIMENTAL

Three-electrode custom-designed PTC (Paint Test Cell) and EG&G K47 assembled corrosion cells were used for corrosion measurements using a saturated calomel electrode and a platinum disk electrode as the reference electrode and counter electrode, respectively. The electrochemical cell setups are described elsewhere [4]. Standard Rapid Electrochemical Assessment of Paint (REAP) procedure was performed to estimate the long term corrosion resistance of conductive polymer coated metals using short term electrochemical tests. A IBM PC featuring author-modified Gamry Instruments CMS 100, 120 and 300 software was used for data acquisition by controlling a potentiostat linked with a lock-in amplifier. The conductive polymer samples of polyaniline-PET blends with/without binder used in this study were supplied by Americhem Inc.

and various coating procedures (curing temperature, coating thickness and mild steel surface treatments) were applied . These different polyaniline formulations were assigned AC1 to AC7, respectively. The average conductive polymer film coating thickness were 25 μm (AC4, AC5 and AC6) and 50 μm (AC1, AC2 and AC3). ASTM standard B117 was used for standard test method of salt spray testing. Polyaniline blend coatings on mild steel coupons overcoated with a high viscosity two-component epoxy resin (e) top-coat were investigated in this study. The coating systems were applied to 2"x2" SAE1010 mild steel panels (s) with appropriate surface pretreatments. The coated coupons were drilled with a 1.18 mm diameter mill (D) to expose a fixed area of bare steel surface. 3.5 % NaCl testing electrolyte was used in immersion and salt fog tests. Sample designations indicate that the sample had a drilled exposed steel surface (D), an epoxy top coat (e), the nature of the conductive polymer coating (ACx), and a mild steel substrate (s). Thus, the designation, D,e/AC4/s, indicates that this sample is a mild steel coupon coated with the polyaniline formulation AC4, overcoated with an epoxy top coat, and contains a drilled exposed bare steel area through the coatings of 1.18 mm diameter.

RESULTS AND DISCUSSION

Visual Inspections

Close-up observations by bare eye and an optical microscope provided valuable corrosion information about passive properties observed in the exposed bare steel areas and protection efficiencies provided by the conductive polymer formulations. Table 1 outlines the corrosion progress of exposed areas in NaCl for various sample configurations. Astonishing, every tested AC formulation avoided the formation of corrosive products at exposure sites for times up to about 3 weeks. After this time the various formulations exhibted different levels of corrosion prevention performance. Some even accelerated corrosion with long exposure times.

Table 1   Visual observations on bare steel holes in NaCl immersion tests

sample	initial stage	intermediate stage	final stage
D,e/AC1/s	shiny until 7 days immersion became 2 separated flat and smooth red-brown colored films at centre of dent (outer: darker & inner:lighter)	red-brown passivated film covered whole drilled hole in two distingished colors and slightly extended out dent edge at week 3	disk center of dent became black but outer ring area still showed red-brown to yelow brown and film thickness increased
D,e/AC1*/s	shiny for the first 3-week	thin light brown film appeared at week 4	voluminous yellow brown rust from week 5
D,e/AC4/s	shiny for first 2-week	thin grayish brown film formed at week 3	thin light brown film covered entire dent since week 4
D,e/AC5/s	shiny through first week	deep brown film at outer edge of dent and light grayish brown film at center of dent	grow-up inner light brown rust & outer deep brown passivated film around the dent edge
D,e/AC6/s	shiny for first 2-week	thin light brown film at week 3 covered whole dent	two deep brown/yellow outer ring-inner disk structure at week 5 uniform thickness of yellow brown rust from week 6

* half thickness of AC1 coating (25 μm) and 2 times of curing temperature (300 °F)

Salt Spray Methods

As pointed out in several articles [5,6], cyclic corrosion weathering tests are assessed as acceptable testing procedures for evaluating coating materials. The 858 and 1500 hour saltfog chamber tests were performed by an independent testing laboratory. The evaluation of cross scribe region and coating blisters describe the blistering size/frequency of samples. Generally, AC1 topcoat with epoxy overcoat achieved less scribe rusts, small scribe blisters and very few topcoat blisters [7].

Electrochemical Test Methods

In deriving quantitative expressions for corrosion rates by the Bulter-Volmer equation, small potential increments were implemented on exposed samples to obtain corrosion information. The 8-week immersion tests indicated that most of Americhem PAni formulation coatings demonstrated apparent protective effects compared to a control (no primer layer of PAni). Corrosion rate data are summarized in Table 2. AC1 containing samples maintained passivation until the end of the immersion test. AC2 did not show any proper protection at all. AC4, AC5 and AC6 established excellent corrosion prevention for the first several weeks testing time possibly caused by more firm polymer-metal adhesion. However, long term corrosion resistance loss for these samples, and the acceleration of corrosion, is puzzling.

Table 2 Corrosion rate comparison of different sample configrations

time\sample	D,e/s	D,e/AC1/s	D,e/AC2/s	D,e/AC3/s	D,e/AC4/s	D,e/AC5/s	D,e/AC6/s
day 1	0.018	0.003	0.016	0.008	0.000	0.000	0.000
day 3	0.015	0.006	0.015	0.01	0.000	0.000	0.000
day 7	0.037	0.008	0.033	0.075	0.000	0.000	0.000
day 14	0.059	0.005	0.065	0.134	0.000	1.687	0.000
day 21	0.106	0.009	0.053	0.195	1.957	3.064	2.056
day 28	0.125	0.008	0.270	0.209			2.618
day 35	0.153	0.012	0.481	0.173	2.207	4.787	3.227
day 42	0.165	0.013	0.136		3.237	4.219	3.597
day 49	0.194	0.009			2.821	2.825	2.810
day 56	0.208	0.009	0.246	0.084	3.111	3.615	2.641

Electrochemical noise analysis

Figure 1 represents the differential domains of electrochemical noise of D,e/AC1/s and the control sample. It is obvious that the monitored pitting activities of the AC1 coated coupon was far less as evidenced by both the difference in current variations and base current values. Most importantly, both ECN and Cyclic Polarization methods are particularly successful in evaluating the pitting corrosion produced at scribed regions, and both methods are in good agreement.

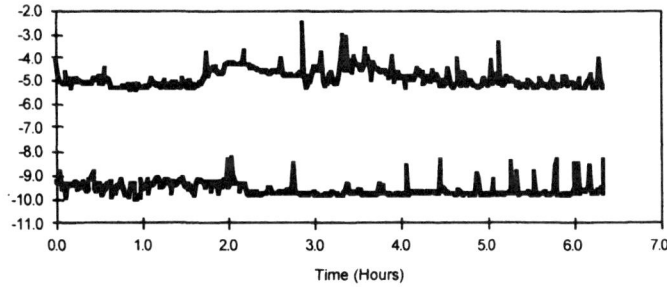

Figure 1. ECN analysis of control sample D,e/s (upper curve) and D,e/AC1/s (bottom curve).

AC Impedance

The low frequency domains of impedance spectra show the resistance response of test coatings. In particular, with AC1 as primer coating, the initial corrosion resistance in the defect area was 129000 ohm compared to 56000 ohm for the control sample (figure 2). After a 7-week immersion, the control sample resistance dropped to 5000 ohm, but AC1 coated sample only dropped to 100000 ohm. Most surprising, with the application of AC1 coating, the corrosion resistance after a 49-day immersion was even higher than the initial immersion resistance of epoxy-topcoat control.

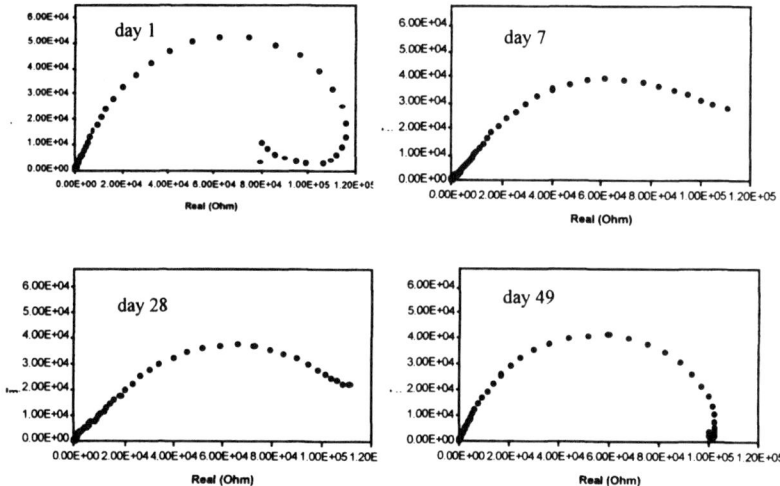

Figure 2. Electrochemical impedance spectra of D,e/AC1/s at different exposure periods.

The nature of a rectifying metal-p type semiconductor (conductive polymer) contact, even under reverse bias conditions to increase the barrier height of interface, can not simply explain why conductive polymers can maintain corrosion protection on a passivating metal surface, especially for an exposed bare metal surface. A further qualitative examination reveals the barrier heights calculated by flat band potentials using the discrete step voltammetry technique under electrostatic conditions are small as compared to film resistances derived from EIS results and potential shifts during immersion tests. Clearly, the metal-semiconductor (MS) contact of rectifying electrical nature under reverse bias should only add little consideration regarding the corrosion protection mechanism [8]. Moreover, a more precise modified metal-oxide-semiconductor (MOS) model structure of electrons and holes accumulated near the contact interface of MO and OS, respectively, should be mentioned which can be viewed as a reasonable explanation for capturing the initial released electrons from the metal surface by holes. Then, the accumulation stage will become the depletion stage, followed by the inversion stage. Tafel slopes of cathodic and anodic polarization curves provide experimental observation corresponding to theoretical accumulation, depletion and inversion stages. The oxide-conductive polymer bilayer will turn into a capacitor as verified by the impedance response of covered films and oxide layers at the interfaces of metals/conducting polymers which form stable adhesive passive films. The outer immersion environment works like a reversed bias to conductive polymer which increases the barrier height blocking electron flow from the metal surface. In addition, all of those stage speculations can be confirmed by the oxidative states of conductive polymer films by XPS at different immersion periods.

No undercutting corrosion near the drilled edge of metal-conductive polymer interface was observed. This provides evidence of no large current concentration effects present at the coating edge with using conductive polymers between barrier coatings and metals. Conductive polymers significantly alter the potential gradient distribution around the drilled defect compared to samples without conductive polymer present. Preliminary scanning electrochemical microscopy identified the anode/cathode distribution on conductive polymer coated metal surfaces containing pinholes [9] and showed that reversed potential distribution is predicted wherein a more negative potential should be observed at the edge and higher at the center, which is opposite of the usual case.

Table 3 XPS core level spectra and stoichiometrics of AC1 coatings under various exposed conditions

sample\valence energy	N(1s)	O(1s)	Fe(2p3/2)	Fe(2p1/2)
AC1/s (top layer)	402.1 (30%), (-N$^+$H=) 400.6 (50%), (-NH-) 398.6 (20%), (-N=)	535.7 (30%), (-OH^{-2}) 534.0 (60%), (-OH$^-$) 531.8 (10%), adsorption-O$^-_2$		
AC1/s (bottom layer)	402.6 (30%) 401.2 (70%)	535.3 (10%) 533.8 (45%) 532.1 (45%)	710.9	724.7
AC1/s, NaCl	402.3 (60%) 400.3 (40%)	535.4 (45%) 533.6 (45%) 530.3 (10%)	712.9	726.2
AC1/glass	401.6 (60%) 400.2 (40%)			
AC1/glass, NaCl	401.6 (40%) 400.2 (50%) 397.9 (5%) 395.8 (5%)			

Surface Analysis of Corrosion Products and Protective Films

Table 3 shows the *ex-situ* XPS determination of oxidative states of nitrogen in AC1 coatings and iron oxides. A step by step diagram of oxide layer production can be seen elsewhere [4]. In addition, according to the shapes and peak locations of Fe(2p$_{3/2}$) and Fe(2p$_{1/2}$), Fe$_2$O$_3$ is the major contributory passive film component at the steel/polymer interface. After immersion, a dual protective oxide layer comprising Fe$_2$O$_3$ and FeOOH forms. Explanation of protection ability of the AC1 coating is noted by the conducting polymer-carbon steel interactions that occur when the coated samples are exposed to aqueous NaCl environments. Upon exposure in the presence of oxygen, the concentration of amine nitrogens (reduced form of PAni) decreases in favor of the imine form. This initial oxidation of Pani in the coating is accompanied by reduction of hydrogen in the bare steel area, thus initially providing a form of cathodic protection in this area. When the

PAni is fully reoxidized to the eneraldine form by oxygen, then the mechanism of protection in the bare steel area gradually changes to anodic polarization, as evidenced by Tafel measurements.

CONCLUSIONS

Polyaniline blend polymers from Americhem were found to provide inhanced corrosion inhibition effciencies in NaCl environments compared to conventional pure polyaniline coatings. These coatings not only protect covered sites but also provide corrosion prohibition to scribed regions under both severe atmospheric and immersed environments. The new formulations of polyaniline stretches the corrosion protection at open defect by redistributing potential gradients around the metal surfaces. By increasing the activation polarization effect within the double layer, the conductive polymers have increased the inhibition capabilities. This sheds new light on the primer coating application and extends application environments to the marine environment.

Impedance data can be used for mechanistic interpretation although conductive polymers do not act like traditional protective barrier coatings, whereas EC voltage and current signatures of other DC corrosion techniques give a more clear indication as to the state of the electroactive conductive polymer coatings. Almost without exception, the corrosion protection by intrinscially electroactive polymers can be considered to setup a polarized electrochemical force field around defects (cathodic protection) at initial and intermediate immersion periods that gradually transfers to anodic protection by generating multi-layer compact metal oxide passive films at later exposure times.

Employed oxygen/inert gas purge experiments suggest that hydrogen reduction is the controlling factor for early to intermediate protection. Oxygen, however, is required for the key control by an anodic protection mechanism. The marked difference between the protection mechanisms of traditional barrier coatings and electroactive polymer coatings should be recognized and emphasized.

REFERENCES

1. J. L. Camalet, J. C. Lacroix, S. Aeiyach, K. Chane-Ching and P. C. Lacaze, J. Electroanal. Chem., **416**, 179 (1996).
2. Naseer Ahmad and A. G. MacDiarmid, Synth. Met., **78**, 103 (1996).
3. N. V. Krstajic, B. N. Grgur, S. M. Jovanovic and M. V. Vojnovic, Electrochim. Acta, **42**, 1685 (1997).
4. W. Lu, S. Bajak and R. Elsenbaumer in Handbook of Conducting Polymers, 2nd ed., edited by T. A. Stotheim (Marcel Dekker, New York, 1997).
5. C. H. Simpson, C. J. Ray and B. S. Skerry, J. Protect. Coat. Linings, **8**, 28 (1991).
6. ASTM B117-90, Annual Book of ASTM Standards, **03.02**, 20 (1990).
7. V. G. Kulkarni, unpublished results.
8. R. F. Pierret, Semiconductor Device Fundamentals, (Addison-Wesley, New York, 1996).
9. P. Kinlen, (private communciation).

ORGANIC METALS AND SUPERCONDUCTORS: FACILE SYNTHETIC METHODOLOGIES BYPASSING COUPLING AND CHROMATOGRAPHY

Ronald L. Meline and Ronald L. Elsenbaumer*
Department of Chemistry and Biochemistry, The University of Texas at Arlington
Arlington, TX 76019 (elsenbaumer@uta.edu)

ABSTRACT

The organic donor tetrathiafulvalene (TTF, 1) and its derivatives are synthesized from tetrathianaphthalene (TTN, 2) in a one pot reaction. The reaction takes advantage of an isomerization of lithiated TTN into lithiated TTF and subsequent substitution with novel disulfide electrophiles yielding substituted TTF donors. The isomerization approach eliminates the need for a chemical coupling reaction of oxones or thiones (a yield limiting procedure), and simple byproducts eliminate the need for tedious purification. The novel methodology allows for a high yield preparation of bis(ethylenedithio)tetrathiafulvalene (BEDT-TTF, 3a, 80%) and bis(propylenedithio)tetrathiafulvalene (BPDT-TTF, 3b), previously unavailable directly from TTF. A high yield of bis(phenylenedithio)tetrathiafulvalene (BPhDT-TTF, 3c, unavailable from dimercaptoisotrithione "DMIT") is also achieved using the synthetic scheme. The method is general, and can be used to prepare a variety of novel organic donors based on TTF. Synthetic procedures along with all relevant characterization are presented.

INTRODUCTION

In addition to thiones and oxones based on dimercaptoisotrithione (DMIT), TTF has been used extensively in recent years as the building block for extended donors owing to the acidity of its vinylic protons as first reported by Green, who demonstrated the proton-lithium exchange by treatment of TTF with n-Buli or LDA.[1] Tetrakis(alkylchalcogeno)-TTF's have been synthesized from tetralithiated TTF with subsequent elemental selenium or tellurium insertion into the carbon-lithium bonds, or with the use of disulfide electrophiles (scheme 1).[2-5] It should be noted that insertion of elemental sulfur into the carbon-lithium bond has been reported as difficult[6] or essentially non-existent[3] Attempted alkylative ring closure of tetrakis(chalcogeno)TTF tetraanions with dibromoalkanes often leads to polymeric products (i.e. intermolecular alkylation) even under high dilution conditions.[2,7] Chalcogen analogs of BEDT-TTF (i.e. capped donors, BEDSe-TTF, BEDTe-TTF) have been prepared by Lee and Williams from TTF through the use of protecting group chemistry and solvent manipulation, respectively.[7,8]

SCHEME 1

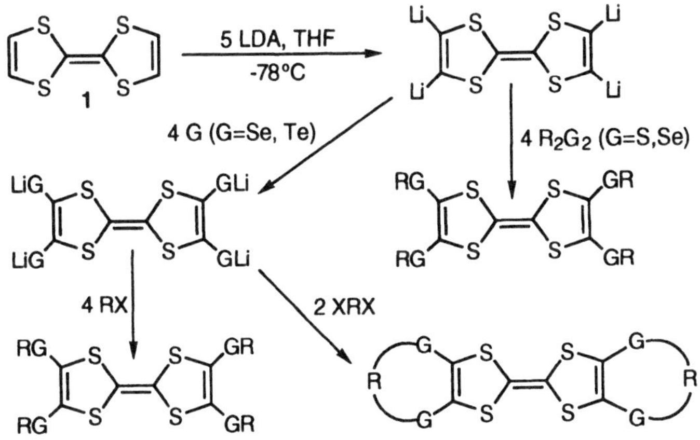

659

Presently, the use of TTF as a precursor to extended donors is hampered by a number of limitations: (1) TTF is expensive and published syntheses are tedious[9] (2) The disulfide electrophiles used must be soluble in THF or Et_2O, and yields are very dependent on the extent of electrophile solubility, (3) As a result of the difficulty of sulfur insertion and subsequent intramolecular alkylation with 1,2-dibromoalkanes, a facile conversion of TTF into BEDT-TTF and other capped tetrathiafulvalene tetrathiols (TTFTT's) has not been realized, (4) The capping of tetrakis(chalcogeno)TTF tetraanions requires the alkyl dihalide to be primary (SN^2 mechanism) thereby excluding aromatic caps, and (5) Tedious purification methods are often employed.[3] A new synthetic method eliminating these limitations accompanied by high yields would invariably be useful for the syntheses of known organic metals and superconductors as well as new materials.

SYNTHETIC METHOD

As has been previously reported, TTN can be used as a precursor to substituted TTF's or it can be functionalized to give substituted TTN.[10,11,12] A simple synthesis of TTN and its conversion to TTF has been reported.[12] Building on prior synthetic routes to alkylthio-TTF's from disulfides, unsymmetric disulfides have been synthesized; each incorporating an ester leaving group. Upon nucleophilic attack of tetralithiated TTF or isomerized tetralithiated TTN onto 4 equivalents of a disulfide, tetrakis(thioalkyl)TTF's and tetrakis(thioaryl)TTF's are produced along with 4 equivalents of ethanol and 4 equivalents of carbon oxysulfide (scheme 2). The simple byproducts facilitate workup. Disulfide formation is based on the chemistry of Brois, who has utilized alkoxycarbonylsulfenyl chlorides and a variety of thiols to produce ester alkyl disulfides (EtOC(O)S-S'R) which selectively yield unsymmetric disulfides (R'-S-S-R") upon reaction with one equivalent of a different thiol (R"SH).[13] The utility of EtOC(O)S-S'R disulfides can be seen when the symmetric disulfide RS'-S'R is not soluble or poorly soluble in THF or Et_2O.

SCHEME 2

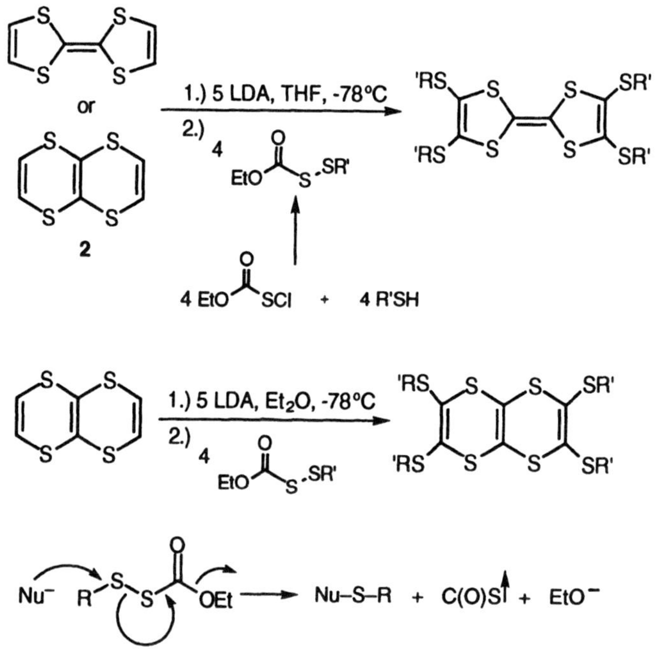

660

The methodology can best be utilized to produce dithioalkyl and dithioaryl capped tetrathiafulvalenes from TTF or TTN. A novel bisdisulfide was synthesized incorporating an ethyl spacing unit and ester leaving groups. Upon nucleophilic attack of tetralithiated TTF or isomerized tetralithiated TTN onto 1,2-di(ethoxycarbonyldithio)ethane (**4a**), BEDT-TTF was produced along with 4 equivalents of ethanol and 4 equivalents of carbon oxysulfide. The simple byproducts facilitated workup, realizing BEDT-TTF in an overall yield of 80%. The bisdisulfide (**4a**) was synthesized from two equivalents of ethoxycarbonylsulfenyl chloride[14] (**5**) and one equivalent of 1,2-ethanedithiol (**6**) in quantitative yield. Crystalline (**4a**) is completely soluble in THF or ether. In this study, the carbanions of tetralithiated TTF easily add two ethane dithiol units from **4a** producing BEDT-TTF (scheme 3). Alkoxycarbonylsulfenyl chloride can be used to prepare a number of alkyl (**4a,4b**) and aryl (**4c**) spaced bisdisulfides in high yield from various dithioles leading to a series of capped TTF's (**3a,3b,3c**), thereby demonstrating the general utility of this approach (scheme 4).

SCHEME 3

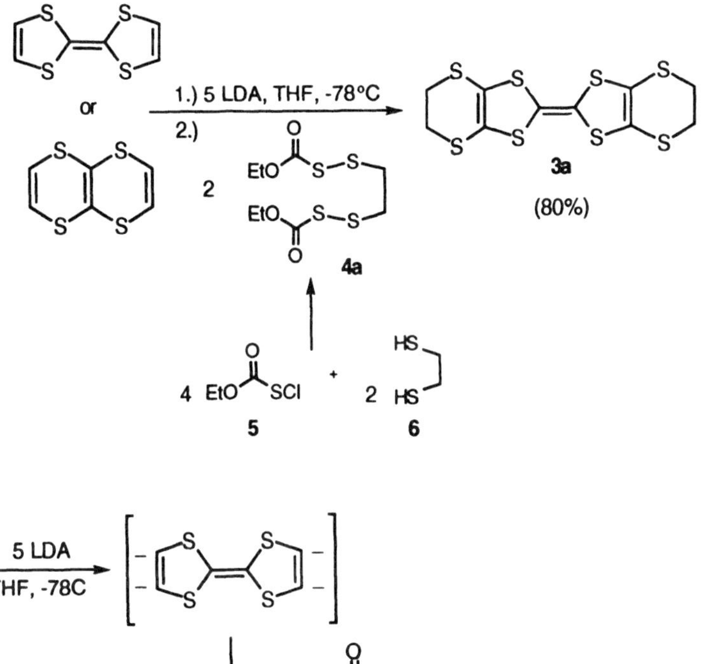
or
1.) 5 LDA, THF, -78°C
2.)
2
EtO
EtO
3a
(80%)
4a
4 EtO
SCl
+
2 HS
HS
5
6

SCHEME 4

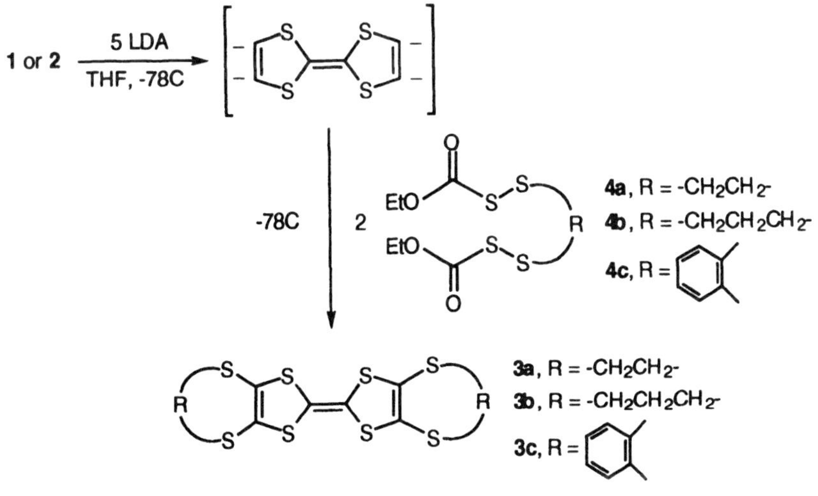
1 or 2
5 LDA
THF, -78C
-78C
2
EtO
EtO
4a, R = -CH2CH2-
4b, R = -CH2CH2CH2-
4c, R =
3a, R = -CH2CH2-
3b, R = -CH2CH2CH2-
3c, R =

SCHEME 5

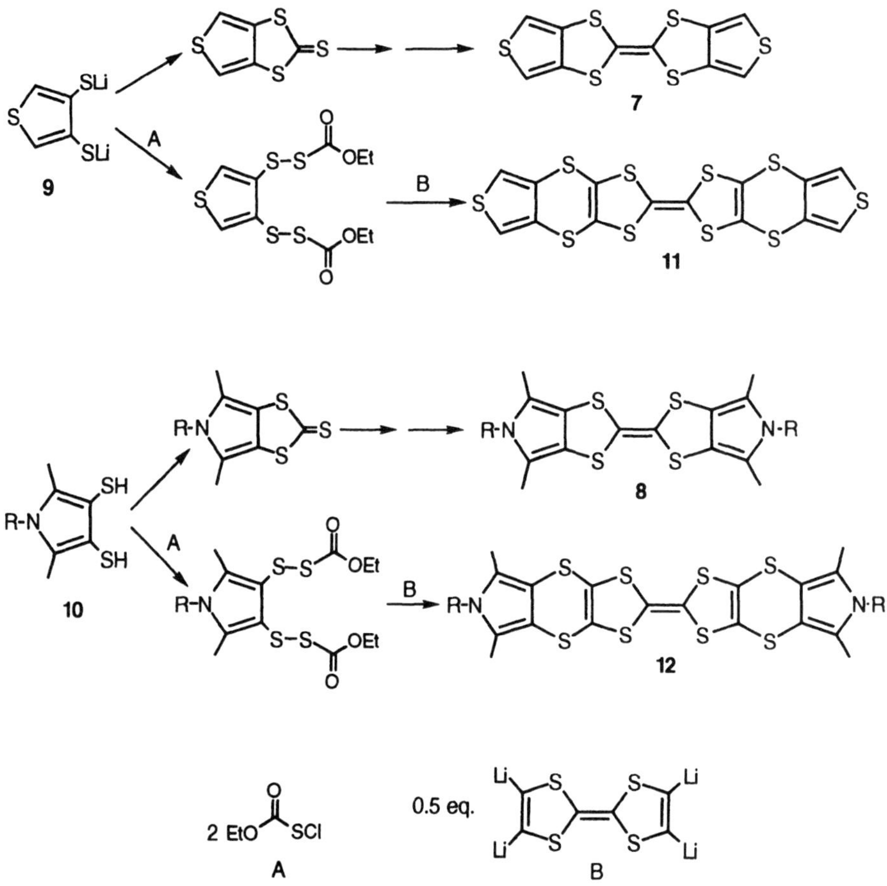

The ability to prepare aryl capped TTF's from dithioles is intriguing in light of recent syntheses of thiophene (**7**),[15-18] selenophene,[19] pyrrole (**8**)[20,21] and furan[22] annelated TTF's. In particular, **8** was shown to have a lower oxidation potential than TTF's.[20] 3,4-dimercaptolithiothiophene (**9**) and 2,5-dimethyl-3,4-dimercapto-*N*-alkylpyrrole (**10**) were used to prepare the respective annelated TTF donors. The dithio heterocycles are also likely precursors to higher dimensional organic metals and potential superconductors (for example **11** and **12**), and such syntheses are presently underway (scheme 5).

CONCLUSIONS

The use of TTF or TTN as a precursor to extended organic metals and superconductors allows for syntheses that bypass coupling of oxones and ketones. This coupling procedure is generally the yield limiting procedure (for example: common syntheses of BEDT-TTF).[23,24] Very soluble disulfide electrophiles with ethyl ester leaving groups allow for the syntheses of extended donors eliminating the need for tedious purification procedures including chromatography. The ability to synthesize numerous sulfur based capped organic metals from TTF or TTN is unprecedented and invariably useful as such annelated TTF's are not available from TTFTT. In general, routes to quasi one-dimensional organic metals that proceed through dithiole precursors[25,26] are potential candidates for the syntheses of quasi two dimensional organic metals and superconductors.

EXPERIMENTAL

1,2-Di(ethoxycarbonyldithio)ethane **4a**
Ethoxycarbonylsufenyl chloride (**5**) was prepared in very high yield according to a published procedure.[14] 12.00g (85 mmol) of **5** was dissolved in 100 ml of dichloromethane, placed in a 250 ml round bottom flask capped with a septum, and was cooled with a dry ice/isopropanol bath to ~-30°C under N_2. 4.00g (42.5 mmol) of **6** dissolved in 20 ml of dichloromethane was syringed into the flask over a 10 min. period turning the yellow solution colorless. The solution was washed with 5% aq. NaOH, water and dried with $MgSO_4$. The filtered solution was condensed to give an oil which crystallizes on standing. Yield: 12.68g = 98.7%. M.P 44-45°C; ^{1}H-NMR ($CDCl_3$/TMS): 1.35 (t, 6H); 3.07 (s, 4H); 4.36 (q, 4H); ^{13}C-NMR ($CDCl_3$/TMS): 14.1, 37.5, 65.2, 168.8; MS (EI) 302(M[+]).

1,2-Di(ethoxycarbonyldithio)propane **4b**
Same procedure as for **4a** from 1,3-propanedithiol in lieu of **6**, producing a yellow liquid. Yield: 97.0%. ^{1}H-NMR ($CDCL_3$/TMS): 1.34 (t, 6H); 2.06 (p, 2H); 2.93 (t, 4H); 4.35 (q, 4H); ^{13}C-NMR ($CDCL_3$/TMS): 14.1, 26.9, 36.8, 65.0, 169.0; MS (EI) 316(M[+]).

1,2-Di(ethoxycarbonyldithio)benzene **4c**
Same procedure as for **4a** from 1,2-benzenedithiol in lieu of **6**, producing crystals. M.P. 49-51°C; Yield: 94.2%. ^{1}H-NMR ($CDCL_3$/TMS): 1.34 (t, 6H); 4.35 (q, 4H); 7.30 (m, 2H); 7.67 (m, 2H); ^{13}C-NMR ($CDCL_3$/TMS): 14.2, 65.5, 129.0, 131.0, 136.5, 168.1; MS (EI) 350(M[+]).

BEDT-TTF **3a**
n-BuLi (40 ml of a 2.5M solution in hexanes, 100 mmol) was syringed through a rubber septum into a solution of diisopropylamine (15 ml, 107 mmol) in dry THF (50 ml) in a 500 ml round bottom flask with a stir bar at -78°C (dry ice/isopropanol) under nitrogen (purge through rubber septum) to form Lithium diisopropylamide (LDA). After stirring the solution for 1 hr, TTF or TTN (4g, 19.6 mmol) was dissolved in dry THF (200 ml) and added to a pressure equalizing funnel and placed on top of the flask. Nitrogen was then transferred to the top of the funnel. The TTF solution was then added dropwise over a 4 hour period to the LDA solution giving a bright yellow solution. The solution was stirred for an additional 4 hours whereupon the funnel was then charged with 12.68 g of **4a** (42 mmol) dissolved in 80 ml of THF. The bisdisulfide solution was added over a 2 hour period, and the reaction flask was allowed to warm to room temperature overnight. The resulting red suspension was filtered on a glass sintered funnel, washed with water (100 ml), methanol (100 ml) and finally ether (200 ml) to give pure BEDT-TTF 6.03g (mp 244-247°C (dec.)), in powder form. BEDT-TTF can be recrystallized from CS_2, sulfolane or thiophene.[12] Similar reaction procedures using **4b** and **4c** in lieu of **4a** were used to prepare **3b** (yield 68%, mp 246-250°C (dec.), lit.[27] mp >250°C (dec.) and **3C** (yield 72%, mp 284-287°C (dec.), lit.[28] mp 286-287°C (dec.) respectively: identical to the known organic metals.[27-30]

ACKNOWLEDGMENTS

The authors would like to thank AFOSR for financial support as well as C.M. Land and Dr. V.V. Sorokin for technical assistance.

REFERENCES:

1 D.C. Green, *J. Org. Chem.* **1979**, *44*, 1476.
2 E. Aharon-Shalom, J.Y. Becker, J. Bernstein, S. Bittner and S. Shaik, *Tetrahedron Lett.* **1985**, 2783.
3 S. Hsu and L.Y. Chiang, *J. Org. Chem.* **1987**, *52*, 3444.
4 N. Okada, H. Yamochi, F. Shinozaki, K. Oshima and G. Saito, *Chem. Lett.* **1986**, 1861.
5 H. Yamochi, N. Iwasawa, H. Urayama and G. Saito, *Chem. Lett.* **1987**, 2265.
6 M.R. Bryce (personal discussions)
7 V.Y. Lee, *Synth. Met.* **1987**, *20*, 161.
8 A.M. Kini, B.D. Gates, M.A. Beno and J.M. Williams, *J.C.S. Chem. Commun.* **1989**, 169.
9 A.J. Moore and M.R. Bryce, *Synthesis* **1997**, *4*, 407.
10 S. Nakatsuji, Y. Amano, H. Kawamura and H. Anzai, *J.C.S. Chem. Commun.* **1994**, 841.
11 R.L. Meline and R.L. Elsenbaumer, *Synthetic Metals.* **1997**, *86*, 1845.
12 R.L. Meline and R.L. Elsenbaumer, *Synthesis* **1997**, *6*, 617.
13 S.J. Brois, J.F. Pilot and H.W. Barnum, *J. Am. Chem. Soc.* **1970**, *92*, 7629.
14 G. Barany, A.L. Schroll, A.W. Mott and D.A. Halsrud, *J. Org. Chem.* **1983**, *48*, 4750.
15 P. Shu, L. Chiang, T. Emge, D. Holt, T. Kistenmacher, M. Lee, J. Stokes, T. Poehler, A. Bloch and D. Cowan, *J.C.S. Chem. Commun.* **1981**, 920.
16 L-Y. Chiang, P. Shu, D. Holt and D. Cowan, *J. Org. Chem.* **1983**, *48*, 4713.
17 N. Santalo, J. Veciana, C Rovira, K. Molins, C. Miravitlles and J. Claret, *Synth. Met.* **1991**, *41-43*, 2205.
18 E.V.K. Suresh Kumar, J.D. Singh, H.B. Singh, K. Das and B. Verghese, *Tetrahedron* **1997**, *53*, 11627.
19 R. Ketcham, A-B. Hörnfeldt and S. Gronowitz, *J. Org. Chem.* **1984**, *49*, 1117.
20 W. Chen, M.P. Cava, M.A. Takassi, and R.M. Metzger, *J. Am. Chem. Soc.* **1988**, *110*, 7903.
21 K. Zong, W. Chen, M.P. Cava and R.D. Rogers, *J. Org. Chem.* **1996**, *61*, 8117.
22 Y. Siquot, P. Frère, T. Nozdryn, J. Cousseau, M. Sallé, M. Jubault, J Orduna J. Garín and A. Gorgues, *Tetrahedron Lett.* **1997**, *38*, 1919.
23 H. Müller and Y. Ueba, *Synthesis* **1993**, 853.
24 N. Svenstrup, K.M. Rasmussen, T.K. Hansen and J. Becher, *Synthesis* **1994**, 809.
25 A. Krief, *Tetrahedron.* **1986**, *42*, 1209.
26 J.M. Williams, J.R. Ferraro, R.J. Thorn, K.D. Carlson, U. Geiser, H.H. Wang, A.M. Kini and H. Whangbo, Organic Superconductors; Prentice Hall: New Jersey, **1992**.
27 M. Mizuno, A.F. Garito and M.P. Cava, *J.C.S. Chem Commun.* **1978**, 18.
28 H. Muller, H.P. Fritz, R Nemetshek, R Hackl, W. Biberacher and C.P. Heidmann, Z *Naturforsch.* **1992**, *47b*, 718.
29 L.C. Porter, A.M. Kini and J.M. Williams, *Acta Cryst.* **1987**, *C43*, 998.
30 J.P. Parakka, A.M. Kini and J.M. Williams, *Tetrahedron Lett.* **1996**, 8085.

SYNTHESIS, CHARACTERIZATION AND PROPERTIES OF CONJUGATED POLY(1-ALKYL-2,5-PYRRYLENE VINYLENE)

In Tae Kim and Ronald L. Elsenbaumer*
Department of Chemistry and Biochemistry, The University of Texas at Arlington, Arlington,
Texas 76019, USA (elsenbaumer@uta.edu)

ABSTRACT

A series of poly(N-alkylpyrrylene vinylenes)(alkyl = methyl(**1**), hexyl(**2**), dodecyl(**3**)) have been synthesized from the monomer, N-alkyl-2,5-bis(thiophenylmethylene)pyrrole, by base induced elimination and polymerization. The resulting deep purple conjugated polymers (**2**, **3**) were soluble in a variety organic solvents. The electrical conductivities of the polymers lie in the range, 10^{-2} to 2.5 S/cm. The highest yield was obtained when the polymers were synthesized in refluxing THF with a monomer/ base mole ratio of 1/4. The optical band gap of the undoped polymers were 1.89 eV(**1**), 1.69 eV(**2**) and 1.65 eV(**3**). Characterization of the polymers includes IR, CV, ^{1}H and ^{13}C NMR, UV-vis spectroscopy, TGA and molecular weight studies.

INTRODUCTION

An attractive feature of the pyrrole system is the ability to readily prepare a number of functionalized polymers by polymerization of functionalized pyrrole monomers. Electrodeposition of poly(pyrrolylene) films on Au and Pt electrodes in an acetonitrile solution of Et_4NBF_4 resulted in polymers with conductivities as high as 100 S/cm. Conductive properties are related to the nature of the counter anions in polypyrrole films[1]. Properties also have been modified by polymerizing pyrrole derivatives with 3-methyl and 3,4-dimethyl substituents. Conductivities of these polymers are 4 and 10 S/cm, respectively[2,3]. On the other hand, the room temperature conductivities of poly(1-alkyl-2,5-pyrrolylene) (alkyl = methyl, hexyl, dodecyl) prepared by chemical oxidation fall in the range, 10^{-3} to 10^{-6} S/cm[4,5]. The large decrease in conductivity of doped poly(pyrrolylenes) with an N-alkylsubstituent and an increase in size of the N-alkyl substituent has been rationalized on a steric basis[6].

Desirable properties of conducting polymers include low band gaps and good processibility. Reduction of band gap (by about 0.4 eV) has been achieved for poly(*p*-phenylene) (PPP) and poly(thienylene) (PT) by the insertion of vinylene linkages between the directly linked aromatic rings to give poly(phenylene vinylene) (PPV) and poly(thiophene vinylene) (PTV), respectively [7-9].

Conducting polymer processibility has been obtained recently by the development of soluble precursor polymers yielding high quality materials which can be easily formed into films for poly(acetylene) [10], poly(*p*-phenylene vinylene) (PPV) [11,12] and poly(2,5-thienylene vinylene) [13]. The common backbone of these polymers is made up of aromatic rings bridged by vinylene linkages which not only reduce steric hindrance between backbone rings and groups attached to them, but also have a beneficial effect on electronic properties as evidenced by experimental and theoretical data on both poly(*p*-phenylene vinylenes) [7] and poly(thienylene vinylenes) [14].

Recently, theoretical calculations using the Valence-Effective Hamiltionian technique were performed on poly(pyrrylene vinylene), poly(3-methylpyrrylene vinylene) and poly(3-methoxy pyrrylene vinylene) [15]. These calculations predict that the electronic properties of poly(pyrrylene vinylene) remain almost unaffected upon methyl substitution, while a noticeable lowering by ~0.4 eV is obtained for the ionization potential and energy gap upon methoxy substitution. Recently, poly(pyrrylene vinylene) synthesized by the thermal elimination of methanol from the precursor polymer poly(1-methoxy-1,2-ethylenediyl-co-2,5-pyrrolylene) exhibited an electrical conductivity on the order of 6.9×10^{-7} S/cm [16].

$$R = -CH_3 \quad (1)$$
$$-(CH_2)_5CH_3 \quad (2)$$
$$-(CH_2)_{11}CH_3 \quad (3)$$

In this paper, we report results for the conductivity and properties of poly(N-alkylpyrrylene vinylenes) (alkyl = methyl(1), hexyl(2), dodecyl(3)), synthesized by chemical means, in comparison with literature data for related polymers.

EXPERIMENTAL

^{1}H- and ^{13}C-NMR spectra were obtained on a Bruker MSL 300 spectrometer. UV-VIS-NIR spectra were obtained on a Varian Cary 5E UV-VIS-NIR spectrometer. FT-IR spectra were recorded on a Digilab FTS-40 FT-IR spectrometer using powdered samples mixed with KBr. Elemental analyses were determined by Quantitative Technologies Inc. TGA and DSC were performed on a DuPont 951 and 910 thermal analysis system, respectively. Conductivity measurements were performed on a Keithley four-in-line probe apparatus. CV studies were carried out on an EG&G PAR Model 273 potentiostat/Galvanostat in a dry box under Ar.

Synthesis of Poly(1-methyl-2,5-pyrrylene vinylene)(1)

To a solution of 4 equivalents(0.748 g) t-BuO$^-$K$^+$ in 20 mL of THF under reflux was added a solution of 2 g (6.12 mmol) of 1-methyl-2,5-bis(phenylthiomethy)pyrrole in 2 mL of THF under nitrogen. Stirring was maintained for 24hr in a round bottomed flask until the light yellow solution turned purple in color. The polymer was separated and purified by repeated washing with hot water, methanol and then extracted (Soxhlet) with acetone for 24 hr. Drying under vacuum at room temperature gave the deep purple blue polymer 1. The overall yield was 75% (0.449 g). Anal. Calcd for C_7H_7N: C, 79.95; H, 6.71; N, 13.32. Found: C, 74.75; H, 6.51; N, 10.22; S, 4.31. λ_{max}: 500nm (2.47 eV) (NMP solution). E_g (band edge): 1.89 eV. IR: 3040, 3010, 2993, 2851, 1604, 1571, 1442, 1380, 1232, 1093, 1043, 932, 749, 692 cm^{-1}. GPC (NMP, against PS standard): $M_n = 7.3 \times 10^3$; $M_w = 9.4 \times 10^3$; Polydispersity (PD): 1.28.

Synthesis of Poly(1-hexyl-2,5-pyrrylene vinylene)(2)

Starting with 1-hexyl-2,5-bis(phenylthiomethy)pyrrole (2 g, 5.06 mmol), the same polymerization procedure as for polymer 1 was followed. The overall yield was 72 % (0.637 g).

^{1}H NMR (in CDCl$_3$): δ 6.8(s,2H), 6.5(s,2H), 3.9(t,2H), 2.3-1.1(m,8H), 0.9(t,3H). ^{13}C NMR (in CDCl$_3$): δ 134.23, 114.56, 107.73, 43.56, 33.38, 31.29, 27.87, 23.45, 14.62. Anal. Calcd for C$_{12}$H$_{17}$N: C, 80.23; H, 9.77; N, 7.99. Found: C, 76.74, H, 8.67; N, 7.15; S, 0.99. IR: 3040, 3010, 2990, 2895, 1600. 1650, 1480, 1300, 965, 750 cm^{-1}. T$_m$: 140 °C and 150 °C. σ = 6.18 × 10^{-7} S/cm (O$_2$), 0.15 S/cm (FeCl$_3$), 2.5 S/cm (AuCl$_3$). Cyclic Voltammogram(CV vs Ag/Ag$^+$): -107 mV (onset oxidation potential), -74 mV(peak oxidation potential). λmax: 563$_{nm}$ (2.20 eV)(film). E$_g$ (Band edge): 1.69 eV. GPC (NMP, against PS standard): M$_n$ = 9.1×10^4; M$_w$ = 1.75×10^5; Polydispersity(PD): 1.92.

Poly(1-dodecyl-2,5-pyrrylene vinylene)(3)

Starting with 1-dodecyl-2,5-bis(phenylthiomethy)pyrrole (2 g, 4.17 mmol), the same polymerization procedure as for polymer 1 was followed. The overall yield was 70% (0.756 g). ^{1}H NMR (in CDCl$_3$): δ 6.72(s, 2H), 6.47(s, 2H), 3.97(s, 2H), 2.43-1.25(m, 20H), 0.84(t, 3H). ^{13}C NMR (in CDCl$_3$): δ 133.36, 114.49, 106.54, 43.38, 31.93-26.91(9C), 22.69, 14.13. IR: 3040, 3010, 2924, 2854, 1646, 1539, 1401, 1261, 1092, 1028, 929, 805, 747 cm^{-1}. λ$_{max}$: 557nm (2.22 eV) (film). E$_g$ (Band edge):1.65 eV. GPC (THF, against PS standard): M$_n$ = 1.68×10^4; M$_w$ = 1.95×10^4; Polydispersity(PD): 1.16. σ = 4.6 × 10^{-1} S/cm (AuCl$_3$). Anal. Calcd for C$_{18}$H$_{29}$N: C, 83.33; H, 11.26; N,5.39. Found: C, 80.50; H, 10.00; N, 5.08; S, 1.15.

RESULTS

Polymer 1 was prepared from 1-methyl-2,5-bis(phenylthiomethyl)pyrrole. It was purified by washing with THF, methanol and acetone. This polymer was soluble in NMP, but not soluble in THF, chloroform or methylene chloride. The conductivity of FeCl$_3$-doped polymer 1 in the form of a pressed powder pellet was measured to be 3×10^{-3} S/cm. After it was doped with AuCl$_3$, polymer 1 showed a conductivity of 2×10^{-2} S/cm. Table 1 show the maximum conductivity values with different dopants for polymer 1.

Polymer 2 was synthesized from 1-hexyl-2,5-bis(phenylthiomethyl)pyrrole and purified by four reprecipitations from THF into methanol and acetone and had a deep blue color with a gold luster. On exposure to air, the polymer became slightly doped reaching a limiting conductivity of 10^{-6} S/cm. On doping with AuCl$_3$, polymer 2 films (cast on quartz) exhibited conductivities of 2.5 S/cm (four-in-line probe) in air. The undoped conjugated polymer 2 was soluble in many common organic solvents. The thermal stabilities of undoped polymers 1, 2 and 3 were evaluated using TGA. The thermogravimetric analysis (TGA; N$_2$; heating rate = 10 °C/min) of polymer 2 indicates the onset of decomposition at 225 °C, and 50 % weight loss by 430 °C. Conversion from the 1-hexyl-2,5-bis(phenylthiomethyl)pyrrole as monomer afforded polymer 2 in 72% yield, which is considerably higher than that typically obtained for poly(phenylene vinylene) prepared from bis-sulfonium salt precursors (20-40%)[17].

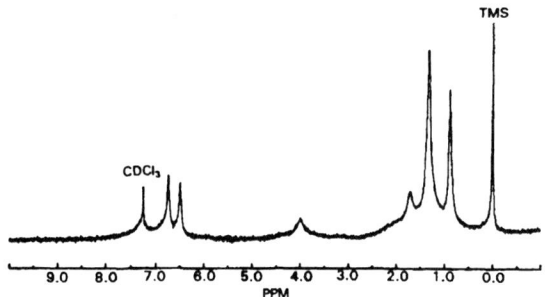

Fig. 1 ¹H-NMR spectrum of polymer **2** at 25 °C in CDCl₃

Fig. 1 shows the ¹H NMR spectrum of polymer **2** in CDCl₃ at 25 °C. The spectrum shows singlet peaks at 6.80 and 6.50 ppm which are assigned to the vinyl protons and β-hydrogens on the ring, respectively. The 1.1, 2.3 and 3.9 ppm peaks are assigned to hexyl protons. The formation of predominant *trans*-polymer **2** was determined by the peaks in the IR spectrum at 950 cm⁻¹, corresponding to the *trans*-vinylene C-H out-of-plane mode and at 3010 cm⁻¹ for the *trans*-vinylene C-H stretching vibration.

DSC analyses of neutral polymer **2** indicated phase transitions (Tm) at about 140°C and 150 °C. Heating and cooling cycles were repeated several times with no significant change in the observed thermal transitions, indicating that the thermal responses are inherent to the polymer and do not result from loss of residual solvent.

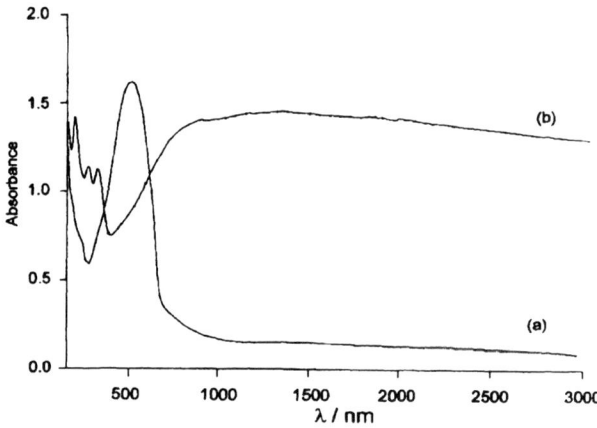

Fig. 2. UV-Vis-NIR absorption spectra of undoped (a) and FeCl₃ doped (b) polymer **2**(cast film on quartz)

As indicated by the UV-vis-NIR spectrum shown in Fig. 2 (a) on an undoped film of cast polymer 2, the onset band gap (low energy absorption edge) was determined to be 1.69 eV for this polymer.

Poly(1-dodecyl-2,5-pyrrylene vinylene) (3), was prepared from 1-dodecyl-2,5-bis(phenyl-thiomethyl)pyrrole. Polymer 3 shows the onset of decomposition at 180°C, and 50 % weight loss by 390 °C. Thus, polymer 2 appears to be slightly more thermally stable than polymer 3. A thin film of polymer 3 was gold in reflectance and greenish purple in transmmitance. A neutral (undoped) polymer film showed an absorption maximum at 563 nm (2.16eV). The absorption edge for the transition occurred at 750 nm, corresponding to an optical band gap (absorption edge) of 1.65 eV which is lower than that of 1.89 eV for polymer 1. This band gap is similar to that observed for polymer 2. Neutral polymers 1 and 2 in THF solution exhibited an absorption maxima at 550 and 548 nm, respectively, as shown in table 1. Upon doping the thin film of polymer 3 with $FeCl_3$ as dopant, a strong absorption appeared at 1400 nm. Both ^1H- and ^{13}C-NMR were taken with $CDCl_3$ as the lock solvent. Singlet peaks at 6.7 and 6.3 ppm are assigned to the vinylene protons and β-hydrogens on the ring, respectively. The 0.8, 1.3 and 3.9 ppm peaks are assigned to dodecyl protons. In the ^{13}C-NMR spectrum, single peaks at 133, 114, 106 ppm are assigned to the carbons of vinylene group, α-carbons on the ring and β-carbons on the ring, respectively.

GPC measurements on polymer 3 (gel permeation chromatography; relative to a polystyrene, with THF as elution solvent) gave a number-average molecular weight of 1.68×10^4, a weight-average molecular weight of 1.95×10^4 and a polydispersity of 1.14. The average degree of polymerization (DP) is at about 65.

A comparison of the properties of polymers 1, 2 and 3 made by the same route are summarized in Table 1.

Table 1. Properties of polymers 1, 2 and 3

Polymer	λmax/nm Soln	λmax/nm Film	Optical gap/ eV	Solubility[a]	Doped conductivity/ S/cm (dopant)
1	500[b]	—	1.89	Insoluble	0.02(AuCl₃)[c], 0.001(FeCl₃)[c]
2	550[d]	569	1.69	Very soluble	2.5(AuCl₃)[e], 0.15(FeCl₃)[e]
3	548[d]	563	1.65	Very soluble	0.48(AuCl₃)[e], 0.11(FeCl₃)[e]

[a] Solubility in moderately polar solvent such as $CHCl_3$, THF, CH_2Cl_2 etc. [b] NMP. [c] Pressed powder pellet [d] THF. [e] Cast films.

CONCLUSIONS

New conducting polymers, poly(1-alkylpyrrylene vinylenes) (alkyl = methyl(1), hexyl(2), dodecyl(3)) were synthesized by polymerization of 1-alkyl-2,5-bis(phenylthiomethyl)pyrrole with an excess of base (t-BuOK) in THF over a temperature range from 0 °C to reflux. The electrical conductivities for polymers (1, 2, 3) are listed in Table 1.

The difference in conductivity between poly(1-alkyl-2,5-pyrrolylene) [$\sigma = 10^{-3}$ (methyl), 4.3 $\times 10^{-5}$ (hexyl), 1.2×10^{-6} S/cm (dodecyl)] and **1, 2** and **3** suggests that the low conductivities of N-substituted polypyrroles as prepared by oxidative polymerization arise from a combination of steric interactions between adjacent rings [6] and from a mixture of α-α', α-β', and β-β' coupled monomers along the polymer chain [18].

We have found the polymerization of 1-alkyl-2,5-bis(phenylthiomethyl)pyrrole by base induced elimination to be general for the electron rich poly(1-alkyl-2,5-pyrrylene vinylenes). Also, we fully expect that this polymerization process will be suitable for the preparation of other electron rich arylene vinylene polymers.

ACKNOWLEDGMENT

We thank the Robert A. Welch Foundation for support of this work. We would also like to thank Professor Michael P. Cava, Professor Martin Pomerantz and Dr. Haitao Cheng for helpful discussions.

REFERENCES

1. Y. Kudoh. S. Tsuchiya, T. Kojima, M. Fukuyama and S. Yoshimura, Synth. Met. **41/43**, 1133 (1991).
2. J. Bargon, S. Mohmand and R. J. Waltman, IBM J. Res. Dev. **27**, 330 (1983).
3. A. Nazzal and G. B. Street, J. Chem. Soc., Chem. Commun. 84 (1983).
4. P. Kovacic, I. Khoury and R. L. Elsenbaumer, Synth. Met. **6**, 31 (1983).
5. I. T. Kim and R. L. Elsenbaumer, Synth. Met., **84**, 157 (1997).
6. A. F. Diaz, J. Castillo, K. K. Kananzwa, A. Logan, M. Salmon and O. Fajardo, J. Electroanal. Chem. **133**, 233 (1982).
7. H. Eckhardt, L. W. Shacklette, K. Y. Jen, and R. L.Elsenbaumer, J. Chem. Phys. **91**, 1303 (1989).
8. K. Y. Jen, C. C. Han and R. L. Elsenbaumer. Mol. Cryst. Liq. Cryst., **186**, 211 (1990).
9. K. Y. Jen, T. R. Jow and R. L. Elsenbaumer, J. Chem. Soc., Chem. Commun. 1113 (1987).
10. D. C. Bott, C. S. Brown, C. K. Chai, N. S. Walker, W. J. Feast, P. J. S. Foot, P. D. Calvert, N. C. Billingham and R. H. Friend, Synth. Met., **14**, 245 (1986).
11. F. E. Karasz, J. D. Capistran, D. R. Gagnon and R. W. Lenz, Mol. Cryst. Liq. Cryst., **118**, 327 (1985).
12. I. Murase, T. Ohnishi, T. Noguchi, M. Hirooka and S. Murakami, Mol. Cryst. liq. Cryst., **118**, 333 (1985).
13. K. Y. Jen, M. Maxfield, L. W. Shacklettle and R. L. Elsenbaumer, J. Chem. Soc., Chem. Commun., 309 (1987).
14. K. Y. Jen, R. Jow, L. W. Shackettke, M. Maxfield, H. Eckhardt and R. L. Elsenbaumer, Mol. Cryst. Liq. Cryst., **160**, 69 (1988).
15. M. C. Piqueras, R. Crespo, E. Ortí and F. Tomás, Synth. Met., **54**, 181 (1993).
16. D. Y. Kim, C. S. Ha and W. J. Cho, Mol. Cryst. Liq. Cryst., **247**, 11(1994).
17. R. W. Lenz, C. C. Han, J. Stenger-Smith and F. E. Karasz J. Polym. Sci. Part A: Polym. Chem. **26**, 3241 (1988).
18. I. T. Kim and R. L. Elsenbaumer, preliminary results.

Conjugated Polymers Containing Pendant Terpyridine Complexes As Photoactive Sensors

Biwang Jiang, Shailesh Sahay, Wayne E. Jones Jr.*
Department of Chemistry and Institute for Materials Research, State University of New York at Binghamton, Binghamton, NY 13902

Abstract

We have synthesized a novel conjugated polymer containing terpyridine receptors. This conjugated, polymer-based fluorescent chemosensory system exhibits unusually high sensitivity toward transition metal ions ($\sim 10^{-9}$ M). The fluorescence quenching response of the terpyridine containing conjugated polymer to transition metal ions is related to a facile energy migration in which energy from absorption of a photon, can migrate through the conjugated polymer system and be quenched by trapping sites. Terpyridine receptors can readily bind to transition metal ions and create a low lying Metal to Ligand Charge Transfer (MLCT) state which can act as an excitation trap.

Introduction

Recently, great efforts have been devoted to the design and construction of highly sensitive and selective molecular sensory systems [1]. One approach to this design effort has been to use flourescent species containing recognition/receptor sites which are capable of quenching the excited state of the fluorophore upon binding the analyte. Conjugated polymer materials are particularly attractive for this application due to the increased reaction volume for interaction with analytes in solution that would lead to quenching of the polymer excited state [2-5]. T. M. Swager et. al. recently demonstrated polyreceptor assemblies, electronically connected by a conjugated polymer (molecular wire) with larger sensitivity gains over more conventional molecule-based fluorescent chemosensors [2]. The origin of this sensitivity enhancement is facile energy migration throughout the conjugated polymer as outlined in Scheme 1. Following excitation, rapid exciton migration along the conjugated polymer chain results in very efficient trapping even at very low quencher concentrations [6]. Based on this concept, we have designed and synthesized a new conjugated polymer-based sensory system that is capable of detecting transition metal ions in extremely low concentration.

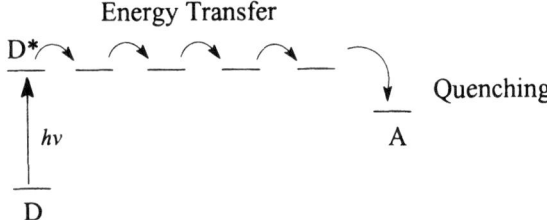

Scheme 1. Representation of energy migration in an extended conjugated polymer system.

Here we present an alternative approach to constructing new transition metal ion sensitive polymers. Our approach as illustrated in Scheme 2, is to prepare, luminescent, conjugated polymers containing terpyridine receptors in which a quenching center is created upon binding of a transition metal ion. The unique of this design is that pendant terpyridine receptors were connected through conjugated bridges to the polymer backbone. Previous designs generally incorporated the receptor units directly into the conjugated backbone. Terpyridine is an ideal receptor for binding transition metals as demonstrated by the well established coordination chemistry that has been observed [7,8]. In addition, transition metal terpyridine complexes contain a photoactive Metal to Ligand Charge Transfer (MLCT) excited state which has previously been shown to undergo photoinduced electron and energy transfer with conjugated polymers [8]. Based on model complexes, the Metal to Ligand Charge Transfer (MLCT) excited states formed between terpyridine and Ni^{2+}, Fe^{2+}, Ru^{2+}, Re^+, and Os^{2+} should all act as excitation traps for the exciton generated in conjugated polymers. Thus, binding of transition metal ions in solution would result in quenching of the polymer emission intensity. Furthermore, the terpyridine receptor groups were expected to have no significant effect on the photophysics of the conjugated polymer backbone compared to that of model polymer 4 shown in Scheme 3.

Scheme 2. Synthesis of terpyridine-containing conjugated polymer and its coordinate reaction with transition metal ions.

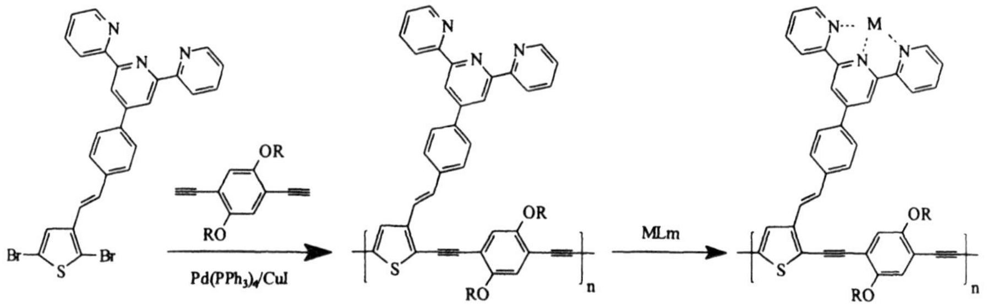

$$R=C_{16}H_{33}$$

1   **2**   **3**

Results and Discussion

Synthesis of polymer precursor 1 is the key step for preparation of polymer 2. The palladium-catalyzed coupling of 1,4-diethynyl-2,5-dihexadecyloxybenzene and 2,5-dibromo-3-terpyridine substituted thiophene 1 leads to polymer 2, as shown in Scheme 2. The polymer 2 is very soluble in common organic solvents such as tetrahydrofuran, chloroform and toluene. The molecular weight was measured by GPC to be Mn = 23,000 against polystyrene. The expected structure of the polymer was confirmed by ^{1}H NMR, IR, UV and fluorescence spectroscopies. Detailed synthetic methods and structure determination of both the polymer 2 and polymer precursor 1 will be described elsewhere [9]. For the comparison, we also synthesized polymer 4 without terpyridine receptor group by using similar palladium-catalyzed coupling reaction.

Scheme 3. Synthesis of conjugated polymer **4**.

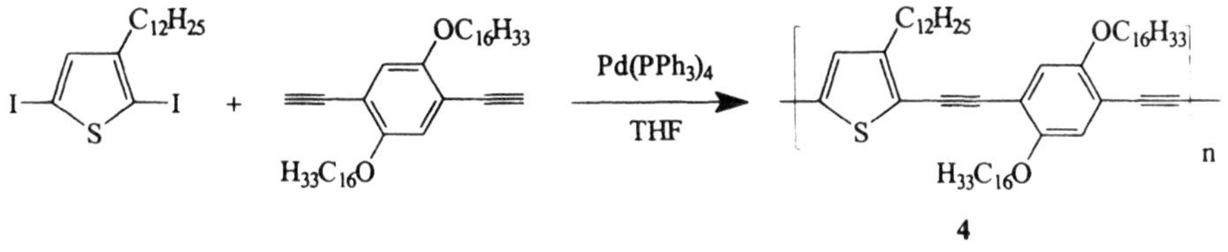

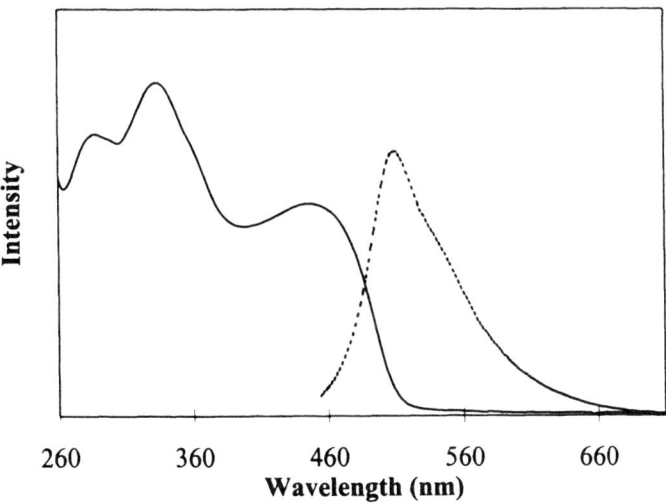

The UV-vis absorption and emission spectra of conjugated polymer **2** in THF solution are shown in Figure 1. Characteristic electronic absorption bands in high energy with λ_{max}=228 and 334 nm from conjugated terpyridine moiety are similar to those of terpyridine monomer **1**. The lower energy absorption maximium at 450 nm is attributed to the extended π-conjugation of polymer main chain and is similar to the polymer **4** (λ_{max} = 445 nm) [10]. The lack of distortion in the absorption bands of the terpyridine and the conjugated polymer main chain suggest limited electronic interaction with the pendant receptor site in the ground state. The polymer exhibits a strong fluorescence at λ_{max}= 495 nm. The emission spectrum is similar to the model polymer **4** in both emission energy and quantum yield.

Figure 1. Absorption (solid line) and emission (dashed line) spectra of polymer **2** in THF solution.

The effect on the emission of the terpyridine-containing polymer **2** in the presence of transition metal ions of THF solutions, $[M^{2+}]$=2.0x10^{-6} M is shown in Figure 2. The emission was observed to be quenched by all three metal dications (Fe, Co and Ni) when dissolved in THF solution as their hydrated salts. The most dramatic shift in emission intensity was observed in the presence of the Ni^{2+} salt. The response is most likely depends on the coordinating ability of the different metals with terpyridine since there is very little difference in the excited state energies of the metal terpyridine complexes [8]. It should also be noted that the coordinated transition metals will not form a closed octahedral shell by binding with a single terpyridine site. Solvent coordination under the dilute conditions described here may also play a significant role in the observed selectivity.

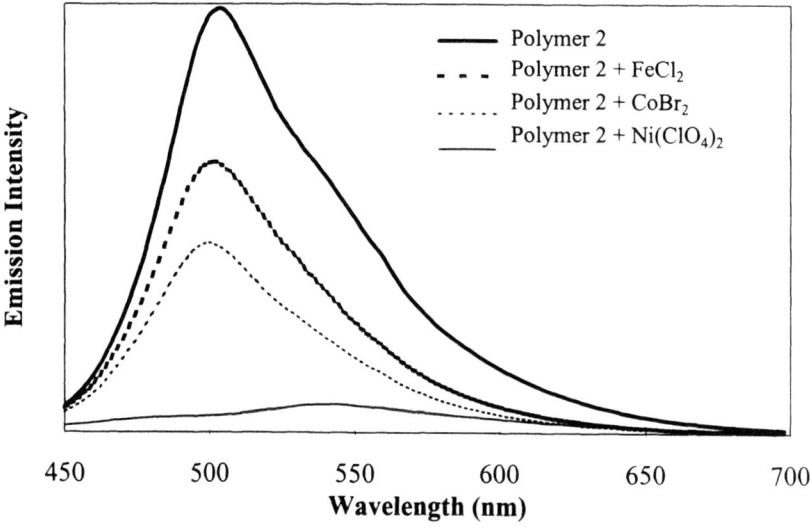

Figure 2. Transition metal ion dependence of emission spectra of polymer **2** in THF solution.

The emission intensity of polymer **2** as a function of concentration of Ni^{2+} over the range 10^{-6} to 10^{-9} M in THF is shown in Figure 3. The emission intensity decreases significantly at concentrations as low as 4.0×10^{-9} M. For the purpose of comparison, the terpyridine units concentration was hold at [Terpy] = 6×10^{-6} M. A comparison of the data in Figure 3 to the emission quenching of the monomer terpyridine **1** solution shows that the monomer requires about three orders of magnitude higher concentration of transition metal ions in order to achieve the same quenching efficiency. This dramatic enhancement of the emission quenching response of polymer based receptors relative to the monomer is consistent with a highly mobile exciton along the conjugated polymer backbone as described previously [2]. No emission quenching was observed for polymer **4** in the presence of a variety of transition metal ions in THF solution, which indicated only terpyridine can interact with the transition metal ions to form an excitation trapping center.

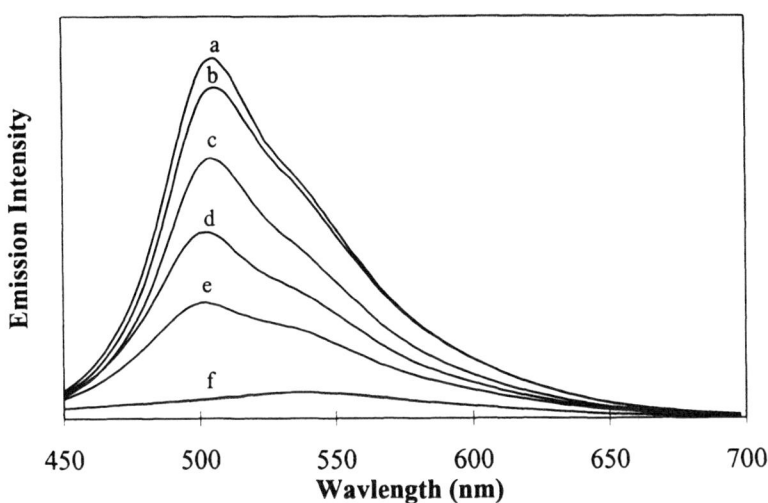

Figure 3. Emission spectra of conjugated polymers as a fuction of Ni^{2+} concentration in THF, (a) $[Ni^{2+}]=0$; (b) $[Ni^{2+}]=4.0\times10^{-9}$; (c) $[Ni^{2+}]=4.0\times10^{-8}$; (d) $[Ni^{2+}]=2.0\times10^{-7}$; (e) $[Ni^{2+}]=8.0\times10^{-7}$; (f) $[Ni^{2+}]=2.4\times10^{-6}$ M.

Although, this terpyridine-containing conjugated polymer shows an unusual sensitivity to transition metal ions. However the selectivity to different transition metals is low, most of the transition metals tested demonstrated the ability to quench the polymer excited state in higher concentration. This factor seriously limits the applicability of the current chemosensor materials. This challenge is currently under investigation.

The solubility of the polymer materials in organic solvents provides an opportunity to prepare solid thin films that can be used to address these two concerns. Thin films of the terpyridine polymer prepared by spin coating of the polymer in THF exhibited substantial emission quenching similar to the solution results in Figure 3. In ethanol solutions of 1 x 10^{-10} M $Ni(ClO_4)_2$ the emission intensity observed at 552 nm was decreased by about 50% as measured by diffuse reflectance spectroscopy at < 45°. Preliminary experiment results indicated the thin films emission intensity could be recovered by immersion in a solution of a competitive metal chelating ligand such as ethylenediamine. This reversibility is critical for the true applicability of this new class of materials in practical sensor applications.

Conclusion

In summary, we have designed and synthesized a soluble, conjugated polymer containing pendant terpyridine receptors as a new class of chemosensor materials. The polymers exhibited strong fluorescent sensitivity toward transition metal ions both in solution and as solid, spin cast, thin films. The fluorescence quenching response observed was significantly stronger than simple monomeric chromophore/receptor assemblies. The enhanced response can be attributed to efficient energy transfer through the conjugated polymer system and coordinating ability of terpyridine to transition metal ions. The conjugated polymer based sensor system was highly sensitive to transition metals to 10^{-9} M in solution and reversibility can be achieved in solid thin films. A modification of receptor structure will be necessary in order to achieve a more highly selective response to different metal ions within a mixture.

Acknowledgements

The authors would like to thank Dr. Omowunmi Sadik for helpful and insightful discussions. This research was funded by the Integrated Electronics and Engineering Center (IEEC) at the State University of New York at Binghamton. The IEEC receives funding from the NY State Science and Technology Foundation, the National Science Foundation and a consortium of industrial members.

Reference

1. (a) A. W. Czarnik, Fluorescent Chemosensory for Ion and Molecular Recognition, Ed. (ACS Symposium Series 538; ACS: Washington, DC, 1993). (b) A. Prasanna de Silva, H. Q. Nimal Gunaratne, T. Gunnlaugsson, A. J. M. Huxley, C. P. McCoy, J. T. Rademacher, and T. E. Rice, Chem. Rev. **97**, 1515 (1997). (c) Omowunmi A. Sadik, Anallytical Methods and Instrumentation, **2**, 293 (1995).
2. (a) Q. Zhou, T. M. Swager, J. Am. Chem. Soc. **117**,12593 (1995);(b) M. J. Marsella, P. J. Carrol, T. M. Swager, ibid. **117**, 9832 (1995); (c) M. J. Marsella, P. J. New land, P. J. Carrol, T. M. Swager, ibid, **117**, 9842 (1995); (d) T. M. Swager, M. J. Marsella, Adv. Mater **6**, 595 (1994).
3. B. Wang, M. R. Wasielewski, J. Am. Chem. Soc. **119**, 12 (1997).
4. T. W. Brockmann, J. S. Tour, J. Am. Chem. Soc. **117**, 4437 (1995).
5. R. D. McCullough, S. P. Williams, Chem. Mater. **7**, 2001 (1995).
6. T. M. Swager, C. J. Gil, M. S. Wrighton, J. Phys. Chem. **99**, 4886 (1995).
7. (a) F. Barigelletti, L. Flamigni, V. Balzani, J. P. Collin, J. P. Sauvage, A. Sour, E. C. Constable, and A. M. W. C. Thompson, J. Am. Chem. Soc. **116**, 7692, (1994). (b) E. C. Constable, P. Harverson and M. Oberholzer, Chem. Commun. 1821, (1996).
8. For reviews see (a) V. Balzani, A. Juris, M. Venturi, Chem. Rev. **96**, 759 (1997). (b) J. P. Sauvage, J. P. Collin, J. C. Chambron, S. Guillerez, and C. Coudret, Chem. Rev. **94**, 993, (1994).
9. Biwang Jiang, Wayne E. Jones Jr. Manuscript in preparation.
10. B. Jiang, S. Yang, W.E. Jones Jr. Chem. Materials, **9**, 2031 (1997).

SELF-ASSEMBLING FUNCTIONALIZABLE POLYDIACETYLENES AND THEIR OPTICAL PROPERTIES

M. Sukwattanasinitt*, X. Wang*, D.-C. Lee*, L. Li**, J. Kumar**, S. K. Tripathy*, and D. J. Sandman*

*Center for Advanced Materials and Department of Chemistry and
**Physics and Applied Physics, University of Massachusetts Lowell,
Lowell, Massachusetts 01854-2881

ABSTRACT

The alkoxycarbonylmethylurethanes of 9-(N-methyl-N-phenylamino)-5,7-nonadiyn-1-ol were synthesized and polymerized by ^{60}Co gamma radiation. The resulting polydiacetylenes (PDA) are soluble, processable, and self-assemble into an acentric array. These PDAs are susceptible to postpolymerization functionalization via its dialkylaniline and ester groups, and hence comprise the most versatile PDA system for systematic chemical modification reported to date. Dialkylaniline functionalization via tetracyanoethylene and diazonium salt coupling led, respectively, to materials with interesting second order nonlinear optical properties and that exhibit directly photofabricated regular surface relief gratings. The latter materials provide the first example of a rigid rod polymer (lacking a detectable glass transition) to exhibit such phenomena. Moreover, these PDAs can form self-assembled multilayer structure in water by ester hydrolysis.

INTRODUCTION

We have recently reported soluble, processable, and the most versatile PDA system reported to date, in terms of chemical richness and diverse optical properties [1].

Over at least the last decade, there has been considerable interest in the introduction of specific side groups into polydiacetylenes to control properties such as polymer chemical reactivity [2,3], electronic structure [2-7], spectroscopy [8-10], second order nonlinear optical (NLO) properties [11-15], and chromic phenomena [16,17]. In recent reports from our laboratories [18-21], the synthesis, acentric self-assembly, and second order properties of the PDA of the n-butoxycarbonylmethylurethane (BCMU) of 1-(5-pyrimidyl)-1,3-octadiyn-8-ol (BPOD) point to a design paradigm that merits elaboration. In particular, we note three design features: 1) a conjugated backbone; 2) one side group as a chromophore; 3) the other side group with both aliphatic and hydrogen-bonded urethane groups to promote solubility and processability. The polymer was found to spontaneously self-assemble into an acentric order with significant second order optical nonlinearity. We now report the synthesis of the alkoxy carbonyl-methyl urethanes of 9-(N-methyl-N-phenylamino)-5,7-nonadiyn-1-ol (1a, b, c) and their polymerization. We find that the polymers are soluble and processable, and that the benzenoid ring group is susceptible to many of the substitution reactions expected for aromatic amines making it a more versatile system for chemical modification than the previously investigated systems. We further describe novel self-assembling and linear and nonlinear optical properties of the modified polymer. These polymers, in addition to possessing the unusual assembly properties of PDA-BPOD, present some unique processing features [7]. The azobenzene functionalized PDA from 1a, for example, was photofabricated

677

to produce regular surface relief gratings. Also, **1b** can be hydrolyzed [22] after polymerization and give water solubility to fabricate self-assembled layer by layer structure [23] which is under investigation.

EXPERIMENTAL

The detailed synthesis of **1a, b**, postpolymerization funtionalization of poly-**1a, -b** , and characterization were reported elsewhere [1]. Cadiot-Chodkiewicz coupling of N-methyl-N-propagylanililne and 5-bromohexyne-1-ol led to 9-(N-methyl-N-phenylamino)-5,7-nonadiyn-1-ol. Reaction of this diacetylene alcohol with the appropriate isocyanate led to **1a, b, c** (Scheme 1); the yield of **1c** was 80%. Unlike **1a, b** which are yellow oils, **1c** crystallized at ambient temperature.

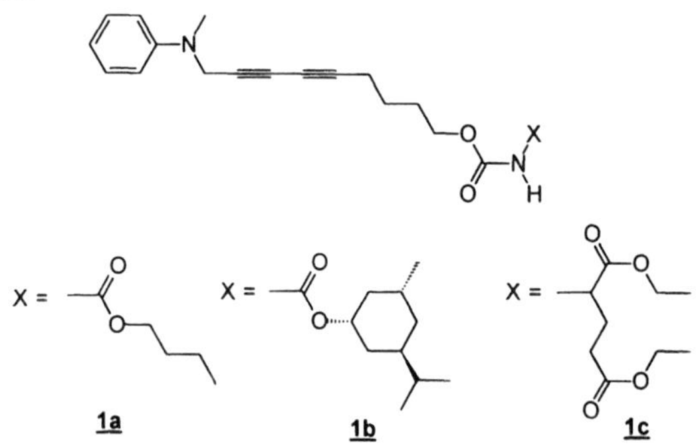

Scheme 1. Structures of diacetylene monomers.

1a, b, c were placed in evacuated ampoules and polymerized in the solid state by exposure to 50 Mrad ^{60}C γ radiation. PDAs from **1a, b, c**, obtained in approximately 40% (PDAs from **1a, b**) and 60% (PDA from **1c**) after monomer removal, were red-colored solids that were found to be soluble in a manner similar to PDA-BPOD and -BCMU in solvents such and chloroform and N,N-dimethylformamide. 100 mg of PDA-**2d** was dissolved in 100 ml of chloroform and 2.6 ml of 0.2 N methanolic potassium hydroxide added to give water soluble polymer. It was further acidified with 2N hydrochloric acid and the red precipitate formed was filtered and washed with water and dried under vacuum for 24 hrs to give PDA-**2e** (Scheme 3).

RESULTS

The ^{13}C NMR and Raman [4,5] spectra of poly-**1a, -b, -c** verified the assignment of the usual en-yne backbone structure [1]. It was found that the PDA of **1a** assembles spontaneously into a noncentrosymmetric film from a frequency doubling experiment that gave a d_{33} value of 0.3 pm/V at 1.064 μm. We have found that the aromatic ring of poly-**1a, -b** is readily substituted. For example, poly-**1a** reacts with excess tetracyanoethylene [24,25] in DMF to give a PDA-**2a** (Scheme 2) in quantitative yield with 40-45% of the aromatic rings tricyanovinylated. This PDA was red in color with a broad visible absorption maximum at 499 nm (CHCl$_3$) and soluble in chloroform and DMF. The presence of the tricyanovinylaniline chromophore in PDA-**2a, -2b** (Scheme 2) invites assessment of second order nonlinear optical properties [26,27].

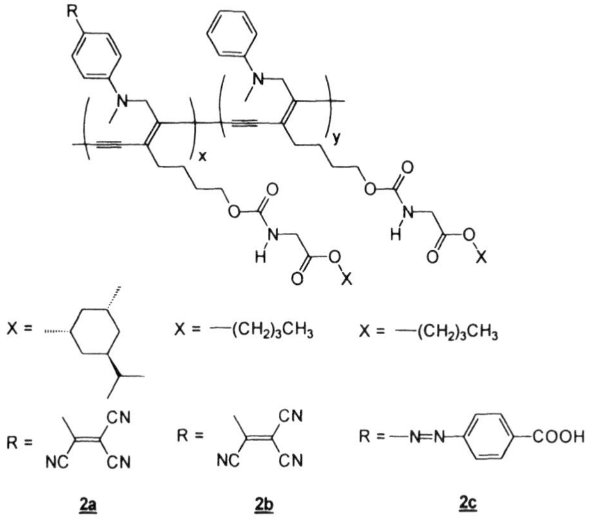

$X = $ $X = —(CH_2)_3CH_3$ $X = —(CH_2)_3CH_3$

2a   **2b**   **2c**

Scheme 2. Structures of postpolymerization functionalized polydiacetylenes.

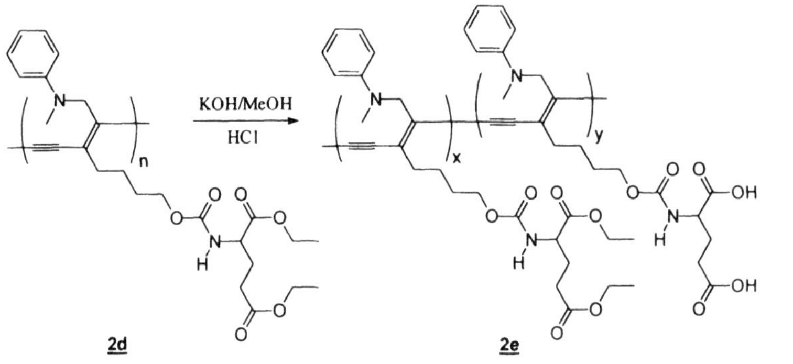

2d   **2e**

Scheme 3. Hydrolysis of PDA-2d.

Spin-coated films of PDA-**2a** gave a d_{33} value of 1 pm/V at 1.064 µm without poling. The nature of the acentric order appears to be similar to that of a poled nonlinear optical polymer film. The d_{33} coefficient increases to 33 pm/V after electric field poling at 130 °C for 30 min. The poling temperature was arbitrarily selected given the fact that no thermal transitions were observed for this polymer until degradation around 200 °C. The PDA backbone is assumed to be rigid with a large persistence length. The side groups, however, are expected to have a sufficient mobility at this temperature. The hydrogen bonded network of the BCMU side groups for example is significantly disrupted at this temperature [28,29]. The PDA from **1a, b** can be further functionalized by a post polymerization azo coupling reaction [30]. The polymer (PDA-**2c**) (Scheme 2) was found by ^{1}H-NMR to have 60% of the aromatic rings functionalized. No glass transition temperature was observed from the DSC study.

Holographic surface relief gratings have been fabricated on azobenzene polymers [31-34]. Exposure of a spin-coated film of an azobenzene polymer to an interference pattern from two polarized argon-ion laser beams at 488 nm led to formation of such a grating [31-34]. An unusual characteristic of this fabrication process is that macromolecules are transported over very large distances (thousands of Å) while the sample was held substantially below its T_g. In the present study, we have fabricated surface relief gratings on azobenzene-functionalized PDAs for the first time. A typical surface relief grating shows very regularly spaced sinusoidal surface structure with depth modulation of over 500 Å. The grating spacing can be adjusted by changing the angle between the two writing beams. This provides the first example of such one step photoinduced grating formation without postprocessing in a rigid rod polymer. Water soluble PDA-2e (Scheme 3) can be self-assembled in basic water with multilayer structure and is expected to have high second nonlinear susceptibility.

CONCLUSIONS

In summary, we have demonstrated a new case of side group manipulation of PDA properties in a versatile self-assembling system. In particular, postpolymerization functionalization is a viable new strategy for the preparation of a whole range of new PDA materials that differs from the traditional approach of monomer synthesis followed by solid state polymerization. In addition to the reactions described above, the PDAs from **1a, b, c** react with other electrophilic reagents. Furthermore, PDA from **1c** can also be postpolymerization functionalized and self-assembled layer by layer in water to give better optical property than PDAs from **1a, b**. Hence, as will be amply demonstrated in future publications, numerous additional novel PDA are at our disposal. Moreover, we have found that the PDA obtained via our postpolymerization functionalization have novel optical properties. Note that the polymer used to form the surface grating is the first example of such grating formation in a rigid rod material.

REFERENCES

1. M. Sukwattanasinitt, X. Wang, L. Li, X. Jiang, J. Kumar, S.K. Tripathy, and D.J. Sandman, Chem. Mater., 1997 (accepted).
2. H. Eckert, J.P. Yesinowski, D.J. Sandman, and C.S. Velazquez, J. Am. Chem. Soc., **109**, 761 (1987).
3. D.J. Sandman, B.S. Elman, G.P. Hamill, J. Hefter, and C.S. Velazquez, in Crystallographically Ordered Polymers, edited by D.J. Sandman (American Chemical Society Symposium Series, **337**, 1987), pp.118-127.
4. S.H.W. Hankin, D.J. Sandman, in Structure-Property Relations in Polymers, edited by M.W. Urban and C.D. Craver (American Chemical Society Advances in Chemistry Series, **236**, 1993), pp243-262.
5. D.N. Batchelder and D. Bloor, in Advances in Infrared and Raman Spectroscopy, edited by R.J.H. Clark and R.E. Hester (Wiley, New York, **11**, 1984), pp. 122-209.
6. B.J. Orchard and S.K. Tripathy, Macromolecules, **19**, 1844 (1986).
7. H. Eckhardt, D.S. Boudreaux, and R.R. Chance, J. Chem. Phys., **85**, 4116 (1986).
8. J.L. Foley and D.J. Sandman, presented at the 1996 MRS Fall meeting, Boston, MA, 1996 (unpublished).

9. C.E. Masse, M. Kamath, N.B. Kodali, W.H. Kim, J. Kumar, and S.K. Tripathy, Macromolecules **26**, 5954 (1993).
10. N.B. Kodali, W.H. Kim, J. Kumar, S.K. Tripathy, and S.S. Talwar, Macromolecules, **27**, 6612 (1994).
11. J.E. Sohn, A.F. Garito, K.N. Desai, and M. Kuzyk, Makromol. Chem., **180**, 2795 (1979).
12. A.F. Garito and K.D. Singer, Laser Focus (1982), pp 59-64.
13. M.S. Paley, D.O. Frazier, H. Abdeldeyem, S. Armstrong, S.P. McManus, J. Am. Chem. Soc., **117**, 4775 (1995).
14. M.S. Paley, D.O. Frazier, S.P. McManus, S.E. Zutaut, and M. Sanghadasa, Chem. Mater., **5**, 1641 (1993).
15. C.E. Masse, K. Vander wiede, W.H. Kim, X.L. Jiang, J. Kumar, and S.K.Tripathy, Chem. Mater., **7**, 904 (1995).
16. K.D. Zemach and M.F. Rubner, in <u>Electrical, Optical, and Magnetic Properties of Organic Solid State Materials</u>, edited by A.F.Garito, A.K.-Y. Jen, C.Y.-C. Lee and L.R. Dalton (Mater. Res. Soc. Proc. **328**, Boston, MA, 1994) pp. 763-768.
17. D.J. Sandman, V. Shivshankar, J.C. Stark, N. Benfaremo, Z.-H. Tsai, C. Sung, S.K. Tripathy, and J. Kumar, Proc. ACS Div. Polym. Mater.: Sci. and Eng., **75**, 171 (1996).
18. W.H. Kim, N.B. Kodali, J. Kumar, and S.K. Tripathy, Macromolecules, **27**, 1819 (1994).
19. S.K. Tripathy, W.H. Kim, B. Bihari, R. Moody, J. Kumar, Synthetic Metals, **71**, 1675 (1995).
20. S.K. Tripathy, W.H. Kim, X. Jiang, and J. Kumar, Nonlinear Optics, **15**, 111 (1996).
21. D.W. Cheong, W.H. Kim, L.A. Samuelson, J. Kumar, and S.K. Tripathy, Macromolecules, **29**, 1416 (1996).
22. G.N. Patel, A.F. Preziosi, and H.R. Bhattacharjee, J. Polym. Sci., **71**, 247 (1984).
23. J.W. Baur, P. Besson, S.A. O'Connor, and M.F. Rubner, in <u>Electrical, Optical, and Magnetic Properties of Organic Solid State Materials III</u>, edited by A.K.-Y. Jen, C.Y.-C. Lee, L.R. Dalton, M.F. Rubner, G.E. Wnek, and Y. Chiang (Mater. Res. Soc. Proc. **413**, Boston, MA, 1996) pp. 583-588.
24. B.C. McCusick, R.E. Heckert, T.L, Cairns, D.D. Coffman, and H.F. Mower, J. Am. Chem. Soc., **80**, 2806 (1958).
25. X. Wang, J.I. Chen, S. Marturunkakul, L. Li, J. Kumar, and S.K. Tripathy, Chem. Mater., **9**, 45 (1997).
26. H.E. Katz, K.D. Singer, J.E. Sohn, C.W. Dirk, L.A. King, H.M. Gordon, J. Am. Chem. Soc., **109**, 6561 (1987).
27. V.P. Rao, A.K.-Y. Jen, K.Y. Wong, and K.J. Drost, J. Chem. Soc., Chem. Commun., 1118 (1993).
28. R.R. Chance, G.N. Patel, and J.D. Witt, J. Chem. Phys., **71**, 206 (1979).
29. G. Walters, P. Painter, P. Ika, and H. Frisch, Macromolecules, **19**, 888 (1986).
30. X. Wang, L. Li, J.-I. Chen, S. Marturunkakul, J. Kumar, and S.K. Tripathy, Macromolecules, **30**, 219 (1997).
31. D.Y. Kim, S.K. Tripathy, L. Li, J. Kumar, Appl. Phys. Lett., **66**, 1166 (1995).
32. D.Y. Kim, L. Li, X.L. Jiang, V. Shivshankar, J. Kumar, and S.K. Tripathy, Macromolecules, **28**, 8835 (1995).
33. X.L. Jiang, L. Li, J. Kumar, D.Y. Kim, V. Shivshankar, and S. K. Tripathy, Appl. Phys. Lett, **68**, 2618 (1996).
34. P. Rochon, E. Batalla, and A. Natansohn, Appl. Phys. Lett., **66**, 136 (1995).

NEW POLYDIACETYLENES WITH VISIBLE CHROMOPHORIC SIDE GROUPS

James L. Foley, Venkataramani Shivshankar, Daniel J. Sandman, Center for Advanced Materials, Department of Chemistry, University of Massachusetts Lowell, Lowell, MA.

ABSTRACT

In the interest of obtaining crystalline polydiacetylene with absorption at longer wavelengths, new derivatives of 2,4-hexadiyne with planar, visible absorbing aromatic chromophores have been designed, synthesized, and thermally polymerized. The resulting polymers have a red shifted maximum with absorption extending to the infrared. Their Raman spectra show the usual polydiacetylene ene-yne structure, and X-ray powder diffraction reveals that the new polymers are crystalline.

INTRODUCTION

In the absence of mechanical strains, the position of the long wavelength absorption maximum of crystalline polydiacetylene (PDA) is determined by interactions between the side groups and the conjugated backbone, and it is suggested that there should be a stronger influence with greater polarity[1]. Following this suggestion four new crystalline diacetylene monomers with side groups having intense visible absorption have been synthesized[2]. The first is the diether 1,6-Bis(p-oxybenzylidinemalononitrile)-2,4-hexadiyne (compound 1), figure 1.

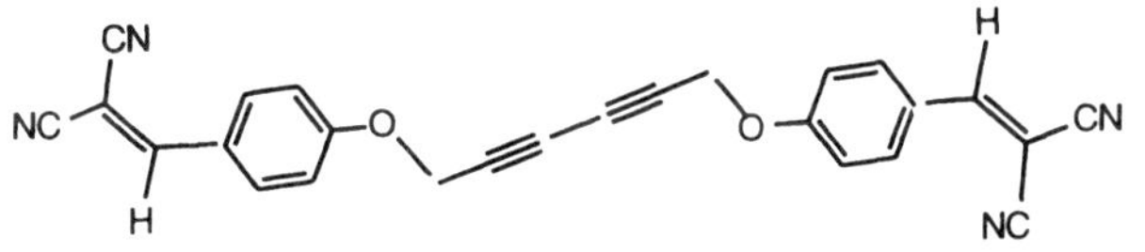

Figure 1. Compound 1.

The three others are the diamines 1,6-bis(N-methyl-N-[p-dicyanovinylphenyl]amino)-2,4-hexadiyne (compound 2), 1,6-bis(N-ethyl-N-[p-tricyanovinylphenyl]amino)-2,4-hexadiyne (compound 3), 1,6-bis(N-ethyl-N-[p-tricyanovinylphenyl]amino)-2,4-hexadiyne (compound 4), figure 2.

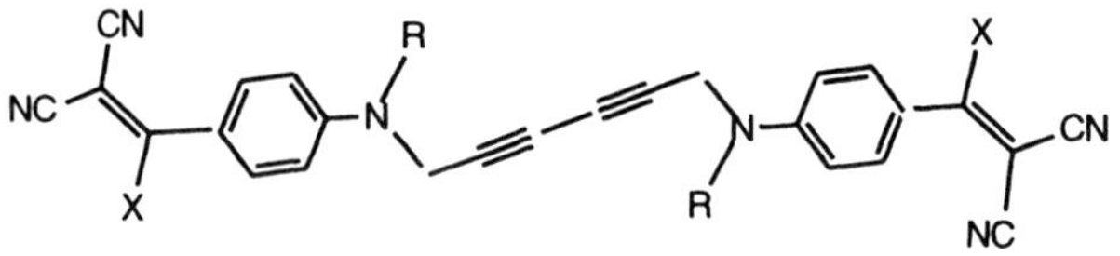

2, X = CN, R = Me

3, X = CN, R = Et

4, X = H, R = Et

Figure 2. Compounds 2 - 4.

EXPERIMENT

Crystallization

All monomers are crystalline solids. Crystallization conditions are presented in Table I.

Table I

Compound	Solvent	Conditions	Description
1	acetic acid	add water	small pale yellow crystals
2	DMF	100 - 25° C	small purple needles
3	acetic acid	118 - 25° C	purple needles
4 A	toluene	111 - 25° C	orange needles
B	...on standing		thin plates

Solid-State Polymerization

The monomers were polymerized by heating the crystals at 120 - 140° C for ten days in a vacuum oven. The other two monomers were heated at the same temperature for two weeks in sealed evacuated ampoules.

Visible Electronic Spectroscopy

For diffuse reflectance electronic spectra, the same crystallization procedures were followed, but before being heated, monomers were first ground in salt, 1% for compound 1, 1% for compound 2, 0.16% for compound 3 and 0.25% for compound 4. Monomers were of the same strength except for compound 2 which was at 0.2%. Sodium chloride was used as the reference.

Raman Spectroscopy

Raman spectra for neat samples of monomers and polymers were obtained at 1064 nm excitation on a Perkin Elmer1760X FT-IR spectrometer fitted with Neodymium YAG laser.

RESULTS

Typical diacetylene polymerization was confirmed by FT-Raman spectroscopy by disappearance of the diacetylene conjugated triple bond peak and the appearance of the PDA triple bond and double bond peaks. The results are summarized in Table II.

Table II

Compound	monomer yne, cm^{-1}	PDA yne, cm^{-1}	PDA ene, cm^{-1}
1	2259	2081	1465
2	not observed	2076	1445
3	2254	2075	1444
4 A	2259	2110	1502

Visible diffuse reflection of the polymers showed peaks beyond the theoretical backbone maximum absorption of 532 nm[3], and tails extending into the infrared. Table III relates the solid state visible maxima of the monomer (chromophores) to the observed backbone absorption of the corresponding polymer backbone.

Table III

Compound	monomer maximum, n m	polymer maximum, n m	polymer absorption edge, n m
1	360	640	990
2	525	660	1060
3	540	665	1120
4 A	420	540	880

The complete spectra are shown in Figure 3

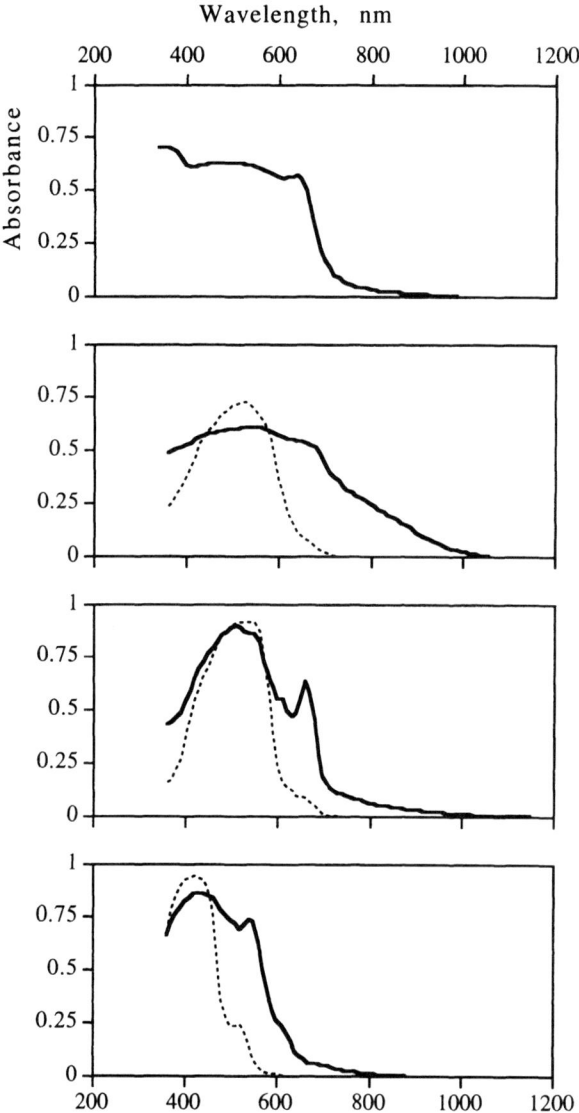

Figure 3. Diffuse reflectance spectra of
compounds 1 to 4A, top to bottom:
-------- monomer, ———— polymer.

X-ray powder patterns show the polymers to be crystalline but not obviously similar in structure to the monomer.

DISCUSSION AND CONCLUSIONS

New hexadiynes with visible chromophoric side groups have been synthesized and thermally polymerized to give new crystalline PDA. The diffuse reflectance results show that the visible absorbing side groups do shift the polydiacetylene backbone absorption to longer wavelengths. Polymers 1 - 3 have spectral maxima at wavelengths as long or longer than PDA-DCH[1], the best defined PDA with single crystal spectrum. PDA-DCH has its maximum in diffuse reflectance, at 635 nm, measured as described above, at 656 nm in single crystal[1]. Hence, the position of the spectral maximum of the PDA backbone is sensitive to the detailed nature of the interactions with the side group. It has been proposed that the exitonic energy of the backbone spectrum is lowered by the effect of the polarizable side groups acting in manner like a solvent[4]. If this effect is operative here, at first glance, it is not linear with side group polarizability or polarity. Additional crystal structure information will be required to further interpret the spectral information.

ACKNOWLEDGMENTS

We are grateful to the Donors of the Petroleum Research Fund, administered by the ACS, for partial financial support, M. J. Downey for furnishing X-ray powder patterns, and Dr. Lian Li for Raman spectra.

REFERENCES

1. R. J. Hood, H. Muller, C. J. Eckhardt, R. R. Chance, K. C. Yee, Chem. Phys. Lett. **54**, 295 (1978).
2. J. L. Foley and D. J. Sandman, to be published.
3. B. E. Kohler and D. E. Schilke, J. Chem. Phys. **86**, 5214 (1987).
4. M. Schott and G. Wegner, Nonlinear Optical Properties of Organic Molecules and Crystals, D. S. Chemla and J. Zyss, eds., (Academic Press, 1987), vol. 2, pp. 3-49.

INVESTIGATION OF ORGANIC-ORGANIC INTERFACES BY TIME-RESOLVED PHOTOCURRENT, ELECTROCHEMICAL, AND PHOTOEMISSION TECHNIQUES

Liang-Bih Lin[*,†], M. Gary Mason[*,‡], Ralph H. Young[*,‡], Deniz E. Schildkraut[‡], Paul M. Borsenberger[*,†], and Samson A. Jenekhe[*,#]

[*]Center for Photoinduced Charge Transfer and [#]Departments of Chemical Engineering and Chemistry, University of Rochester, Rochester, NY 14627
[†]Office Imaging Division and [‡]Imaging Research and Advanced Development, Eastman Kodak Company, Rochester, NY 14650

ABSTRACT

Hole transporting properties and energy barriers at organic-organic interfaces relevant to electrophotographic and organic electroluminescent (EL) devices are described. Three well-known hole transporting molecules, 1,1-bis(di-4-tolylaminophenyl)cyclohexane (TAPC), N,N'-diphenyl-N,N'-bis(1-naphthyl)-(1,1'-biphenyl)-4,4'-diamine (NPB), and N,N,N',N'-tetrakis(4-tolyl)-(1,1'-biphenyl)-4,4'-diamine (TTB) are used in this study. The ionization potentials (IP) and oxidation potentials (E_{ox}) of these materials are determined by photoemission spectroscopy and electrochemical measurements, from which a conversion formula is obtained (IP ~ 4.5 eV + eE_{ox}). Hole transport across organic-organic interfaces is investigated by time-of-flight transient photocurrent techniques. The efficiencies of hole injection are consistent with the energy barriers, when present, at these interfaces.

INTRODUCTION

Amorphous organic solids are used in the electrophotography industry[1] and have potential use in display, optoelectronics, and microelectronic industries. Recently, much attention has been given to organic electronic devices such as organic EL displays,[2,3] photorefractive polymers,[4] and thin film transistors.[5]

One of the less well understood properties of these organic electronic devices is the dynamics of charge injection between organic layers,[6,7] which is critical to the operation of the devices. Hole injection at an organic-organic interface with favorable energy levels is shown in Fig. 1(a). In principle, holes arriving at the interface should be readily transported across. In contrast, in Fig. 1(b), hole injection is expected to be impeded at the interface. Here, we present a study of transient hole injection dynamics and the effects of energetic barriers in organic bilayers.

EXPERIMENT

A typical bilayer cell consists of a 300 Å Se hole-generating layer coated on a poly(ethylene terephthalate) substrate pre-coated with a 300 Å Ni layer, an organic bilayer from sequential vacuum sublimation onto the Se, and finally a 300 Å Au layer deposited onto the free surface (Fig. 1(c)). Thickness of each layer is monitored by a quartz crystal, which is calibrated by ellipsometry, profilometry, and absorption spectroscopy. Photocurrents are measured with conventional time-of-flight photocurrent techniques.[1] Molecular structures of NPB, TAPC, and TTB are shown in Fig. 2. We denote, by HTL1|HTL2, a bilayer with holes moving left to right through the cell (Fig. 1(c)). We prepare four types of bilayers: TAPC|TTB, NPB|TAPC, NPB|TTB, and TTB|NPB.

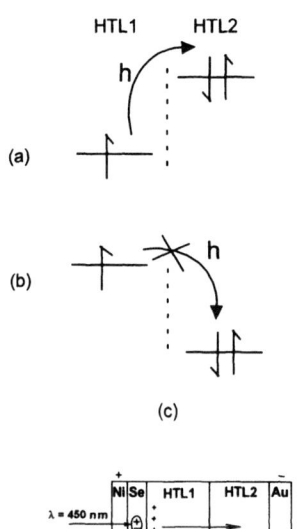

Figure 1. Hole transport across organic-organic interfaces (a) with favorable energies (b) with an energy barrier. (c) A typical configuration of an organic bilayer cell for transient photocurrent measurements.

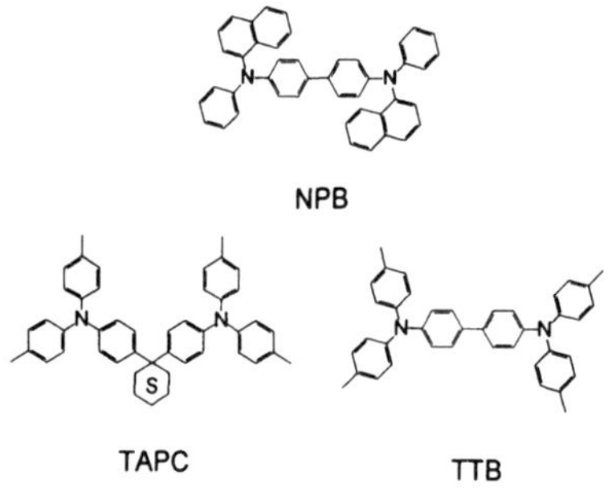

NPB

TAPC   TTB

Figure 2. Molecular structures of the materials used in this work.

Ultraviolet photoemission spectroscopy (UPS) is performed in an analysis chamber with a base pressure ~4 x 10^{-11} Torr. We prepare single layers of NPB or TAPC on indium tin oxide (ITO)-coated glass and a bilayer of TAPC and NPB on a silver substrate in a deposition chamber. The deposition chamber (~8 x 10^{-10} Torr) connects to the analysis chamber via a turntable transport chamber (~1 x 10^{-10} Torr).

Redox potentials are measured using Osteryoung Square Wave Voltammetry at 5 kHz at a 25 μm Pt ultramicro disc electrode in methylene chloride containing 0.1 M tetrabutylammonium fluoroborate as the supporting electrolyte. A three-electrode setup with a Pt wire as the counter electrode and a saturated calomel electrode (SCE) as the reference electrode is used.

RESULTS and DISCUSSION

Typical photocurrent transients of TAPC (6 μm)|TTB (2 μm) and NPB (0.9 μM)|TAPC (4.9 μm) bilayers are displayed in Fig. 3. In the former case, the first and second plateaus represent hole motion through the TAPC and TTB layers, respectively. The photocurrent increases when holes inject from TAPC into TTB. The ends of the plateaus represent the transit times for each layer. The magnitudes of the photocurrent are proportional to the hole mobilities in the two layers, which indicates a complete hole injection in the time scale of the measurement. This is also substantiated by comparing number of carriers in each layer (from integrating the photocurrent-time curve). For the latter case, the photocurrent increases when holes inject from NPB into TAPC. Note that the photocurrent does not reach a plateau in TAPC. Apparently, the trailing tail of the transient charge sheet is still in the NPB layer when the leading edge has started to exit from the TAPC layer into the Au electrode.

To obtain a second flat plateau in the photocurrent transients, the transit time for the second layer cannot be much smaller than that of the first layer. Figure 4 displays photocurrent transients of a NPB (0.9 μm)|TAPC (9.6 μm) bilayer and a NPB (0.9 μm) single layer. A simulation based on convolution of photocurrents from the NPB and a TAPC (9.6 μm) single

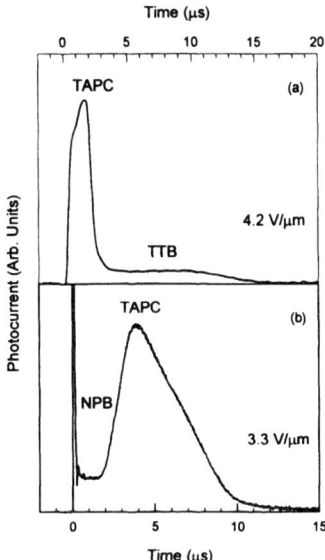

Figure 3. Photocurrent transients of
TAPC (6 μm)|TTB (2 μm) and NPB
(0.9 μm)| TAPC (4.9 μm) bilayers at
296 K

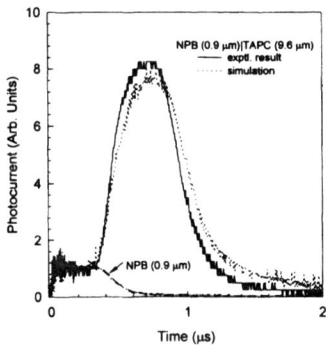

Figure 4. Photocurrent transients of a
NPB (0.9 μm)|TAPC (9.6 μm) bilayer
and a NPB (0.9 μm) single layer. A
simulation based on convolution of
photocurrents from the NPB and a
TAPC (9.6 μm) single layer is also
shown. The field strength is 28 V/μm
and temperature 296 K.

layer is also shown. The rise and decay of the
photocurrent in the bilayer are similar, but
somewhat faster than those of the simulation.
The discrepancy could be due to uncertainty in
the thicknesses of different samples or the
dielectric constants of NPB and TAPC.

Figure 5 displays UPS spectra of NPB (4
nm), TAPC (4 nm), and 4 nm-TAPC on 25 nm-
NPB. The low end high energy edges of NPB
and TAPC is essentially identical, which
indicates that the IP of these materials are very
similar. However, when a thin layer of TAPC
covers NPB, photoelectrons shift to higher
binding energies (ca. 0.5 eV). We attribute the
effect to surface charging as a result of trapped
holes at the TAPC|NPB interface.

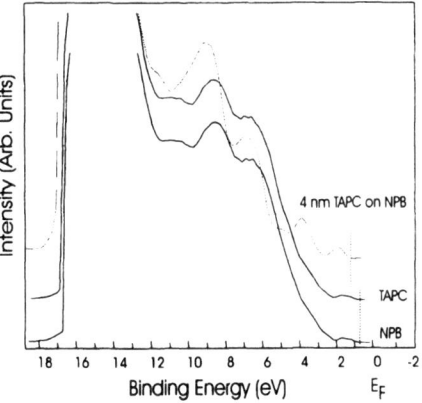

Figure 5. HeI (hv=21.2 eV)
photoemission spectra of NPB, TAPC,
and TAPC on NPB. The horizontal scale
is the binding energy with respect to the
Fermi level of Au.

Square-wave voltammograms of NPB,
TAPC, and TTB are shown in Fig. 6. In general,
these molecules show thermodynamically
meaningful redox properties. Reversible one-
and two-electron E_{ox} are observed. Since these
materials have similar molecular structures,
based on the UPS and electrochemical results,

we obtain a conversion formula: IP ~ 4.5 eV + eE_{ox}, for the first electron transfer E_{ox} (measured in methylene chloride) vs. SCE. The E_{ox} and IP of these molecules are listed in Table I.

Table I. Ionization and oxidation potentials.

	IP (eV)	E_{ox} vs. SCE (V)
TAPC	5.40 ± 0.05	0.91
NPB	5.41 ± 0.02	0.88
TTB		0.76

Figure 7 displays photocurrent transients of NPB|TTB and TTB|NPB bilayers with the same thickness in each layer. The former represents hole injection favorable energetically, whereas, in the latter, holes are expected to be impeded by a ~0.15 eV barrier. The relative magnitudes of the NPB and TTB photocurrents in the two cases are consistent with complete injection from NPB into TTB and partial injection from TTB into NPB.

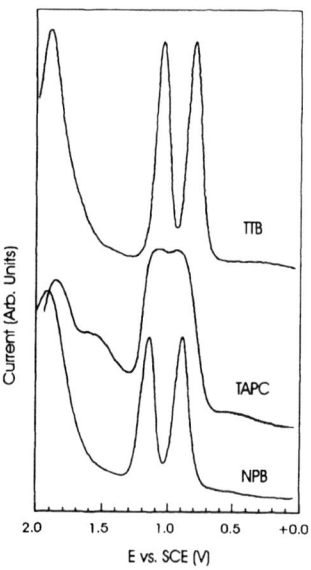

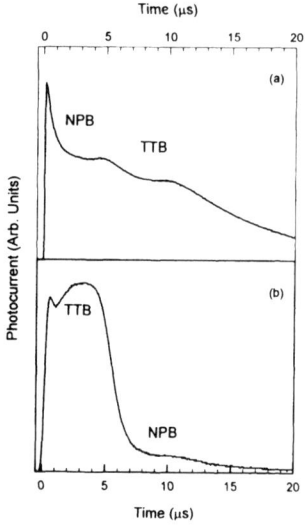

Figure 6. Square-wave voltammograms of NPB, TAPC, and TTB.

Figure 7. Photocurrent transients of NPB (2 μm)|TTB (2 μm) and TTB (2 μm)|NPB (2 μm). The field strength is 5.1 V/μm and the temperature 296 K.

CONCLUSIONS

There are increasing research interests in heterojunctions involving organic materials. We demonstrate here that transient photocurrent measurements provide valuable information on charge injection properties of organic-organic interfaces. The results are consistent with the relative energy levels obtained from photoemission and electrochemical measurements.

ACKNOWLEDGMENTS

Portions of this work were supported by National Science Foundation (CHE-9120001 and CTS-9311741).

REFERENCES

[1] P. M. Borsenberger and D. S. Weiss, *Organic Photoreceptors for Imaging Science and Technologies*, Marcel Dekker, New York, 1992.

[2] S. A. Van Slyke, C. H. Chen, and C. W. Tang, Appl. Phys. Lett. **69**, 2160 (1996).

[3] J. R. Sheats, H. Antoniadis, M. Hueschen, W. Leonard, J. Miller, R. Moon, D. Roitman, and A. Stocking, Science **273**, 884 (1996).

[4] S. Ducharme, J. C. Scott, R. J. Twieg, and W. E. Moerner, Phys. Rev. Lett. **66**, 1846 (1991).

[5] A. Dodabalapur, L. Torsi, and H. E. Katz, Science **268**, 270 (1995).

[6] L. -B. Lin, S. A. Jenekhe, R. H. Young, and P. M. Borsenberger, Appl. Phys. Lett. **70**, 2052 (1997).

[7] L. -B. Lin, R. H. Young, M. G. Mason, P. M. Borsenberger, and S. A. Jenekhe, Appl. Phys. Lett. submitted.

OBSERVATION OF THE TEMPERATURE DEPENDENCE OF THE DYNAMICS OF PHOTOEXCITED STATES IN PRISTINE TRIS(8-HYDROXYQUINOLINE) ALUMINUM (ALQ₃)

Richard Priestley[a,c], Ardie D. Walser[a], and Roger Dorsinville[a,b,c]

[a] Department of Electrical Engineering, The City College and Graduate Center of the City University of New York, 140 Street and Convent Ave., New York, NY 10031.

[b] New York State Center for Advanced Technology for Ultrafast Photonic Materials and Applications at CUNY.

[c] Center for Analysis of Structures and Interfaces(CASI); The City College of New York.

ABSTRACT

We have investigated the temperature dependence of the dynamics of radiative excited states in pristine thin films of tris(8-hydroxyquinoline) aluminum (Alq₃). By measuring the transient photoluminescence (PL) response with subnanosecond resolution, our results revealed an increase in the radiative excited state lifetime and fluorescence quantum yield with decreasing temperature from 300K to 77K. At low temperature we observed a decrease in the bimolecular recombination rate constant, singlet exciton diffusion coefficient and diffusion length. A singlet exciton trapping model is used to explain these results.

INTRODUCTION

Since the first observation of its efficient luminescence[1], tris(8-hydroxyquinoline) aluminum (Alq₃) has been widely used as the emitting layer in organic light emitting diodes (OLED's).[2,3] In fact, it has become the prototypical material among numerous organic electroluminescent systems. The luminescence from this material originates from the radiative decay of singlet excitons formed by photoexcitation, or in the case of electroluminescence (EL), by charge carrier injection. Prior to decay, diffusion and interaction of these excitons with trapping sites can occur and result in a reduction in the PL quantum efficiency or EL quantum yield of Alq₃. Nevertheless, very few publications have been reported regarding the temperature dependence on trapping of singlet excitons.

In previous work, Abe *et al.*[4] observed an increase in the PL and EL intensity at low temperatures in Alq₃ based devices. They associated this increase with a reduction in the nonradiative transition probability typical of molecular crystals at low temperatures.[5] However, no analysis on the dependence of temperature on the excited state lifetime was performed.

An investigation into the nature of trapping sites in Alq₃ has been reported by Burrows *et al.*[6] where they suggested that charge transport in Alq₃ based OLED's is limited by a large density of traps. They determined that a distribution of trapping sites located at 0.15 eV below the LUMO level exists in Alq₃ and the EL observed is

695

produced by the recombination of trapped electrons with holes injected by the hole transport layer. However, they do not address the changes we observed in the behavior of singlet excited states at various temperatures.

In this letter we have performed an investigation on the temperature dependence of the time resolved luminescence of Alq$_3$. We have observed for both low temperature and low excitation, an increase in the singlet excited state lifetime which we associate with a reduction in singlet exciton trapping. In addition, we will demonstrate that for low temperature and high excitation, the mutual annihilation of singlet excitons is lowered. For both low and high excitation, an increase in the PL quantum efficiency was observed with decreasing temperature.

EXPERIMENT

The metal chelate compound Alq$_3$ was synthesized at the Eastman Kodak Co. and purified by the train sublimation technique. Experiments were performed on 1-2 μm thick films thermally evaporated onto quartz slides. Samples were mounted in a liquid nitrogen pour-fill Dewar and investigated within the temperature range of 300 - 77K. An actively-passively mode locked Nd:YAG laser producing 25 psec pulses at 355nm was used for excitation. The excitation density ranged from 4x10^{13} - 2x10^{15} photons/cm^2, allowing us to observe the intensity dependence of the transient behavior. The time resolution for our experimental setup was 500 psec.

RESULTS

For low excitation intensity (< 4x10^{13} photons/cm^2) and under room temperature conditions, the photoluminescence response of Alq$_3$ was observed to have a single exponential decay with a lifetime of 15-16 nsec. This is consistent with our previous measurements[7] and others made by Tang et al.[2] in thin films and toluene solutions. Reducing the temperature to 77K resulted in an increase in the PL decay time to about 20 nsec.

For the case of high excitation intensity (2x10^{15} photons/cm^2), figure 1 shows the transient PL response of Alq$_3$ measured at 300K and 77K. At 300K, we observed a PL decay which consisted of a fast and slow component. Such response was not observed at low excitation. In Alq$_3$, we have previously associated a change in the PL decay from single exponential to nonexponential with exciton-exciton annihilation which occurs at high excitation intensities.[7] At low temperature (77K) an increase in both PL decay components was observed.

Along with an increase in the singlet excited state lifetime, an increase in the PL quantum efficiency was also observed as the temperature decreased from 300K to 77K (Figure 2). This occured under both high and low excitation intensities. At a temperature of 77K the PL quantum efficiency increased by a factor of 4.

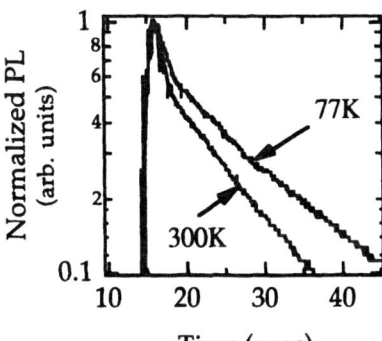

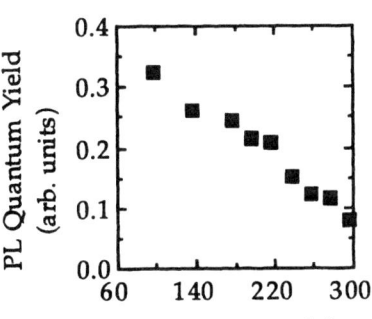

Figure 1. Normalized PL response of Alq$_3$ thin film at 300K and 77K for high excitation.

Figure 2. Dependence of PL quantum yield on temperature for low photoexcitation.

We have interpreted this temperature dependence as a decrease in singlet exciton diffusion at low temperatures which leads to a reduction in exciton trapping. In Alq$_3$, bimolecular recombination processes can be used to determine diffusion parameters. Quantitatively, we can estimate the bimolecular recombination rate constant of singlet excitons by using the following kinetic equation describing the concentration of singlet molecular excitons [S$_1$]: [8,9]

$$d[S_1]/dt = \alpha\, I(t) - k[S_1] - f\, \gamma_{ss}[S_1]^2 \tag{1}$$

where α is the absorption coefficient, I is the excitation intensity, k is the rate of the monomolecular decay of singlet excitons, γ_{ss} is the rate constant for bimolecular recombination, and f is a parameter that takes into account losses due to autoionization and direct nonradiative relaxation to the ground state (S$_0$) from states above the singlet exciton state.

Using equation (1) to fit the experimental PL data shown in figure 1, we extracted the bimolecular recombination rate constant at both 300K and 77K. Theoretical fits provided estimates of $\gamma_{ss} = (8\pm2.5)\times10^{-12}$ cm^3 sec^{-1} and $\gamma_{ss} = (4\pm2)\times10^{-12}$ cm^3 sec^{-1} at 300K and 77K, respectively. These results indicate a reduction in the bimolecular recombination rate constant by a factor of two at low temperature (77K).

From the values obtained for γ_{ss}, we can estimate the diffusion coefficient for singlet excitons, D$_s$, from the equation:

$$\gamma_{ss} = 8\pi D_s R + 8\pi R^2 (D_s k)^{1/2} \tag{2}$$

where R is the spherical reaction radius. [10,11] Taking R = 10Å, we obtained values for D$_s$ within the range of $(3\pm1)\times10^{-6}$ cm^2 sec^{-1} at 300K and $(1.6\pm0.8)\times10^{-6}$ cm^2 sec^{-1} at

77K. The diffusion length of singlet excitons, L, was estimated from the following equation:

$$L \cong \sqrt{Z\tau D_S} \tag{3}$$

where τ is the singlet lifetime and Z is equal to 6, 4, or 2 for three-, two-, or one-dimensional diffusion, respectively.[8] Assuming that exciton diffusion is three-dimensional in isotropic Alq_3 films, one obtains L=(53 ± 9)Å at 300K and L=(37 ± 10)Å at 77K.

Our experimental findings indicate that the diffusion of singlet excitons is reduced at low temperatures. This prevents mobile excitons from reaching trapping sites within the Alq_3 sample. These trapping sites can be defects caused by dislocations within the lattice or guest molecules which act as impurities.[12] Our results are consistent with the assumption that exciton diffusion to randomly distributed traps is the deactivation channel involved in the nonradiative recombination process.

CONCLUSION

In conclusion, we have demonstrated that the excited state lifetime of Alq_3 thin films is temperature dependent. An increase in the excited state lifetime suggests that exciton diffusion to trapping sites is the primary channel for nonradiative energy dissipation. The reduction in the exciton diffusion coefficient and diffusion length at low temperatures supports the exciton trapping model and explains the increase in the PL quantum efficiency we observed at 77K.

ACKNOWLEDGEMENTS

We are grateful to Dr. Ching Tang of Eastman Kodak for providing us with the Alq_3 samples. Author R.P. would like to thank the Center for Analysis of Structures and Interfaces (CASI) for its support. This work was sponsored by NSF, the New York State Technology Foundation, and by the City College Faculty Award Program.

REFERENCES

1 C.W. Tang and S.A. VanSlyke, Appl Phys. Lett. 51, 913 (1987).

2 C. W. Tang, S. A. VanSlyke and C. H. Chen, J. Appl. Phys. 65, 3610 (1989).

3 P. E. Burrows, L. S. Sapochak, D. M. McCarty, S. R. Forrest and M. E. Thompson, Appl. Phys. Lett. 64, (1994).

4 Y. Abe, K.I. Onisawa, S. Aratani and M. Hanazono, J. Electrochem. Soc., 139, 641 (1992).

5   K.C. Kao and W. Hwang in Electrical Transport in Solids , (Pergamon, New York, 1981) , p. 467.

6   P.E.Burrows, Z.Shen, V. Bulovic, D.M. McCarty and S.R.Forrest, J. Appl. Phys. **79**, 7991 (1996).

7   I. Sokolik, R. Priestley, A.D. Walser, R. Dorsinville and C.W. Tang, Appl. Phys. Lett. **69**, 4168 (1996).

8   M. Pope and C.E. Swenberg in Electronic Processes in Organic Crystals , (Oxford University Press, New York, 1982), p. 158.

9   C.E. Swenberg and N.E. Geacinotv in Organic Molecular Photophysics , edited by J.B. Birks (Wiley-Interscience, London, 1973), p. 495.

10   A.J. Campillo, R.C. Hyer, S.L. Shapiro and C.E. Swenberg, Chem. Phys. Lett. **48**, 495 (1977).

11   U. Goesele, Chem. Phys. Lett. **43**, 61 (1976); **46**, 196 (1977).

12   J.B. Birks in Photophysics of Aromatic Molecules , (Wiley-Interscience, London, 1970), p. 533.

TEMPERATURE DEPENDENCE OF THE ELECTRICAL CONDUCTIVITY OF POLY(BENZO[1,2-b:4,5-b'] DITHIOPHENE-4,8-DIYL VINYLENE) AND POLY(DODECYLTHIOPHENE)

R. C. Hyer, R. G. Pethe, T. Yogi, S. C. Sharma, Dept. of Physics, Martin Pomerantz, J. Wang, R. L. Meline, and R. L. Elsenbaumer, Dept. of Chemistry & Biochemistry, University of Texas at Arlington, Arlington, TX 76019, and R. C. McCullough, Dept. of Chemistry, Carnegie Mellon University, Pittsburg, PA.

ABSTRACT

We present results for the electrical conductivity (σ) of thin films of poly(benzo[1,2-b:4,5-b']dithiophene-4,8-diyl vinylene) (PBDV) and poly (dodecylthiophene) (PDDT) as a function of temperature in the range 15-295K. The polymers were doped with $FeCl_3$ and PF_6 which resulted in electrical conductivities differing by two orders of magnitude at room temperature. We examine three sets of $\sigma(T)$-data by using the variable-range hopping (VRH) model that predicts a linear relationship between $\ln(T^{1/2}\sigma)$ and $T^{-1/4}$. We observe a change in the slope of the $\ln(T^{1/2}\sigma)$ vs $T^{-1/4}$ relationship in all three samples at low temperatures. We also analyze the temperature dependence of the resistivity of PBDV by using the thermal fluctuation-induced tunneling model.

INTRODUCTION

Conjugated polymers are known to exhibit remarkable electrical properties. There has been a substantial interest in understanding physical processes that are responsible for the mobility of the charge carriers and related temperature dependence of the electrical conductivity in conducting polymers.[1-9] Unlike saturated polymers, many conjugated polymers are known to exhibit conducting behavior on redox doping, with *dc* electrical conductivities as high as $10^5 \Omega^{-1}$ cm^{-1} measured for polyacetylene. A linear chain structure makes them additionally interesting from the point of view of reduced dimensionality which is expected to influence their electronic properties. The electrical conductivity of the conjugated polymers results from mobile charge carriers introduced into their π electronic configuration. At sufficiently high doping levels, the transport of charges is believed to occur mainly along the conjugated chains and *via* inter-chain hopping processes. The temperature dependence of the electrical conductivity of such materials is usually described by using the Variable Range Hopping (VRH) model. Whereas this model qualitatively describes the temperature behavior of the conductivity, the resulting values of $N(E_f)$ and α^{-1} have been found to be unphysical in several cases.[10-12] The intrinsic electron-phonon scattering, scattering of the charge carriers off defects, dimensionality and microstructure of the film are among important parameters that could significantly affect transport properties of these materials.[13,14]

In an earlier paper, we had discussed $\sigma(T)$ and microscopic structure of poly(pyrrole tosylate) and poly(pyrrole fluoride).[10] These early data had provided useful information about what appeared to be a difficulty in applying the VRH model to $\sigma(T)$ in poly(pyrrole tosylate) and poly(pyrrole fluoride). Specifically, the temperature dependence of σ in both of these conducting polymers could not be explained satisfactorily by the VRH model. For example, a fit of the VRH model to the temperature dependence of the conductivity had provided results for the bipolaron localization lengths of 0.02 Å and 6 x 10^{-8} Å for poly(pyrrole tosylate) and poly(pyrrole fluoride), respectively. The results for the density of states at the Fermi energy were 1 x 10^{27} and 3 x 10^{29} eV^{-1} cm^{-3}, respectively. Obviously, these values have no physical significance. Similar results

Mat. Res. Soc. Symp. Proc. Vol. 488 © 1998 Materials Research Society

3×10^{29} eV^{-1} cm^{-3}, respectively. Obviously, these values have no physical significance. Similar results have also been reported by others.[11, 12] For example, in the cases of oxidized and substituted poly(pyrrole), α^{-1} has been determined to fall within a range from 10^{-4} Å to 3×10^{-9} Å. The corresponding values for $N(E_f)$ range from 7×10^{33} to 3×10^{46} eV^{-1} cm^{-3}. Here we report on the $\sigma(T)$ measurements made on FeCl$_3$ and NOPF$_6$ doped thin films of PBDV and PDDT for temperatures in the range 15-295K. These $\sigma(T)$ data on three different samples, covering a wide range of temperatures, enable a more detailed analysis than was possible in our earlier work.[10]

EXPERIMENTAL

Thin films (thickness ≤ 2 μm) of poly(benzo[1,2-b:4,5-b']dithiophene-4,8-diyl vinylene) (PBDV) were prepared by the precursor polymer route starting with a soluble precursor polymer and heating it at 573 K for two to three hours in a vacuum oven containing a residual reducing atmosphere of 10% H$_2$ and 90% Ar.[20] The conversion of the precursor polymer sulfonium salts provides improved conjugated material by reducing or eliminating oxygen defects (C=O or OH) in the polymer chains.[22, 23] In this conversion the extent of the reaction was followed by UV-vis spectroscopy (λ_{max} = 503 nm, 2.465 eV) and the optimum conditions were as stated above.The band gap, obtained from the band-edge was 1.89 eV and thermogravimetric analysis (N$_2$ atmosphere) showed the onset of decomposition at about 643 K with only 10% weight loss at 858 K and 20% weight loss at 973 K. The precursor polymer could be cast into films from aqueous solution, converted thermally to PBDV, and then doped with various doping (oxidizing) agents. When the burgundy films of PBDV were doped with FeCl3, light blue-green films resulted with conductivities up to 60 Ω^{-1} cm^{-1}.

The dc electrical conductivity measurements were made by using the standard $four$-$probe$ technique. The electrical contacts to the polymer films were made by using a colloidal solution of silver particles as an adhesive layer between the film and lead wires. Upon drying, the sample was mounted into a Displex closed-cycle helium cryostat. The temperature of the samples was computer controlled with an uncertainty ≤ 1 K and was varied over a range of 15-295K. A Keithley Model 236 Source Measure Unit (SMU), interfaced to PC, was used to collect the current-voltage (I-V) data by operating the SMU in the $Source$-$Current$ and $Measure$-$Voltage$ mode in the $four$-$probe$ configuration. Ample time (≥ 1 hr) was allowed at each new setting of the temperature to ensure that the sample had reached thermal equilibrium prior to making conductivity measurements. The I-V data exhibited Ohmic behavior for currents between 100 - 500 μA at temperatures > 150 K and for currents in the range 10 - 90 μA at lower temperatures. The electrical conductivity was calculated by using the thin film approximation,

$$\rho = \sigma^{-1} = (\pi t / \ln 2)(V / I) \tag{1}$$

where ρ is the resistivity (Ω-cm) and t is the thickness of the film (t $< <$ $four$-$probe$ spacing). The measurements, made for both cooling and heating cycles, showed no hysteresis effects.

RESULTS AND DISCUSSION

The conductivities of FeCl3 and NOPF$_6$ doped PBDV and PDDT are shown as a function of temperature in Fig. 1.

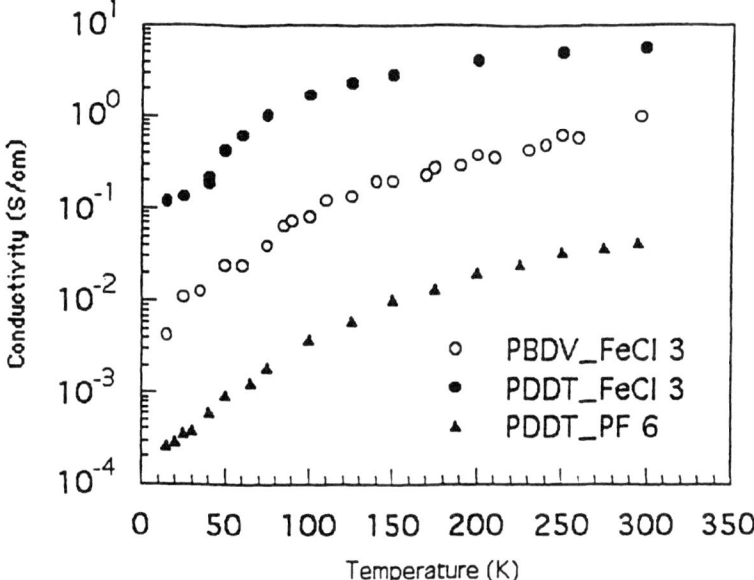

Figure 1. Temperature dependence of the electrical conductivity for FeCl$_3$-doped PBDV and PDDT and for PF$_6$-doped PDDT polymers.

Over the range of temperatures studied the conductivity of FeCl$_3$-doped PBDV increases with temperature by almost three orders of magnitude; we measure σ ~ 4x10^{-3} and 1 Ω^{-1} cm^{-1} at 15 K and 295 K, respectively. In the case of FeCl$_3$-doped PDDT, the conductivity changes from about 0.1 Ω^{-1} cm^{-1} at 15K to 5 Ω^{-1} cm^{-1} at 295K. In PF$_6$-doped PDDT, we obtain 2.5x10^{-4} Ω^{-1} cm^{-1} at 15 K and 3x10^{-2} Ω^{-1} cm^{-1} at 295K. The σ(T) data in doped semiconducting materials are usually explained by using the concept of variable range hopping of the bipolarons.[13-14] In the simplest application of the VRH model, a phonon-assisted hopping rate of the charge carriers is considered between two localized states; one filled at or slightly below the Fermi level, E$_f$, and another empty above E$_f$. These energy states are assumed to be separated in energy by Δ and in space by R. The hopping rate of the electrons between neighboring states is determined by the following three factors: i) exp(-Δ/kT), which is proportional to the probability of finding a phonon of energy Δ at a temperature T, ii) exp(-2αR), which measures overlap between electron wavefunctions of the form exp(-αr), and iii) an attempt frequency, ν$_0$, which depends on the electron-phonon coupling and phonon density of states. The temperature dependence of the conductivity is given by,

$$\sigma = \sigma_0 \exp(-A / T^{1/4}),$$ (2)

where, $\sigma_0 = e^2 \nu_0 [N(E_F) / 32 \pi \alpha k T]^{1/2}$, (3) and $A = 2.1[\alpha^3 / kN(E_F)]^{1/4}$. (4)

From equations (2) - (3), $\sigma T^{1/2} = e^2 v_0 [N(E_F) / 32\pi\alpha k]^{1/2} \exp(-A / T^{1/4})$. (5)

Following equation (5), we plot $\ln(T^{1/2} \sigma)$ vs $T^{-1/4}$ for all three samples in Fig. 2. This figure shows that in each case, the conductivity data do not follow the same linear relationship over the entire range of temperatures. In each case, we observe a change in the slope of the linear relationship at around a temperature that varies with the sample. The "low-temperature" and the "high-temperature" lines meet at a temperature of about 57 K in the case of PBDV. The slopes obtained from the least-squares fits made to data on either side of 57K are significantly different. This results in significantly different values of N(Ef) and α^{-1}. For example, in the case of PBDV we obtain: 1) N(Ef) = 10^{17} eV^{-1} cm^{-3} and α^{-1} = 329 Å for T < 57K and 2) N(Ef) = 3.7 x 10^{27} eV^{-1} cm^{-3} and α^{-1} = 0.034 Å for T > 57K. Similar results are obtained for PDDT.

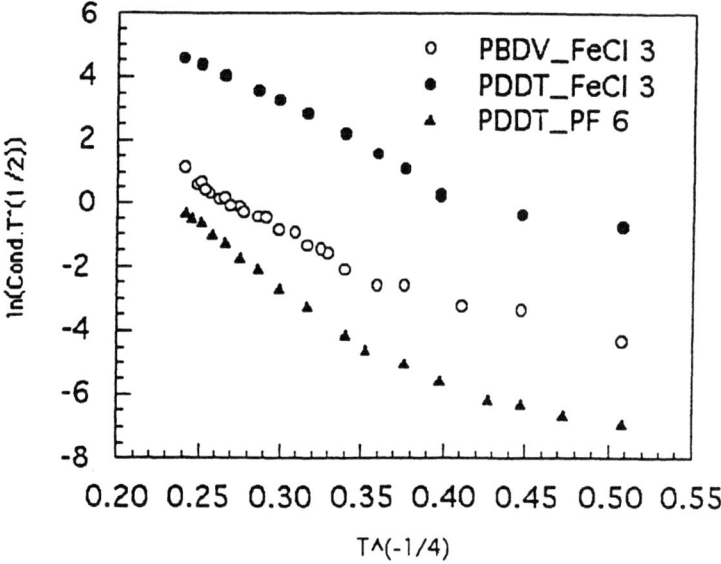

Figure 2. $\ln(T^{1/2} \sigma)$ vs $T^{-1/4}$ for the conductivities of PBDV and PDDT and shown in Fig.1.

Whereas N(Ef) and α^{-1} values at high temperature (T > 57 K) are unphysical, the low temperature results, N(Ef) = 3.4 x 10^{17} eV^{-1} cm^{-3} and α^{-1} = 200 Å, are acceptable. A value for α^{-1}=200 Å may appear much too high from the view point of the range of the electron wavefunction, it compares favorably well with the conjugation length of the polymer. In the case of polyaniline, Wang et. al.[25] have estimated a hopping distance of 110 Å at 25 K and have calculated the most probable hopping distance to be 130 Å. These authors have also estimated about 1.6 states per eV per two rings.[25] From this, $N(E_f) \sim 10^{21}$ states eV^{-1} cm^{-3}. Our low temperature result, $N(E_f) = 3.4 \times 10^{17}$ eV^{-1} cm^{-3} is in reasonable agreement with the $N(E_f)$ value obtained for polyaniline.[25]

In applying the VRH model to $\sigma(T)$ data, it is assumed that transport occurs along a continuous chain of the polymer. However, this may be an over simplification in films in which the growth conditions yield a structure consisting of conducting islands. Under these conditions, the conductivity may be better described as resulting from hopping between islands embedded among insulating barriers. Sheng[24] has developed a model for disordered materials in which thermally activated voltage fluctuations across insulating gaps play an important role in determining the temperature dependence of the conductivity. We have used this model to fit the PBDV data by using the parabolic and "small" barrier approximations. The results of this fit are shown in Fig.3.

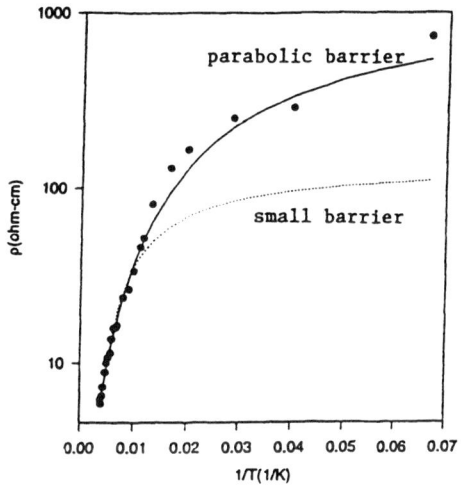

Figure 3. Thermal fluctuation tunneling model fits made to the resistivity data for PBDV.

The parabolic barrier approximation provides a good fit to the resistivity data at all temperatures with reasonable values of the parameters, barrier height ~ 2.8 eV and barrier width ~ 6 Å. We are further investigating these models and the results will be published elsewhere.[25]

CONCLUSIONS

We have studied the temperature dependence of the electrical conductivities of $FeCl_3$ and $NOPF_6$ doped PBDV and PDDT between 15-295K. The VRH model provides reasonable values of $N(E_F)$ and α^{-1} at low temperatures. $N(E_F)$ and α^{-1} values obtained at higher temperatures are unphysical. The thermal fluctuation-induced tunneling model provides a fit to the PBDV data with reasonable values of the barrier parameters.

ACKNOWLEDGMENTS
This research was supported by a grant from the Air Force Office of Scientific Research.

REFERENCES
1. A. J. Heeger, S. Kivelson, J. R. Schrieffer, and W. P. Su, Rev. Mod. Phys., **60**, 781 (1988).
2. J. R. Reynolds, J. Mol. Electronics, **2**, 1 (1986).
3. J. L. Bredas and G. B. Street, Accounts Chem. Res. **18**, 309 (1985).
4. P. Pfluger and G. B. Street, J. Chem. Phys., **80**, 544 (1984).
5. P. Pfluger, U. M. Gubler, and G. B. Street, Solid State. Commun. **49**, 911 (1984).
6. J. C. Scott, J. L. Bredas, K. Yakushi, P. Pfluger, and G. B. Street, Synth. Met., **9**, 165 (1984).
7. K. K. Kanazawa, A. F. Diaz, M. T. Krounbi, and G. B. Street, Synth. Met., **4**, 119 (1982).
8. G. B. Street, T. C. Clarke, M. Krounbi, K. Kanazawa, V. Lee, P. Pfluger, J. C. Scott, and G. Weiser, Mol. Cryst. Liq. Cryst., **83**, 253 (1982).
9. R. Menon, C. O. Yoon, D. Moses, A. J. Heeger, and Y. Cao, Phys. Rev. B **48**, 17685 (1993).
10. S. C. Sharma, S. Krishnamoorthy, S. V. Naidu, C. I. Eom, S. Krichene, and J. R. Reynolds, Phys. Rev. B **41**, 5258 (1990).
11. J. P. Travers, P. Audebert, and G. Bidan, Mol. Cryst. Liq. Cryst., **118**, 149 (1985).
12. A. J. Epstein, H. Rommenlmann, R. Bigelow, H. W. Gibson, D. M., Hoffman and D. B. Tanner, Phys. Rev. Letts., **50**, 1866 (1983).
13. S. R. Elliott, *Physics of Amorphous Materials*, Longman Inc., N. Y., (1983).
14. N. F. Mott & E. A. Davis, *Electronic Processes in Non-Crystalline Materials* Oxford, 1979.
15. M. Pomerantz, J. Wang, S. Seong, K. P. Starkey, L. Nguyen, D. S. Marynick, in *Electrical, Optical, and Magnetic Properties of Organic Solid State Materials*; eds., A. F. Garito, A. K.-Y. Jen, C. Y.-C. Lee, and L. R. Dalton; Proceedings of the MRS: Pittsburgh, PA, **328**, 227 (1994).
16. M. Pomerantz, B. Chaloner-Gill, L. O. Harding, J. J. Tseng, W. J. Pomerantz, Synth. Met., **55**, 960 (1993).
17. M. Pomerantz; B. Chaloner-Gill, L. O. Harding, J. J. Tseng, W. J. Pomerantz, J. Chem. Soc., Chem. Commun., 1672 (1992).
18. M. Pomerantz, J. J. Tseng, H. Zhu, S. J. Sproull, J. R. Reynolds, R. Uitz, H. J. Arnott, M. I. Haider, Synth. Met., **41**, 825 (1991).
19. J. M. Williams, J. R. Ferraro, R. J. Thorn, K. D. Carlson, U. Geiser, H. H. Wang, A. M. Kini, M.-H. Whangbo, *Organic Superconductors (including Fullerenes): Synthesis, Structure, Properties, and Theory*; Prentice Hall, Englewood Cliffs, NJ, 1992.
20. M. Pomerantz, J. Wang, S. Seong, K. P. Starkey, L. Nguyen, D. S. Marynick, *Macromolecules*, **27**, 7478 (1994).
21. W. A. Wampler,K. Rajeshwar,R. G. Pethe,R. C. Hyer,S. C. Sharma, J. M. R. **10**,1811, 1995.
22. F. Papadimitrakopoulos, K. Konstandinidis, T. M. Miller, R. Opila, E. A. Chandross, and M. E. Galvin, Chem. Mater, **6** 1563 (1994).
23. F. Papadimitrakopoulos, T. M. Miller, E. A. Chandross, M. E. Galvin, Polym. Prepr., Am. Chem. Soc. Div. Polym. Chem., **35**, 215 (1994).
24. P. Sheng, Phys. Rev. B **21**, 2180 (1980).
25. T. Yogi, *"Temperature Dependence of the Electrical Conductivity of Polymers"*, M. S. Thesis, University of Texas at Arlington, 1996 (unpublished).

ELECTRON TRANSFER AT THE n-InP | POLY(PYRROLE) INTERFACE

Mark C. Lonergan[*], Christopher T. Cooney, and James A. Myers
Department of Chemistry and The Materials Science Institute, University of Oregon, Eugene, OR 97403-1253

ABSTRACT

Measurements of the barrier height by capacitance-voltage techniques and of the equilibrium exchange current density by current-voltage techniques are performed on the rectifying interface between n-InP and poly(pyrrole) (chemically polymerized and characterized by an electrochemical potential of ≈ 0.2V vs. SCE). The current-voltage data yielded a quality factor of 1.2 ± 0.1 and an equilibrium exchange current density of $(1.2 \pm 0.6) \times 10^{-9}$ A cm^{-2}. The capacitance-voltage data yielded a barrier height of 0.73 ± 0.02 V and measured dopant densities within 15% of the expected value. These data, taken together, are inconsistent with thermionic emission theories developed to describe inorganic semiconductor | metal interfaces and often applied to inorganic semiconductor | doped conjugated polymer interfaces. In particular, the ratio of the rate constant for majority carrier electron capture (surface recombination velocity) at the n-InP | poly(pyrrole) interface to that at n-InP | metal interfaces is found to be $(6 \pm 5) \times 10^{-3}$.

INTRODUCTION

A wide variety of inorganic semiconductor (IS) | doped conjugated polymer (DCP) interfaces have been studied as potential complements to inorganic semiconductor | metal Schottky junctions [1-8]. Qualitatively, the two types of interfaces are similar with examples from both classes exhibiting rectifying current-voltage behavior. Generally, however, the electrical properties of the IS | DCP interfaces are inferior, and consequently, from a purely applications viewpoint, little advantage over IS | metal interfaces has been noted.

Recently, it was reported that the rectifying characteristics of the n-InP | poly(pyrrole) interface can be tuned over a wide range by manipulating the electrochemical potential (Fermi level) of the poly(pyrrole) [9]. Such control was achieved in an operational device resulting in a novel tunable diode with no direct analogue among purely inorganic devices. This single device offers a "turn-on" voltage (more precisely, the forward bias voltage required to pass a particular current) that is tunable over a range of more than 0.6 V. Such control is greater than can be achieved with an entire series of n-InP | metal interfaces and again is achieved within the context of a single device.

Herein, we report a preliminary account of a more detailed study of the electrical characteristics of the n-InP | poly(pyrrole) interface for one particular electrochemical potential of poly(pyrrole) and in the "dry" state (the experiments described above were carried out with the interface immersed in a non-aqueous electrolyte to enable electrochemical manipulation of the conjugated polymer). The specific goal of the work is to test the validity of thermionic emission theories derived for IS | metal interfaces to the n-InP | poly(pyrrole) interface by directly measuring the rate constant for majority carrier electron capture. Because conjugated polymers prepared from different synthetic protocols can exhibit drastically different electrical properties, it is noted that in both the work described above and reported herein, poly(pyrrole) was synthesized using a solution cast route involving the oxidative polymerization of pyrrole by phosphomolybdic acid as described by Freund *et al.* [10].

EXPERIMENTAL

Diode fabrication

The n-InP I poly(pyrrole) diodes were fabricated as follows. A section of n-InP (dopant density: 6.2 x 10^{15} cm^{-3}, thickness = 270µm, typical area ≈ 0.1 cm^2) was ohmically back contacted using Ga/In eutectic. The n-InP was then etched (30s 0.05% Br$_2$/methanol, methanol rinse, 30s 30% NH$_4$OH$_{(aq)}$, water rinse) and blown dry with N$_2$. Immediately following the etch, poly(pyrrole) was solution cast on the n-InP sample using the aforementioned method of Freund and coworkers [10]. Poly(pyrrole) was also cast onto a piece of platinum foil using the same polymerization solution. The electrochemical potential of the poly(pyrrole) fabricated in this manner was measured to be 0.2V vs. SCE in a 0.1M Bu$_4$NBF$_4$ (Bu = butyl) / CH$_3$CN solution using a non-aqueous Ag/Ag$^+$ reference electrode (+0.3V vs. SCE) and its conductivity to be 10-20 Ω^{-1} cm^{-1}. After rinsing with methanol and allowing to dry, the poly(pyrrole) side of the n-InP I poly(pyrrole) diode was ohmically contacted by press contacting it with the poly(pyrrole) covered platinum foil.

Materials

Methanol (J.T. Baker), tetrahydrofuran (Mallinckrodt), phosphomolybdic acid hydrate (Aldrich), and n-InP (Crystacomm) were used as received. Pyrrole (Aldrich) was vacuum distilled prior to use.

Measurements

Current voltage measurements were performed using a Solartron 1287 electrochemical interface and capacitance-voltage measurements were performed using a Solartron 1260 impedance analyzer over the frequency range 1Hz – 1MHz using a 10mV waveform amplitude.

RESULTS

Figure 1 shows the current density (*J*) – voltage (V_{app}) behavior of a typical n-InP I poly(pyrrole) diode. As shown in Fig. 1b, the forward bias current (V_{app}< 0) is modeled well by the diode equation [11] (with the sign convention as described in Fig. 1):

$$J = J_o \left[1 - \exp\left(\frac{-qV_{app}}{nkT} \right) \right] \qquad (1)$$

where J_o is the equilibrium exchange current density, *n* is the diode quality factor, *T* is temperature, *q* is the elementary charge, and *k* is the Boltzmann constant. The very slight "tailing off" at high applied potential is due the series resistance of the junction. Table I summarizes the best fit parameters of the forward bias (V_{app}< 0) data to Eq. 1 for three different n-InP I poly(pyrrole) diodes.

The barrier height of these interfaces were measured using capacitance-voltage (Mott-Schottky) techniques as described by the following relation for an n-type semiconductor [11]:

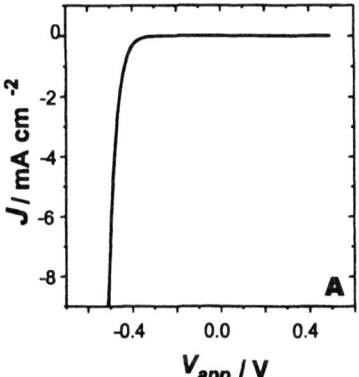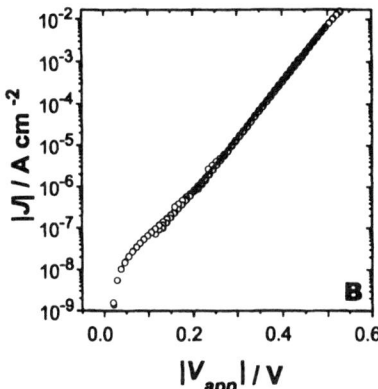

Figure 1: (A) Current density-voltage behavior for an n-InP I poly(pyrrole) interface (sample ml40 in Table I). Negative applied potentials correspond to higher electron energy in the n-InP relative to the PP, and negative currents correspond to net electron flow from the n-InP to the PP. The scan rate for these measurements was 20 mV s^{-1} **(B)** Semi-logarithmic representation of the forward bias data in (A) (note absolute values). The solid line is a fit to the diode Eq. 1 with $n = 1.28$ and $J_o = 1.6 \times 10^{-9}$ A cm^{-2}.

$$C_{sc}^{-2} = \frac{2}{q\varepsilon_s N_D A^2}\left(V_{app} - V_{fb} + \frac{kT}{q}\right) \tag{2}$$

where C_{sc} is the capacitance of the semiconductor depletion layer, ε_s is the static dielectric constant, N_D is the semiconductor dopant density, A is the active device area, and V_{fb} is the flat-band potential, which is related to the barrier height ϕ_b by $\phi_b = V_{fb} + V_n$ where qV_n is the difference between the conduction band edge energy and the Fermi-level .

Measurements were performed over a wide frequency range to identify where the depletion layer capacitance dominates the impedance and to check the validity of the equivalent circuit used to extract C_{sc}. Figure 2a shows the frequency dependent impedance and phase angle for the same interface as in Figure 1. The impedance at semiconductor interfaces can be modeled using the equivalent circuit shown in the inset to Figure 2a where R_{sc} is the interfacial charge transfer resistance at the semiconductor interface, C_{sc} is the semiconductor depletion layer capacitance, and R_s is the series resistance. As shown in Figure 2a, the data for the n-InP I poly(pyrrole) interface are in good agreement with this equivalent circuit. Note that R_{sc} is ideally the intrinsic interfacial charge transfer resistance as embodied by J_o. For the interfaces studied herein, however, undesirable leakage current, although small, dominates R_{sc}. The origin of the undesirable leakage current is uncertain; its magnitude varies from sample to sample and may be due to surface defects or particulate.

A Mott-Schottky plot is shown in Figure 2b where C_{sc}^{-2} is plotted as a function of V_{app} with C_{sc} being calculated according to:

$$\omega C_{sc} = \frac{1 + \left[1 - 4(Z_{im}/R_{sc})^2\right]^{1/2}}{2Z_{im}} \tag{3}$$

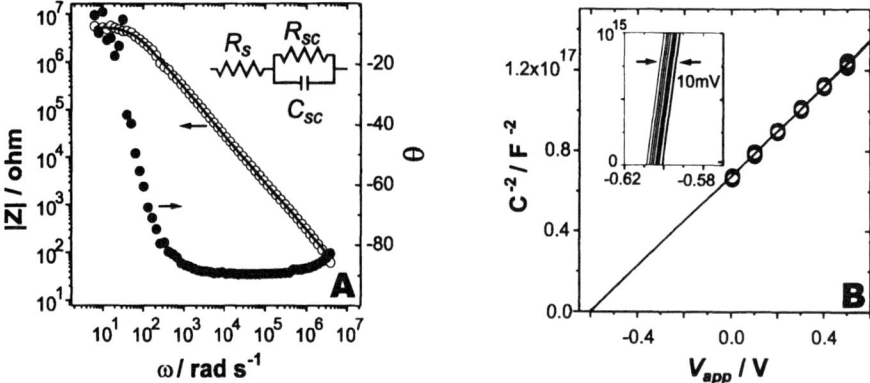

Figure 2. (A) Variable frequency impedance (open circles, left axis) and phase angle (solid circles, right axis) for an n-InP | poly(pyrrole) interface (ml40 in Table I). The solid line represents a fit of the impedance data to the equivalent circuit shown inset with $R_{sc} = 5.2 \times 10^6$ Ω, $C_{sc} = 3.43 \times 10^{-9}$ F, and R_s = 3.0 Ω. **(B)** Mott-Schottky plot for the same interface as in (A). The plot includes data (open circles) for ten frequencies between 1×10^4 rad s^{-1} and 1×10^5 rad s^{-1}. The solid lines represent fits to Eq. 2 and the inset shows a expanded view of the x-intercepts highlighting the minimal frequency dispersion observed.

where $Z_{im} = (|Z| \sin \theta)$ is the imaginary component of the impedance. The frequency range shown in Figure 2b is that where the impedance is most sensitive to the depletion layer capacitance as evidenced by phase angles near $-90°$. In this region, good linearity and very little frequency dispersion (<10mV over an order of magnitude in frequency for the sample in Figure 2b) are observed. Table I summarizes the barrier height and dopant densities extracted for three samples. In calculating the barrier height from the flat-band potential, a value of $V_n = 0.113$ V was used (corresponding to an effective conduction band density-of-states $N_c = 5.36 \times 10^{17}$ cm^{-3}). The slopes of the Mott-Schottky plots yielded dopant densities within 15% of the expected value $(N_D = 6.2 \times 10^{15}$ cm$^{-3})$ further validating these results.

DISCUSSION

A strong analogy has often been drawn between inorganic semiconductor | metal Schottky junctions and inorganic semiconductor | doped (reasonably conducting) conjugated polymer junctions. With regard to the nature of depletion at these interfaces, the analogy is likely good. Both may be considered one-sided in that a substantial depletion layer only exists in the inorganic semiconductor. For the n-InP | poly(pyrrole) interfaces studied herein, support for this assertion comes from the observation that the slope of the Mott-Schottky plot gives the dopant density of the n-InP and also from the fact that interfaces between poly(pyrrole) and degenerately doped $(N_d \approx N_c)$ n-InP do not yield rectifying contacts.

With regard to the kinetics for majority carrier electron transfer, the analogy between inorganic semiconductor | metal and inorganic semiconductor | doped conjugated polymer interfaces may be inappropriate. For majority carrier electron transfer over an interfacial potential barrier, J_o is given by [11, 12]:

Table I: Current-voltage and capacitance voltage data for three n-InP I poly(pyrrole) interfaces.[a]

Sample	Area / cm^2	n	J_o / A cm^{-2}	ϕ_b / V	N_d[b] / cm^{-3}
ml35	0.09(1)	1.13(1)	6.6(9) x 10^{-10}	0.73(1)	5.3(1)x10^{15}
ml38	0.15(2)	1.15(1)	1.5(2) x 10^{-9}	0.71(1)	4.0(1)x10^{15}
ml40	0.15(2)	1.28(1)	1.6(4) x 10^{-9}	0.74(1)	4.5(1)x10^{15}

[a] Numbers in parenthesis indicate error in last digit. For n and J_o, the error estimate is derived from a standard statistical analysis of the non-linear curve fit to the diode equation. For ϕ_b and N_d, the error estimates are the standard deviation of ten measurements covering an order of magnitude in frequency.
[b] From slope of Mott-Schottky plot. Manufacturer's specification 6.22 x 10^{15} cm^{-3}.

$$J_o = qk_n n_s = qk_n N_c \exp\left[\frac{-q\phi_b}{kT}\right] \tag{4}$$

where k_n is the rate constant for majority carrier capture by the contacting material (surface recombination velocity), and n_s is the surface electron concentration. For many inorganic semiconductor I metal interfaces, Bethe's thermionic emission theory is valid [13, 14]. In this case, $k_n = k_n^{TE} = \bar{v} / 4$ where $\bar{v}$ is the average thermal velocity of the electrons in the n-type semiconductor, *independent* of metal. For Schottky junctions, Eq. 4 is generally rewritten in terms of the Richardson constant $A^* = qk_n^{TE}N_c / T^2$ yielding the more familiar expression [13, 14]:

$$J_o = A^*T^2 \exp\left[\frac{-q\phi_b}{kT}\right] \tag{5}$$

For n-InP, $A^* = 9.25$ A cm^{-2} K^{-2} at room temperature [11]. Although Eq. 5 has often been assumed for inorganic semiconductor I doped conjugated polymer interfaces, its validity has been questioned. In particular, unlike k_n^{TE}, the surface recombination velocity at IS I DCP interfaces may depend on the DOS profile of the conjugated polymer at the band edge energy relevant for majority carrier transfer [15] – in much the same way as observed for certain semiconductor I liquid redox couple contacts [12].

With an independent measure of the barrier height from capacitance-voltage techniques, k_n for the n-InP I poly(pyrrole) interface can be calculated directly and compared with k_n^{TE} thereby assessing the validity of Eq. 5. Such an analysis based on the data in Table I yields $k_n / k_n^{TE} = (6 \pm 5)$ x 10^{-3}. Hence, thermionic emission theory with k_n given by the thermal velocity of electrons in the n-type semiconductor does not apply to the n-InP I poly(pyrrole) interface as fabricated herein. Perhaps, the most obvious explanation for the slower kinetics at the n-InP I poly(pyrrole) interface relative to n-InP I metal interface is the presence of a insulating interfacial layer. In particular, the oxidative polymerization carried out on the surface of the n-InP in the presence of oxygen and water may lead to surface oxidation of the n-InP and hence to a tunneling barrier. One of the characteristic signatures of a thin oxide layer is deviation of the diode quality factor from the ideal value of 1 due to a potential drop across the oxide layer. Although the quality factors observed herein are excellent for IS I DCP interfaces, they do not rule out the presence of an oxide layer contributing to the slow kinetics. For instance, Card and Rhoderick measured the effects of thin oxide layers on n-Si I metal interfaces [16]. They observed that an oxide layer 19Å thick can result in $k_n / k_n^{TE} = 5$ x 10^{-3} and a quality factor of 1.2. Similar results have been reported for oxide layers on n-InP [17, 18]. As mentioned above, a second possibility for the slower kinetics observed at the n-InP I poly(pyrrole) interface relative to n-InP I metal interfaces is the much lower density of states (at the conduction band edge energy of n-InP) expected for poly(pyrrole) relative to metals. X-ray photoelectron spectroscopic characterization of the n-InP surface after exposure to the polymerization solution is currently being pursued to help distinguish between the two possibilities mentioned above.

In summary, the n-InP I poly(pyrrole) interface as fabricated herein exhibits good rectification properties and much slower electron transfer kinetics than observed at n-InP I metal interfaces. These results indicate that the application of thermionic emission theories where k_n is given by the thermal velocity of majority carriers in the semiconductor cannot, in general, be applied to IS I DCP interfaces.

ACKNOWLEDGEMENTS

We are grateful to the Dreyfus Foundation and the University of Oregon for support of this work. M.C. Lonergan is grateful to the National Science Foundation (DMR) for a CAREER award. C.T. Cooney and J.A. Myers are also grateful to the National Science Foundation for support through the Research Experience for Undergraduates Program.

REFERENCES

1.   M. Ozaki, et al., *Appl. Phys. Lett.* **35**, 83-85 (1979).
2.   O. Inganas, T. Skotheim, I. Lundstrom, *Phys. Scripta* **25**, 863-867 (1982).
3.   O. Inganas, T. Skotheim, I. Lundstrom, *J. Appl. Phys.* **54**, 3636 (1983).
4.   A. J. Frank, S. Glenis, A. J. Nelson, *J. Phys. Chem.* **93**, 3818-3825 (1989).
5.   A. Watanabe, S. Murakami, K. Mori, Y. Kashiwaba, *Macromolecules* **22**, 4231-4235 (1989).
6.   Y. Renkuan, Y. Hong, Z. Zheng, Z. Youdou, *Synth. Metals* **41-43** (1991).
7.   A. Turut, F. Koleli, *J. Appl. Phys.* **72**, 818-819 (1992).
8.   M. J. Sailor, F. L. Klavetter, R. H. Grubbs, N. S. Lewis, *Nature* **346**, 155-157 (1990).
9.   M. C. Lonergan, *Science* **278**, 2103.
10.   M. S. Freund, C. Karp, N. S. Lewis, *Current Separations* **13**, 66-69 (1994).
11.   S. M. Sze, *Physics of Semiconductor Devices* (Wiley, New York, 1981).
12.   N. S. Lewis, *Annu. Rev. Phys. Chem.* **42**, 543-580 (1991).
13.   H. A. Bethe, "MIT Radiation Lab. Rep" (1942).
14.   E. H. Rhoderick, R. H. Williams, *Metal-Semiconductor Contacts*. P. Hammond, R. L. Grimsdale, Eds., Monographs in Electrical and Electronic Engineering (Oxford University Press, Oxford, ed. 2nd, 1988), vol. 19.
15.   A. Kumar, W. C. A. Wilisch, N. S. Lewis, *Crit. Rev. Solid State Mater.* **18**, 327-353 (1993).
16.   H. C. Card, E. H. Rhoderick, *J. Phys. D* **4**, 1589 (1971).
17.   G. Eftekhari, B. Tuck, D. de Cogan, *J. Phys. D* **16**, 1099 (1983).
18.   Y. S. Lee, W. A. Anerson, *J. Appl. Phys.* **65**, 4051 (1989).

THE DIELECTRIC RESPONSE OF 8-HYDROXYQUINOLINE ALUMINUM USED IN MULTILAYER ORGANIC DEVICES

A. NIKO, C. HOCHFILZER, T. JOST, W. GRAUPNER, G. LEISING
Institut für Festkörperphysik, Technische Universität Graz, Petersgasse 16, A8010 Graz, Austria

ABSTRACT

We present optical measurements of 8-hydroxyquinoline aluminum (Alq_3) a metal-chelate complex which is an important material used for multilayer light emitting diodes. Thin films of Alq_3 produced by vacuum deposition were investigated by transmission and reflectance spectroscopy, and the dielectric coefficients in the near IR to UV-Vis energy range were calculated. Kramers-Kronig analysis was used to determine the real and imaginary dielectric spectral dependence of the thin films. Values obtained by internal field distribution measurements in organic double layer LED's are compared to those determined from optical data. A near-IR value of epsilon (real component) close to 3 was obtained from optical measurements.

INTRODUCTION

8-hydroxyquinoline aluminum (Alq_3) is an important material used for device applications in various multilayer structures. It has been both used as an active layer for light emitting diodes (LED's) [1], and is a useful material as a charge transporting layer in LED's [2]. In particular the combination with conjugated organic materials allows a large scope for device improvement and color tuning required for successful applications in this rapidly expanding field. The methylated ladder type polymer (m-LPPP) [3] is an ideal candidate for multilayer devices made by evaporating organic molecules on top of the m-LPPP layer. The optical characterization of the dielectric function of materials for LED devices is of fundamental importance, as the emission characteristics of the emitting material depend on the intrinsic dielectric response [4]. In regions where the absorption is stronger, the emission color itself can be altered by self-absorption effects. This is even more important in multilayer structures where the position of the emitting zone in a device controls the emission color [5].

For Alq_3, thin films, such as those incorporated into LED's, were investigated in the UV-Vis to near-IR spectral region in transmission and reflectance in order to determine the dielectric function (ε) & complex refractive index ($\tilde{n} = n + iK$) of the material. We also present internal field distribution measurements of multilayers of Alq_3 and m-LPPP devices, which also allows us to determine the dielectric coefficients of these two materials.

EXPERIMENTAL

UV-VIS spectroscopy

In order to determine the complex dielectric function of Alq_3, thin films produced by vacuum deposition onto quartz substrates were investigated by optical spectroscopy. Quartz is an ideal substrate for this purpose due to its transparency in the full UV-Vis region. UV-Vis spectroscopy was done using a Lambda-9 spectrophotometer, in the wavelength range 200-2500 nm at normal incidence for transmission (T), and near normal (5°) in reflectance (R) using a specially designed reflectance sample holder. This requires the calibration of the mirrors making up the sample holder beforehand. Both measurements were done on the same portion of the thin film, so that the spectra are *homologous*. For the characterization, a bare quartz spectrum was also measured, as we require the optical constants of the substrate for the calculations

Shown below are the two spectra for a thin film (160 nm thickness) of Alq$_3$ on quartz. What is immediately apparent is the clear interference pattern in both R and T spectra, where the minima and maxima in R & T show the same spacing. This is a good indication that the film surface is homogeneous, indicating good film quality. Below 250nm, the films do not transmit any light, which means any characterization below this wavelength is problematic.

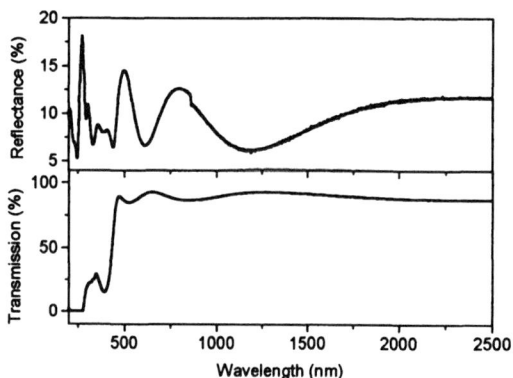

Figure 1: UV-Vis transmission and reflectance spectra of a 160 nm thick Alq$_3$ film

With the two measurements, we can solve Fresnel equations for a thin film on substrate model in order to determine the optical coefficients of a layer, as long as we have knowledge of the film thickness and the refractive index of the substrate [6, 7]. The substrate data was determined from solving Fresnel formulae for a film substrate (quartz has negligible absorption, and a spectrally nearly constant refractive index $n_{subs} = 1.48$). The film thickness was determined in-situ during the vacuum deposition process, and confirmed by the Tolansky method.

Internal Field distribution measurements

The internal field distribution of LED's in a multilayer structure of ITO/m-LPPPP/Alq$_3$/Al:Mg on glass was investigated by electromodulation transmission/absorption spectroscopy [4]. Devices with the same thickness of the m-LPPP layer but varying thickness of transport layer Alq$_3$ were characterized. These devices show a strong dependency of onset voltage, emission color and quantum efficiency with Alq$_3$ film thickness [8].

By applying a suitable model for the dielectric response of the multilayer device, we can arrive at the relationship between the dielectric coefficients of the two materials, with the knowledge of the internal field distribution of the multilayer and the film thickness of each layer.

Taking the electromodulation signal at a particular energy (460nm, the position of the first peak in the electromodulation spectrum of m-LPPP) as a function of the applied electric field in forward and reverse bias directions [9], the relationship between the internal field distribution in the m-LPPP layer and the thickness of Alq$_3$ was investigated. There is no contribution to the electromodulation signal from the Alq$_3$ layer at 460nm, so we directly access the field in m-LPPP.

Considering a model of two series capacitors to describe the multilayer device, we can write that the ratios of potential differences across each layer (U_1 & U_2) are inversely proportional to the capacitance (C_1 & C_2) in each layer. This model can be applied when operating the device in forward bias direction, where we have the system of two dielectric layers separated by a potential barrier which hinders movement of charges. In reverse bias direction, an entirely different effect must be considered, as the m-LPPP/Alq$_3$ boundary is no longer a barrier for charge carriers, which does not allow a simple model of this type to be applied [4]. Hence the field in the m-LPPP layer (E$_2$) can be written as follows for forward bias (ITO is anode, Mg:Al is cathode):

$$E_1 = \frac{U_1}{d_1} = \frac{1}{d_1} \frac{U}{\left(1 + \dfrac{\varepsilon_1 d_2}{\varepsilon_2 d_1}\right)} \qquad \text{—}\!\!\bigg|\bigg|\!\!\text{—}\!\!\bigg|\bigg|\!\!\text{—} \atop C_1 \qquad\quad C_2 \qquad\qquad (1)$$

- considering a layer of m-LPPP with thickness d_1 and an Alq$_3$ layer of thickness d_2, we can determine the relationship between their respective dielectric coefficients. The static dielectric coefficient of m-LPPP layers has been determined by spectral ellipsometry [10], but we have used the same UV-Vis characterization as for ALq$_3$ layers to confirm this experimental data.

RESULTS AND DISCUSSION

UV-VIS spectroscopy

Equations for a film on a substrate were constructed for normal transmission and near normal reflectance, considering multiple reflections and interference in the thin film (no interferences were included in the substrate, thickness 1mm). The equations for reflection and transmission are functions of: wavelength (λ), n_{film}, K_{film}, N_{subs}, K_{subs}, d_{film}, d_{subs} [7]. In a mathematical routine these equations were solved numerically, as they are not explicitly solvable, in order to extract the two variables which are required: the real and imaginary component of the refractive index. This process is done over the entire spectral range measured in order to get the spectral optical coefficients for Alq$_3$. A Kramers Kronig integration of the imaginary component was used to overcome the problems of multiplicity in the real component values determined, originating from the interferences in the spectra. With Kramers Kronig integration we require an integration from zero to infinite energies, so a suitable extrapolation outside the spectral range was used [6]. For low energy, the value of K goes to zero, while for higher energy, the extrapolation can only be reasonably done for a smaller range with a drop to zero, while further out the significance of the extrapolation becomes less important.

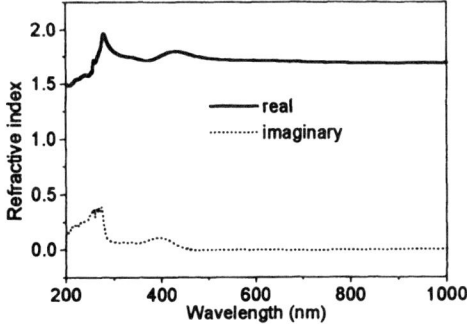

Figure 2: Real and imaginary refractive index of Alq$_3$ film determined from inverting the reflectance and transmission spectra shown in Figure 1.

Shown above is the result of the refractive index inversion of the optical spectra for the Alq_3 thin films using the methods described above. Above 480nm, the imaginary component becomes vanishingly small, and the real component approaches a constant value ($n = 1.65$). Close to the maximum absorption ($\lambda = 400$nm), we see a peak in n at 1.8. Below 250nm, where absorption is highest, the true value for K is difficult to determine accurately, and is best found for thinner films. From the refractive index, the imaginary dielectric function can easily be determined by the relationships. $\varepsilon_1 = n^2 - K^2$, and $\varepsilon_2 = 2nK$. In the following figure is shown the dielectric function of Alq_3 as found by this substitution. We see that for λ going to infinity, a value of $\varepsilon_1 = 3$ is approached.

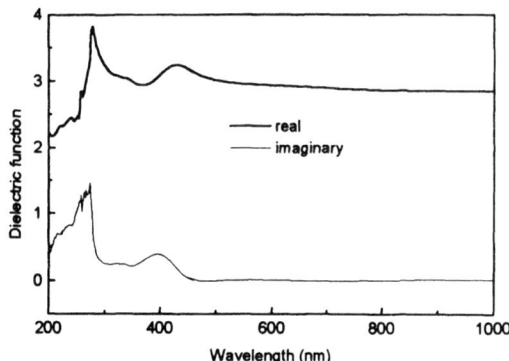

Figure 3: Dielectric function of Alq_3 film determined by inversion of reflectance & transmission spectra in figure 1.

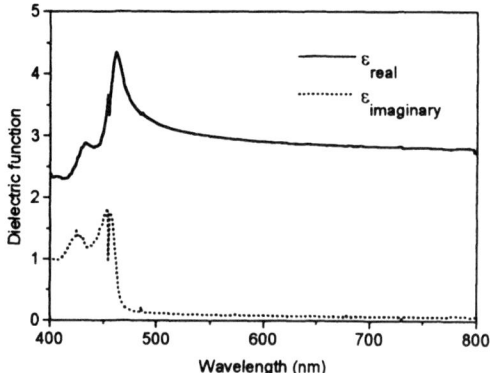

Figure 4: Dielectric function of m-LPPP film determined by inversion of reflectance & transmission spectra.

Using the same UV-Vis spectroscopy methods, single layers of m-LPPP on quartz were investigated to determine the dielectric function, which is illustrated in the figure above. The m-LPPP film investigated here was of similar thickness (200nm), and Kramers-Kronig integration was used to determine the real from the imaginary component. The dielectric functions are in good agreement with values determined by ellipsometry with variable wavelength [10]. Once again here we see a value for ε_1 approaching 3 to higher wavelengths when extrapolating the curve in figure 4 to infinity.

Internal field distribution measurements

m-LPPP/Alq$_3$ devices with a 90nm thick m-LPPP layer and 0, 2, 5, 15, 30 and 50 nm thick Alq3 layers were used for electromodulation measurements [4].The room temperature 460nm electromodulation signal at 10 V voltage amplitude is displayed below in figure 5. Shown is the signal in forward direction determined to access the ratio of epsilons for m-LPPP to Alq$_3$ from the fit through the points shown with equation 1. The field-dependent electroabsorption signal was found to be symmetric for a single layer m-LPPP device which indicates that the field is of the same magnitude in both directions, however the addition of Alq$_3$ layers of different thickness results in a loss of symmetry.

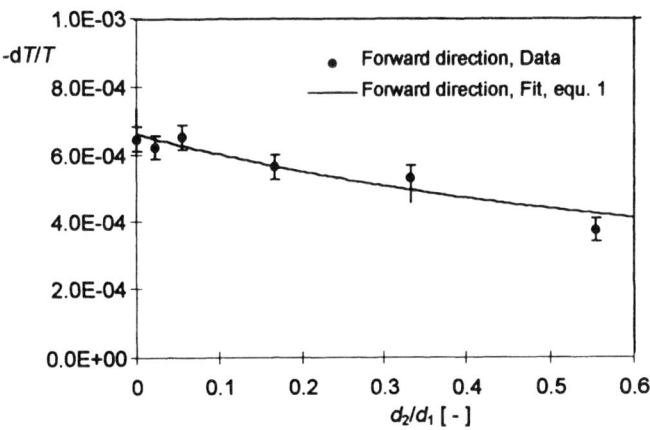

Figure 5: Relationship between the 460nm electromodulation signal, under forward bias of 10V, of a m-LPPP/Alq$_3$ multilayer with varying thickness ratio of two layers d_1 & d_2, shown is a fit using equation 1 in the forward direction.

The fit through the data in forward direction shown in the figure, using the function in equation 1, allows us to determine the ratio of static epsilons between m-LPPP and Alq$_3$ to be 1.0±0.1. This is in good agreement with the values determined by optical spectroscopy methods, where for both m-LPPP and Alq$_3$ a value of ε_1 for higher wavelengths approaching 3 was determined.

It should be noted that when applying reverse bias to the bi-layer device, a different electromodulation signal is seen, which does not follow the same form as that in forward bias when the device emits light. In the reverse bias case, we have to consider that the zone near to the Alq$_3$/m-LPPP interface will be depleted from charge carriers, which is not the case under forward bias, as this is a region where the recombination and luminescence occurs. We therefore have no symmetry in forward and reverse bias electromodulation signals, such as is observed with a single layer m-LPPP device [9].

For thinnest Alq₃ layers this difference in symmetry is not so apparent, as the fields are similar in each direction. Possibly this can be explained by the proximity of the Alq_3/m-LPPP interface with the cathode which may lead to increased reverse injection of electrons into the cathode.

CONCLUSIONS

We have shown the correlation between UV-Vis spectroscopy methods, and electromodulation of a multilayer device determining the dielectric function of Alq_3 and m-LPPP. [1]By applying a simple model of two series capacitors in forward bias direction, we have related the internal field strength of the m-LPPP layer to the electroabsorption signal in forward direction, in order to determine the dielectric response of the bi-layer system. Both materials have a static epsilon value near to 3, which would allow good coupling of layers for multilayer LED applications.

ACKNOWLEDGEMENTS

We are grateful for the financial support from the Spezialforschungsbereich (SFB), *Elektroaktiver Stoffe*.

REFERENCES

1. C. W. Tang, S. A. VanSlyke, Appl. Phys. Lett. **51**, p. 1221 (1987).

2. R. H. Jordan, H. Dodabalapur, M. Strukelj, T. M. Miller, Appl. Phys. Lett. **68**, p. 1192 (1996).

3. U. Scherf, K. Müllen, Makromol. Chem. , Rapid Commun. **12**, p. 489 (1991).

4. C. Hochfilzer, T. Jost, A. Niko, W. Graupner, G. Leising, C. W. Tang, E. Forsythe, Y. Gao, MRS Fall meeting 1997: this Proceedings.

5. S. Tasch, A. Niko, G. Leising, U. Scherf, Appl. Phys. Lett. **68**, p. 1090 (1996).

6. A.Niko, R.Resel, F.Meghdadi, R.Turcu, S.Pruneanu, G.Leising, Synth.Met. **84**, p. 955 (1997).

7. E. Born, Wolf, <u>Fundamentals of optics</u>, Pergamon press, Oxford (1988).

8. C. Hochfilzer, G. Leising, Y. Gao, E. Forsythe, C. W. Tang, Appl. Phys. Lett. (submitted).

9. T. Jost, Diploma thesis, TU Graz 1997.

10. J.Sturm, S.Tasch, A.Niko, G.Leising, E.Toussaere, J.Zyss, T.C.Kowalszyk, K.D.Singer, U.Scherf, J.Huber, Thin Solid Films (1997).

UV Photoemission Study of Interfaces Related to Organic EL Devices

Kiyoshi SUGIYAMA, Kazuhiko SEKI, Eisuke ITO, Yukio OUCHI, and Hisao ISHII
Department of Chemistry, Graduate School of Science, Nagoya University, Furocho,
Chikusa-ku, Nagoya 464-01, JAPAN, seki@chem1.chem.nagoya-u.ac.jp

ABSTRACT

Interfacial electronic structures related to organic electroluminescent (EL) devices were studied by UV photoemission spectroscopy (UPS). The two classes of interfaces studied were: (1) interfaces in a typical multilayer device Al/Alq$_3$/TPD/ITO, where Alq$_3$ is tris(8-hydroxyquinolino)- aluminum, TPD is N,N'-diphenyl-N,N'-(3-methyl-phenyl)-1,1'-biphenyl-4,4'-diamine, and ITO is indium tin oxide, and (2) TTN/metals and TCNQ/metals interfaces, where TTN is tetrathianaphthacene and TCNQ is tetra-cyanoquinodimethane. The UPS studies of the specimen formed by the successive deposition of TPD, Alq$_3$, and Al on ITO revealed interfacial energy diagrams, with the vacuum level shift of - 0.25 eV (downward) and - 0.1 eV (downward) at the TPD / ITO and the Alq$_3$ / TPD interfaces, respectively. The deposition of TTN and TCNQ on metals showed opposite direction of the shift of the vacuum level, with the positive and negative charge at the vacuum side. This can be explained by considering the charge-transfer between the metal and the organic molecule, with these directions being consistent with the electron donating and accepting ability of these molecules.

INTRODUCTION

Recent development of organic electroluminescent (EL) devices has stimulated much research activities [1]. The electronic structure of various interfaces in EL devices is one of the key factors affecting their performance. UV photoemission spectroscopy is a powerful tool to directly examine such interfacial electronic structures [2]. We have been studying various systems formed by depositing organic materials onto metals [3-13], and found that a large shift of vacuum level (change of work function) can occur at the organic/metal interfaces. Such shifts occurs even at the interface between a metal and a long chain alkane n-C$_{44}$H$_{90}$ [10], which is a typical organic insulator. Three possible origins of such shifts were pointed out: i.e. (1) charge transfer between the organics and the metal, (2) electronic polarization of organic molecules due to the image effect by the metal, and (3) other chemical interactions. Similar shift was also found by Park et al. [14] at polymer / metal interfaces.

Other workers also studied the systems formed by depositing metals on organic molecular[15] and polymeric[16] materials, and found that often real chemistry such as covalent formation can occur between the metal atoms and the organic materials.

In this work we report out extension of such studies to two other types of interfaces. One is the model interfaces in the most typical setup of layers, i.e. Al/Alq$_3$/TPD/ITO, where Alq$_3$ is tris(8-hydroxyquinolino)-aluminum, TPD is N,N'-diphenyl-N,N'-(3-methylphenyl)-1,1'-biphenyl-4,4'-diamine, and ITO is indium tin oxide. The other is organic/organic interfaces. The Alq$_3$/TPD interface is an example, and we also examined TCNQ/TTN interface, where TTN is tetrathianaphthacene and TCNQ is N,N,N',N'-tetracyanoquinodimethane. The molecular structures of these organic materials are summarized in Fig. 1.

METHOD

In Fig. 2 we show the principle of the determination of interfacial electronic structures

719

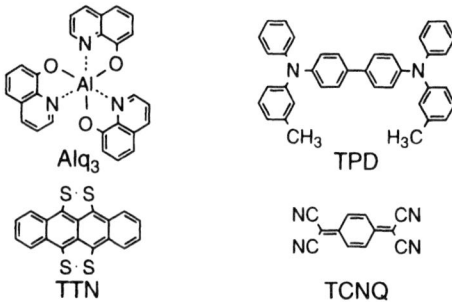

Fig. 1   Organic materials used in this study.

by UPS. At first we measure the UPS spectrum of the clean metal substrate (Fig. 2(a)). Electrons below the Fermi level are excited by UV light, and emitted into vacuum. The kinetic energy distribution of photoelectrons (UPS spectrum) reflects the density of the valence states of the metal. The high energy cutoff corresponds to the emission from the Fermi level, and the low energy cutoff gives the energy of the vacuum level. From these data, we can determine the work function of the metal Φ_m. Next we deposit the organic layer, and measure the UPS spectrum again (Fig. 2(b)). The fastest electron now comes from the highest occupied molecular orbital (HOMO) of the organic molecule. By comparing with the UPS spectrum of the metal, we can deduce the energy separation between the Fermi level of the metal and the HOMO (ε_v^F). Also we can observe the change of the energy of the vacuum level (Δ), and also deduce the energy between the vacuum level and the Fermi level of the metal (ε_{vac}^F). For other interfaces (e.g. organic / organic), we can similarly deduce the energy alignment at the interface.

EXPERIMENTAL

The sample materials of TPD and Alq_3 were supplied from Toshiba R&D Center, and was purified by vacuum sublimation. TTN was supplied by Prof. T. Nogami of University of Electrocommunications, and was used without further purification. TCNQ was commercially supplied from Tokyo Chemical Industry, and was purified by sublimation.

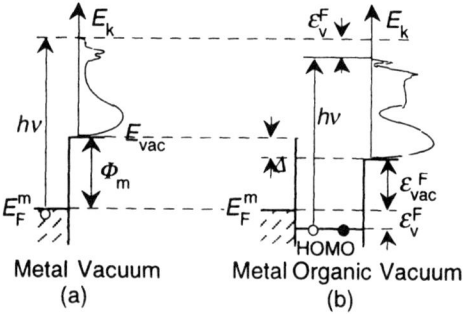

Fig. 2 Principle of the UPS study of an organic/metal interface. (a) Photoemission from the metal. (b) Photoemission from the organic layer deposited on the metal substrate.

The ITO substrate was supplied from Balzers Company (ITO thickness 45 nm, resistance 40 - 50 Ω The ITO substrate was ultrasonically washed with acetone and *i*-propanol, and also by UV irradiation using a lamp (SEN Engineering SUV-40GS) for 5 to 10 minutes under air leading to the cleaning by ozone formation. The work function of ITO cleaned only by organic solvents was about 4.43 eV, while the value after UV cleaning was about 4.75 V. This change was accompanied by the removal of the carbon contamination as observed by XPS.

The UPS spectra were measured on a retarding-field type electron spectrometer with a rare gas discharge lamp emitting HeI light of 21.2 eV. The energy analysis of photoelectrons was performed by ac modulation technique [16] with ac modulation of 0.2 eV peak to peak. The resolution was about 0.2 eV as judged from the spectrum of Fermi edge of Au. The base pressure of the vacuum chamber was 1 x 10⁻⁸ Torr.

Since the UPS signal of the Fermi edge region of ITO was very weak, the location of the Fermi level was determined by measuring the Fermi edge of Au which was electrically connected with the ITO substrate.

RESULTS AND DISCUSSION

In Fig. 3 we show the UPS spectra of TPD with increasing thickness deposited on UV-treated ITO. The origin of the binding energy in the abscissa is the Fermi level of ITO determined as described above. We see that the spectrum changes gradually by TPD deposition, and stop changing at 1 - 1.5 nm. From the shift of the right-hand cutoff, we can estimate the gap between the highest occupied states of ITO and TPD, which corresponds to the hole-injection barrier ε_v^F. From the left-hand cutoff, on the other hand, we can examine the shift of the vacuum level. In Fig. 4, we plot the energies of the vacuum level ε_{vac}^F and the HOMO of TPD relative to the Fermi level of ITO as functions of average film thickness. Most of the downward shift of the vacuum level Δ of about - 0.25 eV occurs abruptly at the interface. The HOMO is at about 0.5 eV below the Fermi level of ITO, and its energy does not depend much on the thickness.

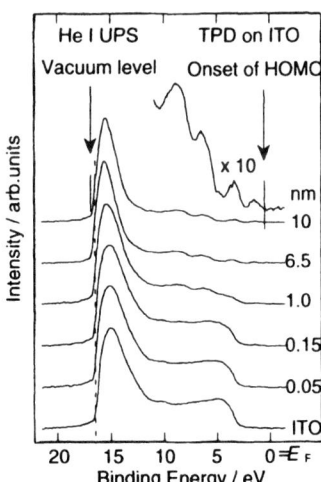
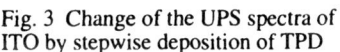
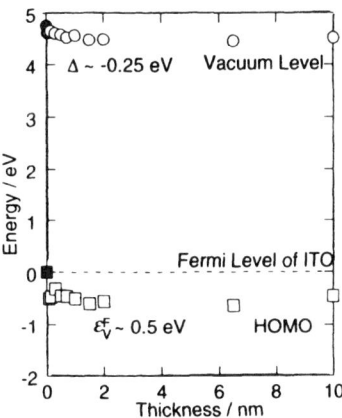

Fig. 3 Change of the UPS spectra of ITO by stepwise deposition of TPD

Fig. 4 The energies of the vacuum level and the HOMO of TPD as functions of the thickness of TPD on ITO.

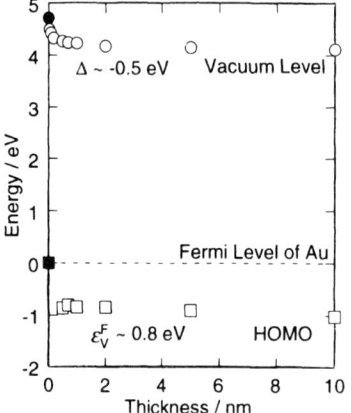

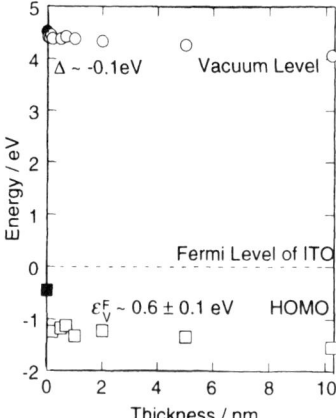

Fig. 5 The energies of the vacuum level and the HOMO of TPD as functions of TPD thickness on Au.

Fig. 6. The energies of the vacuum level and the HOMO of TPD and Alq₃ as functions of Alq₃ on TPD (10 nm) on ITO.

For comparison, we also performed similar experiments for TPD deposited on in-situ evaporated Au film. The results are shown in Fig. 5. Both Δ (- 0.5 eV) and the hole-injection barrier (0.8 eV) are larger than in the case of TPD/ITO interface.

From subsequent UPS measurements at the deposition of Alq3 on TPD (10 nm) formed on ITO, the energy diagram of the Alq₃/TPD interface in Fig. 6 was obtained. We see that the shift of the vacuum level is very small (not exceeding 0.1 eV), while the energy separation between the HOMOs of TPD and Alq₃ is ca. 0.6 eV (within 0.1 eV).

By borrowing the interfacial energy diagram of the Alq₃/Al interface formed by depositing Alq₃ on evaporated Al, we obtain the energy diagram of the whole Al/Alq₃/TPD/ITO system shown in Fig. 7. The injection of hole from ITO to Alq₃ thus occurs through two steps of 0.5 eV and 0.6 eV. We should note, however, these values may be somewhat modified by (1) the exposure to air, as observed at the exposure of porphyrin/metal interfaces to oxygen [6], or (2) the chemical reaction by depositing

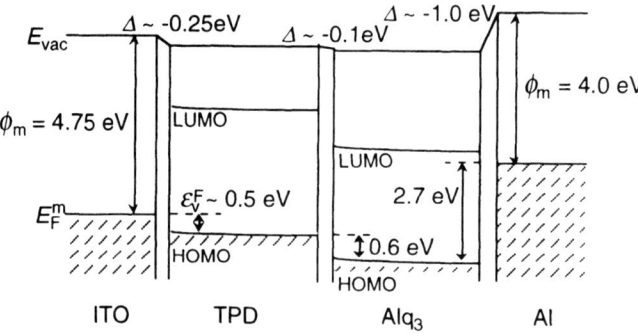

Fig. 7 The estimated energy diagram of an Al/Alq₃/TPD/ITO device. Care must be taken for a direct comparison with the real EL device, in particular for the Al/Alq₃ interface. For details, see the text.

metals on organic materials [15,16]. Thus the results, in particular those for the Al/Alq₃ interface, should be viewed with care.

Next we examine the trend of the vacuum level shift Δ in some more detail. The downward shift of the vacuum level has been observed in many interfaces formed by depositing organic materials on metals such as Au, Ag, Cu, Al, Pb, and Mg. This corresponds to the positive charging of the vacuum side, and was ascribed to the polarization of the electron cloud in the organic molecule probably interacting with the image in the metal substrate or the charge transfer from the molecule to the metal. Exceptional cases of upward shift has only been observed for strong electron donors, where electron is transferred from the metal to the organic molecule to make the vacuum side negative. The present results of downward shift at the TPD/ITO interface is reasonable in view of the moderately electron donating character of TPD.

The negligible shift of the vacuum level at the Alq₃/TPD interface suggests that large interfacial dipole is not formed between organic materials. For examining this point further, we also performed UPS studies on the interface formed by depositing TCNQ on TTN and TTN on TCNQ. The results for TCNQ on TTN are shown in Fig. 8. We now see a small but finite (ca. 0.2 eV) *upward* shift of the vacuum level. This can be ascribed to the electron transfer from TTN, a strong electron donor, to TCNQ, a strong electron acceptor. The negative charge at the TCNQ side is consistent with the upward shift of the vacuum level. For the TTN/TCNQ interface, where the order of deposition was reversed (detailed results not shown), we observed an downward shift, which confirms this interpretation.

CONCLUSION

In this paper we described the UPS studies of organic/ITO and organic/organic interfaces. We observed a shift of the vacuum level also at these interfaces, although the magnitude is not so large as those observed for the interfaces formed among organic materials and common metals we reported previously. The deposition of TPD on ITO showed a downward shift as in most organic/metal interfaces, while almost negligible shift was observed at Alq₃/TPD interface. From these results, the energy diagrams of Al/Alq₃/TPD/ITO device was estimated.

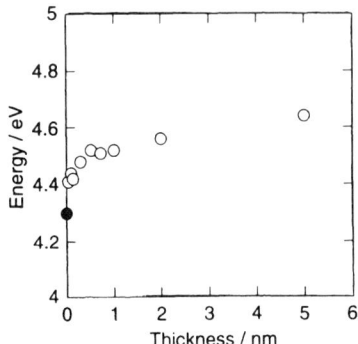

Fig. 8. The energies of the vacuum level of TCNQ as functions of TCNQ thickness deposited on TTN.

ACKNOWLEDGMENTS

We thank Drs. Shun Egusa and Takashi Sasaki of Advanced Research Laboratory of Toshiba R&D Center, Professor Takashi Nogami of the University of Electrocommunications, and Drs. W. Bruetting and M. Meier of Bayreuth University for the generous gift of the samples and ITO substrates. This work was supported in part by the Grant-in-Aid from the Ministry of Education, Science, Sports, and Culture of Japan (Nos. 07NP0303 and 07CE2004) and by the Venture Business Laboratory Program "Advanced Nanoprocess Technologies" of Nagoya University, and the Interdiciplinary Joint Studies Program "Preparation and Characterization of Supramolecular Systems with Combined Functionality " of the Center for Integrated Research in Science and Engineering of Nagoya University.

REFERENCES

[1] For example, N.C. Greenham and R. H. Friend, Solid State Phys., **49**, 1 (1995).
[2] For a review about UPS, *e.g.* Photoemission in Solids, vols. 1 and 2, edited by M. Cardona and L. Ley, (Springer, 1978 , 1979).
[3] S. Narioka, H. Ishii, D. Yoshimura, M. Sei, Y. Ouchi, K. Seki, S. Hasegawa, T. Miyazaki, Y. Harima, and K. Yamashita, Appl. Phys. Lett., **67**, 1899 (1995).
[4] K. Seki, T. Tani, and H. Ishii, Thin Solid Films, **273**, 20 (1996).
[5] K. Sugiyama, D. Yoshimura, E. Ito, T. Miyazaki, Y. Hamatani, I. Kawamoto, Y. Ouchi, and K. Seki, Mol. Cryst. Liq. Cryst., **286**, 561 (1996).
[6] D. Yoshimra, H. Ishii, S. Narioka, M. Sei, T. Miyazaki, Y. Ouchi, S. Hasegawa, Y. Harima, K. Yamashita, and K. Seki, J. Electron Spectrosc., **78**, 359 (1996).
[7] H. Ishii, S. Hasegawa, D. Yoshimura, K. Sugiyama, S. Narioka, M. Sei, Y. Ouchi, K. Seki, Y. Harima, and K. Yamashita, Mol. Cryst. Liq. Cryst., **296**, 427 (1997).
[8] D. Yoshimura, H. Ishii, S. Narioka, M. Sei, T. Miyazaki, Y. Ouchi, S. Hasegawa, Y. Harima, K. Yamashita, and K. Seki, Synth. Metals, **86**, 2399 (1997).
[9] K. Sugiyama, D. Yoshimura, E. Ito, T. Miyazaki, Y. Hamatani, I. Kawamoto, . Ishii, Y. Ouchi, and K. Seki, Synth. Metals, **86**, 2425 (1997).
[10] H. Ishii and K. Seki, IEEE Trans. Electron Devices, **44**, 1295 (1997).
[11] K. Seki, E. Ito, and H. Ishii, Syn. Metals, in press.
[12] H. Ishii, K. Sugiyama, D. Yoshimura, E. Ito, Y. Ouchi, and K. Seki, IEEE J. Selected Topics Quant. Electronics., in press.
[13] K. Seki and H. Ishii, J. Electron Spectrosc., in press.
[14] Y. Park, V. Choong, E. Ettedgui, Y. Gao, B. R. Hsieh, T. Wehrmeister, and K. Muellen, Appl. Phys. Lett., **69**, 1080 (1996).
[15] Y. Hirose, A. Kahn, V. Aristov, P. Soukiassian, Appl. Phys. Lett., **68**, 217 (1996).
[16] W. R. Salaneck, S. Stafstroem, and J. -L. Bredas, Conjugated Polymer Surfaces and Interfaces (Cambridge Univ. Press, 1996).
[17] R.C. Eden, Rev. Sci. Instrum., **41**, 252 (1970).

POLY(2-BUTYNE-1,4-DIYL)S AS PRECURSORS TO

NOVEL SUBSTITUTED POLYACETYLENES

STEVEN K. POLLACK, ASHEBIR FISEHA

Department of Chemistry and the Polymer Science and Engineering Program, Howard
University, Washington, DC 20059, spollack@howard.edu

ABSTRACT

Thermally-induced polymerization of 1,4-diphenyl-1,2,3-butatriene affords poly(1,4-diphenyl-
2-butyne-1,4-diyl) as a soluble high molecular weight material. Structure is characterized by ^{13}C
and vibrational spectroscopy. When treated with base, this material rearranges to a soluble red
poly(acetylene) with a unique substitution pattern.

INTRODUCTION

The polymerization of simple butatrienes has long frustrated spectroscopists in their attempts to
isolate and study the free monomeric molecule[1-6]. Carothers observed this phenomenon and
obtained a patent on the potential use of butatriene as a reactive monomer, not understanding the
nature of the polymer formed[7]. Recently, we have been studying the structure and potential
applications of butatrienes as monomers in addition polymerization[8-10]. In reviewing of the
literature concerning 1,2,3-butatriene, there were reports that in some situations, the normally
white solid formed from its polymerization would convert to a yellow, blue or even black
material[1]. This intrigued us as it implied the formation of a conjugated system. Based on our
previous and ongoing studies, 1,2,3-butatrienes polymerize in the presence or absence of free-
radical initiators via a 1,4 addition to produce a poly(2-butyne-1,4-diyl) structure.

For the case R=H, this creates a polymer which is formally $(C_4H_4)_n$. This is isomeric with the
conjugated polymer poly(acetylene). We reasoned that the formation of colored materials could
be due to rearrangement of the unconjugated, acetylene containing system to the
thermodynamically more stable conjugated polyene. If one could design the butatriene monomer
appropriately, such a rearrangement of the resulting substituted poly(2-butyne-1,4-diyl) could
lead to conjugated polymers with unique electronic and physical properties. In this paper, we
report on a first example of such a rearrangement .

EXPERIMENT

Unless otherwise noted reagents were used as received. All solvents were purified using standard procedures. Melting points were obtained using a Fisher-Johns melting point apparatus and are reported uncorrected. NMR data was obtained on a General Electric QE-300 FT-NMR system equipped with a Tecmag™ data system and were referenced either to an internal TMS standard or to solvent peaks. ESR spectra were obtained both in the solid state and in solution in chloroform using a Bruker ER 2000-SRC spectrometer operating at 9.74 GHz at an RF power of 10 mW, modulation frequency of 100 Khz and a field modulation intensity 2.8 G. UV spectra were obtained using a Perkin Lambda 2 spectrometer. Fluorescence spectra were obtained using a Spex Fluorimax-2 spectrofluorimeter.

Synthesis

1,4-Diphenyl-2-butyn-1,4-ol[11, 12] Under a dry nitrogen atmosphere, 1.57g (16mmol) of trimethyl silyl acetylene was added quickly to a stirred mixture of 1.24g (11.6mmol) of freshly distilled benzaldehyde and 0.208g(0.8 mmol) of tetrabutyl ammonium fluoride in 25 mL of THF at -78°C. The cooling bath was removed after 5 minutes and the mixture was stirred at room temperature until all the benzaldehyde reacts (observation of the C=O band by IR). 7.2 mL(11.6mmol) of 1.6M n-BuLi in hexane was added dropwise over 10 minutes followed immediately with 1.24g(11.6mmol) of benzaldehyde. After complete disappearance of benzaldehyde, 2.4g of potassium fluoride in 25 mL of methanol was added and the mixture was stirred for 6 days. 50 mL of 5% w/v sodium bicarbonate was added and the product mixture was stirred for 15 minutes. The product was extracted several times with ether and the solvent was removed by rotary evaporator leaving the yellow oil. The product mixture was purified by column chromatography using hexane/ethyl acetate as a solvent, giving a white solid, m.p. 142-144°C (yield 25.2%). IR(thin film cast from CH_2Cl_2) 3446(OH), 3055(CH, aromatic), 2986(CH stretch), 2306(acetylene), 1600,1492,1450(aromatic,characterestic), 1423(CH bend) 1285cm^{-1}(CO) 749,698 cm^{-1}(mono-substituted benzene); ^{1}H NMR(CDCL$_3$) δ 2.16-3.29(b, 2H, OH), 4.65,5.5(both s, diastereomeric CH, Ph C̲H(OH)C:::CCH(OH)Ph),7-8ppm(m, 8.95H, 2*Ph); ^{13}C NMR(CDCl$_3$) δ 64.6,65.3(diastereomeric CH, Ph C̲H(OH)C:::CCH(OH)Ph), 86.5(acetylene), 126.6, 126.9, 127.0, 128.2, 128.4, 128.6, 128.99, 140.4 ppm(phenyl).

2,3-Diiodo-1,4-diphenyl-4-Butadiene[13, 14] To a solution of 1.82g (12.9 mmol) of 1-4-diphenyl-2-butyn-1,4-diol in 250 mL of dichloromethane was added a solution of 12.89g (64.4 mmol) of iodotrimethyl silane in 20 mL of dichloromethane at -78°C over a period of 30 minutes, under nitrogen atmosphere. The mixture was stirred for 2 h at -78°C. A saturated solution of sodium bisulfite was added and the mixture was extracted with dichloromethane, dried over magnesium sulfate and the solvent was removed by rotary evaporator. The dark oily product was chromatographed on silica gel using hexane/ethyl acetate yielding a flaky greenish/yellow solid, m.p.=70-72°C (yield 17.2%). ^{1}H NMR(CDCl$_3$) δ 6.95-7.0(s,2.05H, PhC̲H=C(I)C(I)=H̲CPh), 7.2-7.51(m, 10.59H, PhCH=C(I)C(I)=HCP̲h); ^{13}C NMR(CDCl$_3$) δ 99.4 (PhCH=C̲(I)C(I)=HCPh), 136.1(PhC̲H=C(I)C(I)=H̲CPh), 128.3,128.4,141.4-141.7 ppm (P̲hCH=C(I)C(I)=HCP̲h), IR(thin film cast from CCl$_4$) 3050,3021(CH, aromatic), 1598,1492,1442(aromatic,characterestic), 1557(-CH=CH(I)-) 749,698 cm^{-1}(mono-substituted benzene).

Diphenyl Butatriene (DPB, 1) Under a dry nitrogen atmosphere, 2.0 mL (3.314 mmol) of 1.5 M n-BuLi in hexane was added to 1g (2.76 mmol) of diiodo diphenyl butadiene in 30 mL of ether, at -78°C. The mixture was stirred for 2 h at -78°C, warmed up to -50°C and 5 mL of pentane was added and stirring continued for 10 minutes. The temperature was raised to -20°C and 1 mL of water was added. The product mixture was allowed to warm up to room temperature and dried with 1g of anhydrous magnesium sulfate and solvent was removed by rotary evaporator. The product(diphenyl butatriene) was purified(under nitrogen) over Florisil column using degassed hexane (yield 91.9%) . ^{1}H NMR δ 6.35-6.8(s,2.0 H, Ph$\underline{CH}$=C=C=$\underline{HC}$Ph), 7.13-7.62 ppm(m,18.52H, $\underline{Ph}$CH=C=C=HCPh). ^{13}C NMR(CDCl$_3$) δ 109.5,109.6,109.7 (Ph$\underline{CH}$=C=C=$\underline{HC}$Ph), 125-127.8 & 136 ($\underline{Ph}$CH=C=C=HC$\underline{Ph}$, several peaks), 155 ppm(PhCH=$\underline{C}$=$\underline{C}$=HCPh), IR(thin film cast from CCl$_4$) 3050,3021(CH, aromatic), 1598,1492,1442(aromatic,characterestic), 1943 1554(-CH=C=C=CH) 748,688 cm^{-1}(mono-substituted benzene).

Poly(DPB) (2) The monomer 1 (0.352g, 1.81mmol) was dissolved in 10 mL of THF and placed in a sealable vessel. The solution was degassed several times (freeze-pump-thaw, 6 cycles) and thermally polymerized at 120° c for 48 h. The homopolymer which was soluble in THF was precipitated from chlorofom/methanol giving a light yellow precipitate. ^{1}H NMR δ 6.35-6.80 ppm(s,2H,-[-(Ph)$\underline{CH}$:::$\underline{C}$CH(Ph)-]-,7.13-7.62(m, 12H, ,-[-($\underline{Ph}$)CHC:::CCH($\underline{Ph}$)-]-, ^{13}C NMR(CDCl$_3$) δ 45.4-45.9(4 peaks, -[-(Ph)$\underline{CH}$:::$\underline{C}$CH(Ph)-]-,84.8, 84.4(acetylenic (two peaks)), 126.6, 128.7, 138.5,138,7,139.4 ppm , -[-($\underline{Ph}$)CHC:::CCH($\underline{Ph}$) -]-,); IR(thin film cast from CH$_2$Cl$_2$) 3069,3029(CH, aromatic), 2246(acetylene), 1600,1492,1450(aromatic), 749,698 cm^{-1}(mono-substituted benzene).

Rearrangement of Poly(DPB) Under a dry nitrogen atmosphere 60 mg of 2 in 50mL of dry benzene was heated to 75°C. Upon addition of a few drops of saturated potassium t-butoxide in t-butanol, the light yellow solution turned brown and ca 8mL of saturated potassium t-butoxide in t-butanol was added. The mixture was heated for 15 minutes and 25 mL of water was added. The product mixture was extracted with ether, washed several times to remove with water residual t-butanol and dried with anhydrous magnesium sulfate and solvent was removed by rotary evaporator leaving a dark brown solid. ^{1}H NMR 7.13-7.62 ppm(b,-(Ph)CH=CH-). ^{13}C NMR(CDCl$_3$) 125-127.8 & 136 (-(Ph)CH=CH-, broad peaks). IR(thin film cast from CH$_2$Cl$_2$) 3069,3029(CH, aromatic), 1650(C=C), 1600,1492,1450(aromatic,characterestic), 749,698 cm^{-1}(mono-substituted benzene).

RESULTS AND DISCUSSION

We chose 1,4-diphenyl-1,2,3-butatriene (1) as our initial 1,4-disubstituted butatriene. We felt that the presence of the aromatic groups would serve to stabilize the cumulene to enough of a degree that its polymerization could be controlled. Our approach beings the formation of 1,4-diphenyl-1,4-dihydroxy-2-butyne via a two-step one-pot reaction of trimethylsilylacetylene with benzaldehyde[15]. The diol is then treated with iodotrimethylsilane to provide the 2,3-diodo-1,4-diphenylbutadiene and reductive elimination with nBuLi affords the trans-1,4-diphenyl-1,2,3-butatriene in good yield.

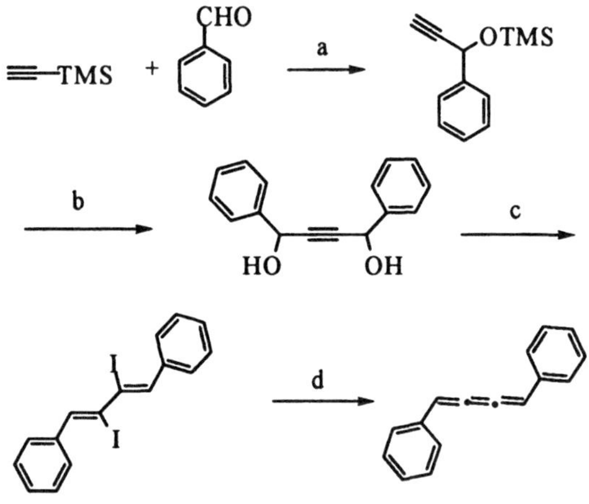

1

a) (n-Bu)₃NF; b)n-BuLi, PhCHO, KF; c) TMSI; d) n-BuLi

When dissolved in THF, degassed and heated to 120°C for 48 hours, a soluble lightly yellow polymer is produced. This solubility is also in contrast to the polymers derived from the monomers mentioned above. Molecular weight as determined by GPC using polystyrene standards is 30,000 g mol⁻¹. Proton NMR analysis shows the expected aromatics peaks due to the benzene rings and a complex multiplet due to the 1,2-diphenylethylidene unit. Note that these two adjacent centers are both chiral, so that we may have a meso or racemic diastereomers at each intermonomer linkage.

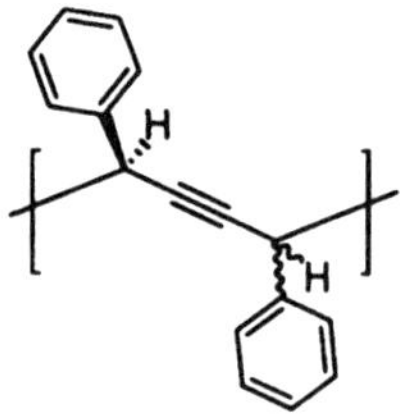

2

This is borne out in the ¹³C spectra (Figures 1 and 2). We note the presence of two acetylenic peaks at 84.8 and 84.4 ppm as well as two closely spaced peaks in the aromatic region at 139.5 and 138.7 ppm, in addition to the other anticipated resonances for the other sp² and sp³ carbons. We attribute these splitting to the proximity of the acetylenic and ipso ring carbon to the chiral center in the meso and racemic diastereomer. The relative heights of these split peaks indicates that one of the two diastereomers is preferred, although at this time, we do not have a definitive assignment. The absence of allenic or vinyl carbons in this spectrum confirms that the polymerization proceeds exclusively via 1,4 addition.

A solution of 2 is light yellow in color. Upon addition of a few drops of saturated potassium t-butoxide in t-butanol, the solution turns immediately dark red. When allowed to react for 15 minutes and then precipitated, the resulting material is dark red. GPC of the isolated red polymer shows little or no change in the molecular weight distribution from the starting polymer **2**. Proton NMR reveals only a broad peak in the aromatic region. Signals due to the benzylic hydrogens are absent. The ^{13}C spectra is equally simplified and give the most dramatic evidence for the nature of the change that has occured.

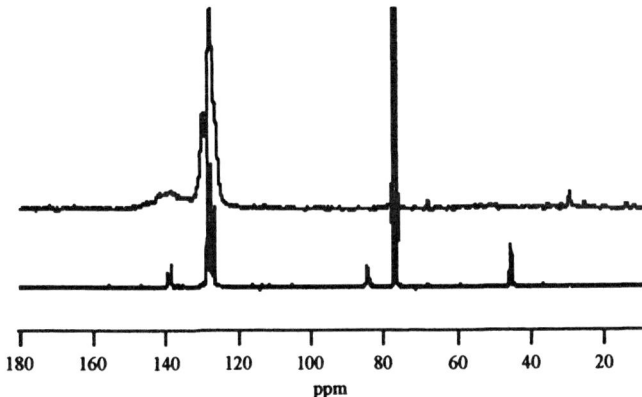

Figure 1. ^{13}C NMR spectra (in CDCl$_3$) of **2** (lower) and **3** (upper)

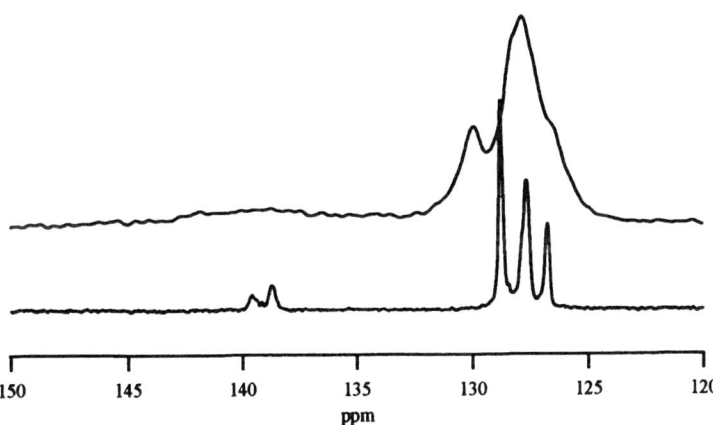

Figure 2. Expansion of aromatic region of Figure 1.

Both the acetylenic carbons and the benzylic carbons are gone and only sp^2 type carbons remain. These latter peaks are somewhat broadened. This and the appearance of the aromatic protons suggested the presence of unpaired spins. The room temperature ESR spectra of **3** exhibits a broad Gaussian peak (both in the solid state and in chloroform solution) with a g ~2.0 and DB$_{pp}$ = 6 G.

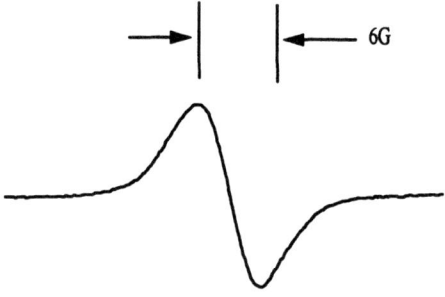

6G

Figure 3.   ESR spectra (in CHCl$_3$) of **3**.

ESR linewidths for undoped poly(acetylene) has been previously observed in the range of 5-10 G at room temperature[16].

UV/vis spectroscopic comparisons of the polymer before and after treatment with base shows the growth of a broad new peak centered at approximately 350 nm in addition to the sharper absorption at 254 nm, due to the aromatic rings.

In addition, fluorescence measurements of the polymers in chloroform with excitation at 350 nm give a broad featureless fluorescence peak at 410 nm for poly(**1**) which shifts to 547 nm after treatment with base.

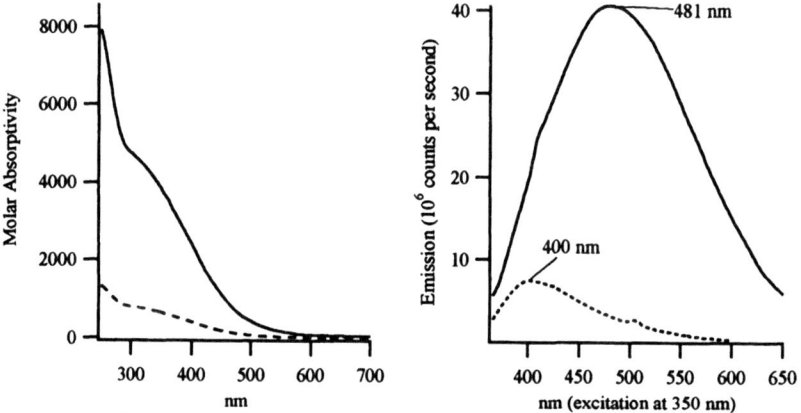

Figure 4.   Absorption and Emmision spectra (excitation at 350 nm) of polymer **2** (dashed) and **3** (solid).

We attribute these observations to a based-catalyzed rearrangement of the non-conjugated, **2** to the conjugated poly(acetylenic) form

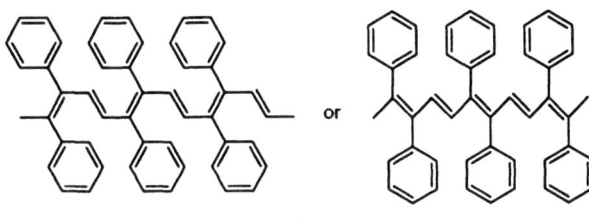

3

the left isomer being the so-called cis-isomer (drawn in the trans-cisoid resonance form) and the right the trans-isomer. Such base-catalyzed rearrangements have precedent for 1,5-dynes and other polyynes, although large excess of base and aggressive conditions must be utilized[17, 18]. It should be noted that the other possible resonance forms for both isomers are not degenerate with those drawn above and may have interesting impact on the electro-optical properties of these polymers.

This polymer can be compared with its isomer, poly(1-phenylacetylene). Since the poly(1-phenylacetylene) is substituted on every third backbone carbon, in the trans-form, there will be severe steric interaction between phenyl rings on adjacent monomers causing this polymer to be non-planar. One can think of the polyacetylene **3** as being the alternating copolymer of acetylene and 1,2-diphenylacetyelene. This substitution pattern for polyacetylene is not readily obtainable by any other route. The ability to utilize benzaldehydes substituted with either electron donating groups or accepting groups in synthesizing the diphenylbutatriene monomer may allow for manipulation of the absorption maximum (the band gap) and potentially, the non-linear optical properties of this conjugated polyene. Work along these lines is currently under way in our laboratories.

CONCLUSIONS

The molecule 1,4-diphenyl-1,2,3-butatriene can be polymerized via thermal polymerization to form a soluble poly(2-butyne-1,4-diyl) of relatively high molecular weight. In turn, using mild conditions, this material can be caused to rearrange to a polyacetylen with a unique substitution pattern. This new route to substituted polyacetylenes may lead to polymers with useful electro-optical properties. Ongoing studies in our laboratories are examining the electro-optical and photophysical properties of these systems.

ACKNOWLEDGMENTS

We wish to acknowledge the Air Force Office of Scientific Research (F49620-95-1-0223) for financial support. We also with to thank Mr. Zhidong Fu for providing GPC analysis and Dr. Neil Blough of the University of Maryland, College Park for aiding us in obtaining ESR spectra of our materials.

REFERENCES

1. W. M. Schubert, T. H. Liddicoet and W. A. Lanka, J. Am. Chem. Soc. **76**, 1929 (1954).

2. E. L. Martin and W. H. Sharkey, J. Am. Chem. Soc. **81**, 5256 (1959).

3. B. Heinrich and A. Roedig, Angew. Chem. Intl. Ed. in English **7**, 375 (1968).

4. E. Kloster-Jensen and J. Wirz, Helvetica Chimica Acta **58**, 162 (1975).

5. F. A. Miller, W. F. Elbert and W. Pingitore, J. of Mol. Struct. **40**, 25 (1977).

6. A. Rogstad, Spectrochemica Acta **36A**, 131 (1980).

7. W. H. Carothers and G. J. Berchet, U.S. Patent No. 2136178 (1937).

8. S. K. Pollack, B. Narayanswamy, R. S. Macomber, D. E. Rardon and I. Constantinides, Macromol. **26**, 856 (1993).

9. V. P. Morris and S. K. Pollack, J. Phys. Chem., B. submitted for publication.

10. S. K. Pollack, A. Fiseha and B. Narayanswamy, Macromol. **29**, 5265 (1997).

11. I. Kuwajima, E. Nakamura and K. Hashimoto, Tetrahedron **39**, 975 (1983).

12. E. W. Colvin, Silicon Reagants in Organic Synthesis, (Academic Press, New York 1988).

13. M. Iyoda, K. Nishioka, M. Nose, S. Tanaka and M. Oda, Chem. Lett. 131 (1984).

14. M. Iyoda, S. Tanaka, H. Otani, M. Nose and M. Oda, J. Am. Chem. Soc. **110**, 8494 (1988).

15. A. Fiseha, PhD. Thesis, Howard University, 1998.

16. P. K. Kahol, G. C. Clark, M. Mehring, in Conjugated Conducting Polymers, edited by H. Keiss (Springer Verlag, New York 1992), p. 217.

17. D. A. Ben-Efraim and F. Sondheimer, Tetrahedron **25**, 2387 (1969).

18. F. Théron, M. Verny and R. Vessière, in The Chemistry of the Carbon-Carbon Triple Bond, Part 1, edited by S. Patai (John Wiley & Sons, New York 1978), p. 381.

POLYMERIC COMPLEXES OF POLYANILINE AS ANTICORROSION COATINGS

R. J. RACICOT*, S. C. YANG**, R. BROWN***
* Department of Chemistry, United States Air Force Academy, Colorado Springs, CO 80840
** Department of Chemistry, University of Rhode Island, Kingston, RI 02881
*** Department of Chemical Engineering, University of Rhode Island, Kingston, RI 02881

ABSTRACT

During the past few years there has been a strong interest in developing conducting polymers as an alternative to the traditional anticorrosion coatings. One of the driving forces for this research comes from the need for an environmentally friendly chromate-free anticorrosion coating for high-strength light weight aluminum alloys. The possibilities for a new scratch-tolerant paint for steel prompted the development of conductive polymer anticorrosion paints.

By molecular engineering, we have synthesized a double-strand polymeric complex of polyaniline that is suitable as an anticorrosion paint on metals in low pH environments. In this article, we will discuss (1) the molecular design for solubility and adhesion, (2) the effectiveness of the electroactive coating under electrochemical impedance tests, and (3) a mechanistic study of the anticorrosion mechanism by examining the polymer/metal interfacial interactions.

INTRODUCTION

A new class of conductive polymer based on polyaniline has been engineered over the past ten years. This new polymer system is a chemical complex consisting of two strands of different polymers intertwined by covalent attraction in a similar manner to a DNA structure [1-3]. The work over the past years with this new system has experimented with a vast array of different polymer strands connected with polyaniline. These double strand complexes have been successful in rendering polyaniline to be soluble in any solvent (polar or nonpolar based) while maintaining it's conductivity. This flexibility in design has opened up new areas of use for conductive polymers that were not easily achievable in the past due to difficulties in processing the polymer. Applications such as rechargeable polymer batteries, signature reduction, static free carpet and tiles and static free electronic parts storage bags are becoming a reality.

One new and exciting area in the use of conductive polymers is as a corrosion resistant coating for steel and aluminum alloys. Past efforts have shown that a conductive polymer coating based on polyaniline can protect steel and aluminum alloy surfaces from corrosion in certain environments [4-15]. The difficulty with some of the single strand based systems of polyaniline, however, is with the processing of the coating and applying it to the surface. The double strand polymer complex offers an advantage over other conductive polymer coatings because the complex can be rendered soluble in any solvent and a paint can be applied to the surface via any typical painting method such as spraying, dipping or brushing. Additionally, the second strand of the complex is a polyacrylic acid. The carboxylic acid groups on this strand provide excellent adhesion properties for the polymer coating on metal surfaces. Furthermore, our testing has shown that the single strand polyaniline coatings are not effective in environments above a pH of 3.0. The polymer turns to a less effective non-conductive blue colored form via the loss of a proton. Our research has shown that the polymer coating must remain conductive to be effective in protecting the metal surface, for both steel and aluminum alloys [14]. The polymer's conductive nature provides for an electroactive interaction with the metal surface and forms a passivation layer between the metal surface and the polymer coating. Therefore, its function is not as a barrier coating, as other polymer type coatings, but as a surface conversion agent, similar to the chromate coatings. Our double strand polymer remained conductive in solutions with pH's as high as 9.5. Although this solution, when casted onto a metal surface, did not provide the best wetting, adhesion or corrosion protection as other coatings we have developed. The best protective coating developed to date has remained conductive at a pH of 6.0, which is right at the edge of the testing solutions pH's, a .5 N NaCl solution. We have observed the slow changing of the polymer from the green state to the blue state in two weeks to one month time. The polymer is shown to protect the metal surface via cyclic, impedance and salt spray testing prior to this color change. We feel confident though, with the right chemical modifications, it is quite feasible to develop a conductive polymer coating that will remain conductive at higher pHs and still offer suitable wetting, adhesion and corrosion protection. When testing in salt environments with a pH below 4.0, our double strand coating offers excellent protection as presented in this paper.

It is well known in the aircraft industry that the Environmental Protection Agency (EPA) has set guidelines for the eventual elimination of chromate conversion coatings for corrosion protection of aluminum alloys [16]. The implications of these guidelines are serious and far reaching. The push is on for the development of a suitable replacement for chromate conversion coatings. Experience with the testing of Alodine 600 and 1200 chromate conversion coatings under cyclic polarization, impedance spectroscopy and ASTM B-117 salt spray tests shows that this is a difficult goal. The chromates are the best and to date nothing seems to be able to perform better or at least as equally as well and have the same ease in applying the coating to the metal's surface. We believe conductive polymer coatings have the potential to perform as well as the chromates with further development work and offer an easy application method to the metal surface. Over the last three years a large amount of data has been collected about the science of applying the double strand conductive polymer as a coating to different aluminum alloys samples and testing under cyclic polarization, electrochemical impedance spectroscopy, ASTM B-117 salt spray exposure, XPS and Auger surface analysis, SEM image analysis and X-ray diffraction analysis. Compelling information has led use to be able to get a firmer understanding of the mechanism of the corrosion protection for conductive polymers.

EXPERIMENT

The electrical conducting polymer was synthesized by a previously reported template-guided polymerization method to obtain a polymeric molecular complex that contains polyaniline and its polymeric dopant poly(acrylic acid)[1]. The dried product was further processed by esterification with methanol to allow conversion of about half of the carboxylic functional groups to methyl ester functional groups. The resulting molecular complex of polyaniline and poly(methylacrylate-co-acrylic acid) is soluble in an organic solvent, such as ethyl acetate. This green colored solution is used as an electroactive coating on metals.

Two types of high-strength aluminum alloys, AA7075-T6 and AA2024-T3, were tested. Test coupons (5 cm x 5 cm square) were first polished with 600 grit silicon carbide paper, then cleaned with a degreaser and de-ionized water. The polyaniline coating solution was casted on the coupon. A uniform film of the conductive polymer id formed after the solvent was evaporated under room temperature.

The chromate conversion coating used in this study was based on the Alodine 600 and 1200 coating process. For the Alodine 600 samples, coupons were placed in a standard solution of chromium trioxide, phosphoric acid and sodium hydrogen fluoride. The alloys were dipped in this solution for 5 minutes to form a uniform green colored sealed oxide coating on the metal surface. The Alodine 1200 samples were commercially prepared by Microfin Corporation of Providence , RI.

EIS has become one of the methods of choice for the study of corrosion on metal surfaces. We use this technique to compare the performance of the double strand conductive polymer coatings against the industry standard chromate conversion coatings in acidic environments on aluminum alloys. Samples were clamped with a 1 cm^2 hollow tube cell and filled with a 0.5N NaCl solution that was buffered to a pH of 3.6. A platinum counter electrode and saturated calomel (SCE) reference electrode were placed in the cell. The electrodes were connected to a EG&G Model 273 potentiostat and a Solartron 1255 frequency response analyzer. A 5.0 mV alternating current (AC) perturbation signal was applied at the open circuit potential. The AC frequency was scanned from 0.003 Hz to 100 kHz. The resulting current and impedance were measured, recorded and displayed in Bode plot format. Equivalent circuit modeling was performed with a non-linear least squares Boukamp software package.

RESULTS

Figure 1 shows Bode plots of impedance and phase angle versus frequency for a 7075-T6 coated alloy and an Alodine 600 chromate conversion coating and an uncoated 7075 as a reference sample at the initial day one test and after 7 days exposure to a 0.5N NaCl solution buffered at pH 3.6.

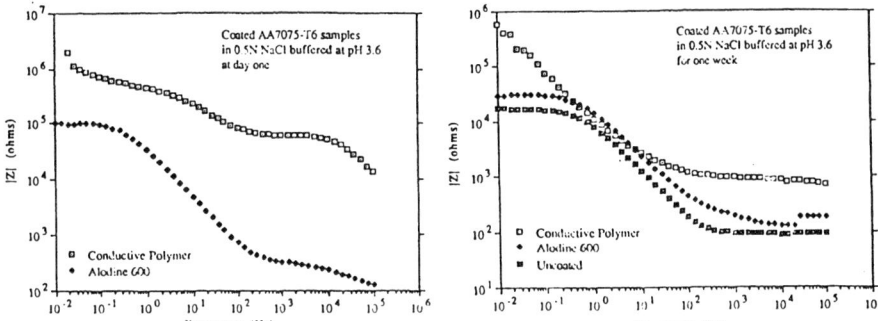

Figure 1. Electrochemical impedance data comparing a PAN/POLYE coated 7075 sample and a chromate conversion coating using the Alodine 600 process tested in an acidic salt solution (0.5N NaCl solution buffered at pH 3.6) The left graph is the initial day one test and the right graph is after one week exposure.

Initially, the Alodine 600 sample originally had a charge transfer resistance close to 10^5, measured as the log impedance at low frequency, yet quickly dropped to the value approaching a typical uncoated aluminum alloy after only one week. This drop in charge transfer resistance is to be expected for a surface converted to an aluminum oxide structure, as the Pourbaix diagram for aluminum shows it's oxide is soluble in solutions with a pH below 4.0. [17] However, the conductive polymer coated sample does not show a drop in charge transfer resistance after exposure to the acidic environment. If fact, continuously run samples have remained above 10^5 for greater than 60 days and continuing. Figure 2 shows the charge transfer resistance fitted from an equivalent circuit model versus time for the polymer coated and chromate coated samples for up to 60 days. The conductive polymer sample maintains a one order of magnitude higher charge transfer resistance than the chromate sample.

Figure 3 shows a similar type EIS Bode plot for the conductive polymer coating on the more corrosive AA2024-T3 alloy versus the Alodine 1200, a better chromate conversion coating, at the initial day one test and after 50 days exposure to the 0.5N NaCl solution buffered at 3.6.

Again, even for the more difficult 2024 alloy, the conductive polymer maintains an order of magnitude higher charge transfer resistance over the better Alodine 1200 sample under long term testing. The removal of the conductive polymer film from the alloy, after testing, reveals no discoloration or evidence of pitting on the surface.

One key issue on conductive polymer coatings concerns the protection mechanism. A question might arise about whether the observed higher impedance in the conducting polymer coated samples is simply due to its ability to serve as a barrier coating and not due to its electroactive nature. There is sufficient evidence that agues against a 'barrier type' protection.

It is well known that polyaniline undergoes a conductor-to-insulator transition when the mobile charge carriers (polarons) are removed by deprotonation of the polymer [18]. For the conducting polymer used in this study, the deprotonation occurs at pH 6.0. A thin film coating will deprotonate to turn into a blue colored coating when exposed to an aqueous environment with a pH higher than 6.0. Upon immersion of the deprotonated sample in a pH 3.6 buffer, the film returns to its green colored conductive state. If the conducting polymer film used in this study was

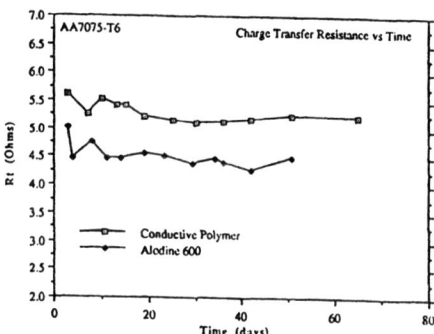

Figure 2. Charge transfer resistance versus time for a double strand conductive polymer coated AA7075-T6 alloy compared with an Alodine 600 coated sample.

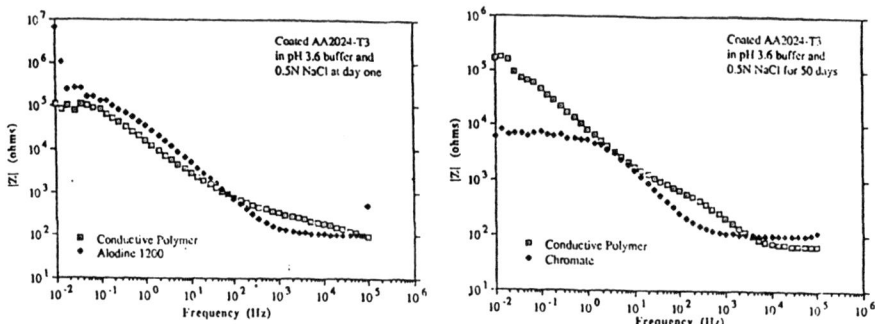

Figure 3. Electrochemical impedance data comparing a double strand conductive polymer coated AA2024-T3 sample and a chromate conversion coating using the Alodine 1200 process tested in an acidic salt solution (0.5N NaCl solution buffered at pH 3.6) The left graph is the initial day one test and the right graph is after exposure to the acidic salt solution for 50 days.

an effective barrier coating, the water and ions would not be able to diffuse into the film to protonate or deprotonate the polymer film. It has been observed that the protonation and deprotonation processes are completed within a short time (two weeks in stagnant solution, two hours in flowing solution) indicating that the coatings used in these tests do not significantly impede the transport of water and ions. The results of these studies suggest that the conductive polymer coating's protective nature is not that of a barrier type coating. Permeability and ionic transport studies have shown that the chemical nature of the conductive polymer coating offers little resistance to the passage of water and ions through the film [19,20].

The reversible conversion property between the conductive and non-conductive states allows for further testing to help answer the mechanism question. If the main reason for observed high impedance originated from its ability to be a transport barrier (despite its ability to allow the transport of protons), the same high impedance should be observed if the coating is converted to its non-conductive form by deprotonation. This contention was tested in an experiment that compared the EIS response of two types of polyaniline coatings of the same thickness. Samples were first prepared in exactly the same manner to form dark green coatings on AA7075-T6 aluminum coupons. A subset of the samples were converted to the blue-colored non-conductive form by washing in de-ionized water or by immersion in dilute (.01M) ammonium hydroxide. These two sets of samples were then immersed in pH neutral 0.5 N NaCl solutions for EIS testing. Figure 4 shows a significant difference between the polymer in its conductive or non-conductive state in the charge transfer impedance. If the high impedance of the green conductive polymer was due to a barrier mechanism, we would expect the same thickness of the blue non-conductive coating to be equally resistant to charge transfer. The experimental results in Figure 4 shows that the non-conductive blue form of the same polymer completely lost its low frequency impedance. This experimental result speaks against the supposition that the observed high impedance of the green conductive state is due a barrier mechanism.

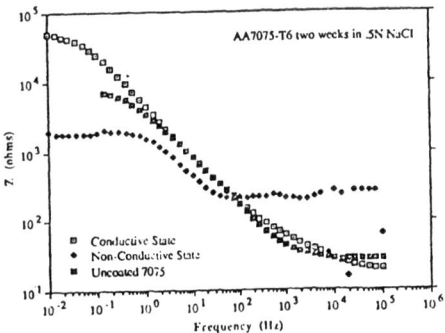

Figure 4. EIS data comparing the conductive green colored state of a double strand polyaniline coating with the blue colored non-conductive state on 7075 samples. The blue colored coating was converted from the conductive state to the non-conductive state buy dipping the sample in a mild basic solution which turned the film blue. Samples have been immersed in a 0.5N NaCl neutral pH solution for two weeks at the time of measurement.

An obvious follow up question to the results presented in figure 4 is what happens to the same conductive polymer coated sample under EIS testing upon reversibly changing the coating from the green form to the blue form and back again. The results of this test are shown in figure 5 for an AA2024-T3 alloy coated sample. Figure 5 is a plot of the log impedance at 0.1Hz, as an estimate for the charge transfer resistance of the coating, versus the pH of the 0.5N NaCl test solution changed with a buffer. The numbers signify the sequence of events for the same coated sample. The sample was coated with a thin coating of the double strand conductive polymer and tested under EIS in the neutral 0.5N NaCl solution, indicated impedance dropped by an order of magnitude (event 2). The test solution was changed to a pH 3.6 buffered solution and the impedance measured at event 3 and left ion this solution for 24 hours. The film had turned back to the green colored form within minutes after exposure to the lower pH solution. It is not expected that the log impedance will change significantly in the low pH environment as evidenced from the testing of other samples whose result are presented in figure 3. The sample is then exposed to a pH buffered at 11.0, which instantly turns the film back to the blue form at event 4. The log impedance is now lower and quickly drops within 3 hours to a value below an uncoated sample, indicating the onset of crevice corrosion as event number 5. The sample is then re-exposed to the pH 3.6 buffered solution and a rapid increase in impedance is observed within the testing time of the EIS equipment to event 6. The results for an uncoated sample exposed to the same pH buffered solutions are also presented to show that the higher impedance under acidic conditions is not due to an interaction of the acid with the underlying metal surface. The uncoated data points are the results after a 24 hour exposure to the sample to the test solutions. These results strongly

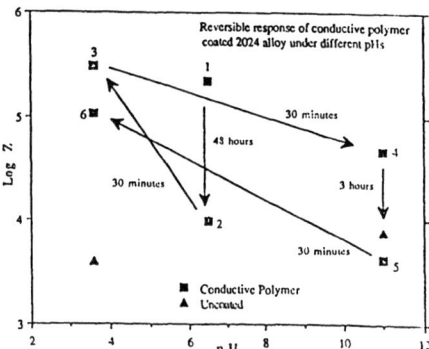

Figure 5. Log impedance at 0.01 Hz versus pH of a conductive polymer coated and an uncoated AA2024-T3 aluminum alloys. The numbers signify the sequence of events for changes in pH environment on the same coasted sample with the length of time labeled for each event.

indicate that the green colored, emeraldine salt form of the conductive polymer complex is directly involved with the formation of passive protective layer on the aluminum alloy surface. This also further demonstrates the permeability of the polymer film to ions and the reversible nature of polyaniline.

The observed EIS response of the conducting polymer coated sample is also characteristically different from that of a barrier type polymer coating. Any barrier type polymer coating tested with EIS, would give extremely high impedance values (approaching 10^{10} ohm) in the initial phase of the EIS test [21]. The green conductive coating of polyaniline does not have as high of an initial impedance (10^5 to 10^6 ohm). Another difference seen in barrier coatings in the slow yet gradual change in capacitance of the film over time as water gets absorbed into the film and changes the dielectric constant of the coating. The conductive polymer coatings impedance value and the capacitance characteristics do not change over time in the pH 3.6 salt solution.

The green conductive state of the polyaniline molecular complex does not impede the transport of ions, water and electrons and it is not expected to impede the transport of oxygen, yet it imparts high impedance for the charge transfer at the interface. The impedance to charge transfer might come from the chemical transformation of the aluminum surface layer by the polyaniline coating. The EIS response presented here shows certain similarity with the chromate coating which is known to chemically convert the aluminum surface. The magnitude of impedance and the capacitance characteristics of the conducting polymer coating in the acidic salt solution is very similar to that of a chromate conversion coating (but tested under neutral condition). The similarity in the EIS responses makes it plausible to propose that the conducting polymer functions as a surface conversion agent to passivate the surface of aluminum and the oxidative power of the conducting electroactive polyaniline may be the reason for maintaining (or repairing) the passive surface layer of aluminum under the aggressive chloride solution. Other evidences for the existence of such a passive layer were found in the analysis of the polymer-metal interface and in the analysis of electrochemical data [22].

Although the conducting polymer and the chromate conversion might be similar in their ability to passivate the aluminum surface, they appear to be effective in different pH ranges. The results presented in figures 1 and 3 show the conducting polymer sustains high impedance in acidic pH 3.6 salt solution while the Alodine 600 chromate conversion coating losses its charge transfer impedance in acidic solutions. The loss of impedance for the chromate conversion coating in pH 3.6 salt solution is understandable because the surface layer structure is aluminum oxide which dissolves in acidic environments [17]. The chromate coatings perform better under neutral environments. The resulted presented here show an interesting contrast between the conducting

738

polymer induced passivation and the chromate induced passivation. If there is indeed a passive layer formed, this passive layer is resistant to acid attack. Preliminary experimental evidences appears to support a non-oxide passive layer [22], but more work is needed to identify the composition of this passive layer.

The difference between the conductive and non-conductive forms of the same polyaniline coating shows that the same polymer in its protonated conductive from shows a higher charge transfer resistance, while the impedance is lost by converting it to the deprotonated non-conductive form [14]. Such a result is inconsistent with the polymer functioning as a barrier to ion or electron transport, but may indicate that the ability for efficient charge transfer at the polymer/metal interface is important for an electroactive polymer to passivate the metal surface under corroding environments. If this polymer is to be used for protection against corrosion, it may be necessary to apply a 'barrier type' polymer overcoat (with or without a primer) on a thin film of the conducting polymer. In this type of coating structure, the conducting polymer will function as a surface conversion agent similar to the role of chromates in the present chromate containing coatings.

CONCLUSION

These developments offer some exciting possibilities. The performance of the conductive polymer coatings, when maintained in the green conductive state, warrants further investigations into these types of coatings. If a double strand complex can be developed that will remain conductive at higher pHs and offer the performance demonstrated at pHs below 4.0, then it is very reasonable to assume that an effective replacement for the chromate conversion coatings on aluminum alloys can be developed. As well, the double strand conductive polymer coating can be applied to the metal surface in an easy and industrially accepted method. Our research team at the University of Rhode Island is actively pursuing several approaches to solve the pH limitations. We are interested in future collaboration efforts with industry partners to bring the conductive polymer coating technology to an end application as a corrosion protection primer coating and a chromate conversion coating replacement.

ACKNOWLEDGEMENTS

The authors wish to gratefully acknowledge the support the University of Rhode Island Foundation and the Alcoa Foundation.

REFERENCES

1. J. Liu and S. C. Yang, J. Chem. Soc., Chem. Comm., 1529 (1991).
2. J. M. Liu, L. Sun, J. H. Hwang and S. C. Yang, Mat. Res. Soc. Symp. Proc., 247, 601 (1992).
3. R. J. Cushman, P. M. McManus and S. C. Yang, J. Electrochem. Soc., 291, 331 (1986).
4. D. W. DeBerry, J. Electrochem. Soc., 132, 1022 (1985).
5. Y. Wei, J. Wang, X. Jia, J. M. Yeh and P. Spellane, ACS Polymer Preprints, 72, 563 (1995).
6. S. Jasty and A. J. Epstein, ACS Polymer Preprints, 72, 565 (1995).
7. S. Sathiyanarayanan, S. K. Dhawan, D. C. Trivedi and K. Balakrishnan, Corrosion Sci., 33 1831 (1992).
8. G. Trochs-Nagels, R. Winard, A. Weymeersch and L. Renard, J. Appl. Electrochem., 22, 756 (1992).
9. S. Ren and D. Barkey, J. Electrochem. Soc., 139, (1992) 1021.
10. D. A. Wrobleski, B. C. Benicewicz, K. G. Thompson, and C. J. Bryan, Polymer Preprints, 35, 265 (1995).
11. B. Wessling, Adv. Mater., 6, 261 (1994).
12. Wei-Kang Lu, R. L. Elsenbaumer and B. Wessling, Synthetic Metals, 71 2163 (1995).

13. R. J. Racicot, R. L. Clark, H-B. Liu, S. C. Yang, M. N. Alias, R. Brown, SPIE Proceedings, Optical and Photonic Applications of Electroactive and Conducting Polymers, Vol. 2528, 198 (1995).

14. R. Racicot, M. N. Alias, R. Brown, R. L. Clark, H-B Liu and S. C. Yang, Proc. Materials Res. Soc., **413**, 529-534 (1996).

15. R. Racicot, R. Brown and S. C. Yang, Proceedings Synthetic Metals Conf., July (1996). (update)

16. S. M. Cohen, Corrosion Eng., **51** (1), 71 (1995).

17. C. Groot and R. M. Peekema, The Potential-pH Diagram for Aluminum, U.S. Atomic Energy Comm. Publ.., H.W. 28556, 13 pages, (1953).

18. S. C. Yang, "Conducting Polymer as Electrochromic Material: Polyaniline", Large-area Chromogenics: Materials and Devices for Transmittance Control, C. M. Lampert and C. G. Granqvist, editors, SPIE Optical Engineering Press, pgs. 335-365, (1990).

19. K. Naoi, M. Lien and W. H. Smyrl, J. Electrochem. Soc., **138**, 440 (1991).

20. G. Schavon, G. Zotti and N. Comisso, J. Phys. Chem., **98**, 4861 (1994).

21. R. M. Latanision, Corrosion Science, **51**, No. 4, 270 (1995).

22. R. J. Racicot, S. C. Yang and R. Brown, Materials Research Society, Fall Meeting, Boston MA, Proceedings Manuscript, (1996). update

NANOCOMPOSITES OF METALLOPHTHALOCYANINES AND CONJUGATED POLYMERS

SHUJIAN YI AND SAMSON A. JENEKHE

Departments of Chemical Engineering and Chemistry
University of Rochester, Rochester, New York 14627-0166

ABSTRACT

Nanocomposites of phenoxy-substituted vanadyl phthalocyanine with a π-conjugated polymer, poly(benzimidazobenzophenanthroline ladder), have been prepared from their Lewis acid complexes in organic solvents. The resulting composite thin films obtained by spin-coating have excellent optical transparency and interesting composition-dependent morphology and photoelectronic properties. Enhanced photoconductivity, compared to the components, was observed in some composites with discrete nanoscale metallophthalocyanine aggregates.

INTRODUCTION

Organic materials, such as metallophthalocyanines (MPc), squaraines, perylenes, and azo compounds, are of interests for applications in photoelectronic devices. Among these materials, metallophthalocyanines are most attractive because of their broad spectral responses and high photosensitivities.[1] For instance, pigments of titanylphthalocyanine have been reported to have one of the best photosensitivities (0.2 $\mu J/cm^2$) among organic photoconductors.[2] However, because most unsubstituted metallophthalocyanines have poor processibility as a result of their very limited solubility, organic photoconductors are usually prepared by dispersing fine particles of metallophthalocyanine pigments in an insulating polymer matrix.[3] The pigment particles, after wet milling, typically have sizes of several hundred nanometers in diameter.[4]

Since the inert polymer matrix is not photoactive nor charge-transporting, it may contain traps that impede charge transport.[5] The use of active polymer matrix may improve the performance of photoconductors. Moreover, there is interest in new photoconductors with sensitivity over a wide spectral range, i.e., with photosensitivity in both the visible and the near-IR region.[6] Approaches to such materials include the combination of pigments with similar or dissimilar structures,[6] and the dispersion of pigment particles in a photoconductive polymer matrix.[5]

Nanoscale size effects on the photoconductivity of nanocomposites and conjugated polymers were recently reported.[7,8] Here, we report the preparation and the photoconductive properties of novel nanocomposites consisting of tetraphenoxy-substituted vanadyl phthalocyanine (VOTPPc) and a conjugated polymer, poly(benzimidazobenzophenanthroline ladder)(BBL) (Figure 1). Our measurements show that some of the composites have an enhanced quantum efficiency of photogeneration and photosensitivity when the dispersed VOTPPc phase is in the nanoscale size range.

EXPERIMENTAL SECTION

Materials. The polymer BBL with an intrinsic viscosity of 32 dL/g in methanesulfonic acid at 30°C was synthesized in this laboratory, using the literature method.[9] Nitromethane (~96%), Gallium (III) chloride (GaCl$_3$, ~99.999%), and vanadyl 2,9,16,23-tetraphenoxy-29H,31H-phthalocyanine (VOTPPc, ~98%) were obtained from Aldrich and were used as received. Binary composites of VOTPPc/BBL with VOTPPc content of 10%, 25%, 50% and 75% molar fraction were prepared by dissolving the two components in nitromethane containing GaCl$_3$ in a vial. All the solutions were prepared in N$_2$ atmosphere dry box. In general, a composite solution contains 0.2-0.4wt% solids. The viscous solutions of all the composites investigated were readily processed into thin films on various substrates (glass, fused silica) by solution spin-coating. Thin films of the composites were prepared by spin coating and the complexed GaCl$_3$ in the thin films was removed by submerging them in deionized water for 4 hours.[10]

Characterization. Optical absorption spectra were obtained with a Perkin-Elmer Model Lambda 9 UV-Vis spectrophotometer. Photoinduced discharge experiments were performed by mounting a bilayer photoreceptor device in a xerographic cycling chamber. The bilayer photoreceptors were composed of a thin layer of composite as the charge generation layer and a thick layer of tri-p-tolylamine (TTA) doped polycarbonate as charge transport layer. The detail methods for device fabrication and evaluation have been discussed elsewhere.[8] For transmission electron microscopy (TEM), the composite thin films on glass substrate were placed in deionized water, peeled off using a blade, and transferred onto copper grids. The TEM images were obtained from a JOEL model JEM2000EX instrument. All measurements were made at room temperature (~22°C).

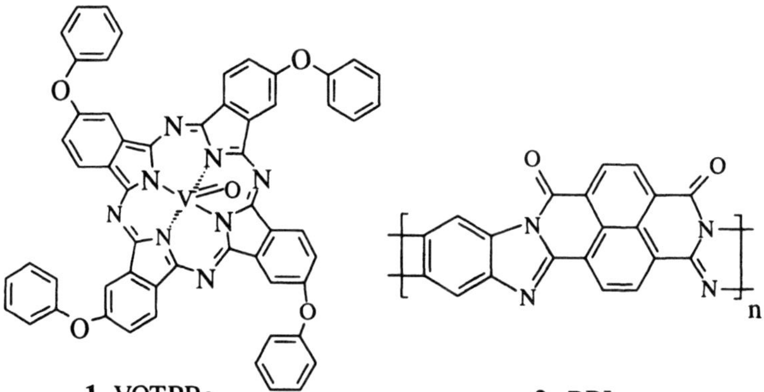

Figure 1. Molecular structures of VOTPPc and BBL

RESULTS AND DISCUSSION

We have exploited the solubility of metallophthalocyanines and conjugated polymers in the same organic solvent to prepare composites of various metallophthalocyanines with different conjugated polymers. We have found that the metallophthalocyanines, such as VOTPPc, H_2Pc, CuPc, and TiOPc, have good solubility in aprotic organic solvents containing metal halide Lewis acids (e.g. $GaCl_3$). The high solubility of the metallophthalocyanines in these solvents is due to their complex formation with Lewis acids at the imine nitrogen sites, similar to the aromatic heterocyclic rigid-rod polymers.[10]

Transmission electron micrographs of the composite thin films showed that VOTPPc molecules formed aggregates in the conjugated polymer matrix (Figure 2). Thin films of pure BBL were structureless, even at higher TEM resolution. The aggregation of VOTPPc molecules in the composites shows that BBL and VOTPPc are thermodynamically incompatible, as expected from the huge disparity in molecular weight and structure. The size of the VOTPPc aggregates was found to depend on the composition. Isolated nanoscale MPc domains with sizes of ~10 nm were observed at low MPc concentration (10% VOTPPc) whereas a largely bicontinuous morphology was observed at higher loading of the metallophthalocyanine (50-75%). At intermediate composition (25%), the composite morphology contains both continuous and isolated domains of the VOTPPc.

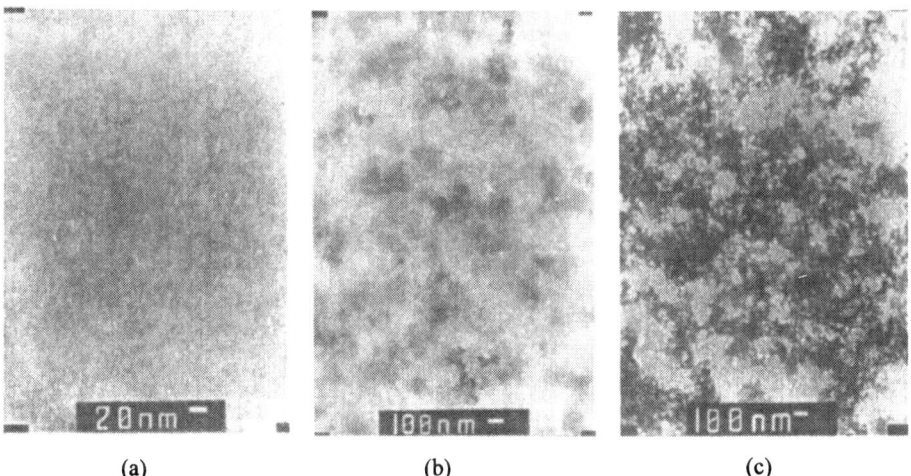

| (a) | (b) | (c) |

Figure 2. Transmission electron micrographs of VOTPPc/BBL composites: (a) 10%; (b) 25%; and (c) 50% VOTPPc

Thin films of the VOTPPc/BBL composites had excellent optical transparency and their color varied depending on the VOTPPc content. As the VOTPPc content varied from 10-75%, the color changed from purple to blue to blue green to green. Figure 3 shows the optical absorption spectra of BBL and the VOTPPc/BBL composite thin films on fused silica substrates. The absorption spectra of the composites were simple compositional averages of those of the pure components. The composites showed absorption bands with λ_{max} at 645 and 720 nm which are characteristic of VOTPPc. Vanadyl phthalocyanine (VOPc) pigment is known to have three different polymorphs: phase I, II, and III.[11] Only phase II has absorption in the near-IR region with λ_{max} around 840 nm, whereas neither phase I nor phase II shows any IR absorption. By comparison with the absorption spectra of the polymorphs of VOPc [11] and t-BuVOPc (phases I and II),[12] we assign the absorptions of the composites at 645 and 720 nm as those of phase I of VOTPPc. Thus, it appears that VOTPPc aggregated primarily in a non-IR absorbing glassy form (phase I) under the spin-coating conditions. Phase I of vanadyl phthalocyanine, substituted or unsubstituted, is metastable and transforms to phase II by heating [11] or by treating with solvent vapor.[12] Our differential scanning calorimetry measurement shows that this phase transformation for VOTPPc starts at 200°C. Thus, all the results presented in this paper are for the phase I polymorph of VOTPPc.

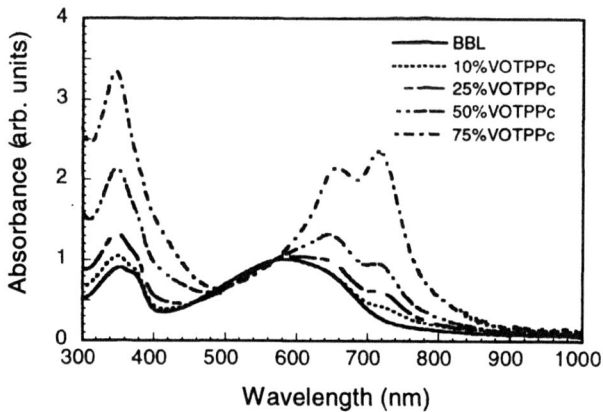

Figure 3. UV-Vis absorption spectra of VOTPPc/BBL composites.

The photoconductivity of the composites was characterized by the photoinduced discharge method.[13] A bilayer device was used with a charge generation layer of 50-nm thick VOTPPc/BBL composite and a 16-μm thick TTA:polycarbonate charge transport layer. The photoinduced discharge curves for the pure components and the VOTPPc/BBL composites are shown in Figure 4. The result for the pure VOTPPc (Figure 4a) in which no difference was observed between dark decay and upon light exposure indicates that phase I aggregates of VOTPPc are not photoactive, in agreement with prior observations on other vanadyl phthalocyanines.[14] The reason why phase I polymorph of VOPc is not photoactive whereas phase II is remains unclear but is speculated to be related to the orientation of V=O bonds in an aggregate.[14] Unidirectional orientation of V=O is believed to enhance intermolecular interactions, favoring efficient charge carrier photogeneration in an ordered crystalline pigment. As observed in t-BuVOPc,[14] disordered orientation of V=O in amorphous aggregates appears to inhibit charge carrier photogeneration.

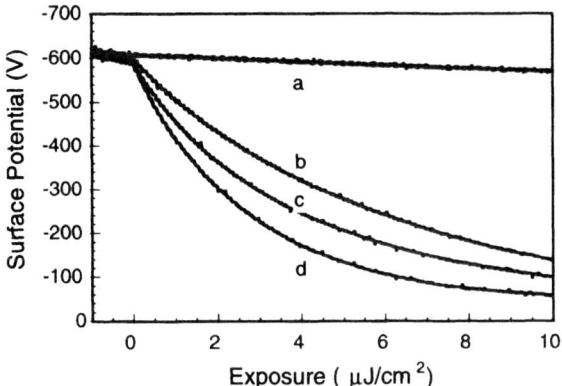

Figure 4. Photoinduced discharge curves of VOTPPc/BBL composite photoreceptors under 570 nm illumination: (a) pure VOTPPc (100%), (b) 25%VOTPPc, (c) pure BBL (0%), and (d) 10%VOTPPc.

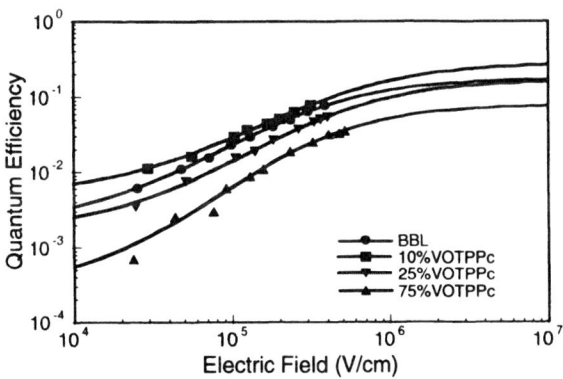

Figure 5. Electric-field dependence of photogeneration quantum efficiency of VOTPPc/BBL composites under 570 nm illumination.

All the VOTPPc/BBL composites were photoactive with photosenitivity that varied with composition (Figure 4). The 10% VOTPPc composite has the best photosensitivity, 1.9 $\mu J/cm^2$, which is significantly better than pure BBL (2.8 $\mu J/cm^2$). The electric field-dependent quantum efficiency for charge photogeneration in pure BBL and three composites is shown in Figure 5. The results show that the 10% composite has enhanced photoconductivity compared to BBL, the matrix material. The main correlation between morphology and the photoelectronic properties of these composites is that enhanced photoconductivity (photosensitivity, charge photogeneration efficiency) is only observed in the composite with discrete nanoscale aggregates of VOTPPc dispersed in the conjugated polymer matrix. Similar nanoscale size effects on photoconductivity of BiI_3-nylon nanocomposites [7] and layered conjugated polymers [8] have been reported. The physical origin of the nanoscale enhancement of photoconductivity in the latter case was elucidated in terms of the *interface-mediated charge separation* and the *small exciton diffusion lengths* in the materials.[8] These mechanisms appear to be the likely explanation of our observed enhancement of photoconductivity in VOTPPc/BBL composites at 10%. However, additional studies are necessary to establish this.

CONCLUSIONS

The solubility of the Lewis acid complexes of metallophthalocyanines and π-conjugated polymers in a common organic solvent has allowed us to prepare well-defined supramolecular materials which combine the electroactive and photoactive properties of both components. The resulting composites have excellent optical transparency, composition-dependent morphology, and enhanced photoconductivity.

ACKNOWLEDGEMENT

This research was supported by the National Science Foundation Center for Photoinduced Charge Transfer (Grant CHE-9120001) and the Office of Naval Research.

REFERENCES

1. K.-Y. Law, Chem. Rev. **93**, 449 (1993).
2. T. Enokida, R. Hirohashi and T. Nakamura, J. Imag. Sci. **34**, 234 (1990).
3. D.-M. Pai and B.E. Springett, Rev. Mod. Phys. **65**, 163 (1993).
4. T. Enokida, R. Hirohashi and S. Mizukami, J. Imag. Sci. **35**, 235 (1991).
5. A. Takimoto, H. Wakemoto and H. Ogawa, J. Appl. Phys. **74**, 1111 (1993).
6. G.D. Hinch and K.A. Voll, SPIE Proceedings. **2850**, 181 (1996).
7. Y. Wang and N. Herron, Science, **273**, 632 (1996).
8. X. Zhang, S.A. Jenekhe and J. Perlstein, Chem. Mater. **8**, 1571 (1996).
9. F.E. Arnold and R.L. Van Deusen, Macromolecules, **2**, 497 (1969).
10. (a) S.A. Jenekhe, P.O. Johnson and A.K. Agrawal, Macromolecules, **22**, 3216 (1989); (b) S.A. Jenekhe and P.O. Johnson, Macromolecules, **23**, 4419 (1990).
11. C.H. Griffiths, M.S. Walker and P. Goldstein, Mol. Cryst. Liq. Cryst. **33**, 149 (1976).
12. K.-Y. Law, J. Phys. Chem. **89**, 2652 (1985).
13. J.A. Osaheni, S.A. Jenekhe and J. Perlstein, J.Phys.Chem. **98**, 12727 (1994).
14. K.-Y. Law, J. Phys. Chem. **92**, 4226 (1988).

LATEX-LIKE WATER-BORNE POLYANILINE FOR COATING APPLICATIONS

Huaibing Liu, Robert Clark, Sze C. Yang, University of Rhode Island, Department of Chemistry, Kingston, RI 02881-0809

ABSTRACT

We report the synthesis of a polymeric complex of polyaniline that is dispersed in water as a stable suspension. The polymer, PAN:PVME-MLA, is a molecular complex of polyaniline and poly(vinylmethylether-co-maleic acid). The synthetic process leads to a stable latex-like suspension in water. The water-borne conducting polymer, once dried as a thin film on a substrate, is not dissolvable by water or other solvents. An example of piece-dyeing process is presented to show its potential for electrostatic dissipation of textile products. Another example illustrates that the material may be used as electroactive thin-films for electrochromic windows and for rechargeable battery applications.

INTRODUCTION

Solution processable polyaniline is useful as a coating material for antielectrostatic, and electrochemical applications. Most coating processes require dispersion in a liquid carrier for easy coating process, but the dried film should not be re-dissolvable by water or other solvents. This requirement makes a latex-like water-borne conducting polymer better suited than the truly water-soluble conducting polymers. Although there were examples in the literature that uses radiation to cross-link a water-soluble polyaniline to render it insoluble after the film is coated, such process is not practical except for application in lithography. A water-bore conducting polymer also has the advantage over an organic soluble conducting polymer because it reduces the use of volatile organic solvent.

In this article we report the synthesis of PAN:PVME-MLA as a latex-like water-borne coating material. PAN:PVME-MLA is a molecular complex of polyaniline and poly(vinylmethylether-co-maleic acid). This polymeric complex is dispersable in water as a stable suspension, and once the coating is dried on a substrate, it is not dissolvable by water. The coating is fixed without the need for radiative or chemical cross-link, and therefore is more generally applicable as a coating material.

The molecular complex PAN:PVME-MLA has another advantage over the conventional polyaniline. The conventional polyaniline easily loses its molecular or ionic dopants to result in a loss of electrical conductivity. Heat and moisture cause the dopants to be either segregated from the conducting polymer or be washed out by water or solvents. This problem is avoided in polymeric complexes of polyaniline because the polymeric dopant is part of the stable molecular complex. The applications of this material as antielectrostatic and electroactive coatings are discussed in this article.

Mat. Res. Soc. Symp. Proc. Vol. 488 © 1998 Materials Research Society

SYNTHESIS OF PAN:PVME-MLA

The soluble PAN:PVME-MLA is a molecular complex of two strands of polymers: (1) a polyaniline molecule, and (2) poly(vinylmethylether-co-maleic acid). These two strands of polymers are bonded by non-covalent intermolecular attractive forces to form a stable molecular complex.[1]

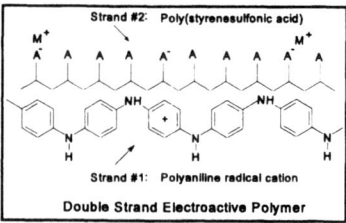

Figure 1 A molecular complex of polyaniline and a polyanion.

The polymeric complex PAN:PVME-MLA was synthesized by a template-guided synthetic method: In the first step, the aniline monomers were adsorbed onto the poly(vinylmethylether-co-maleic acid) chains in aqueous solution. The resulting adduct, PVME-MLA:(Aniline)$_x$, has signatures that can be monitored and verified. In the second step, the attached aniline monomers were oxidatively polymerized to form the polymeric complex.[2] In order to produce latex-like dispersion of the polymeric complex, a new method of synthesis was used that has a subtle but important difference from the other template-guided syntheses reported previously[1,2] . In this new method, the polyanion serves not only as a template for forming the complex, but it also serves as an emulsifier that stabilizes the suspension of the reaction product. The resulting polymer is stably dispersed in water but is not a true solution.

An example of the synthesis of a water-borne molecular complex PAN:PVME-MLA is described as the following: Dissolve 1.92 gm of poly(vinylmethylether-co-maleic acid), PVME-MLA, (containing 0.022 moles of carboxylic functional groups, Aldrich, M.W. = 67,000) in 25 ml of distilled water. Add 5 ml of methanol and slowly add 2 gram of aniline (0.022 mole of aniline) to this solution and stir for one hour. At this stage, aniline is adsorbed on PVME-MLA to form the adduct [poly(vinylmehtylether-co-maleic acid):(An)$_n$]. Slowly add 25 ml 3 M HCl and 6.0 x 10^{-4} mole of ferric chloride to the solution. Stir for 30 minutes. At this stage, the micro emulsion of the adduct [poly(vinylmehtylether-co-maleic acid):(An)$_n$] is stabilized to an appropriate size in the acidic solution. Slowly add 2.5 ml of 30% hydrogen peroxide (containing 0.022 mole H$_2$O$_2$) to initiate the polymerization of the adduct of aniline and PVME-MLA. The reaction mixture soon becomes green in color. After vigorous stirring for 2 hours, the reaction mixture is poured through a filter paper to remove small amount of particles. The filtrate is a dark green homogeneous aqueous dispersion of the reaction product.

The reaction product was purified before performing chemical analysis. The product solutions may contain free polyelectrolyte, uncomplexed PAN, unreacted aniline, low-molecular weight oligomers and inorganic ions. In order to be certain that all the characterization and elemental analysis are performed on samples free of the above-mentioned impurities, we perform a purification process that involves filtration, ion exchange, extraction and dialysis to obtain a single component solution.

Samples were dried in an oven at 70 °C for 72 hours before sealing in airtight sample vials. Elemental analyses were performed by M-H-W Laboratories, Phoenix, Arizona. The purified sample has an elemental content of C: 60.15%, H: 5.87% and N: 7.39%, giving an empirical formula of (C$_7$H$_{10}$O$_5$)$_{0.50}$: (C$_6$H$_4$NH)$_{1.00}$: H$_{1.12}$O$_{0.65}$ for PAN:PVME-MLA. The average molecular weight of PVME-MLA is 67,000 giving an average degree of polymerization of 385 for PVME-MLA. The result of the elemental analysis implies an average chemical formula of $(C_7H_{10}O_5)_{385}$: $(C_6H_4NH)_{770}$ if one assumes that the polyaniline in the complex is a single polymer chain. It is more likely that the complex consists of several shorter chain segments.

The infrared spectrum of PAN:PVME-MLA shows that the reaction product contains functional groups from both the PVME-MLA part (-COOH) and from the polyaniline strand (C-N and aromatic rings). A band at 1718 cm^{-1} is attributed to the stretch mode of carbonyl group of carboxylic acid on PVME-MLA. A strong band at 1160 cm^{-1} indicates that the conductive form of polyaniline is present. Two bands at 1580 cm^{-1} and 1450 cm^{-1} are attributed to the ring stretching combined with C-N stretching. The band at 1263 cm^{-1} is assigned to C-N stretching mixed with C-H bending. Thus the IR spectrum clearly shows the IR features of a molecular complex, i.e., co-presence of unique features from PVME-MLA and polyaniline.

PROPERTY OF PVME-MLA

The suspension stability: The as-synthesized solution remained homogeneous for more than one year without precipitation. The dispersed product does not flocculate in salt solutions such as 0.37 M of sodium sulfate indicating good stability against salting out. By dialysis, the water in the solution may be replaced by isopropanol. The isopropanol suspension is also stable and remained homogeneous for more than one year. The green-colored liquid can be used for coating applications. The coated film is insoluble in water and other solvents. The thin films are electrically conductive with conductivity of about 5 x 10^{-2} Siemens/cm. A film casted on ITO shows the characteristic electrochromism of polyaniline when the electrochemical potential of the sample was cycled..

WATER-BORNE DYEING SYSTEM FOR CONDUCTIVE FABRICS

Electrically conductive polymers are useful as anitielectrostatic coatings for packaging materials, and as electrostatic dissipation materials for fabrics and rugs. The common antielectrostatic materials are ionic salts or resins filled with metals, semiconducting oxides or carbon fibers. These traditional antistatic materials have some shortcomings. The ionic material such as the quarternary ammonium salts loses electrical conductivity when the humidity of the ambient environment is low. The metal, semiconductor, or carbon containing antistatic fiber is difficult to process. The inherently conductive polymer may be useful as antielectrostatic materials for textile products in certain niche applications.

Polypyrrole has been explored as a coating material for textile products. Since polypyrrole is not soluble in water, a conventional dyeing process was not possible. In the previous coating processes, the fabric is first saturated with either an oxidant[3] or the pyrrole monomers.[4] It is then immersed in a chemically reactive bath to produce insoluble polypyrrole. The fabrics produced from these processes are electrically conductive, but their conductivity depends upon a powdery deposition of conductive polymers in the interstices of the fabric. The conductive fabrics produced by this process tends to lose conductivity during usage because the conducting polymer is brittle and the powders are not adherent to the fabric.

The water-borne conducting polymer PAN:PVME-MLA is suitable to be used in the conventional dyeing process. We use a piece-dyeing process to obtain conductive nylon fabrics. A piece of nylon fabric (Monsanto) is soaked in the aqueous dispersion of PAN:PVME-MLA for 20 minutes, taken out of the solution, rinsed with distilled water and air-dried. This process is repeated twice and uniform green colored fabrics are obtained.

No visible solid particles are deposited on the surface. The fabric is still soft and flexible. No mechanical property change has been observed. The surface resistivity of the dyed nylon fabrics is measured as low as 5×10^3 ohm/square, much lower than 10^9 ohm/square which is the surface resistivity of the current best antielectrostatic coating.

This piece-dyeing process is an industrially feasible process to deposit a uniform, smooth, coherent film of a conductive polymer onto individual fibers of the nylon fabric. The conducting polymer adheres to the nylon fabric probably mainly due to hydrogen bonding between PVME-MLA strand and the nylon.

The conventional single-strand polyaniline PAN:HCl is dedoped when it is in contact with water with pH higher than 4. PAN:PVME-MLA is more resistant to dedoping because the polymeric dopant is part of the polymeric complex. An acid/base titration curve shows that the conducing polymer loses it protons at pH 8-8.5 range. A fabric that is washed in pH 7 detergent does not lose electrical conductivity. An alkaline detergent with pH greater than 8.5, however, will deprotonate the polymeric complex. A deprotonated polymeric complex may be easily reprotonated in a slightly acidic rinse (pH 6).

The dyed double-strand polyaniline fabrics were tested in a low humidity chamber (14%) while the electrical conductivity was monitored. The electrical conductivity remains constant (with small changes) over an extended period of time. Under the same condition, an ionic antielectrostatic material such as the quarternary ammonia salt loses its effectiveness because its ionic conductivity requires moisture.

A COATING MATERIAL FOR ELECTROCHROMIC AND RECHARGEABLE BATTERY APPLICATIONS

The aqueous dispersion of PAN:PVME-MLA can be used to prepare adherent electroactive film on different substrates, such as Pt, Au, indium/tin-oxide (ITO) glass. Being a latex-like material, the dried film is not dissolvable in water. This property allows the film to be used as an electrode material in aqueous electrolytes.

PAN:PVME-MLA in isopropanol or in water was spin coated onto fluoride doped tin oxide glass to form a uniform green colored film. Figure 1 shows an example of the cyclic voltammogram of the film immersed in 1 M HCl solution. The film is stable to the electrochemical conditions, and the voltammogram could be repeated for many cycles. The cyclic voltammogram of PANI/PVME-MLA shows electrochemical response similar to a polyaniline film[5] electrochemically synthesized on an electrode surface. The deposited film shows electrochromic effect during an electrochemical potential cycle. The color of the thin film changes from transparent (reduced form) to green, and then to blue color as the film is anodically polarized. The sequence of color changes is reversed when the oxidized film is cathodically polarized.

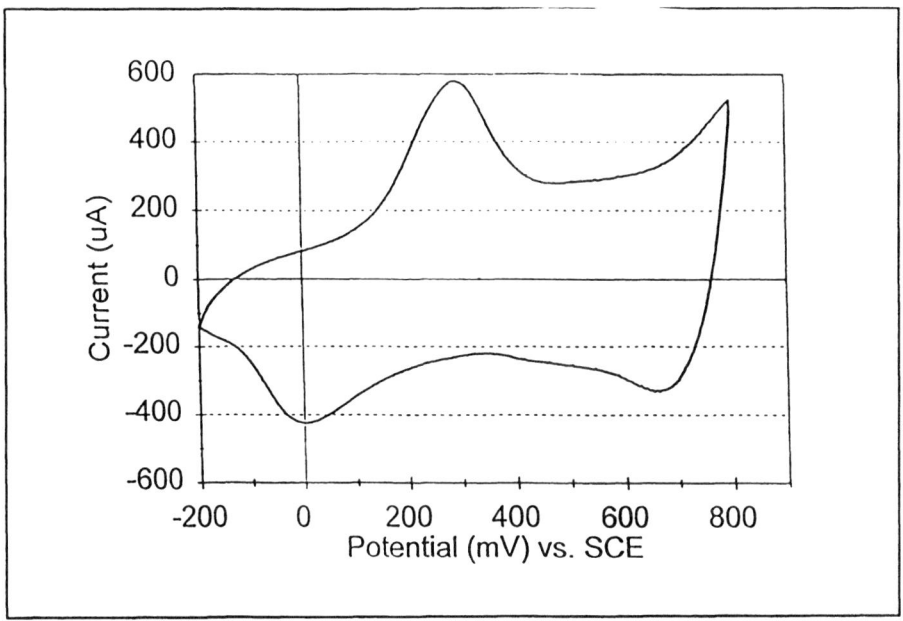

Figure 2 Cyclic voltamogram of PVME-MLA spin coated on tin oxide transparent electrode. Electrolyte: 1M HCl. Scan rate: 50 mV/sec. Reference electrode: Calomel.

The electroactivity of the coated thin film shown in figure 2 suggests that this material is useful as coating of large-area electrochromic smart windows or electrochromic displays.[6, 7] It may also be useful for coating of electrodes for polyaniline batteries. The conventional coating process involves electrochemical synthesis and plating of aniline monomers in acidic electrolysis baths. This method can be used only for small electrochromic windows of small surface area because of the limited size of the electrolysis bath. In addition, the electrochemical deposition results in non-uniform film thickness because the electrochemical deposition involves nucleation and growth of the polymer on the electrode surface. The latex-like PAN:PVME-MLA provides a practical process for coating large-area electrochromic windows and electrochromic displays.

The polyaniline rechargeable batteries are commonly made by electrolytic deposition of polyaniline on electrodes. The electrolytic production process is cumbersome and expensive. A coating process is likely to be less expensive for manufacturing conducting polymer electrodes.

CONCLUSION:

The synthesis and the properties of a water-borne latex-like electrically conducting polymer PAN:PVME-MLA is reported in this paper. This polymer is advantageous for coating applications. It is a water-borne dispersion that is easy to use in a coating process. The dried film is stable in water or solvents. This property makes this material more suitable as a coating material than the water-soluble or solvent-soluble conducting polymers which are either swelled or dissolved by water or solvents. The electrical conductivity of PAN:PVME-MLA is stable under heat, or aqueous environment because the polymeric dopant is part of the molecular complex and is not lost by heat and washing. This material demonstrates properties suitable for applications as antielectrostatics for textiles and as coatings for electrochromic devices, and rechargeable batteries.

ACKNOWLEDGMENT

This research was partially supported by an Alcoa Foundation grant and by a URI Foundation grant.

REFERENCES

1.   L. Sun, H. Liu, R. Clark, S. C. Yang, Syn Met **84**, 67 (1997).

2.   L. Sun, S. C. Yang, Mat. Res. Soc. Symp. Proc. **328**, 209 (1994).

3.   Bjorklund, R.B. and Lundstroem. I.J., J. Electron. Mat. 13, 211 (1984).

4.   Newman, P.R., Warren, L.F., Jr., and Witucki, E.F. U.S. Pat. 4, 617, 228 (1986).

5.   Kobayashi, T., Yaneyama, H., and Tamura, H., J. Electroanal. Chem., 161, 419 (1984).

6.   W.-R. Shieh, S. C. Yang, C. Marzzacco and J.-H. Hwang, MRS Symposium Proceedings, 173, 329-334 (1990).

7.   "Conducting Polymer as Electrochromic Material: Polyaniline", S. C. Yang, pp.335-365 in "Large-area Chromogenics: Materials and Devices for Transmittance Control", C. M. Lampert and C. G. Granqvist, Editor, (SPIE Publishing, 1989).

THERMALLY STIMULATED LUMINESCENCE OF NAPHTHYL-SUBSTITUTED BENZIDINE DERIVATIVE AND TRIS-8-(HYDROXYQUINOLINE) ALUMINUM WITH AND WITHOUT METAL LAYERS

E.W. Forsythe[†], D.C. Morton[††], and Y. Gao[†]
[†]Dept. of Physics and Astronomy, University of Rochester, Rochester, NY 14627-0171
[††] Army Research Laboratories, Adelphi, MD 20783-1197

ABSTRACT

Multilayer organic light emitting devices based on tris-8-(hydroxyquinoline) aluminum (Alq_3) and a naphthyl-substituted benzidine derivative (NPB) have demonstrated practical electroluminescence with a wide range of color. Trap states in these materials play an important role in the carrier transport as well as the light emission process. We report the first observation of thermally stimulated luminescence (TSL) from Alq_3 and NPB. The TSL spectra from 8 K to 300 K were used to determine the trap states in Alq_3 and NPB. The results for Alq_3 show a significant trap distribution at 156 K, which corresponds to a mean trap depth ranging from 0.18 to 0.12 eV, whereas the trap states in NPB are centered from 0.15 eV to 0.01 eV. We have used TSL to study the trap state properties of thin metal layers in the Alq_3 films. In addition, we report photoluminescence as a function of temperature for Alq_3. TSL spectroscopy provides a technique to study the trap states in a specific layer of the device structure.

INTRODUCTION

In recent years, organic materials have been studied for electroluminescent applications [1-6]. Organic light emitting devices (OLEDs) fabricated from small molecule materials, such as naphthyl-substituted benzidine derivative (NPB) and 8-hydroxyquinoline aluminum (Alq_3), have demonstrated high brightness and efficiencies suitable for commercial applications. The NPB layer is the hole transporting material in a typical OLED, and Alq_3 is both the electron transporting material and the emissive layer [4]. Recently, the carrier transport has been modeled as a trap charge limited (TCL) process [6] and later extended for two carrier TCL transport [7]. TCL model for carrier transport requires trap states deeper than the room temperature [8,9]. For Alq_3 devices, the temperature dependent I-V data was fit to the TCL model and estimated the trap depths in bulk Alq_3 of 0.15 eV assuming an exponential distribution in the trap energies in the band gap [6]. Further, the trap density in the bulk Alq_3 film is almost as high as the density of states in the lowest unoccupied molecular orbital. Thus, the physical nature of traps in the materials are important to the carrier transport in the organic layers as well as light emission in the OLEDs.

Thermally stimulated luminescence (TSL) has been shown as a useful technique to study the physical nature of traps in organic [10] and inorganic materials [11]. From, the TSL spectra, the mean trap depths can be estimated, as well as the nature of the trapping mechanism [12-14]. In this letter, we report the TSL spectra for NPB and Alq_3, which is used to estimate the mean trap depths in these materials. In addition, we report the temperature dependent photoluminescence (PL) from Alq_3 at the temperatures corresponding to the observed traps. The TSL results for NPB and Alq_3 each show several states with depths sufficient to support the TCL model for the carrier transport.

EXPERIMENTAL SETUP

The samples shown in Fig. 1a were prepared by thermal evaporation at a base pressure of 5×10^{-7} Torr. A 1500 Å organic layer was deposited onto a quartz substrate and capped with 1000 Å of Ag. A second Alq_3 film was prepared where three 8 Å metal layers where deposited between 150 Å of Alq_3, starting at the quartz substrate followed by 1200 Å of Alq_3 and the 1000 Å Ag cap. These samples were intended to test if the metal layers induce trap states in the Alq_3 layer.

Fig. 1b shows the typical experimental setup for the TSL, similar to Ref. 12. Here, the sample was cooled to 10 K with a CTI cryo-pump in a 1×10^{-3} Torr vacuum. The samples were excited with 100 mW of 363 nm UV radiation from an Ar laser and a 4 cm^2 spot size. The radiation was turned off and the sample heated up at a linear rate of 20 °C/min. A Pritchard model 1980B collected the TSL emission. Then, the peaks were separated by thermal cleansing to more accurately determine the peak shape [15]. Here, the sample was cooled to 10 K and the excited with the UV radiation. The sample temperature was increased at a linear rate and the TSL spectrum measured up to a temperature above the first peak. The sample was then cooled back to 10 K followed by re-heating without the UV excitation. For Alq_3, each thermally cleansed TSL peak was an average of three measurements due to the low signal to noise. All TSL spectra were taken at least three times to ensure the accuracy of the features and to test if the incident radiation result in damage to the molecules. In all cases, the TSL spectra were reproducible and did not show degradation after repeated exposure or thermal cycling. After the TSL measurements, the PL for Alq_3 was measured as a function of sample temperature from 10 K to 300 K in a SPEX Fluorolog, using 363 nm radiation from a xenon lamp.

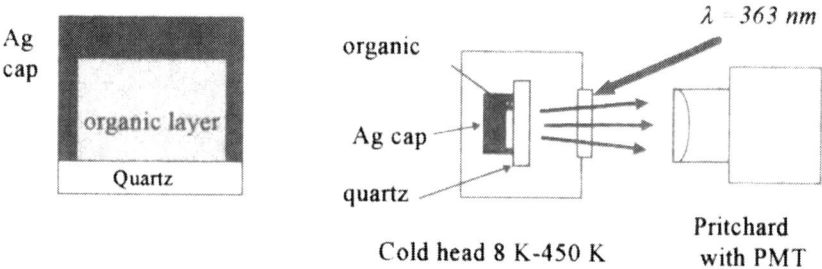

Fig. 1 Schematic illustration of the device structure (a) and of the experimental setup used to measure the thermally stimulated luminescence (b).

EXPERIMENTAL RESULTS

Shown in Fig. 2a is the TSL for NPB and Alq_3. The TSL spectrum for the NPB film has four distinct peaks centered at 40, 65, 110, and 150 K. Conversely, the TSL spectrum for Alq_3 had one broad peak centered at 156 K and smaller peaks at 67 and 60 K. The peaks were isolated by the thermal cleansing method and the results for NPB and Alq_3 are shown in Fig. 3a and 3b, respectively. Fig. 2b is the TSL spectrum for Alq_3 with the metal interlayers. The overall shape of the TSL peaks are similar to the undoped Alq_3,. However, the overall intensity is lowered, which indicates the metal layers induce non-radiative pathways for the TSL without creating distinct trap states, measurable by TSL. The amplitude of the TSL spectrum in Fig. 2b has been scaled by the amplitude of the TSL for undoped Alq$_3$, which demonstrates the reduced intensity due the presence of metal layers.

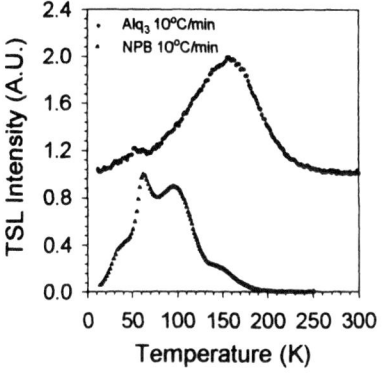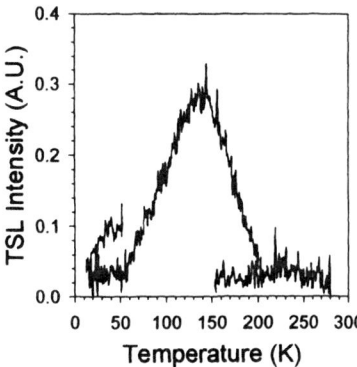

Fig. 2a Thermally stimulated luminescence spectra for NPB (a) and Alq$_3$ and the thermally stimulated luminescence for Alq$_3$ with metal interlayers (b).

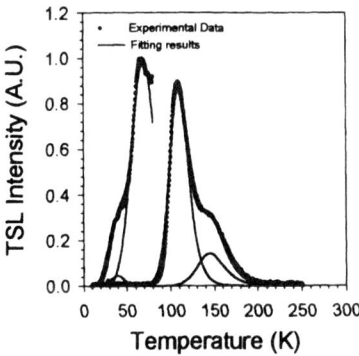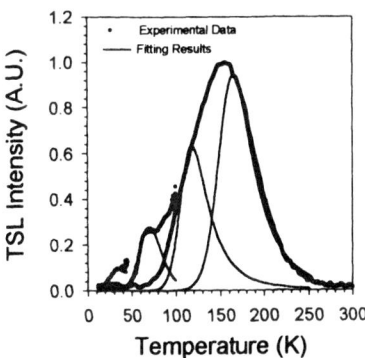

Fig. 3 Thermally stimulated luminescence spectra for NPB, with peaks isolated by thermal cleansing (a) and the Alq$_3$ spectra also with thermal cleansing (b). Fitting results are also shown.

The thermally cleansed TSL peaks were fit to the usual general order kinetics expression shown in Eq. 1 below [13],

$$I = n_o s \exp(-E_t / kT) \times \left(\frac{(l-1)s}{\beta} \int_{T_o}^{T} \exp(-E_t / kT')dT' + 1 \right)^{-\frac{l}{l-1}}, \tag{1}$$

where E_t is the trap energy T is the temperature, and n_0 is the average number of trapped carriers. β is the rate of heating, s is the escape frequency. l is the order parameter, which indicates the order of the trapping process. Where, l equal to 1 is direct recombination and l

equal to 2 indicates re-trapping of the carrier. The expansion in Eq. 2 for the integral in Eq. 1 simplifies the fitting, as described in Ref. 14.

$$\int_0^T \exp\left(-\frac{E_t}{kT'}\right)dT' = T\exp\left(-\frac{E_t}{kT}\right)\sum_{j=1}\left(\frac{kT}{E_t}\right)^j (-1)^{j-1}j!. \tag{2}$$

Using the expansion in Eq. 2, Eq. 1 was fit to the experimental data by vary the parameters, E, l, and s and minimizing the squared difference of the experimental data, y_{exp}, and the model, $(y_{exp} - I)^2$. n_0 was then set equal to the ratio of the experimental data intensity at the peak and the model at the same temperature. For the Alq$_3$ peak at 67 K, a linear background was first subtracted from the data. This background is the leading edge of the next trap peak.

Table 1 summarizes the parameters that characterize the TSL peaks for NPB and Alq$_3$. The four trap states in NPB could be fit to Eq. 1 as described and the results show trap states of 0.148, 0.118, 0.025, and 0.011 eV, with order parameters, l, of 2.0, 1.8, 1.4, 1.6, indicating some re-trapping of the carriers. Since holes are the main charge carriers in NPB, we speculate the trap states observed from the TSL spectra are associated with hole traps. However, the TSL results cannot be used to determine if electrons or holes are trapped.

For Alq3, the fitting was more difficult than NPB as the main peak centered at 156 K was significantly more broad than the peaks in NPB and other inorganic materials [11,13]. When the 156 K peak was fit to a single trap state, the escape frequency, s, is 10 sec^{-1}, which is more than 3 orders of magnitude smaller than typical values reported for TSL peaks [11,13]. In addition, the single TSL peak could not fit the low temperature side of the experimental curve. Therefore, we fit the high and low temperature sides of the 156 K peak separately using equations 1 and 2. This approach is consistent with previous methods for approximating the trap energies from TSL peaks [11]. When this approach is applied to the 156 K peak in Alq$_3$, the trap states are no deeper than 0.180 eV, as this is the last observed TSL peak. Fitting the low temperature side gives an approximate lower limit for the trap energies of 0.12eV. The TSL peak at 67 K is modeled as a single trap state at 0.054 eV. A third state observed at 50 K could not be fit. We assume the trap states observed in the Alq$_3$ TSL spectra are associated with electron traps due the to the electron transporting properties of Alq$_3$.

Organic	T_{max}	E_t	s	l
NPB	150 K	0.148 eV	4.59×10^4 sec^{-1}	2.0
	110 K	0.118 eV	1.41×10^5 sec^{-1}	1.8
	65 K	0.025 eV	12.0 sec^{-1}	1.4
	40 K	0.011 eV	8.90 sec^{-1}	1.6
Alq$_3$	156 K	0.180 eV	9.14×10^4 sec^{-1}	2.9
	156 K	0.120 eV	4.87×10^4 sec^{-1}	3.1
	67 K	0.054 eV	1.12×10^4 sec^{-1}	1.9

Table 1 Summary of the thermally stimulated luminescence results from the fitting.

The temperature dependent PL for Alq₃ is shown in Fig. 4 along with the photoluminescence excitation spectroscopy (PLE) as a function of temperature. A solid line is drawn vertically in Fig. 4 to indicate the wavelength of radiation used for the TSL measurements, which is blue shifted from the peak absorption for Alq₃. Recall, the Alq₃ layer is both the electron transporting material and the emissive layer. The temperature dependent PL shows the typical emission spectra for Alq₃. At the trap release temperatures observed in Fig. 3b, the PL emission does not show additional features, only a less than 5 nm blue shift at low temperature as compared to room temperature with an over all lower emission.

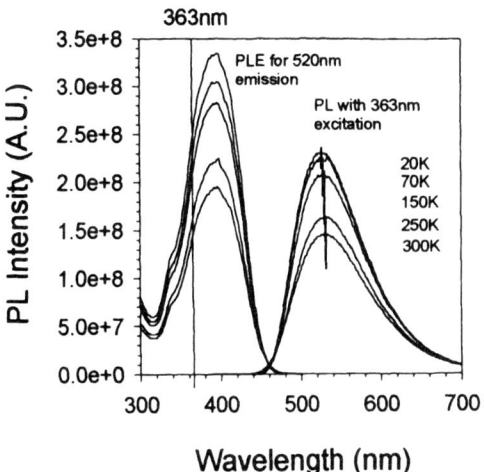

CONCLUSION

The nature of the traps in organic crystals have been attributed to impurities in the molecular films [11]. Here, the molecule with the impurity has a dipole moment and the interaction between the carrier and dipole moment can give rise to trap states. In addition, structural defects can give rise to trap states [16]. Several models have been proposed for the trap states in organic crystals based on local compression of the molecular lattice. Here, the local regions with compressed intermolecular distances have higher polarization energies for the charge carriers. Thus, these local region may act as trap states for electrons and holes [17,18].

In summary, we have measured the trap states in bulk NPB and Alq₃ using thermally stimulated luminescence. The trap states were thermally isolated and fit to a general order TSL expression. NPB has four distinct states centered from 0.148 eV to 0.011 eV, whereas Alq₃ has three states, two of which could be isolated. The trap states in Alq₃ are 0.18, 0.12 and 0.054 eV. For both Alq₃ and NPB, the trap states are sufficiently larger than room temperature thermal energy, which indicates these states can contribute to the carrier transport properties through these layers. In addition, we measured the TSL emission for Alq₃ with five 8 Å metal interlayers to study if the metal induces trap states in the Alq₃ films. We observed the similar TSL spectrum for Alq₃ with and without the metal. However, the sample with the metal layers has an overall lower TSL emission than the film without the layers, indicating the metal induces a more continuous non-radiative states and not additional isolated trap states. The temperature dependent PL for Alq$_3$ was measured at the temperatures corresponding to the emission from the trap states and shows a blue shift less than 5 nm and an overall lower intensity at room temperature than 10 K. In conclusion, TSL measurements have demonstrated an improved understanding of the physical nature of trap states in NPB and Alq$_3$, which may be used to understand the transport mechanisms in these material vital to the development of OLEDs.

The authors would like to thank Dr. C.W. Tang and Eastman Kodak for supplying the material and useful discussions. This work was supported in part by DARPA DAAL01-96-K-0086 and NSF DMR-9612370.

REFERENCES

1. C.W. Tang and S.A. VanSlyke, Appl. Phys. Lett. **51**, p. 913 (1987)
2. C.W. Tang, S.A. VanSlyke, and C.H. Chen, J. Appl. Phys. **65**, p. 3610 (1989)
3. J.H. Burroughs, D.D.C. Bradley, A.R. Brown, R.N. Marks, K. MacKay, R.H. Friend, P.L. Burns, and A.B. Holmes, Nature (London) **347**, p. 539 (1990)
4. L.S. Hung, C.W. Tang, and M.G. Mason, Appl. Phys. Lett. **70**, p. 152 (1997)
5. P.E. Burrows and S.R. Forrest, Appl. Phys, Lett. **64**, p. 2285 (1993)
6. P.E. Burrows, Z. Shen, V. Sulovic, D.M. McCarty, S.R. Forrest, J.A. Cronin, and M.E. Thompson, J. Appl. Phys. **79**, p. 7991 (1996)
7. F. So, J.-H. Xu, M. Kim, H.C. Lee, and J. Shen, Int. Displays Res. Conf. Proc., p. F-23 (1997)
8. M.A. Lampert, Phys. Rev. **103**, p. 1648 (1956)
9. A. Rose, Phys. Rev. **97**, p. 1538 (1955)
10. N.I. Ostapenko and M.T. Shpak, Opt. Spectrosc. (USSR) **68**, p. 511 (1990)
11. D.W. Cooke, B.L. Bennett, W.L. Hults, R.E. Muenchausen, and J.L. Smith, Appl. Phys. Lett. **70**, p. 3594 (1997)
12. J.T. Randell and M.H.F. Wilkins, Proc. Roy. Soc. **A184**, p. 347 (1945)
13. R. Chen, J. Electrochem. Soc. **116**, p. 1254 (1969)
14. R. Chen and Y. Kirsh, <u>Analysis of Thermally Stimulated Processes</u> (Pergamon, Oxford, 1981) pp. 159-165
15. K.H. Nicholas and J. Woods, Br. J. Appl. Phys., **15**, p. 783 (1964)
16. E.A. Silinsh, Phys. Stat. Sol. **3**, p. 817 (1970)
17. M. Batley and L.E. Lyons, Austral. J. Chem. **19**, p. 345 (1966)
18. H. Baessler, G. Herrmann, N. Riehl, and G. Vaubel, J. Phys. Chem. Solids **30**, p. 1579 (1969)

Structural Properties of Self-Assembled Poly(zinc-bisquinoline)

D.L. Thomsen, III, T. Phely-Bobin, and F. Papadimitrakopoulos[*]

Department of Chemistry, Polymer Science Program
Nanomaterials Optoelectronics Laboratory, Institute of Materials Science,
University of Connecticut, Storrs, CT 06269-3136

ABSTRACT

Poly(zinc-bisquinoline) synthesized via a reactive self-assembly technique has been used as an active transport and emitting layer in light emitting diodes. This small molecule self-assembly, pioneered in our laboratory, was shown for the first time to produce pinhole-free films as thin as 400 Å. The ellipsometric trajectory and the refractive index of zinc bisquinoline self-assembled films on silicon were determined using spectroscopic ellipsometry. Psi and Del values were extrapolated at 633nm and 70 deg incidence for films ranging in thicknesses from 0 nm to 125 nm and plotted with model trajectories of 1.65, 1.70, and 1.75 refractive indices. Thin films over 20 nm followed a trajectory having similiar refractive indices between 1.65 and 1.70, indicative of homogenous films. A refractive index of 1.68 ± 0.01 at 633 nm was found to be among the highest reported values for a self-assembled metallorganic multilayer. Atomic force microscopy of a 90 Å film on silicon substrates with mean square roughness of c.a. 12 Å showed an approximate roughness of 14 - 17 Å.

INTRODUCTION

The determination of refractive index by ellipsometry for smooth and transparent films has already been shown to provide information proportional to their densities. Self assembled ordered monolayers of alkyl silanes[1] and thiols[2] have demonstrated increased refractive indexes as a result of alkyl chain alignment.[3-6] Multilayer self-assemblies of zirconium phosphonates[7-11] have reported densities in the range of 1.2 g/cm^3 for films with refractive indexes between 1.52 - 1.58. The density range of self-assembled polyelectrolytes[12-14], and proteins[15], lies between 1.2 and 1.3 g/cm^3. An assembly of greater density would require a better packing, produced by the reduction of hydration sphere, restricted conformational freedom, and/or ordering. This work presents the spectroscopic ellipsometry at multiple incident angles[16] of polymeric chelates of zinc and 8,8'-5,5'-dihyrdoxyquinoline (bisquinoline) and atomic force microscopy (AFM) of a representative thin film. These films were solution self-assembled onto Si from diethyl zinc and 8,8'-5,5'-dihyrdoxyquinoline (bisquinoline) under anhydrous conditions, see Scheme 1. Such films were reported to produce uniform, pin-hole free films over large areas, suitable for electroluminescence applications.[17] AFM and ellipsometry can provide further information about uniformity and refractive index of such films.

---

[*] To whom correspondence should be addressed

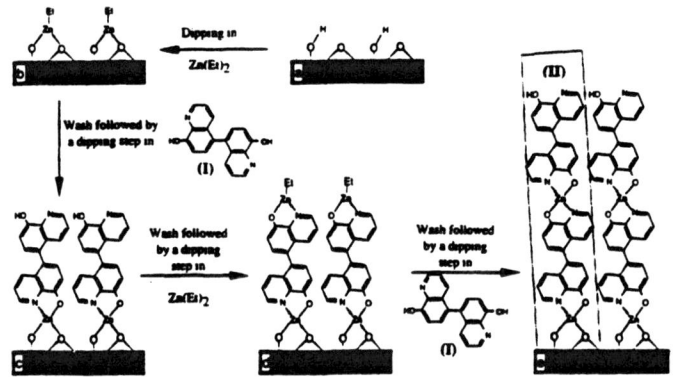

Scheme 1: Schematic representation of poly(zinc bisquinoline) (II) growth via a reactive self-assembly.

EXPERIMENTAL

Self-Assembly of Zinc Bisquinoline: The preparation of bisquinoline and the self-assembly has been described elsewhere.[18-20] The concentrations of diethyl zinc in THF was 1.5×10^{-5}M and the bisquinoline in THF was 3×10^{-4}M. Single polished silicon wafers with native oxide were cleaned in 1/1 H_2SO_4/30%H_2O_2 for 20 minutes followed by washing with distilled H_2O, and rinsing with MeOH. Films were prepared in thicknesses from 0 nm to 125 nm.

Spectroscopic Ellipsometry: Measurements were performed on a WVASE32 Spectroscopic Ellipsometer manufactured by J. A. Woollman Co. Films were scanned between 600-1000 nm at incident angles of 70, 72, 74, 76, and 78 degrees at room temperature in air. Data were collected every 10 nm at 30 revs./meas. for films under 30 nm and at 10 revs./meas. for films over 30nm. Extrapolation for Psi and Del values at 633 nm at 70 degrees incidence had a calculated error due to finite monochromator slit width. For films under 2, 5, 20 and 30 as well as for over 30 nm Psi errors were ± 0.04, ± 0.04, ± 0.04, ± 0.04 and ± 0.22 while Del errors were ± 1.14, ± 0.5, ± 0.3, ± 0.15 and ± 0.50 respectively.

Atomic Force Microscopy (AFM): Images were obtained via a Topometrix TMX 2000 Explorer equipped with a dry scanner of 2x2 μm lateral and 0.8 μm vertical range. Micrographs were recorded from 1000 nm x 1000 nm scan at a scan rate of 4 μm/s and a resolution of 300x300 pixels on a 90 Å zinc bisquinoline film on a single sided polished silicon wafer.

RESULTS AND DISCUSSION

The optical uniformity of zinc bisquinoline was demonstrated by plotting the ellipsometric trajectory of films with thicknesses from 0 nm to 125 nm, grown on silicon wafers. These data correspond to the extrapolated Psi and Del values at 633 nm and 70 degrees incidence, and comparing to model trajectories of 1.65, 1.70, and 1.75 refractive indices (see Figure 1).[21] The

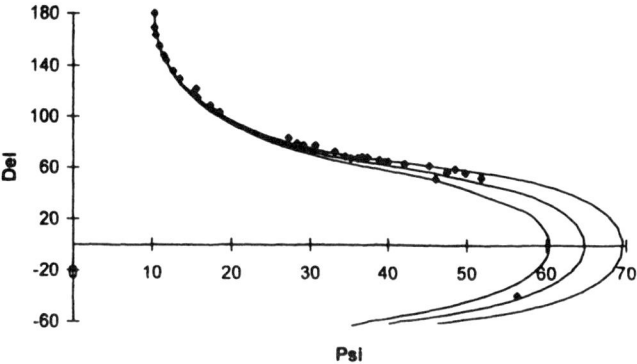

Figure 1: Ellipsometric trajectory of 0 nm to 125 nm poly(zinc-bisquinoline) films extrapolated at 633 nm and 70° incidence and from left to right model trajectories of 1.75, 1.70, and 1.65 refractive indices.

colors of these films ranged from light pink, brown, gold, purple, navy blue, royal blue, and blue-grey as the films increased from 20 nm - 125 nm. The isotropic index of refraction was calculated from the ellipsometric Psi and Del values (see Figure 2) at multiple angles of incidence of seven films of zinc-bisquinoline, between 80 nm and 95 nm respectively, where Psi is near its maximum,[22] based on the Cauchy fit model and its constants given in equation 1:

$$n(\lambda) = A + B/\lambda^2 + C/\lambda^4 \qquad (1)$$

with: $A = 1.62 \pm .01$, $B = 0.02 \pm .01$, $C = 0.002 \pm .002$

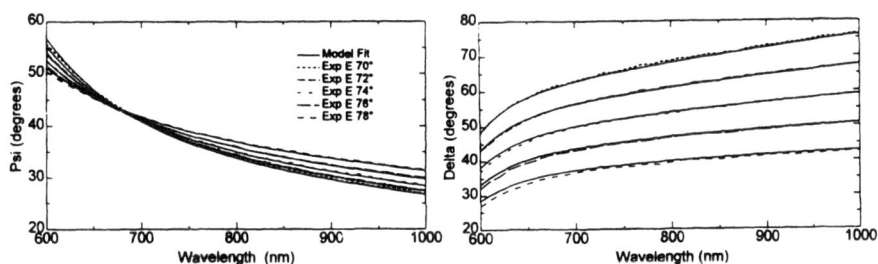

Figure 2: PSI and DEL values of a 93 nm poly(zinc bisquinoline) film and Cauchy model fit.

Figure 3 illustrates the isotropic index of refraction for zinc bisquinoline in its transparent region using the Cauchy fit. When this is extrapolated at 633nm a refractive index of 1.68 ± 0.01 is obtained. The resulting isotropic refractive index and subsequently the density of these films are higher in comparison to reported values of multilayer self-assembled films of zirconium phosphonates[7-11], and polyelectrolytes.[15] The high refractive index of zinc bisquinoline can be attributed to the rigidity of zinc-bisquinoline and its reduced hydration during the self-assembly.

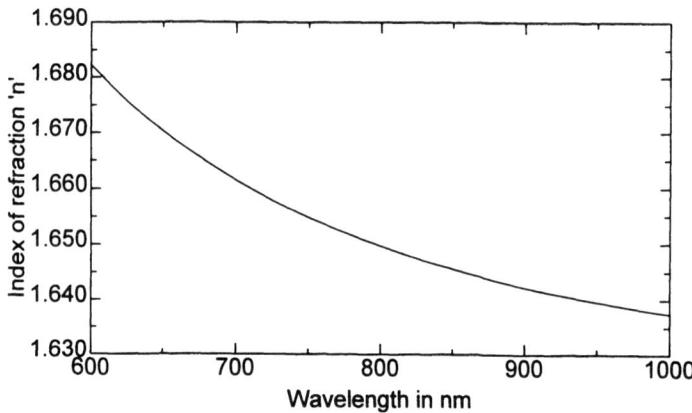

Figure 3: Index of refraction of poly(zinc bisquinoline) from the Cauchy model.

The single crystal of zinc quinoline dihydrate exhibits a density of 1.68 g/cm^3 and refractive indices, as determined by immersion methods to be, $n_\alpha = 1.65$, $n_\beta = 1.78$ and $n_\gamma > 1.82$.[23] Zinc bisquinoline powders have been shown to be semicrystalline in nature,[24] and therefore packing may be a contributing factor to the enhanced refractive index of these self-assembled films.

Zinc bisquinoline films were shown to be capable of forming uniform, pin-hole free films over large areas, capable of producing working electroluminescent devices of thicknesses as thin as 400 Å. The lack of porosity in such films have been crucial in avoiding short circuit formation during the subsequent metallization process of single layered thin films.[17] Figure 1 indicates that films below 20 nm follow a trajectory, although precise refractive indices are difficult to obtain at these thicknesses.[25,26] AFM images on a 90 Å film (Figure 4), during self-assembly, have shown a surface roughness of about 14 - 17 Å. The uniformity of the film becomes more apparent at these thicknesses because the roughness of the substrate is in the order of 12 Å. The initial surface roughness has been attributed to initial surface coverage effects. These films under quartz crystal microbalance (QCM) measurements have shown that the film weight increases nonlinearly during the first four sequential layering steps and reaching a stable growth rate values over the sixth layering step.[27] This initial surface coverage effect has been seen with a number of polyanion/polycation self-assemblies, where at early stages the growth on unfunctionalized

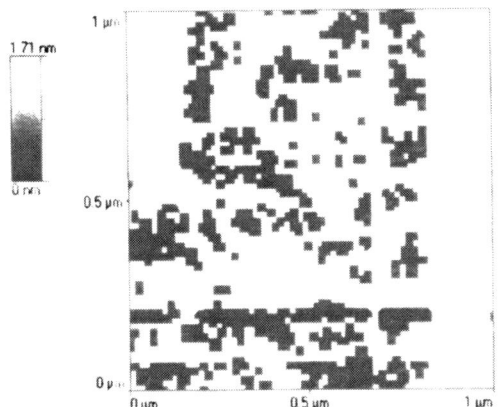

Figure 4: AFM image of a 90 Å poly(zinc bisquinoline) film.

surfaces was reduced.[15,28] The achieved film uniformity makes this a suitable method for thin-film growth of organic semiconductors.

CONCLUSION

Zinc bisquinoline films were determined to have a refractive index of 1.68 ±0.01 at 633 nm. The uniformity of the films was demonstrated by the ellipsometric trajectory of films of varying thicknesses, and the AFM showed a roughness of only 14 - 17 Å for a 90 Å film on silicon of approximate roughness of 12 Å. These films in comparison to spun coated polymer films exhibit increased uniformity and higher index of refraction than other self-assembled multilayer systems. A better understanding of the underlying structure and orientation of zinc bisquinoline self-assemblies will provide the necessary insight for improving the photoluminescence and electroluminescence behavior of these films.

ACKNOWLEDGEMENTS

The authors wish to thank Dr. Y. Lvov, and Mr. T. Fabian for helpful discussions. Financial support from NSF CAREER Grant DMR-9702220 and the Critical Technologies Program through the Institute of Materials Science, University of Connecticut, are greatly appreciated.

BIBLIOGRAPHY

1)J. Sagiv, J. Am. Chem. Soc., **102**, p. 92 (1980).
2)R.G. Nuzzo and D.L. Allara, J. Am. Chem. Soc., **105**, p. 4481 (1983).

3)D.L. Allara and R.G. Nuzzo, Langmuir, 1, p. 45 (1985).

4)N. Tillman, A. Ulman, J.S. Schildkraut and T.L. Penner, J. Am. Chem. Soc., 110, p. 6136 (1988).

5)N. Tillman, A. Ulman and T.L. Penner, Langmuir, 5, p. 101 (1989).

6)M.D. Porter, T.B. Bright, D.L. Allara and C.E.D. Chidsey, J. Am. Chem. Soc., 109, p. 3559 (1987).

7)H. Lee, L.J. Kepley, H.-G. Hong and T.E. Mallouk, J. Am. Chem. Soc., 110, p. 617 (1988).

8)H. Lee, L.J. Kepley, H.-G. Hong, S. Akhter and T.E. Mallouk, J. Phys. Chem., 94, p. 8869 (1990).

9)H.E. Katz, M.L. Schilling, C.E.D. Chidsey, T.M. Putvinski and R.S. Hutton, Chem. Mater., 3, p. 699 (1991).

10)H.E. Katz, G. Scheller, T.M. Putvinski, M.L. Schilling, W.L. Wilson and C.E.D. Chidsey, Science, 254, p. 1485 (1991).

11)H.E. Katz, Chem. Mater., 6, p. 2227 (1994).

12)G. Decher, J.D. Hong and J. Schmitt, Thin Solid Films, 210/ 211, p. 831 (1992).

13)G. Decher, B. Lehr, K. Lowack, Y. Lvov and J. Schmitt, Biosensors & Bioelectronics, 9, p. 677 (1994).

14)G. Decher, Y. Lvov and J. Schmitt, Thin Solid Films, 244, p. 772 (1994).

15)Y. Lvov, K. Ariga, I. Ichinose and T. Kunitake, J. Am. Chem. Soc., 117, p. 6117 (1995).

16)D.E. Aspnes *The Accurate Determination of Optical Properties by Ellipsometry*; Academic Press: New York, 1985.

17)D.L. Thomsen, K.A. Higginson and F. Papadimitrakopoulos, ACS Polymer Preprints, 38, p. 353 (1997).

18)J.P. Phillips, J.F. Deye and T. Leach, Anal. Chim. Acta, 23, p. 131 (1960).

19)R.D. Archer, C.J. Hardiman, K.S. Kim, E.R. Grandbois and M. Goldstein *Metal-Chelate Polymers: Structural/property relationships as a function of the metal ion*; R.D. Archer, C.J. Hardiman, K.S. Kim, E.R. Grandbois and M. Goldstein, Ed.; Plenum Press: NY, 1985, pp 355.

20)D.L. Thomsen and F. Papadimitrakopoulos, A.C.S. Polymer Preprints, 38, p. 398 (1997).

21)"Applied Materials Instruction Manual: Ellipsometer II," Applied Materials, Inc, 1976.

22)H.G. Tompkins *A User's Guide to Ellipsometry*; Academic Press: San Diego, CA, 1993.

23)L.L. Merritt, R.T. Cady and B.W. Mundy, Acta. Cryst., 7, p. 473 (1954).

24)E.W. Berg and A. Alam, Anal. Chim. Acta, 27, p. 454 (1962).

25)P. Drude, Ann. Phys. Chem., 36, p. 865 (1889).

26)A.N. Saxena, J. Opt. Soc. Am., 55, p. 1061 (1965).

27)F. Papadimitrakopoulos, D. L. Thomsen, and T. Phely-Bobin, ACS Polymer Preprints, Spring 1998.

28)Y.M. Lvov and G. Decher, Crystallography Reports, 39, p. 628 (1994).

PREPARATION OF CHROMOPHORIC DIPOLAR DENDRONS
AS SECOND-ORDER NONLINEAR OPTICAL MATERIAL

S. YOKOYAMA AND S. MASHIKO
Communications Research Laboratory, syoko@crl.go.jp,
588-2, Iwaoka, Nishi-ku, Kobe 651-24, JAPAN

ABSTRACT

Dendritic macromolecules, called "dendrons," having a branching structure modified with an electron donor/acceptor functionalized azobenzene chromophore have been synthesized. The structurally different dendrons possess different numbers, 1, 3, 7, and 15, of chromophore units. The second harmonic generation (SHG) of dendrons was measured in the molecular oriented thin films prepared by the Langmuir-Blodgett film transfer technique. In such structurally organized thin films, individual chromophores coherently contribute to the increase in the molecular hyperpolarizability of dendrons. The SHG activity became large as the numbers of branching units of dendrons increased. The highest molecular hyperpolarizability was found to be $>8000 \times 10^{-30}$ esu for dendron containing 15 chromophoric units. The SHG results suggest that the structure of dendrons synthesized is uniaxially dipolar, and that there is a particular merit in this structure for the generation of SHG.

INTRODUCTION

A new synthetic approach to the fabrication of topologically complex materials for nonlinear optical ultrathin films has been reported (Scheme 1). The synthetic method involves repeating coupling reactions of high-β chromophore as the dendritic branching unit, and introduction of hydrophobic alkyl chains at exterior positions. The preparation of thin films was implemented by using the Langmuir-Blodgett (LB) film transfer technique, where aliphatic and polar structure of the products synthesized paved the way to construct a molecular oriented structure in these films. The chromophores were chosen for the high nonlinear optical properties of azobenzene derivatives having a π-electron system coupled with electron donor/acceptor groups [1,2]. The starting compound, **1**, formed the LB monolayer film on glass substrate and exhibited SHG efficiency with χ_{zzz} value of 7.0×10^{-8} esu (29 pm/V) at 1064 nm. This level of SHG response was comparable to or higher than that of the well studied nonlinear optical active LB films [3-5]. This film could form a stable molecular orientation without an external electric field, where molecules were highly organized and their molecular axis was oriented noncentrosymmetrically with respect to the surface normal. In addition, the SHG signal from the film was quite stable for long time. These facts represent an important advance in the fabrication of structurally stabilized organic ultrathin films for optoelectronic applications. A novel preparation technique to extend such small organic compounds to polymeric materials is still challenging [6]. This study achieves the increase in molecular hyperpolarizability of dendritic macromolecules by introducing the electron donor/acceptor azobenzene chromophores as branching units. The key structure of dendron for optoelectronic application is a rod-shaped conformation, where individual chromophores coherently contribute to the molecular hyperpolarizability of the dendron. Here, we show that nonlinear optical chromophore can be organized in a fixed noncentrosymmertic dendritic structure to become dipolar, and present experimental evidence of the second-order nonlinear optical property in molecular oriented thin films.

EXPERIMENT

All the compounds have been synthesized as shown in Scheme 1. The starting azobenzene chromophores are **1** and **2**. They are modified with the strong electron acceptor of the nitro group and expected to have high nonlinear optical response. Synthesis details have been previously reported elsewhere [7]. Structure of the products were fully characterized by ^{1}H and

765

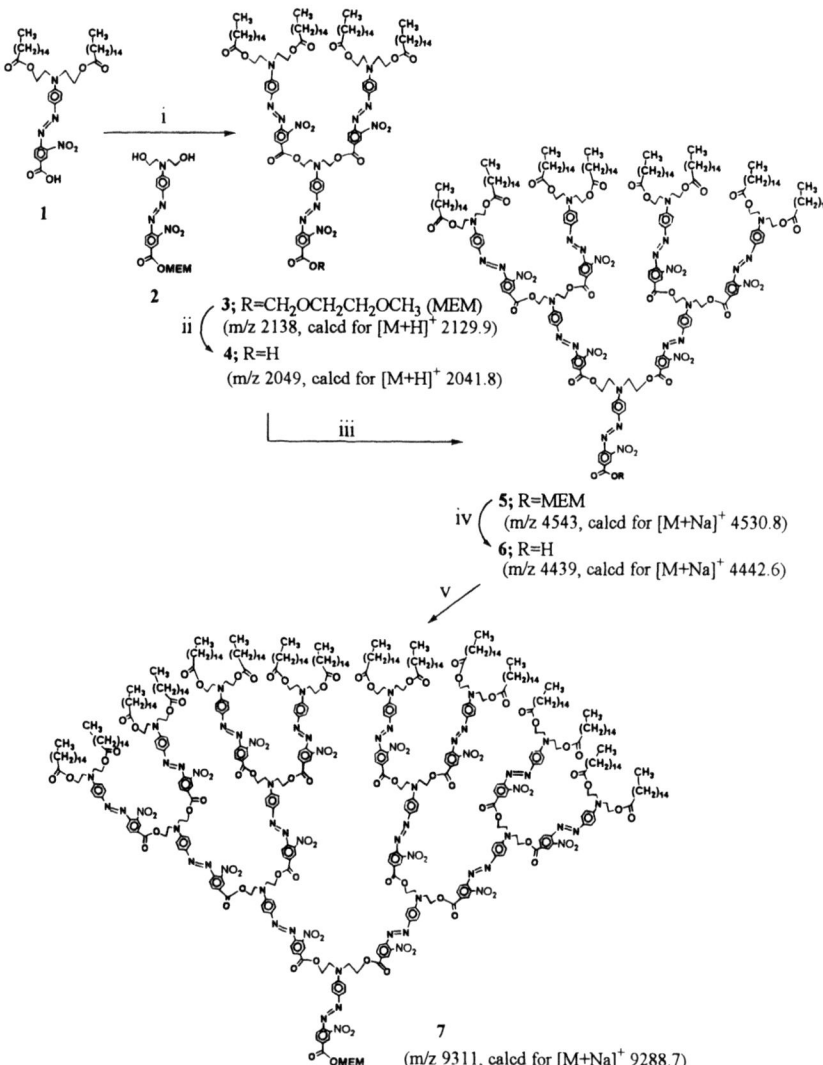

Scheme 1 Preparation of dendrons. *Reagents and conditions*: i, iii, and v, DCC, DMAP, CH₂Cl₂, 20°C; ii and iv, HCl, THF, 0-5°C. The *m/z* values were measure by MALDI TOFMS.

The following labels appear within the scheme:

1

2

3; R=CH₂OCH₂CH₂OCH₃ (MEM)
(m/z 2138, calcd for [M+H]⁺ 2129.9)
4; R=H
(m/z 2049, calcd for [M+H]⁺ 2041.8)

5; R=MEM
(m/z 4543, calcd for [M+Na]⁺ 4530.8)
6; R=H
(m/z 4439, calcd for [M+Na]⁺ 4442.6)

7
(m/z 9311, calcd for [M+Na]⁺ 9288.7)

766

^{13}C NMR spectroscopy after purification by column chromatography. SEC showed the highly purity levels and narrow monodispersity of molecular weights, $M_w/M_n < 1.05$. The m/z values shown in Scheme 1 are exact mass measured by MALDI TOFMS. They were identical to the molecular weights as expected for the structures.

Thin films of **1, 3, 5**, and **7** were prepared by using the LB film transfer technique. Dendrons were dissolved in chloroform with a concentration of 1-0.05 mmol/l, and spread on a water surface in a Langmuir trough. In order to determine the theoretical nonlinear optical response of the films ($\sqrt{I^{2\omega}} (\propto \chi^{(2)}) \propto N\beta$), surface density of dendrons in the films was diluted by mixing with an excessive amount of arachidic acid (AA) in chloroform solution. The single layer films were transferred onto glass substrate from water surface at a constant surface pressure of 20 mN/m.

Transmission SHG measurements were carried out by using a Q-switched Nd-YAG laser (λ=1064 nm, 10 Hz, 8 ns). Fundamental light was passed through polarization optics before being focused onto the sample on a rotating stage. Pulse energies of about 1.0 mJ were used for all the measurements on the thin films. The second harmonic light generated by the sample was separated from the fundamental beam using a filter and a monochromator and subsequently detected by a photomultimeter tube. The resulting signal was processed using a boxcar averager and computer. In all measurements, the p-polarized second harmonic light for p and s-polarized fundamental was determined. This is consistent with $C_{\infty v}$ symmetry for the molecular orientation with respect to the surface normal [8]. A y-cut quartz crystal (d_{11}=0.32 pm/V) was used as a reference.

RESULTS AND DISCUSSION

In previous SHG studies, functional LB monolayer films formed on water surface and solid substrate have been widely used for the model of organic adsorbates formed at interfaces, where the functional unit is oriented in the surface normal direction without a center of inversion by means of alkyl chains [9]. In such a model, it is assumed that the molecule has one main

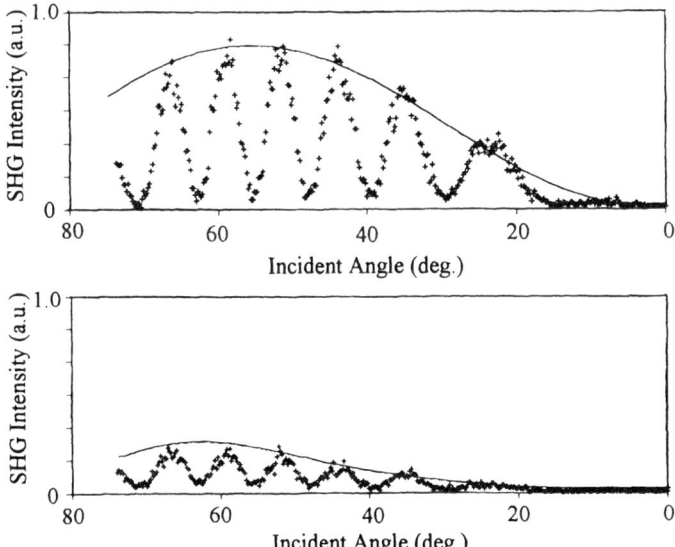

Figure 1. SHG intensity as a function of fundamental beam incident angle from a glass substrate covered with the thin film of **7**. (a) p-Polarized SHG fringe measured by p-polarized fundamental beam. (b) p-Polarized SHG fringe measured by s-polarized fundamental beam. The solid lines are theoretical envelopes fitted to the data.

component of molecular hyperpolarizability, its direction is parallel to the molecular axis, and the molecular orientation is inclined at an averaged tilt angle to the surface normal with a random distribution of the azimuthal angle. We first show that dendrons synthesized can form the LB-type thin films and their film structure determined by SHG measurements is analogous to the conventional SHG active LB monolayer film. Secondly, we determine molecular hyperpolarizability of dendrons, and show an increase in β values as the number of chromophore units becomes large.

Figure 1 shows typical SHG fringes obtained in the film of dendron **7** prepared on a glass substrate. The fringes show the p-polarized SHG signal intensity by p and s-polarized fundamental beams as a function of the incident angles relative to the sample plane. The fringes resulted from the phase difference between two SHG waves generated at either side of the film on a glass substrate. The complementary constructive and destructive interference indicates that the SHG is fully coherent, which is characteristic of the SHG active LB films with uniform morphology and polar ordering [5, 10, 11]. This result is supported by the fact that the surface structure measured by the atomic force microscopy was observed to be uniform within the area larger than the square of a few hundred micrometers (not shown). The sample shows no inplane anisotropy of SHG intensity in the film plane. This indicates that the film has a uniaxial symmetry about the surface normal and that the distribution of the molecular orientation is random. The observation is consistent with the spectroscopy result that the polarized absorption spectra of the films indicated a dichroic ratio for A_s/A_p of 1.05. The films, therefore, have only two independent tensor element of $\chi^{(2)}$, that are χ_{zzz} and $\chi_{zxx}=\chi_{zyy}$ [4, 12, 13]. The theoretical curves shown in Figure 4 are envelopes fitted to the SHG data by using an experimentally determined ratio χ_{zzz}/χ_{zxx}. The best fit between the experimental data and theoretical envelope gave second-order nonlinear coefficients of $\chi_{zzz}=5.0\times10^{-8}$ esu (21 pm/V) and $\chi_{zxx}=3.1\times10^{-8}$ esu (13 pm/V). Similar SHG fringes were also obtained for other films of **1**, **3**, and **5**.

Figure 2 shows the absorption spectra of the diluted films of 7/AA (=1/0-1/80) on glass substrates at an incident angle of 45°. The maximum band at 467 nm is due to π-π^* transition of the azobenzene unit. It can be seen that the chromophore concentration in these films can be diluted by mixing **7** and AA with various mixing ratios. In these diluted films, the number of surface densities of the dendron could be estimated to be 0.47-2.36×10^{14} cm^{-2} from the surface pressure and area curves measured on water surface.

Figure 3a shows the plot of maximum absorbance of the diluted films, **1**, **3**, **5**, and **7**, versus

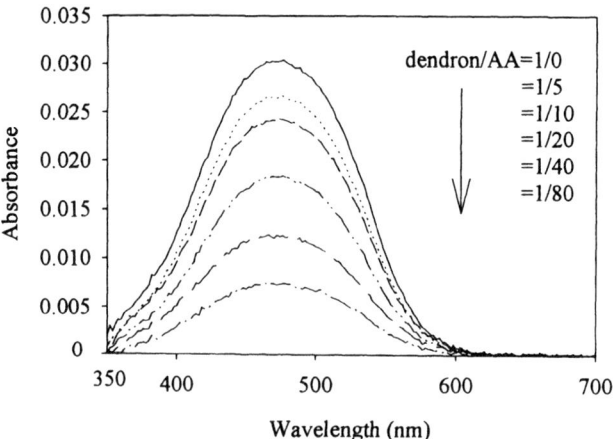

Figure 2. Absorption spectra of the diluted thin films of dendron **7**. Films were prepared by mixing dendron and AA with various ratios of 1/0-1/80 (in mol).

the number of surface densities of dendrons. The linearity of the plots indicates that the surface density of dendrons could be systematically diluted by mixing with AA in the films. For diluted chromophoric thin films the SHG intensity generated should show a quadratic decrease with decreasing the number of surface density, $\sqrt{I^{2\omega}} \propto N$ [14]. The SHG results of the diluted films are shown in Figure 3b in a plot of SHG intensities, relative to a quartz reference, versus the number of surface density. The theoretical relation of $\sqrt{I^{2\omega}} \propto N$ is clearly observed in Figure 3b. For all the diluted films, the plots are linear, as expected for uniform polar films having a constant structure and molecular orientation. In Figure 3b, the slope of the straight lines becomes large as the number of branching unitsof dendron increases. Since the square root of SHG intensity is generally proportional to $\chi^{(2)}$ and the slope, $\sqrt{I^{2\omega}}(\propto \chi^{(2)})/N$, represents the normalized SHG activity, the SHG results suggest that the polar ordering of dendrons becomes large as the number of the chromopholic units in dendrons increases. Furthermore, such an increase in the SHG activities suggests that the chromophores are organized in a noncentrosymmetric arrangement in the dendritic structure and that the individual chromophores should coherently contribute to generate SHG. The normalized SHG activity discussed here includes molecular hyperpolarizability, local field factor, and molecular orientation of dendrons in the thin films [3,9].

In order to determine how well the nonlinear optical activity of the dendrons could be contributed to by that of the individual chromophores, we estimated the molecular hyperpolarizability of the dendrons. Table 1 summarizes the properties of the films of individual chromophore, **1**, and the dendrons, **3**, **5**, and **7**. The film thickness was measured by using

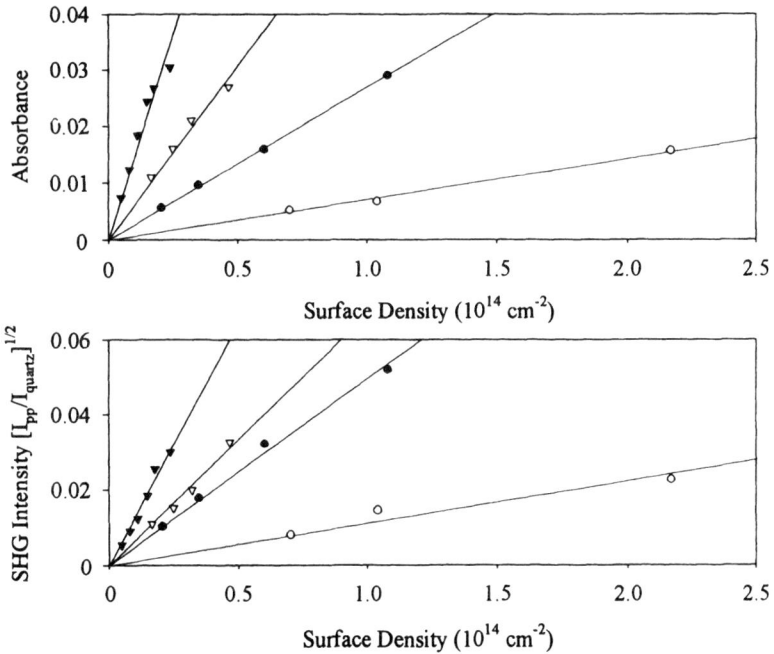

Figure 3. Absorbance and SHG intensity of the diluted thin films of **1** (○), **3** (●), **5** (▽), and **7** (▼) as a function of the number of surface density. (a) Absorbance due to the band of azobenzene unit (see Table 1). (b) Square root of the *p*-polarized SHG intensity by *p*-polarized fundamental beam, relative to SHG intensity of a quartz reference, at an incident angle of 44°. This angle is one of the angles that gives complementary constructive interference.

Table 1. Properties of the azobenzene chromophore and dendrons. M_n, molecular weight; n_c, the number of the chromophore unit; l_f, film thickness; N, number of surface density; λ_{max}, wavelength of the absorption maximum; β, molecular hyperpolarizability.

Compound	M_n	n_c	l_f (nm)	N ($\times 10^{14}$ cm^{-2})	λ_{max} (nm)	β (10^{-30} esu)
1	851	1	2.7	2.2	430	870
3	2129	3	3.7	1.1	463	2630
5	4508	7	4.8	0.47	466	3960
7	9626	15	5.4	0.24	467	8260

atomic force microscope. The β value of compound 1 is estimated to be ~800×10^{-30} esu, a value which is equal to the literature value [6]. Although molecular ordering of the branching structure in dendrons seems imperfect, the molecular hyperpolarizabilities of dendrons were much larger than those of the individual chromophore. The increase in the molecular hyperpolarizability was coherently contributed to the number of chromophores in dendrons, n_c. The highest molecular hyperpolarizability was found to be >8000×10^{-30} esu for dendron 7. The dendron/chromophore hyperpolarizability ratio was found to be 9.5. These SHG experimental results supported our synthesis strategy that chromophores were organized in the dendritic structure to form a rod-shaped conformation in these films rather than the spreading conformation, so products predicted a large increase in the nonlinear optical activity compared with the individual chromophore. The SHG results presented here provided a conclusive demonstration that the rod-shaped dendritic material enhanced its molecular hyperpolarizability through a noncentrosymmetric arrangement of functional units in a branching structure.

REFERENCES

1. A. Ulman, S. C. S. Willand, W. Köhler, D. R. Robello, D. J. Williams, and L. Handley, J. Am. Chem. Soc., 112, p. 7083 (1990).
2. P. N. Prasad and D. J. Williams, Introduction to Nonlinear Optical Effects in Molecules and Polymers, Wiley, New York, 1991.
3. G. Marowsky and R Steinhoff, Opt. Lett. 13, p. 707 (1988).
4. G. Marowsky, L. F. Chi, D. Möbuis, R. Steinhoff, Y. R. Shen, D. Dorsch, and B. Rieger, Chem. Phys. Lett., 147, p. 420 (1988).
5. D. Lupo, W. Prass, U. Scheunemann, A. Laschewsky, Helmut Ringsdorf, and I. Ledoux, J. Opt. Soc. Am. B, 5, p. 300 (1988).
6. M. Kauranen, T. Verbiest, C. Boutton, M. N. Teerenstra, K. Clays, A. J. Schouten, R. J. M. Nolte, and A. Persoons, Science, 270, p. 966 (1995).
7. S. Yokoyama, T. Nakahama, A. Otomo, ans S. Mashiko, Chem. Lett., p. 1,137 (1997).
8. M. A. Carpenter, C. S. Willand, T. L. Penner, D. J. Williams, and S. Mukamel, J. Phys. Chem., 96, p. 2801 (1992).
9. Y. R. Shen, Nature, 337, p. 519 (1989).
10. H. E. Katz, G. Scheller, T. M. Putvinski, M. L. Schilling, W. L. Wilson, and C. E. D. Chidsey, Science, 254, p. 1485 (1991).
11. D. Li, B. I. Swanson, J. M. Robinson, and M. T. Hoffbauer, J. Am. Chem. Soc., 115, p. 6975 (1993).
12. G. Marowsky, G. Lüpke, R. Steinhoff, L. F. Chi, and D. Möbius, Phys. Rev. B., 41, 4480 (1990).
13. T. G. Zhang, C. H. Zhang, and G. K. Wong, J. Opt. Soc. Am. B, 6, p. 902 (1990).
14. I. R. Girling, N. A. Cade, P. V. Kolinsky, R. J. Jones, I. R. Peterson, M. M. Ahmad, D. B. Neal, M. C. Petty, G. G. Roberts, and W. F. Feast, J. Opt. Soc. Am. B, 4, p. 6 (1987).

MODIFICATION OF SPONTANEOUS EMISSION FROM PERIODIC ARRAY OF SPHERICAL POLYSTYRENE PARTICLES CONTAINING FLUORESCENT MOLECULES

T. YAMASAKI, T. TSUTSUI
Graduate School of Engineering Sciences, Kyushu University, Fukuoka, JAPAN 816

ABSTRACT

Emissive behavior of dye molecules embedded in the systems with the size of visible wavelength were studied. The structures consisted of close-packed arrays of spherical dielectric particles with diameter of submicrometer were fabricated. Photoluminescence from self-assembled arrays of dye-doped polystyrene microspheres were examined. In this systems, partial suppression of spontaneous emission by ordered structures were observed. Electrically-pumped emission from organic light emitting diodes with highly ordered arrays of silica microspheres were attempted.

INTRODUCTION

Thin-film light emitting diodes (LED's) made of organic emissive materials have gathered huge interest in recent years. Surface emissive displays made of organic LED's show high quantum efficiency, excellent stability and reliability [1, 2]. Today, experiments of electrically-pumped lasing from organic thin films have been challenged [3, 4]. We have proposed the introduction of the Fabry-Perot microcavity structures to organic LED's and demonstrated that both spectral narrowing and specially directed emission were accomplished [5]. The total thickness of organic layers in LED's are about 100 nm, which are less than the wavelength of emissive light. Thus the quantum optical effects due to the confinement of electromagnetic fields in a small space are expected to be observed by using microcavity structures [6].

On the other hand, we found that a simple planar microcavity structure is insufficient for complete confinement of photons within it, which will be one of the promising starting points for electrically-pumped lasing from organic emissive materials. If two or three dimensional cavity with the size of wavelength of emissive radiation is introduced into organic LED's, one can expect to observe novel quantum phenomena based on more effective confinement of photons. However, it is quite difficult to produce three-dimensional (3D) systems with the size of submicrometer by use of modern technologies, such as lithography. Two dimensional (2D) or 3D periodic structures are expected to be easier to construct rather than 2D or 3D microcavities. Then we direct our attention to the self-organization of organic materials, which are expected to provide 3D periodic structures, in other words, so-called photonic crystals.

Photonic crystals have been attracted much attention as a stage of optical investigations recently [7-11]. Periodic modulations of dielectric constants in a photonic crystal behave essentially like 3D dielectric mirrors and produce the common stop band in all directions, which are called photonic band gap (PBG). Thus one can control the coupling between excited molecules and radiation fields far more effectively by using it than the case of an one-dimensional microcavity. Introducing the structure of photonic crystal in organic LED's, improvement of their emissive characteristics on the basis of quantum optical effects may be expected [9]. However, because of difficulties in fabricating 3D periodic structures with the smaller size, experimental studies on the photonic crystals have been limited in the wavelength region from microwave [7] to near-infrared [10, 11]. In a visible light region, a few pioneering experimental works using cholesteric liquid crystals as a one dimensional periodic structure have been reported [12, 13].

Colloidal crystals made from microsphere latex, such as polystyrene or silica, are used for achieving 3D periodic dielectric structures and exhibit the strong Bragg scattering of visible light [14]. Although these crystals do not show complete PBG, face-centered cubic (fcc) structures of colloidal crystals have the advantage of self-organization in the order of submicrometer and have been used for the study on PBG in the visible light region [15, 16]. Solid self-organized structures consisted of microspheres are formed by evaporating water from the suspensions [17, 18]. The solid crystals are easy to handle and have higher refractive index contrast of 1.60 than the case of 1.20 for colloidal crystals in the water.

We have started the study on the construction of the solid crystals, as an example of photonic crystals in visible regions, and tried to examine spontaneous emission properties by both optical and electrical pumping.

EXPERIMENT

Polystyrene particles doped with a fluorescent dye (YG) (average diameter of 213±8 nm) were used. The ordered arrays of polystyrene spheres were prepared on a glass substrate by controlling temperature and humidity during casting processes [18]. We have succeeded in preparing highly ordered solid films with the average thickness of 30 μm from aqueous suspensions of particle concentration of 2.5 wt %. The other film of dye doped polystyrene without an ordered structure was also prepared for reference by casting onto a glass substrate from the dichromethane solutions of the same particles. The average thickness of this film was 17 μm.

The surface of ordered arrays were observed using an optical microscope and scanning electron microscope (SEM). Fig. 1 shows a typical SEM image from ordered film of dye-doped polystyrene microspheres used. The transmission and reflection spectra of the films were measured with Hitati 330 spectro-photometer. Photoluminescence (PL) spectra of films were obtained with multichannel analyzer (PMA-11, Hamamatsu Photonics K. K.) from the films which were fixed on a rotational stage and excited by the light of xenon lamp through 250 mm monochromator.

Self-organized 2D arrays of silica spheres (average diameter of 550 nm) were prepared on ITO coated glass substrate using the improved method described eleswhere [19]. The device consisted of a glass substrate with a patterned ITO electrode, silica sphere layer, poly(9-vinylcarbazole) layer molecularly doped with 2,5-bis(4-naphtyl)-1,3,4-oxadiazole and Coumarine-6 for EML, an Alq layer for ETL and a top MgAg alloy cathode, in which EML and ETL represent emissive and electron-transporting layers, respectively. This device structure is similar to a standard-type polymer-dispersed dye LED's [20].

RESULTS

We confirmed from optical microscope observations for the surfaces that the ordered films of polystyrene particles were composed of aggregates of microcrystals with specified orientations. The SEM image shown in Fig. 1 clearly demonstrates the presence of the regularly packed arrangement of polystyrene spheres corresponded to either (111) surface of the fcc structure or (001) surface of the hexagonal close packed (hcp) one. At present, we

Fig.1 SEM micrograph of the surface of polystyrene microsphere arrays; scale marker is 500 nm.

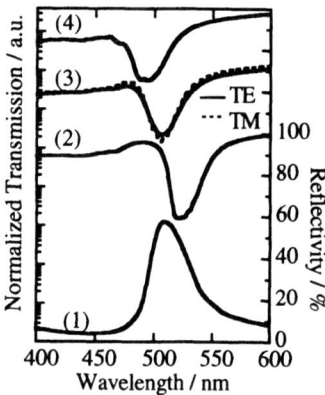

Fig. 2 Reflection and transmission spectra for ordered arrays of poly-styrene microspheres. (1): reflection sprctrum of ordred film at incident angle of 0°, (2) to (4): Normalized transmission spectra taken along incident angles, 0°, 24° and 32° from bottom to top. Dashed (solid) line indicates the spectrum for TM (TE) polarized incident light at 24°.

have no evidence to differentiate between fcc and hcp lattice packing.

Transmission measurements were performed at different incident angle θ with respect to the surface normal. The results obtained for the ordered arrays of polystyrene microspheres are shown in Fig. 2. Attenuation bands, about two order of magnitude deep, due to the Bragg diffraction were observed. With increasing θ, the opacity band shifted toward the shorter wavelength. Bragg reflection peak was appeared in the reflection spectrum for the same film. No attenuation was observed in the transmission spectrum for the polystyrene structure-less film cast from a dichromethane solution. A little differences were found between TE and TM polarized light, as shown in Fig 2 (3). Evidently, the periodic dielectric structure consisted of close-packed polystyrene micro-spheres provides the optical stop band in the visible region.

Photoluminescence from ordered arrays of polystyrene spheres as a function of an observation angle θ are shown in Fig 3. In the same figure, a typical PL spectrum from fluorescent dyes within the structure-less film is shown. One should note that the full width at half maximum of the stop band was narrower than that of the intrinsic emission band of the dyes shown in Fig. 3 (5). Therefore the opacity bands were observed as deep dips in the PL spectra. With increasing observation angle, the dip also shifted toward the shorter wavelength. The dip was observed no longer at $\theta = 50°$.

These behavior are summarized in Fig. 4, where both the center wavelength of the dips in emission spectra and that of the stop band are plotted against observation angle θ (incident angle for the case of transmission experiments). Both curves showed good coincidence and are well explained by using the Bragg low and Snell's low;

$$2 \cdot n_f \cdot d \cdot cos\beta = \lambda \qquad (1)$$

$$n_f \cdot sin\beta = n_2 \cdot sin\theta \qquad (2)$$

where n_f and n_2 are the effective index of refraction for the polystyrene/air composite and the

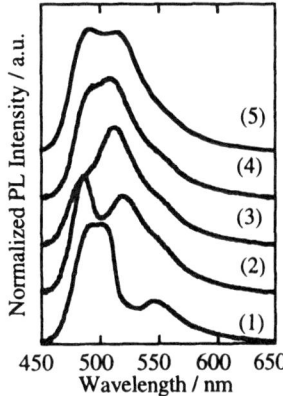

Fig. 3 Photoluminescence spectra for the orderd film of dye doped polystyrene microspheres observed along specific directions. The curves (1) to (4) show for the angle of 0°, 24°, 32° and 50° from bottom to top. The curve (5) indicates photoluminescence spectrum for the structure-less dye-doped polystyrene cast film.

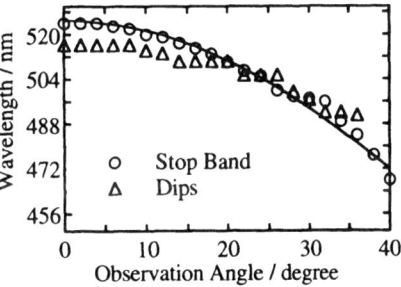

Fig. 4 The angular dependence of the wavelengths of the stop bands and the dips in emission spectra for the ordered array of polystyrene particles. The symbols are the experimental points. The solid line is the curve for the best fit using Bragg condition (R = 219 nm).

refractive index of the air respectively, and β is the Bragg diffraction angle within the structures. The interplaner spacing for the close-packed planes (either (111) of fcc or (001) of hcp), d is estimated from the relation, $d = 0.816 \cdot R$ in which R is an effective particle diameter. For this sample, we used the value of n_f obtained averaging dielectric constants,

$$(n_f)^2 = (n_1)^2 \cdot \phi + (n_2)^2 \cdot (1 - \phi) \qquad (3)$$

where n_1 and ϕ are the reflective index of the polystyrene and a filling factor ($\phi = 0.74$ for close-packed array) respectively. Assuming $R = 219$ nm, the calculated curve gave the best fit to the experimental curves, as indicated in Fig. 4 using a solid line. Although the value of R which gave the best fit was a little longer than the diameter of polystyrene spheres, this difference is within the standard deviation of diameter distribution. As mentioned above, the alteration in emission spectra were explained in terms of the partial suppression of radiation modes within the stop band by ordered structures.

We also confirmed from SEM observations of the surface for the film on a ITO substrate that the ordered 2D array was composed of close-packed arrangement of silica microspheres. After the standard procedure to prepare LED's, it was confirmed also from SEM images that the close-packed arrangement of silica spheres was kept. Fig. 5 shows the device with the 2D periodic structure used. Electroluminescence (EL) were observed from this type of devices. The profiles of EL spectra were modified by the presence of an ordered 2D structure, as shown in Fig. 6.

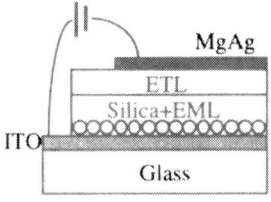

Fig. 5 The device structure of organic LED's with orderd array of silica microspheres.

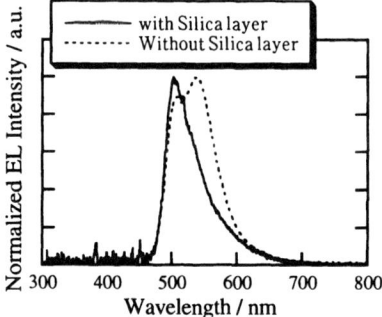

Fig. 6 Electroluminescence spectra from LED's with and without ordered array of silica microspheres .

CONCLUSION

Using the self-organization of materials, we fabricated highly ordered both 3D and 2D structures consisted of close-packed arrays of polystyrene and silica microspheres. The periodic dielectric structures of the prepared arrays provided pseudogaps or nonoverlapping gaps because of low refractive index contrast of 1.60 for polystyrene/air composite and a filling factor of 74%. The partial suppression of radiation modes in the structures, however, resulted in the decrease of the spontaneous emission intensity from the dye embedded in the photonic crystals at the wavelength region of the stop band. Use of specified materials with higher refractive indices are required for the purpose of obtaining a full photonic band gap, and these structures are expected to be useful for both basic and applied researches on the confinement of photon within the structure of wavelength-size scale. We demonstrated the preparation of LED's with self-organized 2D arrays of silica spheres, as a first step for the organic LED's with a photonic crystal structure and the controlling electrically-pumped emission perfectly.

ACKNOWLEDGMENT

The authors would like to thank SHOKUBAI KASEI INDUSTRY Co. for providing silica suspensions. This work has been partly supported by the CREST program from Japan Science and Technology Corporation.

REFERENCES

1. T. Tsutsui, MRS Bulletein **22,** 39 (1997).

2. J. R. Sheats, H. Antoniadis, M. Hueschen, W. Leonard, J. Miller, R. Moon, D. Roitman and A. Stocking, Science **273**, 884 (1996).

3. F.Hide, M. A. Diaz-Garcia, B. J. Schwartz, M. R. Anderson, Q. Pei and A. J. Heeger, Science **273**, 1833 (1996).

4. V. G. Kozlov, V. Bulovic, P. E. Burrows and S. R. Forrest, Nature **389**, 362 (1997); M. Berggren, A. Dodabalapur, R. E. Slusher and Z. Bao, ibid., **389**, 468 (1997).

5. N. Takada, T. Tsutsui, and S. Saito, Appl. Phys. Lett. **63**, 594 (1993); T. Tsutsui, N. Takada, S. Saito and E. Ogino, ibid., **65**, 1868 (1994).

6. Y. Yamamoto and R. E. Slusher, Phys. Today **46**, 66 (1993).

7. E. Yablonovitch, Phys. Rev. Lett. **58**, 2059 (1987); E. Yablonovitch, T. J. Gmitter, and K. M. Leung, ibid., **67**, 2295 (1991).

8. K. M. Ho, C. T. Chan, and C. M. Soukoulis, Phys. Rev. Lett. **65**, 3152 (1990).

9. J. D. Joannopoulos, P. R. Villeneuve, and Shanhui Fan, Nature (London) **386**, 143 (1997).

10. K. Inoue, M. Wada, K. Sakoda, A. Yamanaka, M.Hayashi, and J. W. Haus, Jpn. J. Appl. Phys. **33**, L1463 (1994).

11. T. F. Krauss, R. M. De La Rue, and S. Brand, Nature (London) **383**, 699 (1996).

12. R. Dreher, and H. Schomburg, Chem. Phys. Lett. **25**, 527 (1974).

13. M. Hara, H. Takezoe, A. Fukuda, E. Kuze, Y. Kaizu, and H. Kobayashi, Mol. Cryst. & Liq. Cryst. **116**, 253 (1985).

14. N. A. Clark, A. J. Hurd, and B. J. Ackerson, Nature (London) **281**, 57 (1979).

15. J. Martorell, and N. M. Lawandy, Phys. Rev. Lett. **65**, 1877 (1990); I. I. Tarhan, and G. H. Watson, ibid., **76**, 315 (1996).

16. I. I. Tarhan, M. P. Zinkin, and G. H. Watson, Opt. Lett. **20**, 1571 (1995).

17. H. Miguez, C. Lopez, F. Meseguer, A. Blanco, L. Vazquez, R. Mayoral, M. Ocana, V. Fornes, and A. Mifsud, Appl. Phys. Lett. **71**, 1148 (1997)

18. C. D. Dushkin, K. Nagayama, T. Miwa, and P. A. Kralchevsky, Langmuir **9**, 3695 (1993).

19. S. Matsushita, T. Miwa and A. Fujishima, Langmuir **13**, 2582 (1997).

20. C. P. Lin, T. Tsutsui and S. Saito, J. Polym. Res. **2(3)**, 133 (1995).

FEMTOSECOND OPTICAL NONLINEARITY IN THE SQUARYLIUM DYE J-AGGREGATES

M. FURUKI*, L. S. PU**, F. SASAKI***, S. KOBAYASHI***, T. TANI***
* FESTA Laboratories, 5-5 Tokodai, Tsukuba, Ibaraki 300-26, Japan, furuki@festa.or.jp
** Foundation Res. Lab., Fuji Xerox Co. Ltd., 430 Sakai, Nakaimachi, Kanagawa 259-01, Japan
*** Electrotechnical Laboratory, 1-1-4 Umezono, Tsukuba, Ibaraki 305, Japan

ABSTRACT

Squarylium dye with two propyl and two hexyl groups forms a mono-molecular layer with domains of J-aggregates by spreading its chloroform solution at an air-water interface. Fluorescence microscopy and an absorbance evaluation which also considers the reflectance suggest that this Langmuir-film has a close packed two-dimensional structure with a complete J-aggregate formation of all dye molecules. Strong and ultrafast nonlinear optical responses, indicating delocalization of excitons over 30 molecules, have been observed at 5 °C from this ideal mono-molecular layer of J-aggregates.

INTRODUCTION

Increasing attention has been paid to the nonlinear optical properties of excitons, not only in inorganic materials, but also in organic materials. Especially, the J-aggregate of dye molecules is one of a most interesting topics in this area [1-6]. High optical nonlinearity the and fast decay of excited states in J-aggregates have been discussed based on the picture of delocalized Frenkel-excitons. However, previous studies on the J-aggregates were made on uncertainly dispersed forms in matrixes. For practical applications, observation of optical nonlinearity by Frenkel exciton in J-aggregates with a well-defined structure is desired. Spreading a mono-molecular layer at the air-water interface, called a Langmuir-film, is a most-promising method to achieve highly ordered and highly oriented molecular films with a clear structure [7].

Through our study of squarylium dye Langmuir-films, we have found that molecular structures with two propyl groups substituted on both ends of the chromophore form J-aggregates [8]. Upon them, a Langmuir film of bis[4-(N-propylhexylamino)phenyl]squarylium (SQ36) showed a most intense J-band at 777 nm.

For the purpose of estimating the proper nonlinear optical properties, characterizations of the morphology observed by fluorescence microscopy and the linear optical properties determined by the transmission and reflection spectra of the Langmuir film are discussed in this paper. An analysis of the femtosecond nonlinear responses suggests delocalization of the Frenkel excitons in these two-dimensional J-aggregates.

EXPERIMENT

A solution of squarylium dye was prepared by dissolving SQ36 (Fig. 1) in chloroform at a concentration of 8.31×10^{-4} M. Langmuir films of squarylium dye J-aggregate were formed by spreading this chloroform solution on a water surface [8]. For samples of fluorescence micrographs, thinly spread water on a glass slide was used as a substrate for Langmuir films.

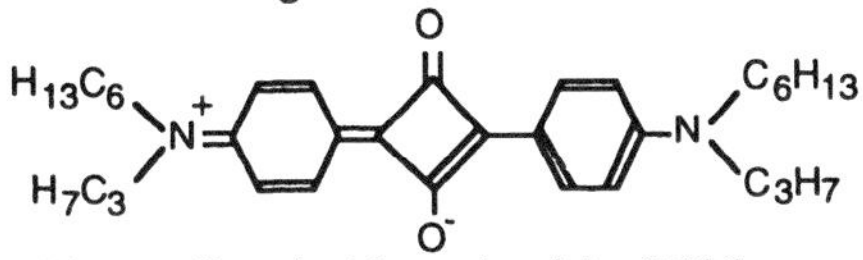

Fig. 1. Chemical formula of the SQ36.

The monolayer was excited by a filtered Xenon lamp through a band-pass filter from 700 to 750 nm. A fluorescence emission image was taken by a CCD camera though a long-pass filter from 770 nm. Polarized images were taken with the polarization of both the excitation and emission light in the same direction.

A transparent Langmuir trough formed on a glass substrate with Teflon side walls having a surface diameter of 20 mm and a depth of 20 mm was used to measure both the linear and nonlinear optical properties. For preparing the subphase of a monolayer, the temperature of pure water was adjusted to 5 °C by a peltier cooler with a water-circulating heat sink. For investigating of the linear optical properties, a spectro-photometer of a linear CCD array with a bundle-type optical fiber was used. The transmission spectra were measured under the illumination of white light from a halogen lump through a facing individual optical fiber. Reflection spectra were measured by using a branched bundle optical fiber connected to a detector and the light source to the blanched ends, with the other end facing normally to the water surface. The transmittance and reflectance of the trough and water without a monolayer was used for a reference of the absorption spectra, respectively.

The femtosecond excited-state dynamics was investigated by measuring the transient absorption change spectra using pump pulses with a center wavelength of 780 nm and probe pulses of a white light continuum with 200 fs duration under a 1 kHz repetition rate. The diameters of both the pump and probe beam were 300 μm. Before conducting pump-probe measurements, the formation of J-aggregates was confirmed by measuring the absorption spectra using femtosecond pulses of the white-light continuum [9].

RESULTS AND DISCUSSIONS

After spreading of a chloroform solution of squarylium dye, the formation of J-aggregates was observed within 30 s due to an increase in the absorption at J-band of 777 nm. From a measurement of the pressure-area isotherm of this monolayer using a Wilhelmy-type Langmuir trough, the occupied area of one dye molecule was determined to be 0.85 nm^2[8]. The microscopy of the fluorescence from the J-aggregate wa applied to characterizing the morphology of the SQ36 Langmuir-film. Figure 2 shows the results of fluorescence microscopy without (a) and with (b) a polarizer, respectively. The flatly bright image without any dark spots in Fig. 2a suggests that the water surface is almost covered with J-aggregates. On the other hand, a mosaic-like view filled with domains of $10^{-5} \sim 10^{-6}$ m size oriented in variable directions was observed by applying a polarizer. Within each single domain, dye molecules were expected to orient in the same direction, because the fluorescence within a single domain shows an almost uniform brightness.

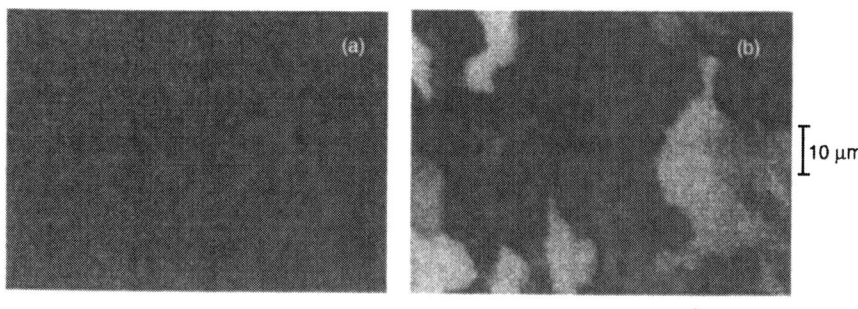

Fig. 2. Fluorescence micrographs of the SQ36 Langmuir-film without a polarizer (a) and with a polarizer (b) (horizontal). Pol.

Figure 3 shows the optical properties of the Langmuir film. The peak absorbance of the J-band, measured with the normal transmission method, reaches almost 0.3 in Fig. 3a. Such a high value, corresponding to absorption of 50%, seems to be suspicious for a single molecular layer. If we apply polarized light for a single domain of the J-aggregate in the same direction as the transition-moment of the J-band, the transmittance should be 0%. We thus checked the reflection spectra of

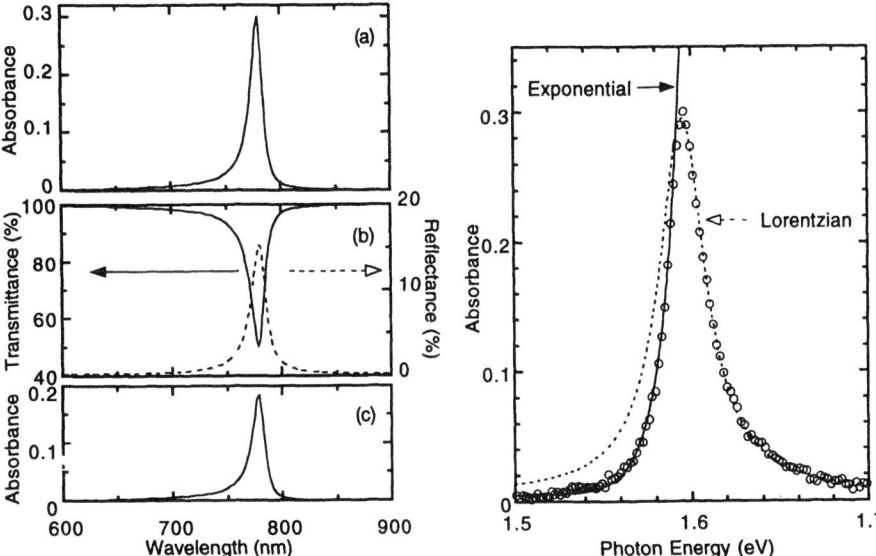

Fig. 3. Optical properties of the Langmuir-film.
(a) Absorption spectrum of transmittion.
(b) Transmittance and reflectance spectra.
(c) Absorption spectrum corrected by
substructing the contribution of reflection.

Fig. 4. Experimental absorbance spectrum
(white circle) with fittings of exponential
(solid line) and Lorentzian (broken line)
curves.

the Langmuir films with the transmission spectrum. At the valley in transmission spectra of 50% the reflectance reaches a peak value of 15% (Fig. 3b). Due to the extremely high optical density of the J-band within such an ultra-thin mono-molecular layer, the reflection enhancement of the air-water interface must have occurred in this Langmuir film [10, 11]. A corrected absorption spectrum produced by subtracting the reflectance contribution from the transmittance is shown in Fig. 3c. The peak absorbance of 0.185 was determined. This absorbance corresponds to molar absorption coefficient of 9.5×10^5 monolayer$^{-1} \cdot$ mol.$^{-1}$ based on the occupied area of one dye molecule for 0.85 nm^2. The oscillator strengths of a absorption band of dye molecules in solutions determined by the integrated area of the molar absorption coefficient as a function of the wave number (cm^{-1}). Using the same procedure for solutions, the oscillator strength of this J-band (700~850 nm) is calculated to be 1.48 based on the integrated area of absorption in Fig. 3c. Generally, the total oscillator strength for one electron transition is unity, never exceed one. The value of 1.48 should be a result of miscount in the orientation factor. If the transition moments of all molecules strictly lie in two-dimensional plane parallel to the water surface perpendicular to the probing light, one half of the incident unpolarized light should be active, where one-third is active for the solutions [11]. The revised oscillator strength of 0.99 (two thirds of 1.48) indicates a J-aggregate formation ratio of almost 100% with the total SQ36 dye molecules on the water. Because there was no change of spectral shape after the correction, the following results use the absorbance determined by only the transmitted light. We also again discuss on the contribution of reflection enhancement at the J-band at the end of this part.

For analyzing of the nonlinear optical properties of this J-aggregate, the nature of the absorption-spectrum shape is important. Curve fittings on the J-band spectrum were applied as shown in [Fig. 4]. Concerning the characteristics of the two-dimensional J-aggregate, a Lorentzian curve fits well with the high-energy side of the absorption band up to the peak[12]. The low-energy side of the

779

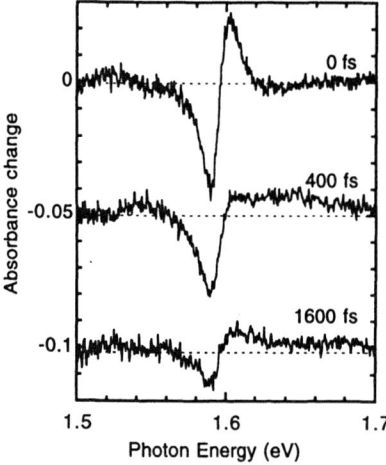

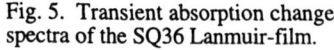

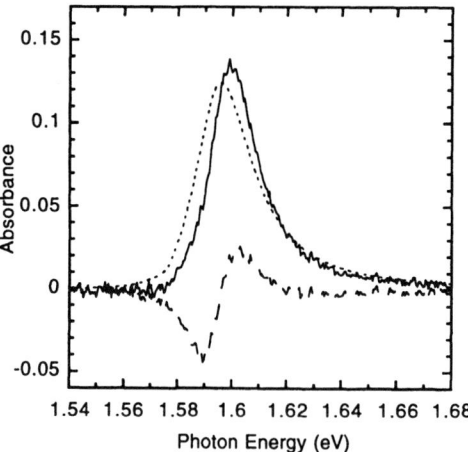

Fig. 5. Transient absorption change spectra of the SQ36 Lanmuir-film.

Fig. 6. Estimated spectra of the induced (solid line) and bleached absorption (dotted line) with the expelimental transient spectrum (broken line).

absorption can be expressed using the Urbach-Martiessen rule, as is also shown in good agreement with the exponential curve. From these results of the curve fittings, we regard the homogeneous band-width to be dominant compared to that of the inhomogeneous one under such a high temperature as 5 °C.

Figure 5 shows the results of pump-probe measurements on different durations measured with a pump energy of 2.5 μJ / cm² · pulse. As in the common cases of the J-aggregates, dispersion-type transient spectra with a decrease in the absorbance at a lower photon energy and an increase at a higher photon energy appeared upon illumination of the pump pulses. Reductions in these absorbance changes were observed within several hundred femtoseconds, suggesting a fast response of the J-aggregates under each pump energy. Based on the Frenkel-type exciton model of J-aggregates [2-6], these transient-absorption spectra must be the summation of a negative-absorption change of the transition between the ground states to the 1-exciton states (bleaching of linear absorption), and a positive-absorption change of the transition from the 1-exciton states to the higher exciton states with slightly larger transition energies (induced absorption).

To analyze these absorption-change spectra, it should be noted that the separation of each contribution from breached absorption and induced absorption is essential. Curve fitting of the experimental results was performed in the following way. Using speculation based on the results of Fig. 4, the total energy of the linear absorption band is expected to be breached at a constant ratio. Thus, the bleached absorption contribution was assumed to have the same shape as that of the linear absorption and the peak height, as determined by the breached ratio of the negative value. We determined the bleached ratio so that substitution from the measured absorbance change would produce an exponential tail on the lower energy side of the residual spectrum, corresponding to the induced absorption. Choosing the exponential tail for the induced absorption is expected to minimize the value of the bleached ratios to avoid over estimations.

The fitted results for the transient spectra at 0 fs with each pump energy are shown in Fig. 6. The spectra of the bleached absorption were reversed to the opposite sign for the purpose of highlighting the shapes of the induced absorption. However, there is less absorbance change in the transient spectra shown in Fig. 6. The calculated bleached ratio reached as high as ~50% of the linear absorption. This distinction should be the result of the large band width of the linear absorption, and also the induced nonlinear absorption overlapping each other under 5 °C. The

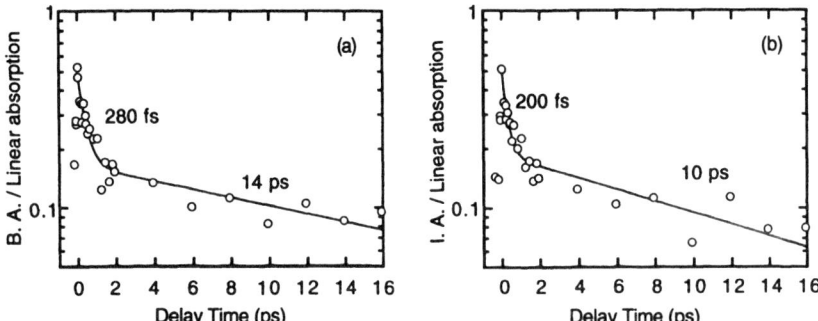

Fig. 7. Temporal change of the bleached absorption (a) and induced absorption (b) on the delay time.

difference in the peak energy between the bleached and induced absorption is 4 meV, which indicates the blue shifted energy of the transition from the one-exciton state to the two-exciton state from that of the stationary transition.

Using this procedure, temporal changes in the absorption saturation and the induced absorption were determined, as shown in Fig. 7. The result indicates that both absorption changes decay while exhibiting the same fast and slow components of 200~300 fs and 10~15 ps. The equality of both absorbance-change ratios also support the correctness of the previous three-level model. As in other J-aggregates, there are two components for the decay time constant for the excited states. The slow time constant can be understood as the decay of excited states by a collective radiative-decay process, called superradiance. The fast time constant of 300 fs must be attributed to coherent effects, such as stimulated emission, re-orientation of excited states, exciton-exciton annihilation or relaxation of multi-exciton states[2-6].

Based on the magnitude of the absorption saturation of 50% at 0 fs, as the result of curve fitting in Fig. 7a, the saturation energy of illumination for this J-band with a pump pulse duration of 200 fs has been determined to be 5 µJ / cm^2 · pulse. Let us remember the contribution of reflection enhancement in Fig. 3. The active ratio of the pump energy absorption is 0.19 based on a calculation of the overlap area in the pump spectrum with the revised absorption spectrum (Fig. 3c). The estimated saturation energy of absorption for 9.5×10^{-7} J / cm^2 · pulse contains 3.7×10^{12} photon/cm^2. The absorption cross section of the one photon for this Langmuir-film of SQ36 was evaluated to be 31.5 nm^2, which is more than 30-times larger than the occupied area of the one dye molecule to be 0.85 nm^2. This estimation can be interpreted into over 30 molecules for coherent cooperation, thus providing strong support of exciton delocalization in these J-aggregates.

CONCLUSIONS

The fluorescence micrographs indicate the uniform and closely packed structure of the SQ36 Langmuir film. The two-dimensional domains of the J-aggregate have a size of 10^{-5}~10^{-6} orders, and are oriented randomly in relation to each other. Concerning the linear optical properties, a large reflectance at the J-band is observed. The almost 100% formations of two-dimensional J-aggregates in this Langmuir film was confirmed by an estimation of the oscillator strength while considering of the reflection. These results concerning the linear optical properties explain that the SQ36 Langmuir film supplies a clear condition of single optical-pass.

The femtosecond nonlinear optical properties exhibit a ultra-fast response of the excited states. Curve fittings of the J-band also enabled estimations of the bleached ratios and speculation concerning the induced nonlinear absorption spectra. Over 30 molecules of coherent excitation in the J-aggregates was estimated by using a bleached ratio of 50% with a pump energy of 2.5 µJ / cm^2 · pulse and a corrected value of the absorbance concerning large reflectivity. The large difference between the estimated coherent size in J-aggregates and the domain size observed in a fluorescence microscope is suggestive in the picture of the Frenkel excitons delocalized in a two-dimensional plane, which is expected to be influenced by a temperature of 5°C and the atmosphere.

An investigation of the nonlinear optical properties with such a well-defined structure should be the proper course, and this was the first observation of the ultra-fast nonlinear properties from a two-dimensional J-aggregate with well defined structures and optical properties consistently.

ACKNOWLEDGMENTS

This work was supported by the New Energy and Industrial Technology Development Organization in the framework of Femtosecond Technology Project. We would like to express our special thanks to Dr. Hitoshi Kawashima, Dr. Tsuyoshi Kato of the Electrotechnical laboratory and Dr. Osamu Wada and other members of the FESTA Laboratories for stimulating and fruitfull discussions.

REFERENCES

1. E. E. Jelly, Nature 138, 1009 (1936).

2. B. Kopainsky, and W. Kaiser, Chem. Phys. Lett. **88**, 357 (1982).

3. S. Kobayashi, and F. Sasaki, Nonlinear Optics **4**, 305 (1993).

4. S. Kobayashi, and F. Sasaki, J. of Lumin. **58**, 113 (1994).

5. K. Minoshima, M. Taiji, K. Misawa, and T. Kobayashi, Chem. Phys. Lett. **218**, 67 (1994).

6. K. Minoshima, M. Taiji, A. Ueki, K. Miyano, and T. Kobayashi, Nonlinear Optics **14**, 39 (1995).

7. H. Kuhn, D. Möbius, and H. Bucher, in A. Weissbreger, and B. W. Rossiter, (Eds.), Physical Methods of Chemistry (Wiley, New York, 1972), vol. 1, Pt. III-B, pp. 577.

8. M. Furuki, S. Kim, L. S. Pu, H. Nakahara, and K. Fukuda, J. Chem. Soc. Japan Chem. Ind. Chem. **10**, 1121 (1990).

9. M. Furuki, L. S. Pu, F. Sasaki, S. Kobayashi, and T. Tani, Abstracts of the 4th International Workshop on Femtosecond Technology, 135 (1997).

10. P. Scherer, B. Kopainsky, and W. Kaiser, Optics Communications **39**(6), 357 (1981).

11. H. Grüniger, D. Möbius, and H. Meyer, J. Chem. Phys. **79**(8), 3701 (1983).

12. A. Nabetani, A. Tomioka, T. Tamaru, and K. Miyano, J. Chem. Phys. **102**, 5109 (1995).

FEMTOSECOND PUMP AND PROBE SPECTROSCOPY ON POLY(*PARA*-PHENYLENES)

W. GRAUPNER*, G. CERULLO**, G. KRANZELBINDER*, G. LANZANI ***, S.STAGIRA**,
M. NISOLI**, S. DE SILVESTRI**, G. LEISING*,
*Institut für Festkörperphysik, Technische Universität Graz, A-8010 Graz, Austria.
**CEQSE - C.N.R., Dipartimento di Fisica, Politecnico, I-20133 Milano, Italy.
***Istituto di Matematica e Fisica, Universita di Sassari, I-07100 Sassari, Italy.

ABSTRACT

We studied the sub-picosecond dynamics of photoexcitations in methyl-substituted poly(*para*-phenylene)-type ladder polymer (m-LPPP) films in a wide excitation density range up to values typical for the regime of narrow-band emission by pump-probe spectroscopy. The singlet excitons relaxation dynamics, monitored at the stimulated emission (SE) peak at 2.53 eV, showed an intensity dependent ultrafast decay component, which occurs on a sub-picosecond time scale, ascribed to the onset of amplified spontaneous emission (ASE). In addition we identified photoinduced absorption (PA) bands of singlet excitons at 1.48 eV and polarons at 1.9 eV respectively. At high excitation density an additional absorption band becomes evident at 2.63 eV.

INTRODUCTION

The interest towards π-conjugated polymers has been constantly growing since their discovery in 1975, when polyacetylene was first doped [1]. Out of many possible applications of conducting polymers which have been envisaged and demonstrated the most promising in optoelectronic devices are light emitting diodes. LEDs with tunable, bright and durable emission are close to be commercial, while the new target is the construction of a "plastic" laser, i.e. a polymer based diode laser. As a consequence most of the fundamental studies, both theoretical and experimental, have been focused on optical properties of luminescent polymers.

Methyl-substituted poly(*para*-phenylene)-type ladder polymer (m-LPPP) [2] is a very appealing candidate for understanding the photophysics of conjugated polymers because the high intrachain order and the narrow conjugation length distribution give rise to very sharp features in the optical spectra. m-LPPP is a stable, soluble and pure fluorescent material [3] which was used to fabricate electroluminescent devices in the blue spectral region with high luminance and efficiency [4] The observation that stimulated emission does not compete with photoinduced absorption [5] pointed to the possibility of using m-LPPP as optical amplifying medium. It was found that at high excitation density or in large gain length geometries a dramatic line-narrowing of the emission spectrum is observed [6]. The physical origin of this behaviour, which is common to other fluorescent conjugated polymers, is presently under debate. Stimulated emission and amplified spontaneous emission, superfluorescence, superradiance and exciton condensation have been proposed as possible explanation. Most of the experimental studies were focussed on photoluminescence experiments carried out using a varity of excitation sources, while transient photoinduced absorption data at high excitation density were so far not reported.

In this work we report on sub-picosecond resolution differential-transmission $\Delta T/T$ experiments on m-LPPP films. Stimulated emission (SE) and photoinduced absorption (PA) are studied both at low excitation and at high excitation density where exciton-exciton interactions are important and line narrowing in the emission spectrum is taking place. The goal is to describe the excited state population dynamics in both regimes.

EXPERIMENTAL

Films of m-LPPP were prepared by dissolving polymer in oxygen-free toluene at 60°C by stirring for 24 hours, which was then dropcast onto infrasil substrates. During experiments samples

Mat. Res. Soc. Symp. Proc. Vol. 488 © 1998 Materials Research Society

were kept in vacuum at 80 K. The laser source was a Ti:sapphire laser system with chirped pulse amplification, which provided 140-fs pulses at 780 nm and 660-μJ energy at a repetition rate of 1 kHz. The excitation pulses at 390 nm were generated by second harmonic of the fundamental beam in a 1 mm thick LiB_3O_5 crystal. The pump beam was focussed to a spot size of 80 μm and the excitation intensity was between 0.3 and 12 mJ/cm^2 per pulse. Pump-probe experiments were carried out in the visible and near infrared spectral range using the white light supercontinuum generated in a thin sapphire plate. Mirrors were used to collect the supercontinuum pulse and to focus it on the sample in order to minimise frequency chirp effects. Two measurements were carried out: i) The whole white light pulse was spectrally analyzed after travelling through the sample using a monochromator and a silicon diode array. Transmission difference spectra were obtained by subtracting pump-on and pump-off data measured in the 2.76-1.65 eV and 1.57-1.31 eV range. ii) Temporal evolution of the differential transmission was recorded at selected wavelengths using a standard lock- in technique. In both experiments the pump beam was linearly polarized at the magic angle with the probe.

Figure 1: Normalized photoluminescence emission spectra of m-LPPP films at 77 K for excitation at 3.2 eV and two different intensities: 0.3 mJ/cm^2 (solid line) and 2.4 mJ/cm^2 (dashed line). The inset shows the room temperature absorption spectrum of m-LPPP.

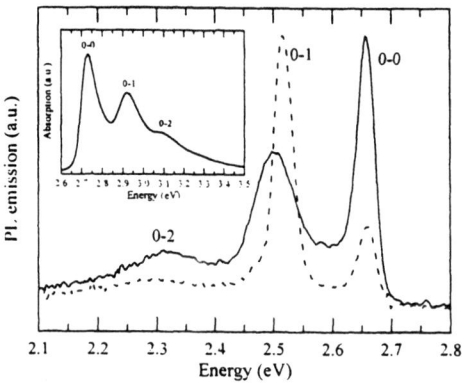

RESULTS

The CW-photoluminescence spectrum of a m-LPPP film, reported in Fig.1, displays a well resolved vibronic structure with the 0-0 transition at 2.67 eV while the contribution in the yellow region is rather small, proving that aggregation is negligible in our material. The 0-0 transition is partially re-absorbed due to the small Stokes shift between emission and absorption. Ground state absorption is reported in the inset of Fig.1 and shows a clear mirror symmetry with the emission spectrum. At high excitation density the emission spectrum collapses into a narrow line located at 2.53 eV, slightly blue-shifted with respect to the low intensity emission. This process has been studied in more detail in [6] and can be assigned to ASE.

$\Delta T/T$ spectra for excitation intensities of 0.3 mJ/cm^2 and different pump-probe delays (δt) are shown in Fig.2. For $\delta t = 0$ ps we measure positive ΔT for probe beam energies higher than 2.3 eV, and negative ΔT for lower energies. Since there is no ground state absorption in the positive ΔT region, we assign it to stimulated emission of the 0-1 and 0-2 lines. In the PA region we see a sharp band at 1.48 eV (Fig.2) which decays on the same time scale of SE and a weak PA band at 1.9 eV. Following the time evolution of the $\Delta T/T$ spectrum up to $\delta t = 400$ ps we see the decay of the main features and finally two PA bands left in the spectrum, at 1.9 and 1.3 eV respectively (see inset of Fig.2).

Upon increasing the excitation densities, up to 12 mJ/cm^2, the shape of the $\Delta T/T$ spectrum changes dramatically in the positive signal region: a dip in the SE spectrum develops around 2.64 eV and at very high power the positive sign of the ΔT signal in the SE region is converted into a negative one – i.e. PA is observed. This intensity–dependent PA is gone in about 4 ps.

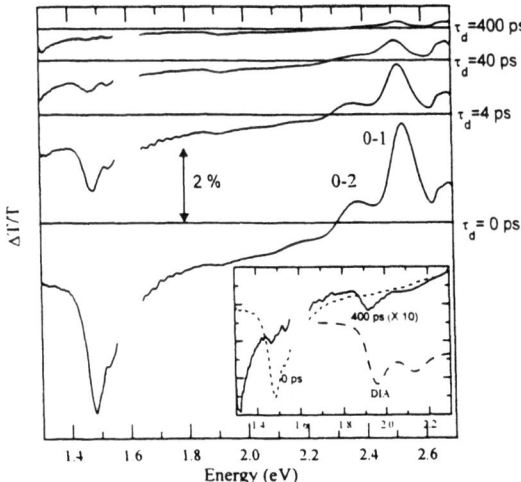

Figure 2: Transmission difference ΔT spectra of m–LPPP films at 77 K excited at 3.2 eV for variuos pump-probe delays. The inset shows the low energy region for 0 ps (solid line) and 400 ps (dashed line) delay. Doping induced absorption (DIA) data are shown for comparison (from [7]).

The dynamics of SE is strongly affected by the excitation density, as shown in Fig.3 where SE decay curves at 2.53 eV for three different excitation conditions are reported. The decay time scale gets shorter for higher excitation densities, reaching about 2 ps for the highest. In the negative (PA) region of the $\Delta T/T$ spectrum we do not detect any change upon increasing the excitation intensity.

DISCUSSION

Within standard Coulomb correlated models for luminescent polymers [8] photoexcitation leads to the generation of singlet excitons with odd parity wavefunction (B_u) which undergoes un ultrafast self–trapping process [9]. These are the radiative states responsible for the spontaneous emission. We assume that photoexcitation of m–LPPP under our experimental conditions generates B_u excitons, thermalized within the time scale of the pulse duration, which are responsible for spontaneous and stimulated emission. The SE band assigned to the 0–1 vibronic transition is initially centered at 2.51 eV, i.e. blueshifted with respect to the time integrated PL spectrum located at 2.48 eV. For longer δt we see a red–shift of the decaying SE spectrum which reaches the spectral position observed in time integrated PL within about 10 ps, as seen in Fig.4. This is consistent with the apparent risetime seen at the lower energy shoulder of the 0–1 SE band, as shown in the inset of Fig.4.

Spectral relaxation of SE is accounted for by exciton migration towards lower energies sites, likely longer conjugation chains with smaller optical gap [10]. In LPPP films containing aggregates this process dominates the $\Delta T/T$ dynamics masking the intrinsic behaviour [10]. In this case photoexcitation produces a broad intra–gap PA band which was assigned to interchain species [10],[5] while the PA of singlet excitons could not be identified. It is important to point out that the initial SE band corresponds to the position where line narrowing is taking place, showing that the apparent blue–shift of the emission line for high excitation can be associated to a fast amplification process which takes place before spectral relaxation occurs. A further clue that a fast exciton depopulation process is taking place is provided by the SE dynamics vs excitation intensity shown in Fig. 3. We conjecture that depletion of exciton population is due to ASE along the film plane. ASE, which is a thresholdless process, acts as intensity dependent mechanism of population depletion. Note that the probe interacts with the exciton population stimulating emission perpendicularly to the film surface. For this reason upon increasing the excitation power we see line narrowing of the spontaneous emission but not of stimulated emission.

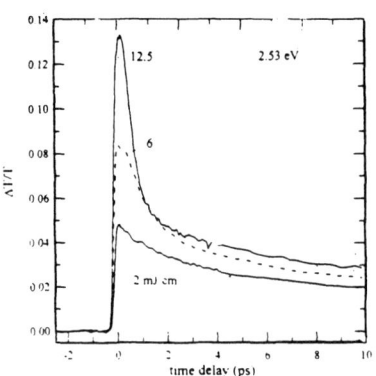

Figure 3: SE decay at 2.53 eV in m–LPPP films at liquid nitrogen temperature versus pump–probe time delay at three different excitation densities.

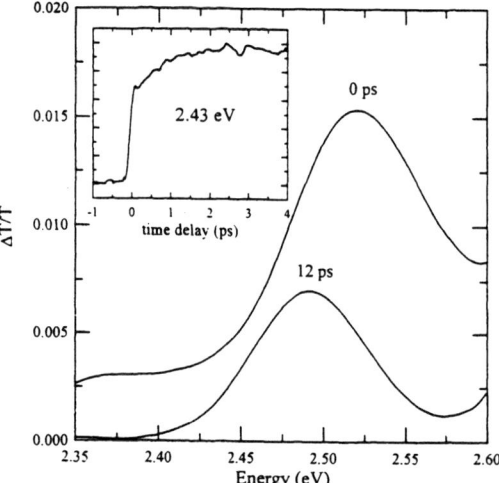

Figure 4: SE spectra in m–LPPP films at 77 K for two pump–probe delays at 0.3 mJ/cm^2 excitation intensity at 3.2 eV. The inset shows ΔT versus pump–probe delay at 2.43 eV.

PA at 1.48 eV appears instantaneously, shows spectral relaxation to the red and decays on the same time scale of SE, as shown in Fig. 5. We assign it to singlet B_u excitons transitions towards higher lying even parity states. A possible candidate for the final state is the intrachain bi–exciton. However its energy level is located below the two exciton state by an amount equal to the binding energy. Since this cannot be larger then the exciton binding energy, estimated to be about 0.5 eV [11], the biexciton level is expected to be around 5 eV. The exciton to biexciton transition would then appear at 2.3 eV and this possibility is thus ruled out. A second candidate is the continuum band state, but electromodulation spectroscopy shows that this is located at 3.3 eV, implying a transition energy of 0.5 eV. The remaining possibility is the so called mA_g state which is known to have a non negligible dipole coupling with the B_u state [8]. The weak shoulder seen at higher energy (around 1.55 eV) is assigned to a vibronic replica of the singlet exciton transition. The shift of 0.07 eV corresponds to the energy of a vibrational mode which is observed in absorption and electro–absorption [11]. We note that the vibronic structure can be detected due to the narrow linewidth of the PA (about 0.1 eV), the most narrow ever reported in conjugated polymers, and this demonstrates indirectly the high quality of m-LPPP.

Since the PA bands at 1.31 and 1.91 eV have different dynamics and longer lifetime they should be assigned to different photoexcitations. Let us first discuss the PA at 1.91 nm. The same PA band has been detected by steady state $\Delta T/T$ experiments on m-LPPP films [12]. In these measurements the pump beam is a mechanically modulated CW laser and the observed photoexcitations have lifetimes in the ms time domain. Based on temperature, excitation intensity and frequency dependence it has been assigned to triplet–triplet transitions. We think that this assignment is also true for our experimental time domain. Due to spectral overlap with the singlet PA band we can not distinguish between slow generation via intersystem crossing or ultrafast generation such as non geminate electron–hole recombination or exciton fission [13]. However the latter is unlikely because the energy for generating a triplet pair is expected to be larger than the pump photon energy [14].

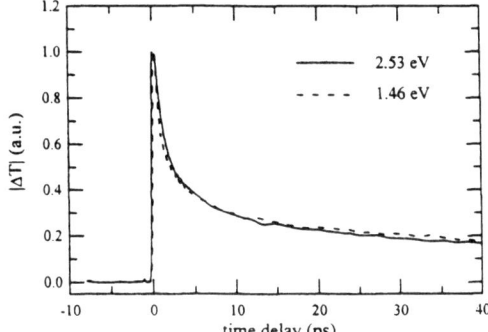

Figure 5: Modulus of ΔT versus pump–probe delay at 1.48 eV and 2.53 eV.

The PA at 1.9 eV and its vibronic replica at 2.14 eV, have been observed in m–LPPP films by various CW techniques: doping induced absorption, charge injection spectroscopy and CW–PA [7]. For these reasons they can be unambiguously assigned to charged species. In the inset Fig.2 we show for comparison the perfect match of the data obtained by chemical doping of m–LPPP films with our transient results. We conclude that charged states are photogenerated within 200 fs, since the PA at 1.9 eV is already present in the $\Delta T/T$ spectrum at $\delta t = 0$, and does not decay within our experimental time range, up to 400 ps. A conclusive statement on the generation mechanism can not be made at this point. Even extrinsic processes cannot be ruled out – we just note that electron transfer reaction between a conjugated polymer and an acceptor can occur in less then 200 fs [15].

CONCLUSIONS

In this work we report on sub–picosecond resolution $\Delta T/T$ experiments on m–LPPP films at different excitation conditions. We identify intrachain B_u singlet excitons which are characterized by SE around 2.53 eV and PA at 1.48 eV and display a non-exponential decay on the 400 ps time scale. Spectral relaxation of SE is detected in the 10 ps time domain and assigned to singlet excitons migration within the inhomogeneously broadened density of states. This process suggests that line narrowing seen at 2.53 eV is due to a fast mechanism which takes place in competition with spectral relaxation and that we identify with ASE along the film plane. Triplet excitons and charged polarons associated to PA at 1.3 eV and 1.9 eV (with replica at 2.14 eV) respectively are also identified. Due to their long lifetime these species appear clearly at 400 ps, when singlet recombination is almost concluded. However PA at 1.9 eV is clearly present at the very beginning, showing that charged species are photogenerated within the pulse duration. At very high excitation a PA band superimposed to SE becomes evident in the ΔT spectrum.

ACKNOWLEDGMENTS

W. Graupner and G. Lanzani wish to acknowledge the support of the Italo/Austrian research programme. This work was supported by CE Human Capital and Mobility (contract CHRX-CT 94-0561). We are indebted to U. Scherf and K. Müllen for the m-LPPP powder.

References

[1] C. K. Chiang, C. R. Fincher, Y. W. Park, A. J. Heeger, H. Shirakawa, E. J. Louis, S. C. Gau, A. G. MacDiarmid, Phys. Rev. Lett. **39**, 1098 (1977).

[2] U. Scherf, K. Müllen, Makromol. Chem., Rapid Commun. **12**, 489 (1991).

[3] J. Stampfl, S. Tasch, G. Leising, U. Scherf, Synth. Met. **71**, 2125 (1995).

[4] S. Tasch, A. Niko, G. Leising, U. Scherf, Appl. Phys. Lett. **68**, 1090 (1996).

[5] W. Graupner, G. Leising, G. Lanzani, M. Nisoli, S. De Silvestri, U. Scherf, Phys. Rev. Lett. **76**, 847 (1996).

[6] C. Zenz, W. Graupner, S. Tasch, G. Leising, K. Müllen, U. Scherf, Appl. Phys. Lett. **71**, 2566 (1997).

[7] W. Graupner, T. Jost, K. Petritsch, S. Tasch, G. Leising, M. Graupner, A. Hermetter, Soc. Plastics Engineers, Technical Papers **XLIII**, 1339 (1997).

[8] J. M. Leng, S. Jeglinski, X. Wei, R. E. Benner, Z. V. Vardeny, F. Guo, S. Mazumdar, Phys. Rev. Lett. **72**, 156 (1994).

[9] X. Weng, Y. Kostoulas, P. M. Fauchet, J. A. Osaheni, S. A. Jenekhe, Phys. Rev. B **51**, 6838 (1995).

[10] R. F. Mahrt, T. Pauck, U. Lemmer, U. Siegner, M. Hopmeier, R. Hennig, H. Bässler, E. O. Göbel, P. Haring Bolivar, G. Wegmann, H. Kurz, U. Scherf, K. Müllen, Phys. Rev. B, **54**, 1759 (1996).

[11] G. Meinhardt, A. Horvath, G. Weiser, G. Leising, Synth. Met. **84**, 669 (1997).

[12] W. Graupner, S. Eder, K. Petritsch, G. Leising, U. Scherf, Synthetic Metals **84**, 507 (1997).

[13] B. Xu, S. Holdcroft, Adv. Mater. **6**, 325 (1994).

[14] S_1 in m-LPPP is located at 2.7 eV. Exciton fission from 3.2 eV would imply a singlet–triplet > 1.1 eV, a value which seems to be rather unlikely. See [13] for comparison with PPV.

[15] N. S. Sariciftci, L. Smilowitz, A. J. Heeger, F. Wudl, Science **258** 1474 (1992).

For further information, please contact Wilhelm Graupner, Institut für Festkörperphysik, Technische Universität Graz, Petersgasse 16, A-8010 Graz, Austria; Tel.: 0043 316 873 8476, FAX: 0043 316 873 8478; e-mail: f513wigr@mbox.tu-graz.ac.at, www.cis.tu-graz.ac.at/if/graupner.htm.

MAGNETIC FIELD EFFECT IN HIGHLY PURE, HIGHLY FLUORESCENT CONJUGATED POLYMERS

W. GRAUPNER *, M. SACHER**, M. GRAUPNER***, C. ZENZ *, G. GRAMPP**,
A. HERMETTER***, G. LEISING*,
*Inst. für Festkörperphysik, Techn. Univ. Graz, A–8010 Graz, Austria.
**Inst. für Physikalische & Theoretische Chemie, Techn. Univ. Graz, A–8010 Graz, Austria.
***Inst. für Biochemie & Lebensmittelchemie, Techn. Univ. Graz, A–8010 Graz, Austria.

ABSTRACT

We report on the magnetic field effect (MFE) on the photoluminescence in films of a highly fluorescent conjugated laddertype poly(paraphenylene) (LPPP). We observe no MFE for the pure material, however, the introduction of a very small content of chemical defects via photo-oxidation leads to a MFE proportional to the amount of defects. The chemically induced increase in MFE is correlated with a change in other properties of the LPPP films, such as photoluminescence emission and excitation spectra, transient photoluminescence and infrared spectra.

INTRODUCTION

Due to their low content of chemical defects [1] the class of conjugated laddertype poly(paraphenylenes) (LPPP) is highly interesting as a model system for basic physical research [2] and attractive for applications. Using LPPP blue and white light emitting devices (LEDs) [3] were demonstrated as well as highly directional, linearly polarized, narrowline ($\approx$ 2 nm) emission from optically pumped polymer waveguides [4]. The effect of weak static magnetic fields on the photoluminescence (PL) quantum yield (QY) [5] and the photoconductivity (PC) [6] of conjugated polymers was attributed to the PL emission being partly due to geminate recombination of charge carriers. In [7] a significant enhancement of this MFE is found for the PC in the presence of oxygen. The aim of our investigation is to study how strong the MFE is in LPPP which is characterized by high intrachain order [8] and a very low defect content [1].

EXPERIMENTAL

The preparation of the LPPP powder is described in [9]. For our investigations we have used the m–LPPP which is distinguished by its methyl substituent (see Fig. 1), increasing the chemical stability as compared to the hydrogen substituted form. The polymer films were prepared by dropcasting from toluene, chloroform and tetrahydrofuran solutions of m–LPPP on glass, sapphire and silicon substrates. All subsequent spectroscopic investigations were either performed on the same or on identically treated samples. Photo–oxidation of the polymer films and solutions is performed by the infrared filtered ultraviolet/visible radiation of a 1000 W high pessure Xe lamp. Photoluminescence excitation (PLEx) and emission (PLEm) spectra were recorded with a Shimadzu spectrofluorophotometer RF–5301PC. The time resolved PL data were obtained with a photomodulation spectrometer I.S.S. GREG200 300 Xe. The infrared spectra were recorded with a Bomem MB 102 Fourier transform infrared spectrometer.

The MFE was investigated with a self–built MARY spectrometer [10], where MARY refers to "MAgnetic field effects on Reaction Yield". A 150 W high pressure Xe lamp (Osram XBO 150/1 OFR) together with appropriate filters was used as excitation source. The sample was placed in a Bruker research magnet B–E 10 and exposed to very homogeneous small magnetic fields B up to 10 mT. In order to detect the small MFE we used a modulation technique. A pair of additional coils, connected in series, was placed onto the pole shoes of the research magnet. This pair of coils was simultaneously fed with a direct current to generate a magnetic field of opposite direction and an alternating current to modulate the magnetic field with a frequency of 190 Hz and a modulation width of 1 mT. The synchronous changes in the PL were detected by a photomultiplier tube RCA

IP28 and picked up with a Princeton Applied Research Model 128A lock–in amplifier. Therefore the recorded MFE spectra represent $\frac{d(PL)}{dB}$ curves.

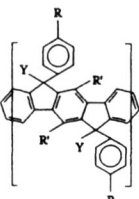

Figure 1: Chemical structure of the PPP-type ladderpolymer, Y=CH$_3$, R= C$_{10}$H$_{21}$, R'= C$_6$H$_{13}$.

RESULTS

Figure 2 shows typical emission and excitation spectra of a thin and a thick m–LPPP film which illustrate that in thick films (1) the PLEm spectrum is distorted by self absorption while (2) the PLEx spectrum is distorted by the *inner filter effect* [11]. The PLEx spectrum of the thin film perfectly matches the absorbance spectrum of the same film. The PLEm and PLEx spectra of the thin film, however, illustrate the small Stokes shift in m–LPPP. It is a consequence of the suppression of torsional degrees of freedom for neighbouring phenyl rings by the methin bridge (see Fig. 1).

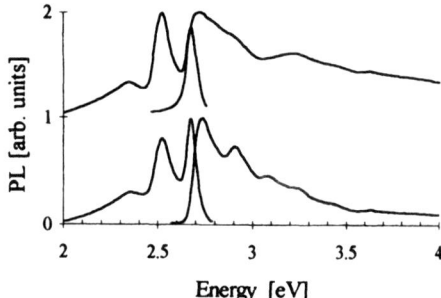

Figure 2: Photoluminescence excitation (PLEx) and emission (PLEm) spectra of a thin (bottom) and a thick (top) m–LPPP film.

Pure m–LPPP films cast on glass were investigated for their MFE. Subsequently these films were photo–oxidized and the MFE was probed again. *Pristine* films made from all the solvents show *no* MFE. In Figure 3 (A) we show selected $\frac{d(PL)}{dB}$ curves as recorded for the same sample at different degrees of photo–oxidation. It is clear that the observed MFE increases with increasing exposition to light and oxygen. Since the spectra are *not* corrected for the absolute PL intensity, which is reduced by photo–oxidation, the $\frac{d(PL)}{dB} \frac{1}{PL}$-values do even increase stronger with increasing defect content. The shape of the $\frac{d(PL)}{dB}$ curves changes slightly with photo–oxidation: the minumum and maximum move closer together for increasing the oxidation time. By numerical integration of the $\frac{d(PL)}{dB}$-curves and correction for the PL intensities we obtain the curves in Fig. 3 (B). The full width of half maximum (FWHM) is found to be about 40 mT for all the curves.

In order to illustrate which physical and chemical changes take place in photo–oxidized m–LPPP samples we investigated PLEx and PLEm spectra (Figs. 4,5) and the infrared spectra (Fig. 6) of these samples, as it was done for poly(*para*–phenylene–vinylene) in [12].

In order to describe the effect of photo–oxidation on the distribution of effective conjugation lengths in m–LPPP we photo–oxidized a solution of m–LPPP in toluene for one hour and examined the resulting PLEm and PLEx spectra at different excitation and emission wavelengths, respectively. In Fig. 5 (A) we have plotted the PLEx spectra for emission energies between 2.53

and 2.95 eV. It is seen that the excitation spectrum changes strongly as the emission wavelength is varied, which can be explained in the following way: *para*–phenylene oligomers of different chain-length have their PLEm maximum at different wavelengths. By setting the emission energy at a certain position we mainly measure the PLEx spectrum of the corresponding oligomer length and the weaker contributions of shorter chains. Therefore Fig. 5 (A) represents the excitation spectra of polymer segments of different conjugation length as a consequence of the introduction of chemical defects (=conjugation breaking) by photo–oxidation. This picture is confirmed by Fig. 5 (B) which shows the PLEm spectra for different excitation wavelengths. In particular, there is strong emission observed at quantum energies above 2.7 eV which is not found at all for a solution of the same concentration consisting of pristine polymer and therefore not containing a significant amount of short oligomers which emit at energies above 2.7 eV.

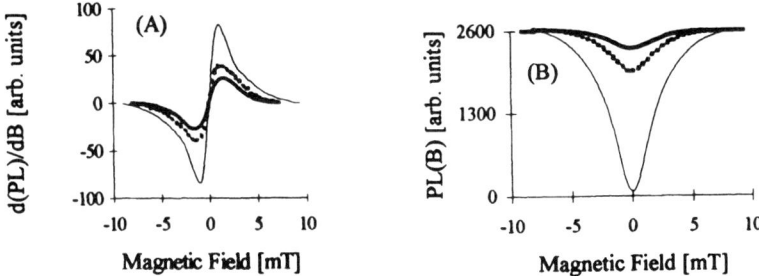

Figure 3: (A) $\frac{d(PL)}{dB}$ spectra of a m–LPPP film cast from a toluene solution after 15, 45 and 105 seconds of photo–oxidation. (B) $\frac{1}{PL} \int (\frac{d(PL)}{dB})dB$ spectra, as obtained by numerical integration of curves (A).

Figure 4: PLEm and PLEx spectra of a pure m–LPPP film (bottom) and after 20 seconds of photo–oxidation.

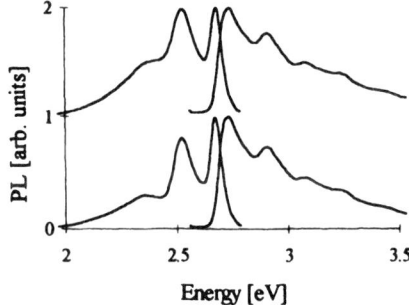

For completeness we want to add that photo–oxidation proceeds much quicker in films than in solutions of the polymer. Also the chemical nature of the solvent influences the process of photo–oxidation. One hour of photo–oxidation completely changes the emission and absorption spectra of m–LPPP in chloroform, while in tetrahydrofuran no change is visible after the same amount of time.

Photo–oxidation drastically reduces the PLQY of m–LPPP films and solutions. Correspondingly, the observed PL decay is expected to become faster with increasing defect content – this was observed for m–LPPP [13].

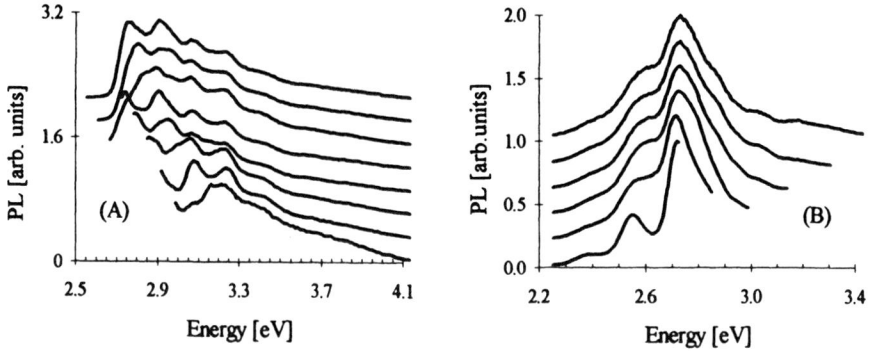

Figure 5: PLEx (A) and PLEm (B) spectra of a m–LPPP solution after 1 hour of photo–oxidation; photon quantum energies for the PLEx spectra are: 2.53, 2.58, 2.64, 2.70, 2.76, 2.82, 2.88 and 2.95 eV (from top to bottom), respectively; photon quantum energies for the PLEm spectra are: 2.76, 2.88, 3.02, 3.18, 3.35 and 3.54 eV (from bottom to top) respectively.

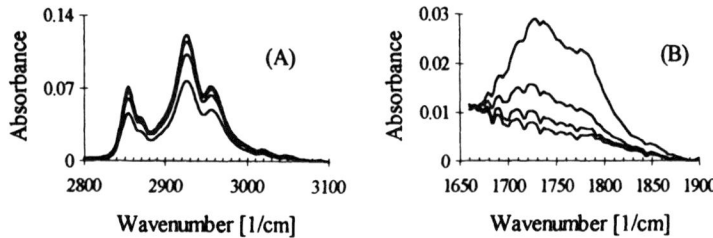

Figure 6: IR spectra of m–LPPP films after 0, 75, 155, and 315 seconds of photo–oxidation. The IR absorbance is decreased in the region of vibrations of CH_3/CH_2–groups (A) and increased in the carbonyl vibration region (B).

The chemical changes in m–LPPP films upon photo–oxidation can be monitored by infrared spectroscopy. In Fig. 6 (A) we show the observed negative change in absorbance after 0, 75, 155, and 315 seconds of photo–oxidation. This region of the IR spectrum is characteristic for the vibrations of CH_3–groups (2960, 2870 cm^{-1}) and CH_2–groups (2920, 2850 cm^{-1}). When normalized to the bands of the CH_2–groups the bands of the CH_3–groups become relatively stronger. The decrease in the mentioned infrared bands corresponds to an increase in the region around 1750 cm^{-1} as shown in Fig. 7 (B). This can be attributed to the formation of carbonyls which are usually detected around 1705–1725 cm^{-1}. However, we have to take into account that aromatic neighbours shift the vibrations of carbonyls by up to 25 cm^{-1} to lower values, conjugation with a double bond leads to a lowering by up to 40 cm^{-1}.

Figure 7: IR Absorbance due to CH_2/CH_3–groups (squares), due to carbonyl groups (triangles) and $\frac{1}{PL}\int(\frac{d(PL)}{dB})dB$ values (crosses); all curves are normalized to the maximum value; lines are shown to guide the eye.

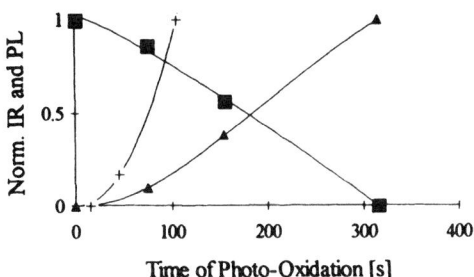

The main question addressed in this paper is whether the MFE detected in conjugated polymers is an intrinsic effect, also observed in ultrapure materials. Therefore we chose the m–LPPP as a model system since its low defect content was experimentally established [1]. As reported by Frankevich and coauthors [7] for substituted poly(para–phenylene vinylene) (PPV), there is a prompt MFE, connected with the existence of coupled charge carriers, and a delayed MFE, connected with the reaction of triplet and singlet excitons with oxygen. We have shown that the degree of oxidation can be well monitored by infrared spectroscopy. In order to determine whether there is any MFE associated with the pure m–LPPP we have plotted the normalized IR absorbance from Fig. 6 and the normalized $\frac{1}{PL}\int(\frac{d(PL)}{dB})dB$ values from Fig. 3 in Figure 7. The former values represent the amount of pure polymer and the number of carbonyl groups in the samples while the latter values represent the MFE. As can be seen in Figure 7, when trying to extrapolate the $\frac{1}{PL}\int(\frac{d(PL)}{dB})dB$ values to zero seconds of photo–oxidation the "residual MFE" is negligible compared to the extrinsic MFE induced by oxygen and air. Therefore we conclude that the proposed mechanism of dissociation of charge carrier pairs as described in [7] is not present in the very pure m–LPPP. This is in excellent agreement with measurements of photoinduced absorption due to trapped charges in m–LPPP [8],[14]. In these studies we found that in the LPPP of the highest quality, namely the m–LPPP, no trapped charges can be detected. Therefore either no pair dissociation takes place or the charge carriers are not trapped in pure m–LPPP. The former scenario is in contrast to recent experiments, where we doped m–LPPP with different buckminster fullerene derivatives [15]. In these studies we saw that fullerene doping leads to high photoconductivity but does not strongly contribute to the formation of trapped charges in m–LPPP while photo–oxidation does [16].

CONCLUSIONS

We have performed optical spectroscopy on samples of a very pure conjugated polymer. Samples which were photo–oxidized show typical changes in their photoluminescence emission and excitation spectra and a lower photoluminescence quantum yield. For the pure samples we found no magnetic field effect on the photoluminescence. The magnetic field effect as expressed by $\frac{1}{PL}\int(\frac{d(PL)}{dB})dB$ increases with increasing exposition to oxygen and air. Extrapolating to zero exposition we conclude that any "intrinsic" MFE in pure m–LPPP is completely negligible compared to the MFE induced by oxygen and light.

ACKNOWLEDGMENTS

We acknowledge the financial support of the Austrian Fonds zur Förderung der wissenschaftlichen Forschung under both *SFB Elektroaktive Stoffe* and *SFB Biokatalyse* (Project No. F0107). We are indebted to U. Scherf and K. Müllen for the m–LPPP powder. We enjoyed the exciting discussions with E. Frankevich, V. Dyakonov, J. Shinar and Z. V. Vardeny.

References

[1] W. Graupner, G. Leditzky, G. Leising, U. Scherf, Phys. Rev. B **54**, 7610 (1996).

[2] R. F. Mahrt, T. Pauck, U. Lemmer, U. Siegner, M. Hopmeier, R. Hennig, H. Bässler, E. O. Göbel, P. Haring Bolivar, G. Wegmann, H. Kurz, U. Scherf, K. Müllen, Phys. Rev. B, **54**, 1759 (1996); M. Samoc, A. Samoc, B. Luther–Davies, U. Scherf, Synth. Met. **87**, 197 (1997); W. Graupner, G. Leising, G. Lanzani, M. Nisoli, S. De Silvestri, U. Scherf, Phys. Rev. Lett. **76**, 847 (1996).

[3] S. Tasch, A. Niko, G. Leising, U. Scherf, Appl. Phys. Lett. **68**, 1090 (1996); S. Tasch, E.J.W. List, O. Ekström, W. Graupner, G. Leising, P. Schlichting, U. Rohr, Y. Geerts, U. Scherf, K. Müllen, Appl. Phys. Lett. **71**, 2883 (1997).

[4] C. Zenz, W. Graupner, S. Tasch, G. Leising, K. Müllen, U. Scherf, Appl. Phys. Lett. **71**, 2566 (1997).

[5] E. L. Frankevich, G. E. Zoriniants, A. N. Chaban, M. M. Triebel, S. Blumstengel, V. M. Kobryanskii, Chem. Phys. Lett. **261**, 545 (1996).

[6] E. L. Frankevich, A. A. Lymarev, I. Sokolik, F. E. Karasz, S. Blumstengel, R. H. Baughman, H. H. Hörhold, Phys. Rev. B **46**, 9320 (1992).

[7] E. Frankevich, A. Zakhidov, K. Yoshino, Y. Maruyama, K. Yakushi, Phys. Rev. B **53**, 4498 (1996).

[8] W. Graupner, S. Eder, M. Mauri, G. Leising, U. Scherf, Synth. Met. **69**, 419 (1995).

[9] U. Scherf, K. Müllen, Makromol. Chem., Rapid Commun. **12**, 489 (1991).

[10] M. Sacher, G. Grampp, Ber. Bunsenges. Phys. Chem. **101**, 971 (1997).

[11] J. R. Lakowicz, *Principles of Fluorescence Spectroscopy*, Plenum Press, New York 1986.

[12] M. Yan, L. J. Rothberg, F. Papadimitrakopoulos, M. E. Galvin, T. M. Miller, Phys. Rev. Lett. **73**, 744 (1994).

[13] W. Graupner, T. Jost, K. Petritsch, S. Tasch, G. Leising, M. Graupner, A. Hermetter, Soc. Plastics Engineers, Technical Papers **XLIII**, 1339 (1997).

[14] K. Petritsch, W. Graupner, G. Leising, U. Scherf, Synthetic Metals **84**, 625 (1997).

[15] C. Waldauf, W. Graupner, S. Tasch, G. Leising, U. Scherf, K. Müllen, Optical Materials in print.

[16] W. Graupner, M. Mauri, J. Stampfl, G. Leising, U. Scherf, K. Müllen, Solid State Communications, **91**, 7 (1994).

For further information, contact Wilhelm Graupner, Institut für Festkörperphysik, Technische Universität Graz, Petersgasse 16, A–8010 Graz, Austria; Tel.: 0043 316 873 8476, FAX: 0043 316 873 8478; e-mail: f513wigr@mbox.tu–graz.ac.at, www.cis.tu-graz.ac.at/if/graupner.htm.

DESIGN, METHODOLOGY AND PREPARATION OF NOVEL POLYMERS FOR NONLINEAR OPTICS

K. G. Chittibabu, L. Li and S. Balasubramanian
Molecular Technologies Inc., Westford, MA 01886

X. Wang,[1] M. Sukwattanasinitt,[1] K. Yang,[2] J. Kumar,[2] D. J. Sandman[1] and S. K. Tripathy[1]*
Center for Advanced Materials, Departments of Chemistry[1] and Physics[2],
University of Massachusetts Lowell, Lowell, MA 01854

ABSTRACT

One-pot post-polymerization modification reactions such as azo-coupling and tricyanovinylation reactions were employed to synthesize a series of polymers containing different nonlinear optical (NLO) chromophoric as well as ionic functionalities. We have extended and established the versatility of our earlier reported post-modification strategy to incorporate various heteroaromatic chromophores as well as ionic functionalities in the polymers, at the final stage of synthesis. The correlation between different heteroaromatic chromophore structures and the NLO properties of the polymers was extensively studied. Polymers containing heteroaromatic chromophores exhibit improved temporal stability and enhanced NLO activity. Polymers with ionic chromophores were employed to fabricate NLO active ultra-thin films using electrostatic self-assembling (ESA) technique. Attempts were also made to synthesize second order NLO active polydiacetylene derivatives using post azo-coupling reaction.

I. INTRODUCTION

Polymeric materials have advantages over inorganic crystals for nonlinear optical (NLO) applications, because of their low dielectric constant, large optical nonlinearity, low cost, and ease of processability. Stable NLO polymeric materials are potential candidates for electro-optic (EO) devices such as high bandwidth electro-optic modulators [1], and optical interconnects [2]. Second-order NLO properties in polymers are present when the chromophores are aligned in a non-centrosymmetric manner, using various approaches including electric field poling [3], formation of acentric Langmuir-Blodgett films [4], and sequential synthesis [5]. Recently, we have reported a new approach using electrostatic self-assembly to achieve acentric alignment of the chromophores in thin films. The process is very simple and involves alternate dipping of appropriate substrates into dilute aqueous solutions of oppositely charged polymers [6].

Chromophores with enhanced NLO susceptibilities can be obtained by increasing electron-donating and/or accepting effects [7], by extending the conjugation length between the donor and acceptor groups [8] and by replacing the phenyl groups in the chromophores with thiophene groups [9]. Efforts were made by our group [10] and various other groups [9, 11] to synthesize and optimize the properties of the chromophore functionalized polymers with high optical nonlinearity. NLO chromophores with large nonlinearity are crucial components in the formulation of photorefractive (PR) materials. PR effect occurs in materials that are photoconducting and in addition exhibit an electro-optic response [12]. This effect results from the formation of light induced space charge fields and consequently the spatial modulation of refractive index through the electrooptic effect. There is a growing interest in the development of fast responding photorefractive polymers with high charge carrier mobilities [13]. Polydiacetylenes are very good photoconductors and exhibit a charge carrier mobility of as high as 100 $cm^2 \cdot V^{-1} \cdot s^{-1}$ [14]. Chromophore functionalized polydiacetylenes may be ideal candidates for PR applications.

We have employed one-pot post-polymerization azo coupling reaction as well as tricyanovinylation, to synthesize a series of NLO polymers containing different heteroaromatic NLO chromophores as well as ionic functionalities, from epoxy and diacetylene based precursor polymers. The effect of the chromophore structure on the bulk NLO and other physical properties was discussed in some detail. The synthesized ionic polymers were assembled as NLO active thin films using ESA technique. Synthesized NLO chromophore functionalized polydiacetylene derivatives are potential candidate materials for PR applications. The synthesis and properties of the synthesized polymers are discussed in this work.

II. EXPERIMENTAL

2.1 Post azo-coupling reaction

The diazonium salts of appropriate chromophoric units were prepared by adding the corresponding amines directly into the nitrososulfuric acid solution at 5 °C, followed by decomposition of excess nitrous acid by addition of suitable amount of sulfamic acid in water. The freshly prepared diazonium salt solution was added dropwise into a solution of appropriate precursor polymer in N,N'-dimethylformamide (DMF) at 0 °C. The solution was stirred at 0 °C for 12 h. Chromophore functionalized polymers were obtained by precipitation in water followed by repeated washing of the polymer with acetone and drying under vacuum at least for 24 h.

2.2 Post tricyanovinylation reaction

Recrystallized tetracyanoethylene was added slowly into the precursor polymer solution in DMF under vigorous stirring. After the addition of tetracyanoethylene, the reaction temperature was raised to 45 °C, at which the reaction solution was stirred for 12h. The solution was subsequently precipitated in water. The precipitate was collected by filtration. The polymer was repeatedly washed with water, dried and further purified by repeated extraction with toluene. The polymer was dried under vacuum at least for 24 h before use.

III. RESULTS AND DISCUSSION

The precursor polymers were functionalized to introduce different chromophores at the final stage of the polymer preparation, using post azo coupling reaction as well as post tricyanovinylation reaction. Heteroaromatic chromophore functionalized epoxy based NLO polymers synthesized by post azo-coupling reaction are shown in Figure 1. Post tricyanovinylation reaction was employed as shown in Figure 2, to synthesize tricyanovinyl functionalized NLO polymers. Second order NLO properties and other physical properties of the synthesized epoxy polymers are listed in Table 1. The first part of the polymer nomenclature is an abbreviation to distinguish between precursor polymers from diglycidyl ether of bisphenol-A (BP) and other epoxide compounds (CH). The following parts refer to the different conjugation bridges and electron acceptor moieties of the chromophores.

Most of the synthesized polymers are solution processable. The glass transition temperatures (Tg's) of the polymers were studied using differential scanning calorimetry (DSC) and are listed in Table 1. Polymer, BP-IM-DC exhibit the highest Tg of 179 °C, among the series and the high Tg is attributed to possible presence of intermolecular hydrogen bonds between imidazole units and hydroxyl groups, causing physical networks. The precursor polymers, BP-AN and BP-TA, have very similar thermal stability as studied by thermogravimetric analysis (TGA) and are stable up to 330 °C under nitrogen atmosphere. Thermal stability of the formed NLO polymers declines, upon functionalization. The magnitude of the decrease depends on the type of heteroaromatic chromophores present in the polymer. BP-TA-TC starts to decompose in the temperature range between 270 and 280 °C, which is as stable as the epoxy based polymer containing 4-tricyanovinylaniline chromophores. Most of the polymers with heteroaromatic azo chromophores studied in this work do not show improved thermal stability over the epoxy based polymers with azobenzene type chromophores. However, BP-TA-TC polymer with highly enhanced NLO activity due to the presence of 2-(4-aminophenyl)-(5-tricyanovinyl)thiophene chromophores was developed, while maintaining thermal stability.

Corona poling was carried out at temperatures 10 to 15 °C higher than Tg of the polymers, to achieve non-centrosymmetric alignment of the chromophores in the polymer films. Second order NLO coefficients (d_{33}) were measured by second harmonic generation (SHG) technique. The measured d_{33} values are enlisted in Table 1. BP-TA-TC with 2-(4'-aminophenyl)-5-(tricyanovinyl) thiophene chromophores exhibited significantly larger optical nonlinearities due to the presence of strong electron-withdrawing tricyanovinyl group. The temporal stabilities of the synthesized polymers are good at 80 °C for 1000 h.

Precursor polymer, CH-AN, was synthesized from 1,4-cyclohexane -dimethanol diglycidyl ether (CH) and aniline (AN) and was functionalized with ionic chromophores using post azo coupling reaction as shown in Figure 3. All the synthesized polymers shown in Figure 3 are highly soluble in water at appropriate pH.

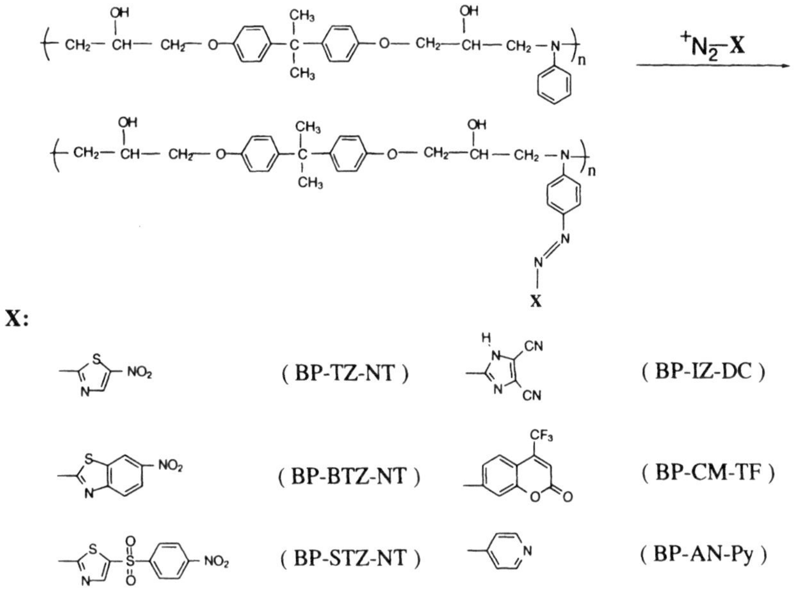

X:

$-\overset{S}{\underset{N}{\bigcirc}}- NO_2$	(BP-TZ-NT)	$-\overset{H}{\underset{N}{\bigcirc}}\overset{CN}{\underset{CN}{}}$	(BP-IZ-DC)
$-\overset{S}{\underset{N}{\bigcirc}}- NO_2$	(BP-BTZ-NT)	$\overset{CF_3}{\bigcirc\bigcirc}$	(BP-CM-TF)
$-\overset{S}{\underset{N}{\bigcirc}}-\overset{O}{\underset{O}{S}}-\bigcirc- NO_2$	(BP-STZ-NT)	$-\bigcirc N$	(BP-AN-Py)

Figure 1. Synthesis of hetero-aromatic chromophore functionalized epoxy polymers by post azo-coupling reaction.

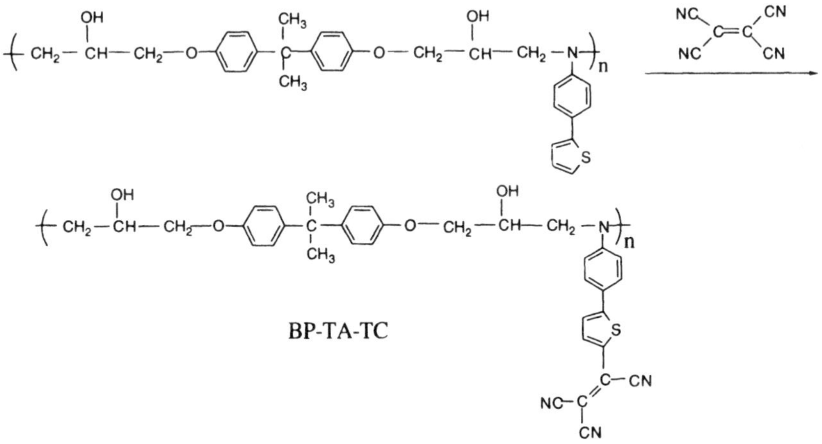

BP-TA-TC

Figure 2. Synthesis of NLO chromophore functionalized epoxy polymers by post tricyanovinylation reaction.

The layer-by-layer deposition of CH-AZ-Ca was achieved by employing poly(diallyl-dimethylammonium chloride) [PDAC] as polycation. UV-Vis spectra of CH-AZ-Ca/PDAC films with increasing number of bilayers are shown in Figure 4. A linear increase of absorbance can be

Table 1. Physical and optical properties of representative NLO polymers.

Polymer	λ_{max} (nm)	DF*	Tg (°C)	Td (°C)	d_{33} at (in pm/V) 1.064 μ	1.550 μ	Absorbance at 532 nm	775 nm
BP-TZ-NT	571	77	141	210	104.5	24.5	0.82	0.010
BP-STZ-NT	545	70	139	222	108.4	14.1	0.97	0.015
BP-BTZ-NT	549	18	128	227	19.0		0.96	0.015
BP-CM-TF	471	85	136	242	59.5		0.64	0.005
BP-IZ-DC	489	72	179	224	24.3		0.70	0.015
BP-TA-TC	615	58	126	273		80.0	0.31	0.120

*** DF - degree of polymerization (±3%) was determined using ^{1}H-NMR technique.**

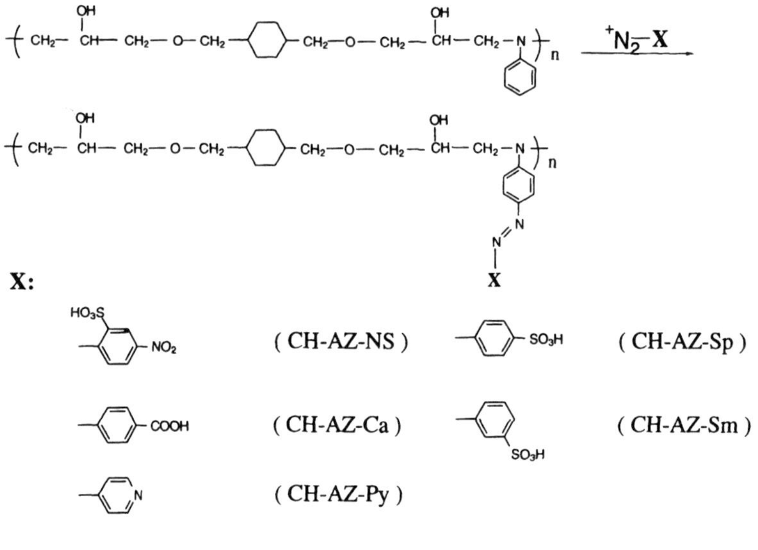

X:
(CH-AZ-NS)
(CH-AZ-Sp)
(CH-AZ-Ca)
(CH-AZ-Sm)
(CH-AZ-Py)

Figure 3. Synthesis of ionic polymers for electrostatic self-assembly (ESA) by post azo-coupling reaction.

seen as the number of bilayers increases. The linear relationship was observed up to at least 15 bilayers. The shape of the absorption curve of the multilayer is similar to that of spun cast film with an absorption maximum of 431 nm. The thickness of a 5 bilayer film was measured to be 72 Å by ellipsometry. The multilayer fabrication process was repeated several times to obtain films of the same exceptional optical quality. The second order NLO coefficients, d_{33}, of the self-assembled multilayers (5 bilayers) was determined to be 19 pm/V. This d_{33} value is in the same range as that of the electric field poled spin coated films of the same polymer. This result indicates that the azo chromophores in both the multilayers and poled spin coated films possess similar order of non-centrosymmetric alignment. Predominant orientation of the chromophores perpendicular to the surface of the substrate was inferred, using the relationship between the transmitted SHG intensity and the incident angle.

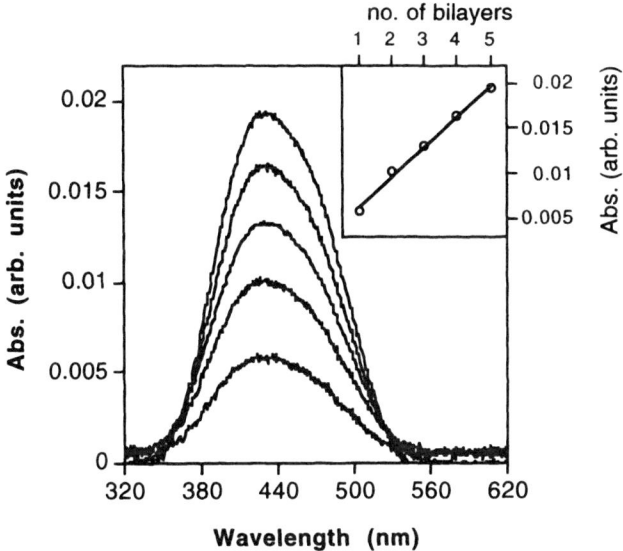

Figure 4. UV-vis spectra of PDAC/CH-AZ-Ca multilayer films; Inset shows increase in absorption with increase in number of bilayers.

We have recently synthesized a new series of diacetylenes, 9-(N-methyl-N-phenylamino)-5,7-nonadiynyl-N-(alkoxycarbonylmethyl)carbamate (MNRC; R = alkyl), whose structures are shown in Figure 5. The diacetylenes contain a functionalizable aniline group and various urethane chains. The presence of the aniline group allows modifications either in the monomer or in the polymer stage. We have carried out one-pot post azo-coupling on the poly(MNRC) [PMNRC such as PMNBC and PMNMC] with different chromophores as shown in Figure 5. Tricyanovinylation reactions can also be carried out on these polydiacetylene derivatives to obtain tricyanovinylated polydiacetylenes (from PMNBC and PMNMC). The tricyanovinylation occurs at the para position of the aniline groups as inferred from visible absorption and ^{1}H-NMR spectra. Large second-order NLO activity was achieved upon corona poling of the tricyanovinylated polymer films at 1.136 μm. The poling was performed at 130 °C for 30 min. A d_{33} value of 33 pm/V was determined from the poled polymer thin films. The unpoled spin-coated films of PMNBC and PMNMC did not show second harmonic generation (SHG) signals.

IV. CONCLUSIONS

We have successfully employed one pot post modification reactions such as post azo-coupling and post tricyanovinylation reactions to prepare NLO chromophore and ionophore functionalized polymers, with various backbone structures including epoxy and conjugated polydiacetylenes. Epoxy polymers containing 2-(4-aminophenyl)-(5-tricyanovinyl)thiophene chromophores, exhibited highest optical nonlinearity with good thermal and temporal stabilities among the series of polymers with variety of heteroaromatic chromophores. Synthesized ionic polymers were fabricated as second order NLO active thin films using electrostatic self assembling technique. NLO chromophore functionalized polydiacetylene derivatives were developed using post azo-coupling and post tricyanovinylation, for studying photorefractive characteristics.

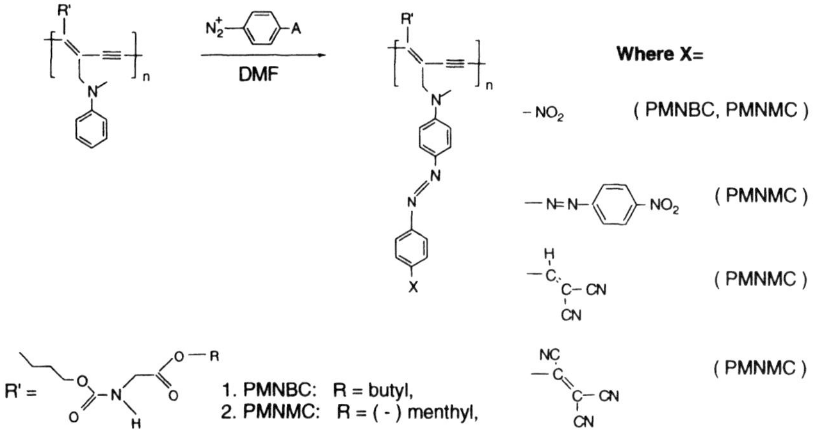

Figure 5. Synthesis of polydiacetylene based photorefractive polymer by post azo-coupling reaction.

ACKOWLEDGMENT

Partial funding from the Office of Naval Research through MURI program is gratefully acknowldged. KGC and LL acknowledge funding from BMDO and NSF through SBIR programs.

REFERENCES

1   D. G. Girton, S. L. Kwiatkowski, G. F. Lipscomb, R. S. Lytel, *Appl. Phys. Lett.*, **58**, 1730 (1991).

2   L. R. Dalton, C. Xu, A. W. Harper, R. Ghosn, B. Wu, Z. Liang, R. Montgomery and A. K-Y. Jen, *Nonlinear Optics*, **10**, 383 (1995).

3   G. A. Meredith, J. G. Van Dusen. D. J. Williams, *Macromolecules* **15**, 1385 (1982).

4   D. W. Kalina, S. G. Grubb, *Thin Solid Films*, **160**, 363 (1988).

5   D. Li, M. A. Ratner, T. J. Marks, C. Zhang, G. K. Wong, *J. Am Chem. Soc.* **112**, 7389 (1990).

6   G. Decher, J. D. Hong and J. Schmitt, *Thin Solid Films* **210/211**, 831 (1992).

7   M. Ahlheim, M. Barzoukas, P. V. Bedworth, A. Fort, Z. Hu, S. R. Marder, J. W. Perry, C. Runser, M. Staelin, B. Zysset, *Science*, **271**, 335 (1996).

8   D. R. Kanis, M. A. Ratner, T. J. Marks, *Chem. Rev.* **94**, 195 (1994).

9   A. K-Y. Jen, V. P. Rao, K. Y. Wong, and K. J. Drost, *J. Chem. Commun.*, 90 (1993).

10   X. Wang, J. Kumar, S. K. Tripathy, L. Li, J. Chen and S. Marturunkakul, *Macromolecules*, **30**, 219 (1997).

11   T. Chen, A. K-Y. Jen, and Y. Cai, *Chem. Mater.*, **8**, 607 (1996).

12   N. V. Kukhtarev, V. B. Markov, S. G. Odulov, M. S. Soskin and V. L. Vinetskii, *Ferroelectrics*, **22**, 949 (1979).

13   W. E. Moerner and S. M. Silence, *Chem. Rev.* , **94**, 127 (1994).

14   Bloor, D., and Chance, R. R., eds, *Polydiacetylenes*, Martinus Nijhoff Publishers, Dordrecht, Netherlands (1985).

CHARACTERIZING THE NLO CHROMOPHORE ORIENTATION OF POLYMERIC FILM BY ELECTROABSORPTION SPECTROSCOPY

Ke Yang *, Xiaogong Wang **, Woohong Kim ***, Aloke Jain *, Lian Li *, Jayant Kumar *
and Sukant Tripathy **,
*Departments of Physics and **Department of Chemistry, Center for Advanced Materials,
University of Massachusetts Lowell, Lowell, MA 01854, U. S. A..
***Samsung Central Research Institute of Chemical Technology, Taejon, Korea 305-380.

Abstract

The dispersion of third-order nonlinear coefficients $\chi_{1133}^{(3)}$ and $\chi_{3333}^{(3)}$ of three different NLO(nonlinear optical) polymer films were determined by electroabsorption spectroscopy. The first material investigated is an epoxy-based polymer BP-2A-NT, with azobenzene NLO chromophore 4-[((4-nitrophenyl)(azo)phenyl)azo)]aniline in its side chain. The other materials are two polydiacetylenes, poly(BPOD) and poly(4-BCMU), in which the delocalized polymer chains contribute to the third-order nonlinearity. The complex spectrum of $\chi_{3333}^{(3)}$ of each material is very similar in shape to corresponding $\chi_{1133}^{(3)}$ spectrum. The ratio of $\chi_{3333}^{(3)}$ to $\chi_{1133}^{(3)}$ is 3.2 for BP-2A-NT, 1.5 for both poly(BPOD) and poly(4-BCMU). These ratios indicate that the distribution of the side-chain NLO chromophores of BP-2A-NT is very close to three-dimensional isotropy, and the distribution of the main-chain chromophores of poly(BPOD) and poly(4-BCMU) is concentrated on the film plane.

I. Introduction

Nonlinear optical (NLO) polymers have received a great deal of attention as promising electro-optic materials over the past decade[1-3]. Various of NLO polymers have been developed as candidate materials for application in high frequency optical modulators and integrated semiconductor-NLO polymer circuits[4-7]. Among these applications, not only the nonlinearity of the material, but also the orientation of the NLO chromophore is of great concern, since a practical application needs to employ a particular or a combination of NLO tensor elements which are sensitive to the orientation of the NLO chromophores.

Electroabsorption spectroscopy can determine the dispersion of each $\chi^{(3)}$ component and can provide information about chromophore orientation in the material[8-10]. The normal incidence of electroabsorption can determine the imaginary part of the change of refractive index $\delta\kappa$ which is due to the applied electric field and is associated with the component $\chi_{1133}^{(3)}$. By using Kramers-Kronig relation , the real part of the change of the refractive index δn can be determined. The complex value of $\chi_{1133}^{(3)}$ is subsequently determined. Similarly, electroabsorption at tilted incidence can determine the complex value $\chi_{eff}^{(3)'}$, which is composed of $\chi_{1133}^{(3)}$ and $\chi_{3333}^{(3)}$. Thus $\chi_{3333}^{(3)}$ can be uniquely determined from these measurements. The obtained ratio of $\chi_{3333}^{(3)}$ to $\chi_{1133}^{(3)}$ can be utilized to offer information about the arrangement of the chromophores which can be treated as a microscopic one-dimensional rod like unit. If the ratio of $\chi_{3333}^{(3)}$ to $\chi_{1133}^{(3)}$ is less than 3, the chromophores are inclined to lie along the film plane. If the ratio of $\chi_{3333}^{(3)}$ to $\chi_{1133}^{(3)}$ equals 3, the chromophores are isotropically oriented in three-dimensional space. If the ratio of $\chi_{3333}^{(3)}$ to $\chi_{1133}^{(3)}$ is larger than 3, the chromophores are inclined to maintain perpendicular to the film plane.

In this paper, we apply this technique to two polydiacetylenes with main-chain chromophore[11], and an epoxy-based polymer BP-2A-NT[12], with side-chain chromophore, to study their third-order nonlinearity and chromophore orientation. Figure 1 shows their molecular structures.

Mat. Res. Soc. Symp. Proc. Vol. 488 © 1998 Materials Research Society

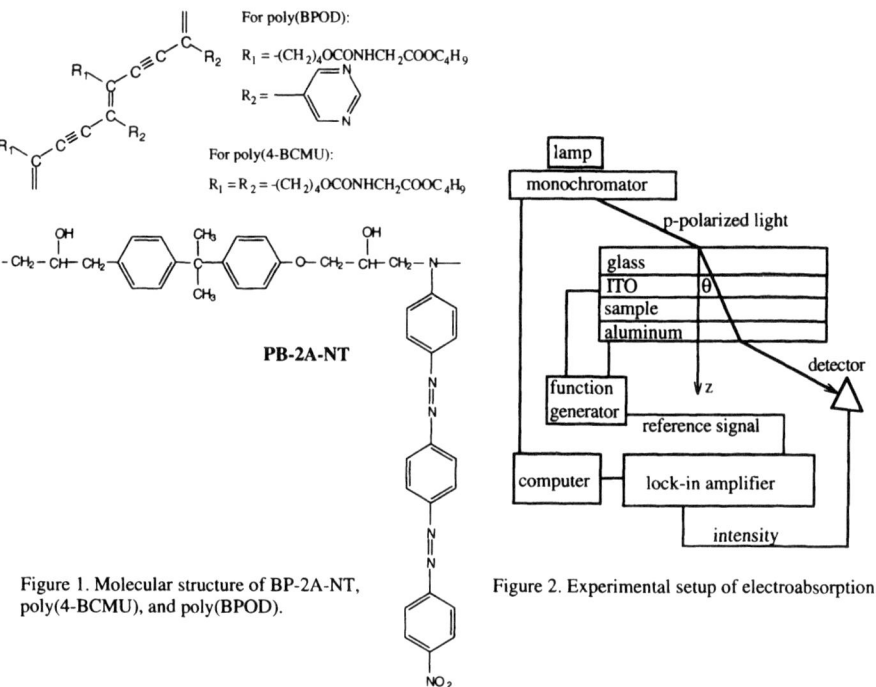

poly(4-BCMU) and poly(BPOD)

For poly(BPOD):

$R_1 = -(CH_2)_4OCONHCH_2COOC_4H_9$

$R_2 = $

For poly(4-BCMU):

$R_1 = R_2 = -(CH_2)_4OCONHCH_2COOC_4H_9$

PB-2A-NT

lamp
monochromator
p-polarized light

glass
ITO   θ
sample
aluminum

function
generator   z   detector

reference signal

computer   lock-in amplifier

intensity

Figure 1. Molecular structure of BP-2A-NT, poly(4-BCMU), and poly(BPOD).

Figure 2. Experimental setup of electroabsorption.

II. Experimental

A layer of 2330 Å poly(BPOD) was spin coated on an indium tin oxide(ITO)-glass substrate. A layer of aluminum film (~220 Å) was deposited as the top electrode on the poly(BPOD) film after thorough drying. A sinusoidal electric field ($E_{ac} = E_{aco} \cos \Omega t$, $\Omega = 2\pi f$, f = 1KHz, V_{p-p} = 25.5 V) was applied to the sample. A beam of p-polarized light coming from a tungsten lamp through a monochromator was incident on the sample with the angles of 53° and 0° ($\theta_0 = 0°$ for normal incidence). The electroabsorption signal $\Delta I_{2\Omega}$, which is defined as the change in the output intensity I at twice of the modulation frequency, was detected by the lock-in amplifier in 2f mode. The sign of $\Delta I_{2\Omega}$ was determined by comparing the amplified signal from lock-in amplifier in 2f mode and the reference signal from the function generator on an oscilloscope. A micro computer was used to control the monochromator and record data from the lock-in amplifier. The experimental setup is shown in Figure 2. For poly(4-BCMU), the experimental conditions were the following: poly(4-BCMU) thickness is 3506 Å; aluminum electrode thickness ~ 300 Å; Other experimental conditions were same as for poly(BPOD). For BP-2A-NT, the experimental conditions were the following: BP-2A-NT thickness is 4246 Å, gold electrode thickness ~ 250 Å. Other experimental conditions were same as those for poly(BPOD). The measurement of poly(BPOD) was performed in the wavelength range of 470nm to 730nm with the angle of incidence 0° and 53°. The measurement of poly(4-BCMU) was performed from wavelength of 400nm to 700nm with the angle of incidence $0°$ and $52°$. The measurement of BP-2A-NT was performed from wavelength of 400nm to 750nm with the angle of incidence $0°$ and $69°$.

The dispersions of the real and imaginary parts of the complex refractive index $\tilde{n} = n + i\kappa$ of poly(BPOD), poly(4-BCMU) and BP-2A-NT were measured by an ellipsometer (Rudolph Research, Type 43603-200E) and a spectrometer (Perkin-Elmer Lambda 9).

III. Data Processing

Let us consider the case of tilted incidence of electroabsorption spectroscopy with p polarization (see Figure 2 for the experimental setup). The expression of $\Delta I_{2\Omega} / I$ is given by reference [8]:

$$\frac{\Delta I_{2\Omega}}{I} = -2\pi \cdot \frac{\omega}{c} t_k E_{aco}^2 \cdot 3 \mathrm{Im}(\frac{\chi_{eff}^{(3)}}{\tilde{n} \cos\theta}) \tag{1}$$

where $\tilde{n}$ is the complex refractive index of the absorptive sample, θ is the complex angle of light propagation inside the material, which is related to the angle of incidence θ_0 by Snell's law: $n_0 \sin\theta_0 = \tilde{n} \sin\theta$ ($n_0 = 1$), t_k is the sample thickness, E_{aco} is the amplitude of the applied electric field. $\chi_{eff}^{(3)}$ is given by:

$$\chi_{eff}^{(3)} = \chi_{1133}^{(3)} \cos^2\theta + \chi_{3333}^{(3)} \sin^2\theta \tag{2}$$

By measuring $\Delta I_{2\Omega} / I$, we can determine the imaginary part of the complex quantity:

$$\chi_{eff}^{(3)'} = \frac{\chi_{eff}^{(3)}}{\tilde{n} \cos\theta} = \frac{\chi_{1133}^{(3)} \cos^2\theta + \chi_{3333}^{(3)} \sin^2\theta}{\tilde{n} \cos\theta} \tag{3}$$

Its real part can be obtained from Kramers-Kronig relation:

$$\mathrm{Re}[\chi_{eff}^{(3)'}(\omega)] = \frac{2}{\pi} \int_{\omega_i}^{\omega_f} \frac{\omega' d\omega'}{\omega'^2 - \omega^2} \mathrm{Im}[\chi_{eff}^{(3)'}(\omega')] \tag{4}$$

where ω_i and ω_f are initial and final optical frequencies, respectively. The complex variable $\chi_{eff}^{(3)'}$ is thus obtained.

For the normal incidence, $\theta = \theta_0 = 0$, then $\chi_{eff}^{(3)'}(\theta = 0) = \frac{1}{n}\chi_{1133}^{(3)}$, and $\chi_{1133}^{(3)}$ is determined. At the tilted incidence, $\chi_{3333}^{(3)}$ is determined from the expression of $\chi_{eff}^{(3)'}$, which gives:

$$\chi_{3333}^{(3)} = (\frac{\tilde{n}}{\cos\theta}\chi_{eff}^{(3)'}\bigg|_{tilted} - \chi_{1133}^{(3)})(\frac{\tilde{n}^2}{\sin^2\theta_0}\bigg|_{tilted} - 1) \tag{5}$$

By measuring $\tilde{n} = n + i\kappa$ and determining $\chi_{eff}^{(3)'}$ through electroabsorption spectroscopy and Kramers-Kronig relation, $\chi_{1133}^{(3)}$ and $\chi_{3333}^{(3)}$ could be determined independently (without using the relation $\chi_{3333}^{(3)} = 3\chi_{1133}^{(3)}$).

IV. Results and Discussion

The dispersions of $\chi_{1133}^{(3)}$ and $\chi_{3333}^{(3)}$ of these material are shown in Figure 3 to 8. We can see that the shape of $\chi_{3333}^{(3)}$ is similar to that of $\chi_{1133}^{(3)}$ for these materials. Their ratio were found to be $\chi_{3333}^{(3)}:\chi_{1133}^{(3)} \approx 1.5$ for both poly (BPOD) and poly(BCMU), $\chi_{3333}^{(3)}:\chi_{1133}^{(3)} \approx 3.2$ for BP-2A-NT. Their chromophores (i.e. the side chain of BP-2A-NT or the main-chain segment of the two

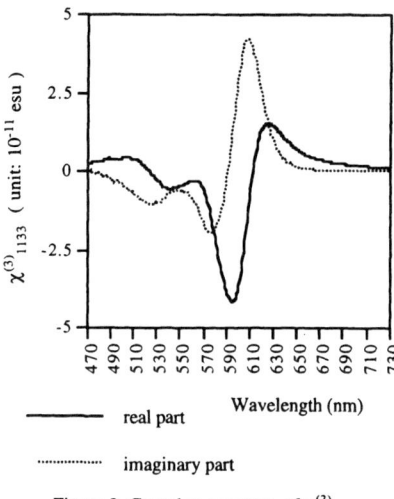

Figure 3. Complex spectrum of $\chi^{(3)}_{1133}$
for poly(BPOD)

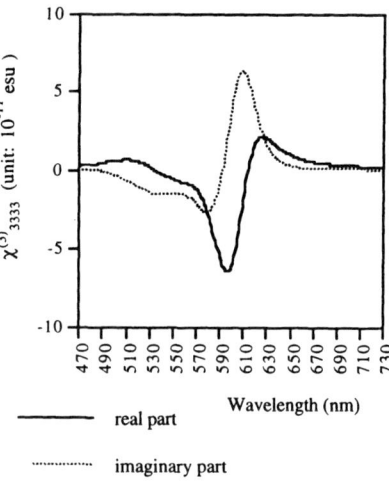

Figure 4. Complex spectrum of $\chi^{(3)}_{3333}$
for poly(BPOD).

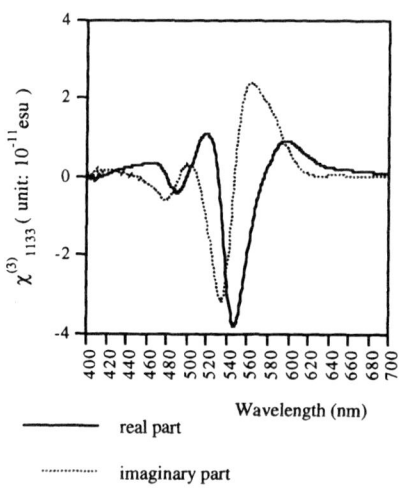

Figure 5. Complex spectrum of $\chi^{(3)}_{1133}$ for
poly(4-BCMU).

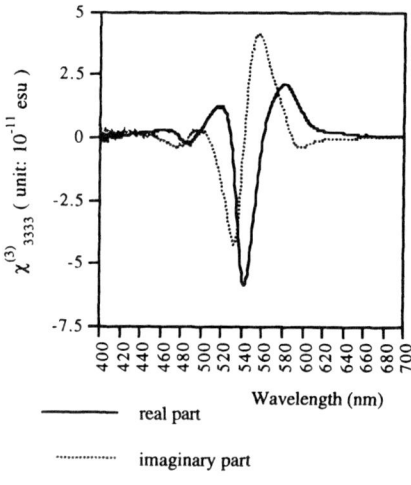

Figure 6. Complex spectrum of $\chi^{(3)}_{3333}$ for
poly(4-BCMU).

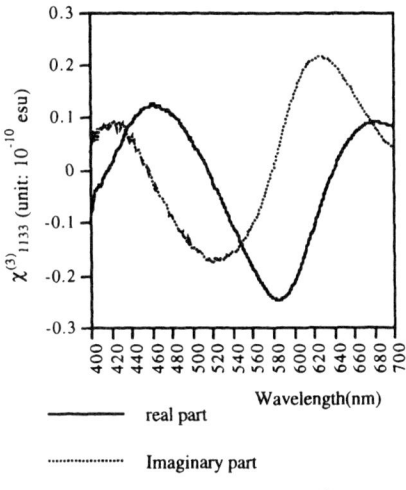

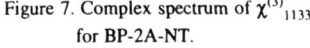

Figure 7. Complex spectrum of $\chi^{(3)}_{1133}$
for BP-2A-NT.

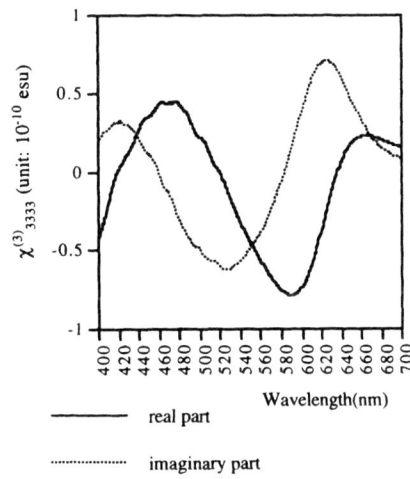

Figure 8. Complex spectrum of $\chi^{(3)}_{3333}$
for BP-2A-NT.

polydiacetylenes) can be treated as one dimensional in shape. It has only one component of microscopic third-order polarizability γ_{zzzz} , where z is along the chromophore direction. We assume the potential energy of the chromophore to be $U = -U_0 \sin\alpha$, where α is the angle between the film normal and the side-chain direction. This potential model represents the interactions among the side chains, main chains and the substrate. The chromophore orientation should satisfy Boltzmann distribution, i.e. the probability of finding a chromophore with angle α to the film normal is proportional to $\exp(-U/kT) = \exp(p \sin\alpha)$, where $p = U_0/kT$. It can be proved that for one-dimensional chromophore, $\chi^{(3)}_{1133} \propto \dfrac{\gamma_{zzzz}}{2}\langle \cos^2\alpha \sin^2\alpha \rangle$, and $\chi^{(3)}_{3333} \propto \gamma_{zzzz}\langle \cos^4\alpha \rangle$. Then we have :

$$\frac{\chi^{(3)}_{3333}}{\chi^{(3)}_{1133}} = \frac{2\langle \cos^4\alpha \rangle}{\langle \cos^2\alpha \sin^2\alpha \rangle} = \frac{2\int_0^\pi \exp(p\sin\alpha)\cos^4\alpha \sin\alpha\, d\alpha}{\int_0^\pi \exp(p\sin\alpha)\cos^2\alpha \sin^3\alpha\, d\alpha} \qquad (6)$$

Computer calculations[8] have shown that when p < 0, the ratio of $\chi^{(3)}_{3333}$ to $\chi^{(3)}_{1133}$ is larger than 3, and the chromophores are preferentially oriented normal to the film plane; when p=0, the ratio of $\chi^{(3)}_{3333}$ to $\chi^{(3)}_{1133}$ equals to 3, and the chromophores are distributed isotropically in three-dimensional space; when p>0, the ratio of $\chi^{(3)}_{3333}$ to $\chi^{(3)}_{1133}$ is less than 3, and the chromophores are inclined to lie along the film plane. For poly(BPOD) and poly(BCMU), the ratio of $\chi^{(3)}_{3333}$ to $\chi^{(3)}_{1133}$ is 1.5, significantly less than 3, and we infer that the distribution of the chromophore is lying along the film plane. For BP-2A-NT, the ratio of $\chi^{(3)}_{3333}$ to $\chi^{(3)}_{1133}$ is close to 3, and we infer that the distribution of the chromophore is close to three-dimensional isotropy.

BP-2A-NT also yields appreciable values of $\chi^{(3)}_{1133}$ and $\chi^{(3)}_{3333}$. The maximum value of

$\chi^{(3)}_{3333}$ is even comparable with those of polydiacetelynes. It dose not mean that the chromophore of BP-2A-NT is as strong in third-order nonlinearity as those of polydiacetelynes. The reason is that the polydiacetelyne chromophore (main chain) is inclined to lie along the film plane, and its $\chi^{(3)}_{3333}$ component is significantly reduced by projection from the chromophore orientation to the film normal. While the BP-2A-NT chromophore (side chain) is oriented isotropically in three-dimensional space. This distribution can convert more third-order nonlinearity of the BP-2A-NT chromophore to the $\chi^{(3)}_{3333}$ component. Due to this reason, the epoxy-based nonlinear optical polymer with side-chain NLO chromophore could be a potential candidate for applications involving the $\chi^{(3)}_{3333}$ component. For the same reason, to exploit large value of $\chi^{(3)}_{3333}$ from NLO polymer with main-chain chromophore such as polydiacetelynes, measures to adjust the main-chain distribution should be considered.

Conclusion

Normal and tilted incidence of electroabsorption spectroscopy was used to study the chromophore orientation in NLO polymers, and the information of polymer chain orientation was inferred. The results indicate that the main-chains of the two polydiacetelynes are inclined to lie along the film plane, while the side-chains of the epoxy-based NLO polymer are distributed close to three-dimensional isotropy.

Acknowledgments: Ke Yang would like to thank a TA support from University of Massachusetts Lowell during this research. Partial funding from ONR is gratefully acknowledged.

References
1. J. Zyss, D. S. Chemla, "Nonlinear Optical Properties of Organic Molecules and Crystals", Academic Press, Inc., Vol. 1 and 2, (1987).
2. P. Prasad, and D. Williams, "Nonlinear Optical Effects in Molecules and Polymers"; John Wiley and Sons: New York, 1991.
3. J. Zyss, "Molecular Nonlinear Optics, Materials, Physics, and Devices", Academic Press, Inc., (1994).
4. D. M. Burland, R. D. Miller, C. A. Walsh, Chemical Reviews, Vol. 94, p31, 1994.
5. L. R. Dalton, A. W. Harper, R. Ghosn, W. H. Steier, M. Ziari, H. Fetterman, Y. Shi, R. V. Mustacich, A. K.-Y. Jen, K. J. Shea, Chem. Mater. Vol. 7, p1060, 1995.
6. L. R. Dalton, A. W. Harper, B. Wu, R. Ghosn, J. Laquindanum, Z. Liang, A. Hubbel, C. Xu, Adv. Mater. Vol. 7, p519, 1995.
7. T. J. Marks, and M. A. Ratner, Angew. Chem. Int. Ed. Engl., Vol. 34, p155, 1995.
8. K. Yang, W. Kim, J. Kumar, L. Li, S. Tripathy. "Dispersion of $\chi^{(3)}$ in polydiacetylene films from electroabsorption spectroscopy", Optics Communications, Vol. 144, pp252-258, 1997.
9. K. Yang, D.-W. Cheong, S. Tripathy, J. Kumar, "Dispersions of Electroabsorption Susceptibilities: Application to a Polymeric Langmuir-Blodgett Film", Optics Communications, Vol. 144, pp259-264, 1997.
10. K. Yang, X. Wang, J. Kumar, A. Jain, L. Li, S. Tripathy, "Electroabsorption Investigation of a Nonlinear Optical Azo Polymer", to be submitted to Nonlinear Optics.
11. W. H. Kim, N. B. Kodali, J. Kumar and S. K. Tripathy, Macromolecules 1994, Vol 27, p1819.
12. X. Wang, J. Kumar, S. K. Tripathy, L. Li, J. Chen, and S. Marturunkakul, "Epoxy-Based Nonlinear Optical Polymers from Post Azo Coupling Reaction", Macromolecules, Vol. 30, pp219-225(1997).

SECOND-ORDER NONLINEAR OPTICAL POLYMER BY DIRECTED LITHIATION OF POLYETHERSULFONE

HAROLD E. McCARRON, CHRISTOPHER J. WALSH, RAJASHREE SEN and BRAJA K. MANDAL*

Department of Biological, Chemical and Physical Sciences, Illinois Institute of Technology, Chicago, IL 60616

ABSTRACT

Directed lithiation of a polyethersulfone and the reactions of the lithiated species have been investigated. This method provides a new route to second-order nonlinear optical polymers.

INTRODUCTION

Polymeric materials with nonlinear optical (NLO) activity have been prepared by several techniques: doping the polymer with an NLO dye, copolymerization of monomers that contain NLO pendant groups and covalent attachment of NLO dyes to polymers.[1-4] Herein, we report, for the first time, a method to prepare NLO polymers by directed lithiation of polyethersulfone (PES).

PES is a tough, strong engineering thermoplastic with excellent strength, flexibility, and creep resistance (Figure 1).[5] The aryl protons *ortho* to the sulfone group of PES are somewhat acidic and can be metalated by using a strong base such as butyllithium.[6,7] These positions are highly favored because the lithiated species forms a five membered ring intermediate with the ortho site and the oxygen of the sulfone group.

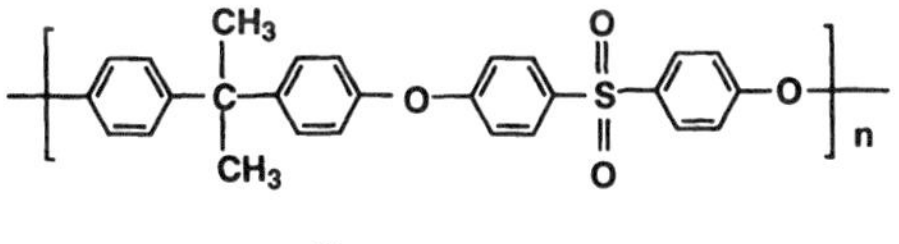

Udel® Polyethersulfone

Figure 1. The structure of polyethersulfone.

In this study, the resulting lithiated polymer reacted with a variety of electrophiles to produce novel PES derivatives. The rapidity and completeness of the substitution at the lithiated sites encouraged an attempt to add NLO functionality by this technique. Particularly attractive was the possibility of adding the N-(4-nitrophenyl)-(S)-prolinoxy moiety, a smaller sized donor-π-acceptor NLO chromophore.

EXPERIMENTAL

1-Bromooctane, ethylene oxide and n-butyl lithium (BuLi) were purchased from Aldrich and used without further purification. Tetrahydrofuran (THF) from Aldrich was distilled from sodium and benzophenone. PES (Aldrich) was purified by dissolving in THF, precipitating from methanol and drying overnight at 70°C. Methanol (histological grade) was purchased from Fisher. Argon (prepurified grade) was supplied by Linox. Carbon dioxide was obtained from dry ice by passing the evolved gas through a 14x3/4 inch tube filled with calcium chloride.

N-(4-nitrophenyl)-(S)-prolinoxy tosylate (NLO-Ts)[8]

This compound was prepared by a minor modification of a previously reported method.[8] Into a single-necked 200 mL round bottomed flask fitted with a stirring bar was weighed 2 g (50 mmoles) of N-(4-nitrophenyl)-(S)-prolinol and 50 mL of pyridine was added. The solution was cooled in an ice-water bath and 3.43 g of p-toluenesulfonyl chloride was added in four equal portions over the course of 45 minutes. The reaction mixture was poured into water, stoppered and refrigerated overnight. The crystals were removed by filtration and redissolved in ethyl acetate for chromatography through an alumina column, eluted with ethyl acetate:hexane::4:1. After reduction of solvent *in vacuo* the product was crystallized from hexane. M.P.: 84°C. Yield: 85%.

General procedure for functionalization of PES

A 100 mL three-necked round bottomed flask was fitted with a gas inlet, a gas outlet and rubber septum. After the flask was flamed and purged with argon for 20 minutes, 50 mL of dry THF and 0.9 g of PES (2.125 mmoles of polymer repeat units) were added to the flask. The flask was stirred under an argon blanket until the polymer had dissolved. The flask was cooled to -78°C with a dry ice/acetone bath. n-BuLi (10 M in hexane, 0.036 mL, 0.636 mmole) was added to the flask with a syringe. The reaction was allowed to proceed for 90 min at -78°C. NLO-TS (0.263 gm, 0.07 mmole) was added to the solution. Argon flow was stopped and the flask allowed to return to room temperature (2 h). The solution was transferred to a single-necked flask and the solvent volume reduced *in vacuo*. The concentrated solution was poured into methanol in a Waring Blender. (Note: For reactions with ethylene oxide and carbon dioxide, the precipitation of PES derivatives was performed in aqueous ammonium chloride solution). The precipitated polymer was filtered immediately, washed with water and again with methanol. The recovered polymer was dried overnight at 70°C.

RESULTS AND DISCUSSION

The directed lithiation of PES was performed in THF at -78°C using n-hexane solution of BuLi. Larger volumes of lower concentration of BuLi (~2.5 M) were found unsuitable due to the precipitation of lithiated PES from the solution. The availability of highly concentrated solutions of BuLi permit addition of quantities of butyllithium significant enough for meaningful metalation. The lithiated PES generated using 10 M BuLi formed a tan colored solution in THF which allowed us to study its reactivity with a variety of electrophiles including an electrophilic compound containing an NLO active chromophore. The reactions are summarized in Scheme-I. A brief discussion for each reaction is described below:

SCHEME-I

Reactions of Lithiated Polyethersulfone

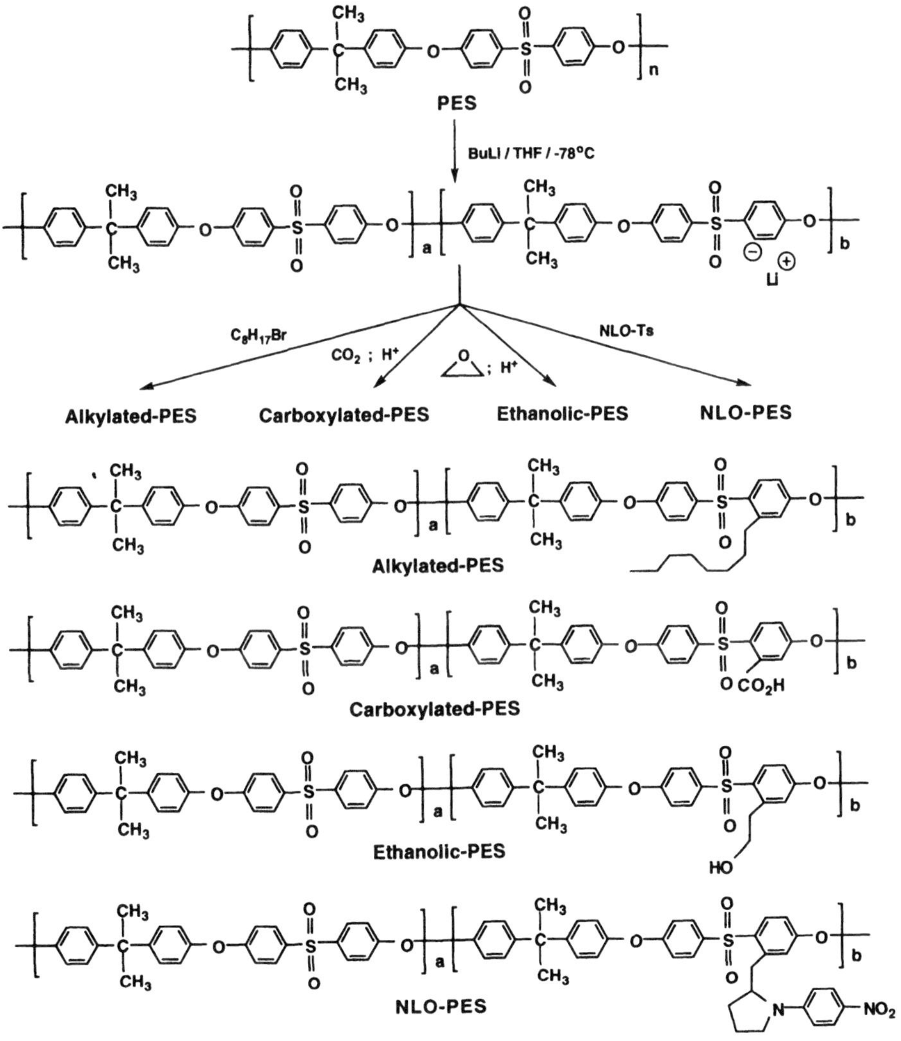

PES
BuLi / THF / -78°C
$C_8H_{17}Br$
CO_2 ; H^+
; H^+
NLO-Ts
Alkylated-PES
Carboxylated-PES
Ethanolic-PES
NLO-PES

Alkylated-PES

Various levels of n-octyl group were attached to the PES. Originally, it was thought that the alkane molecule would disturb the regularity and association of the polymer chain, but high levels of alkane substitution reduce solubility. Modeling studies suggested that the alkane chain lies in the plane of and stretched along the polymer backbone not randomly oriented, as might be expected. This orientation could reduce the freedom of rotation about the sulfone group, shielding it from the polar groups of the solvent. Then again, if every repeat unit contains an alkyl chain, the result of the functionalization is to increase the molecular weight by 26%. The molecular weight of 35,000 becomes 44,000, a significant increase for a difficultly soluble polymer.

Carboxylated-PES

In general, the solubility of the product decreased as the amount of butyllithium increased. Other properties indicated that the polymer morphology had been modified due to CO_2H substitution. Molecular mechanics modeling of the product indicates that in the substituted product, the CO_2H group positions itself as far away from the sulfone group as possible. It is postulated that this significantly changes the coils of the polymer from 28_3 to 28_9 permitting closer fitting of the backbone chain. Both FTIR and NMR data support the substitution mode. Modifications of this polymer are being evaluated as electrolytic membranes.

Ethanolic-PES

Ethylene oxide was added (like carbon dioxide) to the lithiated polymer solution as a gas. This polymer was substituted at the rate of one pendant group for each 1.5 repeat units. FTIR and NMR indicate the presence of the pendant ethanolic groups. The polymer gelled when treated with isophorone diisocyanate.

NLO-PES

As a part of our ongoing investigation on polymeric NLO materials, we attempted to attach NLO dyes using a tosylate group containing NLO compound, NLO-Ts. It seemed that the properties of polyethersulfone were consistent with requirements of NLO usage. The NLO-Ts reacted smoothly with the lithiated polymer yielding a bright yellow polymer which is soluble in THF and forms films readily. FTIR and NMR both show that the dye is attached. NLO studies are in progress.

CONCLUSIONS

Polyethersulfone can be substituted with many functionalizing groups, expanding the use of this polymer which brings its excellent properties to the functionalized polymer. Solubility of the modified polymer can be a problem, and judicious control of the level of functionalization must be maintained. Numerous other modifications suggest themselves and many are under investigation in this laboratory.

ACKNOWLEDGMENTS

This work is supported in part by the Office of Naval Research under the grant # N00014-94-1-1150. One of the authors, R. S., wishes to thank the Howard-Hughes Foundation for support.

REFERENCES

1. H. S. Nalwa, S. Miyata, Eds., Nonlinear Optics of Organic Molecules and Polymers, CRC, New York, 1997.
2. P. N. Prasad and D. J. Williams, Eds., Nonlinear Optical Effects in Molecules and Polymers, Wiley, New York, 1991.
3. T. Chen, A. K. Jen and Y. Cai, Macromolecules, **29**, p.535 (1996).
4. A. K-Y. Jen, K. J. Drost, Y. Cai, V. P. Rao and L. R. Dalton, J. Chem. Soc., Chem. Commun., 965 (1994).
5. J. I. Kroschwitz, Ed., Kirk-Othmer Encyclopedia of Chemical Technology, 4th Edn., Wiley, Vol. 19, p. 945, 1996.
6. M. D. Guiver and J. W. Apsimon, J. Poly. Sci. Part C, Polymer Letter, **26**, p.123 (1988).
7. W. Kerres, W. Cui and S. Reichle, J. Poly. Sci., Part A, Polymer Chemistry, **34**, p.2421 (1996)
8. M. A. Firestone, J. Park, N. Minami, M. A. Ratner, T. J. Marks, W. Lin and G. K. Wong, Macromolecules, **28**, p. 2247 (1995).

SYNTHESIS OF HIGH-Tg AZO POLYMER AND THE OPTIMIZATION OF ITS POLING CONDITION FOR STABLE EO SYSTEM

Takashi Fukuda*°, Hiro Matsuda*, Takao Shiraga[†], Masao Kato[†] and Hachiro Nakanishi[‡]
* National Institute of Materials and Chemical Research (NIMC), Tsukuba 305 Japan
[†] Science University of Tokyo, 2641 Yamazaki, Noda 278 Japan
[‡] Institute for Chemical Reaction Science, Tohoku Univ., Sendai 980-77 Japan

ABSTRACT

A new material for second-order nonlinear optics was synthesized, which was a copolymer of N-phenylmaleimide, 4-isopropenylphenol and 4'-[N-ethyl-N-(4-isopropenylphenoxyethyl) amino]-4''-nitroazobenzene (PMPD). PMPD films were poled by corona-poling technique. The optical nonlinearity of poled PMPD was measured by second harmonic generation (SHG) and electro-optic (EO) effect, and it was demonstrated that this polymer had large optical nonlinearity and a very long-time stability, as was expected. These properties were thought to be sufficient enough for practical EO devices. On the other hand, from the viewpoint of the sample preparation technique, poling conditions were investigated in order to achieve the highest possible dipolar orientation. As a result, it was found that the relationship between the electric resistance of polymer film and substrate was a critical factor for corona-poling efficiency. From a simple model, it was suggested that the poled PMPD film prepared onto the glass substrate with a resistance of ~0.8 GΩ (at 160 °C) exhibits large SHG and EO coefficients, more than ~500 × 10^9 e.s.u (d_{33} at $\lambda = 1.064$ μm) and ~70 pm/V(r_{33} at $\lambda = 632.8$ nm), respectively. It should be noted that this expected values are approximately twice as much as obtained under conventional corona-poling conditions.

INTRODUCTION

Poled polymer systems have attracted much attention as ultrafast and low voltage electro-optic (EO) devices. In recent years, many kinds of polymeric materials with high EO coefficients, good long-time stability, low optical propagation loss, and good processability have been reported. In particular, since the polymers with high glass transition temperature (Tg) can afford to avoid the relaxation of non-centrosymmetric orientation, maleimide-based polymers with NLO-active azo-dye side chain [1]-[3] are thought to be good candidates for practical use. Therefore, we have directed our efforts to synthesize new maleimide-based polymers that are more stable and can be synthesized more easily. Previously, we have reported that the copolymer of 4-[N-(2-maleimidoethyl)-N-methylamino]-4'-nitroazobenzene (MENA) and α-metylstyrene had a large optical nonlinearity and good structural and chemical stability [4]. However, since we recently succeeded in synthesizing a new kind of maleimide-based polymer with larger optical nonlinearity and better long-term stability, we will report on the nonlinear optical properties.

On the other hand, the study on the optimization of poling conditions has been so far directed

---

° E-mail: tfukuda@nimc.go.jp

Mat. Res. Soc. Symp. Proc. Vol. 488 © 1998 Materials Research Society

to poling temperature [5]-[7], applied voltage [5],[6], atmosphere [5],[8], annealing procedure and so on [9]. However, in order to get higher dipolar orientation by the corona-poling method, we came to believe that we should pay attention to the electric resistance of the substrate. Here, we will show experimental data regarding the influence of the electric resistance (related to the substrate thickness and the electric conductivity) of the glass substrate on the poling efficiency, and will point out optimized conditions for the preparation of a poled polymer.

EXPERIMENTAL

The precursor polymer was prepared by radical polymerization of N-phenylmaleimide and isopropenylphenol in THF at 65 °C. The copolymerization ratio of N-phenylmaleimide to isopropenylphenol was 48% to 52% estimated by NMR spectrum. It was converted to a copolymer of 4'-[N-ethyl-N-(4-isopropenylphenoxyethyl) amino]-4''-nitroazobenzene and N-phenyl maleimide (PMPD) by a polymer reaction of Mitsunobu coupling with Disperse Red 1. The chemical structure of PMPD is shown in figure 1. The azo-dye contents of obtained PMPD could be changed by the amount of Disperse Red 1 and reaction time. The azo-dye content of PMPD used in this study was 78%, which was the ratio to the number of hydrogen groups that existed in precursor polymer, estimated from the absorption spectrum. The number average molecular weight (M_n) and the glass transition temperature (T_g) was 7.0×10^4 and 190 °C, respectively.

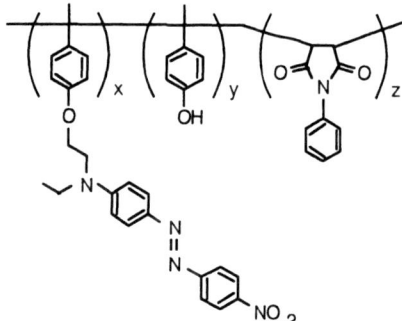

FIGURE 1   Chemical structure of PMPD.

This copolymer is soluble in dichloromethane and chloroform. Thin films were prepared on glass substrates by spin-coating. Approximately 1000Å-thick and 5000Å-thick films were prepared for SHG and EO measurement, respectively. The film thickness was measured with a mechanical stylus profiler. Corona-poling was carried out in the conditions detailed below; a tungsten needle was kept at 5 kV, the electrode gap was 1 cm, poling temperature was 160 °C, poling time was 20 min, and after that, the polymer film was cooled to room temperature at the rate of 3 °C/min. In this report, seven glass substrates that differ the electric resistance in the range from 0.14 to 4.3 GΩ were employed. The temperature dependence of the electric resistance of the substrate was measured in the range from 60 °C to 200 °C. SHG of poled polymers were measured by Maker-fringe method in transmission geometry using a Q-switched Nd:YAG laser (λ = 1.064 μm, pulse duration = 8 ns, repetition frequency = 10 Hz, pulse power = 10 mJ / pulse) as the fundamental light source. The SHG coefficients (d_{33}) were evaluated according to the reference [10] by use of a quartz crystal plate whose d_{11} was 8×10^{-10} e.s.u. EO coefficients (r_{33}) were determined by a simple reflection technique [11] using He-Ne laser ($\lambda = 632.8$ nm, laser power = 0.2 mW). Sample dimensions were 5×4 mm². The angle of incidence of the laser beam was fixed to 45°. In the data analysis, the refractive index of PMPD was assumed as 1.6.

RESULTS AND DISCUSSION

Figure 2 shows a stability of poled PMPD monitored by SHG coefficient (d_{33}). The measurements were performed for both annealed and non-annealed sample at 80 °C. The annealing procedure was done before corona-poling at 110 °C for 7 hours. As is shown in the figure, this poled polymer was quite stable, maintaining ~ 245×10^{9} e.s.u. even after heating treatment at 80 °C for more than 40 days, even though there was ~10% of a quick initial decay. For the annealed sample, the quick initial decay was reduced to only 5%.

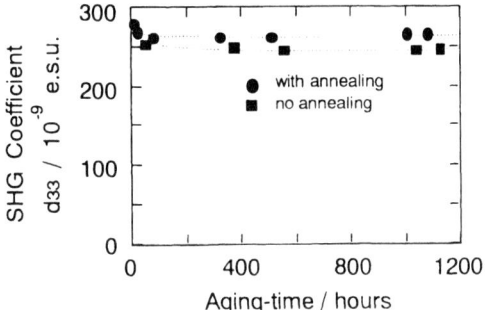

FIGURE 2   The dependence of SHG coefficient (d_{33}) on the aging-time measured at 80 °C. These samples were prepared onto the 0.8 mm-thick soda glass substrate.

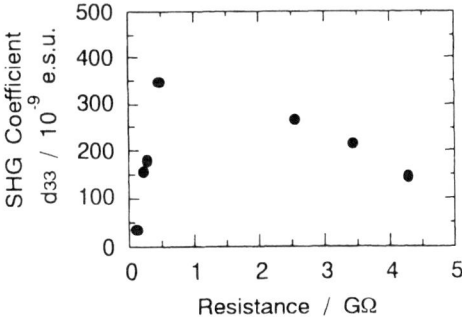

FIGURE 3   The dependence of SHG coefficient (d_{33}) on the electric resistance. The electric resistance shown in this figure is the value at 160 °C.

Figure 3 shows the dependence of SHG coefficient (d_{33}) of poled PMPD on the electric resistance. As is clear in the figure, it was found that SHG coefficient (d_{33}) strongly depended on the electric resistance of the substrate which is characterized by its thickness and electric conductivity. So far, a 0.8 ~ 1.0 mm-thick slide glass was usually employed for the preparation of the poled polymer. From our measurements, the electric resistance of 0.8 and 1.0 mm-thick slide glass measured at 160 °C were 2.6 GΩ and 3.3 GΩ, and the SHG coefficient (d_{33}) obtained with those glass substrates were 270×10^{-9} e.s.u. and 215×10^{-9} e.s.u., respectively. On the other hand, a SHG coefficient (d_{33}) of 350×10^{-9} e.s.u. was obtained for the soda-lime glass substrate with an electric resistance of 0.45 GΩ.

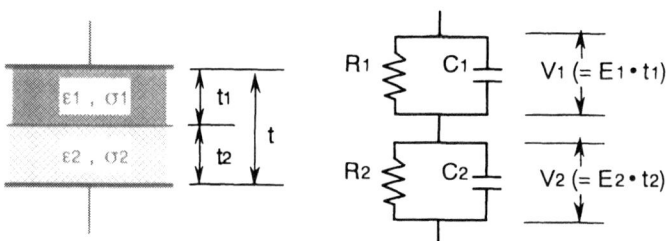

FIGURE 4   Maxwell-Wagner two-layer capacitor and its equivalent circuit.

In order to analyze these results, the Maxwell-Wagner two-layer capacitor model was employed. The schematic of the model and its equivalent circuit were shown in figure 4.

The dielectric consists of a polymer thin film and a glass substrate which are characterized by their dielectric constant, conductivity and thickness, (ε_1, σ_1, t_1) and (ε_2, σ_2, t_2) respectively. When a DC field is suddenly applied, the final distribution follows from the condition of electric current density continuity: $J_1 = J_2$ which can be written as $E_1 / E_2 = \sigma_2 / \sigma_1$, where E_1 and E_2 are the electric field applied at each dielectric, respectively. If we assume that the applied total voltage ($V = V_1 + V_2$) is constant in corona poling, the electric field applied to a poled polymer is given by

$$E_1 = \frac{\sigma_2 V}{\sigma_2 t_1 + \sigma_1 t_2} \quad . \tag{1}$$

The SHG coefficient (d_{33}) of the poled polymer is given by

$$d_{33} = N f_\omega f_\omega f_{2\omega} L(a) \beta_{xxx} \quad , \tag{2}$$

where N is the number of NLO active group in unit volume, f_ω and $f_{2\omega}$ are the local field factors, $L(a)$ is the third-order Langevin function and β_{xxx} is the first-order hyperpolarizability of NLO active group.

$$L(a) = \frac{\mu E_1}{5 k_B T} \quad , \tag{3}$$

where μ is the dipole moment of NLO active group, k_B is the Boltzmann constant and T is the absolute temperature.

From the Eq.(1)-(3), the SHG coefficient (d_{33}) of poled polymer is found to be a monotonically decreasing function of the thickness of glass substrate (t_2). If we assume the case of $\sigma_1 = 10^{-13}$,

$\sigma_2 = 10^{-9}$ S/cm (at 160 °C) and $t_1 = 1000$ Å, the predicted relation between d_{33} and t_2 is shown in Fig. 5. The experimental data of the right slope shown in Fig. 3 is well characterized by this relation. The observed quick drop of d_{33} in the low resistance region in Fig. 3 is thought to be based on the electric breakdown of dielectric. As a result, from this simple model, a large SHG coefficient (d_{33}) of more than ~500 × 10^{-9} e.s.u (at $\lambda = 1.064$ μm) was expected. This value is approximately twice as much as obtained by conventional corona-poling condition.

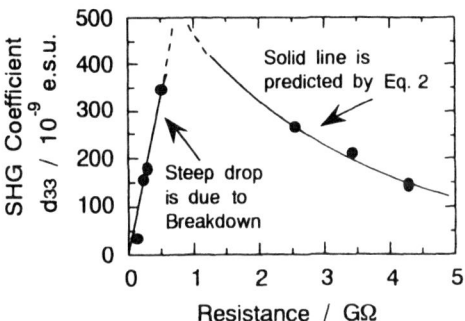

FIGURE 5 The predicted relation between the SHG coefficient (d_{33}) of poled polymer and the thickness of glass substrate (t_2) according to the model above. The scale of the vertical axis of calculated curve (Solid line) was adjusted to fit the experimental data (Black circles).

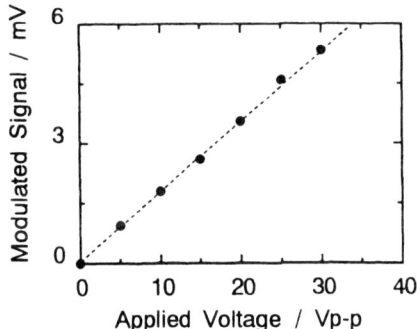

FIGURE 6 Voltage dependence of modulated EO signal. Sinusoidal wave was applied for the modulation.

Figure 6 shows a plot of modulated beam intensity I_m as a function of applied voltage for EO experiment. Since the observed modulated signal is proportional to the applied voltage, this signal is thought to be originated only in linear electro-optic effect (Pockels effect).

If we assume that $n_e \sim n_o \sim n$ and $r_{33} = 3\, r_{31}$ from symmetry conditions, the EO coefficient (r_{33}) is given by following equation, where the subscripts 3 and 1 represent directions, respectively.

$$r_{33} = \frac{3\lambda I_m}{4\pi V_m I_c n^2} \frac{(n^2 - \sin^2\theta)^{3/2}}{(n^2 - 2\sin^2\theta)} \frac{1}{\sin^2\theta} \qquad ,, \qquad (4)$$

where I_m is the amplitude of the modulated signal, V_m is the amplitude of the applied voltage, I_c is the input beam intensity, n is the refractive index, and λ is the wavelength and θ is the angle of incidence of the input beam, respectively.

According to the Eq. (4), the EO coefficient (r_{33}) of PMPD is estimated as 51 pm/V. This value is fairly large, even if we compare it with a typical EO material such as $LiNbO_3$ whose EO coefficient (r_{33}) is approximately 31 pm/V at 632.8 nm. The EO coefficient (r_{33}) of PMPD is expected to be rised up to ~70 pm/V at the optimised poling conditions suggested above.

CONCLUSION

On the preparation of the poled polymer using corona-poling technique, it was demonstrated that the electric conductivity of both polymer film and substrate were critical factors, and it was suggested that the highest possible dipolar orientation could be achieved by optimizing the electric resistance of the substrate.

Since PMPD has a fairly large EO coefficient (r_{33}) and quite stable dipolar ordering induced by corona-poling, this polymer system is quite attractive for practical EO devices such as real-time spatial light modulator.

REFERENCES

[1] M. Ahlheim and F. Lehr, Macromol.Chem., **195**, 361 (1994).
[2] P. Pretre, P. Kaatz, A. Bohren, P. Günter, B. Zysset, M. Ahlheim, M. Stahelin and F. Lehr, Macromolecules, **27**, 5476 (1994).
[3] H. Matsuda, S. Yamada, M. Kato, C. Ishii, T. Miyoshi and H. Nakanishi, *Proceedings of Third International Conference on Organic Nonlinear Optics (ICONO'3)*, 128 (1997).
[4] T. Fukuda, H. Matsuda, H. Someno, M. Kato and H. Nakanishi, *Proceedings of The 8th International Conference on Unconventional Photoactive Systems (UPS-8). that will appear in the special issue of Mol. Cryst. Liq. Cryst.* , in press.
[5] T. A. Pasmore, J. Talbot and H. S. Lackritz, *IEEE 1996 Annual Report. Conference on Electrical Insulation and Dielectric Phenomena*, vol. 2, 688 (1996).
[6] Jong-Ha Lee, Y. K. Kim, Y. H. Won, S. H. An and S. Y. Kim, Ungyong Mulli, **9**, 700 (1996). (in Korean)
[7] R. Gerhard-Multhaupt, S. Bauer, V. Ren, S. Yilmaz and W. Wirges, *Nonlinear Optics, Principles, Materials Phenomena and Devices (Proceedings of 3rd French-Israeli Symposium on Nonlinear-Optics)*, **11**, 309 (1994).
[8] S. J. Bethke, S. G. Grubb, H. L. Hampsh and J. M. Torkelson, *Proc. SPIE - Int. Soc. Opt. Eng.*, **1216**, 260 (1990).
[9] M. A. Mortazavi, A. Knoesen, S. T. Kowel, B. G. Higgins and A. Dienes, J. Opt. Soc. Am. B, Opt. Phys., **6**, 733 (1988).
[10] D. S. Chemla and J. Zyss (eds.), *Nonlinear Optical Properties of Organic Molecules and Crystals, Academic Press*, 1987.
[11] C. C. Teng and H. T. Man, Appl. Phys. Lett., **56**, 1734 (1990).

VERY LARGE SECOND ORDER NON-LINEAR OPTICAL ACTIVITY SHOWN BY HETEROCYCLE-BASED DICYANOMETHANIDE ZWITTERIONS

A. ABBOTTO *, S. BRADAMANTE *, A. FACCHETTI *, G. A. PAGANI *,
I. LEDOUX **, J. ZYSS **
* Department of Materials Science, University of Milano, via Emanueli 15, I-20126 Milano, Italy,
pagani@icil64.cilea.it
** France Telecom – CNET, 196 av. H. Ravera, BP107, 92225, Bagneux Cedex, France

ABSTRACT

The new series of recently synthesized chromophores **2 - 6** in which a positively charged N-alkylpyridinium acceptor and a negatively charged dicyanomethanide donor are spaced by thiophene-based and/or ethylenic bridging units show dramatically enhanced first molecular hyperpolarizabilities ($\mu\beta$ as high as 27000 x 10^{-48} esu), in comparison with some of the best recently reported neutral systems **7** and **8**.

INTRODUCTION

A wide range of applications are responsible for the increasingly greater interest being shown in non linear optic (NLO) materials [1,2], including telecommunications, optical data processing, and data storage. Some of these materials have already been marketed [3], but manufacturing companies feel that substantial improvements are needed in order to consolidate and possibly extend their production. 4-Dimethylamino-4'-nitrostilbene DANS (**1**) is the prototype of a large family of second order NLO active systems in which the acceptor and donor units represent primary organic functionalities. Instead, we have exploited π-excessive and π-deficient heterocycles as donors and acceptors, respectively.

Two approaches in the molecular design of second order NLO materials have been reported in the literature. In the first, and conventional approach, where DANS is often referred as a standard, the NLO activity originates from an efficient long-range charge transfer between the donor and the acceptor groups. In these molecules the ground state is described by an efficient resonance between two limit formula ("the cyanine limit" or "bond length alternation" approach), one aromatic, which slightly predominates, and one quinoid. The aromatic limit formula is always neutral and, by contrast, the quinoid structure is dipolar. The alternative approach draws the attention to the fact that, according to the two-state model, β increases as the difference between the dipole moments of the ground- and excited-state increases. One way to achieve this is to design molecules with a high dipolar character in the ground state, which necessarily strongly decreases in the excited state. However so far this could be obtained in heterocyclic pyridinium betaines with no conjugation between the positively- and negatively-charged fragment, resulting in low values of hyperpolarizability [4].

EXPERIMENT AND RESULTS

Our approach presented here, described by compounds **2 - 6**, implies the design of an heterocyclic analogue of DANS, with the ground state presenting a highly charge-separated zwitterionic structure. In this set the azinium ion is the acceptor and the donor a dicyanomethanide unit: spacer groups are either ethylenic or 2,5-thienylene fragments. The aromatic nature of the dipolar structure, that is exactly the reverse of the more popular class of *DANS-like* push-pull systems where the aromatic form is neutral, provides stability to the high charge-separation structure. The presence of a variable π-conjugation pathway between the donor and the acceptor ensures a more efficient decrease of the charge separation, and therefore of the dipole moment, in the excited state, and a better control on NLO, thermal and optical properties.

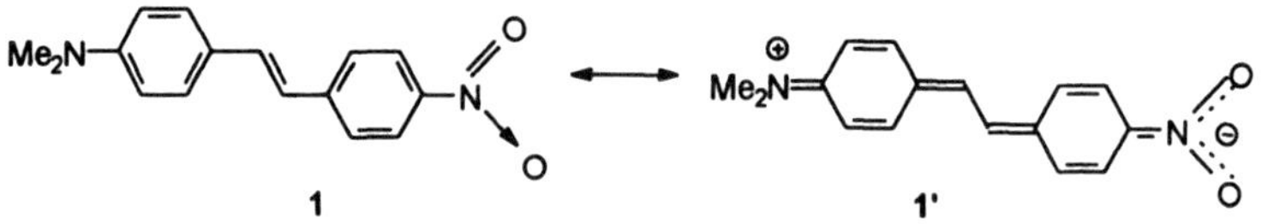

1
ground state: neutral, aromatic

1'
excited state: dipolar, quinoid

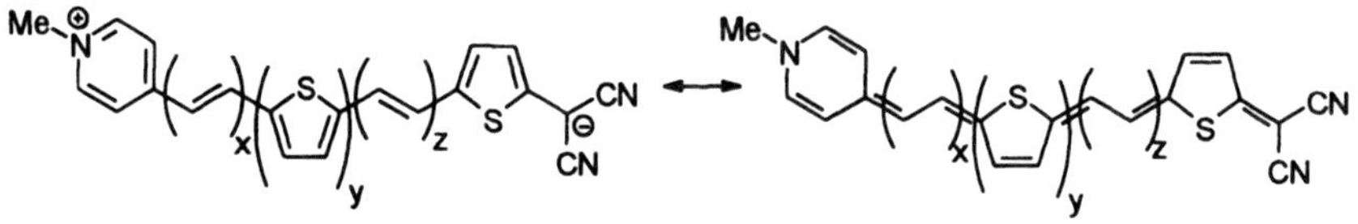

ground state: dipolar, aromatic

excited state: neutral, quinoid

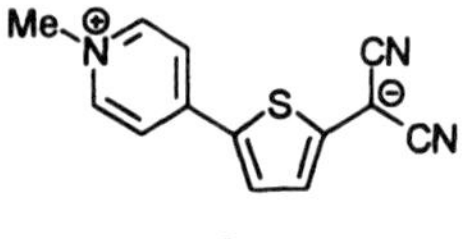

2

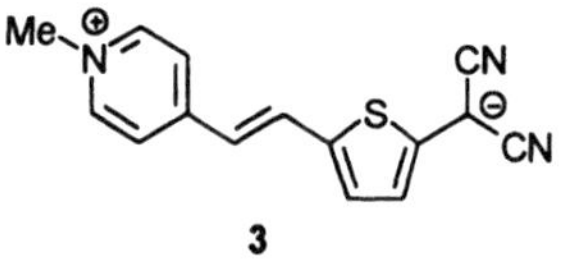

3

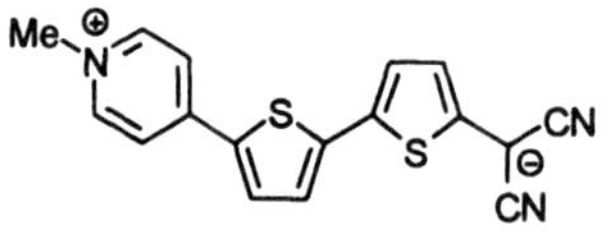

4

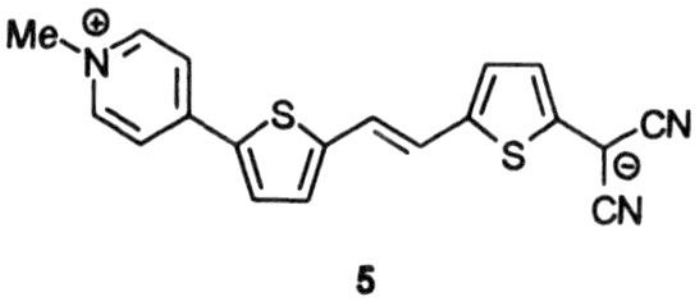

5

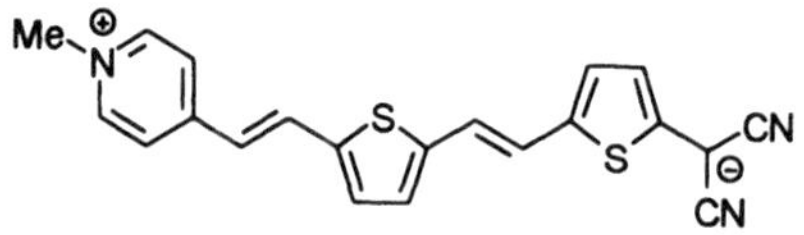

6

The nature of the ground state of the molecules is confirmed by their negative solvatochromism [5] and some NMR indicators [5] that monitor the contribution of the dipolar vs the quinoid structures. Compounds **2 - 6** show a strong negative solvatochromism. For instance, the maximum charge-transfer absorption peak of **2** is at 600 nm in $CHCl_3$ and at 537 nm in MeOH, corresponding to a negative shift of 63 nm with increasing the polarity of the solvent. Compound **3** shows a corresponding blue-shift of 93 nm.

Table. Electronic absorption (charge-transfer band) and first molecular hyperpolarizability data of compounds **2 - 6** in DMF[a] and CHCl$_3$[b] (data for compounds **1**, **7**, and **8** are reported for comparison).

Compd	DMF			CHCl$_3$		
	λ_{max} (nm)	$\mu\beta$ (10^{-48} esu) at $\lambda = 1.9\ \mu m$	$\mu\beta(0)$ (10^{-48} esu)	λ_{max} (nm)	$\mu\beta$ (10^{-48} esu) at $\lambda = 1.9\ \mu m$	$\mu\beta(0)$ (10^{-48} esu)
2	559	2100	1200	600	6740	4190
3	639	4200	2000	702	13450	6990
4	665	8300	3700			
5	692	26900	11000			
6	705	26900	11000			
1[c]						370
7[d]	468	1100	780			
8[d]	522	1300	840			

[a] This work. [b] Ref. 8. [c] Ref. 6. [d] Ref. 7 (data measured in 1,4-dioxane).

The measured $\mu\beta$ values of **2** and **3** in CHCl$_3$ solutions and of **2 – 6** in DMF solutions are listed in the Table, together with those of DANS **1** [6] and of the recently described chromophores **7** and **8** [7]. Our values were obtained using the electric field induced second-harmonic generation (EFISH). Data reported in the Table allow the comparison of the $\mu\beta$ values in CHCl$_3$ [8] and DMF. Because of the insolubility of compounds **4 – 6** in CHCl$_3$, the use of DMF as a solvent was mandatory. Levelling off of the $\mu\beta$ and $\mu\beta(0)$ values, as well as of λ_{max} and the solvatochromic response [5], reached for compounds **5** and **6** discouraged further developments of the series to higher vinylog terms. The introduction of two additional π electrons, out of 18 π electrons present in **5** that directly contribute to the NLO response, is likely not enough to significantly increase the already very high $\mu\beta$ value of **5**. Solutions of these molecules present a significant conductivity, especially in highly polar solvents like DMF (even using freshly prepared solutions). Upon application of the DC-electric filed, conductivity strongly reduces (by a factor of 3 or more) the value of the effective orienting field E applied to the solution, and therefore reduces the second harmonic signal by almost one order of magnitude. Since the conductivity increases with concentration, the acceptable concentration is limited to values as low as 10^{-4} mol l^{-1}. For these reasons values for compounds **2** and **3** in DMF may be affected by large errors. discourage

CONCLUSIONS

The found second order non-linear optical activities of systems **2 – 6** are among the highest so far reported in the literature. Values are particularly high for **5** and **6**, for which the $\mu\beta(0)$ values are not only nearly 30 times larger than that of the commonly used NLO standard DANS [6], but even greater than those obtained very recently either for similar, highly conjugated, systems [6], or for comparable derivatives containing strong sophisticated heterocyclic acceptors [9].

Comparison between **3** and systems **7** and **8** [7] of similar conjugation length shows that the high hyperpolarizability value of **3** cannot be simply attributed to the introduction of the dicyanomethanide moiety or the thiophene ring, but rather to a very fruitful and synergic combination of the thiophene-spaced conjugated bridge with the charged donor and acceptor groups here presented.

REFERENCES

1. P. N. Prasad and J. Williams, *Introduction to Nonlinear Optical effects in Molecules and Polymers* (Wiley, New York, 1991).

2. D. S. Chemla and J. Zyss, *Nonlinear Optical Properties of Organic Molecules and Crystals* (Academic Press, New York, 1987).

3. R. Dagani, C&En 24 (1996).

4. J. Abe, Y. Shirai, N. Nemoto, F. Miyata and Y. Nagase, J. Phys. Chem. B **101**, 576-582 (1997).

5. A. Abbotto, S. Bradamante, A. Facchetti and G. A. Pagani, J. Org. Chem. **62**, 5755-5765 (1997).

6. C.-F. Shu, W. J. Tsai, J.-Y. Chen, A. K.-Y. Jen, Y. Zhang and T.-A. Chen, J. Chem. Soc., Chem. Commun. 2279 (1996).

7. K. Y. Wong, A. K.-Y. Jen, V. P. Rao and K. J. Drost, J. Chem. Phys. **100**, 6818 (1994).

8. A. Abbotto, A. Facchetti, G. A. Pagani, L. Yuan and P. N. Prasad, Gazz. Chim. Ital. **127**, 165-166 (1997).

9. M. Ahlheim, M. Barzoukas, P. V. Bedworth, M. Blanchard-Desce, A. Fort, Z.-Y. Hu, S. R. Marder, J. W. Perry, C. Runser, M. Staehelin and B. Zysset, Science (London) **271**, 335 (1996).

MAGNETIC PROPERTIES OF A BINARY MIXTURE
OF PARAMAGNETIC METALLOMESOGENS

Y. SHIMIZU, S. TANTRAWONG† M. YOSHIDA††, T. SUGINO

Department of Organic Materials, Osaka National Research Institute (*ONRI*), AIST-MITI
Midorigaoka 1-8-31, Ikeda, Osaka 563, Japan

† Joined under STA(Science and Technology Agency) fellowship program. The present
address: Department of Chemistry, Thammasat University, Bangkok, Thailand.
†† Joined from Graduate School of Solid State Electronics, Osaka Electro-Communication
University, Hatsucho, Neyagawa, Osaka 572, Japan.

ABSTRACT

Magnetic properties of a binary mixture of mesogenic tetraphenylporphyrin Cu(II) and
VO(IV) complexes are reported. These mesogens show two lamellar mesophases (D_L and
D_{LC}) and a binary phase diagram revealed that each mesophase is miscible. The magnetic
properties were investigated by magnetization measurements using SQUID. It was found
that the magnetization of 1:1 mixture (mole% base) was larger than those of the pure
compounds below *ca.* 190 K, whilst 1:3 and 3:1 mixtures showed the decrease of
magnetization in the full range of temperature in which the measurements were carried out. It
was shown that χ^{-1} vs. T plots did not obey Curie-Weiss law. For the VO(IV) complex and
1:1 mixture, stepwise changes could be recognized at the temperatures corresponding to the
phase transitions. 1:1 mixture as well as Cu(II) and VO(IV) complexes totally involve antiferromagnetic interaction. However, ferromagnetic behavior in χ^{-1} vs. T plots was seen
about D_L - D_{LC} phase transition.

INTRODUCTION

Molecular magnetic materials have been intensively studied and metal complexes form
an interesting class of magnetic molecules [1]. The molecular geometry and the structure of
molecular condensed phases which could specify magnetic interactions leading to ferro-,
antiferro- and ferri-magnetic properties, have been designed and testified so far [2-4].
However, it is quite difficult to predict the crystal structure only by the knowledge of
molecular structures if one can design the molecular structure possessing an appropriate
electronic structure for required magnetic properties. Usually such molecules have rather
complicated chemical structure and are tend to show some degree of difficulty on synthesis.

On the other hand, it is well known that the molecular orientation could be easily
controlled by an external field in mesomorphic phases. Additionally, as one of the novel
classes of liquid crystals, metallomesogens have been extensively studied in these two
decades and variety of metal complexes containing a paramagnetic metal ion have been
evidenced to show mesomorphic properties [5-7].

Mesomorphic states have several characteristics for their properties as molecular
aggregations. "*Miscibility*" between two mesomorphic compounds which mutually show
thermodynamically identical liquid crystalline phases is one of the special properties of
mesomorphic states. Thus, *miscibility* may enable us to prepare some magnetic materials
consisting of several metallomesogens.

Though it is quite reasonable to believe that liquid crystals are not suitable for magnetic
materials because of the inefficiency of spin-spin interaction due to the molecular fluctuation,
it is considered on one hand to be one concept that mesomorphic states will be utilized as a
field in which molecular alignment is controlled forward the required magnetic interaction
and often it may be required to be fixed into the solid state. In fact, ferromagnetic interaction

Mat. Res. Soc. Symp. Proc. Vol. 488 © 1998 Materials Research Society

a mesogenic tetraphenylporphyrin manganese(III) complex doped with TCNE was recently reported [8].

In this communication, magnetic behavior for a binary mixture consisting of two paramagnetic metallomesogens is reported.

In 1987, Kamachi *et al.* reported ferromagnetic interaction in a polyacrylate possessing tetraphenylporphinatocopper(II) and tetraphenylporphinatooxovanadium(IV) as their pendant groups [9]. In this system, the paramagnetic moieties are sterically forced to stay close together along a polymer chain. In this work, discotic porphyrin metallomesogens, Cu(II) and V=O(IV) complexes of tetradodecylphenylporphyrin (**1** and **2**, respectively) as shown below and the 1:1 mixture were investigated on the magnetic behavior.

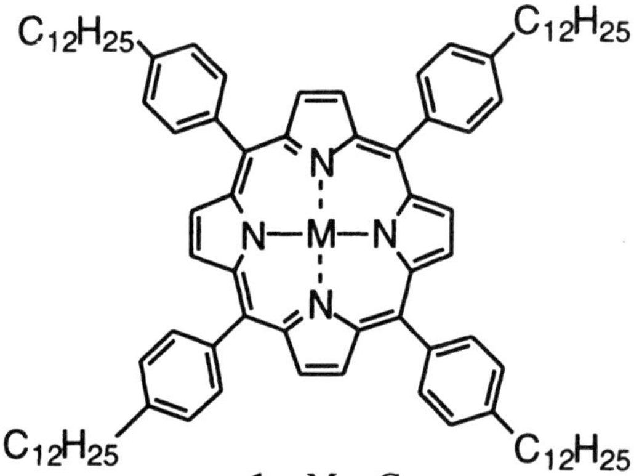

1   M = Cu
2   M = VO
3   M = H$_2$

5,10,15,20-Tetrakis(4-*n*-dodecylphenyl)porphyrin (C$_{12}$TPPH$_2$, **3**) and the metal complexes incorporating metal ions (Co^{2+}, Ni^{2+}, Cu^{2+}, Zn^{2+} and Pd^{2+}), have been reported to exhibit two kinds of lamellar discotic mesophases [10,11]. The results of DSC and X-ray diffraction measurements indicated C$_{12}$TPPH$_2$ exhibits two different discotic lamellar phases (D$_{LC}$ and D$_L$ phases) at the temperature ranges of 52-31°C and 155-52°C, respectively. The recent spectroscopic studies suggest the low-temperature lamellar mesophase (D$_{LC}$ phase) is similar to plastic crystalline phase, whilst the high-temperature mesophase (D$_L$ phase) is liquid crystal and the former has a molecular stacking columnar structure [12]. Additionally, the copper(II) complex has also been reported to exhibit the similar phases with comparable melting temperature but considerably higher clearing temperature, D$_{LC}$ phase: 46-27°C and D$_L$ phase: 185-46°C. On the other hand, the oxovanadium(IV) complex was synthesized in this work and was revealed to show the similar mesomorphic behavior to those of the metal-free and Cu(II) complex, D$_{LC}$ phase: 68-22°C and D$_L$ phase: 153-68°C.

EXPERIMENTAL

Synthesis of the compounds were carried out according to ref.10. The miscibility property between Cu(II) and VO(IV) complexes of the ligand, C$_{12}$TPPH$_2$ was investigated by DSC measurements (TA Instrument, DSC2920). The mixtures with different concentrations were prepared using solution method and magnetization of the mixtures was measured by SQUID (Quantum Design, MPMS-2).

RESULTS AND DISCUSSIONS

Transition temperatures and phase behavior of compounds, **1, 2** and **3** are shown in Table 1. The crystal-mesophase transition temperatures of the ligand and the copper complex are comparable, while that of the oxovanadium complex is somewhat several degree lower.

Table 1   Transition temperatures (T / °C) and enthalpies (ΔH / kJ mol^{-1}) of compounds **1, 2** and **3**. ΔH values are in parenthesis.

Compound	C	C_1	D_L'	D_L	Iso
1	-28	27	46	185	
	(-5.1)	(54)	(12)	(31)	
2	-28.7	22	68	153	
	(-26)	(45)	(16)	(25)	
3	-	31	52	155	
		(46)	(14)	(23)	

VO(IV) complex shows the widest temperature range as for the D_{LC} phase, while that of the ligand was the narrowest and that of Cu(II) complex was amidst. This may also be explained as a consequence of intermolecular interaction derived. The ligand molecules, without any metal ions, possess only organic hydrocarbon parts and hence display weaker intermolecular interaction. As a result, the higher order D_{LC} phase of the ligand is thermally less stable which, in turn, reduces the temperature range of the phase.

Furthermore, due to the fact that the VO complex molecules have an extra V=O bond orthogonal to the molecular plane, it could, on one hand, raise the intermolecular interaction [13], while on the other hand, reduce the molecular symmetry. It is reasonable to think that as a consequence of the high intermolecular interaction derived from the orthogonal V=O bond, the VO complex showed the widest D_{LC} phase range of the three compounds. In contrast,, the V=O bond reduces molecular symmetry which could lower the clearing temperature of the VO complex. In this manner, the clearing temperature of the VO complex was comparable with that of the ligand. The enthalpy data apparently support this interpretation.

Transition temperatures determined by DSC measurements of the mixtures showed normal properties of uninteracted two-component mixtures as seen in Figure 1. The transition temperatures of the mixtures were in-between those of the pure compounds. Furthermore, for the mixtures with lower concentration than 50 mole% of the Cu complex after cooling from the isotropic phase at 5 °C/min, the endothermic peaks on heating around 60 °C of the D_{LC}-D_L transitions became doublets. This may possibly be a result of that the two complexes, in a certain degree, crystallized separately during slow cooling from the isotropic liquid, thus forming two different micro polydomains.

The endothermic peaks at the similar position were all singlet on the first heating. This means a homogeneously dispersed mixture was obtained by the solution method.

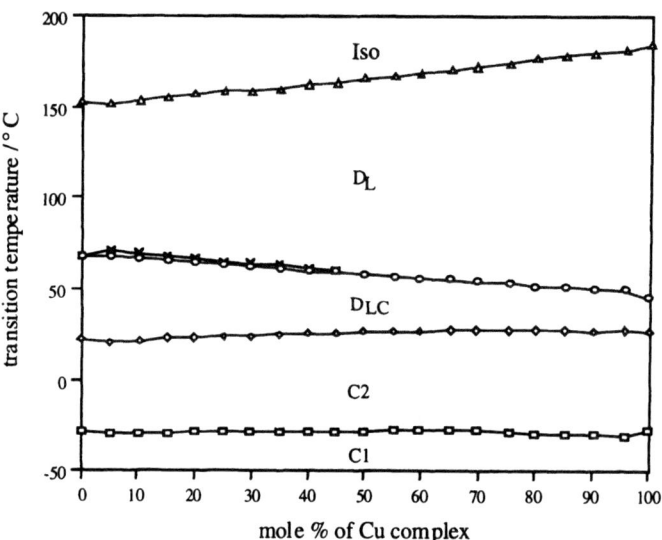

Figure 1　A binary phase diagram of compounds **1** and **2**. The phase transition temperatures were determined by DSC measurements at heating rate, 5 °C/min.

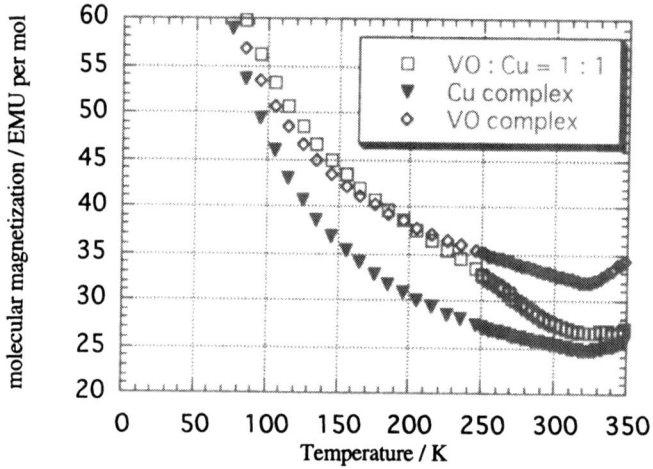

Figure 2　Temperature dependence of the molar magnetizations for the Cu (II) and VO (IV) complexes and the 1 : 1 mixture. The measurements were carried out on heating from 5 K.

The magnetizations of the pure complexes and mixture were measured from 5 K up to 350 K and were calculated the molar magnetization (Figure 2), after the subtraction of ligand diamagnetism (Eq.1). There is no obvious modification of molar magnetizations around the transition temperatures, only a slight reduction of the values at the temperatures was observed. This observation implies that the molecular orientations could affect the intermolecular magnetic interaction, but the interaction is so weak that it has negligible effect on phase behavior.

$$\chi(\text{total}) = \chi(\text{paramag}) + \chi(\text{diamag}) \qquad (1)$$

For the pure complexes the molar magnetization of the VO complex was expected to be similar to that of the Cu complex. However, the former was observed to be lower than the latter. This raised the awareness of molecular orientation of the VO complex and the possibility of intermolecular magnetic interaction. Due to the orthogonal V=O bond, the VO complex molecules could magnetically orient themselves in either a ferromagnetic or an antiferromagnetic manner. The ferromagnetic order reduces the molar magnetization while the antiferromagnetic does the otherwise. The lower magnetization of the VO complex than that of the Cu complex indicates a possibility of antiferromagnetic alignment of the molecules.

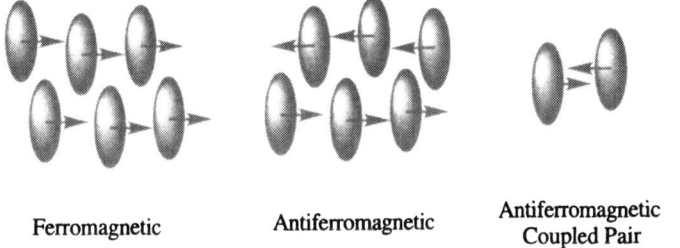

| Ferromagnetic | Antiferromagnetic | Antiferromagnetic Coupled Pair |

On the other hand, the values of magnetization of the mixtures were expected to be in-between those of the pure complexes, in the case that no intermolecular magnetic interaction occurred. However, in the experimental molar magnetization data, it was surprisingly observed that at a specific temperature there was a certain degree of intermolecular magnetic interaction causing the larger magnetization than that derived from the summation of magnetizations of both complexes. This result strongly indicates some intermolecular magnetic interaction generates in the 1:1 mixture. Disappointedly, there was no obvious interrelation between the molar magnetization values of the mixtures with different components, *i.e.*. 1:3 and 3:1 mixtures.

χ^{-1} vs. T plots for the Cu and VO complexes and 1:1 mixture are shown in Figure 3. They all show totally antiferromagnetic behavior. However, ferromagnetic or ferrimagnetic behavior can be seen in some temperature ranges for VO complex and 1:1 mixture. These indicate mesomorphic states could provide a field for constructing magnetic interaction.

CONCLUSIONS

The Cu(II) and VO(IV) complexes of a long-chain tetraphenylporphyrin were shown to exhibit the similar mesomorphism to that of the ligand. The temperature range of mesophase for the complexes was wider than that of the ligand, being indicative of metal-metal and metal-oxygen intermolecular interactions. Both complexes were fully miscible.

The results of magnetic measurements showed possible intermolecular magnetic interactions between two paramagnetic molecules possessing different metal ions.

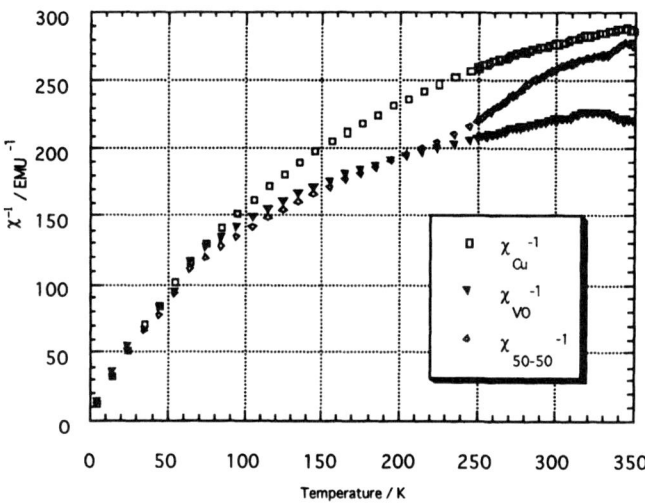

Figure 3 χ^{-1} vs. T plots for the Cu and VO complexes and 1:1 mixture.

ACKNOWLEDGMENTS

The authors would like to thank *Science and Technology Agency* for providing *STA fellowship* to Dr. Sukrit Tantrawong and to thank Dr. I. Matsubara and Dr. R. Funahashi in ONRI for their kind help on SQUID measurements. Agency of Industrial Science and Technology(*AIST*), Ministry of International Trade and Industry(*MITI*) was greatly acknowledged for the financial support under the national R&D project, "*Harmonized Molecular Materials*" operated in Industrial Science and Technology Frontier Program (ISTF).

REFERENCES

1. O. Kahn, <u>Molecular Magnetism</u>, (VCH, Weinheim, 1995).
2. J.S. Milller and A.J. Epstein, Angew.Chem.Int.Ed.Engl., **33**, 385 (1994).
3. D. Gatteschi, Adv.Mater., **6**, 635 (1994).
4. O. Kahn, Angew.Chem. Int.Ed.Engl., **24**, 834 (1985).
5. J.L. Serrano, <u>Metallomesogens: Synthesis, Properties and Applications,</u> (VCH, Weinheim, 1996).
6. D.W.Bruce, <u>Inorganic Materials,</u> 2nd ed., edited by D.W. Bruce and D. O'Hare, (John Wiley & Sons, Chichester, 1996), pp429-522.
7. S.A. Hudson and P.M. Maitlis, Chem.Rev., **93**, 861 (1993).
8. K. Griesar, M.A. Athanassopoulou, E.A. Soto-Bustamante, Z. Tomkowicz, A. Zaleski,   W. Haase, Adv.Mater., **9**, 45 (1997).
9. M. Kamachi, X.S. Cheng, H. Aota, W. Mori, M. Kishita, Chem.Lett., <u>1987</u>, 2331.
10. Y. Shimizu, M. Miya, A. Nagata, K. Ohta, I. Yamamoto and S. Kusabayashi, Liq.Cryst., **14**, 795 (1993).
11. Y. Shimizu, M. Miya, A. Nagata, A. Matsumura, K. Ohta, I. Yamamoto and S. Kusabayashi, Chem.Lett., <u>1991</u>, 25.
12. Y.Shimizu and T.Higashiyama, <u>Proc.SPIE</u>, in press.
13. S. Tantrawong, P. Styring and J.W. Goodby, J.Mater.Chem., **3**, 1209 (1993).

CONTROL OF THE LUMINESCENCE PROPERTIES IN POLYTHIOPHENES SOLUTIONS BY INTERRING ROTATIONAL DISORDER

S. LUZZATI*, G. BONGIOVANNI**, M. CATELLANI*, M. A. LOI**, A. MILANI*, A. MURA**, A. PIAGGI***,
* Istituto di Chimica delle Macromolecole and MITER, CNR, Italy
** Dipartimento di Scienze Fisiche, Università di Cagliari, Italy
*** Istituto di Matematica e Fisica and INFM, Università di Sassari, Sassari, Italy

ABSTRACT

The photoluminescence properties of thiophene-based polymers obtained by the random copolymerization of 3,4-dibutylthiophene and 3-butylthiophene are reported. Optical absorption spectra, cw photoluminescence (PL) spectra, PL quantum efficiencies and PL decays have been measured in dilute solutions. By varying the copolymer chemical composition it is possible to tune the intensity and the color of the luminescence spectra in the whole visible range. We evidence the role of disorder, controlled by interring rotations, as the tuning factor for the disexcitation pathway of the photoexcitations in these materials.

INTRODUCTION

The control of the luminescence properties of conjugated polymers are of considerable interest because of the potential application of these materials as active layers in light emitting diodes [1,2]. A large amount of synthetic and photophysical studies have been devoted to the basic understanding of the radiative decay processes, with the aim to improve the emissive features of organic materials.

Luminescence, in conjugated polymers, is believed to be the result of radiative decay of singlet excitons. The fluorescence quantum yields are far from unity in these materials, as competing nonradiative processes provide additional means of decay. The disexcitation pathway of the excited state is strongly affected by interchain coupling and by the electronic and conformational structure of the isolated chain. The degree of structural order/disorder of the conjugated chain has a strong influence on the luminescence of the material. In poly-3-alkylthiophenes it has been observed that the chain stiffness, varied through the number of head to head enchainment defects, is affecting the emission efficiency [3,4].

Thiophene-based polymers are particularly interesting because of their great chemical versatility. The substitution of the thiophene ring with different groups can be used to modify the band gap and therefore to obtain light emitting devices of controlled colour [5]. We have shown that the random copolymerization of 3-butylthiophene and 3,4-dibutylthiophene monomers in different ratio is an easy route to obtain materials with tuneable optical properties [6]. The steric hindrance of 3,4-dibutylthiophenes induces a twisting of the backbone with a reduction of the conjugation length with a consequent blue shift of the absorption-emission spectra. By increasing the monomer fraction f of 3,4-dibutylthiophene in the copolymer, the colour of the photoluminescence sweeps from red ($f=0$) to blue ($f=1$) [7].

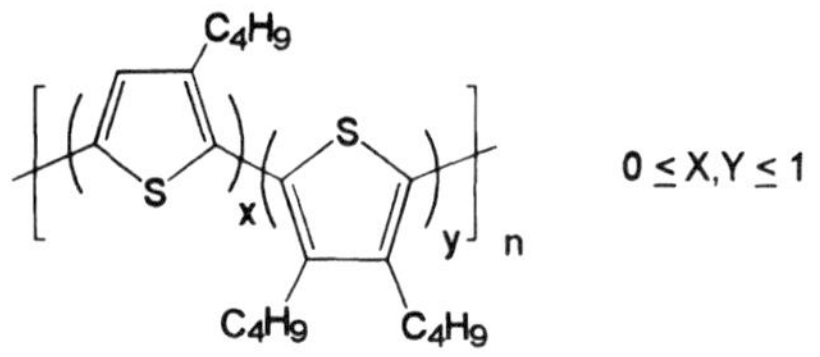

$0 \leq X,Y \leq 1$

Scheme 1

In a previous study we have shown that, for these copolymers, the fluorescence quantum yields in solution are decreasing by increasing f [8]. The aim of this paper is to provide an exhaustive picture of the photoluminescence properties of the above copolymers through a study of the cw and time resolved photoluminescence. In particular we show that the chain disorder, controlled by interring rotations, is the tuning factor for the emission properties of these materials. We display that rotational disorder is strongly affecting the disexcitation pathway of the photoexcitations in these materials.

EXPERIMENT

The copolymer series has been produced via chemical oxidation with $FeCl_3$ starting from mixtures of 3-butyl and 3,4-dibutylthiophene monomers in well defined ratios. The synthesis and macromolecular characterisation are reported elsewhere [6]. The absorption and emission spectra have been performed in anhydrous tetrahydrofuran (THF) at concentrations lower than 10^{-4} M. The absorption spectra have been measured with a Cary 2400 Varian spectrophotometer. Cw photoluminescence spectra have been obtained with a Xe lamp coupled to a monochromator to select the excitation energy and a flat field spectrograph equipped with a N_2 cooled CCD detector. The spectra have been corrected for the instrumental spectral response. The cw and the time resolved photoluminescence have been measured exciting at 370 nm. The time resolved experimental apparatus is described elsewhere [9]. The quantum yields of the solutions have been obtained using quinine sulphate solutions as a standard [10]. Cw and time resolved photoluminescence spectra have been measured on the same solutions.

RESULTS

The absorption and the emission spectra of the copolymer dilute solutions are reported in fig. 1 for various monomer fraction f of 3,4-dibutylthiophene in the copolymer. By increasing f, there is a blue shift of both absorption and emission spectra. This spectral feature is due to the steric hindrance of 3-4 disubstituted units that induces a rotation between adjacent thiophenes with a reduction of the mean conjugation length. The broadening of the absorption spectra , displays a maximum width for intermediate f and then decreases. This implies that the copolymers with these chemical compositions exhibit the highest degree of disorder with the wider distribution of conjugation lengths.

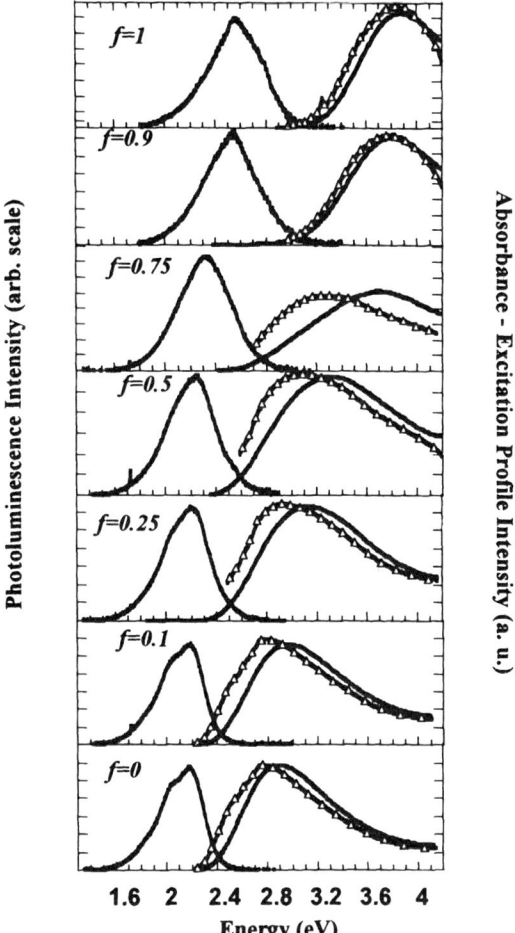

Photoluminescence Intensity (arb. scale)

Absorbance - Excitation Profile Intensity (a. u.)

Energy (eV)

Fig. 1 Absorption, cw PL spectra and excitation profiles (triangles), measured at the PL peak, of diluted copolymer solutions in THF for different copolymer chemical composition f.

The fluorescence quantum yields (Q.Y.) and the photoluminescence (PL) lifetimes of the diluted solutions are reported for the various copolymers in Fig. 2. The efficiency displays a reduction by increasing the number of disubstituted monomers in the chain. The PL lifetimes τ_{PL} and the quantum yields show the same changes with f. If radiative and non radiative decay are monomolecular processes with rates τ_r^{-1} and τ_{nr}^{-1} respectively, the photoluminescence efficiency is given by: Q.Y. $= b\tau_{pl}/\tau_r$ [11]. Here b represents the fraction of absorbed photons leading to the generation of singlet excitons. By reasonably taking $b = 1$ [11], Fig. 2 thus indicates that τ_r is essentially independent on the number of disubstituted monomers in the backbone while the nonradiative decay channels become more relevant by increasing the content of 3,4-

dibutylthiophenes. The relevance of non radiative decay processes in thiophenes-based materials has been already pointed out. In oligothiophenes, by decreasing the conjugation length, there is a reduction of the emission efficiency due to intersystem crossing which is becoming more relevant for shorter oligomers [12]. In poly-3-alkylthiophenes the reduction of chain stiffness and the conjugation length through head to head enchainment defects, leads to a reduction of the emission efficiency [3,4]. It has been suggested that internal conversion through rotational vibrations accounts for this feature [3]. Moreover triplet formation is also favoured by rotational vibrations [13]. Since the presence of disubstituted monomers in the copolymer chain is indeed affecting the conjugation length and chain stiffness, we can infer that by increasing f there is an enhancement of the non radiative decay channels through triplet formation and internal conversion.

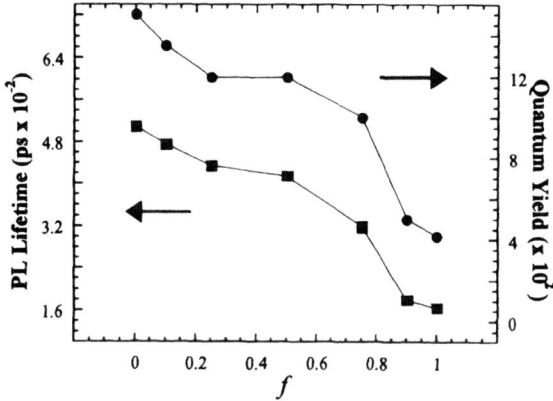

Fig. 2 PL quantum yield (dots) and spectrally integrated photoluminescence lifetime (squares) of diluted copolymer solutions in THF, for different chemical composition f. Excitation wavelength: 370 nm.

Another interesting feature resulting from Fig. 1 is that the apparent Stokes shift, defined as the difference in energy between the absorption and the emission maxima, is increasing with f. Part of this shift is due to the planarization of the excited state which occurs after photoexcitation [14,15]. From the mono to the disubstituted homopolymer the Stokes shift is almost doubled, in agreement with the increasing distortion of the backbone as f increases.
There are however other contributions which may lead to such large Stokes shifts. The PL peak position (see Fig. 1) exhibits almost the same variation with f as the onset of the absorption spectra. This is a common feature for conjugated polymers in the solid state where the emission originates from the tail of the longest conjugations because the excitons migrate to the longest conjugations prior to the radiative recombination [16]. In solution the spectral migration is drastically hindered because the chains are isolated by the solvent. Anyhow some spectral

migration may also occur in solution [17] and will be discussed in the following. The PL spectra could be red shifted also because of the different emission efficiencies coexisting within the copolymer distribution of conjugation lengths. We have reported above that the longest conjugations exhibit an emission efficiency which is much higher than the efficiency of the shorter conjugations. The excitation profiles reported in Fig. 1 are useful to clarify these aspects. It can be seen that the excitation profiles of the copolymers are slightly red shifted but are not far from reproducing the absorption spectra when the copolymer composition has a prevalence of either mono- or di-substituted monomers. The excitation profiles for the copolymers at intermediate f are much more red shifted from the absorption spectra. This indicates that the PL spectra of the copolymers with the broader distribution of conjugation lengths are indeed affected by the coexistence within the copolymer of chain segments with different emission efficiencies. If in such disordered copolymers an efficient migration of the excitons from the shorter to the longest conjugation could occur prior to the radiative recombination, the excitation profiles should reproduce the absorption spectra.

Fig. 3 displays the peak position as a function of the copolymer composition of the absorption, of the cw PL spectra and of the time resolved PL spectra at 0-delay from the excitation (i.e. at delays t_0 shorter than the experimental time resolution, 30 ps). This latter spectrum is blue shifted relative to the cw PL spectrum but exhibits similar variations with f.

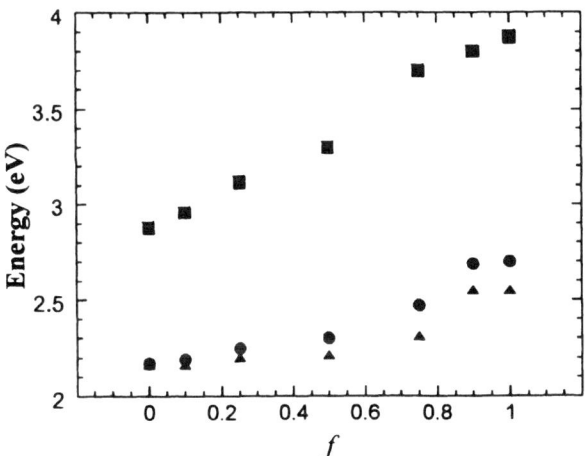

Fig. 3 Peak absorption (squares), 0-delay PL peak (circles),cw PL peak (triangles) for different copolymer chemical composition f.

The 0-delay PL spectra, unlike the cw spectra, do not depend on the quantum yield of the emitting sites, but only on the natural radiative lifetime ($t_0 \ll \tau_{nr}$, τ_r). Since the radiative lifetime τ_r is essentially not varying within the different conjugations (see above), the 0-delay PL peak position is expected to have the same dependence on the copolymer composition as the absorption. The different observed behaviour implies that a relatively fast spectral migration to longer conjugation segments occurs prior to the radiative recombination. At later times, spectral diffusion is mostly inhibited and the cw PL is further red shifted because the emission from shorter conjugations decay faster than from the longest ones. In agreement with the excitation profiles, the red shift is more pronounced for the most disordered copolymers at intermediate f.

CONCLUSION

The photoluminescence properties of a series of thiophene-based polymers obtained by the random copolymerization of 3,4-dibutylthiophene and 3-butylthiophene are reported. By varying the chemical composition it is possible to tune the intensity and the colour of the luminescence spectra in the whole visible range. We provide a coherent picture of the radiative decay mechanism of these materials. In particular we show that disorder, controlled by interring rotations, is the tuning factor for the emission properties of these materials. We display that rotational disorder is strongly affecting the non-radiative decay processes. The disexcitation pathway of the emitting excitons is discussed in terms of spectral migration toward longer conjugation chain segments.

REFERENCES

1. J.H. Burroughes, D.D.C. Bradley, A.R. Brown, R.N. Marks, K. Mackay, R.H. Friend, P.L. Burn and A.B. Holmes, Nature, **347**, 539 (1990).
2. G. Gustafsson, Y. Cao, G.M. Treacy, F. Klavetter, N. Colaneri and A.J. Heeger, Nature, **357**, 477 (1992).
3. B. Xu and S. Holdcroft, Macromolecules, **26**, 4457 (1993).
4. G. Rumbles, I.D.W. Samuel, L. Magnani, K.A. Murray, A.J. DeMello, B. Crystall, S.C. Moratti, B.M. Stone, A.B. Holmes, R.H. Friend, Synth. Met., **76**, 47 (1996).
5. M. R. Andersson, M. Berggren, O. Inganas, G. Gustafsson, J.C. Gustafsson-Carlberg, D. Selse, T. Hjertberg and W. Wennerstrom, Macromolecules, **28**, 7525 (1995).
6. M. Catellani, S. Luzzati, R. Mendichi, A. Giacometti Schieroni, Polymer, **37**, 1059 (1996).
7. M. Catellani, S. Luzzati, C. Botta. P.C. Stein, Mat. Res. Soc. Symp. Proc., **413**, 451 (1996).
8. S. Luzzati, E. Prevosti, M. Catellani, Synth. Met., **84**, 551 (1997).
9. L. Rossi, G. Bongiovanni ,J. Kalinowski, G. Lanzani, A. Mura, M. Nisoli, and R. Tubino, Chem. Phys. Lett. **257**, 545, (1996).
10. J.N. Demas, G.A. Crosby, J. Phys. Chem., **75**, 991 (1971).
11. N.C. Greenham, I.D.W. Samuel, G.R. Hayes, R.T. Phillips, Y.A.R.R. Kessener, S.C. Moratti, A.B. Holmes and R.H. Friend, Chem. Phys. Lett., **241**, 89 (1995).
12. R.S. Becker, J. Seixas de Melo, A.L. Macanita and F. Elisei, Pure & Appl. Chem., **67**, 9 (1995).
13. D. Oelkrug, H.J. Egelhaaf, J. Gierschner, A. Tompert, Synt. Met., **76**, 249 (1996).
14. J.L. Bredas, B. Themans, J.G. Fripiat, J.M. André and R.R. Chance, Phys. Rev. B, **29**, 6761 (1984).
15. G. Lanzani, M. Nisoli, M. Magni, S. De Silvestri, G. Barbarella, G. Zambianchi, R. Tubino, Phys. Rev. B, **51**, 13770 (1995).
16. S. Heun, R.F. Mahrt, A. Grenier, U. Lemmer, H. Bassler, D.A. Halliday, D.D.C. Bradley, P.L. Burn and H. Holmes, J. Phys.:Condens. Matter, **5**, 247 (1993).
17. A. Watanabe, T. Koidara and O. Ito, Chem. Phys. Lett. **273**, 227 (1997).

NEW BASE-DOPABLE WELL-DEFINED POLY-2,7-FLUORENE DERIVATIVES

M. Ranger and M. Leclerc
Université de Montréal, Département de Chimie, C.P. 6128, Succ. Centre-Ville, Montréal (Qc), Canada, H3C 3J7, leclerma@ere.umontreal.ca

ABSTRACT

Well-defined poly(2,7-fluorene) derivatives have been prepared through palladium-catalyzed couplings; using this versatile synthetic method, processable polyfluorenes have been obtained in good yields. In solution, these yellow polymers exhibit blue emission around 410 nm with high quantum yields. Moreover, novel acidic polyfluorene derivatives have been synthesized (e.g. poly(2,7'-(pentyl 9,9-dioctyl-7,2'-bifluorene-9'-carbonyl))) which show, upon base-doping, electrical conductivities of 10^{-2} - 10^{-3} S/cm. This new doping method for conjugated polymers could open the way to the preparation of air-stable electron-injecting electrodes. Polyfluorene derivatives show their potential to play a double function in the fabrication of LEDs, i.e. as n-type electrode and luminescent polymeric materials.

INTRODUCTION

A lot of research activities have been devoted to conjugated polymers due to their unusual optical and electrical properties. Since a few years, these polymers have been used in the fabrication of light-emitting-devices (LEDs and LECs) [1,2]. There is a great interest to produce efficient and stable blue-light emitting materials to get access to the three primary colors. Yoshino and al. reported new class of blue-emitting materials based on poly(9-alkylfluorene)s and poly(9,9-dialkylfluorene)s [3]. These polymers have been obtained from a simple chemical oxidation of the monomers using $FeCl_3$; a similar procedure to that developed for the preparation of poly(3-alkylthiophene)s [4]. However, this non-specific oxidation reaction produces some partially cross-linked materials and the soluble fraction of these polyfluorenes shows some evidence of irregular couplings along the backbone [5]. Recently, Pei and Yang reported efficient photoluminescence and electroluminescence properties with a well-defined soluble polyfluorene derivative, poly(2,7-(9,9-bis(3,6-dioxaheptyl)fluorene)), obtained from a nickel-catalyzed reductive polymerization [6]. We wish to report here another method for the preparation of well-defined polyfluorenes by using palladium-catalyzed Suzuki couplings. With this method, it is possible to build perfectly alternating copolymers and novel acidic polymers with base-doping properties [7,8].

EXPERIMENTAL SECTION

Pentyl 2,7-dibromofluorene-9-carbonyl was synthesized by the treatment of hexanoic anhydride on 2,7-dibromofluorenyl-9-lithium; hexyl 2,7-dibromofluorene-9-carboxylate use in the same way hexyl chloroformate. The synthesis of all other monomeric derivatives will be reported in another publication [7]. The copolymers were prepared using the Suzuki reaction (Scheme 1) [8]. UV-visible absorption measurements were performed in DMF solutions. Fluorescence spectroscopy was carried out in $CHCl_3$ solutions having polymer concentrations around 10^{-6} M. Fluorescence quantum yields were evaluated with 9,10-diphenylanthracene as standard (ϕ_f = 0.9 in hexane). The base-doped form of the polymers was obtained by the treatment of potassium *tert*-butoxide in THF. Ex-situ conductivities were measured by the four-probe technique.

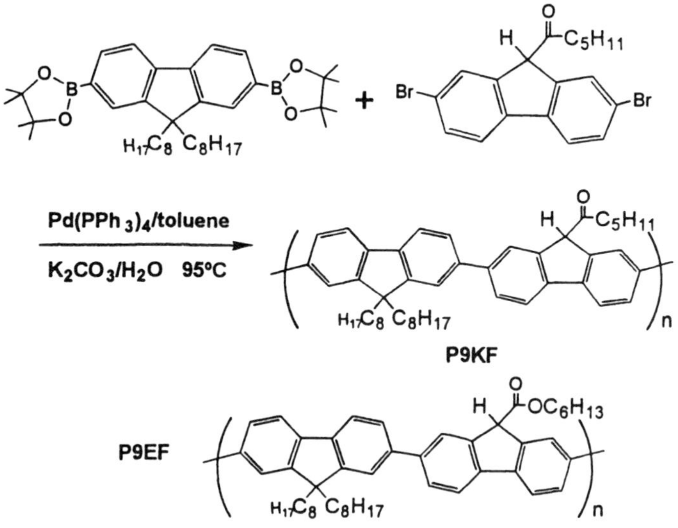

Scheme 1. Preparation of alternating polyfluorene derivatives by Suzuki couplings.

RESULTS

The results of the polymerization reactions are summarized in Table 1. P9EF and P9KF (Scheme 1) are soluble in common organic solvents such as THF, toluene and CHCl₃.

Table 1. Results of the polymerization reactions

Polymers	Rx Time(h)	Yield(%)	M_n	M_w	DP
P9EF	48	75	7800	14000	12
P9KF	48	72	10000	21000	15

DP=degree of polymerization, M_n=Number-average molecular weight, M_w=Weight-average molecular weight. All data were evaluated by GPC against polystyrene standards in THF.

The chemical structure of the copolymers was identified by NMR spectroscopy (Figures 1 and 2). As an example, the ¹H NMR spectrum of P9KF shows an acidic proton at 5.05 ppm. The evaluation of the degree of polymerization (DP) by NMR analysis from terminal protons at 7.49 and 7.38 ppm gave the same value as that estimated by GPC. Figure 2 shows 12 peaks associated to the 12 different carbon atoms in P9KF and confirms that the copolymer has a well-defined structure. This regioregularity should contribute to a better intramolecular transport of the charge carriers (by allowing the resonance between benzoid and quinoid forms). By 2D-NMR analysis, the two peaks (having an asterisk), at 128.68 and 127.11 ppm, were associated to carbon atoms of end groups.

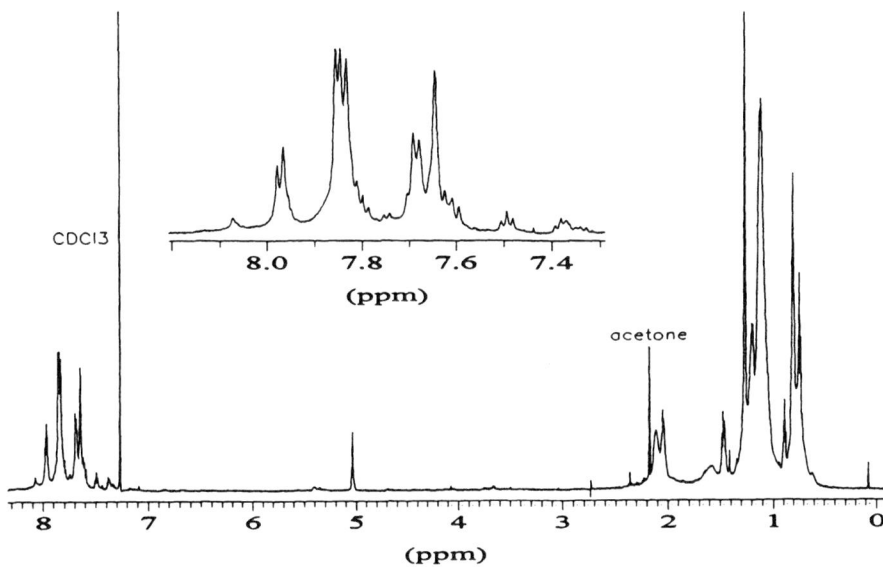

Figure 1. The ^{1}H NMR spectrum of P9KF in CDCl$_3$.

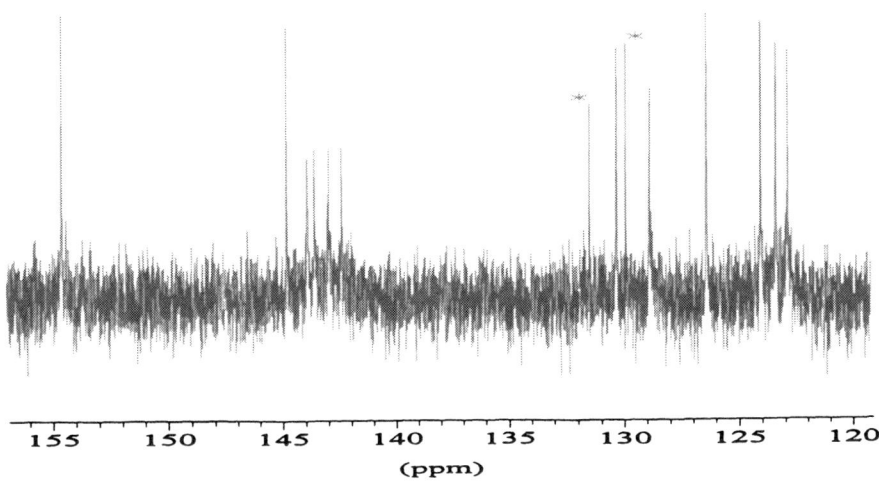

Figure 2. Aromatic region of the ^{13}C NMR of P9KF, in CDCl$_3$.

Photoluminescence measurements have revealed that these copolymers emit strongly in the blue range. Excitation fluorescence spectra correspond perfectly with absorption spectra (Figure 3). P9KF shows an emission maximum at 412 nm and a fluorescence quantum yield of 0.20. P9EF exhibits the same emission spectrum with a quantum yield of 0.50 . The apparent Stokes shift of polyfluorene was small with a average value of 2000 cm^{-1}. This value of Stokes shift is associated to a rigid polymer and could indicate that similar geometries exist in both the ground and excited states. The emission fluorescence spectra showed a vibronic structure associated to the coupling of a C=C vibration mode.

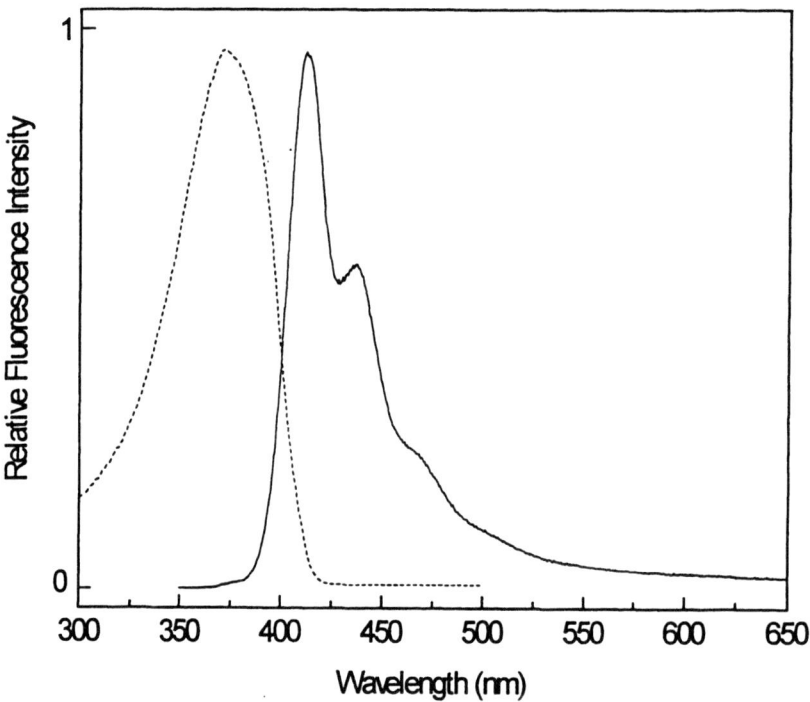

Figure 3. Relative excitation and emission spectra for P9KF in chloroform at room temperature.

Moreover, upon deprotonation, P9KF and P9KF show interesting conductivities of 10^{-5}-10^{-6} and 10^{-2}-10^{-3} S/cm respectively. These conductivities are significant if one compares those obtained with oxidized polyfluorenes [3,4], which are around of 10^{-3} S/cm and considering the fact that all measurements were carried out in air. The higher conductivity of P9KF can be explained by a better stability of the generated anion since the ketone group is a better electron-withdrawing group than ester. At this point, it is difficult to know if there is a contribution from a cationic (ionic) conductivity. UV-visible absorption measurements allow also a certain characterization of the base-doped state (Figure 4). Indeed, this yellow polymer changes to pale green color upon base-doping using potassium *tert*-butoxide in DMF. The acid-base process is reversible and the polymers can recover their initial color. Other electron-withdrawing group (nitrile and benzoyl) are presently studied to optimize the base-doping properties. The nitrile group will also bring more information about the contribution of the enolate form in this process.

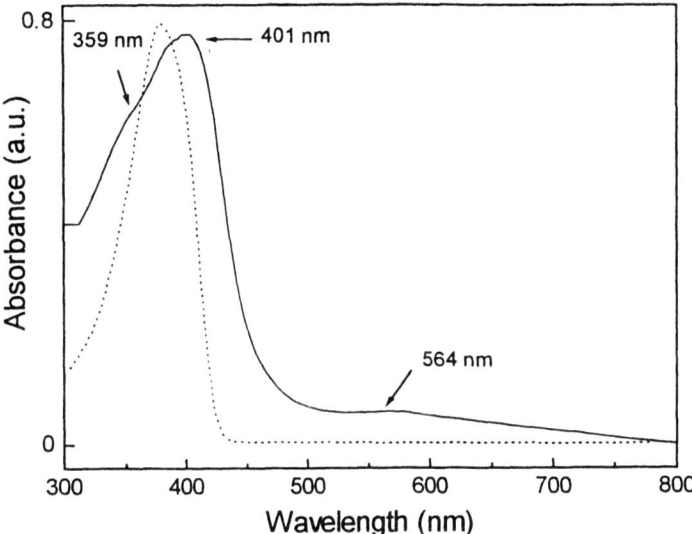

Figure 4: UV-visible Spectra of P9KF in DMF. The dotted line (----) represents the absorption of neutral polymer and the continuous line (——) represents the absorption of the deprotonated form.

CONCLUSIONS

Polyfluorene derivatives demonstrated their potential for a new type of doping reaction, i.e. base-doping. This novel doping process could give another way to prepare n-type conducting materials. Furthermore, these polymers exhibit interesting fluorescence properties for the fabrication of light-emitting diodes and could therefore play a double role in these devices.

ACKNOWLEDGMENTS

This work was supported by operating and collaborative research grants from Natural Sciences and Engineering Research Council (NSERC) of Canada. We thank Dr. M. Belletête and Prof. G. Durocher for their assistance during the fluorescence measurements and Prof. L. Breau and Prof. D. Belanger for many useful advices.

REFERENCES

[1]   J.H. Burroughes, D.D.C. Bradley, A.R. Brown, R.N. Marks, K. Mackay, R.H. Friend, P.L.Burns, A.B. Holmes, *Nature* **347**, 539, (1990).

[2]   Q. Pei, G. Yu, C. Zhang, Y. Yang, A.J. Heeger, *Science*, **269**, 1086, (1995).

[3]   Y. Ohmori, M. Uchida, K. Muro, K. Yoshino, *Jpn. J. of Appl. Phys.*, **30**, L1941, (1991).

[4]   M. Fukuda, M. Sawada, K. Yoshino, *J. of Polym. Sci.: Part A, Polym. Chem.*, **31**, 246, (1993).

[5]   M. Fukuda, M. Sawada, K. Yoshino, *J. of Polym. Sci. : Part A, Polym. Chem.*, 31, 2465, (1993).

[6]   Q. Pei, Y. Yang, *J. Am. Chem. Soc.*, **118**, 7416, (1996).

[7]   M. Ranger, D. Rondeau, M. Leclerc, *Macromolecules*, in press.

[8]   M. Ranger, M. Leclerc, *J. Chem. Soc., Chem. Commun.*, 1597, (1997).

MAGNETIC SUPRAMOLECULAR GRID STRUCTURES: INTRAMOLECULAR COUPLING OF FOUR SEPARATE SPINS

O. WALDMANN[*], J HASSMANN[*], R. KOCH[*], P. MÜLLER[*], G. S. HANAN[**],
D. VOLKMER[**], U. S. SCHUBERT[**], J.-M. LEHN[**]
[*]Physikalisches Institut III, Universität Erlangen-Nürnberg, 91058 Erlangen, Germany,
waldmann@physik.uni-erlangen.de
[**]ISIS, 4 rue Blaise Pascal, Université Louis Pasteur, 67000 Strasbourg, France

ABSTRACT

The magnetization of novel tetranuclear supramolecular grid structures and their mononuclear analogues was measured. In the tetranuclear complexes with Co^{2+} or Ni^{2+} ions an intramolecular antiferromagnetic coupling of the four metal centers is observed. The isotropic coupling strength was determined to be $J = -8$ K for the Ni^{2+} grids, and $J = -2$ K for the Co^{2+} grids.

INTRODUCTION

Supramolecular chemistry is a unique tool to produce, by self-assembly, metallorganic complexes with a defined number of magnetic metal centers [1]. Recently, it has been shown that a novel self-assembled supramolecular grid structure with four Co^{2+} metal centers, the Co-[2×2] grid, represents an almost ideal model system for investigating magnetic interactions in a discrete entity [2]. The Co^{2+} spins exhibit an antiferromagnetic interaction which is confined to within a grid. Thus, each grid forms an antiferromagnetic domain.

In this work we present recent results on the same grid structure but containing four Ni^{2+} ions, denoted as Ni-[2×2] grids. We estimate the coupling constants for both Ni-[2×2] grids and Co-[2×2] grids and compare them with each other. In order to demonstrate the effects of the antiferromagnetic coupling, we also performed measurements of the mononuclear analogues.

THEORY

Within the crystal field approximation the Hamiltonian of the lowest LS-multiplett appropriate for the mononuclear complexes may be written as [3]

$$H_1 = V + T + \lambda L \cdot S + \mu_B (L + 2S) \cdot H , \qquad (1)$$

where λ denotes the spin-orbit coupling constant, μ_B the Bohr magneton, V the octahedral ligand field and T fields of lower symmetry. V and T may be expressed using Stevens operators [3].

In the case of Ni^{2+}, the octahedral field V splits the ^3F multiplett into $^3T_1$, $^3T_2$ and $^3A_2$, $^3A_2$ being lowest. Applying the perturbation theory for an orbital singlet [3] to the Hamiltonian H_1, the effective spin Hamiltonian for the $^3A_2$ level ($S = 1$) is written as

$$H_2 = DS_z^2 + E(S_x^2 - S_y^2) + \mu_B g_x H_x S_x + \mu_B g_y H_y S_y + \mu_B g_z H_z S_z . \qquad (2)$$

For Co^{2+}, the splitting of the ^4F multiplet by the octahedral ligand field leads to $^4T_1$ as lowest level. Two cases will be distinguished. If the splitting of the levels due to lower symmetric fields is larger than that due to the spin-orbit coupling, i.e. $T \gg \lambda$, then again the perturbation theory

Mat. Res. Soc. Symp. Proc. Vol. 488 © 1998 Materials Research Society

for an orbital singlet can be applied leading to the same Hamiltonian as in Eq. 2, but with $S = 3/2$. However, in the case $T \approx \lambda$ the effective Hamiltonian for the $^4T_1$ level is obtained by replacing $L \rightarrow -\frac{3}{2}l$, where $l = 1$ [4]. On restricting ourselves to tetragonal symmetry, this yields [4]

$$H_3 = \delta\, l_z^2 - \tfrac{3}{2}\alpha\lambda l_z S_z - \tfrac{3}{2}\alpha'\,\lambda(l_x S_x + l_y S_y) + \mu_B(-\tfrac{3}{2}\beta l_x + 2S_x)H_x + \mu_B(-\tfrac{3}{2}\beta' l_z + 2S_z)H_z\,, \quad (3)$$

where we distinguished between the α's and β's in order to allow for orbital reduction.

For the tetranuclear complexes, the Hamiltonian within the lowest LS-multiplett may be written as [5]

$$H_{Ex} = -J(S_1 \cdot S_2 + S_2 \cdot S_3 + S_3 \cdot S_4 + S_4 \cdot S_1) + \sum_{i=1}^{4}[V_i + T_i + \lambda L_i \cdot S_i + \mu_B(L_i + 2S_i)\cdot H]\,, \quad (4)$$

where i numbers the metal centers. The term enclosed by the sum will be abbreviated by H_i, the single ion Hamiltonian for the ith spin. As will be seen below, the coupling constant J is much lower than λ and T (and of course V). Thus, the exchange Hamiltonian for the ground state multiplet obtained by a perturbation treatment is similar to H_{Ex} in Eq. 4, but with H_i replaced by one of the single ion Hamiltonians discussed above. The exchange term will not be altered in first order, and we will neglect higher order terms as they are on the order of either λ/V or T/V. In the following, the notation $H_{Ex}(H_1)$ e.g. refers to the Hamiltonian H_{Ex} with H_i replaced by H_1.

EXPERIMENT

The structure of the [2×2] grids consists of four bis(bipyridyl)-pyrimidine ligands and four metal centers. The crystal structure for the Co-[2×2] grid is shown in Fig. 1 [6]. Each metal center is situated in the crossing point of two ligands and is enclosed by six N donor atoms in an almost octahedral geometry. The positive charges are countered by eight PF_6^- ions. The grids were formed by self-assembly in methanol solution [6]. The distance between the metal centers is about 6.5 Å. We also performed measurements on the mononuclear analogue of the [2×2] grids,

(a)   (b)

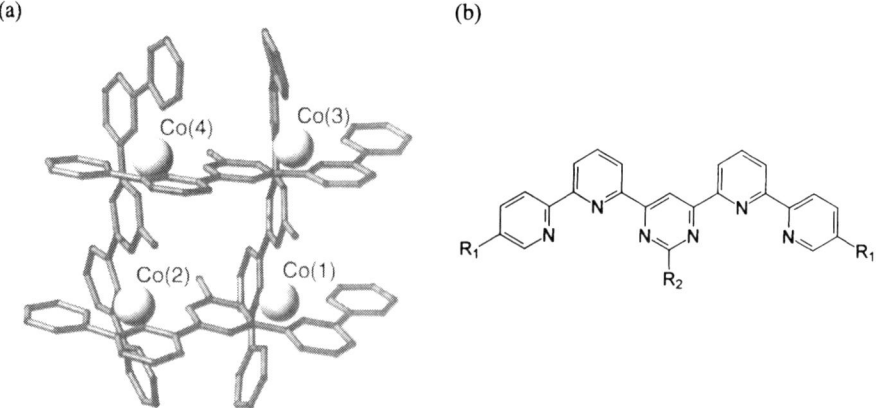

Fig. 1: (a) Crystal structure of the Co-[2×2] grid. (b) Bis(bipyridyl)-pyrimidine ligand with $R_1 = H$ and $R_2 = CH_3$.

the M(terpyridyl)$_2$(PF$_6$)$_2$ complexes with M = Ni, Co, denoted here as Ni-[1×1] and Co-[1×1], respectively, obtained by a method analogous to that used for the [2×2]-grids [7]. For all complexes, the weight of the powder samples was typically 1 - 3 mg.

The temperature and magnetic field dependence of the magnetic moment, $m(T,B)$, was measured with a commercial SQUID magnetometer (Quantum Design). The temperature range was 1.9 - 250 K and the maximum field 5.5 T. The powder samples were fixed with apiezon grease on a plastic straw. The background signal of the plastic straw was found to be below the resolution of the magnetometer. The signal of the grease was measured independently and then subtracted from the data. The diamagnetic contribution of the ligands was determined from measurements of Cd-[2×2] and Cd(terpyridyl)$_2$(PF$_6$)$_2$. For both complexes we obtained χ_D = -(0.6±0.1)×10^{-3} μ_B/(spin T) in fair agreement with an estimation using Pascal's constants. We didn't correct the data for it, but took it into account for the calculations. In this work, we use μ_B/spin, the number of Bohr magnetons per metal ion, as unit for the magnetic moment [2].

RESULTS

In Fig. 2 we present the temperature dependence of the magnetic moment of Ni-[1×1] powder measured at several magnetic fields. Above 50 K the temperature dependence is well described by $\chi T = C + \chi_0 T$ (see inset of Fig. 2), where $\chi = m/B$ and $C = 1.97 \mu_B$/spin T^{-1} K, χ_0 = -0.3×10^{-3} μ_B/(spin T). From the Curie constant C the g-factor and the effective magnetic moment were determined to be g = 2.1 and μ_{eff} = 2.97. Here $g_x = g_y = g_z \equiv g$ was assumed. At lower temperatures a deviation from the Curie-behaviour was observed (inset of Fig. 2) indicating a zero field splitting (ZFS). Fitting the data to the Hamiltonian H_2 yielded $D \approx$ -9 K, $E \approx 0$ where g and χ_0 were fixed to the values given above. The calculated curves using this parameters are shown as solid lines in Fig. 2.

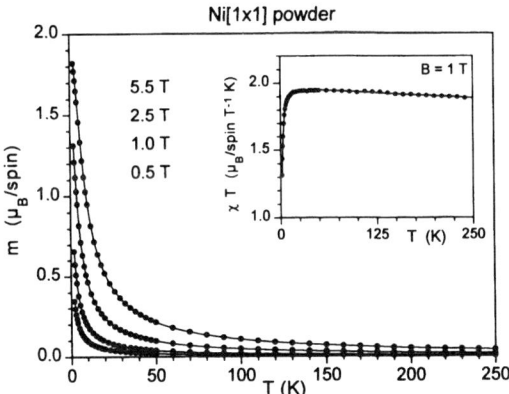

Fig. 2: Temperature dependence of the magnetic moment of Ni-[1×1] powder. Magnetic fields were 0.5 T, 1 T, 2.5 T and 5.5 T from bottom to top. The inset shows χT obtained from the magnetic moments at 1 T. The solid lines represent a fit to the data (see text).

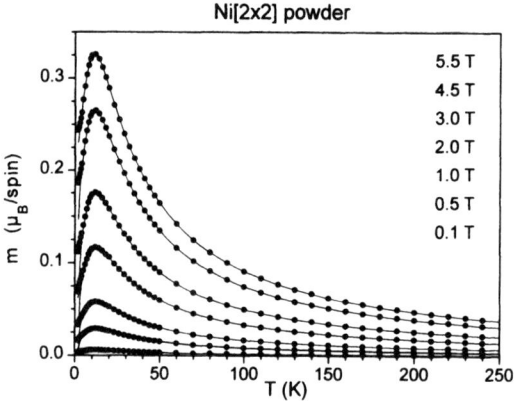

Fig. 3: Temperature dependence of the magnetic moment of Ni-[2×2] powder. Magnetic fields were 0.1 T, 0.5 T, 1 T, 2 T, 3 T, 4.5 T and 5.5 T from bottom to top. The solid lines represent a fit to the data (see text).

In contrast to Ni-[1×1], the temperature dependence of the magnetic moment of Ni-[2×2] powder reveals a maximum at ≈ 13 K (Fig. 3) indicating an antiferromagnetic coupling of the four Ni^{2+} ions within a grid. The appropriate Hamiltonian describing Ni-[2×2] is $H_{Ex}(H_2)$ with $S = 1$. In Fig. 3 the calculated curves using $J = -8$ K, $D = E = 0$, $g = 2.05$ and $\chi_0 = -0.3 \times 10^{-3}$ μ_B/(spin T) are depicted as solid lines.

In Fig. 4 the temperature dependence of Co-[1×1] is reproduced. In Ref. [2] it has been shown that the magnetization at temperatures below ≈ 50 K is governed by the lowest Kramers doublet. For temperatures above ≈ 100 K, indications for a thermal excitation of higher levels were found.

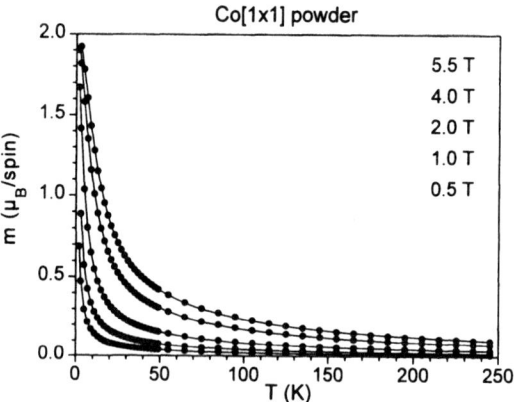

Fig. 4: Temperature dependence of the magnetic moment of Co-[1×1] powder. Magnetic fields were 0.5 T, 1 T, 2 T, 4 T and 5.5 T from bottom to top. The solid lines represent a fit to the data (see text).

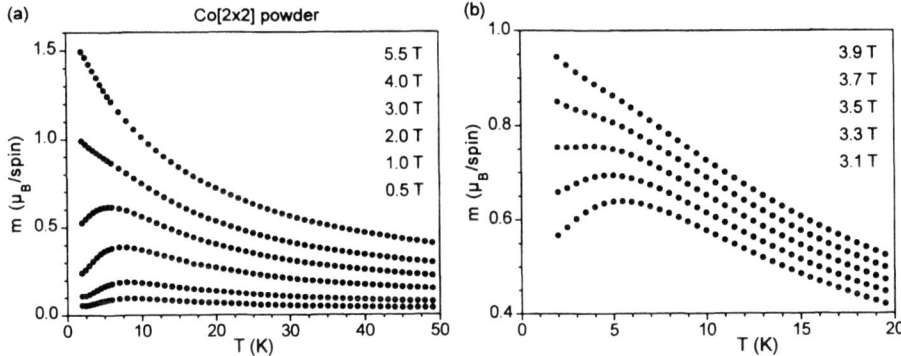

Fig. 5. Temperature dependence of the magnetic moment of Co-[2×2] powder. Magnetic fields are given in the panels.

This suggests that the Co^{2+} ions are in the high-spin state which is further confirmed by the effective magnetic moment at 250 K of $\mu_{eff} = 4.54$. Thus the magnetization in the full temperature regime should better be accounted for by the single ion Hamiltonians H_2 with $S = 3/2$ or H_3, depending on the relative strength of T and λ. A fit using H_2 reproduces the data very well in the whole temperature and magnetic field regime, as demonstrated in Fig. 4. The parameters were $D = -86.7$ K, $E = -19.2$ K, $g_x = g_y = 2.31$ and $g_z = 2.40$. However, using H_3 instead did not lead to better results and the parameters could not be determined unambiguously. In particular, both positive and negative δ yielded reasonable fits. In addition, the parameters α, α', β, and β' took unreasonable values.

Fig. 5(a) shows the temperature dependence of the magnetic moment of Co-[2×2] powder. The low-field magnetization curves exhibit maxima at $T^* \approx 7.5$ K, which has been shown to be a signature of an antiferromagnetic coupling of the four Co^{2+} ions of a grid [2]. At the lowest temperatures a transition to the paramagnetic state occurs at $B^* \approx 3.5$ T as is clearly seen in Fig. 5(b). The results obtained for Co-[1×1] suggest that Co-[2×2] can be described by $H_{Ex}(H_2)$ with $S = 3/2$. Let us assume $D = 0$ and $E = 0$ for the moment. Then, from T^* the coupling constant is estimated to $J = -2.93$ K. The magnetization curves at very low temperatures should exhibit six magnetization steps at fields $B_n = nk_B|J|/\mu_B g$, where $n = 1 - 6$ and k_B denotes the Boltzmann constant. Identifying B^* with B_2 yields $J = -2.75$ K in good agreement with the coupling constant obtained from T^*. Here, $g = 2.35$ was used. However, two objections have to be made. First, a weak indication of the first magnetization step was actually observed in Ref. [2], however it arises at ≈ 0.35 T and not at $B^*/2 = 1.75$ T as expected, indicating a significant anisotropy of the exchange interaction. Furthermore, the large ZFS observed in Co-[1×1] strongly contradicts the neglection of the ZFS and, in contrary, $H_{Ex}(H_2)$ should be evaluated in the limit $|D| >> |J|$, where we assume $E = 0$ for the moment. If only temperatures $T << |D|$ are considered, first-order perturbation theory leads to a Hamiltonian for the ground state multiplet which can be obtained more easily by formally replacing the $S = 3/2$ spin operators by spin-half operators s [8]. For $D < 0$, one has $S_x = 0$, $S_y = 0$, $S_z = 3s_z$, and the Hamiltonian valid at low temperatures is derived to be:

$$H_5 = -9J(s_{1z}s_{2z} + s_{2z}s_{3z} + s_{3z}s_{4z} + s_{4z}s_{1z}) + \mu_B\sum_{i=1}^{4}3g_zH_zs_{iz}. \qquad (5)$$

Thus, in this approximation the exchange interaction is maximally anisotropic, i.e. of the Ising type. From T^*, J is evaluated to be -1.38 K now. The magnetization curves at low temperatures reveal only one step at a field $B_c = 3k_B|J|/\mu_B g$ yielding $J = -1.84$ K. The agreement of these two estimates for J is only fair. We attribute this to the fact that the exchange anisotropy certainly was overestimated, since H_S cannot reproduce the weak step observed at ≈ 0.35 T. Furthermore, the effects of E cannot be neglected, and indeed its inclusion would lead to a reduction of the exchange anisotropy and therewith to the appearance of a second magnetization step well below B_c. We suggest that the coupling constant lies between the two extreme situations considered and therefore should be about $J \approx -2$ K.

In summary, we observed an intramolecular antiferromagnetic coupling of the four spins in both Ni-[2×2] and Co-[2×2]. The isotropic coupling strength was determined to $J = -8$ K for Ni-[2×2] grids and $J = -2$ K for the Co-[2×2] grids.

CONCLUSION

The isotropic coupling constant J used in Eq. 4 is related to the exchange interactions of the magnetic orbitals according to

$$J = 1/N^2 \sum_{\mu=1}^{N} \sum_{\nu=1}^{N} J_{\mu\nu} \qquad (6)$$

where N is the number of magnetic orbitals and $J_{\mu\nu}$ denotes the coupling strength between orbital $|\mu\rangle$ and orbital $|\nu\rangle$ [9]. For Ni^{2+}, we have $N = 2$ and $J = -8$ K, while for Co^{2+} we have $N = 3$ and $J = -2$ K. Since $N^2|J|$ is significantly higher for Ni-[2×2] than for Co-[2×2], it must be concluded that the additional magnetic orbital of Co^{2+} leads to considerable additional ferromagnetic contributions to J. However, since the interaction paths are neither of 90° nor of 180° type, a phenomenological estimation of the $J_{\mu\nu}$ is rather doubtful. We hope that this work will stimulate corresponding theoretical work.

REFERENCES

1. J.-M. Lehn, Supramolecular Chemistry – Concepts and Perspectives, VCH, Wienheim, 1995.

2. O. Waldmann, J. Hassmann, P. Müller, G. S. Hanan, D. Volkmer, U. S. Schubert and J.-M. Lehn, Phys. Rev. Lett. 78, 3390 (1997).

3. A. Abragam and B. Bleaney, Electron Paramagnetic Resonance of Transition Ions, Clarendon Press, Oxford, 1970.

4. A. Abragam and M. H. L. Pryce, Proc. Roy. Soc. London A206, 173 (1950).

5. B. Bleaney, F. R. S. Bowers, and K. D. Bowers, Proc. Roy. Soc. London A214, 451 (1952); M. E. Lines, J. Chem. Phys. 55, 2977 (1971).

6. G. S. Hanan, U. S. Schubert, D. Volkmer, J.-M. Lehn, G. Baum and D. Fenske, Angew. Chem. 109, 1929 (1997); G. S. Hanan, U. S. Schubert, D. Volkmer, E. Riviere, J.-M. Lehn, N. Kyritsakas and J. Fischer, Can. J. Chem. 75, 169 (1997).

7. U. S. Schubert, C. H. Weidl, M. Eigner, C. Eschbaum and J.-M. Lehn, to be published.

8. M. E. Lines, Phys. Rev. 131, 546 (1963).

9. A. Bencini and D. Gatteschi, EPR of Exchange Coupled Systems, Springer, Berlin, 1990, p. 7.

PHYSICAL CHARACTERIZATION AND ELECTRICAL TRANSPORT IN END-GROUP FUNCTIONALIZED SELF-ASSEMBLED MONOLAYERS

J. Collet*, D. Vuillaume*, M. Bonnier**, O. Bouloussa**, F. Rondelez**, J.M. Gay⁺, K. Kham#, C. Chevrot#
* IEMN-CNRS, ISEN Dept. of Physics, BP 69, F-59652 Villeneuve d'Ascq, France.
vuillaume@isen.iemn.univ-lille1.fr
** Institut Pierre et Marie Curie, CNRS, 11 Avenue P. et M. Curie, F-75 Paris, France.
⁺ CRMC2-CNRS, Campus de Luminy, Case 913, F-13988 Marseille, France.
LRPME, Univ. Cergy-Pontoise, F-95031 Cergy-Pontoise, France.

ABSTRACT

We have previously demonstrated that a self-assembled monolayer (SAM) of alkyl-trichlorosilane molecules, covalently bonded to the native oxide of a silicon substrate, allows the fabrication of MIS (Metal-Insulator-Semiconductor) devices with excellent electrical properties [1]. Here we demonstrate that we can functionalize, with π-electron moieties, the end-groups of the molecules, once grafted to the substrate, without disturbing the low conductivity through the SAM ($\sigma_\perp$), while strongly increasing the in-plane conductivity ($\sigma_{//}$). The functionalization of the monolayer is monitored by FTIR, ellipsometry and X ray reflectivity. Conductivity, photoconductivity and capacitance measurements are carried out to study the effect of these chemical functionalizations on the electrical properties of the monolayers. We demonstrate that the low perpendicular conductivity ($\sigma_\perp \approx 10^{-15}$ S/cm) is not altered by the end-group functionalities while high anisotropic conductivities ($\sigma_{//}/\sigma_\perp \approx 10^7$-$10^8$) are obtained at a monolayer level. Interfacial (Si/SAM and SAM/metal) energy barriers are investigated as a function of the end-group functionalities. We also show that conducting monomers substituted by alkyl chains (3,6-dibromocarbazole-N-octyltrimethoxysilane), directly grafted to the substrate, exhibit similar behavior.

INTRODUCTION

Organic self-assembled monolayers (SAM) of saturated alkyl chains present a very good insulating behavior, as demonstrated in our previous works [1, 2], and thus they can be an interesting alternative to thin oxide films for the fabrication of devices at a nanometer scale.

Using a chemical functionalization of the end-groups of such organic SAM's, we aim to fabricate insulator/conductor heterostructures at a molecular level by attaching π-electron moieties at the top of the previously formed insulating monolayers. Our purpose is to perform such a functionalization without disturbing the low conductivity through the SAM ($\sigma_\perp$) while strongly increasing the in-plane conductivity ($\sigma_{//}$).

In this work, we studied the chemical modification of vinyl-terminated monolayers by several routes such as oxidation, hydroboration, esterification and so on. We have obtained monolayers terminated by various chemical groups like -COOH, -CH$_2$OH, -COOCH$_2$-Phenyl, -COO-Retinol and -COOCH$_2$-Pyrene. We also directly fabricated SAM's of 3,6-dibromocarbazole-N-octyltrimethoxysilane (DOMSC). Once modified, we have carried out structural and electrical measurements to analyze the effects of these chemical reactions on molecular ordering and density, electrical conductivity, interfacial energy barrier and dielectric permittivity of the functionalized monolayers.

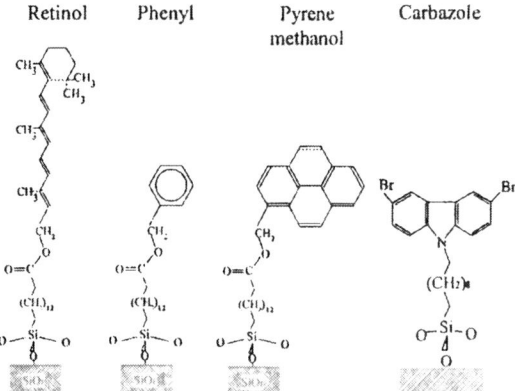

Retinol   Phenyl   Pyrene   Carbazole
                     methanol

Figure 1: Schematic representation of the functionalized monolayers

SAMPLE PREPARATION

Monolayers were prepared in two steps, onto three substrates (glass substrate, <100> silicon substrate phosphorus doped to 10^{19} cm^{-3} and silicon ATR crystals). First, we deposited a vinyl-terminated self-assembled monolayer or a DOMSC monolayer on the substrate, following a procedure described elsewhere [3]. The molecules of interest, $SiCl_3$-$(CH_2)_{12}$-$CH=CH_2$ (tetradecyl-1-enyl trichlorosilane, TETS) and $Si(OCH_3)_3$-$(CH_2)_8$-$NC_{12}Br_2H_6$ (DOMSC, carbazole terminated), were dissolved in an organic bath (C_6H_{14}/CCl_4) at low concentration (ca. 10^{-2} M). They were spontaneously grafted onto the immersed substrate by a chemical reaction between the surface silanol groups of the glass surface or of the native oxide (silicon substrates) and the trichlorosilane or trimethoxysilane head-groups of the molecules.

To obtain a densely packed and well-ordered monolayer, that is the mandatory condition for good insulating properties [1], we always worked below a critical temperature, that depends on the molecule length [3]. We assumed that the critical temperature of TETS and DOMSC are equal to the critical temperature of a 14 carbon atoms alkyl chain (about 12 °C) and a 11 carbon atoms alkyl chain (about 5 °C), respectively.

Once the monolayer of TETS is grafted, we modified the vinyl end-groups by hydroboration and hydrolysis to obtain -CH_2OH terminated monolayers or by oxidation to obtain -COOH terminated monolayers [4]. The yield of these chemical surface modifications is estimated to be in the range 70 % to 90 % [4]. These modifications are the first steps towards the construction of a more complex layer. Using esterification reactions between the -COOH end-groups and different alcohols (benzyl alcohol, retinol, pyrene methanol), we attached highly conjugated moieties onto the previously formed SAM's as shown in Fig. 1. Esterifications were carried out in the presence of a water trap to enhance the yield of the reaction.

PHYSICAL CHARACTERIZATIONS

Infrared Spectroscopy

Fourier Transform Infrared Spectroscopy was used in ATR mode (Attenuated Total Reflection), with p and s polarization to check the efficiency of the functionalization by

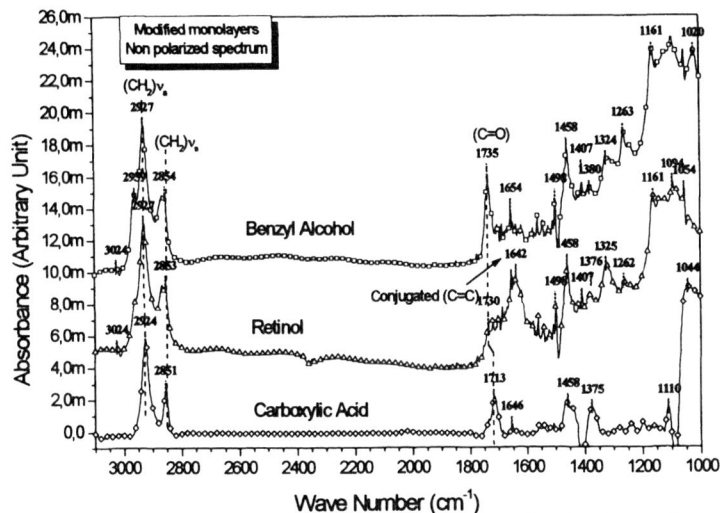

Figure 2: FTIR Spectrum of some functionalized monolayers.

detecting the vibration modes associated to the carbon-carbon double bonds and carbon-oxygen double bonds (Fig. 2) in the modified SAM's. These measurements point out that the π-electron moieties were effectively attached on the previously formed SAM's, even if the transformation yield is not easily obtained from these FTIR data. However, we can obtain the yield of the oxidation reaction (-CH=CH$_2$ $\rightarrow$-COOH) from the decrease in the area of the peak at 1640 cm^{-1}, associated to the carbon-carbon double bonds and the increase in the peak at 1711 cm^{-1} (associated to C=O), before and after the oxidation. The value obtained by this way (about 77%) is in agreement with that obtained by Wasserman et al. from contact angle measurements [4]. By monitoring the displacement of the vibration mode associated to methylene groups at 2922 and 2851 cm^{-1} in the TETS monolayer (2920 and 2850 cm^{-1} in a crystalline structure, a shift towards high wave numbers indicating a more liquid-like structure [5]), and from the change in dichroic ratio of these peaks, we deduced that the order and the density of the monolayer was not disturbed by the chemical modification. We also deduced, by comparison of the methylene peak amplitude in DOMSC monolayer with the methylene peak amplitude of a well-ordered and dense monolayer of alkyl-trichlorosilane, that the monolayer of DOMSC is well ordered and densely packed with a density of about 1 molecule per 50 Å^2. This is comparable to the theoretical value of 39.8 Å^2 obtained from geometry optimization with MOPAC 6.0.

X ray reflectivity

X-ray reflectivity measurements were carried out on a three axes spectrometer with a Ge(111) analyzer and a monochromator at 1.5405 Å wavelength. Thicknesses of the monolayers (Figure 3) were obtained from the first minimum of the signal in the specular reflectivity ($q_{min}=\pi/d$) and by fitting the whole reflectivity curves with one layer model (except for the retinol terminated monolayer for which two and three layer models were used) [6]. From the electronic density and thickness obtained by the fitting procedure we also calculated the area per

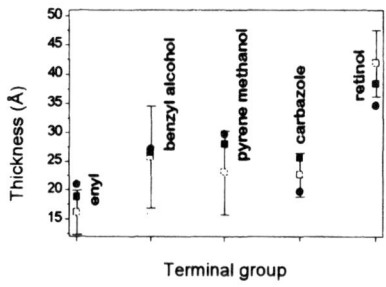

Figure 3: Thickness of the monolayers obtained by X-ray reflectivity measurements (■ from q_{min} □ from fitting) and theoretical values (●). Error bars represent the roughness of the SAM/air interface.

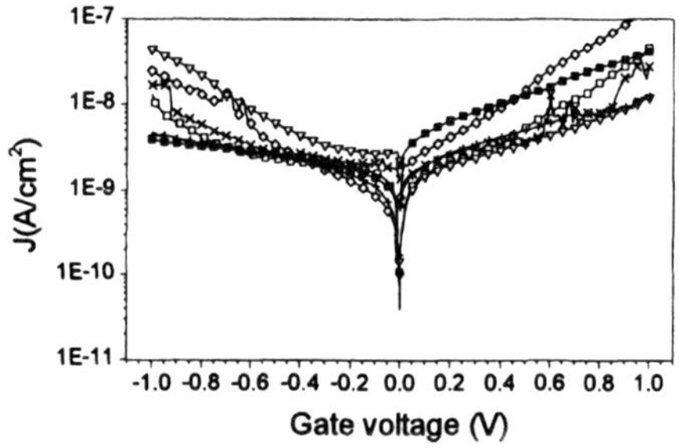

Figure 4 : Current densities through the monolayers for different end-group functionalities : □ enyl, ◊ carboxylic acid, ∇ benzyl alcohol, × pyrene methanol, ■ carbazole, + retinol.

Table I: Thickness of monolayer and area (Σ) per molecules or per end groups (Σ_p).

Terminal end group	Thickness (Å)	Σ (Å²)	Σ_{theo} (Å²)	Σ_p (Å²)	$\Sigma_{p\,theo}$ (Å²)
enyl (TETS)	16.2	22.6	20		
carbazole(DOMSC)	22.7	38	39.8		
pyrene methanol	23.7	23.6		26	26.8-34.3
benzyl alcohol	25.8	19.5		17.8	18.4

molecules (or per end groups) in each monolayer [6]. The values are summarized in Table I. The thicknesses, as well as the area per molecules, compare well to the theoretical values obtained from geometry optimization with MOPAC 6.0 and with those reported in earlier publications [6]. Thus we conclude that monolayers (and π moieties) were well ordered and densely packed. These results were confirmed by ellipsometry and capacity measurements (see below).

ELECTRICAL CHARACTERIZATIONS

We evaporated aluminum counter-electrodes through a shadow mask to carry out electrical measurements on the modified SAM's embedded in a Metal-SAM-Semiconductor device. Starting substrates were degenerated, heavily doped ($>10^{19}$ cm^{-3}), n type silicon wafers

Perpendicular conductivity

For the different functionalized monolayers and by comparison with the precursor layer, -CH=CH$_2$- terminated, we can see that the insulating properties of the monolayers were not altered by the chemical modifications (Figure 4). All monolayers exhibited a low leakage current through the SAM, 10^{-8}-10^{-7} A/cm^2 at 1V, a value that was also reported on our previous studies on -CH3 terminated SAM's [1,2] (Table II). According to our previous studies on -CH$_3$ terminated SAM's [1], this low leakage current is the fingerprint of a well ordered and densely packed monolayer of alkyl chains. Thus we conclude that the grafting of π-moieties has been achieved without introducing too much disorder in the precursor monolayer of alkyl chains.

Table II: Summary of electrical properties (values in brackets are theoretical ones)

End Group	Current @1V (A/cm^2)	Barrier Height (eV)	Monolayer Capacitance $(\mu F/cm^2)$	ellipsometry n.d	Monolayer permittivity	Monolayer thickness(nm)
-CH$_3$	10^{-8}	4.3-4.5	1.07	2.75	2.2	1.84-(1.86)
-CH=CH$_2$	3×10^{-8}	4.1-4.2	0.78	2.81	1.82	2.1-(2.1)
-COOH	8×10^{-8}	4.1-4.2	0.95	3.79	2.55	2.4-(2.1)
-COOCH$_2$- Phenyl	5×10^{-8}	3.6-3.8	0.85	5.03	2.85	2.98-(2.73)
-COO- Retinol	2×10^{-8}	3.6-3.7.	0.75	6.12	3	3.53-(3.48)
-COOCH$_2$- Pyrene	10^{-7}	3.9	0.89	4.92	2.9	2.9-(2.98)
Carbazole	10^{-7}	3.9-3.8	1.016		2.94	(1.98)

Internal photoemission

These measurements [1,7] give us the energy barrier height at the metal-SAM interface or at the semiconductor-SAM interface depending on the applied polarization. Values of these energy barriers are so high, 4 to 4.5 eV (Table II), that carrier tunneling through the monolayer is totally suppressed [1]. This also explains the low leakage current reported above. Again, the chemical modifications do not strongly affect the values of this energy barrier height. Only a little decrease was observed for the highly conjugated systems (phenyl and pyrene), presumably due to dipole effects at the interface (Table II). In these $\sigma-\pi$ type monolayers, the perpendicular electrical transport is thus governed by the blocking behavior of the alkyl chains. No enhanced perpendicular conduction through the molecular orbital of the π species was observed.

Capacitance measurements

Capacitance measurements were made at 1MHz. We deduced the capacitance of the monolayer taking into account: i) the oxide capacitance (assuming a 1 nm thick oxide, as measured by ellipsometry, and a relative permittivity of 3.9) and ii) the Debye capacitance of the semiconductor.

These capacitance measurements were coupled with ellipsometry (at 636 nm) to determine both thickness and dielectric permittivity of the SAM's. From the capacitance measurements we extracted the ratio ε/d, where ε is the dielectric constant and d the SAM thickness and from the ellipsometry we extracted the product nd, where n is the optical index of the monolayer. Assuming that $\varepsilon=n^2$ we combine both results to calculate d and ε (table II). This assumption neglects any possible dipolar contribution in the low frequency dielectric function (ε). This is validated a posteriori by the fact that the measured thicknesses are in close agreement with theoretical ones, calculated for molecules in their all-trans conformation, nearly perpendicular to the substrate (values in brackets in Table II) and with X-ray reflectivity measurements. We can see that the π-moieties slightly increase the relative dielectric permittivity of SAM's from 2 to 3.

In-plane conductivity

For the $\sigma-\pi$ type SAM's (i.e. phenyl, pyrene, retinol and carbazole terminated) we measured their in-plane conductivity. We used glass substrate to avoid any contribution of leakage currents through the substrate. Aluminum electrodes regularly spaced (40 μm) were evaporated onto functionalized SAM's through a shadow mask. Current density flowing between two adjacent electrodes was calculated assuming that only the π moieties contribute to the in-plane conduction. This assumption was validated by measuring the in-plane conductivity of a vinyl terminated SAM ($\sigma_{//}\approx10^{-14}$ S/cm, similar to the perpendicular conductivity). Using a scale of distances between the two contacts, the resistivity measurements gave us the in-plane conductivity of 2.4×10^{-7}, 3.7×10^{-8}, 4.5×10^{-8} S/cm for the phenyl, retinol and carbazole terminated SAM's, respectively. Thus, the anisotropy ($\sigma_{//}/\sigma_{\perp}$) for these functionalized monolayers is of the order of 10^7-10^8. Such values have been obtained for Langmuir-Blodgett multilayers of anthracene derivative [7], but to our knowledge, it's the first time that such a value is obtained at a single monolayer level.

CONCLUSIONS

We have demonstrated that the chemical functionalization (i.e. grafting of π-moiety end-groups) of self-assembled monolayers of alkyl chains allow the fabrication of a insulator/conductor heterostructure at a molecular level with a very high anisotropic conductivity ($\sigma_{//}/\sigma_{\perp}$~$10^7$-$10^8$). Such insulator/conductor heterostructure could be the building block of monomolecularly thick organic Field Effect Transistor [9].

ACKNOWLEDGMENTS

We thank financial support from the research program ULTIMATECH-CNRS.

REFERENCES

1. C. Boulas, J. Davidovits, F. Rondelez, D. Vuillaume, Phys. Rev. Lett. **76**, 4797 (1996)

2. D. Vuillaume et al., Appl. Phys. Lett. **69**, 1646 (1996)

3. J.B. Brzoska, N. Shahidzadeh and F. Rondelez, Nature **360**, 719 (1992)

4. S.R. Wasserman, Y.T. Tao, G.M. Whitesides, Langmuir **5**, 1074 (1989)

5. M.D. Porter, T.B. Bright, D.L. Allara, C.E.D. Chidsey, J. Am. Chem. Soc. **109**, 3559 (1987)

6. I.M Tidswell, B.M. Ocko, P.S. Pershan, S.R. Wasserman, G.M. Whithesides, J.D. Axe, Phys. Rev. B **41**, 1111 (1990)

7. R.J. Powell, J. Appl. Phys. **41**, 2424 (1970)

8. G.G. Roberts, T.M. Ginnity, W.A. Barlow, P.S. Vincett, Thin Solid Films **68**, 223 (1980)

9. C.Zhou, D.M. Newns, J.A. Misewich, P.C. Pattnaik, Appl. Phys. Lett. **70**, 598 (1997)

Ab Initio Calculations on Porphyrins in the Condensed Phase

P.N. DAY, Z. WANG, R. PACHTER, Air Force Research Laboratory, Materials Directorate, Wright-Patterson AFB, OH 45433, daypaul@biotech.ml.wpafb.af.mil, wangz@biotech.ml.wpafb.af.mil, pachterr@ml.wpafb.af.mil

ABSTRACT

Porphyrins are a promising class of materials for optical limiting applications, and in the condensed phase solvent effects have been shown to be significant. We report results with a method designed to simulate the effects of discrete solvent molecules, namely the effective fragment potential (EFP) approach which has been implemented for use in ab initio calculations. Further, a simulated annealing (SA) method has been implemented with the EFP solvation model in an attempt to solve the problem of multiple minima in clusters of molecules. The results with this method indicate some success on models of aqueous formamide and aqueous glutamic acid. Ab initio calculations can now be carried out on porphyrins, and the solvation methods are being updated for their use on these systems.

INTRODUCTION

While most ab initio quantum chemistry calculations are carried out on individual, isolated molecules, the prediction of properties of organic solid-state materials from ab initio calculations requires the inclusion of the effects of neighboring molecules. These may be included explicitly in the calculation or implicitly through a dielectric continuum model,[1] the latter being the most common method for including condensed phase effects in quantum chemistry calculations. The dielectric continuum model can give a reasonable description of the effects of the bulk liquid or solid on the molecule, and has the advantage of requiring only a small amount of computer time beyond that required for the calculation on the isolated molecule. However, continuum methods cannot describe the local effects of the intermolecular interactions that occur in the condensed phase. These effects can be included by the explicit inclusion of neighboring molecules in the ab initio calculation, but this strategy quickly leads to a calculation that is unfeasible. The effective fragment potential (EFP) model[2] was developed to solve this problem by including the effects of spectator molecules through semi-classical one-electron terms, which interact with the molecule (or system of molecules) of interest, treated fully ab initio. The EFP is designed to mimic the interaction obtained from a full ab initio calculation on the entire system, but, like the continuum models, it requires only a small fraction more computer time beyond that needed for the calculation on the isolated molecule.

Porphyrins show promise as a class of optical limiting materials, and would most likely be utilized in a dilute solution in the liquid or solid-phase. Thus the local environment of the porphyrin molecule is a shell of solvent molecules, and in our model these solvent molecules are represented by fragments, which interact with the porphyrin molecule via the EFP. The fragments interact with the nuclei and the electron density of the ab initio molecule through three terms, (1) distributed multipolar expansions, (2) distributed polarizable points, and (3) one of two types of repulsive potentials. The first type of repulsive potential is a fitted repulsive potential which includes the exchange repulsion, charge transfer, and smaller terms in the interaction which have not been included in the first two EFP terms. A fitted EFP for water is built-in to the GAMESS[3] program, and has been shown to quite accurately mimic full ab initio calculations.[4,5] The second type of repulsive potential, which we call the transferable exchange repulsion, has been developed recently by Jensen and Gordon,[6] and has the advantage that it can be quickly generated for any molecule. The fragments also interact with each other through these three types of terms, and the calculation of these interactions does not involve any integration over the electron density. A recently implemented option in the program allows for all the molecules in a system to be replaced with fragments, thus only utilizing the interfragment interactions and completely eliminating the ab initio part of the calculation. These all-fragment calculations can rival empirical molecular mechanics (MM) potentials in computational speed, but should be superior to MM in reproducing the results of ab initio calculations.

The geometry optimization of a cluster of molecules, such as a porphyrin molecule surrounded by many solvent molecules, is difficult due to the many minima on the associated potential energy surface. Several methods for finding the global minimum, such as simulated annealing(SA)[7] and genetic algorithms,[8] have been previously utilized on molecules with many degrees of freedom. However, because these methods require the evaluation of the energy at a large number of geometries, only MM potentials have been used, as ab initio or even semi-empirical energies have been considered too computationally demanding for these methods. The implementation of the all-fragment option in GAMESS has made it possible to carry out these types of calculations at the EFP level of theory.

In this study, the use of SA is demonstrated in the optimization of formamide with 3 water molecules and of glutamic acid (both the neutral and zwitterionic forms) with 10 water molecules. Technical difficulties have prevented us from producing an EFP for a molecule the size of a porphyrin, so that global optimization methods have not yet been utilized on solvated porphyrins. Previously, we reported[9] geometry optimizations on isolated zinc and free-base octobromotetraphenyl porphyrin (OBrTPP), octobromo porphyrin (OBrP), and tetraphenyl porphyrin (TPP), as well as on the ZnOBrTPP surrounded by 10 (EFP) water molecules. For this study we have found the minimum energy geometry for the water-soluble octobromo-tetrapyridinal porphyrin(+4) cation (OBrTPyP^{+4}), and compare it to the structures reported in Ref. 9. Such calculations will be used as a reference when we carry out the SA on the solvated porphyrins.

METHODS

The GAMESS program has been modified to carry out the SA calculations. The method implemented is similar to that suggested by Parks.[10] In SA, a series of Monte Carlo simulations are carried out, starting at a high temperature, and then decreasing the temperature in small increments. In a Monte Carlo simulation, a trial step is generated by incrementing each variable by the product of the current maximum for that variable and a random number between 0 and 1. If the energy at the new geometry is lower than the energy at the current geometry, the step is accepted. If the energy at the new geometry is higher than the energy at the current geometry, the step is accepted only if,

$$\exp\big((E_{old} - E_{new})/kT\big) \ge r \qquad (1)$$

where E_{old} and E_{new} are the current energy and the trial energy, respectively, T is the temperature, k is Boltzmann's constant, and r is another random number between 0 and 1. Otherwise, the step is rejected and a new trial step is generated. Thus, at the high initial temperature, the probability of accepting a step is high, allowing the system to sample a large volume of the variable space. This should prevent the system from getting trapped in a relatively high energy local minimum, and increases the probability that the more selective low-temperature steps will locate the global minimum.

A two-step method has been used in this work to optimize solute-solvent systems: (1) all molecules in the system were replaced with fragments that use the transferable exchange-repulsion potential, and a SA was carried out on the solvent molecules, (2) starting from the minimum-energy geometry or possibly from the final geometry in the SA, a Newton-Raphson geometry optimization was carried out with the solute molecule treated ab initio and the solvent molecules as fragments.

RESULTS

The SA method was first tested on the system of formamide with three water molecules. This system was used by Chen and Gordon[4] to test the fitted EFP for water which is now built into the GAMESS program. The minimum energy geometry they found for this system is accepted to be the global minimum at this level of theory. The interaction energy at this geometry is -27.07 kcal/mol. The two-step optimization process described above was used on this system, and the key structures in this process are shown in Figure 1. Optimization parameters for five different

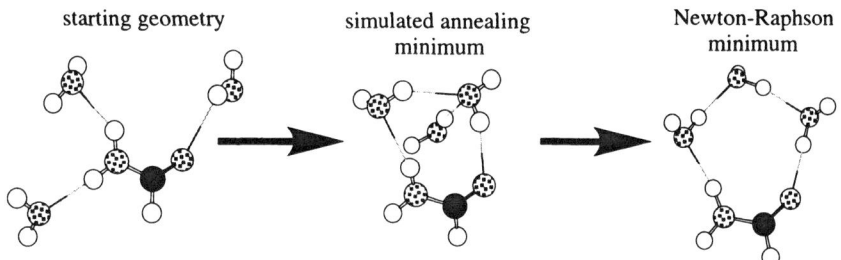

starting geometry   simulated annealing minimum   Newton-Raphson minimum

Figure 1. The geometries involved in finding the global minimum energy for the system of formamide with three water molecules. The SA was carried out at the all-fragment level of theory, while in the Newton-Raphson optimization, the formamide molecule was treated ab initio and the water molecules were represented as fragments.

Table 1. Global optimization data for the formamide with three water molecules system. N_{geo} is the number of geometries tried at each temperature, N_{temp} is the number of different temperatures completed, T_{init} is the initial temperature, T_{fin} is the final temperature, x_{max} is the maximum x coordinate allowed for the fragments, Δx is the initial step size for the translational coordinates, in angstroms, $\Delta\theta$ is the initial step size for the rotational coordinates, in π-radians, α is the rate at which information from successful steps is folded into the current step size, R_{min} is the minimum allowed separation between any two atoms (solute or solvent) that are not in the same molecule, N_E is the number of geometries where the energy was evaluated, r_A is the fraction of those geometries which were accepted, $E_{int}(SA)$ is the interaction energy at the minimum energy geometry found in the SA, $E_{int}(NR)$ is the interaction energy at the minimum energy geometry found in the subsequent Newton-Raphson optimization.

	Run 1	Run 2	Run 3	Run 4	Run 5
N_{geo}	300	300	600	500	300
N_{temp}	105	100	102	71	108
T_{init}	25000	25000	25000	25000	25000
T_{fin}	121	156	141	690	103
x_{max}	4	4	4	4	4
Δx	0.6	0.6	0.6	0.6	0.6
$\Delta\theta$	0.2	0.2	0.2	0.2	0.2
α	0.4	0.4	0.4	0.4	0.4
R_{min}	1.5	1.5	1.5	1.5	1.5
N_E	30529	29042	59480	35008	31504
r_A	0.586	0.614	0.597	0.408	0.591
$E_{int}(SA)$	-16.62	-17.72	-19.11	-18.48	-15.10
$E_{int}(NR)$	-27.07	-24.98	-27.07	-25.79	-26.55

simulations are given in Table 1. The starting geometry shown in Figure 1 was used for all five optimizations. Since each fragment has three translational coordinates and three rotational coordinates, and only the solvent molecules were moved in the SA, this global optimization problem has 18 independent coordinates, and this is not considered to be a trivial problem in global optimization. As is shown in Table 1, two of the optimization runs, labeled Run 1 and Run 3, were successful at finding the accepted global minimum for this system. The other three runs yielded minimum energies just 0.5, 1.3, and 2.1 kcal/mol higher. The structures shown in Figure 1 are from Run 1. Table 1 shows that these five runs used very similar starting parameters. Runs 1, 2, and 5, used identical parameters, while in runs 3 and 4 the number of geometries evaluated at each temperature was increased from 300 to 600 and to 500, respectively. The results of runs 1,

Table 2. Global optimization data for glutamic acid with ten water molecules. The labels are as described for Table 1. In the run labels, "n" is for neutral and "z" is for zwitterion. The two energies listed are both from Newton-Raphson optimizations using an ab initio solute molecule and fragment water molecules, with the "f" indicating that the optimization started from the final geometry in the SA, and the "m" label indicating that the optimization started from the minimum-energy geometry in the SA. The energies are relative to the overall minimum found in Ref. 5.

	N_{geo}	T_{init}	T_{fin}	Δx	$\Delta\theta$	α	R_{min}	x_{max}	E(f)	E(m)
Run n1	30	25000	103	0.3	0.1	0.4	1.5	7	3.59	10.53
Run n2	300	25000	103	0.3	0.1	0.4	1.5	7	6.27	9.04
Run n3	400	25000	103	0.3	0.1	0.4	1.5	5	6.76	6.76
Run n4	400	25000	938	0.6	0.2	0.4	1.6	5	3.95	4.11
Run n5	400	25000	319	0.6	0.2	0.7	1.6	5	6.58	11.19
Run n6	800	25000	5098	0.6	0.2	0.7	1.6	5	6.34	9.72
Run n7	400	25000	201	0.6	0.2	0.7	1.6	4	3.00	3.05
Run n8	400	15000	164	0.6	0.2	0.7	1.6	5	10.02	3.61
Run n9	400	25000	4371	0.9	0.2	0.7	1.6	5	12.95	6.30
Run n10	400	25000	804	0.6	0.4	0.7	1.6	5	10.41	9.20
Run n11	400	25000	1094	0.9	0.4	0.7	1.6	5	5.78	5.78
Run n12	400	15000	104	0.6	0.2	0.4	1.6	5	-1.31	7.23
Run n13	400	25000	804	0.6	0.2	0.4	1.6	5	4.70	11.42
Run z1	400	25000	534	0.3	0.1	0.4	1.5	5	5.91	3.43
Run z2	400	25000	507	0.3	0.1	0.4	1.7	5	6.77	6.77
Run z3	400	25000	457	0.3	0.1	0.4	1.9	5	5.72	3.92
Run z4	400	25000	373	0.6	0.2	0.4	1.9	5	5.84	5.84
Run z5	400	15000	806	0.6	0.2	0.4	1.9	5	1.22	1.22
Run z6	400	15000	483	0.4	0.2	0.4	1.9	5	5.73	5.73
Run z7	400	15000	940	0.8	0.2	0.4	1.9	5	4.50	6.52
Run z8	400	10000	1661	0.4	0.2	0.4	1.9	5	2.65	2.65
Run z9	400	10000	994	0.8	0.2	0.4	1.9	5	5.67	5.67
Run z10	400	15000	848	0.6	0.2	0.4	1.9	5	8.57	4.18
Run z11	600	25000	247	0.3	0.1	0.4	1.9	5	6.91	7.09
Run z12	600	15000	765.7	0.6	0.2	0.4	1.9	5	3.64	12.16
Run z13	600	15000	1215	0.6	0.2	0.7	1.9	5	-1.39	7.96

2, and 5 show that, because of the use of a random number generator in the procedure, simulations with exactly the same starting parameters are not identical. A geometric cooling schedule was used, with each temperature a factor of 0.95 its predecessor. The simulations were designed to stop when no geometries were accepted at a given temperature or when the preset final temperature was completed. Only Run 5 reached the final temperature of 103 K.

Table 2 shows the results of the optimizations on aqueous glutamic acid. This study was done as a comparison to our previous work[5] on glutamic acid surrounded by 10 water molecules, where we found the water molecules to be sufficient to make the zwitterion the equilibrium structure, as is expected in aqueous solution, rather than the neutral form, which is favored in the gas-phase. Thus it seemed that 10 water molecules were enough to at least qualitatively model the effects of aqueous solvation on glutamic acid. In this study, 26 simulated annealing runs were carried out, 13 using the neutral glutamic acid, and 13 using the zwitterion. From the results of each SA, two Newton-Raphson optimizations were carried out, one from the final SA geometry, and one from the minimum-energy SA geometry. As before, for this final optimization step, the solute was treated ab initio, while the water molecules were treated as fragments. The energies given are relative to the lowest energy found previously[5] for this system, which was -548.8146 H for a zwitterionic geometry. In order to surpass the results of the previous study, we needed to find a minimum for the neutral system that was lower than 3.49 kcal/mol (using the energy scale of Table 2) and a minimum for the zwitterion lower than zero. Run n7 was successful, as the "f" result of 3.00 is 0.49 kcal/mol lower than the previous neutral minimum and the "m" result of 3.05 is 0.44 kcal/mol lower than the previous neutral minimum, but the "f" result from run n12 far surpassed these results with an energy of -1.31, which is 4.79 kcal/mol lower than the previous neutral minimum. One of the optimizations on the zwitterionic system surpassed the results of the previous study, as the "f" optimization from run z13 yielded an energy of -1.39 kcal/mol. Since the lowest energy found up to this point for the zwitterionic system is only 0.08 kcal/mol lower than the lowest energy found for the neutral system, the conclusion of Ref. 5 for this system could be questioned, as this is a very small energy difference, and the inclusion of vibrational energy might reverse the energy ordering. Thus, it may require more than 10 water molecules to model aqueous glutamic acid. Since this method does not guarantee the location of a global minimum, we recommend further simulations or a more robust algorithm for global optimization. The dependence of the SA results on the initial positioning of the solvent fragments should also be investigated.

A water soluble porphyrin, octobromo-tetra(pyridinal-chloride) porphyrin (OBrTPyClP) has been synthesized[11] and is of interest as an optical limiting material. The structure of the isolated free-base OBrTPyP^{+4} cation has been optimized at the Hartree-Fock level of theory using the SBK basis, and this structure is shown in Figure 2, for comparison to the calculated structures for OBrP, TPP, and OBrTPP, shown in Figure 3. While OBrP and TPP have a planar structure to the porphyrin ring, the steric hindrance of both the bromine substituents and the phenyl groups in the OBrTPP forces the porphyrin ring to distort into a saddle like structure. The OBrTPP has been shown to be particularly promising as an optical limiting material because its excited state absorption cross-section is much larger than its ground-state absorption cross-section. This could be because the distortion of the ground-state geometry hinders pi-conjugation in the porphyrin ring, thus decreasing the absorption cross-section. The excited state structure may be expanded enough to be planar or much closer to planar, thus allowing for larger pi-conjugation and a much

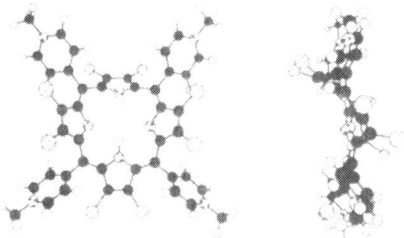

Figure 2. The optimized geometry of the isolated free-base octobromo-tetrapyridinal porphyrin(+4) cation, from two different view points.

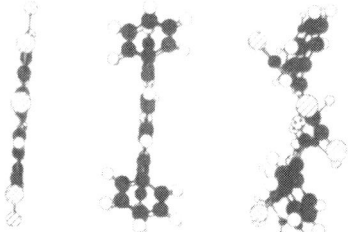

Figure 3. Side views of the free-base optimized geometries of octobromo-porphyrin, tetraphenyl-porphyrin, and octobromo-tetraphenyl-porphyrin.

larger absorption cross-section. As expected, the OBrTPyP^{+4} cation also has this distorted, saddle-like ground-state structure, and so it should be a promising optical limiting material. An EFP for OBrTPyP^{+4} cation will be set up, and the SA procedure we described used to find the global minimum-energy geometry for a system consisting of this molecule and a number of water molecules. Ideally, the number of water molecules will be progressively increased until convergence of the solute structure is reached. This should correctly model the effect of solvation on the structure of this molecule.

CONCLUSIONS

A simulated annealing method has been implemented for use with the EFP solvation model. This method was successful at finding the accepted global minimum structure for the system of formamide with three EFP water molecules. For glutamic acid surrounded by ten EFP water molecules, the method was successful at finding lower energy geometries than had been found previously for both the neutral and zwitterionic isomers. The geometry of a water soluble porphyrin ion, the OBrTPyP^{+4} has been calculated ab initio, and the simulated annealing optimization procedure described here will be used to model the effects of solvation on this ion.

REFERENCES

1. J. G. Kirkwood, J. Chem. Phys. **2**, 351 (1934); L. Onsager,. J. Am. Chem. Soc. **58**, 1486 (1936); O. Tapia, O. Goscinski, Mol. Phys. **29**, 1653 (1975); M. M. Karelson, A. R. Katritzky, M. C. Zerner, Int. J. Quantum Chem. Symp. **20**, 521 (1986); K. V. Mikkelsen, H. Aagren, H. J. A. Jensen, T. Helgaker, J. Chem. Phys. **89**, 3086 (1988); M. W. Wong, M. J. Frisch, K. B. Wiberg, J. Am. Chem. Soc. **113**, 4776 (1991); M. Szafran, M. Karelson, A. R. Katritzky, J. Koput, M. C. Zerner, J. Comput. Chem. **14**, 371 (1993); M. Karelson, T. Tamm, M. C. Zerner, J. Phys. Chem. **97**, 11901 (1993).

2. J. H. Jensen, P. N. Day, M. S. Gordon, H. Basch, ; D. Cohen, D. R. Garmer, M. Kraus, W. J. Stevens, *Modeling the Hydrogen Bond*, edited by D. A. Smith, ACS Symp. Series 569, 1994, p139; P. N. Day, J. H. Jensen, M. S. Gordon, S. P. Webb, W. J. Stevens, M. Krauss, D. Garmer, H. Basch, and D. Cohen; J. Chem. Phys. **105**, 1968 (1996).

3. M. W. Schmidt, K. K. Baldridge, J. A. Boatz, S. T. Elbert, M. S. Gordon, J. H. Jensen, S. Koseki, N. Matsunaga, K. A. Nguyen, S. Su, T. L. Windus, M. Dupuis, and J. A. Montgomery, J. Comput. Chem. **14**, 1347 (1993).

4 . W. Chen, M. S. Gordon, J. Chem. Phys. **105**, 11081 (1996).

5 . P. N. Day, R. Pachter, J. Chem. Phys. **107**, 2990 (1997).

6.. J. H. Jensen and M. S. Gordon, Mol. Phys. **5**, 1313 (1996).

7. S. Kirkpatrick, C. D. Gelatt, Jr., and M. P. Vecchi, Science **220**, 671 (1983).

8. D. Goldberg, *Genetic algorithms in search, optimization, and machine learning*, Addison-Wesley, MA, 1989.

9 . P. N. Day, Z. Wang, and R. Pachter in *Materials for Optical Limiting II*, edited by P. Hood, R. Pachter, K. Lewis, J.W. Perry, D. Hagan, R. Sutherland (Mater. Res. Soc. Proc. 479, Pittsburgh, PA 1997), p. 307-312.

10 . G. T. Parks, Nucl. Technol. **89**, 233 (1990).

11 W. Su, private correspondence.

DEFORMED HELIX FERROELECTRIC LIQUID CRYSTALS WITH LARGE TILT ANGLES IN OPTICALLY ADDRESSED SPATIAL LIGHT MODULATORS FOR DYNAMICAL HOLOGRAPHY APPLICATIONS

L.A. BERESNEV, W. HAASE, A.P. ONOKHOV*, W. DULTZ#, M.V. ISAEV*,
N.A. FEOKTISTOV§, N.L. IVANOVA*, E.A. KONSHINA*, A.N. CHAIKA&,
V.A. BERENBERG& and T. WEYRAUCH

Institute of Physical Chemistry, Darmstadt University of Technology, Petersenstr. 20, 64287
Darmstadt, Germany; *All-Russian S.I.Vavilov Research Center "GOI", Birgevaya Liniya 12,
St. Petersburg, Russia; §A.F.Ioffe Physical Technical Institute, St. Petersburg, Russia;
#Deutsche Telekom AG, Am Kavalleriesand 3, 64276 Darmstadt, Germany; &Institute of Laser
Physics "GOI", Birgevaya Liniya 12, St.Petersburg, Russia.

ABSTRACT

Optically addressed spatial light modulators (OASLMs) based on deformed helix ferroelectric liquid crystals (DHFLC) with high tilt angles on order of 40° and helical pitches less than 0.2μm were developed. The diffraction efficiency reached the order of 20%. The light induced deviation of the optical axis of the DHFLC layer was measured in sandwich structures consisting of photoconductors and liquid crystals. The photoelectric parameters of photoconductive amorphous silicon carbide a-SiC:H and a photoconductive polymeric films were measured with and without light blocking and reflecting layers. The application of the developed OASLMs in a holographic image corrector was demonstrated.

INTRODUCTION

OASLMs composed of photoconductive films and liquid crystal layers embedded between transparent electrodes [1] are considered as key elements for different schemes of optical information treatment. Last years, ferroelectric liquid crystals [2] attracted strong attention [3] as light modulating media owing to a fast response in the range of microseconds. For optical phase conjugation [4] and real time holography [5] the high spatial resolution of the OASLM should be accompanied with an enhanced diffraction efficiency in order to get sufficient conversion of the read-out light into the reconstructed image.

OASLMs using the SSFLC (surface stabilized ferroelectric liquid crystal) [6] geometry, provided so far diffraction efficiencies not more than 10% [4,5]. This restriction is related to relatively small switchable tilt angles of the used SSFLC or DHF (deformed helix ferroelectric) materials on order of $22-30^\circ$. In phase diffraction gratings with a rectangular profile a maximum diffraction efficiency (DE) of 40% can be obtained when the switchable tilt angle of the FLC material is equal to 45° [6,7]. A DE of 34% may be reached when the diffraction grating has sinusiodal phase profile, produced for instance in the photoconducting layer of an OASLM [6] by two interfering laser beams. A significant advantage of FLCs in comparison with nematic liquid crystals as well as with other electrooptical media is the polarization independence of the DE for square shape phase diffraction gratings built from interchanging orientations of the optical axes of the FLC [7,8].

In this paper we describe OASLMs utilizing FLC materials with a high tilt angle of 40° and a very short pitch of less than 0.2μm. The photoelectric parameters of the developed photoconductive films, based on amorphous silicon carbide (a-SiC:H) as well as on photoconductive polyimide films, were measured. A comparable high diffraction efficiency on the order of 20% was obtained. The operation of the developed OASLMs for the holographic correction of images is described.

EXPERIMENT

Deformed helix ferroelectric (DHF) materials

The DHF effect is considered as the lowest voltage effect in liquid crystals [9]. Appreciable electrooptical response takes place for voltages on the order of tenths of a volt. A deviation of the averaged optical indicatrix on the angle of $\pm 22.5^o$ is usually obtained applying less than $\pm$ 5 Volt, and it allows for a gray scale modulation. The response time of DHF material depends on the pitch of the helix and not on the voltage. For applications in OASLMs with high DE we developed DHF materials with a response time on the order of $200 \mu s$ and a spontaneous polarization of about $100 nC/cm^2$. The molecular tilt angle Θ in the totally untwisted state is 39-40° at room temperature. The pitch of the helix p_0 is $0.2 \mu m$ or less. In Table 1 the temperature intervals of the SmC* phase and other characteristic data are listed for some developed FLC materials.

Table 1. Parameters of DHFLC materials.

Parameter \\ FLC material	FLC-459	FLC-461	FLC-464	FLC-471
Interval of SmC* phase, °C	-10...+65	0...63.5	+5...+62	+2...+62.5
Tilt angle Θ (25°C)	41°	41.5^o	40°	39.5^o
Spontaneous polarization, nC/cm^2 (25°C)	18	93	120	115
Pitch of helix p_0, μm (25°C)	0.35	0.22	0.18	0.19

Photoelectric characteristics of photoconductive films

We developed photoconductive films on the basis of photodiode a-SiC:H thin layers in p-i-n configuration [10,11] and of polymer (polyimide) films [12]. These films are transparent for red light and the radiation of He-Ne lasers (633nm) can be used for read-out without an induction of spurious photocurrents in the photosensors. Hence, a transmissive operation can be simply realized. As it was shown [13] the sensitivity and the response time of silicon carbide photoconductors are of the same order as of the well known amorphous silicon a-Si:H films, but the spatial resolution of the a-SiC:H film can be improved owing to the decrease of the dark conductivity of the a-SiC:H layer with the increase of carbide content.

The a-Si$_{1-x}$C$_x$:H p-i-n structure was deposited on glass or quartz substrates with a transparent ITO (indium-tin oxide) electrode by the magnetron sputtering method. The individual layers were deposited by high frequency decomposition of a mixture of gases CH$_4$ and SiH$_4$ in a glow-discharge plasma. Doping was carried out by adding a small amount 0.1-1% of phosphine PH$_3$ (n-type layer) or diborane B$_2$H$_6$ (p-type layer). All layers were grown in a single cycle without exposure to air between intermediate procedures. The thickness of the p-type layer was 150Å, of the i-type layer (with intrinsic conductivity) 5000Å, and of the n-type layer less than 300Å. Though the photosensitivity of silicon-carbide films decreases with the increase of carbide concentration, they have sufficient sensitivity for most applications. The spectral dependence of films containing x=30% of carbide was similar to data presented in [13], with transmittance in the red region of about 60% for a thickness of the structure on the order of 1μm. The dye-sensitized photoconducting polyimide films for blue-green and red light are described in [12,14]. The basic advantage of polymer photoconducting films is the very small dark conductivity and very high spatial resolution obtained in OASLMs utilizing liquid crystals [14].

The measurements of photoelectric properties of the photoconductive films with and without additional blocking and/or reflecting layers were carried out using a simple set-up, where the photoconductive layer was illuminated through the transparent ITO electrode. A small drop of mercury (Hg) was the second electrode in contact with the photoconductor or blocking and/or reflecting layer. Typical photosensitivity curves for some prepared layers are presented in Fig. 1. Continuous illumination by white light with an intensity of $300\mu W/cm^2$ was used. These measurements were carried out for a preliminary selection of the photoconductive films with the best photosensitivity before the assembly of FLC OASLMs. This was especially important in the case of using light-blocking layers on the basis of a-C:H [15] and reflective mirrors on the basis of dielectric stacks and pixelized Al films.

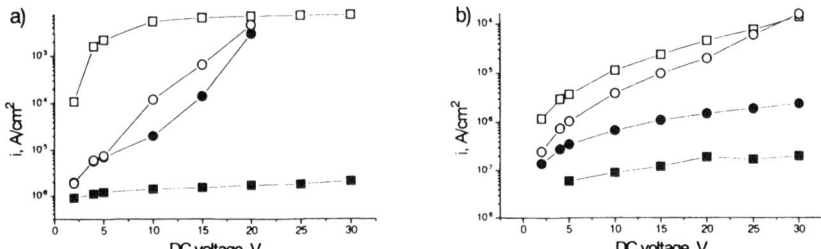

Fig. 1. The current-voltage dependencies of the a-SiC:H film (p-i-n diode structure) with pixelized Al mirror (a) and the photoconductive dye-sensibilized polyimide film (b) in dark (filled symbols) and light (open symbols) condition (squares - positive, circles - negative voltage on Hg).

Electrooptical characterisation of the designed OASLMs

The switching angle of the FLC director was measured between extreme states, corresponding to opposite voltage polarities ("rotation angle" of LC director) in darkness and under illumination with white light (intensity $300\mu W/cm^2$) by means of minimization of the oscillogram response [16]. A square wave voltage (meander) was applied with a peak-to-peak amplitude AC_{pp} and a direct curent bias voltage V_b. As an example, the "rotation angle" is plotted in Fig. 2 in dependence on the bias field V_b at AC_{pp}=30V for an OASLM utilizing an a-SiC:H photolayer in intrinsic conductivity configuration and a 12μm thick FLC layer.

The difference of the "rotation angles" between illuminated and dark states reaches 56 ^{O} at 10 Hz. The phase difference between neighbour (odd and even) stripes of the diffraction grating written in the OASLM at such conditions will reach 112°. If the cell thickness d corresponds to the condition "half of lambda" [17], the maximum diffraction efficiency $\eta_{max} = [2/\pi \sin(2\Theta)]^2$ can be obtained. In our case $\Theta = 28^O$ and thus $\eta_{max} = 27.8\%$. These values are not so far from the ideal case $\Theta = 45^0$ and $\eta_{max} = 40.5\%$. However, taking into account the more complicated profile of the effective refractive index in the DHF phase diffraction grating and periodical deviations of the phase retardation from the condition "half of lambda", the expected value of η will be less than 27.8%.

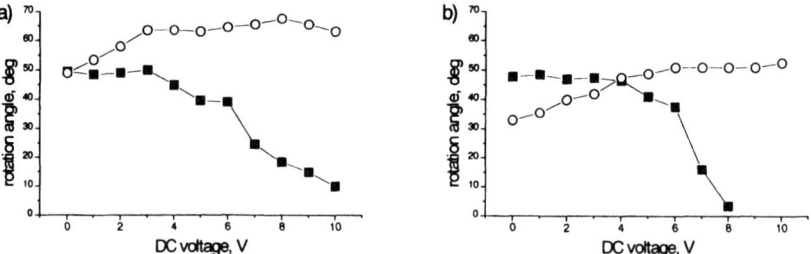

Fig. 2. The "rotation angle" of the FLC director in a OASLM versus the bias voltage in dark (■) and illuminated state (o). a) 10Hz, b) 100Hz.

In Fig. 3 the kinetics of the optical response of an OASLM (with an a-SiC:H film in intrinsic conductivity and a 12μm thick layer of the FLC-471) under pulse light illumination is presented. Response times on the order of 1ms are obvious and thus operation of OASLMs at frequencies in the range of hundreds Hertz is possible. All investigated OASLMs in transmissive mode showed a spatial resolution better than 66 lp/mm measured with a target projection technique[18].

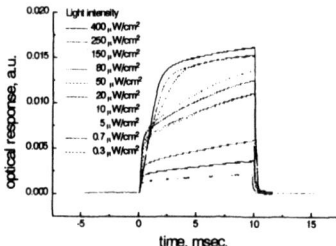

Fig. 3. Kinetics of the response of an OASLM at different intensities of the light pulses, synchronized with the negative polarity of the square wave voltage relative to the photosensor. The peak-to-peak amplitude was 20V, the frequency 50Hz, and the bias field V_b = 0

Measurement of the diffraction efficiency and the quality of image correction

In Fig. 4 the set-up for the measurement of the quality of images and the DE is presented. Lens 2 formed the image of lamp 1 as a light source in infinity relative to the telescope, composed from lenses 3 and 4. The angle scale of the test image was about 0.01 rad. The holographic corrector utilizing the OASLM 5 included the phase diffraction grating 6 for the correction of the chromatism of a two-dimensional hologram, written in the OASLM 5. The correcting hologram was written by means of light pulses of the laser 7 (wavelength 0.53 μm, duration 20-30ns). The signal beam passed through the telescope system and the reference beam was formed using the beam splitter cube 11. The OASLM was placed in the plane, where the eye-piece 4 constructed the image of the objective 3. The spatial frequency of the written hologram and the space frequency of the phase diffraction grating 6 were equal to 95lp/mm. This frequency was sufficient to receive a high diffraction efficiency and to resolve the images of the light source in the neighbour orders of diffraction. The CCD camera 13 recorded the diffracted image, formed by means of the objective 12. The spectral range could be selected by means of color or interference filters (0.53 μm) 14. The noncorrected image formed by the telescope system was recorded by means of the CCD camera 16.

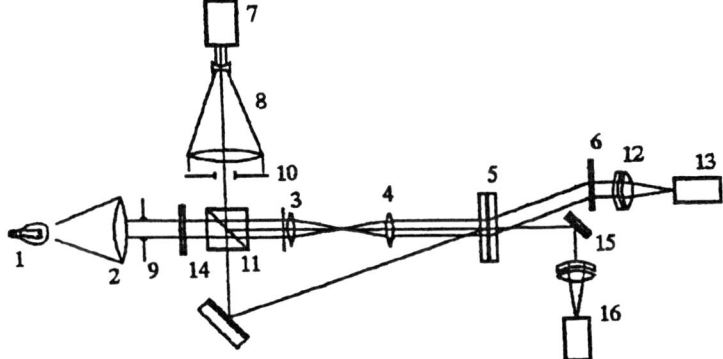

Fig. 4. Experimental set-up for the measurement of the diffraction efficiency and the quality of image correction using OASLMs.

For the measurement of the value and the response time of the DE a HeNe laser (0.63μm) or a LED (0.82μm) were used. The signal was recorded by a photodiode placed instead of the CCD camera. The energy of the writing beam was measured by a photoelectric sensor. The DE was measured under application of a square-wave voltage (peak-to-peak amplitude V_{pp}) and a bias field V_b with a time delay τ_0 between the change of the polarity of the square-wave voltage and the writing laser pulse. The dependence of the DE on the polarization orientation relative to the vector of the written grating was relatively small, the maximum DE corresponds to the direction in coincidence with the grating vector and normal to the smectic layers.

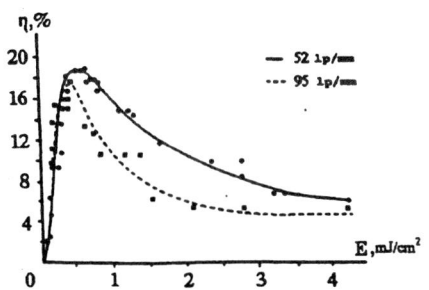

Fig. 5. Dependence of the diffraction efficiency η on the energy density E of the writing laser pulse (duration about 25ns) for two different phase gratings. $V_{pp}=40V$, $V_b=23V$ (52lp/mm), $V_b=21V$ (95lp/mm). Duration of the voltage pulse: 1s, repetition rate: 0.1Hz, $\tau_0=2ms$.

Polymer photosensor: The response time for the OASLM using the polymer photosensor decreased when the delay τ_0 increased to 10-20ms. Further increase of τ_0 didn't change the response time and the amplitude of the DE. The maximum DE for polymer based OASLM was obtained at $V_{pp}=40$-50V. For a certain amplitude V_{pp} a corresponding value of V_b was required to obtain the maximum DE. With polymer based OASLMs repetition rates up to 1Hz were possible with the amplitude of the DE being almost proportional to the laser pulse energy. This operation rate is considerably faster than in polymer based OASLMs using nematic liquid crystals. This is probably connected with a compensating action of the spontaneous polarization on ionic relaxation processes in the polymer photosensor. In Fig. 5 the maximum DE is plotted for gratings with two spatial frequencies 52lp/mm and 95lp/mm. The DE is smaller than the value estimated above because of a non-optimized FLC layer thickness.

a-SiC:H photosensor: Three OASLMs were investigated with the same photoconductive layer (intrinsic conductivity) and with a FLC-471 layer of variable thickness (2μm, 5μm and 12μm, resp.). The dependence of the DE on the energy density E of the writing pulses is similar to the behaviour presented in Fig. 5, but the sensitivity of devices using a-SiC:H films is higher, than of the polymer based device (the maximum of $\eta(E)$ corresponds to 2-5 μJ/cm^2 for a-SiC:H instead of 500-600μJ/cm^2 for polymer based OASLM) The value of DE reaches 20%, that is two times more than that obtained in papers [4,5], where binary OASLMs are investigated with SSFLC materials, having appreciably smaller optically switchable tilt angle on order of 22-30 ^{0}. A significant decrease (2.5 times) of DE at an increase of the spatial frequency from 52 to 100lp/mm was measured, which is connected first of all with a higher dark conductivity of the a-SiC:H layer in comparison to the polymer photosensor.

At pulse laser writing, repetition rates up 6-8Hz were possible for a maximum diffraction efficiency on the level of 20%. At higher frequencies, e.g 100Hz, the DE was not less than 5-6%. In this case, light pulses with a duration of 50ms, an intensity of 200-500μW/cm^2 and the wavelength 0.63μm were used for writing the hologram with a spatial frequency of 85lp/mm. The optimum value of the light intensity was depending on the FLC layer thickness and the amplitude V_{pp} and V_b.

In Fig. 6 the image of the test target is shown which is recorded by the CCD camera 16 without correction of the aberration introduced into the telescope system 3, 4 in the set-up, shown in Fig. 4. The aberration was formed by a totally opaque glass plate placed in front of the objective 3. In Fig. 6 the same test target is recorded by the CCD camera 13 with correction using the laser pulse technique (duration of pulses 20-30ns) and a polymer based DHF OASLM. Green light within the wavelength range 0.5-0,6μm was used for read-out, that was close to the wavelength of the writing light (0.53μm) and correcting hologram 6 is mostly adjusted for green read-out light. The visible contrast of the image was good enough for white light also and only a remarkable decrease of the contrast was observed for red read-out light with the maximum wavelength 100nm far from the wavelength of the writing light.

CONCLUSION

The use of deformed helix ferroelectric liquid crystals with large tilt angles provided high diffraction efficiencies in OASLMs on the order of 20%, which is about 60% of the theoretical limit for phase diffraction gratings with sinusoidal profile. OASLMs based on amorphous silicon carbide allowed to design holographic image correctors with operation rates in the range of hundreds Hz. With photoconductive polymers dynamical holography devices with very high spatial resolution and with real time operation were obtained.

ACKNOWLEDGEMENTS

The work was done in frame of the Volkswagen-Stiftung Projects I/70668 and I/72818. The support (L.A.B.) by the Deutsche Telekom (Contract 4160/75028) is gratefully acknowledged.

a) 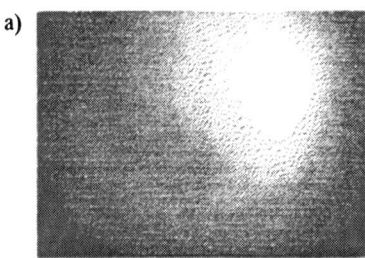 b)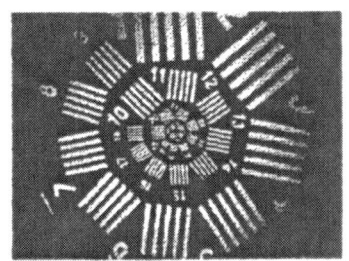

Fig. 6. Image of the test target. a) without correction, b) with correction.

REFERENCES

1. J. Grinberg, A. Jacobson, W. Bleha, L. Miller, L. Fraas, D. Boswell, and G. Myer, Opt. Eng. **14**, 217 (1975)
2. R.B. Meyer, L. Liebert, L. Strzelecki and P. Keller, J. de Phys.-Lett. **36**, L69-L71 (1975).
3. G. Moddel, in <u>Spatial Light Modulator Technology</u>, edited by U. Efron (Dekker, New York, 1994) p. 287.
4. K.M. Johnson, C.C. Mao, G. Moddel, M.A. Handschy, and K. Arnett, Opt. Lett., **15**, 1114 (1990).
5 S. Fukushima, T. Kurokawa, and M. Ohno, Appl. Phys. Lett. **58**, 787 (1991).
6. M.T. Grüneisen, and J.M. Wilkes, "Compensated Imaging by Real-Time Holography with Optically Addressed Spatial Light Modulators", OSA Trends in Optics and Photonics Series (1997), presented at the Spring Topical Meeting on Spatial Light Modulators, Lake Tahoe, NV, March 17-19, 1997, Paper STuB5.
7. M.J. O'Callagan, M.A. Handschy, Optics Letters, **16** (10), 770 (1991); U.S. Patent No. 5 181 665 (26 January 1993).
8. L.A. Beresnev, J. Hossfeld, W. Dultz, A.P. Onokhov, and W. Haase in Electrical, Optical, and Magnetic Properties of Organic Solid State Materials III, edited by A.K.-Y. Jen, C.Y.-C. Lee, L.R. Dalton, M.F. Rubner, G.E. Wnek, and L.Y. Chiang (Mat. Res. Soc. Proc. **413**, Boston, MA, 1996) pp. 363-370.
9. L.A. Beresnev,V.G. Chigrinov, D.I. Dergachev, E.P. Pozhidayev, J. Fünfschilling, and M. Schadt, Liquid Crystals **5**, 1171 (1989).
10. A.V. Zherzdev, V.G. Karpov, A.B. Pevtsov, A.G. Pilatov, and N.A. Feoktistov, Sov. Phys. Semicond., **26** (4), 421 (1992); Fiz. Tekh. Poluprovdn. **26**, 747 (1992).
11. N.L. Ivanova, L.E. Morozova, A.P. Onokhov, A.B. Pevtsov, N.A. Feoktistov, Pis'ma Zh. Tekhn. Fiz. (Russ. J. Tech. Phys.) **22** (4), 7 (1996).
12. V.S. Mylnikov, M.A. Grosnov, N.A. Vasilenko, B.V. Kotov, and L.N. Soms, Sov. J. of Techn. Phys.-Lett. (Pis'ma Zh. Tekhn. Fiz.) **11** (1), 38 (1985).
13. K. Akiyama, A. Takimoto, and H. Ogawa, Applied Optics **32**, 6493 (1993).
14. V. Mylnikov, Mol. Cryst. Liq. Cryst. **152**, 597 (1987).
15. E.A. Konshina, A.P. Onokhov, N.A. Feoktistov, L.A. Beresnev, and W. Haase, to be published.
16. V.A. Baikalov, L.A. Beresnev, and L.M. Blinov, Mol. Cryst. Liq. Cryst. **127**, 397 (1985).
17. V.A. Berenberg et al, to be published.
18. L. Beresnev, A. Onokhov, W. Dultz, and W. Haase, Mol. Cryst. Liq. Cryst. **304**, 285 (1997).

STRUCTURAL PROPERTIES OF HEXAPHENYL POWDER UNDER HIGH PRESSURE

S. Guha*, W. Graupner**, R. Resel**, M. Chandrasekhar*, H.R. Chandrasekhar*, and G. Leising**
*Department of Physics and Astronomy, University of Missouri, Columbia, MO 65211, U.S.A.
**Institut für Festköperphysik, Technische Universität Graz, A-8010 Graz, Austria.

ABSTRACT

We present X-ray and Raman studies of the conjugated oligomer, hexaphenyl, which forms polycrystalline layers. The Raman studies have been carried out under hydrostatic pressures up to 70 kbar. The relative intensity of the two modes at 1220 cm^{-1} and 1280 cm^{-1} shows the influence of planarization under hydrostatic pressure. These results are interpreted within the framework of theoretical calculations and compared with X-ray studies.

INTRODUCTION

a)The crystal structure of oligomers of poly (*para*-paraphenylene)

There have been extensive discussions on the crystal structure of oligomers of poly (*para*-phenylene) (PPP), starting with the structure determination of biphenyl in the late twenties. Currently the crystal structures of all oligomers from biphenyl to septiphenyl have been studied in detail by single crystal studies. Within this series of oligomers the packing of the molecules in the crystal structure is similar. Owing to the sp^2 hybridization of the carbon atoms the molecules are rigid rod-like. In the crystalline state the molecules are arranged in layers. The long axes of the molecules point in the same direction. The thickness of a layer is approximately the length of the molecule, so that the direction of the long axis of the molecule is approximately perpendicular to the layer. The phenyl rings of the oligomers are approximately planar while the planes of two neighboring molecules are tilted relative to each other. Within the classification of polynuclear aromatic hydrocarbons, this type of structure is of the herringbone type and this structural arrangement shows outstanding optical properties[1,2].

The crystal structures are monoclinic with space group P2$_1$/a. The lattice constants are determined to be a≈ 8.1Å, and b ≈5.6Å, with the lattice constant c depending on the length of the oligomer ranging from 9.5Å for biphenyl to 30.6Å for septiphenyl[3,4]. The direction of the chains is chosen as the c-axis.

b)The crystal structure of hexaphenyl

Early attempts of structure determination of hexaphenyl (C$_{36}$H$_{26}$) were done by X-ray powder patterns and by Weissenberg photographs[5,6]. Their results are in reasonable agreement with single crystal investigations [4]: monoclinic unit cell, space group P2$_1$/a with a=8.091Å, b=5.568Å, c=26.24Å, and the monoclinic angle β =98.17°. The refinement factor of the structure solution was 0.062; since the molecule is centrosymmetric, the positions of only 18 C-atoms and 13 H-atoms have to be refined. The distances between the C-atoms vary between 1.322 - 1.507Å , a value of 1.41 Å is expected for aromatic bonds of carbon. The distances between a C- and the adjacent H-atom are determined as 0.95Å in comparison to the expected distance of 1.08Å; it is known that single crystal refinements slightly underestimate the C-H distances[7].

A sketch of the crystal structure of hexaphenyl is shown in Fig. 1. The positions of the atoms are taken from ref. [4]. The large circles represent carbon atoms and the small circles represent hydrogen. The long axis of the unit cell (c-axis) connects two layers of hexaphenyl molecules. Each layer is represented by two molecules: one in projection, the other stretched (Fig.1 (a)). The plane of the two short crystallographic axes of the unit cell (ab-plane) is parallel to the hexaphenyl layers.

Mat. Res. Soc. Symp. Proc. Vol. 488 © 1998 Materials Research Society

The monoclinic angle β is determined by the relative position of two hexaphenyl molecules of neighboring layers which are in the same orientation, that is, the angle of the ab-plane with the c-axis. The centrosymmetry of the space group $P2_1/a$ is expressed by the symmetry of the hexaphenyl molecule passing the origin of the unit cell with the center of the molecule (top-right molecule in Fig.1 (a)). The symmetry element $2_1/a$ of the space group generates the second hexaphenyl molecule of the layer (bottom-right molecule). The arrangement of the molecules within one layer is shown in Fig. 1(b). The molecules are projected along their long axes; within one molecule the larger inner circles represent carbon atoms, the smaller outer circles are hydrogen atoms. The molecules are arranged in a herringbone pattern within one layer.

Figures 1 (a) and (b) show that within the crystal structure solution the molecules are flat. However, molecular simulations on free hexaphenyl molecules reveal a force that tilts the phenyl rings relative to each other. It is caused by the repulsion of the ortho-hydrogens. On the other hand the π-electron system tends to planarize the molecule. An average tilt angle of 45° is determined for free biphenyl molecules[8]. It appears that the close crystallographic packing of the hexaphenyl molecules tends to planarize the molecule.

The phenyl rings of the hexaphenyl molecule have considerable rotational freedom, which indicates torsional librations of the rings. These librations cause large Debye-Waller factors of the refined carbon atoms which are not situated along the long axis of the molecule[4]. No influence of the repulsion of the ortho-hydrogen atoms to the librations is expected: simulations of crystalline packed hexaphenyl show a movement of the phenyl rings in a symmetrical potential where the amplitude of the librations is approximately ±20°[9]. It should be noted that hexaphenyl shows an unrefined low temperature phase at 110K. A metastable high temperature polymorph was observed in thin films of hexaphenyl [4,10].

In case of quaterphenyl, a low temperature phase is observed below 200K. Within this phase the librations are frozen and the two internal phenyl rings are tilted relative to each other by 23°[11]. Single crystal investigations performed at room temperature reveal an average flat molecule for quaterphenyl. However, a double well potential with the minimum at 11° on either side of the mean position of the phenyl rings is observed[12].

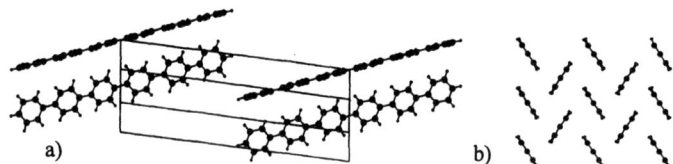

Fig. 1 Sketch of the crystal structure of hexaphenyl. a) The unit cell and the position of the hexaphenyl molecules relative to the unit cell - the unit cell connects two layers of hexaphenyl; b) the herringbone arrangement of the hexaphenyl molecules within one hexaphenyl layer

High pressure vibrational spectroscopy is a useful tool for examining molecular interactions in crystalline and amorphous solids. Since the vibrational frequency of a harmonic solid is independent of compression, pressure-induced changes in the Raman spectrum provide insight into the anharmonicity of the solid-state potential[13]. Furthermore, the vibrational spectra of conjugated polymers under high pressure can provide unique structural information such as the effect of planarization, inter- and intra-chain interactions. The Raman spectrum of the oligomers of poly (*para*-phenylene) is mainly characterized by four intense modes (A_g symmetry). It has been observed that the Raman intensity ratio of the inter-ring C-C stretch mode (1280 cm⁻¹) to the C-H in plane bending mode (1220 cm⁻¹) is a good indication for the number of π conjugated phenyl rings in the polymer chain[14]. Upon doping hexaphenyl, along with a blue shift of the Raman modes there is also a decrease in intensity of the 1280 cm⁻¹ mode relative to the 1220 cm⁻¹ mode[15]. We have studied the intensity ratio of these two modes as a function of pressure and temperature to obtain insight into the structural changes of the material.

EXPERIMENTAL SETUP

The hexaphenyl material was delivered by Tokio Chemical Industries Ltd. The X-ray diffraction pattern was measured with a SIEMENS D501 powder diffractometer. Ni-filtered CuKα radiation was used and the hexaphenyl powder was investigated on a monocrystalline silicon substrate. Raman measurements were carried out in a back-scattering configuration, using the 514.5 nm line of an Ar⁺ laser. The scattered light was detected with a SPEX triple monochromator equipped with a CCD array detector and a holographic supernotch filter. Pressure studies were conducted in a Merrill Bassett type diamond anvil cell with cryogenically loaded argon as the pressure medium and ruby fluorescence as the manometer.

RESULTS

a) X-ray

An X-ray powder pattern of the material is shown on the top panel of Fig. 2. A part of the powder pattern is enlarged in the top panel of Fig. 3. The lower panel of Fig. 2 shows the calculated powder pattern giving information about position and intensity of the X-ray reflections. The calculation is based on single crystal data [4]. The agreement of the experiment with the calculation and other experimental powder patterns is quite good[16], although 001 reflections are overemphasised. This may be explained by a preferred orientation in the powder, since the crystallites are disc shaped and the plane of the discs is (001).

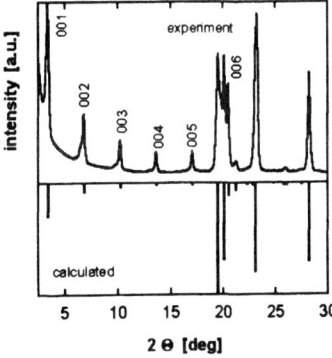 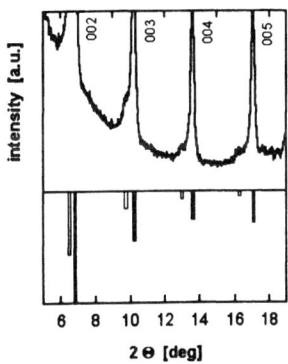

Fig. 2 Experimental X-ray powder pattern of hexaphenyl (top) and calculated pattern (bottom). The full bars give the position as well as the relative intensity of the X-ray reflections.

Fig. 3 Phase analysis of the experimental powder pattern in the range of 5...19° (top) reveals traces of a high temperature polymorph of hexaphenyl which is designated by open bars (bottom).

Small traces of another crystalline phase are found in the powder spectrum. We could identify this impurity as the metastable high temperature polymorph of hexaphenyl. The position as well as intensity of the X-ray reflections from the refined phase as well as from the high temperature polymorph is given by filled and open bars at the bottom of Fig. 3. The packing of the hexaphenyl molecules in the high temperature phase is similar to the structure of the refined phase: the molecules are packed in a herringbone pattern within layers and the layers form a crystalline structure. The inclination of the molecules within layers is smaller, so the thickness of the layer is higher [12].

b) Vibrational Spectra

Figure 4 shows the Raman spectrum of hexaphenyl at 1 bar and two different temperatures. The high energy mode at 1600 cm^{-1} is the ring C-C stretch mode. It is observed that the integrated intensity of the inter-ring C-C stretch mode at 1280 cm^{-1} increases with respect to the C-H in plane bending mode at 1220 cm^{-1} mode with decreasing temperatures. The two modes have almost the same intensity around 150K. Upon further lowering the temperature, the intensity of the 1280 cm^{-1} mode increases as compared to the 1220 cm^{-1} mode. At 300K, $I_{1280}/I_{1220} \approx 0.84$ and at 12K the ratio of the intensity is 1.11. This is indicative of a low temperature phase where the librations are frozen and neighboring phenyl rings are tilted at a fixed angle with respect to each other. This is consistent with X-ray analysis where an unrefined low temperature phase is observed at 110K. Since the 1280 cm^{-1} mode is the inter-ring stretch vibration, the relative orientation of the phenyl rings has a stronger influence on its intensity.

We measured the Raman spectra of hexaphenyl under hydrostatic pressure over a range of 70 kbar. Figure 5 shows the Raman spectra of the 1220 cm^{-1} and 1280 cm^{-1} modes at two different pressures. The frequencies of all the modes increase linearly with pressure and can be fit to $\omega(P)=\omega(0)+(d\omega/dP)$, where P is in kbar. From the pressure derivatives, we can determine the mode Grüneisen parameter γ_i, which is proportional to $(1/\omega_i)(d\omega_i/dP)$. A Lorentzian curve fitting routine was used to determine the frequencies of the Raman modes. The mode frequencies as a function of pressure are shown in Fig. 6 (a) and (b). These measurements were all performed at 300K. The Raman modes have similar pressure coefficients, though the 1600 cm^{-1} has a slightly different Grüneisen parameter compared to the modes at 1220 cm^{-1} and 1280 cm^{-1}. This is quite common in molecular solids since the weak intermolecular bonds are affected more strongly by high pressure compared to the intramolecular bonds, which may lead to vibrational modes with different Grüneisen parameters. On increasing the pressure beyond 60 kbar, the 1280 cm^{-1} mode interferes with the Raman mode from diamond at 1333 cm^{-1}.

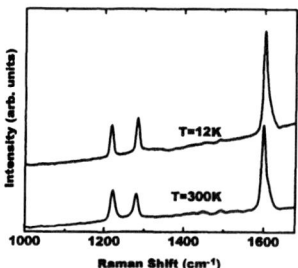

Fig. 4 Raman spectra of hexaphenyl at 1 bar. The top panel is at 12K and the bottom panel is at 300K.

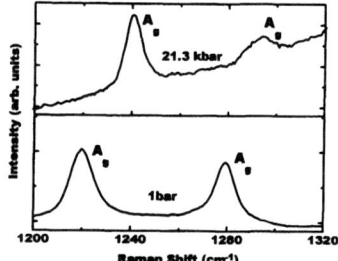

Fig. 5 Raman spectra of hexaphenyl at 21.3kbar (top panel) and at 1 bar (bottom panel). The temperature is 300K for both.

The most striking feature of the Raman spectrum of hexaphenyl under high pressure is the ratio of the intensity of the 1280 cm^{-1} peak to the 1220 cm^{-1} peak, as shown in Fig. 5. Figure 7 shows that at high pressures, the average value of $I_{1280}/I_{1220} \approx 0.3$ and at 1 bar the ratio of the intensities is 0.84. This decrease in intensity of the inter-ring stretch mode indicates that under high pressure the oligomer is planarized. The onset apparently occurs between atmospheric pressure and 20 kbar. A similar decrease of the inter-ring stretch mode under high pressure is observed in (p-phenylene vinylene)[17]. We have also looked at the Raman spectrum of the oligomer under high pressure and low temperatures. At 25 kbar $I_{1280}/I_{1220} \approx 0.3$ and 0.19 at 300K and 12K, respectively. At higher pressures, lowering the temperature further decreases the 1280 cm^{-1} peak, which may indicate stiffening of the bonds. This implies that pressure and temperature influence the oligomer very differently. At low temperatures along with stiffening of the bonds, the torsional librations are frozen and the neighboring phenyl rings are tilted relative to each other by a fixed angle. From *ab initio* calculations of the vibrational spectra of an infinite PPP chain[18], the predicted intensities of the 1220 cm^{-1} and the 1280 cm^{-1} mode agree quite well with experimental results. A planar structure results in the intensity of the 1220 cm^{-1} mode being substantially higher than the inter-ring C-C stretch mode. A helical structure of PPP on the other hand causes the intensity of the 1280 cm^{-1} to be stronger than the 1220 cm^{-1} mode. Our results on the hexaphenyl under high pressure qualitatively agree very well with the above picture. Planarization of the hexaphenyl molecule under high pressure is reflected by the decreased intensity of the inter-ring C-C stretch mode.

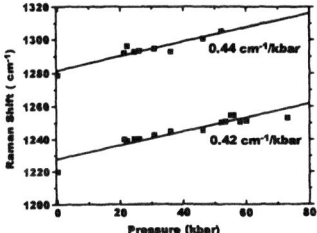

Fig. 6 (a) Frequency of the inter-ring C-C stretch mode and the C-H bending mode as a function of pressure.

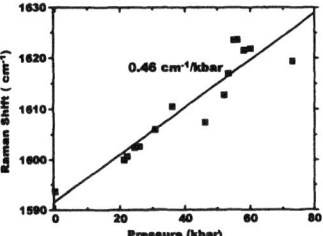

Fig. 6 (b) Frequency of the ring C-C stretch mode as a function of pressure.

CONCLUSIONS

X-ray results on the oligomers of poly (*para*-phenylene) show that within one layer, the molecules are arranged in a herringbone pattern. On increasing the oligomer length the lattice constants decrease, which results in a closer crystallographic packing. In the hexaphenyl molecules this close packing tends to planarize the molecule. Planarization of the molecule is also observed in the Raman spectra of hexaphenyl under high pressure. The intensity of the inter-ring stretch mode at 1280 cm^{-1} decreases compared to the bending mode at 1220 cm^{-1} under high pressures, showing the influence of planarization. The onset occurs between 0 and 20 kbar.

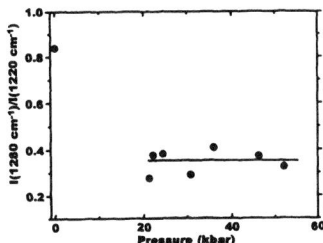

Fig. 7 Ratio of the intensity of the 1280 cm^{-1} mode to the 1220 cm^{-1} mode as a function of pressure.

ACKNOWLEDGMENTS

This project was supported in part by FWF (project 9747-CHE), ÖNB (project 6608) and *Österreichische Industriellenvereinigung.* Work at the University of Missouri was supported by the U.S. Army through Grant DAAL03-92-G0367 (EPSCoR) and DAAL-03-92-0381. We thank Chris Martin for his invaluable help in the laboratory.

REFERENCES

1. G. R. Desiraju, A. Gavezotti, Acta Cryst. B**45**, 473 (1989).

2. B. Stevens, Spectrochim Acta. **18**, 439 (1962).

3. K. D. Aichholzer, Ph. D. thesis, Technical University Graz, Austria (1993).

4. K. N. Baker, A. V. Fratini, T. Resch, H. C. Knachel, W. W. Adams, E. P. Socci, B. L. Farmer, Polymer **34**, 1571 (1993).

5. C. Toussaint, G. Vos, J. Chem. Soc. (B) 813 (1966).

6. C. Toussaint, Acta Cryst. **21**, 1002 (1966).

7. Kitaigorodsky, <u>Molecular crystals and molecules</u>, Academic Press, New York, (1973)

8. O. Bastiansen, Acta Chem. Scand. **3**, 408 (1949).

9. E. P. Socci, B. L. Farmer, W. W. Adams, J. Polymer Sci. B **31**, 1975 (1993).

10. R. Resel, N. Koch, G. Leising, F. Hofer, L. Athouël, Acta Crystallographica B, submitted

11. J. L. Baudour, Y. Delugeard, P. Rivet, Acta Cryst. B **34**, 625 (1978).

12. Y. Delugeard, J. Desuche, J. L. Baudour, Acta Cryst. B **32** 702 (1976).

13. J.R. Ferraro, <u>Vibrational Spectroscopy at High Pressures</u>, pp 2-3. Academic Press, New-York (1984)

14. G. Leising, T. Verdon, G. Louran, and S. Lefrant, Synth. Met. **41**, 279 (1991).

15. G. Louran, L. Athouël, G. Foyer, J.P. Buisson, and S. Lefrant, Synth. Met. **55-57**, 4762 (1993).

16. L. Athouël, G. Froyer, M. T. Riou, M. Scott, Thin Solid Films **274**, 35 (1996).

17. S. Webster and D.N. Batchelder, Polymer **37**, 4961 (1996).

18. L. Cuff and M. Kertesz, Macromolecules **27** 762 (1994).

ELECTRONIC PROPERTIES OF POLY(*PARA*-PHENYLENES) UNDER HIGH PRESSURE

W. GRAUPNER*, S. GUHA**, S. YANG**, M. CHANDRASEKHAR**,
H. R. CHANDRASEKHAR**, G. LEISING*, U. SCHERF***, K. MÜLLEN***

*Institut für Festkörperphysik, Technische Universität Graz, A–8010 Graz, Austria.
** Department of Physics & Astronomy, University of Missouri, MO 65211 Columbia , USA
*** Max–Planck–Institut für Polymerforschung, D–55128 Mainz, Germany.

ABSTRACT

We have studied the photoluminescence emission spectra of polycrystalline and amorphous powder samples of phenylene–type polymers and oligomers. In order to modulate both intrachain and interchain interactions these investigations were performed at hydrostatic pressures up to 70 kbar and in a temperature range of 20 to 300 K. We report characteristic changes of the emission spectrum which depend on both the molecular environment as well as the intramolecular bond configuration.

INTRODUCTION

Although the photophysical properties of conjugated polymers have been investigated since the discovery that polyacetylene can be doped to reach high conductivity [1] such fundamental properties as the nature of the lowest excited states and the decay to the ground state are still not fully understood [2],[3]. Two of the obstacles are (1) the availability of pure materials with a low content of chemical defects and the (2) incomplete understanding of intermolecular interactions. In this study we want to address the latter question with highly defined raw materials. As pointed out in [4] interchain coupling can dramatically alter the electronic properties of isolated chains.

We have chosen materials of (1) technological importance, (2) high chemical purity, and very distinct (3) *inter*chain and (4) *intra*chain interaction, namely one oligo and one polyphenyl (see Fig. 1). Both materials were already used as active layers in polymer light emitting devices (LEDs) [5],[6] and show a photoluminescence quantum yield (PLQY) of about 30 percent in solid films. The oligomer, *para*-hexaphenyl (PHP), is known to form monoclinic crystallites of spacegroup P2₁/a [7], and is characterized by a torsional degree of freedom between the neighbouring phenyl rings [8] (see Fig. 1 (a)). The ladder-type poly(*para*-phenylene (LPPP), called m-LPPP due to the methyl-group in the Y-position in Fig. 1 (b), does not form crystallites due to the bulky sidegroups and shows no torsional degree of freedom between the neighbouring phenyl rings due to the methin-bridge between the rings. Both materials are known to show stimulated emission [9],[10]. For the m-LPPP highly directional, polarized emission with a linewidth of about 2 nm is observed from optically excited waveguides [11]. In addition, for the m-LPPP picosecond PL–spectroscopy revealed the existence of two clearly different emission contributions – a fast *intra*chain and a slower *inter*chain component [12]. We have also shown that (1) the intensity of the fast *intra*chain component shows a strong temperature dependence in contrast to the *inter*chain component [13] and (2) that the *inter*chain component is the main contributor to PL detected magnetic resonance spectra [14].

With this background in mind, we invstigated the PL of these two materials under different hydrostatic pressures up to 70 kbar and at temperatures between 20 and 300 K.

EXPERIMENTAL

The raw materials were provided in powder form by Tokio Chemical Industries Ltd. and the MPI for Polymerforschung. Small powder grains were used for the investigations of the materials. In order to investigate the different samples under high hydrostatic pressures diamond anvil cells [15] with preindented stainless steel gaskets [16] were used for pressure studies of the materials. The diamond surfaces and gasket walls form a micro sample chamber which contains the sample, a ruby particle for pressure calibration and Ar as a pressure medium. The pressure range used in this investigation was 0 to 70 kbar. The pressure cells were mounted in an optically accessible closed cylce He cryostat. Excitation was delivered by the UV lines of an Ar+ ion laser. The emission spectra were recorded with a SPEX double monochromator and a photomultiplier tube attached to a photon counting system.

Figure 1: Chemical structure of (a) PHP and (b) m–LPPP; Y=CH$_3$, R= C$_{10}$H$_{21}$, R'= C$_6$H$_{13}$.

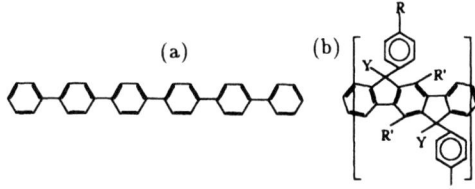

RESULTS

m–LPPP Powder

In Fig. 2 we show the PL spectra of m-LPPP powder at a few selected values of hydrostatic pressure. The emission spectrum of m LPPP is governed by the vibronic progression of the order of 0.2 eV, which is indicative for the coupling of the =C-C=C-C= stretch vibration of the conjugated backbone. Consequently the emissive transition, highest in energy, is called the 0–0 transition to express that it takes place between the zeroth vibronic level in the excited state to the zeroth vibronic level in the ground state. The 0–1 transition is therefore involving the creation of one phonon since the lower state is the first vibronic level in the ground state. As can be seen in Fig. 2 the spectra shift towards lower energy with increasing pressure and also change their overall appearance similar to that reported for poly(*para*phenylenevinylene) (PPV) recently [17].

Figure 2: Normalized PL emission spectra at roomtemperature of m-LPPP at different hydrostatic pressures: the top spectrum is taken from a very thin film at 1 atmosphere, then from top to bottom: m-LPPP powder at 13, 20, 35, 45, 55, 69 kbar.

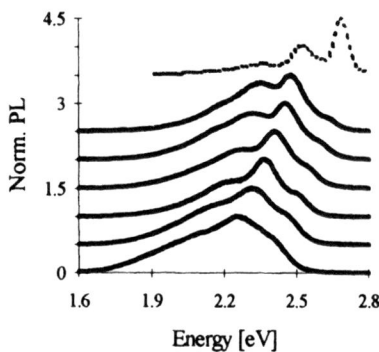

In order to quantify this shift we have displayed parameters, qualifying the spectra, in Figure 3. These parameters are the 0–1 peak position as well as the high and low energy edge (HE and LE respectively) of the spectra taken at 50 or 10 percent of the peak value. The coefficients of the linear regression curves (see Fig. 3), describing the PL parameters of the curves in Figure 2 are summarized in Table 1.

Since it was shown in previous work [18],[19],[20] that the effect of increased pressure on the emission and absorption spectra of conjugated polymers is very similar to what is observed for decreasing the sample temperature, we have also probed the sample's PL spectra at different temperatures (see Figure 4).

Table 1: Coefficients of the linear regression curves, describing the parameters in Figure 3: $E(P) = E(0) + \alpha_P \times$ pressure.

Parameter	E(0) [eV]	α_P [μeV/bar]
HE (10%)	2.727	- 2.95
HE (50%)	2.581	- 2.23
0-1 peak	2.535	- 3.96
LE (50%)	2.208	- 2.74
LE (10%)	1.984	- 2.67

When interpreting the change of the PL emission spectra with increasing pressure one has to take into account that the samples were of high optical density. Therefore the high energy edge of the photoluminescence emission is stronlgy affected by self absorption – an effect which is particularly strong in m-LPPP due to its small Stokes shift [5]. This effect is responsible for the fact that the 0–1 transition is observed as the dominant one in our experiments (see Fig. 2), while sufficiently thin films reveal the 0–0 transition as the dominant one (Fig. 2, top).

In a pressure range up to 2 kbar the absorbance edge of m-LPPP was shown to shift by -7.2 and -7.7 μeV/bar for sample temperatures of 200 and 300 K respectively [18]. In the same experiments the 0-0 and 0-1 transition of the PL undergo a completely identical shift with respect to pressure corresponding to -8.0 μeV/bar.

Figure 3: Parameters of the roomtemperature PL–spectrum of m-LPPP vs. pressure; from top to bottom HE (10%), HE (50%), 0-1 peak, LE (50%), LE (10%).

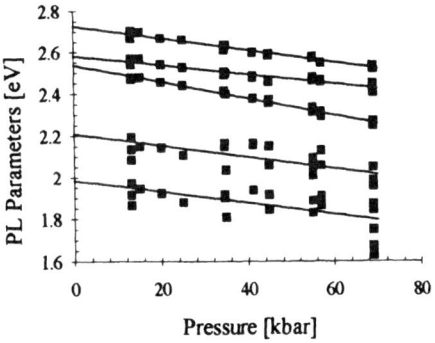

Figure 4: Normalized PL of m-LPPP at 69 kbar at 300, 200, 100 and 16 K (from bottom to top) respectively; curves are vertically displaced for clarity.

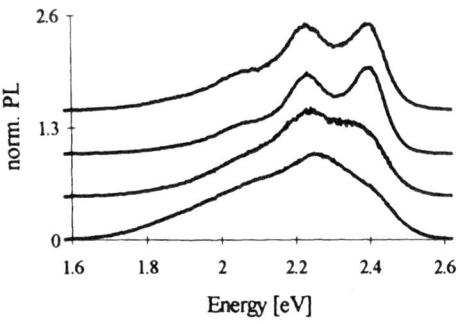

These numerical values have to be compared to the shifts we have determined under pressures up to 69 kbar, i.e. in a pressure range, almost extended by two orders of magnitude. Obviously the strong initial shift of electronic transitions at low pressures is replaced by a weaker dependence at elevated pressure regimes as discussed in [17],[21]. Typical values for the shift with pressure are -5.4 μeV/bar for PPV [17], -10 μeV/bar for poly-3-hexyl-thiophene (P3HT) [22], and -13 μeV/bar [23] for polyacetylene.

In accordance with the results of P3HT we find no significant broadening of either the PL or the absorbance spectrum in the low pressure regime up to 2 kbar [18] and also no broadening in the higher pressure regime up to 69 kbar (see Fig. 3), where we only measured the PL spectrum.

This indicates that interchain coupling between two *conjugated backbones* does not play a strong role in this pressure region. Therefore we propose that the explanation given in ref. [22] for P3HT also applies to low pressure experiments: dipole–dipole interactions with the surroundings of the conjugated backbone lower the excited state energy relative to the ground state ("solvent" or "gas–to–crystal" effect [24],[25],[26],[27]).

PHP Powder

As for m-LPPP we measured the PL of PHP under different hydrostatic pressures up to 73 kbar. In order to quantify the observed shift we have displayed the respective parameters in Table 2. The 0–1 peak shifts with -3.6 μeV/bar which is significantly weaker then the effect seen in m-LPPP. However the values of the parameters displayed in Table 2 indicate that the PHP PL emission spectrum is becoming broader at high pressures – low pressure coefficients at the high energy side of the spectrum are accompanied by high pressure coefficients at the low energy side. Therefore it seems that PHP shows the broadening expected for an increased three dimensional interaction between conjugated molecules [23]. This fact is not surprising considering that PHP forms a polycrystalline material in contrast to m–LPPP where bulky sidegroups prevent a strong interaction between the polymer backbones.

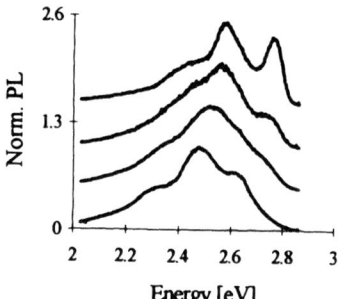

Figure 5: Normalized PL of PHP at 73 kbar at 300, 200, 100 and 21 K (from bottom to top) respectively; curves are vertically displaced for clarity.

Table 2: Coefficients of the linear regression curves, describing the parameters of the PHP emission spectra: $E(P) = E(0) + \alpha_P \times$ pressure.

Parameter	E(0) [eV]	α_P [μeV/bar]
HE (10%)	2.928	-2.09
HE (50%)	2.811	-1.86
0–1 peak	2.729	-3.60
LE (50%)	2.564	-4.41
LE (10%)	2.361	-5.82

As for the m–LPPP powder we have also probed the sample's PL spectra at different temperatures (see Figure 5). The shift of the PL emission center of gravity at 73 kbar occurs with -394 μeV/K as compared to -185 μeV/K in the m-LPPP. As can be seen, comparing Tables 1 and 2 the significantly larger effect in the polycrystalline material with the torsional degree of freedom between the phenyl rings stems mainly from a stronger shift of the whole spectrum and not from a stronger spectral redistribution. From a precise analysis of the PL emission spectra we can also see that the center of gravity shifts stronger than the PL emission edges, which show shifts of around $-300~\mu eV/K$. This effect is very similar to the observed changes under pressure. Also under pressure we observe a stonger shift for the PL center than for the edges.

CONCLUSIONS

In conlusion we have presented the photoluminescence emission spectra of polycrystalline and amorphous powder samples of phenylene–type polymers and oligomers at hydrostatic pressures up to 70 kbar and in a temperature range of 20 to 300 K. In the polycrystalline powder the 0–1 peak shifts with $-3.6~\mu eV/bar$ which is weaker then the effect in the amorphous m-LPPP ($-3.96~\mu eV/bar$). In addition we observe no significant broadening of the PL emission spectrum of m–LPPP while PHP shows this broadening. Upon varying the temperature at elevated pressures the shift of the PL emission center of gravity at 73 kbar occurs with $-394~\mu eV/K$ for PHP as compared to $-185~\mu eV/K$ in the m-LPPP. The significantly larger effect in the polycrystalline material stems mainly from a stronger shift of the whole spectrum and not from a stronger spectral redistribution.

ACKNOWLEDGMENTS

The work in Columbia is supported by U.S. Army under grants DAAL03-92-G0367, DAAL03-92-0381. This work was partly supported by *Österreichische Industriellenvereinigung* and *Österreichische Nationalbank* P 6608.

References

[1] C. K. Chiang, C. R. Fincher, Y. W. Park, A. J. Heeger, H. Shirakawa, E. J. Louis, S. C. Gau, A. G. MacDiarmid, *Phys. Rev. Lett.* **39**, 1098 (1977).

[2] *Primary Photoexcitations in Conjugated Polymers: Molecular Exciton versus Semiconductor Band Model*, ed. by N. S. Sariciftci, World Scientific, Singapore 1996.

[3] J. Cornil, A. J. Heeger, J. L. Bredas, *Chem. Phys. Lett.* **272**, 463 (1997).

[4] refs. [4]-[10] in [3]; E. J. W. List, W. Graupner, M. Wohlgenannt, G. Leising, J. Partee, J. Shinar, P. Schlichting,, U. Rohr, Y. Geerts, U. Scherf, K. Müllen, *Optical Materials in print*; J. Partee, J. Shinar, W. Graupner, G. Leising, S. W. Jessen, A. J. Epstein, Y. Ding, T. J. Barton, *SPIE Proceedings* **3145** *in print.*;

[5] S. Tasch, A. Niko, G. Leising, U. Scherf, *Appl. Phys. Lett.* **68**, 1090 (1996).

[6] S. Tasch, C. Brandstätter, F. Meghdadi, G. Leising, L. Athouel, G. Froyer, Adv. Mat. **9**, 33 (1997).

[7] R. Resel, F. Meghdadi, N. Koch, G. Leising, *Thin Solid Films in print.*

[8] E. P. Socci, B. L. Farmer, W. W. Adams, *J. Polym. Sci. Polym. Phys. Ed.* **31**, 1975 (1993).

[9] W. Graupner, G. Leising, G. Lanzani, M. Nisoli, S. de Silvestri, U. Scherf, *Phys. Rev. Lett.* **76**, 847 (1996).

[10] A. Piaggi, G. Lanzani, G. Bongiovanni, A. Mura, W. Graupner, F. Meghdadi, G. Leising, *Phys. Rev. B* **56**, 10133 (1997).

[11] C. Zenz, W. Graupner, S. Tasch, G. Leising, K. Müllen, U. Scherf, *Appl. Phys. Lett.* **71**, 2566 (1997).

[12] G. Kranzelbinder, H. J. Byrne, S. Hallstein, S. Roth, G. Leising, U. Scherf, *Phys. Rev. B* **56**, 1632 (1997).

[13] W. Graupner, S. Eder, S. Tasch, G. Leising, G. Lanzani, M. Nisoli, S. De Silvestri, U. Scherf, *Journal of Fluorescence* **7**, 195s (1997).

[14] W. Graupner, J. Partee, J. Shinar, G. Leising, U. Scherf, *Phys. Rev. Lett.* **77**, 2033 (1996).

[15] A. Jayaraman, *Rev. Mod. Phys.* **55**, 65 (1983); A. Jayaraman, *Rev. Sci. Instrum.* **57**, 1013 (1986).

[16] D. J. Dunstan, *Rev. Sci. Instrum.* **60**, 3789 (1989).

[17] S. Webster, D. N. Batchelder, *Polymer* **37**, 4961 (1996).

[18] W. Graupner, S. Eder, K. Petritsch, G. Leising, U. Scherf, *Synthetic Metals* **84**, 507 (1997).

[19] K. Yoshino, K. Nakao, M. Onoda, R. Sugimoto, *J. Phys. Cond. Matt.* **1**, 1009 (1989).

[20] L. Rossi, W. Graupner, R. Resel, F. Meghdadi, G. Leising, F. Sannicolo, R. Tubino, *Materials Research Society Meeting, Boston, Dec 1-5, 1997*

[21] M. Hanfland, A. Brillante, K. Syassen, M. Stamm, J. Fink, *J. Chem. Phys.* **90**, 1930 (1989)

[22] B. C. Hess, G. S. Kanner, Z. Vardeny, *Phys. Rev. B* **47**, 1407 (1993).

[23] D. Moses, A. Feldblum, E. Ehrenfreund, A. J. Heeger, T. C. Chung, A. G. MacDiarmid, *Phys. Rev. B* **26**, 3361 (1982).

[24] R. J. Lacey, D. N. Batchelder, G. D. Pitt, *J. Phys. C: Solid State Phys.* **17**, 4529 (1984).

[25] S. A. Rice, J. Jortner in *Physics of Solids at High Pressure* ed. by C. . Tomizuka, R. M. Emrick, Academic Press, London 1965.

[26] S. Webster, D. N. Batchelder, *Macromol. Symp.* **87**, 177 (1994).

[27] K. Lochner, H. Bässler, H. Sowa, H. Ahgbahs, *Chem. Phys.* **52**, 179 (1980).

For further information, contact Wilhelm Graupner, Institut für Festkörperphysik, Technische Universität Graz, Petersgasse 16, A-8010 Graz, Austria; Tel.: 0043 316 873 8476, FAX: 0043 316 873 8478; e-mail: f513wigr@mbox.tu-graz.ac.at, www.cis.tu-graz.ac.at/if/graupner.htm.

DOPING OF POLYANILINE BY CORONA DISCHARGE

ALDO E. JOB[+], JOSÉ A. GIACOMETTI[+] AND LUIZ H. C. MATTOSO[*]

[+]Instituto de Física de São Carlos, Universidade de São Paulo, C.P. 369, São Carlos, 13560-970, SP, Brazil – giacometti@ifqsc.sc.usp.br
[*]Centro Nacional de Pesquisa e Desenvolvimento de Instrumentação Agropecuária, CNPDIA/EMBRAPA, C.P. 741, São Carlos, 13560-970, SP, Brazil

ABSTRACT

It is well known that conductive polyaniline (PANI) films are usually doped by immersing dedoped PANI films in HCl solution. This paper shows that a corona discharge can be successfully employed to dope thin films of polyaniline coated on poly (ethylene terephthalate) films. Similarly to the conventional doping with aqueous HCl the process is accompanied by a color change from blue to green and the conductivity can be tuned in the range from 10^{-10} up to 0.3 Scm^{-1}. Such new doping method presents several advantages over the conventional one namely, dry process, use of no chemicals, rapidity and no dopant migration. Measurements also showed that the conductivity persists for a long time as observed for films prepared in chemical solution doping. It is believed that this novel technique could be employed in a continuous doping process aiming to produce films with large area for anti electrostatic packing applications.

INTRODUCTION

Doping is one of the most important process in conducting polymers, since it makes the transition from the insulating to the conducting regime in these materials. It has been an enormous amount of research [1-6] done to more efficiently and properly dope these class of materials in order to optimize electrical conductivity by a feasible and practical method. Several doping approaches have been proposed which are more or less attractive depending on the final application of the product. The use of the aqueous acidic solution conventionally used to chemically or electrochemically [1] dope polyanilines presents some problems for using these polymer in several technological applications [2], particularly in the micro electronic industry [3,4]. To solve this drawback attempts were made of doping polyaniline - PANI - with high energy radiation such as e-beam [4], X-rays [5] and γ-radiation [6]. The high cost, low efficiency, radiation damage of the polymers and danger of such processes have very drastically limited its use.

Corona discharges can be obtained by applying a high voltage to metallic electrode such as a metallic point. It produces ions and activated neutral species in a region very close to the corona point. Neutral species and/or charged ions can be carried to a sample surface by the corona wind and the electrostatic attraction, respectively. Corona discharge is a technique which has been successfully used for charging polymers [7], in electrophotocopying machines [8], in electrostatic filters [9], etc.. It can be also employed on surface treatments of polymers to modify the polymer surface [10] and to improve adhesion between two different materials [11]. However, for the best of our knowledge corona has not yet been applied to conducting polymer as a doping method.

Mat. Res. Soc. Symp. Proc. Vol. 488 © 1998 Materials Research Society

This work reports the use of the corona discharge as a new method to efficiently dope parent polyaniline to the conductive regime [12]. The corona technique was used to dope thin layers of PANI coated on poly (ethylene terephthalate) - PET films. The effect of experimental conditions of the corona discharge on the electrical conductivity and properties of the conducting polymer is discussed.

EXPERIMENT

The coating of the PET films with PANI, namely PET/PANI composite, was made by chemically synthesizing parent polyaniline having a PET film immersed into the polymerization solution. The synthesis was performed by the conventional chemical method [13] at room temperature using aniline and ammonium peroxydisulfate in a solution of HCl 1.0 molar during around 60 min. The PET/PANI composite was deprotonated in a 0.1 molar ammonium hydroxide solution for 24 h and then dried under vacuum, in order to obtain PANI in the undoped emeraldine base (EB) oxidation state.

The UV-Vis absorption spectra from the PET/PANI composite was obtained using a Perkin-Elmer model Lambda-9 Spectrophotometer. The corona discharge was generated using a corona triode system consisting of a metallic point, a stainless steel grid and a sample holder, as described elsewhere [7,12]. Corona treatments were performed in air at room temperature (~25°C) and relative humidity of about 60%. The electrical conductivity of the PANI coating was measured with the four-point probe technique.

RESULTS

The UV-Vis electronic absorption spectra of polyaniline is a measure of the doping state of the polymer, since it is related to the electron transition energy within the polymeric structure. In reference to Fig. 1, for corona untreated PET/PANI composite with the PANI in the undoped EB state, one main peak can be observed in the UV-Vis spectrum at around 600 nm, assigned to an excitonic transition. As such PET/PANI is exposed to a positive corona discharge the maximum wavelength (λ_{max}) of such peak is increased due to a decrease in the energy associated to the electronic transition and a shoulder appears at around 400nm. Such behavior is characteristic of an doping effect of the polymer, similarly to what occurs for the PANI doped by other methods, as described in the literature [1,4,13].

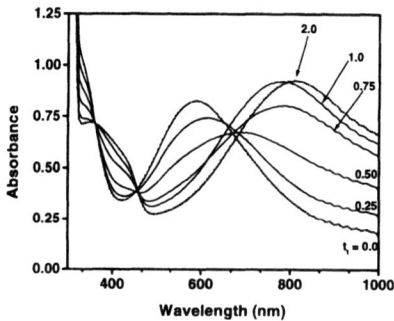

Figure 1 - UV-Vis spectra for PANI coated onto PET films treated with positive corona during different periods of time, t_t. The corona current was 15 μA. The treatment time is indicated for each curve and $t_t = 0$ corresponds to the PANI-EB.

Figure 2 shows the dependence of λ_{max}, corresponding to the 600 nm peak of undoped PANI, on the corona discharge treatment. It is shown that for PET/PANI composite films treated with positive and negative corona polarities, the doping degree of the polymer increases (λ_{max} increases) with the increase of the treatment time, reaching c.a. 850 nm (1.46 eV) for the positive polarity.

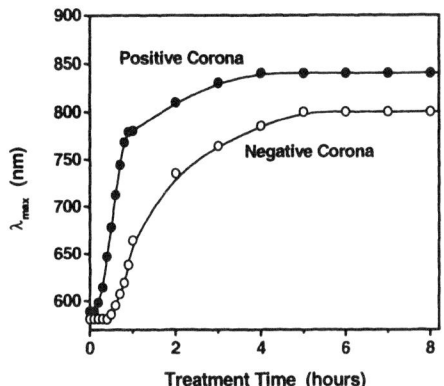

Figure 2 - Wavelength maximum λ_{max} for PANI coated onto PET films treated with positive and negative corona during different periods of time. The corona current was ±15µA.

Very surprisingly, the electrical conductivity of the PANI coating on the composite can be not only increased from 10^{-10} to 0.3 Scm^{-1}, upon corona treatment, but also tuned in this conductivity range, as presented in Figure 3. These conductivity values are stable persisting for long periods of time (more than 5 months, so far). Such behavior confirms that corona discharge can be used as a new doping method for polyanilines. Further, from Figs. 2 and 3 it can be seen that positive corona is more effective since it gives larger values for conductivity and λ_{max} and, in a shorter time as compared with the negative polarity. Preliminary results indicates that such doping increases with the corona current being also dependent on the distances between point-grid-sample. Detailed results on that will be published elsewhere.

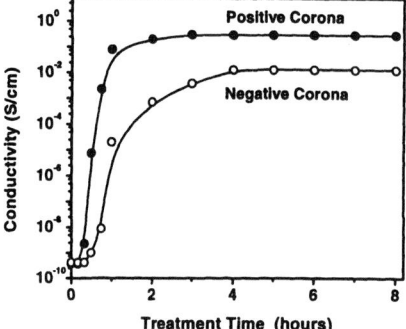

Figure 3 - Dependence of the electrical conductivity for PANI coated onto PET films with positive and negative corona during different period of time. The corona current was ±15 μA.

Fig. 4 shows the dependence of the conductivity on the relative humidity for positive and negative corona . Results indicate that the negative corona is less effective for doping the polyaniline since the conductivity obtained with negative polarity is smaller than the positive in agreement with the results shown in Fig. 3.

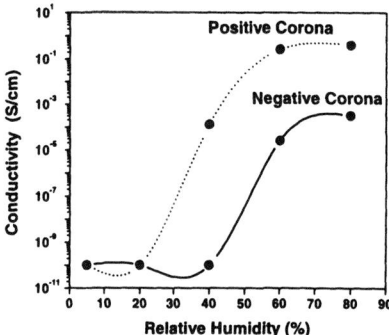

Figure 4 – Dependence of the electric conductivity on the relative humidity for different polarities of the corona discharge. Samples were treated during 1 hour using a corona current of ±15 μA.

For polyanilines, contrary to other conducting polymer, doping is commonly carried out by protonation. It seems likely then that the corona discharge must be dissociating the molecules present so that one can obtain protonic ion which penetrate within the polyaniline, promoting the doping observed. Such hypothesis is also corroborated by the more efficient doping with positive corona. Positive corona discharge produces mainly clusters of $(H_2O)_nH^+$ (n is an integer which increases with moisture) while the negative corona produces ions O_3^-, NO_3^- and CO_3^-, as well as minor quantities of $(H_2O)_nCO_3^-$ ions [14]. It is believed that these ions dissociate in such a way that H^+ ions are produced and which then penetrates into the polymer, being the doping process more efficient for positive corona. Other studies are currently being done to provide more insight into this doping mechanism.

Fig. 5 shows results for the decay of the electric conductivity at different temperatures. Samples were first doped with positive corona during 2 hours using 15 μA. For the conductivity measurements samples were removed from the oven and the conductivity measured at room temperature. Results of decays on Fig. 5 shows that the conductivity persists for a long time at moderate temperature, similarly to what occurs for samples prepared in chemical solution doping. A detailed comparison of these results will be shown elsewhere[15].

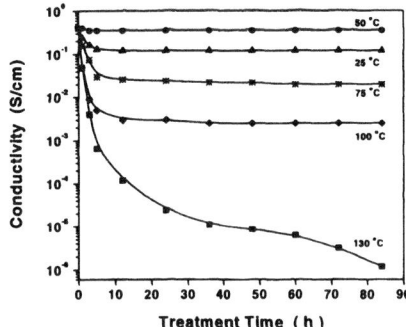

Figure 5 – Decay of the electric conductivity for different temperatures. Samples were doped employing positive corona during 2 hours and corona current of 15 μA.

CONCLUSION

We have demonstrated that the electrical conductivity of PANI coated onto PET films can increase in the range from 10^{-10} to 0.3 Scm^{-1} by treatment with corona discharge. The method gives very reproducible results, the process being easily controlled. In comparison with the usual aqueous acidic solution doping process such new method eliminates the use of chemicals and enables the use of a continuous, rapid and dry doping method. Such characteristics are extremely important on technological applications of conducting polymer in microelectronics, such as antiestatic coating of microchips, sensors, transistors and other electronic devices.

ACKNOWLEDGEMENTS

The financial support given by CNPq, FINEP/PADCT program and FAPESP (Brazil) is gratefully acknowledged. The authors also acknowledge Rhodia-Ster for supplying the PET samples.

REFERENCES

1. W.S. Huang, B.D. Humphrey, and A.G. Macdiarmid, J. Chem. Soc. Faraday Trans. 1, **82**, 2385(1986).

2. S. Roth and W. Grauper, Synth. Met. **55-57**, 3623(1993).

3. W. S. Huang, M. Angelopoulos, J. R. White, and J. M. Park, Mol. Cryst. Liq. Cryst., **189**, 227(1990).

4. M. Angelopoulos, J.M. Shaw, W. S. Huang, and R. D. Kaplan, Mol. Cryst. Liq. Cryst., **189**, 221(1990).

5. J.A. Malmonge and L.H.C. Mattoso, Synth. Met. **84**, 779(1997).

6. M. Wolszczak, J. Kroh, and M. M Abdel-Hamid, Radiat. Phys. Chem., **45**, 71(1995).

7. J. A. Giacometti and O. N. Oliveira Jr., IEEE Trans. Elec. Insul. **27**, 924(1992).

8. R. M. Schaffert, *"Electrophotographic - Part II "*, Focal, 1975.

9. C.D. Hendricks, In *"Electrostatics and its applications"*, Edited by A.D. Moore, Wiley, Chapter 4, (1973), 57-85

10. C. J. Dias, J. N. Marat Mendes, and J. A. Giacometti, J. Phys. D: Appl. Phys. **22**, 663(1989).

11. I. Novak and. Florian, J. Mat. Sci. Lett., **14**, 1021(1995).

12. L. H. C. Mattoso, A. E. Job and J. A. Giacometti, Pending Patent INPI – Brazil (1997).

13. L. H. C. Mattoso, A. G. MacDiarmid, and A. J. Epstein, Synth. Met., **68**, 1(1994).

14. A. Goldman and J. Amoroux, in *"Electrical Breakdown and Discharge in Gases, Macroscopic Processes and Discharges"*, edited by E. E. Kunhardt and L. H. Luessen, Plenum, New York, 1983.

15. In preparation.

Polarized Light-Induced Anisotropy in Polymer Films Doped with Az Dyes in the Photostationary State Studied by IR Spectroscopy

K. TAWA*, K. KAMADA**, T. SAKAGUCHI**, and K. OHTA**

* Fields and Reactions, PRESTO, JST and Osaka National Research Institute, 1-8-31 Midorigaoka, Ikeda, Osaka 563, JAPAN, keiko-o@onri.go.jp
** Osaka National Research Institute, 1-8-31 Midorigaoka, Ikeda, Osaka 563, JAPAN

ABSTRACT

The anisotropy induced in a series of methacrylate polymers doped with Disperse Orange 3 (DO3, NO_2-phenyl-N=N-phenyl-NH_2) was investigated by polarized FTIR spectroscopy. Three kinds of methacrylate polymers with different glass transition temperature (Tg), poly(methyl methacrylate) (PMMA), poly(ethyl methacrylate) (PEMA), and poly(butyl methacrylate) (PBMA), were used as polymer matrices. Observed infrared absorption bands of DO3 in the polymer films were assigned to symmetric (NO_2s) and antisymmetric (NO_2as) stretching modes of NO_2, and the C-N stretching mode of C-NH_2 (C-N). The infrared dichroism was observed in these bands under the irradiation. The orientation factors, K_{fn} (f=x, y, z, n=trans, cis), for both isomers of DO3 were determined in the photostationary state for every polymer matrix. The factors indicate that it is difficult for NO_2 group to move in the PMMA during trans-trans isomerization and that it is difficult for NH_2 group to move in the PBMA. The dynamical behavior of DO3 molecules which depends on the kind of polymers could be interpreted from orientation factors. This study shows that polarized FTIR spectroscopy is one of the most powerful methods available to analyze the physical mechanisms of photoinduced anisotropy.

INTRODUCTION

A polarized light-induced anisotropy [1, 2] has been previously observed in polymer films doped with photoresponsive molecules. In the polymer system containing photoresponsive molecules, the response behavior of the anisotropy can be controlled by the kind of combination between polymer and photoresponsive molecule. Recently, new techniques of optical switching [3] and holography [4] utilizing the polarized light-induced anisotropy in polymer films have been also studied. Todorov et al. [5, 6] studied photoinduced anisotropy in polymer films doped with azo dyes for transient polarization holography which can be utilized as a real-time recording. However, the physical mechanisms of photoinduced anisotropy, has not been clarified yet. In order to clarify the mechanisms, it is necessary to study the molecular dynamics under the polarized light irradiation.

In this study, the photoinduced anisotropy of azo dyes in the polymer films was investigated by using polarized FTIR spectroscopy. Azo dyes have two spatial forms, trans and cis isomers. The isomerization from the trans to cis form is induced by irradiation. However, due to the instability of the cis form, it can isomerize back to the trans form thermally. When the irradiation to azo dyes is continued, azo dyes take the photostationary state where the change of absorbance is not observed. In the polymer films doped with azo dyes, the isomerization rates and the photoequilibrium ratio depend on the kind of azo dyes and polymer matrices. A series of methacrylate polymers with different glass transition temperature (Tg) were used as polymer matrices in order to clarify how DO3 molecules behave in the matrices with the different free volume, since Tg is considered to be the guide value representing the size of the free volume. The infrared absorbance of each functional group changed due to the photoisomerization and the dichroism, i.e., photoinduced anisotropy of the azo dyes, was observed in the photostationary state. A close relationship between the isomerization and the anisotropy will be discussed in terms of the orientation factors [7] obtained from analysis of several vibrational bands in polarized FTIR spectra.

EXPERIMENT

Disperse Orange 3 (DO3, Aldrich, NO_2-phenyl-N=N-phenyl-NH_2) was selected among azo dyes as the photoresponsive molecule. In a series of methacrylate polymers, poly(methyl methacrylate) (PMMA, Aldrich, M_w=120,000, Tg~105℃), poly(ethyl methacrylate) (PEMA, Aldrich, M_w=515,000, Tg~65℃), and poly(butyl methacrylate) (PBMA, Aldrich, M_w=350,000, Tg~20℃) were used as matrices. The sample films were prepared by casting chloroform solution containing a polymer and DO3. The thickness of the films were less than 20 μm. The dye concentration was about 1 wt % in the PMMA films.

A Nicolet 60SXR FTIR spectrometer was used for measuring infrared spectra. The FTIR spectra of the DO3/polymer films were measured before irradiation and under irradiation of linearly polarized light, which is the light from a CW Ar ion laser with a wavelength of 488 nm and an optical power of 20 mW/cm². The IR spectra of PMMA, PEMA, and PBMA were also estimated as a standard sample. All the spectra were measured at a wavenumber resolution of 4 cm⁻¹. Polarized FTIR spectra were observed using the polarizer (KRS-5) in the parallel (Z) and perpendicular (Y) directions to the direction (Z) of polarized light of an Ar ion laser. (Figure 1)

RESULTS

Subtracting the spectrum of the polymer without DO3 from that of DO3 doped in the polymer clarifies the infrared absorption bands of DO3 in the polymer films as shown in Figure 2, where the FTIR spectra were measured in the initial state (solid line: before irradiation) and in the photostationary state (broken line: under irradiation). Our interests were focused on the

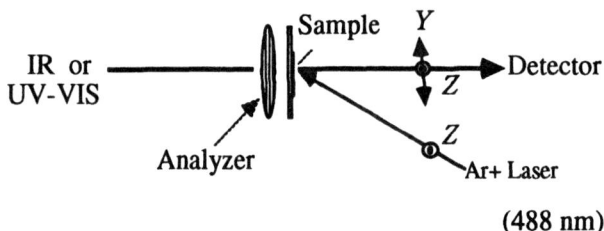

Figure 1 Optical alignment

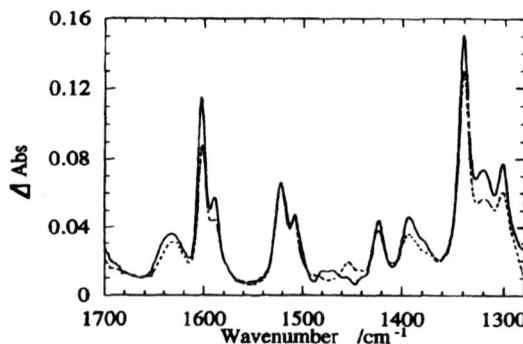

Figure 2 FTIR spectrum of DO3/PMMA in the: ——; initial state, ----; photostationary state.

886

three vibrational bands which were assigned [8, 9] to symmetric (NO_2s, 1341cm^{-1}) and antisymmetric (NO_2as, 1523cm^{-1}) stretching modes of NO_2, and the C-N stretching mode of $C-NH_2$ (C-N, 1303cm^{-1}). By measuring the polarized IR spectra of DO3/polymer in the photostationary state, the infrared dichroism was observed in above three bands. Table I shows the absorbances of NO_2s, NO_2as, and C-N bands in the photostationary state, which are normalized to their initial absorbances before irradiation in each band.

Table I Normalized absorbance of three vibrational bands in PMMA, PEMA, and PBMA matrices in Z (parallel) and Y (perpendicular) directions measured by polarized FTIR method in the photostationary state.

polymer	NO_2s		NO_2as		C-N	
	Z	Y	Z	Y	Z	Y
PMMA	0.910	0.985	1.01	0.960	0.700	0.865
PEMA	0.917	0.951	0.969	0.938	0.791	0.848
PBMA	0.963	0.975	0.976	0.940	0.838	0.855

DISCUSSION

The mechanism of polarized light-induced anisotropy of DO3 in polymer matrix cannot be clarified without considering the process of isomerization. Therefore, it is important to study the dynamics of DO3 molecules in the photoisomerization process from trans to cis and in the thermal isomerization process from cis to trans. In this study, the degree of orientation of the trans and cis isomers will be considered in the photostationary state under irradiation of polarized light. As shown in Figure 2, abovementioned each vibrational band for the cis isomer appears at the same wavenumbers as that for the trans isomer. Therefore, the absorbance of an a-vibrational band can be expressed as the sum of the absorbance of the trans and cis isomers.

$$E(a) = \alpha E(a_t) + (1 - \alpha)E(a_c).$$ (1)

$E(a)$ is the absorbance vector in the laboratory-fixed axes, X, Y, Z, and it is composed of $E_X(a)$, $E_Y(a)$, and $E_Z(a)$. α is the molar fraction of the trans isomer to the total number of molecules. $E(a_n)$ can be represented as equation (2) by the multiplication of the matrix of orientation factors, $K_Z^{(n)}$, and the absorbance vector, $A(a_n)$, in an arbitrary molecular framework, x, y, z. The symbol n means the type of isomers, i.e., trans or cis isomer. The matrix of orientation factors is composed of $K_{Z f_n}$ ($f = x, y, z$, $n = t$ (trans), c (cis)).

$$E(a_n) = K_Z^{(n)}A(a_n).$$ (2)

In this study, the sample can be treated as an uniaxial orientation, $E_X(a) = E_Y(a)$, since the sample was irradiated by the linearly polarized light along the Z axis. Therefore, the matrix of the orientation factor, $K_Z^{(n)}$, can be described as

$$K_Z^{(n)} = \begin{pmatrix} \dfrac{1-K_{Zx_n}}{2} & \dfrac{1-K_{Zy_n}}{2} & \dfrac{1-K_{Zz_n}}{2} \\[2mm] \dfrac{1-K_{Zx_n}}{2} & \dfrac{1-K_{Zy_n}}{2} & \dfrac{1-K_{Zz_n}}{2} \\[2mm] K_{Zx_n} & K_{Zy_n} & K_{Zz_n} \end{pmatrix}.$$ (3)

The physical meaning of matrix elements, K_{Zf_n}, can be understood by introducing the absorption probability, which is represented by the dot product of the Z-polarized light, e_Z (unit vector), and an electric dipole transition moment, $M(a_n)$, as described in equation (4). The $M(a_n)$ and the absorbance, $A(a_n)$, in equation (2) have the relationship as $|M(a_n)|^2 \propto |A(a_n)|$.

$$< (e_Z \cdot M(a_n))^2 > = M_x^2(a_n) < \cos^2\theta_{x_n} > + M_y^2(a_n) < \cos^2\theta_{y_n} > + M_z^2(a_n) < \cos^2\theta_{z_n} >$$
$$+ 2\{ M_x(a_n)M_y(a_n) < \cos\theta_{x_n}\cos\theta_{y_n} > + M_y(a_n)M_z(a_n) < \cos\theta_{y_n}\cos\theta_{z_n} > + M_z(a_n)M_x(a_n) < \cos\theta_{z_n}\cos\theta_{x_n} > \}$$
$$(4)$$

where the off-diagonal terms, such as $M_x(a_n)M_y(a_n) < \cos\theta_{x_n}\cos\theta_{y_n} >$, can vanish by selecting the arbitrary system of molecular axes which coincide with the molecular orientation axes. The terms of $M_f^2(a_n) < \cos^2\theta_{f_n} >$ corresponds to the square of the projection of the f-component of the transition moment, $M_f(a_n)$, onto the Z axis. The square of the cosines are defined as the orientation factors, K_{Zf_n}.

$$K_{Zf_n} = < \cos^2\theta_{f_n} >, \qquad f : x, y, z, \qquad (5)$$
$$K_{Zx_n} + K_{Zy_n} + K_{Zt_n} = 1 . \qquad (6)$$

We can describe the absorption probability simply as equation (7) using the definition of K_{Zf_n}.

$$< (e_Z \cdot M(a_n))^2 > = M_x^2(a_n)K_{Zx_n} + M_y^2(a_n)K_{Zy_n} + M_z^2(a_n)K_{Zz_n} . \qquad (7)$$

In the case of random distribution, such as the sample in its initial state before irradiation, the values of K_{Zx_n}, K_{Zy_n}, and K_{Zz_n} are equivalent to each other, and equal to 1/3.

We defined the molecular axes for the trans and cis isomers respectively as described below. The molecular axes for the trans isomer were chosen as the molecule takes a planar structure in y_t-z_t plane, and the direction of $\pi\pi^*$ transition moment can be assumed to be parallel to z_t axis [10, 11]. In the trans isomer, the vibrational transition moments of NO_2s and C-N bands are almost parallel to the $\pi\pi^*$ transition moment and that of NO_2as is perpendicular to $\pi\pi^*$. The relationship $K_{Zx_t} = K_{Zy_t}$ can be assumed in this case since the $\pi\pi^*$ transition moment is assumed to be parallel to the z_t axis. In the molecular axes for the cis isomer, cis molecule is also assumed to take a planar structure in the y_c-z_c plane. The C-NO_2 bond is assumed to be oriented in the direction of z_c axis and the C-NH_2 bond is oriented in the direction of y_c axis.

The K_{Zf_t} values for both two isomers in each polymer matrix were determined by solving the several equations which were represented for the three bands, NO_2s, NO_2as, and C-N by using the results in the photostationary state shown in Table I, where the α value was determined by UV-VIS spectra and they were 0.79 (PMMA), 0.80 (PEMA), and 0.84 (PBMA).

The determined values of K_{Zf_t} are shown in Figures 3 (a)-(f). In the trans isomer, the K_{Zz_t} value is smaller than the K_{Zx_t} and K_{Zy_t} values in every polymer. This means that the trans isomers were excited by the photoselection of the linearly Z-polarized light. In the case of cis isomer, the K_{Zf_c} values are found to depend on the kind of polymers. In PMMA, the K_{Zz_c} value is larger than the K_{Zx_c} and K_{Zy_c} values. The larger K_{Zz_c} value indicates that there are a lot of cis molecules with a small angle between the z_c and the Z axis, or that more C-NO_2 bonds align to the Z axis. It is considered to be difficult for the NO_2 group to move in the PMMA films during trans-cis-trans isomerization. In PEMA, the K_{Zz_c} value is found to be almost the same as the K_{Zy_c} value. In PBMA, the K_{Zy_c} value is larger than the K_{Zz_c} value. The larger K_{Zy_c} value indicates that there are a lot of cis molecules with a small angle between the y_c and the Z axis, or that more C-NH_2 bonds align to the Z axis in PBMA.

The physical mechanism of photoinduced anisotropy has been suggested to be induced by two processes, the photoselection and the reorientation. The latter means that the configuration of trans molecules rotates by 90° in the back isomerization process from cis to trans. However, this

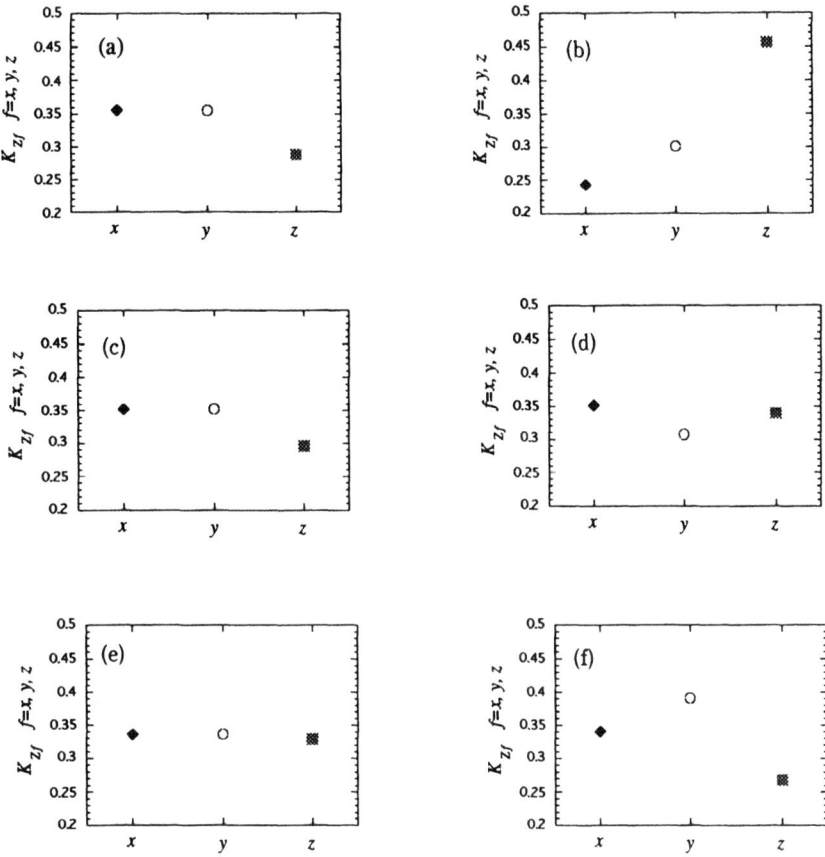

Figure 3 The values of orientation factors K_{zf} for: (a); trans isomer in PMMA, (b); cis in PMMA, (c); trans in PEMA, (d); cis in PEMA, (e); trans in PBMA, and (f); cis in PBMA.

study indicated that the NO_2 group is difficult to move in the PMMA matrix or the entire DO3 molecule cannot rotate in PMMA during a series of isomerization process. Therefore, the physical mechanism of polarized light-induced anisotropy in PMMA doped with DO3 is mainly due to photoselection by the irradiation with linearly polarized light, and it is little induced by the reorientation of DO3 molecules. The free volume of PMMA is considered to be not so large. It can be found from the fact that the glass transition temperature (Tg) is higher, 105 °C. Therefore, the dynamical behavior of DO3 molecules is restricted in the free volume of PMMA. The side of NO_2 group is difficult to move in PMMA possibly due to the larger volume compared to the side of NH_2 group. However, as Tg is lower, the free volume of polymers increases and DO3 molecules become easy to move in matrix. In PBMA, it is easier to move than in PMMA and PEMA with high Tg. DO3 has the amino group and the group can interact with the ester group of a polymer side chain. DO3 in the matrix with high mobility interacts strongly with such polymers as PBMA, therefore, we consider that it may be difficult for NH_2 group to move in the process of photoisomerization.

CONCLUSIONS

We investigated the anisotropy induced in the polymer films with different Tg doped with DO3 using polarized FTIR spectroscopy. In order to analyze the data from the dichroism, the orientation factors for each isomer in each polymer were determined. The factors indicate that it is difficult for the NO_2 group to move in the PMMA matrix. The free volume of PMMA is considered to be not so large, therefore, the dynamical behavior of DO3 molecules is restricted in the free volume of PMMA. The side of NO_2 group is difficult to move in PMMA probably because of the larger volume compared to the side of NH_2 group. However, DO3 molecules become easy to move into the matrix as the free volume is large. In PBMA, it is easier to move or rotate than in PMMA and PEMA with higher Tg. NH_2 group of DO3 in PBMA with a low Tg interacts more strongly with the ester group of a polymer side chain, therefore we consider that it is difficult for NH_2 group to move in the process of photoisomerization.

In this study, it was successfully indicated for the first time that the polarized FTIR spectroscopy is one of the most powerful methods available to find a clue to clarify the physical mechanism of photoinduced anisotropy which depends on the kind of polymers. We are studying the behavior of azo dyes in a series of much more methacrylate polymers under the irradiation of polarized light in order to clarify on which interactions between dyes and polymer matrices the photoinduced anisotropy depends.

REFERENCES

1. S. Xie, A. Natansohn, and P. Rochon, *Chem. Mater.* **5**, 403- 411 (1993).

2. G. S. Kumar and D. C. Neckers, *Chem. Rev.* **89**, 1915-1925 (1989).

3. T. Ikeda, T. Sasaki, and K. Ichimura, *Nature* **361**, 428-430 (1993); T. Ikeda and O. Tsutsumi, *Science* **268**, 1873-1875 (1995).

4. T. Todorov, L. Nikolova, N. Tomova, and V. Dragostinova, *IEEE J. Quantum Electron.* QE-**22**, 1262-1267 (1986).

5. T. Todorov, L. Nikolova, N. Tomova, and V. Dragostinova, *Opt. Quantum Electronics* **13**, 209-215 (1981).

6. V. Mateev, P. Markovsky, L. Nikolova, and T. Todorov, *J. Phys. Chem.* **96**, 3055-3058 (1992).

7. J. Michel and E. W. Thulstrup, *Spectroscopy with Polarized Light*, New York: VCH Publishers, 1986.

8. A. Natansohn, P. Rochon, M. Pezolet, P. Audet, D. Brown, and S. To, *Macromolecules* **27**, 2580-2585 (1994).

9. A. V. Zemskov, G. N. Rodionova, Yu. G. Tuchin, and V.V. Karpov, *Zh. Prikl. Spektrosk.* **49**, 581-586 (1988).

10. D. L. Beveridge and H. H. Jaffé, *J. Am. Chem. Soc.* **88**, 1948 -1953 (1966).

11. M. B. Robin and W. T. Simpson, *J. Chem. Phys.* **36**, 580-589 (1962).

THIN FILM OF CONJUGATED SCHIFF-BASE AS ULTRAHIGH DENSITY DATA STORAGE MATERIAL

W.J.YANG[1], Q.C.YANG[1], H.G.LU[1], H.Y.CHEN[1]*, S.M.HOU[2], L.P.MA[3], H.X.ZHANG[3], Z.Q.XUE[2], S.J.PANG[3]
1 College of Chemistry, Peking Univ., Beijing 100871, P.R.China
2 Department of Electronics, Peking Univ. Beijing 100871, P.R.China
3 Beijing Laboratory of Vacuum Physics, Chinese Academy of Science, Beijing 100080, P.R.China

ABSTRACT

N-(3-nitrobenzylidene)-p-phenylenediamine (NBPDA) was used as ultrahigh density data storage medium by scanning tunneling microscope (STM) technique. Data marks of 1.4nm in diameter were written by applying voltage pulses between the STM tip and the substrate. Structures of single crystal and thin films were characterized by IR, UV-Vis, XRD, STM and verified by DFT quantum chemical calculation.

INTRODUCTION

Conjugated organic and polymeric systems with delocalized electrons have attracted increasing interests in their great potential for high-technology applications, especially ultrahigh density memory and information storage materials [1-3]. The highly polarizable electronic cloud is accompanied by a capability for ultrafast responses originating in the electron mobility, so that the conjugated systems are characterized by optimal electric activities as compared with inorganic crystals.

In recent years, much attention has been focused on conjugated organic charge transfer complexes, such as cyanuric acid melamine-7,7,8,8,tetracyanoquinodimethane (TCNQ) complex or Bis[2-butene-2,3-dithiolato(2-)-S,S']nickel (BBDN)-TCNQ complex, achieving data storage density of $4 \times 10^7 bits/cm^2$ [4,5]. In our laboratory, thin films from a complex of an electron acceptor, m-nitrobenzal malononitrile and a donor, 1,2-benzenediamine, were used as recording media [7,8]. Marks of 1.3nm in size were obtained with STM under ambient conditions. I-V characteristics of the film suggested that the recording mechanism be due to the conductance change of the films from insulating to conducing. In order to further clarify the mechanism and functionality process of such kind of recording media, we prepared a conjugated schiff-base N-(3-nitrobenzylidene)-p-phenylenediamine (NBPDA) with both electron donor and acceptor group located in a single molecule at opposite sites simultaneously. Recording process was successfully carried out to make marks of 1.4nm in diameter by applying voltage pulses between the STM tip and the substrate film. NBPDA crystal and film structure were characterized by IR, UV-Vis, XRD, STM and verified by DFT quantum chemical calculation. The possible mechanism of the recording process were tentatively discussed.

EXPERIMENTS

Materials and Thin Film Preparation

---

* To whom the correspondence should be addressed.

Mat. Res. Soc. Symp. Proc. Vol. 488 © 1998 Materials Research Society

N-(3-nitrobenzylidene)-p-phenylenediamine (NBPDA) was prepared by condensation reaction of equal molar amounts of 3-nitrobenzaldehyde with 1,4-benzenediamine, according to literature [9]. Thus, a solution of 0.1mmol 1,4-benzene-diamine in 10ml ethyl acetate was added to a solution of 0.1mmol 3-nitrobenzaldehyde in 10ml ethyl acetate. After standing at 0°C for 6 days, deep red crystals were obtained with melting point 146~147°C.

Ultrathin films of NBPDA were obtained by a thermal vacuum deposition method. The NBPDA crystals were placed in a glass crucible and heated under 9×10^{-4}Pa and deposited on a substrate kept at room temperature.

Recording Experiments by STM

Experiments were performed with a CSTM-9100 STM in a constant height mode. Ultrathin films were deposited on freshly cleaved highly ordered pyrolytic graphite (HOPG). The STM tips were 0.25mm in diameter Pt/Ir (80/20) wires snipped with a wire cutter. The recording experiments were carried out by applying voltage pulses between the tip and the substrate.

Strcture Characterizations and Quantum Calculation

Crystal structure analysis data were collected with an ENTSF-NONLUS CAD-4 diffractometer. FT-IR spectrum for NBPDA crystal pilled with KBr, and micro FT-IR spectrum for NBPDA thin film on HOPG substrate were both taken by a Nicolet Magna-IR 750. UV-Vis spectra were taken by a Shimadzo UV 250. The crystallinity of the thin film was examined by using a Riguku max/2400 X-ray diffractometer employing Ni-filtered CuKα radiation. Sample film was deposited on a glass substrate.

Quantum chemistry calculations were performed based on density functional theory (DFT) using Amsterdam Density Functional (ADF) 2.3.

RESULTS AND DISCUSSION

Fig.1 shows the molecular structure and the unit cell of NBPDA crystal. The crystal was monoclinic with lattice constants a=0.9601nm, b=1.0229nm, c=1.2697nm, and β=110.33°. Element analysis for $C_{13}H_{11}N_3O_2$: %Calcd(Found): C:64.73(64.51), H:4.56(4.79), N:17.43(17.35).

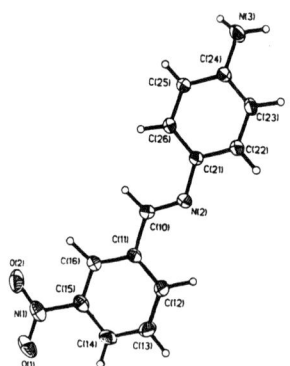 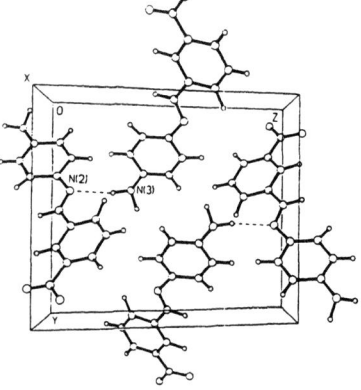

Fig.1(a) Molecular structure of NBPDA.   **(b) Unit cell of NBPDA crystal.**

Fig.2 shows the FT-IR spectrum of NBPDA crystal and Micro FT-IR spectrum of NBPDA film. It seems that the composition of the film is about the same as the original crystal. No significant change of the characteristic absorption bands were observed as shown in Fig.1 and Table I.

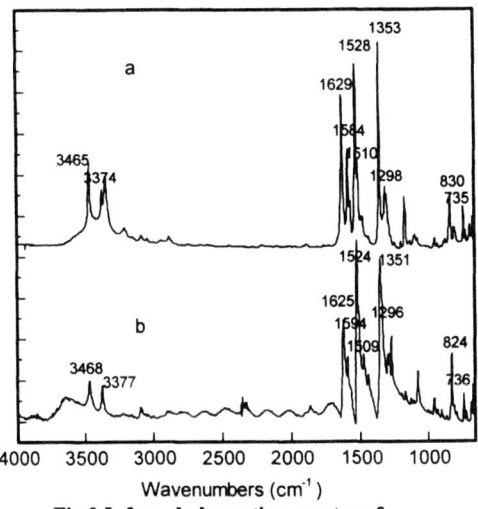

Fig.2 Infrared absorption spectra of (a) NBPDA crystal, (b) NBPDA film.

Table I. FT-IR characteristic absorption bands of NBPDA crystal and film.

Attributions	Wavenumbers (cm^{-1})	
	crystal	film
$v_{as}(NH_2)$	3465	3468
$v_s(NH_2)$	3374	3377
$v(C=N)$	1629	1625
$v_\Phi(C=C)$	1584	1594
	1510	1509
$v_{as}(NO_2)$	1528	1524
$v_s(NO_2)$	1353	1351
$v(C-N)$	1298	1296

Fig.3 shows the UV-Vis spectra of the NBPDA solution and NBPDA film. The red-shift and broadening of the peaks implied the possibility of intermolecular complexing. X-ray diffraction spectrum indicated that the thin films were highly stereo-regularly packed as shown in Fig.4.

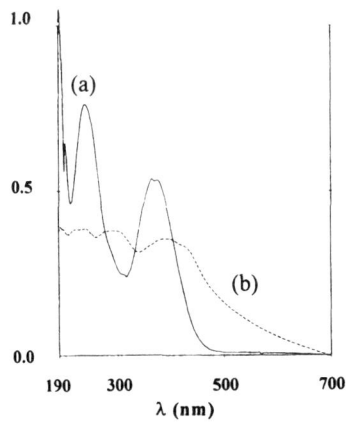

Fig.3 UV-Vis spectra of (a) NBPDA in ethyl alcohol; and (b) NBPDA film deposited on quatz.

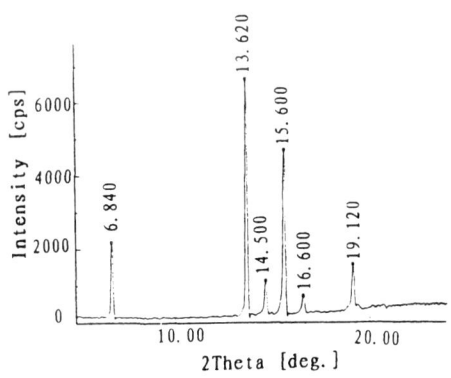

Fig.4 X-ray diffraction spectrum of NBPDA film.

A flat region without visible defects was selected for recording. Recording process was carried out by applying pulse voltage of 4V for 2ms. A typical 28×28nm² STM image of the NBPDA thin film with recorded marks is shown in Fig.5. The average size of recorded marks was 1.4nm in diameter. The distance between two neighboring marks could be 9nm or less, which corresponds to a data storage density of about 10^{12}bits/cm² or above.

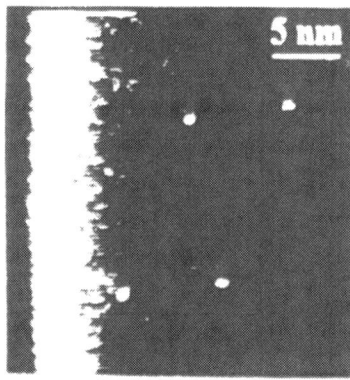

Fig.5 A 28×28nm² STM image on NBPDA thin film with recorded marks by applying pulse voltage of 4V for 2ms. V_b=0.10V, I_t=0.19 nA.

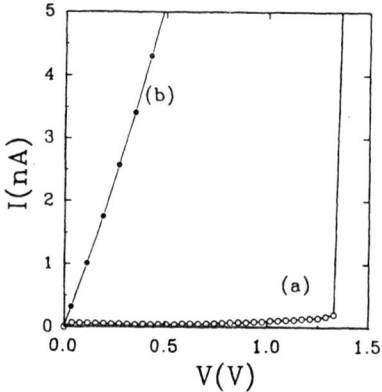

Fig.6 Typical I-V characteristics of the NBPDA film of (a) the unrecorded region and (b) the recorded region by STM.

I-V characteristics of the unrecorded region and recorded region were measured by STM as shown in Fig.6. Curve (a) shows that the medium is insulating when the applied voltage pulse is below the threshold 1.3V, and becomes conductive when above it. Curve (b) shows that the recorded region of the film is conductive.

Fig.7 gives a view of the nearest-neighbor packing of the NBPDA molecules. The molecules packed in a columnar manner, which was similar to the ring-double bond overlap of molecular planes in organic cation-TCNQ structures [10]. The zigzag linkage of C-C-N-C in the molecule was sandwiched between two phenyl groups of its two nearest-neighbor molecules with the interplanar spacings of 0.352nm and 0.345nm, respectively.

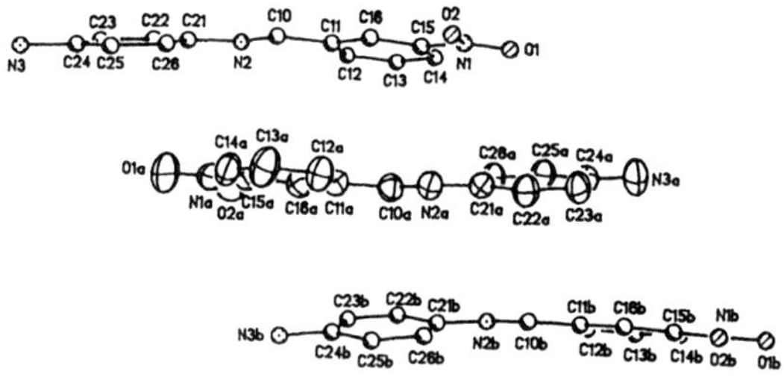

Fig.7 The nearst-neighbor packing of NBPDA molecules. Direction of view is along the molecular plane.

We calculated the three-nearest-neighboring-molecule system by DFT. Calculated Mulliken population for the $^1A_{1g}$ ground state of the molecules are summarized in Table II.

The Mulliken population analysis of the HOMO and LUMO suggested that the electron transfer would occur between the phenylenediamino group of the upper molecule and the nitro-

Table II. Mulliken population analysis for three-nearest-neighboring-molecule system of NBPDA.

	Energy levels ε (eV)	Mulliken			population					
		upper			middle			lower		
		a*	b*	c*	a*	b*	c*	a*	b*	c*
HOMO	-4.882	0.16	0.28	1.32	0.02	0	0.02	0	0	0.06
LUMO	-3.295	0	0	0	1.92	0.02	0	0	0	0

*a: nitrobenzyl group;   b: -C=N-;   c: phenylamino group.

benzylidene group of the middle one. The Mulliken population of the lowest molecule of the three was comparatively small, which indicated that the electron transfer was constricted in two-nearest-neighboring-molecule system. So when appropriate amount of energy was added to the substance by voltage pulses using extremely thin STM tip, electron transfer would occur in certain constricted molecules from the HOMOs of certain amount of NBPDA molecules to the LUMOs of their neighboring ones. Usually, the transferred electrons would settle back shortly after the removal of applied energy, and the substance remained insulating again. In case of experiments performed by STM, the influence of the extremely short distance between the tip and the film (0.3~0.7nm), and the effect of conductive HOPG substrate should not be neglected [11]. The strong tip-sample interaction might stabilize the partial transition of electrons resulting in the expected conductance change. Further work dealing with the detailed electron transfer process and analysis of quantitative calculations are in progress.

CONCLUSIONS

Nanometer-scale recording on thin film of conjugated schiff-base N-[(3-nitrophenyl) methylene]-1,4-Benzenediamine (NBPDA) with STM under ambient conditions was presented. The films were fine enough to show the reproducible electrical bistability. The average size of recorded marks was 1.4nm in diameter. The distance between two neighboring marks could be 9nm or less, which corresponded to a data storage denstiny of about 10^{12}bits/cm^2 or above. Characterizations of the crystal and thin films indicated that the composition of thermal vacuum deposited films was almost the same as the original NBPDA crystal. The existence of intermolecular charge transfer complexing was verified by UV-Vis. Quantum chemistry calculation of the three-nearest-neighboring-molecule system by density functional theory showed the theoretical feasibility of the electron transfer between two neighboring NBPDA molecules, which would result in the conductance change of thin films after applying voltage pulses by STM tip and consequently, the formation of recorded marks.

ACKNOWLEDGMENTS

This work was supported by the National Natural Science Foundation of China. We also should thank Dr. Wang Zheming for his kindness to help us with single crystal analysis and preparation for this manuscript .

REFERENCES

1. W.R.Salaneck, I.Lundstrom and B.Ranly ed. <u>Conjugated Polymers and Related Materials—the Interconnection of Chemical and Electronic Structure, Proceedings, 81st Nobel Symposium,</u> 13 (June 18, 1991, Lulca), (Oxford University Press, Oxford, 1993), pp. 502.

2. M.H.Lyons ed., <u>Materials for Non-Linear and Electro-optico,</u> (IOP publishing Ltd., Bristol and N.Y., 1989).

3. R.S.Potember, T.O.Poehler, R.C.Hoffman, in <u>Molecular Elctronic Devices II,</u> (edited by F.L.Cartex, Marcel Dekker Inc. New York & Basel, 1988), pp. 97-110.

4. W.Xu, G.R.Chen and Z.Y.Hua, Vac.Sci.Tech.(China), **15** (6), 363-367 (1995).

5. W.Xu, G.R.Chen, and Z.Y.Hua, Appl.Phys.Lett. **67**(15), 2241-2242 (1995).

6. K.Yano, R.Kuroda, Y.Shimada, S.Shido, M.Kyogaku, H.Matsuda, K.Takimoto, K.Eguchi and T.Nakagiri, J.Vac.Sci.Technol. **B14**(2),1353-1355 (1996).

7. L.P.Ma, Y.L.Song. H.J.Gao, W.B.Zhao, H.Y.Chen, Z.Q.Xue, S.J.Pang, Appl.Phys. Lett. **69**(24), 1-2 (1996).

8. H.J.Gao, L.P.Ma, H.X.Zhang, H.Y.Chen, Z.Q.Xue and S.J.Pang, J.Vac.Sci.Technol. **B 15**(4), 1581-1583 (1997).

9. J.H.Hodgkin and J.Heller,   J.Polym.Sci.: Part C, **29**, 37-46 (1970).

10. C.J.Fritchie, Acta.Cryst. **20**, 892-898 (1966).

11. C.Julian Chen, <u>Introduction to scanning tunneling microscope,</u> (Chinese Light Industry Press, 1996), pp. 189-205.

PRESSURE DEPENDENCE OF PHOTOLUMINESCENCE OF OLIGOTHIOPHENES

L. ROSSI [a,b], W. GRAUPNER [c], R. RESEL [c], F. MEGHDADI [c], G. LEISING [c], F. SANNICOLO' [d], T. BENINCORI [d], G. LANZANI [a,e], R. TUBINO [a,f]

(a) Istituto Nazionale per la Fisica della Materia (Italy).
(b) Dipartimento di Fisica "A. Volta," Università di Pavia, Pavia (Italy).
(c) Institut für Festkörperphysik, Technische Universität Graz, Graz (Austria)
(d) Dipartimento di Chimica Organica ed Industriale, Università di Milano, Milano (Italy).
(e) Istituto di Matematica e Fisica, Università di Sassari, Sassari (Italy).
(f) Dipartimento di Scienze dei Materiali, Università di Milano, Milano (Italy).

ABSTRACT

We report on the nature of emitting states in two different quaterthiophenes: the unsubstituted quaterthiophenes (T4) and a bridged T4 (T4B), where the bridging have been realized by introducing a second bond between the two central thiophene rings. The effect of the chemical bridging on the photophysics of these compounds have been studied via photoluminescence measurements under hydrostatic pressure, which permit to clarify the influence of interchain coupling, conformational effects and intrachain structural changes on the optical properties.

INTRODUCTION

The class of thiophene oligomers has been widely studied because they possess extensive π-electron delocalization along the molecular backbone which makes them interesting for various optoelectronic applications as active layers in field effect transistors [1-3] and light emitting diodes [4]. Moreover, they can be used as model compounds for the parent polymer: polythiophene. While optical properties of oligothiophenes in solution have been studied in details, open problems still exist on the electronic properties of oligothiophenes in the solid phase such as single crystals and crystalline thin films, where optical properties strongly depend upon supramolecular organization.

T4

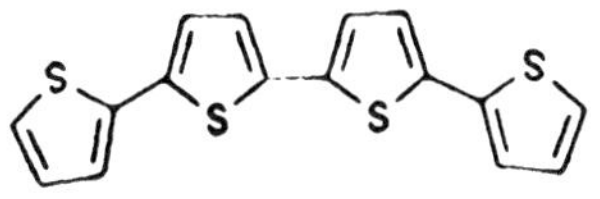

T4B

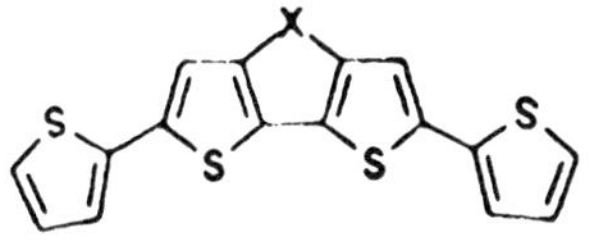

Fig. 1 Chemical structure of T4 and T4B: X=CH_2

Due to the flexibility of molecular design the optical properties of the oligothiophenes can be tuned by changing the length of the conjugated chain or, for a given chain length, by substituting hydrogen atoms with proper chemical groups. With the intent of controlling the molecular conformations and also to induce a controlled supramolecular ordering, we have synthesized a series of quaterthiophenes in which one or two inter-ring torsional degree of freedom have been blocked by chemical bridging [5].

Mat. Res. Soc. Symp. Proc. Vol. 488 © 1998 Materials Research Society

Here we report on the photophysics of a new synthesized quaterthiophene, namely T4B, compared with the unsubstituted T4. The chemical structure of T4B is sketched in Fig. 1, where X=CH$_2$. The bridge between the two central rings in T4B inhibits the inter ring rotation, thus reducing the efficincy of the non radiative decay channels of the excited state as pointed out by the higher photoluminescence (PL) quantum yield value (Φ_F=0.24) measured in T4B in solution with respect to T4 (Φ_F=0.17).

EXPERIMENT

We have evaporated thin films (d≈3500 Å) of T4 and T4B under high vacuum condition (2x10^{-6} mbar) and we have dispersed T4 and T4B molecules in a polymethylmethacrylate (PMMA) matrix. Absorbance and X-ray diffraction measurements were realized by conventional intrumentations. PL measurements under pressure were obtained by exciting with the 351.1/363 nm lines of an Ar$^+$ laser. The detection was realized via a combination of an optical fibre and an Ocean Optics CCD spectrometer. The sample was held in a pressure cell. The pressure was created via a Nova Swiss pressure amplifier, using oil on the low pressure side and He on the high pressure side, and was varied continuously from 0 to 4 kbar.

RESULTS

A. Evaporated Films

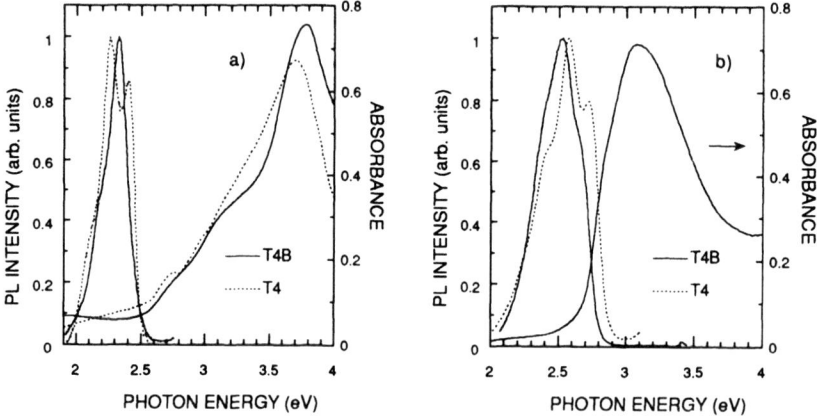

Fig. 2 Absorption and PL spectra at room temperature of T4 and T4B a) evaporated films and b) molecules dispersed in PMMA.

In Fig. 2 a) absorption and PL spectra of T4 and T4B evaporated films are reported. The main absorption peak of T4B is slightly blue shifted and sharpened with respect to that of T4. This peak can also be observed in T5 and T6 polycrystalline films and was assigned to the formation of H aggregates [6], where parallel-orientated transition dipoles result in the formation of an exciton band and only the transition to the higher energy state is allowed. The narrowing of this structure points to a high packing efficiency, driven by the presence of the chemical bridge, also proved by the lack of the contribution to absorption by isolated molecules at ~2.75 eV, clearly present in T4 absorption spectrum.

As expected, the PL spectrum of T4 evaporated films shows a vibronic progression, which is due to the coupling with a vibrational mode of approximately 0.14 eV, and previously assigned to isolated molecules contribution [7]. In contrast, the PL spectrum of T4B consists of two structures, a peak at 2.3 eV and its vibronic replica at 2.12 eV. The energy separation between the replicas is 180 meV, which is very close to typical vibrational energies of aromatic molecules and can be assigned to C-C stretching vibrations [8]. Since the relative intensity of the PL peaks depends on the overlapping of the wavefunctions of the ground and the excited states, the prominence of the high energy peak corresponding to the 0-0 transition is indicative of a small variation of the electronic configuration between those states [9]. This fact suggests that the chemical bridging induces a high degree of intramolecular order in the oligomer.

B. Oligomer Doped Matrix Films

The concentration of oligomers in PMMA was kept very low to prevent aggregation and formation of microcrystallites. X-ray diffraction measurements confirmed that molecules were completely isolated in the PMMA matrix.

The absorption spectrum of T4B in PMMA matrix (PMMA:T4B) shows a structureless peak at 3.1 eV (Fig. 2 b). The main absorption peak is usually associated to the contribution of the conjugated backbone, supposing that the oligothiophenes are torsionally mobile around the inter ring bonds in the ground state. In contrast, the PL spectra show well resolved emission peaks. The reason for this difference is that the structure of the ground state is benzoic so we expect that at room temperature several non planar conformers are present: the inter ring rotational vibration give rise to loss of $\pi-\pi*$ conjugation decreasing the effective conjugation length. Since, the structure of the first excited state is more quinoid, so with double bond character of the ring connecting bonds, it leads to a more planar configuration. The double bond linkage in the quinoid structure does inhibit free inter-ring torsions narrowing the distributions of rotational conformers [10]. Besides the PL spectrum of PMMA:T4B is slightly red shifted with respect to that of PMMA:T4. It is well known that the energy positions of the main absorption and emission peaks of unsubstituted oligothiophenes depend on the number n of thiophene rings [11]: in particular the peak energies shift bathochromic with increasing n. A simple one dimensional free electron model can account quite well for this effect. In this model the main parameter is the effective conjugation length, which describes the π-electron delocalization along the molecular backbone. Up to about six-eight repeating units this length coincides with the physical length of the molecule. Single bridging seems to slightly increase the effective conjugation length (cl_{eff}) resulting in a higher cl_{eff} for bridged molecules when compared to unbridged molecules of the same physical length.

The PL emission in T4B and in PMMA:T4B originates probably from different emitting species, since the PL spectra are completely different in shape. The emission in PMMA is very similar to that of T4, thus can be assigned to emission from isolated molecules.

C. Pressure Studies

In Fig. 3 a) we show the PL spectra of T4B evaporated films for two selected values of hydrostatic pressure. Upon increasing the pressure we observe three concomitant effects: 1) a linear red shift of the main peak, 2) a sharpening of the structures and 3) a slight intensity increase. All the effects are completely reversible.

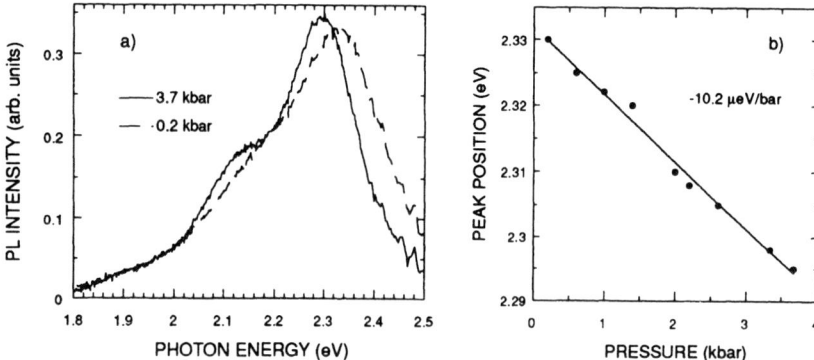

Fig. 3 a) PL spectra at room temperature of T4B evaporated film at two selected values of pressure: 0.2 and 3.7 kbar, b) pressure dependence of the energy position of the main PL peak.

In order to quantify the shift in Fig. 3 b) the position in energy of the main absorption peak is reported as a function of the pressure applied up to 4 kbar. For the whole examined range the shift is linear and its value is -10.2 μeV/bar. This value is quite different with respect to typical values for other conjugated polymers, in particular ladder type poly(para-phenylene) (LPPP) (-7.7 μeV/bar) [12], polyparaphenylenevinylene (PPV) (-5.4 μeV/bar) [13], and polyacetylene (-13 μeV/bar) [14]. Our results are in spite very similar to that obtained in poly-3-hexyl-thiophene (P3HT) (-10 μeV/bar) [15] thus suggesting that the chemistry of the molecular backbone has a determining influence on this value.

The fact that we do not observe broadening of the PL indicates that interchain coupling between two conjugated backbones does not play a significant role in the pressure region that have been analyzed. The rigid shift of the whole spectrum have been usually explained via the so called "solvent" or "gas-to-crystal" effect familiar in the study of molecular crystals [16]: dipole-dipole interactions with the surroundings of the conjugated backbone lower the excited state energy relative to the ground state. In contrast, in our case the behaviour of the PL spectrum under pressure, in particular the sharpening and the enhancement seems to be very similar to PL behaviour upon cooling. The emission in T4B film comes probably from self trapped exciton or from other unknown states which under pressure are in a more planar and hence rigid conformation. In the first case the red shift of PL bands can be explained by a high pressure induced increasing of the Davydov splitting of excitonic bands. The extension of the explored pressure range could allow to see the transition from the two different states of aggregation of the molecules clearly distinguishable in the two T4B samples.

In PMMA:T4B we observe a completely different behaviour: upon increasing the pressure the PL intensity is irreversibly quenched. In Fig. 4 we report PL spectra for three representative pressure values during the rising pressure run. The PL spectra during the decreasing pressure run do not change at all both in intensity and in shape. Similar results are obtained in PMMA:T4. The quenching is associated to a very weak broadening of the PL spectrum, probably due to a higher interaction between molecules under pressure. We have checked for the light induced degradation of the sample and we can exclude this as a dominant effect on the emission. Thus we can infer that the PL quenching is probably due to effects of partial aggregation of molecules in the matrix.

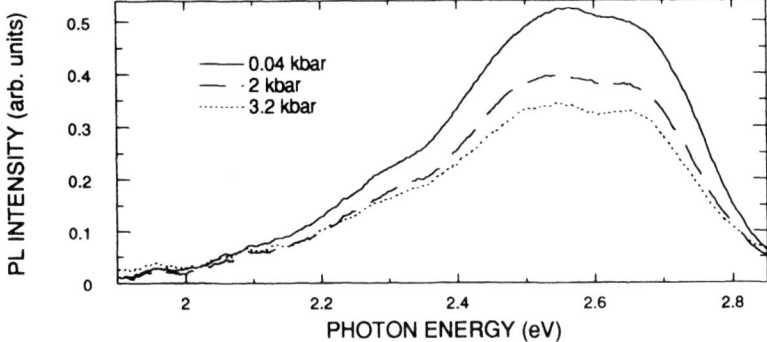

Fig. 4 PL spectra at room temperature of T4B molecules in PMMA at three selected values of pressure: 0.04, 2 and 3.2 kbar.

CONCLUSIONS

We have studied the photophysics of two quaterthiophenes, T4 and T4B, in different forms: thin evaporated films and oligomers doped PMMA matrix films. The results show that emission in T4 always originates from isolated molecules contribution, while in T4B it depends on the aggregation properties which are different in the evaporated film and in the PMMA:T4B sample. From PL measurements under pressure of T4B films we were able to deduce a value of -10.2 μeV/bar for the peak energy shift which is reversible and is accompanied by a slight increase of the intensity. In PMMA:T4B the most relevant effect is an irreversible quenching of the PL intensity probably due to a partial aggregation of the molecules.

Moreover, T4B shows high packing efficiency of molecules in the solid state and a high degree of intramolecular order. The effects of the bridging are an increase in the quantum yield of PL in solution and peculiar self assembly properties in the solid state.

ACKNOWLEDGMENTS

This work was partially funded by a bilateral programme for Scientific-Technological Collaboration Austria-Italy 1997-1999 and OeNB project 6608.

REFERENCES

1. H. Koezura, A. Tsumura, H. Fuchigami and K. Kuramoto, Appl. Phys. Lett. **62**, p. 1794 (1992).

2. F. Garnier, R. Hajlaoui, A. Yassar and P. Srivastava, Science **265**, p. 1684 (1994).

3. A. Dodabalapur, L. Torsi and H.E. Katz, Science **268**, p. 270 (1995).

4. D.D.C. Bradley, Synth. Met. **54**, p. 401 (1993).

5. M.E. Brenna, PhD Thesis, University of Milan, Italy (1993).

6. H.-J. Egelhaaf and D. Oelkrug, Proc. SPIE **2363**, 403 (1995).

7. Y. Kanemitsu, N. Shimizu, K. Suzuki, Y. Shiraishi, and M. Kuroda, Phys. Rev. B **54**, p. 2198 (1996).

8. A. Bree, S. Katagiri, and R. Suart, J. Chem. Phys. **44**, p. 1788 (1966).

9. M. Pope and C.E. Swenberg, <u>Electronic Processes in Organic Crystals</u>, Clarendon, Oxford, 1982, pp. 25-27.

10. H. Akimichi, K. Waragai, S. Hotta, H. Kano, and H. Sakaki, Appl. Phys. Lett. **58**, p. 1500 (1991).

11. D. Fichou, G. Horovitz, B. Xu, and F. Garnier, Synth. Met **39**, p. 243 (1990).

12. W. Graupner, S. Eder, K. Petritsch, G. Leising, and U. Scherf, Synth. Met. **84**, p.507 (1997).

13. S. Webster and D.N. Batchelder, Polymer **37**, p. 4961 (1996).

14. D. Moses, A. Feldblum, E. Ehrenfreund, A.J. Heeger, T.C.Chung, and A.G. MacDiarmid, Phys. Rev. B **26**, p. 3361 (1982).

15. B.C. Hess, G.S. Kanner, and Z. Vardeny, Phys. Rev. B **47**, p. 1407 (1993).

16. R.J. Lacey, D.N. Batchelder, and G.D. Pitt, J. Phys. C: Solid State Phys. **17**, p. 4529 (1984).

SUPERCONDUCTIVITY OF κ-(BEDT-TTF)$_2$ Cu[N(CN)$_2$]Br: ISOTOPE EFFECT REVISITED

M. TOKUMOTO*, N. KINOSHITA*, Y. TANAKA*, H. ANZAI**
* Electrotechnical Laboratory, Tsukuba, Ibaraki 305, Japan, madoka@etl.go.jp
** Himeji Institute of Technology, Kamigori, Hyogo 678-12, Japan

ABSTRACT

In 1991, we reported a normal isotope effect in organic supercon-ductor κ-(BEDT-TTF)$_2$Cu[N(CN)$_2$]Br, when all the hydrogen atoms of BEDT-TTF were replaced with deuterium. In other words, T_c was depressed by as much as 0.9 K, in contrast to the "inverse isotope effect" commonly ob-served in 10K-class organic superconductors. Recently, it was reported that the deuterated κ-(BEDT-TTF)$_2$Cu[N(CN)$_2$]Br shows an insulating nature when cooled very rapidly. Therefore, it is necessary for us to reexamine the superconducting transition of both deuterated and undeuterated κ-(BEDT-TTF)$_2$Cu[N(CN)$_2$]Br by SQUID measurements with special attention to the cooling speed. We studied the effect of cooling rate ranging from 10 K/min down to as slow as 0.02 K/min, and observed a significant ef-fect not only on the superconducting transition temperature T_c but also on the superconducting volume fraction.

INTRODUCTION

κ-(BEDT-TTF)$_2$Cu[N(CN)$_2$]Br, where BEDT-TTF is bis(ethylenedithio) tetrathiafulvalene, is an organic superconductor with the highest T_c of 11.6 K at ambient pressure.[1] We reported a normal "isotope effect", i.e. when all the hydrogen atoms of BEDT-TTF are replaced with deuterium. In other words, T_c was depressed by as much as 0.9 K [2] in contrast to the "inverse isotope effect" observed in κ-(BEDT-TTF)$_2$Cu(NCS)$_2$[3]. Similar reports also confirmed the normal isotope shift in the same material[4-6]. Although, the origin of inverse isotope effect still remains unsolved[7], it has been puzzling why this salt, κ-(BEDT-TTF)$_2$Cu[N(CN)$_2$]Br, is the only ex-ception among the 10 K class organic superconductors which commonly show the "inverse isotope effect", in which T_c increases by substitution with heavier isotope, i.e. hydrogen to deuterium or ^{12}C to ^{13}C.[8] Recently, it was reported that the deuterated κ-(BEDT-TTF)$_2$Cu[N(CN)$_2$]Br is of an in-sulating nature similar to κ-(BEDT-TTF)$_2$Cu[N(CN)$_2$]Cl when cooled very rapidly.[9] Actually, a closer look into the result of our previous report,

Mat. Res. Soc. Symp. Proc. Vol. 488 © 1998 Materials Research Society

reproduced in Fig. 1, clearly indicates that the superconducting fraction for deuterated κ-(BEDT-TTF)$_2$Cu[N(CN)$_2$]Br is almost 40 times smaller than that for undeuterated one, suggesting that something unusual is hidden behind the difference.   On the other hand, a strong depression of T_c, as large as 1.2 K was reported, when the sample of undeuterated κ-(BEDT-TTF)$_2$Cu[N(CN)$_2$]Br was rapidly cooled.[10]   We did not have an opportunity to study further in detail until recently, partly because the facility (old SQUID magnetometer) did not allow us to study the effect of cooling rate in a controlled manner.

However, it has now become possible due to the innovation of the temperature control sysytem of the SQUID magnetometer.   In addition, it is clear that we should reexamine the superconducting properties of both deuterated and undeuterated κ-(BEDT-TTF)$_2$Cu[N(CN)$_2$]Br by SQUID measurements with special attention to the cooling speed of our samples, because we did not care about the cooling rate at all in our previous experiments.   In other word, we did not know the cooling rate, nor did we control it.

In this paper, we present the result of reexamination of the isotope effect on the superconductivity in κ-(BEDT-TTF)$_2$Cu[N(CN)$_2$]Br. We observed a significant effect of cooling rate not only on the superconducting transition temperature T_c but also on the superconducting volume fraction, especially in the case of deuterated salt.   In particular, it was quite unexpected that an extremely slow cooling rate as low as 0.02 K/min was found to have still a significant effect on the superconducting state.

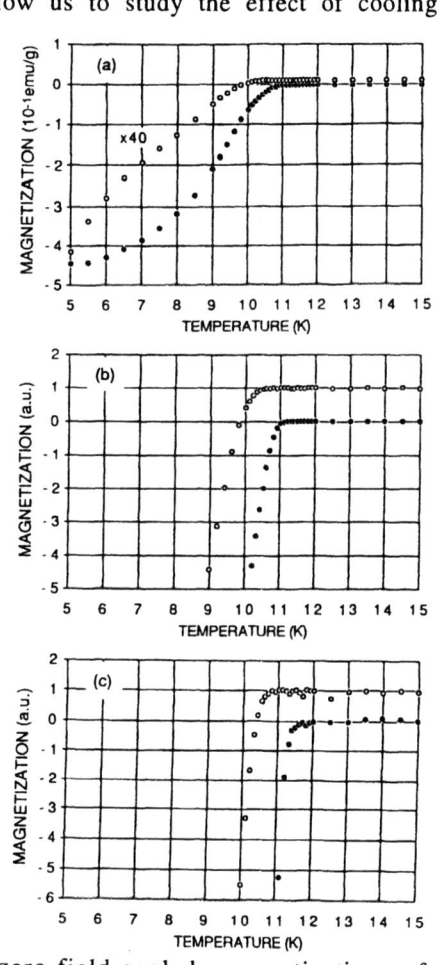

Fig. 1. Temperature dependence of zero-field-cooled magnetization of deuterated (o) and undeuterated (•) κ-(BEDT-TTF)$_2$Cu[N(CN)$_2$]Br in the magnetic field of 10 Oe.   (b) and (c) shows successive blowups of ordinate of Fig. (a) (from ref. 2)

EXPERIMENTAL

Single crystals of κ-(BEDT-TTF)$_2$Cu[N(CN)$_2$]Br were grown by electro-chemical oxidation of BEDT-TTF in the mixture of organic solvent 1,1,2-Trichloroethane with 10% ethanol in the presence of CuBr, NaN(CN)$_2$ and 18-crown-6 ether by application of a constant current of 1μA under nitro-gen gas atmosphere.   Deuterated samples were obtained starting from BEDT-TTFd8, where eight hydrogen atoms of ethylene groups at the end of BEDT-TTF molecule are replaced by deuterium, in a similar way.[2]   The magnetization and its temperature dependence was measured by a QUANTUM DESIGN SQUID magnetometer MPMS-5S equipped with a new temperature controlling system, Continuous Low Temperature Control (CLTC) option, in which operation of Temperature Sweep Mode enables us to change the temperature of our sample with a constant rate ranging between 10 K/min and 0.001 K/min.   This system does not allow us to quench, i.e. rapidly cool our sample, but enabled us to study the effect of extremely slow cooling rate as shown below.   The data were collected by a SQUID magnetometer with Reciprocating Sample Option (RSO) for about 20 mg of non-oriented single crystals.   The sample was first cooled from 100 K to 5 K under zero magnetic field.   For cooling from 100 K to 60 K, cooling rate was kept at a specified constant value ranging from 10 K/min down to 0.02 K/min for each measurement.   From 60 K down to 5 K, the sample was cooled at a rate of 10 K/min, and then the magnetic fields up to 10 Oe was applied at 5 K.

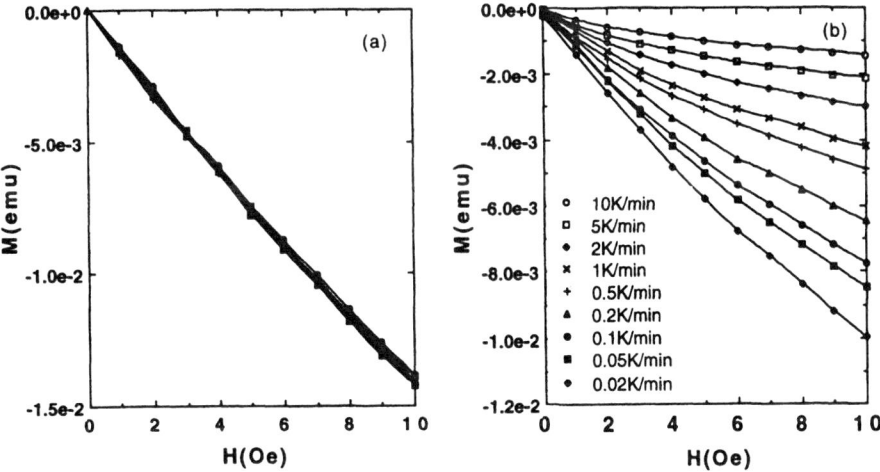

Fig. 2. Effect of cooling speed on the field dependence of the zero-field-cooled magnetization at 5 K. (a) (left) undeuterated, and (b) (right) deuterated κ-(BEDT-TTF)$_2$Cu[N(CN)$_2$]Br.

RESULTS AND DISCUSSION

In order to study the effect of the cooling rate, we must specify the temperature range in which we control the cooling rate. In the case of deuterated κ-(BEDT-TTF)$_2$Cu[N(CN)$_2$]Br, a rapid cooling through 80 K was reported to drive the superconducting phase into a magnetic insulating phase.[11]   A kink in resistivity and an anomaly in the thermal expansion coefficient at 80 K was reported.[12]   A possible candidate for this anomaly is the thermally induced order-disorder transition in the terminal ethylene conformation of BEDT-TTF molecules.

Therefore, when the sample was cooled from 100 K to 5 K under zero magnetic field, cooling rate was controlled at a specified constant value ranging from 10 K/min down to 0.02 K/min .from 100 K to 60 K for each measurement.   From 60 K down to 5 K, the sample was cooled at a rate of 10 K/min, and then the magnetic fields up to 10 Oe was applied. Figure 2 shows the effect of cooling speed on the field dependence of the zero-field-cooled magnetization at 5 K for (a) undeuterated, and (b) deuterated κ-(BEDT-TTF)$_2$Cu[N(CN)$_2$]Br. The effect of cooling rate is almost negligible for undeuterated sample, but is tremendously large for deuterated one.   The superconducting fraction in deuterated sample keeps increasing with decreasing cooling rate, without any indication of saturation even at an extremely slow cooling rate of 0.02 K/min.

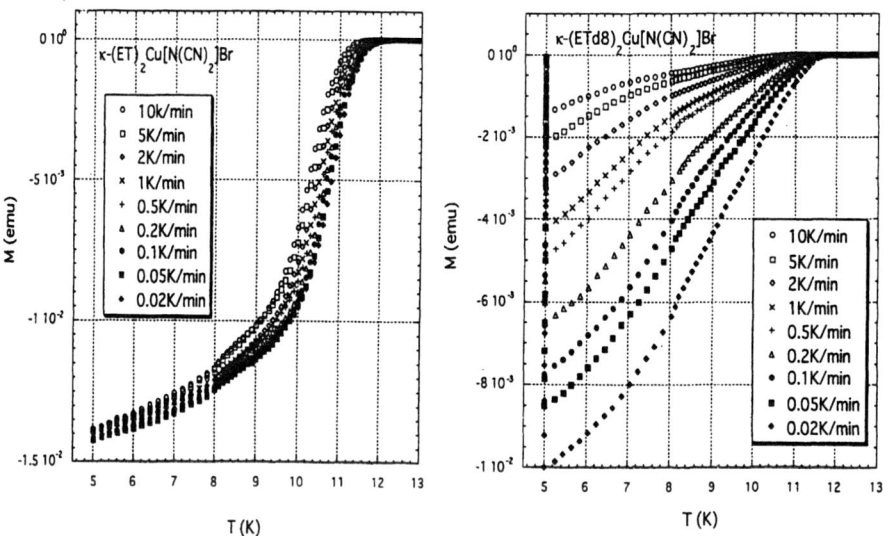

Fig. 3. Effect of cooling speed on the temperature dependence of the zero-field-cooled magnetization (H=10 Oe).   (a) (left) undeuterated, and (b) (right) deuterated κ-(BEDT-TTF)$_2$Cu[N(CN)$_2$]Br.

After the field of 10 Oe was applied at the lowest temperature, the temperature dependence of magnetization M_{ZFC} was measured with increasing temperature. Figure 3 shows the effect of cooling speed on the temperature dependence of zero-field-cooled magnetization between 5 K and 13 K under the magnetic field of 10 Oe for (a) undeuterated, and (b) deuterated κ-(BEDT-TTF)$_2$Cu[N(CN)$_2$]Br. The effect of cooling rate on the temperature dependence is not negligible even for undeuterated samples, in contrast to the case of field dependence shown in Fig. 2(a). This fact is consistent with the previous report.[10] The effect is tremendously large for deuterated samples. Again, the superconducting fraction in deuterated samples keeps increasing with decreasing cooling rate at all temperatures, without any indication of saturation even at an extremely slow cooling rate of 0.02 K/min.

Figure 4 shows a blow up of the temperature dependence of magnetization (Fig. 3) in the vicinity of T_c for (a) undeuterated, and (b) deuterated κ-(BEDT-TTF)$_2$Cu[N(CN)$_2$]Br. In the case of undeuterated sample, T_c increase as much as 0.5 K was observed by decreasing the cooling rate from 10 K/min down to 0.02 K/min, without indication of saturation. On the other hand, in the case of deuterated sample, T_c increase as much as 1.0 K was observed by decreasing the cooling rate from 10 K/min down to 0.02 K/min, also without any indication of saturation. An additional feature in the latter case is the enhancement of paramagnetic susceptibility just above T_c, which becomes more prominent with decreasing the cooling rate.

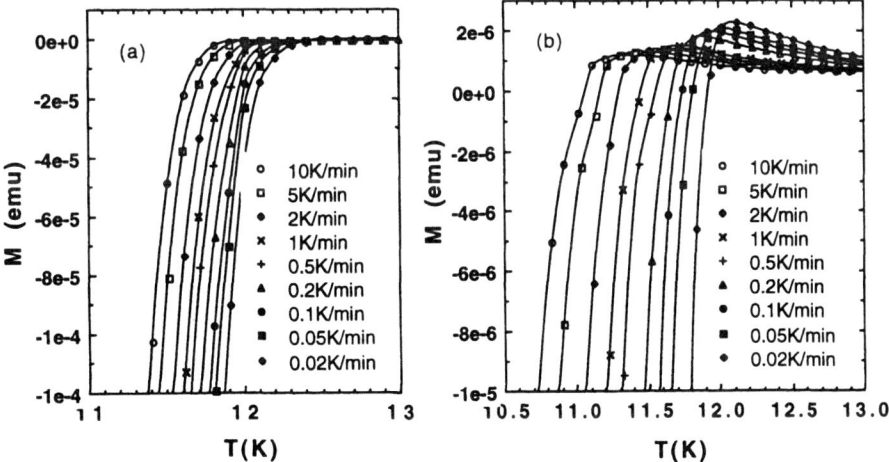

Fig. 4. A blowup of the effect of cooling speed on the temperature dependence of magnetization (Fig. 3) in the vicinity of T_c. (a) (left) undeuterated, (b) (right) deuterated κ-(BEDT-TTF)$_2$Cu[N(CN)$_2$]Br.

CONCLUSIONS

We have reexamined the isotope effect on superconductivity in κ-(BEDT-TTF)$_2$Cu[N(CN)$_2$]Br, when all the hydrogen atoms of BEDT-TTF were replaced with deuterium. In the case of deuterated salt, not only the superconducting transition temperature T_c but also the superconducting volume fraction showed a significant increase by decreasing the cooling rate from 10 K/min down to 0.02 K/min. The effect is much less in the case of undeuterated salts. The normal isotope shift in T_c as much as 0.7 K for the cooling rate of 10 K/min was found to become less than 0.1 K for the slowest cooling rate of 0.02 K/min. Obviously, we need further study, since there is no indication of saturation of the effect of the cooling rate even at the slowest rate of 0.02 K/min in the present study.

REFERENCES

1. A. M. Kini, U. Geiser, H. H. Wang, K. D. Carlson, J. M. Williams, W. K. Kwok, K. G. Vandervoort, J. E. Thompson, D. L. Stupka, D. Jung and M. -H. Wangbo, *Inorg. Chem.* **29**, 2555 (1990)
2. M. Tokumoto, N. Kinoshita, Y. Tanaka and H. Anzai, *J. Phys. Soc. Jpn.* **60**, 1426 (1991)
3. K. Oshima, H. Urayama, H. Yamochi and G. Saito, *Synth. Metals* **27**, A473 (1988)
4. H. H. Wang, K. D. Carlson, U. Geiser, A. M. Kini, A. J. Schultz, J. M. Williams, L. K. Montgomery, W. K. Kwok, U. Welp, K. G. Vandervoort, S. J. Boryschuk, A. V. Strieby Crouchi, J. M. Kommers and D. M. Watkins, *Synth. Metals* **41-43**, 1983 (1991)
5. H. Ito, M. Watanabe, Y. Nogami, T. Ishiguro, T. Komatsu, G. Saito and N. Hosoito, *J. Phys. Soc. Jpn.* **60**, 3230 (1991)
6. T. Komatsu, N. Matsuoka, T. Nakamura, H. Yamochi, G. Saito, H. Ito and T. Ishiguro, *Phosphorus, Sulfur, and Silicon* **67**, 295 (1992)
7. M. Tokumoto, K. Murata, N. Kinoshita, K. Yamaji, H. Anzai, Y. Tanaka, Y. Hayakawa, K. Nagasaka and Y. Sugawara, *Mol. Cryst. Liq. Cryst.* **181**, 285 (1990)
8. G. Saito, A. Otsuka and A. A. Zakhidov, *Mol. Cryst. Liq. Cryst.* **284**, 3 (1996)
9. Y. Nakazawa and K. Kanoda, *Phys. Rev. B* **53**, R8875 (1996)
10. W. K. Kwok, U. Welp, K. D. Carlson, G. W. Crabtree, K. G. Vandervoort, H. H. Wang, A. M. Kini, J. M. Williams, D. L. Stupka, L. K. Montgomery and J. E. Thompson, *Phys. Rev. B* **42**, 8686 (1990)
11. A. Kawamoto, K. Miyagawa and K. Kanoda, *Phys. Rev. B* **55**, 14140 (1997)
12. M. Kund, H. Müller, W. Biberacher, K. Andres and G. Saito, *Physica B* **191**, 274 (1993)

MORPHOLOGY DEPENDENCE OF THE OPTICAL PROPERTIES OF DALM RELATED MATERIALS

K. B. WAGNER-BROWN, K. F. FERRIS[*], J. L. KIEL[**]and R. A. ALBANESE[**], Conceptual MindWorks Inc., San Antonio,TX 78228, [*] Pacific Northwest National Laboratory, Materials and Chemical Sciences Department, Richland, WA 99352, [**] Air Force Research Laboratory, Human Effectiveness Directorate, Brooks AFB, San Antonio, TX 78235.

ABSTRACT

Diazoluminomelanin (DALM) is an electroluminescent polymer which has shown significant optical activity in response to perturbing fields. The current model for this process features optical excitation of a polymer backbone containing conducting conjugation, with subsequent energy transfer to a luminescent group. In this paper we have performed electronic structure calculations using the AM1 Hamiltonian with configuration interaction to estimate the electronic properties of two potential models for the DALM backbone. Contrary to the conventional picture of conjugation, the phenyl groups in the DALM backbone show significant twist angles (42°-55°) depending on substitutional group, resulting in localized electronic excitations.

INTRODUCTION

Diazoluminomelanin (DALM) is a luminescent conjugated polymer discovered and further developed by Kiel et al[1]. The molecule is thermochemiluminescent and can be formed chemically and biologically, the latter in both bacteria and mammalian cells[2-5]. The optical activity of the DALM system is affected by electromagnetic radiation and temperature[1-3]. The sensitivity of the DALM to microwave radiation [3] and its ability to be formed within cells, suggest that when the mechanisms of its luminescence are better understood, it may potentially be useful as a cellular dosimeter in measuring microwave absorption and as a molecular temperature probe[6].

Work by Wright[7] suggests that the backbone of DALM is a poly(m-phenylene) structure, more specifically, a poly-tyrosine as shown in figure 1a . Luminol (5-amino-2,3-dihydro-1,4-phthalazinedione) is also present in the polymer but its positions of attachment along the chain are unknown. Physical characterization of DALM has suggested that the luminescence is due to an excitation of the backbone with subsequent energy transfer to the luminol.

To investigate the luminescence of the DALM system, we have selected two models for the DALM backbone; the phenol oligomers and tyrosine oligomers shown in figure 1. We have examined and compared the electronic transitions and HOMO-LUMO energies of both of these representative backbones. To determine whether it may be necessary to study these properties in a time-averaged manner, we have also calculated the energy barrier for rotation between neighboring rings in a biphenol system and examined the dependence of the electronic properties on the torsion angle.

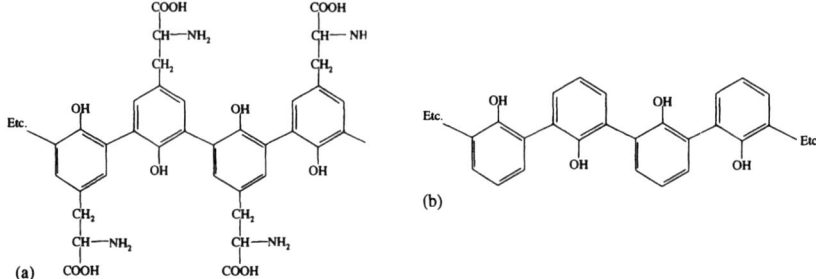

(a)

(b)

Figure 1. A schematic illustration of the (a) poly-tyrosine and (b) poly-phenol system representing a simplified DALM backbone.

COMPUTATIONAL METHODS

We have used the AM1 Hamiltonian [8] as implemented in MOPAC93 [9], to calculate both the optimized geometries and the electronic transitions. The AM1 Hamiltonian has been shown to provide accurate descriptions of the optical properties in aromatic molecules [10]. All bond lengths and angles were optimized by force relaxation methods [11]. The twist angle (ϕ) between the phenol rings was the only constraint for mapping the conformational space of the biphenol molecule. The minimum energy conformation occurs for a twist angle between the phenyl groups of biphenyl at 42°, in good agreement with the experimental values.

The excitation energies have been determined using configuration interaction (CI) methods. The use of CI for semi-empirical Hamiltonians entails some care as inclusion of more configurations can often increase the excitation energy. Excitation energies were obtained by performing configuration interaction (CI) calculations using the 4 highest occupied and lowest 4 unoccupied orbitals. The interaction space consisted of the 100 lowest energy configurations that could be created by excitations from the 4 highest filled molecular orbital to the lowest 4 vacant molecular orbitals. Diagonalization of the resulting matrix yielded the excitation energies and the oscillator strengths of the transitions. There was a better correspondence between the predictions of the calculations and the known phenol band at 270nm when multiple excitations of the system were allowed (as opposed to single excitations) so the electronic calculations were limited to singlet transitions but multiple excitations.

RESULTS

The oligomers of phenols were studied as a first approximation to the backbone of the DALM molecule (figure 1b). Figure 2 shows the HOMO-LUMO energies for the phenol systems as a function of 1/n. Making use of the reciprocal length rule for polymers, we find that for large values of n the energy gap approaches 7.75eV. Calculations were also performed using optimized force field geometries for up to 6 monomers and yielded approximately the same HOMO-LUMO energy for large n.

Orbital amplitude plots for the HOMO's and LUMO's of the phenols show that the HOMO is a π-like orbital and the LUMO is a π* orbital as is seen in most conjugated polymers[12]. When the molecular orbitals are examined as a function of energy, it is noticeable that as the oligomer length increases the unoccupied orbitals become more band-like. Additionally, the orbital amplitude plots show that a HOMO-LUMO transition in the biphenol system would create a charge transfer across the rings

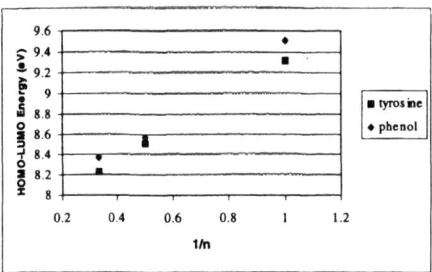

Figure 2. Variation of the HOMO-LUMO energy difference for tyrosine and phenol oligomers of length n.

since the orbitals are localized on different rings. If the neighboring rings in the system were coplanar this localization would be lessened and there would be more efficient conjugation in the system.

In figure 3 the oscillator strengths and excitation energies for phenol oligomers of different length are shown. Transitions with oscillator strengths lower than 0.5 are not shown but only one of these excluded points has an energy lower than the 3.8eV cut-off of the figure (for a single phenol system there is an excited state at 3.7eV with an oscillator strength of 0.14). There is a general trend in the excited states that as the oligomer length grows there are more lower energy transitions with significant oscillator strength, indicating that the longer oligomers could be more easily excited.

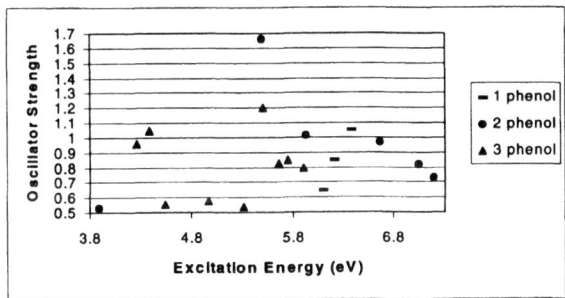

Figure 3. Variation in the oscillator strengths and energies of the transitions of phenol oligomers.

The electronic properties of oligomers of tyrosine (shown in figure 1a) were also examined since this system is the suspected DALM backbone. The HOMO-LUMO energies for these oligomers are also shown in Figure 2. The energy approaches 7.69eV for large n. Figure 4 shows the excited state energies for the tyrosine systems. As in the case of the phenols, increases in the oligomer length yield a greater number of lower energy states with large oscillator strengths. The figure only shows transitions with oscillator strengths above 0.5 such that many states with small oscillator strengths are not

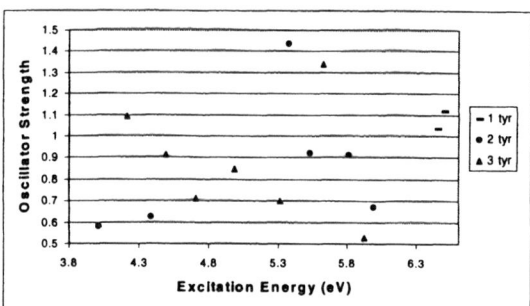

Figure 4. Variation in the oscillator strengths and energies of the transitions of tyrosine oligomers.

shown. There are, however, no states with energies lower than the 4eV state of the 2 tyrosine system. The similarities between the energies for the tyrosine systems and the phenol systems suggests that there are similar electronic processes occurring in both systems and the phenols may be used as a simpler but reliable model for the DALM backbone.

The torsion angle between two neighboring rings in both the phenol systems and the tyrosine systems is 40°-55° at the energy optimized geometries. To determine whether time averaged properties are needed, we calculated the heat of formation for a biphenol system as a function of torsion angle using both the AM1 Hamiltonian and the AMBER force field method, and performed a subsequent molecular dynamics simulation for a three phenol system. The torsion angle between the two phenols, shown in figure 5, was varied through 360° for the AM1 calculations and 180° for the AMBER calculations. The resulting dependence of the heat of formation on the angle is shown in figure 6 for both methods. The AM1 calculations show absolute minima in the biphenol energy surface at about 57° and 307° with local minima at 137° and 227°. The AMBER results are similar but show both the 57° and the 137° minima to be absolute. Both methods show large peaks at 0°, where the OH's are both on the same side of the biphenol, and 180° where the OH's lie opposite one another. The height of the barriers for the geometries occurring at 0° and 180°, approximately 8

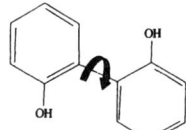

Figure 5. Schematic of the biphenol system. The torsion angle about which the rings were rotated is shown by the arrow.

kcal/mole, are large enough that the rings will not tend to be coplanar, and it is unlikely, at room temperature, that the system will often pass through these configurations. Molecular dynamics simulations support this hypothesis by showing that in the course of 30ps the torsion angle never goes below about 20° and seldom approaches 180°; the system spends most of its time in the geometry having a 50° or 130° torsion angle. Smaller peaks in the energy surface also occur at 90° and 270° when the rings are perpendicular. Although these peaks present a smaller barrier to the rotation, the molecular dynamics simulations show that the system will spend little time in the perpendicular conformations appearing to only pass through the 90° torsion angle when changing between the preferred conformations at 50° and 130°.

Figure 7 shows the transitions of the biphenol system as a function of torsion angle. It is noticeable that the transition energies have their smallest values when the two rings are co-planar at 0° with the two hydroxyl groups on the same side. It should, however, be recalled that the system spends little if any time in this conformation. When the two rings are co-planar at 180° the energies of the excited states are similar to those at the system's optimal geometry. The transition energies increase and are the largest for both perpendicular geometries.

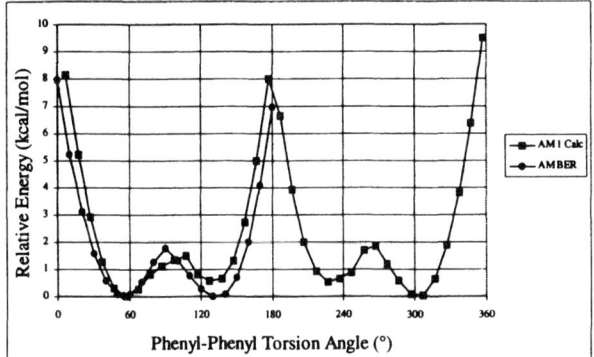

Figure 6. The heat of formation on the biphenol system as a function of the torsion angle between the two rings, relative to the energy occurring at the optimal geometry.

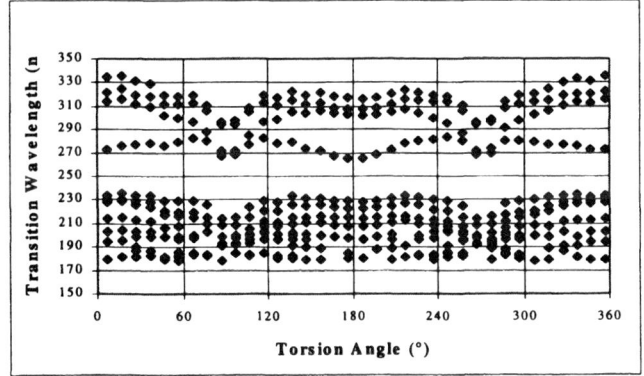

Figure 7. The transitions of the biphenol system as a function of the torsion angle between the rings. States of all oscillator strengths are shown.

CONCLUSIONS

The studies of the electronic properties of the phenol and tyrosine oligomers have shown that in both systems the molecule behaves more like a conducting polymer as its length increases. Since the optimal energy conformation of the systems do not contain co-planar neighboring rings and the electronic transitions are localized on the backbone, the conjugation of the system is broken up. Examinations of the geometries of the excited states should reveal whether large changes in the structure occur that allow for more conjugation along the backbone when the system is excited.

Both the molecular dynamics simulations and the torsional dependence of the heat of formation in the biphenol system indicate that time averages of the electronic processes need only include the two favored geometries having the torsion angles of 50° and 137°. The strengths and energies of the transitions are, however, very dependent upon the torsion angle which may offer a physical explanation for the temperature dependence of the DALM luminescence.

DALM is known to have long lived luminescence (on the order of hours) introducing the strong possibility that there is triplet involvement electronic processes. Future work will include examination of the triplet contribution to the luminescence and consider the probability of singlet-triplet interactions.

ACKNOWLEDGEMENTS

This work was supported in part by the United States Air Force Office of Scientific Research and the U. S. Army Medical Research and Material Command under contract DAMD17-94-C-4069, and the U. S. Department of Energy, Office of Basic Energy Sciences, Materials Sciences Division under contract DE-AC06-76RLO1830.

REFERENCES

1. J. Kiel, G. O'Brien, J. Dillon and J. Wright, Free Rad. Res. Comms. **8**, 115 (1990).
2. J. Bruno and J. Kiel, in Electricity and Magnetism in Biology and Medicine, Ed. M. Blank, San Francisco Press, San Francisco, 1993.
3. J. Bruno and J. Kiel, Bioelectromags. **15**, 315 (1994).
4. J. Kiel, C. Gabriel, D. Simmons, D. Erwin and E. Grant, Proc. IEEE Eng. Med. Biol. Soc. **12**, 1689 (1990).
5. J. Kiel, J. Parker, J. Alls and R. Weber, Proc. IEEE Eng. Med. Biol. Soc. **13**, 1689 (1991).
6. J. L. Kiel, J. G. Bruno and W. D. Hurt, in Radiofrequency Standards, Ed. B. J. Klaunberg et al., Plenum Press, New York, 1994.
7. R. Wright, unpublished work, Southeastern Oklahoma State University.
8. M.J.S. Dewar, E.G. Zoebisch, E.F. Healy, and J.J. Stewart, J. Am. Chem. Soc. **107**, 3902 (1985).
9. MOPAC 93.00, J.J.P. Stewart, Fujitsu Limited, Tokyo, Japan (1993).
10. H.A. Kurtz, J.J.P. Stewart, and K.M. Dieter, J. Comp. Chem. **11**, 82 (1990).
11. D.F. Shanno, J. of Optimization Theory and Applications **46**, 87 (1985).
12. A. J. Heeger and J. Long, Jr., Optics & Photonics News **8**, 24 (1996).

Thermally stable iridescent colors of a new low-molecular-weight cholesteric compound in the solid state and the application to full-color recording

Nobuyuki Tamaoki, Alexander V. Parfenov, Atsushi Masaki and Hiro Matsuda
National Institute of Materials and Chemical Research, 1-1 Higashi, Tsukuba, Ibaraki 305, Japan

ABSTRACT

We synthesized a series of new di-cholesteryl esters of several diacids. One of the compounds containing both of a diyne unit and long alkylene units, in the form of a thin solid film, showed stable iridescent colours in the visible region after rapid cooling process from the temperature of the cholesteric liquid crystalline phase. Measurement of selective reflectance of visible light, X-ray diffraction, optical rotation dispersion of the solid films proved that the films remained the cholesteric ordering of the molecules. The cholesteric structure in the solid films were completely stable at room temperature. The colour of the solid film was totally, partially or in imagewise manner changed into selected colours by adjusting the temperature of the surface just before the rapid cooling process. The property of this compound can be applicable to memory media with high density using the spectrum of light, or to thermo-sensitive full-colour recording materials.

INTRODUCTION

A full-colour recording is technologically important not only because colour information can be stored and reproduced but also because it can be possible to enhance recording density without decreasing the size of one bit by utilizing the spectrum of light for recording and reading memories, as in photochemical hole burning. Although recording media with the property of full-colour "write-once" (like photographs and photocopiers) or rewritable "on-and-off" (like magnetic, opto-magnetic and thermo-optic recordings) are well developed, a rewritable full-colour recording has still not been realized. This is mainly attributable to the difficulty in finding thermally stable materials which reversibly change colours by introduction of external stimuli.

Cholesteric colour has fascinated people since the discovery of liquid crystals[1]. Change of the cholesteric colours in the whole visible region by stimuli like temperature, pressure, electric field and etc. is important due to not only a scientific but also a practical view point. On the other hand, fixing specific colours, namely colours insensitive to stimuli, also attracts much attention in connection with the iridescent colours of a jewel beetle[2] and for some applications like optical filters for selective reflection or large optical rotation. Polymeric materials have already been used to obtain thermally insensitive solid films with cholesteric structure[3, 4, 5]. High viscosity or three dimensional networks of polymer chains practically fix pitch of the configuration of polymers in the helical sense, which determines the colours of cholesteric liquid crystalline compounds, providing a thermally insensitive cholesteric colour. Although it was also possible to tune colour, by changing the monomer ratio in copolymers[4, 6], by changing the temperature during photo polymerization of cholesteric monomers[7] or by changing the ratio between polymerizing solution and liotropic polymers[8], reversible and fast switching of the cholesteric colour, which is desired for the application to rewritable memory or recording media, was impossible to achieve with these polymeric materials. Here we present a low-molecular-weight cholesteric compound which, in the solid state, shows changeable iridescent colours selected from a whole visible region by thermal treatment.

EXPERIMENT

Synthesis and characterization: Uv/vis: Hewlett-Packard HP 8452 Diode Array Spectrophotometer and Shimadzu UV-3100S Spectrophotometer. ^{1}H-NMR: Jeol GSX 270; internal standard: TMS. DSC: Seiko Instruments DSC 120. ORD: Horiba SEPA-200

Kogaku Kogyo Xenon Lamp YX-150 as a light source. X-ray diffraction: Rigaku RAD-C powder diffractometer using graphite-monochromated CuKα radiation (40 KV, 30 mA); The spectra were measured at room temperature between 2° and 50° in the 2θ/θ-scan mode with steps of 0.01° in 2θ and 0.6-s measurement time per step. Melting Points: DSC and Yanaco Micro Melting Point Apparatus MP-500V.

Dicholesteryl 10,12-docosadiynedioate (1)

To a solution of 2 g of cholesterol (Wako Pure Chemical Industries L.T.D.) in 20 ml of dichloromethane, was a solution of 0.96 g of 10,12-docosadiynedioic acid (Farchan Laboratories Inc.), 1.06 g of dicyclohexyl carbodiimide and 0.063 g of 4-dimethylamino-pyridine in 10 ml of dichloromethane added. The mixture was stirred at room temperature for 12 hours. After the removal of the precipitate by filtration, the solution was concentrated under vacuum. The remained white solid was subjected to column chromatography (silica gel/ dichloromethane). Solvent of the first eluent was removed and the product was recrystallized from dichloromethane. 1.58 g of 1 was obtained as a colorless solid (56%). m.p. 119 °C. ^{1}H-NMR (CDCl$_3$) δ = 0.62-2.02 (102 H, m), 2.20-2.38 (16 H, m), 4.60 (2 H, m), 5.39 (2 H, m). Anal. calc. for C$_{77}$H$_{126}$O$_3$: C = 84.08 %; H = 11.55 %. Found: C = 83.15 %; H = 11.23 %.

RESULTS

A new type of di-cholesteryl derivative 1 in scheme 1 was synthesized by condensation of cholesterol and corresponding diacid in the presence of dicyclohexyl carbodiimide and 4-dimethylaminopyridine in dichloromethane. Although it is well-known that many of diacetylene derivatives polymerize in the crystalline phase to provide conjugated polymers, the compound 1 in the crystalline state is not polymerized at all under UV light or γ-ray. This compound, a white powder at room temperature, melts at 119 °C to an isotropic phase during heating process and shows a cholesteric phase between 87 and 115 °C during the cooling process. The cholesteric phase is only the meso phase for the compound, which reverts to crystalline phase again below 87 °C.

Between two glass plates, the compound in its cholesteric phase shows typical iridescent colors ranging from blue to red depending on the temperature. When the compound showing the iridescent colors was quickly cooled down by, for example, dipping in the ice-water, it solidified while keeping the iridescent color observed at the higher temperature at which the quick cooling started. The colors are stable for at least one week at 78 °C, and for at least 10 months at room temperature. These solid films selectively reflect light of narrow spectral band (fig. 1). The wavelength of the peak of the reflection bands increases as the temperature at which quick cooling process started is raised (fig. 2). Band widths of the selective reflection were 50 - 60 nm, and reflectivity was almost 50 %, which coincides with the theoretically predicted value for cholesteric structures. Large optical rotatory dispersion, up to 15 deg / μm in the positive sense, was also observed around the peak of the reflection band (fig. 3). X-ray diffraction measurements showed that the compounds in the colored solid phase did not contain any crystalline ordering of molecules, while the compound in white crystalline phase obtained after recrystallization from solvents or slow cooling from melting state showed sharp peaks with a common pattern verifying crystalline structure (fig. 4). All of these properties verify that cholesteric ordering of the molecules in the liquid crystalline state remained in the solid films with iridescent colors.

DSC measurement for the colored solids showed a glass transition temperature around 80 °C and above this temperature the colored compound gradually turned into a mixture of two kinds of white crystals melting at 114 and 119 °C, respectively. This means that this compound shows polymorphism in crystalline phase, dependent upon the history of the heat treatment. The fraction of the crystals with the melting point of 114 °C increases as with the increment of the temperature at which quick cooling process started on the preparations of the colored solid.

It is also possible to produce images on the film by controlling the surface temperature. Figure 5 shows the images of green letters of alphabet and a dog on the blue background, respectively. The surface were partially cooled down by contact with a flat aluminum or rubber

Scheme 1. Chemical structure of cholesteryl derivative 1.

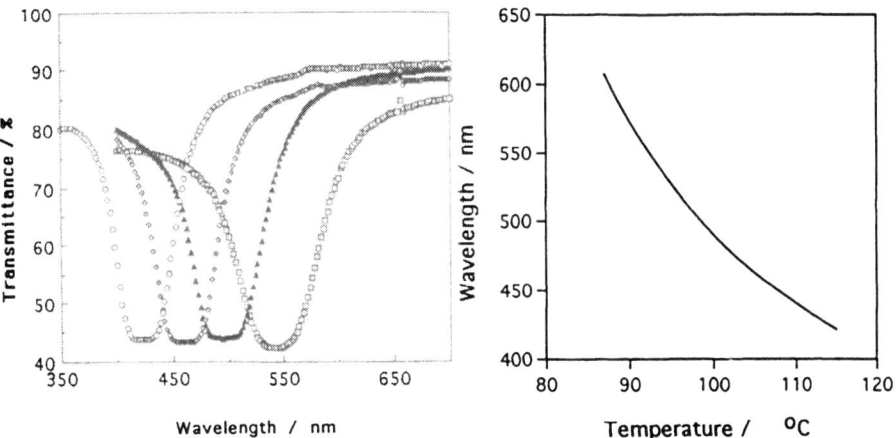

Wavelength / nm

Figure 1. Transmission spectra of colored solid films obtained by changing the temperature at which quick cooling process started. circle: 115 °C; diamond: 105 °C; triangle: 100 °C; square: 95 °C. Negative peak for each film in the measurement of transmission are not due to absorption of the film but due to reflection, which is verified by the coincidence of the position of the positive peaks in separate experiments of reflection spectra.

Figure 2. Relationship between temperature at which quick cooling process started and the wavelength of peaks in transmission band for the solid cholesteric films. The temperature was controlled and measured with a hot stage (Metler FP-80 and FP-82).

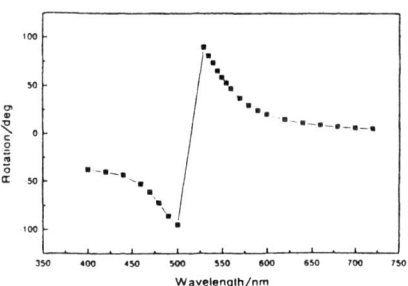

Figure 3. Optical rotation dispersion of the film with a green color.

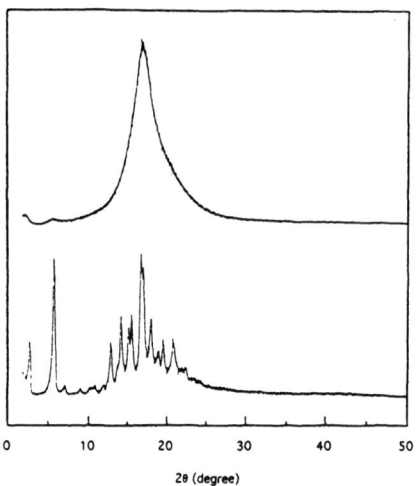

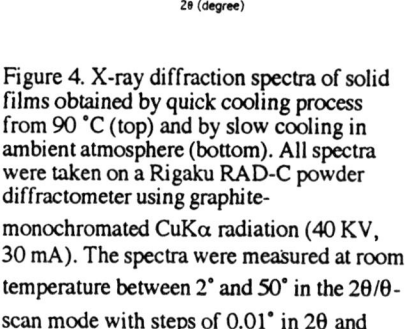

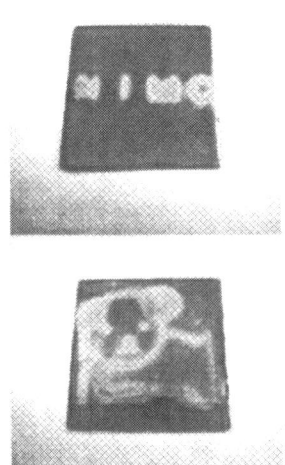

Figure 4. X-ray diffraction spectra of solid films obtained by quick cooling process from 90 °C (top) and by slow cooling in ambient atmosphere (bottom). All spectra were taken on a Rigaku RAD-C powder diffractometer using graphite-monochromated CuKα radiation (40 KV, 30 mA). The spectra were measured at room temperature between 2° and 50° in the 2θ/θ-scan mode with steps of 0.01° in 2θ and 0.6-s measurement time per step.

Figure 5. Photographs of letters of alphabet and a dog recorded on cholesteric solid films. All of these photographs are taken at room temperature. The size and the thickness of the glass plates are 18 x 18 mm and 0.15 mm, respectively. The thickness of the cholesteric films, controlled by the amount of the compound and the pressure during the preparation of the sample, was 10 - 20 μm. Cholesteric films between two glass plates maintained at 115 °C on the hot stage were in imagewise manner cooled down to the temperature, at which cholesteric solid shows green colour, by contact with image plates like stamps made with rubber. Cooling time, main part of which is needed to transfer heat from the cholesteric films to the rubber plates, was less than three seconds. Then the film was quickly immersed to ice-water from the hot stage to solidify the cholesteric compound.

image plate after heating whole surface to 115 °C before the quick cooling process started. The processes of erasing and recording color images can be repeated many times without any degradation of the compound, by heating it above 119 °C first and then cooling quickly from certain temperatures selected between 87 and 115 °C, respectively.

Stable and rapidly changeable colors of these cholesteric films in the solid state was first accomplished in the present work by modifying the molecular structure and the molecular weight. This allowed us to tune the glass transition temperature and the response time needed for the change of the cholesteric pitch to the change of temperature in the cholesteric phase. Polymeric cholesteric films with high and distributed molecular weight takes several hours to stabilize regular cholesteric ordering under the condition of fixed temperature[9]. Dicholesteryl dodecadioate, which has a similar molecular weight and structure to those of 1 without a diyne unit, and cholesteryl 10,12-pentacosadiynate, which contains just one cholesteryl group in each molecular unit, showed too sensitive response of the color to the change of temperature to be frozen into a glassy solid with a selected color. Although the requirement for the molecular structure of the compounds with a stable cholesteric structure in the solid state is unclear, it is already known that many low-molecular-weight compounds with a stable glass state shows polymorphism in the crystalline phase[10]. Polymorphism can be considered as a phenomenon for compounds which have more than two stable molecular structures. Compound 1, which showed two kinds of crystalline phase, would have two stable molecular structures in the solid state, and for this reason it showed a stable glass phase with cholesteric structure.

CONCLUSIONS

We introduced a new type of cholesteric solid film whose iridescent colors selected from a whole visible region can be recorded and erased repeatedly in an imagewise manner by thermal treatment. This thermo-optical effect can be applied to rewritable full-color thermal recording, and also to high density memory media using the full spectrum of light. One unit area can have a reflection spectrum selected from a continuous range of shifted spectra. By controlling the laser power, the temperature of the sample heated by absorption of the laser light would be changed continuously, and the fixed spectrum of one spot after quick cooling process would be controlled as the result. If we accomplish a spectral resolution of 10 nm between 420 and 610 nm for monitoring the reflected light under illumination with white light, 20 different states could be recorded and read in one spot instead of only 2 states as in conventional optical memory.

REFERENCES

1. F. Reinitzer, *Monatsh. Chem.* 9, 421-441 (1888).
2. A. C. Neville and S. Caveney, *Biol. Rev.* 44, 531-562 (1969).
3. E. T. Samulski and A. V. Tobolsky, *Nature* 216, 997 (1967).
4. H. Finkelmann, J. Koldehoff and H. Ringsdorf, *Angew. Chem. Int. Ed. Engl.* 17, 935-936 (1978).
5. It has also been reported that some mixtures of cholesteryl hydrogen phthalate and some cholesteryl derivatives show stable cholesteric solids at 25 °C: W. Mahler, M. Panar, *J. Am. Chem. Soc.* 94, 7195-7197 (1972).
6. D. J. Broer, J. Lub and G. N. Mol, *Nature* 378, 467-469 (1995).
7. P. J. Shannon, *Macromolecules* 17, 1873-1876 (1984).
8. T. Tsutsui and R. Tanaka, *Polymer* 21, 1351-1352 (1980).
9. J. Watanabe, T. Nagase, H. Itoh, T. Ishi and T. Satoh, *Mol. Cryst. Liq. Cryst.* 164, 135-143 (1988).
10. W. Ishikawa, H. Inada, H. Nakano and Y. Shirota, *J. Phys. D: Appl. Phys.* 26, B94-B99 (1993).

Structure and Physical Properties of EDT-TTF Salts

Tetsuo Kondo[*,***], Lyudmila A. Kushch[*,**], Hideki Yamochi[*,***], Gunzi Saito*
*Division of Chemistry, Graduate School of Science, Kyoto University,
Kitashirakawa, Sakyo-ku, Kyoto, 606-01, Japan ,
**Institute of Chemical Physics, Russian Academy of Science, 142 432 Chernogolovka, Russia
***CREST, Japan Science and Technology Corporation (JST)

ABSTRACT

Charge-transfer salts of EDT-TTF with tetrahedral anions MX_4 (M=Fe, Ga ; X=Cl, Br) were prepared. They are isostructural to each other (monoclinic $P2_1/m$) with the stoichiometry of 2:1. All of them are semiconductors having the conductivity at room temperature of 5 to 9 Scm^{-1} and activation energy of 40 to 170 meV. Since the calculated band structure of them indicate that they should be quasi one-dimensional metals, they were discussed from the point of structural deformation and strong electron correlation.

INTRODUCTION

EDT-TTF(ethylenedithio-tetrathiafulvalene) is one of the most typical hybrid TTF-based donor molecule between BEDT-TTF and TTF(Fig. 1)[1]. The studies on the EDT-TTF salts were reported for octahedral, tetrahedral and linear anions [2,3] and intensive examinations were performed for superconducting α-(EDT-TTF)[Ni(dmit)₂] [4,5,6], in which both the donor (EDT-TTF) and anion columns ([Ni(dmit)₂]) construct the metallic (superconducting) Fermi surfaces.

Because of the poor ability of EDT-TTF to yield the stable metallic or superconducting salt in comparison with BEDT-TTF, an intensive study on the EDT-TTF complexes has not been performed except for the reported papers listed above. An elimination of one ethylenedithio moiety from a BEDT-TTF induces a relatively large on-site coulomb repulsion for EDT-TTF. With respect to this, it is of interest to develop new conducting materials having strong electronic correlation. Furthermore, the magnetic exchange interactions between the conducting π-electrons on the donor molecules and localized magnetic moments on the anions is also of interest if they exist. With these scopes, we synthesized new charge-transfer salts of EDT-TTF, mainly for anions with tetrahedral structure, solved the crystal structures, and measured transport and magnetic properties.

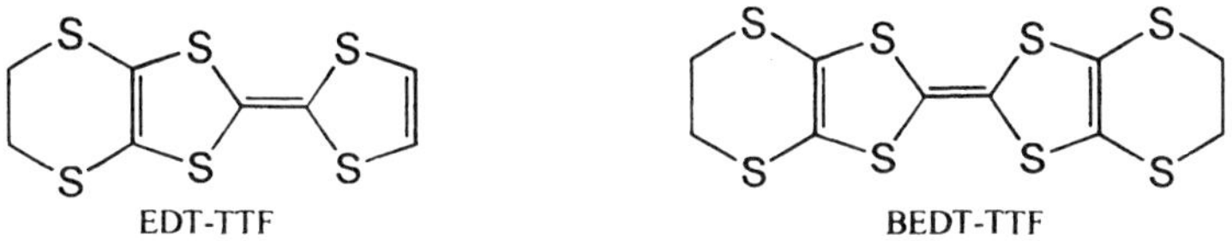

EDT-TTF   BEDT-TTF

Figure 1 Chemical structure of EDT-TTF and BEDT-TTF

EXPERIMENT

Preparation of single crystals was performed by the galvanostatic electro-crystallization method with a current of 0.5μA. The intensity data of the structural analysis were collected on MAC SCIENCE MXC^x four circle diffractometer. X-ray oscillation photographs were taken on MAC SCIENCE DIP-2000K Imaging Plate System at room temperature and 105K using Oxford low temperature cryostat MOCS-2. The d.c. conductivities were measured based on a standard four-probe method. Measurements of the magnetization were performed by the aid of Quantum Design MPMS-2 and MPMS-5S magnetometers in the temperature range of 2.0 to 300K and magnetic field of 1000Gauss.

RESULTS AND DISCUSSION

Single crystals of good quality were obtained for the tetrahedral anions (FeX$_4$ and GaX$_4$; X=Cl, Br) which enable us both the single crystal structure determination and measurement of the transport and magnetic properties with good accuracy. Crystals with clear edges were also obtained for linear (I$_3$) and polymeric type anions (Cu(NCS)$_2$), though no structural determination was possible for these salts at this stage because of the poor quality.

Overview of results of (EDT-TTF)$_2$MX$_4$ (M=Fe, Ga; X=Cl, Br)

Table 1 summarizes crystallographic parameters and electrical conductivity results of (EDT-TTF)$_2$MX$_4$ (M=Fe, Ga; X=Cl, Br). These four salts are crystallographically isostructural (monoclinic P2$_1$/m). A half of the formula unit is crystallographically unique. Figure 2 presents the typical view of (EDT-TTF)$_2$MX$_4$ along the short-axis of EDT-TTF molecule. In these salts, the donor molecules show head-to-tail stacking with weak dimerization along the c-axis, while these donor-columns are alternately separated by the anion layer along the b-axis. The electrical transport of these salts is semiconducting, with a conductivity of 5.0 to 9.4 Scm^{-1} at room temperature with activation energy of 42 to 170 meV. The similarity of the transport properties among them is considered to arise from the fact that these salts are isostructural and have

Table 1 Crystallographic data, final R-value, room temperature conductivity $\sigma_{r.t}$, and activation energy E_a for (EDT-TTF)$_2$MX$_4$.

MX$_4$	GaCl$_4$	GaBr$_4$	FeCl$_4$	FeBr$_4$
Symmetry	Monoclinic	Monoclinic	Monoclinic	Monoclinic
Space Group	P2$_1$/m	P2$_1$/m	P2$_1$/m	P2$_1$/m
a (Å)	7.087(2)	7.100(1)	7.094(1)	7.114(1)
b (Å)	30.440(7)	31.168(3)	30.458(2)	31.161(3)
c (Å)	6.582(2)	6.669(1)	6.601(1)	6.664(1)
β(deg)	100.11(2)	98.913(7)	99.944(4)	98.862(7)
V(Å^3)	1398.0(6)	1457.981(4)	1404.630(3)	1459.636(4)
Z	2	2	2	2
R(%)	6.1	6.3	5.0	5.0
$\sigma_{r.t}$ (S·cm^{-1})	5.0	6.0	7.8	9.4
E_a (meV)	42	57	170	44

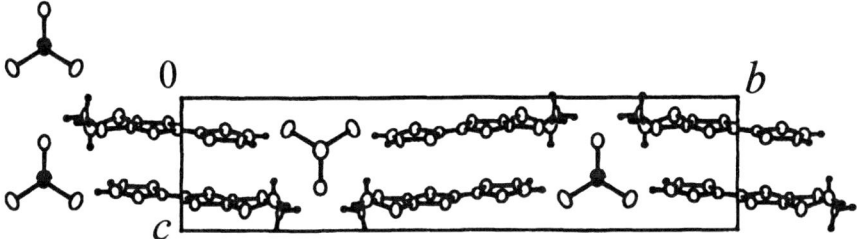

Figure 2 Crystal Structure of (EDT-TTF)$_2$MX$_4$ along the short-axis of EDT-TTF molecule.

similar electronic structures. It should be mentioned here that no singularity is observed in the electrical transport in all of these salts down to 20K.

Electronic Structure of (EDT-TTF)$_2$MX$_4$

The electronic band structure for (EDT-TTF)$_2$MX$_4$ was calculated according to the tight binding method using the structural parameters collected at room temperature. Figure 3 shows the donor column arrangement of (EDT-TTF)$_2$MX$_4$ with the assignment of transfer integrals of these salts. The transfer integrals consist of intra-column ones (p1, p2) and inter-column ones (q1, q2). The calculated transfer integrals are summarized in table 2. Reflecting the weak dimerization of the donor molecules within a column, p1 and p2 are slightly different to each other. A quasi one dimensional electronic structure is suggested since the inter-column transfer integrals (q1, q2) are about 1/5 to 1/6 of those of intra-column ones (p1, p2). Figure 4 presents the energy band dispersion and Fermi surface of (EDT-TTF)$_2$MX$_4$. We confirm that the band structures of these salts are similar to each other except for the small differences in band parameters. As expected, quasi one-dimensional Fermi surfaces are open along the a^*-axis.

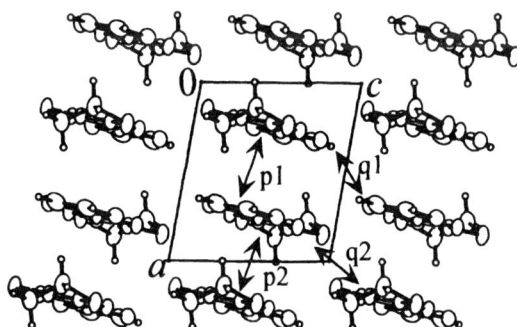

Figure 3 Structure of the EDT-TTF column within the *ac*-plane indicating the transfer integrals p1, p2, q1 and q2 by arrows.

Table 2 Transfer integrals S of (EDT-TTF)$_2$MX$_4$. The values are expressed as $S \times 10^{-3}$.

	GaCl$_4$	GaBr$_4$	FeCl$_4$	FeBr$_4$
p1	30.34	31.23	29.71	30.29
p2	25.45	25.91	24.77	25.70
q1	5.29	4.28	4.30	5.32
q2	6.28	5.20	5.40	6.03

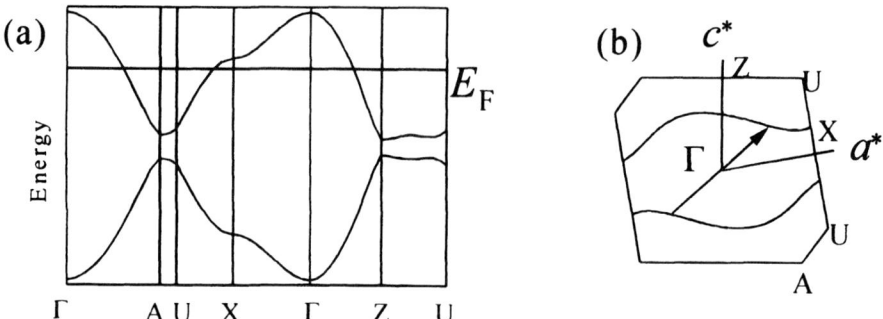

Figure 4 Energy band dispersion (a) and Fermi surface (b) of (EDT-TTF)$_2$MX$_4$. A possible nesting vector is also indicated by an arrow in Fig. 3(b).

The calculated band structures are inconsistent with the observed semiconducting electrical transport. Since an introduction of a nesting vector is possible for the calculated Fermi surface as indicated by an arrow in Fig. 4(b), it is a good assumption that this inconsistency is due to the Fermi surface nesting which is a popular phenomenon in such a quasi one-dimensional system.

The direction of the nesting vector is $a^* + c^*$, which is related to the inter-column direction. Therefore, an appearance of supperlattices along the corresponding direction is expected to be observed. In order to examine the existence of expected superlattice, an X-ray oscillation photograph by an Imaging Plate System was measured for the GaCl$_4$ salt.

Figure 5(a) exhibits the oscillation photograph around the c-axis at room temperature. All diffraction spots observed can be assigned as the Bragg spots according to the determined unit cell described above, while no diffuse scattering was observed, although the intensities of the

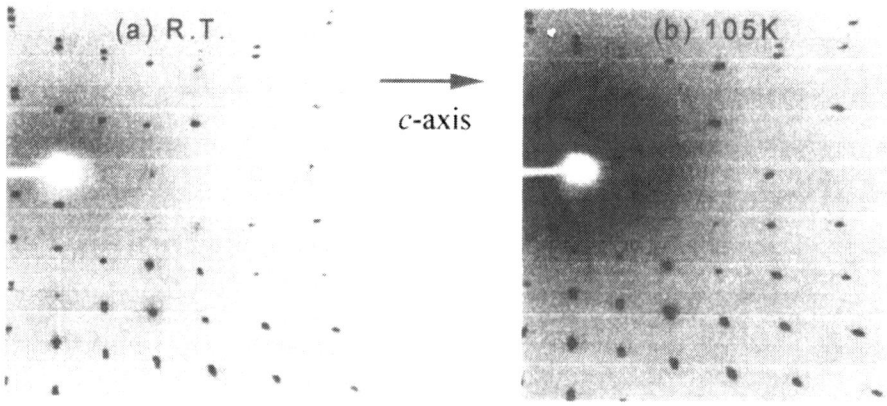

Figure 5 X-ray oscillation photograph of (EDT-TTF)$_2$GaCl$_4$ rotating around the c-axis (a) at room temperature (R.T.), (b) at 105K taken by Imaging Plate. The horizontal is the c-axis.

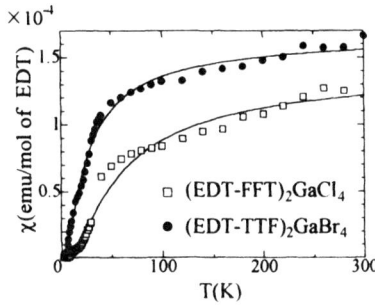

$\times 10^{-4}$

χ(emu/mol of EDT)

□ (EDT-FFT)₂GaCl₄
● (EDT-TTF)₂GaBr₄

100   200   300

T(K)

Figure 6 Temperature dependence of the magnetic susceptibility of (EDT-TTF)₂GaX₆ (X=Cl; squares: X=Br; closed circles). The Pascal diamagnetism and the contribution from paramagnetic impurity are sub-tracted. The fitting curves according to the activation type equation $\chi(T)=A+B\exp(-\Delta/T)/T$) are also presented.

Bragg spots exceed the saturation value of the Imaging Plate. It is expected that the diffuse scatterings would be clearly observed at low temperatures since the superlattice may grow steady as temperature decreases. Figure 5(b) presents the oscillation photograph at 105K. Inconsistent with the above expectation, no diffuse scattering is observed at this temperature aside from a small difference in the photograph arose from the thermal contraction. In the oscillation photograph around the a-axis, diffuse scatterings were not also observed.

Although a non-metallic behavior is observed at room temperature, the existence of lattice distortion was not observed down to 105K. Therefore, another origin that places the non-metallic behavior in this system must be considered. Since the 2:1 composition in these salts leads to a half-filled state in the upper HOMO level, it is not bad assumption that these materials are Mott insulators. In Fig. 6, we present temperature dependence of the magnetic susceptibility of (EDT-TTF)₂GaX₄ (X=Cl, Br), in which the anions do not contain the magnetic species. The susceptibility at room temperature is 1.2×10^{-4} and 1.6×10^{-4} (emu/mol of donor molecule) for the GaCl₄ and GaBr₄ salts, respectively, which is one order of magnitude smaller than those which contain one localized spin on one donor-molecule[7]. Consequently, the possibility of the Mott insulating state for these salts is eliminated at present. The origin of the non-metallic behavior of this system remains as an open question that should be solved by a more detailed investigation for the crystal structure and the physical properties.

Finally, we comment for the temperature dependence of the magnetic susceptibility of this system. The whole behavior of the observed magnetic susceptibility can be fitted by the activation type equation ($\chi=A+B\exp(-\Delta/T)/T$) as is often the case of organic conductors with a Peierls transition in which the energy gap open at the Fermi Energy. The activation energy is estimated at $\Delta\sim10K$ both for the GaCl₄ and GaBr₄ salts, which is at least 50 times smaller than the activation energy obtained from the Arrhenius plot of the electrical conductivity. This is also inconsistent with the suggestion that a Peierls transition occurs in this system.

Magnetic Properties of (EDT-TTF)₂MX₄ (M=Fe, X=Cl, Br)

The magnetic exchange interaction between EDT-TTF molecules and anions containing magnetic ions (FeCl₄ and FeBr₄) is described. According to the crystal structure, no van der Waals atomic contacts are observed among the anion molecules, while atomic contacts between the X(=Cl, Br) atoms and sulfur and ethylene hydrogen atoms of an EDT-TTF molecule are observed. These atomic contacts are expected to give rise to a magnetic exchange interaction between the EDT-TTF molecules and magnetic moments on FeX₄, while the magnetic susceptibilities of (EDT-TTF)₂FeX₄ are found to be paramagnetic one down to 2K as presented in Fig. 7. More strong interaction between EDT-TTF and FeX₄ molecules is necessary in order

925

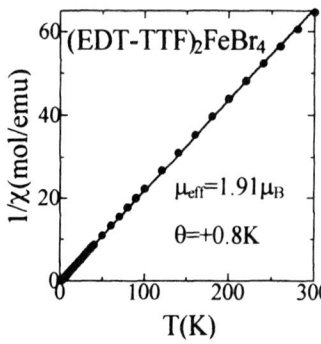

Figure 7 Temperature vs. inverse susceptibility $(1/\chi)$ for $(EDT-TTF)_2FeBr_4$ under a magnetic field of 1000Gauss. The fitting-curve for Curie-Weiss low is also presented by a straight line. The effective moments are estimated at $1.88\mu_B$ and $1.91\mu_B$ for the $FeCl_4$ and $FeBr_4$ salts, respectively. The Weiss temperatures are estimated at $+0.8K$ and $+0.7K$ for the $FeCl_4$ and $FeBr_4$ salts, respectively, using the susceptibilities below 20K.

to place a magnetic phase transition in this system by means of a further development of the appropriate anions.

CONCLUSION

Crystal structures, electronic and magnetic properties were examined for the charge transfer complexes of EDT-TTF with anions having a tetrahedral symmetry MX_4 (M=Fe, Ga: X=Cl, Br). Crystal structures were found to be isostructural to each other. The electrical transports were semiconducting, which were inconsistent with the calculated band structure suggesting the existence of metallic Fermi surfaces. Any sufficient explanations were not obtained for the observed non-metallic behavior at this stage because of the absence of diffuse scatterings caused by the Peierls transition and the lack of localized spins induced by Mott insulating state. Paramagnetic behavior of the $FeCl_4$ and $FeBr_4$ salts suggested the weak exchange interaction between EDT-TTF molecules and magnetic moments within Fe atoms.

ACKNOWLEDGEMENTS

This work is supported by a fund for "Research for Future", from Japan Society for Promotion of Science, and Grant in Aid for Scientific Research, No. 06243105, Ministry of Education, Science, Culture and Sports, Japan.

REFERENCES

1. G. C. Papavassiliou, G. A. Mousdis, J. S. Zambounis, A. Terzis, A. Hountas, B. Hilti, C. W. Mayer and J. Pfeiffer,   Synth. Metals, **B27**, p. 379 (1988)
2. R. Kato, H. Kobayahi and A. Kobayashi,   Chem. Lett. p. 781 (1989)
3. R. Kato, H. Kobayahi and A. Kobayashi,   Chem. Lett. p. 1235 (1989)
4. R. Kato, H. Kobayashi, A. Kobayashi, T. Naito, M. Tamura, H. Tajima and H. Kuroda, Chem. Lett. p. 1839 (1989)
5. H. Tajima, S. Ikeda, A. Kobayashi, H. Kuroda, R. Kato, H. Kobayashi, Solic State Commmun., 86 p. 7 (1993)
6. M. L. Doublet, E. Canadell, B. Garreau, J. P. Legros, J. P. Pouget, and L. Brossard, Synth. Metals **70**, p.1063 (1995)
7. N. F. Mott,   Metal-Insulator Transitions,   Taylor & Francis, (1990) pp. 123-144

A STUDY ON LANGMUIR MONOLAYERS OF METHACRYLATE HOMO- AND COPOLYMERS DERIVATIZED WITH DISPERSE RED DYES

ANANTHARAMAN DHANABALAN, DÉBORA BALOGH, CARLOS JOSÉ LEOPOLDO CONSTANTINO, ANTONIO RIUL JR., OSVALDO N. OLIVEIRA JR. AND JOSÉ A. GIACOMETTI
Instituto de Física de São Carlos, Universidade de São Paulo, CP 369, 13560-970, Brazil
Giacometti@ifqsc.sc.usp.br

ABSTRACT

We report on the organization of different disperse red dye derivatized methacrylate homo- and copolymers as Langmuir monolayers at the air-water interface. The monolayers were investigated using surface pressure and surface potential isotherms. Methacrylic homopolymers containing disperse red-1 (HPDR1) and disperse red-13 (HPDR13) and methacrylic copolymers with different mole percentages of the dye and hydroxyethyl spacer groups were synthesized and characterized. A comparison of the monolayer characteristics of HPDR1 and HPDR13 revealed the influence of the chlorine substitution in the aromatic ring of the dye. Studies with copolymers indicated a clear transition in the monolayer behavior with the change of mole percentage of the dye incorporated in the polymer. While copolymers with low dye content (up to about 5 mole%) presented an expanded monolayer, the copolymers containing higher dye content (12% and above) formed a condensed monolayer similar to that of the homopolymer. These results indicated the critical role of the dye component in the polymer chain in forming the monolayer at the air-water interface.

INTRODUCTION

The amphiphilicity of a molecule seems no longer an essential requirement for producing Langmuir-Blodgett (LB) films, as evidenced from a number of recent reports on the preparation of LB films of semi- and nonamphiphilic molecules [1,2], including polymers [3]. However, it should be mentioned that these non-amphiphilic macromolecules probably tend to form a multilayer stack instead of a true monolayer at the air-water interface [4]. In the case of long alkyl chain derivatized polymers and mixed monolayers (with amphiphilic molecules), the packing arrangement is obviously dictated by the long alkyl chains rather than the main polymer chain [5,6]. Smaller substituents may also be attached to the polymer main chain for improving the processibility and for incorporating interesting functional groups in the LB film structure. Thin LB films of azobenzene functionalized polymeric molecules are of special interest as they can be employed in various device applications, owing to their reversible *cis-trans* isomerization and non-linear optical characteristics [7,8]. In a previous work [10], we have shown that a stable monolayer can be obtained with a methacrylic homopolymer (HPDR13) in which the disperse red-13 (DR-13) is co-valently attached to the main polymer chain with only two methylene groups. The poor transferability associated with the HPDR13 monolayer could greatly be enhanced by using the mixed LB film approach in which HPDR13 was transferred along with cadmium stearate [11]. In the present work, a systematic monolayer study has been carried with homo- and copolymers which contained different percentages of the dye along the methacrylic polymer chain for two different dyes (DR1 and DR13). The structures of the different polymers studied are shown in Fig.1.

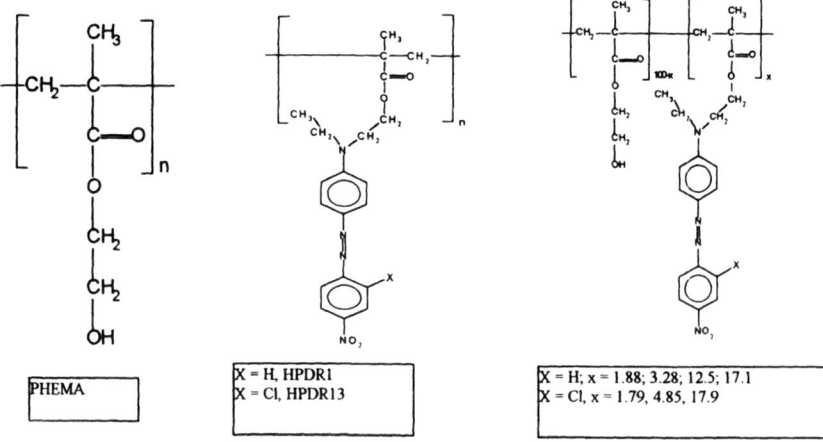

PHEMA

X = H, HPDR1
X = Cl, HPDR13

X = H; x = 1.88; 3.28; 12.5; 17.1
X = Cl, x = 1.79, 4.85, 17.9

Figure 1 –Structures of PHEMA, homopolymers and copolymers.

EXPERIMENTAL

The synthesis of HPDR13 has been carried out as reported elsewhere [10] and the polymer has been characterized by UV-vis, FTIR and NMR spectroscopies. The Differential Scanning Calorimetry (DSC) analysis of the homopolymer showed a clear glass transition at around 48°C. The co-polymerization reaction was performed in a stepwise mode. The DR13 monomer (0.12 g, 2.87×10^{-4} mol) and 0.0043 g of 2,2'-azobisisobutyronitrile (AIBN) in a solvent mixture consisting of dimethylformamide(DMF)/methylethyl ketone (MEK) 1/1 v/v (2 mL) were heated to 70° C. After an interval of 4 hours, 0.25 mL of 2-hydroxyethylmethacrylate (HEMA) and 0.005g of AIBN in 2 mL of solvent mixture were added to the above reaction mixture. After the intervals of 7, 10 and 17 hours, 0.5 mL of HEMA, 2 mL of the solvent mixture and 0.005g of AIBN were added each time. The mixture was kept at 70° C for further 28 hours after the last addition of reactants. The reaction mixture was then poured into n-hexane and the gel-like viscous mass thus separated was washed with n-hexane and 200 mL of ethyl acetate and solidified. The co-polymer separated was then washed with ethyl acetate to remove unreacted monomers and dried under vacuum. The analysis of the polymer with Thin Layer Chromatography (TLC) with dichloromethane/n-hexane 70/30 as eluent confirmed the absence of the monomer. The DSC analysis of the copolymer revealed a glass transition temperature of about 85 °C similar to that observed with the HEMA homopolymer.

For the Langmuir monolayer experiments, the subphase water was supplied by a Milli-RO coupled to a Milli-Q water purification system from Millipore, (resistivity = 18.2 MΩ/cm). Chloroform and a mixture containing DMF and chloroform (2:98 by volume) have been used as spreading solvents (typical concentration was about 0.1 mg/mL) for homopolymers and copolymers respectively. All the experiments were carried out in the dark, at a subphase temperature of 22 ± 0.5° C. Both surface pressure and surface potential isotherms studies were carried out with a KSV-5000 LB system mounted on anti-vibration table in a class

10,000 clean room. The monolayer was compressed with a barrier speed of 10 mm/min. In the case of homopolymers, the mean molecular areas (mMa) were calculated on the basis of the molecular weight of the monomer repeat units (380 for HPDR1, 416 for HPDR13 and 130 for PHEMA). For copolymers, the average molecular weight was calculated based on the percentage of dye substitution in the polymer and used in mMa calculations.

RESULTS AND DISCUSSIONS

Langmuir Monolayers of Homopolymers

The surface pressure isotherm of HPDR13 displays an initial liquid expanded region which is followed by a solid condensed region with a limiting mean molecular area (mMa) of about 20 Å^2 [10]. Such an area was obtained by the extrapolation of the steep part of the isotherm curve. For the sake of comparison, the HPDR13 isotherm is reproduced from ref. 10 in Fig.2. Also shown in the figure is the surface pressure isotherm of HPDR1, which indicates that the nature of the isotherm did not change appreciably with the substitution of chlorine in the azobenzene ring (refer to Fig. 1).

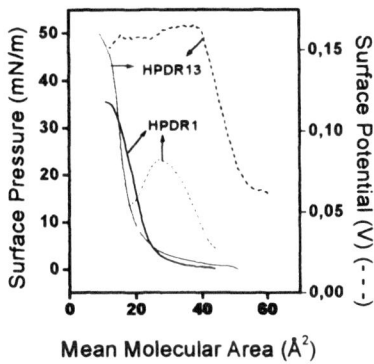

Figure 2 – Surface pressure and surface potential isotherms of HPDR1 and HPDR13 homo-polymers.

The HPDR1 monolayer also exhibited an initial liquid expanded region and a solid condensed region. However, there are noticeable differences in the mMa and the collapse pressure. The HPDR1 monolayer displayed a higher mMa (about 25 Å^2) and a lower collapse pressure (35 mN/m) in comparison to those of HPDR13. Though the overall monolayer organization is mostly dominated by the packing of polymer chains at the air-water interface, as implied by the similarity of the isotherms, the packing arrangement in the condensed region seems to have been dictated by the substituent chlorine in the azobenzene group. The stability of the monolayers was inferred by monitoring the change in the mMa while holding the monolayer in the compressed state (at 20 mN/m). Despite an initial small decrease in the area, no significant decrease could be observed even while holding the HPDR1 monolayer for an hour or so, indicating the formation of a reasonably stable monolayer. Attempts have also been made to obtain surface pressure isotherms of PHEMA in which there is no dye substitution.

However, the compression after spreading PHEMA on the water surface has not resulted in any increase in the surface pressure, even after reaching a very small mean molecular area (2 $\mathring{A}^2$). The surface potential was nevertheless non-zero immediately after the spreading. These results may point to the critical role of the dye substitution in the polymer chain for forming a stable polymeric monolayer at the air-water interface.

Surface potential isotherms were also obtained for HPDR1 and HPDR13, as also shown in Fig. 2. As it has been shown for a number of compounds [6], the surface potential increases in a relatively sharp manner when a critical area, A_c, is reached. Such a critical area is always much higher than the molecular area for the condensed region, and is associated with the decrease in the local dielectric constant of the film/water interface when the monolayer becomes structured [12]. The coming together of polymeric molecules in a non-specific manner, possibly as aggregates, is therefore the reason for the observed non-zero surface potential values immediately after the spreading. Indeed, the presence of J-type aggregation of chromophores in the transferred LB film of HPDR13 has been established by UV-vis spectral studies [10] and the formation of aggregates at areas higher than the limiting mean molecular area has also been visualized through Brewster Angle Microscopy (results not shown). A_c is higher for the HPDR13 monolayer than for the HPDR1 one. This trend is in contrast to that observed with the mMa. The higher A_c for the HPDR13 monolayer may indicate that the chlorine attached to the azobenzene group facilitates the formation of a 2D network at the air-water interface, possibly via hydrogen bonding, at earlier stages of compression. Also observed was the higher maximum surface potential (ΔV_{max}) for HPDR13 monolayer (about 160 mV) in comparison to that of HPDR1 (about 80 mV). This underlines the influence of the chlorine substitution in determining the monolayer characteristics and also the preferred orientation of azobenzene rings within the monolayer structure. Had both HPDR1 and HPDR13 polymers acquired a completely random orientation within the monolayer, one would not observe any difference in the ΔV_{max} values for these two polymers. In the case of PHEMA, a non-zero surface potential (about 40 mV) has been observed immediately after the spreading. There was no change in surface potential during the compression, which is consistent with the surface pressure isotherm in that no monolayer is being formed at the air-water interface.

<u>Langmuir monolayers of copolymers</u>

Having established the critical role of the dye substitution in forming a stable monolayer, it is of interest to investigate the extent of dye substitution in the polymethacrylate chain which is needed to form a stable monolayer. For this purpose, copolymers with different mole percentages of the DR1/DR13 dyes have been studied by surface pressure and surface potential isotherms. The surface pressure data are shown for DR1 in Fig. 3, where two types of isotherm curves can be visualized. With low percentages of dye substitution (up to about 5%), a highly expanded monolayer without any clear phase transition was observed.

The change in the slope at about 10 mN/m may be considered as the collapse of the monolayer. The poor stability as inferred through hysteresis experiments pointed to the formation of meta stable monolayers which possibly comprised extended aggregation of molecules. However, for mole percentages of 12 and above, a condensed monolayer is formed, similar to that observed with HPDR1.

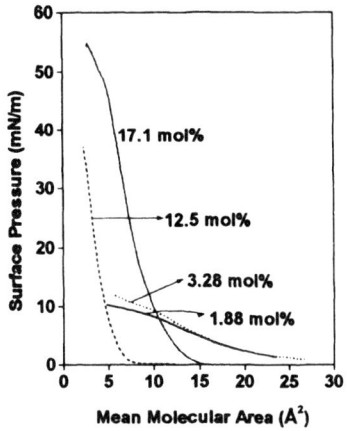

Figure 3 – Surface pressure-mean molecular area isotherms of the copolymers with different mole percentage of DR1.

A noticeable difference in the nature of the isotherm, in comparison to the HPDR1 monolayer, however, is the absence of a distinct liquid expanded region during the initial stages of compression and the higher collapse pressure. Stability experiments have revealed that these monolayers are comparatively less stable in comparison to the HPDR1 monolayer. The mean molecular areas for these copolymers are lower than that of HPDR1 (with 100% dye incorporation) and they increased with increasing mole percentage of the dye (about 5 $Å^2$ for 12.5% and about 10 $Å^2$ for 17.1%). With the assumption, based on the results from the homopolymers in the last section, that the dye molecules are responsible for the monolayer formation, one may infer that the molecular area is directly dependent on the packing of the dyes in the monolayer structure. This trend is expected since no monolayer could be obtained in the absence of the dyes (PHEMA).

The conclusions drawn above for the DR1 copolymers are also applicable to the DR13 dye copolymers, and therefore we omitted the data for the latter copolymers. The mean molecular areas are higher for the DR1 copolymers, which is consistent with the results obtained for the homopolymers.

The surface potential-mean molecular area isotherms of co-polymers containing different mole percentages of DR1 are presented in Fig. 4. For the copolymers with low dye content, the surface potential increased monotonically with the monolayer compression, indicating the formation of expanded-type aggregates with no specific interaction between the polymer molecules. In contrast, the copolymers with 12 mole% and above exhibited a sharp increase in surface potential during the compression, similar to that with the HPDR1 monolayer. The critical area increased with increasing dye content in the polymer, a trend similar to that observed with the limiting mean molecular area. A higher ΔV_{max} of about 150 mV has been measured for the copolymers in comparison with the HPDR1 monolayer (about 80 mV). This probably results from a preferred orientation of the azobenzene functionality in the monolayers of the copolymer.

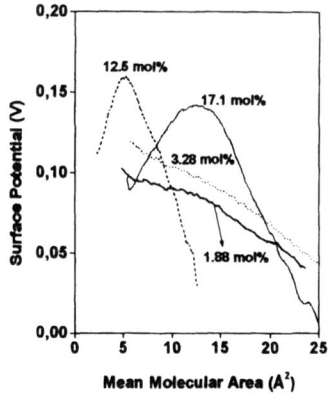

Figure 4 – Surface potential-mean molecular area of the copolymer with different percentage of DR1.

CONCLUSIONS

Surface pressure and surface potential isotherms have been reported for monolayers of disperse dye (DR) derivatized homo and copolymers with different mole percentages of the dye. The results for the homopolymers revealed the influence of the chlorine substitution on the monolayer characteristics. As for the copolymers, with around 12 mole% of dye incorporation the monolayers possess similar characteristics to the homopolymers.

ACKNOWLEDGMENTS

Authors thanks to FAPESP, CNPq, CAPES and FINEP/PADCT for the financial support.

REFERENCES

1. A. Tronin, T. Dubrovsky and C. Nicolini, Thin Solid Films, **284-285,** 894 (1996).
2. C. Gu, L. Sun, T. Zhang and T. Li, Thin Solid Films, **284-285,** 863 (1996).
3. Y.K. Kim, M.H. Sohn, B.C. Sohn, E. Kim and S.D. Jung, Thin Solid Films, **284-285,** 53 (1996).
4. J.H. Cheung and M.F. Rubner, Thin Solid Films, **244,** 990 (1994).
5. I. Watanabe, K. Hong and M.F. Rubner, Langmuir, **6,** 1164 (1990).
6. A. Dhanabalan, A. Riul Jr., L.H.C. Mattoso and O.N. Oliveira Jr., Langmuir, **13,** 4882 (1997).
7. Z.F. Liu, K. Hashimoto and A. Fujishima, Nature, **347,** 658 (1990).
8. A. Nahata, J. Shan, J.T. Yardley and C. Wu, J. Opt. Soc., Am. B. **10,** 1553 (1993).
9. J.B. Peng, G.T. Barnes, A. Schuster and H. Ringsdorf, Thin Solid Films, **210/211,** 16 (1992).
10. A. Dhanabalan, D.T. Balogh, A. Riul Jr., A. Giacometti and O.N. Oliveira Jr., Thin Solid Films , in press.
11. A. Dhanabalan, D.T. Balogh, C.R. Mendonca, A. Riul Jr., C.J.L. Constantino, A. Giacometti, S.C. Zilio and O.N. Oliveira Jr., J. Appl. Phys. (submitted).
12. D.M. Taylor, O.N. Oliveira Jr. and H. Morgan, J. Coll. Interfac. Sci., **139,** 508 (1990).

OPTICAL CHARACTERIZATION OF C_{60}
ORGANIC SEMICONDUCTOR AND BILAYERS

J.P. RAINHO, L. SANTOS, A.A. KHARLAMOV
Physics Department, University of Aveiro 3810 Aveiro, Portugal

ABSTRACT

Preparation and characterization either by optical absorption, photoluminescence and micro-Raman spectroscopy of individual components as well as bilayers consisting of organic dye semiconductor Zinc Phthalocyanine (ZnPc) and fullerene, C_{60}, thin films are reported. The layers and structures were deposited in vacuum and some fullerene films were also prepared by casting the C_{60} solution in benzene. The optical absorption and photoluminescence dependencies on film thickness in bilayers C_{60}/ZnPc were observed and may be discussed in a context of interface induced simmetry reduction of C_{60} molecules.

INTRODUCTION

Efficient synthesis of C_{60} and related fullerenes [1] has led to intense research of structural, electronic and optical properties of these forms of carbon. Noteworthy that films of fullerenes are of interest because they exhibit such properties as photoconductivity [2-4] and luminescence [5,9]. In the case of composites of C_{60} and materials with mild donors such as conducting polymers, spontaneous charge transfer does not occur at the ground state. Upon photoexcitation, it is demonstrated that the photoinduced electrons in the polymer migrate to C_{60}, resulting in a considerable enhancement of photovoltaic effects [10-12].

Metallophthalocyanines are well known for their thermal and chemical stabilities. The phthalocyanine molecule is planar and symmetrical, and provides opportunities for further research: organic dyes like phthalocyanine can be used in photovoltaic energy conversion [13].

This work presents a study of the Raman and Photoluminescence spectra, and optical absorption spectra of individual components, as well as bilayers consisting of ZnPc and C_{60} thin films. Particular attention is devoted to the preparation methods and thickness dependencies of these films and some structural information which they provide.

EXPERIMENT

The C_{60} (99.9 %) powder, obtained from MER Corporation (Tucson,USA), and ZnPc derivative of tetra-*tert*-butil phthalocyanine (Zn-Pc[C(CH$_3$)$_3$]$_4$) from Kodak Corporation were used. The films and structures were fabricated on transparent glasses, and on Au or Al coated glasses by evaporation of compounds in vacuum chamber (2×10^{-5} Torr). The C_{60} and ZnPc powders in the crucibles were heated at 320°C and 210°C, respectively, to deposit the films.

A Alpha Step 200 (Tencor Instruments) profilometer was used to estimate the thickness profile (within $\pm$ 10 nm) and the average roughness value of the films.

Stokes-shifted Raman spectra were measured at room temperature with the Renishaw Raman Microscope-2000 model, at 1 cm^{-1} resolution, using a Spectra-Physics-127 He-Ne laser for excitation at 632.8 nm, with a power of 2.5-25 mW at the source.

Mat. Res. Soc. Symp. Proc. Vol. 488 © 1998 Materials Research Society

The sample temperature was estimated by the Stokes/anti-Stokes scattering intensity ratio of the 1469 cm^{-1} peak of the Raman spectrum of C_{60} film. The temperature obtained for the region illuminated by the laser beam was 31°C for 25 mW input laser power at 632.8 nm. The room temperature photoluminescence spectra were collected with the same Raman spectrometer and the same excitation. The optical absorption was measured at room temperature, with the Shimadzu 2100 spectrophotometer, between 300-900 nm, at 2 nm resolution.

RESULTS AND DISCUSSION

UV-visible absorption in individual films of C_{60} and ZnPc are shown in the insert in Fig.1 The absorption spectrum of C_{60} film shows main features found previously (e.g. [2-4] and [14]). In solid, when compared with solution, the ground state absorption of C_{60} have been attributed to interaction between C_{60} molecules, and, as a result, enhance the forbidden bands around 450 nm and 620 nm.

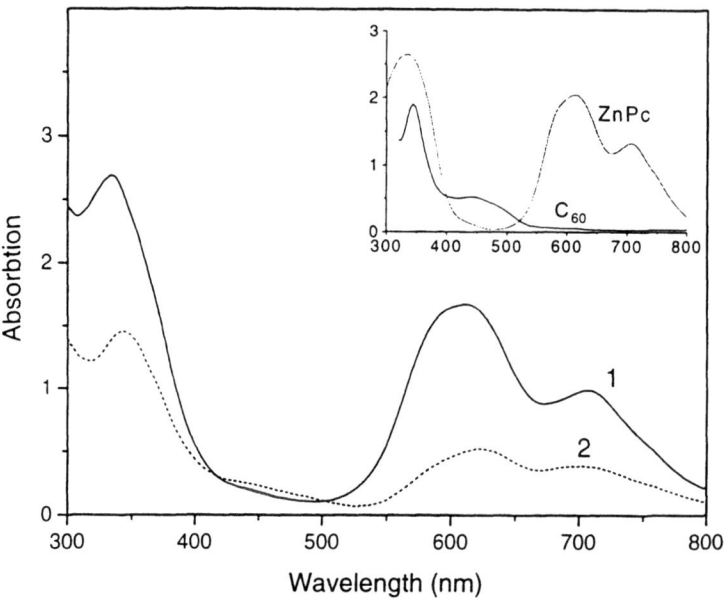

Fig.1 - Optical absorption of C_{60}/ZnPc bilayers: C_{60} (≈ 40 nm) deposited on thick ZnPc (≈ 450 nm) film (curve – 1), and on thin ZnPc (≈ 20 nm) film (curve -2). In the insert: optical absorption of C_{60} and on ZnPc films.

Raman spectra and photoluminescence contributions of C_{60} and ZnPc films are presented in Fig.2.The most obvious feature of the C_{60} photoluminescence spectrum in Fig.2 is that the main peak at 732 nm has a large fraction of the total emission intensity and relatively small vibronic contribution. The photoluminescence spectrum of ZnPc (Fig.2) is dominated by bands at 774, 810, and 853 nm.

Strong Raman bands observed in ZnPc film (see Fig.2) are assigned to a resonance excitation condition (632.8 nm) that is not so far from the ZnPc resonance absorption band at 620 nm.

The laser excitation proved to be a good one because it provides a resonance conditon for ZnPc and, due to its low power, it does not induce polimerization effects in C_{60}.

Fig.1 gives the absorption spectra of the C_{60}/ZnPc bilayers: C_{60} films ($\approx$ 40 nm) were deposited in vacuum onto ZnPc thick ($\approx$ 450 nm) and very thin ($\approx$ 20 nm) films (Fig.1, curves-1 and -2, respectively). The appearance of clear absorption bands at 347, 620, and 710 nm in all spectra of C_{60}/ZnPc bilayers was interpreted as a result of a superposition of the components, while the C_{60} forbidden band intensity around 450 nm is essentially smaller in spectra of the C_{60} films prepared on ZnPc film surface, and can be associated with some simmetry reduction of C_{60}.

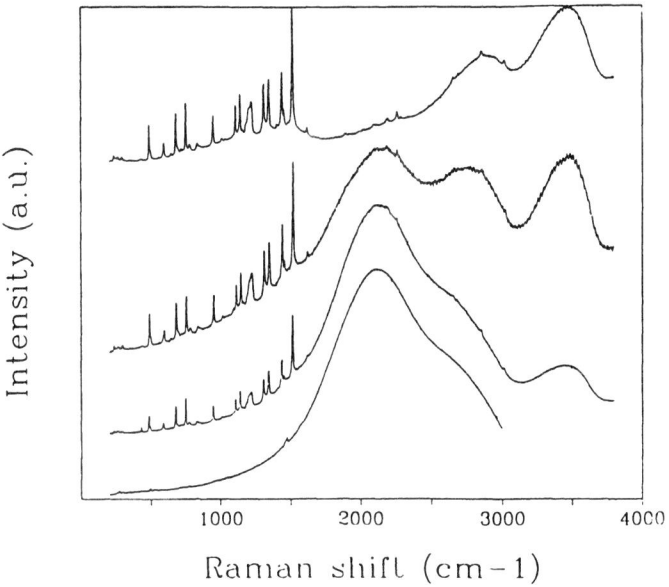

Fig. 2 - Raman spectra and PL contributions of. from bottom to top, C_{60} – film, C_{60}/ZnPc bilayers(two midle curves) and ZnPc – film. Bilayer curves correspond to curves 1 and 2 (fig. 1) respectively.

The photoluminescence and Raman spectra in Fig.2 obtained on vacuum deposited C_{60}/ZnPc bilayers, shown emission and Raman bands at the same locations as for bands of single layers. The main changes are observed in the shape of the photoluminescence bands and may be also associated with the superposition of the PL contributions of C_{60} and ZnPc.

CONCLUSION

The optical characterization of fullerene films, ZnPc films and bilayers can be described as follows: there are some changes in the shape of the PL bands, when we move from single layers to bilayers, although there are no changes in their location.

Strong Raman bands in ZnPc film are found and assigned to a resonance excitation condition.

The results of the absorption measurements are discussed in terms of a simmetry reduction of C_{60}, that would provide a stronger enhancement of the forbidden absorption band around 450 nm, than observed.

ACKNOWLEDGMENTS

L. Santos gratefully acknowledges a funding by FLAD.

REFERENCES

1.   W.Kratschmer, D.L.Lamb, K.Fostiropoulous, D.R.Huffman, Nature **347** p. 354 (1990)

2.   N.Minami, Chem. Lett. Chem. Soc. of Japan **10** p. 1791 (1991)

3.   M. Kiser, J.Reichenbach, H.J. Byrne, J.Andres, W.Maser, S.Roth, A. Zahab, P.Bernier, Solid State Commun. **81** p. 261 (1992)

4.   S. Kazaoui, R. Ross and N.Minami, Solid State Comm. **90**, n° 10, p. 623 (1994)

5.   A.Sinha, J.Menendez, R.C.Hanson, G.B.Adams, J.B.Page, O.F. Sankey, L.D.Lamb, D.R. Huffman, Chem. Phys. Letters **186** p. 2 (1991)

6.   C.Reber, L.Yee, J.McKiernan, J.I.Zink,   R.S. Williams, W.M.Tong, D.A.Ohlberg, R.L.Whetten, F.Diederich, J. Phys. Chem. **95 p.** 2127 (1991)

7.   J.L.Sauvajol, Z.Hricha, N.Coustel, A.Zahab, R.Aznar, J. Phys.: Condens. Matter **5** p. 2045 (1993)

8.   M.Uchida, Y.Ohmori, K.Yoshino, Jpn. J. Appl. Phys., **30** p. 2104 (1991)

9.   M.Uchida, Y.Ohmori, K.Yoshino, Techn. Rep. of the Osaka University **42** p. 145 (1992).

10.   S.Saito, A.Oshiyama, Phys. Rev. Letters **66** p. 2637 (1991)

11.   N.S.Sariciftici, L.Smilowitz, A.J. Heeger, F.Wudl, Science **258** p. 1474 (1992)

12.   L.Smilowitz, N.S.Sariciftici, R.Wu, C.Gettinger, A.J.Heeger, F. Wudl, Phys. Rev. B **47** p. 13835 (1993)

13.   S.Siebentritt, S. Günster and D. Meissner, Synthetic Metals, **41-43** p. 1173 (1991).

14.   L.L.Larina, N.N.Melnik, V.P.Poponin, O.I.Shevaleevskii and A.A.Kalachev, Materials Science Forum, **73-174** p. 231 (1995).

INFRARED STUDIES ON C$_{60}$ POLYMERS

K. KAMARÁS*, L. FORRÓ**, Y. IWASA***
* Research Institute for Solid State Physics, P. O. Box 49, H 1525 Budapest, Hungary
** Département de Physique, Ecole Polytechnique Féderale de Lausanne,
CH 1015 Lausanne, Switzerland
*** JAIST, Tatsunokuchi, Ishikawa 923-12, Japan

ABSTRACT

We present infrared spectra of linear and planar polymeric structures consisting of C$_{60}$ balls. The splittings of infrared lines can be explained fairly well on the basis of symmetry reduction, indicating that the structures are ordered. Increasing the temperature results in breaking of intermolecular bonds. This process is reversible in the alkali salts but irreversible in the neutral rh-C$_{60}$ polymer; we conclude that charged C$_{60}$ balls are more stable in polymer form than as monomers, while for neutral C$_{60}$ the situation is reversed.

INTRODUCTION

Among the many intriguing features of C$_{60}$ is the ability to form several one- and two-dimensional polymers. These can be obtained by photoreaction [1], at high temperature and high pressure [2], or following ionization by alkali metals [3]. Infrared spectroscopy is intimately related to the symmetry changes on polymerization and is therefore a very good indicator of the nature of the bonds forming [4]. In this paper, we compare the infrared spectra of the one- and two-dimensional polymers RbC$_{60}$ and rh-C$_{60}$, respectively, with the expected changes based on reduction of symmetry from C$_{60}$. We also follow the thermal breaking of the intermolecular bonds and the reversibility of these processes.

EXPERIMENTAL

RbC$_{60}$ and $rh-$C$_{60}$ have been prepared as discussed previously [2, 5]. Infrared spectra were taken in KBr pellets on a Bruker IFS 28 infrared spectrometer with 1 cm^{-1} resolution. For the temperature dependence, we employed a cold finger cryostat with liquid nitrogen as coolant and KBr windows.

RESULTS AND DISCUSSION

Room temperature spectra are shown in Fig. 1 and the positions of the lines are listed in Table 1. For the explanation of the splittings we have to apply group theory and construct a correlation table as in Table 2. According to group theory, the principal F$_{1u}$ lines allowed in I$_h$ symmetry should show a twofold splitting in rh-C$_{60}$ (symmetry D_{3d}) and a threefold

Table 1: Infrared lines of C_{60}, rh-C_{60} and RbC_{60} at room temperature.

C_{60}	rh-C_{60}	RbC_{60}	C_{60}	rh-C_{60}	RbC_{60}
526	509	509			
	524	517		775	774
		527			
				965	
576	550	541		997	
	555	554			1016
		570			
			1182	1121	1128
	609	607		1206	1195
					1210
		697			
		701		1306	1290
	707	722	1429	1383	1340
	718	726		1406	1386
		732			1405
	743	747			
	763	755			
		760			

splitting in RbC_{60} (symmetry D_{2h}), respectively. The modes activated by reduced symmetry are also expected to appear as doublets in rh-C_{60} and triplets in RbC_{60}. Indeed, from Table 1 it is apparent that most lines behave according to these predictions, indicating little or no disorder in both materials.

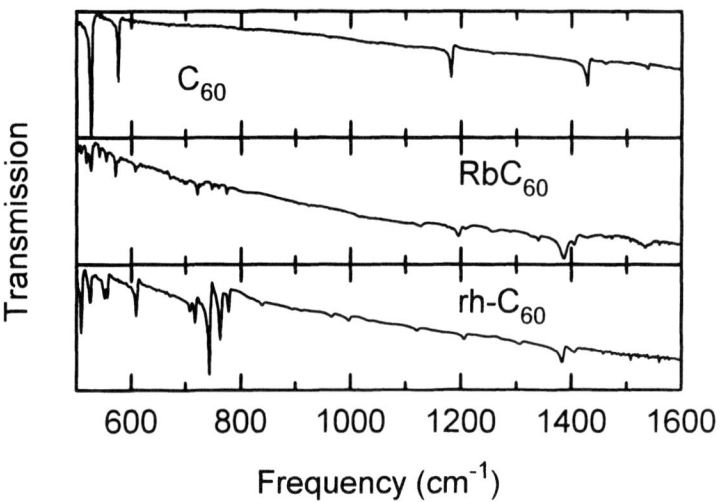

Figure 1. Infrared spectra of C_{60}, rh-C_{60} and RbC_{60} at room temperature.

Table 2: Correlation table for vibrational normal modes of C_{60}, rh-C_{60} and RbC_{60}.

I_h	D_{2h}	Splitting	D_{3d}	Splitting
$2A_g$	$2A_g$	$R\ 1 \to 1$	$2A_{1g}$	$R\ 1 \to 1$
$3F_{1g}$	$3B_{1g} + 3B_{2g} + 3B_{3g}$	$R\ 0 \to 3$	$3A_{2g} + 3E_g$	$R\ 0 \to 1$
$4F_{2g}$	$4B_{1g} + 4B_{2g} + 4B_{3g}$	$R\ 0 \to 3$	$4A_{2g} + 4E_g$	$R\ 0 \to 1$
$6G_g$	$6A_g + 6B_{1g} + 6B_{2g} + 6B_{3g}$	$R\ 0 \to 4$	$6A_{1g} + 6A_{2g} + 6E_g$	$R\ 0 \to 2$
$8H_g$	$16A_g + 8B_{1g} + 8B_{2g} + 8B_{3g}$	$R\ 1 \to 5$	$8A_{1g} + 16E_g$	$R\ 1 \to 3$
$1A_u$	$1A_u$	IR $0 \to 0$	$1A_{1u}$	IR $0 \to 0$
$4F_{1u}$	$4B_{1u} + 4B_{2u} + 4B_{3u}$	IR $1 \to 3$	$4A_{2u} + 4E_u$	IR $1 \to 2$
$5F_{2u}$	$5B_{1u} + 5B_{2u} + 5B_{3u}$	IR $0 \to 3$	$5A_{2u} + 5E_u$	IR $0 \to 2$
$6G_u$	$6A_u + 6B_{1u} + 6B_{2u} + 6B_{3u}$	IR $0 \to 3$	$6A_{1u} + 6A_{2u} + 6E_u$	IR $0 \to 2$
$7H_u$	$14A_u + 7B_{1u} + 7B_{2u}$	IR $0 \to 3$	$7A_{2u} + 14E_u$	IR $0 \to 2$
$R\ 10\ (A_g, H_g)$	$R\ 87\ (A_g, B_{1g}, B_{2g}, B_{3g})$		$R\ 45\ (A_{1g}, E_g)$	
IR $4\ (F_{1u})$	IR $66\ (B_{1u}, B_{2u}, B_{3u})$		IR $44\ (A_{2u}, E_u)$	

In Figures 2 and 3, we show the effect of thermal treatment on the two C_{60} polymers. Heating rh-C_{60} to 500 K (Figure 2), the typical polymer features disappear within an hour and C_{60} lines emerge instead. The process is irreversible, the polymer does not reappear after cooling.

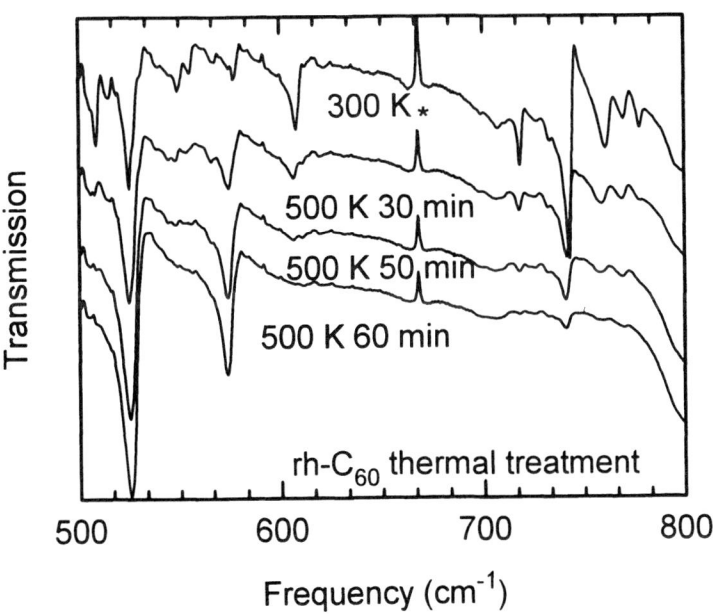

Figure 2. Infrared spectra of rh-C_{60} heated from room temperature to 500 K in the 500 - 800 cm^{-1} range. (At higher frequencies, extrinsic features appear due to outgassing of the cryostat. The line denoted by an asterisk is a CO_2 feature originating in the reference spectrum.)

In contrast, RbC$_{60}$ shows a much more complicated phase diagram depending on temperature and thermal history [7, 8]. Above 450 K, the intermolecular bonds break and the balls exist as monomers in an fcc structure, but this process is reversible so long as no oxygen is present. Figure 3 shows the slow decomposition of RbC$_{60}$ on repeated thermal cycling in vacuum; again, lines of C$_{60}$ emerge, typical of gradual oxidation and depolymerization of the C$_{60}^-$ chains. It has to be stressed, however, that in this case the reversible and irreversible processes compete, the latter being caused by another chemical reaction than the phase transition of the polymer itself. (The oxidation of C$_{60}^-$ anions is probably caused by adsorbed oxygen present on the surface of the crystallites or in the KBr pellet.)

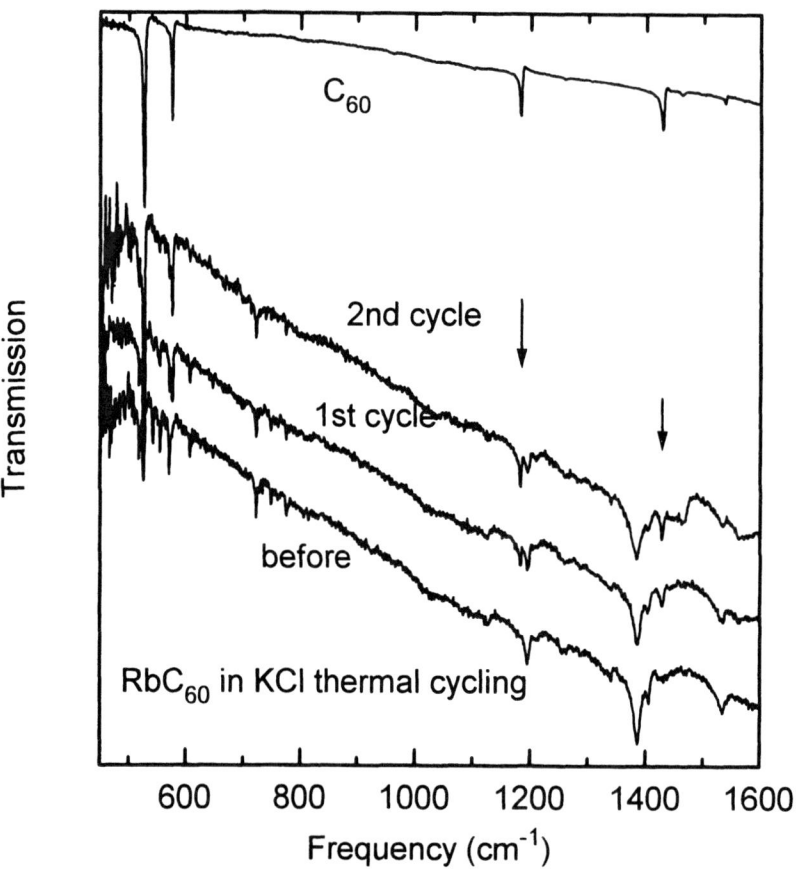

Figure 3. Infrared spectra of RbC$_{60}$ on repeated thermal cycling. One cycle consisted of heating to 500 K, then quenching to 77 K and subsequent slow warmup to room temperature.

CONCLUSIONS

Neutral rh-C_{60} and singly charged RbC_{60} both are capable of forming stable polymers at room temperature. In both cases, the infrared spectra indicate fully ordered structures, thus a high degree of polymerization. The thermal stability of the two substances, is, however, quite different: while for C_{60}^- anions the polymer form seems to be the thermodynamically stable phase, into which they return as long as they retain their extra electron, in rh-C_{60}, obtained under extreme conditions, the decomposition to C_{60} is irreversible. Thus we conclude that the charge on the balls, which causes many remarkable solid-state properties (like one-dimensional metallic behavior [9]) is also responsible for the stability of the polymeric structure.

ACKNOWLEDGEMENTS

Financial support for this research was provided by the Hungarian Academy of Sciences under contract number 96/2-462 and by a joint NSF - Hungarian Academy of Sciences grant.

REFERENCES

[1] A.M. Rao, P.Zhou, K. Wang, G.T. Hager, J.M. Holden, Y. Wang, W. Lee, X. Bi, P.C. Eklund, D.S. Cornett, M.A. Duncan, and I.J. Amster, Science **259**, 955 (1993).

[2] Y. Iwasa, T. Arima, R.M. Fleming, T.Siegrist, O. Zhou, R.C. Haddon, L.J. Rothberg, K.B. Lyons, H.L. Carter, Jr., A.F. Hebard, R. Tycko, G. Dabbagh, J.J. Krajewski, G.A. Thomas, and T. Yagi, Science **264**, 1570 (1994).

[3] S. Pekker, L. Forró, L. Mihály, and A. Jánossy, Solid State Commun. **90**, 349 (1994).

[4] A.M. Rao, P.C. Eklund, J.-L. Hodeau, L. Marques, and M. Nunez-Regueiro, Phys. Rev. B **55**, 4766 (1997).

[5] P.W. Stephens, G. Bortel, G. Faigel, M. Tegze, A. Jánossy, S. Pekker, G. Oszlányi, and L. Forró, Nature (London) **370**, 636 (1994).

[6] K. Kamarás, Y. Iwasa, and L. Forró, Phys. Rev. B **55**, 10999 (1997).

[7] P.W. Stephens in *Physics and Chemistry of Fullerenes and Derivatives*, edited by H. Kuzmany, J. Fink, M. Mehring, and S. Roth (World Scientific, 1995), p. 291.

[8] K. Kamarás, L. Gránásy, D. B. Tanner, and L. Forró, Phys. Rev. B **52**, 11488 (1995).

[9] F. Bommeli, L. Degiorgi, P. Wachter, Ö. Legeza, A. Jánossy, G. Oszlányi, O. Chauvet, and L. Forró, Phys. Rev. B **51**, 14794 (1995).

AUTHOR INDEX

944

SUBJECT INDEX

devices, 563
dialkoxy substituents, 87
diamine derivative, 575
dicyanomethanide, 819
differential scanning calorimetry, 365
dinonylnaphthalene sulfonic acid, 305
dipolar dendron, 765
dipole moment, 165
directed lithiation, 807
disorder, 829
distribution of chromophores, 801
distyrylbenzene, 533
dopant phase, 623
doping, 447, 557
    method, 879
double-strand, 359
    polymer, 733
dyeing, 747
dynamical
    holography, 859
    response, 237

EDT-TTF, 921
8-hydroxyquinoline aluminum, 575, 713
EL, 689
electric
    field, 121
    resistance, 813
electrical
    conductivity, 701, 879
    properties of polymers, 629
    transport properties, 365
electrically(-)
    generated intramolecular proton
        transfer, 545
    pumped, 545
electroabsorption, 121, 413
    spectroscopy, 801
electroactive, 551
electrochemical, 747
    devices, 317
    noise, 653
electrochemically-based solid-state light-
    emitting devices, 63
electrochromic, 747
electrode interfaces, 15
electroluminescence, 3, 27, 51, 87, 527,
    539, 545, 557, 599
        detected magnetic resonance, 99
electroluminescent, 39, 385, 503, 533, 569
    conjugated polymers, 293
    (EL) diodes, 611
electron(-)
    injecting electrodes, 835
    injection, 21, 69
    transfer, 707
electronic
    coupling, 323
    properties, 317
    structure, 489, 909
electro-optic, 193, 243, 413, 441
    activity, 199

chromophores, 199
    polymers, 199
electrostatic self-assembling, 795
ellipsometry, 847
emission, 293
emulsion, 305
    polymerization, 623
energy migration, 671
EO, 813
ESIPT, 545
ESR, 725
excitation energy transfer, 99
exciton-phonon coupling in PTCDA stacks,
    171

fabrics, 747
failure, 521
femtosecond, 777
ferromagnet, 477
ferromagnetism, 471
first molecular hyperpolarizabilities, 819
flat-panel displays (FPDs), 593, 605
fluorescence, 323, 533, 725, 835
    quenching, 671
fluorescent dyes, 569, 575
FTIR, 847
full-color
    displays, 605
    recording, 915
fullerene(s), 237
    C_{60}, 933
functionalization, 847
functionalized pyrrole monomers, 665
fusibility, 377

glass transition temperature, 885
growth mechanism, 105

HCDAH, 641
heat treatment, 515
heterocycle, 819
heterojunctions, 539
hexadecylquinolinium
    tricyanoquinodimethanide, 335
hexadiyne, 683
hexaphenyl, 867
high(-)
    $\mu\beta$ chromophore APII, 151
    pressure, 873
hole
    injection, 21, 689
    transport layer, 617
Holstein model, 171
hopping systems, 635
hydrofluoric acid-treated, 329
hydrolysis, 521
hydrostatic pressure, 867
hyperpolarizability, 211
hyper-Rayleigh scattering, 255
hyperstructured molecules, 255
hysteresis, 359

CPSIA information can be obtained at www.ICGtesting.com
Printed in the USA
LVOW06s2020230514

386805LV00021BA/649/P